The Condensed Encyclopedia of Surfactants

(With Chemical Component and Manufacturer Cross Reference)

Compiled by

Michael and Irene Ash

Contains over 12,000 entries of currently marketed surfactant trademark products

Chemical Publishing Co., Inc.
New York, N.Y.

The Condensed Encyclopedia of Surfactants

ISBN: 978-0-8206-0335-3

Chemical Publishing Company:
www.chemical-publishing.com
www.chemicalpublishing.net

First Edition:

© Chemical Publishing Company, Inc. - New York 1989

Second Impression:

Chemical Publishing Company, Inc. - 2011

Printed in the United States of America

PREFACE

This single-volume compendium fulfills the demand for a compact version of the widely acclaimed Encyclopedia of Surfactants and contains over 12,000 entries. The surfactant products included have been compiled with the cooperation of the major national and international surfactant manufacturers. The first section of this reference is a Trademark Dictionary that contains short entries for all tradename surfactants that appear in Volumes I—IV of the Encyclopedia plus all the new surfactant products that have become available since the publication of this set. Each tradename entry is followed by a number in brackets which refers to the manufacturer that is listed in Part III of this encyclopedia.

What makes this reference unique and especially useful for anyone doing chemical research is the addition of the Chemical Component-to-Tradename Cross Reference section. This provides the user with the capability of finding tradename surfactant products that contain specific chemical compounds. For example, by locating alphabetically the chemical *ammonium lauryl sulfate*, the user is able to retrieve a list of all the tradename products that contain that chemical. The user can then refer back to the first section for detailed descriptions of the specific tradename products listed here.

This section also includes the chemical compound synonyms for the surfactant components which are cross referenced in most cases to the alternate chemical names that contain the tradename product listings. Primary chemical component references are made to the CTFA (Cosmetic, Toiletry, and Fragrance Association) adopted names where possible. For many chemical entries the Chemical Abstract Service Registry Numbers (CAS) are given. This number ensures the unique identification of a chemical regardless of its multiplicity of synonyms. RD (Recognized Disclosure) reference numbers are included in some cases either in addition to the CAS number or because no CAS number has been assigned as yet.

The third section presents the names and addresses of the manufacturers of all of the products included in this compendium with their associated tradenames.

This book is a product of many years of research and we are particularly grateful to Roberta Dakan for her editorial suggestions, ideas, and hours dedicated to the project. We are also indebted to Nancy Roser for her ability to accurately enter relevant data. Their tireless efforts and energy have helped make this reference work possible.

M. & I. Ash

NOTE

The information contained in this series is accurate to the best of our knowledge; however, no liability will be assumed by the publisher for the correctness or comprehensiveness of such information. The determination of the suitability of any of the products for prospective use is the responsibility of the user. It is herewith recommended that those who plan to use any of the products referenced seek the manufacturer's instructions for the handling of that particular chemical.

Unless otherwise specified, when the temperatures are not given for properties such as viscosity, density, solubility, etc., a standard temperature of 25 C is to be assumed.

OTHER BOOKS BY MICHAEL AND IRENE ASH

A Formulary of Paints and Other Coatings, Volumes I and II
A Formulary of Detergents and Other Cleaning Agents
A Formulary of Adhesives and Sealants
A Formulary of Cosmetic Preparations
The Thesaurus of Chemical Products, Volumes I and II
Encyclopedia of Industrial Chemical Additives, Volumes I—IV
Encyclopedia of Surfactants, Volumes I—IV
Encyclopedia of Plastics, Polymers and Resins, Volumes I—IV
Vol. I, What Every Chemical Technologist Wants to Know About...Emulsifiers and Wetting Agents
Vol. II, What Every Chemical Technologist Wants to Know About...Dispersants, Solvents, and Solubilizers
Vol. III, What Every Chemical Technologist Wants to Know About...Plasticizers, Stabilizers, and Thickeners

ABBREVIATIONS

@	at
act.	active
agric.	agricultural
anhyd.	anhydrous
aq.	aqueous
aux.	auxiliary
avail.	available
avg.	average
BHA	butylated hydroxyanisole
BHT	butylated hydroxytoluene
biodeg.	biodegradable
blk.	black
b.p.	boiling point
br.	brown
brnsh.	brownish
C	degrees Centigrade
CC	Closed cup
char.	characteristic
cm	centimeter(s)
COC	Cleveland Open Cup
compd.	compound
conc.	concentration, concentrated
cps	centipoise(s)
cs of cst	centistoke(s)
cryst.	crystalline
DEA	diethanolamine
decomp.	decomposes
dens.	density
deriv.	derivative(s)
disp.	dispersible, dispersion
dist.	distilled
dk.	dark
EO	ethylene oxide
EP	extreme pressure
esp.	especially
F	degrees Fahrenheit
flam.	flammable
f.p.	freezing point
ft^3	cubic feet
g	gram(s)
gal	gallon(s)
gr.	green

HCl .... hydrochloric acid
HLB .... hydrophilic lipophilic balance
hydrog. .... hydrogenated
ingred. .... ingredient(s)
insol. .... insoluble
IPA .... isopropyl alcohol, isopropanol
IPM .... isopropyl myristate
IPP .... isopropyl palmitate
kg .... kilogram(s)
l .... liter(s)
lb .... pound(s)
liq. .... liquid
lt. .... light
max. .... maximum
MEK .... methyl ethyl ketone
mfg. .... manufacture
MIBK .... methyl isobutyl ketone
min. .... mineral, minimum
misc. .... miscible
mixt. .... mixture(s)
ml .... milliliter(s)
mm .... millimeter(s)
mN .... millinewton
m.p. .... melting point
m.w. .... molecular weight
no. .... number
nonflam. .... nonflammable
OC .... Open cup
o/w .... oil-in-water
org. .... organic
PEG .... polyethylene glycol
pH .... hydrogen ion concentration
pkg. .... packaging
PM .... Pensky-Martens
PMCC .... Pensky-Martens closed cup
PO .... propylene oxide
POE .... polyoxyethylene, polyoxyethylated
POP .... polyoxypropylene
pt. .... point
powd. .... powder
quat. .... quaternary
redsh. .... reddish
ref. .... refractive

resp. .... respectively
R.T. .... room temperature
sapon. .... saponification
SBR .... styrene/butadiene resin
sat. .... saturated
SE .... self-emulsifying
sec. .... secondary
sol. .... soluble, solubility
solid. .... solidification
sol'n. .... solution
solv(s). .... solvent(s)
sp.gr. .... specific gravity
Stod. .... Stoddard solvent
SUS .... Saybolt Universal Seconds
syn. .... synthetic
TCC .... Taggart closed cup
TEA .... triethanolamine
tech. .... technical
tert. .... tertiary
TOC .... Taggart open cup
typ. .... typical
unsat. .... unsaturated
USP .... United States Pharmacopeia
veg. .... vegetable
visc. .... viscous, viscosity
wh. .... white
w/o .... water-in-oil
wt. .... weight
yel. .... yellow
ylsh. .... yellowish
% .... percent
≈ .... approximately
< .... less than
> .... greater than
.... number

TABLE OF CONTENTS

A

AA-10. [140] Detergent, wetting agent, emulsifier; wh. liq., flake; odorless; water sol.; dens. 2 lb/gal; sp.gr. 1/5 H_2O; 98% act.

Abex 18-S. [20] Anionic; emulsifier; pale yel. liq.; dens. 8.5 lb/gal; visc. 200 cps (15 C); surf. tens. 48 dynes/cm (1.0%); 35% act.

Abex 22-S. [20] Anionic; emulsifier; pale yel. liq.; dens. 8.7 lb/gal; visc. 1000; surf. tens. 40 dynes/cm; 25% act.

Abex 23 S. [20] Used in emulsion polymerization; paints, paper coatings, textiles, adhesives; liq.; 60% act.

Abex 26-S. [20] Anionic; emulsifier; pale yel. liq.; dens. 8.8 lb/gal; visc. 100 cps; surf. tens. 55 dynes/cm; 33% act.

Abex 33 S. [20] See Abex 23 S; liq.; 30% conc.

Abex 36A. [20] See Abex 23 S; liq.; 58% conc.

Abex 1404. [20] See Abex 23 S; liq.; 42% conc.

Abex JKB. [20] Anionic; emulsifier, surfactant for highly carboxylated acrylic monomers, latex thickeners, alkali-sol. resins; clear liq.; 30% act.

Abex VA-40. [20] Anionic; emulsifier, surfactant for highly carboxylated acrylic monomers, high solids vinyl acetates, latex thickeners, alkali-sol. resins; pale yel. liq.; visc. 600 cps; 40% act.

Abex VA-50. [20] Anionic; emulsifier, surfactant for high solids vinyl acetates; pale-yel. liq.; visc. 400 cps; surf. tens. 47.4 dynes/cm (0.01%); 46% act.

Abil 281. [258; 259] Polysiloxane polyether copolymer; nonionic; silicone surfactant used in personal care products; liq.; sol. in water; 100% conc.

Abil B 8842, B 8843, B 8847, B 8851, B 8852, B 8873. [258; 259] Polysiloxane polyether copolymer; nonionic; surfactant used in personal care products; liq.; water-sol.; 100% conc.

Ablunol 200ML. [588] PEG monolaurate; nonionic; emulsifiers, lubricants, dispersing and leveling agents used in cosmetic, textile, paint and other industrial uses; liq.; HLB 9.5; 100% conc.

Ablunol 200MO. [588] PEG 200 monooleate; nonionic; see Ablunol 200ML; liq.; HLB 7.9; 100% conc.

Ablunol 200MS. [588] PEG 200 monostearate; nonionic; see Ablunol 200ML; solid; HLB 8.0; 100% conc.

Ablunol 400ML. [588] PEG 400 monolaurate; nonionic; see Ablunol 200ML; liq.; HLB 13.1; 100% conc.

Ablunol 400MO. [588] PEG 400 monooleate; nonionic; see Ablunol 200ML; liq.; HLB 11.5; 100% conc.

Ablunol 400MS. [588] PEG 400 monostearate; nonionic; see Ablunol 200ML; solid; HLB 11.6; 100% conc.

Ablunol 600ML. [588] PEG 600 monolaurate; nonionic; see Ablunol 200ML; liq.; HLB 1.48; 100% conc.

Ablunol 600MO. [588] PEG 600 monooleate; nonionic; see Ablunol 200ML; liq.; HLB 13.5; 100% conc.

Ablunol 600MS. [588] PEG 600 monostearate; nonionic; see Ablunol 200ML; solid; HLB 13.6; 100% conc.

Ablunol 1000MO. [588] PEG 1000 monooleate; nonionic; see Ablunol 200ML; solid; HLB 15.2; 100% conc.

Ablunol 1000MS. [588] PEG 1000 monostearate; nonionic; see Ablunol 200ML; solid; HLB 15.2; 100% conc.

Ablunol CO5. [588] Castor oil, ethoxylated; nonionic; emulsifier, lubricant, antistat; liq.; HLB 4.0.

Ablunol CO10. [588] Castor oil, ethoxylated; nonionic; emulsifier, lubricant, antistat; liq.; HLB 6.4.

Ablunol CO15. [588] Castor oil, ethoxylated; nonionic; emulsifier, lubricant, antistat; liq.; HLB 8.5.

Ablunol CO30. [588] Castor oil, ethoxylated; nonionic; emulsifier, lubricant, antistat; liq.; HLB 11.8.

Ablunol CO45. [588] Castor oil, ethoxylated; nonionic; emulsifier, lubricant, antistat; liq.; HLB 13.6.

Ablunol NP4. [588] Nonoxynol-4; nonionic; detergent and dispersant for use in petroleum oils; intermediate for mfg. of surfactants and antistatic agents; co-emulsifier for fats, oils and waxes; liq.; oil-sol.; HLB 8.8; 100% conc.

Ablunol NP6. [588] Nonoxynol-6; nonionic; emulsifier or coemulsifier; coupling agent; intermediate for mfg. of surfactants; emulsifier for silicone and min. oil; liq.; HLB 10.8.

Ablunol NP8. [588] Nonoxynol-8; nonionic; detergent for formulating household and industrial cleaning compounds; liq.; water-sol.; HLB 12.2; 100% conc.

Ablunol NP9. [588] Nonoxynol-9; nonionic; detergent used in textile, paper, and leather industries; metal processing as a wetting agent with mineral acid; liq,; water-sol.; HLB 13.0; 100% conc.

Ablunol NP15. [588] Nonoxynol-15; nonionic; surfactant used in alkaline metal cleaners and bottle washing formulations; liq.; HLB 15.0; 100% conc.

Ablunol S-20. [588] Sorbitan monolaurate; nonionic; emulsifier for oils and fats in cosmetic and industrial oil products; liq.; water-disp.; HLB 8.6; 100% conc.

Ablunol S-60. [588] Sorbitan monostearate; nonionic; preparation of silicone defoamer emulsions; textile process lubricant; solid; HLB 4.7; 100% conc.

Ablunol S-80. [588] Sorbitan monooleate; nonionic; emulsifier for oils, fats and waxes; wet processing of syn. PU leather; liq.; HLB 4.3; 100% conc.

Abluphat LP. [588] Phosphate ester; anionic; antist for syn. fibers; liq.; 30% conc.

Ablusol C-78. [588] Sodium dioctyl sulfosuccinate; anionic; wetting agent for textile, PU leather; emulsion polymerization, agric. emulsion, and other industrial applications; liq.; 70% conc.

Ablusol DBC. [588] Calcium alkylaryl sulfonate; anionic; emulsifier; liq.; 70% conc.

Ablusol NL. [588] Sodium naphthalene sulfonate formaldehyde condensate; anionic; dye dispersant and antiagglomerant; used in the process of polyester and acetate fabrics; liq.; 40% conc.

A.B.S. 87%. [610] Alkylbenzene sulfonic acid; detergent intermediate; liq.; 87% conc.

Accobetaine CL. [66] Complex coco betaine; detergent, wetting agent, emulsifier, high foaming agent, solubilizer, household and cosmetic uses; yel. liq.; pleasant odor; water-sol.; f.p. 0 C; 35% act.

Accomeen C2. [66] PEG-2 cocamine; nonionic; emulsifier, antistat, surfactant; Gardner 4 liq., amine odor; sol. in org. solv., water-insol.; dens. 7.25 lb/gal.; sp.gr. 0.87; 99% act.

Accomeen C5. [66] PEG-5 cocamine; nonionic; emulsifier, surfactant; antistat; Gardner 4 liq., amine odor; sol. in org. solv., water; dens. 8.15 lb/gal.; sp.gr. 0.98; 99% act.

Accomeen C10. [66] PEG-10 cocamine; nonionic; emulsifier, antistat, surfactant; Gardner 6 liq., amine odor; sol. in org. solv., water; dens. 8.3 lb/gal.; sp.gr. 1.0; 99% act.

Accomeen C15. [66] PEG-15 cocamine; nonionic; emulsifier, antistat, surfactant; Gardner 6 liq., mild amine odor; dens. 8.7 lb/gal.; sp.gr. 1.04; 99% act.

Accomeen S2. [66] PEG-2 soyamine; nonionic; emulsifier, antistat, surfactant; Gardner 5 liq.; amine-type odor; sol. in org. solv.; water insol.; dens. 7.6 lb/gal; sp.gr. 0.91; surf. tens. 31.3 (0.1%); 99% act.

Accomeen S5. [66] PEG-5 soyamine; nonionic; emulsifier, antistat, surfactant; Gardner 10 liq.; amine-type odor; sol. in org. solv., water disp.; dens. 7.9 lb/gal; sp.gr. 0.95; 99% act.

Accomeen S10. [66] PEG-10 soyamine; nonionic; emulsifier, surfactant, antistat; Gardner 10 liq.; amine-type odor; sol. in water, org. solv.; dens. 8.5 lb/gal; sp.gr. 1.02; surf. tens. 43 (0.1%); 99% act.

Accomeen T2. [66] PEG-2 tallow amine; nonionic; emulsifier, antistat, surfactant; Gardner 3 paste.; amine-type odor; sol. in org. solv., insol. in water; dens. 7.7 lb/gal;

sp.gr. 0.92; 99% act.

Accomeen T5. [66] PEG-5 tallow amine; nonionic; emulsifier, dispersant; paste; 100% conc.

Accomeen T15. [66] PEG-15 tallow amine; nonionic; emulsifier, dispersant, surfactant; Gardner 5 liq.; amine-type odor; sol. in org. solv., insol. in water; dens. 8.1 lb/gal; sp.gr. 0.97; 99% act.

Accomid C. [66] Coconut oil diethanolamide; nonionic; detergent, stabilizer, visc. improver, foam booster for shampoos and dishwashes; Gardner 4 liq.; mild odor; completely sol. in water at low conc.; dens. 8.3 lb/gal; sp.gr. 0.99; 98% act.

Acconon 300 MO. [66] PEG 300 monooleate; nonionic; emulsifier, lubricant, chemical intermediate; for cosmetics, food, agriculture, plastics; Gardner 6 liq.; sol. in org. solv.; dens. 8.3 lb/gal; sp.gr. 0.99; m.p. < –5 C; 99% act.

Acconon 400 MO. [66] PEG 400 monooleate; nonionic; see Acconon 300 MO; Gardner 6 liq.; sol. in org. solv.; dens. 8.4 lb/gal; sp.gr. 1.01; m.p. < 10 C; 99% act.

Acconon 1300. [66; 579] PPG-3-laureth-9; lt. yel., Gardner 3 max. liq.; dens. 8.35–8.45 lb/gal; sp.gr. 1.016–1.019; cloud pt. 135–145 F.

Acconon TGH. [66; 579] PEG-20-PPG-10 glyceryl stearate; nonionic; wetting agent, foaming agent, surfactant; yel. liq., mild odor; 100% conc.

Accoquat 2C-75. [66] Dicoco dimethyl ammonium chloride; cationic; emulsifier, coupling agent; amber liq., mild odor; dens. 7.2 lb/gal; sp.gr. 0.86; 75% act. in IPA.

Accoquat 2C-75H. [66] Dicoco dimethyl ammonium chloride, high flash; cationic; emulsifier, coupling agent; amber liq., mild odor; 75% act. in hexylene glycol.

Accosoft 550, 620. [66] Tallow amido amine quat. compd.; cationic surfactant used in household and industrial fabric softener, fabric antistat; Gardner 6 liq., gel; flam.; dens. 8.0; sp.gr. 0.96; 75 or 90% act. in IPA.

Accosoft 707. [66] Dimethyl dihydrog. tallow ammonium chloride; cationic surfactant used in household and industrial fabric softener, fabric antistat; Gardner 3 paste; flam.; dens. 7.3; sp.gr. 8.7; 75% act.

Accosoft 748. [66] Dimethyl dihydrog. tallow ammonium methyl sulfate; cationic surfactant used in household and industrial fabric softener or antistat esp. suited to dryer use; Gardner 4 paste, solid; m.p. 125–135 C; biodeg.; flam.; 90% act. in IPA.

Accosoft 808. [66] Tallow imidazoline quat.; cationic; fabric softener and antistat for household and commerical laundries; liq.; 75% conc.

Accosoft A-155. [66] Tallow amido amine quat. compd.; cationic surfactant, household fabric softener base, dispersion, antistat agent; Gardner 7 gel; biodeg.; 75 or 95% act.

Accosoft M 1154. [66] Tallow amine quat. deriv.; cationic; fabric softener for detergent softener systems; liq.; 98% conc.

Accosperse 20. [66] PEG-20 sorbitan monolaurate; nonionic; emulsifier, solubilizer; liq.; 100% conc.

Accosperse 60. [66] PEG-20 sorbitan monostearate; nonionic; emulsifier, solubilizer; amber liq.; sp.gr. 1.1; visc. 500–600 cps; HLB 14.9–15.0; sapon. no. 45–55; 98% act.

Accosperse 80. [66] PEG-20 sorbitan monooleate; nonionic; emulsifier, solubilizer; amber liq.; sp.gr. 1.1; visc. 400–450 cps; HLB 15.0; sapon. no. 45–55; 98% act.

Accosperse TGH. [66] Alkoxylated mono-, diglyceride; nonionic; emulsifier, foaming and wetting agent; yel. liq.; mild odor; f.p. 30 F; cloud pt. 140–146 F (1% aq.); biodeg.; 98–100% act.

Acetamide MEA. [650] Acetamide MEA; nonionic; solvent, humectant used in personal care products; liq.; 70–100% conc.

Acetamide MEA-70. [515] Acetamide MEA; humectant for hand and body lotions; liq.; dens. 9.1 lb/gal; visc. 30 cps; pH 7.0; 70% solids.

Acetamide MEA-100. [515] Acetamide MEA; humectant for hand and body lotions; liq.; dens. 9.3 lb/gal; visc. 150 cps; pH 7.0; 95% solids.

Acetamin 24, 86. [341] *n*-Alkylamine acetate; cationic; surface coating agent for pigments, anticaking agent for fertilizer; emulsifier, dispersant, and softening agent for textiles; mineral flotation reagent; solid; 100% conc.

Acetamin C. [561] *n*-Cocoamine acetate; cationic; flotation of minerals, anticaking agents, emulsifier bactericide; paste, solid; 100% conc.

Acetamin HT. [561] *n*-Hyd. tallow amine acetate; cationic; see Acetamin C; solid.

Acetamin T. [561] *n*-Tallow amine acetate; cationic; see Acetamin C; solid.

Acetoquat CPB. [4] Cetyl pyridinium bromide; cationic; germicide, sanitizing agent; powd.; 95% act.

Acetoquat CPC. [4] Cetyl pyridinium chloride; cationic; germicide, sanitizing agent; powd.; 100% act.

Acetoquat CTAB. [4] Cetyl trimethyl ammonium bromide; cationic; germicide, sanitizing agent; powd.; 95% act.

Acidan. [273] Monoglyceride citric acid ester; anionic; emulsifier, surfactant; used in food industry; powd., flakes; HLB 11.0; 100% act.

Acid Felt Scour. [283] Nonionic; felt cleaner for batch and continuous applications; liq.; 40% act.

Acid Foamer. [608] Quat. ammonium chloride; cationic; aluminum trailer cleaner, brightener, acid inhibitors and cleaners, chrome plating baths; liq.; 100% conc.

Acofor. [505] Dist. tall oil fatty acids; latex stabilizer, dispersant (as soap) for pigments and fillers; Gardner 3 color; sp.gr. 0.903; flash pt. 400 C (COC).

Aconol X6. [283] POE monotallate; nonionic; emulsifier for min. oil; liq.; 100% act.

Aconol X10. [283] Tall oil ethoxylate; nonionic, emulsifier, low-foaming surfactant for built detergent systems; clear liq.; clear sol.; dens. 1.039; visc. 140 cps; 90% act.

Acra-500. [608] Fatty amine complex; cationic; asphalt wetting and antistripping agent; paste; 100% conc.

Acrilev AM. [237] Phosphate ester, potassium salt; anionic; detergent, wetter, dye leveler used in textiles; liq.; 35% conc.

Acrilev AM-Special. [237] Phosphate ester, potassium salt; anionic; detergent, wetter, dye leveler used in textiles; liq.; 65% conc.

Actiflo 68, 70. [121] Natural lecithin; amphoteric; emulsifier, wetting agent, dispersant; used in the food industry; liq.; 66–72% act.

Acto 450. [230] Alkylaryl sodium sulfonate; anionic; detergent, wetting agent, emulsifier, rust preventative; reddish-amber liq.; alcohol odor; flam.; m.w. 470; oil sol., slight sol. in water but readily disp.; dens. 8.24 lb/gal; sp.gr. 0.99; 44% act.

Acto 500. [230] Alkylaryl sodium sulfonate; anionic; see Acto 450; amber, visc. liq.; petroleum odor; m.w. 460; oil sol., slight sol. in water but readily disp.; dens. 8.33 lb/gal; sp.gr. 1.00; 50% act.

Acto 630. [230] Alkylaryl sodium sulfonate; anionic; see Acto 450; amber visc. liq.; bland, petroleum odor; flam.; m.w. 470; oil sol., slight sol. in water but readily disp.; dens. 8.41 lb/gal; sp.gr. 1.01; 63% act.

Acto 632. [230] Alkylaryl sodium sulfonate; anionic; see Acto 450; reddish-amber liq.; alcohol odor; flam.; m.w. 430; oil sol., slight sol. in water but readily disp.; dens. 8.41 lb/gal; sp.gr. 1.01; 63% act.

Acto 636. [230] Alkylaryl sodium sulfonate; anionic; see Acto 450; orange, visc. liq.; alcohol odor; flam.; m.w. 470; oil sol., slight sol. in water but readily disp.; dens. 8.5 lb/gal; sp.gr. 1.02; 63% act.

Acto 639. [230] Alkylaryl sodium sulfonate; anionic; see Acto 450; reddish-amber liq.; alcohol odor; flam.; m.w. 520; oil sol., slight sol. in water but readily disp.; dens. 8.5 lb/gal; sp.gr. 1.02; 63% act.

Actrabase 31-A. [568] Soap sulfonate; anionic; primary emulsifier for oil systems; liq.; HLB > 10; 100% conc.

Actrabase 163. [568] Soap sulfonate; anionic; emulsifier base for sol. oils; liq.; 100% conc.

Actrabase PS-470. [568] Petroleum sulfonate; anionic; emulsifier and rust inhibitor for cutting and lube oils; visc. liq.; 100% conc.

Actrafos 161. [568] Phosphate ester; anionic; lubricant and emulsifier for use in cutting oils; liq.; sol. in water and oil; 100% conc.

Actrafos 186. [568] Aliphatic alcohol phosphate ester; anionic; lubricant and emulsifier for use in cutting oils; liq.; oil and water sol.; 100% conc.

Actrafos 822. [568] Complex aliphatic alcohol phosphate ester; anionic; lubricant and emulsifier for use in cutting oils; liq.; oil and water sol.; 100% conc.

Actrafos S-104, S-109. [568] Phosphate ester; anionic; used in syn. and semisynthetics; solubilizer for nonionic surfactants in alkaline cleaning systems; liq.; 100% conc.

Actrafos SA-208. [568] Linear alcohol phosphate ester; anionic; lubricant and emulsifier in cutting oils; liq.; oil and water sol.; 100% conc.

Actrafos SA-216. [568] Phosphate ester; anionic; lubricant and emulsifier for cutting oils; liq.; sol. in oil and water; 100% conc.

Actrafos SN-306. [568] Phosphate ester; anionic; used in liq. cleaners; liq.; water-sol.; 100% conc.

Actrafos SN-314. [568] Phosphate ester; anionic; used in waterless hand cleaners and solv. cleaners; liq.; 100% conc.

Actrafos SN-315. [568] Phosphate ester; anionic; lubricant, emulsifier for metal working; liq.; 100% conc.

Actralube SOS. [568] Ester; nonionic; emulsifier and lubricity additives; substitute for sperm oil; liq.; 100% conc.

Actramide 189. [568] Alkanolamide; nonionic; foam stabilizer for cleaners and shampoos; liq.

Actramide 202. [568] Alkanolamide; nonionic; emulsifier for sol. oils and metalworking fluids and emulsion cleaners; liq.

Actramide 410, 420, 499. [568] Alkanolamide; nonionic; sec. emulsifier with lubricating properties; liq.; 100% conc.

Actramide 840. [568] Alkanolamide; nonionic; foam booster and thickener; liq.; 100% conc.

Actramide 933. [568] Alkanolamide; nonionic; visc. builder and foam stabilizer; liq.; 100% conc.

Actramide 938. [568] Alkanolamide; nonionic; foam stabilizer for cleaners and shampoos; liq.

Actrasol 167A. [568] Sulfated castor oil, sodium neutralized; anionic; acid etching additive; liq.; 80% conc.

Actrasol C50. [568] Sulfated castor oil; anionic; pigment wetting and dispersion; liq.; 50% conc.

Actrasol C75. [568] Sulfated castor oil; anionic; pigment wetting and dispersion; liq.; 70% conc.

Actrasol C85. [568] Sulfated castor oil; anionic; pigment wetting and dispersion; liq.; 75% conc.

Actrasol CS75. [568] Sulfonated soya bean oil; anionic; liq.; water disp.; 65% conc.

Actrasol EO. [568] Sulfated glycerol trioleate; anionic; shampoos, metal working; liq.; 75% conc.

Actrasol KAP. [568] Sulfated sperm oil substitute; anionic; fat liquor for leather, metal working; liq.; 75% conc.

Actrasol MY75. [568] Sulfated soyabean oil; anionic; liq.; water-disp.; 75% conc.

Actrasol OY75. [568] Sulfated soya bean; anionic; liq.; water-disp.; 75% conc.

Actrasol PSR. [568] Sulfated castor oil; anionic; pigment wetting and dispersion; liq.; 75% conc.

Actrasol SP. [568] Sulfonated TOFA; wet process phosphoric acid defoamer; liq.; 50% conc.

Actrasol SP175K. [568] Sulfated tall oil; anionic; liq.; water-disp.; 75% conc.

Actrasol SR 75, SRK 75. [568] Sulfated oleic acid; anionic; liq.; water-disp.; 75% conc.

Actrol 4DP. [568] PEG diester; nonionic; emulsifier and lubricity additive for metalworking fluids; liq.; HLB < 5; 95% conc.

Actrol 4MP, 628. [568] PEG ester; nonionic; see Actrol 4DP; liq.; HLB > 10; 95% conc.

Acylglutamate CS-11. [9] Monosodium N-cocoyl-L-glutamate; anionic; detergent, emollient for personal care products; bacteriostatic effect; powd.; biodeg.; 100% act.

Acylglutamate CT-12. [9] Monotriethanolamine N-cocoyl-L-glutamate; see Acylglutamate CS-11, liq.; 20% act.

Acylglutamate GS-11. [9] Sodium hydrog. tallow glutamate and sodium cocoyl glutamate; see Acylglutamate CS-11; powd.; 100% conc.

Acylglutamate HS-11. [9] Monosodium N-hydrog. tallowyl-L-glutamate; see Acylglutamate CS-11; powd. 100% conc.

Acylglutamate LS-11. [9] Monosodium N-lauroyl-L-glutamate; see Acylglutamate CS-11; powd.; 100% conc.

Acylglutamate MS-11. [9] Sodium myristoyl glutamate; anionic; see Acylglutamate CS-11; powd.; 100% conc.

Adinol CT95. [159] Sodium N-methyl-N-cocoyl taurate; anionic; cosmetic and pharmaceutical detergent/emulsifier; wh. powd.; 95% min. conc.

Adinol T. [159] Sodium N-methyl-N-oleoyl taurate; detergent, wetting agent, emulsifier, dispersant, foaming agent; liq., paste, gel, powd.; good sol. in hard water.

Adinol T35. [159] Sodium N-methyl-N-oleoyl taurate; anionic; surfactant having detergent, wetting, emulsifying, foaming and dispersant properties used in cosmetic and industrial applications esp. textile processing; liq.; 16% min. conc.

Adma 2. [227] Dodecyldimethylamine; cationic; intermediate for quat. ammonium compds., amine oxides, and betaines; liq.;100% conc.

Adma 4. [227] Tetradecyldimethylamine; cationic; see Adma 2; liq.; 100% conc.

Adma 6. [227] Hexadecyldimethylamine; cationic; see Adma 2; liq.; 100% conc.

Adma 8. [227] Octadecyldimethylamine; cationic; see Adma 2; liq.; 100% conc.

Adma 24. [227] C_{12}/C_{14} alkyldimethylamine; cationic; see Adma 2; liq.; 100% conc.

Adma 46. [227] C_{14}/C_{16} alkyldimethylamine; cationic; see Adma 2; liq.; 100% conc.

Adma 246-451, 246-621. [227] $C_{12}/C_{14}/C_{16}$ alkyldimethylamine; cationic; see Adma 2; liq.; 100% conc.

Adma C8. [227] Octyldimethylamine; cationic; see Adma 2; liq.; 100% conc.

Adma C10. [227] Decyldimethylamine; cationic; see Adma 2; liq.; 100% conc.

Admul 1405. [482] Sat. polyglycerol ester; nonionic; emulsifier for margarines and shortenings; powd.; 100% conc.

Admul 1411. [482] Unsat. polyglycerol ester; nonionic; aerating agent in syn. creams; paste; 100% conc.

Admul CSL. [482] Stearoyl-2-lactylates; anionic; food emulsifier for starch-complexing agent; dough conditioner in breads; flake, powd.; HLB 8.0; 100% conc.

Admul CSL 2007. [482] Calcium stearoyl lactylate; anionic; bread improver; flake.; 100% conc.

Admul CSL-2008. [482] Calcium stearoyl lactylate; anionic; antistaling agent; powd.; 100% conc.

Admul Data. [482] Monoglyceride diacetylated tartaric acid esters; food emulsifier used as dough conditioners in baked products; paste, powd., microbead; HLB 8.0; 100% conc.

Admul GLP, GLS. [482] Monoglyceride lactic acid ester; nonionic; food emulsifier for shortenings, toppings and desserts; flake; HLB 3.5; 100% conc.

Admul MG 4103, 4123, 4143, 4163, 4203, 4223, 4304, 4404, 5103, 6103, 6404, 6504. [482] Monodiglycerides; nonionic; used in bakery products, margarines, shortenings, creams, desserts, ice cream; microbead; HLB 4.0; 100% conc.

Admul PGMP, PGMS. [482] Palmitic and stearic acids propylene glycol ester acids; nonionic; food emulsifier for toppings, desserts, cake improvers; solid; HLB 2.0; 100% conc.

Admul SSL. [482] Stearoyl-2-lactylates; anionic; food emulsifier for starch complexing agent and dough conditioning in bread products; flake, powd.; HLB 8.0; 100% conc.

Admul SSL 2003. [482] Sodium stearoyl lactylate; anionic; o/w emulsifier; bread improver; flake; 100% conc.

Admul SSL 2004. [482] Sodium stearoyl lactylate; anionic; antistaling agent; powd; 100% conc.

Admul WOL 1403. [482] Polyglycerol polyricinoleate; nonionic; visc. modifier in chocolate; liq.; 100% conc.

Admul WOL 1405, 1411. [482] Polyglycerol ester; nonionic; food emulsifier; microbead; HLB 5–6; 100% conc.

Adogen 415. [559] Trimethyl soya quat. ammonium chloride; cationic; emulsifier, dispersant; liq.; 50% act.

Adogen 432. [559] Dialkyl dimethyl ammonium chloride; cationic; softener; used in retail, commerical laundry and textile mill formulations; liq.; 68% conc.

Adogen 432 CG. [559] Dicetyl dimonium chloride, cationic; conditioner, cream rinse base; lt.-colored clear to hazy liq.; disp. in water; pour pt. < 50 F; 74% conc. in aq. IPA.

Adogen 441. [559] Hydrog. tallow trimethyl ammonium chloride; specialty quat.; liq.; Gardner 5 max.; m.w. 341; flash pt. 58 F (PM); 49–52% quat.

Adogen 442. [559] Quaternium-18; cationic; softening agent for textiles, commerical laundries; paste; 75% conc.

Adogen 442 H. [559] Quaternium-18; viscosifier for Bentonite clays for oil-based muds in petroleum industry; paste; flash pt. 65 F (CC); 75% conc.

Adogen 444. [559] Palmityl trimethyl ammonium chloride; cationic; specialty quat.; emulsifier, dispersant, creme rinse, fermentation aid; liq.; Gardner 6 max.; m.w. 319; flash pt. 58 F (PM); 49–52% quat.

Adogen 461. [559] Trimethyl coco quat. ammonium chloride; cationic; emulsifier, dispersant; liq.; 50% conc.

Adogen 462. [559] Dicocodimonium chloride; cationic; emulsifier, dispersant for oilfield chemicals; paste; 75% conc.

Adogen 464. [559] Trimethyl C_8–C_{10} quat. ammonium chloride; cationic; flotation agent, emulsifier for solv. extraction of metal; liq.; 90% conc.

Adogen 470. [559] Ditallow dimethyl ammonium chloride; specialty quat.; liq.; Gardner 6 max.; m.w. 564; flash pt. 65 F (PM); 74–77% quat.

Adogen 471. [559] Tallow trimethyl ammonium chloride, cationic; dispersant, antistat, emulsifier; Gardner 6 max. liq.; m.w. 339; flash pt. 58 F (PM); 49–52% quat.

Adogen 477. [559] N-tallow pentamethyl propane diammonium dichloride; cationic; emulsifier, dispersant; liq.; Gardner 11 max.; m.w. 463; flash pt. 47 F (PM); 49–52% quat.

Adogen 560, 570-S, 572. [559] Fatty aminopropyl amine; cationic; emulsifier; intermediate for corrosion inhibitors; emulsifier for asphalt; liq.; 100% conc.

Adogen R-6. [559] 1:1 blend tallow trimethyl and dicoco dimethyl ammonium chloride; specialty quat.; liq.; Gardner 6 max.; m.w. 389; flash pt. 58 F (PM); 49–52% quat.

Adol 42. [559] Unsat. fatty alcohol; coemulsifier, lubricant, foam control agent, cosolvent, plasticizer, stabilizer, emollient, intermediate; used in metal rolling oils and lubricants, gas scrubbing, printing inks, auto antifreeze, textile dyeing and finishing, emulsion systems, paper pulping, cosmetics, cationic surfactants, min. processing/oil field chemicals, fabric softeners; Lovibond 5Y/0.5R color; m.w. 258; sol. in fatty alcohols, IPA, benzene, trichlorethylene, acetone, turpentine, VMP naphtha, kerosene, lt. min. oil; sp.gr. 0.814 (60/25 C); cloud pt. 36 C max.; sapon. no. 3.5 max.

Adol 52 NF. [559] Cetyl alcohol; see Adol 42; Lovibond 5Y/0.5R max.; m.w. 247; sol. see Adol 42; sp.gr. 0.815 (60/25 C); m.p. 45–50 C; sapon. no. 3.0 max.

Adol 54. [559] Cetyl alcohol; see Adol 42; L C); sol. see Adol 42; solid. pt. 46–50 C; sapon no. 3.0 max.

Adol 61 NF. [559] Stearyl alcohol; used in emulsifiers, surfactants, cosmetics, wax formulations; see also Adol 42; wh. cryst. solid; odorless; sol. in IPA, acetone, naphtha, lt. min. oil; m.w. 272; sp.gr. 0.817 (60/25 C); m.p. 56–60 C; sapon. no. 1.0 max.

Adol 62 NF. [559] Stearyl alcohol; emollient, glass frit binders, waxes, emulsion stabilizers, esters, tertiary amines, surfactants, polymers, chemical intermediate; cosmetic formulations; see also Adol 42; Lovibond 5Y/0.54 max; odorless; m.w. 272; sol. see Adol 61NF; sp.gr. 0.817 (60/25 C); visc. 42 SSU (210 F); m.p. 56–60 C; b.p. 337–360 C (760 mm, 90%); sapon. no. 3.0 max.

Adol 63. [559] Stearyl alcohol; prime base for detergents; used in plasticizers, tert. amines, lube oil additives, textile auxiliaries, mold lubricants, polymers, org. synthesis, chemical intermediates; see also Adol 42; Lovibond 5Y/0.5R max.; m.w. 268; sp.gr. 0.816 (60/25 C); visc. 44 SSU (210 F); solid pt. 48–53 C; b.p. 312–344 C (760 mm, 90%); sapon. no. 3.0 max.

Adol 64. [559] Stearyl alcohol; see Adol 63, Adol 42; Lovibond 5Y/0.5R max.; odorless; m.w. 269; sol. in IPA, acetone, naphtha, lt. min. oil; sp.gr. 0.817 (60/25 C); visc. 44 SSU (210 F); solid. pt. 51–54 C; sapon. no. 3.0 max.

Adol 66. [559] Isostearyl alcohol; see Adol 42; Lovibond 5Y/0.5R max.; m.w. 295; sol. see Adol 42; sp.gr. 0.861; cloud pt. 8.0 C max.; sapon. no. 2.0 max.

Adol 80. [559] Oleyl alcohol; see Adol 42; Lovibond 5Y/0.5R max.; m.w. 263; sol. see Adol 42; sp.gr. 0.840; cloud pt. 13 C max.; sapon. no. 3.0 max.

Adol 85 NF. [559] Oleyl alcohol; see Adol 42; Lovibond 3Y/O.3R max.; m.w. 267; sol. see Adol 42; sp.gr. 0.840; cloud pt. 10 C max.; sapon. no. 2.0 ,max.

Adol 320. [559] Oleyl alcohol; see Adol 42; Lovibond 5Y/0.5R max.; m.w. 261; sp.gr. 0.845; cloud pt. 19 C max.; sapon. no. 3.0 max.

Adol 330. [559] Oleyl alcohol; see Adol 42; Lovibond 5Y/0.5R max.; m.w. 261; sp.gr. 0.845; cloud pt. (AOCS) 22 C max.; sapon. no. 3.0 max.

Adol 340. [559] Oleyl alcohol; see Adol 42; Lovibond 5Y/0.5R max.; m.w. 261; sp.gr. 0.845; cloud pt. (AOCS) 28 C max.; sapon. no. 3.0 max.

Adol 520 NF. [559] Cetyl alcohol; see Adol 42; APHA 50 max.; m.w. 246; sol. see Adol 42; sp.gr. 0.815 (60/25 C); m.p. 45–50; solid pt. 48–53 C; sapon. no. 2.0 max.

Adol 620 NF. [559] Sat. fatty alcohol; see Adol 42; APHA 50 max.; m.w. 267; sol. see Adol 42; sp.gr. 0.817 (60/25 C); m.p. 55–60 C; sapon. no. 2.0 max.

Adol 630. [559] Sat. fatty alcohol; see Adol 42; APHA 100 max.; m.w. 265; sol. see Adol 42; sp.gr. 0.816 (60/25 C); solid pt. 48–53 C; sapon no. 2.0 max.

Adsee 775. [650] POE ethers and special resins; nonionic; spreader sticker and penetrant for agric. spray; surfactant for monosodium methane arsonate formulations; liq.; 100% conc.

Adsee 799. [650] Alkyl POE ether; nonionic; agric. surfactant; soil penetrant; liq.; 100% conc.

Adsee 801. [650] Nonionic; agric. surfactant; liq.

Adsee 815. [650] Nonionic; agric. surfactant, spray tank adjuvant; penetrant; wetting agent; liq.

Adsee 2096. [650] Nonionic; agric. surfactant, spreader penetrant for wetting of pesticide formulations, adjuvant for spray-tank mixts.

Adsee 2141. [650] Anionic/nonionic; agric.

surfactant, premix surfactant for monosodium methyl arsonate formulations; liq.; 100% conc.

Adsee AC62-38. [650] Anionic/nonionic; agric. surfactant, premix surfactant for monosodium methylarsonate formulations.

AE-3. [489] Ethoxylated fatty alcohol; nonionic; detergent, emulsifier; liq.; 100% conc.

AE-7. [489] Ethoxylated fatty alcohol; nonionic; textile lubricants, textile wetting; solid; 100% conc.

AE-1214/3. [489] Laureth-3; nonionic; detergent, emulsifier; liq.; 100% act.

AE-1214/6. [489] Ethoxylated linear alcohols; nonionic; lubricant, wetting agent, used in textiles; solid; 100% act.

AEPD. [52; 323] 2-Amino-2-ethyl-1,3-propanediol; anionic; emulsifier, catalyst; Gardner 2; m.w. 119.17; dens. 9.15 lb/gal; sp.gr. 1.101 (40/40 C); m.p. 38 C; b.p. 152 C; 100% conc.

Aerosol 18. [43; 167] Disodium stearyl sulfosuccinamate; anionic; emulsifying, dispersing, foaming power, detergency and solubilizing on soaps and surface active agents; emulsion polymerization of vinyl chloride and SBRs; foamer for foamed latexes and plastics; Gardner 10 creamy paste; water sol.; dens. 8.9 lb/gal; sp.gr. 1.07; f.p. solid paste < 21 C; surf. tens. 41 dynes/cm; biodeg.; 35% act.

Aerosol 19. [43] Disodium N-alkyl sulfosuccinamate; emulsifier, foaming and wetting agent; 10 Gardner clear liq.; water sol.; dens. 8.9 lb/gal; sp.gr. 1.07; surf. tens. 41 dynes/cm; biodeg.; 35% act.

Aerosol 22. [43] Tetrasodium N-(1,2-dicarboxyethyl)-N-octadecyl sulfosuccinamate; anionic; emulsifier, dispersant, surfactant; Gardner 8, clear to cloudy liq.; water sol.; dens. 9.4 lb/gal; sp.gr. 1.12; visc. 53 cps; f.p. separates < 10 C; m.p. > 200 C; surf. tens. 41 dynes/cm; biodeg.; 35% act.

Aerosol 22 N. [167] Tetrasodium N-(1,2-dicarboxyethyl)-N-octadecyl sulfosuccinamate; anionic; emulsion polymerization, latex foam stabilization; antigelling agent; industrial, household and cosmetic cleaners; metal cleaners; liq.; 35% conc.

Aerosol 30. [43] Cocoamidopropyl betaine; amphoteric surfactant, foam booster, visc. builder; Gardner 4 max., clear yel. liq.; water sol.; sp.gr. 1.04; f.p. 0 C; surf. tens. 35 dynes/cm; partially biodeg.; 35% act.

Aerosol 200. [43] Disodium (ethoxylated alkylamide) sulfosuccinate; emulsifier, foaming agent, dispersant; Gardner 7 max. clear yel. liq.; mild odor; water sol.; dens. 8.93 lb/gal; sp.gr. 1.07; visc. 200 cps max.; cloud pt. < 0 C; surf. tens. 35 dynes/cm; biodeg.; 30% act.

Aerosol 413. [43] Disodium ethanol alkyl (C_{18}) amide sulfosuccinate; emulsifier, foaming agent, dispersant, wetting agent; Gardner 5–7 clear yel. liq.; water sol.; dens. 8.95 lb/gal; visc. 70–90 cps; surf. tens. 38 dynes/cm; biodeg.; 35% act.

Aerosol 501. [43] Disodium alkyl sulfosuccinate; anionic-nonionic; emulsifier, wetting agent, dispersant, foaming agent; APHA 60 clear liq.; water sol.; dens. 9.66 lb/gal; sp.gr. 1.16; visc. 260 cps; f.p. –3 C; surf. tens. 28 dynes/cm; 50% act.

Aerosol A-102. [43; 167] Disodium deceth-6 sulfosuccinate; anionic; emulsifier, solubilizer, foamer, and dispersant for emulsion polymerization of vinyl acetate and acrylic monomers; colorless to lt-yel. clear liq.; water sol.; dens. 9.01 lb/gal; sp.gr. 1.08; visc. 40 cps; f.p. –4 C; surf. tens. 29 dynes/cm; 30% act.

Aerosol A-103. [43; 167] Disodium nonoxynol-10 sulfosuccinate; anionic; emulsifier, wetting agent; emulsion polymerization of acrylic monomers; colorless to lt-yel. clear liq.; water sol.; dens. 9.1 lb/gal; sp.gr. 1.09; visc. 170–190 cps; f.p. –9 C; surf. tens. 34 dynes/cm; 35% act.

Aerosol A-196, A-196-40. [43; 167] Dicyclohexyl sodium sulfosuccinate; anionic; emulsifier, dispersant, stabilizer; for use in styrene butadiene emulsion polymerization; wh. pellets, clear to cloudy liq.; surf. tens. 39 dynes/cm; biodeg.; 85% act. pellets, 40% act. aq. sol'n. resp.

Aerosol A-268. [43; 167] Disodium isodecyl sulfosuccinate; anionic surfactant, sole emulsifier for PVC latexes-vinyl, vinylidene chloride, acrylics, surf. tens. depressant, solubilizer; clear liq.; water sol.; dens. 9.96 lb/gal; sp.gr. 1.19; visc. 150 cps; f.p. –5 C; surf. tens. 28 dynes/cm; 50% act.

Aerosol AY. [43] Sodium diamyl sulfosuccinate; anionic; emulsifier in emulsion polymerization; wetting agent; antigelling agent; used in electroplating, filtration, leaching ores, paint pigments, paper processing, soldering fluxes; APHA 100 max. hard, waxy solid; m.w. 360; sol. in polar org. solv.; sp.gr.

1.2; m.p. 222–227 C; acid no. 2.5 max.; surf. tens. 50.2 dynes/cm (0.01% aq.); 100% solids.

Aerosol AY-65. [43; 167] Sodium diamyl sulfosuccinate; anionic; wetting agent, dispersant, surfactant; sol. in water/ethanol.; dens. 9 lb/gal; sp.gr. 1.081; f.p. < 18 C; surf. tens. 29 dynes/cm; biodeg.; 65% act.

Aerosol AY-100. [43; 167] Sodium diamyl sulfosuccinate; anionic; wetting agent, dispersant, surfactant; sol. in water, org. solv.; dens. 10 lb/gal; sp.gr. 1.2; f.p. < 18 C; surf. tens. 29 dynes/cm; biodeg.; 100% act.

Aerosol C-61. [43] Ethoxylated alkyl guanidine-amine complex; cationic; wetting agent, dispersant, settling agent, textile softener, pigment dispersant, demulsifying agent, antistat for plastics, foaming agent, acid cleaner formulations; lt. tan creamy paste or semiliq.; dens. 8 lb/gal; sp.gr. 1.0; f.p. loses fluidity < 16 C; surf. tens. 34 dynes/cm; partially biodeg.; 70% act.

Aerosol GPG. [43; 167] Sodium dioctyl sulfosuccinate; anionic; wetting agent, surf. tens. depressant, emulsifier; clear, slightly visc. liq.; limited sol. in water, good org. solv.; visc. 200 cps; f.p. –40 C; surf. tens. 26 dynes/cm; biodeg.; 70% act. in water/alcohol.

Aerosol IB-45. [43; 167] Sodium diisobutyl sulfosuccinate; anionic; emulsifier, wetting and dispersing agent; emulsion polymerization of styrene, butadiene and copolymers; for leaching, electroplating; clear liq.; water sol.; extremely hydrophilic; f.p. 20–21 C; surf. tens. 49 dynes/cm; biodeg.; 45% act.

Aerosol MA-80. [43; 167] Sodium dihexyl sulfosuccinate; anionic; wetting agent, emulsifier, dispersant, solubilizing agent, surfactant; Gardner 50 max. clear, slightly visc. liq.; sol. in water, alcohol, org. solv.; dens. 9.4 lb/gal; sp.gr. 1.13; f.p. –28 C; m.p. 199–292 C; surf. tens. 28 dynes/cm; slowly biodeg.; 80% act.

Aerosol OS. [43; 167] Sodium isopropyl naphthalene sulfonate; anionic; emulsifier, wetting and dispersing agent; pigment dispersions, wettable powds., general cleaning formulations; wh. to lt. cream powd.; 50% sol. in water; dens. 11.93 lb/gal; sp.gr. 1.43; m.p. 205 C; surf. tens. 37 dynes/cm; slowly biodeg.; 75% act., 25% sodium sulfate.

Aerosol OT-70 PG. [43; 167] Sodium dioctyl sulfosuccinate; anionic; wetting agent, surf. tens. depressant, emulsifier, surfactant; for use where high flash required; clear to slightly yel. liq.; limited water sol., good org. solv.; sp.gr. 1.09; visc. 200–400 cps; surf. tens. 26 dynes/cm; biodeg.; 70% act.

Aerosol OT-75. [43; 167] Sodium dioctyl sulfosuccinate; anionic; wetting agent, surf. tens. depressant, emulsifier; for textile, rubber, petroleum, paper, metal, paint, plastics, agriculture; APHA 100 max., clear, slightly visc. liq.; m.w. 444; limited water sol., good org. solv.; sp.gr. 1.09; visc. 200 cps; flash pt. 85 C (OC); surf. tens. 28.7 dynes/cm (0.1% aq.); biodeg.; 75 ± 2% solids in water/alcohol.

Aerosol OT-100. [43; 167] Sodium dioctyl sulfosuccinate, anionic; see Aerosol OT-75; waxy solid, APHA 100 max. (50% solids); m.w. 444; sol. in polar and nonpolar solv.; sol. in oils, fats, waxes at 75 C; disp. in water; sp.gr. 1.1; m.p. 153–157 C; surf. tens. 28.7 dynes/cm (0.1% aq.); 100% conc.

Aerosol OT-B. [43; 167] Dioctyl sodium sulfosuccinate and sodium benzoate; anionic; wetting agent, pigment dispersant; wettable powds. incl. agric. chemicals; wh. powd.; m.w. 444; disp. in water; sp.gr. 1.1; m.p. < 300 C; surf. tens. 28.7 dynes/cm (0.1% aq.); 85% act.

Aerosol OT-S. [43; 167] Sodium dioctyl sulfosuccinate; anionic; wetting agent, surf. tens. depressant, emulsifier for plastics, organosols, lacquers, varnishes, all org. media, dry cleaning, corrosion resistant lubricants, agric. emulsions; lt. amber, transparent liq.; limited water sol., good org. solv.; sp.gr. 1.0; visc. 200–300 cps; surf. tens. 26 dynes/cm; biodeg.; 70% act.

Aerosol TR-70. [43; 167] Sodium bistridecyl sulfosuccinate; anionic; emulsifier, surfactant, rust preventative; emulsion and suspension polymerization; dispersant for pigments, dyes, polymers in hydrocarbon and other org. systems; clear sol'n.; high sol. in org. media, limited water sol.; dens. 8.3; sp.gr. 0.995; visc. 110 cps; f.p. –40 C; surf. tens. 26 dynes/cm; biodeg.; 70% act. in water/alcohol.

Aethoxal B. [291] Fatty alcohol polyalkylene glycol ether; nonionic; skin care agent used in personal care products; liq.; 99–100% conc.

AF 10 FG. [603] Silicone antifoam; nonionic; antifoam agent used for general food, poultry, and meat processing applications; wh. liq.; disp. in water; sp.gr. 1.00; dens. 8.3 lb/

gal; flash pt. > 212 F (PMCC); 10% act.

AF 10 IND. [603] Silicone antifoam; nonionic; antifoam for agric., cutting oils, drilling muds, effluent, inks, chemicals, detergents and textiles; wh. liq.; water-disp.; sp.gr. 1.00; dens. 8.3 lb/gal; flash pt. > 212 F (PMCC); pH 4–5 (1% aq.); 10% act.

AF 10 TF. [603] Silicone antifoam; nonionic; antifoam for drilling muds, effluent, cutting oils, agric. formulations; wh. liq.; water-disp.; sp.gr. 1.00; dens. 8.3 lb/gal; flash pt. > 212 F (PMCC); 10% act.

AF 30 FG. [603] Silicone antifoam; nonionic; antifoam for general food, meat, and poultry processing; wh. liq.; water-disp.; sp.gr. 1.01; dens. 8.4 lb/gal; flash pt. > 212 F (PMCC); pH 4–5 (1% aq.); 30% act.

AF 30 IND. [603] Silicone antifoam; nonionic; antifoam for effluent, agric. formulations, antifreeze, detergents; wh. liq.; water-disp.; sp.gr. 1.01; dens. 8.4 lb/gal; flash pt. > 212 F (PMCC); pH 4–5 (1% aq.); 30% act.

AF-60. [249] Dimethyl polysiloxane antifoam emulsion in water; defoaming agent used in adhesive, ink, latex, soap, starch, and paint mfg.; wh. fluid, slight absorbic acid odor; good sol. in water; dens. 8.4 lb/gal; sp.gr. 1.01; visc. 300 max cps; 30% act.

AF-66. [249] Filled polydimethylsiloxane; antifoam agent; grayish wh. fluid; odorless; sol. in aliphatic, aromatic, chlorinated solvs.; dens. 8.4 lb/gal; sp.gr. 1.01; visc. 500 cstks max.; flash pt. 600 F min. (OC); 100% act.

AF-72. [249] PEG-40 stearate, sorbitan stearate, and silica; nonionic; antifoam agent, surfactant; wh. fluid; slight sorbic acid odor; sol. in water; dens. 8.4 lb/gal; sp.gr. 1.01; visc. 1000 cps; 30% act.

AF-75. [249] PEG-40 stearate, sorbitan stearate, and silica; nonionic; antifoam agent, surfactant; wh. fluid; slight sorbic acid odor; sol. in water; dens. 8.4 lb/gal; sp.gr. 1.02; visc. 2500 cps; 10% act.

AF 100 FG. [603] Silicone antifoam; antifoam for food, edible oils, meat and poultry processing; translucent liq.; insol. in water; sp.gr. 1.01; dens. 8.4 lb/gal; pour pt. < 0 F; flash pt. > 600 F (PMCC); 100% conc.

AF 100 IND. [603] Silicone antifoam; antifoam for agric. formulations, drilling muds, vacuum distillations; translucent liq.; insol. in water; sp.gr. 1.01; dens. 8.4 lb/gal; pour pt. < 0 F; flash pt. > 600 F (PMCC); 100% conc.

AF 112. [603] Nonsilicone antifoam; used in solvs., adhesives, latex paints, inks; creamy yel. liq.; water-insol.; sp.gr. 0.92; dens. 7.7 lb/gal; pour pt. 40 F; flash pt. > 200 F (PMCC); 100% act.

AF 6035. [603] Silicone antifoam; used in delayed coker units; clear liq.; insol. in water; sp.gr. 0.84; dens. 7.0 lb/gal; pour pt. < 0 F; flash pt. 108 F (PMCC); 35% act.

AF CM Conc. [603] Silicone antifoam; nonionic; antifoam for carpet extraction machines; wh. visc. liq.; water-disp.; sp.gr. 0.97; dens. 8.1 lb/gal; flash pt. > 212 F (PMCC); pH 6–8 (1% aq.); 20% act.

AF GN-11-P. [603] Nonsilicone antifoam; used for fermentation, drilling muds, effluent, adhesives, gas treating; clear liq.; water-disp; sp.gr. 1.01; dens. 8.4 lb/gal; pour pt. –50 F; flash pt. > 200 F (PMCC); pH 5–7 (1% aq.); 100% act.

AF GN-46. [603] Nonsilicone antifoam; anionic; antifoam for solv. recovery, alcohol extraction; dk. liq.; oil-sol.; sp.gr. 0.90; dens. 7.5 lb/gal.; pour pt. 30 F; flash pt. > 250 F (PMCC); pH 4–6 (1% in 15% IPA/water); 100% act.

AF HL-21. [603] Nonsilicone antifoam; nonionic; for latex paints, adhesives, inks, solvs.; opaque liq.; water-insol.; sp.gr. 0.83; dens. 6.9 lb/gal; flash pt. (PMCC) 270 F; 100% act.

AF HL-23. [603] Nonsilicone antifoam; for cutting oil, boiler water additives, detergents, cleaning sol'ns, syn. lubricants; clear liq.; water-insol.; sp.gr. 1.00; dens. 8.3 lb/gal; pour pt. –20 F; flash pt. > 300 F (PMCC); pH 4–6 (1% in 15% IPA/water); 100% conc.

AF HL-26. [603] Nonsilicone antifoam; nonionic; antifoam for waste water treatment, electroplating, coatings, cleaning aids, dyes, inks; yel. clear liq.; sol. in oil; sp.gr. 1.00; dens. 8.3 lb/gal; pour pt. < 30 F; flash pt. > 220 F (PMCC); 50% act.

AF HL-27. [603] Nonsilicone antifoam; for urea and phenolic resins, latex paints, inks, adhesives; yel. liq.; water-insol.; sp.gr. 0.95; dens. 7.9 lb/gal; pour pt. 32 F; flash pt. > 300 F (PMCC); 100% act.

AF HL-28. [603] Nonsilicone antifoam; for cutting oils, effluents, detergents; clear liq.; water-insol.; sp.gr. 1.00; dens. 8.3 lb/gal; pour pt. –22 F; flash pt. > 400 F (PMCC); 100% act.

Agent 613-95. [575] Sodium alkyl sulfate; dispersant or flocculant in pesticide suspensions; pale yel. clear liq.; sp.gr. 1.073; dens.

8.9 lb/gal; pH 7.0–8.5; 38% act. in water.

Agent 721-47. [575] Formulated liq. dishwash conc.; yel. clear liq.; visc. 1000–2000 cps; pH 6.9–7.4; 47% min. solids.

Agent 768-82. [575] Aq. sodium alkylbenzene sulfonate; aux. surfactant for pesticide suspensions; lt. liq., slurry; dens. 8.7 lb/gal; pH 7.0–8.0; 40% act.

Agent 804. [575] Sodium alpha olefin sulfonate, sodium laureth sulfate, and lauramide DEA; base for personal care product formulations; Gardner 3+ liq.; disp. in water; dens. 8.9 lb/gal; visc. 750 cps; cloud pt. < 10 C; pH 6.5; 52% solids.

Agent X-601-52N. [575] Alkoxylated alkylphenol; wetting and dispersing agent for pesticides; pale-lt. br. solid; sol. in water; sp.gr. 1.036; dens. 8.64 lb/gal; m.p. 40–50 C; flash pt. > 200 F (Seta); pH 5–7.

Agent X-646-12. [575] Alkoxylated alkylphenol; wetting and dispersing agent for pesticide suspensions; straw liq.; sp.gr. 1.028; dens. 8.57 lb/gal; visc. 40 cps; pH 6.0; 30% conc. in water.

Agrilan A. [360] Blend of oil-sol. sulfonates and POE ethers; anionic/nonionic; emulsifier; br. visc. liq.; faint odor; turbid water-sol.; sp.gr. 1.062; visc. 2400 cs; flash pt. > 300 F (COC); pour pt. < 0 C; pH 4.5–6.5 (1% aq.); 100% act.

Agrilan AC. [360] Blend of oil-sol. sulfonates and POE ethers; anionic/nonionic; emulsifier; amber visc. liq.; alcoholic odor; water-sol.; sp.gr. 1.052; visc. 500 cs; flash pt. > 120 F (Abel CC); pour pt. 17 C; pH 4.5–6.5 (1% aq.); 90% act.

Agrilan AEC123. [360] Alkoxylate; nonionic; coemulsifier for agric. toxicants; liq.; 49% conc.

Agrilan AEC145. [360] Polyaromatic ethoxylate; nonionic; coemulsifier used in agrochemical toxicant-emulsion concs. and oil flowables, herbicides and pesticides; liq.; 100% conc.

Agrilan AEC156. [360] Fatty acid ester; nonionic; emulsifier for citrus and other agrochemical spray oil systems; liq.; HLB 9.0; 100% conc.

Agrilan AEC167, AEC178. [360] EO PO copolymer complex; nonionic; coemulsifier for toxicants; liq.; 100% conc.

Agrilan AEC189, AEC200. [360] EO PO copolymer complex; anionic/nonionic; blended emulsifier for production of pesticides; liq.; 83 and 85% conc. resp.

Agrilan BA. [360] Blend of oil-sol. sulfonates and POE ethers; anionic/nonionic; emulsifier; br. visc. liq.; alcoholic odor; turbid water-sol.; sp.gr. 1.034; visc. 3100 cs; flash pt. > 90 F (Abel CC); pour pt. < 0 C; pH 4.5–6.5 (1% aq.); 80% act.

Agrilan BM. [360] Blend of oil-sol. sulfonates and POE ethers; anionic/nonionic; emulsifier; amber liq.; char. odor; turbid water-sol.; sp.gr. 1.022; visc. 196 cs; flash pt. 78 F (Abel CC); pour pt. < 0 C; pH 5.0–6.5 (1% aq.); 70% act.

Agrilan C91D. [360] Phosphate ester; anionic; compatibility agent for agrochemical toxicants; liq.; 70% conc.

Agrilan D54. [360] Alkyl phenol ethoxylate condensate; nonionic; emulsifier for emulsifiable concs. of agrochemical toxicants; liq.; 90% conc.

Agrilan DG102. [360] Alkylaryl ether sulfate; anionic; wetting agents for production of agrochemical toxicants; liq.; 90% conc.

Agrilan DG113. [360] Alcohol alkoxylate; nonionic; wetting agent for production of agrochemical toxicants; liq.; HLB 13.0; 96% conc.

Agrilan EA Range. [360] Alcohol alkoxylate; nonionic; activity optimizers for agrochemical toxicant formulations; liq.; various conc.

Agrilan F460. [360] Alcohol alkoxylate; nonionic/anionic; adjuvants for production of agric. suspension conc. flowables; liq.; 60% conc.

Agrilan F491. [360] Complex phosphate ester; anionic; wetter/dispersant for agric. flowables; liq.

Agrilan F502. [360] Alkylene oxide copolymer; nonionic; wetting and dispersing agent for agric. flowable systems; wh. soft solid; 100% conc.

Agrilan F513. [360] Polyaromatic ethoxylate phosphate ester; anionic; wetting and dispersing agent for agric. toxicants; liq.; 100% conc.

Agrilan F524. [360] Complex sulfonate; anionic; adjuvant for production of suspension conc./flowables; liq.; 55% conc.

Agrilan F535. [360] Phosphate ester; anionic; adjuvant for production of suspension conc./flowables; liq.; 100% conc.

Agrilan TKA 103. [360] Quat. ammonium compd.; cationic; wetting and compatibility agent for quat. herbicides; liq.; 100% conc.

Agrilan TKA125. [360] Complex phosphate ester; anionic; compatibility agent for formulated pesticides; liq.; 70% conc.

Agrilan X98. [360] Calcium dodecylbenzene sulfonate; anionic; emulsifier for herbicide and pesticide formulations; liq.; 60% conc.

Agrilan X 109. [360] Calcium alkylbenzene sulfonate; anionic; coemulsifier for conc. of agrochemical toxicants; liq.

Agrimul 70-A. [187] Alkylaryl polyether alcohol; nonionic; agric. emulsifier for chlorinated hydrocarbons; amber liq.; dens. 8.8 lb/gal; visc. 8850 cps; pour pt. 0 C; 100% conc.

Agrimul A-300. [187] Aromatic sulfonate-oxide condensate blend; anionic/nonionic; agric. emulsifier for chlorinated and phosphated insecticide concs.; amber liq.; dens. 8.9 lb/gal; visc. 1050 cps; pour pt. 5 C; 100% conc.

Agrimul N-300. [187] Aromatic sulfonate-oxide condensate blend; anionic; see Agrimul A-300; liq.; 100% conc.

Agrimul S-300. [187] Aromatic sulfonate-oxide condensate blend; anionic; agric. emulsifier for phosphated toxicants; liq.; 100% conc.

Agrisol PX401, 413. [360] Aromatic alkoxylates; nonionic; cosolvs. for agrochemical toxicants; liq.; 100% conc.

Agriwet CA, C91, C92. [187] Phosphate ester; anionic; wetting agents, detergents, emulsifiers and lubricants; compatibility and resuspension aid for pesticides; coupling agent; liq.; 100% conc.

Agriwet FOA. [187] Phosphate ester; anionic; resuspension aid in pesticide formulations; liq.; 100% conc.

Ahco 759. [320] Sorbitan monolaurate; nonionic; emulsifier, solubilizer for textile use; liq.; HLB 8.6; 100% conc.

Ahco 832. [320] Sorbitan monooleate; nonionic; emulsifier, solubilizer for textile use; liq.; 100% conc.

Ahco 909. [320] Sorbitan monostearate; nonionic; emulsifier, solubilizer for textile use; solid; HLB 4.7; 100% conc.

Ahco 944. [320] Sorbitan monooleate; nonionic; emulsifier, solubilizer for textile use; liq.; 100% conc.

Ahco 3998. [320] POE (20) oleyl ether; nonionic; dye leveling agent and emulsifier; liq.; 100% conc.

Ahco 7166T. [320] POE (20) sorbitan tristearate; nonionic; dye leveling agent and emulsifier; solid; 100% conc.

Ahco 7596D. [320] POE (4) sorbitan monolaurate; nonionic; dye leveling agent and emulsifier; liq.; 100% conc.

Ahco 7596T. [320] POE (20) sorbitan monolaurate; nonionic; dye leveling agent and emulsifier; liq.; 100% conc.

Ahco A-117. [320] Alkylaryl sulfonate; anionic; emulsifier, surfactant, used as dispersant, detergent, wetting agent; amber liq.; faint odor; sol. IPA; sp.gr. 1.0; visc. 7000 cps; 100% act.

Ahco AB-100. [320] Anionic/nonionic blend; emulsifier; amber liq.; solv. odor; sol. IPA; sp.gr. 1.04; visc. 1900 cs; flash pt. 64 F; pour pt. –15 F; flam.

Ahco AB-118. [320] Anionic emulsifier; amber liq.; solv. odor; water sol.; sp.gr. 1.00; visc. 300 cs; flash pt. 100 F; pour pt. 20 F.

Ahco AB-120. [320] Anionic/nonionic emulsifier; amber liq.; solv. odor; sol. IPA; sp.gr. 1.00; visc. 600 cs; flash pt. 80 F; pour pt. 40 F.

Ahco AB-135. [320] Anionic/nonionic emulsifier; amber liq.; solv. odor; sol. IPA; sp.gr. 1.00; visc. 100 cs; flash pt. 90 F; pour pt. 25 F.

Ahco AB-146. [320] Anionic emulsifier; amber liq.; solv. odor; sol. IPA; sp.gr. 1.00; visc. 1000 cs.

Ahco AB-160. [320] Anionic emulsifier; amber liq.; solv. odor; sol. IPA; sp.gr. 1.00; visc. 750 cs.

Ahco AB-228. [320] Anionic emulsifier; amber liq.; solv. odor; water sol.; sp.gr. 1.00; visc. 2600 cs; 100% act.

Ahco AJ-110. [320] Sulfated castor oil; anionic; wetting agent, lubricant, penetrant, dispersant, emulsifier, surfactant; dyeing assistant for textiles; amber liq.; castor oil odor; water sol.; sp.gr. 1.05; HLB 11.0; 50% act. in water.

Ahco AY 2200. [320] Phosphate ester; anionic; low-soiling textile antistat, lubricant; liq.; 45% conc.

Ahco C330. [320] Quat. ammonium compd.; cationic; heat-stable textile fiber antistat; liq.; 100% conc.

Ahco DFO-100. [320] POE (5) sorbitan monooleate; nonionic; emulsifier, solubilizer for textile use; liq.; HLB 10.0; 100% conc.

Ahco DFO-110. [320] POE (20) sorbitan trioleate; nonionic; emulsifier, solubilizer for textile use; liq.; HLB 11.0; 100% conc.

Ahco DFO-150. [320] POE (20) sorbitan monooleate; nonionic; emulsifier, solubilizer for textile use; liq.; HLB 15.0; 100% conc.

Ahco DFP-156. [320] POE (20) sorbitan monopalmitate; nonionic; emulsifier, solubilizer for textile use; liq.; 100% conc.

Ahco DFS-96. [320] POE (4) sorbitan monostearate; nonionic; emulsifier, solubilizer for textile use; solid; HLB 9.6; 100% conc.

Ahco DFS-149. [320] POE (20) sorbitan monostearate; nonionic; emulsifier, solubilizer for textile use; solid; HLB 14.9; 100% conc.

Ahco DHS-111. [320] POE stearate; nonionic; textile lubricant and all-purpose emulsifier; solid; HLB 11.1; 100% conc.

Ahco EO-102. [320] POE sorbitol polyoleate; nonionic; emulsifier, solubilizer for textile use; liq.; HLB 10.2; 100% conc.

Ahco EO-114. [320] POE sorbital polyoleate; nonionic; emulsifier, solubilizer for textile use; liq.; HLB 11.4; 100% conc.

Ahco FO-18. [320] Sorbitan trioleate; nonionic; emulsifier, solubilizer for textile use; liq.; 100% conc.

Ahco FP-67. [320] Sorbitan monopalmitate; nonionic; emulsifier, solubilizer for textile use; solid; 100% conc.

Ahco FS-21. [320] Sorbitan tristearate; nonionic; emulsifier, solubilizer for textile use; solid; HLB 2.1; 100% conc.

Ahcovel Base N-15. [320] Glycerol monostearate, emulsifiable; nonionic; textile softener; solid; 100% conc.

Ahcowet DQ-114. [320] Trideceth-6; nonionic; wetting agent, dispersant, detergent in textiles; liq.; HLB 11.4; 100% conc.

Ahcowet DQ-145. [320] Trideceth-12; nonionic; scour and dye leveling agent in textiles; liq.; HLB 14.5; 100% conc.

Ahcowet RS. [320] Fatty acid ester, sulfated; anionic; wetting agent for textile processing; liq.; 65% conc.

Airrol CT-1. [606] Dialkyl sulfosuccinate; anionic; wetting agent and dyeing assistant; liq.; 70% conc.

Ajicoat SPG. [9] Sodium-poly-L-glutamate; anionic; surface modifier, coemulsifier, codispersant; solid; 100% conc.

Aktaflo-E. [447] Oxyethylated alkyl phenols; nonionic; emulsifier, oil well drilling mud; liq; 100% conc.

Aktaflo-S. [447] Oxyethylated phenols, nonionic; surfactant, oil well drilling mud; liq.; 100% conc.

Akypo AD 100 SPC. [131] Special alkyl ether carboxylate; anionic; surfactant for mild cosmetic products; solid; 100% conc.

Akypo KA 150 CNV, KA 450 CNV. [131] Special alkyl ether carboxylate; anionic; surfactant for mild cosmetic products; liq.; 30% conc.

Akypo MB 1585, 1614-1, 1614-2. [131] Fatty alcohol polyglycol ether carboxylic acid; anionic; surfactant for industrial cleaning; liq.; HLB 15.6, 11.1, and 12.5 resp.; 92, 90, and 93% conc. resp.

Akypo MB 2528-S, 2621-S, 2705-S. [131] Fatty alcohol polyglycol ether carboxylic acid; anionic; surfactant for industrial cleaning; liq.; HLB 12.2 (2528-S); 10.2 (2621-S, 2705-S); 93% conc., 86% conc. (2705-S)

Akypo MB 2717-S. [131] Fatty alcohol polyglycol ether carboxylic acid; anionic; surfactant for industrial cleaning; liq.; HLB 14.2; 83% conc.

Akypo NP 70. [131] Fatty alcohol polyglycol ether carboxylic acid; anionic; emulsifier; liq.; 70% conc.

Akypo NTS. [131] Fatty alcohol polyglycol ether carboxylic acid, sodium salt; anionic; detergent for carpet cleaners esp. aerosols; liq.; 22% conc.

Akypo OP 80, 115, 190. [131] Alkylphenol polyglycol ether carboxylic acids; detergent, emulsifier; used in aq. sol'ns.

Akypo RLM 25. [131] Fatty alcohol polyglycol ether carboxylic acid; nonionic; emulsifier, dispersant for emulsion and dispersion use; liq.; 92% conc.

Akypo RLM 38 NV. [131] Fatty alcohol polyglycol ether carboxylic acid, sodium salt; anionic; detergent for shampoo, liq. soap and foam bath; emulsifier and wetting agent; liq.; 22% conc.

Akypo RLM 45. [20; 131] Laureth-5 carboxylic acid; nonionic; emulsifier, dispersant for emulsion and detergent use; liq.; HLB 11.0; 90% conc.

Akypo RLM 45 N. [131] Fatty alcohol polyglycol ether carboxylic acid, sodium salt; anionic; detergent for shampoos and bubble baths; paste; 83–84% conc.

Akypo RLM 45 NV. [20; 131] Fatty alcohol polyglycol ether carboxylate, sodium salt; anionic; surfactant and additive for personal care use; liq.; HLB 28.0; 22% conc.

Akypo RLM 100. [20; 131] Laureth-10 carboxylic acid; nonionic; see Akypo RLM-Q38; liq.; HLB 14.8; 90% conc.

Akypo RLM 100 MGV. [131] Fatty alcohol polyglycol ether carboxylic acid, magnesium salt; anionic; detergent for shampoos, liq. soap and foam bath; emulsifier, wetting agent; liq.; 22% conc.

Akypo RLM 100 NV. [20; 131] Fatty alcohol polyglycol ether carboxylate, sodium salt; anionic; surfactant and additive for personal care products; liq.; HLB 31.8; 22% conc.

Akypo RLM 130. [20; 131] Fatty alcohol polyglycol ether carboxylic acid; nonionic; see Akypo RLM-Q38; liq.; HLB 15.9; 90% conc.

Akypo RLM 130 NV. [20; 131] Fatty alcohol polyglycol ether carboxylate, Na salt; anionic; surfactant and additive for personal care use; liq.; HLB 32.9; 22% conc.

Akypo RLM 160. [131] Fatty alcohol polyglycol ether carboxylic acids; anionic; emulsifier.

Akypo RLM 160 NV. [20; 131] Fatty alcohol polyglycol ether carboxylate, sodium salt; anionic; detergent for shampoo, liq. soap, and foam bath; emulsifier and wetting agent; liq.; HLB 33.6; 22% conc.

Akypo RLM Q38. [20; 131] Fatty alcohol polyglycol ether carboxylic acid; nonionic; emulsifier, dispersant, additive for personal care products, household and industrial formulas; primary emulsifier for syn. latex; liq.; HLB 10.2; 90% conc.

Akypo RLML 160. [20] Fatty alcohol polyglycol ether carboxylic acid; nonionic; see Akypo RLM-Q38; liq.; HLB 16.6; 90% conc.

Akypo RLML Q38NV. [20] Fatty alcohol polyglycol ether carboxylate, sodium salt; anionic; surfactant and additive for personal care use; liq.; HLB 27.2; 22% conc.

Akypo RO 50. [131] Fatty alcohol polyglycol ether carboxylic acids; anionic; emulsifier for cleaning agents and metal cooling liqs.; liq.; 90% conc.

Akypo RO 90. [131] Fatty alcohol polyglycol ether carboxylic acids; emulsifier.

Akypogene KTS. [131] Fatty alcohol polyglycol ether carboxylic acid and polymer salt blend; anionic; base for rug shampoo and upholstery cleaner with antistatic and anticorrosive properties; liq.; 31% conc.

Akypomine AT. [131] Alkylphenol polyglycol ether carboxylic acid; anionic; mineral flotation; liq.; 90% conc.

Akypomine BC 50, 100. [131] Fatty alcohol ether sulfate; anionic; flotation of minerals; paste; 50 and 95% conc. resp.

Akyporox NP 15. [131] Alkylphenol polyglycol ether; emulsifier; HLB 7.9; 100% conc.

Akyporox NP 40. [131] Alkylphenol polyglycol ether; nonionic; emulsifier, wetting agent used in textile products; liq.; 100% conc.

Akyporox NP 90. [131] Alkylphenol polyglycol ether; nonionic; see Akyporox NP 40; liq.; HLB 12.9; 100% conc.

Akyporox NP 105. [131] Alkylphenol polyglycol ether; nonionic; see Akyporox NP 40; liq.; HLB 13.5; 100% conc.

Akyporox NP 200. [131] Alkylphenol polyglycol ether; nonionic; see Akyporox NP 40; solid.; HLB 20.0; 100% conc.

Akyporox NP 300V, NP 1200V. [131] Alkylphenol polyglycol ether; emulsifier.

Akyporox OP 100. [131] Alkylphenol polyglycol ether; nonionic; emulsifier; 100% conc.

Akyporox OP 115. [131] Alkylphenol polyglycol ether; nonionic; emulsifier, wetting agent; liq.; 70% conc.

Akyporox OP 250. [131] Alkylphenol polyglycol ether; emulsifier; solid; 100% conc.

Akyporox OP 400V. [131] Alkylphenol polyglycol ether; emulsifier; liq.; 70% conc.

Akyporox RLM 22. [131] Fatty alcohol polyglycol ether; nonionic; liq.; emulsifier, oil-sol.; 100% conc.

Akyporox RLM 40. [131] Fatty alcohol polyglycol ether; emulsifier; liq.

Akyporox RLM 80. [131] Fatty alcohol polyglycol ether; emulsifier.

Akyporox RO 90. [131] Fatty alcohol polyglycol ether; nonionic; emulsifier, wetting agent for textile products; liq.; HLB 12.1; 100% conc.

Akyposal 23 ST70. [131] Sodium laureth sulfate; anionic; detergent; shampoo base; paste; water-sol.; 70% conc.

Akyposal 40 DAL. [131] Lauryl ether sulfate, amine salt; anionic; surfactant for cosmetic products; liq.; 40% conc.

Akyposal 100 DAL. [131] Lauryl ether sulfate; anionic; emulsifier for bath oils; liq.; 100% conc.

Akyposal ALS 33. [131] Ammonium lauryl sulfate; anionic; detergent, emulsifier; shampoo base; used in emulsion polymerization; liq.; 33% conc.

Akyposal BA 28. [131] Fatty alcohol ether sulfate; anionic; detergent for personal care, dishwashing, and textile products; liq.; 28% conc.

Akyposal BA 56. [131] Fatty alcohol ether sulfate; anionic; detergent for personal care, dishwashing, and textile products; paste; 56% conc.

Akyposal BA 70. [131] Fatty alcohol ether sulfate; anionic; detergent for personal care, dishwashing, and textile products; paste; 70% conc.

Akyposal DS 28. [131] Sodium lauryl ether sulfate; anionic; detergent; shampoo base; liq.; 28% conc.

Akyposal DS 56. [131] Sodium pareth-23 sulfate; anionic; detergent; shampoo base; liq.; water-sol.; 56% conc.

Akyposal EO 20. [131] Sodium laureth sulfate; anionic; base for shampoos and bubble baths; liq.; 28% conc.

Akyposal MGLS. [131] Fatty alcohol ether sulfate, magnesium salt; anionic; detergent for shampoos, foam bath; liq.; 28% conc.

Akyposal MLES 35. [131] MEA laureth sulfate; anionic; used in personal care products; liq.; 35% conc.

Akyposal NLS. [131] Sodium lauryl sulfate; anionic; detergent; liq.; 31% conc.

Akyposal NPS 100, NPS 250. [131] Alkyl phenol polyglycol ether sulfate; anionic; emulsifier for polymerization; liq.; 34% conc.

Akyposal RLM 56 S. [131] Fatty alcohol ether sulfate; anionic; detergent for personal care products; liq.; 56% conc.

Akyposal RLM 70. [131] Fatty alcohol ether sulfate; anionic; detergent; shampoo base; liq.; water-sol.; 70% conc.

Akyposal TLS 42. [131] TEA lauryl sulfate; anionic; detergent; base for personal care products and car shampoos; liq.; 42% conc.

Alacsan 7LUF. [20] Modified alkyl dimethyl benzyl ammonium chloride; cationic; detergent; Gardner 5 max. visc. liq.; 50% act.

Alacsan QNA. [20] Cationic; surfactant; used for preparation of emulsion polymers; liq.; 60% act.

Albalan. [639] De-oiled lanolin; nonionic; emollient, emulsifier; forms stable w/o emulsions; soft wax; HLB 5.0; 100% conc.

Alcamate FA 82. [21] Disodium N-alkyl sulfosuccinamate; anionic; foam booster and stabilizer for carpet backing; paste; 35% conc.

Alcojet. [22] Sodium metasilicate, sodium carbonate, POE ester of mixed fatty and resin acids; nonionic; detergent used in mechanical washers for stain removal; powd.; water-sol.; biodeg.

Alcolan. [37] Petrolatum, sorbitan sesquiolate, and lanolin alcohol; nonionic; w/o emulsifier, emollient base for cosmetic and pharmaceutical emulsions; paste; 100% conc.

Alcolan 36W. [37] Lanolin, petrolatum, and sorbitan sesquioleate; nonionic; see Alcolan; paste; 100% conc.

Alcolan 40. [37] Petrolatum, lanolin, beeswax, sorbitan sesquiolate, and polysorbate 81; see Alcolan; paste; 100% conc.

Alcolec Series. [45] Phosphatides; nonionic; wetting agent, emulsifier, release agent; amber liq., powd.; sol. in most fat solv. except acetone; sp.gr. 1.03; visc. 10,000 cps; 60–100% act.

Alcolec 439-C. [45] Lecithin; nonionic; wetting agent, emulsifier; for water-based paints; amber liq.; bland odor; sol. in most fat solv. except acetone; water-disp.; sp.gr. 1.03; visc. 7000 cps; 100% act.

Alcolec 495, 619-B, 628G. [45] Lecithin; nonionic; food emulsifier for aq. and oil-base systems; liq.; 100% conc.

Alcolec 621. [45] Lecithin; nonionic; emulsifier, more hydrophilic; liq.

Alcolec 634. [45] Cocamidopropylamine oxide and lecithin; emulsifier; liq.; water-sol.

Alcolec 638. [45] Sodium lauryl sulfate and lecithin; anionic; emulsifier; liq.

Alcolec 650. [45] Lecithin; nonionic; food emulsifier for aq. and oil-base systems; liq.; 100% conc.

Alcolec 658. [45] Lecithin; nonionic; emulsifier; liq.; sol. in alcohol; liq.; 100% conc.

Alcolec 662. [45] Lecithin; nonionic; food emulsifier for aq. and oil-base systems; liq.; 100% conc.

Alcolec 4135. [45] Lecithin, polysorbate 20, sorbitan laurate, propylene glycol stearate, and propylene glycol laurate; nonionic; emulsifier and conditioner for cosmetics; liq.; 100% conc.

Alcolec DS. [45] Lecithin; nonionic; emulsifier for shampoos, cosmetics, and pharmaceuticals; paste liq.

Alcolec Extra A. [45] Lecithin; nonionic; food emulsifier for aq. and oil-base systems; higher phosphatide content; liq.; 100% conc.

Alcolec Granules. [45] Lecithin; nonionic; emulsifier, stabilizer, diet supplement, in cosmetics, pharmaceuticals; amber beads; bland odor; sp.gr. 0.5; 95% act.

Alcolec HS-2. [45] Phosphatide; nonionic; emulsifier for shampoos, cosmetics, and pharmaceuticals; paste liq.

Alcolec HS-3. [45] Sulfated castor oil, lecithin, propylene glycol, and PEG-12 oleate; anionic; emulsifier, stabilizer for cosmetics, pharmaceuticals; powd.; 97% conc.

Alcolec MS-4, MS-10. [45] Phosphatides; nonionic; emulsifier for shampoos, cosmetics, and pharmaceuticals; paste liq.

Alcolec Powd. [45] Lecithin; nonionic; emulsifier, stabilizer, wetting agent, diet supplement, in cosmetics, pharmaceuticals, foods; powd.; sp.gr. 0.6; 100% act.

Alcolec RCX-1, S. [45] Lecithin; nonionic; food emulsifier, wetting and dispersing agent; waxy solids, liq.

Alcolec Z-3. [45] Hydroxylated lecithin; nonionic; wetting agent, emulsifier for personal care products and pharmaceuticals; liq.; sol. in most fat solv. except acetone; visc. 8000 cps.

Alconate 2CH. [20] Dicyclohexyl sodium sulfosuccinate; anionic; emulsifier in styrene butadiene emulsion polymerization; paste; 43% act.

Alconate 2CSA. [20] Tetrasodium dicarboxyethyl stearyl sulfosuccinamate; anionic; surfactant in caustic industrial and household cleaners; emulsifier for emulsion polymers, pesticides; liq.; sol. in calcium, alkaline, and electrolyte sol'ns; 35% conc.

Alconate 80. [650] Emulsifier, wetting agent; dk. br. visc. liq.; sp.gr. 1.1 (60 F); 77% act.

Alconate CPA. [20] Disodium cocamido MIPA sulfosuccinate; anionic; surfactant for personal care products, rug and upholstery shampoos; Gardner 3 clear liq.; dens. 8.9 lb/gal; cloud pt. -5 C; pH 6.5 (5%); 40% solids.

Alconate D-6. [20] Disodium deceth-6 sulfosuccinate; anionic; solubilizer, dispersant, emulsifier used in emulsion polymers; wetting agent for dry powds.; liq.; 31% act.

Alconate L3. [20] Disodium lauryl sulfosuccinate; anionic/nonionic; emulsifier, foaming agent used in personal care products, household detergents, industrial cleaners, and emulsion polymerization; Gardner 3 clear liq.; dens. 9.1 lb/gal; cloud pt. < -5 C; pH 6.5 (5%); 40% solids.

Alconate LEA. [20] Disodium lauramido MEA sulfosuccinate; anionic; detergent, dispersant for rug, upholstery, and fabric cleaners; Gardner 3 opaque liq.; dens. 9.0 lb/gal; cloud pt. 20 C; pH 6.5 (5%); 40% solids.

Alconox. [22] Blend of alkylaryl sulfonates, lauryl alcohol sulfates, phosphates, carbonates; anionic; detergent with wetting, sequestering and synergistic agents; for manual cleaning of laboratory and hospital glassware and instruments; colorless powd.; odorless; sol. in water; biodeg.

Alcopol AH New. [28] Monoethanolamine salt of fatty ester; dispersant esp. useful in nonaq. systems, dry cleaning aid; liq.; 99% conc.

Alcopol FA. [28] Disodium N-octadecyl sulfosuccinamate; anionic; foaming agent for latices, emulsifying and solubilizing agent for soaps; liq.; 35% conc.

Alcopol FL. [28] Tetrasodium N(1,2 dicarboxyethyl)-N octadecyl sulfosuccinamate; anionic; foaming and emulsifying agent; liq.; 35% conc.

Alcopol O 60%. [28] Sodium dioctyl sulfosuccinate; anionic; used in vinyl acetate polymerization; wax; 60% conc.

Alcopol O 70 PG. [28] Sodium dioctyl sulfosuccinate; anionic; dispersant, wetting agent, emulsifying agent; liq.; 70% conc.

Alcopol OB. [28] Sodium diisobutyl sulfosuccinate; anionic; used in vinyl acetate polymerization; liq.; 45% conc.

Alcopol OD. [28] Sodium ditridecyl sulfosuccinate; anionic; dispersant, wetting agent, emulsifier; liq.; 60% conc.

Alcopol OS. [28] Sodium dihexyl sulfosuccinate; anionic; used in vinyl acetate polymerization; liq.; 60% conc.

Alcopol T. [28] Fatty acid ester, sodium salt; anionic; dispersant, wetting and emulsifying agent; liq.; 43% conc.

Alcotab. [22] Blend of alkylaryl sulfonates, lauryl alcohol sulfates, phosphates, carbonates; anionic; detergent with wetting, sequestering and synergistic agents; tablet; biodeg.

Aldo DC. [256] Propylene glycol dicaprylate/dicaprate; coupling agent for mixed solv. systems; emulsifier for food, cosmetic, industrial use; Gardner 2 max. liq.; sol. in ethanol, min. and veg. oil; insol. in water; sp.gr. 0.92; HLB 2 ± 1; iodine no. 0.5 max.; sapon. no. 315–335.

Aldo GLP FG. [256] Glyceryl lactopalmitate; nonionic; food emulsifier; bead; HLB 2.5; 100% conc.

Aldo HMO. [256] Glyceryl monooleate, high mono.; yel. semiliq.; m.p. < 20 C; HLB 3.0; sapon. no. 145–155.

Aldo HMO FG. [256; 372] Glyceryl mono and dioleate (high mono); nonionic; food emulsifier; liq.; HLB 3.0; 100% conc.

Aldo HMS. [256] Glyceryl stearate; nonionic; emulsifier for w/o systems; stabilizer, consistency builder, emollient in personal care products; wh. beads; sol. in ethanol, min. and veg. oil; disp. in water; HLB 2.8; m.p. 61–68 C; sapon. no. 165–175; 52–56% mono content.

Aldo HMS FG. [256] Glyceryl mono- and distearate (high mono); nonionic; food emulsifier; bead; HLB 3.0; 100% conc.

Aldo LS. [256] Glyceryl lactostearate; wh. solid; m.p. 47–54 C; HLB 9.0; sapon. no. 245–260.

Aldo MC. [256; 372] Glyceryl cocoate; nonionic; emulsifier; cream-colored, soft solid; sol. in ethanol, ethyl acetate, toluol, naphtha, min. and veg. oils, disp. in water; sp.gr. 0.97; m.p. 20–26 C; HLB 6.8; sapon. 180–190; pH 7.5–8.5 (5% aq); 100% conc.

Aldo ML. [372] Glyceryl monodilaurate; nonionic; see Aldo DC; liq.; HLB 5.2; 100% conc.

Aldo MLD. [256] Glyceryl laurate, dispersible; nonionic; emulsifier; see Aldo DC; cream soft solid; sol. in ethanol, ethyl acetate, toluol, naphtha, min. and veg. oils, disp. in water; sp.gr. 0.97; m.p. 21–26 C; HLB 6.8; sapon. no. 185–195; pH 7.5–8.5 (5% aq.); 100% conc.

Aldo MLD FG. [256] Glyceryl monodilaurate; nonionic; food emulsifier; liq.; HLB 6.8; 100% conc.

Aldo MO. [256] Glyceryl oleate; nonionic; emulsifier; yel., soft solid; sol. in ethyl acetate; sp.gr. 0.95; m.p. < 25 C; HLB 3.4; sapon. no. 170–180; pH 4.5–6.5; 100% conc.

Aldo MO FG. [256] Glyceryl mono- and dioleate; nonionic; liq. emulsifier and antifoam for foods; emulsifier, solubilizer for cosmetic, pharmaceutical, and industrial applications; liq.; HLB 3.0; 100% conc.

Aldo MO Tech. [256] Glyceryl oleate, tech. grade; nonionic; emulsifier; amber soft liq.; sol. in ethyl acetate; sp.gr. 0.95; m.p. < 25 C; HLB 3.4; sapon. no. 170–180; pH 4.5–6.5; 100% conc.

Aldo MOD. [256] Glyceryl oleate SE, avail. in animal, veg., food, and Kosher grades; nonionic; emulsifier; yel. liq.; sol. in toluol, naphtha, min. and veg. oils, disp. in water; sp.gr. 0.95; m.p. < 0 C; HLB 5.0; sapon. no. 141–147; pH 8.5–9.5 (5% aq.); 100% conc.

Aldo MOD FG. [256] Glyceryl monodioleate, dispersible; nonionic; liq. emulsifier and antifoam for foods; emulsifier, solubilizer for cosmetic, pharmaceutical, and industrial applications; liq.; HLB 4.0; 100% conc.

Aldo MR. [256] Glyceryl ricinoleate, dispersible; nonionic; emulsifier; yel. liq.; sol. in methanol, ethanol, ethyl acetate, toluol, veg. oil, disp. in water; sp.gr. 1.02; m.p. < –8 C; HLB 5.6; sapon. no. 66–69; pH 8.3–9.3; 100% conc.

Aldo MS. [256] Glyceryl stearate; nonionic; emulsifier; wh. beads; sol. hot in methanol, ethanol, toluol, naphtha, min. and veg. oils, disp. in hot water; sp.gr. 0.97; m.p. 57–61 C; HLB 3.6; sapon. no. 158–165; pH 7.6–8.6 (3% aq.); 100% conc.

Aldo MS-20 FG. [372] POE (20) glycerol monostearate; nonionic; emulsifier, antifoam for foods; emulsifier, solubilizer for cosmetic, pharmaceutical, and industrial applications; solid; HLB 13.0; 100% conc.

Aldo MS FG. [372] Glyceryl monodistearate; nonionic; emulsifier for general use in foods; bead, flakes; HLB 3.0; 100% conc.

Aldo MS Industrial. [256] Glyceryl stearate, industrial grade; nonionic; emulsifier; tan flakes; sol. hot in methanol, ethanol, toluol, naphtha, min. and veg. oils, disp. in hot water; sp.gr. 0.97; m.p. 55–62 C; HLB 3.6; sapon. no. 157–177; pH 6.5–8.5 (5% aq.) 100% conc.

Aldo MS LG. [256] Glyceryl stearate; nonionic; see Aldo HMS; wh. beads; sol. in ethanol, min. and veg. oil; disp. in water; HLB 3.3; m.p. 58–62 C; sapon. no. 160–170; 40–45% mono content.

Aldo MS LG FG. [372] Glyceryl monodistearate; nonionic; emulsifier for general use in foods; bead, flakes; HLB 3.0; 100% conc.

Aldo MSA. [256] Glyceryl stearate, PEG-100 stearate; nonionic; emulsifier for o/w personal care products; wh. beads; sol. in ethanol, min. and veg. oil; disp. in water; HLB 11 ± 1; m.p. 56–60 C; sapon. no. 90–100; 17.5% min. mono content.

Aldo MSC. [256] Glyceryl stearate, cosmetic; nonionic; emulsifier; wh. flakes; sol. hot in methanol, ethanol, toluol, naphtha, min. and veg. oils, disp. in hot water; sp.gr. 0.97; m.p. 55–58 C; HLB 3.5; sapon. no. 168–176; pH 4.0–7.0 (5% aq.) 100% conc.

Aldo MSD. [256] Glyceryl stearate SE; nonionic; emulsifier; wh. beads; sol. hot in methanol, ethanol, toluol, naphtha, min. and veg. oils, disp. in hot water; sp.gr. 0.97; m.p. 56–60 C; HLB 5.8; sapon. no. 140–150; pH 9.2–10.2 (3% aq.) 100% conc.

Aldo MSD FG. [372] Glyceryl monodistearate; nonionic; emulsifier for general use in foods; bead; HLB 4.0; 100% conc.

Aldo PGHMS. [372] Propylene glycol stearate;

nonionic; aerating, whipping agent in toppings, cake mixes, shortenings; bead; HLB 3.0; 100% conc.

Aldo PME. [256] Propylene glycol monostearate/glyceryl monostearate; nonionic; emulsifier in cakes, shortenings, toppings; solid; HLB 3.0; 100% conc.

Aldo PMS. [256] Propylene glycol stearate; nonionic; emulsifier, stabilizer, solubilizer, emollient, lubricant, thickener, plasticizing agent, antiirritant, and conditioner in pharmaceutical, cosmetic, and toiletry preparations; foam modifier for detergent systems; wh. flakes; sol. in ethanol, min. and veg. oils; insol. in water; HLB 3 ± 1; iodine no. 1.0 max.; sapon. no. 170–185.

Aldo TO. [256] Glyceryl trioleate; nonionic; yel. liq.; m.p. < 0 C; HLB 0.8; sapon. no. 188–195; 100% conc.

Aldosperse 40/60. [256] 40% POE (20) glyceryl stearate, 60% glyceryl stearate; ice cream emulsifier, dough conditioner for food industry; beads, Gardner 2; HLB 7 ± 1; m.p. 57.

Aldosperse 40/60 FG. [256; 372] 40% POE (20) glyceryl stearate, 60% glyceryl stearate (high mono); nonionic; bakery and food emulsifier; bead; HLB 7.0; 100% conc.

Aldosperse 712 FG. [256] Monodiglycerides and ethoxylated monodiglycerides blend; nonionic; food emulsifier; bead; 100% conc.

Aldosperse AA 40/60. [256] 40% PEG-20 glyceryl stearate, 60% glyceryl stearate; wh. beads; m.p. 40–60 C; HLB 8.0; sapon. no. 120–130.

Aldosperse ML-23. [256] PEG-23 glyceryl laurate; nonionic; emulsifier, solubilizer, suspending and dispersing agent used in personal care products; straw liq.; sol. in ethanol, veg. oil; sp.gr. 1.09; HLB 17 ± 1; sapon. no. 42–50.

Aldosperse ML 230. [256] PEG-23 glyceryl laurate; conditioner for textiles; liq.; HLB 17 ± 1.

Aldosperse MS-20. [256] PEG-20 glyceryl stearate; nonionic; emulsifier, solubilizer, suspending and dispersing agent used in personal care products, food industry, textiles; wh. soft solid; sol. in ethanol; disp. in water; HLB 13 ± 1; m.p. 32–38 C; sapon. no. 65–75.

Aldosperse MS-20 FG. [256; 372] PEG-20 glyceryl stearate; nonionic; bakery and food emulsifier; solid; HLB 13.0; 100% conc.

Aldosperse O-20. [256] 80% glyceryl stearate, 20% polysorbate 80; ice cream emulsifier, dough conditioner for foods; straw beads; m.p. 55–65 C; HLB 5.2; sapon. no. 140–150.

Aldosperse O-20 FG. [256] 80% glyceryl stearate, 20% polysorbate 80; nonionic; emulsifier, solubilizer for frozen desserts, cosmetic, pharmaceutical, and industrial applications; bead; HLB 5.0; 100% conc.

Aldosperse TS-20. [256] 80% glyceryl stearate, 20% polysorbate 65; ice cream emulsifier, dough conditioner for food industry; straw beads; m.p. 55–65 C; HLB 4.3; sapon. no. 150–160.

Aldosperse TS-20 FG. [256] 80% glyceryl stearate, 20% polysorbate 65; nonionic; frozen dessert emulsifier; bead; HLB 4.0; 100% conc.

Aldosperse TS-40. [256] 60% glyceryl stearate, 40% polysorbate 65; ice cream emulsifier and dough conditioner for food industry; straw beads; m.p. 40–60 C; HLB 8.0; sapon. no. 120–130.

Aldosperse TS-40 FG. [256] 60% glyceryl stearate, 40% polysorbate 65; nonionic; frozen dessert emulsifier; bead; HLB 6.0; 100% conc.

Alfol 4. [152; 634] *n*-Butyl alcohol; intermediate for amines, plasticizers, fatty acid esters, detergents; colorless liq.; char. penetrating odor; sol. in water; dens. 6.76; sp.gr. 0.809; visc. 3.4; m.p. –128 C; b.p. 117.7; 99.3% act.

Alfol 6. [152; 634] Hexyl alcohol; detergent intermediate; colorless liq.; dens. 6.83; sp.gr. 0.823; visc. 5.5; m.p. –49 C; b.p. 315 C; 98.5% act.

Alfol 8. [152; 634] Caprylic alcohol; wetting agent, emulsifier, surfactant intermediate, raw material; colorless liq.; mild aromatic odor; dens. 6.86; sp.gr. 0.823; m.p. 5 C; b.p. 383 C; 99.2% act.

Alfol 10. [152; 634] Decyl alcohol; emulsifier, surfactant intermediate; colorless liq., wh. waxy solid; sweet, fat-like odor; sol. in alcohol, ether, benzene, glacial acetic acid; water insol.; dens. 6.90; sp.gr. 0.828; m.p. 44 C; b.p. 448 C; 99% act.

Alfol 12. [152; 634] Lauryl alcohol; detergent intermediate; colorless liq.; sp.gr. 0.831; m.p. 75 C; b.p. 498 C; 99.1% act.

Alfol 14. [152; 634] Myristyl alcohol; detergent intermediate; wh. solid, cryst.; typ. fatty alcohol odor; sp.gr. 0.824; m.p. 99 C; b.p. 552 C; 99% act.

Alfol 16. [152; 634] Cetyl alcohol; detergent intermediate; wh. waxy solid; typ. fatty alcohol odor; dens. 6.77; sp.gr. 0.813; visc. 6.77; m.p. 117 C; b.p. 604 C; 98.9% act.

Alfol 18. [152; 634] Stearyl alcohol; surfactant intermediate; wh. waxy solid; typ. fatty alcohol odor; dens. 6.71; sp.gr. 0.8075; visc. 13.5; m.p. 135 C; b.p. 640 C; 98.7% act.

Alfol 20. [152] Blend of C_{20} and higher even-carbon-number primary linear alcohols; lubricant, defoamer; off-wh. solid; dens. 6.80; sp.gr. 0.817; m.p. 131.

Alfol 610 Alcohol. [152; 634] C_6-C_{10} linear primary alochol; intermediate for surfactants, plasticizers; colorless liq.; mild, sweet odor; sol. in alcohol, ether, benzene, insol. in water; dens. 6.92; sp.gr. 0.829; visc. 11.0; m.p. –8 C; b.p. 350 C; 99% act.

Alfol 810 Alcohol. [152; 634] C_8–C_{10} linear primary alcohol; intermediate for plasticizers and surfactants; colorless liq.; mild, sweet odor; sol. in alcohol, ether, benzene; dens. 6.93; sp.gr. 0.831; visc. 8.9; m.p. 7 C; b.p. 400 C; 99% act.

Alfol 1012 Alcohol. [152] C_{10}–C_{12} linear primary alcohol; intermediate for surfactants; colorless liq.; sweet odor; dens. 7.0; sp.gr. 0.834; visc. 15; m.p. 35–40 C; b.p. 427 C; 99% act.

Alfol 1214 Alcohol. [152; 634] C_{12}–C_{14} linear primary alcohol; intermediate for surfactants; colorless liq.; sweet, typ. fatty alcohol odor; dens. 7.0; sp.gr. 0.838; visc. 14.3; m.p. 73–75 C; b.p. 518 C; 99% act.

Alfol 1216 Alcohol. [152; 634] C_{12}-C_{16} alcohols; intermediate for surfactants, plastics; clear, colorless liq.; sweet, typ. fatty alcohol odor; dens. 7.0; sp.gr. 0.830; visc. 15.5; m.p. 64–70 C; b.p. 514 C; 98.5% act.

Alfol 1218 Alcohol. [152] C_{12}–C_{18} linear primary alcohol; intermediate for surfactants; wh. solid; dens. 6.85; sp.gr. 0.825; visc. 15.5; m.p. 99–101 C; b.p. 524 C; 100% act.

Alfol 1412 Alcohol. [152; 634] Myristyl alcohol; intermediate for surfactants, plastics; wh., soft waxy solid; typ. fatty alcohol odor; dens. 7.0; sp.gr. 0.838; visc. 16.7; m.p. 72–75 C; b.p. 560 C; 99.4% act.

Alfol 1618 Alcohol. [152] C_{16}–C_{18} linear primary alcohol; intermediate for surfactants; wh. waxy solid; typ. fatty alcohol odor; sol. in water, alcohol, chloroform, ether, benzene, glacial acetic acid; dens. 6.81; sp.gr. 0.820; visc. 22; m.p. 110–115 C; b.p. 609 C; 98.7% act.

Alfol 1620 Alcohol. [152] C_{16}–C_{20} linear primary alcohol; intermediate for surfactants; wh. waxy solid; typ. fatty alcohol odor; dens. 6.81; sp.gr. 0.817; visc. 13.5; m.p. 110–113 C; b.p. 605 C; 98.5% act.

Alfol C22-Alcohol. [152] Blend of C_{22} and higher even-carbon-number primary linear alcohols; lubricant, defoamer; off-wh. solid; dens. 6.81; sp.gr. 0.817; m.p. 132.

Alfonic 1012-60. [152] Ethoxylated linear alcohols (60% EO); nonionic; detergent, wetting agent, emulsifier, foaming agent, intermediate, dispersant; APHA 20 liq.; sweet odor; dens. 8.2; sp.gr. 0.98; visc. 20.5 cps; m.p. 36; biodeg.; 99.8% act.

Alfonic 1412-40. [152; 634] Ethoxylated linear alcohols (40% EO); nonionic; detergent intermediate, emulsifier, dispersant; clear liq.; slight odor; oil-sol.; dens. 7.64; sp.gr. 0.91; visc. 52 cps; m.p. 41; biodeg.; 100% act.

Alfonic 1412-60. [152; 634] Ethoxylated linear alcohols (60% E.O.); nonionic; detergent, wetting agent, emulsifier, foaming agent, surfactant; cloudy liq.; sweet, mild odor; dens. 8.2; sp.gr. 0.98; visc. 33 cps; biodeg.; 100% act.

Alfonic 1412-A Ether Sulfate. [152; 634] Sulfated ethoxylated alcohol deriv., ammonium salt; anionic; detergent, wetting agent, emulsifier, foaming agent for detergent formulations, shampoos, emulsion polymerization, oil well drilling, agriculture; clear liq.; mild odor; dens. 8.4; visc. 76 cps; biodeg.; 58% act.

Alfonic 1412-S Ether Sulfate. [152; 634] Sulfated ethoxylated alcohol deriv.; anionic; detergent, wetting agent, emulsifier, foaming agent; clear liq.; mild odor; dens. 8.8; visc. 64 cps; biodeg.; 38% act.

Alfonic 1618-65. [152] Ethoxylated linear alcohols (65% E.O.); nonionic; detergent, emulsifier, foaming and wetting agent; near wh. solid; sol. in water, ethanol, monobutyl glycol ether, IPA, xylene, perchlorethylene, slight in kerosene, insol. in min. oil; dens. 8.2; sp.gr. 0.98; visc. 81 cps; m.p. 88; b.p. 234; biodeg.; 100% act.

Algepon AK. [536] Fatty quat. ammonium compd.; cationic; detergent stripping compd.; paste.

Alipal CD-128. [245] Alcohol ethoxylate sulfate, ammonium salt; anionic; surfactant for aq. systems; air-entraining agent, foamer, frothing agent for concrete, cements, gypsum wallboard; solubilizer and wetting agent; straw liq.; good sol. in electrolyte systems; sp.gr. 1.056; pour pt. –5 C; 56% act.

Alipal CO-433. [245] Sodium nonoxynol-4 sulfate; anionic; detergent, emulsifier, lime-soap dispersant, wetting agent, foamer; dishwashing formulations, scrub soaps, car washes, rug and hair shampoos, emulsion polymerization, petroleum waxes; antisat for plastics and syn. fibers; Varnish 4 max. color, clear liq.; mild aromatic odor; water sol.; dens. 8.9 lb/gal; sp.gr. 1.065; visc. 2500 cps; surf. tens. 32 dynes/cm (1%); 28% act.

Alipal CO-436. [245] Ammonium nonoxynol-4 sulfate; anionic; see Alipal CO-433; Varnish 5 max, clear liq.; alcoholic odor; water sol.; dens. 8.9 lb/gal; sp.gr. 1.065; visc. 100 cps; surf. tens. 34 dynes/cm (1%); 58% act.

Alipal CO-526. [245] Sodium salt of a sulfated alkylphenoxypoly (ethyleneoxy)-ethanol; anionic; detergent, wetting, dispersing, emulsifying and foaming agent; Varnish 10 max. color, hazy liq.; alcoholic odor; water sol.; dens. 9.3 lb/gal; sp.gr. 1.115; visc. 400 cps; surf. tens. 35 dynes/cm (1%); 58% act.

Alipal EP-110. [245] Sulfated nonylphenoxypoly (ethyleneoxy) ethanol, ammonium salt; anionic; emulsifier for polymerization of monomer systems; pale yel. liq.; sp.gr. 1.04; visc. 91.0 cks; pour pt. 0 C; surf. tens. 38.3 dynes/cm (1%); 30% act.

Alipal EP-115. [245] Sulfated nonylphenoxypoly (ethyleneoxy) ethanol, ammonium salt; anionic; emulsifier for latices; liq.; 30% conc.

Alipal EP-120. [245] Sulfated nonylphenoxypoly (ethyleneoxy) ethanol, ammonium salt; anionic; emulsifier for emulsion polymerization of vinyl acetate; pale yel. liq.; sp.gr. 1.06; visc. 111 cks; pour pt. 8 C; surf. tens. 40 dynes/cm (1%).

Alkadet 15. [316] Nonionic wetting agent used in agric. formulations; liq.; 70% conc.

Alkadet LD-50. [25] Blend of surfactants; anionic; detergent; clear, amber liq.; visc. 500 cps; cloud pt. < 30 F; 48% act.

Alkadet LD50-Q. [25] Blend of surfactants; anionic; liq. detergent conc.; clear amber liq.; visc. 500 cps; cloud pt. < 30 F; 48% act.

Alkadet LD-60. [25] Blend of surfactants; anionic; detergent; clear amber liq.; sp.gr. 1.05; visc. 100 cps; cloud pt. 7 F; 58–60% act.

Alkali Surfactant. [230] Amphoteric; surfactant used for coupling nonionic detergents into alkaline cleaners; used in textile formulations; lt. amber liq.; sp.gr. 1.04; pH 6–8 (5%); 35% min. act.

Alkameen C-5, C-10. [25] Tert. amines, ethoxylates of primary tallow amines; cationic; emulsifier, detergent, wetting agent; amber liq.; typ. odor; insol. in water; 100% act.

Alkameen HT-2. [25] Tert. amines, ethoxylates of hydrog. tallow amines; cationic; emulsifier; amber solid; typ. odor; insol. in water; 100% act.

Alkameen HT-5. [25] Tert. amines, ethoxylates of hydrog. tallow amines; cationic; emulsifier, wetting agent, detergent; amber solid; typ. odor; insol. in water; 100% act.

Alkameen HT-12. [25] Tert. amines, ethoxylates of hydrog. tallow amines; cationic; emulsifier, wetting agent, detergent; amber solid; typ. odor; insol. in water; 100% act.

Alkameen HT-15. [25] Tert. amines, ethoxylates of hydrog. tallow amines; cationic; wetting agent, detergent; amber solid; typ. odor; insol. in water; 100% act.

Alkameen HT-25. [25] Tert. amines, ethoxylates of hydrog. tallow amines; cationic; emulsifier, wetting agent, detergent; yel. solid; 100% act.

Alkameen HT-30. [25] Tert. amines, ethoxylates of hydrog. tallow amines; cationic; emulsifier, wetting agent; tan solid; insol. in water; 100% act.

Alkamide 1002. [24; 25] 2:1 Coconut diethanolamide; nonionic-anionic surfactant, detergent; hair shampoo, bubble bath, specialty detergents; straw-color liq.; water sol.; 100% act.

Alkamide 1129. [24] Mixed fatty acid diethanolamide; nonionic/anionic; detergent for hard surface cleaners, floor cleaners, and wax strippers; liq.; 100% conc.

Alkamide 1182. [24] Lauric diethanolamide, modified; nonionic; foam stabilizer and booster, visc. builder in personal care preparations; liq.; 100% conc.

Alkamide 1423. [24] Coconut fatty acid diethanolamide; nonionic/anionic; detergent, wetting, thickening, foaming and foam stabilizing agent in liq. soap formulations; liq.; 100% conc.

Alkamide 1509. [24] Capric diethanolamide; nonionic/anionic; foam booster/stabilizer for personal care products; liq.; 100% conc.

Alkamide 2104. [23; 24; 25] 2:1 Coconut diethanolamide; nonionic-anionic; detergent, emulsifier, foam booster; base for floor and general purpose cleaners; amber, visc. liq.; low odor; water sol.; 100% act.

Alkamide 2106. [24; 25] 2:1 Coconut diethanolamide; nonionic-anionic; detergent base for floor and hard surface cleaners; amber, visc. liq.; water sol.; 100% act.

Alkamide 2110. [24] Coconut 2:1 alkanolamide; nonionic; detergent base to solubilize high silicate concs. in liq. steam cleaners and wax strippers; amber liq.; sol. in water, aromatic solv., perchloroethylene; dens. 1.0 g/ml; pH 8–11 (1% DW); 100% act.

Alkamide 2112. [24; 25] 2:1 Fatty acid alkanolamide, sulfonate blend; nonionic-anionic; detergent base for floor cleaners, alkaline strippers, and degreasers; br., visc. liq.; low odor; water sol.; cloud pt. 45 F max.; 100% act.

Alkamide 2122. [24; 25] 2:1 alkanolamide sulfonate conc.; nonionic-anionic; detergent, wetting, dispersing and air entertaining agent for construction materials industry; amber visc. liq.; water sol.; cloud pt. 45 F max.; 100% act.

Alkamide 2124. [24; 25] 2:1 Lauric diethanolamide; nonionic-anionic; detergent, wetting agent, emulsifier, foaming agent, foam stabilizer, thickener for shampoos; amber visc. liq.; water sol.; sp.gr. 1.0–1.05; 100% act.

Alkamide 2204. [23; 24; 25] 2:1 Coconut diethanolamide; nonionic-anionic; detergent, rust inhibitor; base for hand soap, floor cleaners; amber visc. liq.; low odor; water sol.; 100% act.

Alkamide 2204A. [24] Coconut 2:1 alkanolamide; detergent base for all-purpose, industrial, and household cleaners, wax strippers; amber liq.; sol. in water, aromatic solv., perchloroethylene; dens. 1.0 g/ml; pH 8–11 (1% DW).

Alkamide CDE. [24; 25] Coconut diethanolamide; nonionic; detergent, emulsifier, stabilizer, thickener, foam stabilizer for personal care and detergent products; lt. amber liq.; low odor; sol. in detergent systems, aromatic and chlorinated aliphatic solv., petroleum solv., min. oils; sp.gr. 1.04; 90% act.

Alkamide CDE-Extra. [24] Coconut diethanolamide (1:1); nonionic; emulsifier, stabilizer, thickener, and foam stabilizer for personal care and detergent products; liq.; 100% conc.

Alkamide CDO. [24; 25] Coconut diethanolamide; nonionic; detergent, emulsifier, foam stabilizer, thickener for personal care and detergent products; lt. amber liq.; low odor; sol. in aromatic and chlorinated aliphatic solv., petroleum solv.; min. oils; sp.gr. 1.04; 82% act.

Alkamide CL63. [24; 25] Coconut-lauric diethanolamine; nonionic; detergent, thickener, emulsifier in toiletry and cleaning preparations; cl. to cloudy liq.; sol. in detergent systems, aromatic and chlorinated aliphatic solv., petroleum solv., min. oils; sp.gr. 1.03–1.05; 92% act.

Alkamide CME. [23; 24; 25] Cocamide MEA; nonionic; detergent, foam boosters, visc. builder, opacifier; wh. waxy solid; sol. in detergent systems, various solv., min. oils; m.p. 56 C; 95% act.

Alkamide CMO. [25] Cocamide MEA; detergent, foam boosters, visc. builder, opacifier; cream-colored, waxy solid; sol. in detergent systems, various solvs., min. oils; m.p. 50 C; 90% act.

Alkamide HTDE. [24; 25] Stearic diethanolamide; nonionic; detergent, thickener, visc. builder for kerosene; cream-colored, waxy solid; low odor; m.p. 55 C; 90% act.

Alkamide HTME. [24; 25] Stearic monoethanolamide; nonionic; detergent, opacifier, thickener for cosmetics; cream waxy solid; low odor; m.p. 85–90 C; 90% act.

Alkamide L7DE. [23; 24; 25] Lauric-myristic diethanolamine; nonionic; detergent, foam booster/stabilizer, superfatting and thickening agent for toiletry and cleaning formulations; fortifier for perfumes in soaps; wh. paste, liq. > 25 C; low odor; sol. in detergent systems, various solvs., min. oils; sp.gr. 1.03–1.05; m.p. 15 C; 92% act.

Alkamide L7DE-BT. [24; 25] Lauric-myristic diethanolamide, modified; nonionic; detergent, foam stabilizer, visc. builder for cosmetic preparations; straw-colored liq.; bland odor; sp.gr. 0.99; visc. 275 cps; f.p. 0 C; cloud pt. 0 C; 84% act.

Alkamide L7DE-PG. [24] Lauric/myristic alkanolamide, modified; nonionic; liq. version of Alkamide L7DE; liq.

Alkamide L7ME. [24] Lauric-myristic monoethanolamide; nonionic; foam booster and visc. for liq.; and powd. detergents; solid; 100% conc.

Alkamide L9DE. [23; 24; 25] Lauric diethanolamide, high purity; nonionic; detergent, foam stabilizer and booster, thickener for toiletry and cleaning applications, industrial and household detergents; wh., waxy solid, liq. > 25 C; low odor; sol. in detergent systems, various solvs., min. oils; sp.gr. 1.04; f.p. 25 C; 92% act.

Alkamide L9ME. [24; 25] Lauric monoethanolamide; nonionic; detergent, foam stabilizer and booster, visc. builder; wh., waxy solid; low odor; sol. in detergent systems, various solvs., min. oils; m.p. 60 C; 95% act.

Alkamide LIPA. [23; 24] Lauric monoisopropanolamide; nonionic; foam booster/stabilizer in detergents and laundry compds.; visc. builder in liq. formulations; lt. yel. solid; disp. in water; m.p. 63–68 C; pH 8–11 (1% DW); 95% amide.

Alkamide RDO. [24; 25] Ricinoleic diethanolamide; nonionic; emulsifier, lubricant, antiblocking and mold release agent used in the making of cutting and sol. oils, cleaners; br. clear liq.; low odor, 100% act.

Alkamide SDO. [24; 25] Modified oleic diethanolamide; nonionic; foam stabilizer, visc. builder, superfatting agent for toiletries, cutting and soluble oils, textiles, household and industrial cleaners, corrosion inhibitor; amber, visc. liq.; low odor; sol. in detergent systems, various solvs., min. oils; sp.gr. 1.04; 80% act.

Alkamide STEDA. [24] Ethylene bis-stearamide; nonionic; additive in pulp and paper defoamer formulations; lubricant, plasticizer, antistat and pigment dispersant in resins and plastics; solid; 100% conc.

Alkamidox C-2. [24] Coconut monoethanolamide, ethoxylated; nonionic; thickener, foam stabilizer, and emulsifier for formulated detergents and cosmetics; tan paste.; sol. in aromatic solv., perchloroethylene; disp. in water; pH 10.0–11.5 (5% DW).

Alkamidox C-5. [24] Coconut monoethanolamide, ethoxylated; nonionic; emulsifier, foam stabilizer, and thickener used in alkaline industrial cleaners; carwash formulas; liq. dish and laundry detergents; cosmetics; yel. liq.; sol. in water, aromatic solv., perchloroethylene; dens. 1.0 g/ml; pH 10.0–11.5 (5% DW).

Alkamidox CME-2. [25] PEG-3 cocamide; nonionic; detergent, foam stabilizer; cream-colored solid; sol. in ethanol, xylene, insol. in kerosene, min. oil; 100% act.

Alkamidox CME-5. [25] PEG-6 cocamide; nonionic; detergent, foam stabilizer; straw liq.; sol. in ethanol, xylene, insol. in kerosene, min. oil; 100% act.

Alkamidox L-2. [24] Lauric monoethanolamide, ethoxylated; nonionic; foam stabilizer used in personal care products; wh. solid; sol. in aromatic solv., perchloroethylene; disp. in water; pH 10.0–11.5 (5% DW).

Alkamidox L-5. [24] Lauric monoethanolamide, ethoxylated; nonionic; see Alkamidox C-5; wh. solid; sol. in water, aromatic solv., perchloroethylene; pH 10.0–11.5 (5% DW).

Alkaminox C2. [23; 24; 25] PEG-2 cocamine; cationic; textile scouring, dyeing assistant, softening, antistatic agent; used as corrosion inhibitor in steam generating and circulating systems; coemulsifier; clear amber liq.; sol. in most org. solv., min. acid, insol. in water; sp.gr. 0.87; 100% conc.

Alkaminox C-5. [23; 24] Coconut ethoxylated amine; cationic; see Alkaminox C-2; liq.; 100% conc.

Alkaminox C-10. [23; 24] Coconut ethoxylated amine; cationic; used in cosmetics; Gardner 12 liq.; sol. in water, aromatic solv., perchloroethylene; dens. 1.02 g/ml; HLB 16; 98% min. tert. amine.

Alkaminox HT-12, -25, -30. [23] Hydrog. primary tallow amine ethoxylate; cationic; emulsifier, dispersant; textile scouring and desizing assistants; softening and antistatic agent; solid; 100% conc.

Alkaminox O-2. [24] Primary oleyl amine ethoxylate; cationic; emulsifier for road paving asphalt seal coat emulsions; liq.; 100% conc.

Alkaminox SO-5. [24] Soya ethoxylated amine; cationic; imparts emulsifying, corrosion resistance, and nonfoaming properties in metal cleaning and formulation of cleaners; Gardner 14 liq.; sol. in perchloroethylene; disp. in water; dens. 0.98 g/ml; HLB 14; 98% min. tert. amine.

Alkaminox T-2. [23; 24; 25] PEG-2 tallow amine; cationic; emulsifier, dispersant, scouring, desizing aid, internal antistat, softener for textiles; Gardner 8 max. paste; sol. in most org. solv., min. acid, insol. in water; sp.gr. 0.92; HLB 4; 100% conc.

Alkaminox T-5. [23; 24; 25] PEG-5 tallow amine; cationic; emulsifier, dispersant, used in textile industry, corrosion inhibitor in steam generating and circulating systems, emulsifier for wax emulsions; Gardner 8 max. clear liq.; sol. in most org. solv., min. acid, insol. in water; sp.gr. 0.97; HLB 9; 98% min. tert. amine.

Alkaminox T-10. [23; 24] PEG-10 tallow

amine; cationic; emulsifier, dispersant, leveling agent for dyeing of nylon with anionic dyes; Gardner 14 liq.-paste; sol. in water, perchloroethylene; dens. 1.01 g/ml; 98% min. tert. amine.

Alkaminox T-12. [23; 24] Tallow ethoxylated amine; cationic; see Alkaminox T-10; Gardner 14 liq.-paste; sol. in water; dens. 1.03 g/ml; HLB 13.5; 98% min. tert. amine.

Alkaminox T-15. [23; 24; 25] PEG-15 tallow amine; cationic; see Alkaminox T-5; liq.; sp.gr. 1.03; HLB 14.2; 100% conc.

Alkaminox T-25, T-30, T-50. [23; 24] PEG-25, -30, -50 tallow amine; cationic; coemulsifier, leveling agent for differential dyeing of nylon; Gardner 23 solid; sol. in water; HLB 16.0, 16.6, and 17.5 resp.; 98% min. tert. amine.

Alkamox C2-0. [23; 24] Bis (2-hydroxyethyl) coco amine oxide; nonionic; foaming agent, stabilizer, thickener, and emollient for personal care products and detergents; corrosion inhibitor for nonferrous metals; used in electroplating systems; clear liq.; sol. in water; dens. 1.0 g/ml; 29–31% amine oxide.

Alkamox C20. [25] Amine oxide of ethoxylated coco amine; detergent, wetting agent, foam booster, used for latex foam for carpets and carpet backing; yel. liq.; faint odor; f.p. < 0 C; 33% act.

Alkamox CAPO. [24; 25] Cocamidopropyl amine oxide; nonionic; detergent, wetting agent, foaming agent and foam stabilizer; colorless liq.; faint odor; water sol.; f.p. < 0 C; 30% act.

Alkamox LO. [23; 24; 25] Lauramine oxide; nonionic-cationic; detergent, wetting agent, foaming agent and foam stabilizer, antistat agent; colorless liq.; faint odor; water sol.; f.p. < 0 C; 30% act.

Alkamox ODM. [24] Oleyl dimethylamine oxide; nonionic/cationic; foaming, wetting and conditioning agent; liq.; 50% conc.

Alkamuls 200-DL. [24] PEG 200 dilaurate; nonionic; w/o emulsifier, dispersant, defoamer, coemulsifier, and lubricant in industrial and textile oils, cosmetics; as paper softener; yel. liq. to paste; sol. in min. oil and spirits, aromatic solv., perchloroethylene; disp. in water; dens. 0.96 g/ml; HLB 7.4; sapon. no. 190–200; 100% conc.

Alkamuls 200-D0. [24] PEG 200 dioleate; nonionic; see Alkamuls 200-DL; amber liq.; sol. in min. oil, min. spirits, aromatic solv., perchlorethylene; disp. in water; dens. 0.95 g/ml; HLB 6.0; sapon. no. 147–157; 100% conc.

Alkamuls 200-DS. [24] PEG 200 distearate; nonionic; lubricant and softener in textile applications; opacifier and emulsifier in cosmetic preparations; cream solid; sol. in min. oil and spirits, aromatic solv., perchloroethylene; disp. in water; HLB 5.0; sapon. no. 160–170; 100% conc.

Alkamuls 200-ML. [24] PEG 200 monolaurate; nonionic; emulsifier, coupling agent, and solubilizer in metal working fluids, industrial and textile lubricants; yel. liq.; disp. in water; dens. 1.00 g/ml; HLB 9.8; sapon. no. 142–152.

Alkamuls 200-MO. [24] PEG 200 monooleate; nonionic; w/o emulsifier for min. oils, fatty acids, and solv.; used in industrial applications such as cleaners, degreasers and pesticide formulations; lubricant-softener in tanning and textile processes; yel. liq.; disp. in water; dens. 0.99 g/ml; HLB 8.0; sapon. no. 111–121.

Alkamuls 200-MS. [24] PEG 200 monostearate; nonionic; emulsifier for fats and oils; softener and lubricant for textiles and leather; wh. solid; disp. in water; HLB 8.0; sapon. no. 120–130.

Alkamuls 400-DL. [23; 24; 25] PEG 400 dilaurate; nonionic; lipophilic emulsifier, solubilizer, dispersing agent used in personal care products and industrial applications, agric. chemical sprays, industrial and textile lubricants; yel. liq., paste; sol. in min. oil and spirits, aromatic solv., perchloroethylene; disp. in water; dens. 1.0 g/ml; HLB 10.0; sapon. no. 132–142; 100% act.

Alkamuls 400-DO. [23; 24; 25] PEG 400 dioleate; nonionic; emulsifier, wetting agent for cosmetic and agric. uses; clear, amber liq.; sp.gr. 0.97–0.99; HLB 7.2; 100% act.

Alkamuls 400-DS. [23; 24; 25] PEG 400 distearate; nonionic; see Alkamuls 400-DL; wh., waxy flakes; HLB 7.8; 100% act.

Alkamuls 400-MC. [25] PEG 400 monococoate; liq.; HLB 13.9; 100% act.

Alkamuls 400-ML. [23; 24; 25] PEG 400 monolaurate; nonionic; defoamer, leveling agent for latex paints; dispersant for pigments, dyes; emulsifier for cosmetics and toiletries; clear, amber liq.; sol. (@ 10%) in aromatic solv., perchlorethylene; disp. in water, min. oil, min. spirits; sp.gr. 1.02–1.04; cloud pt. 50 F; HLB 12.8; sapon. no. 191–201; 100% act.

Alkamuls 400-MO. [23; 25] PEG-9 oleate; nonionic; emulsifier, wetting agent, dispersant used in dairy industry, cosmetic and industrial applications; amber, clear liquid; sp.gr. 1.01–1.03; HLB 11.0; 100% act.

Alkamuls 400-MS. [23; 25] PEG 400 monostearate; nonionic; see Alkamuls 400-DL; paste; HLB 11.2; 100% act.

Alkamuls 600-DL. [24] PEG 600 dilaurate; nonionic; see Alkamuls 400-DL; yel. liq., paste; disp. in water; dens. 1.03 g/ml; HLB 11.5; sapon. no. 106–116; 100% conc.

Alkamuls 600-DO. [24; 25] PEG 600 dioleate; nonionic; see Alkamuls 400-DL; amber liq.; sp.gr. 1.01–1.03; HLB 10.0; 100% act.

Alkamuls 600-DS. [24] PEG 600 distearate; nonionic; see Alkamuls 200-DS; cream solid; disp. in water; HLB 10.6; sapon. no. 96–106; 100% conc.

Alkamuls 600-ML. [23; 24; 25] PEG 600 monolaurate; nonionic; defoamer, leveling agent for latex paints; dispersant for pigments, dyes; emulsifier for cosmetics and toiletries; yel. liq. to paste; sol. (@ 10%) in water; disp. in min. oil, min. spirits, aromatic solv., perchlorethylene; dens. 1.05 g/ml; HLB 14.6; sapon. no. 67–77; 100% act.

Alkamuls 600-MO. [24; 25] PEG 600 monooleate; nonionic; dispersant, o/w emulsifier for cosmetic and industrial applications; liq.; HLB 13.0; 100% act.

Alkamuls 600-MS. [23; 25] PEG 600 monostearate; nonionic; defoamer, leveling agent for latex paints; dispersant and dye leveling agent for textiles; wh. solid; sol. @ 10% in water, perchlorethylene; disp. in min. oil, min. spirits; HLB 13.2; sapon. no. 60–70; 100% act.

Alkamuls 6000-DS. [24] PEG 6000 distearate; nonionic; hydrophilic emulsifier for cosmetics; cream solid; sol. in water; HLB 18.4; sapon. no. 14–20; 100% conc.

Alkamuls AG-100,-200 Series. [24] Calcium sulfonate, nonionic ethoxylate blend; agric. emulsifier matched pairs for insecticides and herbicides; liq., paste.

Alkamuls AG-200,-400 Series. [24] Amine sulfonate nonionic ethoxylate blend; agric. emulsifiers for herbicides; liq., paste.

Alkamuls AG-700 Series. [24] Nonionic ethoxylate blend; agric. emulsifiers for toxicants; liq.

Alkamuls AG-821, -822. [24] Ethoxylated ester nonionic blend; agric. emulsifiers and adjuvants for crop oil/surfactant concs.; amber clear liq.; pH 7.5–8.5; 91% act.

Alkamuls AG-900, -902, -906. [24] Nonionic ethoxylate blend; spreading and wetting agents for aq. pesticide systems; clear liq.; HLB 10.0–14.0; pH 7.0–7.5 (5% DW); 92% act.

Alkamuls AG-903. [24] Anionic-nonionic blend; foaming adjuvant with wetting, spreading, and sticking properties for agric. applications; clear liq.; pH 6.0–8.0 (5% DW); 38% act.

Alkamuls AG-904, -905, -907. [24] Anionic-nonionic blend; agric. spreading and wetting agent; clear liq.; pH 6.0–8.0 (5% DW); 84, 75, and 83% act. resp.

Alkamuls AG-CA. [24] Calcium sulfonate; agric. emulsifier component for insecticides and herbicides; amber clear liq.; pH 6.0–8.0 (5% DW).

Alkamuls AG-IPAM. [24] Amine sulfonate; agric. emulsifier component for phenoxy esters; amber clear liq.; pH 6.0–8.0 (5% DW).

Alkamuls DEG-DO. [24] DEG dioleate; nonionic; w/o emulsifier, dispersant, defoamer, coemulsifier, and lubricant in industrial and textile oils; as paper softener; amber liq.; sol. in min. oil, min. spirits, aromatic solv., perchlorethylene; disp. in water; dens. 0.93 g/ml; HLB 2.9; sapon. no. 174–184; 100% conc.

Alkamuls DEG-DT. [23; 25] DEG ditallate; nonionic; dispersing agent, emulsifier; br. liq.; faint odor; 100% act.

Alkamuls DEG-ML. [23; 25] DEG monolaurate; nonionic; defoamer, dispersant, emulsifier, lubricant for industrial and textile oils; coemulsifier; yel. liq. to paste; disp. (@ 10%) in min. oil, min. spirits, aromatic solv., perchlorethylene; insol. water; dens. 0.96 g/ml; HLB 6.0; sapon. no. 190–200; 100% act.

Alkamuls DEG-MO. [25] DEG monooleate; nonionic; lubricant; yel. liq.; sol. (10%) in min. spirits; disp. in min. oil, aromatic solv., perchloroethylene; insol. water; dens. 0.95 g/ml; HLB 3.5; sapon. no. 140–150; 100% act.

Alkamuls DEG-MS. [25] DEG monostearate; nonionic; opacifier, emulsifier, plasticizer; br. liq.; faint odor; HLB 4.7; 100% act.

Alkamuls EGMS. [24] PEG monostearate; nonionic; opacifying and pearlescing agent for liq. cosmetic and detergent compds.; wh. solid; insol. in water; HLB 2.9; sapon. no. 174–184.

Alkamuls GML. [23; 25] Glyceryl laurate; nonionic; emulsifier for textiles, leather, foods; paste; HLB 3.0; 100% act.

Alkamuls GML-45. [24] Glyceryl laurate; nonionic; emulsifier, lubricant, antistat in cosmetic and industrial applications, mold release agents; liq., paste; insol. in water; dens. 0.97 g/ml; HLB 3; 40% min. monoglyceride.

Alkamuls GMO. [23; 25] Glyceryl oleate; nonionic; emulsifier, wetting agent, in cosmetic, agric., textile industries; lt. yel. liq.; m.p. 25 C; HLB 3.4; 100% act.

Alkamuls GMO-45. [24] Glyceryl oleate; nonionic; mold release agent and rust preventive additive for compounded oils; lubricant in syn. fiber spin finishes; processing aid for PVC film; emulsifier, antistat in cosmetic and industrial applications; amber liq., paste; sol. in aromatic solv., perchloroethylene; dens. 0.93–0.98 g/ml; HLB 3; 40% min. monoglyceride.

Alkamuls GMO-45LG. [24] Glyceryl oleate; mold release agent and rust preventive additive for compounded oils; lubricant in syn. fiber spin finishes; emulsifier, antistat in cosmetic and industrial applications; amber liq., paste; sol. in aromatic solv., perchloroethylene; insol. in waer; dens. 0.93 g/ml; HLB 3; 40% min. monoglyceride.

Alkamuls GMO-55LG. [24] Glyceryl oleate; processing aid for PVC film functioning as an internal lubricant, antistat, and antifogging agent; emulsifier for cosmetic and industrial applications; amber liq., paste; sol. in aromatic solv., perchloroethylene; dens. 0.96 gl/ml; HLB 3; 50% min. monoglyceride.

Alkamuls GMR. [23; 25] Glyceryl ricinoleate; nonionic; emulsifier, stabilizer, opacifier for textiles, leather, foods; liq.; 100% act.

Alkamuls GMS. [23; 25] Glyceryl stearate; nonionic; emulsifier, wetting agent for cosmetic, agric., textile industries; solid; HLB 3.4; 100% conc.

Alkamuls GMS-45. [24] Glyceryl stearate; nonionic; emulsifier, lubricant, antistat for personal care and industrial products; textile lubricant-softener; off-wh. flakes; insol. in water; HLB 3; 40% min. monoglyceride.

Alkamuls GTO. [24] Glyceryl trioleate; nonionic; lubricant for textiles, leather, and metals; amber liq.; insol. in water; dens. 0.95 g/ml; HLB 0.8; 100% conc.

Alkamuls NP Series. [24] Nonionic ethoxylates; agric. adjuvant additives for toxicant concs.; liq., solids; pH 7.0–8.0 (5% DW); 100% act.

Alkamuls PSML-4. [24] POE (4) sorbitan monolaurate; nonionic; emulsifier and coupling agent; PVC emulsion polymerization; yel. liq.; disp. in water; dens. 1.0 g/ml; HLB 13.3; sapon. no. 100–115; hyd. no. 215–255; 100% conc.

Alkamuls PSML-20. [23; 24; 25] POE (20) sorbitan monolaurate; nonionic; emulsifier, solubilizer, antistat, lubricant for textiles; yel. liq.; HLB 16.7; 97% act.

Alkamuls PSMO-5. [23; 24; 25] POE (5) sorbitan monooleate; nonionic; emulsifier, solubilizer, antistat, lubricant for paint, food, cosmetic, insecticides, herbicides, fungicides, textiles; amber liq.; HLB 10; 98.5% act.

Alkamuls PSMO-20. [23; 24; 25] POE (20) sorbitan monooleate; nonionic; emulsifier, wetting agent for cosmetic and food applications; yel. liq.; HLB 15; 97% act.

Alkamuls PSMO-30. [24] POE (30) sorbitan monooleate; nonionic; hydrophilic coemulsifier for agric. chemical emulsions; yel. liq.; sol. in water; dens. 1.1 g/ml; HLB 17.2; sapon. no. 34–40; hyd. no. 50–60; 100% conc.

Alkamuls PSMS-4. [23; 24; 25] POE (4) sorbitan monostearate; nonionic; emulsifier; tan solid; typ. odor; water insol.; HLB 9.6; 97% act.

Alkamuls PSMS-20. [23; 24; 25] POE (20) sorbitan monostearate; nonionic; wetting agent, emulsifier for cosmetic and food applications; yel. visc. liq., gel on standing; typ. odor; water sol.; HLB 14.9; 97% act.

Alkamuls PSTO-20. [23; 24; 25] POE (20) sorbitan trioleate; nonionic; emulsifier for cosmetic and food applications; amber liq.; typ. odor; water insol.; HLB 11; 97% act.

Alkamuls PSTS-20. [23; 24; 25] POE (20) sorbitan tristearate; nonionic; wetting agent, emulsifier for cosmetic and food formulations; tan solid; typ. odor; water insol.; HLB 10.5; 97% act.

Alkamuls SML. [23; 24; 25] Sorbitan monolaurate; nonionic; emulsifier for oils and fats in cosmetic and industrial oil products; corrosion inhibitor; amber liq.; typ. odor; disp. in water; moderately sol. most alcohols, veg. and min. oils; sp.gr. 1.05 (60 F); HLB 8.6; sapon. no. 160–170; hyd. no. 320–350; 97% act.

Alkamuls SMO. [23; 24; 25] Sorbitan oleate; nonionic; emulsifier, coupling agent, wetting agent for medicants, petroleum oils, fats, and waxes in the industrial, textile, and cosmetic industries; textile and leather lubricant and softener; corrosion inhibitor; amber liq.; moderately sol. in most veg., min. oils; insol. in water; sp.gr.. 1.0; HLB 4.3; sapon. no. 145–160; hyd. no. 193–210; 97% act.

Alkamuls SMS. [23; 24; 25] Sorbitan stearate; nonionic; emulsifier and coupling agent; used to prepare silicone defoamer emulsions for industrial applications, paraffin wax emulsions for processing paper coatings; textile process lubricant; cosmetics; foods; cream flakes; sol. (10%) in aromatic solv., perchloroethylene; dens. 1.0 g/ml; HLB 4.7; sapon. no 147–157; hyd. no. 235–260; 98.5% act.

Alkamuls STO. [24; 25] Sorbitan trioleate; nonionic; emulsifier and coupling agent; used to compd. textile and leather softener finishes; amber liq.; typ. odor; insol. in water; dens. 1.0 g/ml; HLB 1.8; sapon. no. 170–190; hyd. no. 55–70; 100% act.

Alkamuls STS. [24; 25] Sorbitan tristearate; hydrophobic emulsifier for use as a fiber-to-metal lubricant for syn. and cotton fibers; cosmetics, foods; cream flakes; typ. odor; sol. (10%) in aromatic solv., perchloroethylene; insol. in water; dens. 1.0 g/ml; HLB 2.1; sapon. no. 176–188; hyd. no. 66–80; 100% act.

Alkanol 189-S. [208] Sodium alkyl sulfonate; anionic; wetting agent, detergent, penetrant; reddish-br. liq.; visc. 30 cps; surf. tens. 31.2 dynes/cm (1%, 90 F); 31.5% act.

Alkanol A-CN. [208] EO condensate; antiprecipitant, dyeing assistant, dye solubilizer, contrast agent, nonionic surfactant; lt. amber liq.; 50% sol. in water; dens. 8.2 lb/gal; surf. tens. 38 dynes/cm (1% conc.).

Alkanol CNR. [208] Amphoteric surfactant; dye assistant; amber-colored clear liq.; misc. with water; dens. 8.61 lb/gal; sp.gr. 1.033.

Alkanol DW. [208] Sodium alkylaryl sulfonate; anionic; emulsifier, wetting agent, detergent; yel. liq.; alcoholic odor; misc. in water; dens. 8.9 lb/ gal; cloud pt. 5 C (max); 28% act.

Alkanol ND. [208] Alkyl diaryl sulfonate; anionic; dyeing assistant, surfactant; yel. liq.; terpene odor; misc. with water; dens. 9.6 lb/ gal; visc. slight; 45% act.

Alkanol S. [208] Sodium tetrahydronaphthalene sulfonate; anionic; dispersant, stabilizer, solubilizing agent; lt. cream-colored, small flakes; water sol. 20%; dens. 2.6 lb/ gal; 98% act.

Alkanol WXN. [208] Sodium alkylaryl sulfonate; anionic; wetting, rewetting agent, dyeing assistant; lt. yel. liq.; alcoholic odor; misc. in water; dens. 8.6 lb/gal; cloud pt. 0 C max.; surf. tens. 26 dynes/cm (1%); 30% act.

Alkanol XC. [208] Sodium alkylnaphthalene sulfonate; anionic; wetting agent, dispersant, penetrant; lt. cream-colored powd.; sol. in ethanol, acetone, benzene, 8–10% in water; sp.gr. 0.55.

Alkaphos 3. [25] Free acid of complex org. phosphate esters; anionic; detergent, wetting agent, emulsifier, solubilizer, antistat, lubricant; pale yel., visc. liq.; low odor; sol. in water and most aromatic and chlorinated solv.; sp.gr. 1.025; visc. 863 cps; 100% act.

Alkaphos 6. [25] Free acid of complex org. phosphate esters; anionic; see Alkaphos 3; sp.gr. 1.055; visc. 1.045 cps; 100% act.

Alkaphos 10. [25] Free acid of complex org. phosphate esters; anionic; see Alkaphos 3; sp.gr. 1.085; visc. 2100 cps; 100% act.

Alkaphos K 380. [25] Potassium salt of org. phosphate ester; anionic; solubilizer, hydrotropic surfactant; clear liq.; cloud pt. < 0 C; 44% act.

Alkaphos L6-36S. [24] Complex phosphate ester sodium salt; anionic; solubilizer for surfactants and alkaline salts in cleaners for consumer and industrial applications; liq.; 41% conc.

Alkaphos QS. [25] Free acid of complex org. phosphate ester; anionic; detergent, wetting agent, emulsifier, hydrotropic surfactant, solubilizer; colorless liq.; bland odor; sol. in water, caustic soda, caustic potash; cloud pt. 68 C (1% in 15% NaOH); 90% act.

Alkapol PEG 200. [24] PEG 200; intermediate for processing nonionic surfactants; used in pharmaceuticals as binders and lubricants for compressed tablets; plasticizers in starch pastes and polyethylene sheet; paper softener; humectant and solv. for dyes in stamp pad inks; solv. in cosmetics and industrial applications; antistat, lubricant in textile and plastic processing; liq., solid; m.w. 190-210; sol. in water @ 10%; dens. 1.13 g/ml; pH 5-8 (5% DW); 100% conc.

Alkapol PEG 300. [24] PEG 300; see Alkapol PEG 200; liq.; m.w. 285-315; dens. 1.13 g/ ml; pH 5-8 (5% DW); 100% conc.

Alkapol PEG 400. [24; 25] PEG-9; intermediate, plasticizer, solv., lubricant; wh. liq.; water sol.; sp.gr. 1.13 (20/15.5 C); f.p. 4-8 C; 100% conc.

Alkapol PEG 600. [25] PEG-14; intermediate, plasticizer, solv., lubricant; wh. liq.; water sol.; sp.gr. 1.13 (30/15.5 C); f.p. 20-25 C; 100% conc.

Alkapol PEG 1000. [24] PEG 1000; solv. or cosolv. in industrial applications; used to make suppositories; solid; m.w. 950-1050; sol. in water @ 10%; pH 5-8 (5% DW); 100% conc.

Alkapol PEG 1500. [24] PEG 1500; rubber industry lubricant for calendering the latex onto a textile substrate; lubricant for rayon hosiery knitting; solv. or cosolv. in industrial applications; solid; m.w. 1400-1600; sol. in water @ 10%; pH 5-8 (5% DW); 100% conc.

Alkapol PEG 3350. [24] PEG 3350; mold release agent, antistat for rubber products; binder for tablets in pharmaceuticals; waxy flake; m.w. 3000-3700; sol. in water @ 10%; pH 5-8 (5% DW); 100% conc.

Alkapol PEG 6000. [24] PEG 6000; binding agent in abrasive formulations; mold release agent for rubber articles; chemical intermediate; waxy flake; m.w. 5400-6600; sol. in water @ 10%; pH 5-8 (5% DW); 100% conc.

Alkapol PEG 8000. [24] PEG 8000; base for water-sol. sticks; spreader and binder in aq. shoe polish formulations, lubricant and binder aid for ceramic extrusion; waxy flake; m.w. 7000-9000; sol. in water @ 10%; pH 5-8 (5% DW); 100% conc.

Alkapol PPG-425. [24] PPG-425; intermediate for nonionic surfactants; lubricant, carrier; solv. or cosolv. for cosmetic, pharmaceutical, and industrial formulations; release agent for plastisols; liq.; m.w. 400-450; sol. in perchloroethylene and water; dens. 1.01 g/ml; pH 6-8; 100% conc.

Alkapol PPG 1000. [24] PPG-1000; see Alkapol PPG-425; liq.; 100% conc.

Alkapol PPG-1200. [24] PPG-1200; intermediate for nonionic surfactants; antifoam agent in emulsions; component in hydraulic fluids and greases; liq.; m.w. 1150-1250; sol. in perchloroethylene; dens. 1.01 g/ml; pH 6-8; 100% conc.

Alkapol PPG-2000. [24] PPG-2000; antifoam in industrial applications; additive for tire lubricants; carrier or cosolvent in cosmetic preparations; liq.; m.w. 1900-2100; sol. in perchloroethylene, aromatic solv.; dens. 1.01 g/ml; pH 6-8; 100% conc.

Alkapol PPG-4000. [24] PPG-4000; see Alkapol PPG-2000; liq.; m.w. 3900-4100; sol. in perchloroethylene; dens. 1.01 g/ml; pH 6-8; 100% conc.

Alkaquat C. [23; 24; 25] Coconut imidazoline quat.; cationic; antistat, fabric softener, dye retardant for textile industry, laundry detergents; amber liq.; sol. in aromatic solv., perchlorethylene; disp. in water, min. oil, min. spirits; dens. 0.99 g/ml; f.p. < 0 C; pH 6.0–6.8; 74–76% act.

Alkaquat DMB-451. [25] N-alkyl (C_{12}-C_{14}-C_{16}) methyl benzyl ammonium chloride; cationic; wetting agent, emulsifier, biocide, disinfectant; pale-yel. liq. (working sol'ns. are odorless and colorless); sol. in water, ethanol, acetone, aliphatic solv.; sp.gr. 0.96; surf. tens. 33 dynes/cm (1%); 50% min. act.in ethanol (10%), water (40%).

Alkaquat DMB-ST, 25%. [24] Alkyl benzyl dimethyl quat., stearic deriv., strongly cationic; antistat, conditioner, softener for fibers, hair, paper products; wh. paste; sol. in water; dens. 0.95 g/ml; pH 3.0–5.0 (0.5% aq.); 24–26% act.

Alkaquat O. [25] Oleic imidazoline quat.; cationic; softening agent, liq. form of Alkaquat T for use in high conc. of fabric softeners, textiles, etc.; amber liq.; solv. odor; sol. in min. oil, min. spirits, aromatic solv., perchlorethylene; disp. in water; sp.gr. 0.95; f.p. < 0 C; pH 6.0–6.8; 75% act.

Alkaquat S. [24] Soya imidazoline quat.; antistat, fabric softener for textile applications; amber liq.; sol. in min. oil, min. spirits, aromatic solv., perchlorethylene; disp. in water; dens. 0.98 g/ml; pH 6.0-6.8; 74–76% act.

Alkaquat ST. [25] Modified dialkyl imidazolinium quat. based on stearyl deriv.; fabric softener and antistat for textile fibers; cream solid; IPA solv. odor; sol. in min. oil, aromatic solv., perchlorethylene; disp. in water, min. spirits; m.p. 45 C; pH 6.0–6.8 (10% in IPA); 74–76% act.

Alkaquat T. [24; 25] Ditallow imidazolinium quat.; cationic; textile surfactant, softening agent in home and commercial laundries; pale-yel., slightly visc. fluid; typ. odor; sol. in min. oil, perchlorethylene; disp. in water, min. spirits, aromatic solv.; sp.gr. 0.95; pH 6.0–6.8; 74-76% act.

Alkasurf 075-5. [24] Oleic acid ethoxylated; nonionic; emulsifier, lubricant, antistat for textile fiber additives; liq.; HLB 8.0; 100% conc.

Alkasurf 075-7. [24] Oleic acid ethoxylated; nonionic; see Alkasurf 075-5;. liq.; water-disp.; HLB 10.0; 100% conc.

Alkasurf 075-9. [24] Oleic acid ethoxylated; nonionic; see Alkasurf 075-5; liq.; water-disp.; HLB 11.0; 100% conc.

Alkasurf 61-4.4. [24] Linear alcohol ethoxylate; nonionic; emulsifier in prespotter formulations; used in household and industrial cleaners; liq.; HLB 10.5; 100% conc.

Alkasurf 1012-6. [24] Linear (C_{10}-C_{12}) alcohol ethoxylate; nonionic; surfactant for household and industrial cleaner formulations; liq.; HLB 12.0; 100% conc.

Alkasurf AG-CA, AG-CAHF. [24] Calcium alkylaryl sulfonate; anionic; emulsifier; liq.; HLB 10.5; 60% conc.

Alkasurf ALS. [25] Ammonium lauryl sulfate; anionic; detergent, household, cosmetic, wetting agent, emulsifier; lt.-colored liq., typ. odor; water sol.; 27% act.

Alkasurf BA-PE 70. [24] Primary aliphatic alcohol alkoxylate, straight chain; nonionic; surfactant for emulsification and dispersion applications in insecticides/herbicides, latex polymerization and in leather finishing; soft solid; HLB 16.1; 100% conc.

Alkasurf BA-PE 80. [24] Primary aliphatic alcohol alkoxylate, straight chain; nonionic; see Alkasurf BA-PE 70; HLB 26.1; 100% conc.

Alkasurf CA. [23; 24; 25] Calcium dodecylbenzene sulfonic acid; anionic; industrial o/w emulsifier; dispersant in polyester yarn dyeing; used in agricultural chemicals; amber liq.; sol. in min. oil, min. spirits, aromatic solv., perchlorethylene; disp. in water; dens. 1.00 g/ml; HLB 10.5; pH 6.0–8.0 (5% DW); 59–61% act.

Alkasurf CBX16. [25] Fatty alcohol polyglycol ether carboxylic salt, acid; anionic; detergent, wetting agent; lt. yel. liq.; odorless; good sol. in alkali; f.p. < 0 C; 70% act.

Alkasurf CBX16 Acid. [25] Fatty alcohol polyglycol ether carboxylic salt, acid; anionic; detergent, wetting agent; lt. yel. liq.; odorless; good sol. in alkali; 100% act.

Alkasurf CO-5. [24] PEG-5 castor oil; nonionic; emulsifier and dispersant with antifoaming properties for chlorinated solv.; pigment dispersant in latex paints; emulsifier in textile processing; Gardner 4 liq.; disp. in water; dens. 0.98 g/ml; HLB 3.8; cloud pt. (CO-10) 45-49 C; sapon. no. 145-150.

Alkasurf CO-10. [24] PEG-10 castor oil; nonionic; see Alkasurf CO-5; Gardner 4 liq.; disp. in water; dens. 1.00 g/ml; HLB 7.2; cloud pt. (CO-10) 45-49 C; sapon. no. 1231-126; 100% conc.

Alkasurf CO-15. [24] PEG-15 castor oil; nonionic; dispersant, detergent, wetting agent, defoamer, antistat; emulsifier in syn. fiber lubricants; vat dyeing assistant in the textile industry; cutting oils and hydraulic fluids; Gardner 4 liq.; disp. in water; dens. 1.02 g/ml; HLB 8.6; cloud pt. 56-60 C; sapon. no. 104-109; 100% conc.

Alkasurf CO-20. [24] PEG-20 castor oil; nonionic; emulsifier, lubricant, softener, coemulsifier, antistat, degreaser for textiles, leather processing; pigment dispersants; Gardner 4 liq.; sol. in water, aromatic solv., perchloroethylene; dens. 1.02 g/ml; HLB 10.3; cloud pt. 65-69 C; sapon no. 91-96; 100% conc.

Alkasurf CO-25. [24; 25] PEG-25 castor oil; nonionic; detergent, wetting agent, emulsifier, lubricant for textiles, cosmetics; yel. liq., typ. odor; 100% act.

Alkasurf CO-30. [24] PEG-30 castor oil; nonionic; emulsifier, lubricant for textiles; Gardner 4 liq.; dens. 1.04 g/ml; HLB 11.7; cloud pt. 48-61 C; sapon. no. 73-78; 100% conc.

Alkasurf CO-40. [24] PEG-40 castor oil; nonionic; emulsifier for oils and waxes; coemulsifier for dye carriers in textiles; lubricant for textile/leather processing; liq.; water-sol.; HLB 13.0; 100% conc.

Alkasurf CO-40M. [24] PEG-40 castor oil; nonionic; lubricant and softener for textiles; coemulsifier for dye carrier systems; Gardner 4 liq.; dens. 1.04 g/ml; HLB 13.0; cloud pt. 65-71 C; sapon no. 58-63; 90% conc.

Alkasurf CO-200. [24] Castor oil ethoxylate; nonionic; emulsifier used as lubricant and antistatic humectant for syn. fiber processing in the textile industry; Gardner 4 solid; water-sol.; HLB 18.1; cloud pt. 78-84 C; sapon. no. 12-18; 50% conc.

Alkasurf DA-3. [24] Decyl alcohol ethoxylate; nonionic; wetting agent and penetrant used in pressure dyeing of fabrics and textile or industrial applications; hazy liq.; water-disp.; HLB 9.0; 100% conc.

Alkasurf DA-4. [24] Decyl alcohol ethoxylate; nonionic; see Alkasurf DA-3; hazy liq.;

water-disp.; HLB 10.5; 100% conc.

Alkasurf DA-6. [24] Decyl alcohol ethoxylate; nonionic; see Alkasurf DA-3; hazy liq.; water-disp.; HLB 12.5; 100% conc.

Alkasurf DLS. [25] DEA-lauryl sulfate; detergent, wetting agent, emulsifier; lt.-colored liq.; typ. odor; water sol.; 35% act.

Alkasurf DNP-10. [24] Nonylphenol ethoxylate; nonionic; emulsifier for cleaners, insecticides, wax emulsions; liq.; sol. in min. oil and spirits, aromatic solv., perchloroethylene; dens. 1.03 g/ml; HLB 10.6; cloud pt. 73 C; pH 5-8 (5% DW); 99% min. act.

Alkasurf EA-60. [25] Ammonium laureth sulfate; anionic; detergent, surfactant, wetting agent, emulsifier; clear, lt. amber-colored, slightly visc. liq.; typ. odor; water sol.; sp.gr. 1.04 @ 20 C; cloud pt. < 40 F; 58% act.

Alkasurf ES-60. [25] Sodium laureth sulfate; anionic; general purpose surfactant; detergent, emulsifier, wetting agent; 58% act.

Alkasurf IPAM. [23; 24; 25] Isopropylamine dodecylbenzene sulfonate; anionic; emulsifier, dry cleaning surfactant; clear, pale yel., visc. liq.; water insol., sol. in most commonly used solv.; sp.gr. 1.06; 95% act.

Alkasurf L-2. [24] Lauric fatty acid ethoxylate; nonionic; coemulsifier, dispersing agent, defoamer, emulsifier, and lubricant for industrial and textile oils; yel. liq., paste; insol. in water; dens. 0.96 g/ml; HLB 6.0; sapon. no. 190-200.

Alkasurf L-5. [24] Lauric fatty acid ethoxylate; nonionic; emulsifier, coupling agent, solubilizer in metal working fluids, industrial lubricants, and textile lubricants; yel. liq.; disp. in water; dens. 1.00 g/ml; HLB 9.8; sapon no. 145-152.

Alkasurf L-9. [24] Lauric fatty acid ethoxylate; nonionic; emulsifier and coemulsifier for cosmetic and toiletry preparations; defoamer and leveling agent for latex paints; dispersant for pigments and dyes; yel. liq., paste; sol. in aromatic solv., perchloroethylene; disp. in water; dens. 1.03 g/ml; HLB 12.8; sapon. no. 91-101.

Alkasurf L-14. [24] Lauric fatty acid ethoxylate; see Alkasurf L-9; yel. liq., paste; sol. in water; dens. 1.05 g/ml; HLB 14.6; sapon. no. 67-77.

Alkasurf LA-1. [24] Fatty alcohol (C_{12}-C_{15}) ethoxylate; nonionic; detergent, intermediate, emulsifier, dispersant, wetting agent for mfg. of shampoos; dye retardant in hair color preparations; sol. in min. oil and spirits, aromatic solv., perchloroethylene; insol. in water; dens. 0.88 g/ml; HLB 3.6; 100% conc.

Alkasurf LA-3. [24; 25] C_{12-15} pareth-3; nonionic; emulsifier, detergent base; translucent- to-opaque liq.; oil-sol.; sp.gr. .9; visc. 19 cps (100 F); HLB 7.8; cloud pt. 58-62 C; 100% act.

Alkasurf LA-7. [24; 25] C_{12-15} pareth-7; nonionic; detergent, emulsifier, wetting agent, dispersant for sanitizers, metal cleaners, hard surface cleaners; lt. liq.; water sol.; sp.gr. 0.96; visc. 32 cps (100 F); m.p. 11-15 C; HLB 12; cloud pt. 44-48 C; flash pt. 440 F (COC); pour pt. 70 F; biodeg.; 100% act.

Alkasurf LA-9. [24; 25] Linear fatty alcohol ethoxylate; nonionic; detergent, emulsifier, wetting agent; lt. semisolid; typ. odor; water sol.; sp.gr. 1.01; HLB 13.1; biodeg.; 100% act.

Alkasurf LA-9 85%. [25] Linear fatty alcohol ethoxylate; see Alkasurf LA-9; liq. form of LA-9.

Alkasurf LA-12. [24; 25] C_{12-15} pareth-12; nonionic; detergent, emulsifier, wetting agent; wh. solid; low odor; water sol.; sp.gr. 1.0; HLB 14.4; 100% act.

Alkasurf LA-15. [25] Linear fatty alcohol ethoxylate; detergent, emulsifier, wetting agent, nonionic; wh. solid; biodeg.; 100% act.

Alkasurf LA-16. [25] Linear fatty alcohol ethoxylate; nonionic; detergent for textiles, skins, papermaking, household, industrial, etc., emulsifier used in making of cutting and sol. oils, wetting agent in textile industry; clear colorless liq.; low odor; water sol.; sp.gr. 1.0; visc. 267 SUS; biodeg.; 100% act.

Alkasurf LA23-3. [24] Fatty alcohol (C_{12}-C_{13}) ethoxylate; nonionic; detergent, emulsifier, intermediate, dispersant for mfg. of dishwashing, shampoo, bubble bath, and specialty industrial preparations; liq.; sol. in min. oil, aromatic solv., perchloroethylene; disp. in water; dens. 0.93 g/ml; HLB 8.1; 100% conc.

Alkasurf LA23-6.5. [24] Fatty alcohol ethoxylate, straight chain; nonionic; detergent; industrial usage, household detergent products; liq.; HLB 12.0; 100% conc.

Alkasurf LA1423-3. [24] Linear alcohol ethoxylate; nonionic; emulsifier for min. oils and chlorinated solv.; coemulsifier for use in industrial and consumer products; liq.; HLB 8.0; 100% conc.

Alkasurf LA Acid. [24; 25] Dodecylbenzene

sulfonic acid; anionic; detergent, wetter, emulsifier, penetrant, foamer, intermediate used in liq. dishwashing detergents, cleaners, personal care products, and degreasers; br. visc. liq.; sol. in water, min. oil, perchloroethylene; dens. 1.06 g/ml; biodeg.; 96% min. act.

Alkasurf LA-EP15. [23; 24] Oxyethylated straight chain alcohol, modified; nonionic; detergent, dispersant, wetting agent, emulsifier used in machine dishwasher formulations, metal cleaners, household and industrial formulations; liq.; sol. in water, perchloroethylene; HLB 11.9; cloud pt. 25-30 C (1% DW); pH 5-8 (5%); biodeg.

Alkasurf LA-EP16. [23; 24] Oxyethylated straight chain alcohol, modified; nonionic; detergent, wetting agent, emulsifier and dispersant for industrial and household cleaners; liq.; sol. in water, aromatic solv.; HLB 13.0; cloud pt. 56-63 C (1% DW); pH 5-8 (5%); 100% conc.

Alkasurf LA-EP25. [23; 24] Oxyethylated straight chain alcohol, modified; nonionic; wetting agent used in industrial and household cleaning products; machine dishwashing and dishwasher rinse aids; liq.; sol. in water, min. spirits, aromatic solv., perchloroethylene; HLB 7; pH 5-8; cloud pt. 24-27 C (1% DW); 100% conc.

Alkasurf LA-EP25LF. [23; 24] Oxyethylated straight chain alcohol, modified; nonionic; see Alkasurf LA-EP25; water-sol.; HLB 7; pH 5-8; cloud pt. 4-7 C; 100% conc.

Alkasurf LA-EP35. [23; 24] Oxyethylated straight chain alcohol, modified; nonionic surfactant; see Alkasurf LA-EP25; liq.; water-sol.; HLB 9; pH 5-8 (5%); cloud pt. 34-37; 100% conc.

Alkasurf LA-EP45. [23; 24] Oxyethylated straight chain alcohol, modified; nonionic; see Alkasurf LA-EP25; liq.; water-sol.; HLB 10; pH 5-8 (5%); cloud pt. 42-45.

Alkasurf LA-EP50. [24] Oxyethylated straight chain alcohol, modified; nonionic; see Alkasurf LA-EP25; liq.; HLB 11; pH 5-8 (5%); cloud pt. 46-50 C.

Alkasurf LA-EP59. [23; 24] Oxyethylated straight chain alcohol, modified; nonionic surfactant; useful in all-purpose cleaners, heavy duty liq. and granular cleaners, and detergent sanitizers; liq.; sol. in water, min. spirits, aromatic solv., perchloroethylene; HLB 10.0; cloud pt. 56-60 C (1% DW); pH 5-8 (5%); 100% conc.

Alkasurf LA-EP65. [23; 24] Oxyethylated straight chain alcohol, modified; nonionic surfactant; see Alkasurf LA-EP59; liq.; sol. in water, aromatic solv., perchloroethylene; HLB 11.5; cloud pt. 64-67 C (1% DW); pH 5-8 (5%); 100% conc.

Alkasurf LA-EP73. [23; 24] Oxyethylated straight chain alcohol, modified; nonionic; see Alkasurf LA-EP59; liq.; sol. in water, aromatic solv., perchloroethylene; HLB 14.0; cloud pt. 71-75 C (1% dw); pH 5-8 (5%); 100% conc.

Alkasurf LAN-1. [24] Laureth-1; nonionic; detergent-intermediate for mfg. of shampoo ethoxylates; liq.; sol. in min. oil and spirits, aromatic solv., perchloroethylene; insol. in water; dens. 0.88 g/ml; HLB 3.7; cloud pt. 37-39 C; 100% conc.

Alkasurf LAN-15. [24] Laureth-15; nonionic; coemulsifier for styrene and styrene/acrylic emulsion polymerization; wh. solid; sol. in water, aromatic solv., perchloroethylene; HLB 15.5; cloud pt. 85-88 C (1% in 5% NaCl); 100% conc.

Alkasurf LAN-23. [24; 25] Laureth-23; nonionic; emulsifier, stabilizer, surfactant for cosmetics; almost colorless solid; characteristic odor; HLB 17; biodeg.; 100% act.

Alkasurf MLS. [25] Monoethanolamine lauryl sulfate; detergent, wetting agent, surfactant; pale yel. liq., low odor; 32% act.

Alkasurf NP-1. [24] Nonoxynol-1; nonionic; emulsifier and dispersing agent for petroleum oils; coemulsifier and retardant in hair care formulations; defoamer in ethoxylates; liq.; sol. in min. oil and spirits, aromatic solv., perchloroethylene; insol. in water; dens. 0.99; HLB 4.6; pH 5-8 (5% DW); 99.0% min. act.

Alkasurf NP-4. [24; 25] Nonoxynol-4; nonionic; emulsifier, detergent, dispersant, intermediate, stabilizer; plasticizer, antistat for plastics; used in petroleum oils; lt. liq.; low odor; water insol.; sp.gr. 1.02; HLB 9; 100% act.

Alkasurf NP-5. [24; 25] Nonoxynol-5; nonionic; emulsifier, coemulsifier, dispersant; lt. liq.; low odor; oil-sol., water insol.; sp.gr. 1.03; HLB 10; 100% act.

Alkasurf NP-6. [24; 25] Nonoxynol-6; nonionic; emulsifier, coemulsifier, oil sol. dispersant; lt. liq.; low odor; oil-sol., water insol.; sp.gr. 1.04; HLB 11; 100% act.

Alkasurf NP-7LM. [24] Nonylphenol ethoxylate; nonionic; used in emulsifiable concs.;

liq.; 100% conc.

Alkasurf NP-8. [24; 25] Nonoxynol-8; nonionic; detergent, wetting agent, emulsifier; lt. liq.; low odor; water sol.; sp.gr. 1.05; HLB 12; 100% act.

Alkasurf NP-9. [24; 25] Nonoxynol-9; nonionic; detergent, wetting agent, emulsifier, dispersant for household and industrial cleaners, textile processing, laundry detergent, pesticides; lt. liq.; low odor; water sol.; sp.gr. 1.06; HLB 13.4; 100% act.

Alkasurf NP-10. [24; 25] Nonoxynol-10; nonionic; detergent, wetting agent, emulsifier; lt. liq.; low odor; water sol.; sp.gr. 1.06; HLB 13.5; 100% act.

Alkasurf NP-11. [24; 25] Nonoxynol-11; nonionic; detergent, wetting agent, emulsifier; lt. liq.; low odor; water sol.; sp.gr. 1.06; HLB 13.8; 100% act.

Alkasurf NP-12. [24] Nonoxynol-12; nonionic; surfactant for household and industrial cleaning formulations; liq.; HLB 13.9; 100% conc.

Alkasurf NP-15. [24; 25] Nonoxynol-15; nonionic; detergent, wetting agent; lt. paste; low odor; water sol.; sp.gr. 1.07; HLB 15; 100% act.

Alkasurf NP-15 80%. [24] Nonoxynol-15; nonionic; surfactant with wetting and penetrating properties; used in bottle washing compds., metal cleaners, and heavy duty alkaline cleaners; liq.; sol. in water; dens. 1.07 g/ml; HLB 15.0; cloud pt. 60-64 C; pH 5-8 (5% DW); 80% min. act.

Alkasurf NP-20. [24] Nonoxynol-20; nonionic; wetting agent, stabilizer, dispersant, penetrant; textile scouring agent; coemulsifier for oils, fats, waxes, and solv.; demulsifier for petroleum oil emulsions; wh. solid; sol. in water; dens. 1.08 g/ml; HLB 16.0; cloud pt. 71 C; pH 5-8 (5% DW); 99% min. act.

Aklasurf NP-20 70%. [24] Nonylphenol ethoxylate; see Alkasurf NP-20; wh. liq.; sol. in water; dens. 1.08 g/ml; HLB 16.0; cloud pt. 71 C; pH 5-8 (5% DW); 70% min. act.

Alkasurf NP-30 70%. [24] Nonylphenol ethoxylate; nonionic; solubilizer and coemulsifier for highly polar substances; liq.; sol. in water; dens. 1.09 g/ml; HLB 17.1; cloud pt. 74 C; pH 5-8 (5% DW); 70% min. act.

Alkasurf NP-35 70%. [24] Nonylphenol ethoxylate; nonionic; emulsifier and stabilizer for emulsion polymerization for latex paint, floor finishes, paper coatings, and textile finishes; liq.; sol. in water; dens. 1.09 g/ml; HLB 17.4; cloud pt. 76 C; pH 5-8 (5% DW); 70% min. act.

Alkasurf NP-40. [25] Nonoxynol-40; wetting agent, stabilizer, penetrant, emulsifier, dispersant, nonionic; lt. liq.; low odor; water sol.; f.p. 50-60F; 70% act.

Alkasurf NP-40, 70%. [24] Nonoxynol-40; nonionic; emulsifier, stabilizer, surfactant; coemulsifier and vinyl/acrylic latex stabilizer; liq.; HLB 17.6; 70% conc.

Alkasurf NP-50 70%. [24] Nonoxynol-50; coemulsifier and vinyl/acrylic latex stabilizer; liq.; sol. in water; dens. 1.09 g/ml; HLB 18.0; cloud pt. 76 C; pH 5-8 (5% DW); 70% conc.

Alkasurf NP-100 70%. [24] Nonoxynol-100; coemulsifier for solv., fats and waxes; emulsifier for emulsion polymerization of syn. latices; paste; sol. in water; dens. 1.09 g/ml; HLB 18.0; cloud pt. 76 C; pH 5-8 (5% DW); 70% conc.

Alkasurf O-2. [24] Oleic fatty acid ethoxylate; nonionic; coemulsifier for w/o and o/w emulsions; lubricant component in textile processing and leather tanning; cutting oils, solv. cleaners, degreasers, pesticides; yel. liq.; sol. in min. spirits; insol. in water; dens. 0.95; HLB 3.5; sapon. no. 140-150; 100% conc.

Alkasurf O-5. [24] Oleic fatty acid ethoxylate; nonionic; w/o emulsifier for min. oils, fatty acids and solv.; used in cutting oils, solv. cleaners, degreasers, pesticide formulations, as a lubricant softener in the tanning process and in lubricants and softeners for the textile industry; yel. liq.; disp. in water; dens. 0.99 g/ml; HLB 8.0; sapon. no. 111-121; 100% conc.

Alkasurf O-7. [24] Oleic acid ethoxylate; nonionic; see Alkasurf O-2; liq.; HLB 10.0; 100% conc.

Alkasurf O-9. [24] Oleic fatty acid ethoxylate; nonionic; surfactant used as a dyeing assistant in the textile industry and as an emulsifier for neats-foot oil fat liquors in leather processing; yel. liq.; sol. in aromatic solv., perchloroethylene; disp. in water; dens. 1.02 g/ml; HLB 11.0; sapon. no. 75-85; 100% conc.

Alkasurf O-14. [24] Oleic fatty acid ethoxylate; nonionic; coemulsifier used with O-5 and O-7 in various industrial applications; yel. liq.; sol. in water, aromatic solv., perchloroethylene; dens. 1.04 g/ml; HLB 13.0; sapon. no. 57-67; 100% conc.

Alkasurf OA-2. [24] Oleth-2; nonionic; emulsifier and solubilizer for topical cosmetics; liq.; sol. in min. oil, aromatic solv., perchloroethylene; insol. in water; dens. 0.90 g/ml; HLB 4.9; cloud pt. 47-50 C; 100% conc.

Alkasurf OA-10. [24] Oleth-10; see Alkasurf OA-2; paste; sol. in water, aromatic solv.; HLB 12.4; cloud pt. 53-60 C (1% DW); 100% conc.

Alkasurf OA-20. [24] Oleth-20; nonionic; emulsifier, dispersant, solubilizer and detergent; dyeing assistant for wool/acrylic blends; wh. solid; sol. in water; HLB 15.3; cloud pt. 73-77 C; 100% conc.

Alkasurf OP-1. [24] Octoxynol-1; nonionic; emulsifier and dispersant for petroleum oils; coemulsifier and retardant in hair color preparations; coupling agent; liq.; sol. in min. oil and spirits, aromatic solv., perchloroethylene; dens. 0.99 g/ml; HLB 3.6; pH 5-8 (5% DW); 100% act.

Alkasurf OP-5. [24; 25] Octoxynol-5; nonionic; emulsifier, dispersant, surfactant, dry cleaning detergents, insecticides and wax emulsions; lt.-colored liq.; char. odor; oil sol.; sp.gr. 1.05 (15/15 C); HLB 10.4; 100% act.

Alkasurf OP-6. [24] Octoxynol-6; nonionic; co-emulsifier for oils, nonpolar solv., solv. emulsion cleaners, and dry cleaning soaps; liq.; HLB 11.4; 100% conc.

Alkasurf OP-8. [24; 25] Octoxynol-8 (7–8 EO); nonionic; detergent, emulsifier, wetting agent for household and industrial laundry detergents; lt.-colored liq.; low odor; water sol.; sp.gr. 1.06 (15/15 C); HLB 12.4; 100% act.

Alkasurf OP-10. [24; 25] Octoxynol-10; nonionic; detergent, emulsifier, wetting agent, dispersant for household and industrial cleaners, textile processing, wool scouring, metal cleaning; lt.-colored liq.; low odor; water sol.; sp.gr. 1.1 (15/15 C); HLB 13.5; 100% act.

Alkasurf OP-12. [24; 25] Octoxynol-12 (12–13 EO); nonionic; detergent, emulsifier, wetting agent for metal cleaning, industrial and household liq. detergents; lt.-colored liq.; low odor; water sol.; sp.gr. 1.07; HLB 14.6; 100% act.

Alkasurf OP-16. [24] Octoxynol-16; nonionic; detergent, solubilizer, coupler, hydrotrope, and foam stabilizer; bottle washing compds.; paste; sol. in water; dens. 1.08 g/ml; HLB 15.8; cloud pt. 67 C; pH 5-8; 100% act.

Alkasurf OP-30. [24] Octoxynol-30; emulsifier for vinyl acetate and acrylate emulsion polymerization; foamer in specialty applications; wh. solid; sol. in water; dens. 1.10; HLB 17.5; cloud pt. 75; pH 5-8 (5% DW); 100% act.

Alkasurf OP-30 70%. [24; 25] Octoxynol-30; nonionic; detergent, emulsifier, wetting agent; coemulsifier and vinyl/acrylic latex stabilizer; lt.-colored liq.; low odor; water sol.; sp.gr. 1.1; HLB 17.3; 70% act.

Alkasurf OP-40. [25] Octoxynol-40; nonionic; detergent, emulsifier, wetting agent; lt.-colored liq.; low odor; water sol.; sp.gr. 1.1; HLB 18; 70% act.

Alkasurf OP-40, 70%. [24] Octoxynol-40; nonionic; see Alkasurf OP-30, 70%; liq.; HLB 18.0; 70% conc.

Alkasurf OP-70-50%. [24] Octoxynol-70; nonionic; see Alkasurf OP-30; wh. liq.; sol. in water; dens. 1.08 g/ml; HLB 18.7; cloud pt. 76 C; pH 5-8 (5% DW); 50% conc.

Alkasurf S-1. [24] Stearic fatty acid ethoxylate; nonionic; coemulsifier, opacifying and pearlescing agents for liq. cosmetic and detergent products; wh. solid; HLB 2.9; sapon. no. 174-184; 100% conc.

Alkasurf S-2. [24] Stearic fatty acid ethoxylate; nonionic; see Alkasurf S-1; wh. solid; HLB 3.7; sapon. no. 154-164; 100% conc.

Alkasurf S-5. [24] Stearic fatty acid ethoxylate; nonionic; emulsifier for fats and oils; softener and lubricant for textiles and leather; wh. solid; disp. in water; HLB 8.0; sapon. no. 120-130; 100% conc.

Alkasurf S-8. [24] Stearic fatty acid ethoxylate; lubricant and softener in textile compositions designed for syn. fibers; wh. solid; sol. in aromatic solv., perchloroethylene; disp. in water; HLB 11.2; sapon. no. 80-90.

Alkasurf S-14. [24] Stearic fatty acid ethoxylate; nonionic; defoamer and leveling agent for latex paints; pigment dispersant and dye leveling agent for textiles; wh. solid; sol. in water, perchloroethylene; HLB 13.2; sapon. no. 60-70; 100% conc.

Alkasurf S-40. [24] Stearic fatty acid ethoxylate; nonionic; lubricant and emulsifier in production of textile lubricants and softeners; cosmetics, pharmaceuticals, silicone preparations; wh. flake; sol. in water; HLB 17.0; sapon. no. 25-35; 100% conc.

Alkasurf S65-8. [24] Stearic acid, ethoxylated;

nonionic; see Alkasurf S-8 and S-40; solid; HLB 11.2; 100% conc.

Alkasurf S65-40. [24] Stearic acid, ethoxylated; see Alkasurf S-8 and S-40; flake; HLB 17.0; 100% conc.

Alkasurf SA-2. [24] Steareth-2; nonionic; emulsifier in topical cosmetic applications; wh. solid; sol. in aromatic solv., perchloroethylene; HLB 4.9; cloud pt. 57-61 C; 100% conc.

Alkasurf SA-10. [24] Steareth-10; nonionic; emulsifier used in topical cosmetic applications; wh. solid; sol. in water; HLB 12.4; cloud pt. 60-64 C; 100% conc.

Alkasurf SA-20. [24] Steareth-20; nonionic; detergent and lubricant for fiber and fabric scouring; emulsifier for topical cosmetics; roll-on deodorants; wh. solid; sol. in water; HLB 15.3; cloud pt. 73-77 C; 100% conc.

Alkasurf SS-CDE. [24] Coconut diethanolamide sodium sulfosuccinate ester; anionic; foamer, foam stabilizer used in personal care products and liq. detergent preparations; amber liq.; sol. in water; dens. 1.1 g/ml; pH 5.0-6.0; 39-41% conc.

Alkasurf SS-DA-3. [23; 24] Decyl alcohol, ethoxylated sodium sulfosuccinate ester; anionic; emulsifier, dispersant used in emulsion polymerization of acrylates; liq. 35% conc.

Alkasurf SS-DA4-HE. [24] Disodium decyl PEG-4 sulfosuccinate; anionic; emulsifier, dispersant used in emulsion polymerization; surfactant in personal care products; liq.; 30% conc.

Alkasurf SS-DA-6. [24; 25] Decyl alcohol, ethoxylated sodium sulfosuccinate ester; anionic; detergent, foaming and wetting agent, emulsifier, solubilizer, sequestrant; pale yel. liq.; bland odor; water sol.; sp.gr. 1.1; f.p. < 0 C; 30% act.

Alkasurf SS-IB-45. [24] Sodium diisobutyl sulfosuccinate; anionic; wetting agent and emulsifier for a wide range of applications; liq.; 45% conc.

Alkasurf SS-L7DE. [24] Lauric diethanolamide sodium sulfosuccinate ester; anionic; see Alkasurf SS-CDE; pale yel. liq.; sol. in water; dens. 1.1 g/ml; pH 5.0-7.5; 39-41% conc.

Alkasurf SS-L9DE. [23; 25] Lauric diethanolamide sodium sulfosuccinate ester; anionic; detergent, low irritation surfactant; lt. yel. liq.; bland odor; water sol.; sp.gr. 1.17; biodeg.; 40% act.

Alkasurf SS-L9ME. [24; 25] Disodium lauramido MEA-sulfosuccinate; anionic; detergent for toiletry and carpet shampoo formulations; cream wh., visc. liq.; bland odor; sp.gr. 1.00; biodeg.; 40% act.

Alkasurf SS-L-HE. [24] Disodium lauryl sulfosuccinate; anionic; wetting and foaming agent for hair and bath preparations; liq.; 40% conc.

Alkasurf SS-LA-3. [23; 24; 25] Disodium laureth sulfosuccinate; anionic; detergent, emulsifier, wetter for shampoo and bubble bath base, skin cleansers; lt., clear liq.; typ. odor; water sol.; sp.gr. 1.10; 30% act.

Alkasurf SS-MA-80. [24] Sodium dihexyl sulfosuccinate; anionic; wetting agent and emulsifier; liq.; 80% conc.

Alkasurf SS-NO. [24] Tetrasodium N-alkyl sulfosuccinamate; anionic; emulsifier, solubilizer, wetting agent; liq.; 35% conc.

Alkasurf SS-O-30. [24] Sodium dioctyl sulfosuccinate; anionic; emulsifier and wetting agent used in industrial applications; liq.; 30% conc.

Alkasurf SS-O-40. [24] Sodium dioctyl sulfosuccinate; anionic; emulsifier and wetting agent for industrial applications; liq.; 40% conc.

Alkasurf SS-O-60. [23; 24; 25] Sodium dioctyl sulfosuccinate; anionic; emulsifier and wetting agent; clear, colorless liq.; 1% water sol., sol. in most org. liq.; sp.gr. 1.07; surf. tens. 29 dynes/cm (0.1% aq.); 60% act.

Alkasurf SS-O-60PG. [24] Sodium dioctyl sulfosuccinate; anionic; emulsifier and wetting agent for industrial applications; liq.; 60% conc.

Alkasurf SS-O-70PG. [24] Sodium dioctyl sulfosuccinate; anionic; wetting agent and emulsifier; liq.; 70% conc.

Alkasurf SS-O-75. [23; 24; 25] Sodium dioctyl sulfosuccinate; anionic; wetting agent, emulsifier; clear, colorless liq.; typ. odor; 1% water sol., sol. in most org. liq.; sp.gr. 1.09; surf. tens. 29 dynes/cm (0.1% aq.); f.p. < –20 C; 75% act.

Alkasurf SS-OA. [25] Disodium N-oleyl sulfosuccinamate; foaming agent; amber liq.; low odor; water sol.; visc. 25-30 cps; f.p. < 0 C; 31% act.

Alkasurf SS-OA-HE. [24] Disodium oleyl amido PEG-2 sulfosuccinate; anionic; foaming and wetting agent for shampoo and bath preparations; liq.; 40% conc.

Alkasurf SS-O-ME. [25] Fatty alkanolamide

sulfosuccinate; anionic; emulsifier; liq.; 40% conc.

Alkasurf SS-TA. [23; 24; 25] Disodium N-octadecyl sulfosuccinamate; anionic; detergent, foaming agent, emulsifier, dispersant, solubilizer; cream-colored paste; low odor; poor sol. in water; 35% act.

Alkasurf SS-TA 125. [23; 24] Sodium N-octadecyl sulfosuccinamate; anionic; see Alkasurf SS-TA; liq.; 35% conc.

Alkasurf T [24; 25] TEA-dodecylbenzene sulfonate, cosmetic grade; anionic; detergent, emulsifier, wetting agent for liq. detergents and bubble baths; clear, straw-colored liq.; typ. odor; complete sol. in water; visc. 2500 cps max.; 60% act.

Alkasurf TA-20. [24] Tallow alcohol ethoxylate; nonionic; detergent and dyeing assistant for fiber blends in textle industry; coemulsifier in topical cosmetrics; wh. solid; sol. in water, aromatic solv., perchloroethylene; HLB 15.4; cloud pt. 73-77 C; 100% conc.

Alkasurf TA-30. [24] Tallow alcohol ethoxylate; nonionic; see Alkasurf TA-20; wh. solid; sol. in water; HLB 16.7; cloud pt. 74-79 C; 100% conc.

Alkasurf TA-40. [24] Tallow alcohol ethoxylate; nonionic; see Alkasurf TA-20; wh. solid; sol. in water; HLB 17.4; cloud pt. 75-79 C; 100% conc.

Alkasurf TA-50. [24] Tallow alcohol ethoxylate; nonionic; see Alkasurf TA-20; wh. solid; sol. in water; HLB 17.8; cloud pt. 75-79 C; 100% conc.

Alkasurf TDA-5. [24] Tridecyl alcohol ethoxylate; nonionic; wetting, emulsifying, and dispersing agent for textile scouring, cutting oils; intermediate for phosphate esters; liq.; sol. in min. oil and spirits, aromatic solv., perchloroethylene; disp. in water; dens. 0.96 g/ml; HLB 10.5; cloud pt. 65-68 C; 100% conc.

Alkasurf TDA-6. [24] Tridecyl alcohol ethoxylate; nonionic; emulsifier, wetting, dispersant; dye leveling agent in textiles; solv. degreasers and industrial cleaning compds.; liq.; HLB 11.4; 100% conc.

Alkasurf TDA-7. [24] Tridecyl alcohol ethoxylate; nonionic; emulsifier, rewetting agent for paper towels; solubilizer and foam stabilizer in liq. detergents; intermediate; liq.; sol. in water, min. oil and spirits, aromatic solv., perchloroethylene; dens. 0.98 g/ml; HLB 12.1; cloud pt. 72-75 C; 100% conc.

Alkasurf TDA-7.5. [25] Tridecyl alcohol ethoxylate; emulsifier, wetting agent, nonionic; almost water-wh., opaque liq.; mild odor; water sol.; sp.gr. 1.00 (20/20 C); 100% act.

Alkasurf TDA-8.5. [24] Fatty alcohol (C_{13}) ethoxylate; emulsifier, rewetting agent for paper towels; solubilizer and foam stabilizer in liq. detergents; intermediate; liq.; sol. in water, min. oil and spirits, aromatic solv., perchloroethylene; dens. 0.98 g/ml; HLB 12.5; cloud pt. 52-56 C.

Alkasurf TDA-12. [25] Tridecyl alcohol ethoxylate; nonionic; emulsifier, detergent; wh. solid; mild odor; water sol.; m.p. 27 C; 100% act.

Alkasurf TDA-15. [25] POE (15) tridecyl ether; nonionic; emulsifier, detergent; wh. solid; mild odor; water sol.; m.p. 30 C; 100% act.

Alkasurf TLS. [25] TEA-lauryl sulfate; anionic; detergent, emulsifier, wetting agent, toiletries where low cloud pt. properties are critica; lt. amber liq.; char. odor; water sol.; sp.gr. 1.02; cloud pt. 0 C; 40% act.

Alkasurf WADX. [25] Modified diethanolamine lauryl sulfate; anionic; detergent; straw-colored, clear to slightly hazy visc. liq.; bland odor; water sol.; 30% act.

Alkasurf WAQ. [25] Sodium lauryl sulfate; anionic; detergent, emulsifier, wetting agent; wh., pearlescent liq.; typ. odor; water sol.; 30% act.

Alkasurf WHC. [25] Blend of emulsifiers suited to compounding waterless hand cleaner; wetting agent; amber-colored, clear visc. liq.; typ. odor; 100% act.

Alkaterge-E. [52; 323] Ethyl hydroxymethyl oleyl oxazoline; amphoteric; detergent, emulsifier, wetting agent; Gardner 15 max., clear liq.; sol. in most org. liq., slight sol. in water; dens. 7.74 lb/gal; sp.gr. 0.9; visc. 155 cp; f.p. –31 C; flash pt. > 200 F; surf. tens. 40 dynes/cm; 79% act.

Alkaterge-T. [52; 323] Oxazoline-type compd.; amphoteric; detergent, emulsifier, dispersant, corrosion inhibitor; buff to br., waxy solid; high sol. in aromatic hydrocarbons; flash pt. > 200 F; surf. tens. 30.4 dynes/cm.

Alkaterge-TIV. [52] Oxazoline deriv., ethoxylated; nonionic; emulsifier, corrosion inhibitor; acid scavenger; dispersant; liq.; HLB 8.0; 60% conc.

Alkateric 2CIB. [24] Cocoamphodiacetate; amphoteric; foaming agent; emulsifier for industrial and liq. detergent formulations; used in personal care products; solubilizer, coupling agent; amber clear liq.; sol. in water; dens. 1.18 g/ml; pH 8-9.5; 49.5-50.5% solids.

Alkateric A2P-LPS, 30%. [24] Disodium N-lauryl iminodipropionate; amphoteric; foamer and wetting agent used in cleaners; liq.; 30% conc.

Alkateric A2P-TS, 30%. [24] Disodium N-tallow iminodipropionate; amphoteric; foamer and wetting agent used in cleaners; liq.; 30% conc.

Alkateric BC [23; 24; 25] Coco betaine; amphoteric; detergent and base for shampoos; pale straw liquid; low odor; water sol; f.p. < 0 C; pH 7.5-8.5; 31% act.

Alkateric CAB. [23; 24; 25] Coco amido betaine; amphoteric; detergent, foaming agent, conditioner, solubilizer, visc. builder for shampoos; amber liq.; bland odor; water sol.; 32% act.

Alkateric CIB. [23; 24; 25] Cocoamphoacetate; amphoteric; detergent, foaming agent, wetting agent; amber liq.; bland odor; sp.gr. 1.1; 40% act.

Alkateric EHCIB. [24] Caprylic imidazoline deriv.; amphoteric; wetting agent and emulsifier for alkaline cleaners; bottle washing compds., wax strippers, degreasers; clear liq.; 34% conc.

Alkateric LAB. [24] Lauric amidopropyl betaine; amphoteric; foaming and wetting agent; liq.; 35% conc.

Alkateric OB. [24] Oleyl betaine; amphoteric; foaming and wetting agent; gel; 50% conc.

Alkateric STB. [24; 25] Stearyl betaine; amphoteric; detergent, surfactant/softener for fabric and hair; pale yel. liq.; 50% act.

Alkateric TB. [24; 25] Tallow betaine; amphoteric; detergent, surfactant/softener for fabric and hair; amber liq.; 70% act.

Alkatronic EDP 8-4. [23] Ethylene diamine POP compd., ethoxylated; nonionic; detergent, dispersant, defoamer, wetting and gelling agent used in emulsion polymerization and dye leveling processes; antistat and acid release agents; liq.; HLB 16.0; 100% conc.

Alkatronic EDP 28-1. [23] Ethylene diamine POP compd., ethoxylated; nonionic; see Alkatronic EDP 8-4; liq.; HLB 3.0; 100% conc.

Alkatronic EDP-28-2. [23] Ethylene diamine POP compd., ethoxylated; nonionic; see Alkatronic EDP 8-4; liq.; HLB 7.0; 100% conc.

Alkatronic EDP 28-7. [23] Ethylene diamine POP compd., ethoxylated; nonionic; see Alkatronic EDP 8-4; flake; HLB 27.0; 100% conc.

Alkatronic EDP 38-1. [23] Ethylene diamine POP compd., ethoxylated; nonionic; see Alkatronic EDP 8-4; liq.; HLB 2.5; 100% conc.

Alkatronic EDP 38-4. [23] Ethylene diamine POP compd., ethoxylated; nonionic; see Alkatronic EDP 8-4; paste; HLB 14.5; 100% conc.

Alkatronic EDP 38-8. [23] Ethylene diamine POP compd., ethoxylated; nonionic; see Alkatronic EDP 8-4; flake; HLB 30.5; 100% conc.

Alkatronic EGE 17-1, 17-2, 17-4, 17-8, 25-1, 25-2, 25-5, 25-8, 25-4. [23; 24] Propoxylated POE glycol; nonionic; surfactants used in defoaming and foam control agents in pulp and paper industries; dye leveling and wetting agent in textile processing; liq.; 100% conc.

Alkatronic PGP 10-1. [23; 24] POP glycol ethoxylate; nonionic; surfactant used as oil demulsifier, visc. control agent, foamer, wetting agent; liq.; sol. in water; dens. 1.02 g/ml; HLB 4.5; cloud pt. 40 C (1%); pH 5.0-7.5 (5% DW); 100% conc.

Alkatronic PGP 12-2. [23; 24] POP glycol ethoxylate; nonionic; surfactant used in dishwasher rinse aids; oil demulsifier; liq.; sol. in water; dens. 1.03 g/ml; HLB 8.0; cloud pt. 38 C (1%); pH 5.0-7.5 (5% DW); 100% conc.

Alkatronic PGP 18-1. [23; 24] POP glycol ethoxylate; nonionic; surfactant with foam control for industrial applications; liq.; sol. in perchloroethylene and water; dens. 1.01 g/ml; HLB 3; cloud pt. 24 C (1%); pH 5.0-7.5 (5% DW); 100% conc.

Alkatronic PGP 18-2. [23; 24] POP glycol ethoxylate; nonionic; surfactant used in dishwasher rinse aids and wetting agent in industrial processes; sol. in water and perchloroethylene; dens. 1.03 g/ml; HLB 7.0; cloud pt. 32 C (1%); pH 5.0-7.5 (5% DW); 100% conc.

Alkatronic PGP 18-2D. [23; 24] POP glycol ethoxylate; nonionic; see Alkatronic PGP 18-2; liq.; sol. in water and perchloroethylene; dens. 1.04; cloud pt. 32 C (1%); pH 5.0-7.5 (5% DW); 100% conc.

Alkatronic PGP 18-2LF. [23; 24] POP glycol ethoxylate; nonionic; see Alkatronic PGP 18-2; liq.; sol. in water and perchloroethylene; dens. 1.03 g/ml; cloud pt. 28 C (1%); pH 5.0-7.5 (5% DW); 100% conc.

Alkatronic PGP 18-4. [23; 24] POP glycol ethoxylate; nonionic; surfactant used in detergent formulations; emulsifier and dispersing agent; paste; sol. see Alkatronic PGP 18-2; dens. 1.05 g/ml; HLB 15.0; cloud pt 62 C (1%); pH 5.0-7.5 (5% DW); 100% conc.

Alkatronic PGP 18-8. [23; 24] POP glycol ethoxylate; nonionic; surfactant used as lime soap dispersant, wetting agent latex stabilizer; flake; sol. in water; dens. 1.06 g/ml; HLB 29; cloud pt. > 100 C (1%); pH 5.0-7.5 (5% DW); 100% conc.

Alkatronic PGP 18-8LF. [24] POP glycol ethoxylate; nonionic; see Alkatronic PGP 18-8; flake; sol. in water; dens. 1.06 g/ml; cloud pt. > 100 C (1%); pH 5.0-7.5 (5% DW).

Alkatronic PGP 23-5. [23; 24] POP glycol ethoxylate; nonionc; surfactant used in syn. detergent formulations; paste; sol. in water, perchloroethylene; dens. 1.04 g/ml; HLB 16.0; cloud pt. 85 C (1%); pH 5.0-7.5 (5% DW); 100% conc.

Alkatronic PGP 23-7. [23; 24] POP glycol ethoxylate; nonionic; surfactants used as dispersants, gel agents, and visc. control agent; flake; sol. in water; dens. 1.04 g/ml; HLB 24.0; cloud pt. > 100 C (1%); pH 5.0-7.5 (5% DW); 100% conc.

Alkatronic PGP 23-8. [23; 24] POP glycol ethoxylate; nonionic; see Alkatronic PGP 23-7; flake; sol. in water; dens. 1.06 g/ml; HLB 28.0; cloud pt. > 100 C (1%); pH 5.0-7.5 (5% DW); 100% conc.

Alkatronic PGP 33-1. [23; 24] POP glycol ethoxylate; nonionic; surfactant with foam control in industrial products and processes such as antifreeze and dry cleaning detergents; liq.; sol. in perchloroethylene; dens. 1.02 g/ml; HLB 1.0; cloud pt. 16 C (1%); pH 5.0-7.5 (5% DW); 100% conc.

Alkatronic PGP 33-8. [23; 24] POP glycol ethoxylate; nonionic; surfactant used as dispersant, gelling agent and visc. control agent; flake; sol. in water; dens. 1.06 g/ml; HLB 27.0; cloud pt. > 100 (1%); pH 5.0-7.5 (5% DW) 100% conc.

Alkatronic PGP 40-7. [23; 24] POP glycol ethoxylate; nonionic; surfactant used as dispersant, gelling agent and visc. control agent; flake; sol. in water; dens. 1.05 g/ml; HLB 22.0; cloud pt. > 100 C (1%); pH 5.0-7.5 (5% DW); 100% conc.

Alkatrope SX-40. [25] Sodium xylene sulfonate; hydrotrope, solubilizer and coupling agent; lt.-colored, clear liq.; 41% act.

Alkawet NP-6. [372] Mono and dialkylarylphenoxy POE acid phosphates; anionic; emulsifier; intermediate surfactant for cleaners; liq.; 100% conc.

Alkazine C. [23; 24; 25] Coco hydroxyethyl imadazoline; cationic; emulsifier, antistat, corrosion inhibitor, softener for textiles, asphalt, tar emulsion breakers; water repellent treatment of cement, concrete, and plaster; br. paste; amine odor; sol. in aq. acid; dens. 0.96 g/ml; f.p. 20 C; 85% act.

Alkazine O. [23; 24; 25] Oleic hydroxyethyl imidazoline; cationic; emulsifier, corrosion inhibitor, antistat agent; br. liq.; amine odor; sol. in aq. acid and orgs.; f.p. 5 C; 85% act.

Alkazine ST. [23; 24; 25] Stearic hydroxyethyl imidazoline; cationic; emulsifier, dispersant; tan, waxy solid; amine odor; 85% act.

Alkazine TO. [23; 24; 25] Tall oil hydroxyethyl imidazoline; cationic; emulsifier, corrosion inhibitor; br. liq.; amine odor; sol. in acid systems; 80% act.

Alkazine TO-A. [24] Tall oil amino ethyl imidazoline; cationic; aggregate wetting agent and emulsifier in asphalt emulsions; amber liq.; sol. in min. oil and spirits, aromatic solv., perchloroethylene; disp. in water; dens. 0.93 g/ml; 100% conc.

Alkylate 215. [419] Linear alkylbenzene; detergent, intermediate used in lt.-duty detergents; liq.; 100% conc.

Alkylate 225. [419] Linear alkylbenzene; detergent, intermediate used in liq. and powd. heavy duty detergent; liq.; 100% conc.

Alkylate 230. [419] Linear alkylbenzene; detergent, intermediate used in lt. and heavy duty detergents; liq.; 100% conc.

Alkylate A215. [419] Alkylbenzene; intermediate, detergent; colorless liq.; m.w. 231-241; dens. 7.2 lb/gal; sp.gr. 0.855-0.870; biodeg.

Alkylate A225. [419] Alkylbenzene; intermediate; colorless liq.; m.w. 238-248; dens. 7.2 lb/gal; sp.gr. 0.855-0.870; biodeg.

Alkylate A230. [419] Alkylbenzene; intermediate; colorless liq.; m.w. 231-266; dens. 7.2 lb/gal; sp.gr. 0.855-0.870; biodeg.

Alkylate H230H. [419] By-product of alkylbenzene mfg.; anionic; intermediate, surfactant; yel. liq.; m.w. 360; sp.gr. 0.890-0.910; visc. 140-220 SUS (100 F); I.B.P. 290 C min.

Alkylate H230L. [419] By-product of alkylbenzene mfg.; anionic; intermediate, surfactant; yel. liq.; m.w. 335; sp.gr. 0.880-0.910; visc. 100-160 SUS (100 F); I.B.P. 290 C min.

Alkylene Series. [284] Phosphated ester; anionic; detergent (except 8,10,11,12), wetting agent (17-24), emulsifier (41-49); colorless liquid; practically odorless; sol. in water and org. solv.; up to 100% act.

Alkylox P1904. [551] Alkylpolyethoxyether; nonionic; textile detergent; liq.; 50% conc.

All Wet. [143] Nonylphenol and polyglycol mixture, ethoxylated; nonionic; detergent, wetting, emulsifier used in pesticides; liq.; 80% conc.

Alox 164. [32] Detergent, corrosion inhibitor; br. liq.; mild odor; sol. nonpolar solv.; visc. 35-45 SUS (100 F); flash pt. 150 F min (OC); pour pt. 60 F; 100% act.

Alox 575. [32] Oxygenated hydrocarbon with barium and a sodium petroleum sulfonate; emulsifier, corrosion inhibitor; tan liq., gel; mild odor; oil sol.; dens. 8.5 lb/gal; sp.gr. 1.02 (15.6 C); flash pt. 400 F; pour pt. 80 F; 100% act.

Alox 1689. [32] Ester of an oxygenated hydrocarbon; emulsifier, nonionic; dk. solid; mild odor; sol. nonpolar solv.; dens. 7.7 lb/gal; sp.gr. 0.93; m.p. 115 F; flash pt. 400 F (O.C.); 100% act.

Alox 1727. [32] Emulsifier, corrosion inhibitor; dk., visc. liq.; mild odor; sol. nonpolar solv.; dens. 8.0 lb/gal; sp.gr. 0.98; flash pt. 350 F; 100% act.

Alpha 500. [460] Dist. monoglyceride, polysorbate 60; emulsifier; dry powd.

Alphadim 90 AB. [462] Monoglyceride, distilled; nonionic; food emulsifier; bead; HLB 3.5; 100% conc.

Alphadim 90 LC. [462] Monoglyceride, distilled; nonionic; food emulsifier; plastic; HLB 3.5; 100% conc.

Alphadim 90 NLK, 90 SBK, 90 VCK. [462] Monoglyceride, kosher distilled; nonionic; bead, plastic; HLB 3.5; 100% conc.

Alphenate GA 65. [188] Alkyl polyglycol ether and esters blend; nonionic; detergent for soaping of textiles after dyeing and printing; liq.; 38% conc.

Alphenate HM. [188] Alkyl polyglycol ether blend; nonionic; detergent; liq.; 60% conc.

Alphenate PE Extra. [188] Alkyl polyglycol ester and sulfate semiester of alkylaryl polyglycol ether blend; anionic/nonionic; dyeing assistant for polyester and acrylic fibers; liq.; 50% conc.

Alphenate TFC-76. [188] Sodium alkyl sulfate, modified; anionic; dye leveling agent; liq.; 50% conc.

Alphenate TH 454. [188] Alkyl sulfosuccinate; anionic; wetting agent, emulsifier for emulsion polymerization; liq.; 50% conc.

Alphoxat S 110, 120. [660] Fatty acid polyglycol ester; nonionic; emulsifier and dispersant, greasing agent for textile and chemical technical industries; waxy; 100% conc.

Alrodyne 6104. [137] Polyoxyalkylene fatty ester polymer; emulsifier, nonionic; dk. amber, visc. liq.

Alrosol B. [137] Cocamide DEA; used in liq. soaps and detergents; amber, visc. liq.

Alrosol C. [137] Capramide DEA; foam booster, wetting agent; lt. amber, visc. liq.

Alrosol Conc. [137] Fatty acid amide; foam stabilizer, dispersing, emulsifying, wetting agent for SR and latexes; clear yel., visc. liq., mild odor; sp.gr. 0.99.

Alrosol O. [137] Oleamide DEA; emulsifier for hydrocarbons, pigments, etc.; amber, visc. liq.

Alrosperse 11P Flake. [137] Fatty acid amide; emulsifier, lubricant, nonionic; off-wh., waxy flake.

Alrosperse 100. [137] Blend of nonionic and cationic surfactants; emulsifier, solubilizing agent; dk. amber liq.

Alrowet D-65. [137] Dioctyl sodium sulfosuccinate; anionic; wetting and rewetting agent for textile applications; dispersant, antistat, wetting agent for NR and SR latexes; clear liq., slight odor; sp.gr. ;1.10; 65% conc.

Alscoap LE-240. [606] Triethanolamine alkyl ether sulfate; anionic; foaming agent, base material for hair shampoos; liq.; 40% conc.

Alscoap LN-40. [606] Sodium lauryl sulfate; anionic; detergent, shampoo base; liq.; 40% conc.

Alscoap LN-90. [606] Sodium lauryl sulfate; anionic; detergent, shampoo base, toothpaste; powd.; 90% conc.

Alscoap SS-90. [606] Alkyl ether sulfate/coconut diethanolamide blend; anionic/nonionic; detergent for formulated hair shampoo; pale yel. paste; 90% act.

Alscoap TAP-30. [606] Sodium lauryl ether sulfate; anionic; liq.; 30% conc.

Aluminum Stearate EA. [650] Aluminum distearate; food grade binder, emulsifier, anticaking agent for food applications; disoap; wh. powd.; sol. in turpentine, benzene, toluene, xylene, carbon tetrachloride, veg. and min. oil, oleic acid, waxes; insol. in water; sp.gr. 1.01.

Amascour. [42] Anionic detergent, wetting agent; clear wh. liq.; 100% water sol.; pH 8.

Amasperse N. [42] Anionic dispersing agent, used in textile industry, in metallurgy, etc.; clear liq.; 100% water sol.; dens. 9.7 lb/gal; 40% act.

Amawet AMR. [42] Potassium salt of a long chain ester; anionic; wetting agent; clear, light amber liq.; 100% water sol.; pH 8.

Amawet DF. [42] Dye assistant, nonfoaming; clear yel., slight visc. liq.; dens. 8.6 lb/gal; sp.gr. 1.03.

Amawet LF. [42] Ethoxylated linear alcohol; slightly anionic; wetting, emulsifying, dispersing and penetrating agents; water wh. liq.; faint fatty alcohol odor; biodeg.

Amawet P. [42] Sulfated fatty ester; anionic; dye assistant; dk. amber liq.; water sol.; pH 6.5.

Amberlig. [502] Calcium lignosulfonate; anionic; fly ash retarder, handling and processing aid; liq.; 50–58% conc.

Ambidex. [138] Cationic; wetting agent, emulsifier; dk. br., visc. liq.; amine odor; oil sol.; dens. 8.14 lb/gal; sp.gr. 0.985 (60 F); visc. 6600 SUS; flash pt. 450 F; pour pt. 40 F; 100% act.

Ambidex-PB. [138] Cationic; wetting agent, emulsifier; dk. br., visc. liq.; amine odor; oil sol.; dens. 8.2 lb/gal; sp.gr. 0.985 (60 F); visc. 660 SUS; flash pt. 370 F; pour pt. 72 F; 100% act.

Amerchol 400. [37; 38] Petrolatum, lanolin alcohol, and cetyl alcohol; nonionic; aux. emulsifier, emulsion stabilizer for o/w and w/o systems, incl. makeup, pharmaceuticals; emollient, lubricant; lt. cream soft solid, practically odorless; HLB 9.0; m.p. 49–50 C; acid no. 1.5; sapon. no. 8; 100% conc.

Amerchol BL. [37; 38] Lanolin, min. oil, and lanolin alcohol; nonionic; absorption base, aux. emulsifier for o/w systems, conditioner, emollient, moisturizer, stabilizer for cosmetics and pharmaceuticals; yel.-amber semisolid, slight char. sterol odor; oil sol.; HLB 8.0; sapon. no. 60–70; 100% conc.

Amerchol C. [37; 38] Petrolatum, lanolin, lanolin alcohol; nonionic; see Amerchol BL; pale yel.-cream soft solid, slight, char. sterol odor; oil sol.; HLB 9.5; m.p. 40–46 C; sapon. no. 10–20; 100% conc.

Amerchol CAB. [37; 38] Petrolatum, lanolin alcohol; nonionic; emollient, emulsifier for therapeutic ointments, burn preparations, dermatological products, hypoallergenic preparations; stabilizer, plasticizer; pale cream soft solid, faint, char. sterol odor; oil sol.; HLB 9.0; m.p. 40–46 C; acid no. 1 max.; sapon. no. 1.0 max.; 100% conc.

Amerchol H-9. [37; 38] Petrolatum, lanolin, lanolin alcohol; nonionic; emollient, emulsifier, penetrant, stabilizer, absorption base for cosmetics, pharmaceutical ointments, burn ointments; stabilizer, plasticizer; pale yel. soft solid, slight, char. sterol odor; oil sol.; HLB 9; m.p. 55–62 C; acid no. 1 max.; sapon. no. 15–27; 100% conc.

Amerchol L-99. [37; 38] Mineral oil, lanolin alcohol; nonionic; emulsifier, stabilizer for creams and lotions, cleansing, moisturizing creams, hair and skin preparations, dermatological specialties; lt. yel. oily liq., faint char. sterol odor; oil miscible; sp.gr. 0.855–0.875; HLB 8; acid no. 2 max.; sapon. no. 2 max.; 100% conc.

Amerchol L-101. [37; 38] Mineral oil, lanolin alcohol; nonionic; emollient, penetrant, emulsifier, moisturizer, softener, stabilizer for cosmetics, creams, makeup, hair dressing, pharmaceuticals, aerosols, baby products, textile finishes; plasticizer for hair sprays; pale yel. oily liq., faint, char. sterol odor; oil sol.; sp.gr. 0.840–0.860; visc. 20–30 cps; HLB 8; acid no. 1 max.; sapon. no. 1 max.; 100% conc.

Amerchol L-500. [37; 38] Mineral oil, lanolin alcohol, octyldodecanol; nonionic; emulsifier, stabilizer, emollient, moisturizer, conditioner for hair and skin products, creams, lotions, makeup, aerosols, pharmaceutical vehicles, baby toiletries; cloudy at R.T.; HLB 6; acid no. 2 max.; sapon. no. 5 max.; 100% conc.

Amerchol RC. [37] Petrolatum, lanolin alcohol, stearyl alcohol, and stearone; aux. emulsifier, stabilizer, emollient for creams, lotions, makeup; aids color definition; pale ivory soft solid, slight char. sterol odor; misc. with common oil phase ingred.; m.p. 55–64 C; acid no. 2 max.; sapon. no. 10 max.

Amerlate LFA. [37; 38] Lanolin acid; anionic; stabilizer, emollient for fatty acid systems, aerosol shave creams, cream shampoos, wax systems, household products; ylsh.-tan firm, waxy solid, mild, waxy odor; m.p. 55–62 C acid no. 130–150; sapon. no. 170–190; 100% conc.

Amerlate P. [37; 38] Isopropyl lanolate; nonionic; conditioner, penetrant, emollient, w/o emulsifier, stabilizer, opacifier for cosmetics and pharmaceuticals; pigment dispersant; yel. buttery solid, faint char. odor; HLB 9; sapon. no. 130–155; 100% act.

Amerlate W. [37; 38] Isopropyl lanolate; nonionic; dispersant/wetting agent for pigments in personal care products, pharmaceuticals; emulsifier, softener, lubricant, emollient; yel. buttery solid; faint char. odor; insol. in water; HLB 9; sapon. no. 135–165; hyd. no. 35–55; 100% act.

Amerlate WFA. [37; 38] Lanolin acid; anionic; see Amerlate LFA; also wets and disperses pigments; ylsh.-tan waxy solid, mild waxy odor; misc. with warm min. oil, IPP, IPM, castor oil; m.p. 60 C; acid no. 120; sapon. no. 165; 100% conc.

Ameroxol LE-4. [37; 38] POE (4) lauryl ether; nonionic; o/w emulsifier and solubilizer; liq.; HLB 9.7; 100% conc.

Ameroxol LE-23. [37; 38] POE (23) lauryl ether; nonionic; o/w emulsifier and solubilizer; solid; HLB 17.0; 100% conc.

Ameroxol OE-2. [37; 38] Oleth-2; nonionic; emulsifier, dispersant, stabilizer, lipophilic cosolvent; pale straw-colored, clear liq.; bland odor; sol. in min. oil, isopropyl esters, anhyd. ethanol; HLB 5.0; 100% conc.

Ameroxol OE-10. [37; 38] Oleth-10; nonionic; emulsifier, stabilizer, solubilizer for cosmetics and toiletries; wh. semisolid; bland odor; sol. in ethanol, water, hydroalcoholics, glycols; HLB 12; cloud pt. 47–55 C; 100% conc.

Ameroxol OE-20. [37; 38] Oleth-20; nonionic; see Ameroxol OE-10; wh. waxy solid; bland odor; sol. in ethanol, water, hydroalcoholics, glycols; HLB 15; cloud pt. 87–93 C; 100% conc.

Amerzyme A100-LC. [43a] Blend of amylase, protease, cellulase, and lipase; wetting agent, dispersant, and emulsifier; enzyme blend used for sludge build-up in drains and sewage treatment in home, industrial, commercial, municipal, and institutional installations; gr. liq.; perfumed odor; sol. in water; dens. 8.7 lb/gal; pH 6.5–7.5.

Amfotex FV-10. [494] Dicarboxylic coconut deriv., sodium salt; amphoteric; foaming and wetting agent used in personal care products; liq.

Amfotex FV-15. [494] Sulfonate coconut deriv.; foamer, corrosion inhibitor used in metal cleaning processes; liq.; 38% conc.

Amfotex FV-16. [494] Monocarboxylic lauric deriv., sodium salt; raw material for detergents, shampoos, dishwashers; textile softeners, paint emulsifiers, all-purpose washing agents; liq.; 38% conc.

Amfotex FV-28. [494] Alkyl amidobetaine; detergent, foaming agent, solubilizer, thickener used in personal care products; liq.; 30% conc.

Amidan. [272] Distilled monoglycerides; nonionic; emulsifier, dough conditioner, starch complexing agent in bread rolls, bread improvers; powd.; HLB 4.3; 100% conc.

Amido Betaine C. [659] Coconut amido alkyl betaine; amphoteric; component of personal care products; industrial foamer; liq.; gel; 30% conc.; avail. as 56% gel (C Conc.)

Amidox C-2. [575; 576] PEG-3 cocamide; nonionic; emulsifier, detergent for dishwashing detergents, shampoos, emulsions; Gardner 5 pasty liq.; sol. in ethanol, xylene; disp. in water; pH 10.0–11.5 (1% aq.); 100% conc.

Amidox C-5. [575; 576] PEG-6 cocamide; nonionic; see Amidox C-2; Gardner 5 max. liq.; sol. in water, ethanol, xylene; pH 10.0–11.5 (1% aq.); 100% conc.

Amidox L-2. [575; 576] PEG-3 lauramide; nonionic; see Amidox C-2; Gardner 3 waxy solid; pH 10.0–11.5 (1% aq.); 100% conc.

Amidox L-5. [575; 576] PEG-6 lauramide; nonionic; see Amidox C-2; Gardner 6 max. clear visc. liq.; sol. in water, ethanol, xylene; pH 10.0–11.5 (1% aq.); 100% conc.

Amiet 102, 105, 110, 115, 202, 205, 210, 215, 302, 305, 310, 315, 402, 405, 410, 415, 502, 505, 510, 515. [561] Tert. amines EO condensation products of coco (100 series), soya (200), hydrog. tallow (300), oleyl (400), tallow (500); cationic/nonionic; antistat, textile dyeing assistant, softening and dispersing agent; liq., paste; 100% conc.

Amifat P-21. [9] DL-Pyrrolidone carboxylic acid monostearin ester; nonionic; surfactant used as emulsifier in personal care products; wax; HLB 7.5; 100% conc.

Amifat P-30. [9] DL-Pyrrolidone carboxylic acid monoolein ester; nonionic; see Amifat P-21; paste; HLB 7.0; 100% conc.

Amihope LL-11. [9] Lauroyl-1-lysine; amphoteric; surface modifier, coemulsifier, codispersant; solid; 100% conc.

Amiladin. [172] PEG alkyl amine ether; nonionic; leveling agent for dyestuffs; paste; 50% conc.

Amiladin C-1802. [172] PEG alkyl amine ether; nonionic; dispersant, washing agent; liq.; 100% conc.

Amine 2M12. [348] Dimethyldodecylamine; cationic; intermediate for detergent; liq.; 97% conc.

Amine 2M14. [348] Dimethyltetradecylamine; cationic; intermediates for detergent; liq.; 97% conc.

Amine 2M16. [348] Dimethylhexadecylamine; cationic; intermediate for detergents; liq.; 97% conc.

Amine 2M18. [348] Dimethyloctadecylamine; cationic; liq.; 97% conc.

Amine 2MHBG. [348] Dimethylhydrogenated tallow amine; cationic; intermediate for detergent; liq.; 97% conc.

Amine 2MKK. [348] Dimethylcocoamine; cationic; intermediate for detergents; liq.; 97% conc.

Amine 8D. [348] *n*-Octylamine; cationic; emulsifier; liq., solid; 98% conc.

Amine 10D. [348] *n*-Decylamine; cationic; emulsifier; liq., solid; 98% conc.

Amine 12D. [348] *n*-Dodecylamine; cationic; emulsifier; liq., solid; 98% conc.

Amine 14D. [348] *n*-Tetradecylamine; cationic; emulsifier; liq., solid; 98% conc.

Amine 16D. [348] *n*-Hexadecylamine; cationic; emulsifier; liq., solid; 98% conc.

Amine 18D. [348] *n*-Octadecylamine; cationic; emulsifier; liq., solid; 98% conc.

Amine B11. [348] Alkylamine; cationic; emulsifier; solid; 95% conc.

Amine BG. [348] Tallow amine; cationic; emulsifier, corrosion inhibitor; paste; 98% conc.

Amine C. [137] Imidazoline deriv.; emulsifier, dispersant, detergent, cationic; yel. liq.; sol. in polar org. solv., hydrocarbons, relatively insol. water; 100% act.

Amine HBG. [348] Tallow amine, hydrog.; emulsifier, corrosion inhibitor; solid; 98% conc.

Amine KK. [348] Cocoamine; cationic; emulsifier, corrosion inhibitor; liq.; 98% conc.

Amine O. [137] Oleyl imidazoline; cationic; emulsifier, dispersant, detergent, antistat; amber liq.; sol. in polar org. solv., hydrocarbons, relatively insol. water; 100% act.

Amine OL. [348] Oleylamine; cationic; emulsifier, corrosion inhibitor; liq.; 98% conc.

Amine S. [137] Stearyl imidazoline; cationic; emulsifier, dispersant, detergent, antistat; tan wax; sol. in polar org. solv., hydrocarbons, relatively insol. water; 100% act.

Amine T. [137] Tall oil fatty acid imidazoline; dk. amber liq.; sol. in polar org. solv., hydrocarbons, relatively insol. water; 100% act.

Aminofoam C. [159] TEA-lauroyl animal collagen amino acids; anionic; detergent for personal care products; liq.; 40% conc.

Aminogen DG Liq. [42] Foaming agent, dye assistant; clear liq.; dens. 8.64 lb/gal; pH 8.8; 100% act.

Aminol CA-2. [237] Ricinoleamide DEA; nonionic; softening agent, emulsifier, cosmetic emulsions; liq.; 100% conc.

Aminol CM, CM Flakes. [237] Cocoamide MEA; nonionic; soap additive and foam stabilizer, thickener, and foam booster; hair shampoos; solid, flakes; 100% conc.

Aminol CM-C Flakes. [237] Cocoamide MEA; nonionic; see Aminol CM; flakes; 100% conc.

Aminol CM-D Flakes. [237] Cocoamide MEA; nonionic; see Aminol CM; flakes; 100% conc.

Aminol COR-2. [237] Cocamide DEA; nonionic; used in industrial applications; liq.; 100% conc.

Aminol COR-2C. [237] Cocamide DEA and diethanolamine; nonionic; used in industrial applications; liq.; 100% conc.

Aminol COR-4, -4C. [237] Lauramide DEA; nonionic; foam booster and thickener; detergent in drycleaning applications; liq.; 100% conc.

Aminol HCA. [237] Cocamide DEA; nonionic; foam stabilizer in shampoos and household detergents; liq.; 100% conc.

Aminol KDE. [131] Coco fatty acid diethanolamide; nonionic; foam stabilizer for personal care products; liq.; 100% conc.

Aminol LM-30C Special. [237] Lauramide DEA; nonionic; foam booster of stabilizer in personal care products; liq.; 100% conc.

Aminol N-1918. [237] Stearamide DEA and diethanolamine; nonionic; softener for textile; solid; 100% conc.

Aminol OF. [237] Oleamide DEA; nonionic; thickening agent for hair colors; liq.; 100% conc.

Aminol VR-14. [237] Mixed acid alkanolamide; nonionic; used in detergent compositions as detergents and foam boosters; liq.; 97% conc.

Aminoxid WS35. [258; 259] Cocamidopropylamine oxide; nonionic; detergent, emulsifier, wetting agent, softener, foam stabilizer for detergent preparations, cosmetic and pharmaceutical emulsions; liq.; 35% conc.

Amisoft C-T-12. [9] Monotriethanolamine *n*-cocoyl-*l*-glutamate; emulsifier for cosmetics; liq.; 100% act.

Amisoft GS-11. [9] Monosodium *n*-cocoyl-*l*-glutamate; emulsifier for cosmetics; powd.; 100% act.

Amisoft HS-11. [9] Monosodium *n*-hydrog.-tallowyl-*l*-glutamate; emulsifier for cosmetics; powd.; 100% act.

Amisoft HT-12. [9] Monotriethanolamine *n*-hydrog.-tallowyl-*l*-glutamate; emulsifier for cosmetics; liq.; 20% act.

Amisoft LS11. [9] Monosodium *n*-lauroyl-*l*-glutamate; emulsifier for cosmetics; powd.; 100% act.

Amisoft LT-12. [9] Monotriethanolamine *n*-lauroyl-*l*-glutamate; emulsifier for cosmetics; liq.; 20% act.

Amiter LGOD. [9] N-lauroyl-L-glutamic acid, 2-octyldodecyl diester; used in personal care products; liq.; HLB 0.1; 100% conc.

Amiter LGOD-2. [9] N-lauroyl-L-glutamic acid dioxyethylene 2-octyldodecyl ether diester; nonionic; emulsifer used in cosmetics; liq.; HLB 1.7; 100% conc.

Amiter LGS-2. [9] N-lauroyl-L-glutamic acid; nonionic; emulsifier used in cosmetics; wh. wax; HLB 2.0; 100% conc.

Amiter LGS-5. [9] N-lauroyl-L-glutamic acid, POE (5) stearyl ether diester; nonionic; emulsifier and emulsion stabilizer in cosmetics; wh. soft wax; HLB 5.4; 100% conc.

Ammonium Stearate 33% Liquid. [283] Ammonium stearate; anionic; wetting agent, foam stabilizer for frothed latex systems; wh. liq.; 33% act.

Ammonyx 4. [456] Stearyl dimethyl benzyl ammonium chloride; cationic; emulsifier, conditioner, softener, emollient; paste; sp.gr. 0.99; flash pt. > 200 F; 24–28% solids.

Ammonyx 4B. [456] Stearyl dimethyl benzyl ammonium chloride; cationic; see Ammonyx 4; paste; sp.gr. 0.99; flash pt. > 200 F; 24–26% solids.

Ammonyx 485. [456] Stearyl dimethyl benzyl ammonium chloride; cationic; see Ammonyx 4; paste; sp.gr. 0.45; flash pt. 122 F; 87% min. solids.

Ammonyx 490. [456] Stearyl dimethyl benzyl ammonium chloride; cationic; see Ammonyx 4; paste; sp.gr. 0.47; flash pt. 130 F; 92% min. solids.

Ammonyx 856. [456] Tallow dimethylbenzyl ammonium chloride; cationic; industrial grade pigment dispersant and suspending agent, demulsifier for solv. extraction processes; paste; sp.gr. 0.90; flash pt. 64 F; 75% conc.

Ammonyx 4002. [456] Stearyl dimethyl benzyl ammonium chloride; cationic; emulsifier, conditioner, softener, emollient; powd.; sp.gr. 0.52; flash pt. 170 F.

Ammonyx CDO. [456] Cocamidopropyl amine oxide; nonionic; wetting, foaming agent, foam stabilizer, conditoner for bubble baths, bath oils, dishwashing, hair color systems, softeners, cleansers; liq.; sp.gr. 1.02; flash pt. > 200 F; 30% amine oxide.

Ammonyx CETAC. [456] Cetrimonium chloride; cationic; emulsifier, conditioner, softener, emollient; liq.; sp.gr. 0.93 (25/20 C); flash pt. > 200 F.

Ammonyx CETAC-30. [456] Cetrimonium chloride; cationic; conditioner, softener, and emollient for use in personal care products; cationic emulsifier; liq.; sp.gr. 0.93; 29% min. quat.; 30% min. solids.

Ammonyx CO. [456] Palmitamine oxide; nonionic; conditioner, detergent, foam stabilizer; liq.; sp.gr. 0.96; flash pt. > 200 F; 30% conc.

Ammonyx DMCD-40. [456] Lauramine oxide; nonionic; wetting, foaming agent, foam stabilizer for cosmetic, home and janitorial products; liq.; sp.gr. 0.91; flash pt. 86 F; 40% conc.

Ammonyx DME. [456] Cetethyldimonium bromide; cationic; emulsifier, wetting agent; pharmaceutical applications; paste; sp.gr. 0.96; flash pt. 52 F; 75% conc.

Ammonyx KP. [456] Olealkonium chloride; cationic; conditioner in clear hair rinses, antistat; liq.; sp.gr. 0.98; flash pt. > 200 F.

Ammonyx LO. [456] Lauramine oxide; nonionic; wetting, foaming agent, foam stabilizer, grease emulsifier for personal care products; liq., sp.gr. 0.96; flash pt. > 200 F; 30% conc.

Ammonyx MCO. [456] Myristamine oxide; nonionic; wetting and foaming agent, foam stabilizer for cosmetics, home and janitorial products; liq.; sp.gr. 0.96; flash pt. > 200 F; 30% conc.

Ammonyx MO. [456] Myristamine oxide; nonionic; wetting and foaming agent, foam stabilizer; liq.; sp.gr. 0.96; flash pt. > 200 F; 30% conc.

Ammonyx SO. [456] Stearamine oxide; nonionic; conditioner, emulsifier; paste; sp.gr. 0.99; flash pt. > 200 F.

Ammonyx T. [456] Cetalkonium chloride; cationic; emulsifier, wetting agent, germicide; latex dispersions; sp.gr. 0.99; flash pt. > 200 F; 25% conc.

Ammonyx TDO. [456] Tallow amidopropylamine oxide; nonionic; wetting, foaming agent, foam stabilizer, conditioner for bubble baths, bath oils, dishwashing, hair color systems, softeners, cleansers; liq.; 30% conc.

Amonyl 265 BA. [551] Coco betaine; amphoteric; detergent for shampoo; liq.; 30% conc.

Amonyl 380 BA. [551] Cocoamidopropyl betaine; amphoteric; detergent for shampoos; liq.; 30% conc.

AMP. [52; 323] 2-Amino-2-methyl-1-propanol; anionic; emulsifier, catalyst; dispersant for pigments and latex paints; corrosion inhibitor; stabilizer; resin solubilizer; APHA 20 solid; m.w. 89.14; dens. 7.78; sp.gr. 0.928; visc. 102 cp (30 C); m.p. 30 C; b.p. 165 C; flash pt. 172 F; 100% act.

AMP-95. [52; 323] 2-Amino-2-methyl-1-propanol; anionic; see AMP; colorless liq.; m.w. 89.14; dens. 7.85; sp.gr. 0.942; visc. 147 cp; f.p. –2 C; flash pt. 182 F (TCC); surf. tens. 36–38 dynes/cm; 95% act.

Ampho B11-34. [66; 111] Complex cocobetaine derived from dimethyl amine propyl amine; amphoteric; high foaming surfactant with foam stability, wetting agent, solubilizer used in cosmetic formulations; lt. yel., clear liq.; mild odor; water sol.; dens. 8 lb/gal; sp.gr. 0.96; biodeg.; 35% act.

Ampho T-35. [66; 111] Potassium salt of fatty amine polyethoxy carboxylate; amphoteric; detergent; amber liq.; mild odor; water and polar solv. sol.; dens. 8.7 lb/gal; sp.gr. 1.05; biodeg.; 35% act.

Amphocerin E. [291] Fatty alcohols and wax esters; w/o emulsifier; solid.

Ampholan B-171. [360] Betaine; amphoteric; foaming agent for shampoos, foam based cleaners, cement, and industrial applications; liq.; 30% conc.

Ampholan D197. [360] Betaine; amphoteric; used in personal care products; liq.; 35% conc.

Ampholan U 203. [187; 360] Amphoteric surfactant with carboxyl and amino groups; amphoteric; detergent, foaming and stabilizing agent, dispersant; pale amber liq.; char. odor; water sol.; sp.gr. 1.035; visc. 132 cs; flash pt. > 200 F (COC); pour pt. < 0 C; pH 6.0–8.0 (1% aq); 30% act.

Amphoram CB A30. [117] Coco-alkyl betaine; amphoteric; detergent, bactericidal, emulsifier; liq.; 30% conc.

Amphoram CP1. [117] N-coco amino propionic acid; amphoteric; detergents, foaming agents; used in cosmetics; liq.; 65% conc.

Amphoram CT 30. [117] Coco-alkyl taurine; amphoteric; detergent, bactericidal, emulsifier; liq.; 30% conc.

Amphosol CA. [575; 576] Cocamidopropyl betaine; amphoteric; visc. builder, base for cosmetics and liq. detergents; straw clear liq.; pH 5.0 (10%); 30% act.

Amphosol CG. [575] Cocamido propyl betaine; foam booster, visc. builder, and lime soap dispersant; used in personal care products; amber clear liq.; pH 4.5–6.5; 29–31% act.

Amphosol DM. [576] Alkyl betaine; amphoteric; base for shampoos and foam baths; bacteriostatic; liq.; 30% conc.

Amphotensid 9 M. [660] Coconut imidazoline deriv., modified; amphoteric/anionic; detergent for personal care products; liq.; 30% conc.

Amphotensid B 4. [660] Fatty acid amido alkyl betaine; amphoteric; cosmetics, shampoos, detergents; liq.; 30% conc.

Amphotensid CT. [660] Coconut imidazoline deriv.; amphoteric; detergent for liq. cleaners; liq., paste; 38% conc.

Amphotensid D 1. [660] Alkylamino acid, alkylolamine salt; amphoteric; wetting agent and detergent; liq.; 40% conc.

Amphotensid GB 2009. [660] Coconut imidazoline deriv.; amphoteric; detergent used in personal care products; liq., paste; 38% conc.

Amphoterge J-2. [372] Substituted imidazoline; amphoteric; wetting agent and detergent; liq.; 50% conc.

Amphoterge K. [372] Cocoamphopropionate; amphoteric; detergents used in shampoos, skin cleansers, dishwashing; liq.; 40% conc.

Amphoterge K2. [372] Cocoamphodipropionate; amphoteric; detergent used in shampoos and skin cleansers, liq. cleaners; liq. 40% conc.

Amphoterge KJ-2. [372] Substituted imidazoline; amphoteric; see Amphoterge J-2; liq.; 40% conc.

Amphoterge S. [372] Stearoamphoacetate; amphoteric; textile softener; used in creme rinses; paste; 25% conc.

Amphoterge SB. [372] Cocoamphohydroxypropyl sulfonate; amphoteric; surfactant used in detergent and cosmetic application; liq.; 45% conc.

Amphoterge W. [372] Cocoamphoacetate; amphoteric; see Amphoterge K; liq.; 46% conc.

Amphoterge W2. [372] Cocoamphodiacetate; amphoteric; see Amphoterge K; gel; 52% conc.

Amphoteric L. [230] Surfactant used in anionic systems for personal care products; foaming agent for auto wash and wax; lt. amber liq.; sp.gr. 1.04; pH 6–8 (5%); 35% min. act.

AMS-33. [287] Ethylene glycol monostearate; nonionic; emulsifier, thickener, clouding agent used in personal care products; flakes; HLB 2.0; 100% conc.

Amsul 70. [603] Dodecylbenzene sulfonic acid, amine salt; anionic; emulsifier used in industrial and agricultural formulations; liq.; 70% conc.

Anchoid. [50] Sodium naphthalene formaldehyde sulfonate; dispersant for latex applications, cement, concrete, ceramics; buff free-flowing powd.; sol. in cold water.

Anfomul 01T. [159] 2-Heptadecenyl-4,4-bishydroxymethyl oxazoline; nonionic; emulsifier for emulsion explosives; waxy solid; 100% conc.

Anfomul PL. [159] Wax acid pentaerythritol ester; nonionic; emulsifier for emulsion explosives; soft solid; 100% conc.

Anfomul PLR. [159] Wax and olefinic fatty acids pentaerythritol ester; nonionic; emulsifier for emulsion explosives; liq.; 100% conc.

Anfomul PO. [159] Olefinic fatty acid pentaerythritol ester; nonionic; emulsifier for emulsion explosives; liq.; 100% conc.

Anfomul S4. [159] Sorbitan monooleate; nonionic; emulsifiers for emulsion explosives; liq.; HLB 4.3; 100% conc.

Anfomul S4L. [159] Polyol ester/steroidal alcohol complex; visc.; emulsifiers for emulsion explosives; 100% conc.

Anfomul S6. [159] Sorbitan monoisostearate; nonionic; emulsifiers for emulsion explosives; liq.; HLB 4.7; 100% conc.

Anfomul S43. [159] Sorbitan sesquioleate; nonionic; emulsifiers for emulsion explosives; liq.; HLB 3.7; 100% conc.

Anfomul S50. [159] Sorbitan monooleate; nonionic; emulsifiers for emulsion explosives; liq.; HLB 4.3; 100% conc.

Anionyx 12S. [456] Disodium monooleamido diglycol sulfosuccinate; anionic; detergent for personal care products; liq.; sp.gr. 1.05; flash pt. 138 F; 20% act.

Antara LB-400. [245] Fatty alcohol polyethyleneoxy phosphate ester acid; anionic; lubricity additive, rust inhibitor, emulsifier; amber, opaque visc. liq.; sol. in oil, water, aliphatic and aromatic solv.; sp.gr. 1.04; 99% act.

Antara LE-500. [245] Free acid of a complex org. phosphate ester; anionic; lubricant, emulsifier; yel., clear visc. liq.; sol. in oil, water, aliphatic and aromatic solv.; sp.gr. 1.11.

Antara LE-700. [245] Free acid of a complex org. phosphate ester; anionic; lubricant; slight cloudy, VCS 8, visc. liq.; 100% act.

Antara LF-200. [245] Complex org. phosphate ester free acid; anionic; lubricant; VCS 8, slight cloudy visc. liquid; 100% act.

Antara LK-500. [245] Complex phosphate ester of an ethoxylated linear alcohol; low-foaming lubricant; pale yel., slightly visc., clear liq.; sol. in xylene, ethanol, ethylene glycol, carbon tetrachloride, butyl cellosolve, water; sp.gr. 1.1; visc. 21 cs (100 C); biodeg.; 100% act.

Antara LM-400. [245] Free acid of a complex org. phosphate ester; anionic; lubricant; yel., clear liq.; sol. in oil, aliphatic and aromatic solv.; sp.gr. 1.05; 100% act.

Antara LM-600. [245] Free acid of a complex org. phosphate ester; anionic; lubricant, emulsifier; yel., clear visc. liq.; sol. in oil, water, aliphatic and aromatic solv.; sp.gr. 1.06; 100% act.

Antara LP-700. [245] Free acid of a complex org. phosphate ester; anionic; lubricant; yel., clear visc. liq.; sol. in water, aliphatic and aromatic solv.; sp.gr. 1.20; 100% act.

Antaron FC-34. [245] Monocarboxyl coco imidazoline compd.; amphoteric; detergent, wetting agent, emulsifier, dispersant, emollient, surfactant; amber, clear semivisc. liq.; water sol. esp. high electrolyte content; surf. tens. 32.0 dynes/cm (0.0155%).

Antaron MC-44. [245] Dicarboxylic coco imidazoline, sodium salt; amphoteric; emulsifier, solubilizer, coupling agent; amber visc. liq.; 38% act.

Antaron PC-37. [245] Ethyl PEG-15 cocamine sulfate; amphoteric; shampoo surfactant; amber, clear, slightly visc. liq.; misc. water, ethanol; sp.gr. 1.15; visc. 1100–1200 cps; surf. tens. 39 dynes/cm (0.1%); biodeg.; 75% act.

Antarox BL-214. [245] Linear aliphatic polyether, modified; nonionic; metal cleaning and textile finishing surfactant; liq.; sol. in water and polar solv.; cloud pt. 59–104 F (1% aq.); 100% act.

Antarox BL-225. [245] Modified linear aliphatic polyether; nonionic; low-foaming surfactant, wetting agent for metal cleaning, rinse aids, textiles, floor cleaners; slightly yel., clear liq.; bland odor; sol. water and org. solv. except paraffins; sp.gr. 0.99; visc. 40 cps; cloud pt. 25–29 C; flash pt. 190 C (COC); pour pt. < –18 C; surf. tens. 30 dynes/cm (0.1%); biodeg.; 100% act.

Antarox BL-236. [245] Linear aliphatic polyether, modified; nonionic; surfactant for high temp. spray; rinse aids for commercial and industrial use; liq.; 100% conc.

Antarox BL-240. [245] Modified linear aliphatic polyether; see Antarox BL-225; slightly yel. liq.; bland odor; water sol.; sp.gr. 0.99; visc. 35 cps; cloud pt. 38–42 C; flash pt. 124 C; pour pt. < 21 C; surf. tens. 28 dynes/cm (0.1%); biodeg.; 100% act.

Antarox BL-330. [245] Linear aliphatic polyether, modified; nonionic; detergent, wetting agent for rinse aids, metal cleaning, mechanical dishwashing; liq.; 95% act.

Antarox BL-344. [245] Linear aliphatic polyether, modified; nonionic; detergent and deduster for caustic powds.; liq.; 90% act.

Antarox G-200. [245] Alkylpoly-(ethyleneoxy)-glycolamide; nonionic; dispersant for metal; spin bath additive used in mfg. of tire cords and textile filaments; mfg. of cellophane and steelmaking; br. opaque, visc. liq.; ref. index 1.48; cloud pt. 64–68 C; 100% act.

Antarox LF-222. [245] Aromatic polyether, modified; nonionic; spray metal cleaning with wetting and rewetting properties; liq.; 100% conc.

Antarox LF-330. [245] Linear aliphatic polyether, modified; nonionic; detergent, wetting agent for rinse aid concs., household and industrial detergents, laundry and dairy cleaners; pale yel., clear liq.; bland odor; sol. water and most org. solv. except paraffins; visc. 29 cs; cloud pt. 30 C; surf. tens. 29 dynes/cm (0.1%); 95% act.

Antarox LF-344. [245] Linear aliphatic polyether, modified; nonionic; detergent, wetting agent; pale yel., clear liq.; bland odor; sol. water and most org. solv. except paraffins; visc. 61 cs; cloud pt. 44 C; surf. tens. 33 dynes/cm (0.1%); 90% act.

Antifoam C-196. [245] Silicon emulsion; nonionic; foam controller in aq. systems; liq.

Antifoam CM Concentrate. [603] Blend of antifoaming and defoaming agents with silicone synergist, o/w emulsion, nonionic emulsifier; used in industrial processing; carpet cleaning, wood pulping, paper coating, and textile mfg.; wh. liq. lt. cream; disp. in water; sp.gr. 0.972; pH 6.5–7.5 (1% aq.).

Antifoam Compound SWS-202. [586] Dimethylpolysiloxane compd.; antifoam for use in industrial applications; indirect food contact; cream visc. liq.; odorless; disp. in water; sp.gr. 1.01; dens. 8.4 lb/gal; visc. 10,000 cps; 100% act.

Antifoam Compound SWS-203. [586] Dimethylpolysiloxane compd. with filler; antifoam agent; food applications; gray translucent liq.; odorless; neg. sol. in water; sp.gr. 1.01; dens. 8.4 lb/gal; visc. 2400 cps; flash pt. 600 F (COC); 100% act.

Antifoam Emulsion Q-75. [586] Silicone fluid aq. sol'n.; antifoam agent for food and industrial applications in aq. systems; wh. milky emulsion, liq.; mild odor; b.p. 212 F; sol. in water; sp.gr. 0.991–1.01; dens. 8.3 lb/gal; visc. 3,000–15,000 cps; 30% act.

Antifoam Emulsion SWS-211. [586] Silicone fluid aq. emulsion; antifoam agent for food applications; wh. liq.; odorless; b.p. 212 F; sol. in water; sp.gre. 1.01; dens. 8.3 lb/gal; visc. 5000 cps; 10% act.

Antifoam Emulsion SWS-213. [586] Silicone fluid aq. emulsion; antifoam agent for food applications; wh. liq.; odorless; b.p. 212 F; sol. in water; sp.gr. 1.01; dens. 8.3 lb/gal; visc. 15,000 cps. 30% act.

Antifoam Emulsion SWS-214. [586] Silicone fluid aq. emulsion; antifoam agent for industrial applications; wh. liq.; odorless; b.p. 212 F; sol. in water; sp.gr. 0.991; dens. 8.3 lb/gal; visc. 3000 cps; 10% act.

AO 728 Special. [608] Amine oxide; foam booster and stabilizer for industrial and household detergents; liq.; 50% conc.

Apex Level Dye #2123. [54] Detergent; liq.;

emulsifies in water.

Apex Mersapex #2251. [54] Detergent; water wh. gel; water sol.

Apex Mersynol #450. [54] Detergent, wetting agent; water sol.

Apex Mersynol #4150. [54] Wetting agent, dispersant; clear liq.

Apexomine #1639. [54] Detergent used in textile industry.

Apexomine #1771. [54] Detergent used in textile industry, slight yel. liq.; emulsifies in water.

Apex Pentrapex #200. [54] Detergent, wetting agent; amber liq.

Apex Pentrapex #1923. [54] Wetting agent, penetrating agent; yel. liq.; misc. with water.

Apexscour #1962. [54] Detergent, continuous scouring agent; liq.; water sol.

Apexscour #2590. [54] Anionic/nonionic; detergent, scouring agent, surfactant; liq.

Apex Tarem #20. [54] Detergent, wetting agent; liq.; water disp.

Apex Tarem #1498. [54] Detergent; liq.; forms emulsion in water.

Aqualose L30. [639] PEG-30 lanolin; nonionic; emollient, emulsifier, plasticizer, solubilizer; wax; HLB 14.0; 100% conc.

Aqualose L75. [639] PEG-75 lanolin USP; nonionic; emollient, emulsifier, plasticizer, solubilizer, conditioner, superfatting agent; germicide; used in personal care products; wax; HLB 14.0–16.0; 100% conc.; also avail. as 50% conc. gel (Aqualose L75/50)

Aqualose LL100. [639] PPG-40-PEG-60 lanolin oil; nonionic; emollient, emulsifier, plasticizer, solubilizer; liq.; HLB 13.0; 100% conc.

Aqualose W20/50. [639] Laneth-20; nonionic; emollient, emulsifier; gel; 50% conc.

Aqualox 232. [32] Org. acid amine salts; surfactant used in metal working formulations; alkaline cleaners; pale amber liq.; mild odor; sol. in water, methanol, ethanol, IPA, glycol, and water/alcohol mixt.; sp.gr. 1.14 (15.6 C); dens. 9.5 lb/gal (15.6 C); visc. 75 ± 15 cs (37.6 C); pour pt. –29 C max.; pH 7.6 ± 0.3; 75% act.

Aramide CDM-2. [381] Alkanolamine condensate, modified; nonionic; alkali tolerant for heavy duty cleaners; liq.; 100% conc.

Aramide CDM-4. [381] Alkanolamine condensate; nonionic; visc. builder for all purpose cleaners, emulsifier, and lubricant; liq.; 100% conc.

Aramide CDM-14. [381] Alkanolamide, modified; nonionic; drilling mud emulsifier and surfactant; used in pesticide and aerosols formulations; liq.; 100% conc.

Aramide CDM-15. [381] Alkanolamide, modified; nonionic; detergent, foam builder, hard surface cleaner base formulations; liq.; 100% conc.

Aramide CDR. [381] Alkanolamine condensate; nonionic; thickener for dishwash detergents and cosmetic use; liq.; 100% conc.

Aramide CDS. [381] Super coconut alkanolamide; nonionic; emulsifier, foam stabilizer, thickener for liq. detergents; liq.; 100% conc.

Aramide CDX. [381] Coconut alkanolamide; nonionic; solubilizer, foam stabilizer, detergent and visc. builder; liq.; 100% conc.

Aramide LDS. [381] Super lauric diethanolamide; nonionic; foam stabilizer, lubricant and visc. builder; liq.; 100% conc.

Aramide LMDS. [381] Lauric/myristic alkanolamide; nonionic; detergent, foam stabilizer in detergents and personal care products; solid; 100% conc.

Arbyl 18/50. [277] Fatty alcohol polyglycol ether; nonionic; precleaning and leveling agent; liq.; 50% conc.

Arbyl ASN. [277] Sulfosuccinate; anionic; wetting agent; liq.; 60% conc.

Arbyl N. [277] Fatty alcohol polyglycol ether; nonionic; washing, wetting and aftersoaping agent used in textile products; paste; 50% conc.

Arbyl R Conc. [277] Polyglycol ether; nonionic; cold wetting agent; liq.; 100% conc.

Arbylen Conc. [277] Alkylphenol polyglycol ether; nonionic; washing and wetting agent used in textile products; liq., paste; 50–100% conc.

Ardet 401. [58] Alkylaryl sulfonate, amine salt; anionic; emulsifier for o/w emulsions; liq.; 95% conc.

Ardet A-25. [58] Sodium and amine neutralized DDBS; detergent and wetting agent; base for liq. cleaning compds.; liq.; 25% conc.

Ardet AN. [58] Alkylaryl sulfonate and nonionic detergent blend; anionic; detergent; liq.; 25% conc.

Ardet AP-60. [58] Anionic/nonionic detergent blend; detergent conc. used as a base for dishwash and lt.-duty detergents; liq.; 60% conc.

Ardet AP-62. [58] Anionic/nonionic conc.; base for heavy-duty and all-purpose detergents; liq.; 60% conc.

Ardet C28. [58] Lauric/coconut diethanolamide, modified; nonionic; visc. builder; liq.; 100% conc.

Ardet CB. [58] Coconut diethanolamide, modified; nonionic; detergent for making cleaners; liq.; 40% conc.

Ardet CEA. [58] Coconut monoethanolamide; nonionic; thickener; flakes; 100% conc.

Ardet DC. [58] Coconut diethanolamide 2:1; anionic; general purpose detergent or detergent additive; foam stabilizer and thickener; liq.; 100% conc.

Ardet DCAE. [58] Coconut diethanolamide; nonionic; visc. builder and foam stabilizer; paste; 100% conc.

Ardet DCX. [58] Coconut diethanolamide 1:1; nonionic; thickener and foam stabilizer; liq.; 100% conc.

Ardet DLA. [58] Lauric diethanolamide 2:1; anionic; thickener and foam stabilizer used in detergents; liq., wax; 100% conc.

Ardet DMA. [58] Coconut diethanolamide, modified; anionic; metal finishing compds.; liq.; 100% conc.

Ardet DMC. [58] Coconut diethanolamide, modified; anionic; modification DMF; liq.; 100% conc.

Ardet DMF. [58] Coconut diethanolamide, modified; anionic; formulated for metal finishing in vibrators or barrel tumbling; liq.; 100% conc.

Ardet F-97. [58] Detergent/alkanolamide blend; anionic; base for detergents; liq.; 100% conc.

Ardet LAS. [58] Linear dodecylbenzene sulfonic acid; anionic; liq.; 98% conc.

Ardet LDA. [58] Lauric diethanolamide 1:1; nonionic; wax; 100% conc.

Ardet LEA. [58] Lauric monoethanolamide; nonionic; thickener; flakes; 100% conc.

Ardet LIPA. [58] Lauric monoisopropanolamide; nonionic; used in alkaline formulas; wax, flake; 100% conc.

Ardet LMC. [58] Lauric diethanolamide 1:1, modified; nonionic; foam stabilizer, thickener; liq.; 100% conc.

Ardet LMD. [58] Lauric/myristic diethanolamide 1:1; nonionic; foam stabilizer thickener; solid; 100% conc.

Ardet SEA. [58] Stearic monoethanolamide; nonionic; antitack or opacifier; flaked wax; 100% conc.

Ardet WB. [58] Diethanolamide, long chain; nonionic; used in w/o emulsions; liq.; 100% conc.

Ardet WO. [58] Coconut diethanolmide; nonionic; w/o emulsions; liq.; 100% conc.

Aremsol A. [527] Cocoamidopropyl betaine, amphoteric; mfg. of personal care products; liq.; 40% conc.

Aremsol P Lotion. [527] Pearly shampoo conc.; anionic; preparation of lotion shampoos; liq.; 35% conc.

Aremul AIS. [67] Detergent, emulsifier; amber liq.; mild odor; sol. in various aliphatic and chlorinated solv.; sp.gr. 1.03; 90–95% act.

Aremul DCL. [67] Blend of anionic sulfonates; detergent, emulsifier; dark amber liquid; mild odor; miscible sol. with water and ODCB, TCB, butyl benzoate and methyl salicylate; sp.gr. 1.04; 55–60% act.

Aremul V-91. [67] Blend of alkylaryl sulfonates, modified nonionics and org. coupling agents; detergent, emulsifier; amber liq.; mild odor; sol. in warm water, aliphatic solv.; 90–95% act.

Aremul X-91. [67] Blend of alkylaryl sulfonates, modified nonionics and org. coupling agents; see Aremul V–91; sp.gr. 1.03; 90–95% act.

Argobase 125. [639] Lanolin alcohols extract; nonionic; emollient, w/o emulsifier used in personal care products; liq.; oil-sol.; HLB 3.0; 15% conc.

Argobase EU. [639] Sterols and sterol esters lanolin extracts; nonionic; emollient, emulsifier, stabilizer; absorption base; paste; HLB 4.0; 6% conc.

Argobase LI. [639] Sterols and sterol esters lanolin extracts; nonionic; emollient, emulsifier, stabilizer; absorption base; liq.; HLB 4.0; 24% conc.

Argobase MS-5. [639] Sterols and sterol esters lanolin extracts; nonionic; w/o emulsifier, emollient, stabilizer; paste; HLB 5.0; 25% conc.

Argobase SI. [639] Sterol and sterol esters lanolin extracts; nonionic; emollient, emulsifier, stabilizer; paste; HLB 4.0; 92% conc.

Argonol 40. [639] Isobutylated lanolin oil; nonionic; w/o emulsifier, emollient used in personal care products; fluid; 100% conc.

Argowax Distilled. [639] Lanolin alcohol; nonionic; gelling agent, emulsifier; lt. wax; slight odor; HLB 2.6; 100% conc.

Argowax Standard. [639] Lanolin alcohol; nonionic; see Argowax Distilled; wax; 100% conc.

Aristonate 430. [478] Synthesized dialkylaryl sulfonate sodium salt; emulsifier and dispersant for liqs.; primary ingred. in sol., textile, leather processing and neat oils, metal degreasers, dry cleaning soaps, and rust preventives; Gardner 10 thick liq.; mild odor; oil-sol.; dens. 8.5 lb/gal; visc. 690 SSU (210 F); pH 9.5; 60–62% act. sulfonate.

Aristonate 460. [478] Synthesized dialkaryl sulfonate sodium salt; emulsifier and dispersant for liqs., dispersant/wetting agent for solids and liq. and solid systemss; rust and corrosion inhibitor; primary ingred. in sol., neat, and textile oils, high water base hydraulic fluids; thick liq.; mild odor; sol. see Aristonate 430; dens. 8.5 lb/gal; visc. 121 cSt (210 F).

Aristonate 500. [478] Synthesized dialkylaryl sulfonate sodium salt; emulsifier and dispersant for liqs., dispersant/wetting agent for solids and liq./solid systems; rust and corrosion inhibitor; used as primary ingred. in sol. and neat oils, metalworking fluids, ore flotation reagents; thick liq.; mild odor; sol. see Aristonate 430; dens. 8.5 lb/gal; visc. 134 cSt (210 F); pH 9.0.

Arizona DR-22. [60] Emulsifier, detergent, wetting agent; used to make Arizona disproportionated tall oil rosin soaps (DRS-40, 42, 43,44); sapon. no. 165; softening pt. 66 C.

Arizona DR Mix-26. [60] Emulsifier-mixed disproportionated rosin and fatty acid; emulsifier for SBR industry; paste; 100% conc.

Arizona DRS-40. [60] Potassium soap of disproportionated tall oil rosin; anionic; emulsifier, detergent, wetting agent; for ABS, SBR, other syn. elastomers; tan paste; bland odor; water disp.; dens. 9.11 lb/gal (75 C); sp.gr. 1.1; visc. 500 cps (75 C); 80% solids.

Arizona DRS-42. [60] Potassium soap of disproportionated tall oil rosin; anionic; see Arizona DRS-40; tan paste; bland odor; water-disp.; dens. 9.11 lb/gal (75 C); sp.gr. 1.1; visc. 800 cps (75 C); 80% solids.

Arizona DRS-43. [60] Sodium salt of disproportionated tall oil rosin; anionic; emulsifier, detergent, wetting agent; emulsifier for ABS, SBR, syn. elastomers; tan paste; bland odor; water disp.; dens. 8.94 lb/gal (75 C); sp.gr. 1.1; visc. 1700 cps (75 C); 71% solids.

Arizona DRS-44. [60] Sodium salt of disproportionated tall oil rosin; anionic; see Arizona DRS–43; dens. 8.99 lb/gal (75 C); sp.gr. 1.1; visc. 750 cps (75 C); 75% solids.

Arizona DRS-50. [60] Disproportioned rosin soap, potassium soap; polymerization emulsifier for the SBR industry; water-disp.; dens. 9.09 lb/gal (70 C); visc. 1800 cps (65 C); 84–86% total solids.

Arizona DRS-51E. [60] Disproportioned rosin soap, potassium soap; anionic; see Arizona DRS-50; water-disp.; dens. 9.08 lb/gal (80 C); visc. 800 cps (65 C).

Arkolene GN. [60] Ammonium alkylaryl sulfonate; nonionic; penetrating wetting agent; liq.; 40% conc.

Arkomon A Conc. [301] Fatty acid sarcoside, sodium salt; anionic; basic material for detergents and cleaning agents; corrosion inhibitor; paste; 60% conc.

Arkopal N040. [44; 301] Nonoxynol-4; nonionic; detergent, wetting agent, emulsifier, surface-active raw material for general industrial use; liq., paste; 100% conc.

Arkopal N060. [44; 301] Nonoxynol-6; nonionic; see Arkopal N040; 100% conc.

Arkopal N080. [44; 301] Nonoxynol-8; nonionic; see Arkopal N040; 100% conc.

Arkopal N090. [44; 301] Nonoxynol-9; nonionic; see Arkopal N040; 100% conc.

Arkopal N100. [44; 301] Nonoxynol-10; nonionic; see Arkopal N040; 100% conc.

Arkopal N110. [44; 301] Nonoxynol-11; nonionic; see Arkopal N040.

Arkopal N130. [44; 301] Nonoxynol-13; nonionic; see Arkopal N040; 100% conc.

Arkopal N150. [44; 301] Nonoxynol-15; nonionic; see Arkopal N040; 100% conc.

Arkopal N-230. [301] Nonoxynol-23; nonionic; surface-active raw material; 100% conc.

Arkopal N-300. [301] Nonoxynol-30; nonionic; surface active raw material; paste; 100% conc.

Arkopon Brands. [301] Fatty acid condensation products; anionic; basic material for mfg. of detergents and cleaning agents; auxiliary for the metal and rubber industries; paste, powd.

Arlacel 20. [319; 320] Sorbitan laurate; nonionic; emulsifier for cosmetics, pharmaceuticals; yel. liq.; sol. in methanol, ethanol, min., cottonseed and corn oils, ethylene glycol; sp.gr. 1.0; visc. 4250 cps; HLB 8.6; flash pt. > 300 F; 100% act.

Arlacel 40. [319; 320] Sorbitan palmitate; nonionic; emulsifier for cosmetics, pharmaceuticals; cream beads; sol. in IPA; sp.gr. 1; HLB 6.7; flash pt. > 300 F; 100% act.

Arlacel 60. [319; 320] Sorbitan stearate; nonionic; emulsifier for cosmetics, pharmaceuticals; cream-colored, waxy beads, solid;

sol. in IPA; HLB 4.7; flash pt. > 300 F; pour pt. 53 C; 100% act.

Arlacel 80. [319; 320] Sorbitan monooleate; nonionic; emulsifier for cosmetics, pharmaceuticals; yel. oil, liq.; sol. in IPA, min. and cottonseed oils; sp.gr. 1; visc. 1900 cps; HLB 4.3; flash pt. > 300 F; 100% act.

Arlacel 83. [319; 320] Sorbitan sesquioleate; nonionic; cosmetic and pharmaceutical grade of Arlacel C; yel., oil liq.; sol. in min. and cottonseed oils, ethanol, isopropanol; sp.gr. 1; visc 1500 cps; HLB 3.7; flash pt. > 300 F; 100% act.

Arlacel 85. [319; 320] Sorbitan trioleate; surfactant, nonionic; yel., oil liq.; sol. in isopropanol, alcohol, min., cottonseed and corn oil; sp.gr. 0.95; visc. 250 cps; HLB 1.8; flash pt. > 300 F; 100% act.

Arlacel 165 [319; 320] Glyceryl stearate, PEG-100 stearate; nonionic; surfactant, emulsifier for cosmetics; wh. beads; bland odor; disp. in water; HLB 11; flash pt. > 300 F; pour pt. 48 C; self-emulsifying; 100% act.

Arlacel 186. [319; 320] Glyceryl oleate, propylene glycol; nonionic; surfactant, emulsifier for personal care products; pale yel., clear liq.; sol. in ethanol, IPA, cottonseed and min. oils; sp.gr. 1; visc. 150 cps; HLB 2.8; flash pt. > 300 F; 100% act.

Arlacel 481. [319; 320] Glyceryl sorbitan fatty acid ester, unsat.; nonionic; emulsifier for cosmetic uses; waxy; HLB 4.5; 100% conc.

Arlacel 581. [319] Glycerol sorbitan oleostearate, ethoxylated; nonionic; emulsifier for w/o creams; solid; HLB 5; 100% conc.

Arlacel 582. [319] Glycerol sorbitan isostearate, ethoxylated; nonionic; emulsifier for w/o creams; solid; HLB 5; 100% conc.

Arlacel 986. [319; 320] Glycerol sorbitan fatty acid ester, sat.; nonionic; emulsifier for w/o creams, and lotions; wax; HLB 4.5; 100% conc.

Arlacel 987. [319; 320] Sorbitan monoisostearate; nonionic; emulsifier for w/o cosmetic, creams and lotions; liq.; HLB 4.3; 100% conc.

Arlacel 988. [319; 320] POE glycerol sorbitan fatty acid ester; nonionic; emulsifier for cosmetic uses; waxy, solid; HLB 4.7; 100% conc.

Arlacel 989. [319; 320] PEG-7 hydrog. castor oil; nonionic; emulsifier and softener for w/o lotions, cosmetic uses; waxy, solid; 100% conc.

Arlacel A. [319] Mannide monooleate; nonionic; surfactant for use in emulsified vaccines of the w/o type; liq.; HLB 4.3; 100% conc.

Arlacel C. [319; 320] Sorbitan sesquioleate; nonionic; surfactant, w/o emulsifier; amber, oily liq.; sol. in min. oil, cottonseed oil, ethanol, IPA; sp.gr. 1.0; visc. 1000 cps; HLB 3.7; flash pt. > 300 F; 100% act.

Arlamol E. [319; 320] PPG-15 stearyl ether; nonionic; emollient, solv. for personal care products; oily liq.; sol. in min. oil, isopropyl esters, cottonseed oil, ethanol, IPA, hexadecyl alcohol; sp.gr. 0.95; visc. 90 cps; pour pt. typ. < 0 C; 100% act.

Arlasolve 200. [319; 320] Isoceteth-20; nonionic; surfactant, emulsifier, solubilizer for cosmetics; soft, waxy wh. solid; sol. in water, propylene glycol, ethanol; sp.gr. 1.0; HLB 15.7; flash pt. > 230 F; 100% act.

Arlatone 285. [319; 320] POE castor oil; nonionic; cosmetic grade surfactant, coupling agent, solubilizer for perfumes, fragrances; semifluid; HLB 14.4; 100% conc.

Arlatone 289. [319; 320] POE hydrog. castor oil; nonionic; coupling agent, solubilizer, emulsifier for o/w creams and lotions; waxy; HLB 14.4; 100% conc.

Arlatone 970. [320] POE sorbitan fatty acid ester; nonionic; coupling agent, solubilizer; liq.; 100% conc.

Arlatone 983. [320] POE fatty acid ester; nonionic; emulsifier; solid; 100% conc.

Arlatone 983S. [319] POE fatty acid ester; nonionic; emulsifier for o/w creams and lotions; powd.; HLB 8.7; 100% conc.

Arlatone B. [320] POE sorbitan fatty acid esters and alkylaryl alcohols; nonionic; cosmetic emulsifier; clear, amber liq.; misc. with min., cottonseed oils, deodorized kerosene, IPM, IPP, hexadecyl, ethyl, IPA; sp.gr. 1; visc. 400 cp; HLB 9, flash pt. 300 F; pour pt. 0 F; 100% act.

Arlatone G. [319; 320] PEG-25 hydrog. castor oil; nonionic; surfactant, solubilizer; formulates clear gels; yel. visc. liq. to soft paste; sol. water, ethanol and IPA; sp.gr. 1.0; HLB 10.8; flash pt. > 300 F; 100% act.

Arlatone T. [319; 320] PEG-40 sorbitan peroleate; nonionic; emulsifier, solubilizer, antistat, lubricant; for bath oils; yel. liq.; sol. in veg., min. oils, IPM, IPP; water disp.; sp.gr. 1; visc. 175 cps; HLB 9; flash pt. > 300F; 100% act.

Armac 18D. [10; 63; 64] Octadecylamine acetate, distilled; cationic; wetting agent, emulsifier, corrosion inhibitor, flotation agent; wh. flake; water sol.; sp.gr. 0.827 (80 C); m.p. 85 C; flash pt. 320 F (COC); 100% act.

Armac C. [10; 11; 63; 64] Cocoamine acetate; cationic; wetting agent, emulsifier, corrosion inhibitor, flotation reagent; yel. solid; m.w. 260; water sol.; sp.gr. 0.852 (80 C); HLB 11.2; m.p. 50 C; flash pt. 285 F (COC); 95% conc.

Armac HT. [10; 11; 63; 64] (Hydrog. tallow) amine acetate; cationic; wetting agent, emulsifier, corrosion inhibitor, flotation reagent; wh. solid; sp.gr. 0.820 (80 C); HLB 10.7; m.p. 60 C; flash pt. 325 F (COC); 95% act.

Armac T. [10; 11] Tallowamine acetate; cationic; flotation agent, corrosion inhibitor, lubricant, emulsifier; Gardner 10 max. solid; sp.gr. 0.87 (60 C); HLB 10.8; m.p. 45–61 C; flash pt. 168 C (COC); 94% min act.

Armeen 2C. [10; 63; 64] Dicocamine (sec. amine); cationic; emulsifier, flotation agent, corrosion inhibitor; solid; sol. in chloroform, slight sol. in IPA, toluene, carbon tetrachloride, kerosene; sp.gr. 0.793 (60/40 C); visc. 49.1 SSU (60 C); m.p. 104–117 F; flash pt. 350 F; pour pt. 80 F; 100% conc.

Armeen 2HT. [10; 63; 64] Di(hydrog. tallow) amine (sec. amine); cationic; emulsifier, flotation agent, corrosion inhibitor; solid; sol. in chloroform, slightly sol. in toluene, carbon tetrachloride, kerosene; sp.gr. 0.803 (60/4 C); visc. 63.9 SSU (70 C); m.p. 140–149 F; flash pt. 435 F; pour pt. 115 F; 100% conc.

Armeen 8D. [63] Octylamine (primary amine); liquid; sol. in methanol, ethanol, acetone, IPA, chloroform, toluene, carbon tetrachloride, kerosene; sp.gr. 0.793; visc. 34.4 SSU; m.p. 10–21 F; flash pt. 145 F; pour pt. 15 F.

Armeen 12D. [10; 11; 63; 64] Lauramine (primary amine); cationic; emulsifier, flotation agent, corrosion inhibitor; liquid; sol. in methanol, ethanol, acetone, IPA, chloroform, toluene, carbon tetrachloride, kerosene; sp.gr. 0.801; visc. 42.2 SSU; m.p. 79–86 F; flash pt. 235 F; pour pt. 80 F; 98% conc.

Armeen 16D. [10; 11; 63; 64] Palmitamine (primary amine); cationic; emulsifier, flotation agent, corrosion inhibitor; solid; sol. in methanol, ethanol, IPA, chloroform, toluene, carbon tetrachloride, slight sol. in acetone, kerosene; sp.gr. 0.789 (60/4 C); visc. 37.5 SSU (55 C); m.p. 100–118 F; flash pt. 285 F; pour pt. 100 F; 100% conc.

Armeen 18. [10; 63; 64] Stearamine (primary amine); cationic; emulsifier, flotation agent, corrosion inhibitor, anticaking agent; solid; sol. in ethanol, IPA, chloroform, toluene, carbon tetrachloride, slight sol. in acetone, kerosene; sp.gr. 0.792 (60/4 C); visc. 45.6 SSU; m.p. 122–133 F; flash pt. 320 F; pour pt. 115 F; 100% conc.

Armeen 18D. [10; 11; 63; 64] Stearamine (primary amine); cationic; emulsifier, flotation agent, corrosion inhibitor, anticaking agent; solid; sol. in ethanol, IPA, chloroform, toluene, carbon tetrachloride, slight sol. in methanol, kerosene; sp.gr. 0.791–0.792 (60/4 C); visc. 43.7 SSU; m.p. 122–133 F; flash pt. 300 F; pour pt. 110 F; 97% conc.

Armeen C. [10; 11; 63; 64] Cocamine (primary amine); cationic; emulsifier, flotation agent, corrosion inhibitor, stripping agent for paints; liq.; sol. in methanol, ethanol, acetone, IPA, chloroform, toluene, carbon tetrachloride, kerosene; sp.gr. 0.805; visc. 44.2 SSU (35 C); m.p. 54–59 F; flash pt. 240 F; pour pt. 45 F; 95% conc.

Armeen CD. [10; 11; 63; 64] Cocamine (primary amine); emulsifier, flotation agent, corrosion inhibitor, stripping agent for paints; liquid; sol. in methanol, ethanol, acetone, IPA, chloroform, toluene, carbon tetrachloride, kerosene; sp.gr. 0.804; visc. 43 SSU (35 C); m.p. 57–63 F; flash pt. 230 F; pour pt. 55 F; 98% conc.

Armeen DM12D. [10; 11] Dimethyl lauramine; cationic; surfactant intermediate; yel. liq.; amine odor; water insol.; sp.gr. 0.78; visc. 3.30 cs; f.p. –15 C; b.p. 80–115 C (3 mm Hg); flash pt. 122 C (COC); 98% act.

Armeen DM14D. [10; 11] Dimethyl myristamine; cationic; surfactant intermediate; yel. liq.; amine odor; water insol.; sp.gr. 0.79; visc. 4.7 cs; f.p. –8 C; b.p. 100–125 C (3 mm Hg); flash pt. 28 C (COC); 98% act.

Armeen DM16D. [10; 11] Dimethyl palmitamine; cationic; chemical intermediate, raw material for surfactants; yel. liq.; amine odor; water insol.; sp.gr. 0.80; visc. 6.75 cs; f.p. 8 C; b.p. 100–136 C (3 mm Hg); flash pt. 150 C (COC); 98% act.

Armeen DM18D. [10] Dimethyl stearamine; cationic; chemical intermediate, raw material for surfactants; yel. liq., amine odor; water insol.; sp.gr. 0.79; f.p. 20 C; b.p. 145–160 C (3 mm Hg); flash pt. 178 C (COC); 98% act.

Armeen DMAD. [10] Alkyl (C_{10}–C_{18}) dimethyl amine; chemical intermediate, raw material for surfactants; yel. liq.; amine odor; water insol.; sp.gr. 0.79; visc. 4.44 cs; b.p. 45–150 C (3 mm Hg); flash pt. 130 C (COC); 98% act.; skin, eye irritant.

Armeen DMCD. [10; 11] Dimethyl cocamine; cationic; chemical intermediate, raw material for surfactants; yel. liq.; amine odor; water insol.; sp.gr. 0.79; visc. 4.20 cs; b.p. 42–150 C (3 mm Hg); flash pt. 108 C (COC); 98% act.

Armeen DMHTD. [10; 11] Dimethyl hydrog. tallow amine; cationic; chemical intermediate, raw material for surfactants; yel. liq.; amine odor; water insol.; sp.gr. 0.80 ; visc. 8.75 cs; f.p. 18 C; b.p. 100–155 C; flash pt. 162 C (COC); 98% act.

Armeen DMMCD. [10; 11] Dimethyl cocamine; nonionic; chemical intermediate, raw material for surfactant; yel. liq.; amine odor; water insol.; sp.gr. 0.79; visc. 3.95 cs; f.p. –19 C; b.p. 90–125 C (3 mm Hg); flash pt. 125 C (COC); 98% act.; skin, eye irritant.

Armeen DMOD, DMSD, DMTD. [10] Dimethyl alkyl tert. amines; cationic; surfactant intermediates; liq.; 100% conc.

Armeen HT. [10; 11; 63; 64] (Hydrog. tallow) amine (primary amine); cationic; emulsifier, flotation agent, corrosion inhibitor; Gardner 9 solid; sol. in methanol, ethanol, IPA, chloroform, toluene, carbon tetrachloride, slight sol. in kerosene; sp.gr. 0.795 (60/4 C); visc. 47.5 SSU (55 C); m.p. 79–136 F; cloud pt. 115 F; flash pt. 320 F; pour pt. 110 F; 95% conc.

Armeen HTD. [10; 11; 63; 64] Hydrog. tallow amine (primary amine); cationic; emulsifier, flotation agent, corrosion inhibitor; 9 Gardner 9 solid; sol. see Armeen HT; sp.gr. 0.794 (60/4 C); visc. 44.1 SSU (55 C); m.p. 70–120 F; cloud pt. 110 F; flash pt. 315 F; pour pt. 100 F; 100% conc.

Armeen L-10. [63] Beta-amine; cationic surfactant; Gardner 1 clear liq.; sol. in IPA, min. oil, kerosene; sp.gr. 0.787; cloud pt. –50 F; flash pt. 85 C; biodeg.

Armeen L-11. [63] Beta-amine; cationic surfactant; Gardner 2 clear liq.; sol. in IPA, min. oil, kerosene; sp.gr. 0.802; m.p. –20 F; cloud pt. –20 F; flash pt. 100 C; biodeg.

Armeen L-15. [63] Beta-amine; cationic surfactant; Gardner 5 clear liq.; sol. in IPA, min. oil, kerosene; sp.gr. 0.813; cloud pt. 54 F; flash pt. 157 C; pour pt. 50 F; biodeg.

Armeen O. [10; 11; 63; 64] Oleamine (primary amine); cationic; emulsifier, wetting agent, corrosion inhibitor, dispersant, chemical intermediate, oil additive; cosmetics; Gardner 8 paste; sol. in acetone, methanol, ethanol, IPA, chloroform; toluene, carbon tetrachloride, kerosene, wh. min. oil; sp.gr. 0.820 (38/4 C); visc. 57.0 SSU; m.p. 50–68 F; flash pt. 320 F; 100% conc.

Armeen OD. [10; 11; 64] Oleamine (primary amine); cationic; wetting agent, lube oil additive; Gardner 2 paste; water insol.; sp.gr. 0.79 (60 C); visc. 56.6 SSU; m.p. 8–18 C; flash pt. 154 C (COC); 98% act.

Armeen SD. [10; 63; 64] Soyamine (primary amine); cationic; emulsifier, flotation agent, corrosion inhibitor; Gardner 3 paste; sol. in methanol, ethanol, acetone, IPA, chloroform, toluene, carbon tetrachloride; sp.gr. 0.81 (38/4 C); visc. 46.2 SSU (35 C); m.p. 81–86 F; cloud pt. 85 F; flash pt. 305 F; pour pt. 70 F; 100% conc.

Armeen SZ. [10] Sodium salt of N-coco beta-amino butyric acid; detergent, wetting agent, corrosion inhibitor; yel. liq.; water sol.; sp.gr. 0.96; visc. 19 cps (55 C); f.p.; flash pt. 40 C; 39–41% act.

Armeen T. [10; 11; 63; 64] Tallowamine (primary amine); cationic; Gardner 11 paste; sol. in IPA, methanol, ethanol, chloroform, toluene, carbon tetrachloride; sp.gr. 0.813 (38/4 C); visc. 47 SSU (35 C); m.p. 64–117 F; cloud pt. 100 F; flash pt. 315 F; pour pt. 70 F; 95% act.

Armeen TD. [10; 11; 63; 64] Tallowamine (primary amine); cationic; emulsifier, flotation agent, corrosion inhibitor; Gardner 2 paste; sol. in IPA, methanol, ethanol, chloroform, toluene, carbon tetrachloride; sp.gr. 0.812 (38/4 C); visc. 45.2 SSU (35 C); m.p. 64–118 F; cloud pt. 102 F; flash pt. 305 F; pour pt. 70 F; 98% act.

Armeen Z. [10; 11; 63; 64] Cocaminobutyric acid; anionic/cationic; cosmetic pigment softening, dispersing agent; antifogging agent, foam booster; wetting agent in alkaline paint strippers, latex emulsions; Gardner 8 pumpable slurry; sol. in water, IPA; sp.gr. 0.98; visc. 247 SSU; flash pt. foamed to 210 F; pour pt. 65 F; 100% conc.

Armeen Z-9. [10] N-coco beta-amine butyric acid; emulsifier, pigment dispersant aid; yel. liq.; water sol.; sp.gr. 0.90; flash pt. 33 C; 46–50% act. in 40% water/10% IPA.

Armid 18. [63] Octadecanamide; antiblock

agent, stabilizer, dispersant; Gardner 7 flake; m.p. 99–109 C; flash pt. 225 C; 90% act.

Armid C. [63] Cocoamide; antiblock agent, stabilizer, dispersant; Gardner 10 flake; m.p. 85 C; flash pt. 174 C; 90% act.

Armid HT. [63] (Hydrog. tallow) amide; antiblock agent, stabilizer, dispersant; Gardner 7 flake; m.p. 98–103 C; flash pt. 225 C; 90% act.

Armid O. [63] Oleamide; antiblock agent, stabilizer, dispersant, release agent in cosmetics, penetrant in paper manufacture; Gardner 7 flake, solid; bland odor; m.p. 68 C; flash pt. 207 C; 90% act.

Armoblen S. [63] Long chain fatty amine derivs.; special cationic blend; pigment wetting agent for paints and coatings; Gardner 14 liq.; sp.gr. 0.889; dens. 7.40 lb/gal.

Armogard F-201. [64] Organic amine-based formulation; multifunctional fuel oil additive for use by refiners, fuel oil distributors, and lg. scale oil consumers; used as corrosion and sludge inhibitor and dispersant; clear amber liq.; sol. in hydrocarbon fuel oils; sp.gr. 0.894 (60 F); visc. 35 cP; pour pt. 30 F; flash pt. 145 F (PM).

Armogard G-101. [64] Aliphatic amidoamine; multifunctional gasoline additive for use as corrosion inhibitor, carburetor detergent, and gasoline deicer; clear amber liq.; sol. in gasoline; sp.gr. 0.855 (60 F); dens. 7.12 lb/gal; visc. 6.5 cP; pour pt. 15 F; flash pt. 66 F (PM).

Armoslip 18. [451a] Stearamide; internal lubricant, mold release agent, antiblock and slip agent for PE, PP, cellophane; flakes, pellets.

Armoslip CPM. [451a] Oleamide; internal lubricant, mold release agent, slip agent in polyolefins; flakes, pellets.

Armoslip HT. [451a] (Hydrog. tallow) amide; see Armoslip CPM; flakes, pellets.

Armosoft L. [63] Ditallow dimethyl ammonium chloride; cationic; fabric softener for household and commercial use; Gardner 6 hazy to clear liq.; dens. 7.09 lb/gal; sp.gr. 0.85 (100 F); visc. 45.5 cps (120 F); pour pt. 70–75 F; 75% act.

Armosoft WA-201. [64] Detergent, softener, antistat for home laundry formulations; Gardner 3 max. clear liq.; sol. in water; sp.gr. 1.038; visc. 132 cps (120 F); cloud pt. 45 F (30% aq.); clear pt. 95 F; flash pt. 145 F (PM); 71–74% act. in aq. ethanol.

Armosoft WA-203. [64] See Armosoft WA-201; clear liq., Gardner 3 max.; sol. in water; sp.gr. 1.010; visc. 110 cps (120 F); cloud pt. 50 F (35%); clear pt. 95 F; flash pt. 105 F (PM); pour pt. 40 F; 83–87% act. in aq. ethanol.

Armosoft WA-204. [64] See Armosoft WA-201; Gardner 3 max. clear liq.; sol. in water; sp.gr. 1.035; visc. 230 cps (120 F); cloud pt. 50 F (35%); clear pt. 95 F; flash pt. 115 F (PM); pour pt. 0 F; 83–87% act. in aq. ethanol.

Armotan ML. [11] Sorbitan laurate; nonionic; w/o emulsifier; liq.; 100% conc.

Armotan MO. [11; 12] Sorbitan oleate; nonionic; w/o emulsifier for cosmetic and pharmaceutical preparations, used to make cutting and sol. oils; Gardner 8 liq.; sp.gr. 1.01; visc. 9.5–11 poise; pour pt. –12 C; 100% conc.

Armotan MP. [11] Sorbitan palmitate; nonionic; w/o emulsifier; waxy substance; 100% conc.

Armotan MS. [11; 12] Sorbitan stearate; nonionic; w/o emulsifier; cream needle-like; m.p. 51–58 C; > 99% act.

Armotan PMD 20. [12] Polysorbate 80; wetting agent, emulsifier, for cosmetic and pharmaceutical preparations; Gardner 7 max. liq.; sp.gr. 1.02; visc. 345–445 cs; pour pt. –16 C.

Armotan PML 20. [11] Polysorbate 20; nonionic; o/w emulsifier, solubilizer for bath oils; liq.; 100% conc.

Armotan PMO 20. [11] Polysorbate 80; nonionic; o/w emulsifier; liq.; 100% conc.

Armoteric LB. [11] Betaine; shampoo formulations; liq.

Armul 03. [181] Polyoxyalkylene ether; nonionic; emulsifier for spray oils for agriculture; liq.; HLB 9.3; 100% conc.

Armul 16. [181] Polyoxyalkylene ether; nonionic; emulsifier for industrial applications; liq.; HLB 11.8; 100% conc.

Armul 21. [181] POE sorbitol monotallate; nonionic; emulsifier as plant regulators in tobacco industry; liq.; HLB 15.4; 100% conc.

Arneel 18 D. [63] Stearyl nitrile; detergent, wetting agent, rust inhibitor; Gardner 6; f.p. 39 C; b.p. 330–360 C.

Arneel C. [63] Coco nitrile; detergent, wetting agent, rust inhibitor; Gardner 6.

Arneel OD. [63] Oleyl nitrile; detergent, wetting agent, rust inhibitor; Gardner 3; sp.gr. 0.83; f.p. 5 C; b.p. > 325 C.

Arneel T. [63] Tallow nitrile; detergent, wetting agent, rust inhibitor; f.p. 4 C; b.p. 330–360 C.

Arnox 500, 515. [181] Primary amine oxide adducts; nonionic; surface active agents used as corrosion inhibitors for industrial acid cleaning and refinery applications; liq.; 100% conc.

Arnox 906. [181] Alkyl phenol PEG ether; nonionic; emulsifier and detergent for styrene-butadine latices; used in leather industry for degreasing and penetration for dyes and finishes; liq.; HLB 10.9; 100% conc.

Arnox 908. [181] Alkyl phenol PEG ether; nonionic; emulsifier, detergent, solubilizer for dry cleaning mixture; wetting agent; liq.; oil-sol., disp. in water; HLB 12.3; 100% conc.

Arnox 910. [181] Alkyl phenol PEG ether; nonionic; wetting agent for wettable powds., clays, dust control modifier; lime soap dispersing agent; detergent and visc. builders; used in the pulp and paper industry; liq.; HLB 12.9; 100% conc.

Arnox 912. [181] Nonyl phenoxy polyethoxyethanol; nonionic; surfactant for textile processing; emulsifier and detergent; used in personal care products; liq.; HLB 14.1; 100% conc.

Arnox 930. [181] Nonylphenol PEG ether; nonionic; emulsifier for pesticide formulations, primary emulsifier for acrylic polymer emulsions; solid; HLB 17.1; 100% conc.

Arnox 940. [181] Nonylphenol PEG ether; nonionic; see Arnox 930; solid; HLB 17.8; 100% conc.

Arnox 950. [181] Nonylphenol ether; nonionic; emulsifier for flowable formulations; mfg. of corrosion inhibitor sticks; solid; water-sol.; HLB 18.1; 100% conc.

Arnox 1003. [181] Fatty alcohols straight chain PEG ether; nonionic; emulsifier, detergent, and defoamer intermediate; liq.; HLB 8.1; 100% conc.

Arnox 1005. [181] Fatty alcohols straight chain PEG ester; nonionic; detergent used in industrial and household applications; liq.; HLB 11.5; 100% conc.

Arnox 1007. [181] Fatty alcohols straight chain PEG ester; nonionic; detergent intermediate, wetting agent; variety of uses for paint, oil-field, and textile industries; liq.; HLB 13.1; 100% conc.

Arnox 1009. [181] Fatty alcohols straight chain PEG ester; nonionic; see Arnox 1007; liq.; HLB 13.7; 100% conc.

Arnox 1100, 1110. [181] Primary amine oxide adducts; nonionic; surface active agents used as corrosion inhibitors for industrial acid cleaning and refinery applications; liq.; 100% conc.

Arnox 2404. [181] Polyalkylene ether; nonionic; used in pesticide formulations; solid; 100% conc.

Arnox LF-11, -12, -13. [181] Polyethoxy adducts, modified; nonionic; surfactants used in machine dishwashing, rinse aids, all-purpose cleaners, textile jet dyeing and scouring, dedusting agents, detergent sanitizers, leather processing, metal cleaners and laundry products; liq.; 100% conc.

Arnox TGE-05, 16, 36, 40, 80. [181] Castor oil ethoxylates; nonionic; emulsifier, dispersing, plasticizing, and lubricating agents; used in metal, leather, textile and coating industries; liq., solid; 100% conc.

Arodet 25 S. [67] Blend of amine condensate and alkylaryl sulfonates; detergent, wetting agent, emulsifier; amber liq.; mild odor; water sol.; sp.gr. 0.98; 100% act.

Arodet 60 T Soft. [67] Triethanolamine salt of linear alkyl sulfonate; detergent, wetting agent, emulsifier; pale yel. liq.; mild odor; water sol.; sp.gr. 1.09; 60% act.

Arodet AA-350. [67] Alkylaryl sulfonate with glycol coupling agents; detergent, wetting agent, dyeing assistant; pale yel. liq.; mild odor; misc. in water at ambient temp.; sp.gr. 1.04; 35% act.

Arodet AN-100. [67] Modified nonionic deriv.; detergent, wetting agent, scouring agent, dye dispersant; water sol.; sp.gr. 1.11; 68–70% act.

Arodet BN-100. [67] Ethoxylated alcohol; detergent, wetting agent, emulsifier, dyeing assistant; clear liq.; mild odor; luke-warm water sol.; 100% act.

Arodet F. [67] Fatty amide deriv.; detergent, wetting agent; pale straw-colored gel; mildly perfumed odor; warm water sol.; sp.gr. 1.07; 14–16% act.

Arodet L-70 M. [67] EO deriv.; nonionic; detergent, wetting agent, emulsifier, dye assistant; water wh. liq.; mildly alcoholic odor; water sol.; sp.gr. 1.03.

Arodet RA. [67] Sulfated modified alcohol deriv.; detergent, scouring agent, penetrant, dyeing assistant; pale amber liq; mild odor; warm water sol.; 32–35% act.

Arodet SC Special. [67] Mixt. of alcohol sulfates, alkylaryl sulfonates, amine condensates and sequesterants; detergent, emulsifier, wetting agent; clear liq.; mild odor; water sol.; sp.gr. 1.09.

Arodet TX-7. [67] Modified nonionic blend; detergent, wetting agent, emulsifier; amber liq.; mildly alcoholic odor; water sol.; sp.gr. 1.01; 50–55% act.

Arofos 200 Conc. [67] Phosphate ester blend; detergent, wetting agent, emulsifier, penetrant, dye leveling agent; water wh. to pale yel. liq.; mild odor; water sol.; sp.gr. 1.08; 85% act.

Arofos 326. [67] Poly phosphorylated surfactant; detergent, emulsifier, wetting agent; pale yel. liq.; mild odor; water sol.; sp.gr. 1.09; 99+% act.

Arofos TD-19 Special. [67] Phosphorylated EO adduct in acid form; detergent, emulsifier, wetting agent; light amber gel; mild odor; sol. in aromatic and chlorinated solv., water; sp.gr. 1.10; 98% act.

Arol Defoamer JT. [67] Complex alkyl ethoxylate blend; defoamer; water wh. liq.; mild odor; water sol.; 100% act.

Arolterge 100M. [67] Blend of fatty acid alkanolamides and phosphate esters; detergent, emulsifier, wetting agent; dk. amber liq.; mild odor; water sol.; sp.gr. 1.0; 100% act.

Aromox 18/12. [63] Bis(2-hydroxyethyl) stearamine oxide; cationic; wetting agent, emulsifier, stabilizer, antistat, foaming agent; Gardner 2 soft paste; sp.gr. 0.927; visc. 54 cps; flash pt. 130 F; pour pt. > 120 F; 50% act. in aq. IPA.

Aromox C/12. [10; 63; 64] Dihydroxyethyl cocamine oxide, IPA; cationic; wetting agent, emulsifier, stabilizer, antistat, foaming agent detergent and shampoo formulations; Gardner 2 clear liq.; sp.gr. 0.949; visc. 52 cp; cloud pt. 18 F; flash pt. 82 F; pour pt. 0 F; surf. tens. 33 dynes/cm; biodeg.; 50% act. in aq. IPA.

Aromox C/12-W. [10; 11; 63; 64] Dihydroxyethyl cocamine oxide; cationic; foam stabilizer for detergents, shampoos; wetting agent, emulsifier, stabilizer, antistat, foaming agent; Gardner 2 clear liq.; sp.gr. 0.997; visc. 2097 cp; HLB 18.4; flash pt. 212 F; pour pt. 35 F; surf. tens. 30.8 dynes/cm; biodeg.; 40% act. in water.

Aromox CD/12. [63] Bis(2-hydroxyethyl) cocoamine oxide; cationic; wetting agent, emulsifier, stabilizer, antistat, foaming agent; Gardner 1 clear liq.; sp.gr. 0.949; visc. 52 cp; cloud pt. 18 F; flash pt. < 80 F; pour pt. 0 F; biodeg.; 50% act. in aq. IPA.

Aromox DM14D-W. [10; 11] Tetradecyldimethylamine oxide; detergent; Gardner 1 max. liq.; sp.gr. 0.96; HLB 12.2; m.p. –3 C; nonflam.; pH 7–8; 30% act.

Aromox DM16. [10; 63; 64] Palmitamine oxide, IPA; cationic; wetting agent, emulsifier, foam stabilizer, antistat, foaming agent for detergents, shampoos; Gardner 1 clear liq.; sp.gr. 0.885; visc. 19 cp; cloud pt. 44 F; flash pt. 80 F; pour pt. 0 F; surf. tens. 31.6 dynes/cm; biodeg.; 40% act. in aq. IPA.

Aromox DM16D-W. [10] Hexadecyldimethylamine oxide; detergent, emollient; Gardner 1 max. liq./paste; sp.gr. 0.94 (60 C); m.p. 33 C; nonflam.; pH 7–8; 25% min act.

Aromox DM18D-W. [10] Octadecyldimethylamine oxide; detergent, emollient; paste; water sol.; pH 7–8; 25% min act.

Aromox DMC. [63; 64] Cocamine oxide, IPA; cationic; wetting agent, emulsifier, foam stabilizer, antistat, foaming agent for detergents, shampoos; Gardner 1 clear liq.; sp.gr. 0.89; visc. 18 cp; cloud pt. 6 F; flash pt. < 80 F; pour pt. < 25 F; surf. tens. 33.1 dynes/cm; biodeg.; 40% act. in aq. IPA.

Aromox DMCD. [10; 11] Cocamine oxide, IPA; detergent; Gardner 1 max. liq.; sp.gr. 0.89; HLB 18.6; flash pt. 21 C (APCC); pH 6–9; 39% min. act.

Aromox DMC-W. [10; 63; 64] Cocamine oxide; cationic; wetting agent, emulsifier, foam stabilizer, antistat, foaming agent for detergents, shampoos; Gardner 1 liq.; sp.gr. 0.971; visc. 17 cp; cloud pt. 34 F; flash pt. > 212 F; pour pt. 35 F; surf. tens. 32.5 dynes/cm; biodeg.; 30% act. in water.

Aromox DMHT. [10; 63; 64] Hydrog. tallow amine oxide, IPA; cationic; wetting agent, emulsifier, foam stabilizer, antistat, foaming agent for detergents and shampoos; Gardner 1 clear liq.; sp.gr. 0.886; visc. 32 cp; cloud pt. 64 F; flash pt. < 85 F; pour pt. 35 F; surf. tens. 36.2 dynes/cm (0.1%); biodeg.; 40% act. in aq. IPA.

Aromox DMHTD. [10] Hydrog. tallow dimethylamine oxide; detergent; Gardner 1 max.; sp.gr. 0.89; flash pt. 29 C (PMCC); 39% min act.

Aromox DMMCD-W. [11] Amine oxide; see Aromox C/12-W; liq.; 29% conc.

Aromox DMMC-W. [10; 11; 63; 64] Lauramine oxide; cationic; wetting agent, emulsifier, foam stabilizer, antistat, foaming agent for detergents, shampoos; Gardner 1 clear liq.; sp.gr. 0.96; visc. 90 cp; cloud pt. 22 F; flash pt. > 212 F; pour pt. 36 F; surf. tens. 31.2 dynes/cm (0.1%); 30% act. in water.

Aromox T/12. [10; 11; 63; 64] Dihydroxyethyl tallow amine oxide, IPA; cationic; wetting agent, emulsifier, foam stabilizer, antistat, foaming agent for detergents and shampoos; Gardner 4 clear liq.; sp.gr. 0.94; visc. 77 cp; cloud pt. 60 F; flash pt. 90 F; pour pt. 55 F; surf. tens. 33.0 dynes/cm (0.1%); 50% act. in aq. IPA.

Arosurf 42-E6. [559] Alkoxylated tallow alcohol; emulsifier, detergent; liq.; 100% conc.

Arosurf 42-PE10. [559] Alkoxylated tallow alcohol; low pour pt. detergent for heavy-duty laundry; liq.; 100% conc.

Arosurf 66-E2. [559] Isosteareth-2; nonionic; emulsifier, emollient for personal care products; o/w and w/o systems; coupling agent, emulsion stabilizer; Gardner 1 liq.; HLB 4.6; m.p. –5 C; pH 7 (1% DW); 100% conc.

Arosurf 66-E10. [559] Isosteareth-10; nonionic; see Arosurf 66-E2; also used as a detergent; Gardner 1 semisolid, HLB 12.0; m.p. 22 C; pH 7 (1% DW); 100% conc.

Arosurf 66-E20. [559] Isosteareth-20; nonionic; see Arosurf 66-E2; Gardner 1 soft solid; HLB 15.0; m.p. 35 C; pH 7 (1% DW); 100% conc.

Arosurf 66-PE12. [559] PPG-3-isosteareth-9; nonionic; see Arosurf 66-E2; Gardner 1 liq.; HLB 12.2; m.p. –10 C; pH 7 (1% DW); 100% conc.

Arosurf AA-22. [559] Cationic asphalt emulsifier for emulsions; liq.; water-disp.; sp.gr. 0.88; dens. 7.4 lb/gal; pour pt. 27 F.

Arosurf AA-23. [559] Diamine type; cationic; asphalt rapid set and mixing grade emulsifier; soft solid; 100% act.

Arosurf AA-27 [559] Diamine-type cationic, modified; asphalt emulsifier; soft solid; sp.gr. 0.83 (95/77 F); dens. 6.9 lb/gal; m.p. 93 F; 100% act.

Arosurf AA-28. [559] Diamine-type cationic, modified; asphalt emulsifier for CRS and CMS emulsions; semisolid; sol. in water; sp.gr. 0.85; dens. 7.1 lb/gal; pour pt. 78 F; flash pt. > 200 F (PMCC); 100% act.

Arosurf AA-30. [559] Cationic; asphalt emulsifier; dk. visc. liq.; disp. in water; sp.gr. 1.043; dens. 8.7 lb/gal; pour pt. 0 C; flash pt. 21 C.

Arosurf AA-36. [559] Cationic; asphalt emulsifier for emulsions; visc. liq.; disp. in water; sp.gr. 1.075; dens. 8.46 lb/gal; pour pt. approx. 65 F; flash pt. > 200 F (PMCC); 100% act.

Arosurf AA-37. [559] Cationic; asphalt emulsifier for emulsions; used to product petroleum oil and resins emulsions; liq.; disp. in water; sp.gr. 0.96; dens. 8.0 lb/gal; 50% act.

Arosurf AA-54-100. [559] Diamine-type, modified, cationic; asphalt emulsifier; soft solid; sp.gr. 0.86; dens. 7.1 lb/gal (95 F); pour pt. 55 F; flash pt. 120 F (PMCC); 100% act.

Arosurf AA-55. [559] Diamine-type cationic; asphalt emulsifier; liq.; dens. 7.25 lb/gal (95 F); pour pt. 68 F; flash pt. > 200 F (PMCC); 100% act.

Arosurf AA-57. [559] Modified diamine type; cationic; asphalt emulsifier; liq.; water-disp.; sp.gr. 0.90; dens. 7.5 lb/gal; pour pt. 27 F.

Arosurf AA-60. [559] Modified diamine type; cationic; asphalt emulsifier.

Arosurf AA-61. [559] Diamine-type, modified; cationic; asphalt emulsifier; liq.; sp.gr. 0.87; dens. 7.2 lb/gal (95 F); pour pt. 50 F; flash pt. > 200 F; 100% act.

Arosurf AA-76. [559] Modified diamine type; cationic; asphalt emulsifier.

Arosurf TA-100. [559] Distearyl dimethyl ammonium chloride; cationic; fabric softener conc. for home and commerical laundry and textile processing; powd., Gardner 3 max.; m.w. 583; flash pt. > 200 F (PM); 93% quat. min.

Arosurf TA-101. [559] Distearyl dimethyl ammonium chloride, modified; specialty quat.; powd.; Gardner 4 max.; flash pt. > 200 F (PM); 100% solids.

Arowet NF. [67] Blend of sulfonated esters; wetting agent, anionic; reddish-br. liq.; water sol.; sp.gr. 1.04.

Arowet S. [67] Blend of sulfonated esters; wetting agent, anionic; amber liq.; water sol.; sp.gr. 1.03.

Arphos DI-403-S. [381] Phosphate ester salt; anionic; coupling agent, detergent, wetting agent; liq.; 100% conc.

Arphos DI-404. [381] Complex phosphate ester, free acid form; anionic; wetting and coupling agent for heavy-duty alkaline cleaners; liq.; 100% conc.

Arphos MD-200. [381] Complex phosphate ester, free acid form; anionic; EP lubricant, coupling agent, low-foam detergent; liq.;

100% conc.

Arphos MH-202. [381] Complex phosphate ester, free acid form; anionic; coupling agent, detergent for heavy-duty detergents; liq.; 100% conc.

Arphos MI-405. [381] Complex phosphate ester, free acid form; anionic; emulsifier for pesticides; liq.; 100% conc.

Arphos MK-203. [381] Complex phosphate ester, free acid form; anionic; solubilizer, coupling agent for nonionic surfactants in presence of alkaline builders; liq.; 100% conc.

Arquad 2C-75. [10; 11; 63; 64] Dicocodimonium chloride, IPA; cationic; emulsifier, foaming, wetting, dispersion agents, corrosion inhibitor, antistat for textiles, cosmetics; industrial, agriculture; bactericide, algicide; Gardner 7 semiliq.; m.w. 447; sp.gr.0.89; HLB 11.4; flash pt. < 80 F; pour pt. 10 F; surf. tens. 30 dynes/cm (0.1%); flam.; 75% act. in aq. IPA.

Arquad 2HT-75. [10; 11; 63; 64] Quaternium-18, IPA; cationic; emulsifier, foaming, wetting, dispersion agents, corrosion inhibitor, antistat, bacteriostat for paper softening, household laundry, hair conditioning, combing; soft, wh. paste; m.w. 573; dens. 7.22 lb/gal; sp.gr. 0.87; visc. 47.5 cps (120 F); f.p. 95 F; HLB 9.7; flash pt. 112 F; pour pt. 90–100 F; surf. tens. 37 dynes/cm (0.1%); 75% act. in aq. IPA.

Arquad 2S-75. [64] Dimethyldisoya ammonium chloride; antistat, softener for plastics, textiles; corrosion inhibitor for petroleum industry; paste, Gardner 16 max.; m.w. 605; water sol.; sp.gr. 0.889; HLB 10.9; pH 5–8; flash pt. (PM) 84 F; 74–77% act. in aq. IPA.

Arquad 12-33. [10; 63; 64] Laurtrimonium chloride, IPA; cationic; emulsifier, foaming, wetting, disp. agents, corrosion inhibitor, antistat, textile softener; hair conditioning, combing aid; Gardner 7 liq.; m.w. (act.) 263; sp.gr. 0.98; f.p. 5 F; HLB 17.1; flash pt. 140 F; biodeg.; 33% act. in aq. IPA.

Arquad 12-50. [10; 11; 63; 64] Laurtrimonium chloride, IPA; cationic; see Arquad 12-33; Gardner 1 liq.; m.w. (act.) 263; sp.gr. 0.89; f.p. 13 F; HLB 17.1; flash pt. < 80 F; surf. tens. 33 dynes/cm; biodeg.; 50% act. in aq. IPA.

Arquad 16-29. [10; 63; 64] Cetrimonium chloride; cationic; emulsifier, foaming, wetting and disp. agents, corrosion inhibitor, antistat; Gardner 6 liq.; m.w. (act.) 319; sp.gr. 0.96; f.p. 61 F; HLB 15.8; flash pt. > 212 F; biodeg.; 29% act. in aq. IPA.

Arquad 16-50. [10; 11; 63; 64] Cetrimonium chloride, IPA; cationic; emulsifier, foaming, wetting and disp. agents, corrosion inhibitor, antistat; Gardner 6 liq.; m.w. (act.) 319; sp.gr. 0.88; f.p. 61 F; HLB 15.8; flash pt. < 80 F; surf. tens. 34 dynes/cm; biodeg.; 50% act. in aq. IPA.

Arquad 18-50. [10; 11; 63; 64] Steartrimonium chloride, IPA; cationic; emulsifier, foaming, wetting, disp. and dye-leveling agents, corrosion inhibitor, antistat, visc. stabilizer, fuel oil additives, in lubricant compounding; Gardner 7 liq.; m.w. (act.) 347; sp.gr. 0.88; HLB 15.7; flash pt. < 80 F; surf. tens. 34 dynes/cm; biodeg.; 50% act. in aq. IPA.

Arquad B-50 USP. [10] Alkyl (C_8–C_{18}) dimethylbenzyl ammonium chloride USP; disinfectant, germicide; Gardner 2 max. liq., visc. mass; sp.gr. 0.98; 50% act.

Arquad B-90 USP. [10] Alkyl (C_8–C_{18}) dimethylbenzyl ammonium chloride USP; disinfectant, germicide; Gardner 6 max. liq., visc. mass; 90% act.

Arquad B-100. [63] Dimethylalkylbenzyl ammonium chloride; bacteriocide, EPA approved; Gardner 2 liq.; m.w. 380; flash pt. < 80 F; 50% act. in aq. IPA.

Arquad C-33. [63] Trimethyl coco ammonium chloride; cationic; emulsifier, foaming, wetting and disp. agents, corrosion inhibitor, antistat; Gardner 7 liquid; m.w. (act.) 278; sp.gr. 0.98; f.p. 25 F; HLB 16.5; flash pt. < 80–140 F; biodeg.; 33% act. in aq. IPA.

Arquad C-33-W. [10; 63; 64] Cocotrimethyl ammonium chloride; cationic; emulsion-break retardant in cosmetics; liq.; sp.gr. 0.96; f.p. –3 C; nonflam.; pH 5–8; 32–35% act.

Arquad C-50. [10; 11; 63; 64] Cocotrimonium chloride, IPA; cationic; emulsifier, foaming, wetting, disp. agents, corrosion inhibitor, antistat; Gardner 7 liq.; m.w. (active) 278; sp.gr. 0.89; f.p. 5 F; HLB 16.5; flash pt. < 80 F; surf. tens. 31 dynes/cm (0.1%); biodeg.; 50% act. in aq. IPA.

Arquad DM14B-90. [10; 11] Myristyl dimethylbenzyl ammonium chloride dihydrate; cationic; bactericide, fungicide, germicide, disinfectant; wh. powd.; 90% act.

Arquad DM18B-90. [10] Octadecyl-dimethylbenzyl ammonium chloride; conditioner, emulsifier, softener; white powder; water sol.; pH 5–8 (1%); 90% min act.

Arquad DMCB. [64] Alkyl dimethyl benzyl ammonium chloride; microbicide for disinfectants, sanitizers, algicides for use in swimming pools, air conditioning towers, bathroom cleaners, petroleum recovery; liq., Gardner 4 max.; m.w. 354; sol. in water and most common org. solv.; sp.gr. 0.935; flash pt. 80 F (PM); 79% act. in aq. IPA.

Arquad DMHTB-75. [10] Hydrog. tallow dimethylbenzyl ammonium chloride; bactericide, disinfectant, softening agent; Gardner 4 max. soft paste; f.p. 60 C; 75% act.

Arquad DMMCB-50. [10; 11] Alkyl (C_{12},C_{14},C_{16}) dimethyl benzyl ammonium chloride; cationic; antistat, flocculant, emulsifier, softener, corrosion inhibitor; liq.; 50% act.

Arquad DMMCB-75. [10] Alkyl (C_{12}–C_{14}, C_{16}) dimethyl benzyl ammonium chloride; disinfectant, bactericide, germicide, fungicide; liq.; cloud pt. –36 F; flash pt. < 80 F; biodeg.; flam.; 50% act.

Arquad L-11. [63] Quarternized beta-amines; cationic; emulsifier; Gardner 2 liq.; sp.gr. 0.885; m.p. –90 F; flash pt. 37 F; 50% act.

Arquad L-15. [63] Quarternized beta-amines; cationic; emulsifier; Gardner 9 liq.; water sol.; sp.gr. 0.877; m.p. 46 F; cloud pt. 42 F; flash pt. 100 F; 50% act.

Arquad S-2C-50. [10; 11] 1:1 mixt. of oleyltrimethyl ammonium chloride and dicocodimethyl ammonium chloride; bactericide, wetting agent, corrosion inhibitor; Gardner 10 max. liq.; sp.gr. 9.87; HLB 13.5; f.p. –15 C; pH 5–8; 50–55% act.

Arquad S-50. [10; 11; 63; 64] Soytrimonium chloride, IPA; cationic; emulsifier, foaming, wetting, dispersion agents, corrosion inhibitor, antistat; Gardner 8 max. liq.; sp.gr. 0.89; HLB 15.6; f.p. 20 C; pH 5–8; 49–52% act.

Arquad T-2C-50. [11; 63] 1:1 mixt. of trimethyltallowammonium chloride and dimethyldicocoammonium chloride; cationic; emulsifier, foaming, wetting, disp. agents, corrosion inhibitor, antistat; Gardner 8 liq.; m.w. (act.) 394; sp.gr. 0.87; HLB 13.0; flash pt. < 80 F; flam.; 50% act. in aq. IPA.

Arquad T-27W. [64] Trimethyl tallow ammonium chloride; see Arquad 2S-75; liq., Gardner 3 max.; m.w. 343; water sol.; HLB 14.2; pH 5–8 (10% aq.); 26–29% act. in aq. IPA.

Arquad T-50. [10; 11; 63; 64] Tallowtrimonium chloride, IPA; cationic; emulsifier, foaming, wetting, disp. agents, corrosion inhibitor, antistat; Gardner 8 max. liq.; m.w. 340; sp.gr. 0.88; HLB 14.2; flash pt. < 80 F (PM); pour pt. 15–48 F; pH 5–8 (10% aq.); biodeg.; 50% act. in aq. IPA.

Arsil 486-S. [381] Silicone sol'n.; nonionic; pesticide formulations, latex processing, glycol formulations, ink processing, soap and detergent mfg., wax and polish emulsions; liq.; 10% conc.

Arsul 458-S. [61; 381] Alcohol ether sulfate; anionic; detergent, foaming agent for dishwash, bubble baths, upholstery/rug cleaners, emulsion polymerization; wallboard; liq.; 60% conc.

Arsul DDB. [61; 381] Branched chain dodecylbenzene sulfonic acid; anionic; chemical intermediate for emulsifiers, dispersants, corrosion inhibitors, emulsion breakers, cleaning compds.; liq.; 90% conc.

Arsul LAS. [61; 381] Linear dodecylbenzene sulfonic acid; anionic; see Arsul DDB; liq.; biodeg.; 98% conc.

Arsul WAQ. [61; 381] Sodium lauryl sulfate; anionic; surfactant for emulsion polymerization, cosmetics; liq.; 30% conc.

Artodan. [272; 273] Sodium stearyl-2-lactylate; anionic; food emulsifier for starch complexing, bread improver, freeze/thaw emulsions; powd.; HLB 9; 100% act.

Arylan CA. [360] Calcium dodecylbenzene sulfonate; anionic; emulsifier for herbicides, pesticides; visc. liq.; alcoholic odor; water insol.; sp.gr. 1.012; visc. 3025 cs; flash pt. 94 F; pour pt. < 0 C; pH 5–8 (1% aq); 70% act.

Arylan DA 36. [360] Blended liq. detergent with nonionic foam stabilizer; anionic/nonionic; detergent; liq.; biodeg.; 36% conc.

Arylan DT. [360] Specially blended liq. detergent conc. based on alkylaryl sulfonates with foam stabilizers; detergent; amber, clear liq.; mild odor; sp.gr. 1.05; visc. 330 cs; flash pt. > 200 F (COC); pour pt.< 0 C; pH 6–7.5 (1% aq.); 30% act.

Arylan HAL. [360] Specially blended liq. detergent conc. based on alkylaryl sulfonates with foam stabilizers; detergent; amber, clear liq.; mild, alcoholic odor; sp.gr. 1.04; visc. 300 cs; flash pt. 90 F (Abel CC); pour pt.< 0 C; pH 6.5–7.5 (1% aq.); 87% act.

Arylan HE Acid. [360] Alkylbenzene sulfonic acid, branched chain; anionic; intermediate for detergent mfg. and production of wetting agents and emulsifiers; liq.; 95% conc.

Arylan LQ. [360] Specially blended liq. detergent conc. based on alkylaryl sulfonates with foam stabilizers; detergent; amber, clear liq.;

mild, alcoholic odor; sp.gr. 1.053; visc. 350 cs; flash pt. > 200 F (COC); pour pt.< 0 C; pH 7–9 (1% aq.); 40% act.

Arylan PWS. [360] Amine alkylaryl sulfonate; anionic; emulsifier; liq.; visc. 4200 cs; 90% act.

Arylan S Acid. [192] Dodecylbenzene sulfonic acid (branched chain); anionic; intermediary; liq.; 96% conc.

Arylan S Flake. [360] Sodium benzene sulfonate; wetting agent, emulsifier; cream-colored flake; 82% conc.

Arylan S90 Flake. [360] Sodium benzene sulfonate (branched chain); anionic; wetting agent, emulsifier; 95% conc.

Arylan SBC 25. [360] Sodium dodecylbenzene sulfonate; anionic; detergent base, emulsifier for emulsion polymerization; liq.; 25% conc.

Arylan SBC Acid. [360] Dodecylbenzene sulfonic acid; anionic; detergent intermediate; biodeg.; liq.; 97% conc.

Arylan SC 30. [360] Sodium dodecylbenzene sulfonate; anionic; detergent, emulsifier; emulsion polymerization; liq.; 30% act.

Arylan SC Acid. [360] Dodecylbenzene sulfonic acid; detergent intermediate; br. liq.; visc. 1800 cs; 97% act.

Arylan SNS. [192] Sodium naphthalene sulfonic acid formaldehyde condensate; dispersing agent; buff-colored powd.; 90% act.

Arylan SO60 Acid. [360] Branched chain tridecyl benzene sulfonic acid; detergent, emulsifier, wetting agent, emulsion polymerization aids; br., visc. liq.; char. odor; water sol.; sp.gr. 1.045; visc. > 25,000 cs; flash pt. > 250 F (COC); pour pt. < 0 C; pH 2.0; > 96% act.

Arylan SP Acid. [360] Dodecylbenzene sulfonic acid; anionic; detergent intermediate; liq.; 97% conc.

Arylan SX Flake. [360] Sodium dodecylbenzene sulfonate; anionic; detergent base, wetting agent; liquid; 75% conc.

Arylan TE/C. [360] Benzene sulfonate, TEA salt; anionic; detergent, emulsifier, wetting agent, emulsion polymerization aids; clear liq.; mild odor; water sol.; sp.gr. 1.060; visc. 900 cs; flash pt. > 200 F (COC); pour pt. < 0 C; pH 7–8.5; biodeg.; 40% act.

Arylene M40. [283] Dioctyl sulfosuccinate; anionic; wetting agent, rewetting, dewatering surfactant; clear liq.; dens. 1.03; visc. 50 cps; 40% act.

Arylene M60. [283] Dioctyl sulfosuccinate; anionic; wetting agent, rewetting, dewatering surfactant; clear liq.; dens. 1.06; visc. 50 cps; 60% act.

Arylene TA. [283] Tallow sulfosuccinate; anionic; foam booster for SBR latex backing on carpet; paste; 35% act.

ASA. [227] Alkenyl succinic anhydride; intermediate for defoamers, demulsifiers, emulsifiers, foam boosters, wetting agents, detergents, dispersants; liq.

Asol. [410] Lecithin fraction; nonionic; antispatter agent, release agent, emulsifier for food, cosmetics, pharmaceuticals; liq.; 40–100% conc.

Astromid 18, 18 LV. [19] N-octadecyl sulfosuccinamate, sodium salt; anionic; foaming, frothing, emulsifying; paste and liq. respectively; 35% conc.

Astromid 22. [19] Tetrasodium N-alkyl, N-(1,2 dicarboxyethyl) sulfosuccinamate; anionic; emulsifier; liq.; 35% conc.

Astromid 25. [19] N-alkyl sulfosuccinamate; anionic; foaming, frothing, emulsifying; liq.; 35% conc.

Astrowet 102. [19] Alcohol half-ester sulfosuccinate, sodium salt, ethoxylated; anionic; polymerization emulsifier; liq.; 30% conc.

Astrowet B-45. [19] Sodium diisobutyl sulfosuccinate; anionic; polymerization emulsifier; liq.; 45% conc.

Astrowet H-80. [19] Sodium dihexyl sulfosuccinate; anionic; polymerization emulsifier; liq.; 80% conc.

Astrowet O-70-PG. [19] Sodium dioctyl sulfosuccinate; anionic; wetting, emulsifying; for high flash point applications; liq.; 70% conc. in propylene glycol/water solv.

Astrowet O-75. [19] Sodium dioctyl sulfosuccinate; anionic; wetting, emulsifying agent; liq.;75% conc.

Atcopen SD. [98] Alkyl phosphate; anionic; wetting agent; liq.; 78% conc.

Atcowet C. [164] Alkyl polyphosphate; anionic; detergent, wetting for scouring and kier boiling; liq.; 78% conc.

Atcowet W. [98; 164] Sodium dioctyl sulfosuccinate; anionic; wetting and rewetting agent for textile applications; liq.; 75% conc.

Atlas EM-2, EM-3, EM-13. [72] Glycol ester; nonionic; fiber lubricant, emulsifier; liq.; 100% conc.

Atlas EM-16. [72] Alkanolamide; nonionic; fiber lubricant, emulsifier, wetting agent; liq.; 100% conc.

Atlas EMJ-2. [72] Nonylphenoxyl polyethoxy

ethanol; nonionic; detergent, wetting agent, emulsifier; liq.; HLB 13.4; 100% conc.

Atlas EMJ-C. [72] Nonylphenoxy polyethoxy ethanol; nonionic; detergent, wetting agent, emulsifier; liq.; HLB 13.4; 50% conc.

Atlas G-263. [319; 320] Cetyl ethyl morpholinium ethosulfate; cationic; antistatic agent; textile specialties; liq.; water-sol.; HLB 30.0; 35% act. in water.

Atlas G-263 Anhydrous. [320] Cetyl ethyl morpholinium ethosulfate; antistatic agent, cationic; tan to br., waxy solid; fruity odor; water, ethanol sol.; 99% act.

Atlas G-271. [319; 320] Soya ethyl morpholinium ethosulfate; cationic; deodorant, cosmetic specialty; textile antistatic agent; liq.; HLB 30.0; 35% act. in water.

Atlas G-695. [320] Mono- and diglycerides of edible fats or oils; food emulsifier; paste; bland odor, taste; sol. in lower alcohols, veg. and min. oils; sp.gr. 0.96; visc. 528 cps; HLB 3; clear pt. 40 C; flash pt., pour pt. > 300 F.

Atlas G-711. [320] Alkylaryl sulfonate; anionic; drycleaning detergent; lt. yel. liq.; sol. in chlorinated solv., ethylene glycol, lower alcohols, aniline and petroleum solv.; sp.gr. 1; visc. 5500 cps; HLB 11.7; flash pt. > 300 F; 100% act.

Atlas G-950. [320] Sorbide dioleate; nonionic; surfactant; orange-red liq.; water disp., sol. in aromatic and aliphatic hydrocarbons, veg. oils; sp.gr. 1; visc. 250 cps; HLB 2.7; flash pt. > 300 F.

Atlas G-991. [320] Mono- and diglycerides; food emulsifier; ivory wh., plastic solid; sol. above its m.p. in veg. oils and min. oil; m.p. 118 F; HLB 2.8; flash pt. > 300 F.

Atlas G-1045. [320] POE sorbitol laurate; nonionic; emulsifier; lt. yel., oily liq.; disp. with haze in water; sol. in toluene, min. spirits, acetone, dioxane, cellosolve, ethyl acetate, aniline; sp.gr. 1; visc. 175 cp; HLB 11.5; flash pt. > 300 F; 100% act.

Atlas G-1045A. [320] POE sorbitol polyoleate-laurate; nonionic; emulsifier for textiles; liq.; HLB 13.2; 100% conc.

Atlas G-1052. [320] POE sorbitol distearate; nonionic; surfactant; cream-colored wax; sol. in IPA; HLB 8.5; flash pt. > 300 F.

Atlas G-1069. [320] POE sorbitol ester of mixed fatty acids; nonionic; surfactant; amber liq.; disp. in water, sol. in IPA, most veg. oils, hydrocarbon solv.; sp.gr. 1.0; visc. 200 cps; HLB 9.4; pour pt. 45.

Atlas G-1086. [319; 320] POE (40) sorbitol hexaoleate; nonionic; emulsifier; yel., oily liq.; sol. in toluene, min. spirits, many veg. oils, acetone, carbon tetrachloride, cellosolve, methanol, lower alcohols, aniline; sp.gr. 1; visc. 200 cps; HLB 10.2; flash pt. > 300 F; 100% act.

Atlas G-1087. [320] POE sorbitol polyoleate; nonionic; emulsifier for textiles; liq.; HLB 9.2; 100% conc.

Atlas G-1096. [319; 320] POE (50) sorbitol hexaoleate; nonionic; emulsifier; amber liq.; sol. in veg. oils, acetone, cellosolve, lower alcohols, some aromatic solv., tetrachloride; sp.gr. 1.0; visc. 220 cps; HLB 11.4; flash pt. > 300 F; 100% act.

Atlas G-1186. [320] POE sorbitol oleate; surfactant, nonionic; amber, oily liq.; sol. in lower alcohols, veg. and min. oils; sp.gr. 1; visc. 230 cps; HLB 10.2; flash pt. > 300 F; 100% act.

Atlas G-1223. [320] POE esters of mixed fatty and resin acids with 0.025% antifoam added; nonionic; detergent; lt. amber, oily liq.; sol. in water, dioxane, cellosolve, ethyl acetate, lower alcohols, toluene, acetone; sp.gr. 1.1; visc. 350 cps; HLB 13.8; flash pt. > 300 F.

Atlas G-1236. [320] Blend of nonionic lubricants, emulsifiers, and rewetters; clear amber liq.; disp. in water, sol. in alcohol, chloroform; visc. 600 cps; flash pt. > 300 F; pour pt. 0 C.

Atlas G-1256. [319] POE sorbitol and tall oil ester; nonionic; emulsifier for agriculture, emulsions, and textile formulations; liq.; HLB 9.7; 100% conc.

Atlas G-1281. [319] POE triglyceride; nonionic; see Atlas G-1256; liq.; HLB 9.7; 100% conc.

Atlas G-1284. [319] POE triglyceride; nonionic; see Atlas G-1256; liq.; HLB 13.1; 100% conc.

Atlas G-1285. [319] POE triglyceride; nonionic; see Atlas G-1256; paste; HLB 14.4; 100% conc.

Atlas G-1288. [319] POE triglyceride; nonionic; see Atlas G-1256; paste; HLB 16.0; 100% conc.

Atlas G-1289. [319] POE triglyceride; nonionic; see Atlas G-1256; solid; HLB 14.4; 100% conc.

Atlas G-1292. [319; 320] POE triglyceride; nonionic; see Atlas G-1256; pale yel. liq. to soft paste; sol. in water, ethanol, IPA; sp.gr. 1.0; HLB 10.8; flash pt. > 300 F; 100% conc.

Atlas G-1295. [319; 320] POE triglyceride; nonionic; see Atlas G-125; wh. to cream-colored, hard, waxy solid; sol. in water, lower alcohols, acetone, chloroform; sp.gr. 1.05; HLB 17.5; flash pt. > 300 F; 100% conc.

Atlas G-1300. [319; 320] POE triglyceride; nonionic; see Atlas G-1256; tan, hard waxy solid; sol. in water, lower alcohols, acetone, chloroform; HLB 18.1; flash pt. > 300 F; 100% conc.

Atlas G-1304. [319] POE triglyceride; nonionic; see Atlas G-1256; solid; HLB 18.7; 100% conc.

Atlas G-1425. [319] POE (20) sorbitol lanolin derivative; nonionic; emulsifier for cosmetic waxes; paste; HLB 8.0; 100% conc.

Atlas G-1441. [319; 320] PEG-40 sorbitan lanolate; nonionic; emulsifier for cosmetic waxes, surfactant for personal care products, pharmaceuticals; red br. solid; sol. in water (hazy), IPA; sp.gr. 1; HLB 14; flash pt. > 300 F; pour pt. 33 C; 100% conc.

Atlas G-1471. [319; 320] POE (75) sorbitol lanolin deriv.; nonionic surfactant, emulsifier, cosmetic use; amber, soft paste; sol. in IPA, ethylene glycol; HLB 16; flash pt. > 300 F; 100% conc.

Atlas G-1649. [320] POE alkyl aryl ether; nonionic; surfactant; cream-colored, waxy solid; sol. in water, toluol, methanol, IPA, carbon tetrachloride; sp.gr. 1.1; HLB 16.0; flash pt. > 300 F.

Atlas G-1702. [319; 320] PEG-6 sorbitan beeswax; nonionic; emulsifier in cosmetic creams; tan waxy solid; slight sol. in veg. oils; HLB 5; flash pt. > 300 F; 100% conc.

Atlas G-1726. [319; 320] PEG-20 sorbitan beeswax; nonionic; cosmetic emulsifier; solid; HLB 5; 100% conc.

Atlas G-1790. [320] POE (20) lanolin deriv.; emulsifier, cosmetic use, nonionic; br. solid; water sol.; HLB 11; flash pt. > 300 F.

Atlas G-1795. [320] POE (50) lanolin deriv.; emulsifier, cosmetic use, nonionic; tan waxy solid; sol. in water, ethanol; sp.gr. 1.15; HLB 17; flash pt. > 300 F.

Atlas G-1821. [319] PEG-150 distearate; nonionic; shampoo thickener; flakes; 100% conc.

Atlas G-2079. [320] POE (20) palmitate; nonionic; surfactant; wh. waxy solid; sol. in water, ethanol; HLB 15.5; flash pt. > 300 F.

Atlas G-2090. [320] Blend of POE fatty amine and POE sorbitol oleate; emulsifier blend, detergent; red-br. liq.; water disp., sol. in IPA, xylene, min. spirits, kerosene; sp.gr. 0.95; visc. 40 cps; HLB 12.5; flash pt. > 300 F; 100% act.

Atlas G-2109. [319; 320] POE laurate; nonionic; lubricant, emulsifier for textiles; liq.; HLB 13.3; 100% conc.

Atlas G-2127. [319; 320] POE laurate; nonionic; surfactant, lubricant, emulsifier for textiles; lt. yel. liq.; sol. in IPA, disp. in water, many veg. oils; sp.gr. 1.0; HLB 12.8; flash pt. > 300 F; 100% conc.

Atlas G-2143. [320] POE (10) monoleate; emulsifier, nonionic; lt. yel. liq.; disp. in water, sol. in most min. oils, lower alcohols; flash pt. > 300 F; pour pt. –10 C.

Atlas G-2151. [320] POE (30) stearate; nonionic; surfactant; ivory, waxy solid; sol. in propylene glycol, ethylene glycol, water; HLB 16; flash pt. > 300 F.

Atlas G-2160. [320] POE (25) propylene glycol monostearate; nonionic; surfactant; lt. tan semisolid; sol. in water, methanol and ethanol; HLB 16; flash pt. > 300 F.

Atlas G-2162. [320] POE (25) propylene glycol stearate; nonionic; surfactant; cream-colored semisolid; sol. in water, methanol, ethanol, HLB 16; flash pt. > 300 F; 100% act.

Atlas G-2183. [320] Propylene glycol monopalmitate; food emulsifier; ivory, waxy solid; sol. in veg. oils, common org. solv.; m.p. 87 F; HLB 3.2.

Atlas G-2185. [320] Propylene glycol monooleate; food emulsifier; yel. liq.; sol. in veg. oils, common org. solv.; HLB 3.2.

Atlas G-2198. [320] POE (40) stearate; nonionic; surfactant; lt. tan, waxy solid; sol. in water, acetone, ether, dioxane, cellosolve, carbon tetrachloride, lower alcohols, ethyl acetate, aniline and many org. solv.; HLB 16.9; flash pt. > 300 F; pour pt. 38 C.

Atlas G-2684. [320] Blend of sorbitan monooleate and POE esters of mixed fatty and rosin acids; nonionic; detergent; amber liq.; disp. in water, sol. in acetone, toulene, dioxane, carbon tetrachloride, aniline, min. oils; sp.gr. 1; visc. 500 cps; HLB 7.8; flash pt. > 300 F; pour pt. 15 F.

Atlas G-3300. [320] Alkylaryl sulfonate; anionic; dispersant for pigments in water-based coatings; surfactant; amber liq.; sol. at low conc. in water and hard water, in acetone, carbon tetrachloride, cellosolve, ethylene glycol, methanol, aniline, min. spirits, kerosene, toluene; sp.gr. 1.0; visc.

6000 cps; HLB 11.7; flash pt. > 300 F; 90% act.

Atlas G-3300B. [319] Alkylaryl sulfonate; nonionic; emulsifier for agric. emulsions; liq.; HLB 11.4; 100% conc.

Atlas G-3684. [320] POE fatty amine; surfactant; amber liq.; sol. in acetone, lower alcohols, 8% sulfuric acid, 6% sodium hydroxide, water; sp.gr. 1.03; visc. 150 cps; flash pt. > 300 F; pour pt. 20 F.

Atlas G-3707. [319] POE lauryl alcohol; nonionic; o/w emulsifiers used in topical cosmetic applications; liq.; HLB 12.8; water-sol./disp.; 100% conc.

Atlas G-3780A. [319; 320] PEG-20 tallow amine; nonionic; surfactant, antistat for textiles; amber liq. (may become hazy); sol. in water, lower alcohols, acetone, ethyl acetate, ethylene glycol; sp.gr. 1.04; visc. 250 cps; HLB 15.5; flash pt. > 300 F; 100% conc.

Atlas G-3801. [320] POE (2) alkyl alcohol; nonionic; emulsifier; wh., pourable liq.; sp.gr. 0.89; visc. 60 cps; HLB 5.3.

Atlas G-3811. [320] POE (10) alkyl alcohol; nonionic; emulsifier; wh. pourable liq.; sp.gr. 0.992; visc. 415 cps; HLB 12.9.

Atlas G-3816. [319] POE (16) cetyl alcohol; nonionic; general purpose emulsifier, lubricant used in cosmetic formulas; solid; HLB 14.9; 100% conc.

Atlas G-4822. [319] Cetostearyl alcohol, ethoxylated, self-emulsifying; nonionic; wax; 100% conc.

Atlas G-4884. [319] Sorbitan monooleate, tech.; nonionic; lipophilic surfactant; liq.; HLB 4.3; 100% conc.

Atlas G-4885. [319] Sorbitan trioleate, tech.; nonionic; liq.; HLB 1.8; 100% conc.

Atlas G-4905. [319] POE (20) sorbitan monooleate, tech.; nonionic; hydrophilic surfactant; liq.; water-sol.; HLB 15.0; 100% conc.

Atlas G-7166P. [320] POE (16) sorbitan tristearate; nonionic; surfactant; tan, waxy solid; disp. in warm water, most org. solv., veg. and min. oils; sp.gr. 1.05; HLB 10.

Atlas G-7596H. [320] POE sorbitan laurate; nonionic; emulsifier; orange, oily liq.; water sol.; sp.gr. 1.1; visc. 400 cps; HLB 14.3; flash pt. > 300 F.

Atlas G-7596-J. [320] POE (10) sorbitan laurate; nonionic; detergent; yel. liq.; sol. in water, acetone, dioxane, cellosolve, lower alcohols, ethyl acetate, aniline; sp.gr. 1.1; visc. 400 cps; HLB 14.9; flash pt. > 300 F.

Atlas G-8916F, G-8936CJ. [319] Mixed fatty and resin acids, POE sorbitan esters; nonionic; emulsifiers for agric. emulsions; liq.; HLB 14.6 and 12.5 respectively; 100% conc.

Atlas G-9046T. [320] POE (20) mannitan monolaurate; surfactant, nonionic; yel., oily liq.; sol. in water, ethanol, methanol, propylene glycol; sp.gr. 1.1; HLB 16.7; flash pt. > 300 F.

Atlas Softener AD. [72] Imidazoline of stearic acid; cationic; fiber lubricant and softener for textile and leather applications; paste; 25% conc.

Atlas Sul. L-2. [72] Sulfated neatsfoot oil; anionic; lubricant, emulsifier; liq.; 75% conc.

Atlas WA-100. [72] Dioctyl sulfosuccinate; wetting agent.

Atlas WA-103. [72] Alkylated aromatic sodium sulfonate; anionic; detergent, wetting agent, dye bath additive in textile wet processing; liq.; 25% conc.

Atlas WA-104. [72] Alkanolamide; nonionic; detergent, fiber lubricant, emulsifier, wetting agent; liq.; 100% conc.

Atlas WA-108. [72] Fatty acid sulfate; anionic; textile wetting and dye leveling agent; liq.; 65% conc.

Atlasoft-C. [72] Alkanolamide and linear primary alcohol blend; nonionic; conc. base for textile softeners; waxy solid; 100% conc.

Atlasol 103. [72] Sulfated alcohol; anionic; emulsifier, fiber lubricant; liq.; 75% conc.

Atlasol 160-S, 6920. [72] Sulfated syn. replacement for sperm; anionic; lubricant, emulsifier; liq.; 75% conc.

Atlox 775. [319; 320] POE sorbitol ester; nonionic; emulsifier for agric. spray oils, pesticides; liq.; 100% act.

Atlox 1045A. [319; 320] POE sorbitol oleate-laurate; nonionic; emulsifier for agric. spray oils, pesticides; yel. liq.; sol. in most org. solv. and pesticides, disp. in water; sp.gr. 1.0; visc. 200 cps; HLB 11.4; flash pt. > 300 F; 100% conc.

Atlox 1087. [320] POE sorbitol septaoleate; nonionic; emulsifier; yel. liq.; sol. in most veg. and min. oils, disp. in water; sp.gr. 1; visc. 175 cps; HLB 9.2; flash pt. > 300 F.

Atlox 1096. [319] POE sorbitol hexaoleate; nonionic; emulsifier and coupling agent used in textile industry; liq.; HLB 11.4; 100% conc.

Atlox 1196. [320] POE sorbitol oleate; nonionic; emulsifier; amber, oily liq.; disp. in water, sol. in xylol, methanol, acetone, most agric. conc.; sp.gr. 1; visc. 450 cp; HLB 11.4; flash pt. 475F.

Atlox 1256. [320] POE sorbitol esters of mixed fatty and resin acids; nonionic; surfactant; dk. amber liq.; sol. in most org. solv. and pesticide formulations, disp. in water; sp.gr. 1.0; visc. 1000 cps; HLB 9.7; flash pt. > 300 F.

Atlox 1285. [319] POE triglyceride; nonionic; emulsifier and coupling agent used in textile industry; semisolid; HLB 14.4; 100% conc.

Atlox 2081 B. [71; 319] POE sorbitan esters of fatty and resin acids-alkylaryl sulfonate blend; anionic; emulsifier, surfactant; liq.; biodeg.

Atlox 3300 B. [71] POE sorbitan esters of fatty and resin acids-alkylaryl sulfonate blend; anionic; surfactant; liq.; biodeg.

Atlox 3335. [320] Alkylaryl sulfonate blended with POE sorbitan ester of tall oil; agric. surfactant; amber liq.; disp. in water, sol. in most aromatic solv. and pesticide formulations; sp.gr. 1.1; visc. 2200 cps; HLB 13; flash pt 160 F.

Atlox 3335 B. [71; 319] POE sorbitan esters of fatty and resin acids, alkylaryl sulfonate blend; anionic; agric. emulsifier; pesticides, herbicides; liq.; biodeg.

Atlox 3386 B. [71; 319] POE sorbitol fatty esters; alkyl aryl sulfonate blend; anionic; agric. emulsifier; pesticides, herbicides; liq.; HLB 9.4; biodeg.

Atlox 3387. [320] Anionic-nonionic surfactant blend; agric. surfactant; amber, visc. liq.; sol. in most aromatic solv. and pesticide formulations; sp.gr. 1.0; visc. 4000 cps.

Atlox 3387 B. [71] POE glyceride ester, alkylaryl sulfonate blend; anionic; agric. surfactant; liq.; biodeg.

Atlox 3400 B. [71; 319] POE alkyl phenol, alkylaryl sulfonate blend; anionic; agric. emulsifier; liq.; biodeg.

Atlox 3403F. [320] Anionic-nonionic blend; pesticide emulsifier; amber liq.; sol. in most aromatic solv. used in pesticide formulations, disp. in water; dens. 8.4 lb/gal; sp.gr. 1.01; visc. 100 cp; flash pt. 58 F; pour pt. 25 F; flam.

Atlox 3404F. [320] Anionic-nonionic blend; emulsifier; amber liq.; sol. in kerosene, most aromatic solv. used in pesticide formulations, disp. in water; dens. 8.7 lb/gal; sp.gr. 1.04; visc. 1400 cp; flash pt. 64 F; pour pt. –15 F.

Atlox 3406F. [320] Anionic-nonionic blend; emulsifier; amber liq.; sol. in most aromatic solv. used in pesticide formulations, disp. in water; dens. 8.6 lb/gal; sp.gr. 1.03; visc. 200 cp; flash pt. 61 F; pour pt. 20 F; flam.

Atlox 3407F. [320] Anionic-nonionic blend; emulsifier; amber liq.; sol. in most aromatic solv. used in pesticide formulations, disp. in water; sp.gr. 1.03; visc. 250 cp; flash pt. 59 F; pour pt.10 F; flam.

Atlox 3408F. [320] Anionic-nonionic blend; emulsifier; amber liq.; sol. in most aromatic solv. used in pesticide formulations, disp. in water; dens. 8.5 lb/gal (20 C); sp.gr. 1.02; visc. 150 cp; flash pt. 57 F; pour pt. 20 F; flam.

Atlox 3409F. [320] Anionic-nonionic blend; emulsifier; amber liq.; sol. in most aromatic solv. used in pesticide formulations, disp. in water; dens. 8.4 lb/gal; sp.gr. 1.01; visc. 150 cp; flash pt. 54 F; pour pt. 40 F; flam.

Atlox 3414F. [320] Anionic-nonionic blend; emulsifier; visc. liq.; sol. in kerosene, fuel oil, xylene, heavy aromatic naptha; sp.gr. 1.03; visc. 100–1400 cp; flash pt. > 240 F.

Atlox 3421F. [320] Anionic-nonionic blend; agric. emulsifier; hazy liq.; sol. in most aromatic solv. used in pesticide formulations, disp. in water; sp.gr. 1.04; visc. 250 cp; flash pt. 48 F; pour pt. 55 F; flam.

Atlox 3422F. [320] Anionic-nonionic blend; agric. emulsifier; hazy liq.; sol. in most aromatic solv. used in pesticide formulations, disp. in water; sp.gr. 1.04; visc. 900 cp; flash pt. 58 F; pour pt. 50 F; flam.

Atlox 3450F. [320] Anionic-nonionic blend; agric. emulsifier; dk. br., hazy liq.; sol. in most aromatic solv. used in pesticide formulations, disp. in water; dens. 8.7 lb/gal; sp.gr. 1.04; visc. 700 cp; flash pt. 120 F; pour pt. 60 F.

Atlox 3453F, 3454F, 3455F. [320] Anionic-nonionic blend; general purpose emulsifier for pesticide conc. where high flash pt. is required; liq.; 100% act.

Atlox 3459F. [320] Anionic-nonionic blend; agric. emulsifier; clear liq.; sol. in most aromatic solv. used in pesticide formulations, sol. in water; sp.gr. 1.05 C; visc. 500 cp; flash pt. 64 F; pour pt. 15 F; flam.

Atlox 4851 B. [71; 319] POE triglyceride alkylaryl sulfonate blend; anionic; emulsifier for pesticides, herbicides; liq.; HLB 13.2; biodeg.

Atlox 4853 B. [71; 319] Blend of ionic and nonionic surfactants; emulsifier, wetter, dispersant for pesticides, herbicides; liq.; HLB 11.9; biodeg.

Atlox 4855 B. [71; 319] POE triglyceride, alkylaryl sulfonate blend; anionic; emulsifier for pesticides, herbicides; liq.; HLB 9.2; biodeg.

Atlox 4856 B. [71; 319] POE alcohol alkyl aryl sulfonate blend; anionic; emulsifier for pesticides, herbicides; liq.; HLB 8.7; biodeg.

Atlox 4857 B. [71; 319] POE alkyl phenol, alkylaryl sulfonate blend; anionic; emulsifier for pesticides, herbicides; liq.; HLB 12.3; biodeg.

Atlox 4858 B. [71; 319] POE sorbitan esters of fatty and resin acids, alkylaryl sulfonate blend; anionic; emulsifier for pesticides, herbicides; liq.; HLB 11.6; biodeg.; 100% conc.

Atlox 4860. [71] POE alkyl ether and fatty acids, alkyl aryl sulfonate blend; powd., biodeg.; 100% conc.

Atlox 4861B. [319] Alkylaryl sulfonate; anionic; emulsifying agent for pesticide and herbicide formulations; liq.; HLB 8.6; 70% conc.

Atlox 4862. [71] Anionic suspending agent; wetting/suspending agent, anionic; powd.; 100% conc.

Atlox 4868B. [71; 319] POE triglyceride/alkyl aryl sulfonate blend; wetting/suspending agent; pesticide emulsifier; liq.; HLB 11.9; biodeg.

Atlox 4870B. [71; 319] Polyethylene sorbitan esters of fatty and resin acids, alkylaryl sulfonate blend; wetting/suspending agent; liq.; HLB 13.0; biodeg.

Atlox 4873. [319] POE alkyl ether; nonionic; wetting agent for pesticides and herbicides; powd.; HLB 13.6.

Atlox 4875. [319] POE ester; wetting and dispersing agent for pesticides and herbicides; liq.; HLB 18.2.

Atlox 4883. [319] POE alkyl ether; wetting agent for pesticides and herbicides; powd.; HLB 13.5.

Atlox 4885. [319] Sorbitan trioleate; emulsifying agent for pesticides and herbicides; liq.; HLB 1.8; 100% conc.

Atlox 4991. [71; 319] POE alkylaryl ether; nonionic; surfactant, wetting agent for pesticides, herbicides; cloudy liq.; HLB 13.6; 100% conc.

Atlox 4995. [71; 319] POE alkyl ether; nonionic; wetting/suspending agent for pesticides, herbicides; powd.; HLB 13.6; 100% conc.

Atlox 4996 B. [71] Nonionic; wetting/suspending agent; powd.

Atlox 8916P. [71; 320] POE (16) sorbitan ester of tall oil; nonionic surfactant; amber liq.; sol. in water and most aromatic solv. and pesticide formulations; sp.gr. 1.1; visc. 1320 cps; HLB 14.6; flash pt. > 300 F; 100% conc.

Atlox 8916PF. [319; 320] Mixed fatty and resin acids POE sorbitan esters; emulsifier, agric. surfactant; pesticides and herbicides; amber liq.; sol. hazy in most aromatic solv. used in pesticide formulations, water sol. after heating; dens. 9.0 lb/gal; sp.gr. 1.08; visc. 500 cp; HLB 14.6; flash pt. > 200 F; pour pt. –20 F; 100% conc.

Atlox 8916TF. [320] Alkoxylated polyol fatty acid ester; emulsifier, agric. surfactant; amber liq.; sol. hazy in most aromatic solv. used in pesticide formulations, water sol. after heating; dens. 8.3 lb/gal; sp.gr. 1.00; visc. 500 cp; flash pt. > 200 F; pour pt. –20 F.

Atmer 507. [320] Aliphatic polyether; nonionic; surfactant; amber liq.; disp. in water; sol. in acetone, ethylene acetate, dioxane, cellosolve, lower alcohols; sp.gr. 1.1; visc. 300 cps; flash pt. > 300 F.

Atmer 508. [320] Aromatic polyether; nonionic; surfactant; lt. liq.; sol. in lower alcohols, cellosolve, many petroleum solv.; sp.gr. 1.0; visc. 220 cps; flash pt. > 300 F.

Atmer 665. [320] Nonionic ester; surfactant; tan, waxy beads; IPA sol.; HLB 2.1; flash pt. > 300 F; pour pt. 53 C.

Atmos 150. [319; 320] Glyceryl stearate; nonionic; food emulsifier; ivory wh. beads; bland odor and taste; sol. above its m.p. in veg. oils, min. oil, IPA; m.p. 140 F; HLB 3.2; flash pt. > 300 F; 100% conc.

Atmos 300. [319; 320] Glyceryl oleate; nonionic; food emulsifier; clear yel. liq.; sol. in cottonseed oil, IPA; sp.gr. 0.96; visc. 130 cps; HLB 2.8; flash pt. > 300 F; 100% act.

Atmos 729. [320] Hydrated mono- and diglycerides, polysorbate 60 (11.3%) and sorbitan monostearate (11.3%); food emulsifier; creamy wh., soft plastic; water disp.; flash pt. > 300 F; 45% act.

Atmos 758. [320] Hydrated sorbitan

monostearate (22.5%), polysorbate 60 (13.5%), and mono- and diglycerides; food emulsifier; creamy wh. soft plastic; water disp.; flash pt. > 300 F; 45% act.

Atmos 2462. [320] Hydrated mono- and diglycerides, polysorbate 60 (13%), sorbitan monostearate (13%) and propylene glycol; food emulsifier; creamy wh. soft plastic; water disp.; HLB 7.4; flash pt. > 300 F; 39% act.

Atmul 80. [320] Mono- and diglycerides; nonionic; food emulsifier; ivory wh. plastic solid; sol. above its m.p. in IPA, veg. oils, min. oil; m.p. 115 F; HLB 2.8; flash pt. > 300 F; 100% act.

Atmul 84. [320] Mono- and diglycerides; food emulsifier; ivory wh. beads or flakes; bland odor and taste; sol. above its m.p. in veg., min. oils; m.p. 140 F; HLB 2.8; flash pt. > 300 F.

Atmul 84K. [320] Mono- and diglycerides; food emulsifier; ivory wh. flakes; bland odor and taste; certified as kosher; sol. above its m.p. in veg., min. oils; m.p. 144 F; HLB 2.8; flash pt. > 300 F.

Atmul 86K. [320] Mono- and diglycerides; nonionic; food emulsifier, kosher; ivory wh. plastic solid; sol. above its m.p. in IPA, veg. oils, min. oil; m.p. 117 F; HLB 2.8; flash pt. > 300 F.

Atmul 122. [320] Mono- and diglycerides; nonionic; food emulsifier; ivory wh. plastic solid; sol. above its m.p. in IPA, veg. oils, min. oil; m.p. 126 F; HLB 3.5; flash pt. > 300 F.

Atmul 124. [320] Mono- and diglycerides; food emulsifier; ivory wh. flakes; bland odor and taste; sol. above its m.p. in IPA, veg., min. oils; m.p. 140 F; HLB 3.5; flash pt. > 300 F.

Atmul 500. [320] Mono- and diglycerides; food emulsifier; ivory wh. plastic solid; sol. above its m.p. in IPA, veg. oils, min. oil; m.p. 132 F; HLB 3.5; flash pt. > 300 F.

Atmul 651K. [320] Mono- and diglycerides; food emulsifier for Kosher use; ivory wh. plastic solid; sol. above its m.p. in IPA, veg. oils, min. oil; m.p. 133 F; HLB 3.5; flash pt. > 300 F.

Atmul 700H. [320] Hydrated mono- and diglycerides with polysorbate 60 (1%); food emulsifier, softener for yeast-raised bakery products; creamy wh., soft plastic; water disp.; HLB 4.0; flash pt. > 300 F.

Atolene RW. [183] Oleic acid sulfated ester; anionic; lubricant, wetting, dye leveling agent used in textile processing; liq.

Atolex LDA/40. [571] Dyebath aux., nonionic; leveling and disp. agent.

Atolex LS/3. [571] Sulfated fatty alcohol; nonionic/anionic; detergent, wetting agent, liq.

Atolex NI/100B. [571] Nonionic; detergent, wetting agent for textiles; liq.

Atpet 80. [320] Sorbitan monooleate; nonionic; rust inhibitor; amber liq.; disp. in water, sol. in most petroleum oils, veg. oils, lower alcohols, chlorinated solv.; sp.gr. 1; visc. 1200 cps; HLB 4.3; flash pt. > 300 F; pour pt. 0 F; 100% act.

Atplus 300F. [320] Nonionic surfactants with coupling agents; agric. surfactant for pesticides; amber liq.; sol. in most aromatic solv. used in pesticide formulations, agric. grade paraffin oil, disp. in water; dens. 8.6 lb/gal; sp.gr. 1.03; visc. 400 cp; flash pt. > 200 F; pour pt. 15 F; 92% solids.

Atplus 400. [320] Agric. surfactant, anionic; pale yel. liq.; water sol.; dens. 9 lb/gal; sp.gr. 1.08; visc. < 100 cps.

Atplus 401. [320] Anionic surfactant with coupling agents; agric. surfactant, wetting agent and adjuvant; colorless or lt. yel. liq.; disp. readily w/out heat; sp.gr. 1.21; visc. 31 cstk; flash pt. 118 F; pour pt. –20 F.

Atplus 411F. [320] Oil/surfactant conc.; agric. surfactant for herbicides and harvest aid chemicals.

Atplus 522. [320] Agric. surfactant for use in sodium chlorate sol'n; amber liq.; sol. in water; dens. 10.0 lb/gal; sp.gr. 1.2; visc. 35 cs; flash pt. 115 F; pour pt. < 32 F.

Atplus 526. [320] Conc. blend of fatty acids, fatty acid esters and alkoxylated polyhydric alcohol fatty acid esters; nonionic; agric. surfactant; specifically developed for repackaging by formulators; liq.

Atplus 535. [320] Anionic; agric. surfactant; stabilizer for liq. fertilizer-pesticide mixtures, for repackaging/resale; lt. amber liq.; complete sol. in liq. fertilizers; dens. 10.5 lb/gal; sp.gr. 1.25; visc. 40 cps; flash pt. > 200 F.

Atplus 540. [320] Alkoxylated alcohol ether; agric. foaming agent, spray tank additive; amber liq.; dens. 8.1 lb/gal; sp.gr. 0.97; visc. 30 cps; flash pt. 65 F; pour pt. 20 F; 75% conc.

Atplus 555. [320] Nonionic surfactant and coupling agent; agric. surfactant, wetting agent.

Atplus 989. [320] Nonionic surfactant and coupling agent; agric. surfactant, shows activity with pre-emergent Caporal®.

Atplus 1034. [320] Aromatic sulfates and polyhydroxy compds.; agric. foaming agent, spray tank additive; amber liq.; dens. 10 lb/gal; sp.gr. 1.2; visc. 140 cps; flash pt. > 200 F; pour pt. –35 F.

Atplus 1364. [320] Nonionic surfactant with sticker and coupler; agric. surfactant used with post-emergent WP pesticides.

Atplus 1399, 1400, 1401, 1402. [320] Nonionic surfactant with coupling agents; agric. surfactant.

Atplus 1403. [320] Anionic surfactant with coupling agents; agric. surfactant, wetting agent.

Atranonic Polymer. [73] Alcohol, ethoxylated; nonionic; dispersing agent for pigments used in textiles; liq.; 100% conc.

Atranonic Polymer 40 + 2. [73] EO alcohol condensation product; nonionic; dispersing agent for pigments used in textiles; liq.; 100% conc.

Atsurf 2801. [320] Polyol oleate; nonionic; industrial grade surfactant, emulsifier, for paper coatings, floor polishes, furniture polishes, fiberboard and particle board; amber liq.; sol. in IPA, perchloroethylene, xylene, cottonseed and paraffinic oil; dens. 1.00 g/ml; visc. 379 cs; HLB 5; pour pt. –39 F; 100% act.

Atsurf 2802. [320] Polyol trioleate; nonionic; industrial grade surfactant, emulsifier, for paper coatings, floor polishes, furniture polishes, fiberboard and particle board prod.; amber liq.; sol. in perchloroethylene, xylene, cottonseed and paraffinic oil; dens. 0.94 g/ml; visc. 204 cs; HLB 2.5; pour pt. < –35 F; 100% act.

Atsurf 2803. [320] Polyol stearate; nonionic; industrial grade surfactant, emulsifier, for paper coatings, floor polishes, furniture polishes, fiberboard and particle board prod.; lt. tan solid; sol. in IPA (1%), perchloroethylene, xylene; dens. 1.06 g/ml; visc. 165 cs; HLB 5; pour pt. 125 F; 100% act.

Atsurf 2821. [320] Ethoxylated polyol oleate; nonionic; industrial grade surfactant, emulsifier, for paper coatings, floor polishes, furniture polishes, fiberboard and particle board prod.; amber liq.; sol. in water, IPA, perchloroethylene (10%), xylene (1%), cottonseed oil (1%); dens. 1.06 g/ml; visc. 846 cs; HLB 16; pour pt. 55 F; 100% act.

Atsurf 2822. [320] Ethoxylated polyol trioleate; nonionic; industrial grade surfactant, emulsifier, for paper coatings, floor polishes, furniture polishes, fiberboard and particle board prod.; amber liq.; sol. in IPA, perchloroethylene, xylene, cottonseed and paraffinic oil; dens. 1.02 g/ml; visc. 349 cs; HLB 11; pour pt. 15 F; 100% act.

Atsurf 2823. [320] Ethoxylated polyol stearate; noninic; industrial grade surfactant, emulsifier, for paper coatings, floor polishes, furniture polishes, fiberboard and particle board prod.; amber semisolid; sol. in water, IPA, xylene (1%), cottonseed oil; dens. 1.08 g/ml; visc. 50 cs (160 F; HLB 16; pour pt. 90 F; 100% act.

Avirol 200. [290] Ammonium lauryl sulfate; foaming agent and suspension aid in rock drilling, fire fighting, dispersant for dry wall mfg., dyes and pigments; lt. amber liq.; typ. odor; visc. 1000 cps; cloud pt. 10 C max; 28% fatty alcohol sulfate.

Avirol 300. [290] TEA lauryl sulfate; base for cleaning and cosmetic preparations; lt. amber liq.; visc. 500 cps max; 39–41% fatty alcohol sulfate.

Avirol AOO-1080. [289; 290] Oxidized oleic acid, ammonium salt; anionic; emulsifier for vinyl chloride and acrylic polymers; liq.; 40% conc.

Avirol SA-4106. [289; 290] Sodium 2-ethylhexyl sulfate; anionic; wetting agent, emulsifier, stabilizer for plastics, rubber, adhesives, food contact paper; amber liq.; sp.gr. 1.1; dens. 9.2 lb/gal; visc. 200 cps; cloud pt. 5 C max.; surf. tens. 38 dynes/cm; pH 7–10 (10%); biodeg.; 43–46% solids.

Avirol SA-4110. [289; 290] Sodium *n*-decyl sulfate; anionic; wetting and emulsifying agent for plastics; amber liq.; sp.gr. 1.065; dens. 8.8 lb/gal; visc. 100 cps. max.; pH 8–10 min. (10%); cloud pt. –5 C; surf. tens. 31 dynes/cm (@ CMC); 31–33% solids.

Avirol SA-4113. [289; 290] Sodium tridecyl sulfate; anionic; emulsifier for S/B and vinyl chloride copolymers; liq.; 30% conc.

Avirol SL-2010. [289; 290] Sodium lauryl sulfate; anionic; dispersing agent and emulsifier for acrylates, styrene acrylic, vinyl chloride, vinylidene chloride and vinyl acetate copolymers; liq.; 30% conc.

Avirol SO-70P. [290] Sodium dioctyl sulfosuccinate; anionic; wetting agent, emulsifier for antifog compositions, cleaners for automobiles, dry cleaning formulations, and glass surfaces; pigment dispersion in paints and inks; latex paint stabilization; used in pesticides; APHA 100 clear liq.; sol. in polar and nonpolar solv.; sp.gr. 1.06; dens. 8.8 lb/gal; flash pt. 280 F (PMCC); cloud pt. –5 C; 70% solids.

Aviscour 98/70. [15] Amine neutralized alkylbenzene sulfonic acid; anionic/nonionic; detergent conc. for household detergents; pale yel. clear-hazy visc. liq.; 95% min. solids; visc. 80 ± 25 s (No. 4 Cup); pH 7.5 ± 0.7.

Aviscour 125. [15] Linear alkylbenzene sulfonate and coactive nonionic surfactant blend; anionic/nonionic; detergent conc. used in detergents, hard surface and upholstery cleaning; amber visc. liq.; sol. in water; pH 7.0±0.5 (5%).

Aviscour HP50. [15] Alkylbenzene sulfonic acid detergent conc.; anionic/nonionic; laundry and lt. duty detergents, hard surface cleaners, car shampoos, and foaming bath additives; orange-yel. hazy visc. liq.; dilutable in water; sp.gr. 1.05; cloud pt. 5 C max.; pH 8.2 ± 0.5; 50.0 ± 1.0% solids.

Axol C62. [258] Citric acid ester of glycerol mono/distearates; anionic; food emulsifier; powd.; HLB 10.0; 100% conc.

Axol C63. [258] Citric acid ester of glycerol mono/distearates; anionic; food emulsifier; powd.; 100% conc.

Axol E41. [258] Acetic acid ester of mono/diglycerides; nonionic; food emulsifier; solid; HLB 2–3; 100% conc.

Axol E61. [258] Acetic acid fatty acid glyceride; nonionic; food emulsifier; solid; HLB 2–3; 100% conc.

Axol E66. [258] Acetic acid ester of mono/diglycerides; nonionic; for foodstuffs; liq.; HLB 2–3; 100% conc.

Axol L61, L62. [258] Lactic acid fatty acid glyceride; nonionic; food emulsifier; solid; HLB 5; 100% conc.

Axol L626. [258] Lactic acid ester of mono/diglycerides; nonionic; food emulsifier; paste; HLB 5; 100% conc.

AXS. [152] Ammonium xylene sulfonate; hydrotrope for detergent formulations; near colorless liq.; slight ammonia odor; dens. 9.5; sp.gr. 1.141; visc. 4.8; m.p. 14; b.p. 212; 42% act.

AZdry 40. [76] Di(2-ethylhexyl) sodium sulfosuccinate; anionic; wetting agent; used with surfactants in textile industry, preparation emulsion polymers, dewatering aid in filtration various ore concs.; liq.; 40% conc.

AZdry 70. [76] Di(2-ethylhexyl) sodium sulfosuccinate; anionic; see AZdry 40; liq.; 70% conc.

B

Babinar-715. [393] Fatty acid polyethylene polyamine condensate deriv.; cationic; textile softener for polyacrylonitrile fiber; liq.

Babinar-801C. [393] Fatty acid polyethylene polyamine condensate deriv.; cationic; antistat, softener; paste.

Bardac LF, LF80. [372] Dioctyl dimethyl ammonium chloride; low foaming microbicide for recirculating cooling water systems, swimming pool algicides; liq.; m.w. 312; sp.gr. 0.93 and 0.897 resp.; dens. 7.72 and 7.48 lb/gal resp.; 50% and 80% act. resp. in ethanol-water sol'n.

Barisol Super BRM. [183] Complex multicarbon alcohol phosphate potassium salt; anionic; textile wetting agent for dye leveling, pectin removal from cottons; dispersant; liq.; 30% conc.

Barium Petronate Basic. [650] Over base barium petroleum sulfonates; anionic; emulsifier; lube oil additive; rust preventive; liq.; 44% conc.

Barium Petronate 50-S Neutral. [650] Barium sulfonate neutral; anionic; emulsifier; fuel oil additive; rust preventive; liq.; 50% conc.

Barlox 10S. [372] Decylamine oxide; nonionic; detergent; visc. builder, emollient; liq.; 30% conc.

Barlox 12. [372] Cocamine oxide; nonionic; detergent; visc. builder, emollient; liq.; 30% conc.

Barlox 14. [372] Myristamine oxide; nonionic; detergent; visc. builder, emollient; liq.; 30% conc.

Barlox C. [372] Cocoamidopropylamine oxide; foam stabilizer and visc. builder for shampoos; 40% conc.

Barquat 1552. [372] *n*-Alkyl dimethyl benzyl ammonium chloride and *n*-dialkyl methyl benzyl ammonium chloride; algicide for swimming pool and recirculating cooling towers; Gardner < 2 clear liq.; m.w. 399; sp.gr. 0.955 (20 C); dens. 7.96 lb/gal (20 C); 50% act.

Barre Common Degras. [519] Degras, wool grease deriv.; nonionic; softener for industrial leather; w/o emulsion for corrosion prevention; paste; 100% conc.

Base 11-T. [114; 138] Alkyl sulfonate; anionic; o/w emulsifier, for metal cutting, rust proofing and quenching; br. liq.; oil sol.; dens. 8.7 lb/gal (60 F); sp.gr. 1.05 (60 F); visc. 190 SUS (210 F); 100% act.

Base 75, 76. [345] Petroleum sulfonate and aux. emulsifier blend; anionic; o/w emulsifier for sol. oils and pastes for metalworking industry; liq.

Base 7800. [345] Petroleum sulfonate and aux. emulsifier blend; anionic; o/w emulsifier for sol. oils and pastes for metalworking industry; liq.

Base ML. [345] Methyl lardate; wetting/oiliness agent for metalworking and lubricating oils; liq.; 100% conc.

Base MT. [345] Methyl tallowate; wetting/oiliness agent for metalworking and lubricating oils; liq.; 100% conc.

Basophor A. [86] Fatty acid ester; nonionic; emulsifier for min. and fatty oils; liq.

Basopon LN. [86] Alkylaryl polyglycol ether; nonionic; detergent with dispersant for wool and syn. fibers; liq.

Berol 02. [90] Nonylphenol ethoxylate; nonionic; emulsifier, solv. cleaner; emulsion polymerization; Hazen < 200 clear liq.; sol. in ethanol, xylene, trichloroethylene, lt. fuel oil; disp. in water, paraffin oil, wh. spirit; sp.gr. 1.04; visc. 350 cps; HLB 10.9; cloud pt. 60–70 C; flash pt. > 100 C; pour pt. < 0 C; surf. tens. 31 dynes/cm; pH 6–7 (1% aq.); biodeg.; 100% act.

Berol 07. [90] Fatty alcohol ethoxylate; nonionic; detergent, equalizing and dispersing agent; wh. wax; sol. in water, ethanol, xylene, trichlorethylene, disp. in paraffin oil, wh. spirit, lt. fuel oil; sp.gr. 1.09; visc. 50 cps (50 C); m.p. 36–38 C; HLB 15; cloud pt. 74–75 C; flash pt. > 100 C; surf. tens. 40 dynes/cm; pH 6–7 (1% aq.); biodeg.; 100% act.

Berol 08 Powd. [90] Fatty alcohol ethoxylate; nonionic; detergent, dispersant; wh. powd.; sol. in water, ethanol, trichlorethylene, disp. in xylene; sp.gr. 0.55–0.65; visc. 200 cps (70 C); m.p. 52–54 C; HLB 18.7; cloud pt. 76–77 C; flash pt. > 100 C; surf. tens. 40–48 dynes/cm; pH 6–7 (1% aq.); biodeg.; 100% act.

Berol 09. [90] Nonylphenol ethoxylate; nonionic; detergent, wetting agent, emulsifier, dispersant; Hazen < 150 clear liq.; sol. in water, ethanol, xylene, trichlorethylene; disp. in paraffin oil, wh. spirit, lt. fuel oil; sp.gr. 1.05; visc. 350 cps; HLB 13.3; cloud pt. 52–58 C; flash pt. > 100 C; pour pt. 5 C; surf. tens. 33 dynes/cm; pH 6–7 (1% aq.); biodeg.; 100% act.

Berol 17. [90] Resin acid EO adduct; nonionic; lubricant, emulsifier for textile industry; Gardner 15, clear liq.; sol. in water, ethanol, trichloroethylene, xylene, disp. in acetone, propylene glycol; sp.gr. 1.11; visc. 50 cps; HLB 14; cloud pt. 45 C; flash pt. > 100 C; pour pt. 15 C; surf. tens. 38 dynes/cm; pH 7.5–8.5 (1% aq.); 100% act.

Berol 26. [90] Nonylphenol ethoxylate; nonionic; emulsifier, detergent additive, base for sulfation; Hazen < 250 clear liq.; sol. in ethanol, xylene, trichloroethylene, paraffin oil, wh. spirit, lt. fuel oil, disp. in water; sp.gr. 1.02; visc. 350 cps; HLB 8.9; cloud pt. 42–46 C; flash pt. > 100 C; pour pt. < 0 C; pH 6–7 (1% aq. disp.); biodeg.; 100% act.

Berol 28. [90] Alkyl amine ethoxylate; nonionic; emulsifier; liq.; 100% conc.

Berol 048. [90] Fatty alcohol EO adduct; nonionic; wetting agent, detergent; Hazen < 100 clear liq.; sol. in water, ethanol, disp. in propylene glycol, trichloroethylene, benzene, paraffin oil, wh. spirit; sp.gr. 1.02; visc. 170 cps; HLB 13.5; cloud pt. 70 C; flash pt. > 100 C; pour pt. 0 C; surf. tens. 27 dynes/cm; pH 6–7 (1% aq.); 85% act.

Berol 049. [90] *n*-Primary alcohol EO adduct; nonionic; detergent, for liq. cleaners; Hazen < 100 clear liq.; sol. in water ethanol, xylene, trichloroethylene, paraffin oil, wh. spirit, lt. fuel oil, disp. in water; sp.gr. 0.99; visc. 40 cps; HLB 12; cloud pt. 55–60 C; flash pt. > 100 C; pour pt. 7 C; surf. tens. 30 dynes/cm; pH 6–7 (1% aq.); biodeg.; 100% act.

Berol 050. [90] Fatty alcohol ethoxylate; nonionic; raw material for shampoos, bubble baths; emulsifier; Hazen < 100 clear liq.; sol. in ethanol, xylene, trichloroethylene, paraffin oil, wh. spirit, lt. fuel oil, disp. in water; sp.gr. 0.93; visc. 30 cps; HLB 8.4; cloud pt. 48–52 C; pH 6–7 (1% aq. disp.); 100% act.

Berol 055. [90] *n*-Primary alcohol EO adduct; sulfonation raw material for shampoos, bubble baths; emulsifier; Hazen < 100 clear liq.; sol. in ethanol, xylene, trichloroethylene, paraffin oil, wh. spirit, lt. fuel oil, disp. in water; sp.gr. 0.90; visc. 30 cps; HLB 6.5; cloud pt. 35–40 C; pH 6–7 (1% aq. disp.); 100% act.

Berol 057. [90] Fatty alcohol ethoxylate; nonionic; emulsifier; Hazen < 300 soft wax; sol. in water, ethanol, xylene, trichloroethylene, disp. in water, paraffin oil, white spirit; sp.gr. 1.00 (60 C); visc. 60 cps (40 C); HLB 15.5; cloud pt. 74–76 C; flash pt. > 100 C; pour pt. 30 C; biodeg.; 100% act.

Berol 059. [90] *n*-Primary alcohol EO adduct; nonionic; detergent for liq. cleaners; Hazen < 100 clear liq.; sol. in water, ethanol, xylene, trichloroethylene, wh. spirit, lt. fuel oil; sp.gr. 0.97; visc. 35 cps (40 C); HLB 11.3; cloud pt. 35–40 C (1% aq.); flash pt. > 100 C; pour pt. 5 C; surf. tens. 29 dynes/cm; pH 6–7 (1% aq.); biodeg.; 100% act.

Berol 065. [90] Fatty alcohol ethoxylate; nonionic; detergent for grease and pigment removal; soft, wh. wax; sol. in water, ethanol, xylene, trichloroethylene, disp. in paraffin oil, wh. spirit, lt. fuel oil; sp.gr. 0.98 (40 C); visc. 30 cps (50 C); HLB 12.6; cloud pt. 78–82 C (1% aq.); flash pt. > 100 C; pour pt. 33 C; surf. tens. 34 dynes/cm; pH 6–7 (1% aq.); biodeg.; 100% act.

Berol 79. [90] Nonionic; emulsifier; clear yel. liq; sol. in ethanol, benzene, trichloroethylene, paraffin oil, light min. oil, disp. in water; sp.gr. 0.97; pour pt. 0 C; 100% act.

Berol 081 Flakes. [90] Fatty alcohol ethoxylate; nonionic; detergent, emulsifier, dispersant, stabilizer; wh. flakes; sol. in water, ethanol, trichloroethylene, disp. in xylene; sp.gr. 0.55–0.65; m.p. 46–50 C; HLB 16.7; cloud pt. 75–76 C (1% in 10% NaCl); flash pt. > 100 C; surf. tens. 43 dynes.cm; pH 6–7 (1% aq.); biodeg.; 100% act.

Berol 081 Powd. [90] Fatty alcohol ethoxylate; nonionic; detergent, emulsifier, dispersant, stabilizer; wh. powd.; sol. in water, ethanol, trichloroethylene, disp. in xylene; sp.gr. 0.55–0.65; m.p. 46–50 C; HLB 16.7; cloud pt. 75–76 C (1% in 10% NaCl); flash pt. > 100 C; surf. tens. 43 dynes.cm; pH 6–7 (1% aq.); biodeg.; 100% act.

Berol 087. [90] *n*-Primary alcohol ethoxylate; nonionic; detergent; machine dishwashing, high pressure cleaning; Hazen 75 clear liq.; sol. in water, ethanol, trichloroethylene, disp. in xylene; sp.gr. 0.997; visc. 85 cps; HLB 11.5; cloud pt. 42 C; flash pt. > 100 C; pour pt. –1 C; surf. tens. 31.9 dynes/cm; pH 5–7 (1% aq.); biodeg.; 100% act.

Berol 106. [90] Alkyl glyceride EO adduct; nonionic; o/w emulsifier for biocides and pesticides; Gardner < 4 cloudy liq.; sol. in water, ethanol, propylene glycol, xylene, trichloroethylene; sp.gr. 1.055; visc. 650 cps; HLB 15; cloud pt. 54–58 C; flash pt. > 100 C; pour pt. 15 C; surf. tens. 44 dynes/cm; pH 5–7 (1% aq.); biodeg.; 100% act.

Berol 121. [90] Fatty alcohol ethoxylate; nonionic; detergent, general purpose high foam cleaners in combination with anionics; sol. in ethanol, wh. spirit, lt. fuel oil; opaquely sol. in water, xylene; disp. in trichloroethylene, paraffin oil; sp.gr. 0.97; HLB 11.0; cloud pt. 75–80 C (in 15% *n*-propanol); flash pt. > 100 C; pour pt. 20 C; surf. tens. 29 dynes/cm; pH 6–7 (1% aq.); biodeg.; 97.5% act.

Berol 151. [90] Lauric/myristic acid diethanolamide; nonionic; foam stabilizer, booster for shampoos, bubble bath, modifier, dispersant; Hazen ≤ 200 clear liq.; sol. in ethanol, propylene glycol, xylene; disp. in water, wh. spirit, paraffin oil; sp.gr. 0.98; visc. 1100 cps; HLB 12; flash pt. >100 C; pour pt. –5 C; surf. tens. 29.5 dynes/cm; pH 9.5 (1% aq.); biodeg.; 100% act.

Berol 160. [90] Fatty alcohol ethoxylate; nonionic; detergent, emulsifier; Hazen < 100 clear liq.; sol. in water, ethanol, xylene, trichloroethylene, light fuel oil; disp. in wh. spirit; sp.gr. 0.98; visc. 50 cps; HLB 11.8; cloud pt. 40–50 C; flash pt. ≥ 100 C; pour pt. 15 C; surf. tens. 29 dynes; pH 6–7 (1% aq.); biodeg.; 100% act.

Berol 173. [90] Fatty alcohol ethoxylate; nonionic; detergent, emulsifier; wh. paste; sol. in water, ethanol, xylene, trichloroethylene, light fuel oil; disp. in wh. spirit; sp.gr. 0.99; visc. 30 cps; HLB 12.5; cloud pt. 65–70 C; flash pt. ≥ 100 C; pour pt. 25 C; surf. tens. 33 dynes; pH 6–7 (1% aq.); biodeg.; 100% act.

Berol 185. [90] Modified fatty alcohol ethoxylate; nonionic; detergent; clear, colorless liq.; less tendency than similar products to unpleasant odor > 30 C; sol. in water, ethanol, disp. in xylene, trichloroethylene, wh. spirit, lt. fuel oil; sp.gr. 1.02; visc. 100 cps; HLB 13.5; cloud pt. 65–70 C; flash pt. ≥ 100 C; pour pt. < 0 C; surf. tens. 31.5 dynes/cm (0.1%); pH 6–7 (1% aq.); biodeg.; 90% act.

Berol 186. [90] Modified fatty alcohol ethoxylate; detergent, nonionic; Hazen < 100 clear/opal liq.; sol. in water, ethanol, xylene, trichloroethylene, wh. spirit, lt. fuel oil; sp.gr. 1.01; visc. 100 cps; HLB 13.5; cloud pt. 65–70 C; flash pt. ≥ 100 C; pour pt. 10 C; surf. tens. 31.5 dynes/cm (0.1%); pH 6–7 (1% aq.); biodeg.; 100% act.

Berol 196. [90] Alkyl glyceride EO adduct; nonionic; emulsifier for foodstuffs, superfatting agent in bath oils, hand cleaners; Gardner ≤ 5, clear liq.; sol. in ethanol, propylene glycol, trichloroethylene, wh. spirit, opaquely sol. in water; sp.gr. 1.020; visc. 500 cps; HLB 9; cloud pt. 58 C; flash pt. > 100 C; pour pt. –15 C; pH 5–7 (1% aq.); biodeg.; 100% act.

Berol 252. [90] Alkyl phenol EO adduct; nonionic; emulsifier; Hazen 200 clear liq.; sol. or disp. in water, ethanol; sp.gr. 1.09; visc. 1000 cps; HLB 5.7–14.7; cloud pt. 77 C; flash pt. > 100 C; pour pt. 12 C; surf. tens. 41; pH 6–7 (1% aq.); biodeg.; 70% act.

Berol 259. [90] Nonylphenol ethoxylate; nonionic; stabilizer, foam depressor; liq. cleaner, wetting agent; Hazen ≤ 500 clear liq.; sol. in ethanol, xylene, trichloroethylene, paraffin oil, wh. spirit, lt. fuel oil; sp.gr. 1.01; visc. 550 cps; HLB 5.7; cloud pt. 50–54 C; flash pt. > 100 C; pour pt. < 0 C; pH 6–7 (1% aq.); biodeg.; 100% act.

Berol 263. [90] Alkyl phenol EO adduct; nonionic; detergent; Hazen < 200 soft wax; sol. in water, ethanol, xylene, trichloroethylene, disp. in paraffin oil, wh. spirit; sp.gr. 1.06; visc. 325 cps; HLB 14.7; cloud pt. 92–94 C; flash pt. > 100 C; pour pt. 20 C; surf. tens. 36 dynes/cm; pH 6–7 (1% aq.); biodeg.; 100% act.

Berol 267. [90] Nonylphenol ethoxylate; nonionic; emulsifier, detergent, wetting agent; Hazen < 200 clear liq.; sol. in ethanol, xylene, trichloroethylene, paraffin oil, wh. spirit, lt. fuel oil; sp.gr. 1.04; visc. 350 cps; HLB 12.3; cloud pt. 10 C; flash pt. > 100 C; pour pt. < 0 C; surf. tens. 31 dynes/cm; pH 6–7 (1% aq.); biodeg.; 100% act.

Berol 269. [90] Dinonylphenol ethoxylate; nonionic; emulsifer in polymerization; detergent; Hazen < 500 opal liq.; sol. in ethanol, xylene, trichloroethylene, lt. fuel oil, disp. in water, wh. spirit; sp.gr. 1.02; visc. 550 cps; HLB 10.5; cloud pt. 60 C; flash pt. > 100 C; pour pt. < 0 C; pH 6–7 (1% aq.); biodeg.; 100% act.

Berol 272. [90] Dinonylphenol ethoxylate; nonionic; low-foaming detergent with good stability for spray drying; lt. yel. paste; sol. in water, ethanol, xylene, trichloroethylene, disp. in paraffin oil; sp.gr. 1.04; visc. 300 cps (35 C); m.p. 34 C; HLB 13.4; cloud pt. 60 C (1% aq.); flash pt. >100 C; pour pt. 23 C; surf. tens. 30 dynes/cm (0.1%); pH 6–7 (1% aq.); biodeg.; 100% act.

Berol 274. [90] Alkyl phenol EO adduct; nonionic; emulsifer in emulsion polymerization; Hazen < 200 soft wax; sol. in water, ethanol, xylene, trichloroethylene; sp.gr. 1.07 (50 C); visc. 200 cps (50 C); HLB 17; cloud pt. 77 C (1% in 10/NaCl); flash pt. > 100 C; pour pt. 41 C; surf. tens. 41 dynes/cm (0.1%); pH 6–7 (1% aq.); biodeg.; 100% act.

Berol 277. [90] Alkyl phenol EO adduct; nonionic; emulsifer in emulsion polymerization; Hazen < 200 clear liq.; sol. in water, ethanol; sp.gr. 1.09; visc. 1000 cps; HLB 17; cloud pt. 77 C (1% in 10/NaCl); flash pt. >100 C; pour pt. 12 C; surf. tens. 41 dynes/cm (0.1%); pH 6–7 (1% aq.); biodeg.; 70% act.

Berol 278. [90] Alkyl phenol EO adduct; nonionic; emulsifer in emulsion polymerization; Hazen < 200 clear liq.; sol. in water, ethanol; sp.gr. 1.08; visc. 625 cps; HLB 15.2; cloud pt. 97–99 C (1% aq.); flash pt. > 100 C; pour pt. 0 C; surf. tens. 37 dynes/cm (0.1%); pH 6–7 (1% aq.); biodeg.; 80% act.

Berol 281. [90] Alkyl phenol EO adduct; nonionic; emulsifer in emulsion polymerization; Hazen < 200 clear liq.; sol. in water, ethanol; sp.gr. 1.09; visc. 625 cps; HLB 16; cloud pt. 73 C (1% in 10/NaCl); flash pt. >100 C; pour pt. 7 C; surf. tens. 39 dynes/cm (0.1%); pH 6–7 (1% aq.); biodeg.; 80% act.

Berol 282. [90] Alkyl phenol EO adduct; nonionic; emulsifer in emulsion polymerization; Hazen < 200 soft paste; sol. in water, ethanol; sp.gr. 1.09 (30 C); visc. 625 cps; HLB 17; cloud pt. 77 C (1% in 10/NaCl); flash pt. > 100 C; pour pt. 21 C; surf. tens. 41 dynes/cm (0.1%); pH 6–7 (1% aq.); biodeg.; 80% act.

Berol 291. [90] Alkyl phenol EO adduct; nonionic; emulsifer in emulsion polymerization; Hazen < 200 hard wax; sol. in water, ethanol, xylene, trichloroethylene; sp.gr. 1.07 (60 C); visc. 300 cps (60 C); HLB 18.2; cloud pt. 77 C (1% in 10/NaCl); flash pt. >100 C; pour pt. 51 C; surf. tens. 45 dynes/cm; pH 6–7 (1% aq.); 100% act.

Berol 292. [90] Nonylphenol ethoxylate; nonionic; emulsifer in emulsion polymerization; Hazen < 200 soft wax; sol. in water, ethanol, xylene, trichloroethylene, disp. in wh. spirit; sp.gr. 1.07 (40 C); visc. 200 cps (40 C); HLB 16.0; cloud pt. 73 C (1% in 10/NaCl); flash pt. > 100 C; pour pt. 33 C; surf. tens. 39 dynes/cm; pH 6–7 (1% aq.); biodeg.; 100% act.

Berol 295. [90] Nonylphenol ethoxylate; nonionic; emulsifer in emulsion polymerization; Hazen < 300 clear liq.; sol. in water, ethanol; sp.gr. 1.10; visc. 1500 cps; HLB 17.8; cloud pt. 77 C (1% in 10/NaCl); flash pt. > 100 C; pour pt. 14 C; surf. tens. 43 dynes/cm; pH 5–7 (1% aq.); 70% act.

Berol 296. [90] Alkyl phenol EO adduct; nonionic; emulsifer in emulsion polymerization; Hazen < 200 soft paste; sol. in water, ethanol, xylene, trichloroethylene, disp. in lt. fuel oil; sp.gr. 1.07 (30 C); visc. 375 cps (30 C); HLB 15.2; cloud pt. 97–99 C (1% aq.); flash pt. > 100 C; pour pt. 26 C; surf. tens. 37 dynes/cm; pH 6–7 (1% aq.); biodeg.; 100% act.

Berol 297. [90] Alkyl phenol EO adduct; nonionic; emulsifer used for waxes, oils, solv.; Hazen < 250 clear liq.; sol. in ethanol, xylene, trichloroethylene, paraffin oil, wh. spirit, lt. fuel oil; sp.gr. 1.03; visc. 350 cps; HLB 9.5; cloud pt. 48–54 C; flash pt. > 100 C; pour pt. < 0 C; pH 6–7 (1% aq.); biodeg.; 100% act.

Berol 305. [90] Amine oxide;nonionic; foam controller; pale yel. clear liq.; sol. in water, ethanol, acetone; sp.gr. 0.96; cloud pt. 95 C (9% NaOH); pour pt. 0 C; surf. tens. 31

dynes/cm; 28% act.

Berol 370. [90] EO/PO block polymer; nonionic; detergent, rinse aid additive; machine dishwashing; Hazen < 150 clear liq.; m.w. 1400; sol. in ethanol, xylene, trichloroethylene, wh. spirit, water; sp.gr. 1.05; visc. 230 cps; cloud pt. 30–34 C (1% aq.); flash pt. > 100 C; pour pt. < –10 C; surf. tens. 41 dynes/cm (0.1%); pH 5–7 (1% aq.); 100% act.

Berol 371. [90] EO/PO block polymer; nonionic; detergent, antistat, emollient, solv.; Hazen < 150 soft paste; m.w. 2900; sol. in water, ethanol, xylene, trichloroethylene; sp.gr. 1.05; visc. 700 cps; cloud pt. 62–66 C (1% aq.); flash pt. > 100 C; pour pt. 28 C; surf. tens. 38 dynes/cm (0.1%); pH 5–7 (1% aq.); 100% act.

Berol 372 Flakes. [90] EO/PO block polymer; nonionic; detergent, rinse aid additive, antistat, emollient, solv., emulsifier, foam depressant/detergent; wh. flakes; m.w. 9000; sol. in water, ethanol, trichloroethylene; sp.gr. 0.55–0.65; cloud pt. 70 C (1% in 10% NaCl); flash pt. > 100 C; surf. tens. 42 dynes/cm (0.1%); pH 5–7 (1% aq.); 100% act.

Berol 374. [90] EO/PO block polymer; nonionic; emollient, emulsifier; emulsion polymerization; wh. flakes; m.w. 2200; sol. in ethanol, xylene, trichloroethylene, wh. spirit; disp. water; sp.gr. 1.05; visc. 450 cps; cloud pt. 24–26 C (1% aq.); flash pt. > 100 C; pour pt. < –10 C; surf. tens. 40 dynes/cm (0.1%); pH 5–7 (1% aq.); 100% act.

Berol 452. [90] Sodium lauryl ether sulfate; anionic; wetting and foaming agent, emulsifier; shampoos, bath preparations; Hazen < 100 clear liq.; sol. in water, ethanol, propylene glycol; disp. in acetone; sp.gr. 1.04; visc. 200 cps; flash pt. > 100 C; surf. tens. 37 dynes/cm (0.1%); pH 6.5–8.0 (1% aq); biodeg.; 27–29% act.

Berol 463. [90] DEA lauryl sulfate; anionic; foaming agent used in hair shampoos, bath foams, hand cleaners; Hazen < 100 clear liq.; sol. in water, ethanol, propylene glycol, acetone; sp.gr. 1.050; visc. 1000 cps; clear pt. < 15 C; flash pt. > 100 C; surf. tens. 30 dynes/cm (0.1%); pH 6.5–7.5 (1% aq.); biodeg.; 33–35% act.

Berol 472. [90] Sodium cetyl sulfate; anionic; emulsifier, dispersant; used for emulsion polymerization of vinyl chloride, vinyl acetate, styrene/butadiene, acrylics; Hazen < 150 cloudy liq.; sol. in propylene glycol; disp. in water, ethanol; sp.gr. 1.055; visc. 20 cps; flash pt. > 100 C; surf. tens. 37 dynes/cm (0.1%); pH 7–8 (1% aq.); biodeg.; 32 ±1% act.

Berol 473. [90] Nonylphenol ether sulfate; anionic; detergent with good fatty soil removal, foaming properties; Hazen < 100 clear liq.; sp.gr. 1.075; visc. 205 cps; clear pt. 2 C; flash pt. > 100 C; pour pt. 2 C; pH 3.9; biodeg.; 30% act.

Berol 474. [90] Sodium lauryl sulfate; anionic; emulsifier for emulsion polymerization of vinyl chloride and styrene/butadiene; shampoos, bath preparations; Hazen < 50 soft paste; sol. in water, ethanol, propylene glycol; sp.gr. 1.030; clear pt. 25 C; flash pt. > 100 C; surf. tens. 25 dynes/cm (0.1%); pH 7–8 (1% aq.); biodeg.; 39–41% act.

Berol 475. [90] Sodium alkyl ether sulfate; anionic; degreaser, dispersant for water-based pigment pastes, cleaner; textile use; Hazen < 300 soft paste; sol. in propylene glycol and water, disp. in ethanol, trichloroethylene, xylene; sp.gr. 0.97; flash pt. > 100 C; surf. tens. 37 dynes/cm (0.1%); pH 6.5–8.0 (1% aq.); biodeg.; 39–41% act.

Berol 480. [90] TEA lauryl sulfate; anionic; detergent, foaming agent, used in hair shampoos, bath foams, hand cleaners; < 100 Hazen clear liq.; sol. in water, ethanol, propylene glycol; sp.gr. 1.04; visc. 50 cps; clear pt. 12 C; flash pt. >100 C; surf. tens. 26 dynes/cm (0.1%); pH 6.5–7.5 (1% aq.); biodeg.; 39–41% act.

Berol 484. [90] MEA lauryl sulfate; anionic; foaming for hair shampoos, bath foams, hand cleaners; Hazen ≤ 200 clear liq.; sol. in water, propylene glycol, acetone; disp. in ethanol; sp.gr. 1.020; visc. 150–350 cps; clear pt. ≤ 15 C; flash pt. > 100 C; surf. tens. 33 dynes/cm; pH 6.5–7.5 (1% aq.); biodeg.; 32–34% act.

Berol 490. [90] Sodium alkylaryl sulfonate; anionic; wetting agent, foaming and emulsifier; used in textile industry; Hazen < 300 clear liq.; sol. in water, ethanol, propylene glycol; disp. in lt. min. oil, wh. spirit; sp.gr. 1.037; visc. 250 cps; clear pt. < 18 C; flash pt. > 100 C; surf. tens. 35 dynes/cm; pH 6–8 (1% aq.); biodeg.; 24–26% act.

Berol 493. [90] Sodium alkylaryl sulfonate; anionic; wetting agent, detergent, used in textile industry; Hazen < 150 clear liq.; sol. in water, ethanol, propylene glycol; disp. in acetone, trichloroethylene; sp.gr. 1.100; visc. 5000 cps; clear pt. < 15 C; flash pt. >

100 C; surf. tens. 35 dynes/cm; pH 6–8 (1% aq.); biodeg.; 34–36% act.

Berol 496. [90] Linear sodium alkylaryl sulfonate; anionic; detergent, foaming, wetting, penetrant, emulsifier for textile industry; Hazen 150 soft paste; sol. in water, ethanol, propylene glycol, acetone; disp. in trichloroethylene, spindle oil, wh. spirit; sp.gr. 1.060; clear pt. 0 C; flash pt. > 100 C; surf. tens. 35 dynes/cm; pH 6–8; biodeg.; 80% act.

Berol 513. [90] Alkyl ether phosphate ester and *n*-primary alcohol ethylene adduct; anionic/nonionic; detergent; Gardner < 5 paste; sol. in ethanol, acetone, xylene, propylene glycol, wh. spirit; disp. in water, trichloroethylene; sp.gr. 1.050 (30 C); visc. 330 cps (30 C); flash pt. >100 C; surf. tens. 34.4 dynes/cm; biodeg.; 100% act.

Berol 516. [90] Mixt. of an alkyl ether phosphate ester and an *n*-primary alcohol EO adduct; anionic/nonionic; detergent, corrosion inhibitor; Gardner < 5 paste; sol. in ethanol, acetone, xylene, propylene glycol, white spirit, disp. in water; sp.gr. 1.050 (30 C); visc. 330 cps (30 C); clear pt. 40 C; flash pt. > 100 C; pour pt. 30 C; surf. tens. 34.4 dynes/cm (0.1%); pH 2.3 (1% aq.); biodeg.; 100% act.

Berol 518. [90] Mixt. of an alkyl ether phosphate ester and an *n*-primary alcohol EO adduct; anionic/nonionic; detergent, foam controller; Gardner < 5 soft wax; sol. in ethanol, acetone, xylene, white spirit, trichloroethylene, light fuel oil; disp. in water; sp.gr. 1.01 (50 C); visc. 1500 cps (35 C); clear pt. 35 C; flash pt. > 100 C; pour pt. 35 C; surf. tens. 38 dynes/cm (0.1%); pH 2.3 (1% aq.); biodeg.; 100% act.

Berol 521. [90] Potassium alkyl phosphate esters; anionic; corrosion inhibitor, solubilizer of nonionic surfactants in presence of high electrolyte conc.; Hazen < 300 clear liq.; sol. in water, propylene glycol; sp.gr. 1.145; visc. 315 cps; clear pt. 25 C; flash pt. > 100 C; pour pt. –10 C; pH 7–8 (1% aq.); biodeg.; 40% act. in water.

Berol 522. [90] Potassium alkyl phosphate esters; anionic; solubilizer of nonionic surfactants in presence of high electrolyte conc.; Hazen < 300 clear liq.; sol. in water, propylene glycol; sp.gr. 1.245; visc. < 125 cps; clear pt. < 0 C; flash pt. > 100 C; pour pt. < –10 C; pH 8–9 (1% aq.); biodeg.; 40–42% act.

Berol 525. [90] Phosphate ester; anionic/nonionic; detergent aux.; paste; 100% conc.

Berol 548. [90] Amphoteric surfactant; industrial detergent, wetting agent; Hazen 50 clear liq.; sol. in 30% caustic sol'ns.; visc. 70 cps; clear pt. –5 C; flash pt. > 100 C; surf. tens. 28 dynes/cm (0.1%); pH 9.3 (1% aq); 35% act.

Berol 563. [90] Alkyl polyglycol ether ammonium methyl sulfate; cationic; detergent, alkaline degreasing and cleaning agent; Gardner ≤ 7, clear liq.; sol. in water, ethanol, propylene glycol, xylene; sp.gr. 1.110; visc. 500 cps; cloud pt. 85 C (1% in 9% NaOH); flash pt. > 100 C; pour pt. 10 C; surf. tens. 43 dynes/cm (0.1%); pH 7–9 (1% aq.); 100% act.

Berol 594. [90] Hydroxyethyl-2-alkyl-imidazoline; strong cationic activity in acid sol'ns., oils, solv.; adhesion aid in solv.-based paints and varnishes, dispersing agent, wetting and foaming agent, corrosion inhibitor; Gardner ≤ 12 clear liq.; m.w. 360; sol. in ethanol, xylene, trichloroethylene, wh. spirit, paraffin oil, and dilute sulfuric, hydrochloric, phosphoric and acetic acids; disp. in water; sp.gr. 0.950; visc. 300 cps; clear pt. < –10 C; flash pt. > 100 C; pour pt. < 0 C; surf. tens. 32 dynes/cm (0.1%); pH 10–11 (1% aq.); 100% act.

Berol 716. [90] Alkylphenol EO adduct; nonionic; low foaming detergent with good pigment and grease removal; Hazen < 500 cloudy liq.; sol. in ethanol, xylene, trichloroethylene, paraffin oil (opaque); disp. in water, wh. spirit, lt. fuel oil; sp.gr. 1.04; visc. 800 cps; HLB 13; cloud pt. 80–85 C (in 10% *n*-propanol); flash pt. > 100 C; pour pt. 20 C; pH 6–7 (1% aq.); 100% act.

Berol 730. [90] Polypropylene glycols; antifoaming agent, lubricant, release agent, hydraulic fluid lubricant, substitute for different oils, dust adhesive in air filters; Hazen < 100 clear liq.; m.w. 1200; sol. in min. spirit, acetone, xylene, methanol, carbontetrachloride (5%), water (< 2%); sp.gr. 1.0; visc. 160 cps; flash pt. > 200 C; pour pt. –40 C; 100% act.

Berol 733. [90] Potassium salt of phosphated alkylphenol EO adduct; anionic; detergent, solubilizes nonionic surfactants in presence of inorganic salts, liquid alkaline industrial cleaners; sol. in water, 6% NaOH, 35% tetrapotassium pyrophosphate, 50% sodium metasilicate, propylene glycol; sp.gr. 1.050; visc. 400 cps; clear pt. –5 C; flash pt. > 100

C; pour pt. –7 C; surf. tens. 41 dynes/cm; pH 9–10 (1% aq.); biodeg.; 39–41% act.

Berol 740. [90] Polypropylene glycols; antifoaming agent, lubricant, release agent, hydraulic fluid lubricant, substitute for different oils, dust adhesive fluid in air filters; Hazen < 100, clear liq.; m.w. 1800; sol. in min. spirit, acetone, xylene, methanol, carbon tetrachloride (5%), water (< 0.1%); sp.gr. 1.0; visc. 200 cps; flash pt. > 200 C; pour pt. –35 C; 100% act..

Berol 752. [90] Anionic; surfactant, foam stabilizer; Gardner < 7 visc. liq.; disp. in water, propylene glycol, dioctylphthalate, tributyl phosphate; sp.gr. 1.070; visc. 10,000 cps; flash pt. > 100 C; pour pt. 25 C; pH 6–8 (1% in 1:1 water/ethanol); 95–100% act.

Berol 755. [90] Linear alkylaryl sulfonate and *n*-primary alcohol EO adduct; anionic/nonionic; detergent with relatively strong and stable foaming; Gardner ≤ 5 clear liq.; sol. in water, ethanol, IPA, xylene; disp. in wh. spirit, lt. fuel oil; sp.gr. 1.050; visc. 150 cps; clear pt. 0 C; flash pt. > 100 C; pour pt. –10 C; surf. tens. 36 dynes/cm; pH 6–7; biodeg.; 60% act.

Berol 767. [90] Sodium salt of an alkyl ether sulfate and an alkyl phenol ethoxylate; anionic/nonionic; emulsifier for emulsion polymerization of vinyl acetate and vinyl acetate/acrylate; Hazen 100 max. clear liq.; sol. in water, ethanol, acetone, propylene glycol; disp. in lt. fuel oil; sp.gr. 1.060; visc. 230 cps.; clear pt. < 0 C; flash pt. > 100 C; pour pt. < 0 C; surf. tens. 45 dynes/cm; pH 6–7 (1% aq.); 34–36% act.

Berol 777. [90] Linear alkylaryl sulfonate and alkyl phenol EO adduct; anionic/nonionic; detergent used in textile industry, good pigment disp. properties; Hazen ≤ 250 clear liq.; sol. in water, propylene glycol; disp. in ethanol, acetone, trichloroethylene, xylene; sp.gr. 1.055; visc. 425 cps; clear pt. 1 C; flash pt. > 100 C; pour pt. 0 C; surf. tens. 38 dynes/cm; pH 6–7 (1% aq.); 34% act.

Berol 779. [90] Anionic surfactant used for solv. degreasing; Gardner < 5 clear liq.; sol. in water (opal sol'n.), ethanol, xylene, trichloroethylene, wh. spirit, lt. fuel oil; sp.gr. 0.98; visc. 150 cps; clear pt. < –20 C; flash pt. 70 C; pour pt. < –20 C; pH 5.5–6.5 (1% aq.); biodeg.; 100% act.

Berol 782. [90] Blended emulsifier; used to dissolve large quantities of grease, oil, asphalt; Gardner 12 clear liq.; sol. in ethyl alcohol, xylene, trichloroethylene, wh. spirit, lt. fuel oil; sp.gr. 0.943; visc. 60 cps; clear pt. –30 C; flash pt. 72 C; pour pt. –30 C; pH 6 ± 0.1; biodeg.; 100% act.

Berol 786. [90] O/w emulsion; emulsifier, corrosion inhibitor; yel.-wh. liq.; sp.gr. 0.994; visc. 100 cps; flash pt. > 100 C; pour pt. 0 C.

Berol 788. [90] Linear alkylaryl sulfonate and an alkylphenol EO adduct; anionic/nonionic; detergent with grease and pigment removing properties and relatively strong foaming; Hazen 500 clear liq.; sol. in water, ethanol, propylene glycol, disp. in acetone, trichloroethylene, wh. spirit; sp.gr. 1.050; visc. 450 cps; clear pt. 5 C; flash pt. > 100 C; pour pt. 0 C; surf. tens. 37 dynes /cm; pH 6–7; biodeg.; 30% act.

Berol 792. [90] Nonionic; emulsifier for oils and solv., high penetrating and degreasing properties; clear yel. liq.; sol. in ethanol, paraffin oil, lt. min. oil; disp. in water, benzene, trichloroethylene; sp.gr. 1.02; pour pt. < –10 C.

Berol 808. [90] Anionic/cationic/nonionic; detergent, textile softener, antistat agent, post additive to powd. formulations; clear, yel. liquid; sol. in ethanol, propylene glycol, wh. spirit, lt. fuel oil; opaquely sol. in water; sp.gr. 1.013; visc. 105 cps; cloud pt. > 90 C (1% aq.); flash pt. 60 C; surf. tens. 32 dynes/cm (0.1%); pH 5.5 (1% aq.); biodeg. under normal sewage treatment cond.; 80% act.

Berol 809. [90] Combined anionic/cationic/nonionic surfactant; detergent, textile softener, antistat agent; lt. yel., clear liq.; sol. in water, ethanol, wh. spirit; sp.gr. 1.04; visc. 170 cps; cloud pt. > 55 C (opal 1% aq.); flash pt. > 100 C; pour pt. –4 C; surf. tens. 31.5 dynes/cm (0.1%); pH 6.3 (1% aq.); biodeg. under normal sewage treatment cond.; 80% act.

Berol 822. [90] Calcium alkylaryl sulfonate; anionic; emulsifier for pesticides; Gardner < 10 clear visc. liq.; sol. in ethanol, trichloroethylene, paraffin oil, xylene, wh. spirit, lt. fuel oil; disp. in water; sp.gr. 1.03; visc. 1200 cps; flash pt. 35 C; pour pt. 5 C; pH 5–8 (1% aq.); biodeg.; 60–62% act. in butanol.

Berol Dinonylphenol. [90] Dinonylphenol; chemical intermediate, stabilizer, plasticizer, hydroprobe for nonionic surfactants, used as component in lubricating oils, corrosion inhibitors; yel., visc. liq.; phenol-like odor; sol. in ethanol, acetone, trichloroethylene, butyl glycol, wh. spirit, min. oil; sp.gr.

0.911; visc. 400 cps (40 C); b.p. 326–334 C (5%); flash pt. > 500 C; pour pt. 0 C.

Berol EGA 07. [90] Fatty alcohol ethoxylate; nonionic; textile dye leveling and disp. agent, protective colloid; Hazen < 150 clear liq.; sol. in water, ethanol, xylene, trichloroethylene; disp. in paraffin oil, wh. spirit, lt. fuel oil; sp.gr. 1.01; visc. 10 cps; HLB 15.0; cloud pt. 74–76 C (5% in 10% NaCl); flash pt. > 100 C; pour pt. < 0 C; surf. tens. 40 dynes/cm (0.1%); pH 6–7 (5% aq.); biodeg.; 20% act.

Berol EGA 192. [90] Alkyl glyceride EO adduct; nonionic; textile dye equalizing and disp. agent; Gardner < 3 clear liq.; sol. in water, ethanol, acetone; disp. in propylene glycol, xylene; sp.gr. 1.080; visc. 120 cps; HLB 18; cloud pt. 62 C (1% in 10% NaCl); flash pt. > 100 C; pour pt. 13 C; surf. tens. 42 dynes/cm; pH 6–7 (1% aq.); biodeg.; 85% act.

Berol Fintex 10. [90] Fatty acid EO adduct; nonionic; textile softener and antistat, equalizing agent, emulsifier for greases and oils; wh. wax; sol. in ethanol, trichloroethylene, xylene; disp. in water, propylene glycol, acetone, wh. spirit; sp.gr. 1.010 (36 C); visc. 80 cps (36 C); HLB 12; cloud pt. 75–80 C; flash pt. > 100 C; pour pt. 35 C; surf. tens. 42 dynes/cm (0.1%); pH 6–7 (1% aq.); biodeg.; 100% act.

Berol Fintex 42. [90] Weakly cationic surfactant; textile softener for natural and syn. fibers, antistat, used in anticrease treatment with resins, adhesive bath additive; lt. yel. liq. disp.; sp.gr. 1.0; visc. 500 cps.; flash pt. > 100C; pour pt. 0 C; pH 9 (1% aq.); 25% act.

Berol Fintex 572. [90] Quat. ammonium compd.; cationic; textile softener; liq.; 70% conc.

Berol Fintex 573. [90] Alkyl polyglycol ether ammonium ethyl sulfate and nonionic wetting agent; cationic; antistatic agent for syn. textiles and plastics, bactericide for gram-positive bacteria; Gardner ≤ 7 clear liq.; sol. in water, ethanol, propylene glycol, acetone, trichloroethylene, xylene, white spirit; sp.gr. 1.050; visc. 750 cps; cloud pt. 70 C (1% in 10% NaCl); flash pt. > 100 C; pour pt. 2 C; surf. tens. 35 dynes/cm (0.1%); pH 7 ± 0.5 (1% aq.); 100% act.

Berol Fintex 577. [90] Quat. ammonium compd.; cationic; softening, antistatic and rewetting agent for natural and syn. textiles, antibacterial properties; Hazen < 300, clear liq.; sol. in ethanol, xylene, propylene glycol, wh. spirit; disp. in hot or cold water; sp.gr. 0.94; visc. 100 cps; clear pt. < 15 C; flash pt. 25 C; pour pt. 5 C; surf. tens. 32 dynes/cm; pH 6–7 (1% aq.); 70% act.

Berol Fintex 755. [90] Mixt. of linear alkylaryl sulfonate *n*-primary alcohol; anionic/nonionic; textile wetting, cleaning, disp. agents, leveling agent in dyeing, detergent; Gardner ≤ 5 clear liq.; sol. in water, ethanol, IPA, trichloroethylene (opal), xylene; sp.gr. 1.050; visc. 150 cps; clear pt. 0 C; flash pt. > 100 C; pour pt. –10 C; surf. tens. 36 dynes/cm (0.1%); pH 6–7 (1% aq.); biodeg.; 70% act.

Berol Fintex 776. [90] Fatty alcohol ethoxylate; nonionic; low foam detergent for textile industry; Hazen < 100 clear liq.; sol. in water, ethanol, xylene, trichloroethylene, wh. spirit, lt. fuel oil; sp.gr. 1.00; visc. 50 cps; cloud pt. 45 C; flash pt. > 100 C; pour pt. < 0 C; surf. tens. 32 dynes/cm (0.1%); pH 6–7 (1% aq.); biodeg.; 90% act.

Berol Lanco. [90] Alkyl phenol EO adduct; nonionic; textile wetting and cleaning agent for natural and syn. fibers, emulsifier for waxes, oils, solv., and paint resins, emulsion polymerization, disp. for paint industry; Hazen < 150 clear liq.; sol. in water, ethanol, xylene, trichloroethylene; sp.gr. 1.05; visc. 350 cps; HLB 13.3; cloud pt. 52–58 C (1% aq.); flash pt. > 100 C; pour pt. 5 C; surf. tens. 33 dynes/cm; pH 6–7 (1% aq.); biodeg.; 100% act.

Berol Nonylphenol. [90] Nonylphenol; raw material for detergents, dispersant, wetting agents, additives for lubricating oils, corrosion inhibitors, chemicals for rubber industry, plasticizers, stabilizer; almost colorless visc. liq.; faint phenol odor; m.w. 220; sol. in ethanol, acetone, trichloroethylene, butyl glycol, wh. spirit, min. oil; sp.gr. 0.949; visc. 2750 cps; flash pt. 145 C; pour pt. < 0 C.

Berol PEG 400. [90] Calcium alkylaryl sulfonate; anionic; stabilizer and conservation of wood; liq.; 100% conc.

Berol PEG 4000. [90] Calcium alkylaryl sulfonate; anionic; stabilizer and conservation of wood; flakes; 100% conc.

Berol WASC. [90] Nonylphenol ethoxylate; nonionic; household and industrial detergent, textile washing agent, disp. agent for paint and varnish industry; Hazen < 150, clear liq.; sol. in water, ethanol; disp. in xylene, trichloroethylene, paraffin oil, wh.

spirit; sp.gr. 1.06; visc. 375 cps; HLB 14.1; cloud pt. 75–80 C (1% aq.); flash pt. > 100 C; pour pt. 0 C; surf. tens. 34 dynes/cm (0.1%); pH 6–7 (1% aq.); biodeg.; 95% act.

Bina QAT-43. [137] Lauramidopropyl acetamidodimonium chloride; conditioner and emulsifier for personal care products.

Bio-Dac 50-20. [92] Mixed dialkyl (C_{8-10}) dimethyl ammonium chloride; cationic; antimicrobial, disinfectant, sanitizer, bacteriostat; pale yel. to water-wh. clear liq.; sp.gr. 0.925; dens. 7.7 lb/gal; pH 6–9 (10%); flash pt. 110 F (Seta); 50% act.

Bio-Dac 50-22. [92] Didecyldimethyl ammonium chloride; cationic; antimicrobial, disinfectant for sanitizing applications; pale yel. to water-wh. clear liq.; sp.gr. 0.927; dens. 7.73 lb/gal; pH 6–9 (10%); flash pt. 109 F (Seta); 50% act.

Bio-Dac 205. [92] Alkyl dimethyl benzyl ammonium chloride and mixed dialkyl (C_{8-10}) dimethyl ammonium chloride; cationic; antimicrobial, disinfectant, sanitizer; pale yel. to water wh. clear liq.; sp.gr. 0.946; dens. 7.89 lb/gal; pH 6–9 (10%); flash pt. 116 F (Seta); 50% act.

Biopal NR-20. [245] Nonoxynol-12 iodine; dk., reddish br., visc. free flowing liq.

Biopal NR-20-W. [245] Nonylphenoxypoly (ethyleneoxy) ethanol-iodine complex; anionic; formulates "no rinse" sanitizing sol'n.; liq.; 20% conc.

Biopal VRO-20. [245] Nonoxynol-9 iodine; iodophor conc., iodine complexed with a nonionic surfactant for formulation of rinses and cleaners, disinfecting agents; dk. br. pourable liq.; water sol.; sp.gr. 1.34; visc. 1700 cps; pour pt. 10.4 C.

Bio-Quat 50-24. [92] Alkyl dimethyl benzyl ammonium chloride; cationic; antimicrobial, germicide, disinfectant, sanitizer, fungicide, deodorant; algicide for swimming pools and cooling waters; pale yel. to water-wh. clear liq.; sp.gr. 0.988; dens. 8.23 lb/gal; visc. 60 cps; pH 7.2–8.0 (10%); 50% act.

Bio-Quat 50-25. [92] Alkyl dimethyl benzyl ammonium chloride; antimicrobial for formulating germicides, algicides; pale yel. to water-wh. clear liq.; sp.gr. 0.984; dens. 8.20 lb/gal; pH 7.2–8.0 (10%); 50% act.

Bio-Quat 50-28. [92] Alkyl dimethyl benzyl ammonium chloride; cationic; antimicrobial, germicide, disinfectant, sanitizer, fungicide, deodorant, algicide for cooling waters and swimming pools; pale yel. to water-wh. clear liq.; sp.gr. 0.979; dens. 8.16 lb/gal; visc. 105 cps; pH 7.2–8.0 (10%); 50% act. in IPA.

Bio-Quat 50-30. [92] Alkyl dimethyl benzyl ammonium chloride; cationic; antimicrobial for formulating germicides, algicides; pale yel. to water-wh. clear liq.; sp.gr. 0.985; dens. 8.21 lb/gal; pH 7.2–8.0 (10%); flash pt. 134 F (PM-CT); 50% act.

Bio-Quat 50-35. [92] Alkyl dimethyl benzyl ammonium chloride and alkyl dimethyl ethyl benzyl ammonium chloride; cationic; antimicrobial, germicide, disinfectant, sanitizer, fungicide, deodorant, algicide for cooling waters and swimming pools; pale yel. to water-wh. clear liq.; sp.gr. 0.986; dens. 8.13 lb/gal; visc. 68 cps; pH 7.2–8.0 (10%); 50% act.

Bio-Quat 50-36. [92] Alkyl dimethyl benzyl ammonium chloride and alkyl dimethyl ethyl ammonium chloride; cationic; antimicrobial, germicide, disinfectant, sanitizer, fungicide, deodorant; pale yel. to water-wh. clear liq.; sp.gr. 0.977; dens. 8.13 lb/gal; visc. 68 cps; pH 7.2–8.0 (10%); 50% act.

Bio-Quat 50-40. [92] Alkyl dimethyl benzyl ammonium chloride; cationic; antimicrobial, germicide, disinfectant, sanitizer, fungicide, deodorant, algicide for cooling waters and swimming pools; pale yel. to water-wh. clear liq.; sp.gr. 0.979; dens. 8.16 lb/gal; visc. 425 cps; pH 7.2–8.0 (10%); 50% act.

Bio-Quat 50-42. [92] Alkyl dimethyl benzyl ammonium chloride; antimicrobial for formulating germicides, algicides; pale yel. to water-wh. clear liq.; sp.gr. 0.98–0.99; dens. 8.24 lb/gal; flash pt. 135 F (PM-CT); pH 7.2–8.0 (10%); 50% act.

Bio-Quat 50-60. [92] Alkyl dimethyl benzyl ammonium chloride; cationic; antimicrobial for formulating germicides, algicides; pale yel. to water-wh. clear liq.; sp.gr. 0.961; dens. 8.01 lb/gal; visc. 105 cps; flash pt. 116.6 F; pH 7.2–8.0 (10%); 50% act.

Bio-Quat 50-65. [92] Alkyl dimethyl benzyl ammonium chloride; antimicrobial for formulating germicides, algicides; pale yel. to water-wh. clear liq.; sp.gr. 0.987; dens. 8.23 lb/gal; flash pt. 135 F (PM-CT); pH 7.2–8.0 (10%); 50% act.

Bio-Quat 50-MAB. [92] Alkyl dimethyl benzyl ammonium bromide, cationic; antimicrobial for germicides, algicides; pale yel. to water-wh. clear liq.; sp.gr. 0.950; dens. 7.92 lb/gal; pH 7.2–8.0 (10%); 50% act.

Bio-Quat 50-MAC. [92] Alkyl dimethyl dichlorobenzyl ammonium chloride; cationic; antimicrobial, algicide, slimicide for industrial water treatment; pale yel. to water-wh. clear liq.; sp.gr. 1.038; dens. 8.65 lb/gal; visc. 75 cps; pH 7.2–8.0 (10%); 50% act.

Bio-Quat 80-24. [92] Alkyl dimethyl benzyl ammonium chloride; cationic; antimicrobial; pale yel. to water-wh. clear liq.; sp.gr. 0.916; dens. 7.63 lb/gal; visc. 350 cps; flash pt. 75 F (PM-CT); pH 7.2–8.0 (10%); 80% act.

Bio-Quat 80-28. [92] Alkyl dimethyl benzyl ammonium chloride; cationic; antimicrobial; pale yel. to water-wh. clear liq. above gel pt.; sp.gr. 0.922; dens. 7.73 lb/gal; visc. 405 cps; gel pt. 4–8 C; flash pt. 75 F (PM-CT); pH 7.2–8.0 (10%); 80% act.

Bio-Quat 80-35. [92] Alkyl dimethyl benzyl ammonium chloride and alkyl ethylbenzyl ammonium chloride; cationic; antimicrobial; pale yel. to water-wh. clear liq.; sp.gr. 0.927; dens. 7.72 lb/gal; visc. 330 cps; flash pt. 75 F (PM-CT); pH 7.2–8.0 (10%); 80% act.

Bio-Quat 80-36. [92] Alkyl dimethyl benzyl ammonium chloride and alkyl dimethyl ethyl benzyl ammonium chloride; cationic; antimicrobial; pale yel. to water-wh. clear liq.; sp.gr. 0.922; dens. 7.72 lb/gal; visc. 330 cps; flash pt. 82 F (PM-CT); pH 7.2–8.0 (10%); 80% act.

Bio-Quat 80-40. [92] Alkyl dimethyl benzyl ammonium chloride; cationic; antimicrobial; pale yel. to water-wh. clear liq.; sp.gr. 0.922; dens. 7.73 lb/gal; flash pt. 75 F (PM-CT); pH 7.2–8.0 (10%); 80% act.

Bio-Quat 80-42. [92] Alkyl dimethyl benzyl ammonium chloride; cationic; antimicrobial; pale yel. to water-wh. clear liq.; sp.gr. 0.95; dens. 8.21 lb/gal; flash pt. 75 F (PM-CT); pH 7.2–8.0 (10%); 80% act.

Bio-Quat T-501. [92] Alkyl dimethyl benzyl ammonium chloride and alkyl ammonium bromide; cationic; antimicrobial; pale yel. to water-wh. clear liq.; sp.gr. 0.972; dens. 8.11 lb/gal; visc. 68 cps; pH 7.2–8.0 (10%); 50% act.

Bio-Quat T-502. [92] Alkyl dimethyl dichlorobenzyl ammonium chloride, alkyl dimethyl ammonium chloride, alkyl ether ammonium bromide; cationic; antimicrobial; pale yel. to water-wh. clear liq.; sp.gr. 0.972; dens. 8.31 lb/gal; visc. 68 cps; pH 7.2–8.0 (10%); 50% act.

Bio Soft D-35X. [575] Sodium dodecylbenzene sulfonate; anionic; detergent with good foaming, wetting, emulsifying properties; detergent base; lt. straw liq.; water sol.; dens. 8.9 lb/gal; cloud pt. 10 C; biodeg.; 34% act.

Bio Soft D-40. [575] Sodium dodecylbenzene sulfonate; anionic; detergent with good foaming, wetting, emulsifying properties; detergent base; clear liq.; bland odor; water sol.; dens. 8.7 lb/gal; cloud pt. 29 C; biodeg.; 40% act.

Bio Soft D-60. [575] Linear sodium dodecylbenzene sulfonate; anionic; detergent, foaming and wetting agent; lt.-colored slurry; water sol.; surf. tens. 37.4 dynes/cm (0.1%); biodeg.; 50% act., 21% water.

Bio Soft D-62. [575] Sodium dodecylbenzene sulfonate; anionic; detergent with good foaming, wetting, emulsifying properties; detergent base; straw, yel. slurry; bland odor; water sol.; dens. 9.1 lb/gal; biodeg.; 60% act.

Bio Soft EA-4. [575] Linear primary alkoxylate; nonionic; surfactant; wh. hazy liq.; sol. in org. solvs.; disp. in water; dens. 7.7 lb/gal (60 C); pour pt. 5 C; cloud pt. < 25 C; 100% act.

Bio Soft EA-8. [575] Linear alkoxypolyoxethylene ethanols; nonionic; chemical specialties, laundry detergent, agric. emulsions, sanitizers, metal cleaning and pickling, used in textile and petroleum industries; hazy, wh. liq.; sol. in org. solv., oil, water; dens. 8.42 lb/gal; sp.gr. 1.01; cloud pt. 46–58.5 C; biodeg.; 100% act.

Bio Soft EA-10. [575] Linear alkoxypolyoxethylene ethanols; see Bio Soft EA–8; liq.; 100% conc.

Bio Soft HD-100. [575] Blend of nonionic and anionic surfactants; intermediate for the formulation of built and unbuilt liq. laundry detergents; lt. yel., clear liq.; water sol.; dens. 8.9 lb/gal; visc. 600 cps; pour pt. 5 C; 100% act.

Bio Soft JN. [576] Dodecylbenzene sulfonic acid; anionic; intermediate for production of liq. and powd. detergents; liq.; 97% conc.

Bio Soft LAS-40S. [575] Linear alkylbenzene sulfonic acid, sodium salt; anionic; detergent for household, industrial, and institutional cleaners; liq.; 40% conc.

Bio Soft LAS-50S. [575] Linear alkylbenzene sulfonic acid, sodium salt; anionic; detergent for household, industrial, and institutional cleaners; liq.; 50% conc.

Bio Soft LAS-60T. [575] Linear alkylbenzene sulfonic acid, triethanolamine salt; anionic; liq. detergents, personal and pet care products; liq.; 59% conc.

Bio Soft LAS-97. [575] Dodecylbenzene sulfonic acid; anionic; intermediate for production of liq. and powd. detergents; liq.; 97% conc.

Bio Soft LD-32. [575] Anionic-nonionic blend with sodium cation; liq. dishwashing detergent conc., used in bubble bath, car wash, rug shampoos, textile scouring, air entraining agent in concrete; pale straw, clear liq.; water sol.; dens. 9.23 lb/gal; visc. 350 cps; cloud pt. 5 C; biodeg.; 60% act.

Bio Soft LD-68. [575] Liq. detergent for dishwashing, car wash, fine fabrics; pale straw, clear liq.; dens. 9.0 lb/gal; visc. 800 cps; biodeg.; 60% solids, contains no phosphates.

Bio Soft LD-70. [575] Sodium olefin sulfonate; detergent, foaming agent, nonionic, used in shampoos, bubble bath, dishwashing, car washing, fine fabric detergent; clear, amber liq.; dens. 8.7 lb./gal; visc. 300 cps; cloud pt. > 40 F; biodeg.; 35% act.

Bio Soft LD-80. [575] Sodium olefin sulfonate; detergent, nonionic, used in dishwashing; amber liq.; water sol.; dens. 8.8 lb./gal; visc. 225 cps; cloud pt. < 5 C; biodeg.; 41% solids.

Bio Soft LD-95. [575] Sodium dodecylbenzene sulfonate, sodium laureth sulfate, lauraide DEA, and urea; anionic-nonionic; liq. dishwashing detergent conc.; pale straw clear liq.; water sol.; dens. 9.11 lb/gal; visc. 600 cps; cloud pt. 5 C; biodeg.; 60% act.

Bio Soft N-21. [575] Cocamide DEA; liq. dishwashing detergent conc.; clear amber liq.; water sol.; dens. 8.6 lb/gal; visc. 220 cps; biodeg.; 60% act.

Bio Soft N-300. [575; 576] TEA-dodecylbenzene sulfonate; anionic; liq. dishwashing, car washing detergent, wetting and foaming agent; pale yel. clear liq.; water sol.; dens. 9.0 lb/gal; sp.gr. 1.08; visc. 3200 cps; surf. tens. 38.8 dynes/cm (0.1%); biodeg.; 60% act.

Bio Soft Ninex 21. [576] Amide sulfonate complex; anionic; detergent; liq. dishwash detergent with flash foam; liq.; 60% conc.

Bio Soft S-100. [575] Linear dodecylbenzene sulfonic acid; anionic; detergent, emulsifier, intermediate for formulation of built detergents; dk. visc. liq.; water sol.; sp.gr. 1.04; biodeg.; 97% act.

Bio-Surf DC-730. [92] Quat. ammonium compd.; cationic; emulsifier and dispersant; liq.; oil-sol.; 75% conc.

Bio-Surf PBC-420. [92] Linear DDBSA, TEA salt; anionic; surfactant and wetting agent; liq.; 60% conc.

Bio-Surf PBC-460. [92] Lauryl myristyl dimethyl amine oxides; nonionic/cationic; detergent aid in stabilizing foam; emollient; liq.; 30% conc.

Bio Terge 804. [575] Alpha olein sulfonate base; anionic; conc. for hand soaps and shampoos; liq.; 52% conc.

Bio Terge AS-40. [575; 576] Sodium C_{14-16} olefin sulfonate; anionic; detergent, foaming agent for personal care, commercial and industrial formulations; yel. liquid; water sol.; dens. 8.9 lb/gal; biodeg.; 40% act. in water; contains no phosphates.

Bio Terge AS-90 Beads. [575] Sodium alpha olefin sulfonate; surfactant for personal care products; off-wh. beads; faint odor; dens. 0.30 g/ml; pH 9.8; surf. tens. 46.2 dynes/cm; 88.2% act.

Bio Terge AS-90F. [575; 576] Sodium alpha olefin; anionic; detergent, wetting agent for powd. products; cream flake; water sol.; biodeg.; 90% act.

Bio Terge LD-100. [575] Alpha olefin sulfonate conc.; liq. dishwash conc; yel. clear liq.; mild odor; visc. 150 cps; pour pt. 3 C; flash pt. > 200 F (Seta CC); pH 6.8–7.8; 41–44% solids.

Bio Terge PAS-85. [575] Primary alkane sulfonate; anionic; surfactant used in specialty cleaners for industrial, institutional and household markets; liq.; 38–42% conc.

Biozan. [295] Xanthan gum; thickener, emulsifier, and suspension aid in slurry explosives, foundry coatings, cleaning comps., cosmetics, pharmaceuticals, oil field chemicals and aq. systems; used in food applications.

Biozan SPX 5423. [295] Xanthan gum; thickener, suspension aid; tan powd.; visc. 1300–1600.

Blancol. [245] Sodium naphthalenesulfonate-formaldehyde condensate; anionic; thinning, disp., dye-leveling agent used in paper industry for slime control, improved retention of fillers or fines, improved sizing; tan to br. free-flowing coarse granules; odorless; sol. in warm or cold water; dens. 0.77–0.82 g/ml; 88–90% act.

Blancol N. [245] Sodium naphthalenesulfonate-formaldehyde condensate; see Blancol; tan to br. gran. powd.; odorless; sol. in water;

dens. 0.65–0.75 g/ml; 88–90% act.

Bohrmittel Hoechst. [44] Alkylsulfamido carboxylic acid, sodium salt; anionic; o/w corrosion inhibitor and emulsifier for metalworking fluids; liq.

Borax. [620] Sodium tetraborate decahydrate; dispersant, wetting agent for NR, SR latexes; mold lubricant for general dry rubber molding; wh. powd., odorless; sol. in water, glycerin; sp.gr. 1.73.

Brij 30, 30 SP. [320] Laureth-4; SP grade with preservatives; nonionic; o/w emulsifier for topical cosmetics; colorless to lt. yel. liq. (Brij 30 may become hazy or form a precipitate); sol. in alcohol, propylene glycol, cottonseed oil; sp.gr. 0.95; visc. 30 cps; HLB 9.7 and 9.5 resp.; flash pt. >300 F; 100% conc.

Brij 35, 35 SP. [320] Laureth-23; SP grade with preservatives; nonionic surfactant; wh. waxy solid; sol. in water, alcohol, propylene glycol; sp.gr. 1.05; HLB 16.9; flash pt. >300 F; pour pt. 33 C.

Brij 52. [320] Ceteth-2 (antioxidants added); nonionic surfactant, emulsifier for topical cosmetics; wh. waxy solid; sol. in alcohol, cottonseed oil, min. oil; HLB 5.3; flash pt. > 300F; pour pt. 33 C; 100% conc.

Brij 56. [320] Ceteth-10 (antioxidants added); nonionic surfactant, emulsifier; wh. waxy solid; sol. in alcohol; HLB 12.9; flash pt. > 300F; pour pt. 31 C.

Brij 58. [320] Ceteth-20 with preservatives; nonionic surfactant, emulsifier; wh., waxy solid; sol. in water, alcohol; HLB 15.7; flash pt. > 300 F; pour pt. 38 C; 100% conc.

Brij 72. [320] Steareth-2 with preservatives; nonionic surfactant, emulsifier; wh. waxy solid; sol. in alcohol, cottonseed oil; HLB 4.9; flash pt. >300 F; pour pt. 43 C; 100% conc.

Brij 76. [320] Steareth-10 with preservatives; nonionic surfactant, o/w emulsifier, solubilizer for topical cosmetics; wh. waxy solid; sol. in propylene glycol, ethanol; HLB 12.4; flash pt. >300 F; pour pt. 38 C; 100% conc.

Brij 78. [320] Steareth-20 with preservatives; nonionic surfactant, emulsifier; wh. waxy solid; sol. in alcohol; HLB 15.3; flash pt. >300 F; pour pt. 38 C; 100% conc.

Brij 92, 93. [320] Oleth-2 with preservatives; nonionic surfactant, o/w emulsifier for cosmetics; yel. liq.; sol. alcohol, propylene glycol, cottonseed and min. oils; visc. 30 C; HLB 4.9; flash pt. >300 F; pour pt. 10 C; 100% conc.

Brij 96. [320] Oleth-10 with antioxidants; nonionic; o/w emulsifier, solubilizer for topical cosmetics; yel. liq. with some solids; sol. in water and alcohol; visc. 100 cps; HLB 12.4; flash pt. > 300 F; pour pt. 16 C; 100% conc.

Brij 97. [320] Oleth-10 with preservatives; nonionic; see Brij 96; pale yel. liq.; sol. in water and alcohol; visc. 100 cps; HLB 12.4; flash pt. > 300 F; pour pt. 16 C; 100% conc.

Brij 98, 99. [320] Oleth-20 with antioxidants; nonionic surfactant, emulsifier; cream-colored soft waxy solid; sol. in water, alcohol, propylene glycol; HLB 15.3; flash pt. > 300 F; pour pt. 30 and 33 C respectively; 100% conc.

Brij 700. [320] Steareth-100; nonionic; emulsifier for cosmetics and pharmaceuticals; oil solubilization; wh. solid; HLB 18.8; pour pt. 55 C; 100% conc.

Brij 721. [320] Steareth-21; nonionic; cosmetic emulsifier; colorless waxy solid; waxy odor; HLB 15.5; pour pt. approx. 45 C; flash pt. > 230 F (PMCC); 100% conc.

Bromat. [297] Cetrimonium bromide; cationic; surfactant used as emulsifier in microbial dissipation of oil; germicide; wh. powd.; char. odor; m.w. 364.44; sol. in alcohol, chloroform; pH 5-8; surf. tens. 40.0 dynes; 98% min. assay.

BTC 50, 50 USP. [456] Benzalkonium chloride; cationic; hard surface disinfection, sanitization and deodorization; liq.; sp.gr. 0.96; flash pt. 126; 50% min. quat.

BTC 65. [456] N-alkyl dimethyl benzyl ammonium chlorides; cationic; hard surface disinfection, sanitization and deodorization; liq.; sp.gr. 0.97; flash pt. 132 F; 50% min. quat.

BTC 65 USP. [456] N-alkyl dimethyl benzyl ammonium chlorides; see BTC 65.

BTC 100. [456] N-alkyl dimethyl benzyl ammonium chlorides; cationic; see BTC 65; solid; sp.gr. 0.95; flash pt. > 200 F; 95% min. quat.

BTC 776. [456] N-alkyl dimethyl benzyl and dialkyl methyl benzyl ammonium chlorides; cationic; disinfectant, algicidal and slimicidal agent for swimming pool and industrial water treatment; liq.; sp.gr. 0.96; flash pt. 102 F; 50% min. quat.

BTC 812. [456] N-octyl dodecyl dimethyl ammonium chlorides; cationic; used as hard surface disinfectant and sanitization fungicide; liq.; sp.gr. 0.93; flash pt. 126 F; 50% min. quat.

BTC 818. [456] Octyldecyl dimethyl, dioctyl dimethyl, and didecyl dimethyl ammonium chlorides, cationic; dialkyl quat used as disinfectant, sanitizer, and fungicide for hard surfaces; liq.; water-sol.; sp.gr. 0.93; flash pt. 86 F; 50.0% min. quat.

BTC 824. [456] Myristyl dimethyl benzyl ammonium chlorides; cationic; hard surface disinfection and sanitization, algicide in swimming pool and industrial water treatment; liq.; sp.gr. 0.96; flash pt. > 120 F; 50% min. quat.

BTC 824 P-100. [456] N-tetradecyl dimethyl benzyl ammonium chloride monohydrate; cationic; nondusting water sol. powd. for manufacturing disinfectants, sanitizers, and deodorizers; sp.gr. 0.40; flash pt. > 200 F; 95% min. quat.

BTC 835. [456] N-alkyl dimethyl benzyl ammonium chlorides; cationic; hard surface disinfection and sanitization, algicide in swimming pool and industrial water treatment; liq.; sp.gr. 0.97; flash pt. 130 F; 50% min. quat.

BTC 885. [456] *n*-Alkyl (50% C_{14}, 40% C_{12}, 10% C_{16}) dimethyl benzyl, octyldecyl dimethyl, dioctyl dimethyl, and didecyl dimethyl ammonium chlorides; cationic; germicide for formulation of disinfectant, sanitizer, and fungicidal products used in hospitals, nursing homes, and public institutions; liq.; sp.gr. 0.95; flash pt. 116 F; 50.0% min. quat.

BTC 1010. [456] Didecyldimonium chloride; cationic; quat. used for hard surface disinfection and sanitization; fungicide; liq.; water-sol.; sp.gr. 0.89; flash pt. 86 F; 50% min. quat.

BTC 1100. [456] N-tetradecyl dimethyl-naphthylmethyl ammonium chlorides; cationic; nondusting powd. used as disinfectant, sanitizer, algicide, and deodorizer; sp.gr. 0.50; flash pt. > 200 F; 99% min. quat.

BTC 1326. [456] *n*-Alkyl (60% C_{14}, 30% C_{16}, 5% C_{12}, 5% C_{18}) dimethyl benzyl and *n*-alkyl (68% C_{12}, 32% C_{14}) dimethyl ethylbenzyl ammonium chlorides; cationic; hard surface disinfectant/surfactant for sanitizers for hospitals, nursing homes, public institutions, and industry; powd.; sp.gr. 0.75; flash pt. > 200 F; 13% min. quat.

BTC 2125. [456] Myristalkonium chloride and quaternium-14; cationic; hard surface disinfection and sanitization, algicide in swimming pool and industrial water treatment; liq.; sp.gr. 0.97; flash pt. > 200 F; 50% min. quat.

BTC 2125 80%. [456] N-alkyl dimethyl benzyl and dimethyl ethyl-benzyl ammonium chlorides; cationic; hard surface disinfection and sanitization, algicide in swimming pool and industrial water treatment; liq.; sp.gr. 0.94; flash pt. 130 F; 80% min. quat.

BTC 2125 M. [456] Myristalkonium chloride and quaternium-14; cationic; disinfection and sanitization; liq.; sp.gr. 0.97; flash pt. > 200 F; 50% min. quat.

BTC 2125 M 80%. [456] N-alkyl dimethyl benzyl and dimethyl ethyl-benzyl ammonium chlorides; cationic; disinfection and sanitization; liq.; sp.gr. 0.94; flash pt. 130 F; 80% min. quat.

BTC 2125 MP-40. [456] *n*-Alkyl (60% C_{14}, 30% C_{16}, 5% C_{12}, 5% C_{18}) dimethyl benzyl and *n*-alkyl (68% C_{12}, 32% C_{14}) dimethyl ethylbenzyl ammonium chloride; cationic; germicidal used in disinfectant and sanitizer formulations for hospitals, nursing homes, public institutions and industry; powd.; sp.gr. 0.55; flash pt. > 200F; 40% min. quat.

BTC 2125 P–40. [456] N-alkyl dimethyl benzyl and dimethyl ethyl-benzyl ammonium chlorides; cationic; hard surface disinfection and sanitization, algicide in swimming pool and industrial water treatment; solid; sp.gr. 0.55; flash pt. > 200 F; 40% min. quat.

BTC 2565. [456] N-alkyl dimethyl benzyl ammonium chlorides; cationic; hard surface disinfection and sanitization, algicide in swimming pool and industrial water treatment, cationic; liq.; sp.gr. 0.96; flash pt. 100 F; 50% min. quat.

BTC 8248. [456] N-alkyl dimethyl benzyl ammonium chlorides; cationic; hard surface disinfection and sanitization, algicide in swimming pool and industrial water treatment; liq.; sp.gr. 0.95; flash pt. 100 F; 80% min. quat.

BTC 8249. [456] N-alkyl dimethyl benzyl ammonium chlorides; cationic; hard surface disinfection and sanitization, algicide in swimming pool and industrial water treatment; liq.; sp.gr. 0.94; flash pt. 118 F; 90% min. quat.

BTC E-8358. [456] N-alkyl dimethyl benzyl ammonium chlorides; cationic; hard surface disinfection and sanitization, algicide in swimming pool and industrial water treatment; liq.; sp.gr. 0.93; flash pt. 100 F; 80% min. quat.

BTCO 1010. [456] N-didecyl dimethyl ammonium chloride; cationic; hard surface disinfection and sanitation fungicide; liq.; sp.gr. 0.89; flash pt. 86 F; 50% min. quat.

Bubble Breaker 259, 260, 613-M, 622, 730, 737, 746, 748, 900, 913, 917. [650] Blend of org., nonsilicone compds.; defoamer used in the mfg. of water-based paper coatings, textile processing formulations, agric. chemical products, paints, adhesives, and inks.

Bubble Breaker 776, 3017-A. [650] Disp. of reacted silica in hydrocarbon solv.; defoamer used in latex mfg. operations, formulation of water-based paints and adhesives.

Bubble Breaker 1840X. [650] Sodium salt of sulfonated fatty acid; defoamer for wet process production of phosphoric acid.

Bubble Breaker 3009-F. [650] Blend of org., nonsilicone compounds; defoamer for wash water of sliced potatoes.

Bubble Breaker 3056-A. [650] Disp. of reacted silica in hydrocarbon solv.; defoamer used in latex mfg. operations, formulation of water-based paints and adhesives.

Bubble Breaker 3073-7, D. [650] Blend of org., nonsilicone compounds; defoamer for water-based drilling for the petroleum industry.

Bubble Breaker PR. [650] Sodium salt of sulfonated fatty acid; defoamer for wet process production of phosphoric acid.

C

CAE. [9] N^2 cocoyl-L-arginine ethyl ester DL-pyrrolidone carboxylic acid salt; cationic; surfactant, foamer, and antistat; preservative for cosmetics; powd.; 100% conc.

Calamide C. [478] Cocamide DEA; nonionic; emollient, foam stabilizer, visc. builder; liq. detergents and personal care products; emulsifier for degreasing; lubricant, rust preventive for metal cleaning and cooling systems; detergent and solubilizer for washing floors, walls, metals; amber clear liq.; water-sol.; pH 9.5; 100% conc.

Calamide CW-100. [478] Cocamide DEA and diethanolamine; nonionic; emollient, lubricant, foam stabilizer, visc. builder, detergent, solubilizer used in bubble baths, shampoos, liq. detergents; heavy-duty cleaners; emulsion systems; yel. clear liq.; water-sol.; dens. 8.5 lb/gal; pH 10.5; 100% conc.

Calamide O. [478] Oleamide DEA; nonionic; emollient, lubricant, foam stabilizer, visc. builder, emulsifier; used in personal care products, industrial applications; amber clear liq.; disp. water and oil; pH 10; 100% conc.

Calcium Petronate. [650] Calcium petroleum sulfonate; anionic; detergent and rust inhibitor in lube oil additives; dk. br. visc. liq.; sol. in org. solv.; sp.gr. 0.98; 44% act.

Calcium Stearate, Regular. [650] Calcium stearate; lubricant for ceramics, coated foundry sands, plastics; plasticizer/lubricant for paper coatings; aux. emulsifier; water repellent for cosmetics; flatting, pigment suspending, and thickener for solv.-based paints and varnishes; wh. powd.; sp.gr. 1.03.

Calfoam AAL. [478] Alkanolamide and surfactant blend; liq. detergent for dishwashing and fine fabric laundering; liq.; dens. 9.0 lb/gal; visc. 300 cps; pH 7.0; 45% solids.

Calfoam ES-30. [478] Sodium laureth sulfate; anionic; foam stabilizer, flash foamer, wetter for detergent systems, personal care products; emulsion polymerization; yel. clear liq.; mild odor; dens. 8.8 lb/gal; pH 8.0; 30% solids.

Calfoam NEL-60. [478] Ammonium alcohol ether sulfate; anionic; flash foamer, foam stabilizer, detergent, emulsifier, wetter for liq. detergents, bubble baths, shampoos, car washing; lime soap dispersant; clear liq.; faint alcohol odor; sol. in aq. systems; dens. 8.58 lb/gal; pH 7.5; 57.5% act.

Calfoam SEL-60. [478] Sodium laureth sulfate; general all-purpose cleaning and wetting for use in bubble baths, shampoos, car washing, liquid detergents; liq.; 60% act.

Calfoam SLS-30. [478] Sodium lauryl sulfate; anionic; foamer for personal care products; rug/upholstery shampoos; emulsifier in emulsion polymerization of latex, SBR rubber, polyacrylates, elastomers; foaming agent for foamed rubber; wh. paste; mild odor; pH 8.0; 28% act.

Calimulse PRS. [478] Isopropylamine dodecylbenzene sulfonate; anionic; emulsifier, solubilizer; dry cleaning; degreasers; latex emulsifier; pigment dispersant; agric. sprays, oil slick emulsifiers; clear amber liq.; dens. 8.5 lb/gal; pH 4.8; 90% amine sulfonate.

Calsoft F-90. [478] Sodium dodecylbenzene

sulfonate; anionic; surfactant for all-purpose and hard surface cleaners, degreasers, laundry powds., scouring aids, emulsion polymers; flake; water-sol.; sp.gr. 0.45; 90% act.

Calsoft L-40. [478] Sodium dodecylbenzene sulfonate; anionic; emulsion stabilizer; wetting and foaming agent, detergent, emulsifier for household and industrial detergents, agric. emulsions, dye bath leveling, rug cleaners, bubble baths, ore flotation, and air entrainment in concrete and gypsum board; washing fruits and vegetables; lt. yel. visc. liq.; water-sol.; sp.gr. 1.07; pH 7.5; 42% solids.

Calsoft L-60. [478] Sodium dodecylbenzene sulfonate; anionic; see Calsoft L-40; water-wh. pasty liq.; odorless; dens. 8.7 lb/gal; pH 7.4; 60% solids.

Calsoft LAS-99. [478] Dodecylbenzene sulfonic acid, linear; detergent, emulsifier, intermediate for liq. and dry detergents; Klett 50 syrupy liq.; water-sol.; sp.gr. 1.06; dens. 8.83 lb/gal; visc. 1100 cps; 97.5% act.

Calsoft T-60. [478] TEA-dodecylbenzene sulfonate; anionic; detergent, wetting agent, flash foamer; liq. detergents, wool wash compds., cosmetics and shampoos, agric. emulsifiers, industrial cleaners, textile scouring, and car wash compds.; yel. clear visc. liq.; mild odor; dens. 9.1 lb/gal; visc. 2300 cps; pH 5.7; 60% solids.

Calsolene Oil HS. [317] Sulfated oil, sodium salt, aq. sol'n.; wetting agent and emulsifier; emulsion breaker for w/o and water-in-tar emulsions; liq.

Calsolene Oil HSA. [320] Sulfated fatty acid ester; anionic; surfactant; gold-yel., clear liq. (may be turbid at room temp.); dissolves in hard and soft water and in salt sol'ns.; flash pt. 200 F (OC).

Calsuds CD-6. [478] Cocamide DEA; nonionic/anionic; foamer, foam builder, wetting agent, visc. modifier, and emulsifier; used in liq. detergents, shampoos, wool-washing compds., hand, felt, and janitorial cleaners, textile scours, agric. sprays; clear visc. liq.; dens. 8.6 lb/gal; visc. 2100 cps; pH 9.5 (10%); 100% conc.

Canamulse 55. [109] Propylene glycol mono fatty acid esters; nonionic; food emulsifier for cakes, whipped toppings, starch complexing; flakes, beads; HLB 3.5; 52–58% conc.

Caplube 8350. [111] Glycerol monodioleate; nonionic; o/w emulsifier for food and industrial applications; liq.; HLB 2.7; 100% conc.

Caplube 8385. [111] Castor oil, ethoxylated; nonionic; emulsifier, dispersant, lubricant used in textile compounding; liq.; HLB 14.5; 100% conc.

Caplube 8410. [111] Triglycerol monooleate; nonionic; emulsifier, lubricant for textiles; liq.; HLB 8.8; 100% conc.

Caplube 8440. [111] Decaglycerol tetraoleate; nonionic; emulsifier, antistat, lubricant, and coupler for textile finishes; liq. HLB 6.7; 100% conc.

Caplube 8442. [111] Decaglycerol decaoleate; nonionic; emulsifier, lubricant; HLB 3.2; 100% conc.

Caplube 8445. [111] Decaglycerol tetracocate; nonionic; emulsifier, lubricant, antistat for textile application; liq.; HLB 8.8; 100% conc.

Caplube 8448. [111] Decaglycerol decastearate; nonionic; lubricant for thread finish compounding; opacifier; solid; HLB 3.5; 100% conc.

Caplube 8508. [111] Coconut ethoxylate; nonionic; emulsifier, lubricant for textile applications; liq.; 100% conc.

Capmul EMG. [111] PEG-20 glyceryl stearate; nonionic; dough conditioner for yeast-raised baked goods; solid; HLB 13.1; 100% conc.

Capmul GMO. [111] Glyceryl oleate; nonionic; food emulsifier and dispersant; liq.; HLB 3.4; 100% conc.

Capmul GMS. [111] Glyceryl stearate; nonionic; food emulsifier used in margarine, yeast-raised baked goods; beads, flakes; HLB 3.2; 100% conc.

Capmul GMVS-K. [111] Glyceryl mono shortening; nonionic; food emulsifier for shortenings for cakes and icings, margarine, whipped topping; plastic, solid; HLB 3.4; 100% conc.

Capmul O. [111] Sorbitan oleate; nonionic; food emulsifier and dispersant; plastic, solid; HLB 4.3; 100% conc.

Capmul PGME. [111] Propylene glycol monoesters and monodiglycerides; nonionic; emulsifier for cake mixes, toppings; solid; HLB 2.2; 100% conc.

Capmul POE-L. [111] Polysorbate 20; food emulsifier and solubilizer for flavors; liq.; HLB 16.7; 100% conc.

Capmul POE-O. [111] Polysorbate 80; nonionic; food emulsifier for frozen desserts; liq.; water-sol.; HLB 15.0; 100% conc.

Capmul POE-S. [111] Polysorbate 60; nonionic; food emulsifier for icings, frozen desserts, whipped toppings, coatings; liq.; HLB 14.9; 100% conc.

Capmul S. [111] Sorbitan stearate; nonionic; food emulsifier for chocolate and confectionary coatings, shortenings; bead; HLB 4.7; 100% conc.

Caprol 2G4S. [111] Diglycerol tetrastearate; nonionic; food emulsifier; solid; HLB 2.5; 100% conc.

Caprol 3GO. [111] Polyglyceryl-3 oleate; nonionic; food emulsifier for frozen desserts, veg. dairy products, diet spreads; liq.; HLB 6.2; 100% conc.

Caprol 3GS. [111] Polyglyceryl-3 stearate; nonionic; food emulsifier, stabilizer and whipping agent used in frozen desserts and fat reduction; solid; HLB 6.2; 100% conc.

Caprol 3GVS. [111] Triglycerol mono shortening; nonionic; food emulsifier for icings, shortenings; semisolid; HLB 6.0; 100% conc.

Caprol 6G2O. [111] Polyglyceryl-6 dioleate; nonionic; food emulsifier for frozen desserts; liq.; HLB 8.5; 100% conc.

Caprol 6G2S. [111] Polyglyceryl-6 distearate; nonionic; food emulsifier for whipped toppings, frozen desserts, coffee whiteners; solid; HLB 8.5; 100% conc.

Caprol 10G2O. [111] Decaglycerol dioleate; nonionic; food emulsifier for frozen desserts; liq.; HLB 10.0; 100% conc.

Caprol 10G4O. [111] Decaglycerol tetraoleate; nonionic; food emulsifier, visc. control, stabilizer; liq.; HLB 6.2; 100% conc.

Caprol 10G10O. [111] Decaglycerol decaoleate; nonionic; food emulsifier, solubilizer, and dispersant; liq.; HLB 2.5; 100% conc.

Caprol ET. [111] Polyglycerol mixed esters; nonionic; food emulsifier and crystal inhibitor in oils; plastic; HLB 2.5; 100% conc.

Caprol JB. [111] Decaglycerol decastearate; nonionic; food emulsifier in confectionary coatings; flakes; HLB 2.5; 100% conc.

Caprol PGE860. [111] Decaglycerol monodioleate; nonionic; food emulsifier and beverage cloud agent; liq.; HLB 11.0; 100% conc.

Carbonox. [447] Altered lignite; anionic; thinner, emulsifier, and dispersant for oil well drilling muds; powd.; 100% conc.

Carbopol 907. [261] Polyacrylic acid; anionic; emulsifier, thickener, stabilizer; solid; 100% conc.

Carbopol 910. [261] Carbomer 910; anionic; emulsifier, thickener, stabilizer; powd.; 100% conc.

Carbopol 934. [261] Carbomer 934; anionic; emulsifier, thickener, stabilizer; powd.; 100% conc.

Carbopol 940. [261] Carbomer 940; anionic; emulsifier, thickener, stabilizer; powd.; 100% conc.

Carbopol 941. [261] Carbomer 941; anionic; emulsifier, thickener, stabilizer; powd.; 100% conc.

Carboset 514H. [261] Acrylic resin; anionic; film former, dispersant, wetting used in industrial coatings, cosmetic formulations, paints, textile sizing, polishes, prinking inks, chemical specialities, leather and vinyl coatings; liq.; 40% conc.

Carboset 515. [261] Acrylic/acrylate copolymer; anionic; see Carboset 514H; 1iq.; 100% conc.

Carboset 525. [261] Acrylic/acrylate copolymer; anionic; see Carboset 514H; dry; 100% conc.

Cardene. [138] Wetting agent, sec. surfactant improving o/w emulsions; lt. amber liq.; oil sol.; sp.gr. 0.979 (60 F); visc. 2200 SUS (100 F); 100% act.

Carnapol 77X. [506] Fatty diamide; nonionic; release agent, high m.p. wax for use as release agents and wax finishes; solid; 100% act.

Carsamide 7644. [113] Modified coconut diethanolamide; nonionic/anionic; emulsifier, thickening agent, controlled suds detergent, textile fulling and scouring agent; Gardner 6 liq.; sol. in water, alcohol, chlorinated and aromatic hydrocarbons; disp. in min. oil, kerosene, natural fats and oils, min. spirits; dens. 8.3 lb/gal.

Carsamide C-3. [113] Cocamide DEA; intermediate for liq. detergent formulation, emulsifier for aromatic solv., improves the foaming, detergency and hard water resistance of soaps.

Carsamide C-7944. [113] Cocamide DEA; amber liq. free of foreign matter.

Carsamide CA. [113; 372] Cocamide DEA; nonionic; detergent, emulsifier, dispersant, wetting agent for industrial, institutional, cosmetic and household cleaners; sol. in water, alcohol, chlorinated hydrocarbons, Cellosolve, Carbitol, natural fats and oils; dens. 8.3 lb/gal; sp.gr. 1.0; biodeg.; 100% act.

Carsamide CMEA. [372] Cocamide MEA; nonionic; foam booster, stabilizer, visc. builder for shampoos and detergents; flake; 100% conc.

Carsamide LE. [113] Cocamide DEA; lt. amber liq.

Carsamide SAC. [113; 372] Cocamide DEA; nonionic; detergent, emulsifier, dispersant, wetting agent, visc. builder, foam stabilizer for industrial, institutional, cosmetic and household cleaners; pale amber visc. liq.; mild, fatty odor; sol. in water, alcohol, chlorinated and aromatic hydrocarbons, Cellosolves, Carbitols, natural fats and oils; dens. 8.2 lb/gal; 100% act.

Carsamide SAL-7. [113; 372] Lauramide DEA; nonionic detergent, emulsifier, foaming agent, foam stabilizer, thickener for household, institutional and industrial uses; Gardner 2 solid; mild, fatty odor; sol. in alcohol, chlorinated and aromatic hydrocarbons; dens. 8.2 lb/gal; m.p. 104–114 F; biodeg.; 100% conc.

Carsamide SAL-9. [113] Lauramide DEA; nonionic; see Carsamide SAL–7; solid; dens. 8.1 lb/gal; biodeg.; 100% conc.

Carsamide SAL-82. [113] Lauramide DEA; clear yel. liq. free from foreign matter; visc. 500 cps.

Carsamine 2414. [113] Mixt. of aliphatic and heterocyclic mono and diamines; completely sol. in water; dens. 9.0 lb/gal; b.p. 242–310 C.

Carsamine CB. [113] Cocamidopropyl dimethylamine; clear, amber liq.; dens. 7.6 lb/gal.

Carsamine Oxide LM. [113] Lauryl/myristyldimethyl amine oxide; mildly cationic under acidic conditions; foaming and conditioning agent for personal, household and industrial uses, foam stabilizer, wetting agent in conc. electrolyte sol'ns., Gardner 2 clear liq., 29% act.

Carsodet TD-7C. [113; 372] Trideceth-7 carboxylic acid; anionic; acidulant, emollient, surfactant; neutralized with other materials to produce mild anionics, topical cleansing products, shampoos; colorless liq.; dens. 8.5 lb/gal; pH 1; 90% conc.

Carsofoam 211. [113] Blend of sulfonates, nonionics, and coconut derived alkanolamides; detergent conc. for all purpose cleaners, paint strippers, aircraft cleaners (base), metal cleaners, etc.; amber liq.; mild odor; dens. 8.8 lb/gal; clear pt. 1 C; pH 8.2 ± 0.2; biodeg.; 100% act.

Carsofoam MS Conc. [113] Detergent, shampoo conc. esp. for mild shampoos; lt. yel. hazy liq.; pH 7.3 (10%); 50% act.

Carsofoam PS-1, PS-3. [113] Shampoo conc. used to manufacture liq. cream and pearlized shampoos; wh. pearlized lotion; mild, pleasant odor; pH 7 (PS-3, 10%); 38% act.

Carsofoam SC. [113] Sodium laureth sulfate, cocamide DEA, DEA lauryl sulfate; detergent conc. used for high quality shampoos and bubble baths; Gardner 4 thick, flowable gel; pH 7 (10%); 65% solids.

Carsofoam T-60-L. [113; 372] TEA dodecylbenzene sulfonate; anionic; high foaming detergent, wetting agent for cosmetic, industrial, and institutional usage; Gardner < 5 clear visc. liq.; bland odor; dens. 9.0 lb/gal; pH 7.0–7.6; biodeg.; 60% solids.

Carsofos 4326 [113] Anionic/nonionic blend; detergent, wetting agent, emulsifier, penetrant, for all-purpose cleaners; pale amber liq.; mild, fatty odor; dens. 8.25 lb/gal; sp.gr. 0.989; pH 7.5–8.5; 92% solids.

Carsofos N. [113] Formulation of aq. degreasers; Gardner 5 clear visc. liq.; dens. 8.39 lb/gal; visc. 300 cps (10% aq.); pH 9.2 (1%).

Carsofos NP-6. [113] Amber (Gardner 5) visc. liq.; pH 2.5 (1% aq.).

Carsofos NP-9. [113] Yel. (Gardner 4) slight hazy visc. liq.; pH 2.0 (1% aq.).

Carsonam 3. [113] Cocamidopropyl betaine; amphoteric surfactant, detergent, wetting agent, emulsifier, foaming and coupling agents for shampoos, lotions, germicides, chemical specialties; Gardner 5 liq.; dens. 8.45 lb/gal; pH 5.5–7.5 (10%); biodeg.; 30% act.

Carsonam 32. [113] Cocamidopropyl betaine (industrial grade); industrial use; amber, clear liq.; pH 6.0–7.0; 58–62% solids in proprietary ethanol.

Carsonam 33-S. [113] Cocamidopropyl betaine; amphoteric; surfactant, detergent, wetting agent, emulsifier, foaming and coupling agent for shampoos, lotions, germicides, chemical specialties; liq.; dens. 8.45 lb/gal; pH 5.5–7.5 (10%); biodeg.; 30% act.

Carsonam 3147. [113] Cocamidopropyl betaine; amphoteric; surfactant, detergent, wetting agent, emulsifier, foaming and coupling agents for shampoos, lotions, germicides, chemical specialties; 5 max. color liq.; dens. 8.45 lb/gal; biodeg.; 33% act.

Carsonam BCW. [113] Coco fatty betaine;

foaming and conditioning agent used in household and idustrial detergents, shampoos, bubble baths; Gardner 2 liq.; dens. 8.5 lb/gal; sp.gr. 1.044; pH 8.0; biodeg.; 44% solids, 56% water, 8% sodium.

Carsonam C. [113] Coco imidazoline betaine; amphoteric; emulsifier, detergent, sequestrant, mild germistat and fungistat used for human, rug, upholstery shampoos, cleaners; Gardner 5 liq.; pH 9.7; biodeg.; 37% solids.

Carsonam C-SF. [113] Coco imidazoline betaine (salt free); see Carsonam C.

Carsonam C-SPCL. [113] Coco imidazoline betaine; amphoteric; emulsifier, detergent, sequestrant, mild germistat and fungistat used for human, rug, upholstery shampoos, cleaners; Gardner 7 liq.; pH 6.0; biodeg.; 48% solids.

Carsonam DC. [113] Coco imidazoline betaine dicarboxylate; amphoteric; emulsifier, detergent, sequestrant, mild germistat and fungistat used for human, rug, upholstery shampoos, cleaners; Gardner 9 gel; pH 8.5 (5%); biodeg.; 50% solids.

Carsonam DC-SF. [113] Coco imidazoline betaine dicarboxylate (salt free); amphoteric; emulsifier, detergent, sequestrant, mild germistat and fungistat used for human, rug, upholstery shampoos, cleaners; Gardner 8 gel; pH 9.6 (10%); biodeg.; 38% solids.

Carsonam DC-70%-SF. [113] Coco imidazoline betaine dicarboxylate (salt free); amphoteric; emulsifier, detergent, sequestrant, mild germistat and fungistat used for human, rug, upholstery shampoos, cleaners; Gardner 11 gel; pH 9.6 (10%); biodeg.; 70% solids.

Carsonam L. [113] Lauryl imidazoline betaine; emulsifier, detergent, sequestrant, mild germistat and fungistat, amphoteric, used for human, rug, upholstery shampoos, cleaners; Gardner 6 liq.; sp.gr. 1.1; pH 9.3; biodeg.; 43% solids.

Carsonol ALS, ALS Special. [113; 372] Ammonium lauryl sulfate; anionic; detergent with high foam, good wetting and emulsifying properties used for cosmetics, chemical specialties; clear visc. liq.; water sol.; dens. 8.4 lb/gal; pH 6.5–7.0 (10%); biodeg.; 30% act.

Carsonol ANS. [372] Sulfated nonylphenol ethoxylate, aluminum salt; anionic; surfactant used in dishwashing, soaps, and shampoos; liq.; 58% conc.

Carsonol BD. [113] Blend of carboxybetaine, sulfate and amine groups; detergent, foaming agent, anionic/amphoteric, cosmetic uses; lt. amber, moderately visc. liq.; excellent water sol.; cloud pt. 50 F; pH 7.5–8.5; 30–32% act.

Carsonol BDM. [372] DEA-lauryl sulfate, DEA-lauraminopropionate, and sodium lauraminopropionate; anionic; mild shampoos with conditioning; liq.; 38% conc.

Carsonol DLS. [113; 372] DEA lauryl sulfate; anionic; detergent with high foam, good wetting and emulsifying properties, used for cosmetics, chemical specialties; clear liq.; low odor; water sol.; dens. 8.49 lb/gal; pH 7.5–8.5 (10%); biodeg.; 35% act.

Carsonol ILS. [113] Mixed isopropanolamines lauryl sulfate; anionic; detergent with high foam, good wetting and emulsifying properties used for cosmetics, chemical specialties; visc. liq.; low odor; water sol.; dens. 8.63 lb/gal; pH 7.0–7.4 (10%); biodeg.; 75% act.

Carsonol MLS. [113; 372] Magnesium lauryl sulfate; anionic; detergent with high foam, good wetting and emulsifying properties used for cosmetics, chemical specialties; visc. liq.; low odor; water sol.; dens. 8.57 lb/gal; pH 6.5–7.5 (10%); biodeg.; 30% act.

Carsonol SES-A. [113; 372] Ammonium laureth sulfate; anionic; surfactant with excellent foaming in hard and soft water, for cosmetic, household, and industrial uses; liq.; dens. 8.4 lb/gal; pH 6.5–7.0 (10%); biodeg.; 60% act.

Carsonol SES-S. [113; 372] Sodium laureth sulfate; anionic; surfactant with excellent foaming in hard and soft water, for cosmetic, household, and industrial uses; liq.; dens. 8.8 lb/gal; pH 7.0-8.0 (10%); biodeg.; 60% act.

Carsonol SHS. [113; 372] Sodium 2-ethylhexyl sulfate; anionic; low foaming detergent, wetting agent, penetrant, emulsifier used in caustic sol'ns. for peeling of fruits and vegetables; Gardner 3 clear liq.; dens. 9.6 lb/gal; sp.gr. 1.15; visc. 35 cps; pH 10.3 (10%); 40% act.

Carsonol SLES. [113; 372] Sodium laureth sulfate; anionic; surfactant with excellent foaming in hard and soft water, for cosmetic, household, and industrial uses; liq.; dens. 8.7 lb/gal; pH 7.5–8.5 (10%); biodeg; 30% act.

Carsonol SLES-2. [113; 372] Sodium laureth sulfate; anionic; surfactant with excellent foaming in hard and soft water, for cosmetic, household, and industrial uses; liq.; dens.

8.63 lb/gal; pH 7.0–8.0 (10%); biodeg; 30% act.

Carsonol SLS. [113; 372] Sodium lauryl sulfate; anionic; detergent with high foam, good wetting and emulsifying properties used for cosmetics, chemical specialties; clear, liq. low color; low odor; water sol.; dens. 8.7 lb/gal; pH 7.5–8.5 (10%); biodeg.; 30% act.

Carsonol SLS Paste B. [113; 372] Sodium lauryl sulfate; anionic; detergent with high foam, good wetting and emulsifying properties used for cosmetics, chemical specialties; paste, low color; low odor; water sol.; dens. 8.8 lb/gal; pH 8.4–8.8 (10%); biodeg.; 30% act.

Carsonol SLS Special. [113; 372] Sodium lauryl sulfate; anionic; detergent with high foam, good wetting and emulsifying properties used for cosmetics, chemical specialties; liq., low color; low odor; water sol.; dens. 8.7 lb/gal; pH 7.5–8.5 (10%); biodeg.; 30% act.

Carsonol TLS. [113; 372] TEA-lauryl sulfate; anionic; detergent with high foam, good wetting and emulsifying properties used for cosmetics, chemical specialties, emulsion polymerization; clear liq.; low color, odor; water sol.; dens. 8.7 lb/gal; pH 7.0–7.5 (10%); biodeg.; 40% act.

Carsonon L-985. [113] Laureth-9 (9-10 EO); nonionic; wetting agent, detergent, emulsifier, disp. used for maintenance and institutional cleaners, in textile, paper, and paint industries; colorless liq.; sol. in acid and alkaline sol'ns.; dens. 8.32 lb/gal; cloud pt. 137 F; pour pt. 39–43 F; pH neutral (3%); biodeg.; 85% act.

Carsonon LF-5. [113] Linear alcohol alkoxylate; nonionic; low foam surfactant for pulp and paper, mechanical dishwashing, metal cleaning, dairy equipment cleaning; APHA 100 color, mild odor; sol. in aromatic solv., polar solv., water below cloud pt. and in common chlorinated solv.; sp.gr. 1.02 (60F/60 F); visc. 125 cts; cloud pt. 43–49 C; flash pt. 460 F (COC); surf. tens. 33.5 dynes/cm (0.1%); pH 5.5–7.0 (1% aq.); biodeg.

Carsonon LF-46. [113] Polyoxyalkalene glycol; nonionic; see Carsonon LF-5; Gardner 1 color; mild odor; sol. in aromatic solv., polar solv., water below cloud pt. and in the common chlorinated solv.; visc. 200 cps.; cloud pt. 44 C; flash pt. > 230 F; pH 5.5–7.0 (1% aq.); biodeg.

Carsonon N-4. [113; 372] Nonoxynol-4; nonionic; emulsifier, detergent, wetting agent, dispersant for household and industrial uses; pale-colored liq.; sol. in kerosene, alcohols, aromatic and chlorinated solv., Stod. solv.; dens. 8.6 lb/gal; sp.gr. 1.02; visc. 175–250 cps; flash pt. 500–600 F; pour pt. –5 ± 2 F; 100% act.

Carsonon N-6. [113; 372] Nonoxynol-6; nonionic; emulsifier, wetting agent for household and industrial uses; pale-colored liq; mild, pleasant odor; sol. in alcohols, aromatic, chlorinated and aliphatic solv., fatty oil, disp. in water; dens. 8.68 lb/gal; visc. 180–240 cps; cloud pt. 90–102 F; flash pt. 500 F; 100% act.

Carsonon N-8. [113] Nonoxynol-8 (8–8.5 mole); nonionic; emulsifier, detergent, wetting agent, dispersant for household and industrial uses; pale-colored liq.; mild, pleasant odor; sol. in water, alcohol, aromatic, chlorinated and aliphatic solv.; dens. 8.75 lb/gal; visc. 200–300 cps; cloud pt. 82–86 F; flash pt. 435 F; pour pt. 45 F; pH neutral (1%); 100% act.

Carsonon N-9. [113; 372] Nonoxynol-9 (9–10 mole); nonionic; emulsifier, detergent, wetting agent, dispersant for household and industrial uses; pale-colored liq.; mild, pleasant odor; sol. in water, alcohol, aromatic, chlorinated and aliphatic solv.; dens. 8.83 lb/gal; visc. 180–250 cps; flash pt. 540 F; pour pt. 35–39 F; pH neutral (1%); 100% act.

Carsonon N-10. [113; 372] Nonoxynol-10 (10–11 mole); nonionic; emulsifier, detergent, wetting agent, dispersant for household and industrial uses; pale-colored liq.; mild, pleasant odor; sol. in water, alcohols, aromatic, chlorinated and aliphatic solv.; dens. 8.84 lb/gal; visc. 180–250 cps; cloud pt. 143–146 F; flash pt. 540 F; pour pt. 35–39 F; pH neutral (1%); 100% act.

Carsonon N-11. [372] Nonoxynol-11; nonionic; household and industrial cleaning formulations; liq.; water-sol.; 100% conc.

Carsonon N-12. [372] Nonoxynol-12; nonionic; household and industrial cleaning formulations; paste; water-sol.; 100% conc.

Carsonon N-30. [113; 372] Nonoxynol-30; nonionic; emulsifier, detergent, wetting agent for household and industrial uses, emulsion polymerization; solid; mild, pleasant odor; dens. 9.75 lb/gal; sp.gr. 1.17; cloud pt. 212 F; pH neutral (1%); 100% act.

Carsonon N-30, 70%. [113; 372] Nonoxynol-30; emulsifier, stabilizer; pale yel. liq. to semisolid; sp.gr. 1.09; cloud pt. 167–171 F (1%); pH 6.0–7.0 (3%); 70% act., 30% water.

Carsonon N-40, 70%. [113] Nonoxynol-40; emulsifier, stabilizer; wh. to yel. semisolid; cloud pt. 212 F (1%); pH 6.0–7.0 (3%); 70% act., 30% water.

Carsonon N-50, 70%. [113] Nonoxynol-50; emulsifier, stabilizer; wh. to yel semisolid; cloud pt. 212 F (1%); pH 6.0–7.0 (3%); 70% act., 30% water.

Carsonon N-100, 70%. [113] Nonoxynol-100; emulsifier, stabilizer; wh. to off-wh. semisolid; water sol.; cloud pt. 212 F (1%); pH 6.0–7.0 (3%); 70% act., 30% water.

Carsonon ND-317. [113; 372] Nonylphenol ethoxylate; nonionic; surfactant, base for solid toilet bowl cleaners; Gardner 2 opaque, visc. liq.; sp.gr. 1.11; cloud pt. 162±2 F (1% in 10% NaCl); pH 7.5–8.0 (3% aq.); 100% conc.

Carsonon TD-10. [113] Trideceth-10; nonionic; emulsifier, detergent, wetting agent, foam builder, solubilizer used for household, industrial cleaners, textile processing, paper industry; Gardner 1 liq.; water sol.; dens. 8.5 lb/gal; HLB 13.7; cloud pt. 162–166 F (1% aq.); flash pt. > 300 F (COC); pour pt. 67 F; pH 5.0–7.0 (3% aq.); 100% act.

Carsonon TD-11, TD-11 70%. [113; 372] Trideceth-11; nonionic; detergent, foam builder, emulsifier, wetting agent for household and industrial cleaners; Gardner 1 waxy solid; dens. 8.5 lb/gal; HLB 14.6; cloud pt. 137–141 F (1% in 10% NaCl); flash pt. > 300 F (COC); pour pt. 92 F; pH 5.0–7.0 (3% aq.); 100% conc. and 70% aq. sol'n. resp..

Carsoquat 621. [113] *n*-Alkyl dimethyl benzyl ammonium chloride; cationic; corrosion inhibitor, surfactant; almost water-wh. to pale yel. liq.; mild odor; misc. with water, lower alcohols, most polar solv.; dens. 8 lb/gal; sp.gr. 0.96; 50% act.

Carsoquat 621 (80%). [113] Alkyl dimethyl benzyl ammonium chloride; cationic; corrosion inhibitor, surfactant; straw-colored, clear liq.; m.w. 361; pH 7–8 (1%); 80% act.

Carsoquat 816-C. [113] Cetearyl alcohol, PEG-40 castor oil, and stearalkonium chloride; creme rinse conc.; wh. waxy flakes; mild, fatty odor; disp. readily in water; pH 6.3 (5% disp.); 96% min act.

Carsoquat CTM-29. [113] Cetyl trimethyl ammonium chloride; cationic surfactant, conditioners in specialty hair preparations, coagulating agent in the mfg. of antibiotics and other pharmaceuticals; almost clear to pale yel. liq.; m.w. 319; dens. 8.1 lb/gal; sp.gr. 0.968; pH 7.0–9.0 (10% act.); 29% act.

Carsoquat CTM-429. [113] Cetyl trimethyl ammonium chloride; see Carsoquat CTM-29; pH 3.5–4.0 (2% act.); 29% act.

Carsoquat SDQ-25. [113] Stearalkonium chloride; softener, creme hair rinse and conditioner for cosmetics and natural fibers; wh., thick creamy paste, mild, sweet odor; pH 3.0–4.0 (1% disp.); 20–21% act.

Carsoquat SDQ-85. [113] Stearalkonium chloride; softener, hair conditioner; wh. waxy flake; mild, sweet odor; m.p. 140 F; pH 5.0–7.0 (0.5%); 85% act.

Carsosoft S-75. [113] Quaternium-27; cationic; fabric softener base, excellent rewet properties and nonyellowing; Gardner 4 liq.; m.w. 720; pH 5–7 (5%); 75% act. in IPA.

Carsosoft S-90. [113] Quaternium-27; yel. semisolid; m.w. 723; pH 5.0–7.0 (5% aq.); 85–88% act.

Carsosoft S-90-M [113] Modified alkylamidoethyl alkyl imidazolinium methyl methosulfate; cationic; fabric softener base, excellent rewet properties and nonyellowing; Gardner 4 liq.; readily disp. in warm and cold water; pH 5.0–7.0 (5%); 88–91% act. in IPA.

Carsosoft T-90. [113] Quat. ditallow diamido; cationic; fabric softener; yel. liq.; pH 5–7 (5%); 88–90% act. in IPA.

Carsosoft V-90. [113] Quaternium-18 methosulfate and IPA; wh., Gardner 4 paste; pH 5.5 (10% in 1:1 IPA/aq.); 91% solids.

Carsosoft V-100. [113] Quaternium-18 methosulfate; wh. flake or powd.; pH 5.5–6.5 (10% in 1:1 IPA/aq.); 95% solids.

Carsosulf SXS-Liq. [113] Sodium xylene sulfonate; anionic; coupling agent, solubilizer, nonflam. coupling agent and solubilizer in liq. detergent systems; liq.; sp.gr. 1.18–1.22; pH 7.5–10.5; 40% act.

Carsosulf UL-100 Acid. [113] Linear dodecylbenzene sulfonic acid; m.w. 316; dens. 8.84 lb/gal; sp.gr. 1.06; biodeg.; 97 ± 1% act.

Cartaretin F-4. [537] Adipic acid/dimethylaminohydroxypropyl diethylenetriamine copolymer; cationic polymer, conditioning agent for hair and skin, as wetting agent for sizing; lt. yel. liq.; natural odor; sol. in water and alcohol; sp.gr. 1.07; visc. 400 cps; f.p. -4 C; b.p. 220 F; 30% act.

Cartaretin F-8. [537] Adipic acid/dimethylaminohydroxypropyl diethylenetriamine copolymer; cationic polymer, conditioning agent for hair and skin, as wetting agent for sizing; amber liq.; natural odor; sol. in water and alcohol; sp.gr. 1.07; visc. 800 cps; f.p. -4 C; b.p. 220 F; 30% act.; US Patent 3,632,559.

Castor Oil Sulfated 50%. [145] Sulfated castor oil; anionic; emulsifier, dyeing assistant, dye dispersant; liq.; 50% min. conc.

Castor Oil Sulfated 68%. [145] Sulfated castor oil; anionic; emulsifier, dyeing assistant, dye dispersant; liq.; 69.5% min. conc.

Casul 70 HF. [456; 603] Calcium dodecylbenzene sulfonate; anionic; coemulsifier for o/w and w/o formulations; amber liq.; dens. 8.5 lb/gal; sp.gr. 1.02; visc. 18,400 cps; HLB 10.5 flash pt. 104 F (PMCC); pour pt. 46 F; pH 5–7; 70% act.

Catinal HTB. [606] Alkyl trimethyl ammonium bromide; cationic; dyeing assistant for polyacrylonitrile fiber dyeing; liq.; 30% conc.

Catinex KB-10. [494] Nonylphenol, ethoxylated; nonionic; antifoaming and intermediate emulsifier; liq.; HLB 6.6; 100% conc.

Catinex KB-11. [494] Nonylphenol, ethoxylated; nonionic; detergent intermediate emulsifier; liq. surface act. agent used in petroleum products; liq.; HLB 8.9; 100% conc.

Catinex KB-13. [494] Nonylphenol, ethoxylated; nonionic; drycleaning agent; emulsifier for insecticides; detergent; liq.; HLB 10; 100% conc.

Catinex KB-15. [494] Nonylphenol, ethoxylated; nonionic; detergent, dispersant, and wetting agent; solv. cleaner formulations; emulsifier and stabilizer for paint additives; de-icing fluid and sludge dispersing additive in petroleum products; liq.; HLB 10.9; 100% conc.

Catinex KB-16. [494] Nonylphenol, ethoxylated; nonionic; detergent, dispersant, emulsifier and wetting agent; surface act. agent; liq.; water-sol.; HLB 12.3; 100% conc.

Catinex KB-18. [494] Nonylphenol, ethoxylated; nonionic; see Catinex KB-16; liq.; HLB 12.8; 100% conc.

Catinex KB-19. [494] Nonylphenol, ethoxylated; nonionic; see Catinex KB-16; liq.; HLB 13.1; 100% conc.

Catinex KB-20. [494] Nonylphenol, ethoxylated; nonionic; see Catinex KB-16; liq.; HLB 13.3; 100% conc.

Catinex KB-22. [494] Nonylphenol, ethoxylated; nonionic; see Catinex KB-16; liq.; HLB 14.1; 100% conc.

Catinex KB-23. [494] Nonylphenol, ethoxylated; nonionic; see Catinex KB-16; liq., paste; HLB 14.7; 100% conc.

Catinex KB-24. [494] Nonylphenol, ethoxylated; nonionic; see Catinex KB-16; solid; HLB 16; 100% conc.

Catinex KB-25. [494] Nonylphenol, ethoxylated; nonionic; see Catinex KB-16; solid; HLB 17.1; 100% conc.

Catinex KB-26. [494] Nonylphenol, ethoxylated; nonionic; see Catinex KB-16; solid; HLB 16.7; 100% conc.

Catinex KB-27. [494] Nonylphenol, ethoxylated; nonionic; detergent, dispersant, and wetting agent; emulsifier for oils, fats, and waxes; metal cleaner and degreaser; solid; HLB 15.0; 100% conc.

Catinex KB-31. [494] Nonylphenol, ethoxylated; nonionic; surfactant for foam control; coemulsifier; dispersant for petroleum oils; liq.; HLB 4.6; 100% conc.

Catinex KB-32. [494] Nonylphenol, ethoxylated; nonionic; see Catinex KB-15; liq.; HLB 11.6; 100% conc.

Catinex KB-40. [494] Octylphenol, ethoxylated; nonionic; detergent, emulsifier, dispersant, surface act. agent used in washing formulations and emulsion cleaners; liq.; HLB 11.2; 100% conc.

Catinex KB-41. [494] Octylphenol, ethoxylated; nonionic; detergent, emulsifier, wetting agent for tank cleaners; additive in acid cleaning compds. for metals and lt. duty detergents; liq.; HLB 12.6; 100% conc.

Catinex KB-42. [494] Octylphenol, ethoxylated; nonionic; wetting agent, detergent, dispersant, emulsifier for household and industrial cleaners, textile processes, wool scouring; emulsifier for insecticides and herbicides; liq.; HLB 13.3; 100% conc.

Catinex KB-43. [494] Octylphenol, ethoxylated; nonionic; used in hot cleaning systems, metal pickling operations and electrolytic cleaning; paste; HLB 14.7; 100% conc.

Catinex KB-44. [494] Octylphenol, ethoxylated; nonionic; detergent, wetting agent, coemulsifier and stabilizer for vinyl-acrylic latex; emulsifier for fats and waxes; solid; HLB 17.3; 100% conc.

Catinex KB-45. [494] Octylphenol, ethoxylated; nonionic; primary emulsifier for vinyl acetate and acrylate polymerization; post stabilizer for syn. latex; dyeing assistant and emulsifier for fats and waxes; solid; HLB

17.9; 100% conc.

Catinex KB-48. [494] Octylphenol, ethoxylated; nonionic; see Catinex KB-43; solid; HLB 15.7; 100% conc.

Catinex KB-49. [494] Octylphenol, ethoxylated; nonionic; emulsifier, detergent, dispersant, surface act. agent used in washing formulations and emulsion cleaners; liq.; HLB 10.2; 100% conc.

Catinex KB-50. [494] Octylphenol, ethoxylated; nonionic; emulsifier, detergent, wetting agent; coemulsifier and stabilizer for vinyl-acrylic latex; emulsifier for fats and waxes; solid; HLB 16.2; 100% conc.

Catinex KB-51. [449] Octylphenol, ethoxylated; nonionic; emulsifier, detergent, dispersant, surface act. agent used in wax based washing formulations and emulsion cleaners; liq.; HLB 9.2; 100% conc.

Cationic Guar C-261. [289] Guar hydroxypropyl trimonium chloride; cationic; thickener, emulsion stabilizer and additive for shampoos; solid; 100% conc.

Catisol AO/100. [576] N-alkyl amine acetate; cationic; flotation, flocculation, corrosion inhibitor, emulsifier, antistat; paste; 100% conc.

CC-16. [447] Sodium salt of a humic acid material; anionic; drilling fluid additive, thinner and emulsifier for drilling muds, gel dispersing agent; powd.; sol. in water-based drilling fluids; 100% act.

CE-55-2. [287] Ceteth-2; nonionic; emulsifier for cosmetic and pharmaceutical creams and lotions; antistat for plastics and textile industry; coemulsifier for leather industry; solid; 100% conc.

CE-55-20. [287] Ceteth-20; nonionic; produces emulsifying waxes; solid; HLB 16.0; 100% conc.

Cedemide AX. [199] Lauric diethanolamide; nonionic; foam stabilizer and visc. modifier in liq. detergents and shampoos; wax; 100% conc.

Cedemide B. [199] Alkanolamide; nonionic; detergent; base for mfg. of floor and general purpose cleaners; liq.; 100% conc.

Cedemide CX. [199] Alkanolamide; nonionic; foam and visc. booster; liq.; 100% conc.

Cedemide DX. [199] Coconut oil diethanolamide; nonionic; foam and visc. booster; detergent; liq.; 100% conc.

Cedemide MX. [199] Lauric monoethanolamide; nonionic; foam stabilizer in liq. detergents; solid; 100% conc.

Cedepal 710. [199] Nonoxynol-10; nonionic; detergent, wetting, emulsifier, water surfactant; liq.; HLB 13.6; 100% conc.

Cedepal CA-210. [199] Octoxynol-1 (1.5 EO); nonionic; emulsifier for nonpolar hydrocarbon; liq.; HLB 3.5; 100% conc.

Cedepal CA-520. [199] Octoxynol-5; nonionic; emulsifier, detergent, dispersant, surfactant; liq.; oil-sol.; HLB 10.0; 100% conc.

Cedepal CA-630. [199] Octoxynol-9; nonionic; deteregent, wetting, emulsifier, surfactant; liq.; water-sol.; HLB 13.0; 100% conc.

Cedepal CA-720. [199] Octoxynol-12; nonionic; detergent; liq.; HLB 14.6; 100% conc.

Cedepal CA-890. [199] Octoxynol-40; nonionic; primary emulsifier for vinyl acetate and acrylate polymerization; wax; HLB 18.0; 100% conc.

Cedepal CA-897. [199] Octoxynol-40; nonionic; emulsifier, stabilizer; liq.; HLB 18.0; 70% conc.

Cedepal CO-210. [199] Nonoxynol-1 (1.5 EO); nonionic; defoaming, dispersant, detergent, emulsifier, stabilizer intermediate; surfactant and chemical intermediate; liq.; oil-sol.; HLB 4.6; 100% conc.

Cedepal CO-430. [199] Nonoxynol-4; nonionic; detergent, emulsifier, dispersant, stabilizer intermediate; surfactant and chemical intermediate; liq.; HLB 8.8; 100% conc.

Cedepal CO-436. [199] Ammonium alkylaryl ether sulfate; anionic; detergent, wetting agent, lime soap dispersant; emulsifier in polymers; liq.; 60% conc.

Cedepal CO-500. [199] Nonylphenol ethoxylate; nonionic; emulsifier, detergent, dispersant, stabilizer, intermediate; surfactant; oil-sol.; liq.; HLB 10.0; 100% conc.

Cedepal CO-530. [199] Nonoxynol-6; nonionic; emulsifier, detergent, dispersant; surfactant and intermediate; liq.; oil-sol.; HLB 10.8; 100% conc.

Cedepal CO-610. [199] Nonoxynol-8; nonionic; detergent, wetting, emulsifier, surfactant for detergency; liq.; water-sol.; HLB 12.2; 100% conc.

Cedepal CO-630. [199] Nonoxynol-9; nonionic; see Cedepal CO-610; liq.; HLB 13.0; 100% conc.

Cedepal CO-730. [199] Nonylphenol ethoxylate; nonionic; see Cedepal CO-610; liq.; HLB 15.0; 100% conc.

Cedepal CO-880. [199] Nonylphenol ethoxylate; nonionic; detergent, wetting, emulsifier;

wax; HLB 17.2; 100% conc.

Cedepal CO-887. [199] Nonylphenol ethoxylate; nonionic; detergent, wetting, emulsifier; liq.; HLB 17.2; 70% conc.

Cedepal CO-890. [199] Nonylphenol ethoxylate; nonionic; emulsifier, stabilizer; wax; HLB 17.8; 100% conc.

Cedepal CO-977. [199] Nonylphenol ethoxylate; nonionic; emulsifier, stabilizer; liq.; HLB 18.2; 70% conc.

Cedepal CO-990. [199] Nonylphenol ethoxylate; nonionic; emulsifier, stabilizer, surfactant; wax; water-sol.; HLB 19.0; 100% conc.

Cedepal CO-997. [199] Nonylphenol ethoxylate; nonionic; emulsifier, stabilizer; liq.; HLB 19.0; 70% conc.

Cedepal E-200. [199] Primary alcohol ethoxylate; nonionic; emulsifier, detergent, wetting and foam stabilizing; liq.; HLB 4.0; 100% conc.

Cedepal E-300. [199] Primary alcohol ethoxylate; nonionic; emulsifier, detergent, wetting and foam stabilizing; liq.; HLB 6.0; 100% conc.

Cedepal E-400. [199] Primary alcohol ethoxylate; nonionic; emulsifier, detergent, wetting and foam stabilizing; liq.; HLB 8.0; 100% conc.

Cedepal E-710. [199] Primary alcohol ethoxylate; nonionic; emulsifier, detergent, wetting and foam stabilizing; wax; 100% conc.

Cedepal EL-719. [199] Veg. oil ethoxylate; nonionic; emulsifier, dispersant, lubricant; liq.; 96% conc.

Cedepal EN 600. [199] Primary alcohol ethoxylate; nonionic; emulsifier, detergent, wetting and foam stabilizing; liq.; 100% conc.

Cedepal ET-630. [199] Primary alcohol ethoxylate; nonionic; emulsifier, detergent, wetting and foam stabilizing; liq.; HLB 12.5; 100% conc.

Cedepal FA-406. [199] Primary alcohol ethoxylate; nonionic; emulsifier, detergent, wetting and foam stabilizing; liq.; HLB 12.8; 100% conc.

Cedepal FS-406. [199] Sodium alkyl ether sulfate; anionic; surfactant for high electrolyte applications; liq.; 60% conc.

Cedepal ON-870. [199] Fatty alcohol ethoxylate; nonionic; emulsifier, dispersant, lubricant; wax; 100% conc.

Cedepal ON-877. [199] Fatty alcohol ethoxylate; nonionic; solubilizer for essential oils; liq.; 70% conc.

Cedepal SA-406. [199] Ammonium laureth sulfate; anionic; detergent, dispersant, and wetting agent used in shampoos, dishwashers, car washers; liq.; 60% conc.

Cedepal SD-409. [199] DEA laureth sulfate; used in applications where water content is carefully controlled; liq.; 90% conc.

Cedepal SS-203. [199] Sodium laureth sulfate; anionic; detergent, dispersant, and wetting agent used in shampoos, dishwashers, car washers; liq.; 25% conc.

Cedepal SS-306. [199] Sodium laureth sulfate; anionic; see Cedepal SS-203; liq.; 60% conc.

Cedepal SS-403. [199] Sodium laureth sulfate; anionic; see Cedepal SS-203; liq.; 25% conc.

Cedepal SS-406. [199] Sodium laureth sulfate; anionic; see Cedepal SS-203; liq.; 60% conc.

Cedepal TD-400. [199] Primary alcohol ethoxylate; anionic; emulsifier; liq.; oil-sol.; HLB 8.0; 90% conc.

Cedepal TD-404. [199] Sodium trideceth sulfate; anionic; detergent, wetting agent; liq.; 34% conc.

Cedepal TD-407. [199] Sodium trideceth sulfate; anionic; detergent, wetting agent; liq.; 75% conc.

Cedepal TD-630. [199] Primary alcohol ethoxylate; nonionic; emulsifier, detergent; liq.; HLB 12.5; 100% conc.

Cedepon A 100. [199] Linear alkyl benzene sulfonic acid; anionic; detergent; liq.; 96% conc.

Cedepon AM. [199] Linear alkyl benzene sulfonic acid, amine salt; anionic; emulsifier, detergent for solv. based systems; liq.; 93% conc.

Cedepon LA-30. [199] Ammonium lauryl sulfate; anionic; detergent, emulsifier, dispersant for personal care products; liq.; 28% conc.

Cedepon LS-30, -30P. [199] Sodium lauryl sulfate; anionic; see Cedepon LA-30; liq.; 30% conc.

Cedepon LT-40. [199] TEA lauryl sulfate; anionic; used in shampoos and detergents; liq.; 40% conc.

Cedepon SX-55. [199] Sodium linear alkyl benzene sulfonate; anionic; used where neutralization of acid is impractical; slurry; 55% conc.

Cedepon T. [199] TEA linear alkyl benzene sulfonate; anionic; detergent; liq.; 60% conc.

Cedepon TS. [199] TEA linear alkyl benzene sulfonate plus an alkanolamide; detergent; liq.; 60% conc.

Cellopal 100. [587] Polyethoxy alkylphenol sulfonate, triethanolamine salt; anionic; emulsifier, detergent, dispersant, wetting, and foaming agent; liq.; 100% conc.

Cenegen 7. [163] Alkylaryl sulfonate; anionic; retarding and leveling agent for nylon dyeing; good barre coverage; liq.

Cenegen NWA. [163] Aryl polyethoxy condensate; dyeing assistant for premetallized dyes on nylon and wool; liq

Cenekol Liquid, NCS Liquid. [163] Sulfonated phenolic condensate; acid dye fixative for nylon and stable to electrolytes; liq

Centrocap. [121] Lecithin; amphoteric; natural, sediment-free food emulsifier; liq.; 100% conc.

Centrol Series. [121] Natural soybean lecithin; amphoteric; wetting agent, emulsifier, emollient, softening agent, lubricant, solubilizer, used in cosmetics, food processing, pharmaceuticals, industrial uses; liq.; 100% act.

Centrolene Series. [121] Lecithin; amphoteric; wetting agent, emulsifier, release agent, disp. agent, lubricant, anticorrosive agent, emollient, used in food processing, pharmaceuticals, industrial uses; liq.; 100% act.

Centrolex Series. [121] Lecithin; amphoteric; wetting agent, emulsifier, dietary supplement, dispersant, emollient, used in cosmetics, food processing, pharmaceuticals, industrial uses; granular; 100% act.

Centrolex P. [121] Soybean lecithin; wetting agent for magnetic media industry; dry granular; sol. in THF, cyclohexanone, toluene, MIBK; 97% act.

Centromix Series. [121] Standard soybean lecithin with selected surfactants; emulsifier, wetting agent, dispersant, penetrant, used in cosmetics, food processing, pharmaceuticals, industrial uses; liq., paste; 100% act.

Centrophase Series. [121] Lecithin; amphoteric; food emulsifier, natural release and wetting agent; liq.; 100% conc.

Centrophase 31. [121] Lecithin; wetting agent for the magnetic media industry; fluid; sol. in THF, cyclohexanone, toluene; 62% act.

Centrophase HR. [121] Lecithin, heat resistant; multifunctional ingredient; food substance; lubricant and release agent for heated surfaces; Gardner 13; sol. in liq. veg. and min oil, aliphatic, aromatic and halogenated hydrocarbons, petroleum ether; dens. 1.01 g/cc; visc. 2000 cP; pour pt. 32 F; 52% acetone insol

Centrophil Series. [121] Refined soybean lecithin with selected carriers; wetting agent, emulsifier, dispersant, release agent, dietary supplement, used for cosmetics, food processing, pharmaceuticals, industrial uses; mostare food approved, some for industrial use only; liq., solid, other; 100% act.

Ceralan. [37] Lanolin alcohols; nonionic; w/o emulsifier; wax; 100% conc.

Ceramol. [4] Partially sulfonated fatty alcohol from hydrog. veg. oil; emulsifier for cosmetics; flake.

Ceranine HC Hi Conc. [537] Emulsifier for skin care products; tan solid; slight odor; disp. in water; m.p. 55–65 C; 100% act.

Ceranine HCA. [537] Fatty amide derivative; emulsifier for skin care products, softener for textiles, cationic; cream-colored granules; slight odor; disp. in water; m.p. 71–81 C; 100% act.

Cerasynt 303. [629] Diethylaminoethyl stearate; nonionic; emulsifier, dispersant, and wetting agent; liq.; 100% conc.

Cerasynt 840. [629] PEG-20 stearate; nonionic; emulsifier for creams and lotions; solid; 100% conc.

Cerasynt 945. [629] Glyceryl stearate and laureth-23; nonionic; emulsifier and visc. controller; flake; 100% conc.

Cerasynt D. [629] Stearamide MEA-stearate; nonionic; emulsifier, thickener and opacifier for shampoos and hydrocarbon propellants; flake; 100% conc.

Cerasynt IP. [629] Glycol stearate and other ingredients; nonionic; emulsifier, opacifier and pearling agent for shampoos; flake; 100% conc.

Cerasynt M. [629] Glyceryl stearate; nonionic; emulsifier for cosmetics and pharmaceuticals; flake; 100% conc.

Cerasynt MN. [629] Glyceryl stearate SE; nonionic; emulsifier for cosmetics and pharmaceuticals; flake; 100% conc.

Cerasynt PA. [629] Propylene glycol monostearate; nonionic; emulsifier; flake; 100% conc.

Cerasynt Q. [629] Glyceryl stearate SE; anionic; emulsifier for creams and lotions; flake; 100% conc.

Cerasynt SD. [629] Glyceryl stearate; nonionic; emulsifier for cosmetic and pharmaceutical; flake; 100% conc.

Cerasynt WM. [629] Glyceryl stearate, stearyl alcohol, and sodium lauryl sulfate; anionic; emulsifier; flake; 100% conc.

Cerfak 1400. [305] Alkyl POE ether; nonionic; detergent, wetting agent used in wet processing operations of textiles; liq.; sol. in water and Stoddard solv.; 85% conc.

Cerfax N-100. [305] Fatty acid polyethanolamine condensate; nonionic; detergent, wetting agent used in textile scouring and dyeing; liq.; 100% conc.

Cetal. [37] Cetyl alcohol, NF; emollient for emulsions, oils, makeup; aux. emulsifier; wh. waxy solid, mild, char. odor; m.p. 45–51 C; acid no. 1 max.; sapon. no. 2 max.

Cetalox 8, 25. [654] Fatty alcohol, ethoxylated; nonionic; detergent; paste

Cetalox 50. [653] Fatty alcohol, ethoxylated; nonionic; nonfoaming detergent; flakes

Cetats. [297] Cetrimonium tosylate; cationic; quat. ammonium compd. used as an additive in toothpaste formulations; wh. powd.; char. odor; m.w. 455.72; pH 5–8; 99% min. assay

Cetina. [521] Cetyl esters and stearamide DEA; nonionic; emulsifier, lubricant, emollient for cosmetics; flake; 100% conc.

Cetodan. [273] Acetylated monoglycerides; nonionic; food emulsifier, aerating agent for shortenings, toppings, cakes; edible coating; plasticizer for chewing gum base; antifoam agent, lubricant; solid, liq.; HLB 1.5; 100% conc.

Cetodan 50-00A. [273] Monoglyceride acetic acid ester; food emulsifier; aerating and foam stabilizing agent in food products; block; m.p. 40 C; sapon. no. 280

Cetodan 70-00A. [273] Monoglyceride acetic acid ester; see Cetodan 50-00A; m.p. 35 C; sapon. no. 315

Cetodan 90-40. [273] Monoglyceride acetic acid ester; see Cetodan 50-00A; m.p. 10 C; sapon. no. 370

Cetomacrogol 1000 BP. [160] Ceteareth-20 BP; nonionic; pharmaceutical emulsifier o/w type; solid; HLB 15.6; m.w. 1000; 100% conc.

Cetomacrogol Wax BP. [160] Cetearyl alcohol and ceteth-20; nonionic; pharmaceutical emulsifying wax; solid; HLB 9.0; 100% conc.

Charlab Condensate-K. [116] Alkanolamide; nonionic; wetting agent, emulsifier, scouring agent; amber liq.; water sol.; sp.gr. 1.01; 100% act.

Charlab Leveler AT Special. [116] Sodium hydrocarbon sulfonate; anionic; detergent, emulsifier, textile leveling agent, softener, lubricant; amber solid; sol. in water (160 F); sp.gr. 1.0; m.p. 110–125; 80% act.

Charlab Leveler DSL. [116] Sodium hydrocarbon sulfonate; anionic; detergent, emulsifier, textile leveling agent, softener, lubricant; lt. br. paste; sol. in water (160 F); dens. 1.03 lb/gal; 40% act.

Charlab LPC. [116] Lauryl pyridinium chloride; cationic surfactant; lt. br. solid; pyridine odor; water sol.; f.p. 45 C; 97–100% act.

Chemadene NA-30. [515] Cocamidopropyl betaine; surfactant, solubilizer, emulsifier, and thickener used in cleansers, cleaners, personal care products, industrial and textile applications; Gardner 3 liq.; dens. 8.7 lb/gal; pH 5–6; 37% solids

Chemadox C. [515] Cocamine oxide; foam stabilizer, visc. builder for cosmetic shampoos, bubble baths, creams and lotions; Klett 50 max. color; dens. 7.99 lb/gal; pH 6.7; 30% solids.

Chemadox L. [515] Lauramine oxide; foam stabilizer, visc. builder for cosmetic shampoos, bubble baths, creams and lotions; Klett 50 max. color; dens. 7.99 lb/gal; pH 6.7; 30% solids.

Chemadox M. [515] Myristamine oxide; foam stabilizer, visc. builder for cosmetic shampoos, bubble baths, creams and lotions; Klett 50 max. color; dens. 7.99 lb/gal; pH 6.7; 30% solids.

Chemal BP-261, -262, -262LF, -263, -264. [122] Difunctional blocked polymers; nonionic; defoamer, emulsifier, demulsifier, dispersant, binder, stabilizer, wetting agent, and chemical intermediate; liq., cloud pt. 24, 32, 28, 34, and 58 C resp.; 100% act.

Chemal BP-268, -268/50. [122] Difunctional blocked polymer; nonionic; see Chemal BP-261; solid and liq. resp.; cloud pt. > 100 C; 100% conc.

Chemal DA-4. [122] POE (4) decyl ether; nonionic; wetting and penetrating agent for textile processing, clay soils, and fire fighting products; emulsifier for polyethylene emulsions; liq.; hyd. no. 165–185; 100% conc.

Chemal DA-6. [122] POE (6) decyl ether; see Chemal DA-4; liq.; hyd. no. 131–145; 100% conc.

Chemal DA-9. [122] POE (9) decyl ether; see Chemal DA-4; liq.; hyd. no. 95–110; 100% conc.

Chemal LA-4. [122] Laureth-4; nonionic; o/w emulsifier, lubricant, detergent for cosmetic, silicone polish, and mold release products;

liq.; hyd. no. 150–165; 100% conc.

Chemal LA-9. [122] Laureth-9; nonionic; see Chemal LA-4; liq.; hyd. no. 90–110; 100% conc.

Chemal LA-12. [122] Laureth-12; nonionic; see Chemal LA-4; liq.; hyd. no. 72–87.

Chemal LA-23. [122] Laureth-23; nonionic; see Chemal LA-4; solid; hyd. no. 40–55.

Chemal LF Series. [122] Alkoxylated linear alcohol; nonionic; rinse aids in mechanical dishwashing, spray metal cleaning formulations and detergent products; liq.; 100% conc.

Chemal LFL-10, -17, -19, -28, -38, -47. [122] Alkoxylated linear alcohol; nonionic; used in rinse aids and mechanical dishwashing detergents, spray metal cleaning formulations and detergent products; cloud pt. 10, 17, 19, 28 38, and 47 C resp.

Chemal OA-4. [122] Oleth-4; nonionic; dispersant, detergent; emulsifier and solubilizer for topical cosmetic applications; stabilizer and anticoagulant for natural and synthetic latices; emulsifier for waxes used in coating citrus fruit; liq., semisolid.

Chemal OA-5. [122] Oleth-5; nonionic; emulsifier, lubricant, and solubilizer; liq.; 100% conc.

Chemal OA-9. [122] Oleth-9; see Chemal OA-4; liq

Chemal OA-20. [122] Oleth-20; see Chemal OA-4; also lubricant; semisolid

Chemal OA-20/70. [122] Oleth-20; nonionic; emulsiifer, lubricant, solubilizer; 70% conc.

Chemal TDA-3. [122] Trideceth-3; nonionic; wetting agent, detergent, emulsifier, dispersant, foam stabilizer; solubilizer, penetrant for scouring and dye leveing in textiles, in cleaning and dishwashing compds.; liq.; 100% conc.

Chemal TDA-6. [122] Trideceth-6; nonionic; see Chemal TDA-3; liq.; 100% conc.

Chemal TDA-9. [122] Trideceth-9; nonionic; see Chemal TDA-3; liq.; 100% conc.

Chemal TDA-12. [122] Trideceth-12; nonionic; see Chemal TDA-3; liq.; 100% conc.

Chemal TDA-15. [122] Trideceth-15; nonionic; see Chemal TDA-3; solid; 100% conc.

Chemal TDA-18. [122] Trideceth-18; nonionic; see Chemal TDA-3; 100% conc.

Chemannate C-60B. [123] Calcium dodecylbenzene sulfonate; anionic; base for emulsifiers; amber visc. liq.; oil sol.; dens. 8.45 lb/gal; flash pt. >100 F; pH 4.0–7.0 (5% in 25% IPA); 60% act.

Chemannate C-70B. [123] Calcium dodecylbenzene sulfonate; anionic; base for emulsifiers, used for agric. chemicals; dk. amber visc. liq.; oil sol.; visc. 3500 cps (140 F); flash pt. >100 F; pH 3.0–6.0 (5% in 25% IPA); 70% act.

Chemax CO-5. [122] POE (5) castor oil; nonionic; emulsifier, lubricant for textiles; pigment dispersant in latex paints; essential oils solubilizer; liq.; oil-sol.; HLB 3.8; 100% conc.

Chemax CO-16. [122] POE (16) castor oil; nonionic; emulsifier for fiber lubricants; cutting oils and hydraulic fluids; liq.; water-sol.; HLB 8.6; 100% conc.

Chemax CO-25. [122] POE (25) castor oil; nonionic; emulsifier; pigment dispersant in textiles; liq.; HLB 10.5

Chemax CO-28. [122] POE (28) castor oil; nonionic; see Chemax CO-25; HLB 11.1

Chemax CO-30. [122] POE (30) castor oil; nonionic; see Chemax CO-25; liq.; HLB 11.7; 100% conc.

Chemax CO-36. [122] POE (36) castor oil; see Chemax CO-25; liq.; HLB 12.6

Chemax CO-40. [122] POE (40) castor oil; nonionic; see Chemax CO-25; liq.; HLB 12.9; 100% conc.

Chemax CO-80. [122] POE (80) castor oil; nonionic; see Chemax CO-25; solid; HLB 15.8; 100% conc.

Chemax CO-200/50. [122] POE (200) castor oil; nonionic; see Chemax CO-80; liq.; HLB 18.1; 50% conc.

Chemax CS-10. [122] Silicone defoamer; industrial, cosmetic, and food applications; emulsion; 10% act.

Chemax CS-30. [122] Silicone defoamer; industrial, cosmetic, and food applications; emulsion; 30% conc.

Chemax CS-100. [122] Silicone compd.; see Chemax CS-10; liq.; 100% act.

Chemax DNP-8. [122] Nonyl nonoxynol-8; nonionic; emulsifier for nonpolar solv. and oils; detergent for cellulosic and synthetic fibers; dispersant for hard surface cleaners and laundry compds.; liq.; pour pt. 32 F; 100% conc.

Chemax DNP-15. [122] Nonyl nonoxynol-15; see Chemax DNP-8; liq.; pour pt. 64 F.

Chemax DNP-18. [122] Nonyl nonoxynol-18; nonionic; emulsifier, detergent, solubilizer; solid; 100% conc.

Chemax DNP-150. [122] Nonyl nonoxynol-150; see Chemax DNP-8; solid.

Chemax DOSS-75E. [122] Dioctyl sodium sulfosuccinate; anionic; wetting and rewetting agent, detergent, emulsifier for emulsion polymerization; liq.; 75% conc.

Chemax E-200 ML. [122] PEG-4 laurate; nonionic; emulsifier for min. and cutting oils; dispersant, detergent, lubricant; coemulsifier and defoamer in water-based coatings; liq.; HLB 9.8; 100% conc.

Chemax E-200 MO. [122] PEG-5 oleate; nonionic; oil-sol. emulsifier for min. and fatty oils, solv.; degreaser, dispersant, detergent, lubricant; liq.; HLB 8.0; 100% conc.

Chemax E-200 MS. [122] PEG-5 stearate; nonionic; emulsifier for min. oils and fats used in polishes and metal buffing compds.; dye assistant, lubricant, softner, antistat; soft solid; HLB 7.9; 100% conc.

Chemax E-400 ML. [122] PEG-8 laurate; nonionic; visc. control agent in plastisol formulations; wetting agent and defoamer in latex paint; see also Chemax E-200 ML; liq.; HLB 13.1; 100% conc.

Chemax E-400 MO. [122] PEG-9 oleate; nonionic; emulsifier and lubricant for solv. and oils in pesticides and metal cleaners; see also Chemax E-200 MO; liq.; HLB 11.4; 100% conc.

Chemax E-400 MS. [122] PEG-9 stearate; nonionic; lubricant and softener for synthetic fibers see Chemax E-200 MS; soft solid; HLB 11.6; 100% conc.

Chemax E-600 ML. [122] PEG-14 laurate; nonionic; water-sol. emulsifier and dispersant; liq.; HLB 14.9; 100% conc.

Chemax E-600 MO. [122] PEG-14 oleate; nonionic; surfactant used as coemulsifier and lubricant in industrial formulations; liq.; HLB 13.5; 100% conc.

Chemax E-600 MS. [122] PEG-14 stearate; nonionic; emulsifier used in cosmetic and textile formulations; soft solid; water-sol.; HLB 13.6; 100% conc.

Chemax E-1000 MO. [122] PEG-20 oleate; nonionic; see Chemax E-200 MO; solid; 100% conc.

Chemax E-1000 MS. [122] PEG-20 stearate; emulsifier used in cosmetic and textile formulations; solid; water-sol.; HLB 15.6; 100% conc.

Chemax HCO-5. [122] PEG-5 hydrog. castor oil; nonionic; emulsifier, lubricant, softener, dispersant; coemulsifier for syn. esters; liq.; oil-sol.; HLB 3.8; 100% conc.

Chemax HCO-16. [122] PEG-16 hydrog. castor oil; nonionic; emulsifier, lubricant, dispersant, softener; liq.; HLB 8.4; 100% conc.

Chemax HCO-25. [122] PEG-25 hydrog. castor oil; nonionic; emulsifier, lubricant, and softener; liq.; HLB 10.8; 100% conc.

Chemax HCO-200/50. [122] PEG-200 hydrog. castor oil; nonionic; emulsifier and lubricant; liq.; water-sol.; HLB 18.1; 50% act.

Chemax NP-1.5. [122] POE (1.5) nonylphenol; nonionic; surfactant, emulsion stabilizer in latex emulsions; detergent and dispersant for petroleum oils; emulsifier for silicone and agric. products; coupling and wetting agent; liq.; pour pt. –3 F.

Chemax NP-4. [122] Nonoxynol-4; nonionic; see Chemax NP-1.5; liq.; water oil sol.; pour pt. –15 F; 100% conc.

Chemax NP-6. [122] Nonoxynol-6; nonionic; see Chemax NP-1.5; liq.; pour pt. –26 F; 100% conc.

Chemax NP-9. [122] Nonoxynol-9; nonionic; see Chemax NP 1.5; liq.; pour pt. 31 F; 100% conc.

Chemax NP-10. [122] Nonoxynol-10; nonionic; see Chemax NP-1.5; liq.; pour pt. 49 F; 100% conc.

Chemax NP-15. [122] Nonoxynol-15; nonionic; surfactant, detergent, wetting and rewetting agent, emulsifier in textile, leather, paper, paint, and metal processing; liq.; pour pt. 71 F; 100% conc.

Chemax NP-20. [122] Nonoxynol-20; nonionic; see Chemax NP-15; solid; pour pt. 91 F.

Chemax NP-30. [122] Nonoxynol-30; nonionic; see Chemax NP-15; solid; pour pt. 109 F; 100% conc.

Chemax NP-30/70. [122] Nonoxynol-30; see Chemax NP-15; liq.; pour pt. 34 F.

Chemax NP-40. [122] Nonoxynol-40; nonionic; polymerization emulsifier for vinyl acetate and acrylic emulsions; stabilizer for synthetic latices; wetting agent in electrolyte sol'ns.; solid; pour pt. 112 F; 100% conc.

Chemax NP-40/70. [122] Nonoxynol-40; see Chemax NP-40; liq.; pour pt. 46 F.

Chemax NP-50. [122] Nonoxynol-50; see Chemax NP-40; solid; pour pt. 114 F.

Chemax NP-50/70. [122] Nonoxynol-50; see Chemax NP-40; liq.; pour pt. 52 F.

Chemax NP-100. [122] Nonoxynol-100; see Chemax NP-40; solid; pour pt. 122 F.

Chemax NP-100/70. [122] Nonoxynol-100; see Chemax NP-40; liq.; pour pt. 68 F.

Chemax OP-3. [122] Octoxynol-3; nonionic;

emulsifier, detergent, stabilizer, dispersant, wetting agent; pesticides and floor finishes; liq.; pour pt. –9 F.

Chemax OP-5. [122] Octoxynol-5; see Chemax OP-3; liq.; oil sol.; pour pt. –15 F; 100% conc.

Chemax OP-7. [122] Octoxynol-7; nonionic; detergent comps.; industrial metal cleaning, acid and waterless hand cleaners, and floor finishes; liq.; pour pt. 15 F.

Chemax OP-9. [122] Octoxynol-9; nonionic; see Chemax OP-7; liq.; pour pt. 45 F; 100% conc.

Chemax OP-40. [122] Octoxynol-40; nonionic; see Chemax OP-5; solid; 100% conc.

Chemax OP-40/70. [122] Octoxynol-40; emulsifier for vinyl acetate and acrylate polymerization; liq.; pour pt. 25 F; 70% conc.

Chemax PEG 200 DO. [122] POE (4) dioleate; surfactant; coemulsifier for oils as mold release agent; liq.; oil-sol.; HLB 5.4.

Chemax PEG 400 DO. [122] POE (8) dioleate; emulsifier and solubilizer for solv., fats, and min. oils; liq.; oil-sol.; HLB 8.5.

Chemax PEG 600 DO. [122] POE (12) dioleate; component of defoamers, softeners, and lubricants; semi-liq.; water-sol.; HLB 10.5.

Chemax SBO. [122] Sulfated butyl oleate; softener, emulsifier, wetting agent in textile and metal working industries; liq.

Chemax SCO. [122] Sulfated castor oil; see Chemax SBO; liq.

Chemax TO-8. [122] PEG-8 tallate; nonionic; emulsifier, lubricant, dye assistant; liq.; 100% conc.

Chemax TO-10. [122] PEG-10 tallate; nonionic; emulsifier in industrial lubricants and degreasers; liq.; HLB 11.8.

Chemax TO-16. [122] PEG-16 tallate; nonionic; foam detergent and emulsifier; lubricant, dye assistant; liq.; HLB 13.3; 100% conc.

Chemazine 18, C, O, TO. [122] Imidazoline; cationic; softener, antistat, emulsifier, corrosion inhibitor, filming agents; solid (18, C), liq. (O, TO)

Chemcol 14-Z. [123] Anionic/nonionic blend; emulsifier for aromatic solv. aquatic weed control esp. for xylene or aromatic hydrocarbon-xylene solv.; lt. amber; sol. in xylene, misc. with water; pour pt. < 25 F.

Chemcol 610. [123] Low foaming, oil sol. emulsifier for agric. sprays, esp. petroleum oils for citrus sprays; Gardner < 1 clear liq.; mild odor; oil sol.; pour pt. < –10 F; pH 6–8 (1%).

Chemcol Agwet 110. [123] Alkylaryl POE ether; nonionic; agric. surfactant, reduces surface tension, improves wetting and plant coverage of herbicide and pesticide sprays; clear liq.; water sol.; dens. 8.2 lb/gal; flash pt. 80 F; 80% act.

Chemcol Agwet 115-M. [123] Alkyl POE glycol ether; nonionic; wetting agent, agric. surfactant to improve wetting properties, reduce surf. tens., improve plant coverage of pesticide and herbicide sprays; water wh. to pale yel. liq.; water sol.; dens. 7.8 lb/gal; flash pt. 108 F; biodeg.; 80% act.

Chemcol CO-28. [123] PEG-28 castor oil; nonionic; emulsifier for textiles and agric. chemical; pale yel. liq.; dens. 8.6 lb/gal; cloud pt. < 25 C; pour pt. <0 C; pH 6–8; 100% act.

Chemcol CO-30. [123] PEG-30 castor oil; nonionic; emulsifier for textiles and agric. chemicals; yel. liq.; dens. 8.7 lb/gal; HLB 11.7 cloud pt. > 100 C; pour pt. < 0 C; pH 6–8; 100% act.

Chemcol CO-40. [123] PEG-40 castor oil; nonionic; emulsifier, lubricant, dispersant, antistat for agric. chemicals, textiles, promotes dye leveling; yel. visc. liq.; dens. 8.8–8.9 lb/gal; HLB 13.1; cloud pt. > 100 C; pour pt. 10 C; pH 6.5–7.5; 97% act.

Chemcol HE-1. [123] Anionic/nonionic; agric. emulsifier; amber visc. liq.; dens. 8.3 lb/gal; visc. 80 cps; flash pt. > 100 F; pour pt. < 5 C.

Chemcol HE-2. [123] Anionic/nonionic; agric. emulsifier, lt. amber visc. liq.; dens. 8.6 lb/gal; visc. 146 cps; flash pt. > 100 F; pour pt. 17 C; pH 4–7 (5% in 25% IPA).

Chemcol HE-8. [123] Anionic/nonionic; agric. emulsifier; lt. amber visc. liq.; dens. 8 lb/gal; visc. 1850 cps; flash pt. > 100 F; pour pt. 15 C; pH 4–7 (5% in 25% IPA).

Chemcol JL-60. [123] POE ether alcohol; nonionic; detergent, dispersant, scouring agent, leveling agent used for cosmetics, textiles; wh. cloudy liq.; dens. 8.2 lb/gal; pour pt. 4.5–7.5 C; pH 6–8; 100% act.

Chemcol JL-120. [123] POE ether alcohol; nonionic; detergent, dispersant, scouring agent, leveling agent used for cosmetics, textiles; wh. semisolid; dens. 8.4 lb/gal; pour pt. 50 C; pH 6–8; 100% act.

Chemcol JL-180. [123] POE ether alcohol; nonionic; detergent, dispersant, scouring agent, leveling agent used for cosmetics, textiles; wh. waxy solid; pour pt. 50 C; pH 6–8; 100% act.

Chemcol MS-910. [123] Nonionic/anionic; agric. chemical; misc. surfactant for org. arsenicals; Gardner < 6 clear liq.; mild, pleasant odor; dens. 9.1 lb/gal; sp.gr. 1.09; pH 5.5–6.5 (10%); 100% act.

Chemcol NPE-40. [123] Nonoxynol-4; nonionic; detergent, stabilizer, coemulsifier, dispersant used in textile, household cleaners, agric. chemicals, petroleum oil disp.; pale yel. liq.; oil sol.; dens. 8.5 lb/gal; HLB 8.8; flash pt. > 400 F (TOC); pour pt. <0 C; pH 6–8 (5% in 25% IPA); 100% act.

Chemcol NPE-60. [123] Nonoxynol-6; nonionic; detergent, stabilizer, coemulsifier, dispersant used in textile, household cleaners, agric. chemicals, petroleum oil disp.; pale yel. liq.; water sol.; dens. 8.7 lb/gal; HLB 10.8; flash pt. > 400 F (TOC); pour pt. < 10 C; pH 6–8 (5% in 25% IPA); 100% act.

Chemcol NPE-100. [123] Nonoxynol-10; nonionic; detergent, stabilizer, coemulsifier, dispersant used in textile, household cleaners, agric. chemicals, petroleum oil disp.; colorless to pale yel. liq.; HLB 13.4; pH 6–8 (5% in 25% IPA); 100% act.

Chemcol NPE-140. [123] Nonoxynol-14; nonionic; detergent, stabilizer, coemulsifier, dispersant used in textile, household cleaners, agric. chemicals, petroleum oil disp.; visc. liq. to off-wh. solid; dens. 8.9 lb/gal; HLB 14.7; flash pt. > 450 F (TOC); pour pt. 24 C; pH 3.5–4.5 (10%); 100% act.

Chemcol NPE-200. [123] Nonoxynol-20; nonionic; detergent, stabilizer, coemulsifier, dispersant used in textile, household cleaners, agric. chemicals, petroleum oil disp.; off-wh. solid; HLB 16; flash pt. > 450 F (TOC); pour pt. 36 C; pH 3.5–4.5 (10%); 100% act.

Chemcol NPE-400. [123] Nonoxynol-40; nonionic; detergent, stabilizer, coemulsifier, dispersant used in textile, household cleaners, agric. chemicals, petroleum oil disp.; off-wh. solid; HLB 17.7; flash pt. > 450 F (TOC); pour pt. 48 C; pH 3.5–4.5 (10%); 100% act.

Chemcol PE-34. [123] Anionic/nonionic; surfactant, emulsifier for agric. chemicals; dk., visc. liq.; mild, aromatic odor; dens. 8.9 lb/gal (40 C); sp.gr. 1.067 (40 C); pour pt. < 0 C; pH 3.5–4.5 (10%); 100% act.

Chemcol PS-160. [123] Complex phosphate ester of an EO adduct; detergent, emulsifier, for argric. chemicals, emulsion polymerization; clear, straw-colored liq.; mild odor; dens. 9.2 lb/gal; sp.gr. 1.12; pour pt. < 0 C; 100% act.

Chemcol T-10. [123] PEG-10 tallow amine; cationic; emulsifier, antistat, disp. for textiles; Gardner 8 liq.; dens. 8.2 lb/gal; cloud pt. 70–90 C; pour pt. 54 F; 100% act.

Chemcol T-15. [123] PEG-15 tallow amine; cationic; emulsifier, antistat, disp. for textiles; amber liq.; dens. 8.5 lb/gal; cloud pt. > 100 C; pour pt. < 5 C; pH 11.74; 100% act.

Chemcol T-20. [123] PEG-20 tallow amine; cationic; emulsifier, antistat, disp. for textiles; amber liq.; dens. 8.7 lb/gal; cloud pt. > 100 C; pour pt. 10 C; pH 11.59; 100% act.

Chemcol T-25. [123] PEG-25 tallow amine; cationic; emulsifier, antistat, disp. for textiles; amber (Gardner 6) solid; dens. 8.8 lb/gal; cloud pt. > 100 C; pour pt. 110 F; pH 11.59; 100% act.

Chemcol TO-10. [123] PEG-10 tallate; nonionic; emulsifier, lubricant, dye assistant for textiles; yel. liq.; sp.gr. 0.872; pour pt. < 50 F; sapon. no. 72–82; 100% act.

Chemcol TO-15. [123] PEG-15 tallate; nonionic; emulsifier, lubricant, dye assistant for textiles; Gardner 8 solid; sp.gr. 0.866; pour pt. 80 F; pH 6–8; 100% act.

Chemcol TO-16. [123] PEG-16 tallate; nonionic; emulsifier, lubricant, dye assistant for textiles; lt. amber liq.; dens. 9 lb/gal; HLB 13.2; pour pt. 68 F; pH 7–8; 97% act.

Chemcol VCC-5. [123] Anionic/nonionic blend of metal sulfonates and polyoxylates; multipurpose pesticide emulsifier; amber liq.; dens. 8.49 lb/gal; sp.gr. 1.019; flash pt. 170 F (TOC); pH 4–6 (5% in 3:1 water:IPA); 100% act.

Chemeen 18-2. [122] POE (2) stearyl amine; emulsifier and antistat in textiles, metal buffing, and rubber compds.; lubricant for fiber glass; solid; m.w. 362

Chemeen 18-5. [122] POE (5) stearyl amine; see Chemeen 18-2; solid; m.w. 540

Chemeen 18-50. [122] POE (50) stearyl amine; see Chemeen 18-2; solid; m.w. 2380

Chemeen C-2. [122] PEG-2 cocamine; emulsifier, antistat, dye leveler, wetting agent, lubricant, dispersant; substantive to metals, fibers, and clays; liq.; m.w. 290; 100% conc.

Chemeen C-5. [122] PEG-5 cocamine; see Chemeen C-2; liq.; m.w. 425; 100% conc.

Chemeen C-10. [122] PEG-10 cocamine; see Chemeen C-2; liq.; m.w. 645; 100% conc.

Chemeen C-15. [122] PEG-15 cocamine; see Chemeen C-2; liq.; m.w. 890; 100% conc.

Chemeen DT-3. [122] POE (3) tallow diamine; emulsifier, textile dyeing assistant, corrosion inhibitor used in preparation of asphalt and agric. chemical emulsions; liq.; m.w. 535.

Chemeen DT-15. [122] PEG-15 tallow aminopropylamine; see Chemeen DT-3; liq.; m.w. 1020.

Chemeen DT-30. [122] POE (30) tallow diamine; see Chemeen DT-3; liq.; m.w. 1665.

Chemeen HT-2. [122] PEG-2 hydrog. tallow amine; emulsifier, antistat, lubricant; solid; 100% conc.

Chemeen HT-5. [122] PEG-5 hydrog. tallow amine; emulsifier and antistat in textiles, metal buffing and rubber compds., lubricant for fiber glass; solid; m.w. 495.

Chemeen HT-15. [122] PEG-15 hydrog. tallow amine; see Chemeen HT-5; solid; m.w. 925

Chemeen HT-50. [122] PEG-50 hydrog. tallow amine; see Chemeen HT-5; solid; m.w. 2470; 100% conc.

Chemeen O-30, O-30/80. [122] POE (30) oleyl amine; emulsifier and textile dyeing assistant; antiprecipitant in cross dyeing; solid; m.w. 1595; 100% and 80% conc.

Chemeen S-2, -5. [122] Soya amine, ethoxylated; emulsifier, antistat, lubricant; liq.; 100% conc.

Chemeen S-30. [122] Soya amine, ethoxylated; see Chemeen S-2; solid; 100% conc.

Chemeen S-30/80. [122] Soya amine, ethoxylated; see Chemeen S-2; liq.; 100% conc.

Chemeen T-2. [122] PEG-2 tallow amine; antistat for carpet shampoos; emulsifier, lubricant, dispersant, softener, antiprecipitant, leveling and migrating agent in textile dyeing process; liq.; m.w. 350; 100% conc.

Chemeen T-5. [122] PEG-5 tallow amine; see Chemeen T-2; liq.; m.w. 490; 100% conc.

Chemeen T-10. [122] PEG-10 tallow amine; see Chemeen T-2; liq.; m.w. 700; 100% conc.

Chemeen T-15. [122] PEG-15 tallow amine; see Chemeen T-2; liq.; m.w. 930; 100% conc.

Chemeen T-20. [122] POE (20) tallow amine; see Chemeen T-2; liq.; m.w. 1120; 100% conc.

Chemfac PA-080. [122] Phosphate ester; anionic; detergent, emulsifier, wetting agent, lubricant, antistat; liq.; 100% conc.

Chemfac PB-082, -104, -106, –109, -133, -135. [122] Phosphate esters; anionic; detergent, wetting, and coupling agent, antistat; emulsifier for detergents in alkali sol'ns.; liq.; 99% min. act.

Chemfac PB-184. [122] Oleth-4 phosphate; anionic; see Chemfac PA-080; liq.; 100% conc.

Chemfac PB-264, -804. [122] Phosphate esters; see Chemfac PA-080; liq.; 100% conc.

Chemfac PC-006. [122] POE-phenol phosphate; hydrotrope, detergent, emulsifier; foaming hydrotrope; liq.; 99% min. act.

Chemfac PC-099. [122] POE-alkyl phenol phosphate; detergent, wetting agent, primary emulsifier in emulsion polymerization; liq.; 99% min. act.

Chemfac PC-188, PD-600. [122] Phosphate esters; see Chemfac PA-080; liq.; 100% conc.

Chemfac PD-990. [122] Phosphate ester; see Chemfac PA-080; 90% conc.

Chemfac PX-322. [122] Phosphate ester; anionic; hydrotrope for solubilizing surfactants in alkali or other electrolytes; liq.; 50% conc.

Chemical Base 6532. [536; 537] Stearamidoethyl diethylamine; cationic; emulsifier, conditioning agent for skin and hair; yel. tan solid; slight odor; m.p. 35–45 C; 100% act.

Chemquat 12-33. [122] Laurtrimonium chloride; corrosion inhibitor, antistat, gel sensitizer in latex foam production; visc. depressant in softener formulations; 33% act.

Chemquat 12-50. [122] Laurtrimonium chloride; see Chemquat 12-33; 50% act.

Chemquat 16-50. [122] Cetrimonium chloride; see Chemquat 12-33; 50% act.

Chemstick 56. [123] Spray sticker designed to increase the effectiveness of wettable powd. and emulsifiable conc. for agric. sprays; amber liq.; dens. 7.56 lb/gal.

Chemsulf 2EH-Na. [122] Sodium 2-ethylhexyl sulfate; anionic; surfactant, wetting agent, emulsifier, detergent; liq.; 40% conc.

Chemsulf 094. [122] Ammonium alkyl phenol ethoxy sulfate; anionic; emulsifier, detergent, foaming agent; liq.; 58% conc.

Chimin AX. [598] Cocamidopropyl betaine; amphoteric; conditioner, emulsifier, detergent for cosmetic use in personal care products; liq.; 30% conc.

Chimin BX. [598] Stearyl betaine; ampholite; dispersant, wetting, antistat; soft paste; 40% conc.

Chimin CB. [598] Alkyl betaine; ampholite;

personal care products; liq.; 30% conc.

Chimin DOS. [598] Sodium 2-ethylhexyl sulfosuccinate; anionic; wetting agent; liq.; 68% conc.

Chimin KF3. [598] Sulfated syn. oil; anionic; fat liquoring, leather processing, emulsifier; liq.; 90% conc.

Chimin KSP. [598] Fatty alcohol ester; nonionic; fat liquoring of leather, substitute for sperm oil, textile lubricant; liq. 100% conc.

Chimin P 10. [598] Alkyl polyphosphate; anionic; wetting and penetrating agent; liq.; 100% conc.

Chimin P40. [598] Sodium alkylpolyglycol ether phosphate; anionic; detergent, wetting agent for products in textile, paper, cleaners, machine washing; liq.; 75% conc.

Chimin P45. [598] Sodium alkylpolyglycol ether phosphate; anionic; detergent, antistat, wetting agent for textile, paper, cleaners, machine washing, and personal care products; liq.; 30% conc.

Chimin P50. [598] Phosphoric acid complex org. ester; nonionic; emulsifier, antistat for use in leather, textile and cosmetic applications; liq.; sol. in water and solv.; 100% conc.

Chimin RI. [598] Complex phosphate ester; anionic; alkaline liq. detergents; liq.; 50% conc.

Chimipal APG 400. [598] Polyglycol laurate; nonionic; pigment dispersant and wetter, emulsion polymerization; liq.; HLB 13.0; 100% conc.

Chimipal DCL. [598] Cocamide DEA; nonionic; thickener and foam stabilizer, detergent, and shampoo additive; liq.; 100% conc.

Chimipal FV. [598] Monoglyceride, ethoxylated; nonionic; superfatting agent in cosmetic preparations; liq.; water-sol.; 100% conc.

Chimipal LDA. [598] Lauramide DEA; nonionic; thickener and foam stabilizer, detergent and shampoo additive; 100% conc.

Chimipal MC. [598] Cocamide MEA; nonionic; thickener and foam stabilizer, detergent and shampoo additive; flakes; 100% conc.

Chimipal NH. [598] Fatty amine ethoxylate; cationic; leveling agent, emulsifier; 100% conc.

Chimipal OLD. [598] Oleamide DEA; nonionic; thickener for shampoos, bubble baths and liq. detergents; liq.; 100% conc.

Chimipal OS 2. [598] Fatty alcohol ethoxylate blend; nonionic; powd. detergent, wax emulsifier, leveling agent; solid; 100% conc.

Chimipal PE 300, 302. [598] Polypropylene glycol ethoxylate; nonionic; detergent, dispersant, emulsifier, foam controller; liq.; 100% conc.

Chimipal PE 400. [598] Straight chain alcohol EO/PO; nonionic; detergent, dispersant, wetting, emulsifier, deduster; liq.; 100% conc.

Chimipal PE 500, 510, 520, 530, 540, 550, 560. [598] Terminated alkylene oxide adducts; nonionic; for metals, bottle, alkaline, strong acid cleaners; liq.; 100% conc.

Chimipal WN. [598] Nonylphenol polyglycol ether; nonionic; detergent, emulsifier; liq.; 100% conc.

Chimipon BAC. [598] Linear dodecylbenzene sulfonate; anionic; detergent base; paste; 70% conc.

Chimipon CA. [598] Linear calcium dodecylbenzene sulfonate; anionic; emulsifier for formulated pesticide; liq.; 70% conc.

Chimipon NA. [598] Alkylphenol ether sulfate, ammonium salt; anionic; liq. detergent base, frothing agent, emulsifier in emulsion polymerization; paste; 60% conc.

Chimipon TSB. [598] Linear dodecylbenzene sulfonate, triethanolamine-sodium salt; anionic; household and industrial liq. detergents; paste; 70% conc.

Chocothin. [410] Standardized lecithin fraction; amphoteric; visc. reducing agent for milk chocolates; liq.; 100% conc.

Chupol C. [590] Polyglycol ether; nonionic; reducing agent for concrete; liq.

Cindet 201. [139] Ethoxylated alcohols and aryl sulfonates; anionic; detergent for scouring wool, cotton, synthetics, dyeing assistant, disp. agent, penetrant; lt. amber liq.; pH 7.5–8.5 (1%.); 25% act.

Cindet 542. [139] POE compd.; nonionic; dye assistant, scouring and soaping off agent, detergent, retarding and leveling agent for textiles; water wh. liq.; pH 6.5–7.5 (1%); 50% act.

Cindet AL. [139] Sodium alcohol ether sulfate; anionic; dye assistant, dispersant, penetrant, scouring agent for wool, cotton, synthetics; lt. amber liq.; pH 7.0–8.0 (1%); 60% act.

Cindet AP. [139] Modified fatty nonionic and sulfonated hydrocarbons; anionic; wetting agent for causticizing and bleach operations, scouring agent, afterscour; transparent, amber liq.; water sol.; pH 7.5–8.5 (1%).

Cindet C-12. [139] Coconut amine condensate;

anionic; detergent, emulsifier, wetting agent, scouring agent, lubricant, fulling agent, dyeing assistant, anionic; amber visc. liq.; pH 8.5–9.5 (1%); 99% act.

Cindet GE. [139] Isopropylamine dodecylbenzene sulfonate; anionic; emulsifier in polyester and triacetate dye carriers, emulsifier; amber visc. liq.; sol. in low conc. in water, sol. in many solv.; pH 7.2–7.8 (1%); 94% act.

Cindet JS. [139] POE compd.; nonionic; nonfoaming surfactant, dye assistant, scouring agent, retarding and leveling agent, soaping off agent, detergent; lt. amber liq.; pH 6.5–7.5 (1%).

Cindet K-1. [139] Syn. detergents, phosphates, and other alkalis; anionic; detergent, scouring agent for cotton, rayon, nylon, polyester, polyolefin fibers; wh. powd.; pH 10–11 (1%); 100% act.

Cindet NAK. [139] Org. compds.; anionic; detergent, fulling, scouring and soaping agent, dye dispersant; lt. amber, visc. liq.; pH 7.5–8.5 (1%); 50% act.

Cindet R. [139] Fatty POE compd.; nonionic; dyeing assistant, scouring agent, retarder, detergent as afterscour, emulsifier; lt. amber liq.; pH 6.5–7.5 (1%); 70% act.

Cinwet A-35. [139] Anionic/nonionic surfactants; detergent with high wetting speed for scouring wool, cotton, synthetics, for oil and grease removal, dyeing assistant, dispersant, penetrant; slight yel., clear liq.; pH 7–8 (1%); 35% act.

Cinwet BOS. [139] Sulfated propyl oleate; anionic; dyeing assistant, dispersant, leveling agent, wetting agent, lubricant; amber liq.; pH 6–7 (1%); 65% act.

Cirrasol AEN-XB. [319] Fatty alcohol ethoxylate; nonionic; fiber processing aid; antistat for PVC belting; emulsifier for paraffin waxes, oleins, castor oil into water; liq.; HLB 10.2; 100% conc.

Cirrasol AEN-XF. [319] Fatty alcohol ethoxylate; nonionic; see Cirrasol AEN-XB; semisolid; HLB 13.8; 100%

Cirrasol AEN-XZ. [319] Alkyl phenol ethoxylate; nonionic; fiber processing aid; detergent base, emulsifier, dispersant, wetting, dedusting agent, and antistat; liq.; HLB 12.3; 100% conc.

Cirrasol ALN-FP. [319] Fatty acid ethoxylate; nonionic; fiber processing aid; antistat, emulsifier, and gelling agent; semisolid; HLB 11.6; 100% conc.

Cirrasol ALN-GM. [319] Polyolester ethoxylate; nonionic; fiber processing aid; antistat for plastics and rubber; wetting agent; liq.; HLB 16.9; 100% conc.

Cirrasol ALN-TF. [319] Fatty acid ethoxylate; nonionic; fiber processing aid; general wetting agent; liq.; HLB 11.5; 100% conc.

Cirrasol ALN-TS. [319] Fatty acid ethoxylate; nonionic; fiber processing aid; general wetting agent; liq.; HLB 13.2; 82.4% conc.

Cirrasol ALN-WF. [319] Fatty alcohol ethoxylate; nonionic; fiber processing and antistat lubricant for syn. fibers; emulsifier for waxes; flakes; HLB 14.9; 100% conc.

Cirrasol ALN-WY. [319] Triglyceride, ethoxylated; nonionic; fiber processing aid; wool lubricant; liq.; 90% conc.

Cirrasol EN-MB. [319] Fatty alcohol ethoxylate; nonionic; fiber processing aid; emulsifier/stabilizer for textile lubricant compositions; liq.; 100% conc.

Cirrasol EN-MP. [319] Fatty alcohol ethoxylate; nonionic; fiber processing aid; emulsifier in mfg. of min. and veg. oils; liq.; HLB 6.0; 100% conc.

Cirrasol LAN-SF. [319] Fatty alcohol ethoxylate; nonionic; see Cirrasol EN-MP; HLB 7.0; 100%

Cirrasol LC-HK. [319] Amine alkyl deriv.; cationic; fiber processing aid; lubricant and softener; paste; 33% conc.

Cirrasol LC-PQ. [319] Diamine fatty acid deriv.; cationic; fiber processing aid; lubricant and softener; paste; 25% conc.

Cirrasol LN-GS. [319] Syn. alcohol fatty acid ester; fiber processing aid; heat stable lubricant; liq.; 100% conc.

Cithrol 2DL. [159] POE (4) dilaurate; nonionic; used in cosmetic and industrial applications; dispersant, w/o emulsifier, wetting agent and co-solv.; paste; 97% conc.

Cithrol 2DO. [159] POE (4) dioleate; nonionic; see Cithrol 2 DL; liq.; 97% conc.

Cithrol 2DS. [159] POE (4) distearate; nonionic; antistat in textile finishing; paste; HLB 5.2; 97% conc.

Cithrol 2ML. [159] PEG-4 laurate; nonionic; wetting agent, emulsifier, detergent, thickening agent, solubilizer, dispersant, softener, lubricant, antistat, dye asistant, penetrant for cosmetics, textiles, glass fiber, metal treatment; liq.; m.w. 200; HLB 8.8; sapon. no. 158–168; 100% conc.

Cithrol 2MO. [159] PEG-4 oleate; nonionic; wetting agent, penetrant, detergent, emulsifier, solubilizer, thickening agent, dispersant, textile aux., lubricant for textiles and cosmetics; m.w. 200; HLB 6.2; sapon. no. 110–112; 100% conc.

Cithrol 2MS. [159] PEG-4 stearate; nonionic; emulsifier for insecticides and cosmetics, detergent, wetting agent, solubilizer and thickening agent for perfumery, antifrothing agent, foaming agent; m.w. 200; m.p. 39–41 C; HLB 6.3; sapon. no. 110–120; 100% conc.

Cithrol 3DS. [159] PEG-6 distearate; nonionic; see Cithrol 2 DS; solid; 97% conc.

Cithrol 3MS. [159] PEG-6 stearate; see Cithrol 2MS; paste; m.w. 300; m.p. 30–33 C; HLB 10.7; sapon. no. 95–105; 100% conc.

Cithrol 4DL. [159] PEG-8 dilaurate; nonionic; see Cithrol 2 DL; liq.; 97% conc.

Cithrol 4DO. [159] PEG-8 dioleate; nonionic; see Cithrol 2 DL; liq.; 97% conc.

Cithrol 4DS. [159] PEG-8 distearate; nonionic; see Cithrol 2 DS; solid; 97% conc.

Cithrol 4ML. [159] PEG-8 laurate; nonionic; see Cithrol 2ML; m.w. 400; HLB 19; sapon. no. 92–98.

Cithrol 4MO. [159] PEG-8 oleate; nonionic; see Cithrol 2MO; liq.; m.w. 400; HLB 11.4; sapon. no. 85–93; 100% conc.

Cithrol 4MS. [159] PEG-8 stearate; nonionic; see Cithrol 2MS; paste; m.w. 400; m.p. 31–34 C; HLB 11; sapon. no. 95–105; 10% conc.

Cithrol 6DL. [159] PEG-12 dilaurate; nonionic; see Cithrol 2 DL; liq.; 97% conc.

Cithrol 6DO. [159] PEG-12 dioleate; nonionic; see Cithrol 2 DL; liq.; 97% conc.

Cithrol 6DS. [159] PEG-12 distearate; nonionic; see Cithrol 2 DS; solid; 97% conc.

Cithrol 6ML. [159] PEG-12 laurate; nonionic; see Cithrol 2ML; m.w. 600; HLB 15.8; sapon. no. 64–74.

Cithrol 6MO. [159] PEG-12 oleate; nonionic; see Cithrol 2MO; liq.; m.w. 600; HLB 13.1; sapon. no. 65–75; 100% conc.

Cithrol 6MS. [159] PEG-12 stearate; nonionic; see Cithrol 2MS; paste; m.w. 600; m.p. 33–35 C; HLB 14; sapon. no. 68–76; 100% conc.

Cithrol 10DL. [159] PEG-20 dilaurate; nonionic; see Cithrol 2 DL; paste; 97% conc.

Cithrol 10DO. [159] PEG-20 dioleate; nonionic; see Cithrol 2 DL; liq.; 97% conc.

Cithrol 10DS. [159] PEG-20 distearate; nonionic; see Cithrol 2 DS; solid; 97% conc.

Cithrol 10ML. [159] PEG-20 laurate; nonionic; see Cithrol 2ML; sol. in water, alcohol, polar solv.; m.w. 1000; HLB 18; sapon. no. 51–55; 100% conc.

Cithrol 10MO. [159] PEG-20 oleate; nonionic; see Cithrol 2MO; liq.; m.w. 1000; HLB 16.2; sapon. no. 35–40; 100% conc.

Cithrol 10MS. [159] PEG-20 stearate; nonionic; see Cithrol 2MS; solid; m.w. 1000; m.p. 34–40 C; HLB 16; sapon. no. 36–50; 100% conc.

Cithrol 15MS. [159] PEG 1500 stearate; nonionic; see Cithrol 2MS; solid; m.w. 1500; m.p. 30-41 C; HLB 17; sapon. no. 30–55; 100% conc.

Cithrol 40MO. [159] PEG-75 oleate; see Cithrol 2MO; m.w. 4000; HLB 18.7; sapon. no. 10–15.

Cithrol 40MS. [159] PEG-75 stearate; nonionic; see Cithrol 2MS; solid; m.w. 4000; m.p. 30–41 C; HLB 18.8; sapon. no. 12–17; 100% conc.

Cithrol 60ML. [159] PEG 6000 laurate; see Cithrol 2ML; m.w. 6000; HLB 19.2; sapon. no. 8–13.

Cithrol 60MO. [159] PEG 6000 monooleate; see Cithrol 2MO; m.w. 6000; HLB 19.0; sapon. no. 5–10.

Cithrol A. [159] PEG-8 oleate; nonionic; w/o emulsifier, dispersant, antistat agent for textile, paper processing, cutting oils, polishes, emulsion cleaners; amber liq.; slight agreeable odor; sol. in wh. spirit and liq. paraffin; completely sol. in xylol, ethyl acetate, methylated spirits, oleic acid; disp. in water; sp.gr. 1.013; HLB 11.4; flash pt. 215 C; sapon. no. 85–93; pH 6.5–7.5 (10%); 97% conc.

Cithrol DGDL N/E. [159] Diethylene glycol dilaurate; nonionic; see Cithrol A; paste; 100% conc.

Cithrol DGDL S/E. [159] Diethylene glycol dilaurate SE; anionic; see Cithrol A; paste; 100% conc.

Cithrol DGDO N/E. [159] Diethylene glycol dioleate; nonionic; see Cithrol A; liq.; 100% conc.

Cithrol DGDO S/E. [159] Diethylene glycol dioleate SE; anionic; see Cithrol A; liq.; 100% conc.

Cithrol DGDS N/E. [159] Diethylene glycol distearate; nonionic; see Cithrol A; solid; 100% conc.

Cithrol DGDS S/E. [159] Diethylene glycol distearate SE; anionic; see Cithrol A; solid; 100% conc.

Cithrol DGML N/E. [159] Diethylene glycol

monolaurate; nonionic; see Cithrol A; paste; 100% conc.

Cithrol DGML S/E. [159] Diethylene glycol monolaurate SE; anionic; see Cithrol A; paste; 100% conc.

Cithrol DGMO N/E. [159] Diethylene glycol monooleate; nonionic; see Cithrol A; liq.; 100% conc.

Cithrol DGMO S/E. [159] Diethylene glycol monooleate SE; anionic; see Cithrol A; liq.; 100% conc.

Cithrol DGMS N/E. [159] Diethylene glycol monostearate; nonionic; see Cithrol A; solid; 100% conc.

Cithrol DGMS S/E. [159] Diethylene glycol monostearate SE; anionic; see Cithrol A; solid; 100% conc.

Cithrol DPGML N/E. [159] Dipropylene glycol monolaurate; nonionic; see Cithrol A; liq.; 100% conc.

Cithrol DPGML S/E. [159] Dipropylene glycol monolaurate SE; anionic; see Cithrol A; liq.; 100% conc.

Cithrol DPGMO N/E. [159] Dipropylene glycol monooleate; nonionic; see Cithrol A; liq.; 100% conc.

Cithrol DPGMO S/E. [159] Dipropylene glycol monooleate SE; anionic; see Cithrol A; liq.; 100% conc.

Cithrol DPGMS N/E. [159] Dipropylene glycol monostearate; nonionic; see Cithrol A; solid; 100% conc.

Cithrol DPGMS S/E. [159] Dipropylene glycol monostearate SE; anionic; see Cithrol A; solid; 100% conc.

Cithrol EGDL N/E. [159] Ethylene glycol dilaurate; nonionic; see Cithrol A; solid; 100% conc.

Cithrol EGDL S/E. [159] Ethylene glycol dilaurate SE; anionic; see Cithrol A; solid; 100% conc.

Cithrol EGDO N/E. [159] Ethylene glycol dioleate; nonionic; see Cithrol A; liq.; 100% conc.

Cithrol EGDO S/E. [159] Ethylene glycol dioleate SE; anionic; see Cithrol A; liq.; 100% conc.

Cithrol EGDS N/E. [159] Ethylene glycol distearate; nonionic; see Cithrol A; solid; 100% conc.

Cithrol EGDS S/E. [159] Ethylene glycol dioleate SE; anionic; see Cithrol A; solid; 100% conc.

Cithrol EGML N/E. [159] Ethylene glycol monolaurate; nonionic; see Cithrol A; liq.; 100% conc.

Cithrol EGML S/E. [159] Ethylene glycol monolaurate SE; anionic; see Cithrol A; liq.; 100% conc.

Cithrol EGMO N/E. [159] Ethylene glycol monooleate; nonionic; see Cithrol A; liq.; 100% conc.

Cithrol EGMO S/E. [159] Ethylene glycol monooleate SE; anionic; see Cithrol A; liq.; 100% conc.

Cithrol EGMR N/E. [159] Ethylene glycol monoricinoleate; nonionic; see Cithrol A; liq.; 100% conc.

Cithrol EGMR S/E. [159] Ethylene glycol monoricinoleate SE; anionic; see Cithrol A; liq.; 100% conc.

Cithrol EGMS N/E. [159] Ethylene glycol monostearate; nonionic; see Cithrol A; solid; 100% conc.

Cithrol EGMS S/E. [159] Ethylene glycol monostearate SE; anionic; see Cithrol A; solid; 100% conc.

Cithrol GDL N/E. [159] Glyceryl dilaurate; nonionic; see Cithrol A; liq.; 100% conc.

Cithrol GDL S/E. [159] Glyceryl dilaurate SE; anionic; see Cithrol A; liq.; 100% conc.

Cithrol GDO N/E. [159] Glyceryl dioleate; nonionic; emulsifier, coemulsifier, stabilizer, wetting agent, lubricant, and antistat; used in cosmetic, pharmaceutical, industrial, food applications; liq.; 100% conc.

Cithrol GDO S/E. [159] Glyceryl dioleate SE; anionic; see Cithrol GDO N/E; liq.; 100% conc.

Cithrol GDS N/E. [159] Glyceryl distearate; nonionic; see Cithrol GDO N/E; solid; 100% conc.

Cithrol GDS S/E. [159] Glyceryl distearate SE; anionic; see Cithrol GDO N/E; solid; 100% conc.

Cithrol GML N/E. [159] Glyceryl monolaurate; nonionic; see Cithrol GDO N/E; liq.; 100% conc.

Cithrol GML S/E. [159] Glyceryl monolaurate SE; anionic; see Cithrol GDO N/E; liq.; 100% conc.

Cithrol GMO N/E. [159] Glyceryl monooleate; nonionic; see Cithrol GDO N/E; liq.; 100% conc.

Cithrol GMO S/E. [159] Glyceryl monooleate SE; anionic; see Cithrol GDO N/E; liq.; 100% conc.

Cithrol GMR N/E. [159] Glyceryl monoricinoleate; nonionic; see Cithrol GDO N/E; liq.; 100% conc.

Cithrol GMR S/E. [159] Glyceryl monoricinoleate SE; anionic; see Cithrol GDO N/E; liq.; 100% conc.

Cithrol GMS Acid Stable. [159] Glyceryl monostearate, S/E; nonionic; see Cithrol GDO N/E; solid; 100% conc.

Cithrol GMS A/S ES 0743. [159] Glyceryl monostearate; emulsifier and opacifier for cosmetics; Gardner 3 color; m.p. 52 C; sapon. no. 96.

Cithrol GMS N/E. [159] Glyceryl monostearate; nonionic; see Cithrol GDO N/E; solid; 100% conc.

Cithrol GMS S/E. [159] Glyceryl monostearate SE; anionic; see Cithrol GDO N/E; solid; 100% conc.

Cithrol L, O Range. [159] Polyglycol esters; cosmetic and industrial hydrophilic systems; liq., solid.

Cithrol PGML N/E. [159] Propylene glycol monolaurate; nonionic; see Cithrol GDO N/E; liq.; 100% conc.

Cithrol PGML S/E. [159] Propylene glycol monolaurate SE; anionic; see Cithrol GDO N/E; liq.; 100% conc.

Cithrol PGMO N/E. [159] Propylene glycol monooleate; nonionic; see Cithrol GDO N/E; liq.; 100% conc.

Cithrol PGMO S/E. [159] Propylene glycol monooleate SE; anionic; see Cithrol GDO N/E; liq.; 100% conc.

Cithrol PGMR N/E. [159] Propylene glycol monoricinoleate; nonionic; see Cithrol GDO N/E; liq.; 100% conc.

Cithrol PGMR S/E. [159] Propylene glycol monoricinoleate SE; anionic; see Cithrol GDO S/E; liq.; 100% conc.

Cithrol PGMS N/E. [159] Propylene glycol monostearate; nonionic; see Cithrol GDO N/E; solid; 100% conc.

Cithrol PGMS S/E. [159] Propylene glycol monostearate SE; anionic; see Cithrol GDO S/E; solid; 100% conc.

Cithrol PR. [159] Polyglycerol polyricinoleate; nonionic; w/o emulsifier for veg. oils; liq.; 100% conc.

Cithrol S Range. [159] Polyglycol esters; cosmetic and industrial hydrophilic systems; liq./solid

Citrest LSU. [168] Laureth-6 citrate; anionic; chelating and foaming agent used in personal care products and lubricants; coupling agent for hot systems; liq.; 100% conc.

Citrest LT. [168] Trilauryl citrate; anionic; foaming ingredient for cosmetic product and lubricant; liq.; 100% conc.

Citrest ST. [168] Tristearyl citrate; anionic; coupling agent for hot systems; flakes; 100% conc.

Citrest TD. [168] Dinonoxynol-9 citrate; anionic; foaming fluid surfactant; liq.; 100% conc.

Clearate LV, WD. [143] Soya lecithin deriv.; nonionic; pigment dispersant, wetting agent, and antiflocculant in paint systems; liq.; 56% conc.

Clearate Special Extra. [143] Lecithin; nonionic; dispersant in inks, paints, foods; wetting agent; liq.

Clearate WDF. [143] Soya lecithin; nonionic; o/w emulsifier in the food industry; liq.; HLB 8.0; 60% conc.

Clearlan K50. [217] Anhydrous lanolin, USP; nonionic; w/o emulsions; emollient for the skin; solid; HLB 10.0; 100% conc.

Clindrol 100MC. [144] Cocamide MEA; nonionic; foam stabilizer in shampoo and liq. dishwash detergent; tan flakes; ammoniacal odor; sol. in ethanol, IPA, aliphatic and aromatic hydrocarbons; sp.gr. 1.0; m.p. 63 C; 100% act.

Clindrol 100MCG. [144] Cocamide MEA; nonionic; foam stabilizer and thickener; tan flakes; sol. in alcohols, glycols, aliphatic, aromatic, and chlorinated hydrocarbons; m.p. 60–66 C; pH 8–10.5; 84% min. amide.

Clindrol 101CG. [144] Cocamide DEA; nonionic; emulsifier of aliphatic hydrocarbons, household detergent, thickener; dk. amber liq.; typ. alkanolamide odor; sol. in alcohols, glycols, ketones, esters, aromatic and aliphatic hydrocarbons; sp.gr. 0.99; f.p. 6 C; 100% act.

Clindrol 101LI. [144] Lauramide MIPA; nonionic; foam stabilizer and thickener in rug shampoo, hand cleaners, hair shampoo; tan, solid flakes; sol. in alcohols, glycols, ketones, esters, aromatic and chlorinated hydrocarbons; f.p. 53–60 C; 100% act.

Clindrol 102LI. [144] Lauramide MIPA; nonionic; foam stabilizer and thickener for detergents, rug shampoos and personal care products; wh. waxy flake; sol. in alcohols, glycols, ketones, esters, aromatic and chlorinated hydrocarbons; insol. in water; m.p. 55 ± 2 C; 100% conc.

Clindrol 200CG. [144] Cocamide DEA and diethanolamine; nonionic; detergent, wetting agent, foam stabilizer, thickener for household and textile use; dk. amber liq.;

typ. alkanolamide odor; sol. in water, alcohols, glycols, ketones, aromatic and chlorinated hydrocarbons; sp.gr. 1.00; f.p. –8 C; 100% act.

Clindrol 200CGN. [144] Cocamide DEA and diethanolamine; nonionic; detergent, wetting agent, emulsifier, foam stabilizer, thickener; hard surface and floor cleaners; degreaser, metal cleaner, metalworking compds.; corrosion inhibitor; dk. amber liq.; typ. odor; sol. in alcohols, water, ketones, aromatic and chlorinated hydrocarbons; sp.gr. 0.99; f.p. 0 C; 100% act.

Clindrol 200HC. [144] Cocamide DEA and diethanolamine; nonionic; detergent, emulsifier, foam stabilizer, thickener used for hair shampoos, metal cleaners, household and cosmetic uses; straw-colored liq.; typ. odor; sol. in alcohols, water, ketones, aromatic and chlorinated hydrocarbons; sp.gr. 1.02; f.p. 14–15 C; 100% act.

Clindrol 200L. [144] Lauramide DEA and diethanolamine; nonionic; detergent, wetting agent, emulsifier, foam stabilizer and thickener, for household and cosmetic uses; perfume solubilizer; wh. paste; typ. odor; sol. in alcohols, water, ketones, aromatic hydrocarbons; sp.gr. 1.02; m.p. 25–30 C; 100% act.

Clindrol 200MS. [144] Stearamide MEA; pearl producing agent used in hair shampoos, hand cleaners, skin care products, household and cosmetic uses; tan flakes; ammoniacal odor; sol. in alcohols, glycols, esters, ketones, aromatic hydrocarbons; sp.gr. 0.998; m.p. 98 C; 100% act.

Clindrol 200-O. [144] Oleamide DEA; nonionic; emulsifier, thickener for household and cosmetic uses, hair preparations, waterless hand cleaners, metalworking compds.; amber liq.; sol. in alcohols, glycols, aliphatic and aromatic hydrocarbons, ketones, chlorinated hydrocarbons; sp.gr. 0.99; f.p. –8 C; 100% act.

Clindrol 200RC. [144] Cocamide DEA and diethanolamine; nonionic; detergent, thickener, wetting agent, emulsifier and solubilizer in insecticides, dye solv. in the printing industry, foam stabilizer; amber liq.; sol. in alcohols, glycols, water, ketones, aromatic and chlorinated hydrocarbons; sp.gr. 1.02; 100% act.

Clindrol 200S. [144] Stearamide DEA and diethanolamine; nonionic; emulsifier for skin and hair products and cosmetics, thickener and dispersing agent; wh. waxy solid; sol. in alcohols, glycols, ketones, esters, aromatic and chlorinated hydrocarbons; sp.gr. 1.00; m.p. 36–40 C; pH 9–10 (1% disp.); 100% act.

Clindrol 201CGN. [144] Cocamide DEA; nonionic/anionic; detergent, emulsifier, wetting agent, foam stabilizer, thickener, for use in cleaners, metal degreasers, burnishing compounds, textile industry; dk. amber liq.; typ. alkanolamide odor; sol. in alcohols, water, ketones, aromatic and chlorinated hydrocarbons; sp.gr. 0.99; f.p. 0 C; 100% act.

Clindrol 202CGN. [144] Cocamide DEA and diethanolamine; nonionic/anionic; detergent, emulsifier, wetting agent, foam stabilizer and thickener for use in cleaners, metal degreasers, burnishing compounds, textile industry; dk. amber liq.; typ. alkanolamide odor; sol. in alcohols, water, ketones, aromatic and chlorinated hydrocarbons; sp.gr. 0.99; f.p. 0 C; 100% act.

Clindrol 203CG. [144] Cocamide DEA and diethanolamine; detergent, wetting agent, foam stabilizer, thickener, used in general purpose and other cleaners, textile, household and cosmetic uses; straw to lt. amber liq. typ. alkanolamide odor; sol. in alcohols, water, glycols, ketones, aromatic and chlorinated hydrocarbons; sp.gr. 1.00; f.p. –8 C; 100% act.

Clindrol 204CGN. [144] Cocamide DEA; nonionic/anionic; detergent, emulsifier, wetting agent, foam stabilizer and thickener, used in general purpose and various other cleaners, textile, household and cosmetic uses; straw to lt. amber liq.; typ. alkanolamide odor; sol. in alcohols, water, ketones, aromatic and chlorinated hydrocarbons; sp.gr. 0.99; f.p. 0 C; 100% act.

Clindrol 206CGN. [144] Cocamide DEA and diethanolamine; anionic; detergent in transport and cosmetic industries, emulsifier, wetting agent for removal of spots, insecticides, herbicides, and fungicides, conveyor chain lubricant, household and industrial cleaners; dk. amber liq.; sol. in water, alcohols, glycols, glycol ethers, esters, ketones, aromatic and chlorinated hydrocarbons; sp.gr. 1.02; f.p. –5 C; 100% act.

Clindrol 207CGN. [144] Cocamide DEA and diethanolamine; see Clindrol 206CGN; sp.gr. 1.02; f.p. –5 C; 100% act.

Clindrol 209CGN. [144] Cocamide DEA and

diethanolamine; see Clindrol 206CGN; sp.gr. 1.06; m.p. –7 to –3 C; 100% act.

Clindrol 210CGN. [144] Coconut alkanolamide; see Clindrol 206CGN; sp.gr. 0.99; f.p. 0–5 C; 100% act.

Clindrol LT 14-8. [144] Stearamide MEA; nonionic; opacifier and pearlescent for personal care products; tan flake; sol. in alcohols, glycols, esters, ketone, and aromatic hydrocarbons; insol. in water; sp.gr. 0.998; m.p. 98 C; pH 9–10; 100% conc.

Clindrol LT 15-73-1. [144] Linoleic diethanolamide; nonionic; thickener for personal care products; straw-amber liq.; sol. in alcohol, alcohol ethers, ethers, esters, aromatic and aliphatic hydrocarbons; disp. in water; sp.gr. 0.98; pH 8.5–11; 100% act.

Clindrol SDG. [144] Diethylene glycol monostearate; nonionic; emulsifier, opacifying agent used in skin and hair products; wh. solid; sol. in alcohols, glycols, chlorinated, aromatic and aliphatic hydrocarbons; sp.gr. 1.01; m.p. 42–45 C; 100% act.

Clindrol SEG. [144] Ethylene glycol monostearate; nonionic; opacifier and pearl producing agent used in hair and skin care products, liq. dishwashing detergent; wh. flakes; slight fatty odor; sol. in alcohols, glycols, esters, ketones, aromatic, aliphatic and chlorinated hydrocarbons; sp.gr. 0.998; m.p. 52–60 C; 100% act.

Clindrol Superamide 100C. [144] Cocamide DEA; nonionic; foam stabilizer, thickener, solubilizer used for hair and skin products, liq. dishwashing detergent; straw liq.; sol. in alcohols, glycols, ketones, aromatic and chlorinated hydrocarbons; sp.gr. 0.99; f.p. 8 C; 100% act.

Clindrol Superamide 100CG. [144] Cocamide DEA; nonionic; foam stabilizer, thickener, solubilizer used for hair and skin products, liq. dishwashing detergent; straw to amber liq.; sol. in alcohols, glycols, water, aromatic and chlorinated hydrocarbons; sp.gr. 0.99; f.p. –3 C; 100% act.

Clindrol Superamide 100L. [144] Lauramide DEA; nonionic; foam stabilizer in aq. systems, thickener in gel shampoos; wh. solid; typ. alkanolamide odor; sol. in alcohols, ketones, chlorinated and aromatic solv.; sp.gr. 0.99; m.p. 42–46; 100% act.

Clindrol Superamide 100LM. [144] Diethanolamide of 70% lauric–30% myristic acid; foam stabilizer in aq. systems, thickener in gel shampoos; wh. paste, solid; sol. in alcohols, ketones, chlorinated and aromatic solv.; sp.gr. 0.99; m.p. 29–33 C; 100% act.

Clindrol Superamide 100S. [144] Stearamide DEA; nonionic; thickener, emulsifier used in personal care products; wh. waxy solid; sol. in alcohols, ketones, chlorinated and aromatic solv.; disp. in water; m.p. 58–63 C; pH 9–10.5; 100% act.

Clindrol Superamide LL 660, 743, 931, 1848, 2141. [144] Lauric diethanolamide; nonionic; foam stabilizer, thickener for use in personal care and detergent products; liq.; 100% conc.

Closyl 30. [145] Sodium-N-cocoyl sarcosinate; anionic; wetting, foaming detergent used in personal care and household products; liq.; 30–31% conc.

CM-88. [252] Amine-lignosulfonate; cationic; modifier; replaces 40–60% of a primary emulsifier; solid; dens. 1.25 g/ml; 46% solid.

CMTC-T. [295] Sodium carboxymethyl-cellulose; used as syn. detergent to prevent redeposition of soil on fabrics; powd.; water sol.

Coconad MT. [341] Caprylic triglyceride; nonionic; food emulsifier; coffee whitener, flavor diluting agent; liq.; 100% conc.

Coconad PL. [341] Caprylic/capric triglyceride; nonionic; food emulsifier; health and sports food; liq.; 100% conc.

Coconad RK. [341] Caprylic/capric triglyceride; nonionic; food emulsifier; coffee whitener, releasing agent for processing, flavor diluting agent; liq.; 100% conc.

Coconad SK. [341] Capric triglyceride; nonionic; food emulsifier; health and sports food; liq.; 100% conc.

Collemul H2. [27] PEG 200 monooleate; nonionic; emulsifier; liq.; HLB 8.3; 100% conc.

Collemul H4. [27] POE (8) monooleate; nonionic; emulsifier; liq.; HLB 11.7; 100% conc.

Collemul H6. [27] PEG 600 monooleate; nonionic; detergent; liq.; HLB 14.1; 100% conc.

Collemul L4. [27] POE (8) monolaurate; nonionic; detergent; liq.; HLB 13.4; 100% conc.

Colorol Series. [410] Phospholipids, modified; nonionic; dispersant, wetting and suspending agent for paints, lacquers, aq. systems, printing inks; liq., solid; 20–100% conc.

Compene 1898. [145] Sulfated fish oil; anionic; fat liquor for leather industry; liq.; 65–66% conc.

Compene Conc. 3816. [145] Sulfated fish oil, anionic; emulsifier, fat liquor for leather industry; 66–68% conc.

Comperlan 100. [289] Cocamide MEA; nonionic; foamer; granule; 100% conc.

Comperlan CD. [289] Capric acid diethanolamide; nonionic; surfactant providing flash foam; amber visc. liq.; 98% conc.

Comperlan COD. [291] Coconut oil diethanolamide; foam and visc. increasing agent with emulsifying properties; conditioner for shampoos and other cosmetic and pharmaceutical applications; solv.; liq.; 85% conc.

Comperlan F. [291] Linoleamide DEA; nonionic; foam stabilizer and thickener for shampoos; vitamin E substitute; liq.; 100% conc.

Comperlan HS. [291] Stearamide MEA; nonionic; consistency-giving factor in personal care products; flakes; 100% conc.

Comperlan ID. [289] 1:1 Superamide isostearamide DEA; hair conditioning agent; amber visc. liq.; 100% conc.

Comperlan KD. [289] Cocamide DEA; nonionic; thickener and foamer for personal care products; hair conditioner; amber visc. liq.; 100% conc.

Comperlan KDO. [289] 1:1 Superamide cocamide DEA; nonionic; thickening and foaming agent for personal care products; amber visc. liq.; 100% conc.

Comperlan KM. [291] Cocamide MEA; booster, foam stabilizer, pearly sheen improving agent for surfactant preparations; paste; 30% conc.

Comperlan LD. [289; 291] Lauramide DEA; nonionic; foam booster, stabilizer, detergency and visc. builder, emulsifier for personal care products; waxy solid; 100% conc.

Comperlan LD 9. [289] Distearyl dimonium chloride; cationic; superamide; foam and visc. builder in personal care products; paste; 75% conc.

Comperlan LDO, LDS. [289] 1:1 Superamide lauramide DEA; nonionic; foam booster, stabilizer, detergency and visc. builder, emulsifier for personal care products; amber visc. liq. 100% conc.

Comperlan LM. [289; 291] Lauramide MEA; nonionic; foam and visc. builder for personal care products and cleansers; solid; 100% conc.

Comperlan LMD. [289; 291] Lauramide DEA; nonionic; booster and thickener for personal care products; liq.; 100% conc.

Comperlan LP. [289; 291] Lauramide MIPA; nonionic; foamer; solid; 100% conc.

Comperlan LS. [291] Cocamide DEA and laureth-12; emulsifier, thickener, and foam stabilizer for personal care products; dissolving intermediates; liq.; 65–75% conc.

Comperlan OD. [291] Oleamide DEA; foam stabilizer, thickener and superfatting agent for shampoos and bubble bath products; liq.; 90% conc.

Comperlan PD. [289; 291] 2:1 cocamide DEA; nonionic; thickener, foam builder/stabilizer for personal care products, dishwashing agents; amber visc. liq.; 98% conc.

Comperlan PVD. [291] Veg. oil polydiethanolamide; nonionic; active for dishwashing agents, coemulsifier for cleaning and metal working agents; liq.

Comperlan SD. [289] Coco diethanolamide; nonionic; foamer; liq.; 100% conc.

Comperlan SDO. [289] Cocamide DEA; nonionic; foam and visc. builder for personal care products; liq.; 100% conc.

Comperlan SM. [289] Cocamide MEA; nonionic; superamide, foam and visc. builder; granule; 100% conc.

Comperlan VOD. [289; 291] Linoleamide DEA; nonionic; foam and visc. builder with superfatting effect for personal care products; liq.; 100% conc.

Complemix 50. [43] Sulfonated ester; anionic; wetting agent, emulsifier, clouding agent used in the food and beverage industry; nearly colorless liq.; sp.gr. 0.93; 50% act.

Complemix 100. [43; 167] Sodium dioctyl sulfosuccinate NF; anionic; emulsifier, wetting or clouding agent agent used in the food and beverage industry; dispersant, solubilizer, modifier; wh. waxy solid; water sol.; sp.gr. 1.1; 100% act.

Compound 8-S. [208] Complex alkylaryl sulfonate; anionic; nonfoaming dispersant for disperse- and vat-type dyes, dispersant for carbon black; liq.; water sol.; dens. 9.6 lb/gal; sp.gr. 1.152; visc. readily pourable.

Compound 9264B. [496] Ethoxylated glycerol monostearate; softener, nonionic; solid.

Compritol 888. [248] Glyceryl di/tribehenate; nonionic; food emulsifier and additive for tablet mfg.; solid; HLB 2.0; 100% conc.

Conco AAS-35. [154] Sodium dodecylbenzene sulfonate; anionic; component in household liq. detergents, industrial cleaning and textile wet processing formulations; liq.; 31.5% conc.

Conco AAS-40F. [154] Sodium dodecylbenzene sulfonate; anionic; component in household and industrial detergents, textile scouring compounds, car wash products; flake; 40% conc.

Conco AAS-40G. [154] Sodium dodecylbenzene sulfonate; anionic; component in household and industrial detergents, textile scouring compounds, car wash products; granule; 40% conc.

Conco AAS-40S. [154] Sodium dodecylbenzene sulfonate; anionic; lt. duty dishwashing detergent component; liq.; 40% conc.

Conco AAS-45S. [154] Sodium dodecylbenzene sulfonate; anionic; surfactant for dishwashing compds., textile processing, liq. soaps, leather products, dairy and automobile cleaners, cosmetic products, wall paper remover, rug shampoos, household detergents; tan-wh. paste, liq.; typical odor; cloud pt. 17 C; 41–43% act., 44–47% solids

Conco AAS-50M. [154] Ammonium dodecylbenzene sulfonate; anionic; component in liq. detergent formulations; liq.; 40% conc.

Conco AAS-60S. [154] TEA dodecylbenzene sulfonate; anionic; detergent and wetting agent for household and dishwashing, industrial cleaners, bubble baths, shampoos, textile dyeing compds.; lt. straw visc. liq.; sol. in water and methanol; dens. 9.3 lb/gal; pH 6.8–7.2; 60% act.

Conco AAS-65S. [154] Sodium dodecylbenzene sulfonate; anionic; detergent used in liq. detergent formulations; slurry; 60% conc.

Conco AAS-75S. [154] Calcium dodecylbenzene sulfonate; anionic; surfactant for emulsifier systems for pesticides, in dye carrier formulations; dk. amber visc. liq.; sol. in kerosene, xylene, aromatic naphtha; insol. in water; sp.gr. 1.02; visc. 100–150 cps; flash pt. 105 F (CC); pH 6.0–9.0.

Conco AAS-90S. [154] Sodium dodecylbenzene sulfonate; anionic; component of household and industrial detergent compds.; personal care products; powd., flake, granule; 85–90% conc.

Conco AAS-98S. [154] Dodecylbenzene sulfonic acid; anionic; detergent intermediate used in mfg. of shampoos, detergent, metal cleaners; br. liq.; water sol.; sp.gr. 1.04; biodeg.; 98+% act.

Conco AAS 3H. [154] Isopropylamine dodecylbenzene sulfonate; anionic; surfactant used in solv. formulations for cleaning engine blocks, garage floors, drycleaning detergents, fuel oil additives; amber; typ. odor; dens. 8.4 lb/gal; pH 6.1–7.5 (10%); 97% min. act.

Conco AAS Special 3. [154] Isopropylamine dodecylbenzene sulfonate; anionic; see Conco AAS 3H; amber; typ. odor; sol. in kerosene, alcohol, pine oil, min. spirits, chlorinated and aromatic solv.; dens. 8.4 lb/gal; pH 6.1–7.5; 97% min. act.

Conco AOS-40. [154] Sodium alpha olefin sulfonate; anionic; surfactant for textile auxiliaries, car wash products, industrial cleaners, dishwashing detergents, personal care products, emulsion polymerization; amber liq.; pH 7.5 (5%); 38.5–40.5% act.

Conco AOS-90F. [154] Sodium alpha olefin sulfonate; anionic; used in detergents and personal care products; powd.; 90% conc.

Conco ATR-98S. [154] Tridecyl benzene sulfonic acid; anionic; detergent intermediate, neutralized with wide range of alkalies; liq.; sp.gr. 1.04; biodeg.; 98+% act.

Conco Emulsifier DCS. [154] Amine condensate, modified; anionic; emulsifier for kerosene, Varsol, and min. spirits; liq.; 100% conc.

Conco Emulsifier K. [154] Coconut diethanolamide; nonionic; emulsifier for Varsol, naphtha, tuluol, xylol; liq.; 100% conc.

Concofac 277. [154] Phosphorylated organic ester, anionic; surfactant; detergent and wetting agent for automotive applications; textile penetrant, scouring and wetting agent; floor, rug and kitchen cleaner, metal finishing chemical; water-wh. to lt. straw liq.; sol. in aq. sol'ns.; pH 4–6; 30–32% solids.

Conco NI-21. [154] Nonoxynol-2; nonionic; dispersant, emulsifier, wetting, foaming and penetrating agent for textile wet processing, insecticide and cosmetic formulations, industrial and household detergents, latex and polymer emulsifier and post polymerization stabilizer, waxes and polishes, bottlewashing comps., polyester resin emulsifier, demulsifier for crude petroleum/oil emulsions; liq.; sp.gr. 0.98–0.99; 100% act

Conco NI-43. [154] Nonoxynol-4; nonionic; see Conco NI-21; liq.; sp.gr. 1.02–1.04; 100% act.

Conco NI-60. [154] Nonoxynol-6; nonionic; see Conco NI-21; liq.; sp.gr. 1.03–1.05; cloud pt. 32 F; 100% act.

Conco NI-90. [154] Nonoxynol-9; nonionic; see Conco NI-21; liq.; sp.gr. 1.04–1.08; cloud pt. 87 F; 100% conc.

Conco NI-100. [154] Nonoxynol-10; nonionic; see Conco NI-21; liq.; sp.gr. 1.05–1.07; cloud pt. 125–133 F; 100% conc.

Conco NI-150. [154] Nonoxynol-15; nonionic; see Conco NI-21; liq.; sp.gr. 1.07–1.08; cloud pt. 203–212 F; 100% act.

Conco NI-185. [154] Nonoxynol-20; nonionic; see Conco NI-21; wax; cloud pt. > 212 F; 100% act.

Conco NI-187. [154] Nonoxynol-30; nonionic; see Conco NI-21; wax; cloud pt. > 212 F; 100% act.

Conco NI-190. [154] Nonoxynol-40; nonionic; see Conco NI-21; wax; cloud pt. > 212 F; 100% act.

Conco NI-197. [154] Nonoxynol-40; nonionic; see Conco NI-21; liq.; sp.gr. 1.10; cloud pt. > 212 F; 100% act.

Conoc NI-2000. [154] Nonoxynol-50; nonionic; see Conco NI-21; wax; cloud pt. 163 F; 100% act.

Concopal A, AS. [154] Sulfated alkyl phenoxy poly (ethyleneoxy) ethanol, ammonium salt; anionic; detergent, wetting agent, emulsifier, foamer used in dishwashing formulations, emulsion polymerization, textile wet processing; liq.; 60% conc.

Conco Sulfate 2A1. [154] Sodium dodecyl diphenyl oxide disulfonate; anionic; detergent, emulsifier, wetting agent, dispersant, coupler for textile wet processing, hard surface and built cleaners, cosmetics, emulsion polymerization; yel./br. liq.; typ. odor; sp.gr. 1.16; visc. 150 cps; 45% act.

Conco Sulfate 2A1 Acid. [154] Dodecyl diphenyl oxide disulfonic acid; anionic; used in acidic sol'ns. or in preparation of other salts; dk. liq.; sol. in water, 15% HCl; sp.gr. 1.1; visc. 500 cps; 70% act.

Conco Sulfate 3B2. [154] Sodium decyl diphenyl oxide disulfonate; anionic; dispersant, solubilizer, coupler for agric. chemicals, herbicides, food contact; amber liq.; typ. odor; sol. in water, 25% NaOH, 20% HCl; surf. tens. 38 dynes/cm (0.1% aq.); 45% act.

Conco Sulfate 3B2 Acid. [154] Decyl dephenyl oxide disulfonic acid; anionic; detergent, emulsifier, wetting agent, dispersant used in formulating cleaning and agric. products, herbicides; dk. liq.; sol. in water, 15% HCl; sp.gr. 1.1; visc. 500 cps; 70% act.

Conco Sulfate 4C3. [154] Sodium hexadecyl diphenyl ether disulfonate; anionic; see Conco Sulfate 2A1; yel./br. liq.; typ. odor; sp.gr. 1.12; visc. 150 cps; 35% act

Conco Sulfate 216. [154] Ammonium laureth sulfate; anionic; detergent; foaming agent for dishwashing systems; textile wet processing; all-purpose cleaners, emulsion polymerization; yel. liq.; typ. odor; 60% act.

Conco Sulfate 219, 219 Special. [154] Sodium laureth sulfate; anionic; see Conco Sulfate 216; pale yel. liq.; typ. odor; 60% and 30% act. respectively.

Conco Sulfate A. [154] Ammonium lauryl sulfate; detergent and wetting agent; used in personal care products, rug shampoos, household detergents, lt. yel. liq.; pleasant odor; visc. 6500 cps max.; cloud pt. 5 C; pH 6.3–7.0 (10%); 29–32.5% act.

Conco Sulfate EP. [154] DEA lauryl sulfate; anionic; detergent, wetting in personal care products, household detergents; lt. yel. liq.; visc. 225 cps; cloud pt. 6 C; pH 7.5–8.5; 36–39.9% act.

Conco Sulfate M. [154] Magnesium lauryl sulfate; anionic; household detergents; cosmetic formulations; liq.; 30% conc.

Conco Sulfate P. [154] Potassium lauryl sulfate; anionic; household detergent; cosmetic formulations; wh. paste; pleasant odor; visc. 225 cps max.; cloud pt. 2.5 C; pH 7.0–8.0 (10%); 29.0–32.5% act

Conco Sulfate RA. [154] Alcohol ether sulfate; detergent used in textile wet processing, heavy-duty laundry and household detergents.

Conco Sulfate TL. [154] TEA lauryl sulfate; anionic; detergent, wetting; shampoo base, pigment dispersant; lt. yel. liq.; pleasant odor; visc. 225 cps max.; cloud pt. 2.5 C; pH 6.4–7.7 (10%); 39.5–44.8% act

Conco Sulfate WA. [154] Sodium lauryl sulfate; anionic; shampoo base, rug shampoos, cleansing lotions and household products; pigment dispersant for personal care products; emulsion polymerization; powd.; typ. odor; pH 6.5–7.5 (10%); 92–98% act.

Conco Sulfate WAG. [154] Sodium lauryl sulfate; see Conco Sulfate WA Dry; wh. needle; typ. odor; 92–98% act.

Conco Sulfate WAS. [154] Sodium lauryl sulfate; see Conco Sulfate WA Dry; colorless liq.; pleasant odor; visc. 1000–5000 cps; cloud pt. 17–23 C; pH 7.4–8.6 (10%); 28.5–33.0% act

Conco Sulfate WA Special. [154] Sodium lauryl sulfate; see Conco Sulfate WA Dry; wh. smooth paste; pleasant odor; visc.

19,000–31,000 cps; cloud pt. 17–21 C; pH 7.4–8.6 (10%); 28.5–33.0% act.

Conco Sulfate WB-45, WBS-45. [154] Sodium lauryl sulfate; surfactant for industrial, cosmetic, and household products, e.g., shampoos, rug shampos, ore flotation, pigment dispersions, emulsion polymers; wh. liq.; pleasant odor; cloud pt. 47 F; pH 8.0–8.5 (10%); 45–47.5% act.

Conco Sulfate WE. [154] Sodium laureth sulfate; anionic; detergent and personal care products; liq.; 30% conc.

Conco Sulfate WM. [154] Ammonium laureth sulfate; anionic; detergent, personal care products, and household products; liq.; 30% conc.

Conco Sulfate WR. [154] Sodium lauryl sulfate; anionic; detergent; dentifrice and foaming agent; powd.; 99% conc.

Conco Sulfate WX. [154] Sodium lauryl sulfate; anionic; detergent, dentifrice and foaming agent; powd.; 90–96% conc.

Conco SXS. [154] Sodium xylene sulfonate; surfactant for liq. detergents, household cleaners, wood pulping, dye mfg.; water-wh. liq.; typ. odor; 40% min. act.

Conco XA-C. [154] Palmitamine oxide; cationic, amphoteric; foamer, foam stabilizer for household and industrial detergents; post-polymerization stabilizer, electroplating brightener, cosmetic emollient, polymerization emulsifier, textile dyeing assistant; sol. in water, IPA, ethanol; biodeg.; 20% conc.

Conco XA-L. [154] Lauramine oxide; cationic, amphoteric; see Conco XA-C; sol. in water, IPA, ethanol; biodeg.; 30% conc.

Conco XA-M. [154] Myristamine oxide; cationic, amphoteric; see Conco XA-C; sol. in water, IPA, ethanol; biodeg.; 30% conc.

Conco XA-MC. [154] Myristyl-cetyl dimethyl amine oxide; cationic, amphoteric; see Conco XA-C; sol. in water, IPA, ethanol; biodeg.; 30% conc.

Conco XA-S. [154] Stearamine oxide; cationic, amphoteric; see Conco XA-C; sol. in IPA, ethanol; forms emulsions in water; biodeg.; 30% conc.

Conco XA-T. [154] Tallow amine oxide; cationic, amphoteric; see Conco XA-C; sol. in IPA, ethanol; forms emulsions in water; biodeg.; 30% conc.

Conco XA-Y. [154] Cocamine oxide; cationic, amphoteric; see Conco XA-C; sol. in water, IPA, ethanol; biodeg.; 30% conc.

Condanol 42/FAS. [212] Surfactant blend; detergent used as shampoo base; pale yel. liq.; faint, typ. odor; sp.gr. 1.0; medium visc.; 25% act.

Condanol B2003. [212] Tetrasodium (1,2-dicarboxyethyl)-N-alkyl sulfosuccinamide; anionic; detergent, emulsifier, foaming agent, foam stabilizer, solubilizer for industrial and household uses; clear amber liq.; sol. in water and electrolytes; 35% solids.

Condanol CMS. [212] Disodium coconut sulfosuccinamate; emulsifier, foaming agent, pale yel. liq.; slight, typ. odor; sp.gr. 1.0; 40% act.

Condanol CS 30. [212] Surfactant blend; detergent for carpet shampoo base; pale yel. liq.; formalin odor; sp.gr. 1.0; low visc.; 80% act.

Condanol DLS 35. [212] Diethanolamine lauryl sulfate; detergent, used as shampoo base; pale yel. liq.; slight typ. odor; sp.gr. 1.0; low visc.; 35% act.

Condanol Emulsifier BW. [507] Modified fatty acid deriv.; nonionic; emulsifier used for cleansing liq., hand cleaners, cleaning of oily and greasy surfaces; oil sol.; 97% solids, 3% water.

Condanol Emulsifier BWA. [212] Modified fatty acid deriv.; nonionic; emulsifier used for metal cleaning formulations, solv. degreasers, paraffin wax, cold cleaners to dewax cars; clear, yel. visc. liq.; 97% solids, 3% water.

Condanol FBR. [212] Anionic; detergent, raw material for weakly alkaline household and general-purpose cleaners; clear visc. liq.; sp.gr. 0.98; visc. 350 cp; 98% solids, 2% water.

Condanol FLM. [212] Raw material for foam fire-extinguisher base; lt. yel., clear liq.; sp.gr. 1.0–1.1; clear pt. –12 C; biodeg.

Condanol M 5047. [212] Detergent combination; surface cleaner preconcentrate; colorless, clear liq.; 45% solids.

Condanol MLS 35. [212] Monoethanolamine lauryl sulfate; detergent shampoo base; pale yel. liq.; slight typ. odor; sp.gr. 1.0; high visc.; 35% act.

Condanol NL2. [212] Sodium laureth sulfate (2 EO); detergent shampoo base; water wh. liq.; slight typ. odor; sp.gr. 1.0; medium visc.; 30% act.

Condanol NL3. [212] Sodium laureth sulfate (3 EO); detergent shampoo base; water wh. liq.; slight typ. odor; sp.gr. 1.0; low visc.; 30% act.

Condanol NL3, 28%. [212] Sodium laureth sulfate; anionic; detergent, foaming agent used in high quality shampoos and foam bath preparations; almost colorless, clear liq.; low odor; visc. 50–150 cp; 29% act.

Condanol NLS 28. [212] Sodium lauryl sulfate; anionic; detergent, emulsifier for cosmetic preparations; lt. yel. liq.; 28% act.

Condanol NLS 30. [212] Sodium lauryl sulfate; anionic; detergent, emulsifier for cosmetic preparations; off wh. paste; 30% act.

Condanol NLS 90. [212] Sodium lauryl sulfate; anionic; detergent, emulsifier for cosmetic preparations; wh. to lt. yel. powd.; 85% act.

Condanol OFR. [212] Surface cleaner; clear, colorless liq.; 16% solids.

Condanol S 1590. [212] Anionic; detergent used in wash-active preconcentrate, dishwashing liq.; clear, yel. visc. liq.; 44% solids.

Condanol S 1605. [212] Anionic; detergent with good wetting and emulsifying, foam stability for household and industrial uses; has skin protecting properties; clear yel. visc. liq.; 48% solids.

Condanol SB30L. [212] Detergent, base for cream shampoos; pale yel. liq.; odorless; sp.gr. 1.0; medium visc.; 30% act.

Condanol SBA/1. [212] Sodium laureth sulfosuccinate; detergent, used in hair shampoos, foam baths, metal cleaning; off wh. paste; slight typ. odor; sp.gr. 1.0; 40% act.

Condanol SBDO60. [212] Dioctyl sodium sulfosuccinate; super wetting agent with low foaming properties; pale yel. liq.; strong typ. odor; sp.gr. 1.0; medium visc.; 60% act.

Condanol SBDO70. [212] Sodium diisooctyl sulfosuccinate; anionic; wetting agent in cosmetic preparations and household cleaners, solubilizer; almost colorless liq.; surf. tens. 30 dynes/cm (0.1%); biodeg.; 64% act.

Condanol SBF 12. [212] Disodium lauryl sulfosuccinate; anionic; detergent, foam-improver used in cosmetic preparations, carpet and upholstery shampoos, syn. bar soaps; wh. paste; biodeg.; 39% solids.

Condanol SBF 12-Powd. [212] Disodium lauryl sulfosuccinate; anionic; see Condanol SBF 12; wh. or lt. yel. powd.; water sol.; biodeg.; 95% solids.

Condanol SBFA/3. [212] Sodium laureth sulfosuccinate; detergent for mild hair shampoos and foam baths; water wh. liq.; sp.gr. 1.0; low visc.; 40% act.

Condanol SBFA 30, 40%. [212] Disodium laureth sulfosuccinate; anionic; detergent raw material, base with good foaming; almost colorless, clear liq.; visc. 100–200 cp; biodeg.; 39% solids.

Condanol SBIL/3. [212] Sodium laureth isopropanolamide sulfosuccinate; detergent used for hair shampoos; pale yel. gel; typ. slight odor; sp.gr. 1.0; high visc.; 50% act.

Condanol SBL. [212] Sodium lauric MEA sulfosuccinate; detergent, wetting agent, used as carpet shampoo base, in fire fighting formulations; pale yel. liq., paste; slight typ. odor; sp.gr. 1.0; low visc.; 35% act.

Condanol SBL 203. [212] Disodium lauramido MEA-sulfosuccinate; anionic; mild detergent, foam improver used for cosmetics and household purposes; yel. turbid liq.; biodeg.; 39% solids.

Condanol SBV. [212] Modified sulfosuccinate; anionic; mild detergent, foaming agent, skin protecting properties; lt. yel. clear liq.; visc. 2500 cp; biodeg; 28% solids.

Condanol SBZ. [212] Sodium salt of the monosulfosuccinate of lauric isopropanolamide polyglycol ether (medium chain); detergent used in hair shampoos; pale yel. gel; slight typ. odor; sp.gr. 1.0; medium visc.; 50% act.

Condanol SCK 2040. [212] Anionic; detergent raw material for liq. cleaners, shampoos; yel., clear visc. liq.; visc. 600 cp; biodeg.; 39% solids.

Condanol TLS 40. [212] TEA lauryl sulfate; detergent, shampoo base; pale yel. liq.; slight typ. odor; sp.gr. 1.0; low visc.; 45% act.

Condanol TMS. [212] Disodium alkyl sulfosuccinate; anionic; emulsifier, wetting agent, foam-additive used in emulsion polymerization; soft creamy paste or liq.; biodeg.; 35% solids.

Condanol TS 25. [212] Sulfosuccinate-base detergent combination; anionic; detergent conc. for foam cleaners; lt. yel. liq.; clear pt. 8 C; biodeg.; 25% solids.

Condasol Sulfonic Acid K. [212] Straight chain docecylbenzene sulfonic acid; anionic surfactant used for washing and cleaning formulations; br. visc. liq.; 90% biodeg.

Condasol TKS 90/F. [212] Triammonium dodecylbenzenesulfonate; anionic; raw material for detergent formulations; used to clean steel ball bearings; dk. amber, visc. liq.; biodeg.; 85% act.

Condensate 128, AOC, COA, COM, RA. [97] Alkanolamide; nonionic; softener, foam stabilizer, thickener and detergent; liq.; 100% conc.

Condensate 640. [154] Alkanolamide; detergent and soil-suspending agent; used in floor and hard surface cleaners; lubricant.

Condensate 4476. [490] Amine condensate of fatty acid; wetting agent, detergent, high foamer, substitute for 100% coconut amine condensate, Condensate PC; Gardner 6–8 liq.; sp.gr. 1.020; 88–90% act.

Condensate CO. [154] Cocamide DEA and diethanolamine; nonionic; emulsifier, wetting agent, and detergent in textile wet processing, leather/fur processing, oil emulsification, insecticides, floor cleaning, heavy duty detergents, household cleaning, car wash, metal cleaning; amber liq.; typ. odor; water-sol.; sp.gr. 1.05 (60 F); pH 9–10; 97% min. act.

Condensate NP. [154] Cocamide DEA and diethanolamine; see Condensate CO; amber liq.; typ. odor; water-sol.; sp.gr. 1.05 (60 F); pH 8.1–9.5; 97% min. act.

Condensate PA. [154] Cocamide DEA; thickener, dispersant, foamer for cosmetic, personal care products, industrial and household cleaners, textile products, buffing and polishing compds.; yel. thick liq.; typ. odor; disp. in water; sp.gr. 0.97; alkali no. 30–50; pH 9.0–9.8; 90–95% amide content.

Condensate PC. [490] Coconut-diethanolamine condensation product; wetting and rewetting agent, detergent, high foamer, thickener, used in the textile industry; amber liq.; 100% act.

Condensate PE. [154] Lauramide DEA; see Condensate PA; disp. in water; wh. solid; typ. odor; sp.gr. 0.98; m.p. 97.5–99.0 F; alkali no. 30–47; pH 9.0–9.7; 90–95% amide.

Condensate PL. [154] Lauramide DEA and diethanolamine; nonionic; wetting agent, dispersant, detergent; foamer in surfactants; thickener and foam stabilizer in personal care products and dishwashing detergents; buffing compounds; lt. amber liq.; typ. odor; sp.gr. 1.02; alkali no. 175–205; pH 9–10; 100% act.

Condensate PM. [154] Myristamide DEA; see Condensate PA; wh. solid; typ. odor; disp. in water; sp.gr. 0.97; m.p. 100–102 F; alkali no. 30–48; pH 9.0–9.7; 90–95% amide.

Condensate PN. [154] Cocamide DEA and diethanolamine; nonionic; see Condensate CO; also used as an emulsifier; lt. amber liq.; typ. odor; water-sol.; sp.gr. 1.04 (60 F); pH 8.0–9.5; 97% min. act.

Condensate PO. [154] Cocamide DEA; see Condensate PA; lt. yel. liq.; typ. odor; disp. in water; sp.gr. 0.98; alkali no. 38–50; pH 9.0–9.7; 85% min. amide.

Condensate PS. [154] Cocamide DEA and diethanolamine; emulsifier, wetting agent, foam builder and stabilizer; used in consumer and commercial cleaners and textile processing applications; lt. amber liq.; characteristic odor; sol. in water, alcohol, carbon tetrachloride; sp.gr. 1.02; pH 8.5–9.5; 97% min. act.

Consamine 15. [153] Alkyl POE ether; nonionic; detergent, scouring, wetting agent for textiles; liq.; 100% conc.

Consamine CA. [153] Fatty alkanolamide; nonionic; detergent, emulsifier, intermediate, detergent base, thickener, fulling agent used in textiles; liq

Consamine K-Gel. [153] Sodium-N-methyl-N-oleyl-taurate; anionic; wetting, scouring, and detergent for textile applications; gel.

Consamine PA. [153] Complex alcohol ester; anionic; wetting agent, and penetrant for textile processes; liq

Consonyl FIX. [153] Alkyl naphthalene sulfonate; anionic; fixative for improving wash fastness of acid dyestuffs on nylon; resist for direct and naphthol dyes on nylon; liq.

Consos Castor Oil. [153] Sulfated castor oil; anionic; retarder and leveling agent for dyeing application in textiles; dispersant and lubricant; liq

Consoscour 47. [153] Alkylaryl sulfonate; anionic; detergent, scouring, and leveling agent for dyeing applications; liq.; 47% conc.

Consostat DKM. [153] POE alkyl amine; cationic; antistat used for wool and syn. fibers; liq.; 88% conc.

Consotard No. 895. [153] Quat. deriv.; cationic; retarder for dyestuffs on acrylic fiber; liq

Cordon AES-65. [237] Aliphatic ester sulfate; anionic; lubricant, dye leveling, prevents chafe marks in textiles; liq.; 65% conc.

Cordon N-400. [237] Complex alkyl-naphthalene sulfonate; anionic; disperse dye leveling; liq.; 22% conc.

Cordon NU 890/75. [237] Sulfated castor oil; anionic; emulsifier for pine oil, plasticizer, lubricant, textile dyeing assistant; liq.; 75% conc.

Cordon PB-870. [237] Sulfated tall oil; anionic; lubricant, emulsifier; liq.; 42% conc.

Corona Lanolin. [159] Anhydrous lanolin; nonionic; conditioning emollient, moisturizing agent, cosolvent, plasticizer, w/o emulsifier, superfatting agent, wetting/dispersing agent, used as pharmaceutical base; paste; 100% conc.

Coronet Lanolin. [159] Superrefined cosmetic grade lanolin BP; nonionic; conditioning emollient, moisturizing agent, co-solv., plasticizer, w/o emulsifier, superfatting agent, wetting/dispersing agent, used as pharmaceutical base; minimum odor, color, batch variation in this grade; slightly higher m.p. than USP and BP standards; paste; 100% conc.

Cosmopon 35. [598] Sodium alkyl sulfate; anionic; spray dried wool detergent powd., hand cleaner paste form; paste; 48% conc.

Cosmopon BL. [598] Fatty alkanolamide sulfosuccinate; anionic; rug shampoo, bubble bath, conditioner; paste; 50% conc.

Cosmopon BLM. [598] Sodium alkyl sulfosuccinate; anionic; cream shampoos; paste; 30% conc.

Cosmopon BN. [598] Sodium oleyl sulfosuccinamate; anionic; emulsifier for emulsion polymerization; foaming agent for latex emulsions; antigelling and cleaning agents for paper mill felts; paste; 35% conc.

Cosmopon BT. [598] Sodium alkyl ethoxy sulfosuccinate; anionic; baby shampoo and bubble paths; liq.; 40% conc.

Cosmopon ESC. [598] Sodium lauryl ether sulfate with amide; anionic; hair shampoos; paste; 40% conc.

Cosmopon LE 50. [598] Sodium lauryl ether sulfate; anionic; personal care products and liq. detergents; paste; 50% conc.

Cosmopon TR. [598] TEA lauryl sulfate; anionic; hair shampoo, bubble bath, car wash, liq. hand cleaner; liq.; 40% conc.

Cosmowax. [159; 160] Stearyl alcohol, steareth-20, and steareth-10; nonionic; wax, self-emulsifying; o/w emulsifier, thickener, stabilizer for cosmetics, pharmaceuticals; solid; low odor; HLB 8.0; m.p. 47–55 C; sapon. no. 1 max.; 100% conc.

Cosmowax J. [159; 160] Cetearyl alcohol and ceteareth-20; nonionic; o/w emulsifier and stabilizer for personal care products, pharmaceutical creams; creamy wh. waxy solid; low odor; HLB 8.5; m.p. 47–55 C; sapon. no. 1 max.; 100% conc.

Cosmowax K. [159; 160] Stearyl alcohol and ceteareth-20; nonionic; emulsifier, stabilizer for personal care products and pharmaceuticals; creamy wh. waxy solid; low odor; HLB 9.0; m.p. 55–63 C; sapon. no. 1 max.; 100% conc.

Cowax-3, -10. [287] Wax, emulsifying; nonionic; o/w and w/o emulsions for cosmetic and pharmaceutical industry; solid

Cowax-13. [287] Wax, emulsifying; nonionic; cosmetic and pharmaceutical industries; solid; HLB 4.5

CPA-Alpha. [196] Anionic/nonionic; detergent, wetting agent, emulsifier for water and drycleaning solv.; very lt. yel. amber liq.; slight odor; complete sol. in drycleaning solv.; 100% act.

CPH-27-N. [280] PEG-4 laurate; nonionic; solubilizer and emulsifier; liq.; 100% conc.

CPH-30-N. [280] PEG-8 laurate; nonionic; solubilizer and emulsifier; liq.; HLB 12.9; 100% conc.

CPH-31-N. [280] Glyceryl oleate; nonionic; antiblocking agent; liq.; HLB 3.7; 100% conc.

CPH-39-N. [280] PEG-4 oleate; nonionic; spreading agent, defoamer; liq.; HLB 7.6; 100% conc.

CPH-41-N. [280] PEG-12 oleate; nonionic; solubilizer and emulsifier; liq.; HLB 13.6; 100% conc.

CPH-43-N. [280] PEG-12 laurate; nonionic; solubilizer and dispersant; solid; HLB 14.6; 100% conc.

CPH-53-N. [280] Glyceryl stearate; nonionic; thickener and emulsifier; flake; HLB 3.7; 100% conc.

CPH-79-N. [280] PEG-8 dilaurate; nonionic; wetting agent, pigment dispersant; liq.; HLB 9.7; 100% conc.

CPH-205-NX. [280] Glyceryl oleate; nonionic; lube additive, rust preventive; liq.; HLB 3.0; 100% conc.

CPH-211-N. [280] PEG-8 dioleate; nonionic; water treatment, lubricant; liq.; HLB 8.1; 100% conc.

CPH-213-N. [280] PEG-12 dioleate; nonionic; lubricant emulsifier; liq.; HLB 10.5% conc.

CPH-250-SE. [280] Glyceryl monostearate; anionic; emulsifier; flake; 100% conc.

CPH-353-SE. [280] Glyceryl oleate; anionic; corrosion inhibitor in oils; liq.; 100% conc.

CPH-362-N. [280] Glyceryl oleate; nonionic; oil additive, rust preventive; liq.; HLB 4.7; 100% conc.

CPH-376-N. [280] POE laurate; nonionic; solubilizer and emulsifier; liq.; 100% conc.

Crafol AP-31, -33. [494] Potassium lauryl phosphate; anionic; mfg. of manmade textile fibers as spinning oil for cut polyester fibers and cotton; antistat; spinning oil for carpets; paste.

Cralane KR-13, -14. [494] Lanolin, ethoxylated; nonionic; wetting, cleaning, and superfatting agents, o/w emulsifier, conditioner, solubilizer, and plasticizer for personal care products and germicides; solid; 100% conc.

Cralane LR-10, -11. [494] Lanolin, ethoxylated; nonionic; see Cralane KR-13; visc. liq.; 50% conc.

Crapol AU-20. [494] Alkyl dimethyl benzyl ammonium chloride; cationic; germicide, disinfectant, algicide for swimming pools, industrial waste water; liq.; 80 ± 2% conc.

Crapol AU-21. [494] Alkyl dimethyl benzyl ammonium chloride, modified; cationic; retarding and leveling agent for acrylic fiber dyeing; liq.; 50 ± 1% conc.

Crapol AU-23. [494] Stearyl dimethyl benzyl ammonium chloride plus additives; cationic; antistat for creams; paste; 25 ± 1% conc.

Crapol AU-24. [494] Alkyl dimethyl benzyl ammonium chloride; cationic; see Crapol AU-20; liq.; 80 ± 2% conc.

Crapol AU-25. [494] Quat. ammonium blend; cationic; migration and leveling agent for acrylic fiber dyeing; liq.; 40 ± 1% conc.

Crapol AU-40. [494] Alkylamine polyglycol ether metasulfate plus additives; cationic; antistat and surface act. agent used in mfg. of heavy duty detergents; liq.; 46 ± 2% conc.

Crapol AV-10. [494] Coconut hydroxyethyl imidazoline; cationic; used as corrosive inhibitor, dispersant and fluidizing agent for pigments, emulsions, and agric. products; acid detergent for food and dairy products; liq., paste; 100% conc.

Crapol AV-11. [494] Lauryl hydroxyethyl imidazoline; cationic; see Crapol AV-10; liq.; 100% conc.

Cremophor A 6. [82] Stearyl alcohol; nonionic; emulsifier for cosmetic and pharmaceutical preparations; wh. wax; sol. in water, alcohol; sp.gr. 0.896–0.906 (60 C); HLB 10–12; sapon. no. < 3; pH 6–7; 100% act.

Cremophor A 11. [82] Ceteareth-11; nonionic; emulsifier for cosmetic and pharmaceutical preparations; wh. wax; sol. in water, alcohol; sp.gr. 0.964–0.968 (60 C); HLB 12–14; sapon. no. < 1; pH 6–7; 100% act.

Cremophor A 25. [82] Ceteareth-25; nonionic; emulsifier for cosmetic and pharmaceutical preparations; wh. powd.; sol. in water, alcohol; sp.gr. 1.020–1.028 (60 C); HLB 15–17; sapon. no. < 3; pH 5–7; 100% act.

Cremophor EL. [82; 86] PEG-36 castor oil; nonionic; solubilizer, emulsifier used for essential oils, pharmaceuticals, cosmetics, veterinary medicine; pale yel. liq.; sol. in water, ethanol, propanol, ethyl acetate, chloroform, carbon tetrachlorethylene, benzene, toluene, and xylene; sp.gr. 1.05–1.06; visc. 700–850 cps; HLB 12–14; sapon no. 65–70; pH 6–8 (10% aq.); 100% act.

Cremophor RH 40. [82; 86] PEG-40 hydrog. castor oil; nonionic; solubilizer for essential oils and perfumery, emulsifier; wh. paste; sol. in water, ethanol, IPA, *n*-propanol, ethyl acetate, benzene, toluene, and xylene; HLB 14–16; pH 6–7 (10% aq.); 100% act.

Cremophor RH 60. [82; 86] PEG-60 hydrog. castor oil; nonionic; See Cremophor RH 40; wh. paste; sol. see Cremophor RH 40; HLB 15–17; pH 6–7 (10% aq.); 100% act.

Cremophor RH 410. [82] PEG-40 hydrog. castor oil; nonionic; solubilizer for essential oils and perfumery synthetics, emulsifier; liq.; sol. see Cremophor RH 40; HLB 14–16; visc. 1500–2000 cps; pH 6–7 (10% aq.); 90% act.

Cremophor RH 455. [82] PEG-40 hydrog. castor oil and propylene glycol; nonionic; solubilizer for essential oils and perfumery synthetics, emulsifier; liq.; sol. see Cremophor RH 40; HLB 14–16; visc. 1000–1500 cps; pH 6–7 (10% aq.); 90% act.

Cremophor S 9. [82] PEG-9 stearate; nonionic; emulsifier for o/w type, thickening agent; ylsh. wh. solid; sol. in water, alcohols, acetone, ethyl acetate, chloroform, benzene, castor oil, and oleic acid; sp.gr. 0.97 (60 C); HLB 12; 100% act.

Cremophor WO7. [82] PEG-7 hydrog. castor oil; nonionic; emulsifier for cosmetic w/o preparations; visc. liq.; HLB 5; 100% conc.

Cremophor WO-A. [82] Fatty alcohol glycerol ether; nonionic; emulsifier for w/o emulsion-based cosmetics and pharmaceuticals; yel. liq.; sol. in min. and native oils and fats; sp.gr. 0.950–0.980; HLB 4.8–5.8; sapon. no. < 5; 100% act.

Crester KZ. [159] Polyglycerol and edible fatty acid mixed esters; nonionic; emulsifier for use in the food industry; solid; 100% conc.

Crester L. [159] Polyglycerol and fatty acid mixed ester; nonionic; emulsifier for margarine and syn. cream; paste; 100% conc.

Crester MG. [159] Mixed fatty acid monoglyceride and polyglyceryl ester blend; nonionic; margarine emulsifier; solid; 100% conc.

Crester PR. [159] Polyglycerol polyricinoleate; nonionic; visc. modifier in the food industry; liq.; 100% conc.

Cresterge AAS Spec. [128] Alkylaryl sulfonate amine salt, modified; anionic; detergent emulsifier for aliphatic, aromatic and chlorinated solvents; liq.; 95% conc.

Crestex Emulsifier S-11. [128] POE fatty ester; nonionic; emulsifier; solid; HLB 12.1; 100% conc.

Crestoil. [128] Sulfonated neatsfoot oil; anionic; emulsifier for new oils, paraffin oils in textiles; liq.; 75% conc.

Crestoil L-35. [128] Fatty acid and PEG ester; nonionic; emulsifier for textile and leather trades; liq.; 100% conc.

Crestol 88. [128] Org. phosphate ester; anionic; detergent, emulsifier; liq.; 88% conc.

Crestol 100, 326. [128] Org. phosphate ester; anionic; detergent, emulsifier; liq.; 100% conc.

Crestol E. [128] Org. phosphate ester; anionic; wetting and scouring agent, dispersant and detergent used in textile applications; liq.; 42% conc.

Crestolan Conc. [128] EO carboxylate; anionic; detergent, emulsifier used in textile, paper, and leather cleaning; gel; 75% conc.

Crestomite 24, 25. [128] Fatty amine condensate; nonionic; detergent, emulsifier used in cosmetic preparations; 80% conc.

Crestomul T. [128] Amine EO condensate; cationic; emulsifier, dispersant, dyeing, retarder, and antistat used in textile processes; liq.; 100% conc.

Crestonol-30C, -60T. [128] Alkylated aromatic sulfonate amine salt; anionic; detergent, wetting agent used in textile and leather processing; liq.; 60% conc.

Crestonol Acids. [128] Alkylated aromatic sulfonic acids; anionic; detergent, intermediate; used in preparation of surfactants; liq.; 88% conc.

Crestonol HCS. [128] Alcohol ethoxylate; nonionic; emulsifier for paraffin wax; liq.; 50% conc.

Crestonol SLR. [128] Alkylated aromatic sodium salt; anionic; detergent and wetting agent, surfactant; liq.; 35% conc.

Crestonon. [128] Alkylated aromatic sodium salt; anionic; detergent, emulsifier; textile, leather, compd.ing, cosmetics, paper; liq.; 100% conc.

Crestonon H, NI, Conc. [128] EO condensate; nonionic; surfactant, detergent, penetrant, emulsifier used in textile, leather, compounding, cosmetics, paper; liq.; HLB 5.0; 35% conc.

Crestopen 5X. [128] Sodium dioctyl sulfosuccinate; anionic; wetting agent; liq.; 60% conc.

Crestopen D Conc. [128] Sulfated syn. ester; anionic; wetting emulsifier; rewetting and softening agent for textile, leather, and paper trade; liq.; 67% conc.

Crestopen M. [128] SO_3 sulfonate; anionic; wetting, emulsifier for use in salt; liq.; 50% conc.

Crill 1. [159] Sorbitan laurate; nonionic; emulsifier, dispersant, cosolv., wetting agent, antifoam, visc. reducer, mold release, antiblock agent; used for cosmetics, food and food pkg., insecticides and herbicides, leather treatment, metalworking fluids, oil slick disp., paints and inks, pharmaceuticals, plastics, polishes, textiles; pale yel. clear visc. liq.; sol. in ethanol, oleyl alcohol, min. oil; HLB 8.6; sapon. no. 158–170; 98% conc.

Crill 2. [159] Sorbitan monopalmitate; see Crill 1; pale tan hard waxy solid; partially sol. in propylene glycol, ethyl and oleyl alcohols, olive oil, oleic acid; m.p. 46 C; HLB 6.7; sapon. no. 140–150; 98% conc.

Crill 3. [159] Sorbitan monostearate; emulsifier, lubricant, antistat; pale tan hard waxy solid; low odor; partially sol. in oleyl alcohol, olive oil, oleic acid; m.p. 51–54 C; HLB 4.7; sapon. no. 147–157; 98% conc.

Crill 4. [159] Sorbitan monooleate; w/o emulsifier, wetting agent, pigment dispersant; amber visc. liq.; sol. in ethyl and oleyl alcohols, min. oil, IPM, olive oil, oleic acid; HLB 4.3; sapon. no. 147–160; 98% conc.

Crill 6. [159] Sorbitan monoisostearate; w/o emulsifier, wetting agent, pigment dispersant; pale yel. visc. liq.; sol. in mineral oil, olive oil, partly sol. in oleyl alcohol, IPM; HLB 4.7; sapon. no. 143–153; 98% conc.

Crill 35. [159] Sorbitan tristearate; emulsifier, lubricant, antistat; pale tan hard waxy solid; partly sol. in oleyl alcohol, min. and olive oil, IPM, oleic acid; m.p. 48 C; HLB 2.1; sapon. no. 176–188; 98% conc.

Crill 43. [159] Sorbitan sesquioleate; w/o emulsifier, wetting agent, pigment dispersant; amber visc. liq.; sol. in oleyl alcohol, min. and olive oil, oleic acid; HLB 3.7;

sapon. no. 149–160; 98% conc.

Crill 45. [159] Sorbitan trioleate; w/o emulsifier, wetting agent, pigment dispersant; amber visc. liq.; sol. in oleyl alcohol, min. and olive oil, IPM, oleic acid; HLB 1.8; sapon. no. 172–186; 98% conc.

Crill 50. [159] Sorbitan monooleate (tech.); see Crill 1; liq., paste; HLB 4.3; 98% conc.

Crillet 1. [159] PEG-20 sorbitan monolaurate; nonionic; solubilizer, emulsifier, dispersant, wetting agent; often combined with a member of the Crill range in emulsification systems; used in cosmetics, food and food pkg., household products, insecticides, herbicides, metalworking fluids, paints and inks, pharmaceuticals, textiles; clear, yel. liq.; low odor; sol. in water, ethyl and oleyl alcohol, oleic acid; HLB 16.7; sapon. no. 40–51; surf. tens. 38.5 dynes/cm (0.1%); 97% conc.

Crillet 2. [159] PEG-20 sorbitan monopalmitate; emulsifier, dispersant, penetrant, leveling agent, lubricant, antistat; yel. amber pasty liq.; sol. in water, ethyl and oleyl alcohol, oleic acid; HLB 15.6; sapon. no. 43–49; surf. tens. 41.5 dynes/cm (0.1%); 97% conc.

Crillet 3. [159] PEG-20 sorbitan monostearate; emulsifier for cosmetics and pharmaceuticals, dispersant for insecticides, herbicides, cattle dyes, penetrant, leveling agent, lubricant, antistat; yel. liq. gels to soft solid on cooling; sol. in ethyl and oleyl alcohol, oleic acid; HLB 14.9; sapon. no. 45–55; surf. tens. 42.5 dynes/cm (0.1%); 97% conc.

Crillet 4. [159] PEG-20 sorbitan monooleate; emulsifier, dispersant, solubilizer, lubricant, detergent, antistat, wetting agent for cosmetics and pharmaceuticals; clear yel. amber liq.; faint char. odor; sol. in water, ethyl and oleyl alcohols, oleic acid; HLB 15.0; sapon. no. 45–55; surf. tens. 42.5 dynes/cm (0.1%); 97% conc.

Crillet 6. [159] PEG-20 sorbitan monoisostearate; see Crillet 1; clear yel. liq.; sol. in water, ethyl and oleyl alcohols, oleic acid, xylene, trichlorethylene; HLB 14.9; sapon. no. 40–50; surf. tens. 38.6 dynes/cm (0.1%); 97% conc.

Crillet 11. [159] PEG-4 sorbitan monolaurate; solubilizer used in cosmetics, emulsifier and dispersant for pharmaceuticals; clear yel. liq.; sol. in ethyl and oleyl alcohols, oleic acid, kerosene; HLB 13.3; sapon. no. 100–115; surf. tens. 34.7 dynes/cm (0.1%); 97% conc.

Crillet 31. [159] PEG-4 sorbitan monostearate; emulsifier for cosmetics and pharmaceuticals, dispersant for insecticides, herbicides, cattle dyes, penetrant, leveling agent, lubricant, antistat; cream pale yel. waxy solid; sol. in oleyl alcohol, oleic acid; HLB 9.6; sapon. no. 98–113; surf. tens. 41.5 dynes/cm (0.1%); 97% conc.

Crillet 35. [159] PEG-20 sorbitan tristearate; see Crillet 31; cream/buff waxy solid; sol. in ethyl and oleyl alcohols, oleic acid, trichlorethylene, partly sol. in water; HLB 10.5; sapon. no. 88–98; surf. tens. 42.5 dynes/cm (0.1%); 97% conc.

Crillet 41. [159] PEG-5 sorbitan monooleate; emulsifier, dispersant, solubilizer, lubricant, detergent, antistat, wetting agent, for cosmetics and pharmaceuticals; clear yel. amber liq.; sol. in ethyl and oleyl alcohol, IPM, oleic acid, kerosene, butyl stearate; HLB 10.0; sapon. no. 96–104; surf. tens. 37 dynes/cm (0.1%); 97% conc.

Crillet 45. [159] PEG-20 sorbitan trioleate; see Crillet 41; clear amber visc. liq.; sol. in ethyl and oleyl alcohols, IPM, oleic acid, kerosene, trichlorethylene, butyl stearate; HLB 11.0; sapon. no. 82–95; surf. tens. 41 dynes/cm (0.1%); 97% conc.

Crillon CDY. [159] Coconut diethanolamide; nonionic; sol. and cutting oils; liq.

Crillon LDE. [159] Lauric diethanolamide; nonionic; detergent and foam stabilizer; o/w emulsifiers, antistat; anticorrosive; liq.

Crillon LME. [159] Lauric monoethanolamide; nonionic; lubrication; skin protection agent; solid.

Crillon ODE. [159] Oleic diethanolamide; nonionic; used in personal care products; liq.

Crodafos Series. [159] Alkyl ether phosphates, available in acid form or neutralized with appropriate base; anionic surfactants, detergents, wetting agents, coupling agents, corrosion inhibitors, foam stabilizers, thickeners, lubricants, antistats, for textile and other industrial uses; many and varied forms; 99% min. act.

Crodafos 25D Series. [159] Alkyl ether phosphates based on syn. C_{12}/C_{15} alcohols; available with 2, 5, and 10 moles EO and as Crodafos 25D TEA salt; anionic; see Crodafos Series; colorless to pale yel. liq. to soft paste; pH 1.8–2.5 (3%); 99% min. act.

Crodafos 25 D2 Acid. [159] Phosphated C_{12}–C_{15} ether (2 EO); anionic; see Crodafos 02; liq.; 99% conc.

Crodafos 25 D5 Acid. [159] Phosphated

C_{12}–C_{15} ether (5 EO); anionic; see Crodafos 02; liq.; 99% conc.

Crodafos 25 D10 Acid. [159] Phosphated C_{12}–C_{15} ether (10 EO); anionic; see Crodafos 02; paste; 99% conc.

Crodafos CAP. [160] PPG-10 cetyl ether phosphate; anionic; w/o emulsifier; antistat used in personal care products; liq.; 100% conc.

Crodafos CDP. [159] Cetyl diethanolamine phosphate; anionic; emulsifier and stabilizer for o/w emulsions; powd.; 100% conc.

Crodafos CS Series. [159] Alkyl ether phosphates based on ceto stearyl alcohol; avail. with 2, 5, and 10 moles EO and as Crodafos CS TEA salt; see Crodafos Series; wh. to buff hard wax; pH 1.8–2.5 (3%); 99% min. act.

Crodafos CS2 Acid. [159] Phosphated ceto/stearyl ether (2 EO); anionic; emulsifier and stabilizer for o/w emulsions; solid; 99% conc.

Crodafos CS5 Acid. [159] Phosphated ceto/stearyl ether (5 EO); anionic; see Crodafos CS2 Acid; solid; 99% conc.

Crodafos CS10 Acid. [159] Phosphated ceto/stearyl ether (10 EO); anionic; see Crodafos CS2 Acid; solid; 99% conc.

Crodafos ID Series. [159] Alkyl ether phosphates based on isodecanol; avail. with 2, 5, and 10 moles EO and as Crodafos ID TEA salts; see Crodafos Series; colorless to pale yel. fluid to paste; pH 1.8–2.5 (3%); 99% min. act.

Crodafos N Series. [159] Phosphated oleyl polyether; anionic; avail. as Crodafos N3 based on Volpo N3 and Crodafos N10 based on Volpo N10; avail. as the free acids and as neutralized (salt) versions; emulsifier esp. for use at low conc., solubilizer, detergent, antistat for use in cosmetics, pharmaceuticals, and toiletries; paste; 100% act.

Crodafos N3 Acid. [159; 160] Oleth-3 phosphate; anionic; surfactant, conditioner, antistat, emulsifier, gelling agt. for cosmetics, pharmaceuticals, toiletries; corrosion inhibitors in aerosols; lubricant in textile and metal processing; Gardner 11 max.; sapon. no. 125–145; pH 1–3 (2% aq.); 100% act.

Crodafos N3 Neutral. [160] DEA-oleth-3 phosphate; anionic; see Crodafos N3 Acid; Gardner 11 max. paste; pH 6–7 (2% aq.); 100% conc.

Crodafos N5 Acid. [159] Oleth-5 phosphate; anionic; see Crodafos N3 Acid; liq.; 99% conc.

Crodafos N10 Acid. [159; 160] Oleth-10 phosphate; anionic; see Crodafos N3 Acid; paste; sapon. no. 75–115; pH 1–3 (2% aq.); 100% act.

Crodafos N10 Neutral. [160] DEA-oleth-10 phosphate; anionic; see Crodafos N3 Acid; Gardner 11 max. paste; pH 5.5–7.0 (2% aq.); 100% conc.

Crodafos O2 Acid. [159] Oleth-2 phosphate; anionic; conditioner for hair and shampoo products; liq.; pH 1.8–2.5 (3%); 99% conc.

Crodafos O2 TEA. [159] TEA-oleth-2 phosphate; pH 1.8–2.5 (3%); 99% min act.

Crodafos O5 Acid. [159] Oleth-5 phosphate; anionic; see Crodafos 02 Acid; liq.; pH 1.8–2.5 (3%); 99% conc.

Crodafos O10 Acid. [159] Oleth-10 phosphate; anionic; see Crodafos 02 Acid; paste; pH 1.8–2.5 (3%); 99% conc.

Crodafos SG. [159; 160] PPG-5 ceteth-10 phosphate; anionic; see Crodafos N3 Acid; esp. for shampoos and cream rinses; Gardner 8 max. liq.; sapon. no. 90–110; pH 1–3 (3%); 100% act.

Crodafos T Series. [159] Alkyl ether phosphates based on tridecanol; avail. with 2, 5, and 10 moles EO and as Crodafos T TEA salts; colorless to pale yel. liq. to visc. liq.; pH 1.8–2.5 (3%); 99% min act.

Crodafos T2 Acid. [159] Phosphated tridecyl ether (2 EO); anionic; see Crodafos SG; liq.; 99% conc.

Crodafos T5 Acid. [159] Phosphated tridecyl ether (5 EO); anionic; see Crodafos SG; liq.; 99% conc.

Crodafos T10 Acid. [159] Phosphated tridecyl ether (10 EO); anionic; see Crodafos SG; paste; 99% conc.

Crodalan 0477. [159] Lanolin alternative; nonionic; emollient, superfatting agent, w/o emulsifier for cosmetic and pharmaceutical systems; paste; 100% conc.

Crodalan AWS. [159; 160] Acetylated ethoxylated lanolin alcohols; nonionic; emollient, superfatting agent, conditioner, o/w emulsifier, dispersant, wetting agent, solubilizer used in cosmetic preparations; golden liq.; faint fatty odor; sol. in alcohol, water; sp.gr. 1.02–1.08; acid no. 3 max.; hyd. no. 55–67; pH 5–7 (10% aq.); 100% act.

Crodalan IPL. [159] Isopropyl lanolate; emollient, wetting agent used in bath oils, skin, hair and makeup cosmetics; pale yel. thin opalescent liq. or soft paste liquifying at skin temp.

Crodalan LA. [159; 160] Acetylated lanolin alcohols; nonionic; emollient, penetrant, plasticizer used in cosmetics, pharmaceuticals; clear, thin mobile liq.; odorless; low visc.; sol. in min. oil; acid no. 2 max.; sapon. no. 180–200; 100% act.

Crodamets. [159] Primary amine, ethoxylated; cationic; emulsifier; liq., paste; 100% conc.

Crodamine 1.16D. [162] Primary cetyl amine; cationic; emulsifier for herbicides, ore flotation, pigment dispersion; aux. for textiles, leather, rubber, plastics, and metal industries; liq., paste; 100% conc.

Crodamine 1.18D. [162] Stearylamine; cationic; see Crodamine 1.16D; solid; 100% conc.

Crodamine 1.HT. [162] Tallow amine, hydrog.; cationic; see Crodamine 1.16D; liq.; 100% conc.

Crodamine 1.O. [162] Oleyl amine; cationic; see Crodamine 1.16D; solid; 100% conc.

Crodamine 1.OD. [162] Oleyl amine; cationic; see Crodamine 1.16D; liq.; 100% conc.

Crodamine 1.T. [162] Tallow amine; cationic; see Crodamine 1.16D; liq.; 100% conc.

Crodamine 3.A16D. [162] Palmityl dimethylamine; cationic; see Crodamine 1.16D; liq.; 100% conc.

Crodamine 3.A18D. [162] Stearyl dimethylamine, dist.; cationic; see Crodamine 1.16D; paste; 100% conc.

Crodamine 3.ABD. [162] Dimethyl behenylamine; cationic; see Crodamine 1.16D; solid; 100% conc.

Crodamine 3.AED. [162] Dimethyl erucylamine; cationic; see Crodamine 1.16D; liq.; 100% conc.

Crodamine 3.AOD. [162] Dimethyl oleylamine; cationic; see Crodamine 1.16D; liq.; 100% conc.

Crodamol BS. [159] Butyl stearate; emollient without oiliness in lotions and creams, plasticizer in hair sprays and nail varnish, wetting agent for pigments and dyes in lipsticks; water wh. liq.

Crodasinic LS30. [159] Sodium N-lauroyl sarcosinate; anionic; foaming, wetting agent and detergent for acidic conditions; corrosion inhibitor; bacteriastat and inhibitor; used in dental care preparations, pharmaceuticals, personal care products, household and industrial applications; clear liq.; water-sol.; 30% act.

Crodasinic LS35. [159] Sodium N-lauroyl sarcosinate; anionic; foaming agent, wetting agent, detergent, lubricant, antistat, corrosion inhibitor, bacteriostat, penetrant used in dental , pharmaceutical, shampoos, depilatories, and shaving preparations, household and industrial uses; clear liq.; water sol.; biodeg.; 35% act.

Crodasinic OS35. [159] Sodium N-oleoyl sarcosinate; anionic; see Crodasinic LS35; lower foaming; liq.; 35% conc.

Crodateric C. [162] Derived from coconut fatty acid; amphoteric surfactant; m.w. 360; 50% act. aq. sol'n.

Crodateric Cy. [162] Derived from caprylic acid; amphoteric surfactant; m.w. 284; 50% act. aq. sol'n.

Crodateric L. [162] Derived from lauric acid; amphoteric surfactant; m.w. 340; 50% act. aq. sol'n.

Crodateric O, O.100. [162] Derived from oleic acid; amphoteric surfactant; m.w. 417; 50% act. aq. sol'n. and 100% act. resp..

Crodateric S. [162] Derived from stearic acid; amphoteric surfactant; m.w. 410; 50% act. aq. sol'n.

Crodazoline C. [162] Fatty acid imidazoline; Gardner 9–10; m.w. 288; m.p. 42 C; surf. tens. 32.5 (1%/0.36%); pH 11.5 (10% aq. disp.); 90% min. imidazoline.

Crodazoline Cy. [162] Caprylic imidazoline; Gardner 10–11; m.w. 212; m.p. 45 C; surf. tens. 27.8 (1%/0.36%); pH 12 (10% aq. disp.); 90% min. imidazoline.

Crodazoline H. [162] Fatty acid imidazoline; Gardner 8–9; m.w. 383; m.p. 57 C; surf. tens. 43 (1%/0.36%); pH 11.5 (10% aq. disp.); 90% min. imidazoline.

Crodazoline L. [162] Fatty acid imidazoline; Gardner 4–5; m.w. 268; m.p. 40 C; surf. tens. 37.2 (1%/0.36%); pH 11.3 (10% aq. disp.); 90% min. imidazoline.

Crodazoline M. [162] Fatty acid imidazoline; Gardner 9–10; m.w. 297; m.p. 51 C; surf. tens. 36.5 (1%/0.36%); pH 11.5 (10% aq. disp.); 90% min. imidazoline.

Crodazoline O. [162] Oleic imidazoline; Gardner 6–7; m.w. 345; m.p. 40–57 C; surf. tens. 26.5 (1%/0.36%); pH 11.1 (10% aq. disp.); 90% min. imidazoline.

Crodazoline S. [162] Stearic imidazoline; Gardner 4–5; m.w. 338; m.p. 52 C; surf. tens. 43.2 (1%/0.36%); pH 11.0 (10% aq. disp.); 90% min. imidazoline.

Crodazoline T. [162] Fatty acid imidazoline; Gardner 10–11; m.w. 350; m.p. 40–57 C; surf. tens. 36 (1%/0.36%); pH 11.2 (10% aq. disp.); 90% min. imidazoline.

Crodazoline TW. [162] Fatty acid imidazoline; Gardner 7–8; m.w. 338; m.p. 49 C; surf. tens. 38 (1%/0.36%); pH 11.1 (10% aq. disp.); 90% min. imidazoline.

Croderol G7000. [162] Glycerin; humectant, moisturizing agent in cosmetics and pharmaceuticals, diluent and plasticizer for many polar materials.

Crodesta A-10. [159; 160] Acetylated sucrose distearate; dispersant, emulsifier, wetting agent, solubilizer and detergent in cosmetics, toiletries, pharmaceuticals; yel./wh. wax; sol. in veg. and min. oils; HLB 0.5; m.p. 43–49 C; sapon. no. 230–290; 100% act.

Crodesta A-20. [160] Acetylated sucrose distearate; see Crodesta A-10; solid; HLB 1.0; 100% conc.

Crodesta F-10. [159; 160] Sucrose distearate; see Crodesta A-10; creamy wh. powd.; HLB 3.0; m.p. 60–68 C; sapon. no. 140–200; 100% act.

Crodesta F-50. [159; 160] Sucrose distearate; see Crodesta A-10; wh. powd.; HLB 6.5; m.p. 74–78 C; sapon. no. 93–153; 100% act.

Crodesta F-110. [159; 160] Sucrose distearate and sucrose stearate; see Crodesta A-10; thickener and suspending agent; wh. powd.; HLB 12.0; m.p. 72–78 C; sapon. no. 85–145; 100% act.

Crodesta F-160. [159; 160] Sucrose stearate; see Crodesta A-10; thickener and suspending agent; wh. powd.; HLB 14.5; m.p. 70–74 C; sapon. no. 75–153; 100% act.

Crodesta SL-40. [159; 160] Sucrose cocoate; nonionic; see Crodesta A-10; amber liq.; HLB 15.0; 55% mono ester.

Crodet L4. [159] POE 4 lauric acid; nonionic; o/w emulsifier for cosmetics and pharmaceutical creams, lotions and ointments, wetting agent, solubilizer for perfumes or aq. alcoholic preparations, dispersant; pale straw liq.; sol. in ethyl, oleyl, and cetearyl alcohols, oleic acid; HLB 9.8; sapon. no. 138–150.

Crodet L8. [159] POE 8 lauric acid; see Crodet L4; pale straw liq.; sol. see Crodet L4.

Crodet L12. [159] POE 12 lauric acid; see Crodet L4; pale straw liq.; sol. in water, ethyl, oleyl, and cetearyl alcohols, oleic acid; HLB 14.5; sapon. no. 72–82.

Crodet L24. [159] POE 24 lauric acid; see Crodet L4; off wh. soft paste; sol. in water, ethyl and cetostearyl alcohol; HLB 16.8; sapon. no. 42–48.

Crodet L40. [159] POE 40 lauric acid; see Crodet L4; off wh. waxy solid; sol. in water, ethyl and cetostearyl alcohol; HLB 17.9; sapon. no. 26–31.

Crodet L100. [159] POE 100 lauric acid; see Crodet L4; pale yel. waxy solid; sol. in water, ethyl and cetostearyl alcohol; HLB 19.1; sapon. no. 11–15.

Crodet S4. [159] POE 4 stearic acid; nonionic; o/w emulsifier for cosmetics and pharmaceutical creams, lotions and ointments, wetting agent, solubilizer for perfumes or aq. alcoholic preparations, dispersant; off wh. soft paste; sol. see Crodet L12; HLB 7.7; sapon. no. 117–129.

Crodet S8. [159] POE 8 stearic acid; see Crodet S4; off wh. soft paste; sol. in ethyl and oleic alcohols, oleic acid, cetostearyl alcohol; HLB 10.8; sapon. no. 84–94.

Crodet S12. [159] POE 12 stearic acid; see Crodet S4; off wh. waxy solid; sol. in ethyl and cetostearyl alcohol, water; HLB 13.4; sapon. no. 65–75.

Crodet S24. [159] POE 24 stearic acid; see Crodet S4; off wh. waxy solid; sol. in ethyl and cetostearyl alcohol, water; HLB 15.8; sapon. no. 38–47.

Crodet S40. [159] POE 40 stearic acid; see Crodet S4; off wh. waxy solid; sol. in ethyl and cetostearyl alcohol, water; HLB 16.7; sapon. no. 23–30.

Crodet S100. [159] POE 100 stearic acid; see Crodet S4; off wh. waxy solid; sol. in ethyl and cetostearyl alcohol, water; HLB 18.8; sapon. no. 10–14.

Crodex A. [159] Cetostearyl alcohol and sodium lauryl sulfate; anionic; emulsifying wax BP for pharmaceuticals and cosmetic uses; almost wh. waxy solid; faint char. odor.

Crodex C. [159] Cetostearyl alcohol and Cetrimide BP; cationic; emulsifying wax BPC, bacteriacides, for pharmaceuticals and cosmetics; almost wh. waxy solid; faint char. odor.

Crodex N. [159] Cetostearyl alcohol and ceteth-20; nonionic; emulsifying wax BP, wetting agent, penetrant, emulsifier for most emollient materials in cosmetics and pharmaceuticals; almost wh. waxy solid; faint char. odor.

Croduret 10. [159] PEG-10 hydrog. castor oil; nonionic; emulsifier, solubilizer, emollient, superfatting agent, detergent used for cosmetics, textiles, metalworking fluids, emulsion polymerization, insecticides, herbicides, household detergents; straw-colored liq.; sol. in oleyl alcohol, naphtha, MEK, trichloroethylene; HLB 6.3; cloud pt. < 20 C; sapon. no. 120–130; surf. tens. 37 dynes/cm (0.1%).

Croduret 30. [159] PEG-30 hydrog. castor oil; see Croduret 10; straw colored liq.; sol. in water, ethanol, naphtha, MEK, trichloroethylene; HLB 11.6; cloud pt. 48 C; sapon. no. 70–80; surf. tens. 46 dynes/cm (0.1%).

Croduret 40. [159] PEG-40 hydrog. castor oil; see Croduret 10; off wh. visc. paste; sol. in water, ethanol, naphtha, MEK, trichloroethylene; HLB 13.0; cloud pt. 62 C; sapon. no. 60–65; surf. tens. 46 dynes/cm (0.1%).

Croduret 60. [159] PEG-60 hydrog. castor oil; see Croduret 10; off wh. stiff paste; sol. in water, ethanol, MEK, trichloroethylene; HLB 14.7; cloud pt. 71 C; sapon. no.45–50; surf. tens. 47.5 dynes/cm (0.1%).

Croduret 100. [159] PEG-100 hydrog. castor oil; see Croduret 10; off wh. waxy solid; sol. in water, ethanol, MEK, trichloroethylene; HLB 16.5; cloud pt. 65 C; sapon. no. 25–35; surf. tens. 46.2 dynes/cm (0.1%).

Crolactil CSL. [159] Acyl lactylates; w/o emulsifiers, ingredients in food products; off-wh. powd.; HLB 5; sapon. no. 175–250.

Crolactil SISL. [159] Acyl lactylates; o/w emulsifiers, ingredients in food products; yel. liq.; HLB 6.5; sapon. no. 210–280.

Crolactil SSL. [159] Acyl lactylates; o/w emulsifiers, ingredients in food products; off-wh. powd.; HLB 6.5; sapon. no. 210–280.

Cutina AGS. [289] Glycol distearate; opacifier; emulsion shampoos; solid.

Cutina CBS. [289] Glyceryl stearate, cetearyl alcohol, cetyl palmitate, and cocoglycerides; nonionic; cream base for mfg. of creams and lotions of the o/w type; flakes; 100% conc.

Cutina E-24. [289; 291] PEG-20 glyceryl stearate; nonionic; emulsifier for creams and emulsions for baby and children's preparations; paste; 100% conc.

Cutina GMS. [289; 291] Glyceryl stearate; nonionic; cream base for o/w and w/o emulsions; beads; 100% conc.

Cutina KD 16. [289; 291] Glyceryl stearate SE; emulsifier for o/w ointments and creams; solid.

Cutina MD. [289; 291] Glyceryl stearate; stabilizer used in ointments, creams, and liq. emulsions; solid.

Cutina MD-A. [289; 291] Glyceryl stearate; nonionic; cream base for cosmetic and pharmaceutical creams and emulsions; granules; 100% conc.

CW-6. [210] Mono- and diglycerides and sodium stearyl lactylate; emulsifier system for freeze-thaw coffee whitener; m.p. 130–140 F.

Cyclochem 326A. [168] Syn. beeswax; nonionic/anionic; emulsifier for w/o and o/w emulsions incl. hair creams, cold creams, lotions and lipsticks; off-wh. flakes; m.p. 51–54 C; 100% conc.

Cyclochem EGDS. [168] Ethylene glycol distearate pure; thickener, opacifier used in shampoos and cosmetic lotions; wh. flakes; m.p. 61 C; sapon. no. 195.

Cyclochem EGMS. [168] Ethylene glycol monostearate pure; adds visc., pearlescence and opacity to lotions and shampoos; wh. flakes; m.p. 56 C; sapon. no. 182.

Cyclochem EM 324. [168] Glycol stearate SE; anionic; emulsifier useful in formulating o/w hair and skin conditioning creams; off-wh. flakes; m.p. 47–50 C.

Cyclochem EM 560. [168] Cetearyl alcohol, sodium lauryl sulfate, cetyl esters, and myristyl alcohol; anionic; emulsifier, self-emulsifying, useful for o/w creams on the acid side; off-wh. color; m.p. 62–65 C.

Cyclochem GMO. [168] Glyceryl monooleate; nonionic; emulsifier, emollient, opacifier, provides lubricity to pre- and after-shave lotions and cosmetic systems; amber liq.; m.p. 9 C; sapon. no. 170; 100% conc.

Cyclochem GMS. [168] Glyceryl monostearate pure; nonionic; strongly lipophilic emulsifier for personal care products; wh. flakes; m.p. 56 C; sapon. no. 170.

Cyclochem GMS 21. [168] Glyceryl monostearate SE, acid stable; emulsifier with electrolyte tolerance for creams, lotions; off-wh. flakes; m.p. 52 C; sapon. no. 157.

Cyclochem GMS 165. [168] Glyceryl stearate and PEG-100 stearate; nonionic; acid stable, self emulsifying; electrolyte-tolerant emulsifier, emollient for creams, lotions and toiletry products; off-wh. flakes; m.p. 55 C; sapon. no. 93; 100% conc.

Cyclochem GMS/SE. [168] Glyceryl stearate SE; anionic; self-emulsifying form of Cyclochem GMS; flakes; HLB 5.5; 100% conc.

Cyclochem GTIS. [168] Triisostearin; emollient, coemulsifier, softener, provides good texture in creams, promotes softness on skin; yel. liq.; m.p. < 0 C; sapon. no. 192.

Cyclochem GTL. [168] Trilaurin; emollient;

promotes an oily feel to cream and lotions; liq./paste; m.p. 27 C; sapon. no. 245.

Cyclochem LIPA. [168] Lauric monoisopropanolamide; visc. builder, foam stabilizer; off-wh. flakes; 100% act.; 95% amide content.

Cyclochem PEG 400 DS. [168] PEG 400 distearate; nonionic; wh. soft solid; m.p. 33 C; sapon. 123; acid no. 10.0 max.

Cyclochem PEG 400 ML. [168] PEG 400 monolaurate; lt. straw liq.; m.p. 10 C; sapon. no. 90.

Cyclochem PEG 400 MO. [168] PEG 400 monooleate; lt. straw liq.; m.p. 10 C; sapon. no. 85.

Cyclochem PGMS. [168] Propylene glycol stearate, pure; nonionic; aux. emulsifier for lotions and makeup preparations, occlusive agent; off-wh. soft solid; m.p. 36 C; sapon. no. 186; 100% conc.

Cyclochem POL [168] Cetearyl alcohol, ceteth-20, and glycol stearate; nonionic; emulsifier for creams and lotions containing high quantities of alkaline ingredients; off-wh. wax; m.p. 48–52 C; 100% conc.

Cyclogol Cetyl Alcohol NF. [168] *n*-Hexadecanol; nonionic; emollient carrier and visc. builder for creams, lotions, other cosmetic emulsions; off-wh. flakes; m.p. 45–50 C.

Cyclogol Cetyl-Stearyl Alcohol. [168] *n*-Hexadecanol and *n*-octadecanol; nonionic; combines the visc. effect of stearyl with the skin feel of cetyl in lotions and cosmetic emulsions; off-wh. flakes; m.p. 48–53 C.

Cyclogol EL. [168] Laureth-12; nonionic; solubilizer in perfume oils, superfatting agents and microbicides; wh. paste; m.p. 28–33 C.

Cyclogol LTH. [168] Ethoxylated linear alcohol; nonionic; detergent, dispersant for emulsifying oils, degreasing, engine cleaners; clear liq.; m.p. 22–24 C.

Cyclogol NI. [168] Cetearyl alcohol and ceteareth-20; nonionic; emulsifier for preparing o/w ointments, creams, and lotions containing hard to solubilize additives; wh. flakes; m.p. 49–50 C; 100% conc.

Cyclogol Stearyl Alcohol. [168] *n*-Octadecanol; nonionic; visc. builder for creams, lotions and other cosmetic emulsions; off-wh. flakes; m.p. 56–60 C.

Cyclomide AC 28. [168] Fatty acid condensate; anticorrosive min. oil additive for cutting fluids and sol. cutting oil compositions; liq.; 100% act.

Cyclomide BWA/M. [168] Fatty alkanolamide blends; nonionic; effective o/w emulsifier system for use as an engine degreaser; liq.; 100% act.

Cyclomide C212. [168] Cocamide MEA; nonionic; foam builder and stabilizer for soap or syn. based washing powds.; cream-colored flakes; 100% act.; 95% amide.

Cyclomide CODI. [168] Cocamidopropyl dimethylamine; cationic; emulsifier; base for emulsions for creams, lotions and hair rinse formulations; yel. liq.; 100% act., 98% amide.

Cyclomide DC212. [168] Cocamide DEA and diethanolamine; nonionic; visc. booster, foaming agent, detergent, used in cosmetic and laundry products; amber liq.; 100% act.

Cyclomide DC212M. [168] 2:1 Modified coconut diethanolamide; nonionic; cleaners, solventless degreasers and wax strippers; amber liq.; 100% act.

Cyclomide DC212/S. [168] 1:1 Cocamide DEA; nonionic; foam stabilizer and visc. booster with low cloud pt.; liq.; 100% act.; 85% amide.

Cyclomide DC212/SE. [168] 1:1 Cocamide DEA; nonionic; detergent, foam stabilizer, thickener, foam booster in shampoos and detergent formulations; liq.; 100% act.; 95% amide.

Cyclomide DIN 295/S. [168] Linoleamide DEA and diethanolamine; nonionic; visc. builder, thickener; conditioning to hair; liq.; 100% conc.

Cyclomide DL203. [168] 2:1 Lauramide DEA; nonionic; lather stabilizer for shampoos and detergents; off-wh. liq., wh. paste; 100% act.

Cyclomide DL203/S. [168] 1:1 Lauramide DEA; nonionic; foam and detergency booster for cosmetic products; superfatting agent; off-wh. cryst. solid; 100% act; 95% amide.

Cyclomide DL207/S. [168] 1:1 Lauramide DEA; nonionic; foam and visc. booster, wetting agent, superfatting agent; off-wh. liq., wh. paste; 100% act.; 95% amide.

Cyclomide DL207/SL. [168] Lauramide DEA; nonionic; wetting action; liq.; 78% conc.

Cyclomide DO280. [168] Oleamide DEA and diethanolamine; nonionic; emulsifier for sol. oils; corrosive inhibitor; liq.; 100% conc.

Cyclomide DO280/S. [168] 1:1 Oleamide DEA; nonionic; thickening and superfatting agent, emulsifier, for lotion shampoos; liq.; 100% act.; 85% amide.

Cyclomide DOTS. [168] 1:1 Linoleamide DEA; visc. builder, foam stabilizer, emulsifier, pigment dispersant; amber liq.; 100% act.; 95% amide.

Cyclomide DP240/S. [168] 1:1 Palm kernelamide DEA; nonionic; visc. builder and foam stabilizer for org. formulations; liq.; 100% act.; 85% amide.

Cyclomide IODI. [168] Isostearamidopropyl dimethylamine; cationic; base for preparing emulsions, promotes substantivity and adhesion of additives to the skin; clear liq.; 100% act.; 98% amide.

Cyclomide KD. [168] 1:1 Cocamide DEA; nonionic; visc. builder, foam stabilizer; pale yel. liq.; 100% act.; 95% amide.

Cyclomide L203. [168] Lauramide MEA; nonionic; foam builder/stabilizer for soap and syn. washing powds.; visc. builder; off-wh. flakes; 100% act.; 95% amide.

Cyclomide LE. [168] Lauramide DEA; nonionic; visc. improvement, foam stabilizing in shampoo; liq.; 100% conc.

Cyclomide LIPA. [168] Lauramide MIPA; nonionic; foam booster, stabilizer and visc. booster; flake; 100% conc.

Cyclomide Pinemulse. [168] Modified coconut polydiethanolamide; nonionic; emulsifier and coupler for preparing stable pine oil cleaners without heat; amber liq.; 100% act.

Cyclomide S280. [168] Stearamide MEA; nonionic; skin protectant in toilet bars, creams, lotions, pastes; off-wh. flakes; 100% act.; 95% amide.

Cyclomide SODI. [168] Stearamidopropyl dimethylamine; cationic; emulsifier; off-wh. solid; 100% act.; 98% amide.

Cyclomide WHC. [168] Coconut DEA; nonionic; emulsifier for preparation of waterless hand cleaners; liq.; 100% conc.

Cyclomox C. [168] Coconut dimethylamine oxide; nonionic; flash foamer and viscosifier in hard water; nonionic; liq.; 30% conc.

Cyclomox CO. [168] Cocoamidopropylamine oxide; cationic; conditioner, emollient with visc. and foam building effects; liq.; 30% conc.

Cyclomox L. [168] Lauramine oxide; cationic; used in dishwash and shampoo formulations; liq.; 27–30% conc.

Cyclomox LO. [168] Lauramidopropyl amine oxide; nonionic; top foam production; liq.; 30% conc.

Cyclomox SO. [168] Stearamidopropylamine oxide; nonionic; foamer and viscosifier; solid; 30% conc.

Cyclophos P4. [168] Aliphatic acid ester; anionic; imparts detergency, moderate foaming and good coupling properties in powd. machine dishwashing products and steam carpet cleaners, surf. tens. depressant; clear, thin liq.; 98% act.

Cyclophos PB12. [168] Aromatic acid ester; anionic; metal cleaning, highly stable under prolonged use in soak tank cleaning; clear liq.; 98% act.

Cyclophos PE5. [168] Aromatic acid ester; anionic; promotes detergency; disp. in oil and water systems; clear liq.; 98% act.

Cyclophos PE9. [168] Aromatic acid ester; anionic; promotes detergency and visc. in heavy duty, all purpose liq. formulations and built-in steam cleaners; clear liq.; 99% act.

Cyclophos PE30. [168] Aromatic acid ester; anionic; useful in emulsion polymerization to promote clarity and stability of cast acrylic and PVA films, latex stabilizer, dyeing assistant; off-wh. paste; 90% act.

Cyclophos PE-41K. [168] Org. phosphate ester, potassium salt; anionic; hydrotrope in liq. concs.; solubilizer; liq.; 50% conc.

Cyclophos PL3. [168] Laureth-3 phosphate; anionic; emulsifier for cosmetic oils, creams and waterless hand cleaners, fiber antistat lubricant; pale yel. liq.; 99% act.

Cyclophos PL6A. [168] Aliphatic acid ester; anionic; coupler for nonionics in liquid alkaline systems, moderate foamer; pale yel. liq.; 98% act.

Cyclophos PL31. [168] Aliphatic acid ester; anionic; coupling agent in extreme alkali systems, improves lubricity and loadbearing properties of oil-based lubricants; emulsification of cutting oils; clear liq.; 98% act.

Cyclophos PO3. [168] Oleth-3 phosphate; anionic; surfactant for shampoos and conditioners; liq.; 100% conc.

Cyclophos TE12. [168] Aromatic acid ester; emulsifier and solubilizer of nonionics in high electrolyte sol'ns.; amber liq.; high water sol.; 98% act.

Cyclopol SB-5. [168] Sodium laneth-5 sulfosuccinate; anionic; moisturizing agent for shampoos, foam baths, skin cleansers, skin creams, hair sprays, dishwashing liq.; yel. liq.; 48–50% act.

Cyclopol SBCA198. [168] Mixed fatty amine sulfosuccinate; anionic; liq. surface active agent in latex systems; yel. liq.; 33–35% act.

Cyclopol SBCP. [168] Disodium cocamido

MIPA-sulfosuccinate; anionic, promotes low irritation in mild shampoos, bubble baths, and skin cleansers; yel. liq.; 39–41% act.

Cyclopol SBDO. [168] Dioctyl sodium sulfosuccinate; anionic; surf. tens. depressant, solubilizer for soaps and surfactants; liq.; 68–70% act.

Cyclopol SBFA30. [168] Disodium laureth sulfosuccinate; anionic; mild, nonirritating raw material for bubble baths, shampoos and general cleansing agents; liq.; 38–40% act.

Cyclopol SBG 212. [168] Disodium monococamide sulfosuccinate; anionic; mild, nonirritating base, foaming agent for personal care products; yel. liq.; 29–31% act.

Cyclopol SBG 280. [168] Disodium oleamido PEG-2 sulfosuccinate; anionic; foaming agent, nonirritating surfactant, visc. control in personal care products; yel. liq.; 29–31% act.

Cyclopol SBG 280/T. [168] TEA-monooleamide sulfosuccinate; anionic; base for skin products and shampoos, antiirritant; yel. liq.; 39–41% act.

Cyclopol SBL203. [168] Disodium lauramido MEA-sulfosuccinate; anionic; detergent raw material for carpet shampoos; yel liq.; 38–40% act.

Cyclopol SBR-3. [168] Disodium ricinoleamido MEA-sulfosuccinate; anionic; skin-protective, anti-irritant surfactant; yel. liq.; 38–40% act.

Cyclopol SBU-185. [168] Disodium undecylenamido MEA-sulfosuccinate; anionic; fungicidal and bacteriostatic additive, antidandruff agent used in shampoos, body lotions, foot preparations; yel. liq.; 49–51% act.

Cycloryl 21. [168] Sodium lauryl sulfate cosmetic grade; low cost base for cream and liq. shampoos and specialty cleaners; pale yel. liq./paste; 28–30% act.

Cycloryl 21 LS. [168] Sodium lauryl sulfate, low salt; anionic; base for shampoos, hand cleaners and detergents; pale yel. liq.; 28–30% act.

Cycloryl 21 SP. [168] Sodium lauryl sulfate low cloud pt; flowable process aid for the preparation of compd.s, frothing agents, and detergents; yel. liq.; 28–30% act.

Cycloryl ABSA. [168] Linear dodecylbenzene sulfonic acid; anionic; detergent used for hard surface cleaners, built detergents and shampoo systems; amber liq.; biodeg.; 97% min. act.

Cycloryl ALC. [168] PEG-80 sorbitan laurate, sodium trideceth sulfate, PEG-150 distearate blend; conc. base for baby shampoos; liq.; 62% conc.

Cycloryl DA. [168] DEA lauryl sulfate; anionic; base for preparation of clear, mild shampoos, anionic; yel. liq.; 34–36% act.

Cycloryl DCA. [168] Modified coconut polydiethanolamide; anionic; conc. detergent for human and pet shampoos, builds visc. and solubilizes difficult ingredients; amber liq.; 98–100% act.

Cycloryl DDB-40. [168] Sodium dodecylbenzene sulfonate; anionic; base for medium-duty detergents, cleaners and strippers; yel. liq.; 38–40% act.

Cycloryl DDB-60T. [168] TEA-dodecylbenzene sulfonate; anionic; base for cleaners and shampoos; yel. liq.; 58–60% act.

Cycloryl JB-8. [168] Cocamide DEA; base for foaming bath gels and liquids, anionic; amber liq.; 33–36% act.

Cycloryl LDC. [168] Ammonium lauryl ether sulfate plus sulfonate and amide; anionic; base for light duty detergent and liq. dishwashing compds., car washing application; amber liq.; 60–62% act.

Cycloryl M1. [168] Sodium laureth sulfate, lauramide MIPA, glycol stearate and cocamide MEA; anionic; built pearlized base for cream shampoos having high flash foam; wh. creamy liq.; 34–36% act.

Cycloryl MA. [168] Ammonium lauryl sulfate; anionic; base for shampoos, bubble baths, and lt. duty detergents; pale yel. liq.; 28–30% act.

Cycloryl MA 330. [168] Ammonium laureth sulfate; anionic; base for lt. duty dish detergents, fine fabric cleaners, shampoo, grease cutters, degreasing agent; pale yel. liq.; 27–29% act.

Cycloryl MA 360. [168] Ammonium laureth sulfate; anionic; base for lt. duty dish detergents, fine fabric cleaners, shampoo, superior grease cutters, degreasing agent; pale yel. liq.; 58–60% act.

Cycloryl ME 60. [168] Sodium myreth sulfate; anionic; conc. base for shampoos, bath oils, bubble baths and cosmetic emulsions; pale yel. liq.; 58–60% act.

Cycloryl ME 361. [168] Ammonium myreth sulfate; anionic; conc. base for acid shampoos and mild high foaming cleansing products; pale yel. liq.; 58–60% act.

Cycloryl MG. [168] Magnesium lauryl sulfate; anionic; carpet, upholstery shampoo base having low cloud pt., forms tack free residue; pale yel. liq.; 28–30% act.

Cycloryl MS. [168] Sodium myristyl sulfate; anionic; mild emulsifier effective at low levels; detergent for cleansing creams, fine fabric detergents; wh. paste; 25–27% act.

Cycloryl N 70. [168] Sodium laureth sulfate; anionic; used in shampoos and cleansers; soft gel; 70% conc.

Cycloryl NA. [168] Sodium laureth sulfate; anionic; base for shampoos, cosmetics and lt. duty and/or high foaming detergents; pale yel. liq.; 27–29% act.

Cycloryl NA-2. [168] Sodium laureth sulfate; anionic; see Cycloryl NA; clear liq.; 24–26% act.

Cycloryl NA 61, NA61 CG. [168] Sodium laureth sulfate; anionic; see Cycloryl NA; pale yel. liq.; 58–60% act.

Cycloryl NWC. [168] Sodium laureth sulfate, cocamide DEA, TEA-lauryl sulfate, and propylene glycol; anionic; base for shampoos and bubble baths; amber liq.; biodeg.; 62–64% act.

Cycloryl SA. [168] MEA-lauryl sulfate; anionic; shampoos and related products; liq.; 28% conc.

Cycloryl SET. [168] Ethoxylated primary alcohol plus additives; anionic; base for superior heavy-duty liq. laundry detergent; liq.; 79–81% act.

Cycloryl TD. [168] TEA laureth sulfate; anionic; base for visc. shampoos with high salt tolerance; pale yel. liq.; 39–41% act.

Cycloryl TEALS. [168] TEA-cocoyl sarcosinate; anionic; high flash foaming detergent for shampoos and bubble baths, good in hard water applications; liq.; 39–41% act.

Cycloryl TS30. [168] Sodium lauryl sulfate plus additives; anionic; detergent for carpet shampoo conc.; amber liq.; 28–39% act.

Cycloryl WAT. [168] TEA lauryl sulfate; anionic; base for lt. colored, clear shampoos, bubble baths; pale yel. liq.; 38–40% act.

Cycloryl WAT-1. [168] TIPA lauryl sulfate; anionic; base for highly conc. foaming toiletry products where water content must be minimized; liq.; 86–90% act.

Cycloryl XL. [168] Amphoteric modified lauryl sulfate; anionic; detergent base for shampoos; pale yel. liq.; 36–38% act.

Cyclosheen 100. [168] Lauramide MIPA, glycol stearate, and cocamide MEA; nonionic; emulsifier, opacifier imparting stable pearl, improved lathering and increased visc. of liq.; shampoos and cleaners; off-wh. soft wax; m.p. 58–60 C.

Cycloteric BET C-30. [168] Cocamidopropyl betaine; amphoteric, foaming, conditioning and thickening agent for use with skin and hair preparations; yel. liq.; pH 4.5–5.5; 29–31% act.

Cycloteric BET-CS. [168] Cocamidopropyl hydroxysultaine; amphoteric; ingredient for personal care products; liq.; 49–51% conc.

Cycloteric BET I-30. [168] Isostearamidopropyl betaine; amphoteric; foaming, conditioning and thickening agent for use with skin and hair preparations; yel. liq.; pH 5.5–6.5; 29–31% act.

Cycloteric BET O-30. [168] Oleamidopropyl betaine; amphoteric; foaming, conditioning and thickening agent for use with skin and hair preparations; yel. liq.; pH 5.5–6.5; 29–31% act.

Cycloteric BET OD-40. [168] Capric/caprylic amidopropyl betaine; amphoteric; surfactant; liq.; 40% conc.

Cycloteric DC-SF. [168] Cocoamphodipropionate; amphoteric; cleansing agent with high tolerance to alkali and other electrolytes; clear liq.; pH 8.0–9.0; 38–40% act.

Cycloteric MV-SF. [168] Cocoamphopropionate; amphoteric; low foaming agent valuable in compd.ing highly acidic and highly alkaline cleansers where fast action and rust inhibition are sought, wetting agent; clear liq.; pH 9–11; 38–40% act.

Cycloton CT 100. [168] Stearyl alcohol and cetrimonium bromide; cationic; emulsifier for creme rinse/conditioner; wh. flakes; m.p. 76–77 C.

Cyncal 80%. [299] Myristalkonium chloride; cationic; antimicrobial agent, antistat, disinfectant, sanitizer for use in food, beverage processing, for industrial and farm use, fabrics; lt. yel. liq.; mild, pleasant odor; m.w. 359; sol. in water, lower alcohols, ketones, glycols; dens. 7.8 lb/gal; sp.gr. 0.94; 80% act.

D

D-9. [462] Veg. monoglycerides, dist.; emulsifier in food products; lt. tan powd.; bland odor; 17–20% protein content; 41–43% solv. extractables.

Dapral. [11] Phosphate derivs.; detergent, wetting agent, emulsifier for making of cutting and sol. oils, for household, cosmetic, industrial uses; liq., paste.

Dapral AS. [11] Phosphoric acid ester, potassium salt; anionic; polyester yarn spinning; paste; 75% conc.

Dariloid Series. [346] Alginates; dairy stabilizers and stabilizer/emulsifiers; milk sol.; lt. ivory granules; pH 4.5–10.2.

Datagel. [159] Monoglyceride activated diacetyl tartaric acid ester; nonionic; food emulsifier; paste, gel; 100% conc.

Datamuls 42. [258] Fatty acid glyceride diacetyl tartaric acid ester; nonionic; emulsifier for yeast-raised bakery goods; pumpable; 100% conc.

Datamuls 43, 47 [258] Diacetyl tartaric acid monoglyceride; nonionic; food emulsifier for yeast-raised bakery goods; paste and powd. resp.; 100% conc.

Datamuls 4720, 4741. [258] Diacetyl tartaric acid monoglyceride; nonionic; food emulsifier for yeast-raised goods; powd.; 80 and 60% conc. resp.

Datamuls 4820. [258] Fatty acid glyceride diacetyl tartaric acid ester; nonionic; food emulsifier for yeast-raised doughs and baking aids; powd.; 80% conc.

Datem esters. [159] Monoglyceride diacetyl tartaric acid ester; nonionic; emulsifier for use in yeast-raised bakery goods; solid; 100% conc.

DDBS 100. [659] Alkylbenzene sulfonic acid, branched; anionic; detergent intermediate; liq.; 100% conc.

DDBS-100 SP. [659] Alkylbenzene and toluene cosulfonic acid; anionic; used in spray drying; liq.; 100% conc.

DDBSA 99-b. [419] Dodecylbenzene sulfonic acid; for use in low salt liq. detergents, general purpose neutralization reactions, spray-dried products; m.w. 316; sp.gr. 1.0; 98% act.

Deatron N. [437] Alkyl phosphate; anionic; antistat for nylon fabrics; liq.; 50% conc.

Deceresol Surfactant NI Conc. [43] Polyether alcohol; nonionic; detergent, emulsifier, wetting agent for textile industry; clear liq.; dissolves readily with stirring in cold water; dens. 8.8 lb/gal; sp.gr. 1.05; pH 6–7; 100% act.

Deceresol Surfactant OT 75%. [43] Sulfonated ester; anionic; wetting agent for textile industry; water-wh. visc. liq.; slight hexanol odor; completely sol. as dilute sol'n.; dens. 8.9 lb/gal; sp.gr. 1.082; 75% act.

Deceresol Surfactant OT Special. [43] Sulfonated ester; anionic; wetting agent for textile industry; colorless thin liq.; slight odor; dissolves with agitation in warm water; dens. 8.8 lb/gal; sp.gr. 1.05; pH 5.7; 70% act.

Dehydag Wax E. [290; 291] Sodium cetearyl sulfate; anionic; o/w emulsifier used in personal care products and powd. cleaners; wh. powd.; water-sol.; dens. 0.180 g/ml; pH 6–8; 87% min. act.

Dehydag Wax N. [291] Cetearyl alcohol and

sodium cetearyl sulfate; anionic; mfg. of o/w creams and liq. emulsions; solid; 100% conc.

Dehydag Wax O. [290; 291] Cetearyl alcohol; consistency modifier used in personal care products and pharmaceuticals; wh. granules; dens. 0.816 g/ml (60 C); solid. pt. 48–52 C; sapon. no. 1.0 max.

Dehydag Wax SX. [290; 291] Cetearyl alcohol, sodium lauryl sulfate (90:10 ratio); anionic; surfactant used in mfg. of o/w emulsions for personal care products; wh. granules; solid. pt. 50–54 C; sapon. no. 2.0 max.; 99.0% min. fatty alcohol.

Dehydag Wax W. [291] Cetostearyl alcohol and sodium lauryl sulfate (90:10 ratio); anionic; base for mfg. of ointments, creams and liniments; granules; 100% conc.

Dehydol 100. [291] Laureth-10; nonionic; detergent, wetting; used in insecticides, cleansing products; paste; 100% conc.

Dehydol 737. [291] Fatty alcohol polyglycol ether; nonionic; raw material for laundry detergents; paste; 100% conc.

Dehydol D 3. [291] Fatty alcohol polyglycol ether; nonionic; emulsifier, solubilizer for oils; liq.; HLB 9.0; 100% conc.

Dehydol LS 2, LS 3, LS 4. [291] Laureth-2, -3, -4 resp.; nonionic; emulsifier, solubilizer for solvents, oils, bases for production of sulfates; raw material for dishwashing, cleansing agent and cold cleaners; liq.; 99–100% conc.

Dehydol LT 2, LT 3, LT 4, LT 5, LT 7. [291] Fatty alcohol polyglycol ether; nonionic; see Dehydol LS 2; liq.; 99–100% conc.

Dehydol O4. [291] Octyl alcohol (4 EO); nonionic; raw material for mfg. of surfactants and cleaning agents; liq.; HLB 11.5; 100% conc.

Dehydol TA 5. [291] Fatty alcohol polyglycol ether; nonionic; raw material for detergents; emulsifier; solid; 100% conc.

Dehydol TA 6. [291] Fatty alcohol polyglycol ether; nonionic; emulsifier for min. oils and paraffin; solid; HLB 10.0; 100% conc.

Dehydol TA 11. [291] Tallow fatty alcohol (11 EO); nonionic; raw material for detergents, toilet cubes; emulsifier; solid; 100% conc.

Dehydol TA 12. [291] Fatty alcohol polyglycol ether; nonionic; emulsifier for solv., waxes, oils; solid; HLB 13.4; 100% conc.

Dehydol TA 14. [291] Fatty alcohol polyglycol ether; nonionic; see Dehydol TA 11; solid; 100% conc.

Dehydol TA 20. [291] Fatty alcohol polyglycol ether; nonionic; emulsifier for paraffins, waxes; raw material for foam-controlled detergents, toilet cubes; solid; HLB 15.5; 100% conc.

Dehydol TA 30. [291] Fatty alcohol polyglycol ether; nonionic; see Dehydol TA 20; solid; HLB 16.5; 100% conc.

Dehydol WM. [291] Fatty alcohol polyglycol ether; nonionic; raw material for laundry detergents; paste; 100% conc.

Dehydol WM 90. [291] Fatty alcohol polyglycol ether; nonionic; see Dehydol WM; liq.; 90% conc.

Dehydrophen 65. [291] Alkylaryl polyglycol ether; nonionic; detergent, wetting, dishwashing agent; liq.; HLB 11.3; 100% conc.

Dehydrophen 100. [291] Alkylaryl polyglycol ether; nonionic; detergent, wetting, emulsifier; liq. dishwashing agent; liq.; HLB 13.3; 100% conc.

Dehydrophen 150. [291] Alkylaryl polyglycol ether; nonionic; emulsifier for various hydrocarbons; paste; HLB 15.0; 100% conc.

Dehymuls E. [289; 291] Sorbitan sesquioleate, beeswax, aluminum stearate, anionic; cosmetics emulsifier; wax-like substance.

Dehymuls F. [291] Fatty acid ester, fatty acid salts mixt.; anionic; w/o emulsifier for creams and emulsions; solid; 100% conc.

Dehymuls K. [291] Petrolatum, decyl oleate, sorbitan sesquioleate, beeswax, min. oil, ceresin, aluminum stearate; self-emulsifying base for mfg. of cosmetic and pharmaceutical preparations of the w/o type; solid.

Dehymuls SML. [291] Sorbitan monolaurate; nonionic; w/o emulsifier and coemulsifier for cosmetic and pharmaceutical applications; liq.

Dehymuls SMO. [291] Sorbitan monooleate liq.; nonionic; emulsifier and coemulsifier for ointments and creams of the w/o type.

Dehymuls SMS. [291] Sorbitan monostearate; nonionic; emulsifier for w/o; used in the cosmetic and pharmaceutical industry; flakes.

Dehymuls SSO. [291] Sorbitan sesquioleate; nonionic; w/o emulsifier and coemulsifier for waxes and oils; liq.

Dehypon Conc. [291] Fatty alcohol polyglycol ether blend; nonionic; wetting agent, emulsifier and dispersant; paste; 99% conc.

Dehypon LS 24, LS 36, LS 45, LS 54. [291] Fatty alcohol EO/PO adducts; nonionic; surfactant, dispersant, wetting agent for industrial cleaners, liq. laundry detergents, dishwashing agents; liq.; 100% conc.

Dehypon LT 104. [291] Fatty alcohol polyglycol ether; nonionic; surfactant, antifoaming agent; paste; 100% conc.

Dehyquart A. [290] Cetrimonium chloride; cationic; softener, conditioner, bactericide, fungicide, and odor inhibitor in personal care products; pale yel. clear liq.; 24–26% act.

Dehyquart C. [290] Laurylpyridinium chloride; cationic; surfactant, fungicide, bactericide, and disinfectant used in cleaning formulations and personal care products; sequestrant for min. oil industry; wh. paste, powd.; 80–82% act.

Dehyquart C Crystals. [290; 291] Laurylpyridinium chloride; cationic; bactericide, fungicide, corrosion inhibitor, sequestrant, conditioner used in personal care products; 90–94% act.

Dehyquart LT. [291] Laurtrimonium chloride; cationic; wetting agent, antistat, bactericide, conditioning component for hair care products; liq.; 34–36% conc.

Dehyquart SP. [290] Quaternium-52; surfactant, softening and antistatic agent used in personal care products; metal corrosion inhibitor; clear visc. liq.; pH 6.8–7.2 (10%); 49.0–51.0% solids.

Dehyquart STC-25. [290] Stearalkonium chloride; cationic; antistat, conditioner for hair and fibers; emulsifier in creams, lotions; wh. slurry; 25% conc.

Dehyton AB-30. [289; 291] Coco-betaine; amphoteric; detergent, foamer used in personal care products; lt. amber liq.; 29–30% conc.

Dehyton G. [289; 291] Cocamphocarboxyglycinate; amphoteric; raw material for personal care products; liq.; 30% conc.

Dehyton K. [291] Cocamidopropyl betaine; amphoteric; raw material for mfg. of surfactant preparations, conditioning agent; liq.; 30% conc.

Delestat P35. [551] Quat.; cationic; resistivity control for painting; liq.; 80% conc.

Delion 342. [590] Alkyl phosphate POE alkyl ether; anionic/nonionic; finishing oil for acrylic fibers; liq.; 60% conc.

Delion 624. [590] Phosphate surfactant; anionic; antistat and lubricant for fibers; paste; 60% conc.

Delion 662. [590] Phosphate surfactant; anionic; antistat for syn. fibers; paste.

Delion 964. [590] Alkyl phosphate; anionic/nonionic; spin finish oil for polyester staple; paste; 80% conc.

Delion 6067. [590] Alkyl phosphate; anionic; staple finish oil for polyester staple; paste; 60% conc.

Delion A-016. [590] Sulfonated surfactant; anionic; antistat for syn. fibers; wetting agent and detergent; liq.; 40% conc.

Delion A-160. [590] Phosphated surfactant; anionic; antistat agent for fibers; liq.

Delvet 68, 70. [187] Nonionic surfactant used as chlorinated paraffin flame retardant additive; fluid; 68% act.

Demelan CB-10, -12, -14. [494] Alkylbenzene sulfonate blend; detergent for liq. dishwashing formulations; gel; sol. in water; 61 ± 1, 20 ± 1, and 35 ± 1% conc. resp.

Demelan CB-28. [494] Alkylbenzene sulfonate blend; detergent product for production of hand cleaners; liq.; 79 ± 1% conc.

Demelan CB-30, -42. [494] Alkylbenzene sulfonate blend; scouring agent for textile industry; liq.; 20 ± 1 and 30 ± 1% conc. resp.

Demelan CB-57. [494] Alkylbenzene sulfonate blend; see Demelan CB-10; gel; 50 ± 1% conc.

Demelan FB-12. [494] Alkylbenzene sulfonate blend; see Demelan CB-28; paste; 65 ± 1% conc.

Demelan FB-16. [494] Alkylbenzene sulfonate blend; see Demelan CB-30; paste; 65 ± 1% conc.

Demulfer Series. [606] Demulsifier for crude oil production; liq.

Denphos P-610. [263] Phosphate ester; anionic; detergent, emulsifier, wetting agent, hydrotrope for household and industrial cleaners; water-wh. to pale yel. liq.; mild odor; sol. in aq. sol'ns. with high conc. of alkaline builders; sp.gr. 1.155; pH 1.5–2.5 (2%); 100% act.

Densol 284. [263] Highly sulfated fatty acid; anionic; stabilizer for natural and syn. latex; protects from coagulation; dk. amber liq.; fatty odor; 70% conc.

Densol 6920. [263] Syn. replacement for sulfated castor oil; anionic; emulsifier for min. oil, disp. for colors in oil systems; liq.; sp.gr. 0.972 (60 F); pH 5.4 (10%); 75% act.

Densol 6930. [263] Syn. replacement for sulfated sperm oil; emulsifier; pH 5.4 (10%).

Densol BP-61, -62. [263] Block polymer; nonionic; defoamer, emulsifier, dispersant, wetting agent, and stabilizer; liq.; 100% conc.

Densol P-82. [263] Sulfated oleic ester; anionic; wetting agent, rewetting agent, emulsifier for paper towels, latex impregnation and dipping, textile dyeing; amber liq.; dens. 8.57 lb/gal; sp.gr. 1.0288; pH 5.5 (2%); 63% act.

Densulf TA-75. [263] Sulfonated tallow; softener for textiles; fatliquoring agent for leathers; plasticizer for cotton finishing; ivory soft paste, mild fatty odor; sol. in water; sp.gr. 1.09.

Denwet CM. [263] Sodium dioctyl sulfosuccinate; anionic; wetting agent, emulsifier for emulsion polymerization, battery separators, glass cleaners; clear liq.; dens. 9.0 lb/gal; pH 7.0 ± 0.5 (2%); 75% act.

Denwet RG-7. [263] Sulfonated alkyl ester; anionic; wetting agent, coupling agent, solubilizer used in drycleaning, herbicides, pesticides; clear liq.; dens. 9.0 lb/gal; pH 7.0 ± 0.5 (2%); 75% act.

Depasol AS-26. [494] Monoalkyl sulfosuccinate; anionic; mild detergent for cosmetic formulations; liq.; 32% conc.

Depasol AS-27. [494] Dialkyl sodium sulfosuccinate; anionic; wetting and penetrating agent; used in drycleaning detergents and emulsion polymerization; rewetting agent for textile and paper industries; liq.; 70 ± 2% conc.

Depasol CM-41. [494] Dialkyl sodium sulfosuccinate; anionic; see Depasol AS-27; liq.; 50 ± 2% conc.

Depuma. [42] Silicone-type nonionic; antifoaming agent; pearly wh., pourable visc. emulsion; water disp.; dens. 8.2 lb/gal.

Depuma OB. [42] Org. nonsilicone-type nonionic; antifoaming agent; wh. liq.; sol. in most org. solvs., emulsifies in warm water; 100% biodeg.; 100% act.

Dergon OM. [536] Coconut alkanolamide; nonionic; detergent, emulsifier; liq.; 90% conc.

Deriphat 130-C. [290] Monosodium N-octyl-beta-iminodipropionic acid; amphoteric; surfactant used in liq. lt.-duty, transportation, alkali, or acid cleaners; wetting agent; pale to colorless liq.; sol. in water, ethanol, IPA, butyl Cellosolve; dens. 8.6 lb/gal; pH 8.6; 49% solids.

Deriphat 151. [290] Sodium-N-coco beta-aminopropionate; amphoteric; wetting agent and emulsifier; wh. powd.; sol. in strong acids, alkali, and ionic systems; dens. 2 lb/gal; pH 12; 98% solids.

Deriphat 151C. [289; 290] N-coco-beta aminopropionic acid; amphoteric; wetting agent, detergent, emulsifier, corrosion inhibitor; hard surface cleaners; Gardner 5 clear liq.; sol. in strong acids, alkalies, ionic systems; sp.gr. 1.03; pH 5.5; 34% solids.

Deriphat 154. [289; 290] Disodium N-tallow-beta iminodipropionate; amphoteric; detergent, solubilizer for hard surface cleaning, textiles, emulsion polymerization; wh. powd.; sol. in strong acids, alkalies, and ionic systems; dens. 2 lb/gal; pH 11; 98% solids.

Deriphat 160. [289; 290] Disodium N-lauryl beta-iminodipropionate; amphoteric; detergent, solubilizer, primary emulsifier used in org. and inorg. compds.; wh. powd.; dens. 2.0 lb/gal; 98% solids.

Deriphat 160C. [289; 290] Monosodium-N-lauryl beta-iminodipropionate; amphoteric; detergent, solubilizer, stabilizer; used in petroleum processing and emulsion polymerization; amber clear liq.; sol. in strong acid, alkali, and ionic systems; sp.gr. 1.04; dens. 8.6 lb/gal; pH 7.5; 30% solids.

Deriphat 170C. [290] N-lauryl, myristyl beta aminopropionic acid; amphoteric; wetting agent; Gardner 3 clear liq.; sol. in alcohols; sp.gr. 1.03; pH 6; 45% solids.

Deriphat BA, BAW. [290] Cocamido betaine; amphoteric; foaming and wetting agent in acid, alkali, or electrolyte sol'ns., and household and industrial detergent formulations; liq.; 45–60% act.

Deriphat BC, BCW. [290] Coco dimethyl ammonium carboxylic acid betaine; amphoteric; foam stabilizer, wetting agent, frothing agent for liq. detergents; lime soap dispersant; liq.; 59 and 43% act. resp.

Desadipol AX Extra. [551] Alkylpolyethoxyether; nonionic; skin detergent; liq.; 100% conc.

DeSomeen TA-2. [181] Ethoxylated tallow amine (2 EO); emulsifier, dispersant, textile scouring, dyeing assistant, desizing assistant, softener, antistat; paste; 99% act.

DeSomeen TA-15. [181] Ethoxylated tallow amine (15 EO); see DeSomeen TA-2; liq.; cloud pt. 172–179 F; 100% act.

DeSomeen TA-20. [181] Ethoxylated tallow amine (20 EO); see DeSomeen TA-2; liq.; cloud pt. 179–181 F (10% NaCl); 100% act.

DeSonate 50-S. [181] Sodium dodecylbenzene sulfonate; anionic; detergent component; liq.; 53% act.

DeSonate 60-S. [181] Sodium dodecylbenzene sulfonate; anionic; detergent component; liq.; 60% act.

DeSonate 60-T. [181] Triethanolamine linear alkylbenzene sulfonate; surfactant for liquid detergents, bubble bath, shampoos; liq.; 60% act.

DeSonate AOS. [181] Sodium alpha olefin sulfonate; anionic; detergent component for shampoos, bubble bath, detergents, liquid soaps; liq.; 40% act.

DeSonate AUS. [181] Sodium C_{14}-C_{16} alpha olefin sulfonate; anionic; surfactant for use in shampoos and detergent products; liq.; 37% conc.

DeSonate SA. [181] Linear dodecylbenzene sulfonic acid; detergent intermediate, emulsifier, base for liquid and powd. detergents; liq.; 97% act.

DeSonate SA-H. [181] Branched dodecylbenzene sulfonic acid; chemical intermediate, emulsifier; liq.; 97% act.

DeSonic 1.5N. [181] Nonyl phenol ethoxylate (1.5 EO); defoamer, detergent, emulsifer; liq.; oil-sol.; 100% act.

DeSonic 3K. [181] Linear alcohol ethoxylate (3 EO); nonionic; detergent, emulsifier, wetting agent, defoamer for industrial detergents, textile scouring; liq.; oil-sol.; 100% act.

DeSonic 4N. [181] Nonyl phenol ethoxylate (4 EO); nonionic; detergent, emulsifier, defoamer for pesticide, paint, paper, and textile industries; liq.; oil-sol.; HLB 8.8; 100% act.

DeSonic 5D. [181] Dodecyl phenol ethoxylate; nonionic; emulsifier, wetter, and surfactant for agric., pesticide, industrial cleaner, and detergent formulations; liq.; HLB 9.1; 100% conc.

DeSonic 5K. [181] Linear alcohol ethoxylate (5 EO); nonionic; detergent, emulsifier, wetting agent for household/industrial cleaners; liq.; cloud pt. 75–79 F; 100% act.

DeSonic 5N. [181] Nonyl phenol ethoxylate (5 EO); detergent, surfactant; liq.; oil-sol.; 100% act.

DeSonic 6C. [181] Castor oil, ethoxylated; nonionic; emulsifier, lubricant, dye leveler, antistat, and dispersant for textile applications; mfg. of polyurethane foams; softening and rewetting agents for paper; liq.; HLB 4.4; 100% conc.

DeSonic 6D. [181] Dodecyl phenol ethoxylate (6 EO); nonionic; emulsifier for solv. and emulsion cleaners; liq.; oil-sol.; HLB 10.0; 100% act.

DeSonic 6N. [181] Nonyl phenol ethoxylate (6 EO); nonionic; emulsifier, detergent; liq.; HLB 10.8; 100% act.

DeSonic 6T. [181] Tridecyl alcohol ethoxylate (6 EO); nonionic; low-foam wetting agent, detergent, foamer for mechanical and spray cleaning; pulp, paper, textile industries; liq.; HLB 11.4; 100% act.

DeSonic 7K. [181] Linear alcohol ethoxylate (7 EO); nonionic; detergent, emulsifier, wetting agent for household/industrial cleaners; liq.; cloud pt. 118–122 F; 100% act.

DeSonic 7N. [181] Nonyl phenol ethoxylate (7 EO); low-foaming surfactant, detergent, emulsifier; liq.; HLB 11.7; 100% act.

DeSonic 9D. [181] Dodecyl phenol ethoxylate (9 EO); emulsifier, detergent for emulsion cleaners; liq.; cloud pt. 61–64 F; 100% act.

DeSonic 9K. [181] Linear alcohol ethoxylate (9 EO); nonionic; detergent, emulsifier, wetting agent for household/industrial cleaners; liq.; cloud pt. 131–138 F; 100% act.

DeSonic 9N. [181] Nonyl phenol ethoxylate (9 EO); nonionic; detergent, wetting agent, emulsifier for textile, paper, metal cleaning; liq.; HLB 12.8; cloud pt. 127–133 F; 100% act.

DeSonic 9T. [181] Tridecyl alcohol ethoxylate (9 EO); surfactant for lt. and heavy-duty detergents, textile leveling and scouring; liq.; cloud pt. 154–170 F; 100% act.

DeSonic 10D. [181] Dodecyl phenol ethoxylate (10 EO); nonionic; detergent/wetting agent for industrial and heavy-duty detergents; liq.; HLB 13.2; cloud pt. 145–153 F; 100% act.

DeSonic 10N. [181] Nonyl phenol ethoxylate (10 EO); detergent, wetting agent, emulsifier for textile, paper, metal cleaning; liq.; cloud pt. 140–149 F; 100% act.

DeSonic 10T. [181] Tridecyl alcohol ethoxylate; nonionic; foam builder and detergent for cleaning formulations; scouring agent in textile industry; liq.; HLB 13.8; 100% conc.

DeSonic 11N. [181] Nonyl phenol ethoxylate (11 EO); nonionic; detergent, wetting agent, emulsifier for textile, paper, metal cleaning; liq.; HLB 13.6; cloud pt. 158–162 F; 100% act.

DeSonic 12D. [181] Dodecyl phenol ethoxylate (12 EO); detergent/wetting agent for industrial and heavy-duty detergents at high temps.; liq.; cloud pt. 165–174 F; 100% act.

DeSonic 12K. [181] Linear alcohol ethoxylate

(12 EO); nonionic; detergent, wetting agent for household/industrial cleaners, high-temp. applications; liq.; cloud pt. 187–192 F; 100% act.

DeSonic 12N. [181] Nonyl phenol ethoxylate (12 EO); wetting agent, detergent used with high temp. and electrolytes; liq.; cloud pt. 176–181 F; 100% act.

DeSonic 12T. [181] Tridecyl alcohol ethoxylate (12 EO); surfactant for lt.- and heavy-duty detergents; leveling and scouring; liq.; cloud pt. 187–201 F; 100% act.

DeSonic 13N. [181] Nonylphenol ethoxylate; nonionic; see DeSonic 4N; liq.; HLB 14.3; 100% conc.

DeSonic 14D. [181] Dodecyl phenol ethoxylate; nonionic; see DeSonic 5D; solid; HLB 14.0; 100% conc.

DeSonic 15N. [181] Nonyl phenol ethoxylate (15 EO); detergent, wetting agent at high temps. and electrolyte; metal cleaning; liq.; cloud pt. 143–149 F (10% NaCl); 100% act.

DeSonic 15T. [181] Tridecyl alcohol ethoxylate (15 EO); surfactant for lt.- and heavy-duty detergents at high temps.; leveling and scouring agent; liq.; cloud pt. 156–167 F (10% NaCl); 100% act.

DeSonic 20N. [181] Nonyl phenol ethoxylate (20 EO); detergent, wetting agent at high temps. and electrolyte; solid; cloud pt. 154–162 F (10% NaCl); 100% act.

DeSonic 30C. [181] Castor oil ethoxylate (30 EO); nonionic; emulsifier, lubricant, dye levelers, antistat, dispersant for textiles; emulsifer for polyurethane foams; softener/rewetter for wet-strength paper; liq.; HLB 11.7; 100% act.

DeSonic 30N. [181] Nonyl phenol ethoxylate (30 EO); nonionic; emulsifier for fats, oils, waxes; solid; HLB 17.1; cloud pt. 159–163 F (10% NaCl); 100% act.

DeSonic 30N70. [181] Nonyl phenol ethoxylate (30 EO); nonionic; emulsifier for fats, oils, waxes; liq.; HLB 17.1; cloud pt. 159–163 F (10% NaCl); 70% act.

DeSonic 36C. [181] Castor oil ethoxylate (36 EO); nonionic; see DeSonic 30C; liq.; HLB 12.6; cloud pt. 122–140 F; 100% act.

DeSonic 40C. [181] Castor oil ethoxylate (40 EO); nonionic; see DeSonic 30C; liq.; HLB 13.1; cloud pt. 173–179 F; 100% act.

DeSonic 40N. [181] Nonyl phenol ethoxylate (40 EO); surfactant for high temps. and electrolytes; emulsifier; solid; cloud pt. 165–176 F (10% NaCl); 100% act.

DeSonic 40N70. [181] Nonyl phenol ethoxylate (40 EO); emulsifier; surfactant for high temps. and electrolytes; liq.; cloud pt. 165–176 F (10% NaCl); 70% act.

DeSonic 50N. [181] Nonyl phenol ethoxylate (50 EO); emulsifier for waxes, polishes; surfactant for high temps. and electrolytes; solid; 100% act.

DeSonic 50N70. [181] Nonyl phenol ethoxylate (50 EO); emulsifier for waxes, polishes; surfactant for high temps. and electrolytes; liq.; 70% act.

DeSonic 54C. [181] Castor oil ethoxylate (54 EO); nonionic; see DeSonic 30C; liq.; HLB 14.4; cloud pt. 136–142 F (10% NaCl); 100% act.

DeSonic 100N. [181] Nonyl phenol ethoxylate (100 EO); emulsifier for waxes, polishes; surfactant for high temps. and electrolytes; solid; 100% act.

DeSonic 100N70. [181] Nonyl phenol ethoxylate (100 EO); emulsifier for waxes, polishes; surfactant for high temps. and electrolytes; liq.; 70% act.

DeSonic AJ-85. [181] Alcohol, ethoxylated; nonionic; detergent and wetting agent for household and industrial cleaners; wetting of carpet fibers for dyeing process; liq.; 85% conc.

DeSonic AJ-100. [181] Alcohol ethoxylate; nonionic; anhyd. version of Flo Mo AJ85; paste; 100% conc.

DeSonic DA-4. [181] Decyl alcohol ethoxylate (4 EO); nonionic; rapid wetter/rewetter with low foaming properties for textile uses; liq.; water-disp.; HLB 10.5; 100% act.

DeSonic DA-6. [181] Decyl alcohol ethoxylate (6 EO); wetting agent for scour systems; liq.; water-sol.; HLB 12.5; cloud pt. 115–120 F; 100% act.

DeSonic S-45. [181] Octyl phenol ethoxylate (4.5 EO); emulsifier, surfactant, emulsion cleaner, dry dishwashing detergent, polish emulsifier; liq.; 100% act.

DeSonic S-100. [181] Octyl phenol ethoxylate (9–10 EO); metal and textile processing surfactant; household and industrial cleaners; emulsifier for vinyl and acrylic polymerization; liq.; cloud pt. 140–158 F; 100% act.

DeSonic S-102. [181] Octyl phenol ethoxylate (12–13 EO); surfactant for hot cleaning and pickling systems, hard surface cleaners, household and industrial cleaners; liq.; cloud pt. 187–194 F; 100% act.

DeSonic S-114. [181] Octyl phenol ethoxylate

(7–8 EO); surfactant for metal and acid cleaners, tank-soak, household, and industrial cleaners; liq.; cloud pt. 72–79 F; 100% act.

DeSonic S-405. [181] Octyl phenol ethoxylate (40 EO); coemulsifier for vinyl and acrylic polymerization; dye assistant; liq.; cloud pt. 165–176 F (10% NaCl); 70% act.

DeSonic SMO. [181] Sorbitan monooleate; nonionic; emulsifier, fiber lubricant and softener; liq.; sol. in oils and org. solvs.; insol. in water; HLB 4.3; 100% conc.

DeSonic SMO-20. [181] POE (20) sorbitan monooleate; nonionic; emulsifier, wetting agent, surfactant exhibiting antistat and lubricating properties; liq.; HLB 15.0; 100% conc.

DeSonic SMT. [181] Sorbitan monotallate; nonionic; see DeSonic SMO; liq.; insol. in water; sol. in oils and org. solvs.; HLB 4.1; 100% conc.

DeSonic SMT-20. [181] POE (20) sorbitan monotallate; nonionic; see DeSonic SMO-20; liq.; HLB 15.0; 100% conc.

DeSonic TA-2, -15. [181] Tallow amine, ethoxylated; cationic; emulsifier, dispersant, textile scouring agent, desizing assistant, softener, and antistat; 80-100% conc.

DeSonic TA-25CWS. [181] Tallow amine, ethoxylated; cationic; see DeSonic TA-2; liq.; 80% conc.

DeSonic TDA-9. [181] Alcohol, ethoxylated; nonionic; wetting agent, degreaser, detergent, and emulsifier; liq.; 100% conc.

DeSonol A. [181] Ammonium lauryl sulfate; anionic; surfactant for liquid detergents, shampoos, bubble bath; liq.; 28–30% act.

DeSonol AE. [181] Ammonium laureth sulfate; anionic; surfactant for shampoos, bubble bath, liquid detergents; liq.; 58–60% act.

DeSonol AES. [181] Ammonium laureth sulfate; anionic; detergent, surfactant for personal care products; liq.; 60% conc.

DeSonol S. [181] Sodium lauryl sulfate; anionic; surfactant for liquid detergents, shampoos, bubble baths; liq.; 28–30% act.

DeSonol SE. [181] Sodium laureth sulfate; anionic; surfactant for shampoos, bubble bath, liquid detergents; liq.; 58–60% act.

DeSonol SE-2. [181] Sodium laureth-2 sulfate; anionic; surfactant for shampoos, bubble bath, liquid detergents; liq.; 27–29% act.

DeSonol T. [181] TEA lauryl sulfate; anionic; surfactant for shampoos, liquid detergents; liq.; 39–40% act.

DeSophos 3 DP. [181] Phosphate ester; textile wetting agent; alkaline scouring; as potassium or sodium salts in dry cleaning detergents; liq.; 100% act.

DeSophos 3 OMP. [181] Acid ester of aliphatic hydrophobic base; emulsifier for paraffinic oils; liq.; 100% act.

DeSophos 3 OMPDEA. [181] Partly neutralized diethanolamine salt of DeSophos 3 OMP; emulsifier for paraffinic oils; liq.; 98% act.

DeSophos 4 CP. [181] Phosphate ester; softener, antistat for textile finishing; in lubricants for filament yarns, syn. fibers, wool; emulsifier for cosmetic oils and creams, polymerization of latexes; liq.; 95% act.

DeSophos 4 NP. [181] Phosphate ester; emulsifier for emulsion polymerization; liq.; most oil-sol. of series; water-sol. when neutralized; 100% act.

DeSophos 5 AP. [181] Phosphate ester; coupling agent for nonionics in liquid alkaline detergents; moderate foamer; liq.; 100% act.

DeSophos 5 BMP. [181] Phosphate ester; coupling agent for industrial uses; liq.; 100% act.

DeSophos 5 BP. [181] Phosphate ester; coupling agent for agric. and industrial uses; liq.; sol. in water, nonpolar aliphatic solv., oils; 100% act.

DeSophos 6 DNP. [181] Phosphate ester; paraffinic oil emulsifier as triethanolamine salt; rust inhibitor; most oil-sol. in DeSophos DNP series; liq.; 100% act.

DeSophos 6 DP. [181] Phosphate ester; dry cleaning detergent, penetrant, antistat; liq.; 100% act.

DeSophos 6 MPNa. [181] Phosphate ester; hard-surface detergent, corrosion retarder, moderate foamer in detergent concentrates; dry cleaning detergent; textile antistat; liq.; 88% act.

DeSophos 6 NP. [181] Phosphate ester; emulsifier, detergent; for emulsion polymerization; liq.; 100% act.

DeSophos 6 NP4. [181] Phosphate ester; polymerization emulsifier for polyvinyl acetate and acrylic films; also in waterless hand cleaners and laundry detergents; liq.; 100% act.

DeSophos 6 OMP. [181] Acid ester; more hydrophilic version of DeSophos 3 OMP; liq.; 100% act.

DeSophos 6 OMPDEA. [181] Partly neutralized diethanolamine salt of DeSophos 6

OMP; liq.; 98% act.

DeSophos 7 DP. [181] Phosphate ester; alkali-stable emulsifier, detergent, wetting agent, dispersant; rewetter in textile applications; liq.; 100% act.

DeSophos 7N. [181] Nonylphenol ethoxylate; nonionic; detergent and emulsifier used in pesticide, paper, paint, and textile industries; liq.; HLB 11.7; 100% conc.

DeSophos 7 OPNa. [181] Phosphate ester; dispersible surfactant in emulsification of min. oils; textile softener and antistat; in lubricants for syn. and wool fibers; liq.; sol. in aromatic solv.; 98% act.

DeSophos 8 DNP. [181] Phosphate ester; emulsifier for phosphated and chlorinated pesticide concs., polyethylene; in industrial cleaners and dry cleaning; liq.; sol. aromatic solv., kerosene; disp. water; 100% act.

DeSophos 9 NP. [181] Phosphate ester; dedusting agent for dry cleaning detergent, alkaline powders, water-repellent fabric finishes; emulsion polymerization of polyvinyl acetate and acrylic films; liq.; 100% act.

DeSophos 14 DNP. [181] Phosphate ester; see DeSophos 8 DNP; liq.; 100% act.

DeSophos 30 NP. [181] Phosphate ester; emulsifer, stabilizer; in preparation of polyvinyl acetate and acrylic copolymers; liq.; 100% act.

DeSotan SMO. [181] Sorbitan monooleate; lipophilic emulsifier, fiber lubricant, softener; liq.; sol. oils, org. solv.; insol. water; HLB 4.3; 100% act.

DeSotan SMO-20. [181] POE (20) sorbitan monooleate; hydrophilic emulsifier, wetting agent; general-purpose surfactant with antistatic and lubricating properties; liq.; cloud pt. 149–158 F (10% NaCl); 100% act.

DeSotan SMT. [181] Sorbitan monotallate; lipophilic emulsifier, fiber lubricant, softener; liq.; sol. oils, org. solv.; insol. water; HLB 4.1; 100% act.

DeSotan SMT-20. [181] POE (20) sorbitan monotallate; hydrophilic emulsifier, wetting agent; general-purpose surfactant with antistatic and lubricating properties; liq.; cloud pt. 149–158 F (10% NaCl); 100% act.

DET. [584] Alkylaryl sulfonate sodium salt with S.T.P.P. and other alkaline builders; detergent suitable for hand washing fabrics; wh. powd.; pH 10.5 (1%).

Detergent ADC. [317] Sodium alkylbenzene sulfonate; anionic; detergent base; wetting agent and scouring agent; emulsifier; liq.; 37% conc.

Detergent Concentrate 840. [418] Alkanolamide, modified; nonionic/anionic; used in car wash, bubble baths, all purpose cleaners; liq.; 82% conc.

Detergent CR. [67] Fatty ethanolamine condensate; detergent, wetting agent, emulsifier, penetrating and leveling agent, for use in personal care, food, and household products; pale amber liq.; mild odor; readily sol. in water; sp.gr. 1.01; 100% act.

Detergent RP. [317] DEA alkylbenzene sulfonate; anionic; emulsifier; detergent for dry cleaning; liq.; 80% conc.

Detergent Slurry. [127] Alkylaryl sulfonate; anionic; base for detergent uses; paste; 40–42% conc.

Detergyl 955FS. [317] Sodium alkylbenzene sulfonate; anionic; wetting/scouring agent; liq.; 20% conc.

Detergyl S. [317] 1:2 Coco diethanolamide; nonionic; detergent; scouring assistant for wool and cotton; penetrating assistant in melange printing; liq.; 100% conc.

DET Liq. [584] Alkylaryl sulfonate, POE alcohol ether sulfate plus additives; household cleanser, detergent; clear liq.; pH 8.0–9.5 (1%).

Det-O-Jet. [22] Low sudsing detergent with good wetting, emulsifying, and penetrating properties, for ultrasonic and mechanical washers; liq.; sol. in water; biodeg.

DET-Washmatic. [584] Sodium alkylaryl sulfonate; detergent for use in washing machines; blue, spray-dried powd.; pH 10.5 (1%).

Dextrol OC-50. [183] Complex org. phosphate ester, partially sodium neutralized; anionic; detergent formulations; liq.; sol. in water and solvs.; 88% conc.

Diamiet 503, 508, 520, AB, C. [561] *n*-Propylene alkyl fatty diamine EO condensation product; cationic; emulsifier, dispersant, corrosion inhibitor, wetting agent; liq., paste, solid; 100% conc.

Diamin DO. [561] Diamine dioleate salts; cationic; dispersant for oil flushing agents for pigments, oil/wax emulsifier; liq./paste; 100% conc.

Diamin HT. [561] *n*-Propylene hydrog. tallow fatty diamine; cationic; asphalt emulsifier, corrosion inhibitor, antistripping agent; solid; 100% conc.

Diamin O. [561] *n*-Propylene oleyl fatty diamine; cationic; see Diamin HT; liq. or paste; 100% conc.

Diamin S. [561] *n*-Propylene soya fatty diamine; cationic; see Diamin HT; liq., paste; 100% conc.

Diamin T. [561] *n*-Propylene tallow fatty diamine; cationic; see Diamin HT; paste; 100% conc.

Diamine B11. [348] Alkyl-1,3-propylene diamine; cationic; emulsifier, corrosion inhibitor; solid; 95% conc.

Diamine BG. [348] Tallow-1,3-propylene diamine; cationic; emulsifier, corrosion inhibitor; paste; 95% conc.

Diamine KKP. [348] Coco-1,3-propylene diamine; cationic; emulsifier, corrosion inhibitor; liq.; 95% conc.

Dianol. [172] Fatty acid diethanolamide; nonionic; emulsifier, dispersant for pigments, wax, solvs.; detergent, solubilizer; liq.; 100% conc.

Dianol 300. [172] Fatty acid diethanolamide; nonionic; detergent, solubilizer, emulsifier, and dispersant solvs.; liq.; 100% conc.

Diazital O Extra Conc. [192] Fatty alcohol polyglycol ethers; nonionic; dyeing assistant; solid; 100% conc.

Diazopon SS-837. [245] POE alkylphenol; nonionic; dispersant, solubilizer, stabilizer, anticrock and soaping agent for naphthol dyeings; used in textile dyeing processes; liq.; pH neutral; 70% conc.

Dibactol. [297] Myristalkonium chloride; biocide used in mfg., formulating, tableting, and dry blending; wh. powd.; m.w. 404.06; pH 5–8; 99–103% assay.

Dilasoft RS. [538] Fatty acid deriv.; amphoteric; hydrophilic softener for cellulosic fibers; liq.; 100% conc.

Dilasoft RW. [538] Fatty acid deriv. modified; cationic/nonionic; softening agent with wetting properties for cellulosic fibers; liq.

Dilasoft TF. [538] Sulfonate; softening/filling agent, anionic; liq.

Dimodan CP. [273] Cottonseed oil dist. monoglyceride, unsat.; food emulsifier; plastic; m.p. 50 C; 90% min. monoester.

Dimodan LS. [273] Sunflower oil dist. monoglyceride, unsat.; food emulsifier; soft plastic; m.p. 40 C; 90% min. monoester.

Dimodan O. [273] Veg. oil dist. monoglyceride, unsat.; food emulsifier; block; m.p. 60 C; 90% min. monoester.

Dimodan P. [273] Lard dist. monoglyceride, sat.; food emulsifier; m.p. 70 C; 90% min. monoester.

Dimodan PM. [273] Lard or tallow dist. monoglyceride, sat.; food emulsifier; beads; m.p. 70 C; 90% min. monoester.

Dimodan PM 300. [273] Lard or tallow dist. monoglyceride, sat.; food emulsifier; fine powd.; m.p. 70 C; 90% min. monoester.

Dimodan PV, PV 300. [273] Soybean oil dist. monoglyceride, unsat.; food emulsifier; beads, powd.; m.p. 72 C; 90% min. monoester.

Dimodan PVP. [273] Veg. oil dist. monoglyceride, sat.; food emulsifier; beads; m.p. 69 C; 90% min. monoester.

Dimodan S. [273] Lard dist. monoglyceride, unsat.; food emulsifier; block; m.p. 55 C; 90% min. monoester.

Dimodan TH. [273] Tallow dist. monoglyceride, sat.; food emulsifier; beads; m.p. 70 C; 90% min. monoester.

Dinoram C. [117] N-coco propylene diamine; cationic; chemical intermediate; paste.

Dinoram O. [117] N-oleyl propylene diamine; cationic; chemical intermediate; paste.

Dinoram S. [117] N-tallow propylene diamine; cationic; emulsifier for bitumen; antistripping agent for road making; paste.

Dinoram SH. [117] N-hydrog. tallow propylene; cationic; chemical intermediate; paste.

Dinoramac C. [117] N-coco propylene diamine acetate; cationic; flotation, bactericide, emulsifier, anticaking, soil stabilization, flocculation, corrosion inhibitor; paste.

Dinoramac O. [117] N-oleyl propylene diamine acetate; cationic; see Dinoramac C; paste.

Dinoramac S. [117] N-tallow propylene diamine acetate; cationic; see Dinoramac C; paste.

Dinoramox S 3 to 12. [117] POE N-tallow propylene diamine; cationic; dispersant, wetting agent, emulsifier, corrosion inhibitor used in paint, agric., chemical, and textile industries; liq.; 100% conc.

Dinoramox SH 3. [117] PEG-3 N-hydrog. tallow propylene diamine; cationic; emulsifier, dispersant, corrosion inhibitor for use in textile and detergent industries; solid; 100% conc.

Dionil OC, SD, SH 100, W 100. [308] Fatty acid amide polyglycol ethers; nonionic; detergent; additive for detergent, dishwashing agents and cosmetic preparations; component for textile auxs.; liq. to wax; 100% conc.

Dispatex G. [437] Aromatic condensated compound; anionic; dispersing agent for polyester dyeing; liq.; 26% conc.

Dispersogen ASN. [301] Higher alcohol, ethoxylated; dispersant for polyester dyeing; liq.; 26% conc.

Dispex A40. [27] Ammonium salt of a polymeric carboxylic acid; deflocculant used in the ceramic industry, for sanitaryware, tableware; pale yel. mobile liq.; misc. in water; sp.gr. 1.16; visc. 3.5–5.5 cs (15%); pH 7.5–8.5; 40% act.

Dispex N40. [27] Sodium salt of a polymeric carboxylic acid; see Dispex A40; pale yel. mobile liq.; sp.gr. 1.30; visc. 15–25 cs (25%); pH 7.0–7.5; 40% act.

Dissolvan Brands. [301] PO/EO block polymers; nonionic; emulsion breaker for w/o emulsions; dehydration and desalting agent for crude oil; liq.

DK Ester F-10, -20, -50, -70, -90, -110, -140, -160. [172] Sucrose fatty acid ester; nonionic; food additive for stabilizing emulsion, dispersant, solubilizer; powd.; HLB 1.0, 2.0, 6.0, 8.0, 9.5, 11.0, 13.0, 15.0 resp.; 100% conc.

DMAMP. [323] 2-Dimethylamino-2-methyl-1-propanol; emulsifier for waxes in combination with an unsat. fatty acid; corrosion inhibitor, solubilizer for resin systems; catalyst for production of polyurethane foams; raw material; liq.; essentially anhyd.; sol. in water, org. solvs.; dens. 7.5 lb/gal; sp.gr. 0.9; Gardner visc. A–A_2; f.p. 19 C; b.p. 160 C (760 mm Hg); flash pt. 135 F (TOC); 100% act.

DMAMP-80. [323] 2-Dimethylamino-2-methyl-1-propanol; see DMAMP; dens. 7.9 lb/gal; sp.gr. 0.95; Gardner visc. A–A_2; f.p. –20; b.p. 98 C (760 mm Hg); flash pt. 150 F (TOC); 80% act.

DMO-33. [287] Diethylene glycol monooleate; nonionic; emulsifier for emollient creams, biocides, detergents, and polishes; gelling agent for latex; antifoamer for min. oil; liq.; HLB 5.0; 100% conc.

DMS-33. [287] Diethylene glycol monostearate; nonionic; overcreasing agent in soaps, consistency regulator in o/w emulsions; emulsifier for creams and ointments; pigment suspension; flakes; HLB 5.0; 100% conc.

DO-33-F. [287] Sorbitan dioleate; nonionic; emulsifier for w/o creams and ointments; pigment stabilizer for paint, anticorrosion and min. oil emulsions; liq.; HLB 3.5; 100% conc.

Dobane 45. [555] Narrow cut linear alkylbenzene; detergent, base material; clear liq.; not. misc. with water; dens. 0.859 kg/l; visc. 13 cs; m.p. –69 C; flash pt. 172 C; 90% biodeg.

Dobane 055. [555] Linear alkylbenzene; anionic after sulfonation; detergent, base material; clear liq.; dens. 0.860 kg/l; visc. 10 cs; m.p. < –70 C; flash pt. 150 C; 90% biodeg.; 100% act.

Dobane 80. [555] Narrow cut linear alkylbenzene; detergent, base material; clear liq.; dens. 0.862 kg/l; visc. 5 cs; m.p. < –70 C; 90% biodeg.

Dobane 83. [555; 556] Linear alkylbenzene; anionic after sulfonation; detergent, base material; clear liq.; m.w. 230; dens. 0.861 kg/l; visc. 7 cs; m.p. < –70 C; flash pt. 138 C; 90% biodeg.; 100% act.

Dobane 102. [555; 556] Narrow cut linear alkylbenzene; anionic; detergent, base material; clear liq.; m.w. 235; dens. 0.861 kg/l; visc. 7 cs; m.p. < –70 C; f.p. 140 C; 90% biodeg; 100% conc.

Dobane 103. [556] Linear alkylbenzene; anionic; see Dobane 83; liq.; m.w. 238; 100% conc.

Dobane 113. [556] Linear alkyl benzene; anionic; see Dobane 83; liq.; m.w. 240; 100% conc.

Dobane 123. [555] Narrow cut linear alkylbenzene; detergent, base material; clear liq.; dens. 0.860 kg/l; visc. 11 cs; m.p. < –70 C; f.p. 160 C; 90% biodeg.

Dobane 945. [555] Broad cut linear alkylbenzene; detergent, base; clear liq.; dens. 0.860 kg/l; visc.11 cs; m.p. –70 C; flash pt. 142 C; 90% biodeg.

Dobane HP. [555] Broad cut linear alkylbenzene; detergent, base; clear liq.; dens. 0.861 kg/l; visc. 9 cs; m.p. < –70 C; flash pt. 140 C; 90% biodeg.

Dobane JN. [555] Linear alkylbenzene; anionic after sulfonation; detergent, base material for controlled foam and high foam household powders and lt. duty liq.; clear liq.; dens. 0.861 kg/l; visc. 8 cs; m.p. < –70 C; flash pt. 144 C; 90% biodeg.; 100% act.

Dobanic Acid 83. [555; 556] Linear alkylbenzene sulfonic acid; anionic; preparation of detergents; liq.; m.w. 310; 97% conc.

Dobanic Acid 102. [555] Linear alkylbenzene sulfonic acid; detergent for dishwashing liqs., household powds.; brown liq.; sulfur dioxide odor; miscible in water up to 40%; dens. 1.06 kg/l; visc. 1800 cs; m.p. –12 to –10

C; > 90% biodeg.; 97% act.

Dobanic Acid 103. [556] Alkylaryl sulfonic acid; anionic; preparation of detergents; liq.; m.w. 318; 97% conc.

Dobanic Acid JN. [555] Linear alkylbenzene sulfonic acid; detergent for dishwashing liq., household powders; br. liq.; sulfur dioxide odor; misc. in water up to 40%; dens. 1.05 kg/l; visc. 1600 cs; m.p. –12 to –10 C; > 90% biodeg.; 97% act.

Dobanol 23. [555; 556] C_{12}–C_{13} linear primary alcohol; nonionic; base material for preparation of alcohol ethoxylates, sulfates, ethoxysulfates; liq.; mild odor; dens. 0.831 kg/l; visc. 14 cs (40 C); m.p. 17–19 C; flash pt. 132 C; > 95% biodeg.; 100% act.

Dobanol 23-2. [555; 556] C_{12}–C_{13} linear primary alcohol ethoxylate; nonionic; detergent, intermediate for production of laundry powds.; clear liq.; dens. 0.91 kg/l; visc. 15 cs (40 C); m.p. 4–5 C; HLB 6.2; flash pt. 148 C; 100% act.

Dobanol 23-2S/70. [556] C_{12}-C_{13} linear primary alcohol ethoxysulfate, sodium salt; anionic; detergent, surfactant used in personal care products; paste; 70% conc.

Dobanol 23-3. [556] C_{12}-C_{13} linear primary alcohol ethoxylate; nonionic; detergent intermediate for cosmetic, specialty and dishwashing formulations; liq.; HLB 8.1; 100% conc.

Dobanol 23-6.5. [555; 556] C_{12}–C_{13} linear primary alcohol ethoxylate (6.5 EO); nonionic; see Dobanol 23-2; wh. slurry; dens. 0.96 kg/l; visc. 26 cs (40 C); m.p. 17–19 C; HLB 11.9; flash pt. 172 C; 100% act.

Dobanol 25. [555; 556] C_{12}–C_{15} linear primary alcohol; base material for preparation of alcohol ethoxylates, sulfates, and ethoxysulfates; liq.; mild odor; dens. 0.831 kg/l; visc. 14 cs (40 C); m.p. 20–22 C; biodeg.; 100% act.

Dobanol 25-3. [555; 556] Pareth-25-3; nonionic; detergent, intermediate for mfg. of cosmetic, specialty industrial, and dishwashing formulations; clear liq.; dens. 0.93 kg/l; visc. 18 cs (40 C); m.p. 5–6 C; HLB 7.8; flash pt. 162 C; > 95% biodeg.; 100% act.

Dobanol 25-3A/27 EU. [555] Ammonium pareth-25 sulfate; detergent, component of higher performance lt. duty liq. formulations; clear liq.; misc. in all proportions; dens. 1.00 kg/l; visc. 30 cs; flash pt. 33 C; pH 7.0–9.0 (5% aq.); > 95% biodeg.; 25.5–27.5% act.

Dobanol 25-3A/60. [555; 556] Ammonium pareth-25 sulfate; anionic; see Dobanol 25-3A/27 EU; liq.; misc. in all proportions; dens. 1.04 kg/l; visc. 60 cs; cloud pt. 8 C; flash pt. 26 C; pH 6.5–8.0 (5% aq.); > 95% biodeg.; 59–61% act.

Dobanol 25-3S/27. [555; 556] Sodium pareth-25 sulfate; anionic; see Dobanol 25-3A/27 EU; clear liq.; misc. in all proportions; dens. 1.05 kg/l; visc. 60 cs; cloud pt. 1 C; flash pt. 26 C; pH 6.5–8.0 (5% aq.); > 95% biodeg.; > 26% act.

Dobanol 25-3S/60. [555; 556] Sodium pareth-25 sulfate; anionic; see Dobanol 25-3A/27 EU; misc. in all proportions; dens. 1.05 kg/l; visc. 60 cs; flash pt. 25 C; pH 6.5–8.0 (5% aq.); > 95% biodeg.; 58–60% act.

Dobanol 25-3S/70. [555; 556] Sodium pareth-25 sulfate; anionic; see Dobanol 25-3A/27 EU; clear gelled liq.; misc. in all proportions; dens. 1.08 kg/l; visc. 5000 cP at shear rate of 10 reciprocal s; pH 7.5–9.5 (5% aq.); > 95% biodeg.; 68–72% act.

Dobanol 25-7. [555; 556] Pareth-25-7; nonionic; detergent, wetting agent in laundry powds.; wh. paste/solid; misc. in all proportions; dens. 0.97 kg/l (40 C); visc. 31 cs (40 C); HLB 12.0; cloud pt. 48–54 C; flash pt. 180 C; pH 6.8 (0.5% aq.); 100% act.

Dobanol 25-9. [555; 556] Pareth-25-9; nonionic; detergent, wetting in laundry powds.; wh. solid; misc. in all proportions; dens. 0.98 kg/l (40 C); visc. 40 cs (40 C); HLB 13.2; cloud pt. 75–83 C; flash pt. 190 C; pH 6.8 (0.5% aq.); 100% act.

Dobanol 45. [555; 556] C_{14}–C_{15} linear primary alcohol; detergent, intermediate, base material; liq.; mild odor; dens. 0.824 kg/l (40 C); visc. 17 cs (40 C); m.p. 29–31 C; flash pt. 152 C; biodeg.; 100% act.

Dobanol 45-7. [555; 556] Pareth-45-7; nonionic; detergent, wetting agent in high performance laundry powds.; wh. solid; misc. in all proportions; dens. 0.96 kg/l (50 C); visc. 33 cs (40 C); HLB 11.6; cloud pt. 43–49 C; flash pt. 185 C; pH 6.8 (0.5% aq.); 100% act.

Dobanol 45-11. [555; 556] Pareth-45-11; nonionic; see Dobanol 45-7; wh. solid; misc. in all proportions; dens. 0.99 kg/l (50 C); visc. 47 cs (40 C); HLB 13.7; cloud pt. 84–92 C; flash pt. 200 C; pH 6.8 (0.5% aq.); 100% act.

Dobanol 45-18. [555; 556] Pareth-45-18; nonionic; see Dobanol 45-7; wh. solid; misc. in all proportions; dens. 1.02 kg/l (50 C); visc. 42 cs (40 C); HLB 15.7; cloud pt. > 100 C;

flash pt. 200 C; pH 6.8 (0.5% aq.); 100% act.

Dobanol 91. [555; 556] C_9–C_{11} linear primary alcohol; base material for preparation of alcohol ethoxylates, sulfates, ethoxysulfates; dens. 0.832 kg/l; visc. 9 cs (40 C); m.p. –10 to –8 C; flash pt. 107 C (PMCC); biodeg.; 100% act.

Dobanol 91-2.5. [555; 556] Pareth-91-3; nonionic; detergent intermediate; clear liq.; dens. 0.93 kg/l; visc. 11 cs (40 C); HLB 8.1; flash pt. 120 C (PMCC); pH 6.8 (0.5% aq.); 95% biodeg.; 100% act.

Dobanol 91-5. [555; 556] Pareth-91-5; nonionic; wetting agent, emulsifier; clear liq.; misc. in all proportions; dens. 0.98 kg/l; visc. 18 cs (40 C); HLB 11.6; flash pt. 146 C (PMCC); pH 6.8 (0.5% aq.); 100% act.

Dobanol 91-6. [555; 556] Pareth-91-6; nonionic; detergent, wetting agent; clear liq.; misc. in all proportions; dens. 0.99 kg/l; visc. 21 cs (40 C); HLB 12.5; flash pt. 150 C (PMCC); pH 6.8 (0.5% aq.); 100% act.

Dobanol 91-8. [555; 556] Pareth-91-8; nonionic; detergent; liq./slurry; misc. in all proportions; dens. 1.00 kg/l; visc. 28 cs (40 C); HLB 13.7; flash pt. 160 C (PMCC); pH 6.8 (0.5% aq.); 100% act.

Dobanol 235. [555] A blend of primary alcohols; base material for preparation of alcohol ethoxylates, sulfates, ethoxysulfates; dens. 0.831 kg/l; visc. 14 cs (40 C); m.p. 18–20 C; flash pt. 134 C (PMCC); biodeg.

Dobanol 235-3. [555] Primary alcohol ethoxylate; chemical intermediate; clear liq.; dens. 0.92 kg/l; visc. 17 cs (40 C); m.p. 4–5 C; HLB 7.8; flash pt. 160 C (PMCC); > 95% biodeg.; 100% act.

Dodiflood Brands. [301] Ether sulfonates, ether carboxylates; anionic; surfactant for enhanced oil recovery and microemulsion flooding; liq.

Doittol 14. [192] Sulfate fatty acid ester; anionic; wetting agent and leveling assistant for dyeing and bleaching; liq.; 45% conc.

Doittol 891. [187; 192] Sodium fatty acid soap, modified; anionic; surfactant; rubber antiblocking agent; liq.; 20% act.

Doittol APS Conc. [192] Alkylaryl polyglycol ether, semiester; anionic; dyeing assistant for dyeing polyester and blends; liq.; 60% conc.

Doittol FL. [192] Alkyl polyglycol ether blend; detergent for soaping after dyeing; semiliq.; 50% conc.

Doittol P690. [192] Complex org. phosphate ester; amphoteric; texile fiber antistat; liq.; 100% conc.

Dousoft BK 5078. [145] Imidazoline deriv.; cationic; softener conc. for textiles, paper and liq. fabric softeners; liq.; 75% min. conc.

Dovanox 23H, 23M, 25N, 231. [370] POE syn. alcohol; nonionic; detergent, penetrant, emulsifier, dispersant; liq., wax; 100% conc.

Dow Corning F1-1630. [203] Dimethyl polysiloxane; surfactant used in flexible urethane foam.

Dow Corning FF-400. [203] Silicone and glycol copolymer; nonionic; lubricant, stabilizer, antistat, fiber finish for textiles, threads; lt. amber fluid; sol. in polar solvs.; water disp.; sp.gr. 1.022; visc. 300 cs; HLB 7.0; flash pt. 177 C (OC); 100% act.

Dow Corning FF-412. [203] Organopolysiloxane; fiber finish for textiles, threads; lt. amber fluid; sol. in low m.w. alcohols, glycol ethers; water disp.; sp.gr. 1.040; visc. 200 cs; flash pt. 177 C (OC); 100% act.

Dow Corning FF-414. [203] Organopolysiloxane; fiber finish for textiles, threads; lt. amber fluid; sol. in water, alcohols, glycol ether, water disp.; sp.gr. 1.070; visc. 465 cs; flash pt. 177 C (OC); 100% act.

Dow Corning 544 Antifoam. [203] Silicone compd.; foam control agent for detergents, textile jet dyeing, cutting oils; med. off-wh. liq.; 100% act.

Dow Corning Antifoam 1400. [203] Silicone compd.; foam control agent for glycol scrubbing, resin mfg., oil refining, asphalts, adhesives/coatings, metalworking, pesticide/fertilizer, and detergents industries; med. off-wh. to wh. liq.; 100% act.

Dow Corning Antifoam 1410. [203] Silicone emulsion, nonionic; foam control agent for inks, textile starching/sizing, cutting oils, adhesives/coatings, waste water treatment, pesticide/fertilizer industries; thin wh. cream; 10% act.

Dow Corning Antifoam 1430. [203] Silicone emulsion, nonionic; foam control agent for detergents, distillation, glycol scrubbing, metalworking, waste water treatment, adhesives/coatings, and textile industries; med. wh. cream; 30% act.

Dow Corning Antifoam 1500. [203] Silicone compd.; foam control agent for foods, pesticides/fertilizers, gas-oil separation, printing inks, asphalt; med. off-wh. liq.; 100% act.

Dow Corning Antifoam 1510-US. [203] Silicone emulsion; nonionic; foam control agent for pesticide/fertilizer, waste water treatment, food industries; med. wh. cream; 20% act.

Dow Corning Antifoam 1520-US. [203] Silicone emulsion; nonionic; foam control agent for food, resin mfg., waste water treatment, adhesives/coatings, metalworking, and textile industries; thin wh. cream; 20% act.

Dow Corning Antifoam A. [203] Simethicone; foam control agent for distillation, resin sizes, textile latex backing; also avail. in food grade; med. off-wh. to gray liq.; 100% act.

Dow Corning Antifoam AF. [203] Simethicone; nonionic; foam control agent for food, pesticide/fertilizer, paper/printing, textile, adhesives/coatings industries; thick wh. cream; 30% act.

Dow Corning Antifoam B. [203] Silicone emulsion; nonionic; foam control agent for adhesives/glues, textile latex backing, inks, detergents, distillation, resin mfg, pesticide/fertilizer industries; thin wh. cream; 10% act.

Dow Corning Antifoam C. [203] Simethicone; nonionic; food grade foam control agent for food industry and paper coatings; med. wh. cream; 30% act.

Dow Corning Antifoam FG-10. [203] Silicone emulsion; nonionic; foam control agent for food industry and waste water treatment; thin wh. cream; 10% act.

Dow Corning Antifoam H-10. [203] Silicone emulsion; nonionic; foam control agent for detergents, waste water treatment, adhesives/coatings textile, metalworking, and paper/printing industries; thin wh. cream; 10% act.

Dow Corning Antifoam Y-30. [203] Silicone emulsion; nonionic; foam control agent for paper/printing industry for resin sizes and inks; med. wh. cream; 30% act.

Dow Corning 200 Fluid. [203] Dimethicone; foam control agent for distillation, resin mfg., asphalt, oil refining, gas-oil separation; clear fluid; visc. 1000–10,000 cSt and 12,500–60,000 cSt grades; 100% act.

Dow Corning 200 Fluid, Food Grade. [203] Dimethyl silicone fluid; foam control agent for food industry, inks; thin clear fluid; visc. 350 cSt; 100% act.

Dow Corning 470A Fluid. [203] Silicone glycol copolymer; nonionic; emulsifier, emulsifier additive, wetting agent, surf. tens. depressant for cosmetic, detergent uses; medium straw liq.; sol. in water @ < 160 F; sp.gr. 1.075; visc. 275 cst.; flash pt. 500 F (OC); 100% solids.

Dow Corning 1248 Fluid. [203] Silicone polycarbinols (graft copolymer type); surfactant for org. systems and urethanes; clear liq.; sol. in hydrocarbons; visc. 160 cs; HLB 1.5; flash pt. 250 F; pour pt. –130 F; 100% conc.

Dow Corning FS-1265 Fluid. [203] Dimethyl silicone fluid; foam control agent for aromatic/chlorinated solvs., dry cleaning, metal cleaning and degreasing, oil refining; clear fluid; visc. 300 cSt and 1000–10,000 cSt grades; 100% act.

Dow Corning Q2-8026 Fluid. [203] Silicone polycarbinols (graft copolymer type); improves properties of orgs. and urethanes; clear straw liq.; sol. in hydrocarbons; visc. 465 cs; flash pt. 400 F; pour pt. 60 F.

Dow Corning Q4-3557 Fluid. [203] Silicone polycarbinols (graft copolymer type); see Dow Corning Q2-8026; clear straw liq.; sol. in hydrocarbons; visc. 40 cs; flash pt. >200 F; pour pt. –80 F.

Dow Corning Q4-3667 Fluid. [203] Silicone polycarbinols (ABA type); nonionic; see Dow Corning Q2-8026; clear straw liq.; sol. in hydrocarbons; visc. 320 cs; HLB 9.4; flash pt. 285 F; pour pt. 65 F; 100% conc.

Dow Corning 3225C Formulation Aid. [203] Silicone glycol copolymer; nonionic; used in personal care products; liq.; HLB 4.0; 10% conc.

Dow Corning 190 Surfactant. [203] Dimethicone copolyol; nonionic; silicone surfactant used for producing flexible slab stock urethane foam; ingredient in personal care products; Gardner 2, low-visc. liq.; sol. in water and ethanol; sp.gr. 1.035; visc. 1500 cst; HLB 14.4; flash pt. 250 F (CCC); 100% conc.

Dow Corning 193 Surfactant. [203] Dimethicone copolyol; nonionic; see Dow Corning 190 Surfactant; Gardner 2, clear vis. liq.; sp.gr. 1.07; visc. 465 cs; HLB 13.6; f.p. 50 F; flash pt. 300 F (COC); pour pt. 52 F.

Dow Corning 196 Surfactant. [203] Silicone glycol copolymer; profoamer used for producing low-dens., flexible-slab urethane foams; Gardner < 4 low-visc. liq.; sol. in water, polyol, fluorocarbon or water-amine streams; sp.gr. 1.035; visc. 2000 cst; flash pt. 225 F (OC).

Dow Corning 197 Surfactant. [203] Nonhydrolyzable silicone; profoamer for high and low dens. rigid urethane foams with good system compatibility; clear yel. thin liq.; sp.gr. 1.05; visc. 300 cst; flash pt. 120 F.

Dow Corning 198 Surfactant. [203] Silicone glycol copolymer; surfactant for high- and low-dens. flexible urethane foams; Gardner < 4, low-visc. liq.; sol. in water, polyol, fluorocarbon or water-amine streams; sp.gr. 1.035; visc. 1500 cst.; flash pt. > 150 F (OC).

Dow Corning 1250, 1251 Surfactant. [203] Silicone in solv. (xylene); profoamer for mechanical frothing; allows foam fusing at low temp.; very lt. straw-colored thin liq.; sp.gr. 1.00; visc. 6 and 4 cst resp.; flash pt. 85 F (OC); 50% silicone.

Dow Corning 1252, 1253 Surfactant. [203] Silicone in solv. (texanol isobutyrate); see Dow Corning 1250; very lt. straw-colored thin liq.; sp.gr. 1.06 and 1.07 resp.; visc. 250 cs; flash pt. 135 F (OC); 50% silicone.

Dow Corning 1315 Surfactant. [203] Silicone glycol copolymer; nonionic; surfactant used in flame-retarded flexible foam systems; Gardner 2–4 liq.; sol. in polyol/water/amine premixes; sp.gr. 1.03; visc. 1000 cs; HLB 15.3; flash pt. 60.6 C (CC).

Dow Corning 5043 Surfactant. [203] Silicone glycol copolymer; surfactant used in molded and slab stock urethane foam; lt. straw low visc. liq.; sol. in polyol/water/amine premixes; sp.gr. 1.0; visc. 300 cSt; flash pt. 60 C (CC).

Dow Corning 5098 Surfactant. [203] Nonhydrolyzable silicone glycol copolymer; surfactant for production of foam; lt. straw clear liq.; sp.gr. 1.08; visc. 250 cSt; flash pt. (CC) 56 C.

Dow Corning Q2-5103 Surfactant. [203] Silicone surfactant; used to produce rigid polyurethane foam and flexible urethane foam; Gardner 2–4, lt. straw clear liq.; sol. in water, polyol, fluorocarbon, and water-amine systems; sp.gr. 1.05; visc. 250 cSt; flash pt. (CC) > 200 C.

Dow Corning Q2-5125 Surfactant. [203] Silicone surfactant; see Dow Corning Q2-5103; Gardner 2–4 clear liq.; sol. in water, polyol, fluorocarbon, and water-amine streams; sp.gr. 1.03; visc. 1000 cs; flash pt. 60.6 C (CC).

Dowfax 2A0. [201; 203] Dodecyl diphenyloxide disulfonic acid; anionic; intermediate used in acidic systems or preparation of salts; black liq.; sp.gr. 1.11; visc. 510 cps; surf. tens. 30 dynes/cm (0.10%); 40% act.

Dowfax 2A1. [201; 203] Sodium dodecyl diphenyloxide disulfonate; anionic; detergent, emulsifier, wetting agent in electroplating, solubilizer, leveling agent for acid dyeing of nylon, dyeing assistant; amber liq.; sol. in aq. sol'ns. of acids, alkalies, electrolytes; sp.gr. 1.16; visc. 145 cps (0.10%); HLB 16.7; surf. tens. 31 dynes/cm (0.10%); 45% act.

Dowfax 2A-Monoethanolamine. [201] MEA dodecyl diphenyloxide disulfonate; anionic; used in agric. formulations; liq.; 45% conc.

Dowfax 2A-Monoisopropylamine. [201] Monoisopropylamine dodecyl diphenyloxide disulfonate; anionic; used in agric. formulations; liq.; 45% conc.

Dowfax 2A-Triethylamine. [201] Triethylamine dodecyl diphenyloxide disulfonate; anionic; used in agric. formulations; liq.; 45% conc.

Dowfax 2A-Triisopropanolamine. [201] Triisopropanolamine dodecyldiphenyloxide disulfonate; anionic; used in agric. formulations; liq.; 45% conc.

Dowfax 2EP. [201] Sodium dodecyl diphenyloxide disulfonate; anionic; see Dowfax 2A1; liq.; HLB 16.7; 45% conc.

Dowfax 3B0. [201] N-decyl diphenyloxide disulfonate; anionic; see Dowfax 2A1; also used in formulating cleaning products and agric. formulations; liq.; 40% conc.

Dowfax 3B2. [201; 203] Sodium *n*-decyl diphenyloxide disulfonate; anionic; detergent, wetting agent, emulsifier, solubilizer, used as pesticide spray adjuvant; red-br. liq.; sp.gr. 1.17; visc. 120 cps; surf. tens. 38 dynes/cm (0.10%); 45% act.

Dowfax 3B-Monoethanolamine. [201] Monoethanolamine *n*-decyldiphenyloxide disulfonate; anionic; surfactant for agric. formulations; liq.; 45% conc.

Dowfax 3B-Monoisopropylamine. [201] Monoisopropylamine *n*-decyldiphenyloxide disulfonate; anionic; surfactant for agric. formulations; liq.; 45% conc.

Dowfax 3B-Triethylamine. [201] Triethylamine n-decyldiphenyloxide disulfonate; anionic; surfactant for agric. formulations; liq.; 45% conc.

Dowfax 3B-Triisopropanolamine. [201] Triisopropanolamine *n*-decyldiphenyloxide disulfonate; anionic; surfactant; used in agric. formulations; liq.; 45% conc.

Dowfax 9N. [202] Nonyl phenol ethoxylate; nonionic; intermediate, detergent, wetting agent, dispersant; used for household and industrial cleaners, wool scouring, textile aux., general industrial usage; liq., waxy solid; 100% conc.

Dowfax 63N10. [202] POE POP block copolymer; nonionic; detergent for textiles, household uses, defoamer for anionic detergent slurries; colorless liq.; slt. water sol.; sp.gr. 1.02; visc. 340 cst; 100% act.

Dowfax 63N20. [202] POE POP glycol copolymer; nonionic; detergent in liq. dishwashing agent, defoamer; liq.; 100% conc.

Dowfax XDS 8292.00. [201; 202] Sodium N-hexyl diphenyloxide disulfonate; anionic; see Dowfax 2A1; liq.; 45% conc.

Dowfax XDS 8390.00. [201; 202] Sodium *n*-hexadecyl diphenyloxide disulfonate; anionic; surfactant for emulsion polymerization; liq.; HLB 14.4; 35% conc.

Dowfax XDS 30599. [201; 202] Sodium dodecyl diphenyloxide disulfonate; anionic; used in emulsion polymerization; liq.; 45% conc.

DRA-1500. [606] POE disproportionated rosin ester; nonionic; emulsifier, dispersant; liq.; HLB 14.0; 100% conc.

Dresinate 81. [295] Rosin sodium soap; anionic; emulsifier ingred. and/or stabilizer in solv. cleaners; pale-colored liq.; dilutable with water and aq. sol'ns. of alcohols and glycols; dens. 8.7 lb/gal; visc. 5.7 poises (140 F); 87% solids.

Dresinate 90. [295] Rosin potassium soap; anionic; see Dresinate 81; pale liq.; dilutable with water and aq. sol'ns. of alcohols and glycols; dens. 9.0 lb/gal; visc. 14.7 poises (140 F); 90% solids.

Dresinate 91. [294; 295] Rosin potassium soap; anionic; see Dresinate 81; emulsifier for metalworking oils, drilling; pale liq.; dilutable with water and aq. sol'ns. of alcohols and glycols; dens. 8.8 lb/gal; visc. 5.3 poises (140 F); 88% solids.

Dresinate 92. [295] Potassium soap; anionic; see Dresinate 81; dk. liq.; dilutable with water and aq. sol'ns. of alcohols and glycols; dens. 9.0 lb/gal; visc. 9.2 poises (140 F); 88% solids.

Dresinate 93. [295] Potassium soap; anionic; see Dresinate 81; dk. liq.; dilutable with water and aq. sol'ns. of alcohols and glycols; dens. 9.5 lb/gal; visc. 7.3 poises (140 F); 87% solids.

Dresinate 94. [295] Potassium soap; anionic; see Dresinate 81; pale liq.; dilutable with water and aq. sol'ns. of alcohols and glycols; dens. 9.3 lb/gal; visc. 12.4 poises (140 F); 87% solids.

Dresinate 95. [294; 295] Potassium soap; anionic; see Dresinate 81; emulsifier for metalworking oils, drilling; dk. liq.; dilutable with water and aq. sol'ns. of alcohols and glycols; dens. 9.1 lb/gal; visc. 4.5 poises (140 F); 88% solids.

Dresinate 214. [294; 295] Potassium soap of a pale modified rosin; anionic; emulsifier, pigment wetting agent and dispersant, foaming agent, used in mfg. of adhesives, polymer emulsion syn. latices; paste; dilutable with water and aq. sol'ns. of alcohols and glycols; dens. 9.22 lb/gal (60 C); visc. 23 poises (140 F); 80% solids.

Dresinate 515. [294; 295] Potassium soap of a pale modified rosin; anionic; see Dresinate 214; paste; dens. 9.24 lb/gal (60 C); visc. 64 poises (140 F); 80% solids.

Dresinate 731. [294; 295] Sodium soap of a pale modified rosin; see Dresinate 214; paste; dilutable with water and aq. sol'ns. of alcohols and glycols; dens. 9.1 lb/gal (60 C); visc. 11 poises (140 F); 70% solids.

Dresinate TX. [295] Tall oil sodium soap; anionic; emulsifier for oils and asphalts, detergent aid, disp. for pigments; stabilizer for syn. rubber latices, used for industrial and household cleaners; dk. br. powd.; sol. in water and aq. sol'ns. of alcohols and glycols; dens. 26 lb/ft^3; 96% min. solids.

Dresinate X. [295] Sodium soap of a pale rosin; anionic; see Dresinate TX; pale cream, powd.; sol. in water and aq. sol'ns. of alcohols and glycols; dens. 26 lb/ft^3; 95% min. solids.

Dresinate XX. [295] Sodium soap of a dk. rosin; see Dresinate TX; lt. br. powd.; sol. in water and aq. sol'ns. of alcohols and glycols; dens. 28 lb/ft^3; 96% min. solids.

Drew BRX-354. [206] Coco amide propyl amine oxide; nonionic; detergent used in bubble bath and acid shampoo formulations; liq.; 30% act.

Drewfax 0007. [205] Sodium dioctyl sulfosuccinate; wetting, penetrating, surf. tens. reducing agent; colorless liq.; dens. 9.0 lb/gal; sol. in water and various solvs.; 70% act.

Drewfax 218. [206] Sodium laureth sulfate; anionic; wetting agent and emulsifier for polymerization; liq.; 26–28% conc.

Drewfax 220. [206] Sodium alkyl sulfate; anionic; wetting agent for pesticides, emulsifier for polymerization; liq.; 29–30% act.

Drewfax 250. [206] Sodium alkyl sulfate; anionic; wetting agent for pesticides, emulsifier for polymerization; powd.; 90% act.

Drewfax 358. [206] Phosphate ester; anionic; emulsifier; liq.; 100% conc.

Drewfax 495. [206] Alkylaryl sulfonate; anionic; emulsifier; liq.; 95% conc.

Drewfax 502. [206] Sulfosuccinate deriv.; anionic; wetting agent and emulsifier for polymerization and flowable pesticides; liq.; 30% conc.

Drewfax 528. [206] Sodium dioctyl sulfosuccinate; anionic; wetting agent and emulsifier for polymerization; paste; 70% act.

Drewfax 539. [206] Sulfosuccinate deriv.; anionic/nonionic; wetting agent and emulsifier for polymerization and flowable pesticides; liq., powd.

Drewfax 549, 550, 560, 577. [206] Sulfosuccinate deriv.; see Drewfax 539; liq. (549), powd. (550, 560); 70% conc. (549), 66–68% (550), 83% (560).

Drewfax 585. [206] Sulfosuccinate deriv.; anionic; wetting agent for pesticide powds.; powd.; 97% min. conc.

Drewfax 1785. [206] Sorbitan ester; nonionic; emulsifier for w/o and o/w emulsions; liq.; 100% conc.

Drewfax 1980. [206] Glycol ester; nonionic; visc. modifier for thermoplastics; liq.; 100% conc.

Drewlate 30. [496] Acetylated tartrated monoglyceride of vegetable fat; food emulsifier; amber syrupy liq.

Drewlene 10. [496] Propylene glycol monostearate; food emulsifier increases whipping ability and aeration; yel. solid, flake, bead; m.p. 110 F; HLB 3.4; sapon. no. 170–190.

Drewmulse 3-1-O. [496] Triglycerol monooleate; nonionic; emulsifier, solubilizer, disp. for w/o and o/w emulsions, creams, lotions, for internal use; Gardner 8 liq.; sol. in IPA, peanut oil, min. oil, water disp.; HLB 7.0; sapon. no. 125–150.

Drewmulse 3-1-S. [496] Triglycerol monostearate; nonionic; see Drewmulse 3-1-O; Gardner 10 liq.; disp. in water, IPA, peanut oil, min. oil, propylene glycol; HLB 7.0; sapon. no. 120–140.

Drewmulse 6-2-S. [496] Hexaglycerol distearate; nonionic; see Drewmulse 3-1-O; Gardner 13 solid; sol. in IPA, peanut oil, min. oil; HLB 8.0; sapon. 105–125.

Drewmulse 10. [496] Glycerol mono fat from soya oil; lipophilic emulsifier used in food industry as dispersing aid, antistaling agent, antistick agent; cream plastic; m.p. 118 F; HLB 2.8; sapon. no. 170–190; 40% alpha mono.

Drewmulse 10-4-O. [496] Decaglycerol tetraoleate; nonionic; see Drewmulse 3-1-O; Gardner 8 liq.; sol. in IPA, peanut oil, min. oil; water disp.; HLB 6.0; sapon. no. 125–145.

Drewmulse 10-8-O. [496] Decaglycerol octaoleate; nonionic; see Drewmulse 3-1-O; Gardner 8 liq.; sol. in peanut oil, min. oil; HLB 4.0; sapon. no. 155–175.

Drewmulse 10-10-O. [496] Decaglycerol decaoleate; nonionic; see Drewmulse 3-1-O; Gardner 8 liq.; sol. in peanut oil, min. oil; HLB 3.0; sapon. no. 165–185.

Drewmulse 10-10-S. [496] Decaglycerol decastearate; nonionic; see Drewmulse 3-1-O; 8 Gardner solid; sol. in peanut oil, min. oil; HLB 3.0; sapon. no. 160–180.

Drewmulse 15. [496] Glycerol mono cottonseed oil; see Drewmulse 10; cream soft solid; m.p. 110 F; 40% alpha mono.

Drewmulse 20/200/V. [496] Glycerol monostearate; see Drewmulse 10; cream beads; m.p. 140 F; 40% alpha mono.

Drewmulse 21. [496] Glycerol monostearate from soya oil; see Drewmulse 10; cream bead, flake; HLB 8.4; sapon. no. 160–180; 30% alpha mono.

Drewmulse 30/900. [496] Glycerol monostearate; see Drewmulse 10; cream beads; m.p. 138 F; 52% alpha mono.

Drewmulse 50. [496] Glycerol mono fat; see Drewmulse 10; cream colored cream; 52% alpha mono.

Drewmulse 55. [496] Glycerol mono shortening from tallow and soya; cream plastic; m.p. 130–135 F (tallow), 116–122 F (soya); HLB 3.4 (tallow), 2.4 (soya); sapon. no. 165–185; 52% alpha mono.

Drewmulse 70. [206] Blend of glycerol monoesters; nonionic; food emulsifier; liq.; 13.5–19% act.

Drewmulse 75. [206; 496] Glyceryl cocoate; nonionic; food emulsifier; paste; 100% act.

Drewmulse 85. [496] Glycerol monooleate from tallow and soya; see Drewmulse 10; lt. yel. semiliq.; HLB 3.4; sapon. no. 160–180; 40% alpha mono.

Drewmulse 200. [496] Glycerol monooleate from tallow and soya; see Drewmulse 10; cream beads; m.p. 135–145; HLB 2.8;

sapon. no. 154–177; 40% alpha mono.

Drewmulse 300. [496] Glycerol monostearate (80%) and polysorbate (20%) blend from soya and tallow oils; see Drewmulse 10; cream beads; m.p. 135 F; HLB 4.3; 30% alpha mono.

Drewmulse 365. [206] Glycerol monostearate (67%) and polysorbate 65 (33%) blend from soya and tallow oils; see Drewmulse 10; cream beads; m.p. 136 F; HLB ~5.3; 25–30% alpha mono.

Drewmulse 700. [496] Glycerol monostearate (80%) and polysorbate 80 (20%) blend from soya and tallow oils; see Drewmulse 10; cream beads; m.p. 135 F; HLB 5.2; 30% alpha mono.

Drewmulse 900. [496] Glycerol monostearate; see Drewmulse 10; cream beads; m.p. 138 F; HLB 3.8; sapon. 160–180; 52% alpha mono.

Drewmulse 1128. [496] Glyceryl stearate, PEG-120 glyceryl stearate, and PEG-120 propylene glycol stearate; nonionic; emulsifier, stabilizer, and visc. builder in cosmetics and pharmaceuticals, opacifier; Gardner 2 flake; HLB 8.4; 20% mono.

Drewmulse 1129. [496] Glyceryl stearate and PEG-90 stearate; see Drewmulse 1128; flake; HLB 11.0; 20% mono.

Drewmulse CNO. [496] Glyceryl cocoate; see Drewmulse 1128; Gardner 4 solid; HLB 3.8; 50% mono.

Drewmulse DGMS. [206] Diethylene glycol monostearate; nonionic; emulsifier for w/o and o/w systems, opacifier and pearlescence agent for shampoos, additive for pigment milling; flakes; 85% act.

Drewmulse EGDS. [206] Ethylene glycol distearate; nonionic; see Drewmulse DGMS; flakes; 48% act.

Drewmulse EGMS. [206] Ethylene glycol monostearate; nonionic; see Drewmulse DGMS; flakes; 48% act.

Drewmulse GMC-8. [496] Glycerol monocaprylate-caprate; stabilizer and visc. builder in cosmetics and pharmaceuticals, opacifier, for food applications; Gardner 3 liq.; HLB 8.3; 75% mono.

Drewmulse GMO. [496] Glyceryl oleate; nonionic; stabilizer and visc. builder in cosmetics and pharmaceuticals, opacifier; Gardner 6 liq.; HLB 3.4; 40–45% conc.

Drewmulse GMRO. [206] Glycerol monoricinoleate; nonionic; see Drewmulse DGMS; liq.

Drewmulse GMS. [206] Glyceryl monostearate; nonionic; see Drewmulse DGMS; flakes; 40–45% act.

Drewmulse GMS-AE. [206] Glyceryl stearate; used in preparation of creams; flakes.

Drewmulse HM-100. [496] Glyceryl stearate and PEG-40 stearate; stabilizer and visc. builder in cosmetics and pharmaceuticals, opacifier, esp. effective in antiperspirant formulations; Gardner 2 flake; HLB 8.4; 24% mono.

Drewmulse M. [206] Blend of glycerol monoesters; nonionic; food emulsifier; paste; 32% act.

Drewmulse PGML. [496] Propylene glycol monolaurate; chemical specialty; liq.

Drewmulse PGMS. [206] Propylene glycol monostearate; nonionic; see Drewmulse DGMS; flakes; 48% act.

Drewmulse POE-SML. [496] Polysorbate 20; nonionic; w/o emulsifier for cosmetics and pharmaceuticals, solubilizer, dispersant, wetting agent, detergent, visc. and control agent; liq.; HLB 16.7; sapon. no. 40–51.

Drewmulse POE-SMO. [496] Polysorbate 80; nonionic; see Drewmulse POE-SML; liq.; HLB 15.0; sapon. no. 45–55.

Drewmulse POE-SMS. [496] Polysorbate 60; nonionic; see Drewmulse POE-SML; solid; HLB 14.9; sapon. no. 45–55.

Drewmulse POE-STS. [496] Polysorbate 65; nonionic; see Drewmulse POE-SML; solid; HLB 10.5; sapon. no. 88–98.

Drewmulse PNO. [206] Blend of glycerol monoesters; nonionic; food emulsifier; paste; 38–42% act.

Drewmulse SML. [496] Sorbitan monolaurate; nonionic; see Drewmulse POE-SML; liq.; HLB 8.6; sapon. no. 159–171.

Drewmulse SMO. [496] Sorbitan monooleate; nonionic; see Drewmulse POE-SML; liq.; HLB 4.3; sapon. no. 144–156.

Drewmulse SMS. [496] Sorbitan monostearate; nonionic; see Drewmulse POE-SML; solid; HLB 4.7; sapon. no. 147–157.

Drewmulse STS. [496] Sorbitan tristearate; nonionic; see Drewmulse POE-SML; solid; HLB 2.1; sapon. no. 170–190.

Drewmulse TP. [496] Glyceryl stearate; nonionic; emulsifier, stabilizer, and visc. builder in cosmetics and pharmaceuticals, opacifier; Gardner 4 flake; HLB 3.8; 40% mono.

Drewmulse V. [496] Glyceryl stearate (veg.); see Drewmulse TP; Gardner 6 bead; HLB 3.8; 40% mono.

Drewmulse V-SE. [496] Glyceryl stearate SE;

see Drewmulse TP; Gardner 6 bead; HLB 8.4; 30% mono.

Drewpol 3-1-O. [496] Triglycerol monooleate; food emulsifier and additive; flavor and color solubilizer; amber liq.; HLB 7.0; sapon. no. 120–160.

Drewpol 3-1-S. [496] Triglycerol monostearate; nonionic; see Drewpol 3-1-O; amber beads; HLB 7.0; sapon. no. 120–140.

Drewpol 3-1-SH. [496] Triglycerol mono shortening; see Drewpol 3-1-O; thick, amber liq.; HLB 7.4; sapon. no. 110–140.

Drewpol 6-2-S. [496] Hexaglycerol distearate; see Drewpol 3-1-O; amber beads; HLB 4.0; sapon. no. 100–130.

Drewpol 10-4-O. [496] Decaglycerol tetraoleate; see Drewpol 3-1-O; amber liq.; HLB 6.0; sapon. no. 125–145.

Drewpol 10-8-O. [496] Decaglycerol octaoleate; see Drewpol 3-1-O; amber liq.; HLB 4.0; sapon. no. 150–180.

Drewpol 10-10-O. [496] Decaglycerol decaoleate; see Drewpol 3-1-O; amber liq.; HLB 3.0; sapon. no. 160–190.

Drewpol 10-10-S. [496] Decaglycerol decastearate; see Drewpol 3-1-O; tan solid; HLB 3; sapon. no. 150–190.

Drewpon 40, 40LS. [206] TEA lauryl sulfate; anionic; emulsifier, detergent, wetting agent, dispersant used for household cleaning products, cosmetics, emulsion polymerization; liq.; 39–42 and 38–40% act. resp.

Drewpon 100. [206] Sodium lauryl sulfate; anionic; detergent, wetting agent for personal care products, pesticides; powd.; 90% act.

Drewpon AC-50. [206] Amphoteric imidazoline; detergent for nonirritating shampoos; liq.; 38% act.

Drewpon AS, CN. [206] Sodium lauryl sulfate; anionic; detergent for nonirritating shampoos; liq./paste; 28–30% act.

Drewpon BP. [206] Imidazoline; amphoteric; detergent for nonirritating shampoos; liq.; 27.5–30.5% conc.

Drewpon EBO. [206] Sodium laureth sulfate; anionic; detergent for nonirritating shampoos; liq.; 26–28% act.

Drewpon ECM. [206] MEA laureth sulfate; anionic; detergent for nonirritating shampoos; liq.; 30–32% act.

Drewpon EDT. [206] TEA laureth sulfate; anionic; detergent for nonirritating shampoos; liq.; 34–36% act.

Drewpon EG. [206] Sodium laureth sulfate; anionic; detergent for pearlescent shampoos; liq.; 22% act.

Drewpon EKZ. [206] Ammonium laureth sulfate; anionic; detergent for nonirritating shampoos; 23–25% act.

Drewpon ELB. [206] Sodium laureth sulfate; anionic; detergent for nonirritating shampoos; 27–28% act.

Drewpon ESG. [206] Magnesium laureth sulfate; anionic; detergent for nonirritating shampoos; 26–28% act.

Drewpon ETH, EVE. [206] Sodium laureth sulfate; anionic; detergent for nonirritating shampoos; 26–28% act.

Drewpon MEA. [206] MEA lauryl sulfate; anionic; detergent for nonirritating shampoos; 32–34% act.

Drewpon MG. [206] Magnesium lauryl sulfate; anionic; detergent for nonirritating shampoos; 28–30% act.

Drewpon NH. [206] Ammonium lauryl sulfate; anionic; detergent for nonirritating shampoos; 27–29% act.

Drewpon NHAC. [206] Ammonium lauryl sulfate; anionic; detergent for nonirritating shampoos; 37.5% act.

Drewpon SLA. [206] Lauryl ether sulfate and sulfosuccinate; anionic; used in shampoos; liq.; 25.5–28% conc.

Drewpone 60. [496] PEG-20 sorbitan monostearate; food emulsifier, foaming agent, dough conditioner; yel. liq. to soft solid; HLB 14.9; sapon. no. 45–55.

Drewpone 65. [496] PEG-20 sorbitan tristearate; food emulsifier, foaming agent, dough conditioner; amber soft solid; HLB 10.5; sapon no. 88–98.

Drewpone 80. [496] PEG-20 sorbitan monooleate; food emulsifier, foaming agent, dough conditioner; yel. liq.; HLB 15; sapon. no. 45–55.

Drewsorb 60. [496] Sorbitan monostearate; food emulsifier; cream beads; water disp.; HLB 4.7; sapon. no. 147–157.

Drewsoft 100. [496] Glyceryl monostearate, modified; nonionic; softener for fibers and blends; flake; 100% conc.

Driltreat. [447] Phosphatide; nonionic; surfactant for oil mud; liq.; 100% conc.

Druspin LO-VIS. [496] Blended product; textile lubricant, antistat, emulsifier; liq.; low visc.; 100% act.

Druspin Supreme. [496] Blend of min. oil, emulsifiers, lubricants, antistats; liq.; 100% act.

Drymet. [573] Sodium metasilicate, anhyd.;

soap builder and detergent; ingredient in industrial cleaning compds.; granules; 100% conc.

Drynol E/20, E/30, E/40. [253] Calcium dodecylbenzene sulfonate with ethoxylated alkylphenols and stabilizing agent blend; nonionic/anionic; emulsifier for pesticide concs.; thick liq.; 65%, 70%, and 75% conc.

Dulectin. [207] De-oiled soybean lecithin; nonionic; surfactant, emulsifier, dispersant, and stabilizer for pharmaceuticals and cosmetics; powd.; 97% conc.

Duofol AS. [284] Sulfated ester; anionic; wetting and rewetting agent used in wet processing cotton goods; liq.; 65% conc.

Duofol T. [283] Sulfated ester; anionic; wetting, disp., finishing agents for textiles; amber liq.; clear sol. 5% water; sp.gr. 1.068; visc. 70 cps; 55% act.

Duomac C. [11] Coco-1,3-diaminopropane diacetate; corrosion inhibitor, bactericide in drilling fluids; sol. in IPA, water; sp.gr. 0.85; m.p. 20–24 C; 89% act.

Duomac T. [10; 64] N-tallow-1,3-propanediamine diacetate; cationic; emulsifier, corrosion inhibitor, flotation reagent; pigment flushing, flocculation; br. paste; vinegar odor; m.w. 480; sol. in water, chloroform, ethanol, IPA; sp.gr. 0.892 (90 C); HLB 10.7; m.p. 181 F; b.p. > 300 C; flash pt. 335 F (COC); 95% conc.

Duomeen C. [11; 64] N-coco-1,3-propanediamine; cationic; chemical intermediate; corrosion inhibitor, fuel oil additive, flotation agent; dk. amber. liq.; ammonia odor; m.w. 276; sol. in naphtha, min. oil, IPA; sp.gr. 0.836; m.p. 71 F; b.p. > 300 C; flash pt. 295 F (COC); 100% conc.

Duomeen CD. [11] Coco-1,3 diamino propane; cationic; intermediate; liq., paste; 89% conc.

Duomeen C Special. [64] N-coco-1,3-propanediamine; see Duomeen C; yel. liq.; ammonia odor; m.w. 276; sol. in naphtha, min. oil, IPA; sp.gr. 0.836; m.p. 80 F; b.p. > 300 C; flash pt. 295 F (COC).

Duomeen CX. [11] Diamine; cationic; chemical intermediate; liq., paste; 100% conc.

Duomeen L-11. [64] N-s-alkyl-1,3-propanediamine; see Duomeen C; yel. liq.; ammonia odor; m.w. 300; sol. in min. oil, IPA; sp.gr. 0.841; m.p. –32 F; b.p. > 300 C; flash pt. 226 F (COC).

Duomeen L-15. [64] N-s-alkyl-1,3-propanediamine; see Duomeen C; yel. liq.; ammonia odor; m.w. 383; sol. in min. oil, IPA; sp.gr. 0.842; m.p. 44 F; b.p. > 300 C; flash pt. 266 F (COC).

Duomeen O. [64] N-oleyl-1,3-propanediamine; cationic; see Duomeen C; yel. paste; ammonia odor; m.w. 350; sol. in naphtha, IPA; sp.gr. 0.841; m.p. 59 F; b.p. > 300 C; flash pt. 370 F (COC).

Duomeen OX. [11] Oleyl-1,3 diaminopropane; cationic; intermediate; liq., paste; 92% conc.

Duomeen T, T Special [10; 64] N-tallow-1,3-propanediamine; see Duomeen C; textile finishing agent, dispersant for inorg pigments in paints, larvacidal oil additive; bitumen emulsifier; dk. amber paste; ammonia odor; m.w. 330 and 350 resp.; sol. in naphtha, IPA; sp.gr. 0.841; m.p. 115 F; b.p. > 300 C; flash pt. 370 F (COC); 100% conc.

Duomeen TDO. [64] N-tallow-1,3-propanediamine dioleate; cationic; corrosion inhibitor, fuel oil additive, flotation agent, metal drawing additive; dk. semisolid; ammonia odor; m.w. 924; sol. in naphtha, min. oil, IPA; sp.gr. 0.865 (65 C); m.p. 77 F; b.p. > 300 C; flash pt. 455 F (COC).

Duomeen TX. [10; 11] Tallow-1,3-diaminopropane; emulsifier for bitumen emulsions, base material; antisettling agent for paints; Gardner 7 solid; ammoniacal odor; water insol.; sp.gr. 0.84; m.p. 46–55 C; flash pt. 180 C; 92% act.

Duoquad T-50. [10; 11] N, N, N′, N′, N′-pentamethyl-*n*-tallow-1,3-propanediammonium dichloride; cationic; detergent, corrosion inhibitor, metal cleaner; emulsifier for sec. oil recovery; Gardner 13 max. liq.; m.w. 480; sp.gr. 0.90; HLB 14.4; f.p. –20 C; flash pt. 13 C; pH 6–8; 48–52% act. in aq. IPA.

Duowet EHS, EHS Conc. [245] Sulfonated aliphatic polyester, anionic; wetting agent used in dyeing, sizing, desizing, raw stock processing; liq.

Du Pont Retarder LAN. [208] Alkyl trimethylammonium halide; cationic; dye retarder and stripping agent with azoic dyes, emulsifier; dk. red-br. liq.; surf. tens. 43.4 dynes/cm (0.1%); 20% act.

Duponol 80. [208] Sodium *n*-octyl sulfate, tech.; anionic; wetting agent, dispersant, penetrant, softening agent, emulsifier used for textile, metal, and leather processing; lt. yel. liq.; m.w. 232; sol. in presence of many multivalent ions; visc. 30 cps; surf. tens. 51.7 dynes/cm (0.1%); 33–35% act.

Duponol C. [208] Sodium lauryl sulfate USP;

anionic; emulsifier, dispersant, detergent, wetting agent for cosmetic, dental and medical preparations; wh. spray-dried beads; mild fatty odor; water sol.; dens. 3.3 lb/gal.

Duponol D Paste. [208] Sodium lauryl oleyl sulfate, tech.; anionic; detergent, emulsifier, wetting agent, dispersant used for textile, metal, and leather processing; lt. yel. paste; m.w. 328; surf. tens. 30.5 dynes/cm (0.1%); 38% act.

Duponol EP. [208] DEA-lauryl sulfate, tech.; anionic; foaming agent, detergent, wetting agent, emulsifier used for cosmetics, personal care products; very pale golden, slightly visc. liq.; bland and clean odor; sol. in water, polar solv. with some electrolyte precipitation; dens. 8.4 lb/gal; visc. 50–150 cps ; cloud pt. 5 C; 33–36% act.

Duponol G. [208] Fatty alcohol amine sulfate; anionic; emulsifier, wetting agent; yel. visc. translucent paste; mild fatty penetrating odor; sol. in hydrocarbons, 50% in water; dens. 8.2 lb/gal; surf. tens. 29.6 dynes/cm (0.1%); 92% act.

Duponol LS Paste. [208] Sodium oleyl sulfate tech.; anionic; scouring agent in textile and leather industries, rewetting agent for paper; lt. tan paste; m.w. 355; surf. tens. 37.8 dynes/cm (0.1%); 26% act.

Duponol ME Dry. [208] Sodium lauryl sulfate, tech.; anionic; detergent, wetting agent, foaming agent, emulsifier for general industrial uses; wh. to med. cream powd.; m.w. 302; 30% sol. in water; dens. 3.3 lb/gal; cloud pt. 70 F (1% aq.); surf. tens. 35 dynes/cm (0.1%); 90–96% act.

Duponol QC. [208] Sodium lauryl sulfate; anionic; shampoo base; water wh. slightly visc. liq.; bland clean, char. odor; sol. in water, polar solv. with some electrolyte precipitation; visc. < 500 cps; cloud pt. 65 F; 29–30% act.

Duponol RA. [208] Fortified sodium ether-alcohol sulfate; anionic; detergent, dispersant, emulsifier, penetrating, leveling, and wetting agent; textile processing; lt. amber clear liq.; misc. with water; dens. 8.7 lb/gal; sp.gr. 1.04; visc. 178 cps (80 F); cloud pt. 68 F; flash pt. > 212 F (TOC); surf. tens. 33 dynes/cm (0.1%).

Duponol WA Dry. [208] Sodium lauryl sulfate; anionic; detergent, emulsifier, wetting agent, penetrant; wh. powd.; water sol., the act. ingred. is sol. in highly polar solv.; dens. 2.6 lb/gal; cloud pt. 37 F (1%); surf. tens. 33 dynes/cm (0.1%); 44% conc.

Duponol WA Paste, WAQ. [208] Sodium lauryl sulfate; shampoo base, superior cleansing and foaming.

Duponol WAQE. [208] Sodium lauryl sulfate, tech.; anionic; emulsifier for emulsion polymerization; pale yel. liq.; m.w. 302; water sol., act. ingred. is sol. in polar solv. with some electrolyte precipitation; dens. 8.6 lb/gal; visc. 500 cps; cloud pt. 65 F; 29–30.5% act.

Duponol WN. [208] Sodium octyl/decyl sulfate, tech.; anionic; wetting and penetrating agent; pale yel. liq; m.w. 246; visc. 30 cps; surf. tens. 52.5 dynes/cm (0.1%); 33–35% act.

Duponol XL. [208] Alkyl alcohol sulfate; anionic-amphoteric; shampoo surfactant; golden moderately visc. liq.; faintly spicy odor; sol. in water, in polar solv. with some electrolyte precipitation; visc. 2500–4500 cps; cloud pt. 50 F; clear pt. 50–55 F, 36% solids.

Dur-Em 104K, 107K. [210] Mono- and diglycerides from hydrog. veg. oil with BHA, BHT, and citric acid; nonionic; food emulsifier in bakery, prepared mixes, convenience foods (avail. in kosher form); cream plastic and wh. bead resp.; m.p. 115–121 F; 100% act.

Dur-Em 114. [210] Glyceryl oleate; nonionic; emulsifier for personal care products; HLB 2.8; m.p. 43–49 C; 40% min. alpha monoglyceride.

Dur-Em 114K. [210] Mono- and diglycerides from hydrog. veg. oil with BHA, BHT and citric acid; see Dur–Em 104K; cream plastic; m.p. 110–120 F; 100% act.

Dur-Em 117. [210] Glyceryl stearate; nonionic; textile lubricant and finishing agent; emulsifier for cosmetic and pharmaceutical creams and lotions; lubricant for thermoplastics; dispersant for inorg. pigments; HLB 2.8; m.p. 62–65 C; 40% min. alpha monoglyceride.

Dur-Em 117K. [210] Mono- and diglycerides from hydrog. veg. oil with citric acid; nonionic; see Dur-Em 104K; wh. bead; m.p. 145–150 F.

Dur-Em 127. [210] Mono- and diglycerides with citric acid; emulsifier; peanut butter stabilizer; m.p. 140–145 F.

Dur-Em 127K. [210] Mono- and diglycerides from hydrog. veg. oil with citric acid; peanut butter stabilizer (avail. in kosher form); wh.

bead; m.p. 140–145 F; 100% act.

Dur-Em 136K. [210] Mono- and diglycerides from hydrog. veg. oil with BHA, BHT and citric acid; nonionic; food emulsifier in bakery, prepared mixes, convenience foods, veg. dairy sys. (avail. in kosher form); plastic; 100% act.

Dur-Em 187. [210] Mono- and diglycerides from hydrog. animal fats with citric acid; nonionic; see Dur-Em 104K; wh. bead or flake; m.p. 140–145 F; 100% act.

Dur-Em 195. [210] Mono- and diglycerides from hydrog. veg. and animal fats with BHA, BHT and citric acid; food emulsifier in dry mix sys., dog foods, prepared mixes; cream plastic; m.p. 123–129 F; 100% act.

Dur-Em 204. [210] Glyceryl oleate; food emulsifier; m.p. 114–121 F.

Dur-Em 204K. [210] Mono- and diglycerides from hydrog. veg. oil with BHA, BHT and citric acid; nonionic; see Dur-Em 104K; cream plastic; m.p. 120–130 F; 100% act.

Dur-Em 207. [210] Mono- and diglycerides with citric acid to protect flavor; emulsifier for food; peanut butter stabilizer; m.p. 140–146 F.

Dur-Em 207-E. [210] Glyceryl monostearate SE; food emulsifier, emulsifier hydrate for breads, for pharmaceutical, cosmetic uses; wh. powd.; m.p. 140–146 F; 100% act.

Dur-Em 207K. [210] Mono- and diglycerides from hydrog. veg. oil with citric acid; see Dur-Em 104K; wh. bead or plastic; m.p. 140–146 F; 100% act.

Dur-Em 227K. [210] Mono- and diglycerides from hydrog. veg. oil with citric acid; see Dur-Em 104K; bead; 100% act.

Dur-Em 287. [210] Mono- and diglycerides from hydrog. animal fat with citric acid; nonionic; food emulsifier in bakery, prepared mixes, convenience foods; plastic; 100% act.

Dur-Em 287E. [210] Mono- and diglycerides from hydrog. animal fat with citric acid; nonionic; see Dur-Em 287; wh. powd.; m.p. 140–146 F; 100% act.

Dur-Em 295. [210] Mono- and diglycerides from hydrog. veg. oil with BHA, BHT and citric acid; nonionic; see Dur-Em 287; cream plastic; m.p. 123–129 F; 100% act.

Dur-Em GMO. [210] Glyceryl oleate; nonionic; lubricant for personal care products, dry cleaning bases, paints, and insecticides; HLB 2.8.

Durfax 20. [210] POE (20) sorbitan monolaurate; nonionic; used in personal care and household products; food emulsifier; fabric antistat; HLB 16.5; sapon. no. 39–52.

Durfax 60. [210] POE (20) sorbitan monostearate; nonionic; food emulsifier; personal care products; preshave beard lubricant and softener products; paste; sol. in water; HLB 14.9; sapon. no. 45–55.

Durfax 60K. [210] POE (20) sorbitan monostearate; nonionic; food emulsifier for o/w emulsions; yel. visc. liq.; sol. in water, cottonseed oil; HLB 14.9; 100% act.

Durfax 65. [210] POE (20) sorbitan tristearate; nonionic; food emulsifier; pesticide dispersant; solid; water-disp.; HLB 10.5; sapon. no. 88–98.

Durfax 65K. [210] POE (20) sorbitan tristearate; nonionic; food emulsifier for o/w emulsions; tan solid; sol. in cottonseed oil, disp. in water; 100% act.

Durfax 80. [210] POE (20) sorbitan monooleate; nonionic; food emulsifier; personal care products; antifog agent in plastics and aerosol furniture polish; liq.; water-sol.; HLB 15.9; sapon. no. 45–55.

Durfax 80K. [210] POE (20) sorbitan monooleate; nonionic; food emulsifier for o/w emulsions; yel. liq.; sol. in water, cottonseed oil; HLB 15.9; 100% act.

Durfax EOM. [210] PEG-20 glyceryl stearate; food emulsifier; dough conditioner; lubricant and fabric softener; HLB 13.5; sapon. no. 65–75.

Durfax EOM K. [210] PEG-20 glyceryl monostearate; nonionic; food emulsifier, dough conditioner; paste; m.p. 80–85 F; HLB 13.1; 100% act.

Durlac 100W. [210] Glyceryl stearate lactate; emulsifier for food; starch gelling agent in industrial processes; HLB 2.4; m.p. 46–54 C.

Durlac 100WK. [210] Glyceryl lactoesters of fatty acids from hydrolyzed veg. oil; nonionic; food emulsifier; flake; m.p. 115–130 F; 100% act.

Durlac 200K. [210] Glyceryl lactoesters of fatty acids from veg. oil; nonionic; food emulsifier, for cosmetic and pharmaceutical use; solid; m.p. 97–103 F; 100% act.

Durlac 300. [210] Glycerol and propylene glycol lactylated fatty acid ester; emulsifier for food, cosmetic, pharmaceutical applications; m.p. 95–105 F.

Durlac 300K. [210] Lactylated fatty acid esters of glycerin and propylene glycol; nonionic;

food emulsifier, cosmetic and pharmaceutical uses; cream flake; m.p. 95–105 F; 100% act.

Dur-Lo. [210] Mono- and diglycerides with BHA and citric acid; food emulsifier.

Durpeg 400MO. [210] PEG-8 oleate; nonionic; food emulsifier, processing aid in animal feeds; cosmetic products; amber semisolid; HLB 11.1; sapon. no. 80–88; 100% act.

Durpro 107. [210] Propylene glycol mono- and diesters of fats and fatty acids with citric acid; emulsifier for cakes and cake mixes; m.p. 115–125 F; 50% min. PGME.

Durpro 107-55. [210] Propylene glycol stearate; nonionic; occlusive agent and emulsifier for personal care products; HLB 2.2; m.p. 46–52 C; 55% min. act.

Durpro 107K. [210] Propylene glycol, mono- and diesters, and mono- and diglycerides with citric acid; nonionic; food emulsifier for cake and cake mixes; cream flake; m.p. 117–125; HLB 2.5; 100% act.

Durpro 187. [210] Propylene glycol, mono- and diesters, and mono- and diglycerides; nonionic; food emulsifier for cake and cake mixes; cream flake; m.p. 117–125; HLB 2.5; 100% act.

Durtan 20. [210] Sorbitan monolaurate; nonionic; used in cosmetic, toiletry, and household products; fabric antistat and lubricant; HLB 7.4; sapon. no. 158–170.

Durtan 60. [210] Sorbitan monostearate; nonionic; food emulsifier; dispersant for inorganics used in thermoplastics; beads; HLB 4.7; sapon. no. 147–157.

Durtan 60K. [210] Sorbitan monostearate; nonionic; food emulsifier; cream beads; m.p. 121–127 F; HLB 4.7; 100% act.

Durtan 80. [210] Sorbitan monooleate; nonionic; used in water softening compds., auto polishes, pesticide formulations; antifog in plastics; rust inhibitor additive; ink pigment dispersant; HLB 4.3; sapon. no. 148–161.

Dusoran MD. [207; 630] Lanolin alcohols, dist.; nonionic; w/o emulsifier, stabilizer, softener, emollient; lt. yel., waxy solid; sol, in min. oil, abs. and 95% alcohols, chloroform, ether, lt. petroleum, toluene; sapon. 8.0; 100% act.

Dusoran NG, NO. [630] Wool and lanolin alcohols; stabilizer, softener, emulsifier for cosmetic and personal care products; tan and lt. br. resp.; solid; sol. in min. oils, abs., alcohols, chloroform, toluene; sapon. no. 10.

Dyasulf 2031. [187] Sulfated oleic acid sodium salt; wetting agent and penetrant in textile operations; detergent and penetrant in fabrics; amber liq.; dens. 1.09 g/ml; pH 5.3 (2% aq.); 60% act.

Dyesperse DC. [283] Naphthalene formaldehyde sulfonate; anionic; dispersant for disp. and acid dyes; amber liq.; clear sol.; sp.gr. 1.138; visc. 10 cps; 33% act.

Dymsol 38-C. [187] Fatty acid sulfate, sodium salt; anionic; emulsifier for polymerization; dk. amber liq.; 70% act.

Dymsol 2031. [187; 192] Sulfonated fatty acid, sodium salt; anionic, primary or sec. emulsifier for emulsion polymerization; dk. amber liq.; biodeg.; 60% act.

Dymsol L. [187; 192] PEG fatty ester; nonionic; stabilizer, wetting agent, visc. controller for latex and resin emulsions; liq.; biodeg.; 100% act.

Dymsol LP. [187] Alkylaryl sulfonate; anionic; emulsifier for large particle size latices; clear yel. liq.; biodeg.; 33% act.

Dymsol S. [187] Sulfated syn. sperm oil; anionic; emulsion breaker, solubilizer, coupling agent; fatliquors for tanning leather; greenish-br. liq.

Dynacet. [213] Acetylated monoglyceride; nonionic; food emulsifier; solid, wax, liq.; 100% conc.

Dyqex. [252] Sodium lignosulfonate; anionic; dye disp. extender; 46% liq.; 100% powd.

E

EC-25, -25K. [210] Propylene glycol mono- and diesters of fats and fatty acids, mono- and diglycerides, partially hydrog. soybean oil, lecithin, BHA, citric acid; nonionic; food emulsifier for cakes, mixes; lt. straw plastic; m.p. 90–100 F; HLB 2.6; 100% act.

EC-117MK. [210] Mono- and diglycerides and lactylic esters of fatty acids; nonionic; food emulsifier; cream to lt. tan beads; m.p. 123–133 F; HLB 4.6; 100% act.

Eccoful FC Conc. [215] Amide ethoxylate, modified; anionic; detergent and scouring agent for wool and wool blends; liq.; 65% conc.

Eccolene OW. [215] Sulfated fatty ester; anionic; wetting aid, dispersant, emulsifier, penetrant and scouring agent for textile processing; liq.; 65% conc.

Ecconol 61. [224] Amine sulfonate complex; anionic; detergent, wetting agent for dish and carwash products; amber liq.; water sol.; biodeg.; 67% act.

Ecconol 66. [224] Fatty amide condensate; nonionic; detergent, thickener, emulsifier used in cleaning compds. and lubricants; amber liq.; biodeg.; 100% act.

Ecconol 606. [224] Org. alkylaryl sulfonate; anionic; detergent, wetting agent for liq. formulations; amber liq.; biodeg.; 60% act.

Ecconol 628. [224] Coconut oil amide; nonionic; detergent, wetting agent, cleaning agent, thickener, foam stabilizer; liq.; biodeg.; 100% act.

Ecconol B. [224] Fatty alkanolamide; nonionic; detergent, wetting agent, cleaning agent for heavy-duty formulations; emulsifier; liq.; biodeg.; 90% act.

Eccoscour CB. [215] Sodium alkylaryl sulfonate; anionic; detergent, wetting agent, and emulsifier for textile scouring and dyeing applications; liq.; 30% conc.

Eccoscour D-7. [215] Sulfate alkyl phenol ethoxylate; amphoteric; surfactant and scouring agent for textile fabric and yarns; liq.; 42% conc.

Eccoscour KG. [215] Complex alcohol, ethoxylated; nonionic; scouring agent for removal of graphite from elasticized textiles; liq.

Eccoterge EO. [215] Phenol, ethoxylated; nonionic; detergent, wetting and dye leveling agent for fabrics; kier boiling assistant; liq.; 100% conc.

Eccoterge MV Conc. [215] Fatty condensate; anionic; detergent, wetting and scouring agent for textile use; liq.; 100% conc.

Eccoterge NF-2. [215] Complexed ester, ethoxylated; anionic; detergent and scouring agent for textile and nontextile applications; liq.; 60% conc.

Eccoterge SCH. [215] Complexed amide, ethoxylated; anionic; wetting, scouring agent, and dyeing assistant for textiles; liq.; 60% conc.

Eccotex P. Conc. [215] Alkylaryl comp., modified; anionic; dispersant, penetrant used in textile industry; wetting of dyestuffs; gel; 60% conc.

Eccowet LF Conc. [215] Sodium alkyl naphthalene sulfonate; anionic; wetting agent, dispersant used in dyeing applications; liq.; 60% conc.

Eccowet W-50. [215] Sodium aliphatic ester

sulfonate; anionic; wetting, penetrant used in textile industry; liq.; 50% conc.

Edenol 302. [289] Propylene glycol dicaprylate/dicaprate; emollient oil and cosolv. for personal care products; yel. low visc. liq.; insol. in water; dens. 0.92 g/ml; HLB 9; solid. pt. < 0 C; cloud pt. < 0 C; sapon. no. 315–335; ref. index 1.435.

Effcol TS-214. [397] Linear hydrocarbon sulfonate; anionic; antistat agent for warp sizing; liq.; 40% conc.

Ekaline F. [538] Aliphatic polyglycol ether; nonionic; agent for washing off dyeings; leveling and dispersing agent; liq., flakes.

Ekaline G-80 Flakes, Liq. [537] Nonionic; detergent, wetting agent, emulsifier, for general purpose scouring; wh. flakes, colorless liq.; water sol.; 100% and 20% act.

Elasdo +. [460] Mono- and diglycerides and ethoxylated mono and diglycerides; food emulsifier; bead; HLB 7.7.

Elasdo Power 70. [460] Ethoxylated monoglyceride and mono- and diglycerides; food emulsifier; plastic; HLB 8.8; 100% conc.

Elecut S-507. [590] Alkylaryl sulfonate salt; anionic; detergent; liq.; 75% conc.

Eleminol ES-70. [539] POE and nonylphenyl ether sulfate; anionic; primary emulsifier for emulsion polymerization; used in textile products; liq.; 80% conc.

Eleminol HA-100. [539] POE nonylphenyl ether; nonionic; emulsifier for emulsion polymerization; flake; HLB 18.2; 100% conc.

Eleminol HA-161. [539] POE nonylphenyl ether; nonionic; see Eleminol HA-100; flake; HLB 18.8; 100% conc.

Eleminol JS-2. [539] Sodium alkylaryl sulfosuccinate; emulsifier for emulsion polymerization; liq.; 40% conc.

Eleminol MON-2. [539] Sodium alkyldiphenyl ether disulfonate; emulsion polymerization of styrene-butadiene and acrylate resins; liq.; 48% conc.

Eleminol MON-7. [539] Sodium dodecyl diphenyl ether disulfonate; anionic; emulsifier for emulsion polymerization; liq.; sol. in acidic, alkaline and electrolyte sol'ns.; 49% conc.

Elfacos C26. [11] Hydroxyoctacosanyl hydroxystearate; nonionic; consistency regulating agent for w/o emulsions, cosmetic applications; pellets; 100% conc.

Elfacos E200. [11] Methoxy PEG-22/dodecyl glycol copolymer; nonionic; w/o emulsifier; solid; 100% conc.

Elfacos ST 9. [11] PEG-45/dodecyl glycol copolymer; nonionic; stabilizer for w/o emulsions and emollient used in cosmetics; paste; HLB 7.0; 100% conc.

Elfacos ST 37. [11] PEG-22/dodecyl glycol copolymer; nonionic; see Elfacos ST 9; liq.; 100% conc.

Elfan 200. [11] Sodium lauryl sulfate; anionic; detergent for cosmetics, toothpaste; wh. powd., beads; dens. 220 g/l; 90% act.

Elfan 240. [11] Sodium lauryl sulfate; anionic; personal care products and lt.-duty detergents; liq.; 30% conc.

Elfan 240M. [11] MEA lauryl sulfate; anionic; shampoo and bubble baths; liq.; 29% conc.

Elfan 240T. [11] TEA lauryl sulfate; anionic; see Elfan 240M; liq.; 40% conc.

Elfan 240 T/S. [11] TEA lauryl sulfate; detergent for cosmetics, shampoos; yel. liq.; water sol.; sp.gr. 1.04; visc. 150 cps; 40% act.

Elfan 260 S. [11] Sodium lauryl sulfate; detergent for textiles, skins, cosmetics, emulsifier for emulsion polymerization; yel. liq., paste; water sol. (30 C); sp.gr. 1.04; visc. 200 cps (30 C); f.p. 15 C; 30% act.

Elfan 280. [11] Coco fatty alcohol sulfate, sodium salt; anionic; general purpose heavy duty detergent, industrial, household, personal care uses; yel. paste; water sol. 700 g/l (50 C); sp.gr. 1.05; visc. 8000 cps (50 C); pH 8.5 ± 1; 42% act.

Elfan 280 Powder. [11] Sulfated coconut fatty alcohol, sodium salt; anionic; lt.-duty detergents and all-purpose washing agents, toothpaste; powd.; 90% conc.

Elfan 680. [11] Oleyl-cetyl alcohol sulfate, sodium salt; anionic; powd. detergent, cleaning agent, textile uses, anionic; yel.-br. paste; water sol. < 10 g/l; sp.gr. 1.05; m.p. 40 C; pH 11 ± 1; 50% act.

Elfan A. [11] Betaine-type surfactant; amphoteric; wetting agent, emulsifier, for cosmetics; liq.

Elfan A 432. [11] Betaine type from lauric-myristic acid; amphoteric; detergent for cosmetics, baby shampoos, bubble baths; yel. liq.; water sol.; sp.gr. 1.04; f.p. < 0 C; pH 5 ± 1; 30% act.

Elfan KM 500, KM 550. [11] Alpha olefin sulfonate and lauryl ether sulfate; anionic; personal care products; dishwashing detergents, car shampoos and lt.-duty liqs.; liq.; 31% conc.

Elfan KM 730. [11] Alpha olefin sulfonate and lauryl ether sulfate mixture; anionic; shampoos, bubble baths; liq.; 33% conc.

Elfan KT 550. [11] Sulfated coco-tallow fatty alcohol, sodium salt; anionic; heavy duty detergent, industrial, household, cosmetic uses; yel. paste; sp.gr. 1.05; visc. 10,000 cps (50 C); pH 8 ± 0.5; 42% act.

Elfan L 310. [11] Ethylene glycol distearate; anionic; additive for shampoos and bubble baths; lt. yel. solid, flakes; water disp.; sp.gr. 0.96; m.p. 64–66 C; sapon. no. 200.

Elfan NS. [11] Lauryl ether sulfates; general purpose detergent, wetting agent; liq., paste; 30 or 70% act.

Elfan NS 242. [11] Sodium lauryl ether sulfate (2.5 EO); anionic; detergent, wetting agent used in fire extinguishers and for industrial, household, and cosmetic uses; yel. paste; water sol. 400 g/l; visc. 30,000 cps; f.p. < 0 C; pH 7–8; 70% act.

Elfan NS 243 S. [11] Sodium lauryl ether sulfate (3 EO); anionic; see Elfan NS 242; clear liq.; water sol.; sp.gr. 1.04; visc. 50 cps; f.p. < 0 C; pH 6.5–7.5; 28% act.

Elfan NS 243 S Conc. [11] Sodium lauryl ether sulfate (3 EO); anionic; see Elfan NS 242; yel. paste; water sol.; sp.gr. 1.12; visc. 30,000 cps; f.p. < 0 C; pH 7.5–9.5; 70% act.

Elfan NS 252 S. [11] Sodium lauryl ether sulfate (2.5 EO); anionic; see Elfan NS 242; yel. liq.; water sol.; sp.gr. 1.05; visc. 150 cps; f.p. < 0 C; pH 6.5–7.5; 28% act.

Elfan NS 252 S Conc. [11] Sodium lauryl ether sulfate (2.5 EO); anionic; see Elfan NS 242; yel. paste; water sol.; sp.gr. 1.11; visc. 30,000 cps; f.p. < 0 C; pH 7.5–9.5; 70% act.

Elfan NS 423 SH. [11] Sodium lauryl ether sulfate (3 EO); anionic; detergent for dishwashing, textiles, household, skins; yel. liq.; water sol.; visc. 100 cps; f.p. < 0 C; 28% act.

Elfan NS 423 SH Conc. [11] Sodium lauryl ether sulfate (3 EO); anionic; see Elfan NS 423 SH; yel. paste; water sol.; sp.gr. 1.10; visc. 30,000 cps; f.p. < 0 C; 70% act.

Elfan NS 682 KS. [11] Potassium tallow ether sulfate; anionic; detergent; paste; 50% conc.

Elfan OS. [11] Olefin sulfonates; detergent, wetting agent, emulsifier used for dishwashing, shampoos, textiles, laundering, household and cosmetics; liq.; 30–40% act.

Elfan OS 46. [11] Sodium alpha-olefin sulfonate (C_{14}/C_{16}); anionic; detergent for shampoos, personal care, skin, household and cosmetic cleaners; yel. liq.; water sol.; sp.gr. 1.05; visc. 250 cps; f.p. < 0 C; pH 7–9; 37% act.

Elfan SG 35. [11] Silver gloss conc. from ether sulfate and ethylene glycol distearate; detergent used for cosmetics, shampoo base, bubble bath; wh. paste; water disp.; sp.gr. 1.04; visc. 10,000 cps; pH 7.5–9.5; 37% act.

Elfan SP 500. [11] Ether sulfate and alkylbenzene sulfonate with a nonionic surfactant; all purpose detergent for household use; yel. liq.; sol. 250 g/l water; sp.gr. 1.05; visc. 5000 cps; f.p. < 0 C; pH 6.5–7.5; 50% act.

Elfan WA. [11] Sodium dodecylbenzene sulfonate; anionic; heavy-duty detergent for textiles, household; emulsifier for emulsion polymerization; yel. paste; sol. 250 g/l water; sp.gr. 1.08; f.p. < 0 C; m.p. 5 C; pH 8–10; 50% act.

Elfan WA Pulver (Powder). [11] Sodium dodecylbenzene sulfonate; anionic; see Elfan WA; lt. yel. beads, powd.; sol. 100 g/l water; dens. 300 g/l; pH 8–9; 80% act.

Elfan WA Sulfosaure (Sulphonic Acid). [11] Dodecylbenzene sulfonic acid; anionic; raw material; dk. br. liq.; SO_3 odor; water sol.; sp.gr. 1.06; visc. 500 cps; flash pt. 150 C (PMCC); 98% act.

Elfan WAT. [11] TEA dodecylbenzene sulfonate; anionic; detergent for textile, household, personal care uses; emulsifier for insecticides; yel.-br. liq.; water sol.; sp.gr. 1.06; visc. 1000 cps; f.p. < 0 C; pH 5.5–6.5; 50% act.

Elfanol 510. [11] Sulfosuccinic acid monoester of tallow fatty acid monoethanolamide, sodium salt; anionic; lt. duty, multipurpose detergent and Syndet bars; lt. yel. paste; poor water sol.; sp.gr. 1.06; pH 5.5–7.5; 50% act.

Elfanol 616. [11] Sulfosuccinic acid monoester of an ethoxylated fatty alcohol, sodium salt (3 EO); anionic; detergent for cosmetics, shampoos, bubble and baby baths; clear liq.; water sol.; sp.gr. 1.05; visc. 100 cps; f.p. 0 C; pH 7; 30% act.

Elfanol 850. [11] Sulfosuccinic acid monoester of an ethoxylated coco fatty acid monoethanolamide, sodium salt; anionic; detergent for shampoos, bubble baths, cosmetics; yel.-br. liq.; poor water sol.; sp.gr. 1.06; f.p. 14 C; pH 7; 45% act.

Elfanol 883. [11] Sodium dioctyl sulfosuccinate; anionic; wetting agent for technical processes; yel. liq.; water sol. 15 g/l; sp.gr. 1.08; visc. 50 cps; f.p. < 0 C; pH 7; 63% act.

Elfapur KA 45. [11] Coco fatty acid monoethanolamide polyglycol ether; nonionic; detergent for textile, household, personal care use; yel. liq.; water sol 300 g/l; sp.gr. 1.05; f.p. 10–15 C; cloud pt. 74–76 C (1% aq.); 88% act.

Elfapur LM 20. [11] $C_{12}C_{14}$ fatty alcohol ethoxylate; nonionic; base material for production of anionics; liq.; 100% conc.

Elfapur LM 25. [11] C_{12}/C_{14} lauryl alcohol polyglycol ether; see Elfapur LM 20; yel. liq.; poor water sol.; sp.gr. 1.01; f.p. 5 C; cloud pt. 56–60 C; flash pt. 180 C (10% in 25% butyl diglycol); 97% act.

Elfapur LM 30 S. [11] Syn. C_{12}/C_{14} fatty alcohol ethoxylate; nonionic; see Elfapur LM 20; yel. liq.; poor water sol.; f.p. 5 C; cloud pt. 60–64 C (10% in 25% butyl diglycol); 97% act.

Elfapur LM 75 S. [11] Syn. C_{12}/C_{14} fatty alcohol ethoxylate; nonionic; see Elfapur LM 20; yel. liq.; water sol.; sp.gr. 1.00; f.p. 10 C; cloud pt. 60–62 C (1% aq.); 87% act.

Elfapur LP 25 S. [11] Syn. C_{12}/C_{15} fatty alcohol ethoxylate; nonionic; see Elfapur LM 20; yel. liq.; poor water sol.; sp.gr. 1.01; f.p. 5 C; cloud pt. 50–54 C (10% in 25% butyl diglycol); flash pt. 180 C; 97% act.

Elfapur ML 30 SH. [11] Lauryl polyglycol ether from syn. C_{12}/C_{14} fatty alcohol; see Elfapur LM 20; yel. liq.; poor water sol.; sp.gr. 1.01; f.p. 5 C; cloud pt. 62–66 C (10% in 25% butyl diglycol); flash pt. 180 C (PMCC); 97% act.

Elfapur N 50. [11] Nonylphenol polyglycol ether; nonionic; detergent, emulsifier for textiles, household uses; yel. liq.; poor water sol.; sp.gr. 1.05; f.p. –5 C; cloud pt. 60–62 C (10% in 25% butyl diglycol); flash pt. 90 C (PMCC); 97% act.

Elfapur N 70. [11] Nonylphenol polyglycol ether; nonionic; see Elfapur N 50; yel. liq., gel at 50–70% act.; water sol.; sp.gr. 1.05; f.p. 0 C; cloud pt. 22–24 C (1% aq.); flash pt. 103 C (PMCC); 97% act.

Elfapur N 90. [11] Nonylphenol polyglycol ether; nonionic; see Elfapur N 50; yel. liq., gel at 50–70% act.; water sol.; sp.gr. 1.06; f.p. 15 C; cloud pt. 60–62 C (1% aq.); flash pt. 114 C (PMCC); 97% act.

Elfapur N 120. [11] Nonylphenol polyglycol ether; nonionic; see Elfapur N 50; yel. liq., gel at 50–70% act.; water sol.; sp.gr. 1.07; cloud pt. 88–90 C (1% aq.); flash pt. 102 C (PMCC); 97% act.

Elfapur N 150. [11] Nonylphenol polyglycol ether; nonionic; see Elfapur N 50; yel. liq., gel at 40–70% act.; water sol.; sp.gr. 1.07; cloud pt. 64–66 C (1% in 10% NaCl); flash pt. 94 C (PMCC); 97% act.

Elfapur O 80. [11] Oleyl-cetyl alcohol ethoxylate; nonionic; heavy-duty detergent used in textile industry; yel.-br. liq., gel at 15–90% act.; water sol. 400 g/l; sp.gr. 1.05; m.p. 25–30 C; cloud pt. 44–46 C (1% aq.); 97% act.

Elfapur O 160. [11] Oleyl-cetyl alcohol polyglycol ether; detergent used in textile industry; water sol. 400 g/l; sp.gr. 1.06; m.p. 29–35 C; cloud pt. 64–66 C (1% in 10% NaCl); 97% act.

Elfapur T 110. [11] Tallow fatty alcohol polyglycol ether; nonionic; foam controlled, heavy-duty detergent powders, for textile use; yel. paste, gel at 30–80% act.; water sol. 300 g/l; sp.gr. 1.05; m.p. 35–40 C; cloud pt. 77–79 C (1% aq.); 97% act.

Elfapur T 250. [11] Tallow fatty alcohol polyglycol ether; nonionic; see Elfapur T 110; wh. powd., beads, gel at > 25% act.; water sol. 250 g/l; cloud pt. 77–79 C (1% in 10% NaCl); 98% act.

Elimina-254. [397] POE fatty alcohol sulfate; anionic; detergent, dispersant, antistat, emulsifier used in detergent and shampoo formulations, latex, dyestuffs, pigments, oil and waxes; used for syn. fibers; paste; 30% conc.

Eltesol ACS 60. [17] Ammonium cumenesulfonate; hydrotrope acting as solubilizer, coupling agent, and visc. modifier in liq. formulations; cloud pt. depressant in detergent formulations; pale pink liq.; dens. 1.10 g/cm^3; pH 7.0–8.0 (10% aq.); 60.0% min. act.

Eltesol AX 40. [17] Ammonium xylene sulfonate; anionic; hydrotrope, cloud pt. depressant used in the detergent mfg.; pale yel. liq.; pH 7–8.5; 41% act.

Eltesol CA 65. [17] Cumene sulfonic acid; anionic; catalyst for foundry resins; descaling agent for metal cleaning; antistress additive and plating aid in electroplating bath; curing aid in the plastics industry; raw material in the mfg. of dyes and pigments; detergents industry; dens. 1.15 g/cm^3; visc. 30 cs; 65 ± 1.0% act. in water.

Eltesol CA 96. [17] Cumene sulfonic acid conc.; anionic; see Eltesol CA65; br. visc. liq.; dens. 1.25 g/cm^3; visc. 1000 cs; 96.0 ± 1.0% act.

Eltesol PSA 65. [17] Phenol sulfonic acid; anionic; see Eltesol CA-65; mfg. pharmaceutical chemicals and disinfectants; dens. 1.3 g/cm^3; visc. 50 cs; 65.0 ± 1.0% act. in water.

Eltesol PT 93. [17] Potassium toluene sulfonate; anionic; see Eltesol AX 40; off wh. powd.; pH 9–10.5; 93% act.

Eltesol PX 40. [17] Potassium xylene sulfonate; see Eltesol ACS 60; straw liq.; dens. 1.10 g/cm^3; pH 7.0–10.5 (10% aq.); 40.0 ± 1.0% act. in water.

Eltesol PX 93. [17] Potassium xylene sulfonate; anionic; see Eltesol AX 40; off wh. powd.; pH 9–10.5; 93% act.

Eltesol ST 34. [17] Sodium toluene sulfonate; anionic; see Eltesol AX 40; pale yel. liq.; pH 7–10; 34% act.

Eltesol ST 40. [17] Sodium toluenesulfonate; see Eltesol ACS-60; straw liq.; dens. 1.10 g/cm^3; pH 7.0–10.5 (10% aq.); 40.0±1.0% act.

Eltesol ST 90. [17] Sodium toluene sulfonate; anionic; see Eltesol AX 40; off wh. powd., pellets; pH 9–10.5; 89% act.

Eltesol ST Pellets. [17] Sodium toluene sulfonate; see Eltesol ACS-60; wh. pellets; dens. 0.55 g/cm^3; pH 9.0–10.5 (3% aq.); 85.0% min. act.

Eltesol SX 30. [17] Sodium xylene sulfonate; anionic; see Eltesol AX 40; pale yel. liq.; pH 7–10; 30% act.

Eltesol SX 93. [17] Sodium xylene sulfonate; anionic; see Eltesol AX 40; off wh. powd., pellets; pH 9–10.5; 93% act.

Eltesol SX Pellets. [17] Sodium xylene sulfonate; see Eltesol ACS-60; wh. pellets; dens. 0.5 g/cm^3; pH 9.0–10.5 (3% aq.); 88.0% min. act.

Eltesol TA. [17] Toluene sulfonic acid; anionic; see Eltesol CA-65; hydrotrope, intermediate; used in mfg. of acrylonitrile; dens. 1.2 g/cm^3; visc. 15 cs; 61.2% min. act. in water.

Eltesol TA 65. [17] Toluene sulfonic acid; anionic; intermediate, catalyst in foundry and chemical industries, hardening agent in plastics, activator for nicotine insecticides; lt. amber clear liq.; dens. 1.2 g/cc; visc. 9–12 cs; 65.5% act.

Eltesol TA 96. [17] Toluene sulfonic acid; see Eltesol TA 65; br. thick crystalline liq.; dens. 1.3 g/cm^3; 97% act.

Eltesol TA/F. [17] Toluene sulfonic acid aq. sol'n.; see Eltesol CA-65 and TA; amber/br. clear liq.; dens. 1.2 g/cm^3; visc. 15 cs; 61.2% min. sulfonic acid.

Eltesol XA. [17] Xylene sulfonic acid aq. sol'n.; see Eltesol CA-65; amber/br. clear liq.; dens. 1.2 g/cm^3; visc. 15 cs; 61.0% min. act.

Eltesol XA65. [17] Xylene sulfonic acid; anionic; intermediate, catalyst in preparation of esters, hardening agent in plastics, activator for nicotine insecticides; lt. amber, clear to slightly hazy liq.; dens. 1.2 g/cc; visc. 9–12 cs; 65% act.

Eltesol XA90. [17] Xylene sulfonic acid conc.; anionic; see CA-65; br. visc. liq.; dens. 1.35 g/cm^3; visc. 3000 cs; 95.0 ± 2.0% act.

Eltesol XA/M65. [17] Xylene sulfonic acid; see Eltesol XA65.

Emal 0. [341] Sodium lauryl sulfate; anionic; foaming agent for dentifrice; powd.; 98.5% conc.

Emal 10. [341] Sodium lauryl sulfate; anionic; emulsifier for PVC polymerization; detergent for personal care products; foaming agent for dentifrice; powd., needle; 90% conc.

Emal 20C. [341] Sodium POE alkyl ether sulfate; anionic; detergent, emulsifier, foam stabilizer, shampoo base; scouring agent for syn. fibers; liq.; 25% conc.

Emal 20T. [341] POE alkyl ether sulfate triethanolamine; anionic; shampoo base; foaming agent for pharmaceuticals and cosmetics; liq.; 40% conc.

Emal 40. [341] Sodium fatty alcohol sulfate; anionic; wetting agent for agric. powd.; scouring for textiles; powd.; 30% conc.

Emal AD-25. [341] Ammonium lauryl sulfate; anionic; shampoo base; foaming agent; liq.; 25% conc.

Emal E-25C, -70C. [341] Sodium POE alkyl ether sulfate; anionic; detergent, emulsifier, foam stabilizer, shampoo base; liq.; 70% conc.

Emal NC-35. [341] Sodium POE alkylaryl ether sulfate; detergent, emulsifier, foam stabilizer, shampoo base; scouring agent for syn. fibers; liq.; 35% conc.

Emal TD. [341] TEA lauryl sulfate; anionic; shampoo base, foaming agent; emulsifier in skin cream; liq.; 40% conc.

Emanon 1112, 3115. [341] PEG fatty acid ester; nonionic; emulsifier for cosmetics and pharmaceuticals; liq., solid; 100% conc.

Emargol KL. [650] Sodium sulfoacetate of mono- and diglycerides; anionic; food processing defoamer and o/w emulsifier; dough modifier; paste; disp. in oil and water.

Emasol L-10. [341] Sorbitan monolaurate; nonionic; food emulsifier; liq.; HLB 8.6;

100% conc.

Emasol L-106. [341] POE (4) sorbitan monolaurate; nonionic; emulsifier and solubilizer; liq.; HLB 13.3; 100% conc.

Emasol L-120. [341] POE (20) sorbitan monolaurate; nonionic; emulsifier for pharmaceuticals and cosmetics; stabilizer for emulsion polymerization; liq.; HLB 16.7; 100% conc.

Emasol O-10. [341] Sorbitan monooleate; nonionic; dispersant for colorants; liq.; HLB 4.3; 100% conc.

Emasol O-15 R. [341] Sorbitan sesquioleate; nonionic; emulsifier and lubricant; liq.; HLB 3.7; 100% conc.

Emasol O-30. [341] Sorbitan trioleate; nonionic; see Emasol O-10; liq.; HLB 1.8; 100% conc.

Emasol O-105 R. [341] POE sorbitan monooleate; nonionic; emulsifier and solubilizer; liq.; HLB 10.0; 100% conc.

Emasol O-106. [341] POE (5) sorbitan monooleate; nonionic; see Emasol O-105 R; liq.; HLB 10.0; 100% conc.

Emasol O-120. [341] POE (20) sorbitan monooleate; nonionic; see Emasol O-105 R; liq.; HLB 15.0; 100% conc.

Emasol O-320. [341] POE (20) sorbitan trioleate; nonionic; see Emasol O-105 R; liq.; HLB 11.0; 100% conc.

Emasol P-10. [341] Sorbitan monopalmitate; nonionic; see Emasol O-10; solid; HLB 6.7; 100% conc.

Emasol P-120. [341] POE (20) sorbitan monopalmitate; nonionic; emulsifier for pharmaceuticals and cosmetics; solubilizer for colorants; stabilizer for emulsion polymerization; liq.; HLB 15.6; 100% conc.

Emasol S-10. [341] Sorbitan monostearate; nonionic; see Emasol O-10; solid; HLB 4.7; 100% conc.

Emasol S-20. [341] Sorbitan distearate; nonionic; emulsifier and lubricant; flake; HLB 4.4; 100% conc.

Emasol S-30. [341] Sorbitan tristearate; nonionic; see Emasol O-10; HLB 2.1; 100% conc.

Emasol S-106. [341] POE (4) sorbitan monostearate; nonionic; see Emasol S-20; solid; HLB 9.6; 100% conc.

Emasol S-120. [341] POE (20) sorbitan monostearate; nonionic; see Emasol S-20; solid; HLB 14.9; 100% conc.

Emasol S-320. [341] POE (20) sorbitan tristearate; nonionic; see Emasol S-20; solid; HLB 10.5; 100% conc.

Emcol 14. [654] Polyglycerol oleate; nonionic; w/o emulsifier; corrosion inhibitor for aerosol; spreader sticker, antifoaming agent; liq.; 100% conc.

Emcol 27-21, 27-21N. [650] Blend of oil sol. sulfonate and POE ethers; anionic/nonionic; emulsifier for Ortho-Dichlorobenzene; liq.; 100% act.

Emcol 150. [650] Isopropyl alcohol and stearyl alcohol; cationic; cosmetics and toiletries surfactant functioning as antistat, conditioner, lubricant, and substantive agent; paste; disp. in water.

Emcol 500B. [652] Sulfonate and POE ether; anionic/nonionic; emulsifier for pesticide emulsifiable concs.; liq.; 100% conc.

Emcol 1655. [650] Cocamidopropyl dimethylamine propionate; cationic; cosmetics and toiletry surfactant used as antistat, conditioner, emollient, foaming and substantive agent; liq.; water-sol.

Emcol 3780. [650] Stearamidopropyl dimethylamine lactate; cationic; see Emcol 1655; water-sol.

Emcol 4072. [650] Disodium hydrog. cottonseed glyceride sulfosuccinate; anionic; cosmetics and toiletries surfactant used as conditioner, emollient, emulsifier; paste; water-sol.

Emcol 4100M. [650] Disodium myristamido MEA-sulfosuccinate; anionic; dispersant, wetting agent, foaming agent, detergent, emulsifier for bubble bath, shampoo, carpet and upholstery cleaners, textiles; wh. creamy semisolid; disp. in water; sp.gr. 1.01; surf. tens. 37.8 dynes/cm (0.05%); 38% solids.

Emcol 4161L. [650; 654] Disodium oleamido MIPA sulfosuccinate; anionic; dispersant, wetting, foaming, detergent, and emulsifying agent for bubble bath, shampoos, cleansers for cosmetics and toiletries; lt. yel. clear liq.; sol. in water; sp.gr. 1.10; flash pt. (PMCC) 93 C; surf. tens. 32.6 dynes/cm (0.5%); 38% solids.

Emcol 4300. [650; 654] Disodium C_{12-15} sulfosuccinate; anionic; dispersant, wetting, foaming, detergent, emulsifying agent for bubble bath, shampoo, cosmetics and toiletries; lt. clear liq.; sol. in water; sp.gr. 1.09; surf. tens. 29.3 dynes/cm (0.05%); 33% solids.

Emcol 4400-1. [650] Disodium lauryl sulfosuccinate; anionic; cosmetics and toiletries surfactant used as emulsifier, cleansing, foaming agent, stabilizer; paste; water-sol.

Emcol 4500. [650] Sodium dioctyl sulfosuccinate; dispersant, detergent, wetting, foaming, emulsifying agent; for cosmetics, toiletries, textiles, industrial processing slurries; clear, lt. visc. liq.; sol. in hydrocarbons, kerosene, Stod.; disp. water and alcohol; sp.gr. 1.10; flash pt. > 93 C; surf. tens. 26.3 dynes/cm (0.05%); 70% solids.

Emcol 4560. [650] Sodium dioctyl sulfosuccinate; dispersant, wetting, foaming, detergent, emulsifying agent, solubilizer; used in drycleaning; clear, lt. visc. liq.; sol. in perchlorethylene, carbon tetrachloride, Stod.; slightly sol. in water; disp. in alcohol; sp.gr. 1.08; pH 5.1 (3% aq. disp.); surf. tens. 27.4 dynes/cm (0.05%); 90% solids.

Emcol 4580PG. [650] Sodium diester sulfosuccinate; anionic; detergent, emulsifier, and water solubilizer in dry cleaning; textile industry detergent and scouring agent for solv. systems; liq.; sol. in perchloroethylene, Stod.; disp. in water.

Emcol 4600. [650] Sodium ditridecyl sulfosuccinate; detergent, foam modifier, wetting agent, dispersant, processing aid for pigments, in drycleaning; clear, lt. visc. liq.; sol. in perchlorethylene, carbon tetrachloride, Stod., kerosene, xylene, IPA; slightly sol. in alcohol; insol. water; sp.gr. 1.00; flash pt. > 93 C; pH 6.6 (3% aq. disp.); 70% solids.

Emcol 4930. [650] Sodium diester sulfosuccinate; anionic; emulsion polymerization surfactant offering latex stabilization, wetting, and emulsification; liq.; water-sol.

Emcol AD 5-06, AD 6-89A. [650] Blend of oil sol. sulfonate and POE ethers; anionic/nonionic; emulsifier used for pesticide emulsifiable conc.; liq.; 100% act.

Emcol AG 4-48N. [650] Blend of oil sol. sulfonate and POE ethers; see Emcol AD 5–06; liq.; 100% act.

Emcol AK 16-97N, 18-35. [650] Blend of oil sol. sulfonate and POE ethers; anionic/nonionic; emulsifier used for pesticide emulsifiable conc.; liq.; 100% act.

Emcol CC-9. [650; 654] PPG-9 diethylmonium chloride, cationic; dispersant, antistat, wetting agent, o/w emulsifier, conditioner, penetrant, lubricant; for cosmetics, toiletries, textiles, industrial slurries; lt. amber clear liq.; sol. in water, IPA; sp.gr. 1.01; flash pt. > 93 C; pH 6.5 (10% aq.); 100% conc.

Emcol CC-36. [650; 654] PPG-25 diethylmonium chloride; cationic; dispersant, o/w emulsifier; plasticizer for hair polymers; antistat; skin cleanser for cosmetics; used in dry cleaning systems; lt. amber clear liq.; sol. in water, IPA; sp.gr. 1.01; flash pt. > 93 C; pH 6.7 (10% in 10:6 IPA:water); 100% conc.

Emcol CC-37-18. [650] Coco-betaine; amphoteric; foaming agent and stabilizer for cosmetics and toiletries, visc. modifier; liq.

Emcol CC-42. [650] PPG-40 diethylmonium chloride, cationic; pigment dispersant, emulsifier, solv., conditioner, antistat, lubricant, corrosion inhibitor for toiletries, cosmetics, germicides, syn. fibers and plastics; textile; ore flotation additive; lt. amber oily liq.; sol. in IPA, acetone, MEK, min. spirits, ethanol; partly sol. in water; sp.gr. 1.01; flash pt. > 200 C (PMCC).

Emcol CC-55. [650] Polypropoxylated quaternary ammonium acetate; cationic surfactant used for industrial, cosmetic, and toiletries applications; lt. amber oily liq.; sol. (25%) in ethanol, IPA, acetone, MEK; disp. in water; sp.gr. 1.02; flash pt. > 93 C (PMCC).

Emcol CC-57. [650] Polypropoxylated quaternary ammonium phosphate; cationic surfactant, detergent, antistat for drycleaning, industrial uses, cosmetics, and toiletries.

Emcol CC-422. [650] Polypropoxy quat. ammonium chloride; cationic; detergent and dry cleaning applications; anticaking and cleansing agent, antistat, dispersant and o/w emulsifier for aerosol formulations; liq.; sol. in halogenated hydrocarbons, perchloroethylene, and water.

Emcol CS-136, -143, -151, -165. [650] Org. phosphate ethers; anionic; detergent, emulsifier, dispersant, solubilizer; liq.

Emcol CS-1361. [650] Sodium salt of complex organo-phosphate ester; used for scale control; clear liq.; sol. in hydrocarbons, min. oil, kerosene, ethanol, water; sp.gr. 1.10; pour pt. 40 F; surf. tens. 31.9 dynes/cm (0.05%); pH 5 (3% aq.); 10% moisture.

Emcol D 24-25. [650] Calcium sulfonate; o/w emulsifier for industrial use.

Emcol D75-13,-33. [650] Alkoxylated acetylenic alcohol; nonionic; corrosion inhibitor for use in metal processing; liq.; water-sol.

Emcol DG. [650] Cocamidopropyl betaine; amphoteric; detergent, emulsifier, foaming and wetting agent, foam stabilizer for detergent industry; liq.; water-sol.

Emcol E-607L. [650; 654] Lapyrium chloride; cationic; emollient, emulsifier, foamer, substantive agent, deodorant for cosmetics, toiletries, industrial applications; wh. cryst.; sol. 37% in water and ethanol, 28% in IPA; surf. tens. 37 dynes/cm (0.1% aq.); pH 3.9 (1% aq.); 97.5% act.

Emcol E-607S. [650; 654] Steapyrium chloride; cationic; see Emcol E-607L; wh. to off-wh. powd.; sol. 2.5% in water; pH 3.4 (1% aq.); 94% act.

Emcol H-30, -31A, -32. [652] Polyglycol fatty acid ester; emulsifier for petroleum products; liq.; 100% conc.

Emcol H-400X. [650; 652] Sulfonate and POE ether; anionic/nonionic; emulsifier for pesticide emulsifiable concs.; clear to dk. amber, slightly visc. liq.; oil-sol.; sp.gr. 1.03; pH 6–7 (3% disp.); 100% conc.

Emcol H-470X. [650] Blend of oil sol. sulfonate and POE ethers; emulsifier for pesticide emulsifiable conc.; liq.; 100% act.

Emcol ISML. [650] Isostearamidopropyl morpholine lactate; cationic; cosmetics and toiletries surfactant used as antistat, conditioner, emollient, foaming and substantive agent; liq.; water-sol.

Emcol K-8300. [650; 654] Alkanolamide disodium sulfosuccinate half ester; anionic; dispersant, wetting and foaming agent, detergent; emulsifier in emulsion polymerization; sol. in water; sp.gr. 1.10; pH 6.5 (3% aq.); surf. tens. 32.6 dynes/cm (0.05%); 38% solids.

Emcol L. [650] Lauramine oxide; cationic; cosmetics and toiletries surfactant used as antistat, cleansing and substantive agent, emollient, lubricant, and visc. modifier; detergent and foam stabilizer for industrial detergents; liq.; water-sol.

Emcol M. [650] Myristamine oxide; cationic; see Emcol L; liq.; water-sol.

Emcol N-83T, N-85T. [650] Blend of oil sol. sulfonate and POE ethers; emulsifier for pesticide emulsifiable conc.

Emcol N-139B, -139BN. [652] Sulfonate and POE ether; anionic/nonionic; emulsifier for Trifluralin and Alachlor emulsifiable concs.; liq.; 100% conc.

Emcol N-140B. [650] Sulfonate and POE ethers; anionic/nonionic; emulsifier for pesticide emulsifiable conc.; dk. amber, hazy liq.; sp.gr. 1.01–1.03; pH 6–7; 100% act.

Emcol N-141B. [650] Blend of oil sol. sulfonate and POE ethers; see Emcol N-140B; liq.; 100% conc.

Emcol N-300B. [650] Blend of oil sol. sulfonate and POE ethers; emulsifier for pesticide emulsifiable conc.; clear to dk. amber, slight visc. liq.; sp.gr. 1.03; pH 6–7; 100% act.

Emcol N-369, N-469. [650] Blend of oil sol. sulfonate and POE ethers; emulsifier for pesticide emulsifiable conc.; liq.; 100% act.

Emcol N-500B. [650] Blend of oil sol. sulfonate and POE ethers; emulsifier for pesticide emulsifiable conc.; clear to dk. amber, slight visc. liq.; sp.gr. 1.03; pH 6–7; 100% act.

Emcol N-710, N-712, N-714. [650] Blend of oil sol. sulfonate and POE ethers; emulsifier for pesticide emulsifiable conc.

Emcol NA-30. [650] Cocamidopropyl betaine; amphoteric; cosmetics/toiletries surfactant used as antistat, cleansing, foaming, spreading agent, conditioner, foam stabilizer, solubilizer and visc. modifier; liq.; water-sol.

Emcol P-1020 BU. [650; 652] Calcium alkylaryl sulfonate; anionic; o/w emulsifier for industrial use; intermediate for agric. emulsifiers; amber visc. liq.; oil sol.; sp.gr. 1.01–1.03; pH 6–7.5 (3% in 20% IPA); 68–70% act.

Emcol P-1045. [650] Amine salt of alkylaryl sulfonic acid; wetting agent in oil-based systems, sludge dispersant in fuel oil; liq.; water disp., oil sol.; 90% act.

Emcol P-1049. [650] Amine salt of alkylaryl sulfonic acid; anionic surfactant; liq.; water sol.; 60% act.

Emcol P-1059. [650] Amine salt of alkylaryl sulfonic acid; anionic surfactant for o/w emulsions, cleaning and degreasing; liq.; 90% act.

Emcol P-1059B. [650] Amine salt of alkylaryl sulfonic acid; anionic surfactant, water solubilizer; liq.; 90% act.

Emcol P-1073. [650] Amine salt of alkylaryl sulfonic acid; anionic surfactant for visc. control; dk. br. visc. cloudy liq.; oil sol.; 90% act.

Emcol PS-331, -400, -415. [650] Org. phosphate ester; anionic surfactant for water-based systems; liq.

Emcol Q. [650] Polyquaternium-5; cationic; cosmetics and toiletries surfactant used as antistat, conditioner, emulsifier, and substantive agent; liq.; water-sol.

Emcol TS-230. [650] Org. phosphate ester; dispersant and lubricant in oil drilling muds, o/w emulsifier, flow control agent; amber, clear liq.; sol. in xylene, ethanol, carbon tetrachloride, water; sp.gr. 1.22; pour pt. < 50 F; surf. tens. 49.6 dynes/cm (0.05%); pH 2 (3% aq.).

Emcol ZM-1117. [650] Blend of oil sol. surfactants and POE ethers; anionic/nonionic; emulsifier for pesticide emulsifiable conc.; liq.; 100% act.

Emcon E. [232] Egg oil; nonionic; emollient, w/o emulsifier for food, cosmetics, household, pharmaceuticals; yel. liq.; bland odor; sol. in min. and veg. oils, most org. solvs.; dens. 9.95; 100% act.

Emcon E-5, E-20. [232] Lecithin; emollient, moisturizer for hair and skin care products; w/o emulsifier; superfatting agent, humectant, mold release agent; yel. liq.; E-5 sol. in min. and veg. oils, slightly disp. in most org. solv., insol. water; E-20 sol. in min. and veg. oils, disp. in most org. solv., water; sp.gr. 0.95; 85.4 and 62.3% triglycerides resp..

Emcon E-L. [232] Egg lecithin; nonionic; lecithin applications; solid; 100% conc.

Emcon W. [232] Wheat germ oil; nonionic; emollient, conditioner, source of vitamin E; liq.; 100% conc.

Emerest 2325. [217] Butyl stearate; nonionic; dye solubilizer in personal care products; Gardner 4 liq.; insol. in water; dens. 7.1 lb/gal; visc. 7 cSt (100 F); pour pt. 18 C; flash pt. 370 F.

Emerest 2326. [217] Butyl stearate; noninic; lubricant used in aluminum foil rolling, fiber and yarn processing; Gardner 1 liq.; sol. in IPA, toluol, min. oil; dens. 7.2 lb/gal; visc. 10 cs; f.p. 20 C; flash pt. 188 C; sapon. no. 172; 100% act.

Emerest 2350. [217] Glycol stearate; nonionic; opacifying and pearlescing agent used in liq. cosmetic and detergent compds.; Gardner 1 beads; sol. in IPA, toluol, min. oil; f.p. 50 C; HLB 2.1; sapon. no. 185; 100% act.

Emerest 2355. [217] Glycol distearate; nonionic; opacifier used in liq. detergent and cosmetic products; Gardner 4 flakes; disp. in min. oil; f.p. 62 C; HLB 1.2; sapon no. 195; 100% act.

Emerest 2381. [217] Propylene glycol stearate SE; anionic; pearlescing and spreading agent, for cosmetic and dishwashing products, hand lotions; Gardner 2 beads; sol in IPA, disp. in water, min. oil, hot toluol; f.p. 40 C; HLB 4.0; sapon. no. 170; 100% act.

Emerest 2400. [217] Glyceryl stearate; nonionic; emulsifier for hand creams, cosmetics, textiles, industrial lubricants; Gardner 1 beads; sol. in IPA, hot toluol, hot min. oil; f.p. 57.5; sapon. no. 153; 100% act.

Emerest 2401. [217] Glyceryl stearate; nonionic; tech. grade of Emerest 2400, emulsifier for hand creams, cosmetics, industrial lubricants; Gardner 2 beads; sol. in IPA, hot toluol, hot min. oil; f.p. 58; sapon. no. 153; 100% act.

Emerest 2407. [217] Glyceryl stearate, SE; modified anionic; emulsifier used in cosmetic creams and lotions; Gardner 3 beads; sol. in IPA, hot toluol, min. oil; f.p. 58 C; sapon. no. 153; 100% act.

Emerest 2421. [217] Glyceryl oleate; nonionic; emulsifier, in mold release agents, anti-icing fuel additive, rust preventative, in textiles as a lubricant component in syn. fiber spin finishes; Gardner 11 liq.; sol. in toluol, min. oil; dens. 7.9 lb/gal; visc. 91 cs (38 C); f.p. 6 C; flash pt. 235 C; sapon no. 170; 100% act.

Emerest 2423. [217] Triolein; nonionic; lubricant for metals, leather, textiles, emulsifier, called syn. olive oil; Gardner 3 liq.; sol. in toluol, min. oil; dens. 7.6 lb/gal; visc. 47 cs; f.p. 9 C; flash pt. 293 C; sapon no. 197; 100% act.

Emerest 2426. [217] Glyceryl dioleate; emulsifier used in w/o emulsions or industrial lubricants; Gardner 8 liq.; insol. in water; dens. 7.7 lb/gal; visc. 55 cSt (100 F); HLB 1.6; pour pt. < 0 C; flash pt. 470 F.

Emerest 2485. [217] Pentaerythritol tetrapelargonate; nonionic; primary lubricant base or modifier in lubricant formulations used in metal working and syn. fiber processing; Gardner 1 liq.; insol. in water; dens. 8.0 lb/gal; visc. 35 cSt (100 F); pour pt. 10 C; flash pt. 555 F.

Emerest 2620. [217] PEG 200 monolaurate; nonionic; emulsifier, coupling agent, defoamer in water base coatings, visc. control additive; Gardner 1 liq.; disp. in water, min. oil, butyl stearate, glycerol trioleate, perchloroethylene, Stod.; dens. 8.2 lb/gal; visc. 40 cs; HLB 9.8; cloud pt. < 25 C; 100% act.

Emerest 2622. [217] PEG 200 dilaurate; nonionic; coemulsifer and lubricant in self-emulsifiable textile and industrial oils, mold release agent, visc. control agent; Gardner 2 liq.; disp. in water, disp. in min. oil, butyl stearate, glycerol trioleate, perchloroethylene, Stod.; dens. 8.0 lb/gal; visc. 35 cs; HLB 6.8; cloud pt. <25 C; 100% act.

Emerest 2624. [217] POE (4) monooleate; emulsifier for min., cutting and fatty oils, and solv.; solv. emulsifier in metal cleaners and degreasers, w/o emulsifier for consumer pesticide aerosols; lubricant component in textile, leather processing; Gardner 3 liq.; sol. in xylene, disp. in water; dens. 8.1 lb/gal; visc. 34 cSt (100 F); HLB 8.3; pour pt. < –15 C; flash pt. 415 F; cloud pt. < 25 C.

Emerest 2625. [217] PEG 200 monoisostearate; emulsifier used as a component in fiber lubricants, processing aids, and in conc. fabric softeners; Gardner 2 liq.; sol. in Stod., disp. in water; dens. 8.2 lb/gal; visc. 50 cSt (100 F); HLB 8.3; pour pt. –8 C; flash pt. 310 F; cloud pt. < 25 C.

Emerest 2630. [217] PEG-6 monolaurate; hydrophilic emulsifier used as a lubricant component and scrooping agent for textile fibers and yarn; visc. control agent for plastisols; Gardner 2 liq.; sol. in xylene, disp. in water; dens. 8.4 lb/gal; visc. 37 cSt (100 F); HLB 12.1; pour pt. 9 C; flash pt. 475 F; cloud pt. < 25 C.

Emerest 2632. [217] PEG 300 monooleate; emulsifier and lubricant in formulating textile softeners; Gardner 3 liq.; sol. in xylene, disp. in water; dens. 8.3 lb/gal; visc. 46 cSt (100 F); HLB 10.4; pour pt. –10 C; flash pt. 425 F; cloud pt. < 25 C.

Emerest 2634. [217] PEG 300 monopelargonate; surfactant for open-end spinning of syn. fibers; Gardner 1 liq.; sol. in water, xylene; dens. 8.6 lb/gal; visc. 25 cSt (100 F); HLB 12.8; pour pt. < –15 C; flash pt. 485 F; cloud pt. < 25 C.

Emerest 2636. [217] PEG 300 monostearate; emulsifier for oils and fats in industrial lubricants; softening and lubricating for textiles and leather; Gardner 1 solid; sol. in xylene, disp. in water; dens. 8.3 lb/gal; visc. 45 cSt (100 F); HLB 10.1; m.p. 29 C; flash pt. 460 F; cloud pt. < 25 C.

Emerest 2640. [217] PEG 400 monostearate; nonionic; emulsifier of oils and fats in industrial lubricants; thickening agent and stabilizer for starch coatings on paper, lubricant for channeling wire through conduit; Gardner < 1 soft, waxy solid; disp. in water, min. oil, butyl stearate, perchloroethylene, Stod.; m.p. 32 C; HLB 11.7; cloud pt. <25 C; 100% act.

Emerest 2642. [217] PEG 400 distearate; nonionic; emulsifier, thickener, for cosmetic and industrial emulsions; Gardner 2 soft solid; sol. in min. oil, butyl stearate, perchloroethylene, Stod., disp. water; m.p. 36 C; HLB 8.1; cloud pt. <25 C; 100% act.

Emerest 2644. [217] PEG 400 monoisostearate; see Emerest 2625; Gardner 1 liq.; disp. in water, min. oil, butyl stearate, xylene; dens. 8.5 lb/gal; visc. 70 cSt (100 F); HLB 11.3; pour pt. 10 C; flash pt. 450 F; cloud pt. < 25 C.

Emerest 2646. [217] PEG 400 monooleate; nonionic; emulsifier for oil-based sol. cutting oils and other specialty industrial lubricants, visc. stabilizer for vinyl plastisols; Gardner 2 liq.; disp. in water, butyl stearate, perchloroethylene, Stod.; dens. 8.5 lb/gal; visc. 100 cs; HLB 11.7; cloud pt. < 25 C; 100% act.

Emerest 2647. [217] PEG 400 sesquioleate; nonionic; oil and fat emulsifier in industrial and textile lubricants; Gardner 2 liq.; sol. in min. oil, butyl stearate, perchloroethylene, Stod.; dens. 8.2 lb/gal; visc. 75 cs; HLB 9.2; cloud pt. < 25 C; 100% act.

Emerest 2648. [217] PEG 400 dioleate; lipohilic emulsifier and solubilizer for min. oils, fats and solv., used in industrial lubricants; 100% act.

Emerest 2650. [217] PEG 400 monolaurate; nonionic; wetting agent, defoamer and leveling agent in latex paints, dispersant, solubilizer for oils and solv.; Gardner 1 liq.; sol. in water, disp. in butyl stearate, perchloroethylene, Stod.; dens. 8.6 lb/gal; visc. 100 cs; HLB 13.1; cloud pt. 33 C; 100% act.

Emerest 2652. [217] PEG 400 dilaurate; emulsifier used as softener, lubricant, release agent for paper, in textile industry; Gardner 2 liq.; sol. in min. oil, butyl stearate, perchloroethylene, Stod., disp. in water; dens. 8.3 lb/gal; visc. 65 cs; HLB 10.4; cloud pt. <25 C; 100% act.

Emerest 2654. [217] PEG 400 monopelargonate; surfactant used as base lubricant for syn. fiber finishes, other textile processing; coemulsifier, coupling agent; Gardner 1 liq.; sol. in water, xylene; dens. 8.7 lb/gal; visc. 34 cSt (100 F); HLB 14.3; pour pt. 5 C; flash pt. 440 F; cloud pt. 40 C (2% saline).

Emerest 2660. [217] PEG 600 monooleate; emulsifier and detergent, used in textile industry, component in systems for the acid washing of printed circuit boards; Gardner 2 liq.; sol. in water, disp. in butyl stearate, Stod.; dens. 8.7 lb/gal; visc. 115 cs; HLB 13.6; cloud pt. 47 C; 100% act.

Emerest 2661. [217] PEG 600 monolaurate; lubricant in processing syn. fibers; Gardner 2 liq.; sol. @ 5% in water, xylene; dens. 8.6 lb/gal; visc. 60 cSt (100 F); HLB 14.8; pour pt. 14 C; flash pt. 525 F; cloud pt. 63 C.

Emerest 2662. [217] PEG 600 monostearate; emulsifier for personal care products; visc. modifier in creams and lotions; component in textile softeners and self-emulsifying lubricants; Gardner 1 solid; sol. in xylene, disp. in water; dens. 8.5 lb/gal; HLB 13.8; m.p. 40 C; flash pt. 440 F; cloud pt. 55 C.

Emerest 2664. [217] PEG 600 monoisostearate; fatty acid ester used in fiber lubricants and processing aids and as an emulsifier; Gardner 5 liq.; disp. in water; dens. 8.6 lb/gal; visc. 90 cSt (100 F); HLB 13.0; pour pt. 10 C; flash pt. 490 F; cloud pt. < 25 C.

Emerest 2665. [217] PEG 600 dioleate; emulsifier, solubilizer used in pesticide and agric. sprays, latex paints, metalworking fluids, solv., specialty and industrial lubricants; Gardner 3 liq.; sol. in butyl stearate, Stod., xylene, disp. in water; dens. 8.3 lb/gal; visc. 64 cSt (100 F); HLB 10.3; pour pt. 19 C; flash pt. 530 F; cloud pt. < 25 C.

Emersal 6400. [217] Sodium lauryl sulfate; foaming agent; detergent for hair and rug shampoo, lt. duty cleaners; emulsifier for latex polymerization; Gardner 1 liq.; sol. @ 5% in water; dens. 8.7 lb/gal; visc. 424 cSt (100 F); pour pt. 15 C; cloud pt. > 100 C (5% saline).

Emersal 6430. [217] Ammonium lauryl sulfate; foaming agent and detergent in hair shampoo, cleaning lotions, and delicate fabric detergents; Gardner 1 liq.; sol. @ 5% in water; dens. 8.6 lb/gal; visc. 2490 cSt (100 F); pour pt. 6 C; cloud pt. > 100 C (5% saline).

Emersal 6434. [217] TEA-lauryl sulfate; foaming agent used in personal care products and fine fabric detergents; Gardner 2 liq.; sol. @ 5% in water; dens. 8.7 lb/gal; visc. 51 cSt (100 F); pour pt. < 0 C; cloud pt. > 100 C (5% saline).

Emersal 6453. [217] Sodium laureth sulfate; foaming agent and detergent in personal care products, rug shampoo, and lt. duty cleaning formulations; Gardner 1 liq.; sol. @ 5% in water; dens. 8.8 lb/gal; visc. 3690 cSt (100 F); pour pt. < 0 C; cloud pt. > 100 C (10% saline).

Emersal 6462. [217] Sodium alkyl sulfate; rewetting agent and detergent for textile, metal cleaning, and automotive formulations; used in drilling operations; Gardner 1 liq.; sol. @ 5% in water; dens. 9.0 lb/gal; visc. 11 cSt (100 F); pour pt. < 0 C; cloud pt. > 100 C (10% saline).

Emersal 6465. [217] Sodium 2-ethylhexylsulfate; wetting agent, penetrant, and cleaner used in hard surface, metal, and tire cleaning formulations, emulsion polymerization, electrolytic plating assistant, and textile mercerization; lye peeling of fruits and vegs.; Gardner 2 liq.; sol. @ 5% in water; dens. 9.1 lb/gal; visc. 11 cSt (100 F); pour pt. < –15 C; cloud pt. > 100 C (10% saline).

Emersol 210, 220. [217] Oleic acid; emulsifier.

Emersol 6321. [217] Oleic acid, food grade; emulsifier.

Emery 515. [217] Coco/tallow fatty acid blend; emulsifier (latex).

Emery 5440. [217] Sodium isethionate; anionic; intermediate raw material for surfactants; liq.; 56% conc.

Emery 6748. [217 Cocamidopropyl betaine; amphoteric surfactant used in hair and bath products, skin cleaners, specialty cosmetics; Gardner 4 liq.; sol. in water; dens. 8.7 lb/gal; 30% act.

Emery 6752. [217] Monocarboxylate coconut imidazolinium deriv.; amphoteric; surfactant used in cleaning compds. and shampoos, visc. improver and foam stabilizer, detergent; Gardner 9 liq.; sol. in water; dens. 8.7 lb/gal; 38% act.

Emery 6779. [217] Methoxy PEG-400 monooleate; surfactant used for foaming stability in industrial applications; Gardner 4 liq.; disp. in water; dens. 8.4 lb/gal; visc. 29 cSt (100 F); pour pt. < 15 C; flash pt. 490 F; cloud pt. 42 C.

Emery 6849. [217] Linear POE alcohol; used in solid built toilet bowl cleaners; Gardner 1 liq.; water sol.; dens. 8.2 lb/gal; cloud pt. > 100 C; 65% act.

Emid 6500. [217] Cocamide MEA; nonionic; foam stabilizer, thickener in shampoos, liq. detergents, rug cleaners; Gardner 4 flake; sol. in min. oil, butyl stearate, perchloroethylene, Stod., disp. in water; m.p. 71 C; 100% act.

Emid 6510. [217] Lauramide DEA; nonionic; foam booster, stabilizer and thickener in shampoos, bubble baths, detergents; Gardner 2 solid; sol. in perchloroethylene, Stod., disp. in water, min. oil, butyl stearate; m.p. 42 C; 100% act.

Emid 6511. [217] Lauramide DEA; nonionic; emulsifier in liq. formulations; Gardner 2 solid; sol. in perchloroethylene, Stod., disp. in water, min. oil, butyl stearate; m.p. 40 C; 100% act.

Emid 6514. [217] Cocamide DEA; nonionic; foam stabilizer, thickener, detergent, used in liq. cleaning compds.; Gardner 3 liq.; sol. in water, butyl stearate, perchloroethylene, Stod., disp. in min. oil; dens. 8.2 lb/gal; 100% act.

Emid 6515. [217] Cocamide DEA; nonionic; foam booster and stabilizer, inhibitor of soil redeposition; Gardner 3 liq.; sol. in water, butyl stearate, perchloroethylene, Stod., disp. in min. oil; dens. 8.3 lb/gal; visc. 390 cSt (100 F); pour pt. 0 C; flash pt. 370 F; 100% act.

Emid 6531. [217] Cocamide DEA and diethanolamine; thickener, detergent, emulsifier, and foam stabilizer for personal care products, liq. detergents, and janitorial cleaners; deduster for detergents; Gardner 3 liq.; sol. @ 5% in water; dens. 8.4 lb/gal; visc. 360 cSt (100 F); pour pt. –10 C; flash pt. 338 F.

Emid 6533. [217] Cocamide DEA and diethanolamine; emulsifier, thickening agent, foamer for detergents, scouring agent for textiles; Gardner 4 liq.; sol. in water, perchloroethylene, Stod., disp. in min. oil, butyl stearate; dens. 8.3 lb/gal; 100% act.

Emid 6534. [217] Cocamide DEA and diethanolamine; nonionic; emulsifier, thickening agent, foam stabilizer, used in shampoos and cosmetic formulations; Gardner 4 liq.; sol. in water, perchloroethylene, Stod., disp. in min. oil, butyl stearate; dens. 8.5 lb/gal; 100% act.

Emid 6538. [217] Modified coco diethanolamide; nonionic; thickener, foam stabilizer for detergents; Gardner 4 liq.; sol. in water, perchloroethylene, Stod., disp. in min. oil, butyl stearate; dens. 8.3 lb/gal; 100% act.

Emid 6541. [217] Lauramide DEA and diethanolamine; nonionic; emulsifier, foam booster, stabilizer, for shampoos, general purpose cleaners; Gardner 2 paste; sol. in water, perchloroethylene, Stod.; dens. 8.3 lb/gal; visc. 275 cSt (100 F); pour pt. –15 C; flash pt. 300 F; 100% act.

Emid 6543. [217] Stearic-oleic diethanolamide; nonionic; hydrophobic coemulsifier for min. oils, fatty esters, various solv.; Gardner 4 liq.; sol. in water, perchloroethylene, Stod., disp. in min. oil, butyl stearate; dens. 8.5 lb/gal; 100% act.

Emid 6544. [217] Capramide DEA and diethanolamine; nonionic; foam booster, wetting agent, detergency builder, used in salt and alkali systems, dedusting agent for alkaline powd.; sol. in butyl stearate, perchloroethylene, Stod., disp. in water, min. oil; m.p. 30 C; 100% act.

Emid 6545. [217] Oleic diethanolamide; nonionic; emulsifier in formulating antistatic fiber processing aids and yarn lubricants, foam suppressant in dye carrier and solv. emulsions; Gardner 7 liq.; sol. in min. oil, butyl stearate, perchloroethylene, Stod., disp. in water; dens. 7.8 lb/gal; 100% act.

Emkabase. [218] Naphthenic, amino sulfonate blend; anionic; emulsifier for min. oil; liq.

Emkafix RXC. [218] Quat. ammonium deriv.; cationic; dispersant and improves fixation of direct dyes; liq.

Emkagen 49. [218] Amino condensate and alkylaryl sulfonate blend; anionic; detergent for print washing in textile applications; liq.

Emkal BNS, BNX Powd. [218] Butyl naphthalene sodium sulfonate; anionic; dispersant, detergent, dyeing assistant, scouring, and wetting agent, emulsifier used in textile industry; liq. and powd. resp.; 35 and 80% conc. resp.

Emkal BNS Acid. [218] Butyl naphthalene sulfonic acid; anionic; see Emkal BNS; liq.; 90% conc.

Emkal NNS. [218] Nonyl naphthalene sodium sulfonate; anionic; see Emkal BNS; liq.; 35% conc.

Emkal NNS Acid. [218] Nonyl naphthalene sulfonic acid; anionic; see Emkal BNS; liq.; 90% conc.

Emkal NOBS. [218] Sodium nonyl benzene sulfonate; anionic; see Emkal BNS; liq.; 40% conc.

Emkalon AVR. [218] Quat. compound; cationic; softener, antistat used in textile applications; paste.

Emkane Acid. [218] Alkylaryl sulfonic acid; anionic; detergent, dyeing assistant, scouring and wetting agent; emulsifier used in the textile industry; liq.

Emkane HAD, HAL, HAX. [218] Alkylaryl sulfonate; anionic; see Emkane Acid; paste, liq., and liq. resp.

Emkapon. [218] Amide sulfonates; detergent blend.

Emkapon 71. [218] Polyglycol ether blend; nonionic; wetting and scouring agent for the textile industry; liq.

Emkapon CT. [218] POE deriv.; anionic; detergent used for textile applications; paste.

Emkapon Jel BS. [218] Fatty amide sulfonate;

anionic; see Emkane Acid; liq.

Emkapon K. [218] Sulfated fatty amide; anionic; detergent, textile agent for scouring and wetting, emulsifier; paste.

Emkapon SS, TS. [218] Amide sulfonate; anionic; general textile detergents; liq.

Emkastat PC. [218] Quat. deriv.; cationic; antistat; used in soil-resistant finishes for textiles; paste.

Emkatex AES. [218] Aliphatic ester sulfate; anionic; lubricant and dyeing assistant used in textile applications.

Emkatex CF 32. [218] Amino condensate with penetrants; anionic; detergent used in textile industry; liq.

Emkatex DX. [218] Phosphated alcohol; anionic; dispersant and textile dye bath stabilizer and leveling agent used in the textile industry; liq.

Emkatex LS. [218] POE deriv.; nonionic; detergent, dye leveling and retarding agent used in the textile industry; paste.

Emkatex PX29. [218] Sulfated amide; anionic; dye assistant, dispersant, emulsifier, and anticrock used in textiles; jelly; 33% conc.

Emkatex RA. [218] Sulfated fatty ester; anionic; emulsifier, wetting and textile scouring agent; liq.; 41% conc.

Emkatex WNX. [218] Sulfated esters; anionic; wetting and leveling agent, lubricant used in textile industry; gel.

Emphos CS-121. [650] Alkylaryl ethoxylate phosphate ester; anionic; detergent, o/w emulsifier, lubricant, solubilizer, and wetting agent for metal cleaning; liq.; sol. in naphthenic and paraffinic oils; disp. in water.

Emphos CS-136. [650] Nonyl phenol ethoxylate phosphate ester, acid form; anionic; lubricant; emulsifier for cutting fluids, polyvinyl acetate and acrylic film formation; drycleaning detergent; waterless hand cleaner component; clear visc. liq.; sol. @ 25% in water, ethanol, carbon tetrachloride, perchloroethylene, xylene, aromatic naphtha; sp.gr. 1.09; pour pt. 18 C; pH 2.0 (3% aq.); surf. tens. 31.2 dynes/cm (0.05% aq.).

Emphos CS-141. [650; 652] Nonyl phenol ethoxylate phosphate ester, acid form; anionic; lubricant for cutting fluids; detergent for heavy-duty liq. formulations; emulsifier in polyvinyl acetate and acrylic film formation; textile wetting agent; water-repellent fabric finishing; clear visc. liq.; sol. in water, ethanol, carbon tetrachloride, perchloroethylene, xylene, naphtha; sp.gr. 1.12; pour pt. > 4 C; pH 2.0 (3% aq.); surf. tens. 30.6 dynes/cm (0.05%).

Emphos CS-143. [650; 652] Phosphoric acid ester; emulsifier for Kelthane; clear liq.; pH 2.5–3 (3% aq.).

Emphos CS-147. [650] Alkylaryl ethoxylate phosphate ester; anionic; industrial detergent, o/w emulsifier, lubricant, solubilizer, foaming and coupling agent; liq.; water-sol.

Emphos CS-151. [650; 652] Org. phosphate ester; anionic; see Emphos CS-141; clear visc. liq.; sol. see Emphos CS-141; sp.gr. 1.11; pour pt. > 50 F; pH 2 (3% aq.); surf. tens. 36.8 dynes/cm (0.05%).

Emphos CS-330. [650] Alkylaryl ethoxylate phosphate ester; anionic; detergent, o/w emulsifier, and lubricant for oils in metal processing.

Emphos CS-733. [650] Alkylaryl ethoxylate phosphate ester; anionic; industrial coupler, o/w emulsifier, lubricant with electrolyte tolerance; liq.; water-sol.

Emphos CS-1361. [650] Sodium nonoxynol-9 phosphate; anionic; antistat; emulsifier for transparent gels; detergent for hard surfaces, metal cleaners, and drycleaning systems; clear liq.; sol. @ 25% in water, ethanol, carbon tetrachloride, perchloroethylene, aromatic naphtha, kerosene; sp.gr. 1.10; pour pt. 4 C; pH 5.0 (3% aq.); surf. tens. 31 dynes/cm (0.05% aq.).

Emphos D70-30C. [650] Sodium glyceryl oleate phosphate; anionic; emollient, emulsifier, substantive agent, and moisture barrier for personal care products; aerosol formulation and food processing surfactant; liq.; oil-sol.

Emphos D70-31. [650] Mono- and diglycerides phosphate ester; anionic; pigment dispersant for oil-base paints; w/o emulsifier, lubricant, release agent, and thickener in food processing; flake; oil-sol.

Emphos F27-85. [650] Hydrog. veg. glycerides phosphate; anionic; mold lubricant for food use; pigment dispersant in oil-based systems; moisture barrier; tan soft solid; insol. in water; sp.gr. 1.01; pour pt. 17 C; pH 6.9 (3% aq.).

Emphos PS-21A. [650] Alcohol ethoxylate phosphate ester; anionic; detergent base, emulsifier, foaming and wetting agent, dispersant, lubricant, used for detergent industry and industrial surfactants; liq.; oil-sol.

Emphos PS-121. [650] Linear alcohol ethoxylate phosphate ester; anionic; emulsifier/lubricant for cutting fluids; drycleaning detergent with antistat properties; textile wetting and scouring agent; clear liq.; sol. in ethanol, carbon tetrachloride, perchloroethylene, xylene, naphtha, kerosene, min. oil; disp. in water; sp.gr. 1.06; pour pt. < 10 C; pH 2.0 (3% aq.); surf. tens. 28.6 dynes/cm (0.05% aq.).

Emphos PS-220. [650] Linear alcohol ethoxylated phosphate ester; anionic; lubricant; emulsifier; polymer particle dispersant; clear liq.; sol. see Emphos PS-121; insol. in water; sp.gr. 1.02; pour pt. 10 C; pH 2.0 (3% aq.); surf. tens. 32.3 dynes/cm.

Emphos PS-222, PS-230. [652] Aliphatic phosphate esters; anionic; emulsifier for pesticide formulations, detergents, hydrotropes and emulsion polymerization; liq.

Emphos PS-236. [650] Alcohol ethoxylate phosphate ester; anionic; detergent base, emulsifier, coupling, foaming and wetting agent, corrosion inhibitor, lubricant, penetrant, and solubilizer used in the detergent industry, industrial surfactants, textile systems; metal cleaning and emulsion polymerization surfactant; liq.; sol. in naphthenic oils and water.

Emphos PS-400. [650; 652] Linear alcohol ethoxylate phosphate ester; anionic; polymer particle dispersant; monomer/water emulsifier; pesticides, detergents, emulsion polymerization; clear liq.; disp. in water; sp.gr. 1.00; pour pt. > 10 C; pH 2.0 (3% aq.); surf. tens. 30.6 dynes/cm (0.05% aq.).

Emphos PS-410. [652] Aliphatic phosphate ester; anionic; emulsifier for pesticide formulations, detergents, hydrotropes and emulsion polymerization; liq.

Emphos PS-415M. [650] Alcohol ethoxylate phosphate ester; anionic; industrial coupling and defoaming agent, lubricant, solubilizer; textile industry detergent, dispersant, scouring and wetting agent for aq. systems; liq.; water-sol.

Emphos PS-440. [650] Alcohol ethoxylate phosphate ester; anionic; industrial detergent, o/w emulsifier, lubricant, penetrant, solubilizer, coupling, foaming and wetting agent; metal processing surfactant; liq.; sol. in naphthenic oils and water.

Emphos PS-810. [650] Linear alcohol ethoxylate phosphate ester; anionic; emulsifier/lubricant for cutting fluids; lube additive; clear liq.; disp. in water; sp.gr. 1.03; pour pt. > 10 C; pH 2.0 (3% aq.); surf. tens. 37.6 dynes/cm (0.05% aq.).

Emphos PS-900. [650; 652] Alcohol ethoxylate phosphate ester; anionic; detergent, o/w emulsifier, and lubricant for use in metal processing; liq.; sol. in naphthenic and paraffinic oils.

Emphos TS-230. [650] Phenol ethoxylate phosphate ester; anionic; industrial corrosion inhibitor, defoamer, lubricant; metal processing surfactant for syn. oils; liq.; water-sol.

Empicol 0031/T. [17] DEA-lauryl sulfate; anionic; detergent used in personal care products; liq.; 34% conc.

Empicol 0045. [17] Sodium lauryl sulfate; anionic; foaming agent used in personal care products; wh. fine powd.; dens. 0.35 g/cm^3; pH 9.5–10.5 (1% aq.); 94.0% min. act.

Empicol 0185. [17] Sodium lauryl sulfate; anionic; detergent, wetting and rewetting agent, emulsifier, dispersant, disintegrator; used in personal care products, pharmaceuticals, dental care, textile printing inks, agric. and horticultural preparations; wh. powd.; dens. 0.35 g/cm^3; pH 9.5–10.5 (1% aq.); 94.0% min. act.

Empicol 0216. [17] Complex org. phosphate ester; anionic; foam stabilizer, conditioner, and antistatic agent for toiletries; detergent used in textile processing; liq.

Empicol 0919. [17] Sodium lauryl sulfate; wetting, foaming, and emulsifying agent used in personal care products, carpet and upholstery shampoos, and specialty cleaners; sp.gr. 1.05 ± 0.005; visc. 120 cP max.; pH 8.5 ± 0.7 (10%); 30.0 ± 1.0% act.

Empicol AL 30. [15; 17] Ammonium lauryl sulfate; anionic; detergent for shampoos and carpet shampoos; amber liq./paste; m.w. 291; visc. 14,000 cs; pH 6–7 (5%); 29% act.

Empicol AL-30T, 70. [17] Ammonium lauryl sulfate; anionic; detergent amd base used in the cosmetic industry; liq.; 30–70% conc.

Empicol BSD. [17] Sodium magnesium lauryl ethoxy sulfonate; anionic; detergent and toiletry raw material; liq.; 26% conc.

Empicol CHC 30. [17] Sodium cetyl oleyl sulfate; anionic; multipurpose textile detergent; cream soft paste; water sol.; pH 8.5 (2%); 30% act.

Empicol DA. [15; 17] DEA-lauryl sulfate; anionic; ingred. in personal care products, automobile cleaners; pale yel. visc. liq.; dilutable in water; sp.gr. 1.05; pH 7.0 ± 0.5

(10% aq.); 34.0 ± 1.0% act.

Empicol DLS. [17] DEA-lauryl sulfate; anionic; detergent used in liq. and lotion shampoos; straw clear liq.; m.w. 379 of act.; dens. 1.05 g/cc; visc. 500 cs; cloud pt. < 0 C; pH 7–8 (1% aq.); 34% act.

Empicol EAA70. [17] Ammonium laureth sulfate; anionic; detergent and toiletry raw material; base for personal care products; liq.; 70% conc.

Empicol EAB. [17] Ammonium laureth sulfate; anionic; detergent and base used in the cosmetic industry; liq. 23% conc.

Empicol EGB. [17] Magnesium laureth sulfate; anionic; detergent and toiletry raw material; liq.; 25% conc.

Empicol EL. [17] MEA-lauryl sulfate; anionic; detergent used in the mfg. of liq. shampoos; amber visc. liq.; sp.gr. 1.025; visc. 1143 cs (50%); cloud pt. –2 C; 50% act.

Empicol EMD. [17] Blend of natural, straight chain fatty alcohol; mild detergent for shampoos, baby products; pale straw liq.; almost odorless; cold water sol.; dens. 1.05 g/cc; visc. < 250 cs; cloud pt. < 0 C; pH 6.5–7.5 (5%); biodeg.; 30% act.

Empicol ESB. [15] Sodium laureth sulfate; anionic; shampoo ingred.; liq.; 28% conc.

Empicol ESB3. [17] Sodium laureth sulfate; anionic; base and foam stabilizer for personal care products; colorless-pale yel. liq.; dens. 1.0 g/cm^3; visc. 2000 ± 800 cP; cloud pt. 0 C; pH 7.0 ± 0.7 (5%); 28.0 ± 1.0% act.

Empicol ESB3GA. [15] Sodium laureth sulfate; foam stabilizer for toiletries; colorless to pale yel. clear visc. liq.; visc. 2750 ± 750 cP; pH 7.0 ± 0.5 (1%); 27.2 ± 1.2% act.

Empicol ESB 3/S. [17] Sodium laureth sulfate; shampoo base; colorless liq.; water sol.; dens. 1.01; visc. 2600 cs; cloud pt. < 0 C; biodeg.; 25% act.

Empicol ESB30. [17] Sodium laureth sulfate; see Empicol ESB3; colorless clear liq.; m.w. (of act.) 384; dens. 1.0 g/cm^3; visc. 20,000 cP; cloud pt. 0 C; pH 6.5–7.5 (5% aq.); 27.5% act.

Empicol ESB50. [17] Sodium laureth sulfate; see Empicol ESB3; pale straw hazy liq.; m.w. (of act.) 384; dens. 1.1 g/cm^3; pH 6.5–7.5 (5% aq.); 50.0 ± 1.0% act.

Empicol ESB70. [17] Sodium laureth sulfate; see Empicol ESB3; pale straw hazy liq.; dens. 1.1 g/cm^3; pH 7.0–9.0 (2% aq.); 68.0 ± 2.0% act.

Empicol ESB70-AU. [15; 17] Sodium laureth sulfate; anionic; see Empicol ESB3; pale yel. hazy liq.; dens. 1.1 g/cm^3; pH 8.0 ± 1.0 (2% aq.); 68.0 ± 2.5% act.

Empicol ESC3. [17] Sodium laureth sulfate; raw material used in the formulation of personal care products; colorless clear liq.; dens. 1.0 g/cm^3; visc. 35 cs; cloud pt. 0 C; pH 6.0–8.0 (5% aq.); 28.0 ± 1.0% act.

Empicol ESC70. [17] Sodium laureth sulfate; anionic; see Empicol ESC3; pale straw visc. hazy liq.; dens. 1.1 g/cm^3; pH 7.0–8.0 (5% aq.); 69.5 ± 2.0% act.

Empicol ESC70-AU. [15; 17] Sodium lauryl triethoxysulfate; anionic; base for personal care products; pale yel. visc. liq.; dilutable in water; sp.gr. 1.1; visc. 210 cP max.; cloud pt. 2 C max.; pH 7.2 ± 0.6 (2%); 68.0 ± 2.5% act.

Empicol ESC-AU. [15; 17] Sodium lauryl triethoxysulfate; anionic; surfactant used as a base in personal care products; pale yel. clear liq.; dilutable in water; sp.gr. 1.05; visc. 140 cP max.; pH 7.2 ± 0.6 (1% aq.); 27.5 ± 1.0% act.

Empicol ETB. [17] TEA-laureth sulfate; anionic; see Empicol EGB; liq.; 30% conc.

Empicol LM. [17] Sodium lauryl sulfate; anionic; see Empicol EGB; powd.; 86% conc.

Empicol LM45. [17] Sodium lauryl sulfate; primary act. ingred., detergent, and foaming agent used in personal care and household products; wh. paste; dens. 0.95 g/cm^3; pH 8.0–10.0 (2% aq.); 45% act.

Empicol LM/T. [17] Sodium lauryl sulfate; see Empicol LM45; wh. powd.; dens. 0.35 g/cm^3; pH 9.0–10.5 (1% aq.); 94% act.

Empicol LMV. [17] Sodium lauryl sulfate; see Empicol EGB; needles; 82% conc.

Empicol LMV/T. [17] Sodium lauryl sulfate; see Empicol LM45; wh. needles; dens. 0.5 g/cm^3; pH 9.0–10.5 (1% aq.); 90% act.

Empicol LQ33. [17] MEA-lauryl sulfate; anionic; surfactant used in the formulation of personal care products; lt. amber clear to opalescent visc. liq.; sp.gr. 1.05; visc. 17 cSt max.; cloud pt. 2 C max.; pH 7.0 ± 0.5 (5% aq.); 33.5% ± 1.0% act.

Empicol LQ33/T. [17] MEA-lauryl sulfate; anionic; surfactant in the mfg. of personal care products; emulsifier in the mfg. of rubber latices and for resins; bactericidal detergents; pale amber visc. liq.; dens. 1.03 g/cm^3 visc. 6000 cs; pH 7.0 ± 0.5 (5% aq.); 33.5 ± 1.0% act.

Empicol LQ 70. [17] MEA-lauryl sulfate; detergent and toiletry raw material; base for personal care products; liq.; 70% conc.

Empicol LS30. [17] Sodium lauryl sulfate; wetting, dispersing, emulsifying, and foaming agent for industrial processes, detergent/cleaner formulations; pale yel. clear visc. liq.; visc. 750 ± 300 cps (35 C); pH 8.5 ± 1.0 (10%); 30 ± 1% act.

Empicol LS30B. [15] Sodium lauryl sulfate; base for personal care products; pale yel. clear visc. liq.; visc. 800 ± 200 cP; pH 8.5 ± 1.0 (10%); 30.0 ± 1.0% act.

Empicol LS30P. [15; 17] Sodium lauryl sulfate; anionic; emulsifier, wetting and foaming agent for emulsion polymerization and carpet and upholstery shampoos; base for personal care shampoos; specialty cleaners and industrial processes; pale yel. clear liq.; sp.gr. 1.05; visc. 120 cP max.; pH 7.5 ± 0.7 (10% aq.); 30.0 ± 1.0% act.

Empicol LX. [17] Sodium lauryl sulfate; emulsifier in the mfg. of plastics, resins, and syn. rubbers; foaming agent for rubber foams, personal care products and carpet and upholstery shampoos; lubricant in mfg. of molded rubber goods; wh. powd.; dens. 0.35 g/cm^3; pH 9.5–10.5 (1% aq.); 90% act.

Empicol LX28. [17] Sodium lauryl sulfate; see Empicol LX; pale yel. liq., paste; dens. 1.05 g/cm^3; pH 8.0–9.5 (5% aq.); 28% act.

Empicol LXS95. [17] Sodium lauryl sulfate; anionic; detergent, foamer, emulsifier used in toothpastes, emulsion polymerization, rubber, pharmaceuticals; wh. powd.; dens 0.35 g/cm^3; pH 9.6–10.5 (1% aq.); 94.0% min. act.

Empicol LXV. [17] Sodium lauryl sulfate; see Empicol LX; wh. needles; dens. 0.5 g/cm^3; pH 9.5–10.5 (1% aq.); 85% act.

Empicol LY28/S. [17] Sodium lauryl sulfate; coprecipitant in mfg. of photographic film; emulsion polymerization in the plastic and rubber industries; foaming agent in carpet processes; pale yel. liq. to paste; dens. 1.05 g/cm^3; visc. 28 cs; cloud pt. –1 C; pH 8.5–9.5 (5% aq.); 29.0 ± 1.0% act.

Empicol LZ. [17] Sodium lauryl sulfate; detergent, wetting and foaming agent in personal care products; emulsifier in mfg. of rubbers, plastics, and resins by emulsion polymerization; foaming agent in mfg. of foam rubber goods; lubricant used in plastic goods; wh. powd.; dens. 0.35 g/cm^3; pH 9.5–10.5 (1% aq.); 89.0% min. act.

Empicol LZ34. [17] Sodium lauryl sulfate; wh. paste; dens. 1.05 g/cm^3; pH 8.5–9.5 (2% aq.); 34.0% min. act.

Empicol LZ/D. [15] Sodium lauryl sulfate; anionic; shampoo ingred., emulsion polymerization; powd.; 90% conc.

Empicol LZ/E. [17] Sodium lauryl sulfate; anionic; emulsifier in mfg. of plastics, resins, and syn. rubbers by emulsion polymerization; emulsifier and dispersant in textile printing pastes; dispersant/lubricant in extrusion of PVC; wh. powd.; dens. 0.35 g/cm^3; 90.0% min. act.

Empicol LZG 30. [17] Sodium lauryl sulfate; anionic; surfactant, foaming and wetting agent used in industrial processes; emulsifier in mfg. of syn. rubbers, plastics, and resins by emulsion polymerization; cream paste; dens. 1.05 g/cm^3; pH 8.0–9.5 (2% aq.); 30% act.

Empicol LZGV. [17] Sodium lauryl sulfate; anionic; see Empicol LZG 30; cream needles; dens. 0.5 g/cm^3; pH 9.5–10.5 (1% aq.); 85% act.

Empicol LZGV/C. [17] Sodium lauryl sulfate; see Empicol LZGV/C; cream needles; dens. 0.5 g/cm^3; pH 9.5–10.5 (1% aq.); 89% act.

Empicol LZP. [17] Sodium lauryl sulfate; see Empicol LZ; wh. powd.; dens. 0.35 g/cm^3; pH 9.5–10.5 (1% aq.); 89.0% min. act.

Empicol LZV. [17] Sodium lauryl sulfate; see Empicol LZ; wh. needles; dens. 0.5 g/cm^3; pH 9.5–10.5 (1% aq.); 85.0% min. act.

Empicol LZV/E. [17] Sodium lauryl sulfate; see Empicol LZ/E; wh. needles; dens. 0.5 g/cm^3; 90.0% min. act.

Empicol MD. [17] Alkyl ether sulfate; anionic; mild surfactant for use in personal care products; pale straw liq., almost odorless; cold water sol.; dens. 1.05 g/cm^3; visc. 50 cs; cloud pt. < 0 C; pH 6.5–7.5 (5% aq.); biodeg.; 30.0 ± 1.0% act.

Empicol MIPA. [15; 17] MIPA-lauryl sulfate; anionic; base for hair shampoos; pale yel. clear visc. liq.; visc. 22 ± 10 s; pH 7.5 ± 0.5 (5%); 30.0 ± 1.0% act.

Empicol ML 26. [17] Magnesium lauryl sulfate; anionic; detergent used in the mfg. of shampoos and toothpaste; pale straw liq.; misc. with water; dens. 1.0 g/cc; visc. 150 cs; 26% act.

Empicol SBB. [17] Disodium alkylolamide sulfosuccinate; anionic; raw material for toiletries and detergent; liq.; 29% conc.

Empicol SCC. [17] Disodium alkylolamide sulfosuccinate; anionic; see Empicol SBB; liq.; 35% conc.

Empicol SDD. [17] Disodium alkyl ethoxy sulfosuccinate; anionic; raw material for toiletries and detergents; liq.; 40% conc.

Empicol SEE. [17] Disodium undecylenic monoethanolamide sulfosuccinate; anionic; see Empicol SDD; liq.; 25% conc.

Empicol SFF. [17] Disodium alkyl ethoxy sulfosuccinate; anionic; see Empicol SDD; liq.; 30% conc.

Empicol SGG. [17] Disodium alkylolamide ethoxy sulfosuccinate; anionic; see Empicol SDD; liq.; 50% conc.

Empicol SHH. [17] Sodium monoethanolomine alkyl ethoxy sulfosuccinate; anionic; see Empicol SDD; liq.; 32% conc.

Empicol SLL, STT. [17] Disodium alkyl sulfosuccinate; anionic; raw material for toiletries and detergents; liq., paste; 32% conc.

Empicol TA40. [15; 17] Triethanolamine lauryl sulfate; ingred. for personal care products; pale yel. clear liq.; pH 7.0 ± 0.5 (5%); 40.0 ± 1.0% act.

Empicol TA40A. [15; 17] Triethanolamine lauryl sulfate; base for personal care products; water-wh. clear liq.; pH 7.2 ± 0.3 (10%); 40.5–42.0% act.

Empicol TAS30. [17] Sodium tallow sulfate; collector in the beneficiation of minerals by ore flotation; detergent ingred.; cofoaming agent for latex foam compds.; wh./straw paste; dens. 1.10 g/cm³; pH 8.0–10.0 (1%); 30–33% act. in water.

Empicol TAS90. [17] Sodium tallow sulfate; fabric scouring agent, sizing additive; cream powd.; pH 10; 90% act.

Empicol TAS80V. [17] Sodium tallow sulfate; heavy-duty detergent, wetting and emulsifying agent; cream needles; pH 10; 80% act.

Empicol TC30/T. [17] Triethanolamine ammonium lauryl sulfate; anionic; detergent, foaming agent for personal care products; pale yel. visc. liq.; dens. 1.05 g/cm³; visc. 26,000 cs; cloud pt. 0 C; pH 6.5 ± 0.5 (5% aq.); 21–28% act.

Empicol TC 34. [17] Ammonium lauryl sulfate and TEA-lauryl sulfate; anionic; detergent used in shampoos, foaming agent; clear, amber liq.; m.w. 317; visc. 27,000 cs; pH 6.2–6.9 (5%); 31.5% act.

Empicol TCR. [17] Ammonium lauryl sulfate, TEA-lauryl sulfate and lauramide MIPA; anionic; detergent used in shampoos, foaming agent; clear golden yel. liq.; m.w. 317; visc. 12,500 cs; pH 6–7 (5%); 21.5% act.

Empicol TCR/T. [17] Lauryl sulfate mixed ammonium/triethanolamine salts; foaming agent for personal care products; pale yel. clear liq.; dens. 1.05 g/cm³; visc. 12,000 cs; cloud pt. 0 C; pH 6.5 ± 0.5 (5% aq.); 21% act.

Empicol TDL 75. [17] TEA-lauryl sulfate; raw material for shampoo, bubble baths and oils; yel. to amber soft paste; m.w. 410 (of act.); water sol.; dens. 1.1 g/cc; pH 7.3–7.7 (2%); 75% act.

Empicol TL 40. [17] TEA-lauryl sulfate; detergent used in liq. and lotion shampoos; pale yel., clear liq.; misc. with water; dens. 1.025 g/cc; visc. 300 cs; pH 6.8–7.1 (5%); 40% act.

Empicol TL40/S. [17] Triethanolamine salt of sulfated fatty alcohol; detergent, foaming agent, used as liq. and lotion shampoo base; pale yel., clear liq.; dens. 1.025 g/cc; 38% act.

Empicol TL40/T. [17] TEA-lauryl sulfate; anionic; surfactant for personal care products; emulsifier in the mfg. of rubber latices; pale yel. clear liq.; dens. 1.08 g/cm³; visc. 45 cs; cloud pt. < 0 C; pH 7.0 ± 0.5 (5% aq.); 41% act. in water.

Empicol TLP. [17] TEA-lauryl sulfate and lauramide MIPA; detergent used for shampoos, as foam booster, stabilizer and visc. modifier; golden yel. clear liq.; misc. with water; dens. 1.025 g/cc; visc. 1000 cs; pH 6.8–7.1 (5%); 40.5% act.

Empicol TLP/T. [17] TEA-lauryl sulfate; anionic; built product used in personal care products; pale yel. clear liq.; dens. 1.07 g/cm³; visc. 450 cs; cloud pt. < 0 C; pH 7.0 ± 0.5 (5% aq.); 39% act. in water.

Empicol TLR. [17] TEA-lauryl sulfate and lauramide MIPA; see Empicol TLP; visc. 4000 cs; pH 6.5–7.2 (5%); 35.5% act.

Empicol TLR/T. [17] TEA-lauryl sulfate; anionic; see Empicol TLP/T; pale yel. clear liq; dens. 1.06 g/cm³; visc. 2200 cs; cloud pt. < 0; pH 7.0 ± 0.5 (5% aq.); 35% act. in water.

Empicol WA. [17] Sodium lauryl sulfate; industrial surfactant used in rubber foams for carpet backing; pale yel. clear liq.; m.w. 292 (of act.); sp.gr. 1.04; pH 8.5-9.5 (5%); 27% act.

Empicol WAK. [15] Sodium lauryl sulfate; anionic; emulsifier in refining veg. oils; base for personal care products; wetting and foaming agent in specialty cleansers and industrial processes; pale yel. clear liq.; sp.gr. 1.050; pH 8.5 ± 0.7 (10% aq.); 35.0 ± 1.0% act.

Empicol XC35. [17] Blend of anionic and

nonionic components; shampoo base, formulations should contain preservative; visc. pearly liq.; almost odorless.

Empicol XC 35/S. [17] Blend of anionic and nonionic components; shampoo base, formulations should contain preservative; wh. liq.; almost odorless.

Empicol XDB. [17] Detergent complex; detergent, foaming and conditioning agent for shampoos; golden clear liq.; visc. 4000 cs; cloud pt. < 0 C; pH 7.5–8.5 (5%); 30% solids.

Empicol XM 17. [17] Anionic; detergent in carpet shampoo double conc.; pale amber liq.; visc. 3000 cs; cloud pt. 5 C; pH 6–7 (5%); 21% anionic act.

Empicol XT 45. [17] MEA-lauryl sulfate; shampoo base, act. ingredient for lacquer removing hair shampoo; pale yel. mobile liq.; sp.gr. 1.05; visc. 100 cs; cloud pt. < 0 C; pH 9.0–9.5 (5%); 26–27% act.

Empicryl APD. [17] Maleic anhydride/diisobutylene copolymer, disodium salt, aq. sol'n.; pigment dispersant for water based paints, pigment stabilizer; pale yel. clear liq.; sp.gr. 1.12 (18 C); visc. 16 cs; f.p. –3 C; pH 12.0–12.7.

Empicryl APD90. [17] Maleic anhydride/diisobutylene copolymer, sodium salt; dispersing agent for pigments and extenders in aq. systems; agric. applications; wh. powd.; dens. 0.35 g/cm^3; pH 10.0–10.5 (25% aq.); 90.0% solids.

Empicryl APD/B. [17] Maleic anhydride/diisobutylene copolymer, disodium salt, aq. sol'n.; see Empicryl APD; pH 10.0–10.5.

Empicryl APD B90. [17] Maleic anhydride/diisobutylene copolymer, sodium salt; see Empicryl APD90; wh. powd.; dens. 0.35 g/cm^3; pH 12.0–12.7 (25% aq.); 90.0% total solids.

Empigen 2125-AU. [15] Soya diethanolamide; nonionic; co-active ingred. in toiletries; liq.; 80% conc.

Empigen AB. [17] Dimethyl lauramine; precursor for mfg. of derivs.; catalyst for polyurethane foam, resin curing agent, corrosion inhibitor, and flotation aid; pale straw liq.; char. odor; dens. 0.79 g/cm^3; 91.0% min. tert. amine.

Empigen AD, AF, AG. [17] Tert. alkyl dimethyl amine; cationic; intermediate for amine deriv.; resin curing agents; corrosion inhibitor; ore flotation chemicals; liq.; 97% conc.

Empigen AH. [17] Alkyl dimethyl amine, dist.; see Empigen AB; pale straw liq.; char. odor; dens. 0.79 g/cm^3; 91.0% min. tert. amine.

Empigen AM. [17] Alkyl dimethyl amine, dist.; cationic; see Empigen AB; pale straw waxy solid; char. odor; dens. 0.84 g/cm^3; 96.0% min. tert. amine.

Empigen AS. [17] Alkyl amido propyl dimethyl amine; cationic; industrial corrosion inhibitor for cutting oil formulations, oil drilling, and refinery operations; chemical intermediate for surfactants; amber liq., paste; char. odor; dens. 0.91 g/cm^3; 95% conc.

Empigen AT. [17] Alkyl amide propyl dimethyl amine; cationic; see Empigen AS; amber liq./paste; char. odor; dens. 0.95 g/cm^3; 94% conc.

Empigen AY. [17] Tert. alkyl ethoxy dimethyl amine; cationic; see Empigen AB; amber liq.; dens. 0.89 g/cm^3; 94.0% min. tert. amine.

Empigen BAC. [17] Benzalkonium chloride; disinfectant for dairy, food processing, restaurant, brewing, and bottling industries, medical, pharmaceutical uses; pale yel. liq.; misc. with water, alcohol, acetone; 50% act.

Empigen BB. [17] Lauryl betaine; amphoteric; foam booster/stabilizer, wetting agent, thickening agent, used for shampoos and detergents; almost water wh. liq.; 30% conc.

Empigen BB-AU. [17] Alkyl betaine; amphoteric; foam booster/stabilizer, thickener, emulsifier, dispersant, wetting and foaming agent; used in personal care products, industrial and institutional cleaners; pale yel. liq.; sp.gr. 1.04; cloud pt. 1 C max.; pH 8.5 ± 1.0 (5% aq.); 30% act. in water.

Empigen BCB50. [17] Alkyl benzyl dimethyl ammonium chloride; cationic; retarder in dyeing of acrylic fibers; pale yel. liq.; dens. 0.98 g/cm^3; visc. 120 cs; cloud pt. < 0 C; pH 7.0–9.5 (5% aq.); 50.0 ± 1.5% act. in water.

Empigen BCF 80. [19] Alkyl benzyl dimethyl ammonium chloride; cationic; used in industrial application; liq.; 80% conc.

Empigen BCJ-50. [17] Oleyl dimethyl benzyl ammonium chloride; basic ingred. in personal care products; yel. clear liq.; sp.gr. 0.977; visc. 450 cP; cloud pt. < 0 C; pH 4.0 ± 1.0 (1% aq.); 50.0 ± 3.0% quat. act. in water.

Empigen BCM75, BCM75/A. [17] Hydrog.-tallow dimethyl benzyl ammonium chloride; cationic; used in mfg. of bentones; flotation aid; wh. waxy solid; m.p. 60 C; flash pt. 40 C

(CC); pH 6.0–8.5 (5%); 75.0–80.0% act.

Empigen BCY40. [17] Alkyl ethoxy benzyl dimethyl ammonium chloride; retarder in the dyeing of acrylic fibers; pale yel. liq.; dens. 0.99 g/cm^3; visc. 400 cs; cloud pt. < 0 C; pH 7.0–9.5 (5% aq.); 40.0 ± 2.0% act. in water.

Empigen BS. [17] Alkyl amido betaine; amphoteric; foam booster and coactive agent for toiletries and detergents; liq.; 30% conc.

Empigen BT. [17] Alkyl amido propyl dimethyl amine betaine; amphoteric; surfactant for personal care products; pale amber liq.; bland odor; dens. 1.0 g/cm^3; pH 4.5–8.0 (5% aq.); 30.0 ± 1.0% act. in water.

Empigen CDL 10. [17] Lauric imidazoline betaine; amphoteric; detergent used in personal care products; liq.

Empigen CDL 30. [17] Lauric imidazoline betaine, modified; amphoteric; see Empigen CDL 10; liq.

Empigen CDR10. [17] Coconut imidazoline betaine; amphoteric; surfactant used in personal care products, household and industrial cleaners; amber visc. liq.; dens. 1.20 g/cm^3; visc. < 500 cs; pH 8.7–9.3 (10% aq.).

Empigen CDR30. [17] Coconut imidazoline betaine, modified; amphoteric; amber visc. liq.; dens. 1.20 g/cm^3; visc. 40,000–180,000 cs; pH 8.2–8.8 (10% aq.).

Empigen CHB. [17] Alkyl trimethyl ammonium chloride; cationic; pharmaceutical material; powd.; 96% conc.

Empigen CM. [17] Alkyl trimethyl ammonium methosulfates; cationic; antistat and conditioning agent for personal care products; retarder for dyeing of acrylic fibers; pale straw cloudy liq.; dens. 0.96 g/cm^3; flash pt. 40 C; cloud pt. 23 C; pH 6.5–8.5 (5% aq.); 30.0 ± 1.5% act.

Empigen FRC75S. [17] Alkyl imidazoline methosulfate; textile conditioning agent, antistat, and lubricant for fibers; pale amber hazy liq.; disp. in water; dens. 0.95 g/cm^3; visc. 300 cp; flash pt. 23 C (CC); pH 4.0–6.5 (8%); 75.0 ± 2.0% act.

Empigen OB. [17] Lauramine oxide; coactive, foam booster/stabilizer and visc. modifier for personal care products, fire fighting foam concs.; pale straw liq.; dens. 0.98 g/cm^3; visc. 25 cs; pH 7.5 ± 0.5 (5% aq.); 30.0 ± 1.5% act.

Empigen OH. [17] *n*-Myristyl dimethyl amine oxide; see Empigen OB; pale straw liq.; dens. 0.99 g/cm^3; visc. 900 cs; pH 7.5 ± 0.5 (5% aq.); 30.0 ± 1.5% act.

Empigen OH25. [17] Alkyl amine oxide; amphoteric; foam booster and coactive agent for toiletries and detergent; bleach additive; liq.; 25% conc.

Empigen OS/A. [17] Alkyl amido amine oxide; amphoteric; foam booster and coactive agent for toiletries and detergents; liq.; 30% conc.

Empigen OY. [17] PEG-3 lauramine oxide; see Empigen OB; pale straw liq.; dens. 1.0 g/cm^3; visc. 23 cs; pH 6.5 ± 0.5 (5% aq.); 25.0 ± 1.0% act.

Empigen XDR 112. [17] Coconut imidazoline betaine binary deriv.; amphoteric; detergent used for personal care products; liq.

Empigen XDR121. [17] Coconut imidazoline betaine with sodium lauryl ethoxy sulfate; used in personal care products; pale straw visc. liq.; dilutable with water; dens. 1.1 g/cm^3; pH 8.2–8.8.

Empigen XDR123. [17] Coconut imidazoline betaine with sodium lauryl ethoxy sulfate; see Empigen XDR121; pale straw liq.; dilutable with water; dens. 1.1 g/cm^3; med. visc.; pH 8.2–8.8.

Empigen XDR302. [17] Coconut imidazoline betaine with sodium lauryl sulfate; see Empigen XDR121; pale straw liq.; dilutable with water; dens. 1.1 g/cm^3; pH 7.4–7.8.

Empilan 0004. [17] Castor oil EO/PO condensate; nonionic; emulsifier and lubricant for formulation of cutting oils, grinding fluids, and textile lubricants; mfg. of brake fluids; pale amber liq.; visc. 550 cs; HLB 12.0; sapon. no. 67.5 ± 3.5.

Empilan 2020. [17] Lauryl alcohol ethoxylate; stabilizer for latices produced by emulsion polymerization; wh./cream solid; dens. 1.05 g/cm^3; m.p. 35 C; HLB 15.7; hyd. no. 64 ± 4.

Empilan 2125-AU. [17] Soya bean diethanolamide; visc. modifier, superfatting agent, and coactive ingred. in personal care products; amber clear visc. liq.; sp.gr. 1.01; pH 10.5 ± 1.0 (10% aq.).

Empilan 2502. [17] Coconut diethanolamide; nonionic; foam booster/stabilizer; liq.; 100% conc.

Empilan BQ 100. [17] PEG-8 oleate; emulsifier for insecticides, spindle oils, industrial wetting agent, scouring agent, antifoam and antistat, in laundry works, glue mfg., paper coating, hand cleaning jellies, turbine oil additive; dk. br. liq; faint, typ. odor; m.w. 400 (of glycol base); disp. in water, yields clear sol'ns. with ethanol, xylene, oleic acid; sp.gr. 1.017; visc. 130 cps; m.p. –1 C; 100% act.

Empilan CDE. [17] Cocamide DEA; nonionic; foam boosting/stabilizing agent for use in personal care and detergent products; antistat in plastics; pale yel. visc. liq.; dens. 1.0 g/cm³; 100% act.

Empilan CDEY. [15; 17] Coconut diethanolamide; nonionic; foam booster and visc. modifier in hair shampoos; pale yel. clear visc. liq.; 78.0% conc.

Empilan CDX. [17] Cocamide DEA; nonionic; foam booster/stabilizer, solubilizer, and visc. modifier for toiletry and detergent formulations; antistatic agent in plastics; pale yel. visc. liq.; dens. 1.0 g/cm³.

Empilan CIS. [17] Cocamide MIPA; nonionic; see Empilan CME; cream waxy flake; dens. 0.4 g/cm³; m.p. 46 C; 100% conc.

Empilan CM. [15; 17] Coconut monoethanolamide; nonionic; foam booster/stabilizer in toiletry and detergent formulations; intermediate; cream wax-like solid.

Empilan CME. [17] Cocamide MEA; nonionic; foam booster/stabilizer in detergent systems; stabilizer for hair and carpet shampoos; base for mfg. of ethoxylated alkylolamides; cream waxy flake; disp. in hot water; dens. 0.4 g/cm³; m.p. 68 C; 94.2% act.

Empilan DL 40. [17] Fatty alcohol ethoxylate; emulsifier for fatty acids and min. oils for fiber lubricants, wetting and stabilizing agent for syn. rubber latices, textile dye leveling and disp. agent in textile industry; pale yel. clear liq.; misc. with water; dens. 1.046 g/cc; cloud pt. > 100 C in dist. water; 40% act.

Empilan DL 100. [17] Fatty alcohol ethoxylate; see Empilan DL 40; stiff waxy solid; warm water sol.; m.p. 36 C; cloud pt. > 100 C (1% aq.); 100% act.

Empilan EGMS. [17] Glycol stearate; opacifier/pearling and emulsifying/stabilizing agent in shampoos; wh. hard, wax-like flake; oil-sol.; m.p. 56 C; 65% diester, 30% monoester.

Empilan FD. [17] Coconut diethanolamide; nonionic; surfactant for mfg. of personal care products, industrial and institutional liq. detergent systems; coactive and visc. modifier for laundry detergents and hard surface cleaners; pale yel. clear visc. liq.; sp.gr. 1.00–1.01; 80% conc.

Empilan FD20. [15; 17] Coconut diethanolamide; nonionic; act. ingred. in liq. detergents; foam booster and visc. modifier; pale yel. clear visc. liq.; 65% conc.

Empilan FE. [15; 17] Coconut diethanolamide; nonionic; foam booster, visc. modifier used as coactive ingred. in toiletries; pale yel. clear visc. liq.; sp.gr. 0.970; visc. 250 cP; flash pt. 28 C (PMCC); 90% conc.

Empilan GMS LSE32. [17] Glyceryl stearate SE; nonionic; food grade emulsifier; wh. wax-like powd.; m.w. 358 (of monoglyceride); m.p. 55–60 C; pH 7–9 (10% aq.); 32.5% min. monoglyceride.

Empilan GMS LSE40. [17] Glyceryl stearate SE; emulsifier in the baking and food industry; lubricant in production of PVC sheets and expanded polystyrene; wh. microbead powd.; dens. 0.5 g/cm³; m.p. 55–65 C; 40.0% min. monoglyceride.

Empilan GMS LSE80. [17] Glyceryl stearate SE; see Empilan GMS LSE40; wh. microbead powd.; dens. 0.5 g/cm³; m.p. 64 C; 80.0% min. monoglyceride.

Empilan GMS MSE40. [17] Glyceryl stearate SE; see Empilan GMS LSE40; wh. microbead powd.; dens. 0.5 g/cm³; m.p. 55–60 C; 36.0% min. monoglyceride.

Empilan GMS NSE32. [17] Glyceryl stearate; nonionic; food grade emulsifier; wh. wax-like powd.; m.p. 54–59 C; pH 6–8 (10% aq.); 32.5% monoglyceride.

Empilan GMS NSE40. [17] Glyceryl stearate SE; nonionic; emulsifier in the baking and food industry; lubricant and antistat in production of PVC sheets and expanded polystyrene; stabilizer; wh. microbead powd.; dens. 0.5 g/cm³; m.p. 55–60 C; pH 6–8 (10% aq.); 40.0% min. monoglyceride.

Empilan GMS NSE90. [17] Glyceryl stearate SE; see Empilan GMS NSE40; wh. microbead powd.; dens. 0.5 g/cm³; m.p. 65 C; 90.0% min. monoglyceride.

Empilan GMS SE32. [17] Glyceryl stearate SE; nonionic; emulsifier; wh. wax-like powd.; m.p. 55–60 C; pH 7–9 (10% aq.); 32.5% min. monoglyceride.

Empilan GMS SE40. [17] Glyceryl stearate SE; see Empilan GMS LSE40; also emulsifier for cosmetic and pharmaceuticals; wh. microbead powd.; dens. 0.5 g/cm³; m.p. 55–60 C; pH 7–9 (10% aq.); 36.0% min. monoglyceride.

Empilan GMS SE70. [17] Glyceryl stearate SE; see Empilan GMS LSE40; wh. microbead powd.; dens. 0.5 g/cm³; m.p. 61 C;

70.0% min. monoglyceride.

Empilan KA3. [17] Fatty alcohol EO condensate; nonionic; scouring and wetting agents for textiles, emulsifier, dye leveling and dispersing agent, detergent, in insecticides and pesticides; pale straw clear liq.; sol. in min. oils and other org. solvs.; HLB 8.9; biodeg.; 100% act.

Empilan KA5. [17] Fatty alcohol EO condensate; see Empilan KA3; HLB 11.5; cloud pt. 37 C; biodeg.; 100% act.

Empilan KA590. [17] Fatty alcohol EO condensate; nonionic; see Empilan KA3; water wh. clear liq.; dens. 9.8 lb/gal; HLB 11.4; cloud pt. 95 F; biodeg.; 90% act.

Empilan KA880. [17] Fatty alcohol EO condensate; see Empilan KA3; HLB 13.6; cloud pt. 81 C; biodeg.; 80% act.

Empilan KA1080. [17] Fatty alcohol EO condensate; see Empilan KA3; HLB 14.6; cloud pt. 95 C; biodeg.; 80% act.

Empilan KB 2. [17] Laureth-2; nonionic; emulsifier, foam booster, superfatting agent, used in detergents and emulsifying systems; almost colorless liq.; insol. in cold water; dens. 0.9; visc. 25 cs; 100% act.

Empilan KB 3. [17] Laureth-3; nonionic; see Empilan KB 2; liq.; 100% conc.

Empilan KC 3. [17] Laureth-3; nonionic; see Empilan KB 2; liq.; 100% conc.

Empilan KCMP 0703/F. [17] Fatty alcohol mixed EO/PO condensate; nonionic; emulsifier, detergent, wetting agent for industrial and domestic applications; rinse aids for dishwashing machines, mfg. of latices, metal cleaning; straw clear liq.; sol. in water; dens. 0.96 g/cm^3; HLB 14.0; cloud pt. 10–15 C (1% aq.); pH 4.0–7.5 (5% aq.).

Empilan KCMP 0705/F. [17] Fatty alcohol mixed EO/PO condensate; see Empilan KCMP 0703/F; pale straw clear liq.; sol. in water; dens. 0.97 g/cm^3; HLB 14.4; cloud pt. 23–27 C (1% aq.); pH 4.0–7.5 (5% aq.).

Empilan KL 6. [17] Oleth-6; emulsifier for min. oils in cosmetics, detergent and wetting agents in textile industry; cream soft paste; cold water disp.; dens. 0.9 g/cc; m.p. 35 C; pH 8–10; 100% act.

Empilan KL 10. [17] Oleth-10; see Empilan KL 6; m.p. 40 C; pH 8–10; 100% act.

Empilan KL 20. [17] Oleth-20; see Empilan KL 6; cream soft solid; cold water disp.; dens. 0.9 g/cc; m.p. 42 C; pH 8–10; 100% act.

Empilan KM 11. [17] Steareth-11; nonionic; emulsifier, foam control agent in syn. heavy duty detergents, soap additive; colorless to pale straw paste; 100% act.

Empilan KM 20. [17] Ceteareth-20; nonionic; wetting agent, detergent for industrial/domestic applications; emulsifier, foam control agent in syn. heavy duty detergents; colorless to pale straw waxy flake/block; 100% act.

Empilan KM 50. [17] Steareth-50; nonionic; see Empilan KM 20; colorless to pale straw waxy flake/block; 100% act.

Empilan LDE. [17] Lauramide DEA; nonionic; foam stabilizer, solubilizer, detergent used in shampoos and liq. detergent formulations; cream solid; dens. 1.0 g/cm^3.

Empilan LDX. [17] Lauramide DEA and diethanolamine; nonionic; see Empilan CDX; cream solid; dens. 1.0 g/cm^3; 100% conc.

Empilan LIS. [17] Lauramide MIPA; nonionic; foam booster/stabilizer for liq. and powd. detergents, visc. modifier, base for mfg. of ethoxylated deriv.; cream waxy flake; hot water disp.; dens. 0.4 g/cc; 87.5% act.

Empilan LME. [17] Lauramide MEA; see Empilan CME; waxy flake; dens. 0.4 g/cm^3; m.p. 85 C.

Empilan LP2. [17] EO condensate of Empilan CME; nonionic; liq. detergent additive as foam stabilizer, detergent booster, solubilizer; cream yel. soft paste; faint char. odor; sapon. no. 14; 100% act.

Empilan LP10. [17] Fatty acid alkylolamide ethoxylate; nonionic; foam and detergent booster; paste; 100% conc.

Empilan MAA. [17] PEG-6 cocamide; nonionic; see Empilan LP2; paste; sapon. no. 15; 100% act.

Empilan NP9. [17] Nonoxynol-9; nonionic; wetting agent, detergent, emulsifier, solubilizer; pale straw soft paste/liq.; dens. 1.0 g/cc; visc. 300 cs; cloud pt. 55 C (1% aq.); 100% act.

Empimin AO-60. [15] Ammonium alkyl ether sulfate; anionic; detergent and wetting agent; liq.

Empimin AQ60. [15; 17] Ammonium alkyl triethoxysulfate; anionic; foaming and wetting agent, dispersant in institutional and industrial cleaning specialties; pale yel. clear liq.; dilutable in water; sp.gr. 1.02; flash pt. 30 C (PMCC); pH 7.0 ± 0.6 (10% aq.); 58.0 ± 3.0% act.

Empimin KSN27. [17] Sodium laureth sulfate;

anionic; raw material for mfg. of liq. detergents and cleaning compositions; straw clear liq.; dens. 1.05 g/cm³; visc. 100 cs; cloud pt. 0 C; pH 6.5–7.5 (5% aq.); biodeg.; 27.0 ± 1.0% act.

Empimin KSN60. [17] Sodium laureth sulfate; anionic; see Empimin KSN27; pale amber translucent liq.; dens. 1.0 g/cm³; visc. mobile liq.; visc. 2500–3000 cs; flash pt. 38 C; pH 7.0–8.0 (5% aq.); biodeg.; 60.0 ± 1.5% act.; 10% ethanol.

Empimin KSN70. [17] Sodium laureth sulfate; see Empimin KSN27; straw opaque liq.; dens. 1.1 g/cm³; visc. liq.; pH 7.0–9.0 (2% aq.); 70.0 ± 2.0% act.

Empimin LAM. [17] Ammonium alkyl ether sulfate; anionic; foaming agent used in mfg. of plasterboard; liq.; 30% conc.

Empimin LR28. [17] Sodium lauryl sulfate; anionic; foaming agent in rubber latex systems; emulsifier for emulsion polymerization; yel./amber liq.; dens. 1.04 g/cm³; cloud pt. < 0 C; pH 7.0–9.0 (5% aq.); 28.0% min. act.

Empimin LSM30. [17] Sodium alkyl ether sulfate; anionic; foaming agent for slurries in plasterboard mfg.; foam-boosting additive in latices for carpet backing; wetting agent; ingred. in alkaline liq. cleaners; pale yel. liq.; dens. 1.06 g/cm³; visc. 40 cs; cloud pt. < 0 C; pH 8.0 ± 1.0 (5% aq.); 30.0 ± 1.0% act. in water.

Empimin LSM30-AU. [15; 17] Sodium alkyl ethoxysulfate; anionic; see Empimin LSM30; liq.; sp.gr. 1.05; visc. 40 cSt; pH 8.0 ± 1.0 (5%); cloud pt. < 0 C; 30.0 ± 1.0% act.

Empimin MA. [17] Sodium dihexyl sufosuccinate; anionic; emulsifier, wetting agent for emulsion polymerization; straw clear liq.; sol. in water; dens. 1.1 g/cm³; visc. 80 cs; flash pt. 30 C (CC); pH 6.0 ± 1.0 (5%); 63.0% min. act.

Empimin MHH. [17] Disodium N-cocoyl sulfosuccinate; foaming agent for rubber latex compds.; emulsifier in the mfg. of polymers by emulsion polymerization; dens. 1.1 g/cm³; pH 8.0 ± 0.5 (5% aq.); 32.5% min. act.

Empimin MKK. [17] Disodium N-stearyl sulfosuccinamate; anionic; foaming agent for rubber latices; pale cream soft paste; pH 8 (5% aq.); 35% act.

Empimin MKK98. [17] Disodium N-cetyl stearyl sulfosuccinate; anionic; see Empimin MHH; pale cream powd.; dens. 0.3 g/cm³; pH 8.5 ± 0.5 (5% aq.); 98% act.

Empimin MKK/L. [17] Disodium N-cetyl stearyl sulfosuccinamate; anionic; see Empimin MHH; pale cream opaque liq.; dens. 1.1 g/cm³; pH 8.25 ± 0.25 (5% aq.); 28.0% min. act.

Empimin MSS. [17] Diammonium N-cocoyl sulfosuccinate; see Empimin MHH; yel./pale cream clear/opaque liq.; dens. 1.1 g/cm³; pH 6.5 ± 0.5 (5% aq.); 31.0% min. act.

Empimin MTT. [17] Disodium N-oleyl sulfosuccinamate; foaming agent for rubber latices; pale cream liq.; pH 8 (5% aq.); 40% act.

Empimin MTT/A. [17] Disodium N-oleyl sulfosuccinate; anionic; see Empimin MHH; yel./pale cream clear/opaque liq.; dens. 1.1 g/cm³; pH 8.0 ± 0.5 (5% aq.); 28.0% min. act.

Empimin OP45. [17] Sodium dioctyl sulfosuccinate; anionic; emulsifier, dispersant, wetting agent used for emulsion polymerization, oil slicks, textiles; straw clear liq.; sol. in water; dens. 1.05 g/cm³; visc. 100 cs; flash pt. > 100 C (CC); pH 6.0 ± 1.0; 44% conc.

Empimin OP70. [17] Sodium dioctyl sulfsuccinate; anionic; see Empimin OP45; straw clear liq.; sol. in water; dens. 1.10 g/cm³; visc. 450 cs; flash pt. > 100 C (CC); pH 6.0 ± 1.0 (5%); 72.0 ± 2.0% act.

Empimin OT. [17] Sodium dioctyl sulfosuccinate; anionic; emulsifier, wetting agent for o/w emulsions, emulsion polymerization, filler and extender dispersions; straw clear liq.; dens. 1.0 g/cm³; visc. 55 cs; flash pt. 25 C (CC); pH 6.0 ± 1.0 (5%); 60.0% min. act.

Empimin SDS. [15; 17] Sodium decyl sulfate; anionic; emulsifier, dispersant, detergent, and wetting agent for industrial and institutional cleansers, mfg. of pigments, alkaline cleansers; dust suppression; pale yel. clear liq.; sp.gr. 1.06; pH 7.0 ± 0.6 (5% aq.); 30.0 ± 1.0% act.

Empimin SQ25. [15; 17] Sodium alkyl triethoxysulfate; dispersant, wetting and foaming agent in institutional and industrial cleaners; pale yel. clear liq.; dilutable in water; sp.gr. 1.043 ± 0.005; visc. 80 ± 45 cP; pH 7.0–0.7 (5% aq.); 25.0 ± 0.5% act.

Empimin SQ70. [15; 17] Sodium alkyl triethoxylsulfate; see Empimin SQ25; off-wh. to pale yel. hazy visc. liq.; dens. 1.1 g/cm³; visc. 550 cP max.; cloud pt. 0 C max.; pH 8.0 ± 0.6 (2% aq.); 68.0 ± 2.0% act.

Empiwax SK. [17] Cetearyl alcohol and sodium lauryl sulfate; self-emulsifying wax as o/w emulsifier for pharmaceutical and toilet preparations and ointments; wh. flake.

Empiwax SK/BP. [17] Cetearyl alcohol and sodium lauryl sulfate; see Empiwax SK; complies with B.P. specs. for emulsifying wax; wh. flake.

Emplex. [462] Sodium stearoyl lactylate; anionic; starch and protein complexing agent for bakery products; emulsifier and conditioning agent for processed foods; lt. tan powd.; mild caramel odor; sol. in water and oil; 100% conc.

Emrite 6003. [217] GMS, food grade; nonionic; lipophilic emulsifier used in food industry; bead; HLB 3.9; 100% conc.

Emrite 6008. [217] GMO, food grade; nonionic; see Emrite 6003; liq.; HLB 3.4; 100% conc.

Emrite 6105. [217] SMS; nonionic; see Emrite 6003; bead; HLB 5.2; 100% conc.

Emrite 6120. [217] Polysorbate 80; nonionic; food emulsifier; liq.; HLB 15.0; 100% conc.

Emrite 6125. [217] Polysorbate 60; nonionic; food emulsifier; HLB 15.2; 100% conc.

Emsorb 2500. [217] Sorbitan oleate; nonionic; emulsifier, coupling agent, lubricant, softener used in textile fibers, leather, cosmetics, and household products; Gardner 8 liq.; sol. in min. oil, butyl stearate, perchloroethylene, Stod.; dens. 8.3 lb/gal; visc. 1000 cs; HLB 4.6; 100% conc.

Emsorb 2502. [217] Sorbitan sesquioleate; nonionic; primary emulsifier for w/o systems, coemulsifier for o/w systems, coupling agent; Gardner 6 liq.; sol. in min. oil, butyl stearate, perchloroethylene, Stod., water insol.; dens. 8.2 lb/gal; visc. 1500 cs; HLB 3.7; 100% conc.

Emsorb 2503. [217] Sorbitan trioleate; nonionic; coupling agent, emulsifier, used for processing and softening of textiles and leather; Gardner 7 liq.; sol. in min. oil, butyl stearate, perchloroethylene, Stod.; dens. 7.9 lb/gal; visc. 200 cs; HLB 2.7; 100% act.

Emsorb 2505. [217] Sorbitan stearate; nonionic; hydrophobic emulsifier, coemulsifier for industrial oils, household products, cosmetics; textile lubricant; Gardner 4 solid; disp. in butyl stearate, perchloroethylene, water insol.; m.p. 54 C; HLB 5.4; 100% conc.

Emsorb 2507. [217] Sorbitan tristearate; hydrophobic emulsifier similar to Emsorb 2503, fiber-to-metal textile lubricant; Gardner 4 waxy solid; sol. in butyl stearate, perchoroethylene, Stod., disp. in min. oil, water insol.; m.p. 54 C; HLB 2.4; 100% conc.

Emsorb 2510. [217] Sorbitan monopalmitate; emulsifier for cosmetics and household products, fiber-to-metal lubricant for syn. and cellulosic fibers; Gardner 8 waxy solid; sol. in oil, Stod.; disp. in butyl stearate; m.p. 48 C; HLB 6.6; 100% conc.

Emsorb 2515. [217] Sorbitan monolaurate; nonionic; emulsifier with antifoam properties for household specialties, industrial oils, cosmetics and emulsion polymerization; Gardner 7 liq.; sol. in min. oil, butyl stearate, perchoroethylene, Stod., disp. in water; dens. 8.8 lb/gal; visc. 4500 cs; HLB 7.6; 100% conc.

Emsorb 6900. [217] POE (20) sorbitan monooleate; nonionic; hydrophobic emulsifier and coemulsifier, dispersant for pigments in coatings; solubilizer for oils and fragrances; Gardner 5 liq.; sol. in water; dens. 9.0 lb/gal; visc. 400 cs; HLB 15.1; cloud pt. > 100 C; 100% conc.

Emsorb 6901. [217] POE (5) sorbitan monooleate; nonionic; o/w emulsifier and lubricant in industrial lubricants, textile lubricants and softeners; Gardner 6 liq.; sol. in Stod., disp. in water and min. oil; dens. 8.5 lb/gal; visc. 450 cs; HLB 10.1; cloud pt. < 25 C; 100% conc.

Emsorb 6903. [217] POE (20) sorbitan trioleate; nonionic; o/w emulsifier, lubricant for metals, textiles, leather; Gardner 6 liq.; sol. in butyl stearate, Stod., disp. in water, min. oil and perchloroethylene; dens. 8.6 lb/gal; visc. 300 cs; HLB 10.9; cloud pt. < 25 C; 95% act.

Emsorb 6905. [217] POE (20) sorbitan monostearate; nonionic; o/w emulsifier, fiber-to-metal textile lubricant; coemulsifier with Emsorb 2505, cosmetic and household uses; Gardner 3 waxy semisolid; sol. in water, Stod., disp. in perchloroethylene; dens. 8.9 lb/gal; visc. 500 cs; HLB 15.2; cloud pt. > 100 C; 100% conc.

Emsorb 6906. [217] POE (4) sorbitan monostearate; nonionic; w/o emulsifier in household formulations, fiber-to-metal lubricant for syn. and cellulosic fibers and yarns; Gardner 3 solid; sol. in Stod., disp. in water, min. oil, butyl stearate, perchloroethylene; m.p. 39 C; HLB 9.0; cloud pt. < 25 C; 100% conc.

Emsorb 6907. [217] POE (20) sorbitan tristearate; nonionic; o/w emulsifier, in textile processing and finishing compds. as lubricant and softener; Gardner 5 waxy solid; sol. in min. oil, butyl stearate, Stod., disp. in water, perchloroethylene; m.p. 34 C; HLB 11.0; cloud pt. < 25 C; 100% conc.

Emsorb 6910. [217] POE (20) sorbitan monopalmitate; nonionic; o/w emulsifier and lubricant; Gardner 6 visc. liq.; sol. in water, disp. in perchloroethylene; dens. 9.2 lb/gal; visc. 600 cs; HLB 15.8; cloud pt. > 100 C; 100% conc.

Emsorb 6915. [217] POE (20) monolaurate; nonionic; o/w emulsifier and solubilizer of petroleum oils, solv. and fats, essential oils, fragrances, visc. modifier in shampoos, used in textile industry; Gardner 6 liq.; sol. in water, disp. in perchloroethylene; dens. 9.2 lb/gal; visc. 400 cs; HLB 16.5; cloud pt. > 100 C; 100% conc.

Emsorb 6916. [217] POE (4) sorbitan monolaurate; o/w emulsifier in emulsion polymerization of PVC; Gardner 7 liq.; disp. in water, butyl stearate, perchloroethylene, Stod.; dens. 9.0 lb/gal; visc. 700 cs; HLB 12.1; cloud pt. < 25 C.

Emthox 5882. [217] Laureth-4; nonionic; dispersant for bath oil; emulsifier for creams and otions; liq.; HLB 9.7; 100% conc.

Emthox 5967. [217] Laureth-12; nonionic; emollient, thickener in shampoos; emulsifier in creams and lotions; solid; HLB 14.0; 100% conc.

Emulan A. [82] Oleic acid oxyethylate; nonionic; o/w emulsifier for oils and metal polishing emulsions; yel. liq.; char. odor; sol. in min. oils, fatty oils, fatty acids, polar org. solvs., emulsifiable in water; sp.gr. 0.89; visc. 120 mPa•s; HLB 11; flash pt. 210 C; sapon. no. 110; pH 6–7.5 (1% aq.); 100% act.

Emulan AF. [82] Fatty alcohol oxyethylate; nonionic; emulsifier for paraffin wax and min. oil emulsions; wh. soft wax; sol. in min. oil and polar org. media; sp.gr. 0.91 (50 C); m.p. 42 C; HLB 11; flash pt. 190 C; sapon. no. 0; pH 6-7.5 (1% aq.); 100% act.

Emulan AT 9. [82] Fatty alcohol + EO; nonionic; emulsifier for paraffin waxes; wax; HLB 13.0; 100% conc.

Emulan EL. [82] Castor oil oxyethylate; nonionic; emulsifier; pale yel. liq.; char. odor; sol. in water, polar org. media; sp.gr. 1.06 (30 C); visc. 200–500 mPa•s (30 C); HLB 14; flash pt. 280 C; sapon. no. 60; pH 6–7.5 (1% aq.); 100% act.

Emulan FJ. [82] Fatty acid alkylolamide; nonionic; w/o emulsifier for impregnation, lubrication, polishing and cleaning, o/w emulsifier in metalworking; br. liq.; faint odor; sol. in min. oils, fatty oils and mixtures, emulsifiable in water; sp.gr. 0.96; visc. 200 mPa•s; HLB 3–5 (w/o type), 8.5 (o/w type); sapon. no. 150; pH 9 (1% aq.); 100% act.

Emulan FM. [82] Triethanolamine monooleic ester; nonionic; see Emulan FJ; br. liq.; char. odor; sol. in min. oils, fatty oils and mixt.; sp.gr. 0.95; visc. 200 mPa•s; HLB 3–5 (w/o type), 9.5 (o/w type); flash pt. 240 C; sapon. no. 135; pH 10 (1% aq.); 100% act.

Emulan NP 2080. [82] Alkylphenol + EO; nonionic; emulsifier for o/w or w/o emulsions; liq.; HLB 16.0; 80% conc.

Emulan NP 3070. [82] Alkylphenol + EO; nonionic; emulsifier for emulsion polymerization; liq.; HLB 17.0; 70% conc.

Emulan OC. [82] Fatty alcohol oxyethylate; nonionic; emulsifier, stabilizer; wh. to pale yel. soft wax; sol. in water, fatty acids, waxes, polar org. solv.; sp.gr. 1.03 (50 C); m.p. 35–40 C; HLB 17; flash pt. 210 C; sapon no. 0; pH 6–7.5 (1% aq.); 100% act.

Emulan OG. [82] Fatty alcohol oxyethylate; nonionic; emulsifier, dispersant, stabilizer; waxy powd. in bead form; sol. in water, polar org. solv.; m.p. 45–50 C; HLB 17; flash pt. 240 C; sapon no. 0; pH 6–7.5 (1% aq.); 100% act.

Emulan OP 25. [82] Alkylphenol + EO; nonionic; emulsifier in emulsion polymerization; wax; HLB 16.5; 100% conc.

Emulan OSN. [82] Fatty alcohol oxyethylate; nonionic; o/w emulsifier for wax emulsions; dispersant, stabilizer; wh. to pale yel. soft wax; sol. in water, fatty acids, waxes, polar org. solv.; sp.gr. 1.03 (50 C); m.p. 35–40 C; HLB 17; flash pt. 230 C; sapon no. 0; pH 6–7 (1% aq.); 100% act.

Emulan OU. [82] Fatty alcohol oxyethylate; nonionic; emulsifier, dispersant, stabilizer; wh. to pale yel. soft wax; sol. in water, fatty acids, waxes, polar org. solv.; sp.gr. 1.02 (50 C); m.p. 40 C; sapon no. 0; pH 6–7.5 (1% aq.); 100% act.

Emulan P. [82] Fatty alcohol oxyethylate; nonionic; emulsifier; colorless clear liq.; sol. in oils; sp.gr. 0.91; HLB 11–12; flash pt. 150 C; sapon no. 0; pH 6–7.5 (1% aq.); 100% act.

Emulan PO. [82] Alkylphenol oxyethylate; nonionic; emulsifier for cleaners, formwork release oils, drilling oils; colorless to pale yel. clear liq.; sol. in min. oils, fatty oils, polar org. solv., emulsifiable in water; sp.gr.

1.04; visc. 400 mPa•s; HLB 10–12; flash pt. 240 C; sapon. no. 0; pH 6–7.5 (1% aq.); 100% act.

Emulbesto. [410] Hydrophilic lecithin; nonionic; o/w emulsifier for calf milk replacement and piglet starters; liq.; 60–65% conc.

Emulbon S-83. [606] Di(glycerol) borate sesquioleate; nonionic (semipolar); emulsifier, dispersant; liq.; 100% conc.

Emulbon T-20. [606] Di(glycerol) borate POE monolaurate; nonionic (semipolar); detergent, emulsifier; liq.; 100% conc.

Emulbon T-80. [606] Di(glycerol) borate POE monooleate; nonionic (semipolar); emulsifier, dispersant; liq.; 100% conc.

Emuldan. [273] Mono- and diglycerides; nonionic; food emulsifier; bead, powd., plastic, liq.; HLB 2.8–4.1; 100% conc.

Emuldan HA 40. [273] Tallow or lard monodiglyceride; food emulsifier and stabilizer; beads; m.p. 60 C; sapon. no. 160–180; 40% min. monoester.

Emuldan HA 52. [273] Tallow or lard monodiglyceride; see Emuldan HA 40; beads; m.p. 62 C; sapon. no. 155–175; 52% min. monoester.

Emuldan HV 40 K. [273] Veg. oil monodiglyceride; see Emuldan HA 40; beads; m.p. 62 C; sapon. no. 155–175; 40% min. monoester.

Emuldan HV 52 K. [273] See Emuldan HA 40; beads; m.p. 65 C; sapon. no. 160–180; 52% min. monoester.

Emuldan HVF 52 K. [273] Veg. oil monodiglyceride; see Emuldan HA 40; powd.; m.p. 65 C; sapon. no. 160–180; 52% min. monoester.

Emulfluid. [410] Hydrophilic lecithin; amphoteric; for production of foodstuffs; liq.; water-disp.; 100% conc.

Emulgade 1000NI. [291] Cetearyl alcohol and ceteareth-20; nonionic; emulsifier for creams and lotions; flakes; 100% conc.

Emulgade A. [291] Cetearyl alcohol and laureth-10; nonionic; base for mfg. of creams and emulsions; granules; 100% conc.

Emulgade F. [289; 291] Cetearyl alcohol, PEG-40 castor oil, and sodium cetearyl sulfate; anionic; self-emulsifying raw material for cosmetic creams and fluid emulsions; flakes; 95% fatty matter.

Emulgade F Special. [289; 291] Cetearyl alcohol and PEG-40 castor oil; nonionic; self-emulsifying raw material for fluid o/w emulsions used in cosmetic creams, in processing cationics; 99% fatty matter.

Emulgane E. [188] POE veg. oil; nonionic; emulsifier for animal, veg. fats and oils; dispersant for pigments; liq.; water-sol.; 100% conc.

Emulgane FL 7. [188] Polyoxyalkylene fatty esters and esters; nonionic; emulsifier for min. oils in jute batching; liq.; 100% conc.

Emulgane O. [188] Alkylpolyglycolether; nonionic; dispersant and emulsifier; dyeing assistant in acid media; paste; 100% conc.

Emulgane OP. [188] Polyoxyalkylene fatty esters; nonionic; emulsifier for min., veg. oils and solv.; liq.; 100% conc.

Emulphogene BC-420. [245] Trideceth-3; nonionic; emulsifier, detergent, and dispersant for petroleum oils, intermediate in mfg. of high-foaming anionic surfactants; VCS 2 (50 C) cloudy, slightly visc. pourable liq.; mild, pleasant odor; sol. in toluene; insol. in water; sp.gr. 0.92–0.94; flash pt. > 200 F (PMCC); pour pt. –7 to –1 C; pH 6.5–8.5 (10%); 100% act.

Emulphogene BC-610. [245] Trideceth-6; detergent and wetting agent for dishwashing formulations, alkaline cleaners, emulsifier for silicones and petroleum oils; VCS 2 (50 C) cloudy, slightly visc., pourable liq.; mild, pleasant odor; sol. in alcohols, acetone, toluene, disp. in water; sp.gr. 0.97–0.99; flash pt. > 200 F (PMCC); pour pt. 4.5–7.5 C; pH 6–8 (10%); 100% act.

Emulphogene BC-630. [245] Tridecyloxypoly (ethyleneoxy) ethanol; see Emulphogene BC–610; VCS 2 (50 C) cloudy, slightly visc. pourable liq.; mild, pleasant odor; sol. in alcohols, acetone, toluene, tetrachloroethylene; sp.gr. 0.99; cloud pt. 50–60 C (1%); flash pt. > 200 F (PMCC); pour pt. 15–20 C; pH 6–8 (10%); 100% act.

Emulphogene BC-720. [245] Trideceth-10; nonionic; foam stabilizer, detergent booster, wetting and rewetting agent, emulsifier, and solubilizer; VCS 2 (50 C) opaque, visc. pourable liq.; mild, pleasant odor; sol. in water, alcohol, acetone, toluene, tetrachloroethylene; sp.gr. 1.00–1.03 (40 C); cloud pt. 58–62 C (1%); flash pt. > 200 F (PMCC); pour pt. 20 C; pH 6–8 (10%); 100% act.

Emulphogene BC-840. [245] Trideceth-15; see Emulphogene BC-720; sp.gr. 1.01–1.04 (40 C); cloud pt. > 95 C (1%); flash pt. > 200 F (PMCC); pour pt. 33–40 C; pH 6–8 (10%); 100% act.

Emulphogene DA-530, DA-630. [245] Isodeceth-4, -6; nonionic; wetting agent for textile industry; intermediates; liq.; 100% act.

Emulphogene DA-639. [245] POE decyl alcohols; nonionic; wetting agent for textile industry; liq.; 90% act.

Emulphogene TB-970. [245] Linear aliphatic ethoxylate; nonionic; detergent, dispersant, textile wet processing aid, used in cleaners and sanitizers; off-wh. waxy solid and flakes; sol. in water, ethanol, carbon tetrachloride, Cellosolve, xylene; cloud pt. > 100 C (1% aq.); surf. tens. 50.8 dynes/cm (0.1%); 100% act.

Emulphor EL-620. [245] PEG-30 castor oil; nonionic; emulsifier for cosmetic and pharmaceutical preparations, dispersant, antistat, lubricant, solubilizer, wetting agent, used in leather, pesticides, herbicides, paper industries; lt. br. clear liq. sol. in water, acetone, carbon tetrachloride, alcohols, veg. oil, ethers, toluene, xylene; dens. 8.7–8.8 lb/gal; sp.gr. 1.04–1.05; visc. 600–1000 cps; cloud pt. 40–45 C (1% aq.); flash pt. 291–295 C; surf. tens. 41 dynes/cm (0.1%); 100% act.

Emulphor EL-719. [245] PEG-40 castor oil; nonionic; see Emulphor EL–620; yel. clear liq.; sol. see Emulphor EL-620; dens. 8.9–9.0 lb/gal; sp.gr. 1.06–1.07; visc. 500–800 cps; cloud pt. 83–88 C (1% aq.); flash pt. 275–279 C; surf. tens. 38 dynes/cm (0.1%); 96% act.

Emulphor EL-980, EL-985. [245] POE veg. oil; nonionic; emulsifier, antistat, lubricant, spin finish additive; VCS 5 max. waxy solid and VCS 4 max. visc. liq. resp.; water sol.; 100 and 50% act. resp.

Emulphor ON-870. [245] Oleth-20; nonionic; emulsifier, dispersant, stabilizer, solubilizer, wetting agent in metal cleaners with min. acids and corrosion inhibitors; used in rubber-based paints, adhesives, in textile, cosmetic, pharmaceutical industries; wh. solid wax; sol. in water, xylene, ethanol, ethylene glycol, butyl Cellosolve; sp.gr. 1.04; cloud pt. > 100 C; flash pt. 93 C (PMCC); pour pt. 46 C; surf. tens. 37 dynes/cm (0.1%); 100% act.

Emulphor ON-877. [245] POE oleyl alcohol; see Empulphor ON–870; VCS 3 max. pourable visc. liq.; sol. in water, xylene, ethanol, ethylene glycol, butyl Cellosolve; sp.gr. 1.06; cloud pt. 95–100 C; flash pt. > 93 C (PMCC); pour pt. 15 C; surf. tens. 40.5 dynes/cm (0.1%); 70% act.

Emulphor TO-530, TO-639. [245] POE (10 and 16) tall oil; leveling agent, dye assistant; TO-639 also as low-foam wetting agent in metal cleaners; liq.; 100 and 97% act. resp.

Emulphor VN-430. [245] PEG-5 oleate; nonionic; w/o emulsifier in cosmetics, lubricant, surfactant used in textile industry; dk. amber clear liq.; disp. in water, sol. in xylene, wh. min. oil, ethanol, butyl Cellosolve; sp.gr. 1.0; cloud pt. < 10 C (1%); flash pt. > 93 C (PMCC); pour pt. 13 C; surf. tens. 32 dynes/cm (0.1%); pH 7–7.5; 100% act.

Emulphor VT-650. [245] PEG-9 stearate; nonionic; emulsifier for silicone and nonsilicone defoamers, lubricant for dye carriers, used in textile processing; VCS 2 soft wax; slight acetic acid odor; sol. in hydrocarbons, ethanol, ether; dens. 8.3 lb/gal; sp.gr. 0.99 (50 C); visc. 9 cs (100 C); flash pt. 162 F (CC); sapon. no. 83–94; 100% act.

Emulpon EL 33, EL 40, HT. [654] Castor oil, ethoxylated; nonionic; textile, wetting agent used in dyeing; liq.; 100% conc.

Emulpur. [410] Lecithin fraction; amphoteric; food industry; powd.; 98% conc.

Emulpur N. [410] Lecithin fraction, oil-free; amphoteric; natural emulsifier for the food industry; powd.; 95% conc.

Emulsan A-67. [506] PEG 400 ester of mixed fatty acids; nonionic; emulsifier; liq.; 97% conc.

Emulsan K. [506] PEG 400 cocoate; nonionic; emulsifier for textile, leather, other industrial applications; liq.; 99%.

Emulsan O. [506] PEG 400 monooleate; nonionic; emulsifier; same as Emulsan K but with better oil sol.; liq.; 99% act.

Emulsifier 30. [478] Sodium dodecylbenzene sulfonate; emulsifier for emulsion polymers; paste; nonbiodeg.; 30% act.

Emulsifier 83. [21] Linear alcohol ethoxylated, modified; nonionic; scouring agent used for wet processing of textiles; liq.; 100% conc.

Emulsifier 99. [478] Dodecylbenzene sulfonic acid; intermediate to produce agric. emulsifiers, emulsion polymer assistants; thick liq.; nonbiodeg.; 98% act.

Emulsifier DMR. [301] Fatty acid polyglycol ester; nonionic; emulsifier for mfg. of o/w emulsions which are thickeners for printing of reactive dyes; waxy flakes.

Emulsifier K-700. [228] Mixt. of wetting agents, conditioning oils, lanolin; clouding agent, wetting agent, conditioner for hair preparations; ylsh. clear oily liq.

Emulsifier WHC. [575] Emulsifier used in

waterless hand cleaner formulations; lt. amber liq.; disp. in water, sol. in kerosene; dens. 8.225 lb/gal; pH 8 (1% aq.).

Emulsifying Oil C. [427] Sulfated fatty acid; anionic; carrier for pine oil, basis for detergent emulsifier for xylene; liq.

Emulsogen B2M. [44] Amine salt of alkyl sulfamidocarboxylic acids with partially chlorinated hydrocarbons; anionic; emulsifier, corrosion inhibitor for metal machining; br. clear oily liq.; dissolves in water to transparent sol'n., in min. oils to clear sol'n.; sp.gr. 0.94–0.95; flash pt. 128 C (Marcusson); pH 7.3 ± 0.2 (1% aq.).

Emulsogen H. [301] Alkyl sulfamidocarboxylic acid with chlorinated hydrocarbons; anionic; base for corrosion inhibitors; dk. br. visc. liq.; dissolves to clear sol'n. in most min. oils, to turbid sol'n. in water.

Emulsogen S. [44] Inorganic/nonionic blend; emulsifier/corrosion inhibitor for metalworking oils; br. clear visc. liq.; sol. in min. oils; dens. 0.97 g/cm³; solid. pt. –9 C; flash pt. 130 C; pH 8.5–9.5 (10% aq.).

Emulsogen STH. [44] Alkylsulfamido carboxylic acid; anionic; antioxidant, emulsifier for min. oil emulsions with good rust protection; golden yel. clear oily liq.; sol. with turbidity in water, clear to turbid in min. oils.

Emulthin M-35. [410] Lecithin; amphoteric; emulsifier, dispersant, wetting agents for all food uses esp. for baking industry; releasing agent for waffles; flour improver; spray-dried powd.; 50% conc.

Emulthin M-501. [410] Lecithin; amphoteric; natural wheat flour improver; spray-dried powd.; 50% conc.

Emulton AA. [553] Fatty acid-alkanolamine condensate; nonionic; emulsifier, detergent, thickener; 100% act.

Emulton G. [553] Fatty acid-glycerol esters; nonionic; emulsifier, stabilizer; solid, liq.; 100% act.

Emulton PG. [553] Fatty acid-polyglycol esters; nonionic; emulsifier, lubricant, plasticizer, wetting agent, softener, defoamer; solid, liq.; 100% act.

Emulton PS. [553] Petroleum sulfonates from sodium, calcium, barium, and ammonium salts; anionic; emulsifier, rust preventative, oil additive; 60% act. in min. oil.

Emulvin S. [416] Polythioether; nonionic; emulsifier and latex stabilizer; ylsh. br. oily liq.; dens. 1.13 g/cm³; pH 6.

Emulvin W. [416] Aromatic polyglycol ether, nonionic; emulsifier, stabilizer, wetting agent for latices; ylsh. br. liq.; sp.gr. 1.13.

Emulvis. [280] PEG-150 distearate; nonionic; visc. builder in cosmetics, sol. retarder for gums and resins; antiblock agent for decals; wh. to cream flake; bland odor; 50% sol. in 70% IPA; 3% sol. in water; m.p. 53–58 C; pH 4–6 (3% aq.).

Endet. [584] Heavy duty detergent powd. containing proteolytic enzyme; household detergent; wh. spray dried powd., colored enzyme beads; pH 9.5 (1% aq.).

Epal 6, 8, 10, 12. [227] C_6, C_8, C_{10}, and C_{12} resp. linear primary alcohols; detergent and emulsifier intermediate; liq.; 100% conc.

Epal 12/70, 12/85. [227] C_{12}-C_{14} linear primary alcohol; see Epal 6; liq.; 100% conc.

Epal 14, 16NF, 18NF. [227] C_{14}, C_{16}, and C_{18} resp. linear primary alcohols; see Epal 6; waxy solid; 100% conc.

Epal 108, 610. [227] C_6-C_{10} linear primary alcohol; see Epal 6; liq.; 100% conc.

Epal 618. [227] C_{16}-C_{18} linear primary alcohol; waxy solid; see Epal 6; 100% conc.

Epal 810. [227] C_8-C_{10} linear primary alcohol; see Epal 6; liq.; 100% conc.

Epal 1012. [227] C_{10}-C_{12} linear primary alcohol; see Epal 6; liq.; 100% conc.

Epal 1214. [227] C_{12}-C_{16} linear primary alcohol; see Epal 6; sl. visc. liq.; 100% conc.

Epal 1218. [227] C_{12}-C_{18} linear primary alcohol; see Epal 6; waxy solid; 100% conc.

Epal 1416. [227] C_{14}-C_{16} linear primary alcohol; see Epal 6; waxy solid; 100% conc.

Epal 1418. [227] C_{14}-C_{18} linear primary alcohol; see Epal 6; waxy solid; 100% conc.

Epal 1618. [227] C_{16}-C_{18} linear primary alcohol; see Epal 6; waxy solid; 100% conc.

Equex AEM. [489] Ammonium laureth-12 sulfate, cocamide DEA and SD alcohol 40-B; anionic; detergent, foamer for foaming baths, shampoos; lt. amber clear liq.; sweet, pleasant odor; dens. 8.2 lb/ gal; sp.gr. 0.98; visc. 60 cps; cloud pt. 40 F; pH 6.3; 37.5% act.

Equex S. [489] Sodium lauryl sulfate; anionic; cosmetic grade detergent, emulsifier, used in shampoos, liq. cleaners, specialty products; liq.; dens. 8.7 lb/gal; sp.gr. 1.04; visc. 50–200 cps; cloud pt. 54 F; pH 7.8 (10%); 29% act.

Equex SP. [489] Sodium lauryl sulfate; anionic; polymerization grade emulsifier; pale yel. clear liq.; clean, pleasant odor; dens. 8.7 lb/ gal; sp.gr. 1.04; visc. 90 cps (80 F); cloud pt.

55 F; pH 8 (10%); 29.8% act.

Equex STM. [489] Sodium lauryl sulfate, cocamide MEA and TEA-lauryl sulfate; anionic; detergent, foamer for shampoos, cleaners, specialty; opaque wh. to straw thick, honey-like fluid; clean, pleasant odor; dens. 8.7 lb/gal; sp.gr. 1.04; visc. 9000 cps; cloud pt. 52–55 F; pH 8; 22.5% act.

Equex SW. [489] Sodium lauryl sulfate; anionic; detergent, wetting agent, emulsifier for cream shampoos, paste cleaners, rug shampoos; pale yel. thick honey-like fluid; clean, pleasant odor; dens. 8.8 lb/gal; sp.gr. 1.05; visc. 40,000 cps (80 F); cloud pt. 77 F; pH 7.8 (10%); 28.5% act.

Equex T. [489] TEA-lauryl sulfate; anionic; detergent for shampoos; water wh. to pale yel. clear liq.; clean, pleasant odor; dens. 8.7 lb/gal; sp.gr. 1.04; visc. 50 cps (80 F); cloud pt. 28 F; pH 7.3 (10%; 40% act.

Erional NW, PA, RN. [137] Aromatic sulfonic acid condensation product; anionic; resist agent for nylon and fixative for acid dyes on nylon; liq.

ES-1239. [115] Dimethylaminopropyl ricinolamide benzyl chloride; cationic; emulsifier with antistatic, softening, bactericidal, wetting, dispersing properties for cosmetic, textile, agric. industries; Gardner 10+ clear liq.; sol. in water, alcohol, polar solv.; sp.gr. 1.022; dens. 8.48 lb/gal; visc. 15.3 stokes; flash pt. 230 F; 70% solids in propylene glycol.

ESI-Terge 10. [220] Coconut diethanolamide; nonionic; foam stabilizer, thickener for household, cosmetic, industrial products; liq.; 100% conc.

ESI-Terge 40% Coconut Oil Soap. [220] Cochin oil soap; anionic; detergent for liq. hand soap; straw liq.; coconut odor; water sol.; dens. 8.5 lb/gal; f.p. 320 F; 40% act.

ESI-Terge 320. [220] Phosphated nonylphenoxy polyethoxy ethanol; anionic; detergent, visc. aid, self-coupling wetting agent in cleaners; lt. straw liq.; mild odor; water sol.; dens. 9.34 lb/gal; 100% act.

ESI-Terge 330. [220] Phosphated glycol ester; anionic; detergent, wetting agent, used in cleaners; lt. straw liq.; mild odor; water sol.; dens. 9.18 lb/gal; 100% act.

ESI-Terge AH 20. [627] Amine condensate, modified; nonionic/anionic; detergent base; visc. builder; liq.; 100% conc.

ESI-Terge B-15. [220] 2:1 coconut diethanolamide; nonionic; detergent for mild all-purpose cleaners, foam builder and thickener, wetting agent; amber liq.; water sol.; dens. 8.4 lb/gal; sp.gr. 1.01; pH 9.8–10.8 (1%); biodeg.; 100% act.

ESI-Terge C-5. [220] Modified coconut diethanolamide; nonionic; see ESI-Terge B-15; amber liq.; water sol.; dens. 8.1 lb/gal; sp.gr. 0.97; pH 8.5–9.5 (1%); biodeg.; 100% act.

ESI-Terge HA-20. [220] Modified amine condensate; anionic/nonionic; self-coupling detergent for cleaners, detergent base; amber liq.; sol. in water and some solv.; dens. 8.64 lb/gal; pH 7.5–8.5 (1%); 100% act.

ESI-Terge P-304. [627] Phosphated alcohol ethoxylate; anionic; detergent base for drycleaning soaps, household cleaners and heavy duty industrials; liq.; 98% conc.

ESI-Terge S-10. [220] 1:1 coconut diethanolamide; nonionic; detergent, emulsifier, foam stabilizer, thickener, for liq. dishwashing and car washing detergents; lt. straw liq.; dens. 8.15 lb/gal; sp.gr. 0.98; pH 9–10; biodeg.; 100% act.

ESI-Terge T-60. [220] TEA dodecylbenzene sulfonate; anionic; detergent, wetting agent, foam stabilizer, for car and dishwashing detergent, syn. hand soap; amber liq.; dens. 9.1 lb/gal; sp.gr. 1.09; f.p. 320 F; pH 6.5–7.5 (1%); biodeg.; 60% act.

Ethoduomeen T/13. [11] PEG-3 tallow aminopropylamine; emulsifier used in making of bitumen emulsions, dispersing waxes; Gardner 18 max. liq.; sp.gr. 0.95; f.p. 20 C; m.p. 17 C; b.p. 150 C; flash pt. 204 C (COC); 95% min. act.

Ethoduomeen T/20. [11] PEG-10 tallow aminopropylamine; wetting agent used in coating preparation on paperboard; Gardner 18 max. liq.; sp.gr. 0.99; f.p. < 25 C; b.p. 150 C; flash pt. 90 C (PMCC); 95% min. act.

Ethoduomeen T/25. [11] PEG-15 tallow aminopropylamine; corrosion inhibitor in water treatment chemicals in sec. oil recovery; Gardner 18 min. liq.; sp.gr. 1.02; m.p. 25 C; flash pt. 238 C (COC); 95% act.

Ethoduomeen TD/13. [64] Ethoxylated diamine from tallow fatty acid; see Ethoduomeen T/13; Gardner 4 max. clear liq.; sp.gr. 0.95; surf. tens. 34.2 dynes/cm (0.1%).

Ethoduoquad T/20. [64] POE quat. ammonium salt from tallow fatty acid; cationic; antistat, emulsifier, dyeing assistant used in textile industry, as electroplating bath additive; clear liq.; pH 6–9 (10% aq.); 72% act.

Ethofat 60/15. [64] PEG-5 stearate; nonionic; emulsifier, detergent, wetting agent, dispersant, suspending agent, for textile, cosmetic, agric., metal and leather treating; Gardner 8 soft paste; sol. in acetone, IPA, carbon tetrachloride, benzene, disp. in water; sp.gr. 1.01; sapon. no. 110–120; surf. tens. 39 dynes/cm (0.1%); pH 6–7.5 (10% aq.).

Ethofat 60/20. [64] PEG-10 stearate; see Ethofat 60/15; Gardner 8 clear liq.; sol. see Ethofat 60/15; sp.gr. 1.02; sapon. no. 70–80; surf. tens. 36 dynes/cm (0.1%); pH 6–7.5 (10% aq.).

Ethofat 60/25. [64] PEG-15 stearate; see Ethofat 60/15; Gardner 11 paste; sapon. no. 55–65; surf. tens. 39.5 dynes/cm (0.1%); pH 6–7.5 (10% aq.).

Ethofat 142/20. [64] EO condensate of 4% rosin fatty acid; see Ethofat 60/15; Gardner 11 clear liq.; sp.gr. 1.03; sapon no. 72–82; surf. tens. 39.5 dynes/cm (0.1%); pH 6–7.5 (10% aq.).

Ethofat 242/25. [64] EO condensate of 50% rosin fatty acid; see Ethofat 60/15; Gardner 12 clear liq.; sp.gr. 1.08; sapon no. 55–65; surf. tens. 42 dynes/cm (0.1%); pH 6–7.5 (10% aq.); 100% conc.

Ethofat C/15. [64] PEG-5 cocoate; nonionic; see Ethofat 60/15; Gardner 9 clear liq.; sol. see Ethofat 60/15; sp.gr. 1.00; HLB 10.6; sapon. no. 120–130; surf. tens. 33 dynes/cm (0.1%); pH 6–7.5 (10% aq.); 100% conc.

Ethofat C/25. [64] PEG-15 cocoate; see Ethofat 60/15; Gardner 8 paste; sol. see Ethofat 60/15; sapon. no. 60–70; pH6–7.5 (10% aq.).

Ethofat O/15. [64] PEG-5 oleate; see Ethofat 60/15; 9 max. Gardner clear liq.; sol. see Ethofat 60/15; sp.gr. 0.99; sapon. no. 110–120; surf. tens. 35 dynes/cm (0.1%); pH 6–7.5 (10% aq.); 100% conc.

Ethofat O/20. [64] PEG-10 oleate; see Ethofat 60/15; 9 max. Gardner clear liq.; sol. see Ethofat 60/15; sp.gr. 1.03; sapon. no. 75–85; surf. tens. 41 dynes/cm (0.1%); pH 6–7.5 (10% aq.).

Ethomeen 18/12. [64] PEG-2 stearamine; cationic; emulsifier, dispersant used in textile processing; Gardner 7 solid; sol. in acetone, IPA, carbon tetrachloride, benzene; sp.gr. 0.96; 95% tert. amine.

Ethomeen 18/15. [64] PEG-5 stearamine; see Ethomeen 18/12; Gardner 8 liq. to paste; sol. in acetone, IPA, carbon tetrachloride, benzene; sp.gr. 0.98; surf. tens. 34 dynes/cm (0.1%).

Ethomeen 18/20. [64] PEG-10 stearamine; see Ethomeen 18/12; Gardner 8 clear liq.; sp.gr. 1.02; surf. tens. 40 dynes/cm (0.1%).

Ethomeen 18/25. [64] PEG-15 stearamine; see Ethomeen 18/12; Gardner 8 clear liq.; sp.gr. 1.04; surf. tens. 43 dynes/cm (0.1%).

Ethomeen 18/60. [64] PEG-50 stearamine; see Ethomeen 18/12; wh. solid; sol. in acetone, IPA, carbon tetrachloride, benzene, water; sp.gr. 1.12; surf. tens. 49 dynes/cm (0.1%).

Ethomeen C/12. [64] PEG-2 cocamine; cationic; emulsifier, dispersant used in textile processing; Gardner 6 max. clear liq.; sol. in acetone, IPA, carbon tetrachloride, Stod., benzene; forms gel in water; sp.gr. 0.87; 95% tert. amine.

Ethomeen C/15. [64] PEG-5 cocamine; see Ethomeen C/12; Gardner 6 max. clear liq.; sol. in acetone, IPA, carbon tetrachloride, Stod., benzene, water (cloudy); sp.gr. 0.98; surf. tens. 33 dynes/cm (0.1%).

Ethomeen C/20. [64] PEG-10 cocamine; see Ethomeen C/12; Gardner 11 max. clear liq.; sol. see Ethomeen C/15; sp.gr. 1.02; surf. tens. 39 dynes/cm (0.1%).

Ethomeen C/25. [64] PEG-15 cocamine; see Ethomeen C/12; Gardner 12 max. clear liq.; sol. in acetone, IPA, carbon tetrachloride, Stod., benzene, water; sp.gr. 1.04; surf. tens. 41 dynes/cm (0.1%).

Ethomeen O/12. [64] PEG-2 oleamine; cationic; emulsifier, dispersant used in textile processing; Gardner 8 max. clear liq.; sp.gr. 0.90; surf tens. 31.5 dynes/cm (0.1%).

Ethomeen O/15. [64] PEG-5 oleamine; see Ethomeen O/12; Gardner 8 max. clear liq.; sp.gr. 0.96; surf tens. 35.3 dynes/cm (0.1%).

Ethomeen O/25. [64] PEG-15 oleamine; see Ethomeen O/12; Gardner 8 max. paste.

Ethomeen S/12. [64] PEG-2 soyamine; cationic; emulsifier, dispersant used in textile processing; Gardner 14 max. clear heavy liq; sol. in acetone, IPA, carbon tetrachloride, Stod., benzene; insol. water; sp.gr. 0.91; surf. tens. 31.4 dynes/cm (0.1%); 95% tert. amine.

Ethomeen S/15. [64] PEG-5 soyamine; see Ethomeen S/12; Gardner 14 max. clear heavy liq; sol. in acetone, IPA, carbon tetrachloride, Stod., benzene; forms gel or disp. in water; sp.gr. 0.95; surf. tens. 33 dynes/cm (0.1%).

Ethomeen S/20. [64] PEG-10 soyamine; see Ethomeen S/12; Gardner 14 max. clear heavy liq; sol. in acetone (cloudy), IPA, carbon tetrachloride, Stod., benzene, water; sp.gr. 1.02; surf. tens. 40 dynes/cm (0.1%).

Ethomeen S/25. [64] PEG-15 soyamine; see Ethomeen S/12; Gardner 14–18 max. clear heavy liq; sol. see Ethomeen S/20; sp.gr. 1.04; surf. tens. 43 dynes/cm (0.1%).

Ethomeen T/12. [64] PEG-2 tallow amine; cationic; emulsifier, dispersant used in textile processing; Gardner 8 paste; sol. in acetone, IPA, carbon tetrachloride, Stod., benzene, water; sp.gr. 0.92; 95% tert. amine.

Ethomeen T/15. [64] PEG-5 tallow amine; see Ethomeen T/12; Gardner 8 max. clear liq.; sol. cloudy in acetone, IPA, carbon tetrachloride, Stod., Benzene; forms gel in water; sp.gr. 0.97; surf. tens. 34 dynes/cm (0.1%).

Ethomeen T/25. [64] PEG-15 tallow amine; see Ethomeen T/12; Gardner 8 max. clear liq.; sol. cloudy in acetone, IPA, carbon tetrachloride, benzene, water; sp.gr. 1.03; surf. tens. 41 dynes/cm (0.1%).

Ethomeen TD/15. [64] Ethoxylated tert. amine from tallow fatty acid; see Ethomeen T/12; Gardner 4 max. clear liq.; sp.gr. 0.97; surf. tens. 35.8 dynes/cm (0.1%).

Ethomeen TD/25. [64] Ethoxylated tert. amine from tallow fatty acid; see Ethomeen T/12; Gardner 4 max. clear liq.; sp.gr. 1.03; surf. tens. 43.5 dynes/cm (0.1%).

Ethomid HT/15. [64] Ethoxylated amide from hyd. tallow acid; nonionic; emulsifier, dispersant, detergent; Gardner 10 max. solid; sol. in IPA; disp. in acetone, carbon tetrachloride, Stod., benzene, water; sp.gr. 1.03; surf. tens. 37 dynes/cm (0.1%); 100% conc.

Ethomid HT/23. [63] POE (12.5) hydrog. tallow amide; nonionic; emulsifier, dispersant, detergent, dye leveling agent; for silicone finishing agents, sizing lubricants; Gardner 8 max. solid; sp.gr. 1.029; cloud pt. 130–170 F (1%); flash pt. (PM) > 400 F; surf. tens. 37 dynes/cm (0.1%).

Ethomid HT/60. [64] PEG-50 tallow amide; see Ethomid HT/15; Gardner 11 max. hard solid; sol. in IPA, carbon tetrachloride, benzene, water; sp.gr. 1.14; surf. tens. 47 dynes/cm (0.1%).

Ethomid O/15. [64] PEG-5 oleamide; nonionic; emulsifier, dispersant, detergent; Gardner 12 max. liq.; sol. in IPA, carbon tetrachloride, gels in water @ 50–80 C; surf. tens. 35 dynes/cm (0.1%).

Ethoquad 18/12. [64] PEG-2 stearmonium chloride and IPA; cationic; antistat, emulsifier, dyeing assistant used in textile industry, as electroplating bath additives; Gardner 6 max. paste; sol. in acetone, IPA, benzene, water; sp.gr. 0.919; surf. tens. 40.2 dynes/cm (0.1%); pH 7–8 (10% aq.); 70% act.

Ethoquad 18/25. [64] PEG-15 stearmonium chloride; see Ethoquad 18/12; Gardner 11 max. clear liq.; sol. in acetone, IPA, benzene, water, carbon tetrachloride; sp.gr. 1.058; surf. tens. 50.1 dynes/cm (0.1%); pH 6–9 (10% aq.); 95% act.

Ethoquad C/12. [64] PEG-2 cocomonium chloride and IPA; see Ethoquad 18/12; Gardner 9 max. clear liq.; sol. see Ethoquad 18/25; sp.gr. 0.969; surf. tens. 35.4 dynes/cm (0.1%); pH 7–8 (10% aq.); 75% act.

Ethoquad C/25. [64] PEG-15 cocomonium chloride; see Ethoquad 18/12; Gardner 11 max. clear liq.; sol. see Ethoquad 18/25; sp.gr. 1.071; surf. tens. 43.4 dynes/cm (0.1%); pH 7–8 (10% aq.); 95% act.

Ethoquad O/12. [64] PEG-2 oleamonium chloride and IPA; see Ethoquad 18/12; Gardner 9 max. clear liq.; sol. see Ethoquad 18/25; sp.gr. 0.932; surf. tens. 40.3 dynes/cm (0.1%); pH 7–8 (10% aq.); 72% act.

Ethoquad O/25. [64] PEG-15 oleamonium chloride; see Ethoquad 18/12; Gardner 11 max. clear liq.; sol. see Ethoquad 18/25; sp.gr. 1.062; surf. tens. 40.8 dynes/cm (0.1%); pH 7–9 (10% aq.); 95% act.

Ethosperse CA-2. [256] Ceteth-2; nonionic; o/w emulsifier, thickener, stabilizer for hair care products, antiperspirants; wh. solid; sol. in ethanol; disp. hot in water; HLB 6 ± 1; m.p. 29–33 C.

Ethosperse G-26. [256] Glycereth-26; nonionic; emulsifier for cosmetic, pharmaceutical and industrial uses; pale straw liq.; sol. in water, methanol, ethanol, acetone, toluol; sp.gr. 1.12 (38 C); visc. 150 cps (38 C).

Ethosperse LA-4. [256] Laureth-4; nonionic; emulsifier for cosmetic, pharmaceutical and industrial uses; pale straw liq.; sol. in methanol, ethanol, acetone, toluol, min. oil, misc. with water, veg. oils; sp.gr. 0.95; visc. 30 cps; HLB 9.5; 100% act.

Ethosperse LA-12. [256] Laureth-12; nonionic; emulsifier for cosmetic, pharmaceutical and industrial uses; turbid liq. to soft paste; sol. see Ethosperse LA-4; sp.gr. 1.10; visc. 1000 cps; HLB 4.8; 100% act.

Ethosperse LA-23. [256] Laureth-23; nonionic; see Ethosperse CA-2; wh. solid; sol. in water, hot ethanol; HLB 17 ± 1; m.p. 30–45 C.

Ethosperse OA-2. [256] Oleth-2; nonionic; see Ethosperse CA-2; also as spreading agent and emollient in bath oils; colorless liq.; sol. in ethanol, min. oil; disp. hot in water; visc. 30 cps; HLB 4 ± 1.

Ethosperse SL-20. [256] Sorbeth-20; nonionic; emulsifier for in cosmetic, pharmaceutical and industrial uses; lt. yel. liq.; sol. in water, methanol, ethanol, acetone; sp.gr. 1.16; visc. 460 cps; HLB 16.6; 100% act.

Ethosperse TDA-6. [256] Trideceth-6; nonionic; emulsifier for in cosmetic, pharmaceutical and industrial uses; pale straw liq.; sol. in methanol, ethanol, acetone, toluol, disp. in water; sp.gr. 0.98; visc. 80 cps; HLB 11.0; 100% act.

Ethylan 20. [360] Nonoxynol-20; nonionic; detergent, wetting agent, stabilizer, emulsifier in waxes, resins, hand cleaning gels; wh. waxy solid; negligible odor; water sol. sp.gr. 1.065 (40 C); visc. 160 cs; HLB 16.0; cloud pt. > 100 C (1% aq.); flash pt. > 400 F (COC); pour pt. 30 C; pH 6–8 (1% aq.); 100% act.

Ethylan 44. [360] Nonyl phenol ethoxylate; nonionic; emulsifier for hydrocarbons, etc.; straw clear liq.; negligible odor; water insol.; sp.gr. 1.023; visc. 340 cs; HLB 9; flash pt. > 400 F (COC); pour pt. < 0 C; pH 6–8 (1% aq.); 100% act.

Ethylan 55. [360] Nonyl phenol ethoxylate; nonionic; emulsifier for hydrocarbons, etc.; straw clear liq.; water insol.; sp.gr. 1.035; visc. 290 cs; HLB 10.5; flash pt. > 400 F (COC); pour pt. < 0 C; pH 6–8 (1% aq.); 100% act.

Ethylan 77. [360] Nonyl phenol ethoxylate; nonionic; emulsifier for hydrocarbons, etc.; straw clear liq.; negligible odor; water disp.; sp.gr. 1.038; visc. 360 cs; HLB 10.9; cloud pt. < 2 C (1% aq.); flash pt. > 400 F (COC); pour pt. < 0 C; pH 6–8 (1% aq.); 100% act.

Ethylan 172. [360] Cetyl/oleyl alcohol ethoxylate; nonionic; emulsifier for min. oils and hydrophobic waxes; lt. amber liq.; HLB 7; 100% act.

Ethylan A2. [360] PEG 200 oleate; nonionic; emulsifier, antifoam agent; lt. amber liq.; sp.gr. 0.976; visc. 67 cs; HLB 7; 100% act.

Ethylan A3. [360] PEG 300 oleate; nonionic; emulsifier, antifoam, dispersant for industrial uses; lt. amber liq.; sp.gr. 1.0; visc. 100 cs; HLB 8.9; 100% act.

Ethylan A4. [360] PEG 400 oleate; nonionic; emulsifier, antifoam agent, dispersing agent for industrial uses; lt. amber liq.; sp.gr. 1.017; visc. 130 cs; HLB 10.3; 100% act.

Ethylan A6. [360] PEG 600 oleate; nonionic; dispersing agent for industrial uses; lt. amber liq.; sp.gr. 1.037; visc. 160 cs; HLB 12.3; 100% act.

Ethylan A15. [360] Coconut diethanolamide; nonionic; emulsifier used in industrial degreasing, metal cleaning; straw clear liq.; sp.gr. 1.010; visc. 880 cs; 99% act.

Ethylan BBC31. [360] Modified linear alcohol ethoxylate; nonionic; low foam iodophor intermediate; liq.; 100% conc.

Ethylan BCP. [360] Nonyl phenol ethoxylate; nonionic; detergent, wetting agent, emulsifier; clear, colorless liq.; water sol.; sp.gr. 1.058; visc. 320; HLB 12.9; cloud pt. 50 C (1% aq.); flash pt. > 400 F (COC); pour pt. 0 C; pH 6–8 (1% aq.); 100% act.

Ethylan BV. [360] Nonyl phenol ethoxylate; nonionic; foam stabilizer, solubilizer, used in pesticides, perfumes; wh. soft paste; water sol.; sp.gr. 1.083; visc. 380; HLB 14.5; cloud pt. 90 C (1% aq.); flash pt. > 400 F (COC); pour pt. 20 C; pH 6–8 (1% aq.); 100% act.

Ethylan BZA. [360] Modified alkylphenol ethoxylate; nonionic; low-foam detergent for mechanical dishwashing, metal cleaning; liq.; cloud pt. 18 C; 100% act.

Ethylan C 40 AH. [360] PEG ester of unsat. fatty acid; nonionic; emulsifier for chlorinated solv., olein and wax, pesticides; yel. hazy liq.; sp.gr. 1.054; visc. 320 cs (40 C); HLB 13.5; 100% act.

Ethylan C 160. [360] PEG ester of unsat. fatty acid; nonionic; emulsifier for olein and wax, dispersing aid, perfumery solubilizer; cream waxy solid; HLB 17.6; 100% act.

Ethylan CD913. [360] Syn. lower fraction primary alcohol EO condensate; nonionic; detergent, o/w and w/o emulsifier, wetting agent, solubilizer; colorless, clear liq.; water insol.; sp.gr. 0.936; visc. 27 cs; HLB 8.8; flash pt. 290 F (COC); pour pt. < 0 C; pH 6–8 (1% aq.); 97% act.

Ethylan CD916. [360] Syn. lower fraction primary alcohol EO condensate; nonionic; see Ethylan CD913; colorless clear liq.; char. fatty alcohol odor; water sol.; sp.gr. 0.993; visc. 53 cs; HLB 12.8; flash pt. 355 F (COC); pour pt. 0 C; pH 6–8 (1% aq.); 96% act.

Ethylan CD919. [360] Syn. lower fraction primary alcohol EO condensate; nonionic; see Ethylan CD913; colorless clear liq.; char. fatty alcohol odor; water sol.; sp.gr.

1.008; visc. 77 cs; HLB 14.4; flash pt. 385 F (COC); pour pt. 7 C; pH 6–8 (1% aq.); 97% act.

Ethylan CD9112 [360] Syn. lower fraction primary alcohol EO condensate; nonionic; see Ethylan CD913; wh. waxy solid; char. fatty alcohol odor; water sol.; sp.gr. 1.031 (40 C); visc. 56 cs (40 C); HLB 15.5; flash pt. > 400 F (COC); pour pt. 23 C; pH 6–8 (1% aq.); 100% act.

Ethylan CD9130. [360] Syn. lower fraction primary alcohol EO condensate; nonionic; o/w emulsifier, wetting agent; wh. waxy solid; char. fatty alcohol odor; water sol.; visc. 62 cs (60 C); HLB 17.8; flash pt. > 400 F (COC); pour pt. 45 C; pH 6–8 (1% aq.); 100% act.

Ethylan CF71. [360] Coconut fatty ester; nonionic; emulsifier for cosmetics, pharmaceuticals, fiber lubricant; pale straw liq.; HLB 14; 100% act.

Ethylan CH. [360] Ethoxylated coconut fatty acid alkylolamide; nonionic; foam stabilizer in liq. detergents, coemulsifier; amber clear liq.; mild odor; water sol.; sp.gr. 1.025; visc. 270 cs; cloud pt. 80 C (1% aq.); flash pt. > 300 F (COC); pour pt. 14 C; pH 7–9 (1% aq.); 100% act.

Ethylan CL. [360] Alkylolamide EO condensate; nonionic; emulsifier, foam stabilizer for shampoos; liq.; cloud pt. 26 C; 100% act.

Ethylan CPG 630. [360] Modified alcohol ethoxylate; nonionic; low foaming wetting agent and detergent for rinse aids, machine dishwashing, metal cleaning; colorless clear liq.; mild odor; water sol.; sp.gr. 0.978; visc. 65 cs; cloud pt. 38 C (1% aq.); flash pt. > 350 F (COC); pH 6–7.5 (1% aq.); 99% act.

Ethylan CPG 660. [360] Modified alcohol ethoxylate; see Ethylan CPG 630; water sol.; sp.gr. 0.984; visc. 77 cs; cloud pt. 29 C (1% aq.); flash pt. > 350 F (COC); pH 6–7.5 (1% aq.); 99% act.

Ethylan CPG 745. [360] Modified alcohol ethoxylate; see Ethylan CPG 630; water sol.; sp.gr. 0.985; visc. 84 cs; cloud pt. 37 C (1% aq.); flash pt. > 350 F (COC); pH 6–7.5 (1% aq.); 99% act.

Ethylan CPG 945. [360] Modified alcohol ethoxylate; see Ethylan CPG 630; water sol.; sp.gr. 1.005; visc. 88 cs; cloud pt. 47 C (1% aq.); flash pt. > 350 F (COC); pH 6–7.5 (1% aq.); 99% act.

Ethylan CRS. [360] Ethoxylated coconut fatty acid alkylolamide; nonionic; foam stabilizer in liq. detergent, general purpose, or hard surface cleaners and shampoos, coemulsifier; amber clear liq.; mild odor; water sol.; sp.gr. 1.026; visc. 250 cs; flash pt. > 300 F (COC); pour pt. 10 C; pH 7–9 (1% aq.); 100% act.

Ethylan D252. [360] Syn. primary alcohol ethoxylate; nonionic; o/w and w/o emulsifier for min. oils and waxes; colorless clear liq.; faint odor; oil sol.; sp.gr. 0.903; visc. 31 cs; HLB 5.6; pour pt. 5 C; pH 6–8 (1% aq.); 100% act.

Ethylan D253. [360] Syn. primary alcohol ethoxylate; nonionic; o/w and w/o emulsifier for min. oils and waxes, visc. depressant; colorless clear liq.; faint odor; oil sol.; sp.gr. 0.930; visc. 36 cs; HLB 7.8; pour pt. 3 C; pH 6–8 (1% aq.); 100% act.

Ethylan D254. [360] Syn. primary alcohol ethoxylate; nonionic; o/w and w/o emulsifier for min. oils and waxes, dry cleaning assistant; colorless clear liq.; faint odor; oil sol.; sp.gr. 0.948; visc. 57 cs; HLB 9.8; pour pt. 5 C; pH 6–8 (1% aq.); 97% act.

Ethylan D256. [360] Syn. primary alcohol ethoxylate; nonionic; o/w emulsifier for waxes, detergent, wetting agent; hazy liq.; faint odor; water sol.; sp.gr. 0.972; visc. 72 cs; HLB 11.4; cloud pt. 43 C (1% aq.); pour pt. 15 C; pH 6–8 (1% aq.); 100% act.

Ethylan D257. [360] Syn. primary alcohol ethoxylate; nonionic; o/w emulsifier for waxes, detergent, wetting agent; hazy liq.; faint odor; water sol.; sp.gr. 0.974 (40 C); visc. 31 cs (40 C); HLB 12.2; cloud pt. 49 C (1% aq.); flash pt. 385 F (COC); pour pt. 21 C; pH 6–8 (1% aq.); 100% act.

Ethylan D259. [360] Syn. primary alcohol ethoxylate; nonionic; detergent, wetting agent, emulsifier, solubilizer for wax, oils; wh. solid; water sol.; sp.gr. 0.988; visc. 36 cs (40 C); cloud pt. 76 C; 100% act.

Ethylan D2512. [360] Syn. primary alcohol ethoxylate; nonionic; o/w emulsifier; detergent, wetting agent, solubilizer; wh. solid; faint odor; water sol.; sp.gr. 1.001 (40 C); visc. 46 cs (40 C); HLB 14.2; cloud pt. 92 C (1% aq.); pour pt. 29 C; pH 6–8 (1% aq.); 100% act.

Ethylan D2560. [360] Syn. primary alcohol ethoxylate; nonionic; emulsifier, wetting agent; wh. flake; faint odor; water sol.; visc. 120 cs (60 C); HLB 18.6; cloud pt. > 100 C (1% aq.); pour pt. 45 C; pH 6–8 (1% aq.); 100% act.

Ethylan DP. [360] Nonyl phenol ethoxylate; nonionic; foam stabilizer and booster, emulsifier, solubilizer; in pesticides; slight hazy visc. liq.; negligible odor; water sol.; sp.gr. 1.068; visc. 400 cs; HLB 14.0; cloud pt. 80 C (1% aq.); flash pt. > 400 F (COC); pour pt. 13 C; pH 6–8 (1% aq.); 100% act.

Ethylan ENTX. [360] Alkylphenol ethoxylate; nonionic; emulsifier for min. oil and wax, mastic plasticizer; pale straw liq.; faint odor; oil sol.; sp.gr. 1.006; visc. 440 cs; HLB 9.0; flash pt. > 350 F (COC); pour pt. 2 C; pH 6–7.5 (1% aq.); 100% act.

Ethylan GD. [360] Special fatty acid diethanolamide; nonionic; emulsifier for oils and metal degreasing agents, antistat, lubricant; clear amber liq.; mild odor; oil sol.; sp.gr. 0.982; visc. 1200 cs; flash pt. > 300 F (COC); pour pt. 10 C; pH 8–11 (1% aq.); 75% act.

Ethylan GEL-2 [360] Polysorbate 20; w/o emulsifier, solubilizer esp. with sorbitan esters; used in agric., perfumes, fiber lubricants, textile antistats, polymer additives; liq.; HLB 16.5; 100% conc.

Ethylan GEO-8 [360] Polysorbate 80; see Ethylan GEL-2; liq.; HLB 15.0; 100% conc.

Ethylan GEP-4 [360] Polysorbate 40; see Ethylan GEL-2; liq.; HLB 15.5; 100% conc.

Ethylan GES-6 [360] Polysorbate 60; see Ethylan GEL-2; paste; HLB 15.0; 100% conc.

Ethylan GL-20. [360] Sorbitan laurate; surfactant for use as antistat and textile fiber lubricant; liq.; HLB 8.0; 100% conc.

Ethylan GLE-21 [360] Polysorbate 21; see Ethylan GEL-2; liq.; HLB 13.3; 100% conc.

Ethylan GMF. [360] Alkylphenol ethoxylate; nonionic; wetting agent, emulsifier for hydrocarbon solv., emulsion polymerization; straw clear liq.; faint odor; water sol.; sp.gr. 1.050; visc. 330 cs; HLB 12.5; cloud pt. 47 C (1% aq.); flash pt. > 350 C (COC); pour pt. 7 C; pH 6–7.5 (1% aq.); 100% act.

Ethylan GO-80 [360] Sorbitan oleate; see Ethylan GL-20; liq.; HLB 4.5; 100% conc.

Ethylan GOE-21 [360] Polysorbate 81; see Ethylan GEL-2; liq.; HLB 10.0; 100% conc.

Ethylan GP-40 [360] Sorbitan palmitate; see Ethylan GL-20; solid; HLB 6.7; 100% conc.

Ethylan GPS-85 [360] Polysorbate 85; see Ethylan GEL-2; liq.; HLB 11.0; 100% conc.

Ethylan GS-60 [360] Sorbitan stearate; see Ethylan GL-20; solid; HLB 5.0; 100% conc.

Ethylan GT-85 [360] Sorbitan trioleate; see Ethylan GL-20; liq.; HLB 1.5; 100% conc.

Ethylan HA. [360] Nonyl phenol ethoxylate; nonionic; detergent, wetting agent, stabilizer, emulsifier in waxes, resins, hand cleaning gels; wh. waxy solid; negligible odor; water sol.; sp.gr. 1.064 (60 C); visc. 120 cs (60 C); HLB 17.4; cloud pt. > 100 C (1% aq.); flash pt. > 400 F (COC); pour pt. 43 C; pH 6–8 (1% aq.); 100% act.

Ethylan HB1. [360] Short chain ethoxylate; nonionic; coalescing agent for surface coatings; liq.; 100% conc.

Ethylan HB4. [360] Aromatic ethoxylate; nonionic; penetrant, hydrotrope, cosolv.; liq.; 100% conc.

Ethylan HB30. [360] Aromatic ethoxylate; nonionic; penetrant for oil well drilling muds; solid; 100% conc.

Ethylan HP. [360] Nonyl phenol ethoxylate; see Ethylan HA; wh. waxy solid; negligible odor; water sol.; sp.gr. 1.072 (40 C); visc. 78 cs (60 C); HLB 16.6; cloud pt. > 100 C (1% aq.); flash pt. > 400 F (COC); pour pt. 36 C; pH 6–8 (1% aq.); 100% act.

Ethylan KELD. [360] Modified fatty alkylolamide; nonionic; emulsifier for wax, kerosene and olein, solv. degreasing; clear amber liq.; mild odor; disp. in water; sp.gr. 1.006; visc. 630 cs; flash pt. > 300 F (COC); pour pt. < 0 C; pH 8–10 (1% aq.); 100% act.

Ethylan KEO. [360] Nonyl phenol ethoxylate; nonionic; detergent, wetting agent, emulsifier, foam stabilizer, solubilizer, used in pesticides, perfumes; clear colorless liq.; negligible odor; water sol.; sp.gr. 1.060; visc. 340 cs; HLB 13; cloud pt. 54 C (1% aq.); flash pt. > 400 F (COC); pour pt. 3 C; pH 6–8 (1% aq.); 100% act.

Ethylan L10. [360] PEG 1000 laurate; emulsifier for cosmetics and pharmaceuticals; wh. solid; sp.gr. 1.056; visc. 76 cs (40 C).

Ethylan LD. [360] Coconut diethanolamide; nonionic; foam stabilizer, emulsifier; clear straw liq.; mild odor; disp. in water; sp.gr. 0.981; visc. 960 cs; flash pt. > 350 F (COC); pour pt. 15 C; pH 8.5–10.5 (1% aq.); 90% act.

Ethylan LDG. [360] Coconut diethanolamide; nonionic; detergent, shampoo foam stabilizer; liq.; 82% conc.

Ethylan LDS. [360] Coconut diethanolamide; see Ethylan LD; sp.gr. 0.990; visc. 810 cs; flash pt. > 350 F (COC); pour pt. 10 C; pH 8.5–10.5 (1% aq.); 87% act.

Ethylan LM. [360] Coconut monoalkanolamide; nonionic; emulsifier, antistat, foam stabilizer for liq. detergents; pale straw flake; mild odor; water insol.; sp.gr. 0.909 (80 C); visc. 40 cs (80 C); flash pt. > 350 F (COC); pour pt. 63 C; pH 8.5–10.5 (1% aq.); 94% act.

Ethylan LM2. [360] Ethoxylated coconut fatty acid alkylolamide; nonionic; foam stabilizer, emulsifier for detergent formulations; off-wh. soft paste; mild odor; water sol.; sp.gr. 0.963 (40 C); visc. 108 cs (40 C); flash pt. > 300 F (COC); pour pt. 2 C; pH 7–9 (1% aq.); 100% act.

Ethylan ME. [360] Cetyl oleyl alcohol ethoxylate; nonionic; emulsifier for org. solv., veg. oils, paraffin waxes; liq.; HLB 10; m.p. 15 C; 100% conc.

Ethylan MLD. [360] Lauric diethanolamide; nonionic; foam booster and stabilizer in toiletries and detergent formulations, antistat for plastics; pale cream waxy flake; negligible odor; disp. in water; visc. 130 cs (60 C); flash pt. > 300 F (COC); pour pt. 40 C; pH 8–9 (1% aq.); 95% act.

Ethylan N30. [360] Nonyl phenol ethoxylate; nonionic; detergent, wetting agent, stabilizer, emulsifier in waxes, resins, hand cleaning gels; wh. waxy solid; negligible odor; water sol.; sp.gr. 1.064 (60 C); visc. 90 cs (60 C); HLB 17.0; cloud pt. > 100 C (1% aq.); flash pt. > 400 F (COC); pour pt. 39 C; pH 6–8 (1% aq.); 100% act.

Ethylan N50. [360] Nonyl phenol ethoxylate; see Ethylan N30; wh. waxy solid; negligible odor; water sol.; sp.gr. 1.073 (60 C); visc. 135 cs (60 C); HLB 18.2; cloud pt. > 100 C (1% aq.); flash pt. > 400 F (COC); pour pt. 43 C; pH 6–8 (1% aq.); 100% act.

Ethylan N92. [360] Nonyl phenol ethoxylate; see Ethylan N30; wh. waxy solid; negligible odor; water sol.; sp.gr. 1.078 (60 C); visc. 340 cs (60 C); HLB 19.0; cloud pt. > 100 C (1% aq.); flash pt. > 400 F (COC); pour pt. 50 C; pH 6–8 (1% aq.); 100% act.

Ethylan NP 1. [360] Nonyl phenol ethoxylate; nonionic; defoamer, oil emulsifier; clear, straw liq.; oil sol.; sp.gr. 0.987; visc. 650 cs; HLB 4.5; flash pt. 300 F (COC); pour pt. < 0 C; pH 6–8 (1% aq.); 100% act.

Ethylan PQ. [360] Modified alkylphenol ethoxylate; nonionic; solubilizer used in the mfg. of iodophor; water-wh. liq (clear @ 30 C); faint odor; water sol.; sp.gr. 1.045 (40 C); HLB 14.2; cloud pt. 83 C (1% aq.); flash pt. > 350 F (COC); pour pt. 15 C; pH 6–7.5 (1% aq.); 98% act.

Ethylan OE. [360] Cetyl/oleyl alcohol ethoxylate; nonionic; emulsifier for fatty acids, alcohols, and waxes, oil and latex stabilizer, dye leveling agent; cream waxy solid; negligible odor; water sol.; sp.gr. 1.009 (40 C); visc. 67 cs (40 C); HLB 14; cloud pt. 90 C (1% aq.); flash pt. > 350 F (COC); pour pt. 31 C; pH 6–8 (1% aq.); 100% act.

Ethylan R. [360] Cetyl/oleyl alcohol ethoxylate; see Ethylan OE; cream waxy solid; negligible odor; water sol.; sp.gr. 1.026 (40 C); visc. 150 cs (40 C); HLB 15.4; cloud pt. > 100 C (1% aq.); flash pt. > 350 F (COC); pour pt. 36 C; pH 6–8 (1% aq.); 100% act.

Ethylan TB345. [360] Amine EO-PO copolymer; nonionic; low-foam surfactant for metal treatment formulations; liq.; 100% conc.

Ethylan TC. [360] Coconut fatty amine ethoxylate; nonionic; wetting agent for metal cleaning and stripping of surface coatings, fiber antistat, used in cosmetics; dk. amber liq.; mild fatty amine odor; water sol.; sp.gr. 1.040; visc. 140 cs; HLB 15; flash pt. > 300 F (COC); pour pt. < 0 C; pH 8.5–10.0 (1% aq.); 100% act.

Ethylan TCO. [360] Complex amine oxide; nonionic; foam and suds stabilizer in light- and heavy-duty detergents, cosmetic formulations, foam booster, emulsifier; straw clear to hazy liq.; mild fatty amine odor; water sol.; sp.gr. 1.008; visc. 159 cs; flash pt. > 300 F (COC); pour pt. 0 C; pH 6–8 (1% aq.); 40% act.

Ethylan TF. [360] Ethoxylated coconut fatty amine; nonionic; oil emulsifier with anticorrosive properties, antistat for syn. fibers with polystyrene; lt. amber liq.; mild fatty amine odor; disp. in water; sp.gr. 0.912; visc. 140 cs; HLB 6; flash pt. > 300 F (COC); pour pt. < 0 C; pH 8.5–10.0 (1% aq.); 100% act.

Ethylan TH-2. [360] Hydrog. fatty amine ethoxylate; nonionic; see Ethylan TF; pale br. solid; mild fatty amine odor; disp. in water; sp.gr. 0.878 (60 C); visc. 33 cs (60 C); HLB 5; flash pt. > 300 F (COC); pour pt. 40 C; pH 8.5–10.0 (1% aq.); 100% act.

Ethylan TN-10. [360] Coconut fatty amine ethoxylate; nonionic; see Ethylan TF; also for cosmetics mfg.; dk. amber liq.; mild fatty amine odor; water sol.; sp.gr. 1.015; visc. 174 cs; HLB 14; flash pt. > 300 F (COC); pour pt. < 0 C; pH 8.5–10.0 (1% aq.); 100% act.

Ethylan TT-15. [360] Tallow amine ethoxylate; nonionic; wetting agent for metal cleaning and stripping of surface coatings, fiber antistat, used in cosmetics; pale br. paste; mild fatty amine odor; water sol.; sp.gr. 1.030; visc. 252 cs; HLB 14; flash pt. > 300 F (COC); pour pt. 0 C; pH 8.5–10.0 (1% aq.); 100% act.

Ethylan TU. [360] Nonyl phenol ethoxylate; nonionic; textile scouring agent, dry cleaning detergent, wetting and emulsifying agent for oils and aromatic solv.; pale straw clear liq.; negligible odor; water sol.; sp.gr. 1.055; visc. 340 cs; HLB 12.2; flash pt. > 400 F (COC); pour pt. < 0 C; pH 6–8 (1% aq.); 100% act.

Ethylan VPK. [360] PEG unsat. fatty acid ester; nonionic; textile oil emulsifier, fiber lubricant; pale clear liq.; sp.gr. 1.043; visc. 120 cs; HLB 12.4; 100% act.

Etocas 10. [159] PEG-10 castor oil; nonionic; cosmetic and essential oil solubilizer, emulsifier, lubricant, emollient, superfatting agent, antistat, softener, detergent; pale yel. liq.; sol. in ethanol, naphtha, MEK, trichlorethylene, disp. in water; HLB 6.3; cloud pt. < 20 C; sapon. no. 120–130; pH 6–7.5; 97% act.

Etocas 35. [159] PEG-35 castor oil; nonionic; see Etocas 10; pale yel. liq.; sol. see Ethocas 35; HLB 12.5; cloud pt. 35–40 C; sapon. no. 62–72; surf. tens. 41.5 dynes/cm (0.1% aq.); pH 6–7.5; 97% act.

Etocas 40. [159] PEG-40 castor oil; nonionic; see Etocas 10; pale yel. liq.; sol. see Etocas 35; HLB 13; cloud pt. 50 C; sapon. no. 60–65; pH 6–7.5; 97% act.

Etocas 60. [159] PEG-60 castor oil; nonionic; see Etocas 10; pale yel. soft paste; sol. see Etocas 35; HLB 14.7; cloud pt. 60 C; sapon. no. 45–50; surf. tens. 43.2 dynes/cm (0.1% aq.); pH 6–7.5.

Etocas 100. [159] PEG-100 castor oil; nonionic; see Etocas 10; pale yel. waxy solid; sol. see Etocas 35; HLB 16.5; cloud pt. 66 C; sapon. no. 25–35; surf. tens. 41.6 dynes/cm (0.1% aq.); pH 6–7.5; 97% act.

Eumulgin B-1. [289] Ceteareth-12; wetting agent and dispersant for paint systems; waxy solid; sp.gr. 0.95; solid. pt. 50–68 F.

Eumulgin B-2. [289] Ceteareth-20; see Eumulgin B-1; waxy solid; sp.gr. 0.95; solid. pt. 50–68 F.

Eumulgin B-3. [289] Ceteareth-30; see Eumulgin B-1; waxy solid; sp.gr. 0.95; solid. pt. 50–68 F.

Eumulgin L. [289] PPG-2-ceteareth-9; nonionic; emulsifier, solubilizer; liq.; 100% conc.

Eumulgin O10. [289] Oleth-10; nonionic; wetting agent and dispersant for paint systems; soft waxy solid; sp.gr. 0.95; solid. pt. 50–68 F.

Eumulgin RO-40. [289] PEG-40 hydrog. castor oil; emulsifier, solubilizer for perfume oils; visc. liq.

Eumulgin ST-8. [289] PEG-8 stearate; nonionic; emulsifier for o/w liq. emulsions and creams; waxy solid.

Euperlan PK 771. [289] Sodium laureth sulfate, glycol distearate, and cocamide MEA; anionic/nonionic; pearlescent base for shampoos and bath products; liq.; 45% conc.

Euperlan PK 776. [289] Sodium laureth sulfate, glycol distearate, and cocamide MEA; anionic/nonionic; see Euperlan PK 771; liq.; 41% conc.

Euperlan PK-810. [289] Sodium laureth sulfate, glycol stearate, and cocamide MEA; anionic/nonionic; pearlescent base for lotion shampoos and bath products; liq.; 37% conc.

Eureka 102. [72] Sulfated castor oil; anionic; emulsifier, detergent, plasticizer in finish coatings; lt. liq.; pH 7.6; 72% act.

Eureka 102-WK. [72] Sulfated castor oil; anionic; used in textile industry and component for shampoo bases and cosmetic formulations; liq.; 50% conc.

Eureka 392. [72] Sulfated fatty acid from veg. oil; anionic; emulsifier, carrier for refined oils, base for solv. fatliquor systems; liq.; pH 5.6; 75% act.

Eureka 400. [72] Sulfonated refined sperm oil; emulsifier, lubricant, softener used in leather processing; lt. liq.; pH 4.5; 87% act.

Eureka E-2. [72] Fatty acid diamine condensate deriv.; cationic; emulsifier for raw oils, lubricant, softener; heavy visc. liq.; dens. 8.32 lb/gal (60 F); sp.gr. 0.999 (60 F); pH 6–6.5 (10%); 100% act.

Examide-CS. [566] Fatty acid deriv.; amphoteric; detergent, emulsifier and scouring agent for use in textiles; liq.; 50% conc.

Examide-DA. [566] EO condensate; nonionic; detergent, wetting, emulsifier; surfactant used as scouring and leveling agent, dyeing assistant in textile industry; liq.; 50% conc.

Excel 84. [341] Monodiglyceride stearic acid base; nonionic; food emulsifier; powd.; HLB 2.8; 100% conc.

Excel 122. [341] Monodiglyceride oleic and stearic acid base; nonionic; food emulsifier;

plastic; HLB 3; 100% conc.

Excel 124. [341] Monodiglyceride stearic acid base; nonionic; food emulsifier; powd.; 100% conc.

Excel 150. [341] Monodiglyceride stearic acid base; nonionic; see Excel 124; powd.; HLB 3.2; 100% conc.

Excel 200. [341] Monoglyceride blend, distilled; nonionic; food emulsifier; powd.; HLB 3.5; 100% conc.

Excel 300. [341] Monodiglyceride propylene glycol blend; nonionic; defoaming agent for beverages; liq.; HLB 2.8; 100% conc.

Excel O-95F. [341] Monoglyceride oleic acid base, distilled; nonionic; food emulsifier; solid; HLB 3.5; 100% conc.

Excel O-95N. [341] Monoglyceride, dist.; nonionic; food emulsifier; plastic; 100% conc.

Excel O-95R. [341] Monoglyceride oleic acid base, dist.; nonionic; food emulsifier; solid; HLB 3.5; 100% conc.

Excel P-40S. [341] Monodiglyceride; nonionic; food emulsifier; powd.; HLB 2.8; 100% conc.

Excel T-95, T-95P, VS-95. [341] Monoglyceride, dist.; food emulsifier; bead, powd.; HLB 3.8; 100% conc.

F

Famodan MS. [273] Fatty acid sorbitan esters; food emulsifier; powd.; m.p. 54 C; sapon. no. 147–157.

Fancol HL. [232] Hydrog. lanolin; emollient, moisturizer, lubricant, plasticizer, chemical intermediate, humectant, mold release agent for pharmaceuticals, cosmetics, industrial applications; wh. solid, trace odor; sol. in IPM, min. oil, castor oil, acetone (@ 75 C); insol. water; sp.gr. 0.855–0.865; m.p. 48–53 C; sapon. no. 6 max.

Fancol LA. [232] Lanolin alcohol; emollient, thickener, emulsifier, stabilizer, plasticizer, superfatting agent, chemical intermediate, lubricant, humectant, mold release agent for cosmetics, pharmaceuticals, soaps, industrial applications; sol. in carbon tetrachloride, chloroform, IPM (@ 75 C), min. oil (@ 75 C), oleyl alcohol; insol. water; m.p. 56 C; acid no. 2 max.; sapon. no. 12 max.

Fancol LAO. [232] Min. oil and lanolin alcohol; nonionic; conditioner, surfactant, stabilizer, moisturizer/humectant, penetrant, emollient, and primary emulsifier for use in cosmetics and pharmaceuticals; plasticizer in aerosol formulas; yel. clear liq.; odorless; sol. in oils; insol. in water; sp.gr. 0.84–0.86; sapon. no. 3.0 max.

Fancol OA 50, 70, 80, 90, 95. [232] Oleyl alcohol; industrial plasticizer, emulsion stabilizer, antifoam and coupling agent, aerosol lubricant, petroleum additive, pigment dispersant; rust preventive; detergent, release agent, cosolvent, softener, tackifier, spreading agent used for metalworking, petrochemicals, pulp and paper, paints and coatings, plastics and polymers, food applications; chemical intermediate; liq.; sapon no. 1 max.; cloud pt. 30–40 C, 18–25 C, 12–18 C, 6 C, and 5 C max. resp.

Fancor Lanwax. [232] Natural lanolin wax ester; plasticizer in wax crayons; personal care and pharmaceutical products; extender and crystallization inhibitor for natural waxes in industrial applications; emulsifier; m.p. 48–52 C; sapon. no. 90–110.

Fancor LFA. [232] Lanolin fatty acids; emollient, stabilizer, emulsifier for personal care and pharmaceutical products; used in industrial leather treating, coatings, polishes, corrosion inhibitors, lubricants; wax-like solid; m.p. 57–65 C; sapon. no. 158–175.

Farmin 2C. [561] Sec. fatty di-*n*-alkyl amine; intermediate for textile finishing; softener; antistat; solid; 100% conc.

Farmin 20, 60, 68, 80, 86, AB, C. [561] Primary fatty amines; emulsifier, anticaking agent, textile additive, corrosion inhibitor, flotation reagent; liq./solid; 100% conc.

Farmin D86. [561] Sec. fatty di-*n*-alkyl amine; see Farmin 2C; solid; 100% conc.

Farmin DM20, 40, 60, 80, 86, DMC. [561] Tert. fatty *n*-alkyl dimethylamine; intermediate for corrosion inhibitors; quat. synthesis, benzalkonium chloride; liq.; 100% conc.

Farmin HT, O, S, T. [561] Primary fatty amines; see Farmin 20; liq.; 100% conc.

Farmin M2C, M2R86. [561] Tert. fatty di-*n*-alkyl methyl amine; see Farmin DM20; paste, solid; 100% conc.

Finsolv TN. [654] C_{12-15} alcohol benzoate; solubilizer, sunscreen vehicle, deoiler for creams and oils; liq.; 100% conc.

Flexricin 9. [446] Propylene glycol ricinoleate; nonionic; wetting agent, dye solv., wax plasticizer for textile, household, and cosmetic applications, rewetting dried skins; Gardner 2+ liq.; sol. in toluene, butyl acetate, MEK, ethanol; sp.gr. 0.96; visc. 3 stokes; 100% act.

Flexricin 13. [446] Glyceryl ricinoleate; nonionic; wetting agent, wax plasticizer, plasticizer and mold release agent for rubber polymers, antifoam agent, household and cosmetic applications, rewetting dried skins; Gardner 2 min. liq.; sol. in toluene, butyl acetate, MEK, ethanol; sp.gr. 0.985; visc. 8.8 stokes; 100% act.

Flexricin 15. [446] Glycol ricinoleate; nonionic; wetting agent, plasticizer, textile, household, and cosmetic applications, rewetting dried skins; Gardner 4 liq.; sol. in toluene, butyl acetate, MEK, ethanol; sp.gr. 0.965; visc. 3.9 stokes.

Flo-Mo AJ-85. [181] Tridecyl alcohol ethoxylate (7.5 EO); detergent, wetting agent; cleaning food handling equipment; liq.; cloud pt. 102–109 F; 85% act.

Flo-Mo AJ-100. [181] Tridecyl alcohol ethoxylate (7.5 EO); see Flo-Mo AJ-85; liq.; cloud pt. 102–109 F; 100% act.

Fluilan. [159] Lanolin oil; nonionic; w/o emulsifier, dispersant for pigments; conditioning agent, emollient, moisturizer, penetrant, superfatting agent; for skin care cosmetics, pharmaceuticals; golden visc. liq.; faint char. lanolin odor; 100% act.

Fluilanol. [159] Lanolin oil and oleth-3; nonionic; emulsifier, emollient, dispersant, wetting agent, superfatting agent used in cosmetics, soaps, detergents; pale liq.; faint sterol odor; 100% act.

Fluorad FC-93. [377] Ammonium perfluoroalkyl sulfonate; anionic; wetting agent in etching sol'ns. in semiconductor devices; clear liq.; sol. in alcohol/water mixts.; sp.gr. 1.1; flash pt. 81 F (PMCC); pH 7 (1% aq.); 25% act.

Fluorad FC-95. [377] Potassium perfluoroalkyl sulfonate; anionic; wetting and foaming agents for coatings, etchants, plating baths, cleaning systems; powd.; low. sol. in water and most inorg. solv.; surf. tens. 22 dynes/cm (0.1% aq.); pH 7–8 (0.1% aq.); 100% act.

Fluorad FC-98. [377] Potassium perfluoroalkyl sulfonate; see Fluorad FC-95; powd.; low. sol. in water and most inorg. solv.; surf. tens. 40 dynes/cm (0.1% aq.); pH 6–8 (0.1% aq.); 100% act.

Fluorad FC-99. [377] Amine perfluoroalkyl sulfonate; anionic; see Fluorad FC-95; clear yel. to amber liq.; misc. with water; surf. tens. 22 dynes/cm (0.4% aq.); pH 6–7; 25% act. in water.

Fluorad FC-100. [377] Fluorinated alkyl mixt.; amphoteric; surfactant used in electrolyte systems; plating bath antimist and leak detector sol'ns.

Fluorad FC-120. [377] Ammonium perfluoroalkyl sulfonate; anionic; wetting, leveling agent, surfactant in aq. coating systems; low visc. liq.; sp.gr. 1.06; pH 8.5–9.5; 25% act.

Fluorad FC-126. [377] Fluorochemical; anionic; surfactant; powd.; sol. in water, acetone, methanol; decomp. @ 120 C surf. tens. 44 dynes/cm (0.001%); pH 4–6; 100% act.

Fluorad FC-128. [377] Potassium fluorinated alkyl carboxylate; anionic; wetting, spreading and leveling agent used in alkaline cleaners, leak detection sol'ns.; powd.; forms gel in water; surf. tens. 17 dynes/cm (0.1–0.2% aq.); pH 7–8; 100% act.

Fluorad FC-129. [377] Potassium fluorinated alkyl carboxylate; anionic; wetting and leveling agent, used in alkaline cleaners, floor polishes; amber liq.; limited sol. in aq. acids and bases; surf. tens. 17 dynes/cm (0.1–0.2% aq.); pH 7–8; 50% act.

Fluorad FC-134. [377] Fluorinated alkyl quat. ammonium iodides; cationic; wetting, spreading and leveling agent, used in cleaning and polishing products, photographic development, leak detector sol'ns.; powd.; surf. tens. 17 dynes/cm (0.1% aq.); pH 3–4; 100% act.

Fluorad FC-135. [377] Fluorinated alkyl quat. ammonium iodides; cationic; wetting, spreading and leveling agent, used in cleaning and polishing products; dk. amber liq.; surf. tens. 17 dynes/cm (0.2% aq.); pH 3–4; 50% act.

Fluorad FC-143. [377] Ammonium perfluoroalkyl carboxylate; anionic; emulsifier for fluoropolymer emulsions, wetting agent in alkaline media; powd.; sol. in water, methanol, acetone, dens. 0.6–0.7 g/cc; decomp. @ 130 C; surf. tens. 44 dynes/cm (0.1% aq.); pH 4–6 (0.5 aq.); 100% act.

Fluorad FC-170-C. [377] Fluorinated alkyl POE ethanol; nonionic; wetting, spreading and leveling agent for floor polish formulations; amber liq.; sol. in HCl, alcohols; sp.gr. 1.32; cloud pt. < 0 C; flash pt. > 220 F

(PMCC); surf. tens. 20 dynes/cm (0.1% aq.); pH 7–8 (1% aq.); 95% act.

Fluorad FC-171. [377] Fluorinated alkyl alkoxylate; nonionic; wetting, spreading and leveling agent for floor polish formulations, coatings; amber liq.; sol. in butyl Cellosolve, toluene, MEK, IPA, dibutyl phthalate; sp.gr. 1.4; flash pt. > 220 F (PMCC); pour pt. 20 F; surf. tens. 20 dynes/cm (0.1% aq.); 100% act.

Fluorad FC-430. [377] Fluorinated alkyl ester; nonionic; wetting, leveling, spreading and flow control agents for water-based coatings and polymer systems, soldering systems; straw to amber liq.; sol. > 20 g/100 ml in water, Cellosolve acetate, MEK, toluene; sp.gr. 1.15; visc. 15,000 cp; flash pt. > 300 F (TOC); surf. tens. 30.3 dynes/cm (0.1 g/100 ml aq.); 100% act.

Fluorad FC-431. [377] Fluorinated alkyl ester; see Fluorad FC-430; straw to amber liq.; sol. > 20 g/100 ml in butyl Cellosolve, Cellosolve acetate, MEK, toluene; sp.gr. 1.05; visc. 200 cp; flash pt. 40 F (TOC); surf. tens. 36.6 dynes/cm (0.1 g/100 ml aq.); 50% act.

Fluorad FC-432. [377] Fluorinated alkyl ester; nonionic; surfactant; low visc. liq.; sol. in hydrocarbons; sp.gr. 0.78; flash pt. 19.4 (TCC); surf. tens. 20.5 (2 g/100 g ethyl acetate); 25% act.

Fluorad FC-750. [377] Fluorinated alkyl quat. ammonium iodides; cationic; well stimulation additive, offers lowest surf. tens. for aq. acids, enhances activity of org. corrosion inhibitors; dk. liq.; sp.gr. 1.2; flash pt. 53 F (PMCC); pour pt. 32–40 F; surf. tens. 17.5 dynes/cm (100 ppm aq.); 50% solids in 2/1 IPA/water.

Fluorad FC-760. [377] Fluorinated alkyl alkoxylates; nonionic; well stimulation additive, useful at low levels to reduce surf. tens. of fluids used for acidizing, fracturing and water flooding; amber liq.; sp.gr. 1.4; flash pt. > 220 F (PMCC); pour pt. 20 F; surf. tens. 19 dynes/cm (100 ppm aq.).

Fluowet Brands. [301] Fluorochemical; anionic/cationic/nonionic; surfactant, intermediate products; liq., solid; 100% conc.

Foamaster 206-A, 267-A, 335-A [188] Nonsilicone type based on chemically modified oils and fats; nonionic; defoamer for textile industry; liq.; water disp.; 100% act.

Foamaster 1407-50. [188] Nonionic; defoamer for fluidized paste; water disp.

Foamaster 1719-A. [188] Nonionic; defoamer for use in gelatine; liq.; 50% act.

Foamaster 8034. [188] Nonionic; defoamer for paints, adhesives, rubber, and latex; liq.; 100% act.

Foamaster B. [188] Nonioic; predispersible defoamer for latex systems; liq.; 100% act.

Foamaster DD-72. [188] Nonionic; defoamer for emulsion paints, wallpaper coatings, inks; liq.; 100% act.

Foamaster DF-177-F, -178. [188] Nonionic; foam suppressor for coating and adhesive formulations; liq.; 100% act.

Foamaster DF-198-L. [188] Defoamer for paints based on fine particle size acrylic emulsions; liq.; 100% act.

Foamaster DNH-1. [188] Nonionic; defoamer for emulsion paints and polymers; liq.; 100% act.

Foamaster G. [188] Disp. latex stripping defoamer; liq.; 100% act.

Foamaster JMY. [188] Nonionic; foam control agent for adhesives, rubber, and latex; liq.; 100% act.

Foamaster KF-99. [188] Nonsilicone type based on chemically modified oils and fats; defoamer for textile industry; liq.; disp. in water; 100% act.

Foamaster NDW. [188] Nonionic; defoamer for water-based paints, adhesive and cement screeds, rubber and latex; liq.; 100% act.

Foamaster NS-20. [188] Nonionic; silicone-free defoamer for coatings or paper stock, air knife coatings; liq.; 100% act.

Foamaster NXZ. [188] Defoamer for syn. latices, paints, adhesives, blade and roll coatings, emulsifiable latex stripping; liq.; 100% act.

Foamaster P. [188] Nonionic; foam controller in polyvinyl alcohol systems in adhesive industry; liq.; 35% act.

Foamaster PD-1. [188] Nonionic; defoamer for dry mixes, adhesives and cement screeds, joint cements, spackles; powd.; 67% act.

Foamaster TDB. [188] Nonionic; foam control agent in dry adhesives, joint cements, spackle, rubber and latex; liq.; 100% act.

Foamaster TBD-1. [188] Defoamer for water-sol. resins; liq.; 100% act.

Foamaster V. [188] Nonionic; disp. latex stripping defoamer; liq.; 100% act.

Foamaster VL. [188] Nonionic; emulsifiable latex monomer stripping defoamer, used in latex and emulsion paints; liq.; 100% act.

Foamer. [283] Fatty acid amide sulfosuccinate ester; anionic; stabilizer for shampoos, detergents, and bubble baths; yel. liq.; 40% act.

Forlanit E. [291] Phosphoric acid hydroxyalkyl ester; o/w emulsifier for cosmetic and pharmaceutical emulsions; flakes; 99% conc.

Forlanit P. [291] Sodium lauryl ether phosphate; anionic; wetting agent, flotation aux. for hot-baths containing cyanides; liq.; 30% conc.

Forlanon. [291] Sodium lauryl ether phosphate; anionic; see Forlanit P; liq.; 10% conc.

Fosfamide N. [289] Fatty amido phosphate complex; anionic; detergent, foaming and wetting agent for textile, all-purpose, institutional, and metal cleaners; liq.; 100% act.

Fosterge BA-14 Acid. [289] Mono and dialkyl phenoxy POE acid phosphate; anionic; detergent intermediate for cleaners and emulsifier for min. oils and pesticides; liq.; dens. 8.50 lb/gal; 98–100% act.

Fosterge H Acid. [289] Mono and dialkyl phenoxy POE acid phosphate; emulsifier; liq.; dens. 8.8 lb/gal; 98–100% act.

Fosterge LF Acid. [289] Alkylphenoxy POE acid phosphate; anionic; intermediate for foaming cleaners and solv. emulsion cleaners; emulsifier for pesticides; corrosion inhibitor; liq.; dens. 9.17 lb/gal; 98–100% act.

Fosterge R Acid. [289] Mono and dialkyl phosphoric acid; anionic; intermediate for emulsifying, penetrating, and anticorrosion compds.; dens. 8.5 lb/gal; liq.; 100% act.

Fosterge RD. [289] Fosterge R Acid diethanolamine salt; anionic; emulsifier, penetrant, and anticorrosion agent used in industrial cleaners; liq.; dens. 9.15 lb/gal; 86–88% act.

Fosterge W Acid. [289] Alkyl phenoxy POE phosphoric acid; anionic; detergent, emulsifier, coupling agent and corrosion inhibitor; dedusting agent for alkaline powds.; liq.; dens. 9.29 lb/gal; 100% act.

Fostex S. [289] Polymeric alkyl phosphonate; anionic; scale inhibitor and dispersant for water treating applications; liq.; 48% act.

Fostex SN. [289] Polymeric alkyl phosphonate sodium salt; anionic; see Fostex S; liq.; 43% act.

G

Gafac BH-650. [245] Free acid of complex org. phosphate ester; anionic; pesticide emulsifier, industrial and household detergent, hard surface cleaners; clear visc. liq.; sol. in butyl Cellosolve, ethanol, water; dens. 11.4 lb/gal; sp.gr. 1.36–1.38 (50 C); pour pt. < 5 C; pH < 3 (10%); 100% act.

Gafac BI-729. [245] Free acid of complex org. phosphate ester; anionic; pesticide emulsifier, industrial and household detergent; clear visc. liq.; sol. in butyl Cellosolve, ethanol, water; dens. 10.3 lb/gal; sp.gr. 1.23–1.25 (50 C); pour pt. < 0 C; pH 2.5 (10%); 89% act.

Gafac BI-750. [245] Free acid of complex org. phosphate ester; anionic; see Gafac BI-729; clear visc. liq.; sol. in butyl Cellosolve, ethanol, water; dens. 10.6 lb/gal; sp.gr. 1.26–1.28 (50 C); pour pt. < 0 C; pH < 2.5 (10%); 100% act.

Gafac GB-520. [245] Partial sodium salt of a complex org. phosphate ester; anionic; textile antistat, lubricant, emulsifier for fiber finishes; hazy visc. liq.; sol. in xylene, butyl Cellosolve, perchloroethylene, ethanol; disp. water; dens. 8.6 lb/gal; sp.gr. 1.03–1.04; pour pt. 9 C; pH 2.5–4.5 (10%); 98% act.

Gafac LO-529. [245] Sodium nonoxynol-6 phosphate; anionic; detergent for household and industrial formulations; clear visc. liq.; sol. in kerosene, Stod., xylene, butyl Cellosolve, perchloroethylene, water; dens. 9.1 lb/gal; sp.gr. 1.05–1.15; pour pt. 6 C; pH 5–6 (10%); 88% act.

Gafac MC-470. [245] Partial sodium salt of a complex org. phosphate ester; anionic; household and industrial detergent, textile antistat, lubricant, emulsifier; clear visc. liq.; sol. in min. oil, kerosene, xylene, perchloroethylene, disp. in water; dens. 8.6 lb/gal; sp.gr. 1.02–1.04; pour pt. 0 C; pH 5–6.5 (10%); 95% act.

Gafac PE-510. [245] Free acid of complex org. phosphate ester; anionic; pesticide emulsifier, household, industrial, drycleaning detergent, emulsion polymerization; clear to slightly hazy visc. liq.; sol. in kerosene, Stod., xylene, butyl Cellosolve, perchloroethylene, ethanol; disp. in water; dens. 9 lb/gal; sp.gr. 1.08–1.09; pour pt. 20 C; pH < 2.5 (10%); 100% act.

Gafac RA-600. [245] Deceth-4 phosphate; anionic; detergent, dispersant, wetting agent for household, industrial, textile wet processing; clear to slightly hazy visc. liq.; sol. in kerosene, xylene, butyl Cellosolve, perchloroethylene, ethanol, water; dens. 8.9 lb/gal; sp.gr. 1.06–1.08; pour pt. < 0 C; pH < 2.5 (10%); 98.50% act.

Gafac RB-400. [245] Free acid of complex org. phosphate ester; anionic; lubricant, antistat, emulsifier for fibers and metal; opaque visc. liq.; sol. in min. oil, kerosene, Stod., xylene, butyl Cellosolve, perchloroethylene, ethanol; disp. in water; dens. 8.6 lb/gal; sp.gr. 1.03–1.04; pour pt. 18 C; pH < 2.5 (10%); 98% act.

Gafac RD-510. [245] Free acid of complex org. phosphate ester; anionic; emulsifier, antistat, lubricant, solubilizer for fibers, metals, cosmetics; clear visc. liq.; sol. in min. oil, kerosene, xylene, butyl Cellosolve, perchloroethylene, ethanol; disp. in water; dens. 8.8 lb/gal; sp.gr. 1.05–1.06; pour pt. 13 C; pH < 2.5 (10%); 98% act.

Gafac RE-410. [245] Free acid of complex org. phosphate ester; anionic; detergent, emulsifier, wetting agent, dispersant, antistat, lubricant for pesticides, drycleaning, textile wet processing, metals; clear visc. liq.; sol. in kerosene, Stod., xylene, butyl Cellosolve, perchloroethylene, ethanol; disp. in water; dens. 9.1 lb/gal; sp.gr. 1.00–1.20; pour pt. 18 C; pH < 2.5 (10%); 100% act.

Gafac RE-610. [245] Nonoxynol-9 phosphate; anionic; see Gafac RE-410; also for household and industrial detergents, fabric finishes, emulsion polymerization; slightly hazy visc. liq.; sol. in xylene, butyl Cellosolve, perchloroethylene, ethanol, water; dens. 9.2 lb/gal; sp.gr. 1.1–1.12; pour pt. < 0 C; 100% act.

Gafac RE-870. [245] Free acid of complex org. phosphate ester; anionic; see Gafac RE-410; also for cosmetics; soft waxy paste; sol. in xylene, butyl Cellosolve, perchloroethylene, ethanol, water; dens. 9.6 lb/gal; sp.gr. 1.1–1.2 (50 C); pour pt. 29 C; pH < 2.5 (10%); 99% act.

Gafac RE-960. [245] Free acid of complex org. phosphate ester; anionic; see Gafac RE-410; also for emulsion polymerization; soft waxy paste; sol. in butyl Cellosolve, ethanol, water; dens. 9.8 lb/gal; sp.gr. 1.17–1.18 (50 C); pour pt. 20 C; pH < 2.5 (10%); 90% act.

Gafac RK-500. [245] Free acid of complex org. phosphate ester; anionic; lubricant, antistat, emulsifier for metal treatment; clear, visc. liq.; sol. in xylene, butyl Cellosolve, perchloroethylene, ethanol, water; dens. 9.1 lb/gal; sp.gr. 1.05–1.15; pour pt. < 0 C; pH < 2.5 (10%); 100% act.

Gafac RM-410. [245] Nonyl nonoxynol-7 phosphate; anionic; detergent, emulsifier for drycleaning, pesticides; slightly hazy, visc. liq.; sol. in min. oil, kerosene, Stod., xylene, butyl Cellosolve, perchloroethylene, ethanol; disp. in water; dens. 8.8 lb/gal; sp.gr. 1.05–1.07; pour pt. 19 C; pH < 2.5 (10%); 100% act.

Gafac RM-510. [245] Nonyl nonoxynol-10 phosphate; anionic; emulsifier, solubilizer for pesticides, cosmetics; hazy, visc. liq.; sol. in kerosene, Stod., xylene, butyl Cellosolve, perchloroethylene, ethanol; disp. in water; dens. 8.8 lb/gal; sp.gr. 1.05–1.07; pour pt. 5 C; pH < 2.5 (10%); 100% act.

Gafac RM-710. [245] Nonyl nonoxynol-15 phosphate; anionic; emulsifier, solubilizer for cosmetics; hazy, visc. liq.; sol. in xylene, butyl Cellosolve, perchloroethylene, ethanol, water; dens. 8.9 lb/gal; sp.gr. 1.06–1.08; pour pt. 15 C; pH < 3 (10%); 100% act.

Gafac RO-310. [245] Free acid of complex org. phosphate ester; anionic; cosmetic emulsifier for lotions, creams, hair grooming; clear visc. liq.; sol. in min. oil, kerosene, xylene, butyl Cellosolve, perchloroethylene, ethanol, disp. in water; dens. 8.4; sp.gr. 1.00–1.02; pour pt. < 0 C; pH < 2.5 (10%); 98% act.

Gafac RO-660. [245] Free acid of complex org. phosphate ester; anionic; see Gafac RO-310; soft waxy paste; sol. in min. oil, kerosene, Stod., xylene, butyl Cellosolve, perchloroethylene, ethanol, water; dens. 8.6; sp.gr. 1.03–1.04; pour pt. 25 C; pH < 2.5 (10%); 98% act.

Gafac RP-710. [245] Free acid of complex org. phosphate ester; anionic; household and industrial detergent; hard surface cleaning; slightly hazy visc. liq.; sol. in xylene, butyl Cellosolve, perchloroethylene, ethanol, water; dens. 10.2; sp.gr. 1.22–1.23; pour pt. –7 C; pH < 2.5 (10%); 100% act.

Gafac RS-410. [245] Free acid of complex org. phosphate ester; anionic; pesticide emulsifier, drycleaning detergent, textile wetting agent, antistat, lubricant for fiber and metal treatment; hazy visc. liq.; sol. in min. oil, kerosene, Stod., xylene, butyl Cellosolve, perchloroethylene, ethanol, disp. in water; dens. 8.6 lb/gal; sp.gr. 1.03–1.04; pour pt. –15 C; pH < 2.5 (10%); 100% act.

Gafac RS-610. [245] Trideceth-6 phosphate; anionic; see Gafac RS-410; haxy, visc. liq.; sol. in kerosene, Stod., xylene, butyl Cellosolve, perchloroethylene, ethanol; disp. in water; dens. 8.7 lb/gal; sp.gr. 1.04–1.06; pour pt. < 0 C; pH < 2.5 (10%); 100% act.

Gafac RS-710. [245] Free acid of complex org. phosphate ester; anionic; see Gafac RS-410; opaque visc. liq.; sol. in Stod., xylene, butyl Cellosolve, perchloroethylene, ethanol, water; dens. 8.7 lb/gal; sp.gr. 1.04–1.06; pour pt. 18 C; pH < 2.5 (10%); 100% act.

Gafamide CDD-518. [245] Cocamide DEA; anionic; foam stabilizer, visc. builder, detergent, wetting agent used in shampoos, bubble baths, laundry, industrial cleaners; clear visc. liq.; sol. in water, ethanol, perchloroethylene butyl Cellosolve; sp.gr. 1.0; visc. 1050 cst (Ostwald-Fenske); pH 10 (1%); 84% act.

Gafterge AW-123. [245] Phosphate ester; anionic; surfactant, wetting agent, penetrant, and dispersant used in textile applications; liq.

Gaftex 288. [245] Sulfated alkyl ester; anionic; penetrant, wetting and leveling agent for textile dyeing; liq.; sol. in water.

Gaftex CD-169. [245] Phosphate ester deriv. complex; anionic; surfactant, penetrant, dispersant, and wetting agent used in textile dyeing; liq.

Gaftex COM-154, -154 Conc. [245] Polyether; nonionic; wetter and dispersant for dyes; emulsifier for residual oil and lubricant in dyebath; liq.

Gaftex DP-100. [245] Phosphate ester deriv., modified; anionic; detergent and wetting agent for use in becks, pkg. machines, rotary tubs used in dyeing; liq.

Gaftex WSP. [245] Sulfonated org. ester; anionic; wetting agent and penetrant used in dyeing; liq.

Ganex V, P Series. [245] Alkylated vinylpyrrolidone polymers; dispersant, solubilizer in aq. agric. chemicals, pigment suspensions, personal care products, dyes; petroleum additive; antiredeposition agent in drycleaning; liq. to waxy solid.

Gantrez AN. [245] PVM/MA copolymer; anionic; dispersant, coupling, stabilizing agent, thickener, emulsifier, solubilizer, corrosion inhibitor, film former, antistat, used in agric., paper and textile industries, chemical processing, industrial products, detergents, cosmetics; wh. fluffy powd.; sol. in water and several org. solvs.; dens. 0.32 g/cc; sp.gr. 1.37; pH 2 (free acid, 5% aq.); 100% conc.

Gantrez AN-119, -139, -149, –169, –179. [245] PVM/MA copolymer; thickener, dispersant and stabilizer used in emulsion polymerization; gelling agent, coupler, and suspending aid; used in adhesives, detergents; powd.; sol. in water, acid, caustic, and org. solv.

Gardilene IPA. [15] Isopropylamine alkylbenzene sulfonate; anionic; emulsifier, drycleaning assistant, dispersant; amber, clear visc. liq.; pH 6.5; flamm.; 85% act. in solv.

Gardilene S25L. [15] Sodium alkylbenzene sulfonate; anionic; wetting, dispersing, and foaming agent, emulsifier in emulsion polymerization; pale yel. clear visc. liq.; visc. 3300 cps; clear pt. < 6 C; pH 7.2 (10%); 25 % act. in water.

Gardinol CX. [15] Sodium cetyl/oleyl alcohol sulfate; anionic; detergent, wetting agent, dispersant, dyeing assistant and emulsifier used in fiber applications, hand cleaners, laundry detergents; off-wh. to cream soft paste, visc. liq.> 30 C; sp.gr. 1.05 (35 C); pH 7.0 ± 0.5 (10% aq.); > 80% biodeg.; 30.0 ± 1.0% act.

Gardinol CXM. [15] Sodium cetyl/oleyl sulfate, modified; textile scouring agent; off-wh. to cream paste, visc. liq. > 30 C; dissolves in water; sp.gr. 1.02 (30 C); pH 7.5 ± 0.5 (10%); > 80% biodeg.; 23.0 ± 0.7% act.

Gardinol WA Paste. [527] Sodium lauryl sulfate; anionic; detergent for shampoos, fabrics; emulsifier for cosmetics; wh. to cream stiff paste; 40% conc.

Gardiquat 12H. [15] C_{12} alkyl benzyl dimethyl ammonium chloride; bactericide in disinfectants and antiseptics; used in food processing industries and oil drilling muds; colorless to pale yel. clear liq.; sp.gr. 0.98; cloud pt. 0 C; pH 7.0 ± 0.5 (1% aq.); 50.0 ± 1.0% act. in water.

Gardiquat 1450. [15] Alkyl dimethyl benzyl ammonium chloride (USP); germicide, deodorant, algicide, slimicide; almost water wh. clear liq.; cloud pt. 1 C; pH 7.0 ± 0.5 (1%); flamm.; 50% act.

Gardiquat 1480. [15] Alkyl dimethyl benzyl ammonium chloride (USP); germicide, deodorant, algicide, slimicide; almost water wh. clear liq.; sp.gr. 0.945; cloud pt. 1 C max.; flash pt. 32 C (PMCC); pH 7 ± 0.5 (1%); 80% act.

Gardiquat SV 480. [15] Alkyl dimethyl benzyl ammonium chloride; germicide, deodorant, algicide, slimicide; water wh. clear liq.; sp.gr. 0.929; visc. 106 s (#4 cup); cloud pt. < 0 C; flash pt. 29 C (PMCC); pH 7.2 (1%); 80% act.

Gardisperse AC. [15] Sodium alkyl phenoxy POE sulfate; anionic; emulsifier, dispersant, wetting agent; amber clear liq.; visc. 120 cP max.; cloud pt. 2 C max.; pH 8.5 ± 1.0 (2%); 30.0 ± 1.0% act.

Geitol RC-100. [607] Mixt. of nonionic agent and specific builders; detergent, reduction cleaning agent for polyester fibers; powd.; 100% conc.

Geleol. [247] Glyceryl stearate; nonionic; emulsifier; solid; 100% conc.

Gelot 64. [247] PEG-75 stearate and glyceryl stearate; nonionic; emulsifier for o/w creams and ointments; solid; HLB 10.0; 100% conc.

Gemtex PA-75. [237] Dioctyl sodium sulfosuccinate; wetting agent for textile uses and emulsion polymerization; dye leveler; bubble baths, bath oils; liq.; 75% conc.

Gemtex PA-85P, PAX-60. [237] Dioctyl sodium sulfosuccinate, anhyd.; see Gemtex PA-75; liq.; 85 and 60% conc. resp.

Gemtex SC-40, -70, -75. [237] Dioctyl sodium sulfosuccinate; wetting/rewetting agent, dye leveler for textiles; liq.; 40, 70, 75% conc. resp.

Gemtex SC-75E, SC Powd. [237] Dioctyl sodium sulfosuccinate; see Gemtex PA-75; liq. and powd. resp.; 75 and 83% conc. resp.

Genagen CA-050. [301] Fatty acid alkylolamide polyglycol ether; nonionic; cleansing skin protective component for detergents and cleaning agents; 100% conc.

Genagen CD. [301] Fatty acid ester; nonionic; lubricant for fiber preparation; liq.

Genamin CC, CS, OL, SH, TA Brands. [301] Fatty amine and diamine, primary, sec., and tert.; cationic; basic material for paints and varnishes, road construction, anticaking agent; 100% conc.

Genamin O, S, T Brands. [301] Fatty amine ethoxylate (2–25 EO); cationic; surface act. basic material for industrial fields; 100% conc.

Genamine C-020. [44] Coconut fatty amine oxethylate; cationic; raw material for min. oil additives, insecticides, pesticides, cosmetic bases, adhesives; clear liq.; sol. in min. oil, turbid in water (10 g/l); sp.gr. 0.895 (50 C); visc. 28.8 cps (50 C); flash pt. 188 C (Marcusson); surf. tens. 26 dynes/cm; pH 9–10; 100% act.

Genamine C-050. [44] Coconut fatty amine oxethylate; cationic; see Genamine C-020; clear liq.; sol. in water, min. oil; sp.gr. 0.95 (50 C); visc. 30.4 cps (50 C); flash pt. 224 C (Marcusson); surf. tens. 32 dynes/cm; pH 9–10; 100% act.

Genamine C-080. [44] Coconut fatty amine oxethylate; cationic; see Genamine C-020; clear liq.; sol. in water, turbid in min. oil; sp.gr. 0.988 (50 C); visc. 35.2 cps (50 C); flash pt. 249 C (Marcusson); surf. tens. 37 dynes/cm; pH 9–10; 100% act.

Genamine C-100. [44] Coconut fatty amine oxethylate; cationic; see Genamine C-020; clear liq.; sol. in water, turbid in min. oil; sp.gr. 1.0 (50 C); visc. 37 cps (50 C); flash pt. 273 C (Marcusson); surf. tens. 40 dynes/cm; pH 9–10; 100% act.

Genamine C-150. [44] Coconut fatty amine oxethylate; cationic; see Genamine C-020; clear liq.; sol. in water; sp.gr. 1.027 (50 C); visc. 45.5 cps (50 C); flash pt. 287 C (Marcusson); surf. tens. 44 dynes/cm; pH 9–10; 100% act.

Genamine C-200. [44] Coconut fatty amine oxethylate; cationic; see Genamine C-020; clear liq.; sol. in water, min. oil; sp.gr. 1.043 (50 C); visc. 55 cps (50 C); flash pt. 295 C (Marcusson); surf. tens. 45 dynes/cm; pH 9–10; 100% act.

Genamine C-250. [44] Coconut fatty amine oxethylate; cationic; see Genamine C-020; turbid liq.; sol. in water; sp.gr. 1.059 (50 C); visc. 65.2 cps (50 C); flash pt. 298 C (Marcusson); surf. tens. 47 dynes/cm; pH 9–10; 100% act.

Genamine O-020. [44] Oleylamine oxethylate; cationic; see Genamine C-020; turbid liq.; sol. in min. oil; sp.gr. 0.883 (50 C); visc. 37.4 cps (50 C); flash pt. 220 C (Marcusson); pH 9–10; 100% act.

Genamine O-050. [44] Oleylamine oxethylate; cationic; see Genamine C-020; clear liq.; sol. in min. oil; sp.gr. 0.956 (50 C); visc. 45.3 cps (50 C); flash pt. 269 C (Marcusson); surf. tens. 42 dynes/cm; pH 9–10; 100% act.

Genamine O-080. [44] Oleylamine oxethylate; cationic; see Genamine C-020; clear liq.; sol. in water min. oil; sp.gr. 0.983 (50 C); visc. 49.6 cps (50 C); flash pt. 282 C (Marcusson); surf. tens. 39 dynes/cm; pH 9–10; 100% act.

Genamine O-100. [44] Oleylamine oxethylate; cationic; see Genamine C-020; turbid liq.; sol. in water; sp.gr. 0.995 (50 C); visc. 52.1 cps (50 C); flash pt. 288 C (Marcusson); surf. tens. 41 dynes/cm; pH 9–10; 100% act.

Genamine O-150. [44] Oleylamine oxethylate; cationic; see Genamine C-020; turbid liq.; sol. in water; sp.gr. 1.024 (50 C); visc. 65.1 cps (50 C); flash pt. 307 C (Marcusson); surf. tens. 45 dynes/cm; pH 9–10; 100% act.

Genamine O-200. [44] Oleylamine oxethylate; cationic; see Genamine C-020; paste; sol. in water; sp.gr. 1.063 (50 C); visc. 75.9 cps (50 C); flash pt. 314 C (Marcusson); surf. tens. 48 dynes/cm; pH 9–10; 100% act.

Genamine O-250. [44] Oleylamine oxethylate; cationic; see Genamine C-020; wax; sol. in water; sp.gr. 1.08 (50 C); visc. 89.2 cps (50 C); flash pt. 317 C (Marcusson); surf. tens.

51 dynes/cm; pH 9–10; 100% act.

Genamine S-020. [44] Stearylamine oxethylate; cationic; see Genamine C-020; wax; sol. in min. oil, turbid in water; sp.gr. 0.889 (50 C); visc. 40.6 cps (50 C); flash pt. 219 C (Marcusson); pH 9–10; 100% act.

Genamine S-050. [44] Stearylamine oxethylate; cationic; see Genamine C-020; clear liq.; sol. in min. oil, turbid in water; sp.gr. 0.943 (50 C); visc. 39 cps (50 C); flash pt. 258 C (Marcusson); surf. tens. 35 dynes/cm; pH 9–10; 100% act.

Genamine S-080. [44] Stearylamine oxethylate; cationic; see Genamine C-020; turbid liq.; sol. in water, min. oil; sp.gr. 0.974 (50 C); visc. 43.3 cps (50 C); flash pt. 273 C (Marcusson); surf. tens. 39 dynes/cm; pH 9–10; 100% act.

Genamine S-100. [44] Stearylamine oxethylate; cationic; see Genamine C-020; turbid liq.; sol. in water, min. oil; sp.gr. 0.987 (50 C); visc. 46.1 cps (50 C); flash pt. 277 C (Marcusson); surf. tens. 42 dynes/cm; pH 9–10; 100% act.

Genamine S-150. [44] Stearylamine oxethylate; cationic; see Genamine C-020; turbid liq.; sol. in water; sp.gr. 1.011 (50 C); visc. 55.6 cps (50 C); flash pt. 280 C (Marcusson); surf. tens. 44 dynes/cm; pH 9–10; 100% act.

Genamine S-200. [44] Stearylamine oxethylate; cationic; see Genamine C-020; turbid liq.; sol. in water, turbid in min. oil; sp.gr. 1.033 (50 C); visc. 67.8 cps (50 C); flash pt. 281 C (Marcusson); surf. tens. 47 dynes/cm; pH 9–10; 100% act.

Genamine S-250. [44] Stearylamine oxethylate; cationic; see Genamine C-020; paste; sol. in water; sp.gr. 1.050 (50 C); visc. 81.8 cps (50 C); flash pt. 282 C (Marcusson); surf. tens. 49 dynes/cm; pH 9–10; 100% act.

Genaminox CS. [301] Coco dimethyl amine oxide; nonionic; foaming agent and stabilizer, thickener for personal care products; liq.; 30% conc.

Genaminox KC. [301] Cocamine oxide; nonionic; see Genaminox CS; liq.; 30% conc.

Genapol AMS. [44] TEA-PEG-3 cocamide sulfate; anionic; detergent, foaming agent used in top-grade cosmetics cleansers; lime soap dispersant; clear yel. low-visc. liq.; weak odor; dens. 1.0 g/cm³; visc. 200 cps max.; pH 6.5–8.0 (1%); biodeg.; 40% act.

Genapol C-020. [44] Coconut fatty alcohol polyglycol ether; nonionic; raw material for mfg. of textile, leather, paper auxs., detergents, emulsifiers, cosmetics; turbid liq.; sol. in min. oil, benzene, turbid in water; sp. gr. 0.892 (50 C); visc. 10.6 cps (50 C); flash pt. 170 C (Marcusson); surf. tens. 34 dynes/cm; pH 9–10; 100% act.

Genapol C-050. [44] Coconut fatty alcohol polyglycol ether; nonionic; see Genapol C-020; turbid liq.; sol. in min. oil, benzene, turbid in water; sp. gr. 0.952 (50 C); visc. 17.6 cps (50 C); flash pt. 201 C (Marcusson); surf. tens. 32 dynes/cm; 100% act.

Genapol C-080. [44] Coconut fatty alcohol polyglycol ether; nonionic; see Genapol C-020; paste; sol. in water, benzene, turbid in min. oil; sp. gr. 0.979 (50 C); visc. 25.4 cps (50 C); flash pt. 246 C (Marcusson); surf. tens. 36 dynes/cm; 100% act.

Genapol C-100. [44] Coconut fatty alcohol polyglycol ether; nonionic; see Genapol C-020; paste; sol. in water, benzene, turbid in min. oil; sp. gr. 0.990 (50 C); visc. 30.4 cps (50 C); flash pt. 251 C (Marcusson); surf. tens. 38 dynes/cm; 100% act.

Genapol C-150. [44] Coconut fatty alcohol polyglycol ether; nonionic; see Genapol C-020; wax; sol. in water, benzene; sp. gr. 1.027 (50 C); visc. 45.8 cps (50 C); flash pt. 260 C (Marcusson); surf. tens. 43 dynes/cm; 100% act.

Genapol C-200. [44] Coconut fatty alcohol polyglycol ether; nonionic; see Genapol C-020; wax; sol. in water, benzene; sp. gr. 1.032 (50 C); visc. 61.5 cps (50 C); flash pt. 264 C (Marcusson); surf. tens. 44 dynes/cm; 100% act.

Genapol C-250. [44] Coconut fatty alcohol polyglycol ether; nonionic; see Genapol C-020; wax; sol. in water, benzene; sp. gr. 1.055 (50 C); visc. 80.6 cps (50 C); flash pt. 267 C (Marcusson); surf. tens. 47 dynes/cm; 100% act.

Genapol CRT 40. [44; 301] Alkyl sulfate triethanolamine ammonium salt; anionic; detergent raw material for cosmetics, shampoos, body cleansing agents; pale yel. clear liq.; slight odor; m.w. 422; sol. in water; dens. 1.0–1.05 g/cm³; visc. 100 cps max.; pH 6.5–7.5 (1%); biodeg.; 40% conc.

Genapol GEV. [301] Fatty alcohol, ethoxylated; nonionic; lubricant for fiber preparation; liq.

Genapol LRO Liq., Paste. [44; 301] Sodium laureth sulfate; anionic; detergent, foaming agent used in cosmetic products, personal care products, lime soap dispersant; pale yel.

clear liq., mobile paste resp.; slight odor; misc. with water; dens. 1.05 and 1.0 g/cm³ resp.; visc. 100 cps max. (liq.); pH 6.5–8.0 (1% aq., liq.), 7.2 ± 0.6 (1% aq., paste); biodeg.; 27 and 68% conc.

Genapol O-020. [44] Oleyl alcohol polyglycol ether; nonionic; raw material for mfg. of textile, leather, paper auxs., detergents, emulsifiers, cosmetics; clear liq.; sol. in min. oil, benzene, turbid in water; sp.gr. 0.894 (50 C); visc. 12.4 cps (50 C); flash pt. 186 C (Marcusson); 100% act.

Genapol O-050. [44] Oleyl alcohol polyglycol ether; nonionic; see Genapol O-020; turbid liq.; sol. in benzene, turbid in water, min. oil; sp.gr. 0.936 (50 C); visc. 18.4 cps (50 C); flash pt. 225 C (Marcusson); surf. tens. 54 dynes/cm; 100% act.

Genapol O-080. [44] Oleyl alcohol polyglycol ether; nonionic; see Genapol O-020; turbid liq.; sol. in water, benzene, turbid in min. oil; sp.gr. 0.960 (50 C); visc. 25.1 cps (50 C); flash pt. 246 C (Marcusson); surf. tens. 44 dynes/cm; 100% act.

Genapol O-100. [44] Oleyl alcohol polyglycol ether; nonionic; see Genapol O-020; paste; sol. in water, benzene, turbid in min. oil; sp.gr. 0.989 (50 C); visc. 33 cps (50 C); flash pt. 260 C (Marcusson); surf. tens. 41 dynes/cm; 100% act.

Genapol O-120. [44] Oleyl alcohol polyglycol ether; nonionic; see Genapol O-020; paste; sol. in water, benzene, turbid in min. oil; sp.gr. 1.0 (50 C); visc. 42.5 cps (50 C); flash pt. 265 C (Marcusson); surf. tens. 42 dynes/cm; 100% act.

Genapol O-150. [44] Oleyl alcohol polyglycol ether; nonionic; see Genapol O-020; wax; sol. in water, benzene, turbid in min. oil; sp.gr. 1.02 (50 C); visc. 49.1 cps (50 C); flash pt. 271 C (Marcusson); surf. tens. 45 dynes/cm; 100% act.

Genapol O-200. [44] Oleyl alcohol polyglycol ether; nonionic; see Genapol O-020; wax; sol. in water, benzene, turbid in min. oil; sp.gr. 1.037 (50 C); visc. 65.9 cps (50 C); flash pt. 278 C (Marcusson); surf. tens. 47 dynes/cm; 100% act.

Genapol O-230. [44] Oleyl alcohol polyglycol ether; nonionic; see Genapol O-020; wax; sol. in water, benzene, turbid in min. oil; sp.gr. 1.042 (50 C); visc. 79.5 cps (50 C); flash pt. 279 C (Marcusson); surf. tens. 47 dynes/cm; 100% act.

Genapol O-250. [44] Oleyl alcohol polyglycol ether; nonionic; see Genapol O-020; wax; sol. in water, benzene, turbid in min. oil; sp.gr. 1.042 (50 C); visc. 88.1 cps (50 C); flash pt. 280 C (Marcusson); surf. tens. 48 dynes/cm; 100% act.

Genapol PAF, PF, PL, PN Brands. [301] Propylene and ethylene oxide polymerization products; nonionic; defoamer, component for detergent, cleaning and antistat; basic material for wide variety of aux.; liq., powd.; 100% conc.

Genapol PGM Conc. [44] Sodium laureth sulfate, glycol distearate, and cocamide MEA; anionic; pearl-luster conc. used in shampoos, bubble baths, cosmetics; wh. fluid paste; slight odor; misc. with water; dens. 1.0 g/cm³; pH 7.2 ± 0.8 (1% act.); biodeg.; 40.0 ± 1.0% act.

Genapol PN 30. [44] EO/PO cross linked block polymer; nonionic; lubricant for formulating water sol. metalworking fluids; controlled foam detergents, wetting agents; yel. liq.; cloudy sol. in water; sp.gr. 1.01 (50 C); cloud pt. 29–31 C (5 g in 25 ml of 24% butyl diglycol); 100% act.

Genapol S-020. [44] Stearyl alcohol polyglycol ether; nonionic; raw material for mfg. of textile, leather, paper auxs., detergents, emulsifiers, cosmetics; wax; sol. in min. oil, benzene, turbid in water; sp.gr. 0.890 (50 C); visc. 15.2 cps (50 C); flash pt. 198 C (Marcusson); surf. tens. 50 dynes/cm; 100% act.

Genapol S-050. [44]. Stearyl alcohol polyglycol ether; nonionic; see Genapol S-020; wax; sol. in min. oil, benzene, turbid in water; sp.gr. 0.934 (50 C); visc. 21.2 cps (50 C); flash pt. 224 C (Marcusson); surf. tens. 50 dynes/cm; 100% act.

Genapol S-080. [44]. Stearyl alcohol polyglycol ether; nonionic; see Genapol S-020; paste; sol. in benzene, turbid in water, min. oil; sp.gr. 0.971 (50 C); visc. 28.4 cps (50 C); flash pt. 228 C (Marcusson); surf. tens. 44 dynes/cm; 100% act.

Genapol S-100. [44]. Stearyl alcohol polyglycol ether; nonionic; see Genapol S-020; paste; sol. in benzene, turbid in water, min. oil; sp.gr. 0.978 (50 C); visc. 34 cps (50 C); flash pt. 239 C (Marcusson); surf. tens. 45 dynes/cm; 100% act.

Genapol S-150. [44]. Stearyl alcohol polyglycol ether; nonionic; see Genapol S-020; wax; sol. in water, benzene, turbid in min. oil; sp.gr. 1.009 (50 C); visc. 50.3 cps (50 C);

flash pt. 266 C (Marcusson); surf. tens. 47 dynes/cm; 100% act.

Genapol S-200. [44]. Stearyl alcohol polyglycol ether; nonionic; see Genapol S-020; wax; sol. in water, benzene, turbid in min. oil; sp.gr. 1.031 (50 C); visc. 67.4 cps (50 C); flash pt. 267 C (Marcusson); surf. tens. 48 dynes/cm; 100% act.

Genapol S-250. [44]. Stearyl alcohol polyglycol ether; nonionic; see Genapol S-020; wax; sol. in water, benzene, turbid in min. oil; sp.gr. 1.049 (50 C); visc. 106 cps (50 C); flash pt. 273 C (Marcusson); surf. tens. 49 dynes/cm; 100% act.

Genapol T. [301] Fatty alcohol polyglycol ether (2–50 EO); nonionic; detergent base and basic material for cosmetic and specialty chemical industries; liq., paste, wax; 100% conc.

Genapol X Brands. [301] Isotridecanol polyglycol ether (3–15 EO); nonionic; basic material for mfg. of detergents, cleaning and rinsing agents; mfg. of leather and paper aux.; 100% conc.

Genapol ZRO Brands. [301] Sodium laureth sulfate; anionic; see Genapol LRO Brands; 70% conc.

Generol 122. [289; 290] Soya sterol; nonionic; emollient, aux. or primary emulsifier, emulsion stabilizer, visc. modifier for cosmetics; wh. waxy flakes; HLB 3.0; 90% conc.

Generol 122 E 5. [289; 290] PEG-5 soya sterol; nonionic; cosmetics emulsifier, emollient, emulsion stabilizer; soft, waxy amorphous solid; HLB 5; 100% conc.

Generol 122 E 10. [289; 290] PEG-10 soya sterol; nonionic; cosmetics emulsifier, emollient, aux. emulsifier, emulsion stabilizer and visc. modifier; lt. amber soft wax; HLB 12; 100% conc.

Generol 122 E 16. [289; 290] PEG-16 soya sterol; nonionic; cosmetics emulsifier, emollient, solubilizer, and pigment wetter; ivory hard wax; HLB 15; 100% conc.

Generol 122 E 25. [289; 290] PEG-25 soya sterol; nonionic; cosmetics emulsifier, solubilizer, emollient, pigment wetter and deflocculating agent; ivory hard wax; HLB 17; 100% conc.

Geronol ACR/4. [253] Disodium laurylethoxy sulfosuccinate; anionic/nonionic; emulsifier for emulsion polymerization, detergent base, foamer, foam stabilizer, dispersant; liq.; 30% conc.

Geronol ACR/9. [253] Sodium salt alkylarylethoxy half ester sulfosuccinic acid; anionic; emulsifier for emulsion polymerization, detergent base, foamer, foam stabilizer, dispersant; liq.; 30% conc.

Geronol Aminox/3. [253] C_{16}-C_{18} alkanolamide; cationic; cationic; adjuvant for cutting oils for metals and rust protector; 100% conc.

Geronol AZ82. [253] POE/POP alkylphenols and calcium dodecylbenzene sulfonate; anionic/nonionic; emulsifier for pesticide emulsifiable conc.; specific for malathion; waxy paste; 85% conc.

Geronol MOE/2/N. [253] Alkylaryl polyglycolic ether and ethoxylated, modified fatty acids; nonionic; emulsifier for agric. o/w emulsions; liq.; 90% conc.

Geronol PRH/4-A, -B. [253] POE alkylphenol phosphate; anionic; emulsifier for various compds.; liq.; 100% conc.

Geronol SC/138. [253] POE/POP block polymer; nonionic; emulsifier for Dimethoate E.C.; waxy paste; 100% conc.

Geronol TZ/14. [253] POE arylphenol phosphate amine salt; anionic; dispersant/suspending agent for pesticides; visc. liq.; 100% conc.

Geronol TZ/A. [253] Org. sulfonate, straight chain and POE/POP block polymer; anionic/nonionic; see Geronol TZ/14; liq.; 90% conc.

Geronol TZ/B. [253] Alkylphenol, ethoxylated; nonionic; see Geronol TZ/14; thick liq.; 100% conc.

Geronol V/87, V/497. [253] Dodecylbenzene sulfonate and POE/POP alkylphenols; nonionic/anionic; emulsifier for pesticide conc.; specific for Alachlor; thick liq.; 80% conc.

Geropon 40/D. [253] Alkylaryl polyglycol ether; nonionic; wetting agent for pesticides; powd.; 40% conc.

Geropon AB/20. [253] Ammonium laureth sulfate; anionic; emulsifier for emulsion polymerization and min. oils; detergent; liq.; 30% conc.

Geropon AY. [253] Sodium diamyl sulfosuccinate; anionic; wetting agent; wax; sol. in water and org. polar solv.; wax; 70% conc.

Geropon BIS/SODICO. [253] Bis-tridecyl sodium sulfosuccinate; anionic; emulsifier for polymerization; liq.; 60% conc.

Geropon CET/50/P. [253] Fatty acids, ethoxylated; nonionic; wetting/suspending agent for pesticides; waxy powd.; 100% conc.

Geropon CYA/45. [253] Sodium diisobutyl

sulfosuccinate; anionic; surfactant for emulsion polymerization; liq.; 45% conc.

Geropon CYA/DEP. [253] Sodium diisooctyl sulfosuccinate; anionic; wetting agent for textile industry, pesticides, emulsion polymerization, printing inks, water paints; liq.; 75% conc.

Geropon FMS. [253] Sodium diphenylmethane sulfonate; anionic; dispersant and suspending agent for pesticide formulations; powd.; 70% conc.

Geropon IN. [253] Sodium isopropyl naphthalene sulfonate; anionic; see Geropon FMS; powd.; 65% conc.

Geropon K/65. [253] Fatty acids with ethoxylated alcohols, modified; anionic/nonionic; wetting and suspending agent for Dodine wettable powds.; powd.; 85% conc.

Geropon K/202. [253] POE fatty alcohol; nonionic; wetting and suspending agent for Dodine wettable powds.; powd.

Geropon LIV/30. [253] Ammonium alkylaryl ether sulfate; anionic; emulsifier for emulsion polymerization; wetting agent in household and industrial detergents; liq.; 30% conc.

Geropon LIV/2330. [253] Ammonium alkylaryl ether sulfate; anionic; primary emulsifier for acrylic, vinyl acetate, vinyl acrylic, styrene acrylic, styrene-butadiene; liq.; 30% conc.

Geropon MLS/A. [253] Sodium methallyl sulfonate; anionic; dye improver reactive comonomer for acrylic fibers polymerization; reactive emulsifier or coemulsifier; powd.; 100% conc.

Geropon NK. [253] Sodium dibutyl naphthalene sulfonate; anionic; wetting agent for pesticides and general purpose; powd.; 65% conc.

Geropon PL/68, PL/87. [253] POE/POP block copolymers; nonionic; wetting/antistat for agric. powds. and general purpose; waxy powd.; 100% conc.

Geropon RM/77. [253] Sodium dinaphthyl methane sulfonate; anionic; dispersant/suspending agent for pesticides and general purpose; powd.; 80% conc.

Geropon SDS. [253] Sodium dioctyl sulfosuccinate; anionic; dispersant and wetting agent for pigments and dyes in plastics; powd.; 100% conc.

Geropon TA/72/S. [253] Polycarboxylic acid, sodium salt; anionic; see Geropon RM/77; powd.; 93% conc.

Geropon TBS. [253] Alkyl polyglycolic ether, modified; anionic; detergent for fibers, cosmetics, leather, paper, and metal industry; liq.

GLS. [159] Glyceryl lactostearate; nonionic; used in the food industry; solid; 100% conc.

Glucam E-10. [38] Methyl gluceth-10; nonionic; fluid polyol for cosmetics and pharmaceuticals; adds gloss, emollience and conditioning, humectant; liq.; 100% conc.

Glucam E-20. [38] Methyl gluceth-20; nonionic; see Glucam E-10; liq.; 100% conc.

Glucam E-20 Distearate. [38] Methyl gluceth-20 distearate; nonionic; aux. o/w emulsifier, moisturizer, emollient and lubricant; soft solid; HLB 12.5; 100% conc.

Glucam P-10. [38] PPG-10 methyl glucose ether; nonionic; see Glucam E-10; liq.; 100% conc.

Glucam P-20. [38] PPG-20 methyl glucose ether; nonionic; see Glucam E-10; liq.; 100% conc.

Glucamate DOE-120. [37] PEG-120 methyl glucoside dioleate; nonionic; thickener, emulsifier, solubilizer for shampoos; wax; 100% conc.

Glucamate SSE-20. [37; 38] PEG-20 methyl glucose sesquistearate; nonionic; o/w emulsifier, solubilizer used with Glucate SS; pale yel. soft solid; sol. in water, IPA, ethanol, castor oil, corn oil; HLB 15.3; cloud pt. 74 C (1% in 5% NaCl); flash pt. 570 F (OC); sapon. no. 47; pH 6.5 (10% aq.); 100% conc.

Glucate DO. [38] Methyl glucoside dioleate; nonionic; w/o emulsifier, aux. emulsifier for o/w systems; emollient, lubricant, and pigment dispersant; liq.; HLB 5.0; 100% conc.

Glucate SS. [37; 38] Methyl glucose sesquistearate; nonionic; w/o emulsifier used with Glucamate SSE-20 to provide visc. stability, mildness; off wh. flakes; sol. in IPA, misc. with common oil phase ingred., water insol.; m.p. 51 C; HLB 6.4; flash pt. 530 F (OC); sapon. no. 136; pH 5.5 (10% in 1:1 IPA:water); 100% conc.

Glycerox L8. [159] POE (8) glyceryl monolaurate; nonionic; emulsifier for cosmetic preparation; liq.; HLB 11.0; 100% conc.

Glycerox L15. [159] POE (15) glyceryl monolaurate; nonionic; see Glycerox L8; liq.; HLB 14.0; 100% conc.

Glycerox L40. [159] POE (40) glyceryl monolaurate; nonionic; see Glycerox L8; soft paste; HLB 17.0; 100% conc.

Glycidol Surfactant 10G. [455] para-Nonylphenoxy-polyglycidol; nonionic; surfactant used in agric. chemical emulsions, leather processing, paint, emulsion polymerization, waste paper deinking, alkaline cleaners, photographic film emulsion and coating formulations; amber liq.; mild odor; sol. in alcohols, polyethers, acids; sp.gr. 1.032–1.114; dens. 8.6–9.3 lb/gal; visc. 110–195 cps; pour pt. –9 to –14 C; b.p. 93–103 C; cloud pt. > 100 C (0.5% aq.); pH 7.1–7.9 (1% aq.); surf. tens. 27–30 dynes/cm (0.1% act.); 50% act.

Glycomul L. [256] Sorbitan laurate; nonionic; emulsifier for edible, cosmetic, industrial, pharmaceutical uses; amber liq.; sol. in methanol, ethanol, naphtha; sp.gr. 1.0; visc. 4500 cps; HLB 8.6; sapon. no. 157–171; 100% conc.

Glycomul LC. [256] Sorbitan laurate, cosmetic grade; nonionic; cosmetic emulsifier; lt. amber liq.; sol. in methanol, ethanol, naphtha, disp. in water; sp.gr. 1.0; visc. 5000 cps; HLB 8.6; sapon. no. 154–163; 100% conc.

Glycomul MA. [256] Sorbitan ester mixed fatty acids; nonionic; see Glycomul L; dk. amber liq.; sol. in methanol, ethanol, naphtha, acetone, toluol, min. and veg. oil; sp.gr. 1.0; visc. 1000 cps; sapon. no. 139–156; 100% conc.

Glycomul O. [256] Sorbitan oleate; nonionic; see Glycomul L; amber liq.; sol. in ethyl acetate, min. and veg. oils, disp. in water; sp.gr. 1.0; visc. 1000 cps.; HLB 4.3; sapon. no. 148–161; 100% conc.

Glycomul P. [256] Sorbitan palmitate; nonionic; see Glycomul L; cream beads; sol. in veg. and min. oil, ethyl acetate, ethanol, acetone, toluol; HLB 6.7; sapon. no. 139–150; 100% conc.

Glycomul S. [256] Sorbitan stearate (also avail. in veg. and kosher grade); nonionic; see Glycomul L; cream beads; sol. in veg. oil; HLB 4.7; sapon. no. 146–158; 100% conc.

Glycomul SOC. [256] Sorbitan sesquioleate; nonionic; see Glycomul L; cream beads; sol. in methanol, ethanol, ethyl acetate; sp.gr. 1.0; visc. 1000 cps; HLB 4.0; sapon. no. 149–166; 100% conc.

Glycomul TAO. [256] Sorbitan ester mixed resin and fatty acids; nonionic; see Glycomul L; amber liq.; sol. in methanol, ethanol, toluol, naphtha, min. and veg. oils, disp. in water; sp.gr. 1.02; visc. 1300 cps.; HLB 4.3; sapon. no. 135–148; 100% conc.

Glycomul TO. [256] Sorbitan trioleate; nonionic; see Glycomul L; amber, oily liq.; sol. in ethyl acetate, toluol, naphtha, min. and veg. oils, disp. in water; sp.gr. 0.95; visc. 200 cps; HLB 1.8; sapon. no. 171–185; 100% conc.

Glycomul TS. [256] Sorbitan tristearate; nonionic; see Glycomul L; lt. tan beads; poorly sol. in ethyl acetate, toluol, disp. in acetone, naphtha, min. and veg. oils; HLB 2.1; sapon. no. 175–190; 100% conc.

Glycosperse L-20, L-20X. [256] Polysorbate 20 (L-20X anhyd.); nonionic; emulsifier for food, cosmetic, pharmaceutical uses; yel. liq.; sol. in water, alcohol, acetone; sp.gr. 1.1; visc. 400 cps; HLB 16.7; sapon. no. 39–52; 100% conc.

Glycosperse O-5. [256] Polysorbate 81; nonionic; see Glycosperse L-20; amber liq.; sol. in alcohol, ethyl acetate, min. oil; disp. in water; sp.gr. 1.0; visc. 450 cps; HLB 10.0; sapon. no. 95–105; 100% conc.

Glycosperse O-20. [256] Polysorbate 80; nonionic; see Glycosperse L-20; yel. liq.; sol. in water, alcohol, ethyl acetate, toluol, veg. oil; sp.gr. 1.0; visc. 400 cps; HLB 15; sapon. no. 44–56; 100% conc.

Glycosperse O-20 Veg. [256] Polysorbate 80, veg. grade; nonionic; see Glycosperse L-20; yel. liq.; sol. see Glycosperse O-20; sp.gr. 1.0; visc. 400 cps; sapon. no. 44–56; 100% conc.

Glycosperse O-20X. [256] Polysorbate 80, anhyd.; nonionic; see Glycosperse L-20; yel. liq.; sol. see Glycosperse O-20; sp.gr. 1.0; visc. 600 cps; sapon. no. 44–56; 100% conc.

Glycosperse P-20. [256] Polysorbate 40; nonionic; see Glycosperse L-20; yel. liq.; sol. in water, methanol, ethanol, acetone, ethyl acetate; sp.gr. 1.0; visc. 550 cps; HLB 15.6; sapon. no. 40–53; 100% conc.

Glycosperse S-20. [256] Polysorbate 60; nonionic; see Glycosperse L-20; yel. liq.; sol. in water, ethyl acetate, toluol; sp.gr. 1.1; HLB 14.9; sapon. no. 44–56; 100% conc.

Glycosperse TO-20. [256] Polysorbate 85; nonionic; see Glycosperse L-20; yel. liq., gels on standing; sol. in ethanol, methanol, ethyl acetate; water disp.; sp.gr. 1.0; visc. 300 cps; HLB 11.0; sapon. no. 82–95; 95% conc.

Glycosperse TS-20. [256] Polysorbate 65; nonionic; see Glycosperse L-20; tan waxy solid; sol. in ethanol, methanol, acetone, ethyl acetate, naphtha, min. and veg. oils; disp. water, toluol; sp.gr. 1.05; HLB 10.5;

sapon. no. 88–98; 100% conc.

GMA-33. [287] Glycerol arachidate; nonionic; emulsifier for cosmetic, pharmaceutical and food industries; solid; HLB 4.0; 100% conc.

GMF-33. [287] Fatty acid glycerides; nonionic; food emulsifier; paste; 100% conc.

GMM-33. [287] Glycerol myristate; nonionic; emulsifier for producing w/o and o/w creams and bases for pharmaceuticals and cosmetics; flakes; HLB 5.5; 100% conc.

GMO-33. [287] Glycerol oleate; nonionic; cosmetic and pharmaceuticals; consistency builder and thickener for w/o emulsions; dispersant for pigments; food emulsifier; liq.; HLB 4.0; 100% conc.

GMR-33. [287] Fatty acid glycerides; nonionic; see GMF-33; liq.; 100% conc.

GMS-33. [287] Glycerol stearate; nonionic; cosmetics and pharmaceuticals; emulsifier for w/o emulsions and food applications; bases; lubricant; flakes; HLB 3.5; 100% conc.

GMS-33-SEN. [287] Glycerol and POE stearate mixture; nonionic; emulsifier for production of various creams; flakes; HLB 11.0; 100% conc.

GMS-33-SES. [287] Glycerol stearate palmitate SE; anionic; emulsifier for food and cosmetic industry; HLB 5.0.

GMS-66, -99. [287] Glycerol stearate palmitate; nonionic; emulsifier, thickener and consistency builder used in the cosmetic, pharmaceutical and food industries; HLB 4.5 and 5.0 resp.; 100% conc.

GMS-99-SES. [287] Glycerol stearate palmitate SE; anionic; see GMS-99; flakes; HLB 6.0; 100% conc.

GMS-333. [287] Polyglycerol stearate; nonionic; see GMS-99; solid; HLB 7.5; 100% conc.

GMS-333-SES. [287] Polyglycerol stearate SE; anionic; see GMS-99; solid; HLB 8.5; 100% conc.

Golden Dawn Grade 1, 2. [639] Anhyd. lanolin; nonionic; emollient, w/o emulsifier, ointment base, hair conditioner, wax crystal inhibitor, lipstick binder; soft wax, Grade 2 slightly darker color.

Golden Dawn Superfine. [639] Anhyd. lanolin; nonionic; see Golden Dawn Grade 1; high quality; pale soft wax; HLB 4.0; 100% conc.

Golden Fleece DF. [639] Extra refined anhyd. lanolin (detergent-free); nonionic; see Golden Dawn Grade 1; soft solid; HLB 4.5; 100% conc.

Golden Fleece P-80. [639] Anhyd. lanolin (trace pesticides reduced by 80%); nonionic; see Golden Dawn Grade 1; pale soft wax.

Golden Fleece P-95. [639] Anhyd. lanolin (trace pesticides reduced by 95%, low in natural free alcohols); nonionic; see Golden Dawn Grade 1; pale soft wax.

Golden Fleece RA. [639] Anhyd. lanolin (detergent-free and low in natural free alcohols); nonionic; see Golden Dawn Grade 1; soft solid; 100% conc.

Gradol 250. [263] Sodium salt of a carboxylated polyelectrolyte; anionic; dispersant for pigments in latex paint, visc. depressant for latex, clay slurries; APHA 40 max. clear liq.; sol. in water systems; dens. 9.6 lb.gal; f.p. < –5 C; surf. tens. 69–71 dynes/cm (1%); pH 9.5–10.5 (1%); 25% act.

Gradol 250A. [263] Ammonium salt of a carboxylated polyelectrolyte; dispersant in ceramics industry where sodium or potassium salts undesirable; amber clear liq.; sol. in water systems; dens. 9.1–9.4 lb/gal; pH 7.5–9.0; 25% act.

Gradol 300. [263] Sodium salt of carboxylated polyelectrolyte; anionic; nonfoaming dispersant for pigments, fillers and other insol. fine particles in water; lt. liq.; sol. in aq. systems; sp.gr. 1.21; pH 10 (5% sol'n.); 30% solids.

Gradonic 400-ML. [263] PEG 400 laurate; nonionic; emulsifier for oils and solvs.; plasticizer, visc. reducer for starches; stabilizer, wetting agent for latex paints; liq.; 100% conc.

Gradonic N-95. [263] Nonoxynol-9.5; nonionic; detergent, wetting agent, dispersant, emulsifier; almost colorless liq.; mild odor; sol. in water, dilute inorg. salt and caustic sol'ns., min. acid sol'ns.; pH 7 (2%); 100% act.

Grilloten LSE87. [519] Sucrose laurate; o/w emulsifier, solubilizer; liq.; HLB 12.5; 100% conc.

Grilloten LSE87K. [519] Sucrose cocoate; see Grilloten LSE87; liq.; HLB 12.5; 100% conc.

Grilloten PSE141G. [519] Sucrose stearate; see Grilloten LSE87; liq.; HLB 15.0; 100% conc.

Grilloten ZT12, ZT40, ZT80. [519] Sucrose ricinoleate; see Grilloten LSE87; liq.; HLB 9.0, 8.5, and 9.5 resp.; 100% conc.

Grindtek CA-P. [273] Hydrog. tallow glyceride citrate; o/w emulsifier; cream powd.; sol. warm in toluene, wh. spirit, propylene glycol; HLB 11.

Grindtek DAT-L. [273] Diacetylated tartaric acid esters of glycerol (oleic/stearic) fatty acid esters; o/w emulsifier; yel.-br. visc. liq.; sol. in ethanol, toluene, peanut oil, disp. in water; HLB 7.

Grindtek DAT-S. [273] Diacetylated tartaric acid ester of glycerol (stearic/palmitic) fatty acid esters; see Grindtek DAT-L; cream flakes; sol. in toluene, wh. spirit, peanut oil; HLB 8.

Grindtek DAT-VO. [273] Diacetylated tartaric acid esters of glycerol (oleic/stearic) fatty acid esters; w/o emulsifier; yel.-br. liq.; sol. in peanut oil, toluene, wh. spirit, disp. in water, propylene glycol; HLB 6.

Grindtek FAL 1. [273] Sodium stearoyl lactylate; o/w emulsifier; cream powd.; sol. warm in ethanol, propylene glycol, toluene, wh. spirit, paraffin oil; HLB 10.

Grindtek FAL 2. [273] Calcium stearoyl lactylate; w/o emulsifier; cream flake; sol. warm in toluene, wh. spirit; HLB 2.

Grindtek FAL 3. [273] Stearoyl lactylic acid; o/w emulsifier; tan block; sol. in toluene; sol. warm in ethanol, propylene glycol, peanut oil; HLB 2.

Grindtek FAL 4. [273] Caprylic fatty acid ester of lactylic acid and its salts; o/w emulsifier with fungistatic properties; yel./br. plastic; sol. in ethanol, propylene glycol, toluene, warm water, wh. spirit; HLB > 13.

Grindtek LA 15. [273] Glyceryl palmitate lactate; component in w/o and o/w creams; wh. flake; sol. warm in peanut oil, toluene, wh. spirit, paraffin oil; HLB 5.1.

Grindtek LA 30. [273] Glyceryl stearate lactate; component in w/o and o/w creams; wh. block; sol. warm in ethanol, peanut oil, toluene, wh. spirit, paraffin oil; HLB 8.2.

Grindtek ML 90. [273] Glyceryl laurate; component in w/o and o/w creams, lubricant anitstat, antifogging agent in plastics; wh. block; sol. in ethanol, warm in propylene glycol, toluene, wh. spirit; HLB 5.3.

Grindtek MM 90. [273] Glyceryl myristate; see Grindtek ML 90; wh. block; sol. warm in ethanol, propylene glycol, toluene; HLB 5.0.

Grindtek MOL 90. [273] Glyceryl linoleate; see Grindtek ML 90; wh. semisolid; sol. in ethanol, toluene, warm in propylene glycol, peanut oil, wh. spirit, paraffin oil; HLB 4.3.

Grindtek MOP 90. [273] Lard glyceride; see Grindtek ML 90; wh. block; sol. warm in ethanol, propylene glycol, tolene, wh. spirit; HLB 4.3.

Grindtek MSP 32-6, 40, 40F, 52, 90. [273] Glyceryl stearate (32-6: SE grade); see Grindtek MLJ 90; wh. powd.; sol. warm in toluene; MSP 32-6 disp. in water; MSP 90 sol. warm in ethanol; HLB 2.8–4.3.

Grindtek PGE 25. [273] Polyglyceryl-3 oleate; o/w emulsifier, antifogging agent for plastics; yel.-br. liq.; sol. in ethanol, warm in peanut oil, toluene, paraffin oil, disp. in water, propylene glycol; HLB 5.5.

Grindtek PGE 55, 55-6. [273] Polyglyceryl-3 stearate (55-6: SE grade); see Grindtek PGE 25; wh. powd., tan block; sol. warm in toluene (55), ethanol (55-6); HLB 6.8, 7.4.

Grindtek PGE-DSO. [273] Oleic/linoleic fatty acid ester of polyglycerol; o/w emulsifier, antifogging agent for plastics, effective for w/o emulsifications with veg. oils; yel.-br. visc. liq.; sol. in toluene, warm in peanut oil, wh. spirit, paraffin oil; HLB 2.8.

Grindtek PGMS 90. [273] Propylene glycol stearate; w/o emulsifier; whitish block; sol. in ethanol, toluene; sol. warm in propylene glycol, peanut oil, wh. spirit, paraffin oil; HLB 3.5.

Grindtek PK 60. [273] Palm kernel glycerides; component in w/o and o/w creams, lubricant, antistat, antifogging agent for plastics; wh. block; sol. in ethanol, warm in peanut oil, toluene, paraffin oil, propylene glycol, tolene; HLB 4.1.

Grindtek SMS. [273] Sorbitan stearate; w/o and o/w emulsifier; tan powd.; sol. warm in toluene; HLB 5.2.

Grindtek STS. [273] Sorbitan tristearate; w/o and o/w emulsifier; tan powd.; sol. warm in toluene, wh. spirit; HLB 2.3.

Grocor 1000, 1200. [276] Triolein; nonionic; wetting agent, emulsifier, sulfated product is a surfactant for textiles, leathers; Gardner 6 max. and amber resp., liqs.; mild odor; sol. in oils and hydrocarbons; sp.gr. 0.9145–0.945; visc. 190–205 SUS (100 F); cloud pt. 5–18 F; 97% min. act.

Grocor 2000. [276] Glycerol monooleate; nonionic; wetting agent, emulsifier, penetrant, textile finishing compds.; pale yel. liq.; mild odor; sol. in oils and hydrocarbons; sp.gr. 0.949; visc. 375 SUS (100 F); cloud pt. 5 F; 97.5% min. act.

Grocor 2500. [276] Glycerol dioleate; nonionic; aux. emulsifier, emulsion stabilizer, thickener for cosmetics and paints; yel. liq.; mild odor; sol. in oils and hydrocarbons; sp.gr. .93; visc. 90 cps; f.p. 0 C; 95% min. act.

Grocor 3000. [276] Propyl oleate; nonionic; wetting agent, stabilizer, used for chemical specialties, textiles, leather finishing; lt. yel. liq.; mild odor; sol. in IPA, oils; sp.gr. 0.89; f.p. 0 C; 97% min. act.

Grocor 4000. [276] Butyl oleate; nonionic; textile wetting agent, plasticizer for chloroprene and vinyl chloride, leather treatment; slightly yel. liq.; sol. in IPA, hydrocarbons, min. oil, sp.gr. 0.86; f.p. –20 C; b.p. 170 C; 97% min. act.

Grocor 5221 SE. [276] Diglycol stearate SE; anionic; textile lubricant and finishing, wetting agent, emulsifier, plasticizer, opacifying agent for cosmetics; wh. solid, beads, flakes; sol. in hydrocarbons, disp. in hot water; sp.gr. 0.96; 100% act.

Grocor 5410. [276] Butyl stearate; nonionic; lubricant and spinning compd. for textiles, paper defoamer, plasticizer; water wh. liq.; mild odor; sol. in IPA, toluol, oils; sp.gr. 0.855; visc. 10 cs; f.p. 20 C; 98% min. act.

Grocor 5500. [276] Glyceryl stearate; nonionic; emulsifier, thickener, opacifier for cosmetics, lubricant softener for textiles, dispersant for dry powd.; wh. solid, powd., beads, flake; mild odor; sol. in IPA; sp.gr. 0.97; m.p. 56–59 C; 100% act.

Grocor 5510. [276] Butyl stearate; nonionic; wetting agent, plasticizer for chlorinated rubbers, lubricant and cutting oils for metals, defoamer for paper and paperboard; water wh. liq.; mild odor; sol. in IPA and toluol; sp.gr. 0.855; visc. 10 cs; f.p. 20 C; 98% min. act.

Grocor 5721. [276] Tridecyl stearate; wetting agent for pigments and coating disp. dry powd., emulsifier, dye carrier and textile lubricant, nonionic; water wh. liq.; sp.gr. 0.86; f.p. 6 C; 100% conc.

Grocor 5820 G. [276] 2-Hexyldecyl stearate; emollient, lubricant and intermediate for cosmetics, textiles, printing and paints, wetting agent, nonionic; water wh. liq.; mild odor; sp.gr. 0.853; visc. 90 SUS (100 F); f.p. –7 C; 98% min. act.

Grocor 6000, 6000 E, 6000 SE. [276] Glyceryl stearate (E: food grade, SE: self emulsifying); nonionic; emulsifier, lubricant, plasticizer, suspending agent, stabilizer, used in textiles, rubber, paper, cosmetic and chemical specialties, 6000 E used as antitack agent for foods; wh. solid, beads; mild odor; sol. in IPA, SE disp. in water, sp.gr. 0.91–0.92; m.p. 58 C; 100% conc.

Grocor 8002, 8008. [276] Methyl oleate; nonionic; wetting agent, lubricant, plasticizer, used in textiles, chemical specialties, cosmetics, leather, plastics, dyeing, sizing; lt. yel. liq.; mild odor; sol. in IPA; sp.gr. 0.872; cloud pt. 0 C; 98% min. act.

Grocor 8041. [276] Methyl tallowate; nonionic; detergent, lubricant, plasticizer, used for textiles, leather and wire drawing; water wh. liq.; f.p. 15 C; 98% min. act.

Grocor 8058. [276] Methyl stearate; nonionic; detergent, lubricant, plasticizer, for textiles, leather, wire drawing; water wh. liq., gel; sol. in IPA; sp.gr. 0.860; f.p. 32 C; 98% min. act.

Grocor PEG 300. [276] PEG 300 monooleate; nonionic; detergent, wetting agent, emulsifier, textile lubricant, used for dyeing and finishing, paper, cosmetics, leather and chemical specialties; lt. yel. liq.; mild odor; sp.gr. 0.97; visc. 170 SUS (100 F); 98% min. act.

Grotex 1120. [276] Nonionic; self-emulsifying softener, detergent, wetting agent, lubricant, thickener, suspending agent for textiles; wh. solid, beads; mild odor; disp. in water; sp.gr. 0.92; m.p. 52 C; 100% conc.

GTS-33. [287] Glycerol tristearate; nonionic; filler, food industry, cosmetic and pharmaceuticals; flakes; HLB 1.0; 100% conc.

H

Hallcomid M-6. [280] N,N-Dimethyl caproamide; detergent, demulsifier, penetrant, solubilizer, plasticizer, softening agent, solv. agriculture, automotive, coatings, cosmetics, drilling muds, textiles, germicides; colorless to pale yel. liq.; low odor; m.w. 143; sol. in common org. solvs., hydrocarbons, veg. oils, esters, water; visc. 7.0 cps; f.p. –40 C; b.p. 83–89 C (3 mm Hg); flash pt. 190 F; 95% conc.

Hallcomid M 8-10. [280] N,N-Dimethyl caprylamide-capramide; nonionic; mutual solv. for polar and nonpolar ingred. in cosmetics, cleaners, corrosion inhibitor; colorless to pale yel. liq.; m.w. 171–199; sol. in water, most common org. solvs., hydrocarbons, plasticizers, fatty acids, veg. oils; sp.gr. 0.88; visc. 10.5 cps; f.p. –21 C; b.p. 115–170 C (3 mm Hg); flash pt. 245 F; 95% conc.

Hallcomid M-12. [280] N,N-Dimethyl lauramide; nonionic; hydrotrope, solubilizer for polar and nonpolar emulsifiers, dyes, perfumes, corrosion inhibitor; colorless to pale yel. liq.; m.w. 227; sol. in water, most common org. solvs., hydrocarbons, plasticizers, fatty acids, veg. oils; sp.gr. 0.872; visc. 14.5 cps; f.p. 14–16 C; b.p. 147–163 C (3 mm Hg); HLB 8.0; flash pt. 292 F; 95% conc.

Hallcomid M-18. [280] N,N-dimethyl stearamide; nonionic; solubilizer for shampoos, corrosion inhibitor; colorless to pale yel. solid; m.w. 312; sol. in water, most common org. solvs., hydrocarbons, plasticizers, fatty acids, veg. oils; sp.gr. 0.851; visc. 13.5 cps (40 C); f.p. 38–40 C; b.p. 194–214 C (3 mm Hg); HLB 7.0; flash pt. 399 F; 95% conc.

Hallcomid M 18-OL. [280] N,N-dimethyl oleamide; nonionic; solubilizer, solv., dispersant, wetting agent; slightly yel. liq.; m.w. 310; sol. in water, most common org. solv., hydrocarbons, plasticizers, fatty acids, veg. oils; sp.gr. 0.875; visc. 18.5 cps; f.p. –8 C; b.p. 196–211 C (3 mm Hg); HLB 7.0; flash pt. 410 F; 95% conc.

Hampfoam 35. [262] Carboxylated fatty amide, sodium salt; anionic; surfactant used in alkaline industrial formulations; pale yel. clear liq.; visc. 100 cps; pH 8.0–9.5; 35.0% min. solids.

Hamposyl C. [262] Cocoyl sarcosine; anionic; detergent, wetting and foaming agent for hair and rug shampoos and cosmetics; Gardner 3 liq.; sol. in most org. solv.; sp.gr. 0.97–0.99; m.p. 23–26 C; biodeg.; 94% act.

Hamposyl C-30. [262] Sodium cocoyl sarcosinate; anionic; see Hamposyl C; APHA 100 liq.; misc. in water; sp.gr. 1.02–1.03; visc. 30 cps; f.p. –1 C; surf. tens. 30 dynes/cm; biodeg.; 30% act.

Hamposyl L. [262] Lauroyl sarcosine; anionic; see Hamposyl C; Gardner 2 solid; sol. in most org. solvs.; sp.gr. 0.97–0.99; m.p. 34–37 C; biodeg.; 94% act.

Hamposyl L-30, L-95 [262] Sodium lauroyl sarcosinate; anionic; see Hamposyl C; APHA 60 liq., wh. powd. resp.; misc. and sol. in water resp.; sp.gr. 1.02–1.03; visc. 30 cps; f.p. –1 C; surf. tens. 30 dynes/cm; biodeg.; 30%, 94% act.

Hamposyl M. [262] Myristoyl sarcosine; anionic; see Hamposyl C; Gardner 2 solid; sol. in most org. solv.; sp.gr. 0.97–0.99; m.p. 48–53 C; biodeg.; 94% act.

Hamposyl M-30. [262] Sodium myristoyl sarcosinate; anionic; see Hamposyl C; APHA 80 liq.; misc. in water; sp.gr. 1.02–1.03; visc. 30 cps; f.p. –1 C; surf. tens. 30 dynes/cm; biodeg.; 30% act.

Hamposyl O. [262] Oleoyl sarcosine; anionic; see Hamposyl C; Gardner 4 liq.; sol. in most org. solv.; sp.gr. 0.95–0.97; visc. 250 cps; biodeg.; 94% act.

Hamposyl S. [262] Stearoyl sarcosine; anionic; see Hamposyl C; Gardner 4 solid; sol. in most org. solv.; sp.gr. 0.96–0.98; m.p. 53–58 C; biodeg.; 94% act.

Haroil SCO-65, -7525. [263] Sulfated castor oil; anionic; superfatting agent for cosmetic creams, emulsifier for cosmetic formulations, plasticizer for adhesives, antisagging and antisettling agent for paints and stains; amber liq.; pH 7.0 ± 0.5 (SCO-65, 2% aq.); dens. 1.025 (SCO-7525); 50%, 75% act.

Harol D. [263] Highly polymerized naphthalene sulfonate; anionic; stabilizer, visc. depressant, dispersant; paper coating compds.; powd.; dens. 47 lb/ft^3; surf. tens. 71 dynes/cm (1%); pH 9 (1%); 83–85% act.

Harol RG-71. [263] Sodium salt of polymerized alkyl naphthalene sulfonic acid; anionic; dye retardant, leveler, pitch dispersant; powd; dens. 35–37 lb/ft^3; surf. tens. 70–72 dynes/cm (1%); pH 7.8–10.4 (1%); 95% act.

Hartaine CB-40. [284] Coco-betaine; detergent for textiles, household, cosmetics, wetting agent for cottons, bleaching, emulsifier for dispersing waxes, pigments, resins, printing; very lt. liq.; water sol.; dens. 8.85 lb/gal; sp.gr. 1.06; 40% act.

Hartamide 9137. [283] Oleic diethanolamide; anionic; coupling agent, emulsion stabilizer, lubricant and antistat; liq.; 80% amide.

Hartamide AD. [284] Coconut alkanolamide; nonionic; detergent for textiles, household, cosmetics, industrial cleaning, wetting agent for dyeing, emulsifier for oils, waxes, pigments, resins, thickener; amber liq.; fatty odor; water sol.; 100% act.

Hartamide CE 80. [284] Lauric-myristic super amide; nonionic; detergent for textiles, household, cosmetics, industrial cleaning, wetting agent for dyeing, emulsifier for oils, waxes, pigments, resins, thickener; lt. amber liq.; fatty odor; water sol.; 100% act.

Hartamide CE-90. [284] High purity coconut fatty acid alkanolamide; nonionic; see Hartamide CE 80; lt. amber liq.; fatty odor; water sol.; 100% act.

Hartamide LDA 70, 90. [283] Lauric diethanolamide; nonionic; detergent, foam stabilizer and visc. regulator for liq. and powd. detergent systems; clear solid; 100% act.

Hartamide LE-90. [284] High purity lauric alkanolamide; nonionic; detergent for textiles, household, cosmetics, industrial cleaning, wetting agent for dyeing, emulsifier for oils, waxes, pigments, resins, thickener; lt. solid; fatty odor; water sol.; 100% act.

Hartamide LM 11D. [284] High purity lauric myristic alkanolamide; nonionic; see Hartamide CE 80; lt. gel/solid; fatty odor; water sol.; 100% act.

Hartamide LMEA-70. [283] Lauric myristic monoethanolamide; anionic; foam stabilizer and visc. modifier for detergents and personal care products; wh. solid; 95% amide.

Hartamide LMEA-90. [283] Lauric monoethanolamide; anionic; foam stabilizer for detergents and bubble bath preparations; wh. solid; 95% amide.

Hartamide OD. [283] Coconut diethanolamide; nonionic; foam stabilizer and visc. regulator for liq. and powd. detergent systems; yel. liq.; dens. 1.02 lb/gal; 100% act.

Hartamide OS. [284] Coco-oleic super amide; nonionic; detergent for textiles, laundering, wetting agent, emulsifier for cutting and sol. oils, disp. waxes, resins, pigments; lt. amber liq.; fatty odor; sol. in water; 100% act.

Hartenol LAS-30. [284] Sodium lauryl sulfate; anionic; detergent for textiles, skins, leather, laundering, wetting agent, emulsifier for dispersing waxes, pigments, resins; very lt. liq.; 30% act.

Hartenol LES-60. [284] Sodium lauryl sulfate; detergent, wetting agent, emulsifier for textiles, dyeing, household, cosmetics and cleaners; lt. liq.; 60% act.

Hartex San Q 50. [283] Lauryl dimethylbenzyl ammonium chloride; germicide, sanitizer, disinfectant; amber liq.; dens. 0.980 lb/gal; visc. 60 cps; 50% act.

Hartex V63, V64. [283] Sulfonated castor oil; anionic; penetrant, emulsifier for terpineols, min. and fatty acids, lubricant; amber liq.; dens. 1.02–1.056 lb/gal; visc. 250–520 cps; 70% and 45% act.

Hartolan, Hartolan Super. [159] Lanolin alcohols; spreading agent, dispersant, stabilizer, plasticizer, o/w emulsifier and emollient for cosmetic and pharmaceutical systems; br. solid wax and pale amber solid wax resp.; m.p. 58 and 60 C resp.; sapon. no. 5 mg. max.

Hartolite. [159] Fraction of lanolin alcohols; nonionic; w/o emulsifier, emollient, skin conditioner, moisturizing agent; pale yel. soft waxy solid; 100% act.

Hartolon NA. [283] Sat. esters and alkyl condensate; nonionic; softener used in textiles; paste.

Hartonyl L531, L535, L537. [283] Ethoxylated fatty amine; cationic; leveling aid, dispersing and retarding agent, dyeing assistant, antiprecipitant for textiles; yel. liq.; dens. 0.999–1.025 lb/gal; 33–60% act.

Hartonyl W504. [283] Modified linear alcohol ethoxylate; nonionic; wetting and foaming aid used in carpet dyeing; clear liq.; 85% act.

Hartopol 25R2, 31R1. [283] Polyoxyalkylene glycol; nonionic; surfactant for rinse aids and windshield washer fluids; liq.; cloud pt. 24–32 and 23–26 C resp.; 100% act.

Hartopol 4270. [283] Polyoxyalkylene glycols; nonionic; detergent, emulsifier for rinse aids, textiles, latex coatings; clear liq.; visc. 425 cps; 100% act.

Hartopol 4277. [283] Polyoxyalkylene glycols; nonionic; detergent, emulsifier, for rinse aids, polyurethane, oil demulsifier; clear liq.; visc. 275 cps; 100% act.

Hartopol 4278. [283] Polyoxyalkylene glycols; nonionic; detergent, emulsifier, for cosmetics, pharmaceuticals, paper, emulsion polymerization; clear liq.; dens. 1.05 lb/gal; visc. 540 cps; 100% act.

Hartopol 4299. [283] Polyoxyalkylene glycols; nonionic; detergent, emulsifier for pulp/paper foam control, water treatment, emulsion polymerzation; clear liq.; visc. 400 cps; 100% act.

Hartopol L42. [283] Polyoxyalkylene glycol; nonionic; low foaming rinse aid; liq.; HLB 8.0; cloud pt. 36–40 C; 100% act.

Hartopol L44. [283] Polyoxyalkylene glycol; nonionic; dispersant and demulsifier; liq.; HLB 16.0; cloud pt. 69–73 C; 100% act.

Hartopol L62, L62LF. [283] Polyoxyalkylene glycol; nonionic; used in rinse aids and industrial formulations; liq.; HLB 7.0 and 6.6 resp.; cloud pt. 31–33 and 27–29 C resp.; 100% act.

Hartopol L64. [283] Polyoxyalkylene glycol; nonionic; dispersant and emulsifier in emulsion polymerization; liq.; HLB 15.0; cloud pt. 58–62 C; 100% act.

Hartopol L81. [283] Polyoxyalkylene glycol; nonionic; industrial defoamer; liq.; HLB 2.0; cloud pt. 18–22 C; 100% act.

Hartopol LF-1, -2, -4. [283] Polyoxalkylene glycols; nonionic; detergent, emulsifier for low-foam dishwash, general household use; clear liq.; sp.gr. 1.01–1.035; visc. 320–400 cps; 100% act.

Hartopol LF-5. [283] Polyoxyalkylene glycol; nonionic; mechanical dishwashing rinse aid conc. blend; liq.; HLB 6.8; cloud pt. 22–26 C; liq.; 87% act.

Hartopol P65, P85. [283] Polyoxyalkylene glycol; nonionic; dispersant and demulsifier; paste; HLB 17.0 and 16.0 resp.; cloud pt. 80–84 and 83–87 C resp.; 100% act.

Hartopon FS. [283] Modified linear alcohol ethoxylate; nonionic; detergent for soaping off of dyestuffs, emulsifying and suspending agent; clear liq.; dens. 1.04 lb/gal; biodeg.; 75% act.

Hartopon TDA. [283] Linear alcohol ethoxylate, modified; scouring and wetting agent, emulsifier used in textiles; liq.; 70% conc.

Hartotrope AXS 40. [283] Ammonium xylene sulfonate; detergent, solubilizer and cloud pt. depressant for lt. duty and built liq. detergent systems; clear liq.; dens. 1.125 lb/ gal; 40% act.

Hartotrope KTS 50. [283] Potassium toluene sulfonate; detergent, solubilizer and cloud pt. depressant for lt. duty and built liq. detergent systems; clear liq.; dens. 1.24 lb/ gal; 50% act.

Hartotrope STS 40, Powd. [283] Sodium toluene sulfonate; anionic; detergent, solubilizer and cloud pt. depressant for lt. duty and built liq. detergent systems; clear liq., wh. powd. resp.; dens. 1.19 lb/gal (liq.); 40%, 93% act.

Hartotrope SXS 40 [283] Sodium xylene sulfonate; detergent, solubilizer and cloud pt. depressant for lt. duty and built liq. detergent systems; clear liq.; dens. 1.19 lb/ gal; 40% act.

Hartowet SLN. [283] Alcohol condensate; nonionic; wetting agent for finishing formulations in textiles; liq.; 60% conc.

Hartowet VSP. [283] Alkoxylate, modified; anionic; detergent and wetting agent effective under alkaline conditions; clear liq.; 50% liq.

Hartox DMCD. [284] Cocodimethylamine oxide; cationic; detergent for textiles, dyeing and cosmetics, wetting agent for bleaching, dyeing; liq.; 40% act.

Hercules AR 150. [295] PEG ester; nonionic; low-foaming detergent; Gardner 10 liq./soft wax; sol. in water, IPA, benzene, toluene, ester solv.; visc. 300 cps (38 C); cloud pt. 60 C (1%); surf. tens. 39 dynes/cm (0.01% aq.); pH 9 (1%).

Hetamide LA. [296] Lauric acid; nonionic; foam stabilizer, emulsifier, visc. builder, detergent for car shampoos, cleaning, cosmetics, dispersants, lubricants; liq.; sp.gr. 1.00; pH 9–10 (1% aq.); 100% act.

Hetamide MC. [296] Cocamide DEA; nonionic; foam stabilizer, emulsifier, visc. builder, detergent for car shampoos, cleaning, cosmetics, dispersants, lubricants; liq.; sp.gr. 0.98; pH 9.5–10.5 (1% aq.); 100% act.

Hetamide ML. [296] Lauramide DEA; nonionic; foam stabilizer, emulsifier, visc. builder, detergent for car shampoos, cleaning, cosmetics, dispersants, lubricants; solid; sp.gr. 0.96; m.p. 35 C; 100% act.

Hetamide MMC, OC. [296] Mixed fatty acids; nonionic; foam stabilizer, emulsifier, visc. builder, detergent, for car shampoos, cleaning, cosmetics, dispersants, lubricants; liq.; sp.gr. 1.01; pH 9–10.5 (1% aq.); 100% act.

Hetamide RC. [296] Cocamide DEA; nonionic; foam stabilizer, emulsifier, visc. builder, detergent for car shampoos, cleaning, cosmetics, dispersants, lubricants; liq.; sp.gr. 1.01; pH 9–10 (1% aq.); 100% act.

Hetester FAO. [296] C12–15 alcohols octanoate; wetting agent; used in personal care products; yel. clear liq.; sol. @ 5% in corn, safflower, and min. oils, octyl palmitate, oleyl alcohol, ethanol; insol. in water; sapon. no. 138–152; pH 6.5–7.5.

Hetester PCA. [296] PPG-1 ceteth-3 acetate; antichalking agent, emollient used in personal care products; colorless clear liq.; sol. @ 5% in water; cloud pt. 15 C; sapon. no. 110–130; pH 6.0–7.0.

Hetester PHA. [296] PPG-1 isoceteth-3 acetate; antichalking agent, emollient used in personal care products; wh.-pale yel. clear liq.; self-emulsifying in water; sapon. no. 110–130; pH 6.0–7.0.

Hetoxamate FA-5. [296] PEG-4 tallate; nonionic; detergent, emulsifier, lubricant, softener for cosmetics, textiles, leather, metal cleaning; liq.; sol. in IPA, min. oil, disp. in water; HLB 8.4; sapon. no. 100–120.

Hetoxamate FA-20. [296] PEG-20 tallate; nonionic; see Hetoxamate FA-5; semisolid; sol. in water, IPA; disp. min. oil; HLB 14.9; sapon. no. 50–60.

Hetoxamate LA-5. [296] PEG-5 laurate; nonionic; se Hetoxamate FA-5; liq.; sol. in IPA; disp. in water; HLB 9.8; sapon. no. 125–145.

Hetoxamate LA-9. [296] PEG-9 laurate; nonionic; se Hetoxamate FA-5; liq.; sol. in water, IPA; HLB 13.1; sapon. no. 89–96.

Hetoxamate MO-2. [296] PEG-2 oleate; detergent, emulsifier for personal care products; softener for leather; Gardner 2 max. liq.; sol. in IPA, min. oil; insol. in water; HLB 5.3; sapon. no. 145–160.

Hetoxamate MO-5. [296] PEG-5 oleate; nonionic; see Hetoxamate FA-5; liq.; sol. in IPA; disp. water; HLB 8.0; sapon. no. 115–125.

Hetoxamate MO-9. [296] PEG-9 oleate; nonionic; see Hetoxamate FA-5; liq.; sol. in IPA; disp. water; HLB 11.4; sapon. no. 80–88.

Hetoxamate MO-15. [296] PEG-15 oleate; nonionic; see Hetoxamate FA-5; liq.; sol. in water, IPA; HLB 13.5; sapon. no. 60–70.

Hetoxamate SA-5. [296] PEG-5 stearate; nonionic; see Hetoxamate FA-5; solid; sol. in IPA; disp. hot water; HLB 8.0; sapon. no. 120–130.

Hetoxamate SA-7. [296] PEG-7 monostearate; see Hetoxamate MO-2; Gardner 1 max. solid; sol. in IPA; disp. in water; HLB 10.5; sapon. no. 90–100.

Hetoxamate SA-9. [296] PEG-9 stearate; nonionic; see Hetoxamate FA-5; solid; sol. in IPA; disp. hot water; HLB 11.5; sapon. no. 84–93.

Hetoxamate SA-13. [296] PEG-12 stearate; see Hetoxamate MO-2; Gardner 2 max. liq.; sol. in water, IPA; HLB 13.4; sapon. no. 60–70.

Hetoxamate SA-23. [296] PEG-20 stearate; see Hetoxamate MO-2; Gardner 2 max. solid; sol. see Hetoxamate SA-13; HLB 15.6; sapon. no. 39–49.

Hetoxamate SA-35. [296] PEG-35 stearate; nonionic; see Hetoxamate FA-5; solid; sol. in water, IPA; HLB 16.9; sapon. no. 24–34.

Hetoxamate SA-40. [296] PEG-40 stearate; see Hetoxamate MO-2; Gardner 2 max. flake; sol. see Hetoxamate SA-13; HLB 17.2; sapon. no. 24–34.

Hetoxamate SA-90. [296] PEG-90 stearate; nonionic; see Hetoxamate FA-5; solid; sol. in water IPA; HLB 18.6; sapon. no. 10–18.

Hetoxamide C-4. [296] PEG-5 coconut amide; emulsifier, oil-loving; dk. br. liq.; pH 8.5–9.5 (10%).

Hetoxamine C-2. [296] PEG-2 cocamine; cationic; emulsifier, softener, antistat, water repellent, desizing agent in agriculture, waxes, oils, textile/leather, metal cleaning; liq.; m.w. 285; sol. IPA, min. oil; gel in water.

Hetoxamine C-5. [296] PEG-5 cocamine; cationic; see Hetoxamine C-2; liq.; m.w. 425; sol. water, IPA, min. oil.

Hetoxamine C-15. [296] PEG-15 cocamine; cationic; see Hetoxamine C-2; liq.; m.w. 860; sol. water, IPA.

Hetoxamine O-2. [296] PEG-2 oleamine; nonionic; see Hetoxamine C-2; liq.; m.w. 350; sol. in IPA, min. oil.

Hetoxamine O-5. [296] PEG-5 oleamine; nonionic; see Hetoxamine C-2; liq.; m.w. 492; sol. in IPA, min. oil.

Hetoxamine O-15. [296] PEG-15 oleamine; nonionic; see Hetoxamine C-2; liq.; m.w. 930; sol. in water, IPA.

Hetoxamine S-2. [296] PEG-2 soyamine; cationic; see Hetoxamine C-2; liq.; m.w. 350; sol. in IPA, min. oil.

Hetoxamine S-5. [296] PEG-5 soyamine; cationic; see Hetoxamine C-2; liq.; m.w. 480; sol. in IPA; partly sol. in min. oil; gel in water.

Hetoxamine S-15. [296] PEG-15 soyamine, cationic; desizing agent, antistat; emulsifier in agriculture, waxes and oils, leather processing, and metal cleaning industries; water repellent and wet spinning assistant in textile industries; Gardner 12 max. liq., solid; sol. see Hetoxamate SA-13; 95.0% min. tert. amine.

Hetoxamine ST-2. [296] PEG-2 stearamine; nonionic; see Hetoxamine C-2; solid; m.w. 388; sol. in IPA, min. oil.

Hetoxamine ST-5. [296] PEG-5 stearamine; nonionic; see Hetoxamine C-2; solid; m.w. 520; sol. in IPA, min. oil.

Hetoxamine ST-15. [296] PEG-15 stearamine; see Hetoxamine S-15; Gardner 8 max. solid; sol. see Hetoxamate SA-13; 95.0% min. tert. amine.

Hetoxamine ST-50. [296] PEG-50 stearamine; see Hetoxamine S-15; Gardner 8 max. solid; sol. see Hetoxamate SA-13.

Hetoxamine T-2. [296] PEG-2 tallow amine; cationic; see Hetoxamine C-2; liq.; m.w. 350; sol. in IPA, min. oil.

Hetoxamine T-5. [296] PEG-5 tallow amine; cationic; see Hetoxamine C-2; liq./paste; m.w. 490; sol. in water, IPA, min. oil.

Hetoxamine T-15. [296] PEG-15 tallow amine; cationic; see Hetoxamine C-2; liq./paste; m.w. 925; sol. in water, IPA.

Hetoxamine T-20. [296] PEG-20 tallow amine; cationic; see Hetoxamine C-2; liq./paste; m.w. 1150; sol. in water, IPA.

Hetoxide BN-13. [296] B-naphthol ethoxylate; nonionic; emollient, emulsifier, used in cosmetics, household, textile industry; paste; sol. in water, IPA, min. oil.

Hetoxide BP-3. [296] Modified butanol ethoxylate; nonionic; emollient, emulsifier, used in cosmetics, household, textile industry; liq.; sol. in water, IPA.

Hetoxide BY-1.8. [296] POE (1.8) butynediol; metal plating emulsifier; liq.; sol. in water, IPA; insol. in min. oil.

Hetoxide BY-3. [296] POE (3) butyndiol; nonionic; emollient, emulsifier used in cosmetics, household, textile industry; liq.; sol. in water, IPA.

Hetoxide C-2. [296] PEG-2 castor oil; nonionic; emollient, emulsifier, used in cosmetics, household, textile industry; liq.; sol. in IPA, min. oil; HLB 4.3; sapon. no. 155–170.

Hetoxide C-9. [296] PEG-9 castor oil; nonionic; see Hetoxide C2; liq.; sol. in IPA, min. oil; HLB 11.2; sapon. no. 120–136.

Hetoxide C-15. [296] PEG-15 castor oil; nonionic; see Hetoxide C2; liq.; sol. in IPA, min. oil; HLB 13.6; sapon. no. 95–105.

Hetoxide C-25. [296] PEG-25 castor oil; nonionic; see Hetoxide C2; liq.; sol. in IPA, min. oil; HLB 15.6; sapon. no. 74–82.

Hetoxide C-30. [296] PEG-30 castor oil; nonionic; perfume solubilizer, emollient, emulsifier, visc. control and scouring agent, lubricant, dispersant for cosmetic formulations; dyeing assistant, dye carrier for textiles; used in household cleaning comps., metal treatment, metal plating, chemical intermediate; Gardner 5 max. liq.; sol. in IPA; disp. in water; HLB 11.7; sapon. no. 65–75.

Hetoxide C-40. [296] PEG-40 castor oil; nonionic; see Hetoxide C2; paste; sol. in water, IPA; HLB 16.9; sapon. no. 55–65.

Hetoxide C-200. [296] PEG-200 castor oil; nonionic; see Hetoxide C2; solid; sol. in water, IPA; sapon. no. 16–18.

Hetoxide C-200-50%. [296] PEG-200 castor oil; see Hetoxide C-30; Gardner 4 max. liq.; sol. in water, IPA; sapon no. 7–10; 49.0–51.0% moisture.

Hetoxide DNP-4, DNP-9.6. [296] Nonyl nonoxynol-4, -9.6; nonionic; see Hetoxide C;

liq.; sol. in IPA, min. oil, disp. in water; HLB 6.6, 10.7; hyd. no. 120–140, 65–75.

Hetoxide G-7. [296] Glycereth-7; nonionic; see Hetoxide C; ; liq.; sol. in IPA, water.

Hetoxide G-26. [296] Glycereth-26; humectant for pressure-sensitive adhesives; Gardner 2 max. liq.; sol. in water, IPA; acid no. 2.0 max.

Hetoxide HC-16. [296] PEG-16 hydrog. castor oil; nonionic; see Hetoxide C; liq.; sol. in IPA, min. oil; HLB 11.3; sapon. no. 85–95.

Hetoxide HC-40. [296] PEG-40 hydrog. castor oil; see Hetoxide C-30; Gardner 4 max. semisolid; sol. see Hetoxamate SA-13; HLB 13.1; sapon. no. 50–60.

Hetoxide HC-60. [296] PEG-60 hydrog. castor oil; see Hetoxide C-30; Gardner 4 max. solid; sol. see Hetoxamate SA-13; HLB 14.8; sapon. no. 41–51.

Hetoxide MTG. [296] Alcohol ethoxylate; nonionic; see Hetoxide C; liq.; sol. in IPA, min. oil, water.

Hetoxide NP-4. [296] Nonoxynol-4; nonionic; see Hetoxide C; liq.; sol. in IPA, min. oil; HLB 7.6; hyd. no. 135–140.

Hetoxide NP-9. [296] Nonoxynol-9; nonionic; see Hetoxide C; liq.; sol. in water, IPA; HLB 13.0; hyd. no. 85–95.

Hetoxide NP-40. [296] Nonoxynol-40; nonionic; see Hetoxide C; liq.; sol. in water, IPA; HLB 17.0.

Hetoxol 15 CSA. [296] POE (15) cetearyl ether; nonionic; detergent, emulsifier, leveling agent, intermediate; used in personal care products, wax, oil and textiles, scouring agents, dyes, household formulations, silicone emulsification, surfactants; Gardner 1 max. solid; sol. see Hetoxamate SA-13; HLB 14.2.

Hetoxol 916P. [296] POE (6) PPG 2.5 C_9–C_{11} alcohols ether; surfactant for industrial applications; Gardner 2 max. liq.; sol. see Hetoxamate SA-13; cloud pt. 42–45 C (1% aq.).

Hetoxol CA-2, CA-10, CA-20. [296] Ceteth-2, -10, –20; nonionic; detergent, emulsifier, leveling agent, intermediate, used for cosmetics, household formulations, silicone emulsification, textile processing; solid; sol. in IPA; CA-2 sol. in min. oil; CA-10 sol. in water, IPA; CA-20 sol. in water; HLB 5.3, 12.7, 15.7; hyd. no. 160–180 (CA-2), 45–60 (CA-20).

Hetoxol CD-4. [296] POE (4) 2-ethyl hexyl ether; nonionic; emulsifier; APHA 100 max. liq.; sol. see Hetoxamate SA-13; HLB 11.5.

Hetoxol CS-4. [296] Ceteareth-4, –5; see Hetoxol 15 CSA; Gardner 1 max. solid; sol. in IPA; disp. in water; HLB 8.2 and 9.2 resp.

Hetoxol CS-9. [296] Ceteareth-9; nonionic; see Hetoxol CA2; solid; sol. in water, IPA; HLB 12.0; hyd. no. 80–90.

Hetoxol CS-15. [296] Ceteareth-15; see Hetoxol 15 CSA; Gardner 1 max. solid; sol. water, IPA; HLB 14.2.

Hetoxol CS-20. [296] Ceteareth-20; nonionic; see Hetoxol CA2; solid; sol. in water, IPA; HLB 15.4; hyd. no.50–70.

Hetoxol CS-30. [296] Ceteareth-30; nonionic; see Hetoxol CA2; solid; sol. in water, IPA; HLB 16.7; hyd. no. 40–52.

Hetoxol CS-50, CS-50 Special. [296] Ceteareth-50; see Hetoxol 15 CSA; Gardner 2 max. flake; sol. in water, IPA.

Hetoxol L-3N. [296] Laureth-3; nonionic; see Hetoxol CA2; liq.; sol. in IPA, min. oil; HLB 7.9; hyd. no. 170–176.

Hetoxol L-4N. [296] Laureth-4; nonionic; see Hetoxol CA2; liq.; sol. in IPA, min. oil; HLB 9.7; hyd. no. 145–165.

Hetoxol L-9N. [296] Laureth-9; nonionic; see Hetoxol CA2; liq.; sol. in water, IPA; HLB 11.8; hyd. no. 90–110.

Hetoxol L-12. [296] Laureth-12; emulsifier for personal care products; Gardner 1 max. semisolid; sol. in water, IPA; HLB 14.6.

Hetoxol L-23N. [296] Laureth-23; nonionic; see Hetoxol CA2; solid; sol. in water, IPA; HLB 16.9; hyd. no. 40–55.

Hetoxol LS-9. [296] Laureth-9; nonionic; see Hetoxol CA2; solid; sol. in water, IPA; HLB 13.3; hyd. no. 90–100.

Hetoxol M-3. [296] Myreth-3; emulsifier and pigment dispersant in makeup; sol. in IPA; disp. in water; HLB 7.6; pH 5.5–7.0.

Hetoxol OA-3 Special. [296] POE (3) oleyl ether; emulsifier and pigment dispersant for cosmetic applications; pale yel. liq.; sol. in IPA, min. oil; insol. in water; HLB 6.4.

Hetoxol OA-5 Special. [296] POE (5) oleyl ether; see Hetoxol OA-3 Special; pale yel. liq.; sol. in IPA, min. oil; disp. in water; HLB 9.0.

Hetoxol OA-10 Special. [296] POE (10) oleyl ether; see Hetoxol OA-3 Special; wh. semisolid; sol. in water, IPA; HLB 12.4.

Hetoxol OA-20 Special. [296] POE (20) oleyl ether; see Hetoxol OA-3 Special; wh. solid; sol. see Hetoxol OA-10 Special; HLB 15.3.

Hetoxol OL-2. [296] Oleth-2; nonionic; see

Hetoxol CA2; liq.; sol. in IPA, min. oil; HLB 4.9; hyd. no. 152–162.

Hetoxol OL-4. [296] Oleth-4; nonionic; see Hetoxol CA2; liq.; sol. in IPA, min. oil; HLB 8.0; hyd. no. 120–135.

Hetoxol OL-5. [296] Oleth-5; nonionic; see Hetoxol CA2; liq.; sol. in water, IPA HLB 9.1; hyd. no. 110–120.

Hetoxol OL-10. [296] Oleth-10; surfactant for cosmetic formulations; Gardner 3 semisolid; sol. in water, IPA; HLB 12.4.

Hetoxol OL-10H. [296] Oleth-10; see Hetoxol OL-10; Gardner 1 max. semisolid; sol. in water, IPA; HLB 12.8; cloud pt. 71–77 C (1% aq.).

Hetoxol OL-20. [296] Oleth-20; see Hetoxol OL-10; Gardner 2 max. solid; sol. in water, IPA; HLB 15.3.

Hetoxol OL-23. [296] Oleth-23; nonionic; see Hetoxol CA2; semisolid; sol. in water, IPA; HLB 15.9; hyd. no. 47–62.

Hetoxol OL-40. [296] Oleth-40; nonionic; see Hetoxol CA2; solid; sol. in water, IPA; HLB 17.4; hyd. no. 30–45.

Hetoxol P. [296] Fatty alcohol; emulsion base; flake; sapon. no. 14.0 max.

Hetoxol PA-1. [296] Ethylene glycol monopropargyl ether; metal plating emulsifier, wetting agent; liq.; sol. in water, IPA.

Hetoxol STA-2. [296] Steareth-2; nonionic; see Hetoxol CA-2; liq.; sol. in IPA; HLB 4.9; hyd. no. 145–165

Hetoxol STA-10. [296] Steareth-10; see Hetoxol 15; Gardner 1 max. solid; sol. in water, IPA; HLB 12.4.

Hetoxol STA-20. [296] Steareth-20; see Hetoxol 15 CSA, STA-10; Gardner 1 max. solid; sol. in water, IPA; insol. in min. oil; HLB 15.3.

Hetoxol STA-30. [296] Steareth-30; see Hetoxol 15 CSA, STA-10; Gardner 2 max. flake; sol. see Hetoxol STA-20; HLB 16.6; m.p. 46–50 C.

Hetoxol TD-3. [296] Trideceth-3; nonionic; see Hetoxol CA2; liq.; sol. in IPA, min. oil; HLB 7.9; hyd. no. 165–175.

Hetoxol TD-6. [296] Trideceth-6; nonionic; see Hetoxol CA2; liq.; sol. in IPA, water; HLB 11.3; hyd. no. 115–125.

Hetoxol TD-9. [296] Trideceth-9; wetting agent, industrial surfactant; Gardner 1 max. liq.; sol. in water, IPA; HLB 13.2%

Hetoxol TD-12. [296] Trideceth-12; nonionic; see Hetoxol CA2; liq.; sol. in IPA, water; HLB 14.5; hyd. no. 72–87.

Hetoxol TD-25. [296] Trideceth-25; see Hetoxol TD-9; Gardner 2 max. semisolid; HLB 16.9.

Hetoxol TDEP-15. [296] POE (10) PPG 15 tridecyl ether; surfactant for industrial use, automatic dishwashing formulations; Gardner 2 max. liq.; sol. in water, IPA; cloud pt. 21–23 C (1% aq.).

Hetoxol TDEP-63. [296] POE (6) PPG 3 tridecyl ether; see Hetoxol TDEP-15; Gardner 2 max. liq.; sol. see Hetoxol TDEP-15; cloud pt. 27–33 C (1% aq.).

Hetsorb L-4. [296] Polysorbate 21; emulsifier; Gardner 8 max. liq.; sol. in IPA; disp. in water; sapon. no. 100–115.

Hetsorb L-10. [296] PEG-10 sorbitan laurate; detergent, emulsifier, lubricant for cosmetics; liq.; sol. in water, IPA; HLB 8.4; sapon. no. 66–76.

Hetsorb L-20. [296] Polysorbate 20; see Hetsorb L-10; liq.; sol. in water, IPA; HLB 16.7; sapon. no. 40–50.

Hetsorb L-80-72%. [296] PEG-80 sorbitan laurate; surfactant for shampoos; used as counterirritant; Gardner 5 clear liq.; sol. in water, IPA; sapon. no. 8–18; 70–74% solids.

Hetsorb O-5. [296] PEG-5 sorbitan monooleate; food emulsifier; Gardner 8 max. liq.; sol. in IPA; disp. in water; sapon. no. 96–104.

Hetsorb O-20. [296] Polysorbate 80; see Hetsorb L-10; liq.; sol. in water, IPA; HLB 15.0; sapon. no. 45–55.

Hetsorb P-20. [296] Polysorbate 40; general purpose emulsifier; Gardner 6 max. liq., semisolid; sol. in water, IPA; sapon. no. 40–53.

Hetsorb S-4. [296] PEG-4 sorbitan monostearate; emulsifier; Gardner 8 max. semisolid; sol. in IPA; disp. in water; sapon. no. 95–115.

Hetsorb S-20. [296] Polysorbate 60; see Hetsorb L-10; liq.; sol. in water, misc. in IPA; HLB 14.9; sapon. no. 45–55.

Hetsorb TO-20. [296] Polysorbate 85; general purpose emulsifier; Gardner 8 max. liq.; sol. in IPA, min. oil; disp. in water; sapon. no. 82–95.

Hetsorb TS-20. [296] Polysorbate 65; general purpose emulsifier; Gardner 7 max. solid; sol. in IPA, hot min. oil; disp. in water; HLB 10.5; sapon. no. 88–98.

Hetsulf 40, 40X. [296] Sodium dodecylbenzene sulfonate; anionic; wetting agent, emulsifier, dispersant, intermediate, detergent, liq. formulation syndet; paste, liq.; pH

7–8.5 (5% aq.); biodeg.; 38% and 35% act.

Hetsulf 50A. [296] Ammonium dodecylbenzene sulfonate; anionic; wetting agent, emulsifier, dispersant, for lt. duty detergent formulations; slurry; pH 6.0–7.0 (5% aq.); biodeg.; 48% act.

Hetsulf 60S. [296] Sodium dodecylbenzene sulfonate; anionic; wetting agent, emulsifier, dispersant, base for formulated products; slurry; pH 7.0–8.0 (5% aq.); biodeg.; 57% act.

Hetsulf 60T. [296] Amine dodecylbenzene sulfonate; anionic; wetting agent, emulsifier, dispersant, for cosmetic, bath and shampoos uses; liq.; pH 7.0–7.5 (5% aq.); biodeg.; 52% act.

Hetsulf Acid. [296] Dodecylbenzene sulfonic acid; anionic; wetting agent, emulsifier, dispersant, intermediate, base for neutralized surfactant; liq.; dens. 1.05; biodeg.; 97% act.

Hetsulf IPA. [296] MIPA-dodecylbenzene sulfonate; anionic; wetting agent, emulsifier, dispersant; liq.; pH 5.0–6.5 (5% aq.); biodeg.

Hexaryl D 60 L, L 30. [654] Alkylaryl sulfate; anionic; multipurpose detergent; liq.

Hi-Brite. [146] Detergent for industrial, institutional cleaners; pink liq.; bland odor; sol. in water; dens. 8.60 lb/gal; sp.gr. 1.030; 17% act.

Hipochem ADN. [298] Quat. ammonium complex; nonionic; dyeing assistant, leveling agent of acid dyes on nylon; amber liq.; mild alcohol odor; sol. in water; dens. 8.73 lb/gal; sp.gr. 1.048; 50% act.

Hipochem AM-99. [298] Super amide condensate; nonionic; multipurpose detergent, wetting agent, emulsifier, foam builder; amber liq.; mild amine odor; water sol.; dens. 8.89 lb/gal; sp.gr. 1.06; 99% act.

Hipochem AMC. [298] Amine condensate; nonionic; see Hipochem AM-99; dk. br. liq.; mild amine odor; water sol.; dens. 8.16 lb/gal; sp.gr. .98; 99–100% act.

Hipochem AR-100. [298] Alkylphenol polyethoxide ether; detergent and textile scouring agent; liq.; 100% conc.

Hipochem C-95. [298] Quat. ammonium compd.; cationic; dyeing assistant, retarder for cationic dyes on acrylics; amber liq.; mild alcohol odor; water sol.; dens. 8.29 lb/gal; sp.gr. 0.995; 50% act.

Hipochem CAD. [298] Aromatic ester, emulsified; amphoteric; dyeing assistant for textiles; liq.; 100% conc.

Hipochem CDL. [298] Fatty alcohol ethoxylate; nonionic; dyeing assistant and scouring agent for textiles; nonionic; liq.; 30% conc.

Hipochem D2. [298] Quat. ammonium compd.; cationic; dyeing assistant, retarder for cationic dyes on acrylics; amber liq.; mild alcohol odor; water sol.; dens. 8.08 lb/gal; sp.gr. 0.97; 40% act.

Hipochem Dispersol GTO. [298] Sulfated glycerol trioleate; anionic; detergent, wetting agent, emulsifier, dyeing assistant; dk. amber liq.; mild sulfate odor; water sol.; dens. 8.66 lb/gal; sp.gr. 1.04; 55% act.

Hipochem Dispersol SB. [298] Sulfated butyl oleate; anionic; see Hipochem Dispersol GTO; dk. amber liq.; mild sulfate odor; water disp.; dens. 8.41 lb/gal; sp.gr. 1.01; 55% act.

Hipochem Dispersol SCO. [298] Sulfated castor oil; anionic; see Hipochem Dispersol GTO; dk. amber liq.; mild sulfate odor; water disp.; dens. 8.33 lb/gal; sp.gr. 1.0; 70% act.

Hipochem Dispersol SP. [298] Sulfated propyl oleate; anionic; dyeing assistant, detergent, emulsifier used in textiles; liq.; 60% conc.

Hipochem EFK. [298] Aliphatic sulfate; detergent, textile lubricant, substantive to cellulosic fibers; wh. paste; water sol.; 40% act.

Hipochem EK-6. [298] Long chain phosphated alcohol; anionic; wetting agent for desizing, dyeing, bleaching; lt. amber liq.; mild alcohol odor; water sol.; dens. 8.70 lb/gal; sp.gr. 1.044; 60% act.

Hipochem EK-18. [298] Sodium dioctyl sulfosuccinate; anionic; wetting, rewetting, desizing, bleaching, scouring for textiles; lt. amber liq.; slight alcohol odor; water sol.; dens. 8.91 lb/gal; sp.gr. 1.07; 75% act.

Hipochem LCA. [298] Sulfonated oleyl alcohol; anionic; detergent, lubricant, textile dye leveling agent; wh. paste; mild sulfate odor; disp. in water (10%); 40% act.

Hipochem LH-Soap. [298] Coconut oil soap; anionic; detergent, wetting agent, emulsifier, developer and brightener; water sol.; dens. 7.77 lb/gal; sp.gr. 0.93; 40% act.

Hipochem No. 3. [298] Sulfated oil plus soap; anionic; detergent and dye leveling agent used in textiles; liq.; 45% conc.

Hipochem No. 40-L. [298] Modified ABS; anionic; detergent, wetting agent, emulsifier, dyeing penetrant; dk. amber liq.; water sol.; dens. 8.33 lb/gal; sp.gr. 1.0; 40% act.

Hipochem No. 641. [298] Alkyldiaryl sulfonate; anionic; dyeing assistant for polyamide fibers; lt. br. liq.; water sol.; dens. 9.7 lb/gal; sp.gr. 1.16; 45% act.

Hi-Tex. [146] Detergent used for laundering and rug shampooing; amber liq.; sol. in water; dens. 8.45 lb/gal; sp.gr. 1.012; 14% act.

Hi-Tex AC. [146] Detergent used for cleaners; amber liq.; mild odor; sol. in water; dens. 8.36 lb/gal; sp.gr. 1.001; 12% act.

Homodan. [273] Special emulsifiers and blends; plasticizer, improves plasticity of margarine for flakier puff pastry, nonionic; flakes, liq., pills, solid; 100% act.

Homotex PS-90. [341] Propylene glycol monostearate; nonionic; cake mix, fluid shortening; powd.; HLB 3.4; 100% conc.

Homotex PT. [341] Monoglyceride, med. chain; nonionic; bacteria control; liq.; HLB 3.2; 50% min.

Honoralin PL. [590] Sulfated oil; anionic; wetting and penetrating agent; dyeing aux. for cotton and wool; liq.

Hostacerin CG. [301] Triceteareth-4 phosphate and sodium C14-17 sec alcohol sulfonate; anionic; base material for mfg. of o/w creams; waxy flakes.

Hostacerin DGO. [44; 301] Polyglyceryl-2 sesquioleate; nonionic; emulsifier for cosmetics and pharmaceuticals; yel. liq.; weakly fatty odor; sol. in most org. solv., paraffin and neutral oil, warm fatty alcohols, fatty acid esters and their mixtures; dens. 0.97 g/cm³; visc. 400–800 cps; HLB 3–4; flash pt. 250 C; sapon. no. 165 ± 15; 100% act.

Hostacerin DGS. [301] Polyglyceryl-2-PEG-4 stearate; nonionic; see Hostacerin DGO; wax; HLB 7–8.

Hostacerin O-5. [44] Oleth-5; nonionic; emulsifier for cosmetics creams and lotions; yel. slightly cloudy liq.; readily sol. in most org. solv., dissolves to cloudy sol'n. in water; sp.gr. 0.953 ± 0.005; visc. 67 ± 3 cps; flash pt. 240 C (Marcusson); sapon. no. 1; 100% act.

Hostacerin T-3. [44; 301] Ceteareth-3; nonionic; emulsifier, superfatting agent, base for ointments, creams, liq. emulsions, shampoo additive; wh. soft, waxlike substance; sol. warm in all hydrocarbons, fatty alcohols; sp.gr. 0.905 (50 C); visc. 15 ± 3 cps (50 C); HLB 7–8; cloud pt. 54 C (in butyl diglycol); flash pt. 220 C; sapon. no. 1 max.; 100% act.

Hostacerin WO. [301] Polyglyceryl-2 sesquiisostearate, beeswax, min. oil, magnesium stearate, and aluminum stearate; nonionic/anionic; emulsifier conc. for cosmetic w/o emulsions; wax; HLB 4–5; 100% conc.

Hostacor BM. [44] Alkyl sulfonamidocarboxylic acid ester 70%; emulsifier and corrosion inhibitor in coolants; yel., br. clear liq.; water sol; sp.gr. 1.1; pour pt. 14 F; pH 9 (1%).

Hostacor BS. [44] Aryl sulfonamidocarboxylic acid ester 70%; emulsifier and corrosion inhibitor in coolants; yel., br. clear liq.; water sol; sp.gr. 1.1; pour pt. 14 F; pH 9 (1%).

Hostacor DT. [44] Fatty acid alkanolamide; corrosion inhibitor used in metalworking fluids; br. oily liq.; dissolves to turbid sol'n. in water, clear sol'n. in min. oils; sp.gr. 1.01; flash pt. 210 C (Marcusson); pH 9 (1% aq.).

Hostacor H Liq. [44] Arylsulfonamidocarboxylic acid; anionic; chemical intermediate for mfg. of corrosion inhibitors for aq. systems; yel. visc. liq.; sol. in ethanol, glycol, water insol.; sp.gr. 1.17 ± 0.05 (122 F).

Hostapal 3634 Highly Conc. [301] Alkylaryl polyglcol ether; nonionic; wetting agent and detergent used in fibers; liq.

Hostapal BV Conc. [44; 301] Alkylaryl polyglycol ether sulfate, sodium salt; anionic; dispersant for pigments, wetting agent, detergent for fibers, emulsifier for emulsion polymerization; yel. br. gelatinous paste; misc. sol. with water; dens. 9.10 lb/gal; 49–51% act.

Hostapal CV Brands. [301] Alkylaryl polyglycol ether; nonionic; wetting agent and detergent for textile industry; stiff-flowing liq.

Hostapal CVS. [44] Alkylphenol polyglycol ether; anionic; wetting agent for org. and inorganic pigment preparations, for use with anionic dispersing agents; slighty yel. visc. liq.; dens. 8.85 lb/gal; cloud pt. 52–56 C (1% aq.); 100% act.

Hostaphat F Brands. [301] Phosphoric acid complex org. ester; anionic; antistat finish for syn. fibers; liq.; 95% conc.

Hostaphat KL340N. [301] Trilaureth-4 phosphate; anionic; o/w emulsifier for cosmetic purposes; liq.; 100% conc.

Hostaphat KO 300. [301] Trioleyl phosphate; nonionic; emulsifier for cosmetic w/o emulsions; liq.; insol. in water; 100% conc.

Hostaphat KO 380. [301] Trioleth-8 phosphate; nonionic; see Hostaphat KO 300 N; liq.; 100% conc.

Hostaphat KW 340 N. [301] Triceteareth-4

phosphate; nonionic; see Hostaphat KO 300; wax; 100% conc.

Hostaphat L Brands. [301] Org. phosphoric acid ester; anionic; antistat for textile industry.

Hostaphat OPS. [301] Alkyl phosphonic acid; anionic; hydrotrope used in electrolyte sol'ns.; solubilizer; paste; 99% conc.

Hostapon CT Paste. [301] Sodium methyl cocoyl taurate; anionic; detergent used in shampoos; paste; 30% conc.

Hostapon KA Special. [301] Fatty acid isethionate, sodium salt; anionic; detergent base for cosmetic industry; powd.; 80% conc.

Hostapon KTW New. [301] Sodium lauroyl taurate; anionic; detergent base for cosmetics; powd.; 50% conc.

Hostapon STT Paste. [301] Fatty acid methyl tauride, sodium salt; anionic; see Hostapon CT Paste; paste; 30% conc.

Hostapon T. [301] Sodium methyl oleoyl taurate; anionic; see Hostapon KA Special; powd.; 63% conc.

Hostapur CX Highly Conc. [301] Alkyl polyglycol ether; wetting agent and detergent for textiles; liq.

Hostapur OS Brands. [301] alpha-Olefin sulfonate, sodium salt; anionic; basic materials for detergents and cleaning agents; carpets; liq., powd.

Hostapur SAS 30, SAS 60, SAS 93. [44; 301] Sodium C14–17 sec alcohol sulfonate; detergent base for mfg. of washing-up liq., detergent, shampoo, wetting agent; colorless to weakly yel. clear liq., yel. soft paste, flake; virtually odorless; m.w. 328; sp.gr. 1.048–1.087; visc. 26–400 cps; biodeg.; 30, 60, 93% act.

Humectol C, C Highly Conc. [301] Sulfonated fatty acid amines; anionic; wetting and leveling agent, dispersant for dyeing and printing; liq.

Hyamine 10-X. [524] Methylbenzethonium chloride; cationic; germicide, disinfectant, sanitizer in restaurant and pharmaceutical uses; wh. crystals; dens. 27.5 lb/ft^3; visc. solid; surf. tens. 40 dynes/cm (0.01% aq.); 100% act.

Hyamine 1622. [524] Benzethonium chloride; cationic; see Hyamine 10-X; lt. amber liq.; dens. 8.6 lb/gal; flash pt. 110 F (TOC); pour pt. 25 F; 50% act.

Hyamine 2389. [524] Dodecylbenzyl trimonium chloride (80%), dodecylxylylditrimonium chloride (20%); cationic; see Hyamine 10-X; lt. amber liq.; dens. 8.3 lb/gal; visc. 360 cps; flash pt. > 200 F (TOC); pour pt. 25 F; surf. tens. 43 dynes/cm (0.01% aq.); 50% act.

Hyamine 3500. [524] *n*-Alkyl (50% C_{14}, 40% C_{12}, 10% C_{16}) dimethyl benzyl ammonium chloride; cationic; see Hyamine 10-X; VCS 2–5 liq.; dens. 8.0 lb/gal; visc. 42 cps; flash pt. 105 F (TOC); pour pt. 15 F; surf. tens. 40 dynes/cm (0.01% aq.); 50% act.

Hybase C-300. [650] Highly basic oil-sol. calcium sulfonate; detergent, rust inhibitor used in marine or stationary diesels; dk. br. liq.; sol. in org. solvs.; dens. 9.4 lb/gal; sp.gr. 1.13 (60 F); visc. 800 SUS (210 F); flash pt. 380 F; 29% act.

Hybase M-300, -400. [650] Highly basic oil-sol. magnesium sulfonate; lubricant, corrosive inhibitor, used in marine or stationary diesels; dk. br. liq.; petroleum odor; sol. in org. solv.; dens. 9.3–9.6 lb/gal; sp.gr. 1.12–1.15 (60 F); visc. 300 cs (210 F); flash pt. 375–380 F; 27–28% act.

Hydagen DEO. [290] Triethyl citrate and BHT; act. ingred. for deodorant systems and personal care products; pale yel. clear oily liq.; sp.gr. 1.11–1.12; 99.0% act.

Hydrofol Acid 1655-CG-NF. [559] 48% stearic/52% palmitic acid (triple pressed stearic); used in cosmetic emulsions; Lovibond 2Y/0.5R max.; sapon. no. 206–214.

Hydrofol Acid 1690. [559] 90% palmitic acid; used in cosmetic emulsions; Lovibond 5Y/0.5R max.; sapon. no. 216–222.

Hydrofol Acid 1855. [559] 55% stearic acid; used in cosmetic emulsions; Lovibond 3-5Y/0.3R max.; sapon. no. 205–213.

Hydrofol Acid 1895. [559] 95% stearic acid; see Hydrofol Acid 1655-CG-NF; Lovibond 5Y/0.5R max.; sapon. no. 194–202.

Hymolon CWC. [283] Coconut diethanolamide; anionic; fulling agent for wool and household cleaners; liq.

Hymolon K-90. [283] Coconut diethanolamide; nonionic; household detergent and emulsifier, scouring agent for wool, cotton, polyester; yel. liq.; dens. 1.04 lb/gal; 91% act.

Hymono 1103, 1123, 1163, 3203, 4404, 7804, 8803, 8903. [482] Monoglyceride, dist.; nonionic; used in food applications; microbead, paste; HLB 4.0; 100% conc.

Hyonic CPG 745. [187] Linear aliphatic alcohol block polymer; nonionic; surfactant and wetting agent used in rinse aids, machine dishwashing, spray cleaning, bottle washing, metal cleaning; clear liq.; sp.gr. 110; dens. 9.2 lb/gal; visc. 59 cP; pour pt. 0 C; cloud pt. 2 C (10%); 99% act.

Hyonic NP-40. [187] Nonoxynol-4; nonionic; emulsifier for solv. cleaning compds., waterless hand cleaners, agric. emulsifiable concs., silicone products, detergent for petroleum oils; visc. reducer for plastisols; intermediate for production of anionic surfactants; metal cleaners; food contact applications; clear liq.; insol. in water; dens. 1.02 g/ml; HLB 9; pour pt. 26 C; pH 7.0 (1% aq.); > 99% act.

Hyonic NP-60. [187] Nonoxynol-6; nonionic; emulsifier, wetting agent, detergent for textile lubricants, agric. emulsifiable concs.; intermediate for textile antistats; color enhancer in latex paint; post-polymerization stabilizer; food applications; clear liq.; disp. in water; dens. 1.04 g/ml; HLB 12; pour pt. –32 C; cloud pt. < 0 C (1% aq.); pH 7.0 (1% aq.); surf. tens. 30 dynes/cm (0.01%); > 99% act.

Hyonic NP-90. [187] Nonoxynol-9; nonionic; detergent and penetrant for household and industrial cleaning compds., textile processing; degreasing compds.; wetting agent for paper toweling or tissue; emulsifier in agric. applications; clear liq.; sol. in water; dens. 1.06 g/ml; HLB 13; pour pt. 2 C; cloud pt. 54 C (1% aq.); pH 7.0 (1% aq.); surf. tens. 31 dynes/cm (0.01%); > 99% act.

Hyonic NP-100. [187] Nonoxynol-10; nonionic; surfactant, wetting agent, emulsifier; base for liq. dishwashing detergents, laundry preparations, household and industrial cleaners; agric. applications; penetrant for corrosion inhibitors; stabilizer for latex; rewetting agent for paper toweling and tissue; food applications; clear liq.; sol. in water; dens. 1.07 g/ml; HLB 13.2; pour pt. 4 C; cloud pt. 68 C (1% aq.); pH 7.0 (1% aq.); surf. tens. 32 dynes/cm (0.01%); > 99% act.

Hyonic NP-110. [187] Nonoxynol-11; nonionic; see Hyonic NP-100; clear liq.; sol. in water; dens. 1.07 g/ml; HLB 13.6; pour pt. 13 C; cloud pt. 72 C (1% aq.); pH 7.0 (1% aq.); surf. tens. 34 dynes/cm (0.01%); > 99% act.

Hyonic NP-120. [187] Nonoxynol-12; nonionic; detergent, wetting agent, penetrant, coemulsifier; food and agric. applications; liq., semisolid; sol. in water; dens. 1.07 g/ml; HLB 14; pour pt. 15 C; cloud pt. 91 C (1% aq.); pH 7.0 (1% aq.); surf. tens. 36 dynes/cm (0.01%); > 99% act.

Hyonic NP-407. [187] Nonoxynol-40; nonionic; stabilizer for water dispersions; primary emulsifier in mfg. of vinyl acetate or acrylic latex; wetting agent for min. acid and alkali sol'ns.; coemulsifier for agric. formulations; food applications; clear liq.; sol. in water; dens. 1.10 g/ml; HLB 17.6; pour pt. –6 C; cloud pt. 100 C (1% aq.); pH 7.0 (1% aq.); 70% act.

Hyonic NP-500. [187] Nonoxynol-50; nonionic; see Hyonic NP-407; clear solid; sol. in water; dens. 1.08 g/ml; HLB 18; pour pt. 24 C; cloud pt. 100 C (1% aq.); pH 7.0 (1% aq.); > 99% act.

Hyonic OP-40. [187] Octoxynol-4; nonionic; emulsifier in solv. cleaning compds., waterless hand cleaners, agric. emulsifiable concs., silicone products; metal cleaners; detergent for petroleum oils; visc. reducer for plastisols; intermediate for production of anionic surfactants; agric. use; clear liq.; disp. in water; dens. 1.02 g/ml; HLB 8; pour pt. –26 C; cloud pt. 53 C (1% aq.); pH 7.0 (1% aq.); surf. tens. 28 dynes/cm (0.01%); > 99% act.

Hyonic OP-70. [187] Octoxynol-7; nonionic; surfactant, emulsifier, wetting agent, intermediate for textile lubricants, agric. emulsifiable concs.; color development enhancer in latex paint; post-polymerization stabilizer; clear liq.; disp. in water; dens. 1.06 g/ml; HLB 12; pour pt. 11 C; cloud pt. 23 C (1% aq.); pH 7.0 (1% aq.); surf. tens. 27 dynes/cm (0.01%); > 99% act.

Hyonic OP-100. [187] Octoxynol-9; wetting agent, coemulsifier; detergent for household and industrial cleaning compds. and textile processing; penetrant; degreasing compds.; wetting agent for paper toweling/tissue; agric. applications; clear liq.; sol. in water; dens. 1.07 g/ml; HLB 13; pour pt. 10 C; cloud pt. 65 C (1% aq.); pH 7.0 (1% aq.); surf. tens. 30 dynes/cm (0.01%); > 99% act.

Hyonic OP-407. [187] Octoxynol-40; nonionic; stabilizer for water dispersions; mfg. and stabilize latexes; wetting agent in min. acid and alkali formulations; coemulsifier and stabilizer in agric. formulations; clear liq.; sol. in water; dens. 0.91 g/ml; HLB 18; pour pt. –2 C; cloud pt. 73 C (1% aq.); pH 7.0 (1% aq.); surf. tens. 48 dynes/cm (0.01%); 70% act.

Hyonic OP-705. [187] Octoxynol-70 see

Hyonic OP-407; clear liq.; sol. in water; dens. 1.10 g/ml; HLB 18.5; pour pt. –12 C; cloud pt. 91 C (1% aq.); pH 7.0 (1% aq.); surf. tens. 49 dynes/cm (0.01%); 50% act.

Hyonic PE-40. [187] Nonyl phenol ethoxylate; nonionic; emulsifier for solv. cleaning compds., waterless hand cleaners, metal cleaners, surfactant intermediate; clear liq.; dens. 8.5 lb/gal; cloud pt. 0 C (1% aq.); pour pt. –29 C; surf. tens. 28 dynes/cm (0.01%); pH 7.0 (1% aq.); > 99% act.

Hyonic PE-90. [187] Nonyl phenol ethoxylate; nonionic; detergent, wetter, emulsifier, stabilizer for household and industrial cleaning, textile, paper processing, latex adhesives; clear liq.; sp.gr. 1.06; dens. 8.8 lb/gal; cloud pt. 54 C (1% aq.); pour pt. 2 C; surf. tens. 31 dynes/cm (0.01%); pH 7.0 (1% aq.); > 99% act.

Hyonic PE-100. [187] Nonoxynol-10; nonionic; detergent, wetter, emulsifier, base, penetrant for household and industrial cleaners, paper toweling; latex stabilizer; clear liq.; sp.gr. 1.068; dens. 8.9 lb/gal; cloud pt. 68 C (1% aq.); pour pt. 4 C; surf. tens. 32 dynes/cm (0.01%); pH 7.0 (1% aq.); > 99% act.

Hyonic PE-120. [187] Nonyl phenol ethoxylate; nonionic; detergent, penetrant, wetting agent, coemulsifier; liq./semisolid; dens. 8.9 lb/gal; cloud pt. 91 C (1% aq.); pour pt. 15 C; surf. tens. 36 dynes/cm (0.01%); pH 7.0 (1% aq.); > 99% act.

Hyonic PE-240, -260. [187] Alkyl phenol POE ethanol; nonionic; wetting and foaming agent; clear liq.; dens. 8.8 and 8.9 lb/gal resp.; cloud pt. 25 and 65 C resp.; pour pt. –9 and 7 C resp.; 100% act.

Hyonic PE-250. [187] Octoxynol-9; nonionic; wetting and foaming agent; clear liq.; dens. 8.9 lb/gal; cloud pt. 52 C; pour pt. 5 C; 100% act.

Hyonic PE-360. [187] Tridecyl alcohol; nonionic; detergent, emulsifier, wetting agent; semisolid; dens. 8.1 lb/gal; cloud pt. 0 C; pour pt. 18 C.

Hystrene 4516. [312] 45% palmitic fatty acid; lubricant, textile aux., emulsifier, plasticizer, intermediate, used in cosmetics, shampoos, pharmaceuticals; solid; sapon. no. 207–212; 100% conc.

Hystrene 5016. [312] Triple pressed stearic acid; see Hystrene 4516; solid; sapon. no. 208–211; 100% conc.

Hystrene 7018. [312] 70% stearic fatty acid; see Hystrene 4516; solid; sapon. no. 200–206; 100% conc.

Hystrene 7022. [312] 70% arachidic and behenic fatty acid; see Hystrene 4516; solid; sapon. no. 172–182; 100% conc.

Hystrene 8016. [312] 80% palmitic fatty acid; see Hystrene 4516; solid; sapon. no. 214–218; 100% conc.

Hystrene 8018. [312] 80% stearic fatty acid; see Hystrene 4516; solid; sapon. no. 199–204; 100% conc.

Hystrene 9008. [312] 90% caprylic fatty acid; see Hystrene 4516; liq.; sapon. no. 379–388 100% conc.

Hystrene 9010. [312] 90% capric fatty acid; see Hystrene 4516; solid; sapon. no. 318–328; 100% conc.

Hystrene 9014. [312] 90% myristic fatty acid; see Hystrene 4516; solid; sapon. no. 241–244; 100% conc.

Hystrene 9016. [312] 90% palmitic fatty acid; see Hystrene 4516; solid; sapon. no. 217–244; 100% conc.

Hystrene 9018. [312] 90% stearic fatty acid; see Hystrene 4516; solid; sapon. no. 197–202; 100% conc.

Hystrene 9022. [312] 90% behenic fatty acid; see Hystrene 4516; solid; sapon. no. 169–179; 100% conc.

Hystrene 9512. [312] 95% lauric fatty acid; see Hystrene 4516; solid; sapon. no. 277–282; 100% conc.

Hystrene 9718. [312] 92% stearic fatty acid; see Hystrene 4516; solid; sapon. no. 196–201; 100% conc.

I

Iberpal B.I.G. [282] Ethoxylated alkyl phenol; nonionic; detergent and scouring agent; liq.; 100% conc.

Iberpenetrant-114. [282] Alkylphenol polyethylene sulfonate; anionic; textile wetting agent, dispersant, foaming agent; liq.

Iberpol Gel. [282] Aminoethyl sulfonate, sodium salt; anionic; detergent and textile scouring acid; liq.; 32% conc.

Iberpon W. [282] Alkylaryl sodium sulfonate; anionic; detergent; textile scouring agent; liq.; 25% conc.

Iberscour AC. [282] Amine condensate; anionic; surface act. agent; wetting agent and detergent; liq.

Iberscour P. [282] Ethoxylated alkyl phenol; nonionic; wetting agent, detergent, emulsifier used in polyester scouring; liq.

Iberscour W Conc. [282] Alkyphenol EO ether; nonionic; detergent; textile wetting agent; liq.; 100% conc.

Iberterge 65. [282] Alcohol straight chain ethoxylate, modified; detergent, emulsifier; textile scouring agent; liq.; 35% conc.

Iberterge CO-40. [282] Sodium alkylaryl sulfonate; anionic; detergent; textile scouring agent; flake, powd.; 40% conc.

Iberwet B.I.G. [282] Alkylaryl sulfonate, modified; anionic; wetting agent; liq.

Iberwet-BO. [282] Butyl oleate and builders; nonionic; rewetting agent for textile use; sanforizing assistant; liq.

Iberwet-E. [282] Sodium alkylaryl sulfonate; detergent; textile scouring aid; liq.

Iberwet W-100. [282] EO nonylphenol; nonionic; detergent; textile wetting agent; liq.; 100% conc.

Ice #2. [210] Glyceryl stearate and polysorbate 80; stabilizer and emulsifier for the food industry; lubricant for textiles and plastics; fabric softener; HLB 5.2; m.p. 59–63 C; 32–38% alpha monoglyceride.

Ice #2FK. [210] 80/20 blend of monoglycerides and polysorbate 80; nonionic; emulsifier used with stabilizers in frozen desserts; cream beads; m.p. 138–144 F; HLB 5.2; 100% conc.

Ice #10K. [210] 80/20 blend of mono- and diglycerides and polysorbate 65; nonionic; emulsifier used with stabilizers in frozen desserts; cream, lt. ivory beads; m.p. 141–144 F F; HLB 5.9; 100% conc.

Ice #12. [210] Mono- and diglycerides and polysorbate 65; see Ice #2; HLB 4.5; m.p. 59–63 C; 32–36% alpha monoglyceride.

Ice #12K. [210] 80/20 blend of mono- and diglycerides and polysorbate 65; nonionic; emulsifier used with stabilizers in frozen desserts; cream, lt. ivory beads; m.p. 140–144 F; HLB 4.1; 100% conc.

Ice #81. [210] Mono- and diglycerides and fatty acids polyglycerol esters; see Ice #2; HLB 4.8; m.p. 58–63 C; 34–40% alpha monoglyceride.

Icodimeen T-30. [86] Tallow diamine adduct (30 EO); cationic/nonionic; wetting agent, penetrant, emulsifier, stabilizer, dispersant, antistat, lubricant; wax; water-sol.; m.w. 1488; sp.gr. 1.07; HLB 17.0; m.p. 28 C; pH 8.0–9.5 (5% aq.); 100% act.

Icomeen 18-5. [86] PEG-5 stearamine; cationic/nonionic; see Icodimeen T-30; also as solubilizer; Gardner 8 max. cast solid; disp. in water; sp.gr. 0.98; HLB 8.9; pH 9 (5% aq.); 100% act.

Icomeen 18-50. [86] PEG-50 stearamine; cationic; surfactant; Gardner 9 max. cast solid; water-sol.; sp.gr. 1.07; HLB 17.7; pH 9 (5%); 100% act.

Icomeen O-30. [86] PEG-30 oleamine; cationic/nonionic; see Icomeen 18-5; Gardner 10 max. paste; m.w. 1590; water-sol.; sp.gr. 1.05; HLB 16.2; pH 8.0–9.5 (5% aq.); 100% act.

Icomeen O-30-80%. [86] PEG-30 oleamine; cationic/nonionic; see Icomeen 18-5; Gardner 10 max. liq.; water-sol.; sp.gr. 1.05; HLB 16.2; pH 7.5–9.0 (5% aq.); 80% act.

Icomeen S-5. [86] PEG-5 soyamine; cationic/nonionic; see Icomeen 18-5; disp. in water; sp.gr. 0.98; HLB 9.0; pH 9 (5% aq.); 100% act.

Icomeen T-2. [86] PEG-2 tallow amine; cationic/nonionic; see Icodimeen T-30; Gardner 8 max. paste; insol. in water; m.w. 350; sp.gr. 0.92; HLB 5.0; pH 10 (5% aq.); 100% act.

Icomeen T-5. [86] PEG-5 tallow amine; cationic/nonionic; see Icodimeen T-30; Gardner 8 max. liq., paste; water-sol.; sp.gr. 0.96; HLB 9.2; cloud pt. 45 C (1% aq.); pH 9 (5% aq.); surf. tens. 33 dynes/cm (0.1% aq.); 100% act.

Icomeen T-7. [86] PEG-7 tallow amine; cationic/nonionic; see Icodimeen T-30; Gardner 8 max. liq.; water-sol.; m.w. 570; sp.gr. 0.98; HLB 10.8; cloud pt. 85 C (1% aq.); pH 9 (5% aq.); surf. tens. 36 dynes cm (0.1% aq.); 100% act.

Icomeen T-15. [86] PEG-15 tallow amine; cationic; surfactant; Gardner 8 max. liq.; water-sol.; m.w. 925; sp.gr. 1.03; HLB 14.1; pH 9 (5% aq.); surf. tens. 41 dynes/cm (0.1% aq.); 100% act.

Icomeen T-20. [86] PEG-20 tallow amine; cationic/nonionic; see Icodimeen T-30; Gardner 8 max. liq.; water-sol.; m.w. 1145; sp.gr. 1.04; HLB 15.4; pH 9 (5% aq.); 100% act.

Icomeen T-25. [86] PEG-25 tallow amine; cationic/nonionic; see Icodimeen T-30; Gardner 8 max. liq.; water-sol.; m.w. 1365; sp.gr. 1.05; HLB 16.0; pH 9 (5% aq.); 100% act.

Icomeen T-25 CWS. [86] Tallow amine ethoxylate, modified; cationic/nonionic; surfactant for gel formation; see also Icodimeen T-30; Gardner 8 max. liq.; water-sol.; sp.gr. 1.05; HLB 16.0; pH 7.5–9.0 (5% aq.); 100% act.

Icomeen T-40. [86] PEG-40 tallow amine; cationic/nonionic; see Icodimeen T-30; Gardner 8 max. cast solid; water-sol.; m.w. 2030; sp.gr. 1.09; HLB 17.2; m.p. 33 C; pH 8.0–9.5 (5% aq.); 100% act.

Icomeen T-40-80%. [86] PEG-40 tallow amine; cationic/nonionic; see Icodimeen T-30; Gardner 8 max. liq.; water-sol.; m.w. 2030; sp.gr. 1.08; HLB 17.2; pH 8.0–9.5 (5% aq.); 80% act.

Icomid HT-50. [86] PEG-50 hydrog. tallow amide; noninic; emulsifier, foam stabilizer, finishing agent and lubricant; Gardner 8 max. wax. solid; water-sol.; m.w. 1200; sp.gr. 1.14; HLB 17.7; pH 8–10 (5% aq.); surf. tens. 47 dynes/cm (0.1% aq.); 100% act.

Icomid O-5. [86] PEG-5 oleamide; nonionic; see Icomid HT-50; Gardner 8 max. liq.; disp. in water; m.w. 510; sp.gr. 1.0; visc. 500 cps; HLB 8.8; pH 5–7 (5% aq.); surf. tens. 35 dynes/cm (0.1% aq.); 100% act.

Iconol 28. [86] Coco diethanolamide; nonionic; see Icomid HT-50; Gardner 12 max. liq.; water-sol.; sp.gr. 1.01; visc. 2300 cps; cloud pt. > 100 C; pH 8.0–10.0 (5% aq.); 100% act.

Iconol CNP-1, CNP-10. [86] Nonylphenol fatty esters, ethoxylated; surfactant; Gardner 8 and 5 max. resp. liqs.; disp. in water; sp.gr. 1.03; visc. 200 and 700 cps resp.; cloud pt. < 25 C; 100% act.

Iconol COA. [86] Coco diethanolamide; nonionic; see Icomid HT-50; Gardner 10 max. liq.; water-sol.; sp.gr. 1.02; visc. 1300 cps; cloud pt. > 100 C; pH 8.5–11.0 (5% aq.); 100% act.

Iconol DA-4. [86] Deceth-4; nonionic; wetting agent, detergent; APHA 70 max. liq.; disp. in water; m.w. 330; sp.gr. 0.96; visc. 30 cps; HLB 10.5; pour pt. < 0 C; cloud pt. < 25 C (1% aq.); pH 6.0–7.5 (5% aq.); 100% act.

Iconol DA-5.5. [86] Deceth-5.5; nonionic; wetting agent, detergent; APHA 70 max. liq.; water-sol.; m.w. 390; sp.gr. 0.98; visc. 20 cps; HLB 12.1; pour pt. 8 C; cloud pt. 26–34 C (1% aq.); pH 6.0–7.5 (5% aq.); 100% act.

Iconol DA-6. [86] Deceth-6; nonionic; wetting agent, detergent; APHA 50 max. liq.; water-sol.; m.w. 410; sp.gr. 0.99; visc. 20 cps; HLB 12.5; pour pt. 8 C; cloud pt. 34–40 C (1% aq.); pH 5.5–6.5 (5% aq.); 100% act.

Iconol DA-6-90%. [86] Deceth-6; nonionic; wetting agent, detergent; APHA 50 max.

liq.; water-sol.; m.w. 410; sp.gr. 0.99; HLB 12.5; pour pt. < 0 C; cloud pt. 34–40 C (1% aq.); pH 5.5–7.5 (5% aq.); 90% act.

Iconol DA-9. [86] Deceth-9; nonionic; wetting agent, detergent; APHA 70 max. paste; water-sol.; m.w. 550; sp.gr. 1.02; HLB 14.2; pour pt. 10 C; cloud pt. 79–84 C (1% aq.); pH 6.0–7.5 (5% aq.); 100% act.

Iconol DDP-10. [86] Dodoxynol-10; nonionic; emulsifier, wetting and cleaning agent, penetrant, detergent, stabilizer, dispersant, coemulsifier; Gardner 1 max. liq.; water-sol.; m.w. 725; sp.gr. 1.04; visc. 400 cps; HLB 12.6; pour pt. 10 C; cloud pt. 39–44 C (1% aq.); pH 6.0–7.5 (5% aq.); 100% act.

Iconol DNP-8. [86] Nonyl nonoxynol-8; nonionic; see Iconol DDP-10; Gardner 4 max. clear liq.; disp. in water; m.w. 700; sp.gr. 1.0; visc. 500 cps; HLB 10.0; pour pt. 9 C; cloud pt. < 25 C (1% aq.); pH 6.0–7.5 (5% aq.); 100% act.

Iconol DNP-24. [86] Nonyl nonoxynol-24; nonionic; see Iconol DDP-10; Gardner 1 max. paste; water-sol.; m.w. 1400; HLB 15.0; m.p. 33 C; cloud pt. > 95 C (1% aq.); pH 6.0–7.5 (5% aq.); 100% act.

Iconol DNP-150. [86] Nonyl nonoxynol-150; nonionic; see Iconol DDP-10; Gardner 1 max. flakes, cast solid; water-sol.; m.w. 6900; sp.gr. 1.05; HLB 19.0; m.p. 55 C; cloud pt. > 100 C (1% aq.); pH 6.0–7.5 (5% aq.); 100% act.

Iconol NP 1.5. [86] Nonoxynol-1.5; nonionic; surfactant; Gardner 2 max. clear liq.; oil-sol.; insol. in water; m.w. 281; sp.gr. 0.99; visc. 350 cps; HLB 4.6; pour pt. –11 C; pH 6.0–7.5 (5% aq.); 1.0% max. water.

Iconol NP-4. [86] Nonoxynol-4; nonionic; surfactant; APHA 100 max. liq.; oil-sol.; insol. in water; m.w. 391; sp.gr. 1.02; visc. 350 cps; HLB 8.9; pour pt. –27 C; pH 6.0–7.5 (5% aq.); 1.0% max. water.

Iconol NP-5. [86] Nonoxynol-5; nonionic; surfactant; APHA 100 max. liq.; disp. in water; m.w. 435; sp.gr. 1.03; visc. 300 cps; HLB 10.0; pour pt. –27 C; pH 6.0–7.5 (5% aq.); surf. tens. 29 dynes/cm (0.1% aq.); 1.0% max. water.

Iconol NP-6. [86] Nonoxynol-6; surfactant; APHA 100 max. liq.; disp. in water; m.w. 479; sp.gr. 1.04; visc. 300 cps; HLB 10.8; pour pt. –28 C; pH 6.0–7.5 (5% aq.); surf. tens. 29 dynes/cm (0.1% aq.); 1.0% max. water.

Iconol NP-7. [86] Nonoxynol-7; nonionic; wetting and cleaning agent, penetrant, detergent, emulsifier, stabilizer, dispersant, coemulsifier; used in food; APHA 100 max. liq.; water-sol.; m.w. 523; sp.gr. 1.05; visc. 300 cps; HLB 11.9; pour pt. 5 C; cloud pt. 22–27 C (1.0% aq.); surf. tens. 30 dynes/cm (0.1% aq.); 100% conc.

Iconol NP-9. [86] Nonoxynol-9; nonionic; surfactant; APHA 70 max. liq.; water-sol.; m.w. 611; sp.gr. 1.06; visc. 300 cps; HLB 13.0; pour pt. 4 C; cloud pt. 52–56 C (1.0% aq.); pH 6.0–7.5 (5% aq.); surf. tens. 32 dynes/cm (0.1% aq.); 1.0% max. water.

Iconol NP-10. [86] Nonoxynol-10; nonionic; surfactant; APHA 70 max. liq.; water-sol.; m.w. 655; sp.gr. 1.06; visc. 300 cps; HLB 13.5; pour pt. 9 C; cloud pt. 60–65 C (1.0 aq.); pH 6.0–7.5 (5% aq.); surf. tens. 34 dynes/cm (0.1% aq.); 1.0% max. water.

Iconol NP-12. [86] Nonoxynol-12; nonionic; see Iconol NP-7; liq.; HLB 14.0; 100% conc.

Iconol NP-15. [86] Nonoxynol-15; nonionic; surfactant; APHA opaque visc. liq.; water-sol.; m.w. 875; sp.gr. 1.07; visc. 700 cps; HLB 15.0; pour pt. 26 C; cloud pt. 95–100 C (1.0% aq.); pH 6.0–7.5 (5% aq.); surf. tens. 39 dynes/cm (0.1% aq.); 1.0% max. water.

Iconol NP-20. [86] Nonoxynol-20; nonionic; surfactant; water-sol.; m.w. 1095; sp.gr. 1.08 (50/25 C); HLB 16.0; m.p. 30 C; cloud pt. > 100 C (1% aq.); pH 6.0–7.5 (5% aq.); surf. tens. 40 dynes/cm (0.1% aq.); 1.0% max. water.

Iconol NP-30. [86] Nonoxynol-30; nonionic; surfactant; wh. wax; water-sol.; m.w. 1535; sp.gr. 1.08 (50/25 C); visc. 30 cps (100 C); HLB 17.2; m.p. 41 C; cloud pt. > 100 C (1% aq.); pH 6.0–7.5 (5% aq.); surf. tens. 42 dynes/cm (0.1% aq.); 1.0% max. water.

Iconol NP-30-70%. [86] Nonoxynol-30; nonionic; surfactant; APHA 100 max. liq.; water-sol.; m.w. 1535 (act.); sp.gr. 1.09; visc. 1100 cps; HLB 17.2; pour pt. 0 C; cloud pt. > 100 C (1% aq.); pH 6.0–7.5 (5% aq.); surf. tens. 42 dynes/cm (0.1% aq.); 28.5–31.5% water.

Iconol NP-40. [86] Nonoxynol-40; nonionic; surfactant; wh. wax; water-sol.; m.w. 1975; sp.gr. 1.09 (50/25 C); visc. 40 cps (100 C); HLB 17.7; m.p. 48 C; cloud pt. > 100 (1% aq.); pH 6.0–7.5 (5% aq.); 1.0% max. water.

Iconol NP-40-70%. [86] Nonoxynol-40; nonionic; surfactant; water-sol.; m.w. 1975 (act.); sp.gr. 1.1; visc. 1400 cps; HLB 17.7; pour pt. 7 C; cloud pt. > 100 C (1% aq.); pH

6.0–7.5 (5% aq.); 28.5–31.5% water.

Iconol NP-50. [86] Nonoxynol-50; nonionic; surfactant; wh. wax; water-sol.; m.w. 2415; sp.gr. 1.1 (50/25 C); visc. 60 cps (100 C); HLB 18.0; m.p. 49 C; cloud pt. > 100 C (1% aq.); pH 6.0–7.5 (5% aq.); 1.0% max. water.

Iconol NP-70. [86] Nonoxynol-70; nonionic; surfactant; wh. wax; water-sol.; m.w. 3300; sp.gr. 1.1 (50/25 C); visc. 80 cps (100 C); HLB 18.6; pour pt. 51 C; cloud pt. > 100 C (1% aq.); pH 6.0–7.5 (5% aq.); 1.0% max. water.

Iconol NP-70-70%. [86] Nonoxynol-70; nonionic; surfactant; APHA 100 max. liq.; water-sol.; m.w. 3300; sp.gr. 1.1; visc. 3600 cps; HLB 18.6; pour pt. 17 C; cloud pt. > 100 C (1% aq.); pH 6.0–7.5 (5% aq.); 28.5–31.5% water.

Iconol NP-100. [86] Nonoxynol-100; nonionic; surfactant; Gardner 1 max. flake, cast solid; water-sol.; m.w. 4315; sp.gr. 1.12 (50/25 C); visc. 150 cps (100 C); HLB 19.0; m.p. 52 C; cloud pt. > 100 C (1% aq.); pH 6.0–7.5 (5% aq.); 1.0% max. water.

Iconol NP-915. [86] Alkylphenoxypoly (ethylenoxy) ethanol; nonionic; surfactant; Gardner 1 liq., paste; water-sol.; sp.gr. 1.06; visc. 400 cps; pour pt. 16 C; cloud pt. 70–78 C (1% aq.); pH 6.0–7.5 (5% aq.); 100% conc.

Iconol OP-5. [86] Octoxynol-5; nonionic; surfactant; APHA 100 max. liq.; disp. in water; m.w. 425; sp.gr. 1.04; visc. 300 cps; HLB 10.0; pour pt. < 0 C; cloud pt. < 25 C (1% aq.); pH 6.0–7.5 (5% aq.); surf. tens. 28 dynes/cm (0.1% aq.); 1.0% max. water.

Iconol OP-7. [86] Octoxynol-7; nonionic; surfactant; APHA 100 max. liq.; water-sol.; m.w. 515; sp.gr. 1.05; visc. 300 cps; HLB 12.4; pour pt. –9 C; cloud pt. 21–25 C (1% aq.); pH 6.0–7.5 (5% aq.); surf. tens. 29 dynes/cm (0.1% aq.); 1.0% max. water.

Iconol OP-10. [86] Octoxynol-10; nonionic; surfactant; APHA 100 max. liq.; water-sol.; m.w. 650; sp.gr. 1.06; visc. 250 cps; HLB 13.5; pour pt. 7 C; cloud pt. 63–67 C (1% aq.); surf. tens. 30 dynes/cm (0.1% aq.); 1.0% max. water.

Iconol OP-30. [86] Octylphenol ethoxylate; nonionic; emulsifier, wetting agent, dispersant, syn. latex stabilizer, and detergent in formulating industrial, institutional, and household cleaning products; primary emulsifier for acrylic and vinyl emulsion polymerization and for asphalt emulsion systems; solid; HLB 17; cloud Pt.: > 100 C (1% aq.); surf. tens. 46.4 dynes/cm (0.1%).

Iconol OP-30-70%. [86] Octylphenol ethoxylate; nonionic; see Iconol OP-30; liq.; HLB 17; cloud pt.: > 100 C (1% aq.).

Iconol OP-40. [86] Octoxynol-40; nonionic; surfactant; Gardner 1 max. cast solid; water-sol.; m.w. 1970; sp.gr. 1.09 (50/25 C); visc. 40 cps (100 C); HLB 17.9; m.p. 50 C; cloud pt. > 100 C (1% aq.); pH 6.0–7.5 (5% aq.); surf. tens. 42 dynes/cm (0.1% aq.); 1.0% max. water.

Iconol OP-40-70%. [86] Octoxynol-40; nonionic; surfactant; APHA 100 max. liq.; water-sol.; m.w. 1970; sp.gr. 1.1; visc. 500 cps; HLB 17.9; pour pt. –4 C; cloud pt. > 100 C (1% aq.); pH 6.0–7.5 (5% aq.); 28.5–31.5% water.

Iconol P-5. [86] Ethoxylated phenol (5 EO); surfactant; Gardner 1 max. liq.; water-sol.; m.w. 314; sp.gr. 1.11; visc. 100 cps; cloud pt. > 100 C (1% aq.); pH 6.0–7.5 (5% aq.); 100% act.

Iconol P-6. [86] Ethoxylated phenol (6 EO); surfactant; Gardner 1 max. liq.; water-sol.; m.w. 358; sp.gr. 1.11; visc. 100 cps; cloud pt. > 100 C (1% aq.); pH 6.0–7.5 (5% aq.); 100% act.

Iconol PD-8-90%. [86] Decyl alcohol alkoxylate; nonionic; surfactant; APHA 70 max. liq.; water-sol.; sp.gr. 1.0; HLB 11.0; pour pt. < –10 C; cloud pt. 45–50 C (1% aq.); pH 6.0–7.5 (5% aq.); 9–11% water.

Iconol TDA-3. [86] Trideceth-3; nonionic; surfactant; APHA 70 max. liq.; disp. in water; m.w. 325; sp.gr. 0.95; visc. 30 cps; HLB 8.0; pour pt. –15 C; cloud pt. < 25 C (1% aq.); pH 6.0–7.5 (5% aq.); 100% act.

Iconol TDA-6. [86] Trideceth-6; nonionic; wetting agent, degreaser, emulsifier, detergent; APHA 70 max. liq.; disp. in water; m.w. 505; sp.gr. 0.98; visc. 50 cps; HLB 11.4; pour pt. 10 C; cloud pt. < 25 C (1% aq.); pH 6.0–7.5 (5% aq.); 100% act.

Iconol TDA-8. [86] Trideceth-8; nonionic; see Iconol TDA-6; APHA 70 max. liq.; water-sol.; m.w. 550; sp.gr. 1.0; HLB 12.7; pour pt. 10 C; cloud pt. 40–45 C (1% aq.); pH 6.0–7.5 (5% aq.); 100% act.

Iconol TDA-8-90%. [86] Trideceth-8; nonionic; see Iconol TDA-6; APHA 70 max. liq.; water-sol.; m.w. 550 (act.); sp.gr. 1.0; visc. 60 cps; HLB 12.7; pour pt. < 0 C; cloud pt. 40–45 C (1% aq.); pH 6.0–7.5 (5% aq.); 9–11% water.

Iconol TDA-9. [86] Trideceth-9; nonionic;

surfactant; APHA 70 max. liq.; water-sol.; m.w. 590; sp.gr. 1.0; visc. 75 cps; HLB 13.2; pour pt. 20 C; cloud pt. 55–62 C (1% aq.); pH 6.0–7.5 (5% aq.); 100% act.

Iconol TDA-10. [86] Trideceth-10; nonionic; surfactant; APHA 70 max. liq., paste; water-sol.; m.w. 640; sp.gr. 1.03; HLB 13.7; pour pt. 20 C; cloud pt. 77–88 C (1% aq.); pH 6.0–7.5 (5% aq.); 100% act.

Iconol TDA-18-80%. [86] Trideceth-18; nonionic; surfactant; APHA 70 max. liq.; water-sol.; m.w. 1000 (act.); sp.gr. 1.07; HLB 16.0; pour pt. 5 C; cloud pt. > 100 C (1% aq.); pH 6.0–7.5 (5% aq.); 19–21% water.

Iconol TDA-29-80%. [86] Trideceth-29; nonionic; surfactant; APHA 70 max. liq.; water-sol.; m.w. 1475 (act.); sp.gr. 1.08; HLB 17.4; pour pt. 7 C; cloud pt. > 100 C (1% aq.); pH 6.0–7.5 (5% aq.); 19–21% water.

Iconol WA-1. [86] Alkoxylated phenolic; nonionic; surfactant; pigment dispersant for aq. systems; Gardner 6 clear liq.; sol. in water, alcohols, toluene; sp.gr. 1.02; visc. 300 cps; HLB 13; cloud pt. 53–58 C (1% aq.); pH 6.0–7.5 (5% aq.); 90% act.

Iconol WA-4. [86] Alkoxylated phenolic; nonionic; surfactant; Gardner 6 clear liq.; sol. see Iconol WA-1; sp.gr. 1.02; visc. 300 cps; HLB 13; cloud pt. 53–58 C (1% aq.); pH 6.0–7.5 (5% aq.); 90% act.

Idet 5L. [584] Sodium salt of alkylaryl sulfonate; anionic; wetting agent, detergent; clear yel. liq.; pH 6.8 (1%); 17% conc.

Idet 5LP. [584] Alkylaryl sulfonate; anionic; dispersant and wetting agent used in paints and emulsion latices; liq.; 25% conc.

Idet 5L SP NF. [584] Specially formulated alkylaryl sulfonate; anionic; detergent and wetting agent for wet processes in textile industry, dispersing and penetrating agent; yel. clear visc. liq.; pH 6.8 (1%); 25% conc.

Idet 10, 20–P. [584] Alkylaryl sulfonate sodium salt; anionic; wetting agent in textile wet processing, in formulations of water disp. powd., detergent; spray–dried powd.; pH 7.0–9.0 (1% aq.); 37%, 76% act.

Igepal CA-210. [245] Octoxynol-1.5; nonionic; emulsifier for solv. cleaners, drycleaning, pesticides, floor polish; yel. liq.; aromatic odor; sol. in wh. min. oil, kerosene, Stod., xylene, butyl Cellosolve, perchloroethylene, ethanol; sp.gr. 0.99; visc. 500–800 cps; HLB 4.8; flash pt. > 200 F (PMCC); pour pt. 23 ± 2 F; 100% act.

Igepal CA-420. [245] Octoxynol-3; nonionic; see Igepal CA-210; pale yel. liq.; aromatic odor; sol. in kerosene, Stod., xylene, butyl Cellosolve, perchloroethylene, ethanol; sp.gr. 1.02; visc. 310–400 cps; HLB 8.0; flash pt. > 200 F (PMCC); pour pt. –10 ± 2 F; 100% act.

Igepal CA-520. [245] Octoxynol-5; nonionic; see Igepal CA-520; also detergent, wetting agent; pale yel. liq.; aromatic odor; sol. in Stod., xylene, butyl Cellosolve, perchloroethylene, ethanol; disp. in water; sp.gr. 1.04; visc. 250–280 cps; HLB 10.0; flash pt. > 200 F (PMCC); pour pt. –15 ± 2 F; surf. tens. 30 dynes/cm (0.01%); 100% act.

Igepal CA-620. [245] Octoxynol-7; nonionic; detergent, wetter, emulsifier for household and industrial detergents, textile, paper; pale yel. liq.; aromatic odor; sol. in xylene, butyl Cellosolve, perchloroethylene, ethanol, water; sp.gr. 1.05; visc. 240–260 cps; HLB 12.0; cloud pt. 70–75 F (1%); flash pt. > 200 F (PMCC); pour pt. 15 ± 2 F; surf. tens. 30 dynes/cm (0.01%); 100% act.

Igepal CA-630. [245] Octoxynol-9; nonionic; see Igepal CA-620; pale yel. liq.; aromatic odor; sol. in xylene, butyl Cellosolve, perchloroethylene, ethanol, water; sp.gr. 1.06; visc. 230–260 cps; HLB 13.0; cloud pt. 146–153 (1%); flash pt. > 200 F (PMCC); pour pt. 45 ± 2 F; surf. tens. 31 dynes/cm (0.01%); 100% act.

Igepal CA-720. [245] Octoxynol-12.5; nonionic; detergent, wetter, emulsifier for industrial cleaning; opaque liq.; aromatic odor; sol. in xylene, butyl Cellosolve, perchloroethylene, ethanol, water; sp.gr. 1.07; visc. 70–76 cps (50 C); HLB 14.6; cloud pt. 187–194 (1%); flash pt. > 200 F (PMCC); pour pt. 69 ± 2 F; surf. tens. 32 dynes/cm (0.01%); 100% act.

Igepal CA-880. [245] Octylphenoxypoly ethyleneoxy ethanol; nonionic; emulsifier for fats and oils, vinyl acetate and acrylate polymerization; post-stabilizer for syn. latices; dyeing assistant; solid; HLB 17.4; 100% conc.

Igepal CA-887. [245] Octoxynol-30; nonionic; emulsifier, stabilizer for plastics; dyeing assistant; pale yel. liq.; aromatic odor; sp.gr. 1.10; HLB 17.4; flash pt. > 200 F (PMCC); pour pt. 36 ± 2 F; surf. tens. 39 dynes/cm (0.01%); 70% act.

Igepal CA-890. [245] Octoxynol-40; nonionic; see Igepal CA-887; off-wh. wax; aromatic odor; sol. in butyl Cellosolve, ethanol, water;

sp.gr. 1.08 (50 C); HLB 18.0; flash pt. > 200 F (PMCC); pour pt. 115 ± 2 F; surf. tens. 42 dynes/cm (0.01%); 100% act.

Igepal CA-897. [245] Octoxynol-40; nonionic; see Igepal CA-887; pale yel. liq.; aromatic odor; sp.gr. 1.10; HLB 18.0; flash pt. > 200 F (PMCC); pour pt. 25 ± 2 F; surf. tens. 48 dynes/cm (0.01%); 70% act.

Igepal CA-950. [245] Octylphenoxypoly ethyleneoxy ethanol; nonionic; see Igepal CA-880; solid; HLB 18.7; 100% conc.

Igepal CO-210. [245] Nonoxynol-1.5; nonionic; foam and emulsion stabilizer, detergent, coemulsifier, intermediate, defoamer, dispersant; yel. liq., aromatic odor; sol. in wh. min. oil, kerosene, Stod., naphtha, xylene, butyl Cellosolve, perchloroethylene, ethanol; sp.gr. 0.99; visc. 300–400 cps; HLB 4.6; flash pt. > 200 F (PMCC); pour pt. –3 ± 2 F; 100% act.

Igepal CO-430. [245] Nonoxynol-4; nonionic; coemulsifier, plasticizer, stabilizer, antistat, detergent, dispersant; pale yel. liq., aromatic odor; sol. in kerosene, Stod., naphtha, xylene, butyl Cellosolve, perchloroethylene, ethanol; sp.gr. 1.02; visc. 250–325 cps; HLB 8.8; flash pt. > 200 F (PMCC); pour pt. –15 ± 2 F; 100% act.

Igepal CO-520. [245] Nonoxynol-5; nonionic; intermediate, emulsifier, coupling agent, dispersant; pale yel. liq., aromatic odor; sol. in Stod., naphtha, xylene, butyl Cellosolve, perchloroethylene, ethanol; sp.gr. 1.03; visc. 240–300 cps; HLB 10.0; flash pt. > 200 F (PMCC); pour pt. –24 ± 2 F; surf. tens. 30 dynes/cm (0.01%); 100% act.

Igepal CO-530. [245] Nonoxynol-6; nonionic; emulsifier, detergent, dispersant for agric., petroleum, paper industries; pale yel. liq., aromatic odor; sol. in Stod., naphtha, xylene, butyl Cellosolve, perchloroethylene, ethanol; sp.gr. 1.04; visc. 230–300 cps; HLB 10.8; flash pt. > 200 F (PMCC); pour pt. –26 ± 2 F; surf. tens. 28 dynes/cm (0.01%); 100% act.

Igepal CO-610. [245] Nonoxynol-8 (7–8 EO); nonionic; detergent, wetting agent, emulsifier; pale yel. liq., aromatic odor; sol. in naphtha, xylene, butyl Cellosolve, perchloroethylene, ethanol, water; sp.gr. 1.05; visc. 230–290 cps; HLB 12.2; cloud pt. 72–82 F (1%); flash pt. > 200 F (PMCC); pour pt. 37 ± 2 F; surf. tens. 30 dynes/cm (0.01%); 100% act.

Igepal CO-620. [245] Nonoxynol-8; nonionic; detergent, wetting agent, emulsifier; surfactant used in liq. detergent compds.; liq.; water-sol.; HLB 12.6; 100% conc.

Igepal CO-630. [245] Nonoxynol-9; nonionic; detergent, wetting agent, penetrant, emulsifier for textile, paper, household/industrial cleaners, paints, metal processing; almost colorless liq., aromatic odor; sol. in naphtha, xylene, butyl Cellosolve, perchloroethylene, ethanol, water; sp.gr. 1.06; visc. 225–300 cps; HLB 13.0; cloud pt. 126–133 F (1%); flash pt. > 200 F (PMCC); pour pt. 31 ± 2 F; surf. tens. 31 dynes/cm (0.01%); 100% act.

Igepal CO-660. [245] Nonoxynol-10; nonionic; see Igepal CO-630; pale yel. liq., aromatic odor; sol. see Igepal CO-630; sp.gr. 1.06; visc. 225–275 cps; HLB 13.2; cloud pt. 140–149 F (1%); flash pt. > 200 F (PMCC); pour pt. 46 ± 2 F; surf. tens. 31 dynes/cm (0.01%); 100% act.

Igepal CO-710. [245] Nonoxynol-10 (10–11 EO); nonionic; see Igepal CO-630; pale yel. liq., aromatic odor; sol. see Igepal CO-630; sp.gr. 1.06; visc. 240–300 cps; HLB 13.6; cloud pt. 158–165 F (1%); flash pt. > 200 F (PMCC); pour pt. 49 ± 2 F; surf. tens. 32 dynes/cm (0.01%); 100% act.

Igepal CO-720. [245] Nonoxynol-12; nonionic; see Igepal CO-630; opaque liq., aromatic odor; sol. see Igepal CO-630; sp.gr. 1.06; visc. 260–340 cps; HLB 14.2; flash pt. > 200 F (PMCC); pour pt. 62 ± 2 F; surf. tens. 34 dynes/cm (0.01%); 100% act.

Igepal CO-730. [245] Nonoxynol-15; nonionic; detergent, dispersant, wetting agent, penetrant, emulsifier for industrial cleaners; yel. liq., aromatic odor; sol. see Igepal CO-630; sp.gr. 1.07; visc. 450–550 cps; HLB 15.0; cloud pt. 203–212 F (1%); flash pt. > 200 F (PMCC); pour pt. 71 ± 2 F; surf. tens. 36 dynes/cm (0.01%); 100% act.

Igepal CO-850. [245] Nonoxynol-20; nonionic; detergent, wetting agent, dispersant, emulsifier, for industrial cleaners; latex stabilizer; demulsifier; pale yel. wax; aromatic odor; sol. see Igepal CO-630; sp.gr. 1.08 (50 C); HLB 16.0; flash pt. > 200 F (PMCC); pour pt. 91 ± 2 F; surf. tens. 39 dynes/cm (0.01%); 100% act.

Igepal CO-880. [245] Nonoxynol-30; nonionic; see Igepal CO-850; also textile scouring, solubilizer; pale yel. wax; aromatic odor; sol. see Igepal CO-630; sp.gr. 1.08 (50 C); HLB 17.2; flash pt. > 200 F (PMCC);

pour pt. 109 ± 2 F; surf. tens. 43 dynes/cm (0.01%); 100% act.

Igepal CO-887. [245] Nonoxynol-30; nonionic; see Igepal CO-880; pale yel. liq.; aromatic odor; sp.gr. 1.09; HLB 17.2; flash pt. > 200 F (PMCC); pour pt. 34 ± 2 F; 70% act.

Igepal CO-890. [245] Nonoxynol-40; nonionic; emulsifier, stabilizer, wetting agent, dyeing assistant for plastics, floor polishes, etc.; off-wh. wax; aromatic odor; sol. in ethanol, water; sp.gr. 1.09 (50 C); HLB 17.8; flash pt. > 200 F (PMCC); pour pt. 112 ± 2 F; 100% act.

Igepal CO-897. [245] Nonoxynol-40; nonionic; see Igepal CO-890; pale yel. liq.; aromatic odor; sp.gr. 1.10; HLB 17.8; flash pt. > 200 F (PMCC); pour pt. 46 ± 2 F; 70% act.

Igepal CO-970, -977. [245] Nonoxynol-50; nonionic; see Igepal CO-890; off-wh. wax and pale yel. liq. resp.; aromatic odor; sol. in ethanol, water; sp.gr. 1.10 (50 C); HLB 18.2; flash pt. > 200 F (PMCC); pour pt. 114 ± 2 F and 52 ± 2 F resp.; 100 and 70% act. resp.

Igepal CO-980. [245] Nonylphenoxypoly ethyleneoxy ethanol; nonionic; stabilizer and dyeing assistant; polymerization emulsifier for vinyl acetate and acrylic emulsions; solid; water-sol.; HLB 18.7; cloud pt. > 212 F (1% aq.); 100% conc.

Igepal CO-985, -987. [245] Nonylphenoxypoly ethyleneoxy ethanol; nonionic; see Igepal CO-980; liq.; water-sol.; HLB 18.7; cloud pt. > 212 F (1% aq., CO-987); 50 and 70% conc. resp.

Igepal CO-990. [245] Nonoxynol-100; nonionic; see Igepal CO-890; off-wh. wax; aromatic odor; sol. in ethanol, water; sp.gr. 1.12 (50 C); HLB 19.0; flash pt. > 200 F (PMCC); pour pt. 122 ± 2 F; 100% act.

Igepal CO-997. [245] Nonoxynol-100; nonionic; see Igepal CO-890; pale yel. liq.; aromatic odor; sp.gr. 1.11; HLB 19.0; flash pt. > 200 F (PMCC); pour pt. 68 ± 2 F; 70% act. in water.

Igepal CTA-639. [245] Modified alkylphenoxypoly (ethyleneoxy) ethanol; nonionic; detergent, wetting agent, emulsifier in rubber latex emulsion paints; slightly visc. liq.; sol. in water, xylene, ethanol, carbon tetrachloride; sp.gr. 1.06; visc. 387 cps; cloud pt. 70–78 C (1% aq.); flash pt. 249 C (COC); surf. tens. 29.8 dynes/cm (0.1% aq.); pH 6 (1%); 100% act.

Igepal CTA-639W. [245] Alkylphenoxypoly ethyleneoxy ethanol; nonionic; detergent, wetting; emulsifier in rubber latex emulsion paints; surfactant; liq.; water-sol; 100% conc.

Igepal DM-430. [245] Dialkylphenoxypoly (ethyleneoxy) ethanol; nonionic; emulsifier for agric., emulsion polymerization, leather industries; yel. slightly hazy liq.; sol. in naphtha, xylene, perchloroethylene, ethanol; dens. 8.3 lb/gal; sp.gr. 0.995; visc. 13 cps (100 C); HLB 9.4; flash pt. 500–520 F (COC); pour pt. –7 C; 100% act.

Igepal DM-530. [245] Nonyl nonoxynol-10; nonionic; emulsifier for cutting oils, agric., textile finishing; yel. slightly hazy liq.; sol. in naphtha, xylene, perchloroethylene, ethanol; partly sol. water; dens. 8.43 lb/gal; sp.gr. 1.010; visc. 15 cps (100 C); HLB 10.6; flash pt. 500–520 F (COC); pour pt. 0 C; surf. tens. 29 dynes/cm (0.01%); 100% act.

Igepal DM-710. [245] Dinonyl phenyl ethoxylate; nonionic; detergent, emulsifier for textiles, leather, metal cleaners, paper, latex, pesticides; pale yel. opaque liq.; aromatic odor; sol. in naphtha, xylene, perchloroethylene, ethanol, water; dens. 8.72 lb/gal; sp.gr. 1.045; visc. 19 cps (100 C); HLB 13.0; cloud pt. 48–52 C (1%); flash pt. 500–520 F (COC); pour pt. 18 C; surf. tens. 29 dynes/cm (0.01%); 100% act.

Igepal DM-730. [245] Dinonyl phenyl ethoxylate; nonionic; detergent, emulsifier for textiles, leather, metal cleaners, paper, latex, pesticides; pale yel. paste; aromatic odor; sol. in xylene, ethanol, water; dens. 8.75 lb/gal; sp.gr. 1.049; visc. 25 cps (100 C); HLB 15.1; cloud pt. > 100 C (1%); flash pt. 500–520 F (COC); pour pt. 25 C; surf. tens. 33 dynes/cm (0.01%); 100% act.

Igepal DM-880. [245] Nonyl nonoxynol-49; nonionic; emulsifier, solubilizer for emulsion polymerization, latex stabilization, essential oils, cleaners; pale yel. wax; aromatic odor; sol. in xylene, ethanol, water; sp.gr. 1.050 (50 C); visc. 55 cps (100 C); HLB 17.2; cloud pt. > 100 C (1%); flash pt. 500–520 F (COC); pour pt. 47 C; surf. tens. 38 dynes/cm (0.01%); 100% act.

Igepal DM-970. [245] Nonyl nonoxynol-150; nonionic; detergent, dispersant, wetter, stabilizer for laundry, household, textile, hard-surface detergents, cosmetics, insecticides, paper, petroleum, paints; wh. flakes; sol. in water, aromatic solv., methanol, ethanol; sp.gr. 1.05 (80 C); HLB 19; cloud pt. > 100 C (1%); 100% act.

Igepal OD-410. [245] Alkyl phenoxypoly ethyleneoxy ethanol; nonionic; solv. for resins; ingred. of metal cleaners, paint strippers and cleaning compds.; ink vehicle; liq.; 100% conc.

Igepal RC-520. [245] Dodecylphenoxypoly (ethyleneoxy) ethanol; nonionic; rewetting agent, emulsifier for paper, textiles, etc.; VCS 2 max., slightly visc. liq.; sol. in most hydrocarbon and polar solv.; sp.gr. 1.03; visc. 360 cps (Ostwald-Fenske); HLB 12.0; pour pt. –10 C; 100% act.

Igepal RC-620. [245] Dodecylphenoxypoly (ethyleneoxy) ethanol; nonionic; detergent, wetting agent, emulsifier, dispersant for agric., household/industrial cleaners; APHA 200 max. liq.; HLB 10.0; cloud pt. 38–42 C (1% aq.); 100% act.

Igepal RC-630. [245] Dodecylphenoxypoly (ethyleneoxy) ethanol; nonionic; detergent, wetting agent, emulsifier, dispersant for household/industrial cleaners; APHA 200 max. liq.; HLB 12.7; cloud pt. 62–66 C (1% aq.); 100% act.

Igepon AC-78. [245] Sodium cocoyl isethionate; anionic; surfactant with good dispersing, foaming for detergent bars, dentifrices, shampoos, bubble baths, other cosmetics; powd.; ≥ 83% act.

Igepon AC-941. [245] Sodium cocoyl isethionate compd.; anionic; mfg. of cleansing bars; solid; 93% conc.

Igepon AM-78. [245] Myristic acid, 2-sulfoethyl ester, sodium salt; anionic; detergent for personal care and cosmetic products; powd.; surf. tens. 32.5 dynes/cm (0.05%); 83% act.

Igepon CN42. [245] Sodium-N-cyclohexyl-N-palmitoyl taurate; anionic; low-foaming detergent and dispersant for mechanical dishwasher and industrial cleaners, stabilizer for syn. rubber emulsions; lt. tan, soft firm paste; fatty odor; m.w. 467.2; water sol.; pH 8.0–9.5 (10%); 23% min. act.

Igepon T-33. [245] Sodium-N-methyl-N-oleoyl taurate; anionic; detergent, wetting agent, dispersant for textiles, industrial detergents, paper industry; pale yel. nearly clear liq., perfumed alcoholic odor; m.w. 425; water sol.; pH 6.5–8.0 (10%); 32% act.

Igepon T-43. [245] Sodium-N-methyl-N-oleoyl taurate; anionic; see Igepon T-33; wh. visc. liq. slurry, fatty odor; m.w. 425; water sol.; pH 6.5–8.0 (5%); 33% act.

Igepon T-51. [245] Sodium-N-methyl-N-oleoyl taurate; anionic; see Igepon T-33; lt. amber gel, perfumed odor; m.w. 425; water sol.; pH 6.5–8.0 (10%); 13.9% act.

Igepon T-77. [245] Sodium-N-methyl-N-oleoyl taurate; anionic; see Igepon T-33; cream flakes, fatty odor; m.w. 425; water sol.; pH 6.5–8.0 (5%); 67% act.

Igepon TC-42. [245] Sodium methyl cocoyl taurate; anionic; detergent, foamer, dispersant in cosmetics; wh. soft, smooth paste; m.w. 363; pH 7.0–8.5 (10%); 24% act.

Igepon TK-32. [245] Sodium-N-methyl-N-tall oil acid taurate; anionic; detergent, suspending/dispersing agent, precipitation inhibitor for petroleum industry; VCS 8 max. clear liq.; m.w. 439; water sol.; pH 8–10 (10%); 20% min. act.

Igepon TM-43. [245] Sodium methyl oleoyl taurate; anionic; detergent for personal care and cosmetic cleansing products; slurry; surf. tens. 36.8 dynes/cm (0.05%); 33% act.

Igepon TN-74. [245] Sodium methyl palmitoyl taurate; anionic; detergent with suspending and dispersing agent; wh. fine powd.; m.w. 406; water sol.; pH 6.5–8.0 (2%); 44% min. act.

Imerol DU. [538] EO adduct; dry cleaning assistant, softener, antistatic agent, nonionic; liq.

Imerol NCP, NCP Liq. [538] Sulfonate containing solvs.; wetting agent, emulsifer, dispersant, used for print paste dyes; sp.gr. 1.09; 57% act.

Imerol SS Liq. [538] Detergent, used for solv. scouring in textile industry; yel. liq.; misc. with water; sp.gr. 0.874; 88% act.

Imerol VLF Liq. [538] Detergent, used for solv. scouring in textile industry; yel. liq.; disp. in water @ 70–100 F; sp.gr. 0.94; 100% act.

Imerol XN. [538] Modified polyglycol ether; weakly anionic; odorless scouring and stain removing agent, liq.; sol. in solv.

Imidrol LC. [28] Imidazoline lauryl deriv.; cationic; dispersant for pigments in printing ink formulation; solid, wax; 100% conc.

Imidrol OC. [28] Imidazoline oleyl deriv.; cationic; dispersant for organic pigments; emulsifier wet adhesion agents; liq.; 100% conc.

Imidrol SC. [28] Imidazoline stearyl deriv.; cationic; see Imidrol OC; solid, wax; 100% conc.

Imwitor 191. [213] Glyceryl stearate; coemulsifier, dispersant for personal care products;

emulsifier in o/w and w/o emulsions; lubricants and binders used in the pharmaceutical industry; Gardner 4 max. powd.; m.p. 56–61 C; solid. pt. 63–68 C; sapon. no. 155–170; 90% 1-monoglycerides.

Imwitor 325. [214; 344] Monodiglyceride; nonionic; food emulsifier; powd.; 100% conc.

Imwitor 333. [214; 344] Monodiglyceride lactic acid ester; nonionic; food emulsifier; solid; 100% conc.

Imwitor 369. [213; 344] Monoglyceride citric ester; nonionic; food emulsifier; solid; 100% conc.

Imwitor 370. [213] Glyceryl stearate citrate; nonionic; food emulsifier; solid; 100% conc.

Imwitor 371. [214; 344] Mono- and diglycerides; nonionic; food emulsifier; powd.; 100% conc.

Imwitor 440. [213; 344] Natural oil monodiglyceride; nonionic; emulsifier for cosmetics and food industries; powd.; HLB 3.7; 100% conc.

Imwitor 460. [213; 344] Natural oil monodiglyceride; nonionic; food emulsifier; powd.; 100% conc.

Imwitor 490. [213; 344] Natural oil monodiglyceride; nonionic; food emulsifier; powd.; 100% conc.

Imwitor 595. [344] Monoglyceride, molecular dist.; nonionic; stabilizer and plasticizer for emulsions; dispersant; powd.; HLB 4.4; 100% conc.

Imwitor 742. [213] Caprylic/capric glycerides; nonionic; plasticizer for hard fats, solv. for lipophilic ingreds.; emollient; paste; HLB 4.0; 100% conc.

Imwitor 780. [213] Glyceryl isostearate; nonionic; w/o emulsifier; liq.; HLB 3.7; 100% conc.

Imwitor 780 K. [213; 344] Isostearyl diglyceryl succinate; nonionic; w/o emulsifier for cosmetic and pharmaceutical preparations; yel., med. visc. liq.; slight char. odor; sol. in chloroform, benzene, 96% alcohol, oils; sp.gr. 0.96–0.98; visc. 700–900 cps; HLB 3.7; sapon. no. 240–260; 100% conc.

Imwitor 845. [344] Lard or tallow monodiglyceride; nonionic; food emulsifier; paste, plastic; 100% conc.

Imwitor 900. [213; 344] Palmitic/stearic acid monodiglyceride; see Imwitor 191; Gardner 4 max. powd.; m.p. 56–61 C; solid. pt. 56–61 C; sapon. no. 162–173; 40–50% 1-monoglyceride.

Imwitor 900 K. [213; 344] Glyceryl stearate; nonionic; stabilizer, emulsifier, thickener for cosmetics; wh. waxy solid; slightly fatty odor; sol. in all fats, oils, waxes, chloroform, ether, ethanol (@ 60 C); m.p. 56–60 C; HLB 3.8; sapon. no. 160–176; 42–48% monoglycerides.

Imwitor 910. [213; 344] Hardened oil monodiglyceride; nonionic; food emulsifier; powd.; 100% conc.

Imwitor 940. [213] Palm oil glycerides; nonionic; stabilizer for cosmetics; flakes, powd.; HLB 3.8; 100% conc.

Imwitor 940 K. [213] Glycerol monostearate/palmitate Ph. Eur.; see Imwitor 191; Gardner 4 max. powd.; m.p. 53–57 C; solid. pt. 54–60; sapon. no. 165–178; 42–48% 1-monoglyceride.

Imwitor 945. [213] Palmitic/stearic acid glycerol monodiester; nonionic; stabilizer for cosmetics; powd.; HLB 3.8% 100 conc.

Imwitor 960. [213] Glyceryl stearate SE; emulsion bases for cosmetics; Gardner 4 max. powd.; m.p. 56–61 C; sapon. no. 155–175; 30–40% 1-monoglyceride.

Imwitor 960K. [213; 344] Glyceryl stearate SE; anionic; emulsifier and ointment base for pharmaceutical and cosmetic creams of the o/w type; wh. hard waxy solid; typical odor; sol. in all fats, oils, waxes, in chloroform, benzene, ether, ethanol; HLB 12.0; m.p. 56–60 C; sapon. no. 155–170; 30–40% monoester; 100% conc.

Imwitor 965. [213] Palm oil glyceride and potassium stearate; anionic; o/w emulsifier for cosmetics; flakes, powd.; 100% conc.

Imwitor 965 K. [213] Glycerol monostearate/palmitate and potassium stearate; see Imwitor 960; Gardner 4 max. flakes; m.p. 53–57 C; solid. pt. 56–62 C; sapon. no. 150–170; 30–42% 1-monoglyceride.

Imwitor 1330. [213; 344] Monoglyceride diacetylated tartaric acid ester; food emulsifier; powd.; 100% conc.

Imwitor 2020. [213; 344] Monoglyceride diacetylated tartaric acid ester; food emulsifier; liq.; 100% conc.

Imwitor 2320. [213; 344] Monoglyceride diacetylated tartaric acid ester; food emulsifier; paste; 100% conc.

Incrocas 10. [161] PEG-10 castor oil; nonionic; emulsifier, solubilizer, emollient, superfatting agent, lubricant for personal care products, detergents, metalworking fluids, insecticides, herbicides, household products;

lubricant, antistat, softener, surfactant, and dye leveling agent for textile; pale yel. liq.; disp. in water; sol. @ 10% in ethanol, naphtha, MEK, oleic acid, trichloroethylene; HLB 6.3; cloud pt. 20 C (1% in 10% brine); sapon. no. 120–130; pH 6.0–7.5 (3% aq.); 100% act.

Incrocas 30. [161] PEG-30 castor oil; nonionic; see Incrocas 10; also lime soap dispersant in alkaline scouring systems; pale yel. liq.; sol. in water, ethanol, oleyl alcohol, naphtha, MEK, oleic acid, trichloroethylene; HLB 12.5; cloud pt. 35–40 C (1% in 10% brine); sapon. no. 72–82; pH 6.0–7.5 (3% aq.); surf. tens. 41.5 dynes/cm (0.1% DW); 100% act.

Incrocas 40. [161] PEG-40 castor oil; nonionic; see Incrocas 10; pale yel. liq.; sol. @ 10% see Incrocas 30; HLB 13.0; cloud pt. 50 C (1% in 10% brine); sapon. no. 60–65; pH 6.0–7.5 (3% aq.); surf. tens. 40.90 dynes/cm (0.1% DW); 100% act.

Incrocas 60. [161] PEG-60 castor oil; nonionic; see Incrocas 30; pale yel. soft paste; sol. see Incrocas 30; HLB 14.7; cloud pt. 60 C (1% in 10% brine); sapon. no. 45–50; pH 6.0–7.5 (3% aq.); surf. tens. 43.20 dynes/cm (0.1% DW); 100% act.

Incrocas 100. [161] PEG-100 castor oil; nonionic; see Incrocas 10; also lime soap dispersant in alkaline scouring systems; pale yel. waxy solid; sol. see Incrocas 30; HLB 16.5; cloud pt. 66 C (1% in 10% brine); sapon. no. 25–35; pH 6.0–7.5 (3% aq.); surf. tens. 41.60 dynes/cm (0.1% DW); 100% act.

Incrochem PPI. [161] Stearamidopropyl dimethylamine lactate, ethylene glycol monostearate, propylene glycol; surfactant for personal care products; paste; 32% act.

Incrochem SBJ. [161] Cetearyl alcohol, stearamidopropyl dimethylamine, propylene glycol, ceteareth-20; conc. surfactant; Gardner 2 max. soft wax; m.p. 39 C.

Incrochem SC1. [161] Disodium oleamido MEA sulfosuccinate, sodium laureth sulfate, lauramide DEA, cocamidopropylamine oxide, propylene glycol, sodium chloride, polysorbate 60, citric acid; conc. for formulation of personal care products; yel. liq.; visc. 2000 cps min.; pH 6.5–7.5 (5%); 60% act.

Incrochem SC4. [161] Sodium laureth sulfate, disodium oleamido MEA sulfosuccinate, lauramide DEA, polysorbate 60, propylene glycol, citric acid; see Incrochem SC1; yel. liq.; pH 6.5–7.5 (5%); 60% act.

Incromate CDL. [161] Cocamidopropyl dimethylamine lactate; cationic; surfactant for personal care products; visc. liq.; 95% act.

Incromate CDP. [161] Cocamidopropyl dimethylamine propionate; cationic; see Incromate CDL; Gardner 4 max. liq.; pH 6.0–7.0; 40% act.

Incromate CPML. [161] Cocamidopropyl morpholine lactate; cationic; see Incromate CDL; yel. visc. liq.; sol. in water; pH 4–5 (5%); 29–31% act.

Incromate IDL. [161] Isostearamidopropyl dimethylamine lactate; cationic; see Incromate CDL; yel. visc. liq.; pH 6–7 (5%); 95–97% act.

Incromate ISML. [161] Isostearamidopropyl morpholine lactate; cationic; surfactant for personal care products and cationic emulsions; Gardner 6 max. liq.; sol. in water; pH 4–5 (5%); 25% min. act.

Incromate LDL. [161] Linoleamidopropyl dimethylamine lactate; cationic; see Incromate CDL; yel. visc. liq.; pH 6–7 (5%); 28–30% act.

Incromate ODL. [161] Oleamidopropyl dimethylamine lactate; cationic; surfactant in personal care products; aids in emulsification of fatty materials; Gardner 4 liq., gel; pH 6–7 (5%); 28–30% act.

Incromate PPI. [161] Stearamidopropyl dimethylamine lactate, ethylene glycol monostearate, propylene glycol emulsion, cationic; conditioner, emulsifier, pearlizing agt., humectant, visc. control agt. for creme rinses, shampoos, hand cleansers and lotions; wh. thick paste; pH 4–5 (5%); 30.5–32.5% solids.

Incromate SDL. [161] Stearamidopropyl dimethylamine lactate; cationic; softener for hair shampos, fabrics; Gardner 3 max. slurry; pH 4–5 (10%); 24–26% conc.

Incromectant AMEA-70. [161] Acetamide MEA; solv., humectant for cosmetics; clarifying agent for shampoos; conditioner for creme rinses; sol'n.; sp.gr. 1.07–1.17; dens. 9.3 lb/gal; pH 6.0–8.5; 70% min. act.

Incromide CA. [161] Cocamide DEA; nonionic; surfactant, foam stabilizer, emulsifier and thickener used in household, cosmetic, and industrial formulations; liq.; 100% act.

Incromide CAL. [161] Coco diethanolamide; nonionic; visc. builder and foam enhancer for bath products and shampoo systems; Gardner 5 max. liq.; bland odor; 100% act.

Incromide CM. [161] Coco monoethanolamide; nonionic; foam stabilizer, thickener, and opacifier for cosmetic, industrial, household, and AOS formulations; Gardner 4 max. flake; 100% act.

Incromide CME. [161] Cocamide MEA; surfactant; Gardner 5 max. flake; m.p. 68 C; 2% max. free amine.

Incromide L-90. [161] Lauramide DEA; nonionic; foam stabilizer, thickener, detergent, and foaming agent in household, industrial and institutional cleaning comps., car washes, rug and upholstery cleaners, and personal care products; Gardner 2 max. solid; mild fatty odor; sol. in alcohol, chlorinated and aromatic hydrocarbons; disp. in water; 100% act.

Incromide LA. [161] Linoleic diethanolamide; nonionic; superfatting agent and thickener for personal care and household products; useful in increasing visc. of various sulfate and ether sulfate dilutions; conditioner and lubricant; Gardner 5 max. liq.; bland odor; 100% act.

Incromide LCL. [161] Lauric diethanolamide; surfactant; yel. liq.

Incromide LEP. [161] Coco diethanolamide; nonionic; see Incromide CA; liq.; 100% act.

Incromide LI. [161] Lauric MIPA; nonionic; visc. builder and foam booster; wh. flake; m.p. 54 C; 100% act.

Incromide LL. [161] Lauric diethanolamide; nonionic; foam booster, stabilizer, visc. builder, solubilizer, coupler used in personal care products and liq. detergent formulations; Gardner 4 max. liq., paste; bland odor; pH 9.7–10.3; 100% conc.

Incromide LLA. [161] Lauramide DEA, linoleamide DEA; nonionic; thickener, foam stabilizer, wetting agent, emulsifier, and emulsion stabilizer used in personal care products; Gardner 5 max. liq.; 100% act.

Incromide LM-70. [161] Lauramide DEA; nonionic; see Incromide L-70; Gardner 2 max. solid; 100% act.

Incromide LMD. [161] Lauric diglycolamide; nonionic; surfactant, thickener, and foam stabilizer; Gardner 3 max. flake; m.p. 52 C; 100% act.

Incromide LMI. [161] Lauric/myristic monoisopropanolamide; nonionic; visc. builder and foam booster; Gardner 4 max. flake; m.p. 48–50 C; 100% act.

Incromide LMM. [161] Lauramide MEA; nonionic; foam enhancer, visc. builder, gelling and wetting agent, thickener, and foam stabilizer used in personal care products; Gardner 5 max. flake; slight amine odor; 100% act.

Incromide LR. [161] Lauric diethanolamide; nonionic; thickener, foam stabilizer used in personal care products, dishwash detergents and industrial cleaners; Gardner 5 max. liq.; bland odor; 100% act.

Incromide OM. [161] Oleic monoethanolamide; nonionic; visc. builder for personal care products; Gardner 6 max. solid; m.p. 50–52 C; 100% act.

Incromide OPD. [161] Oleic diethanolamide; nonionic; visc. builder, conditioner, emulsifier, foam stabilizer, and thickener for personal care products; solvent degreasers; yel. paste, liq.; bland odor; 100% act.

Incromide OPM. [161] Oleic monoethanolamide, modified; nonionic; visc. builder used in personal care products; Gardner 5 max. solid; m.p. 49–50 C; 100% act.

Incromide SM. [161] Stearamide MEA; nonionic; softener and dye carrier in textile industry; pearlizier, opacifier and thickener in cosmetic preparations; flake; 100% conc.

Incromide UM. [161] Undecylenamide MEA; nonionic; foam stabilizer for antidandruff and fungicidal shampoos; Gardner 8 max. flake; m.p. 60 C; 100% act.

Incromine BB. [161] Behenamidopropyl dimethylamine; nonionic/cationic; emollient conditioner, lubricant, and moisturizer for personal care products; yel. flakes; m.w. 420–450; m.p. 70–72; 100% act.

Incromine CB. [161] Cocamidopropyl dimethylamine; nonionic/cationic; foaming agent and surfactant used in personal care products; yel. liq., solid; 100% act.

Incromine CPM. [161] Cocamidopropyl morpholine; nonionic/cationic; intermediate used in personal care products; wh.-cream soft solid; m.p. 32 C; 100% act.

Incromine IB. [161] Isostearamidopropyl dimethylamine; nonionic/cationic; o/w emulsifier and lubricant for personal care products; yel. visc. liq., soft paste; m.w. 364–384; 100% act.

Incromine ISM. [161] Isostearamidopropyl morpholine; nonionic/cationic; hair conditioner in shampoos; yel. paste, liq.; 100% act.

Incromine L-40. [161] Lauramine oxide; nonionic; foam stabilizer, visc. builder, booster, softener and conditioner used in cosmetic and multifunctional applications; wetting agent in electrolyte sol'ns.; liq.; 40% conc.

Incromine LSB. [161] Linoleamidopropyl dimethylamine; nonionic/cationic; see Incromine CPM; yel. visc. liq., soft solid; 100% act.

Incromine OPB. [161] Oleamidopropyl dimethylamine; nonionic/cationic; emollient conditioner, lubricant and moisturizer for personal care products; yel. soft paste; 100% act.

Incromine OPM. [161] Oleamidopropyl dimethylamine; nonionic/cationic; see Incromne BB; soft solid; 100% conc.

Incromine Oxide B-30P. [161] Behenamine oxide; nonionic; visc. builder, emulsifier, lubricant, wetting, and foam stabilizer used in personal care products; wh. paste; pH 6.5–8.0 (3%); 29–31% act.

Incromine Oxide C. [161] Coco amidopropyl dimethylamine oxide, aq. sol'n.; nonionic; visc. builder, foam stabilizer, conditioner used in personal care products; lt. straw liq.; pH 6.5–7.5 (5%); 29.5–31.5% act.

Incromine Oxide I. [161] Isostearamidopropyl amine oxide; nonionic; foam stabilizer, thickener, lubricant, and visc. builder used in cosmetic, household, and janitorial products; wetting agent for conc. electrolyte sol'ns.; Gardner 1 max. gel; pH 6.5–7.5; 24.5–26.5% act.

Incromine Oxide ISMO. [161] Isostearamidopropyl morpholine oxide; nonionic; see Incromine I; Gardner 2 max. liq.; pH 6.5–7.5; 24.5–26.5% amine oxide.

Incromine Oxide L. [161] Lauryl dimethylamine oxide; nonionic; surfactant for dishwashing comps., household products, hair conditioner used in personal care products; colorless clear liq.; pH 7.0–8.0 (5% aq.); 29.0–31.0% amine oxide.

Incromine Oxide M. [161] Myristyl dimethylamine oxide; nonionic; surfactant, emulsifier, emollient, conditioner used in personal care products; visc. liq.; pH 7.0–8.0 (5%); 29.5–30.5% amine oxide.

Incromine Oxide MC. [161] Myristyl cetear yl amine oxide; nonionic; foam stabilizer and visc. builder used in personal, household and industrial applications; wetting agent in electrolyte sol'ns.; liq.

Incromine Oxide O. [161] Oleamidopropyl dimethylamine oxide; nonionic; see Incromine I; colorless gel; pH 6.5–7.5; 24.5–26.5% amine oxide.

Incromine Oxide OD-50. [161] Oleamine oxide; nonionic; see Incromine I; Gardner 1 max. liq.; pH 7–8 (10%); 50–52% amine oxide.

Incromine Oxide S. [161] Stearamine oxide; nonionic; conditioner and cosmetic emulsifier for personal care products; wh. paste; 24.5–26.5% amine oxide.

Incromine PB. [161] Palmitamidopropyl dimethylamine; nonionic/cationic; see Incromine CPM; substantive conditioner and emulsifier used in personal care products; off-wh. flakes; m.w. 340–350; m.p. 56–57 C; 100% act.

Incromine SB. [161] Stearamidopropyl dimethylamine; nonionic/cationic; see Incromine CPM; cr. waxy flakes; faint char. odor; m.p. 63 C; 100% act.

Incromine TB. [161] Tallowamidopropyl dimethylamine; nonionic/cationic; see Incromine CPM; soft solid; 100% act.

Incronam 30. [161; 162] Cocamidopropyl betaine; amphoteric; surfactant, emulsifier, coupling agent, visc. builder, foam detergent for personal care products, chemical specialities, rug and upholstery shampoos, dishwashing compds.; Gardner 5 max. clear liq.; sol. in water; pH 5.5–7.5 (10% aq.); 30% act.

Incronam B-40. [161; 162] Behenyl betaine; amphoteric; foaming surfactant, conditioner, lubricant used in personal care products; wh. paste; pH 5.5–7.0 (3%); 40% act.

Incronam I-30. [161; 162] Isostearamidopropyl betaine; amphoteric; see Incronam B-40; yel. clear visc. liq.; pH 6–7.5 (5%); 29.5–31% act.

Incronam ISM-30. [161] Isostearamidopropyl morpholine betaine; amphoteric; conditioner, visc. builder used in personal care products; yel. liq.; pH 5–7 (5%); 29.5–31.0% act.

Incronam OP-30. [161] Oleamidopropyl betaine; amphoteric; see Incronam B-40; yel. clear liq.; pH 6.0–7.2 (5%); 30% act.

Incronam P-30. [161] Palmitamidopropyl betaine; amphoteric; thickener and conditioner used in personal care products; yel. visc. liq., slurry; pH 6–7.2 (5%); 30% min. act.

Incronol ALS. [161] Ammonium lauryl sulfate; anionic; detergent dispersant, suspending, and foaming agent used in personal care products; pH 6.3–6.8 (10%); 27–30% act.

Incronol ILS. [161] Monoisopropanolamine lauryl sulfate; surfactant for personal care products; lt. straw visc. liq.; pH 7.0–7.4 (10%); 75% min. act.

Incronol SLS. [161] Sodium lauryl sulfate;

anionic; shampoo base; also used in rug and upholstery shampoos; industrial cleaners, emulsion polymerization systems; pH 8.0–8.5 (10% aq.); 28–30% act.

Incronol TLS. [161] Triethanolamine lauryl sulfate; surfactant for personal care products; lt. straw liq.; pH 7.0–7.5 (10% aq.); 40.0% min. act.

Incropearl Liquid. [161] Sodium laureth sulfate and hydroxyethyl MIPA; anionic; pearling agt. for shampoos, dishwashing, bubble baths, shower cleansers, rug cleaners; creamy, pearly paste, mild char. odor; pH 8.0–8.6; 35–40% solids.

Incropearl Liquid NI. [161] Hydroxyethyl stearamide MIPA and PPG-5 ceteth-20; nonionic; see Incropearl Liquid; creamy, pearly paste, mild char. odor; pH 7–8; 18–22% solids.

Incropol 233. [161] Propoxylated alcohol; nonionic; surfactant used as emulsifier, lubricant, detergent for industrial and general household products; coupling agent, antistat, fiber lubricant and solubilizer in personal care products; colorless liq.; pH 5.5–7.5 (3%).

Incropol 290. [161] Alcohol, ethoxylated; nonionic; see Incropol 233; colorless liq.

Incropol CS-12. [161] Ceteareth-12; nonionic; see Incropol 233; Gardner 1 max. solid; pH 5.5–7.5 (3%); 100% act.

Incropol CS-20. [161] Ceteareth-20; nonionic; see Incropol 233; Gardner 1 max. solid; pH 5.5–7.5 (3%).

Incropol CS-30. [161] Ceteareth-30; nonionic; see Incropol 233; Gardner 1 max. solid; pH 5.5–7.5 (3%); 100% act.

Incropol CS-40. [161] Ceteareth-40; nonionic; see Incropol 233; wh. flake; pH 5.5–7.5 (1%).

Incropol CS-50. [161] Ceteareth-50; nonionic; see Incropol 233; Gardner 1 max. flake; pH 6–8 (3%); 100% act.

Incropol CS-60. [161] Ceteareth-60; nonionic; see Incropol 233; wh. flake; pH 5.5–7.5 (1%).

Incropol L-2, L-7. [161] Laureth-2, –7 resp.; nonionic; see Incropol 233; Gardner 1 max. liq.; pH 5.5–7.5 (3%); 100% act.

Incropol L-7-90. [161] Laureth-7; nonionic; see Incropol 233; Gardner 1 max. liq.; pH 5.5–7.5 (3%); 90% act.

Incropol L-12. [161] Laureth-12; nonionic; see Incropol 233; Gardner 1 max. paste; pH 5.5–7.5 (3%); 100% act.

Incropol L-23. [161] Laureth-23; nonionic; see Incropol 233; solid; 100% act.

Incropol L-30. [161] Laureth-30; nonionic; see Incropol 233; Gardner 1 max. paste; pH 5.5–7.5 (3%); 100% act.

Incroquat Behenyl BDQ/P. [162] Behenyl dimethyl benzyl ammonium chloride; cationic; o/w emulsifier and conditioner; paste; 25% conc.

Incroquat Behenyl TMC/P. [162] Behenyl trimethyl ammonium chloride; cationic; o/w emulsifier and conditioner; paste; 25% conc.

Incroquat Behenyl TMS. [161] Behenyl trimethyl ammonium methosulfate, cetearyl alcohol; cationic; conditioner, softener, emollient, o/w emulsifier used in personal care products; flake; 25% act.

Incroquat CR Conc. [161] Cetearyl alcohol, PEG-40, castor oil, stearalkonium chloride; cationic; see Incroquat Behenyl TMS; wh. flake; mild fatty odor; pH 6.3 (5% disp.); 96% min. act.

Incroquat CTC-25. [161] Cetyl trimethyl ammonium chloride, cationic; conditioner in specialty hair preparations; see also Incroquat Behenyl TMS; APHA 100 max. liq.; pH 3.5–4.0 (2% aq.); 24.0–26.0% min. act.

Incroquat CTC-30. [161] Cetyl trimethyl ammonium chloride; cationic; see Incroquat CTC-25; Gardner 2 max. liq.; pH 3.5–4.0 (5%); 29% act.

Incroquat S-75 CG. [161] Tallow imidazolinium methosulfate; cationic; conditioner, antistat, and ingred. used in personal care products, fibers, and synthetics; textile and paper softener; dispersant for pigments and dyestuffs; Gardner 5 max. liq.; pH 5–7 (5%); 74–76% act.

Incroquat S-85. [161] Stearalkonium chloride; cationic; conditioner, softener; see also Incroquat Behenyl TMS; wh. waxy flake; m.p. 140 F; pH 5–7 (0.5%); 85.0% min. act.

Incroquat SDQ-25. [161] Stearalkonium chloride; cationic; surfactant used as ingred. in personal care products, textile and paper; dispersant for pigments and dyestuffs; antistat for fibers and synthetics; paste; m.p. 140 F; pH 3.0–4.0 (1%).

Incrosoft CFI. [161] 1-Methyl-L-oleylamidoethyl-2-oleyl-imidazolinium methosulfate; cationic; fluid disp. for softening fabrics and tissue or as paper debonding agents; Gardner 7 clear liq.; sp.gr. 0.95; flash pt. (PM) 73 F; pH 5.0–6.5; 74–76% solids.

Incrosoft CFI-75. [161] 1-Methyl-L-oleylamidoethyl-2-oleyl-imidazolinium methosulfate; cationic; Gardner 7 max. clear liq.; pH 5–7.5 (5% in 50% IPA); 74% min. act.

Incrosoft O-90. [161] Complex difatty quat. sulfate; fabric softener for clear softening products; used in home and commercial laundry products; yel. clear visc. liq.; pH 4–5 (3%); 88–90% act.

Incrosoft S-75. [161] Tallow imidazolinium methosulfate; softener base, lubricant, antistat and rewetting agent for fabrics and syns.; Gardner 5 max. liq.; water-sol.; pH 4–7 (5%); 74–76% act.

Incrosoft S-90. [161] Tallow imidazolinium methosulfate; see Incrosoft S-75; also used in textile maintenance and finishing; yel. semisolid; disp. in water; pH 5.0–7.0 (5% in DW); 85–88% act.

Incrosoft S-90M. [161] Tallow imidazolinium methosulfate; see Incrosoft S-75; Gardner 4 max. liq.; water-sol.; pH 5–7 (5%); 89–90% act.

Incrosoft T-75. [161] Ditallow diamido methosulfate; fabric softener, lubricant and antistat for home and commercial laundry products; liq.; disp. in water; 75% act.

Incrosoft T-90. [161] Ditallow diamido methosulfate; see Incrosoft T-75; Gardner 6 max. visc. liq.; disp. in water; pH 4–7 (3%); 88–90% act.

Incrosorb S-60. [161] POE (20) sorbitan monostearate; nonionic; o/w emulsifier and solubilizer for alkaline and acid pH systems; used in personal care products; pale yel. liq. gel; sapon. no. 45–55.

Incrosul LAFS. [161] Disodium laneth-5 sulfosuccinate; anionic; conditioning surfactant used in personal care products; yel. slurry/liq.; 40% act.

Incrosul LMA. [161] Diammonium lauramido MEA sulfosuccinate; anionic; surfactant for cosmetic products; yel. liq./slurry; pH 6–7 (50%); 39–41% act.

Incrosul LMS. [161] Disodium lauramido MEA-sulfosuccinate; anionic; foaming agent and cleanser used in personal care products and carpet shampoos; Gardner 4 max. liq./slurry; pH 6–7 (3%); 39–41% act.

Incrosul LMS-Special. [161] Disodium lauramide MEA sulfosuccinate; anionic; see Incrosul LMS; yel. clear liq.; pH 6–7.5 (5%); 38–40% solids.

Incrosul LS. [161] Disodium lauryl sulfosuccinate; anionic; foaming surfactant for cosmetic products; wh. paste; pH 6–7 (5%); 39–41% act.

Incrosul LSA. [161] Diammonium lauryl sulfosuccinate; anionic; surfactant used in personal care products; Gardner 1 max. liq.; mild char. odor; sol. in water; pH 6–7 (3%); 39–41% act.

Incrosul LTS. [161] Disodium laureth sulfosuccinate; anionic; surfactant used in shampoos and general cleansing preparations; Gardner 1 max. clear liq.; mild char. odor; pH 6–7 (1%); 38–43% solids

Incrosul OMS. [161] Disodium oleamido MEA sulfosuccinate; anionic; see Incrosul LSA; yel. clear visc. liq.; mild char. odor; pH 6–7 (5%); 34–36% act.

Incrosul TS. [161] Disodium tridecyl sulfosuccinate; anionic; see Incrosul LSA; Gardner 1 max. liq.; mild char. odor; sol. in water; pH 6–7 (3%); 39–41% act.

Incrosul UMS, UMS-45. [161] Disodium undecylenamido MEA sulfosuccinate; anionic; surfactant used in personal care products; lime soap dispersant and fungicidal activity; yel. clear liq.; faint char. odor; pH 6–7 (3%) and 5.25–6.75 (5%) resp.; 49–51 and 43.5–46% act. resp.

Indoil PDA. [86] Phosphate ester, neutralized; surfactant; Gardner 1 max. liq.; water-sol.; sp.gr. 1.1; visc. 1000 cps; cloud pt. > 100 C (1% aq.); pH 7.0–8.5 (5% aq.)

Indulin AQS, AQS-1M. [642] Tall oil-based carboxylic acid compd.; anionic; emulsifier; liq.; water-sol

Indulin AS-1, -Special. [642] Amine deriv.; cationic; asphalt antistripping additive; liq.; 100% conc.

Indulin AT. [642] Kraft pine lignin; used in polymeric applications where dispersant and adsorption properties are required; br. powd.; sol. in ethylene glycol, dixoane, monoethanolamine, dimethyl formamide, Cellosolve; sp.gr. 1.3; dens. 26 lb/ft^3; flash pt. 349 F; pH 6.5 (2% aq.); surf. tens. 43 dynes/cm; 97% lignin content

Indulin C. [642] Kraft pine lignin, sodium salt; anionic; stabilizer with dispersant properties for asphalt emulsions; primary emulsifier; industrial emulsions used in dispersion of fillers; br. powd.; dens. 30 lb/ft^3; pH 10.2 (2% aq.); surf. tens. 43 dynes/cm (1% aq.)

Indulin HF. [642] Tall oil deriv.; anionic; asphalt emulsifier; liq.; 100% conc.

Indulin RK-2, -4. [642] Amine deriv.; cationic; asphalt emulsifier for cationic emulsions;

liq.; 100% conc.

Indulin SAL. [642] Kraft lignin salt, formulated; anionic; emulsifier and stabilizer for asphalt; liq.; 40% conc.

Indulin W-1. [642] Amine deriv. of pine lignin; cationic; emulsifier in asphalt emulsions, retarding agent in cement; br. powd.; water sol. ≥ pH 8.3 and ≤ pH 4.2; dens. 22 lb/ft^3 (loose); visc. 100 cps @ 20% solids; surf. tens. 47 dynes (1% aq.); pH 10.5 (2% aq.); 100% conc.

Indulin W-2. [642] Lignin, modified, amine deriv.; cationic; emulsifier and stabilizer for o/w emulsions; powd.; 100% conc.

Indulin W-3. [642] Lignin amine; cationic; emulsifier for asphalt emulsions, slurry seal applications; primary emulsifier, stabilizer, and retarder in emulsions; br. liq.; amine odor; water-sol.; sp.gr. 1.090; dens. 9.1 lb/gal; visc. 450 cps; pH 12.0 (1%); 35% solids.

Indu-Sol 238. [25] Wetting agent, suspending agent, lubricant, rust inhibitor, used in grinding and general machining operations; greenish-red, translucent liq.; dens. 11 lb/gal (Imp.); pH 9.2–9.6 (5%); biodeg.

Industrene 105. [310; 312] Oleic fatty acid; used in alkyd resins, rubber compounding, water repellents, polishes, soaps, abrasives, cutting oils, candles, crayons, emulsifiers; FG grades as lubricant, release agent, binder, defoamer in foods, intermediate for food emulsifiers; Gardner 6 liq.; sapon. no. 197–205; 100% conc.

Industrene 138. [312] Dist. cotton fatty acid; see Industrene 105; Gardner 7 liq.; sapon. no. 197–205; 100% conc.

Industrene 143. [310; 312] Dist. animal fatty acid; see Industrene 105; Gardner 5 paste; sapon. no. 203–207; 100% conc.

Industrene 154-BG. [312] Hydrog. buffing grade fatty acid; see Industrene 105; Gardner 9 solid; sapon. no. 191–203; 100% conc.

Industrene 205. [310; 312] Wh. oleic fatty acid; see Industrene 105; liq.; sapon. no. 197–205; 100% conc.

Industrene 206. [310] Oleic acid; see Industrene 105; liq.; 100% conc.

Industrene 225. [312] Dist. soya fatty acid; see Industrene 105; Gardner 3–4 liq.; sapon. no. 197–204; 100% conc.

Industrene 225 FG. [310] Soya fatty acid; see Industrene 105; liq.; 100% conc.

Industrene 226. [310; 312] Dist. soya fatty acid; see Industrene 105; Gardner 3–4 liq.; sapon. no. 197–204; 100% conc.

Industrene 226 FG. [310] Soya fatty acid; lubricant, release agent, binder, defoaming agent and intermediate for food; liq.; 100% conc.

Industrene 232. [312] Dist. corn-type fatty acid; see Industrene 105; Gardner 6 liq.; sapon. no. 197–205; 100% conc.

Industrene 239. [312] Dist. cotton fatty acid; see Industrene 105; Gardner 5 liq.; sapon. no. 200–206; 100% conc.

Industrene 325. [310; 312] Dist. coconut fatty acid; see Industrene 105; paste; sapon. no. 260–280; 100% conc.

Industrene 328. [310; 312] Stripped coconut fatty acid; see Industrene 105; paste; sapon. no. 254–261; 100% conc.

Industrene 333. [312] Hydrog. stripped coconut fatty acid; see Industrene 105; paste; sapon. no. 251–265; 100% conc.

Industrene 365. [312] 50% caprylic acid; see Industrene 105; sapon. no. 345–360; 100% conc.

Industrene 3022. [312] Hydrog. fish fatty acid; see Industrene 105; solid; sapon. no. 195–203; 100% conc.

Industrene 4022. [312] 40% arachidic and behenic fatty acid; see Industrene 105; solid; sapon. no. 191–199; 100% conc.

Industrene 4516. [310; 312] 45% palmitic fatty acid; see Industrene 105; solid; sapon. no. 206–212; 100% conc.

Industrene 4518. [310; 312] Single pressed grade fatty acid; see Industrene 105; 3 Gardner solid; sapon. no. 205–212; 100% conc.

Industrene 5016. [310; 312] Double pressed stearic acid; see Industrene 105; solid; sapon. no. 208–211; 100% conc.

Industrene 5016 FG. [310] Stearic acid; see Industrene 105; solid; 100% conc.

Industrene 5022. [312] 50% arachidic and behenic fatty acid; see Industrene 105; 7 Gardner solid; sapon. no. 178–188; 100% conc.

Industrene 6018. [310] Hydrog. buffing grade fatty acid; see Industrene 105; solid; 100% conc.

Industrene 7018. [310; 312] 70% stearic acid; see Industrene 105; solid; sapon. no. 201–208; 100% conc.

Industrene 7018 FG. [310] 90% stearic acid; see Industrene 105; solid; 100% conc.

Industrene 8518. [312] 85% stearic fatty acid; see Industrene 105; solid; sapon. no. 198–205; 100% conc.

Industrene 8718 FG. [310] Stearic acid; see

Industrene 105; solid; 100% conc.

Industrene 9018. [310; 312] 90% stearic fatty acid; see Industrene 105; solid; sapon. no. 197–202; 100% conc.

Industrene B. [310; 312] Hydrog. mixed fatty acid; see Industrene 105; solid; sapon. no. 200–208; 100% conc.

Industrene R. [310; 312] Hydrog. rubber grade fatty acid; see Industrene 105; 10 Gardner solid; sapon. no. 194–214; 100% conc.

Industrol 68. [86] Sorbitol stearate ethoxylate; emulsifier; wax; sp.gr. 1.1; m.p. 35 C; sapon. no. 60–75; 100% act.

Industrol 400-MOT. [86] POE (8) monotallate; nonionic; surfactant; Gardner 10 max. liq.; disp. in water; sp.gr. 1.04; visc. 400 cps; HLB 11.0; pour pt. –5 C; cloud pt. < 25 C (1% aq.); pH 6.0–7.5 (5% aq.); 100% act.

Industrol 1025-DT. [86] POP 1000 ditallate; nonionic; surfactant; Gardner 6 max. liq.; insol. in water; sp.gr. 0.98; pour pt. < –10 C; cloud pt. < 25 C (1% aq.); 100% act.

Industrol 1186. [86] Sorbitol oleate ethoxylate; emulsifier; Gardner 7 max. liq.; disp. in water; sp.gr. 1.0; visc. 450 cps; cloud pt. < 25 C (1% aq.); sapon. no. 75–95; pH 5.5–7.0 (5% aq.); 100% act.

Industrol CO-25. [86] Castor oil ethoxylate (25 EO); nonionic; surfactant; Gardner 5 max. liq.; water-sol.; m.w. 2100 (theoret.); sp.gr. 1.04; HLB 10.8; pour pt. –5 C; pH 6.0–7.5 (5% aq.); 100% act.

Industrol CO-30. [86] Castor oil ethoxylate (30 EO); nonionic; surfactant; Gardner 5 max. liq.; water-sol.; m.w. 2350 (theoret.); sp.gr. 1.04; visc. 800 cps; HLB 11.7; pour pt. 9 C; pH 6.0–7.5 (5% aq.); 100% act.

Industrol CO-36. [86] Castor oil ethoxylate (36 EO); nonionic; surfactant; Gardner 5 max. liq.; water-sol.; m.w. 2600 (theoret.); sp.gr. 1.05; visc. 700 cps; HLB 12.5; pour pt. 18 C; pH 5.5–7.0 (5% aq.); 100% act.

Industrol CO-40. [86] Castor oil ethoxylate (40 EO); nonionic; surfactant; Gardner 5 max. liq.; water-sol.; m.w. 2780 (theoret.); sp.gr. 1.05; visc. 600 cps; HLB 12.9; pour pt. 18 C; pH 6.0–7.5 (5% aq.); 100% act.

Industrol CO-80-80%. [86] Castor oil ethoxylate (80 EO); nonionic; surfactant; water-sol.; m.w. 4500 (theoret.); sp.gr. 1.08; HLB 15.5; pH 6.0–7.5 (5% aq.); 80% act.

Industrol CO-200. [86] Castor oil ethoxylate (200 EO); nonionic; surfactant; wh. wax; water-sol.; m.w. 9800 (theoret.); sp.gr. 1.08; HLB 18.1; m.p. 40 C; pH 6.0–7.5 (5% aq.); 100% act.

Industrol CO-200-50%. [86] Castor oil ethoxylate (200 EO); nonionic; surfactant; APHA 200 max. liq.; water-sol.; sp.gr. 1.04; visc. 2000 cps; HLB 18.1; pour pt. 7 C; pH 5.0–7.0 (2% aq.); 50% act.

Industrol COH-25. [86] POE (25) hydrog. castor oil; nonionic; surfactant; Gardner 3 max. liq.; water-sol.; m.w. 2140 (theoret.); sp.gr. 1.03; HLB 11.0; pour pt. 5 C; pH 6.0–7.5 (5% aq.); 100% act.

Industrol COH-200. [86] POE (200) hydrog. castor oil; nonionic; surfactant; Gardner 2 max. wax; water-sol.; m.w. 9800 (theoret.); HLB 18.1; m.p. 45 C; pH 6.0–7.5 (5% aq.); 100% act.

Industrol CSS-25. [86] Ethoxylated cetyl-stearyl alcohol (25 EO); nonionic; surfactant; Gardner 1 max. flake/wax; water-sol.; m.w. 1340; HLB 16.1; m.p. 41 C; cloud pt. > 100 C (1% aq.); pH 5.0–6.5 (5% aq.); 100% act.

Industrol DL-9. [86] POE (8) dilaurate; nonionic; surfactant; Gardner 3 max. liq.; insol. in water; m.w. 815; sp.gr. 0.99; HLB 10.8; pour pt. 8 C; cloud pt. < 25 C (1% aq.); pH 6.0–7.5 (5% aq.); 100% act.

Industrol DO-9. [86] PEG 400 dioleate; nonionic; surfactant; Gardner 4 max. liq.; disp. in water; m.w. 960; sp.gr. 0.97; HLB 8.5; pour pt. 6 C; cloud pt. < 25 C; pH 5.0–7.0 (5% aq.); 100% act.

Industrol DO-13. [86] PEG 600 dioleate; nonionic; surfactant; Gardner 4 max. liq.; disp. in water; m.w. 1160; sp.gr. 1.0; HLB 10.6; pour pt. 20 C; cloud pt. < 25 C (1% aq.); pH 5.0–7.0 (5% aq.); 100% act.

Industrol DT-13. [86] PEG 600 ditallate; nonionic; surfactant; Gardner 4 max. liq.; disp. in water; sp.gr. 1.00; HLB 10.5; pour pt. 10 C; cloud pt. < 25 C (1% aq.); pH 6.0–7.5 (5% aq.); 100% act.

Industrol L-20-S. [86] Polysorbate 20; nonionic; surfactant; liq.; water-sol.; sp.gr. 1.13; visc. 400 cps; HLB 16.7; sapon. no. 40–50; 3.0% max. water

Industrol LG-70. [86] Oleyl alcohol (20 EO) ethoxylate, modified; nonionic; surfactant to prevent gelling in aq. sol'ns.; Gardner 1 max. liq.; water-sol.; sp.gr. 1.07; HLB 16.0; pour pt. 18 C; cloud pt. > 100 C (1% aq.); pH 5.0–7.5 (5% aq.); 70% act.

Industrol LG-100. [86] Oleyl alcohol (20 EO) ethoxylate, modified; see Industrol LG-70; Gardner 1 max. wax; water-sol.; HLB 16.0;

m.p. 32 C; cloud pt. > 100 C; pH 6.0–7.5 (5% aq.); 100% act.

Industrol MIS-9. [86] PEG 400 monoisostearate (9 EO); nonionic; surfactant; Gardner 7 max. liq.; disp. in water; sp.gr. 1.03; visc. 100 cps; HLB 11.8; pour pt. 10 C; pH 6.0–7.5 (5% aq.); 100% act.

Industrol ML-5. [86] PEG 200 monolaurate (5 EO); nonionic; surfactant; Gardner 4 max. liq.; disp. in water; m.w. 425; sp.gr. 1.02; visc. 100 cps; HLB 9.8; pour pt. 9 C; cloud pt. < 25 C (1% aq.); pH 6.0–7.5 (5% aq.); 100% act.

Industrol ML-9. [86] PEG-9 laurate; nonionic; surfactant; Gardner 4 max. liq.; water-sol.; m.w. 600; sp.gr. 1.04; HLB 13.1; pour pt. 12 C; cloud pt. 40 C (1% aq.); pH 6.0–7.5 (5% aq.); 100% act.

Industrol ML-14. [86] PEG-14 laurate; nonionic; surfactant; Gardner 4 max. liq.; water-sol.; sp.gr. 1.05; HLB 14.9; pour pt. 14 C; cloud pt. 64 C; pH 6.0–7.5 (5% aq.); 100% act.

Industrol MO-5. [86] PEG 200 monooleate (5 EO); nonionic; surfactant; Gardner 6 max. liq.; disp. in water; m.w. 485; sp.gr. 0.99; HLB 8.3; pour pt. –10 C; pH 6.0–7.5 (5% aq.); 100% act.

Industrol MO-6. [86] PEG 250 monooleate (6 EO); nonionic; surfactant; Gardner 6 max. liq.; disp. in water; m.w. 544; sp.gr. 0.99; HLB 9.7; pour pt. –10 C; pH 6.0–7.5 (5% aq.); 100% act.

Industrol MO-7. [86] PEG-7 monooleate; surfactant; Gardner 6 max. liq.; disp. in water; m.w. 585; sp.gr. 1.0; HLB 10.4; pour pt. –10 C; pH 6.0–7.5 (5% aq.); 100% act.

Industrol MO-9. [86] PEG-9 oleate; surfactant; Gardner 6 max. liq.; disp. in water; m.w. 675; sp.gr. 1.01; HLB 11.7; pour pt. 5 C; pH 5.5–7.0 (5% aq.); 100% act.

Industrol MO-13. [86] POE (12) monooleate; nonionic; surfactant; Gardner 7 max. liq.; water-sol.; m.w. 850; sp.gr. 1.04; HLB 13.6; pour pt. 18 C; pH 6.0–7.5 (5% aq.); 100% act.

Industrol MS-5. [86] PEG-5 stearate; nonionic; surfactant; Gardner 3 max. wax; insol. in water; m.w. 500; HLB 9.1; m.p. 28 C; pH 5.5–7.0 (5% aq.); 100% act.

Industrol MS-7. [86] PEG 300 monostearate (7 EO); nonionic; surfactant; Gardner 2 max. wax; disp. in water; m.w. 575; HLB 10.1; m.p. 29 C; pH 5.5–7.0 (5% aq.); 100% act.

Industrol MS-8. [86] PEG 350 monostearate (8 EO); nonionic; surfactant; Gardner 2 max. wax; disp. in water; m.w. 620; HLB 11.4; m.p. 30 C; pH 5.5–7.0 (5% aq.); 100% act.

Industrol MS-9. [86] PEG-9 stearate; nonionic; surfactant; Gardner 2 max. wax; disp. in water; m.w. 665; HLB 12.0; m.p. 30 C; pH 5.5–7.0 (5% aq.); 100% act.

Industrol MS-23. [86] PEG 1000 monostearate (23 EO); nonionic; surfactant; Gardner 1 max. wax; water-sol.; m.w. 1280; HLB 15.7; m.p. 42 C; pH 5.5–7.0 (5% aq.); 100% act.

Industrol MS-40. [86] PEG-40 stearate; nonionic; surfactant; Gardner 1 max. wax.; water-sol.; m.w. 2030; HLB 16.9; m.p. 48 C; pH 5.5–7.0 (5% aq.); 100% act.

Industrol MS-100. [86] PEG 4400 monostearate (100 EO); nonionic; surfactant; Gardner 1 max. flake/wax; water-sol.; m.w. 4670; HLB 18.8; m.p. 57 C; pH 5.5–7.0 (5% aq.); 100% act.

Industrol O-20-S. [86] Polysorbate 80; nonionic; surfactant; liq.; water-sol.; sp.gr. 1.08; visc. 400 cps; HLB 15; sapon. no. 45–55; 3.0% max. water

Industrol OAL-20. [86] Oleyl alcohol ethoxylate (20 EO); surfactant; Gardner 1 max. wax; m.w. 1150; sp.gr. 1.03; HLB 15.8; m.p. 32 C; cloud pt. > 100 C (1% aq.); pH 6.0–7.5 (5% aq.); 100% act.

Industrol S-20-S. [86] Polysorbate 60; nonionic; surfactant; water-sol.; sp.gr. 1.13; visc. 600 cps; HLB 14.9; sapon. no. 45–55; 3.0% max. water

Industrol SO-13. [86] PEG 600 sesquioleate; nonionic; surfactant; Gardner 4 max. liq.; disp. in water; m.w. 1020; sp.gr. 1.0; HLB 12.1; pour pt. 19 C; cloud pt. < 25 C (1% aq.); 100% act.

Industrol STS-20-S. [86] Polysorbate 65; nonionic; surfactant; tan solid; disp. in water; sp.gr. 1.05; HLB 10.5; sapon. no. 88–98; 3.0% max. water

Industrol TFA-8. [86] Fatty acid ethoxylate (8 EO); nonionic; surfactant; Gardner 8 max. liq.; disp. in water; sp.gr. 1.02; HLB 10.0; pour pt. 0 C; pH 7.0–8.5 (5% aq.); 100% act.

Industrol TFA-15, TFA-15-80%. [86] Fatty acid ethoxylate (15 EO); nonionic; surfactant; Gardner 7 max. solid and liq. resp.; water-sol.; sp.gr. 1.08 and 1.06 resp.; HLB 14.0; pH 6.0–7.5 (5% aq.); 100 and 80% act. resp.

Industrol TO-10. [86] Tall oil fatty acid ethoxylate (10 EO); nonionic; surfactant; Gardner 10 max. liq.; disp. in water; m.w. 660; sp.gr.

1.03; HLB 11.2; pour pt. 0 C; pH 6.0–7.5 (5% aq.); 100% act.

Industrol TO-16. [86] Tall oil ethoxylate (16 EO); nonionic; surfactant; Gardner 10 max. liq.; water-sol.; m.w. 900; sp.gr. 1.05; HLB 13.5; pour pt. 16 C; pH 6.0–7.5 (5% aq.); 100% act.

Infrasan C. [506] Coconut fatty acids alkanolamide (2:1); nonionic; used for detergency and wetting; liq.; 100% conc.

Intermediate 300. [310] Coco/tallow monoethanolamide; nonionic; thickener, foam booster, superfatting agent; solid; 100% conc.

Intermediate 325. [310] Coco monoethanolamide; nonionic; see Intermediate 300; solid; 100% conc.

Intermediate 512. [310] Lauric diethanolamide; nonionic; see Intermediate 300; solid; 100% conc.

Interwet 33. [329] Nonionic; emulsifier, penetrant for coatings, adhesives, insecticides, wax dispersions, bitumen emulsions; yel. liq., mild fatty odor; sol. in alcohols, acetones, toluol, disp. in water; sp.gr. 1.01; visc. 85 cps; f.p. 2 C; b.p. 290 C; HLB 11.5; 100% act.

Intex Scour 707. [330] Phosphate; anionic; detergent, emulsifier, fabric preparations, scour after dyeing; liq.; 90% conc.

Intex Scour Base 706. [330] Alkanolamide; nonionic; foam additive, scour, pulling soap, shampoo base; liq.; 100% conc.

Intralan Salt HA. [163] Neutralized naphthalene condensate; anionic; dispersing and leveling agent for disperse dyes and dyeing of syn. fibers, dyeing auxiliary; powd.; water sol.

Intralan Salt N. [163] EO condensate; nonionic; dye assistant, leveling agent, penetrant for textiles; yel. liq.; sol. in boiling water.

Intraphasol COP. [163] Hydrocarbons, solubilized, aliphatic and sulfonic acid salts; anionic; emulsifier, wetting agent assistant for dyeing of fabrics and carpets; liq.

Intraphasol PC. [163] Sulfuric acid ester salt of polyglycol ether compd.; anionic; wetting agent, emulsifier, dispersant used in printing of polyamide and polyester carpets; amber liq.; alcoholic odor; water sol.; dens. 8.7 lb/gal.

Intrapol 1014. [188] Solvent and alkylpolyglycol ether; nonionic; foaming detergent and dispersant; kier boiling assistant; liq.; 80% conc.

Intratex A. [163] Complex amino condensate; amphoteric; nonfoaming leveling agent for dyeing wool; tan liq.; water sol.; dens. 9 lb/gal.

Intratex AN. [163] EO condensate; nonionic; leveling agent for wool dyeing, dye stripping; liq.; water sol.

Intratex B. [163] Polyglycol ether deriv.; amphoteric; leveling agent, detergent, for dye leveling of wool; tan liq.; water sol.; dens. 8.7 lb/gal.

Intratex C. [163] Complex amino condensate; amphoteric; leveling agent for dyeing wool; clear liq.; dens. 8 lb/gal.

Intratex OR. [163] Low molecular polyamide, nonionic; leveling agent, vat dyeing assistant; yel. br. liq.; water sol.; dens. 8.6 lb/gal; sp.gr. 1.04.

Intratex POK. [163] Sulfonated benzimidazole deriv.; anionic; dispersant, leveling agent for vat and disperse dyes; yel. br. liq.; water sol.

Intratex SCS. [163] Metallic salt; inorganic; tannic acid after-treatment of acid dyed nylon

Intratex W New. [163] EO condensate; nonionic; leveling agent, penetrant, for wool dyeing; yel. liq.; water sol.; dens. 8.5 lb/gal.

Intravon AN. [163] Fatty acid deriv.; scouring and emulsifying agent for wool, cotton, synthetics and blends, detergent in laundering, dyeing, household and cosmetic applications; amber liq.; water sol.; dens. 8.4 lb/gal; sp.gr. 1.01.

Intravon JET. [163] Nonionic/anionic blend; detergent, emulsifier, low foaming scouring agent for textiles; yel. br. liq.; water sol.

Intravon JF, JU. [163] EO condensate; nonionic; detergent, emulsifier, dispersant, wetting agent, penetrant for textile, household and cosmetic applications; yel. liq.; water sol.; dens. 8.4–8.5 lb/gal.

Intravon SO. [163] Anionic/nonionic blend; detergent for laundering and household uses, wetting agent, emulsifier for scouring and dyeing textiles; yel. liq.; water sol.; dens. 8.6 lb/gal.

Invermul. [447] Modified calcium salts of higher org. acids; emulsifier for w/o emulsion drilling fluids, reduces filtration; powd.; 100% act.

Inversol 140, 170, 190. [345] Complex polyglycol ester; nonionic; lubricity agent for metalworking liqs.; liq.; 100% conc.

Ionet 300. [539] POE alkyl ether type surfactant; anionic; scouring agent for fibers; leveling agent; penetrant; dispersant; liq.

Ionet DL-200. [539] POE dilaurate; nonionic; emulsifier for emulsion polymerization, metal processing lubricant and personal care products; liq.; HLB 6.6; 100% conc.

Ionet DO-200, -400, -600, -1000. [539] POE dioleate; nonionic; see Ionet DL-200; liq. except DO-1000 (solid); HLB 5.3, 8.4, 10.4, and 12.9 resp.; 100% conc.

Ionet DS-300, -400. [539] POE distearate; nonionic; see Ionet DL-200; solid; HLB 7.3 and 8.5 resp.; 100% conc.

Ionet MO-200, -400. [539] POE monooleate; nonionic; emulsifier for min. oil; pigment dispersions; base material for detergents; liq.; HLB 8.4 and 11.8 resp.; 100% conc.

Ionet MS-400, -600, -1000. [539] POE monostearate; nonionic; see Ionet MO-200; solid; HLB 11.9 (MS-400), 15.7 (MS-1000); 100% conc.

Ionet S-20. [539] Sorbitan monolaurate; nonionic; emulsifier for personal care products; lubricant, rust inhibitor; pigment dispersant; spreading agent for agricultural pesticides; base for textile lubricants; liq.; HLB 8.6; 100% conc.

Ionet S-60 C. [539] Sorbitan monostearate; nonionic; see Ionet S-20; solid; HLB 4.7; 100% conc.

Ionet S-80. [539] Sorbitan monooleate; nonionic; see Ionet S-20; liq.; HLB 4.3; 100% conc.

Ionet S-85. [539] Sorbitan trioleate; nonionic; see Ionet S-20; liq.; HLB 1.8; 100% conc.

Ionet T-20 C. [539] POE (20) sorbitan monolaurate; nonionic; base and emulsifier for personal care products, metal processing, lubricant and rust inhibitor; pigment dispersant; spreader sticker for agricultural pesticides; base for textile lubricants; liq.; HLB 16.7; 100% conc.

Ionet T-60 C. [539] POE sorbitan monostearate; nonionic; see Ionet T-20 C; liq.; HLB 14.9; 100% conc.

Ionet T-80 C. [539] POE sorbitan monooleate; nonionic; see Ionet T-80 C; liq.; HLB 15.0; 100% conc.

Irgalube 53. [137] POE alkyl ester and fatty ester; nonionic; dyeing assistant for fabrics; dye bath lubricant; paste

Irgasol N. [137] Sodium lignosulfonate; anionic; dye dispersant; liq.

Iscolan. [159] Lanolin ester; nonionic; emollient, wetting and spreading agent, used for skin, make-up and hair care cosmetics; pale yel., slightly opalescent liq.

Isokleen. [331] Detergent, alkaline conc. cleaner, dispersant, solubilizer for natural or syn. latices, organisols, etc.; dry powd.; water sol.; 100% act.

Isomul Extra. [331] Alkanolamide; nonionic; detergent, emulsifier, base for solubilizing waxes, solvents, oils, etc., base for formulated metal and textile cleaners; liq.; biodeg.; 100% act.

Isoplast D. [331] Ethoxylated tall; nonionic; emulsifier, disperser, stabilizer; liq.; 100% conc.

Isorezoff. [331] Neutral alkanolamide; nonionic; emulsifier, disperser, stabilizer; liq.; 100% conc.

Isoslushoff. [331] Alkaline alkanolamide; nonionic; detergent, cleaner, suspending agent, resin emulsifier; liq.; 88% conc.

Isoslushoff X-5. [331] Phenolic base; detergent, high strength cleaner where biodeg. not important.

Isoslushoff X-M-50. [331] Alkaline alkanolamide; detergent, solubilizer, nonionic; liq.; odorless; water dilutable; biodeg., nonvolatile, nonflam.

Isoslushoff X-NP-25. [331] Alkaline alkanolamide; nonionic; detergent for paper, resins, polyethylenes; odorless liq.; water dilutable; biodeg.

Isoslushoff X-NP-30. [331] Alkaline-free type Isoslushoff X-NP-25; nonionic; detergent used wherever alkali is objectionable; biodeg., nonflam.

Isothan Q-75. [456] *n*-Alkyl isoquinolinium bromides; cationic; fungicide in agric. products, shampoos, hair rinses; liq.; sp.gr. 1.08; flash pt. 92 F.

J

Jet Amine D-C, D-O, D-T. [336] Diamine; cationic; emulsifier of bitumen, fuel oil and gasoline additive, corrosion inhibitor, mineral flotation, bactericide; liq., paste; 100% conc.

Jet Amine DE-13, -810, -1215. [336] Ether diamine R-O-$(CH_2)_3$-N-$(CH_2)_3$-NH_2; emulsifier, corrosion inhibitor; liq.

Jet Amine PC, PHT, P-O, P-S, PT. [336] Primary amine; cationic; corrosion inhibitor, ore flotation agent, emulsifier, paint pigment dispersant, mold release agent, lube oil additive, fertilizer anti-cake; liq., solid; 100% conc.

Jet Amine PE-13, -810, -1215. [336] Primary amine R-O-$(CH_2)_3$-NH_2; ore flotation, emulsifier, corrosion inhibitor, fuel oil additive; liq.

Jetco-AEA. [336] Linear alkyl benzyl sulfonate; anionic; asphalt emulsifier; liq.; 60% conc.

Jet Quat 2C-75. [336] Dicoco dimethyl ammonium chloride; cationic; bactericide, textile softener, asphalt emulsifier, petroleum processing; liq.; 75% conc.

Jet Quat 2HT-75. [336] Dihydrog. tallow dimethyl ammonium chloride; cationic; see Jet Quat 2C-75; liq.; 75% conc.

Jet Quat C-50. [336] Coco trimethyl ammonium chloride; cationic; see Jet Quat 2C-75; liq.; 50% conc.

Jet Quat S-2C-50. [336] Soya dicoco quat. ammonium chloride; cationic; see Jet Quat 2C-75; liq.; 50% conc.

Jet Quat S-50. [336] Soya trimethyl ammonium chloride; cationic; see Jet Quat 2C-75; liq.; 50% conc.

Jet Quat T-2C-50. [336] Tallow dicoco quat. ammonium chloride; cationic; see Jet Quat 2C-75; liq.; 50% conc.

Jet Quat T-50. [336] Tallow trimethyl ammonium chloride; cationic; see Jet Quat 2C-75; liq.; 50% conc.

Jorchem PRM-98. [339] Phosphate ester; anionic; wetting and scouring agent, penetrant for all fibers; liq.; 90% conc.

Jordamide 22. [339] Coconut alkanolamide, modified; nonionic; emulsifier, low foam thickener for hard surface cleaners, waterless hand cleaners; brown liq.; acid no. 50–56; 100% conc.

Jordamide 29-78. [339] Modified 2:1 alkanolamide; see Jordamide 22; brown liq.; acid no. 22–32.

Jordamide 201. [339] Oleamide DEA (2:1); nonionic; emulsifier and wetting agent in cleaning applications; liq.; 100% conc.

Jordamide 1214. [339] Lauramide DEA (2:1); nonionic; emulsifier, wetting agent, foam stabilizer for cleaning applictions, personal care products; brown liq.; acid no. 10–17; 100% conc.

Jordamide 1281. [339] Coconut alkanolamide, modified; nonionic; foam booster and stabilizer, visc. builder, detergent and emulsifier; liq.; 100% conc.

Jordamide AXS. [339] Coconut alkanolamide, modified; nonionic/anionic; visc. and foamer for liq. detergents; liq.; 100% conc.

Jordamide CCO. [339] Cocamide DEA (2:1); nonionic; emulsifier and wetting agent in cleaning applications; liq.; 100% conc.

Jordamide CFAM. [339] Cocamide MEA (1:1); nonionic; foam stabilizer, visc. builder for household, industrial, and cosmetic products; lt. yel. flake; 100% conc.

Jordamide CLD. [339] Cocamide DEA (1:1); nonionic; foam stabilizer, visc. builder for household, industrial, cosmetic products; amber liq.; 100% conc.

Jordamide CLM. [339] Lauramide DEA (1:1); nonionic; see Jordamide CLD; yel. liq.; 100% conc.

Jordamide CMEA, CMEA Extra. [339] Cocamide MEA (1:1); foam stabilizer, visc. builder, thickener, and emulsifier for personal and household products; lt. yel. flake; solid. pt. 68 C (CMEA Extra); 100% conc.

Jordamide CP. [339] Coconut alkanolamide (2:1), modified; nonionic; foam booster and stabilizer, visc. builder, detergent, and emulsifier for hard surface cleaners; liq.; acid no. 45–55; 100% conc.

Jordamide J-10. [339] Coconut alkanolamide, modified; nonionic; emulsifier and wetting agent used in cleaning applications; amber liq.; acid no. 70–75; 100% conc.

Jordamide JR-100. [339] Coconut alkanolamide (2:1), modified; see Jordamide J-10; brown liq.; acid no. 42–53; 100% conc.

Jordamide JT-128. [339] Cocamide DEA (1:1); nonionic; foam stabilizer, thickener and emulsifier in personal and household products; yel. liq.; 100% conc.

Jordamide LLD. [339] Linoleamide DEA (1:1); nonionic; see Jordamide JT-128; amber liq.; 100% conc.

Jordamide PCS. [339] Coconut alkanolamide, modified; nonionic/anionic; foam booster, stabilizer, visc. builder, detergent and emulsifier; liq.; 100% conc.

Jordamide RO. [339] Modified 2:1 alkanolamide; emulsifier, low foam thickener for hard surface cleaners, waterless hand cleaners; brown liq.; acid no. 75–85.

Jordamide SCD. [339] 1:1 linoleamide/cocamide DEA; foam stabilizer and visc. builder for household, industrial, and cosmetic products; amber liq.

Jordamide TC. [339] Tallow diethanolamide, modified; nonionic; wool fulling and scouring agent, thickener; liq.; 100% conc.

Jordamide WC Conc. [339] Cocoamide DEA (1:1); nonionic; see Jordamide PCS; yel. liq.; 100% conc.

Jordamine 1281. [339] Modified 2:1 alkanolamide; visc. builder for hard surface cleaners; amber liq.; acid no. 50–56.

Jordamine DAPI. [339] Isostearamidopropyl dimethylamine; cationic; intermediate used in quats., amine oxides; oil and grease additive; emollient, conditioner for skin and hair care products; solid; 99% conc.

Jordamine DAPL, DMCAPA. [339] Cocamidopropyl dimethylamine; cationic; see Jordamine DAPI; solid and liq. resp.; 99 and 90% conc. resp.

Jordamine DAPSA. [339] Stearamidopropyl dimethylamine; cationic; see Jordamine DAPI; solid; 99% conc.

Jordamine DDBSA. [339] Dodecylbenzene sulfonic acid; anionic; formulation of household and industrial detergents; liq.; 97% conc.

Jordamine S-13. [339] Stearamidopropyl dimethylamine; cationic; see Jordamine DAPI; liq.; 99% conc.

Jordamine S-13 Lactate. [339] Stearamidopropyl dimethylamine lactate; hair conditioner, shampoos; liq.; 23% aq.

Jordamine SHCFA. [339] Cocamidopropyl dimethylamine; cationic; see Jordamine DAPI; solid; 99% conc.

Jordamox CAPA. [339] Cocamidopropylamine oxide; nonionic; wetting and scouring agent, stabilizer, foam booster, conditioner, visc. builder, detergent used in cosmetic, household and industrial formulations; liq.; 29.5–31.5% conc.

Jordamox CDA-40. [339] Palmitamine oxide; nonionic; see Jordamox CAPA; liq.; 29–31% conc.

Jordamox LDA. [339] Lauramine oxide; nonionic; see Jordamox CAPA; liq.; 29–31% conc.

Jordamox MDA. [339] Myristamine oxide; nonionic; wetting agent, foam booster/stabilizer, visc. builder for cosmetics; liq.; 29–31% conc.

Jordamox ODA. [339] Oleamine oxide; hair conditioner; liq.; 50–52% act.

Jordamox SDA. [339] Stearamine oxide; nonionic; conditioner, emulsifier; paste; 24.5–26.5% conc.

Jordanol AL 300. [339] Ammonium lauryl sulfate; anionic; used in personal care products and dishwash detergents; liq.; 30% conc.

Jordanol DDBSA. [339] Dodecylbenzene sulfonic acid; formulation of household and industrial detergents; liq.; 97% conc.

Jordanol NP-95. [339] Nonylphenol, ethoxylated; nonionic; formulation of household and industrial detergents; liq.; 100% conc.

Jordanol SL 300. [339] Sodium lauryl sulfate; anionic; see Jordanol AL 300; liq.; 30% conc.

Jordanol TL-400. [339] Triethanolamine lauryl sulfate; anionic; see Jordanol AL 300; liq.; 40% conc.

Jordaphos 151. [339] Aromatic phosphate ester; anionic; emulsifier for chlorinated solvs. and aerosol propellents; detergent surfactant; liq.; 100% conc.

Jordaphos 236. [339] Aliphatic phosphate ester; anionic; emulsifier for aliphatic solvs.; coupling agent; Gardner 3 max. clear liq.; 100% conc.

Jordaphos DT. [339] Aliphatic phosphate ester; anionic; surfactant for hard surface cleaning, textile scouring; liq.; 100% conc.

Jordaphos FDEO. [339] Aliphatic phosphate ester; anionic; surfactant; coupling agent; Gardner 8 max. clear liq.; 100% conc.

Jordaphos JA-60. [339] Aliphatic phosphate ester; anionic; detergent surfactant and coupling agent; Gardner 6 max. clear visc. liq.; 100% conc.

Jordaphos JB-40. [339] Aliphatic phosphate ester; anionic; lubricity additive, rust inhibitor, emulsifier; Gardner 9 max. opaque visc. liq.; oil and water-sol.; 100% conc.

Jordaphos JE-41. [339] Aromatic phosphate ester; anionic; emulsifier for aromatic solvs.; detergent, solubilizer; Gardner 6 max. clear visc. liq.; 100% conc.

Jordaphos JE-51, -61. [339] Aromatic phosphate ester; anionic; detergent for emulsion polymerization; liq.; 100% conc.

Jordaphos JM-51. [339] Aromatic phosphate ester; emulsifier for aliphatic and chlorinated solvs. and aerosol propellents used in cutting oils and lubricants; liq.; 100% conc.

Jordaphos JM-71. [339] Aromatic phosphate ester; emulsifier for cutting oils; liq.; water-sol.; 100% conc.

Jordaphos JP-70. [339] Aromatic phosphate ester; lubricant; liq.; water-sol.; 100% conc.

Jordaphos JS-41. [339] Aliphatic phosphate ester; anionic; textile wetting agent used for scouring; drycleaning; liq.; 100% conc.

Jordaphos JS-61. [339] Aliphatic phosphate ester; anionic; lubricant, cutting oil emulsifier; Gardner 6 max. hazy visc. liq.; 100% conc.

Jordaphos JS-71. [339] Aliphatic phosphate ester; anionic; detergent, emulsifier, wetting agent, dispersant; Gardner 6 max. opaque visc. liq.; 100% conc.

Jordaphos RA-60. [339] Aliphatic phosphate ester; anionic; emulsifier for aliphatic solvs.; coupling agent; Gardner 4 max. clear liq.; 100% conc.

Jordapon CI. [339] Sodium cocoyl isethionate; anionic; detergent; lime soap dispersant; powd.; 80% conc.

Jordapon CI Paste. [339] Sodium cocoyl isethionate; anionic; foaming surfactant used in personal care products; paste; 40% conc.

Jordaquat 522. [339] Isostearamidopropyl ethyldimonium ethosulfate; quat. for hair conditoners and shampoos; clear visc. liq.; exc. water sol.; 100% act.

Jordaquat 1033. [339] Soya ethyldimonium ethosulfate; quat. for hair conditioners and shampoos, rug shampoos; low foaming; antistat; clear yel. liq.; 58% act.

Jordaquat Dimer 18. [339] Hydroxypropyl bisstearyldimonium chloride; quat. for hair and skin conditioners, perms, mousses; wh. paste; 50% act.

Jordaquat JO-50. [339] Olealkonium chloride; quat. for clear hair rinses; conditioner, antistat; clear yel. liq.; 50% act.

Jordaquat JS-25. [339] Stearalkonium chloride; quat. for hair conditioning; softener; wh. paste; 25% act.

Jordawet DMDS. [339] Disodium oleamido PEG-2 sulfosuccinate; anionic; wetting and penetrating agent for hair, bath and home products; liq.; 30% conc.

Jordawet DSLES. [339] Disodium laureth sulfosuccinate; anionic; used in personal care and home products; liq.; 40% conc.

Jordawet LA-40. [339] Disodium lauryl sulfosuccinate; anionic; used in hair, bath, and home products; 40% conc.

Jorscour TR. [339] Alkyl phenol, ethoxylated; nonionic; scouring agent; liq.; 100% conc.

Jortaine C, CAB-35. [339] Cocamidopropyl betaine; amphoteric; wetting agent, stabilizer, foamer and visc. builder in personal care products; liq.; 35% conc.

Jortaine CB-40. [339] Coco betaine; amphoteric; wetting agent, stabilizer, foam booster, and visc. builder in personal care products; liq.; 40% conc.

Jortaine CFA-35. [339] Cocoamidopropyl betaine; amphoteric; see Jortaine CB-40; liq.; 35% conc.

Jortaine COSB. [339] Cocamidopropyl hydroxysultaine; amphoteric; wetting agent, stabilizer, foamer, visc. builder used in detergents and personal care products; liq.; 50% conc.

Jortaine CSB, CSB-50. [339] Cocamidopropyl hydroxysultaine; amphoteric; visc. builder used in personal care products and cleaners; liq.; 35 and 50% conc. resp.

Jortaine LMAB. [339] Lauramidopropyl betaine; amphoteric; see Jortaine COSB; liq.; 35% conc.

Jortaine TM. [339] Dihydroxyethyl tallow glycinate; thickener, conditioner, foamer; for high or low pH formulations, shampoos; visc. liq. to gel; 39% act.

Juniorlan. [217] Lanolin wax and triolein; nonionic; emulsifying properties; solid; 100% conc.

K

Kadif 50 Flakes. [653] Sodium dodecylbenzene sulfonate; general purpose wetting and penetrating agent, detergent for industrial, institutional and household uses; lt. flakes; practically odorless; water sol.; pH 8.0 (1% aq.); 50% act.

Kalcohl 08H. [341] *n*-Octyl alcohol; raw material for plasticizer; tobacco-offshoots controller; liq.; 100% conc.

Kalcohl 5-24, 6-24, 7-24. [341] Lauryl myristyl alcohol; raw material for sodium lauryl sulfate; metal rolling agent; liq.; 100% conc.

Kalcohl 10H. [341] *n*-Decyl alcohol; tobacco-offshoots controller; raw material for antioxidant, alkyl phosphate, ethoxylate, chloride; liq.; 100% conc.

Kalcohl 20. [341] Lauryl alcohol; see Kalcohl 10H; liq.; 100% conc.

Kalcohl 40. [341] Myristyl alcohol; see Kalcohl 10H; solid; 100% conc.

Kalcohl 60. [341] Cetyl alcohol; see Kalcohl 10H; beads or solid; 100% conc.

Kalcohl 68. [341] Cetyl stearyl alcohol; base for ointment and cream; beads, solid; 100% conc.

Kalcohl 80. [341] Stearyl alcohol; see Kalcohl 68; beads, solid; 100% conc.

Kalcohl 86. [341] Stearyl cetyl alcohol; lubricant for plastic processing; beads, solid; 100% conc.

Kaorich Beads. [210] Veg. oil (cottonseed, soybean), hydrog.; emulsifier for food processing; 2.5R max. color; m.p. 124–130 F.

Karawet CS. [376] Org. phosphate ester, complex salt; anionic; emulsifier; detergent, and wetting agent; liq.; 40% conc.

Karawet DOSS. [376] Sodium dioctyl sulfosuccinate; anionic; wetting and rewetting agent for textile applications; liq.; 70% conc.

Karawet SB. [376] Org. phosphate ester, complex salt; alkaline stable emulsifier detergent and wetter; liq.; 50% conc.

Kar-Kote. [146] Detergent used for washing painted surfaces on cars, trucks and buses; tan powd.; bland odor.

Katapol OA-860. [245] PEG-30 oleamine; hydrophilic emulsifier, leveling agent; textile dyeing assistant; antiprecipitant for dyeing processes; solid; water-sol.; 99% act.

Katapol OA-910. [245] POE (30) oleyl amine; cationic; emulsifier and textile dyeing assistant; antiprecipitant to prevent cross-staining; stripping agent and dye leveler for acid dyes; liq.; 98% conc.

Katapol PN-430. [245] PEG-5 tallow amine; cationic; emulsifier for min. oils, acid corrosion inhibitor for ferrous alloys; br. liq.; sol. in kerosene with slight haze; pH 8.0–10.0 (10% aq. disp.); 99% act.

Katapol PN-730, PN-810. [245] PEG-15, -20 tallow amine; cationic; emulsifier, dye assistant, softener, antistat, lubricant for textiles and leathers; amber clear liq.; water sol.; 99% min. act.

Katioran AF. [82] Fatty acid hydroxylalkylamide and fatty alcohol, ethoxylated; nonionic; emulsifier and thickener for cosmetic preparations; waxy.

Kelacid. [346] Alginic acid; used as gelling agent, emulsifier and stabilizer in food and industrial applications; stabilizer in paper and textile industry; wh. fibrous particles;

pH 2.9 (1% aq.); surf. tens. 53 dynes/cm; 7% moisture.

Kelcoloid D. [346] Propylene glycol alginate; see Kelacid; cream fibrous particles; sp.gr. 1.46; dens. 33.71 lb/ft³; visc. 170 cps; ref. index 1.3343; pH 4.4; surf. tens. 58 dynes/cm; 13% max. moisture

Kelcoloid DH, DO, DSF. [346] Propylene glycol alginate; see Kelacid; cream agglomerated; sp.gr. 1.46; dens. 33.71 lb/ft³; visc. 400, 25, 20 cps resp.; ref. index 1.3343; pH 4.0, 4.3, 4.0 resp.; surf. tens. 58 dynes/cm.

Kelcoloid HVF, LVF, O, S. [346] Propylene glycol alginate; see Kelacid; cream fibrous particles; sp.gr. 1.46; dens. 33.71 lb/ft³; visc. 400, 120, 25, 20 cps resp.; ref. index 1.3343; pH 4.0, 4.0, 4.3, 4.0 resp.; surf. tens. 58 dynes/cm.

Kelco-Pac. [346] Sodium alginate; see Kelacid; ivory granular particles; sp.gr. 1.64; dens. 43.38 lb/ft³; visc. 55 cps; pH 7.2; surf. tens. 70 dynes/cm; 9% moisture.

Kelcosol. [346] Sodium alginate; see Kelacid; cream fibrous particles; sp.gr. 1.64; dens. 43.38 lb/ft³; visc. 1300 cps; pH 7.2; surf. tens. 70 dynes/cm; 9% moisture.

Kelgin F. [346] Sodium alginate, refined; see Kelacid; ivory granular particles; sp.gr. 1.59; dens. 54.62 lb/ft³; visc. 300 cps; ref. index 1.3343; surf. tens. 62 dynes/cm; 13% moisture.

Kelgin HV, LV, MV. [346] Sodium alginate; see Kelacid; ivory granular particles; sp.gr. 1.59; dens. 54.62 lb/ft³; visc. 800, 60, 400 cps resp.; ref. index 1.3343; pH 7.5; surf. tens. 62 dynes/cm; 13% moisture

Kelgin QH, QL, QM. [346] Sodium alginate; see Kelacid; ivory granular particles; visc. 400, 30, 180 cps resp.; pH neutral.

Kelgin RL, XL. [346] Sodium alginate; see Kelacid; ivory granular particles; sp.gr. 1.59; dens. 54.62 lb/ft³; visc. 10 and 30 cps resp.; ref. index 1.3343; pH 7.5; surf. tens. 62 dynes/cm; 13% moisture.

Kelig 32. [40] Lignosulfonate and modified sugar acids; anionic; sequestrant, dispersant, lubricant removal from zinc phosphate coatings, oil well cement retardant; powd; 100% conc.

Kelig 100. [502] Sodium lignosulfonate; anionic; metal cleaning component for industrial cleaning applications; powd., liq.; 55–100% conc.

Kelmar, Kelmar Improved. [346] Potassium alginate; see Kelacid; cream granular and fibrous particles resp.; visc. 270 and 400 cps resp.; pH neutral.

Kelset. [346] Alginate; see Kelacid; lt. ivory fibrous particles; pH neutral.

Keltex, Keltex P, S. [346] Industrial sodium alginate; see Kelacid; tan granular particles; visc. 800, 765, and 1300 cps resp.; pH neutral.

Keltone. [346] Sodium alginate; see Kelacid; cream fibrous particles; sp.gr. 1.64; dens. 43.38 lb/ft³; visc. 400 cps; pH 7.2; surf. tens. 70 dynes/cm; 9% moisture.

Keltose. [346] Calcium alginate and ammonium alginate; see Kelacid; ivory granular particles; visc. 250 cps; pH neutral.

Kelvis. [346] Sodium alginate, refined; see Kelacid; ivory granular particles; sp.gr. 1.59; dens. 54.62 lb/ft³; visc. 760 cps; ref. index 1.3343; pH 7.5; surf. tens. 62 dynes/cm. 13% moisture.

Kemamide B. [310; 312] Behenamide; slip and antiblock agent in plastic films; foam stabilizer, release agent, thickener, pigment dispersant; Gardner 5 max. solid; m.w. 312; water insol., slightly sol. in org. solv.; m.p. 98–108 C; 100% conc.

Kemamide E. [310; 312] Erucamide; see Kemamide B; Gardner 5 max solid; m.w. 335; water insol., slightly sol. in org. solv.; m.p. 76–86 C; 100% conc.

Kemamide E-180. [310] Stearyl erucamide; see Kemamide B; wh. powd.; dens. 0.8074 (110 C); m.p. 72–75 C; flash pt. (CC) 258 C.

Kemamide E-221. [310] Erucyl erucamide; see Kemamide B; wh. powd.; dens. 0.8165 (110 C); m.p. 55–58 C; flash pt. (CC) 260 C.

Kemamide O. [310] Oleamide; see Kemamide B; solid; 100% conc.

Kemamide P-181. [310] Oleyl palmitamide; see Kemamide B; wh. powd.; m.p. 69–72 C; flash pt. (CC) 262 C.

Kemamide S. [310; 312] Stearamide; see Kemamide B; Gardner 4 max. powd., pellet, waxy solid; m.w. 278; water insol., slightly sol. in org. solv. at R.T.; dens. 0.809 g/ml (130 C); visc. 5.8 cP (130 C); m.p. 98–108 C; flash pt. (COC) 246 C; 100% conc.

Kemamide S-65. [310] Tallow amide; see Kemamide B; solid; 100% conc.

Kemamide S-180. [310] Stearyl stearamide; see Kemamide B; wh. powd.; dens. 0.8042 g/ml (110 C); m.p. 92–95 C; flash pt. (CC) 246 C.

Kemamide S-221. [310] Erucyl stearamide; see Kemamide B; dens. 0.7877 g/ml (110 C);

m.p. 72–75 C; flash pt. (CC) 268 C.

Kemamide U. [312] Oleamide; see Kemamide B; Gardner 5 max solid; m.w. 275; water insol., slightly sol. in org. solv. at R.T.; m.p. 68–78; 100% conc.

Kemamine A650. [311] Coconut amine acetate; cationic; pigment dispersant, fungicide, petroleum additive, collectors in ore flotation and boiler water treatment; solid; 100% conc.

Kemamine A970. [311] Hydrog. tallow amine acetate; cationic; see Kemamine A650; solid; 100% conc.

Kemamine A974. [311] Tallow amine acetate; cationic; see Kemamine A650; solid; 100% conc.

Kemamine A990. [311] Octadecyl amine acetate; cationic; pigment dispersant, fungicide, petroleum additive; solid; 100% conc.

Kemamine AD 650. [311] N-coco 1,3-propylene diamine diacetate; cationic; see Kemamine A990; solid; 100% conc.

Kemamine AD 974. [311] N-tallow 1,3 propylene diamine diacetate; cationic; see Kemamine A990; solid; 100% conc.

Kemamine BQ-2802C. [310] Dimethyl 80%-behenyl benzyl ammonium chloride; cationic; antistat, textile softener, dyeing aid, corrosion inhibitor, emulsifier; used in personal care products, e.g., creams, lotions, shampoos, hair conditioners; Gardner 4 max.; m.w. 475; pH 9 max. (5%); 75% min. act.

Kemamine BQ-9702C. [310; 312] Dimethyl hydrog. tallow benzyl quat. ammonium chloride; cationic; emulsifier, corrosion inhibitor, germicide, sanitizer, slimicide, antistat, textile softening agent, dyeing aid; also for personal care products; Gardner 4 max. liq.; m.w. 420; disp. in water; pH 9 max. (5%); 75% act.

Kemamine BQ-9742C. [310; 312] Tallowalkonium chloride; cationic; see Kemamine BQ-9702C; Gardner 6 max. paste; m.w. 420; disp. in water; pH 9 max (5%); 75% act.

Kemamine D-150, -190. [312] Arachidyl-behenyl 1,3-propylenediamine; cationic; gasoline detergent, bactericide, corrosion inhibitor in petroleum production, epoxy hardener; Gardner 5 max. solid; 88% conc

Kemamine D-650. [312] N-coconut 1,3-propylenediamine; cationic; see Kemamine D-150; Gardner 6 max. paste; 88% conc.

Kemamine D-970. [312] N-hydrog. tallow 1,3-propylenediamine; cationic; see Kemamine D-150; Gardner 6 max. solid; 88% conc.

Kemamine D-974. [312] Tallow aminopropylamine; cationic; see Kemamine D-150; Gardner 12 max. solid; 88% conc.

Kemamine D-989. [312] N-oleyl 1,3-propylenediamine; cationic; see Kemamine D-150; Gardner 7 max. liq.; 88% conc.

Kemamine P-150, P-150D. [312] 50% arachidyl-behenyl primary amine (P-150D—dist.); cationic; emulsifier, flotation agent, dispersing and flushing agent, intermediate, used in metalworking oils, as fuel oil additive; Gardner 3 and 1 resp., solid; sol. in common org. solv.; 93 and 97% conc.

Kemamine P-190, P-190D. [311; 312] 90% arachidyl-behenyl primary amine (P-190D—dist.); cationic; see Kemamine P-150; Gardner 3 and 1 resp., solid; sol. in common org. solv.; 93, 97% conc.

Kemamine P-650, P-650D. [311; 312] Coco amine (P-650D—dist.); cationic; see Kemamine P-150; Gardner 3 and 1 resp., liq.; sol. in common org. solv.; 93%, 97% conc.

Kemamine P-690, P-690 D. [311] Lauramine (P-690 D—dist.); cationic; see Kemamine P-190; liq.; 100% conc.

Kemamine P-790, P-790 D. [311] Tetradecyl primary amine (P-790 D—dist.); cationic; see Kemamine P-190; solid; 100% conc.

Kemamine P-880, P-880D. [311; 312] Palmityl primary amine (tech. and dist. resp.); cationic; see Kemamine P-150; Gardner 3 and 1 resp., solid; sol. in common org. solv.; 93, 97% conc.

Kemamine P-970, P-970D. [310; 311; 312] Hydrog. tallow amine (tech. and dist. resp.); cationic; see Kemamine P-150; Gardner 3 and 1 resp., solid; sol. in common org. solv.; 93, 97% conc.

Kemamine P-974, P-974D. [311; 312] Tallow amine (tech. and dist. resp.); cationic; see Kemamine P-150; Gardner 3 and 1 resp., paste; sol. in common org. solv.; 93, 97% conc.

Kemamine P-989, P-989D. [310; 311; 312] Oleamine (tech. and dist. resp.); cationic; see Kemamine P-150; Gardner 3 and 1 resp., liq.; sol. in common org. solv.; 93, 97% conc.

Kemamine P-990, P-990D. [311; 312] Stearamine (tech. and dist. resp.); cationic; see Kemamine P-150; Gardner 3 and 1 resp., solid; sol. in common org. solv.; 93, 97% conc.

Kemamine P-997, P-997D. [311; 312] Soyamine (tech. and dist. resp.); cationic; see

Kemamine P-150; Gardner 3 and 2 resp., paste; sol. in common org. solv.; 93, 97% conc.

Kemamine Q-1302C. [310] Dimethyl di-30% arachidyl-behenyl quat. ammonium chloride; cationic; emulsifier, corrosion inhibitor, textile and paper softener, household laundry uses; solid; 75% conc.

Kemamine Q-1902C. [311; 312] Dibehenyl/diarachidyl dimonium chloride; cationic; germicide, sanitizer, slimicide, antistat, textile softening agent, dyeing aid, corrosion inhibitor, emulsifier; Gardner 4 max. solid; m.w. 680; disp. in water; pH 9 max (5%); 75% act.

Kemamine Q-2802C. [310] Dimethyl di-80%-behenyl benzyl ammonium chloride, cationic; see Kemamine BQ-2802C; Gardner 4 max.; m.w. 690; pH 9 max. (5%); 75% min. act.

Kemamine Q-6502C. [310; 311; 312] Dimethyl dicoconut quat. ammonium chloride; cationic; see Kemamine Q-1902C; Gardner 4 max. liq.; m.w. 465; disp. in water; pH 9 max (5%); 75% act.

Kemamine Q-6503 B. [311] Dicoco dimethyl ammonium chloride; cationic; see Kemamine Q 1902 C; liq.; 50% conc.

Kemamine Q-6903 B. [311] Laurtrimonium chloride; cationic; see Kemamine Q 1902 C; liq.; 50% conc.

Kemamine Q-9702C. [310; 312] Quaternium-18; cationic; see Kemamine Q-1902C; Gardner 2 solid; m.w. 575; disp. in water; pH 9 max (5%); 75% act.

Kemamine S-190. [310; 311] Di-90% arachidyl-behenyl sec. amine; cationic; industrial use, grease, corrosion inhibitor; solid; 90% conc.

Kemamine S-650. [311] Dicoconut sec. amine; cationic; see Kemamine S-190; solid; 100% conc.

Kemamine S-970. [311] Hydrog. di-tallow amine; cationic; see Kemamine S-190; solid; 100% conc.

Kemamine T-1902D. [310; 312] Dist. dimethyl-90% arachidyl-behenyl tert. amine; cationic; chemical intermediate for quat. ammonium derivs., epoxy hardener, catalyst in mfg. of flexible polyurethane foams; Gardner 1 max. liq.; 95% conc.

Kemamine T-6501. [310; 312] Methyl dicoconut tert. amine; cationic; see Kemamine T-1902D; 3 max Gardner liq.; 95% conc.

Kemamine T-6502D. [310; 312] Dist. dimethyl cocamine; cationic; see Kemamine T-1902D; Gardner 1 max.; 95% conc.

Kemamine T-9701. [310; 312] Dihydrog. tallow methyl amine; cationic; see Kemamine T-1902D; Gardner 3 max. liq.; 95% conc.

Kemamine T-9702D. [310; 312] Dist. dimethyl hydrog. tallow amine; cationic; see Kemamine T-1902D; Gardner 1 max. liq.; 95% conc.

Kemamine T-9742D. [310; 312] Dist. dimethyl tallow amine; cationic; see Kemamine T-1902D; Gardner 1 max. liq.; 95% conc.

Kemamine T-9902D. [310; 312] Dist. dimethyl stearamine; cationic; see Kemamine T-1902D; Gardner 1 max. liq.; 95% conc.

Kemamine T-9972D. [310; 312] Dist. dimethyl soya amine; cationic; see Kemamine T-1902D; Gardner 1 max. liq.; 95% conc.

Kemester 104. [310] Methyl oleate; nonionic; emulsifier, emollient for cosmetics; lubricant for leather; liq.; 100% conc

Kemester 105, 205. [310; 312] Methyl oleate (205—white); intermediate in production of superamides, in metalworking lubricants, specialized solv.; opacifier, visc. control agent; Gardner 2 and 1 resp., liq.; sapon. no. 194–203; 100% conc.

Kemester 226. [310; 312] Methyl soyate; see Kemester 105; Gardner 4 max. liq.; sapon. no. 195–200; 100% conc.

Kemester 325. [310; 312] Methyl cocoate; see Kemester 105; Gardner 1 max. liq.; sapon. no. 246–256; 100% conc.

Kemester 1000. [310] Glycerol trioleate; nonionic; base for surfactants for textiles and leather; liq.; 100% conc.

Kemester 2000. [310] Glycerol monooleate; nonionic; emulsifier, stabilizer, wetting agent for textiles; liq.; 100% conc.

Kemester 4000. [310] Butyl oleate; nonionic; wetting agent; liq.; 100% conc.

Kemester 5221SE. [310] Diglycol stearate, SE; anionic; emulsifier, plasticizer, lubricant for cosmetics, rubber, textiles; flake; 100% conc.

Kemester 5500. [310] Glycerol monostearate; nonionic; emulsifier; stabilizer, plasticizer, lubricant for cosmetic, paper, textile, and industrial uses; bead, flake; 100% conc.

Kemester 6000. [310] Glycerol monostearate; nonionic; cosmetic and industrial emulsifier; bead; 100% conc.

Kemester 6000E. [310] Glycerol monostearate; nonionic; emulsifier, stabilizer, plasticizer; bulk, beads; 100% conc.

Kemester 6000SE. [310] Glycerol monostearate SE; anionic; emulsifier, lubricant, plasticizer for cosmetic, paper and textiles; bulk, beads; 100% conc.

Kemester EGDS. [310; 312] Glycol distearate; see Kemester 105; Gardner 2 max. solid; sapon. no. 190–200; 100% conc.

Kemester EGMS. [310] Ethylene glycol monostearate; see Kemester 105.

Ke-Mul 181. [252] Lignosulfonate; anionic; asphalt emulsifier and stabilizer; liq.; 45% conc.

Ke-Mul A97. [252] Lignosulfonate; anionic; emulsifier for o/w emulsions; liq.; 45% conc.

Kerapol 791. [192] Coconut alkylolamide; nonionic; detergent; foam booster; liq.; 100% conc.

Kerasol 1014. [192] Solv. and alkylolamide; nonionic; detergent; used in textile processing; kier boiling assistant; liq.; 100% conc.

Kerinol 2012 F. [192] Fatty acid alkanolamide; nonionic; detergent, foam stabilizer and thickener; liq.; 100% conc.

Kerinol AA 62. [192] Fatty acid alkanolamide; nonionic; see Kerinol 2012 F; paste; 100% conc.

Kerinol C 109. [192] Alkanolamide, modified; nonionic; synergist and scouring agent for wool; liq.; 100% conc.

Keripon NC. [536] Fatty acid ester, glycol; nonionic; wetting, rewetting; antistat; leather degreaser and penetrant with fat liquors; used in paper mfg.; liq.; 50% conc.

L

LA-55-3. [287] Laureth-3; nonionic; emulsifier for cosmetics; technical applications for min. oil; mfg. of emulsions; liq.; HLB 7.5; 100% conc.

LA-55-4. [287] Laureth-4; nonionic; emulsifier for personal care products; latex stabilizer for emulsion polymerization for polishes and fabrics; liq.; HLB 9.5; 100% conc.

LA-55-23. [287] Laureth-23; nonionic; emulsifier for personal care products; leather dressing agent; stabilizing of photo emulsions for mfg. of polishes; solid; HLB 17.0; 100% conc.

Labrafac Hydro. [247; 248] Caprylocapric glycerides, ethoxylated; nonionic; hydrophilic oil; liq.; 100% conc.

Labrafil ISO. [247] Isostearic glycerides, ethoxylated; nonionic; hydrophilic oil; liq.; 100% conc.

Labrafil M 1944 CS. [248] Apricot kernel oil PEG-6 complex; nonionic; hydrophilic oil; liq.; 100% conc.

Labrafil M 1969 CS, M 1980 CS. [247; 248] Oleic olive glycerides, ethoxylated; nonionic; hydrophilic oil; liq.; 100% conc.

Labrafil M 2125 CS. [248] Corn oil PEG-6 complex; nonionic; hydrophilic oil; liq.; 100% conc.

Labrafil M 2130 BS, M 2130 CS. [248] Hydrog. palm/palm kernel oil PEG-6 complex; hydrophilic hydrog. oil; solid; 100% conc.

Labrafil M 2130 CS. [248] Palm kernel oil, palm oil, and PEG-6; hydrophilic hydrog. oil; solid; 100% conc.

Labrafil M 2735 CS. [248] Triolein PEG-6 complex; nonionic; hydrophilic hydrog. oil; liq.; 100% conc.

Labrafil WL 1905 CS. [247] Linoleic-erucic colza glycerides, ethoxylated; nonionic; hydrophilic oil; liq.; 100% conc.

Labrafil WL 1958 CS. [247] Ricinoleic glycerides, ethoxylated; liq.; hydrophilic wax; liq.; 100% conc.

Labrasol. [247; 248] Glyceryl caprylate/caprate and PEG-8 caprylate/caprate; nonionic; hydrophilic oil; liq.; 100% conc.

Labs-100. [659] Linear alkylbenzene sulfonic acid; anionic; detergent intermediate; liq.; 100% conc.

Labs-100 SP. [659] Alkylbenzene and toluene, cosulfonic acid; anionic; spray drying; liq.; 100% conc.

Lactodan. [273] Monoglyceride lactic acid esters; nonionic; food emulsifier, improves aeration and foam stabilization; solid, flakes; HLB 3.5; 100% act.

Lactomul 461. [277] Monodistearate POE glycol; nonionic; emulsifier for calf-milk replacer; paste; 100% conc.

Lactomul 463. [277] Glycerol polyethylene tallow fatty acid ester; nonionic; see Lactomul 461; liq.; 100% conc.

Lactomul 843. [277] POE glycol monodioleate; nonionic; see Lactomul 461; liq.; 100% conc.

Lactomul 925. [277] Hydrog. lard monoglyceride, distilled; nonionic; see Lactomul 461; powd.; 100% conc.

Lakeway 100-CA. [95] Coco diethanoloamide; nonionic; foam booster, thickener, stabilizer, used in shampoos, bubble baths, lt. duty detergents; lt. straw liq.; dens. 8.3 lb/

gal; cloud pt. –3 C; pH 10 (10%); 90% conc.

Lakeway 101-10. [95] Sodium lauryl sulfate; anionic; detergent used in personal care products, hair shampoo, bubble bath, rug shampoos; wh. to lt. yel. liq.; dens. 8.5 lb/gal; visc. 350 cps; cloud pt. 18 C; pH 8.0 (10%); 29% conc.

Lakeway 101-10P, N. [95] Dry sodium lauryl sulfate; used for emulsion polymerization, ink disp., personal care products, lubricant; wh. powd. or needle; pH 10 (1%); 90% conc.

Lakeway 101-11, 101-19. [95] Sodium lauryl sulfate; anionic; detergent for emulsion polymerization; wh. to lt. yel. liq.; dens. 8.5–8.6 lb/gal; visc. 250 and 100 cps resp.; cloud pt. 15 and 19 C resp.; pH 8.0 (10%); 29 and 45% conc.

Lakeway 101-20. [95] Ammonium lauryl sulfate; detergent used in personal care products, bubble baths, industrial processing, fire fighting foams; wh. to lt. yel. liq.; dens. 8.5 lb/gal; visc. 1000 cps; cloud pt. 14 C; pH 7.0 (10%); 29% conc.

Lakeway 101-30. [95] Triethanolamine lauryl sulfate; anionic; detergent with low cloud pt., low irritancy for personal care products; wh. to lt. yel. liq.; dens. 8.6 lb/gal; visc. 120 cps; cloud pt. 3 C; pH 8.0 (10%); 40% conc.

Lakeway 106R, 120. [95] Formulated conc.; anionic; rug shampoo conc.; wh. to lt. yel. liq.; dens. 8.6–8.9 lb/gal; visc. 150–265 cps; pH 7.0–7.5 (10%); 30% conc.

Lakeway 201-10, 201-11. [95] Sodium lauryl polyether sulfate; anionic; high actives, low cloud pt. detergent for shampoos, personal care products; straw liq.; dens. 8.8 and 8.5 lb/gal resp.; visc. 285 and 100 cps resp.; cloud pt. 13 and 0 C resp.; pH 7.5 and 8.0 resp. (10%); 58 and 29% conc.

Lakeway 201-20, 201-29. [95] Ammonium lauryl polyether sulfate; anionic; detergent for dishwash, laundry, shampoos, personal care products; straw liq.; dens. 8.4 lb/gal; visc. 75 and 60 cps resp.; cloud pt. 0 C; pH 7.2 and 6.5 resp. (10%); 58 and 29% conc.

Lakeway 301-10. [95] Sodium alpha olefin sulfonate; anionic; high sudsing detergent for hair shampoos, bubble bath, personal care products, dishwash, rug shampoos, industrial processing products; lt. yel. liq.; dens. 8.9 lb/gal; visc. 75 cps; cloud pt. 2 C; pH 8.5 (10%); 39% conc.

Lakeway 301-10F. [95] Dry sodium alpha olefin sulfonate; anionic; detergent, high actives flake form for dry bubble baths, metal cleaning; wh. to off-wh. flake, powd.; dens. 2.9 lb/gal; pH 7.0 (10%); 90% conc.

Lakeway 302, 303. [95] Formulated conc.; anionic; detergent used in liq. hand dishwash, car wash base, industrial hand soap; straw, pink liq.; dens. 9.3 lb/gal; visc. 710 and 640 cps resp.; cloud pt. < –6 C; pH 6.8 (10%); 60% conc.

Lakeway 401-M. [95] Formulated conc.; anionic; detergent base conc. for hair shampoo, bubble bath; straw liq.; dens. 8.6 lb/gal; visc. 5000 cps; cloud pt. 6 C; pH 9.4 (10%); 39% conc.

Lakeway 402. [95] Formulated anionic blend; used in personal care products, hair shampoo, bubble bath, rug shampoo; wh. to yel. liq.; dens. 8.7 lb/gal; visc. 100 cps; cloud pt. 10 C; pH 8.0 (10%); 40% conc.

Lakeway 500 Series. [95] Formulated conc.; anionic; foaming agent, various types available for gypsum board production; liq.; dens. 8.5 lb/gal; pH 7.0 (10%); 28–60% conc.

Lakeway I Series. [95] Formulated conc.; anionic; upholstery and rug shampoo conc.; wh. to lt. yel. liq.; dens. 8.5–8.6 lb/gal; visc. 65–240 cps; cloud pt. 2–18 C; pH 7.2–7.8 (10%); 21–31% conc.

Lakeway SA. [95] Linear dodecylbenzene sulfonic acid; anionic; low cost sudsing detergent base for washes and cleaners; dk. visc. liq. fluid; dens. 8.7 lb/gal; visc. 1440 cps; 98% conc.

Lakeway SXS. [95] Sodium xylene sulfonate; anionic; hydrotrope for solubilizing; lt. yel., low visc. liq.; dens. 10 cps; cloud pt. < –5 C; pH 8.5; 41% conc.

Lamacit AP 6. [277] Fatty acid and POE derivs.; anionic; emulsifier for paraffins used in particle and fiber boards; paper mfg.; solid; 100% conc.

Lamacit ER. [277] PEG-20 glyceryl ricinoleate and ricinoleamide DEA; nonionic; solubilizer and emulsifier for cosmetics; liq.; 100% conc.

Lamacit GML 12. [277] PEG-12 glyceryl laurate; nonionic; see Lamacit ER; liq.; 100% conc.

Lamacit GMO 25. [277] PEG-25 glyceryl oleate; nonionic; see Lamacit ER; liq.; 100% conc.

Lamecerin 50-80. [277] POE glycol lanolin derivate; nonionic; refatting agent for personal care products; liq.; water-sol.; 50% conc.

Lamecreme SA 7. [277] POE stearoylether; nonionic; emulsifier for cosmetics; solid; 100% conc.

Lamefix 680. [277] Fatty acid polyglycol ether and PEG; nonionic; accelerator for HT-fixation of polyester and triacetate; liq.; 98% conc.

Lameform C. [277] Lanolin derivs., polyglycerol esters, and monoglycerides; w/o emulsifier for cosmetics; paste.

Lameform GE 2. [277] Polyglyceryl-3 diisostearate, isostearamide DEA; nonionic; w/o emulsifier for cosmetics; liq.; 90% conc.

Lamegin CSL. [277] Calcium stearoyl lactylate; anionic; lipophilic emulsifier and antistaling agent for food industry; powd.; 100% conc.

Lamegin DWF, DWH, DWP. [277] Monodiglyceride diacetyl tartaric acid ester; anionic; hydrophilic emulsifier, bread improver; liq., flakes, paste respectively; 100% conc.

Lamegin EE. [277] Acetylated hydrog. tallow glyceride; nonionic; emulsifier and plasticizer for cosmetic, food, and edible coatings; solid; 100% conc.

Lamegin GLP 10, 20. [277] Hydrog. tallow glyceride lactate; nonionic; emulsifier and plasticizer for cosmetics, foods, and drugs; solid; 100% conc.

Lamegin NSL. [277] Sodium stearoyl lactylate; anionic; emulsifier for cosmetics, foods, and drugs; powd.; 100% conc.

Lamegin ZE 30, 60. [277] Hydrog. tallow glyceride citrate; anionic; emulsifier for cosmetic, margarine and meat industry; solid; 100% conc.

Lamepon 287 SF. [277] Sulfonic acid; anionic; dispersant, wetting agent, protective colloid for dyestuffs; liq.; 40% conc.

Lamepon A. [277] Protein fatty acid; anionic; dispersant, protective colloid for dyes; stabilizer for peroxide bleaching; liq.; 45% conc.

Lamepon N. [277] Sulfonic acid; anionic; dispersant and protective colloid for dyes; liq.; 40% conc.

Lamepon PA-TR. [277] TEA abietoyl hydrolyzed animal protein; anionic; detergent for shampoos; liq.; 32% conc.

Lamepon RE. [277] Sulfonic acid; anionic; dispersant and leveling agent for dyes; liq.; 40% conc.

Lamepon S, S 2. [277] Potassium coco-hydrolyzed animal protein; anionic; detergent for personal care products; liq.; 32% conc.

Lamepon ST40, S-TR. [277] TEA-coco-hydrolyzed animal protein; anionic; see Lamepon S; liq.; 40 and 32% conc. resp.

Lamepon UD. [277] Potassium undecylenoyl hydrolyzed animal protein; anionic; detergent for hair preparations and shampoos; liq.; 32% conc.

Lamesoft LMG. [277] Glycerol monolaurate, dispersed; anionic; refatting agent and thickener for foam bath and shampoos; liq.; 25% conc.

Lamigen ES-30, -60, -100. [172] PEG lanolin fatty acid ester; nonionic; softener and antistat for textile; oiling agent for leather; emulsifier for emulsion polymerization; additive to lubricating oil and water sol. paint; paste, solid; HLB 11.0, 13.0, 14.5 resp.; 100% conc.

Lamigen ES-180, ET-20, -70, -90, -180. [172] PEG lanolin fatty alcohol ether; nonionic; see Lamigen ES-30; solid, paste; HLB 16.0, 12,0, 14.0, 15.0, 16.0 resp.; 100% conc.

LAN-401. [437] Min. oil and surfactant; nonionic; coning oil for syn. fibers; liq.; 100% conc.

Lanacet 1705. [217] Acetylated lanolin; nonionic; used in personal care products; solid; 100% conc.

Lanamine. [37] Mixed isopropanolamines myristate; anionic; gel; 50% conc.

Lanapol CT. [551] Lanolin deriv.; anionic; cosmetic assistant; 50% conc.

Lan-Aqua-Sol. [232] Ethoxylated lanolins; anionic; emulsifier for cosmetic and pharmaceutical emulsions, adds emollience to skin and hair care products, household detergent applications; yel. gel, solid; trace odor; sol. in water, alcohol; sp.gr. 1.00–1.03; m.p. 45–51 C; 50 and 100% act.

Lan-Aqua-Sol Hydrophilic 50, 100. [232] PEG-40 lanolin; nonionic; primary o/w emulsifier; emollient, wetting agent, dispersing aid, conditioner and superfatting agent; used in personal care products, cleaners, dishwashing detergents; Gardner 10–11 max. resp.; faint odor; water-sol.; sp.gr. 1.00–1.03 and 1.02–1.06 resp.; m.p. 45–51 C; sapon. no. 5–25; pH 4.0–6.0 and 4.5–6.5 resp.; 50 and 100% act. resp.

Lan-Aqua-Sol Hydrophilic-Plus 50, 100. [232] PEG-50 lanolin; nonionic; see Lan-Aqua-Sol Hydrophilic 50; Gardner 10 and 11 max. resp.; faint odor; sol. in water; sp.gr. 1.00–1.04 and 1.02–1.06 resp. (50/4 C); m.p. 45–51 C; sapon no. 7–10 and 14–20 resp.;

pH 5.0–8.0 and 4.0–7.0 resp.; 50 and 100% act. resp.

Lan-Aqua-Sol Super-Hydrophilic 50, 100. [232] PEG-85 lanolin; nonionic; see Lan-Aqua-Sol Hydrophilic 50; Gardner 9 and 11 max. resp.; faint odor; water-sol.; sp.gr. 1.00–1.05 and 1.03–1.07 resp. (50/4 C); m.p. 46–52 C; sapon. no. 5–12 and 14–22 resp.; pH 4.0–6.0; 50 and 100% act. resp.

Lan-Aqua-Sol xtra-Hydrophilic 50, 100. [232] PEG-75 lanolin; nonionic; see Lan-Aqua-Sol Hydrophilic 50; Gardner 10 and 11 max. resp.; faint odor; water-sol.; sp.gr. 1.00–1.05 and 1.03–1.07 resp. (50/4 C); m.p. 46–52 C; sapon. no. 6.5–15 and 14–22 resp.; pH 4.0–6.0; 50 and 100% act. resp.

Lanbritol Wax N21. [527] Cetearyl alcohol, ceteth-12, and oleth-12; nonionic; emulsifier for pharmaceuticals, cosmetics and hair care preparations; wh. waxy solid; m.p. 45–53 C; biodeg.; 100% conc.

Lancare. [188] Anionic/nonionic surfactant; used for a range of formulated shampoos, bubble baths, mild abrasive cleaners; liq.

Landemul DCC Series. [360] EO/PO copolymers; nonionic; demulsifier; liq.

Lanethyl. [159] Alcohol extract of lanolin alcohols; nonionic; plasticizer or film modifier for hair setting resins, w/o emulsifier used in alcohol-based systems, conditioning agent; amber brittle wax; sol. in ethanol; lipophilic; 100% conc.

Laneto 40. [519] PEG-40 lanolin; nonionic; moisturizer, lubricant and solubilizer surfactant for soap and detergent systems; emollient, resin modifier and solubilizer for personal care products; Gardner 12 max.; char. odor; 50 ± 1% water.

Laneto 50. [519] PEG-75 lanolin; see Laneto 40; Gardner 12 max. liq.; char. odor; water-sol.; sapon. no. 1.00–1.00; 50 ± 1% water.

Laneto 60. [519] PEG-60 lanolin; see Laneto 40; m.p. 48–52 C; sapon. no. 10–16.

Laneto 100. [519] PEG-75 lanolin; nonionic; emollient, lubricant, solubilizer, emulsifier, plasticizer for cosmetics and pharmaceuticals; wax; water sol.; 100% conc.

Lanette 14. [289] Myristyl alcohol; nonionic; consistency agent for cosmetic and pharmaceutical o/w and w/o creams, emulsions, sticks; flakes; 100% conc.

Lanette 16. [289] Cetyl alcohol; nonionic; see Lanette 14; flakes; 100% conc.

Lanette 18. [289] Stearyl alcohol; nonionic; see Lanette 14; flakes; 100% conc.

Lanette 18-22. [289] Fatty alcohol C_{18}-C_{22}; nonionic; see Lanette 14; flakes; 100% conc.

Lanette 22. [289] Behenyl alcohol; nonionic; see Lanette 14; fused flakes; 100% conc.

Lanette E. [289] Sodium cetearyl sulfate; anionic; emulsifier and wetting agent for o/w emulsions, personal care products; wh. powd.; 88% act.

Lanette N. [289; 291] Cetyl stearyl alcohol and sodium cetyl stearyl sulfate; anionic; emulsifying raw material for o/w creams and emulsions; granules; 100% conc.

Lanette O. [289; 291] Cetearyl alcohol; nonionic; consistency factor for ointments, creams, emulsions; granules; 100% conc.

Lanette SX. [289; 291] 90% cetyl stearyl alcohol and 10% sodium lauryl sulfate; anionic; emulsifier; basic material for mfg. of o/w ointment, creams and emulsions; granules.

Lanette W. [291] Cetearyl alcohol and sodium lauryl sulfate; anionic; base for mfg. of ointments, creams and liniments; granules; 100% conc.

Lanette Wax CAT. [527] Cetearyl alcohol, cetrimonium bromide and laureth-12; cationic; self-emulsifying wax for use in toiletry, pharmaceutical preparations, creams and lotions; waxy solid; 100% conc.

Lanette Wax SX, SXBP. [527] Cetearyl alcohol and sodium C12–15 alcohols sulfate (SXBP complies to B.P. specifications); anionic; o/w emulsifier for use in toiletry, pharmaceutical preparations, creams, ointments and lotions; cream to pale yel. waxy flakes; faint char. odor; partially sol. in alcohol, almost insol. in water; m.p. 50 C; 100% conc.

Lanex. [159] Alcohol extract of pure lanolin; nonionic; emollient, plasticizer and conditoner in cosmetics, hair sprays and preshave lotions, superfatting agent, film modifier; amber visc. liq.; sol. in ethanol; lipophilic; 100% conc.

Lanexol AWS. [159] PPG-12-PEG-50 lanolin; nonionic; emollient and lubricant for alcoholic and aq. compositions, plasticizer for hair sprays, superfatting agent in solv. systems based on alcohol or acetone, o/w emulsifier; amber liq.; sol. in water, ethanol and mixts.; cloud pt. 65–80 C (1% aq.); pour pt. 13 C max.; sapon. no. 10–20; 97% conc.

Lanfrax. [217] Lanolin wax; nonionic; cosmetic emulsion stabilizer, w/o emulsifier and waxing agent, o/w aux. emulsifier, slip reducing agent; wax; 100% conc.

Lanfrax 1776. [217] USP lanolin, wax fraction; nonionic; emulsifier, emollient; waxing agent, o/w aux. emulsifier; slip reducing agent in floor finishing compds.; w/o emulsion stabilizer and thickener; wax; 100% conc.

Lanfrax WS 55. [217] PEG-75 lanolin wax; nonionic; cosmetic o/w emulsifier in hair tonics, hair sprays; wax; 100% conc.

Lankrolan SHR-3. [188] Org. thiosulfate; anionic; shrink resist agent for wool; liq.; 42% conc.

Lankro Mud-Aids. [188] Series of general workability aids for drilling muds incl. mud surfactants and defoamers.

Lankro Mud Detergents. [188] Anionic or nonionic detergent for well drilling mud applications.

Lankro Mud-Emuls. [188] Series of w/o and o/w emulsifiers for water-based and invert muds.

Lankromul OSD. [360] Emulsifier in aliphatic hydrocarbon solv.; emulsifier and dispersant for oil spills; straw clear liq.; mild, fatty odor; disp. in water; sp.gr. 0.808 (10 C); visc. 3.5 cs; flash pt. 88 C (COC); pour pt. < –20 C.

Lankropearl T. [188] Blend of pearlizing agents in an anionic base; anionic/nonionic; surfactant, pearlizing agent for toiletries; liq.; 40% conc.

Lankropol ADF. [360] Modified sulfosuccinamate; anionic; foaming agent used in foamed aq. polymer systems, textile, leather, upholstry industries; lt. br. clear liq.; char. odor; sol. in water; sp.gr. 1.078; visc. 11 cs; flash pt. > 200 F (COC); pour pt. –5 C; pH 6–8 (1% aq.); 35% act.

Lankropol ATE. [360] Tetrasodium N-(1,2-dicarboxyethyl)-N-octadecyl sulfosuccinamate; primary emulsifier and mechanical stabilizer for emulsion polymers, aux. foaming agent, solubilizing agent; amber slightly hazy liq.; ethanolic odor; sp.gr. 1.119; visc. 36 cs; flash pt. 73 F (Abel CC); pour pt. 3 C; pH 7.0–8.5 (1% aq.); 35% act.

Lankropol KMA. [360] Sodium dihexyl sulfosuccinate; anionic; wetting agent esp. in sol'ns. of electrolytes; solubilizer for soaps, emulsion polymerization aid; pale straw hazy liq.; ethanolic odor; sol. in water; sp.gr. 1.082; visc. 31 cs; flash pt. 91 F (Abel CC); pour pt. < 0 C; pH 6.0–7.5 (1% aq.); 60% act. in ethanol.

Lankropol KNB 22. [360] Monoalkyl sulfosuccinate; foaming agent for personal care products; liq.; 29% conc.

Lankropol KO2. [360] Sodium dioctyl sulfosuccinate; anionic; wetting agent, emulsion polymerization aid; pale straw clear liq.; ethanolic odor; sol. up to 0.5% in water; sp.gr. 1.050; visc. 35 cs; flash pt. 81 F (Abel CC); pour pt. < 0 C; pH 6–7 (1% aq.); 60% act. in ethanol.

Lankropol KO Special. [360] Sodium dioctyl sulfosuccinate; anionic; solv. emulsifier, water carrier in dry cleaning formulations, dewatering aid, emulsifier for min. oil with nonionics; pale straw liq.; mild odor; disp. in water; sp.gr. 1.050; visc. 1500 cs; flash pt. > 200 F (COC); pour pt. < 0 C; pH 6–7 (1% aq.); 60% act. in min. oil.

Lankropol ODS/LS, ODS/PT. [360] Disodium N-octadecyl sulfosuccinamate; anionic; foaming agent for aq. polymer dispersions; lt. amber hazy liq., paste; mild odor; sol. in water; sp.gr. 1.066–1.082; visc. 12 and 5400 cs resp.; flash pt. > 200 F (COC); pour pt. –5 to 4 C; pH 7–9 (1% aq.); 35% act.

Lankropol OPA. [360] Potassium salt of a fatty acid sulfonate; anionic; surfactant used in metal industry and for household use, wetting agent, detergent, dispersant; dk. amber clear liq.; mild fatty odor; sol. in water; sp.gr. 1.125; visc. 270 cs; flash pt. > 200 F (COC); pour pt. < 0 C; pH 5.5–6.0 (1% aq.); biodeg.; 50% act.

Lankropol WA. [360] Ammonium salt of a sulfated monoester of fatty acid; anionic; low foaming wetting agent in textile industry, dispersion aid in paint industry; solv. emulsifier in degreasing formulations and herbicides, detergent; dk. amber clear liq.; char. odor; sol. in water; sp.gr. 1.030; visc. 120 cs; flash pt. > 200 F (COC); pour pt. < 0 C; pH 6–7 (1% aq.); biodeg.; 50% act.

Lankropol WN. [360] Sodium salt of a sulfated monoester of fatty acid; anionic; see Lankropol WA; dk. amber clear liq.; char. odor; sol. in water; sp.gr. 1.050; visc. 75 cs; flash pt. > 200 F (COC); pour pt. < 0 C; pH 6–7 (1% aq.); biodeg.; 50% act.

Lankrosol SXS-30. [360] Sodium xylene sulfonate; anionic; hydrotrope and visc. modifier for high act. liq. detergents, solubilizer for anionic surfactants; pale straw clear liq.; negligible odor; sol. in water; sp.gr. 1.128; visc. 2.7 cs; flash pt. > 200 F (COC); pour pt. < 0 C; pH 7 (1% aq.); 30% act.

Lanogel 21, 31, 41, 61. [37; 522] PEG-27, -40, -75, -85 lanolin; nonionic; emollient, emulsifier, dispersant, wetting agent, solubilizer, foam stabilizer, used in cosmetics, personal care products, pharmaceuticals, and facial tissues; ASTM 3 max. gel; disp. to sol. in water; sapon. no. 4–24; 50% act.

Lanogene. [37] Lanolin oil; emollient, moisturizer, and emulsifier which imparts oil sol. and spreading properties; yel.-amber liq.; slight char. odor; sol. in oils, esters, hydrocarbons, and IPA; iodine no. 18–36; sapon. no. 85–105.

Lanol C. [551] Cetyl alcohol; nonionic; cosmetic assistant; wax; 100% conc.

Lanol CS. [551] Cetearyl alcohol; nonionic; see Lanol C; wax; 100% conc.

Lanol CT, ST. [551] Fatty alcohol, emulsifiable; nonionic; see Lanol C; wax; 100% conc.

Lanol S. [551] Stearyl alcohol; nonionic; see Lanol C; wax; 100% conc.

Lanolin Alcohols LG. [630] Nonionic; emulsifier, stabilizer, emollient, used in textile, wood and paper industries; tan, soft, bleached wax; sol. in hydrocarbons, chlorinated hydrocarbons, veg. and min. oils, insol. in water; sapon. no. 20–40; 100% conc.

Lanolin Alcohols LO. [630] Nonionic; see Lanolin Alcohols LG; brown soft solid; sol. in hydrocarbons, chlorinated hydrocarbons, veg. and min. oils, insol. in water; sapon. no. 20–40; 100% conc.

Lanolin Alcohols THG. [630] Nonionic; emulsifier, stabilizer, emollient, used in cosmetic and tech. applications; tan bleached solid; sol. in hydrocarbons, chlorinated hydrocarbons, veg. and min. oils, insol. in water; sapon. no. 20 max.; 100% conc.

Lanolin Alcohols THO. [630] Nonionic; see Lanolin Alcohols THG; sol. in hydrocarbons, chlorinated hydrocarbons, veg. and min. oils, insol. in water; sapon. no. 20 max.; 100% conc.

Lanolin Fatty Acids G, O. [207; 630] Anionic; emulsifier, used in lubricating greases, polishing compositions, anticorrosion compds.; wax, brown and bleached resp.; sol. in min. oils and many org. solv.; sapon. no. 140–165; 100% conc.

Lanoquat 1756. [217] Quaternium-33, ethyl hexanediol; cationic; conditioner; emulsifier for skin moisturizers; liq.; 50% conc.

Lanotein AWS 30. [232] PEG-75 lanolin oil and hydrolyzed animal protein; conditioner, film former, and emollient used in hair products; lt. amber liq.; bland odor; water-sol.; sp.gr. 1.05–1.08; pH 5.0–6.0.

Lanoxal 75. [551] Lanolin, ethoxylated; nonionic; cosmetic assistant; wax; 100% conc.

Lanoxyl 30. [654] Lanolin, ethoxylated; nonionic; emulsifier; paste; 100% conc.

Lanpol 5. [159] POE (5) lanolin acids, dist.; nonionic; o/w emulsifier, solv. for bromo acid dyes in lipsticks, skin care and makeup products, solubilizer, wetting agent; pale yel. pasty solid; HLB 7.5; sapon. no. 105–120; pH 4–7 (3%); 97% conc.

Lanpol 10. [159] POE (10) lanolin acids, dist.; nonionic; see Lanpol 5; pale yel. pasty solid; HLB 10.9; sapon. no. 60–80; pH 4–7 (3%); 97% conc.

Lanpol 20. [159] POE (20) lanolin acids, dist.; nonionic; see Lanpol 5; solid; 97% conc.

Lanpol 520 [159] POE lanolin acid; nonionic; see Lanpol 5; pale yel. pasty solid; HLB 14.1; sapon. no. 45–60; pH 4–7 (3%); 97% conc.

Lanpolamide 5. [160] PEG-5 lanolinamide; nonionic; w/o emulsifier for hydrocarbon sprays, insecticides, room deodorants; solid, liq.; HLB 3.6; 50–100% conc.

Lanquell 206, 217. [360] Polyglycol-based, low-toxicity antifoam for food industry, used in mfg. of paper for food pkg.; pale yel. liq.; faint odor; disp. in water; sp.gr. 0.984 and 1.005 resp.; visc. 180 cs and 355 cs resp.; flash pt. > 150 and > 200 F resp. (Abel CC); pour pt. –28 and –13 C resp.; pH 7 (1% aq.); 92 and 98% act.

Lantrol. [217] Lanolin oil; nonionic; skin and hair emollient which emulsifies with water, used in cosmetics, personal care products; visc. liq.; HLB 9.0; 100% conc.

Lantrol 1692 AWS. [217] Alkoxylated lanolin; nonionic; aux. o/w emulsifier; liq.; sol. in water and alcohol; HLB 18.0; 100% conc.

Lantrol AWS. [217] Ethoxylated lanolin; nonionic; aux. o/w emulsifier for personal care products, emollient; visc. liq.; sol. in water and alcohol; HLB 18.0; 100% conc.

Larosol NRL-40. [400] Sodium alkyl diphenyl oxide disulfonate; anionic; leveling, transfer and streak coverage; nylon carpet dyeing; emulsifier for dye carriers; liq.; 45% conc.

Larostat 264-A, 264-A Anhyd. [400] Soyadimethylethyl ammonium ethosulfate; cationic; applications in the textile industry; liq., gel, wax; 35–100% conc.

Lathanol LAL Flake, Powd. [575] Sodium lauryl sulfoacetate; anionic; emulsifier, wetting agent, detergent, foaming agent, used in cosmetics and personal care products; wh. flake, powd.; faint lauryl alcohol odor; slightly acrid taste; water sol. 3.5 g/100 ml; dens. 0.38 g/ml (untamped) and 0.85 g/ml (tamped) resp.; surf. tens. 30.1 dynes/cm (0.1%) pH 5.0–7.5 (5% conc.); 70% act.

Lauramide 11, D, ME. [659] Coconut diethanolamide; nonionic; foam booster, thickener, and superfatting agent; liq. (11, D); liq. to paste (ME); 80% conc. (11), 90% conc. (ME).

Lauramide R. [659] Fatty acid diethanolamide; nonionic; see Lauramide 11; liq.; 85% conc.

Lauramide S. [659] Unsat. fatty polyalkanolamide; nonionic; additive in metal cleaning compounds; liq.

Laurex 16/18, 16/18D. [16] Primary fatty alcohol derived from naturally occurring oils and fats; (16/18D—dist. grade); raw material for mfg. of sulfated derivs. and additives for lubricating oils; wh. solid; m.p. 50–54 C; sapon. no. 1.5 and 0.8 resp.

Laurex 810. [17] Octyl-decyl alcohol fraction; raw material in the mfg. of alkyl phthalates, phosphoro diethionates, alkyl methacrylate monomers; defoaming agent in drilling muds and fermentation broths; colorless clear liq.; dens. 0.83 g/cm^3; flash pt. 92 C; sapon. no. 1.0 max.

Laurex CH. [16] Coconut alcohol; superfatting agent in shampoos, raw material for sulfation, ethoxylation; wh. soft solid; m.p. 18–23 C; sapon. no.1.5.

Laurex CS. [16] Cetearyl alcohol; see Laurex CH; wh. flake or solid; m.p. 50–54 C; sapon. no. 1.5.

Laurex CS/D. [17] Cetyl stearyl alcohol; used in mfg. of pharmaceuticals and personal care products; wh. waxy flake; dens. 0.4 g/cm^3; m.p. 48–53 C; flash pt. 150 C; sapon. no. 2.0 max.

Laurex L1, NC. [16] Lauryl alcohol; see Laurex CH; wh. soft solid; m.p. 20–23 C; sapon. no. 0.5.

Laurex PKH. [16] Palm kernel alcohol; see Laurex CH; wh. soft solid; m.p. 18–23 C; sapon. no. 1.5.

Lauridit KD, KDG. [11] Coco fatty acid diethanolamide; nonionic; detergent, emulsifier for cosmetic and household applications, visc. increasing additive; yel. liq.; water sol. 40 and 800 g/l resp.; sp.gr. 1.02; flash pt. 104 and 114 C resp. (PMCC); pH 9.0–9.5; 90 and 85% act.

Lauridit KM. [11] Coco fatty acid monoethanoloamide; nonionic; detergent for textile, household and cosmetic applications, foam stabilizer for detergents, shampoos, bubble baths; yel. flakes; poor water sol.; sp.gr. 0.97; m.p. 65–70 C; sapon. no. 14 max.; 93% act.

Lauridit LM. [11] Lauric acid monoethanolamide; nonionic; see Lauridit KM; ylsh. flakes; poor water sol.; sp.gr. 1.01 (80 C); m.p. 80–84 C; sapon. no. 20 max; pH 9; 93% act.

Lauridit OD. [11] Oleic acid diethanolamide; nonionic; detergent, emulsifier for cosmetic and household applications, visc. increasing additive; yel. liq.; water sol. 100 g/l; sp.gr. 0.97; f.p. 0 C; flash pt. 100 C (PMCC); pH 9; 85% act.

Lauridit PD. [11] Fatty acid polydiethanolamide/fatty acid diethanolamide 1:2; household detergent, corrosion inhibitor for cutting oils, additive for acid cleaners, emulsifier; br. liq.; water sol.; sp.gr. 1.02; flash pt. 105 C (PMCC); 65% act.

Lauridit PPD. [11] Palm kernel fatty acid polydiethanolamide; nonionic; detergent for personal care products, dishwashing agent, dissolving intermediary for perfumes and liq. hand cleaners; liq.; 100% conc.

Lauropal 2. [654] Fatty alcohol, ethoxylated; nonionic; base for liq. detergents and shampoos; liq.

Lauropal 4. [654] Fatty alcohol, ethoxylated; nonionic; emulsifier for min. oils and cosmetics; liq.

Lauropal 9. [654] Fatty alcohol, ethoxylated; nonionic; wetting agent, emulsifier for textiles; wax.

Lauropal 11. [654] Fatty alcohol, ethoxylated; nonionic; wetting agent and emulsifier for textiles, metal cleaning; wax.

Lauropal 0205. [654] Fatty alcohol, ethoxylated; nonionic; detergent for textiles; liq.

Lauropal 0227. [654] Fatty alcohol, ethoxylated; nonionic; wetting agent and detergent for industrial use; liq.

Lauropal 950. [654] Fatty alcohol, ethoxylated; nonionic; see Lauropal 9; gel.

Lauropal 1150. [654] Fatty alcohol, ethoxylated; nonionic; see Lauropal 11; liq.

Lauroxal 3. [654] Fatty alcohol, ethoxylated; nonionic; wetting agent and detergent for industrial use; liq.

Lauroxal 6. [654] Fatty alcohol, ethoxylated; nonionic; wetting agent and detergent for textiles; liq.

Lauroxal 8. [654] Fatty alcohol, ethoxylated; nonionic; wetting agent and detergent for metal cleaning and leather; wax.

Laventin CW. [82] Alkanol polyglycol ether; nonionic; water-free wetting agent and detergent used in textile industry; colorless to slightly yel. clear, low visc. liq.; water sol.; dens. 0.99 g/ml; pH 6.5–7.5 (1%); biodeg.

Laventin W. [82] Oxyethylated fatty acid deriv.; nonionic; water-free detergent used in textile industry; yel.-br., visc. liq. or semisolid paste; solidifies at low temps.; water sol.

Lebon 101H, 105. [539] Alkylimidazoline type surfactant; shampoo base; lt.-duty detergents; liq.; 35 and 36% conc. resp.

Lebon 2000. [539] Cocoamidopropyl betaine; amphoteric; see Lebon 101H; liq.; 30% conc.

Lecidan. [273] Blends of monoglycerides and special soya lecithin; nonionic; food emulsifier; pills; HLB 5; 100% act.

Lecithin W.D. [611] Lecithin product; wetting agent, dispersant for pigments in water-based paints; 90% conc.

Lenetol 527. [317] EO condensate and detergent; anionic/nonionic; detergent and wetting agent for textiles; liq.; 29% conc.

Lenetol 9130 AH. [317] C_9-C_{11} alcohol (3 EO); nonionic; emulsifier, solubilizer, stabilizer for emulsion polymerization; paste; 100% conc.

Lenetol ASC. [317] Phosphate ester, modified; anionic; detergent used in textiles; liq.; 70% conc.

Lenetol B Conc. [317] Linear fatty acid, ethoxylated; nonionic; detergent/emulsifier; liq.; 84% conc.

Lenetol NS. [317] Alkyl phenol EO condensate; nonionic; wetting agent and detergent for dyes; liq.; 28% conc.

Lenetol WLF. [317] Phosphate ester, modified; anionic/nonionic; wetting agent and detergent for textiles; liq.; 55% conc.

Leomin OR. [301] Fatty acid polyglycol ester; nonionic; dispersant and preparation agent for textile applications; liq.

Leomin TR. [301] Fatty amine ethoxylate; cationic; cleaning agent for HT apparatus; liq.

Leonil DB Powder. [301] Dialkyl naphthalene sulfonate; anionic; wetting agent and dyeing auxiliaries for textiles; powd.

Leophen BN. [82] Alcohol sulfonate; foam-free wetting agent in mercerizing textiles; yel.-br. clear low-visc. liq.; misc. with water and dilute caustic soda; dens. 1.09 g/ml; biodeg.

Leophen LG. [82] Mixt. of sodium salts of aliphatic sulfonic acids; wetting agent used in caustic baths; yel. almost clear slightly visc. liq.

Leophen M. [82] Neutral phosphoric acid ester with nonionic emulsifiers; wetting agent and detergent in textile industry, antifoam; colorless to weak yel. liq.; misc. in water; dens. 1.0 g/ml.

Leophen ML. [82] Specific alkylsulfate; wetting agent for yarns, fabrics, knitgoods with min. foam formation; yel.-br. liq.; dens. 1 g/ml; pour pt. –3 C.

Leophen RA. [82] Sodium salt of a sulfosuccinate; anionic; wetting/rewetting agent for fibers and textiles; yel. liq.; water sol.; pH 5.5–7.0 (1% aq.).

Leophen RBD. [82] Dialkyl sulfimide; wetting and rewetting agent used in bleaching and dyeing fibers; golden yel. clear visc. liq.; water sol.; dens. 1.1 g/ml.

Leophen U. [82] Nonionic/anionic surfactant blend; wetting agent, detergent and dispersant used in desizing, treating with alkali and bleaching; yel. paste; water sol.; pH 7.5–8.0 (10 g/l).

Letocil LP 80, SP 1000. [131] Fatty acid polyglycol ester; nonionic; emulsifier for cosmetic industry; liq., solid resp.; 100% conc.

Levelan AO192. [360] Fatty alcohol ethoxylate; nonionic; leveling agent for direct dyestuffs; clear liq.; faint odor; water sol.; sp.gr. 1.018; visc. 13 cs; cloud pt. > 100 C (1% aq.); flash pt. > 200 F (COC); pour pt. < 0 C; pH 6–8 (1% aq.); 20% act.

Levelan NKS. [393] Amino carboxylic acid; amphoteric; leveling agent for textiles; paste

Levelan P148. [360] Nonyl phenol ethoxylate; nonionic; primary emulsifier in emulsion polymer industry, stabilizer in latex stabilization; clear liq.; faint odor; water sol.; sp.gr. 1.069; visc. 490 cs; cloud pt. > 100 C (1% aq.); flash pt. > 200 F (COC); pour pt. < 0 C; pH 6–8 (1% aq.); 80% act.

Levelan P208. [360] Nonyl phenol ethoxylate; nonionic; wetting agent in textile industry, latex stabilizer and emulsifier in emulsion polymer industry; clear liq.; faint odor; water sol.; sp.gr. 1.086; visc. 460 cs; cloud pt. > 100 C (1% aq.); flash pt. > 200 F (COC); pour pt. < 0 C; pH 6–8 (1% aq.); 80% act.

Levelan P307. [360] Alkylphenol ethoxylate; wetting agent for high temp. applications; liq.; 70% act.

Levelan P357. [360] Nonyl phenol EO condensate; nonionic; see Levelan P208; clear liq.; faint odor; water sol.; sp.gr. 1.090; visc. 960 cs; cloud pt. > 100 C (1% aq.); flash pt. > 200 F (COC); pour pt. 2 C; pH 6–8 (1% aq.); 70% act.

Levelan R-15, -200. [393] Quat. ammonium compd.; cationic; leveling agent for dyestuffs; liq.

Levelan WS. [393] Fatty amine condensate, modified; amphoteric; see Levelan R-15; liq.; 50% conc.

Levelene. [137] POE alkyl ether, modified; nonionic; rewetting and leveling agent, penetrant and stripper used in texile applications; liq.

Levenol A Conc. [341] POE alkyl amine; nonionic; retarding and stripping agent for textiles; paste; 100% conc.

Levenol DS-1. [341] POE alkyl amine; nonionic; stripping agent for textiles; liq.; 100% conc.

Levenol PW. [341] POE alkyl ether; nonionic; leveling agent; flake; 100% conc.

Levenol RK. [341] Quat. ammonium halide; cationic; retarding agent for dyeing of acrylics; liq.; 41% conc.

Levenol TD-326. [341] Quat. ammonium halide; nonionic; leveling and dispersant for polyester dyeing; liq.

Levenol WX, WZ. [341] POE alkyl or alkylaryl ether sulfate; nonionic; leveling agent and dye coagulation preventing agent for dyeing applications; liq.; 25% conc.

Lexaine C. [325] Cocamidopropyl betaine; amphoteric; visc. builder, foam booster, thickener in conditioners, specialty shampoos, personal care products; Gardner 4 max. liq.; sp.gr. 1.044; f.p. 0 C; pH 4.5–5.5; 30% act.

Lexaine CG-30. [325] Cocamidopropyl betaine; amphoteric surfactant used in light duty liq. detergent systems, personal care products; straw to lt. amber clear nonvisc. liq.; bland odor; pH 6.5–8.0; 29% act.

Lexaine CG-50. [325] Cocamidopropyl betaine, propylene glycol and glycerin; amphoteric; see Lexaine CG-30; straw to lt. amber clear nonvisc. liq.; bland odor; pH 7.5 (10% aq.); 50% act.

Lexaine CS. [325] Cocamidopropyl betaine; amphoteric; surfactant, foam booster, and visc. builder used in personal care products; liq.; 35% conc.

Lexaine CSB-35, -50. [325] Cocamidopropyl hydroxysultaine; amphoteric; surfactant, foaming and wetting agent used in personal care products; liq.; 35 and 48% conc. resp.

Lexaine IBC-70. [325] Cocamidopropyl betaine, DEA-lauryl sulfate, propylene glycol, and glycerin; amphoteric/anionic; shampoo conc.; amber, slightly hazy liq.; dens. 8.8 lb/gal; visc. 1850 cps; f.p. –4 C; pH 6–7 (10% aq.); 70% act.

Lexaine IBC-70S. [325] Cocamidopropyl betaine, DEA lauryl sulfate, propylene glycol, glycerin; surfactant used in personal care products; liq.; 65% betaine.

Lexaine O. [325] Oleamidopropyl betaine and propylene glycol; amphoteric; surfactant used in personal care products; liq.; 25% conc.

Lexamine 22. [325] Stearamidoethyl diethylamine; cationic; emulsifier used in hair care products; solid; 100% conc.

Lexamine B-13. [325] Behenamidopropyl dimethylamine; cationic; see Lexamine 22; flakes; 100% conc.

Lexamine C-13. [325] Cocamidopropyl dimethylamine; cationic; see Lexamine 22; soft solid; 100% conc.

Lexamine L-13. [325] Lauramidopropyl dimethylamine; cationic; see Lexamine 22; solid; 100% conc.

Lexamine O-13. [325] Oleamidopropyl dimethylamine; cationic; see Lexamine 22; liq.; 100% conc.

Lexamine P-13. [325] Palmitamidopropyl dimethylamine; cationic; see Lexamine 22; solid; 100% conc.

Lexamine R-13. [325] Ricinoleamidopropyl dimethylamine and glycerin; cationic; see Lexamine 22; liq.; 100% conc.

Lexamine S-13. [325] Stearamidopropyl dimethylamine; cationic; see Lexamine 22; flakes; 100% conc.

Lexamine S-13 Lactate. [325] Stearamidopropyl dimethylamine lactate; conditioner for hair care products; liq.; 23% conc.

Lexate CL-60. [325] Propylene glycol, oleamidopropyl dimethylamine, cocamidopropyl betaine, and oleamide DEA; hair conditioner conc.; Gardner 7 liq.; mild, fatty odor; sp.gr. 1.01; visc. 150 cps; cloud pt. < 5 C; 75% act.

Lexate CRC. [325] Stearamidopropyl dimethylamine, glycol stearate, and ceteth-2; cationic; cream rinse conc. and conditioner, emulsifier; cream to tan, waxy flake; mild, fatty amino odor; easily melted and dispersed in water; m.p. 49–53 C; 100% act.

Lexate PX. [325] Petrolatum, lanolin, and ozokerite; o/w and aux. emulsifier, lanolin cream base, conditioning additive in soaps and shaving cream, emollient; faintly yel. semisolid; m.p. 41 C; sapon. no. 1.0.

Lexate TA. [325] Glyceryl stearate, isopropyl myristate, and stearyl stearate; aux. lipophilic emulsifier in w/o emulsions, emollient in cosmetic creams and lotions; wh. waxy solid; odorless to bland, fatty odor; m.p. 45–50 C; sapon. no. 159–169.

Lexate TL. [325] Glyceryl stearate, butyl stearate, and stearyl stearate; superfatting agent in milled bar soap, aux. lipophilic emulsifier in w/o systems, emollient; wh. waxy solid; mild, fatty odor; m.p. 50–55 C; sapon. no. 145–152.

Lexein S620. [325] Potassium coco-hydrolyzed animal protein; anionic; used in personal care products; liq.; 30% conc.

Lexein S620S. [325] TEA-coco-hydrolyzed animal protein and sorbitol; anionic; cleansing agent and a humectant used in personal care products; water sol'n.; HLB 60.0

Lexein S620TA. [325] TEA-coco-hydrolyzed animal protein; anionic; visc. builder and foam modifier; water sol'n.; HLB 40.0

Lexemul 55G. [325] Glyceryl stearate; nonionic surfactant used in cosmetics and topical pharmaceuticals; wh. flakes; mild fatty odor; m.p. 55–59 C; sapon. no. 160–170; 100% conc.

Lexemul 400 MS. [325] PEG 400 monostearate; nonionic; emulsifier; solid; 100% conc.

Lexemul 503, 515. [325] Glyceryl stearate; emulsifier, stabilizer, thickener, opacifier in emulsions or surfactant systems; flakes; HLB 3.2 and 3.8 resp.; 100% conc.

Lexemul 530. [325] Glyceryl stearate SE; anionic modified; primary emulsifier in o/w systems; wh. to cream flakes; low odor; sol. @ 60 C in carbon tetrachloride, ethyl acetate, IPA, min. oil; disp. in water; HLB 5.0; m.p. 56 C; sapon. no. 146–154; pH 8.5 (3% aq. disp.); 100% conc.

Lexemul 561. [325] Glyceryl stearate and PEG 100 stearate; nonionic; cosmetic grade emulsifier for o/w cream or lotion systems; wh., brittle flakes; mild ethoxylate odor; disp. in water (60 C); HLB 10.0; sapon. no. 90–100; pH 4.3 (3% aq. disp.); 100% conc.

Lexemul AR. [325] Glyceryl stearate and stearamidoethyl diethylamine; cationic; emulsifier, stabilizer, opacifier, and emollient for cationic systems in personal care products; wh. to cream flakes; mild fatty odor; disp. in water (60 C); HLB 4.1; m.p. 60 C; sapon. no. 166–174; 100% conc.

Lexemul AS. [325] Glyceryl stearate, sodium lauryl sulfate; anionic; self-emulsifying emulsifier, stabilizer, opacifier and emollient in personal care products; wh to cream flakes; HLB 4.9; m.p. 60 C; sapon. no. 153–162; 100% conc.

Lexemul CS-20. [325] Cetearyl alcohol, ceteareth-20; nonionic; emulsifier used in personal care products; flakes; HLB 5.4; 100% conc.

Lexemul EGDS. [325] Glycol distearate; opacifier and pearling agent for cosmetic surfactant systems; flakes; HLB 1.5; 100% conc.

Lexemul EGMS. [325] Glycol stearate; opacifier and pearling agent for personal care products; flakes; HLB 2.3; 100% conc.

Lexemul LPM. [325] Hydrog. palm glycerides, cetyl alcohol, TEA-isostearoyl hydrolyzed animal protein; cosmetic emulsifier; pellet; HLB 14.9; 100% conc.

Lexemul P. [325] Propylene glycol stearate SE; for low visc. emulsions for personal care products; flakes; HLB 4.0; 100% conc.

Lexemul T. [325] Glyceryl stearate SE; for use as emulsifier, opacifier, stabilizer, and emollient in alkaline anionic systems; flakes; HLB 5.5; 100% conc.

Lexin K. [45] Lecithin; nonionic; emulsifier; semi-solid.

Lexomul PEG 200 DL. [325] Glycol ester; nonionic; emulsifier and lubricant for metalworking fluids, paints, polishes and misc. industrial formulations; liq.; HLB 5.9.

Lexomul PEG 400 DL. [325] Glycol ester; nonionic; emulsifier and lubricant in textile softeners and lubricants, metalworking oils, misc. industrial formulations; liq.; HLB 9.8.

Lexomul PEG 400 DO. [325] Glycol ester; nonionic; emulsifier for metalworking fluids, wetting and release agent in molding operations; liq.; HLB 8.5.

Lexomul PEG 400 ML. [325] Glycol ester; nonionic; emulsifier and wetting agent in metalworking fluids, cleaners, detergent polishes and misc. industrial formulations; liq.; HLB 13.1.

Lexomul PEG 400 MO. [325] Glycol ester; nonionic; emulsifier and wetting agent in metalworking fluids, cleaners, detergent polishes and misc. industrial formulations; liq.; HLB 11.4.

Lexoteric 2. [325] Cocoamphodiacetate; amphoteric; foam boosting surfactant used in hair care products; liq.; 49% conc.

Lignosite. [252] Calcium lignosulfonate; anionic; dispersant, emulsifier and emulsion stabilizer; 50% liq., 100% powd.

Lignosite 17. [252] Ammonium lignosulfonate; anionic; 48% liq., 100% powd.

Lignosite 260. [252] Modified lignosulfonate; anionic; emulsifier, dispersant for pigments, insecticides; o/w emulsifier; tan fine powd.; 100% conc.

Lignosite 401. [252] Calcium lignosulfonate; anionic; 40% liq., 100% powd.

Lignosite 431. [252] Sodium lignosulfonate; anionic; dispersant for pigments, wettable powds., wax emulsions, industrial cleaners; 100% conc. br. fine powd., 50% sol'n.; sol. in water; dens. 10.5 lb/gal; pH 4 (10%).

Lignosite 458. [252] Sodium lignosulfonate; anionic; dispersant for pigments, mfg. of concrete admixtures, wax emulsions, wettable powds., industrial cleaners; 100% conc. br. fine powd., 46–48% sol'n.; sol. in hot or cold water; dens. 23 lb/ft^3 (powd.), 10.3 lb/gal (liq.).

Lignosite 704. [252] Calcium lignosulfonate; anionic; 40% liq., 100% powd.

Lignosite 823. [252] Sodium lignosulfonate; anionic; dispersant for dyestuffs and pigments, mfg. of wax emulsions, wettable powds.; sol. in cold or hot water; dens. 23 lb/ft^3; pH 7–8 (104% aq.); 100% powd., 47% liq.

Lignosite 889. [252] Sodium lignosulfonate; anionic; 100% powd., 47% liq.

Lignosite 1840. [252] Calcium lignosulfonate; extender for adhesive and binder systems, dispersant for mfg. of wax emulsions; 100% conc. lt. fine powd., 50% sol'n.; sol. in hot or cold water; dens. 23 lb/ft^3 (powd.), 10.5 lb/gal (liq.); pH 4.5 (10% aq.).

Lignosite AC. [252] Lignosulfonate, modified; anionic; 44% liq., 100% powd.

Lignosite L. [252] Lignosulfonate formulation; liq.; 50% conc.

Lignosite FML. [252] Ferromagnetic iron lignosulfonate; ferromagnetic fluid used in separation processes; liq.; dens. 10.1 lb/gal; visc. 60 cps; 35% solids.

Lignosite GPM-38. [252] Modified sodium lignosulfonate; 100% powd.; 38–40% liq.; sol. in hot or cold water; pH 10 (10% aq.).

Lignosol AXD. [503] Sodium lignosulfonate; anionic; wetting agent for insecticides, herbicides and fungicides, emulsifier, dispersant; br. powd.; water sol.; dens. 28–32 lb/ft^3; pH 4.9 (27%); 95% act.

Lignosol B. [503] Calcium lignosulfonate; anionic; wetting agent, emulsifier, dispersant, binder in construction, insecticides, herbicides, fungicides; black liq.; sp.gr. 1.25; visc. 200 cps (80 F); f.p. –10 C; b.p. 105 C; 50% act.

Lignosol BD. [503] Calcium lignosulfonate; anionic; see Lignosol B; yel. powd.; water sol.; dens. 28–32 lb/ft^3; pH 4.5 (27%); 95% act.

Lignosol D-10, D-30. [503] Sodium lignosulfonate; anionic; wetting agent, dispersant for disperse dyes; br. powd.; water sol.; 95% act.

Lignosol DXD. [503] Sodium lignosulfonate; anionic; wetting agent and dispersant in industrial cleaners, insecticides, herbicides and fungicides, emulsifier, emulsion stabilizer for asphalt emulsions; black powd.; water sol.; dens. 30–32 lb/ft^3; pH 9.0 (27%); 95% act.

Lignosol FTA. [502; 503] Sodium lignosulfonate; anionic; wetting agent, primary dispersant for disperse dye; br. powd.; water sol.; dens. 30–32 lb/ft^3; pH 9.5 (27%); 95% act.

Lignosol HCX. [503] Sodium lignosulfonate; anionic; wetting agent, dispersant, used in insecticides, herbicides, fungicides; br. powd.; water sol.; 95% act.

Lignosol NSX 120. [502] Sodium lignosulfonate, modified; see Lignosol FTA; powd.

Lignosol R30. [503] Calcium sodium lignosulfonate; slime control agent in mining; powd.

Lignosol SF. [503] Calcium lignosulfonate; anionic; dispersant for Portland cement and concrete, retanning agent; br. powd.; water sol.; dens. 28–32 lb/ft^3; pH 6.0 (27%); 95% act.

Lignosol SFL. [503] Calcium lignosulfonate; anionic; dispersant for Portland cement and concrete; black liq.; water sol.; sp.gr. 1.24; f.p. –10 C; b.p. 100 C; 48% act.

Lignosol SFX. [502; 503] Sodium lignosulfonate; anionic; dispersant for concrete admixtures, dyestuffs, sludge conditioner; br. powd.; dens. 28–32 lb/ft^3; pH 6.8 (27%); 95% act.

Lignosol SFX-65. [503] Lignosulfonate; yel. powd.; dens. 28–32 lb/ft^3; pH 6.0 (27%); 95% act.

Lignosol TS. [503] Ammonium lignosulfonate; anionic; wetting agent, emulsifier, slurry water reducer and grinding aid in cement mfg.; black liq.; water sol.; sp.gr. 1.23; f.p. –10C; 47% act.

Lignosol TSD. [503] Ammonium lignosulfonate; anionic; wetting agent, leather retanning agent, emulsifier, dispersant, slurry water reducer and grinding aid in cement mfg.; yel. powd.; water sol.; dens. 28–32 lb/ft^3; pH 4.5 (27%); 95% act.

Lignosol TSF. [503] Lignosulfonate; dispersant for insecticides, fungicides, herbicides; br. powd.; dens. 28–32 lb/ft^3; pH 4.3 (27%); 95% act.

Lignosol WT. [503] Low m.w. lignosulfonate; anionic; wetting agent, dispersant, sludge conditioner; yel. powd.; water sol.; 95% act.

Lignosol X. [503] Sodium lignosulfonate; anionic; wetting agent, emulsifier, tanning and retanning agent, used for dye leveling, water reducer, grinding aid; black liq.; water sol.; sp.gr. 1.24; f.p. –5 C; b.p. 105 C; 47% act.

Lignosol XD. [503] Sodium lignosulfonate; anionic; wetting agent, emulsifier for wax emulsions, used for retanning leather, water treatment; yel. powd.; water sol.; dens. 28–32 lb/ft^3; pH 6.5 (27%); 95% act.

Lignosol XD-65. [503] Sodium lignosulfonate; slurry explosive and flotation aid in mining industry; yel. powd.; dens. 28–32 lb/ft^3; pH 6.8 (27%); 95% act.

Lilamac 140, 160, 170. [368] Acetate amine salt of Lilamin 140, 160, 170 resp.; cationic; flotation aid for crude oil extraction, soil stabilization, wax emulsions, ore flotation; Gardner 8 max. liq.; m.w. 280–332; m.p. 38–60 C; 95% act.

Lilamac S1. [348] Amine acetate; cationic; flotation agent; flakes.

Lilamin 101 D. [368] Arachidyl behenyl primary amine (dist.); cationic; emulsifier, anticorrosive, lubricant, bactericidal, pigment dispersants in natural and syn. rubbers, mold release agent for hard rubber, additives for crude oil extraction, lubricants; Gardner 3 max. flakes; m.w. 295; 97% act.

Lilamin 115, 115 D. [368] Soyabean primary amine (115 D, dist. grade); cationic; see Lilamin 101 D; Gardner 5 max. flakes; m.w. 270; 95% act.

Lilamin 140, 140 D. [368] Hydrog. tallow primary amine (140 D, dist. grade); cationic; see Lilamin 101 D; Gardner 3 max. solid; m.w. 266; 95% act.

Lilamin 142, 142 D [368] Stearyl primary amine (142 D, dist. grade); cationic; see Lilamin 101 D; Gardner 3 max. flakes; m.w. 273; 95% act.

Lilamin 151. [368] Tall primary amine; cationic; see Lilamin 101 D; Gardner 5 max. flakes; m.w. 270; 92% act.

Lilamin 160, 160 D. [368] Coconut primary amine (160 D, dist. grade); cationic; see Lilamin 101 D; Gardner 3 max. flakes; m.w. 205; sp.gr. 0.781 (60 C); visc. 2.75 cps (60 C); m.p. 16 C; flash pt. 112 C (OC); 95% act.

Lilamin 163, 163 D. [368] Lauryl primary amine (163 D, dist. grade); cationic; see Lilamin 101 D; Gardner 3 max. flakes; m.w. 190; sp.gr. 0.777 (60 C); visc. 1.80 cps (60 C); m.p. 25 C; flash pt. 110 C (OC); 95% act.

Lilamin 170, 170 D. [368] Tallow primary amine, dist. grade (170 D); cationic; see Lilamin 101 D; Gardner 3 max. paste; m.w. 266; 95% act.

Lilamin 172, 172 D. [368] Oleyl primary amine, also avail. in dist. grade (172 D); cationic; see Lilamin 101 D; Gardner 3 max. flakes; m.w. 270; sp.gr. 0.802 (60 C); visc. 3.20 cps (60 C); m.p. 19 C; flash pt. 155 C (OC); 95% act.

Lilamin 308 D. [368] Caprylyl dimethyl tert. amine (dist.); cationic; emulsifier, anticorrosive, lubricant, bactericidal, pigment and filler dispersants in natural and syn. rubbers, additives for crude oil extraction, polymerization catalysts in epoxy resins and polyurethane foams; Gardner 2 max. liq.; m.w. 165; 95% act.

Lilamin 310 D. [368] Capryl dimethyl tert. amine (dist.); cationic; see Lilamin 308 D; Gardner 2 max. liq.; m.w. 195; 95% act.

Lilamin 312 D. [368] Lauryl dimethyl tert. amine (dist.); cationic; see Lilamin 308 D; Gardner 2 max. liq.; m.w. 220; 95% act.

Lilamin 314 D. [368] Myristyl dimethyl tert. amine (dist.); cationic; see Lilamin 308 D; Gardner 2 max. liq.; m.w. 250; 95% act.

Lilamin 316 D. [368] Palmityl dimethyl tert. amine (dist.); cationic; see Lilamin 308 D; Gardner 2 max. liq.; m.w. 280; 95% act.

Lilamin 342 D. [368] Stearyl dimethyl tert. amine (dist.); cationic; see Lilamin 308 D; Gardner 2 max. liq.; m.w. 305; flash pt. 164 C (OC); 95% act.

Lilamin 343. [368] Dihydrog. tallow methyl tert. amine; cationic; see Lilamin 308 D; Gardner 2 max. solid; m.w. 525; m.p. 30 C; 95% act.

Lilamin 345 D. [368] Hydrog. tallow-dimethyl tert. amine (dist.); cationic; see Lilamin 308 D; Gardner 2 max. liq.; m.w. 295; flash pt. 160 C (OC); 95% act.

Lilamin 363. [368] Trilauryl tert. amine; cationic; see Lilamin 308 D; Gardner 2 max. liq.; m.w. 529; 95% act.

Lilamin 364. [368] Tri (caprylyl capryl) tert. amine; cationic; see Lilamin 308 D; Gardner 2 max. liq.; m.w. 390; 95% act.

Lilamin 367 D. [368] Coconut dimethyl tert. amine (dist.); cationic; see Lilamin 308 D; Gardner 2 max. liq.; m.w. 230; flash pt. 109 C (OC); 95% act.

Lilamin 368. [368] Tri (caprylyl capryl/lauryl) tert. amine; cationic; see Lilamin 308 D; Gardner 2 max. liq.; m.w. 436; 95% act.

Lilamin 369. [368] Dicoco methyl tert. amine; cationic; see Lilamin 308 D; Gardner 2 max. liq.; m.w. 390; m.p. 30 C; 95% act.

Lilamin 372 D. [368] Oleyl dimethyl tert. amine (dist.); cationic; see Lilamin 308 D; Gardner 2 max. liq. m.w. 300; 95% act.

Lilamin 381. [368] Tri isooctyl tert. amine; cationic; see Lilamin 308 D; Gardner 2 max. liq.; m.w. 362; 95% act.

Lilamin 382. [368] Tri isodecyl tert. amine; cationic; see Lilamin 308 D; Gardner 2 max. liq.; m.w. 450; 95% act.

Lilamin 383. [368] Tri tridecyl tert. amine; cationic; see Lilamin 308 D; Gardner 2 max. liq.; m.w. 567; 95% act.

Lilamin 540. [368] Hydrog. tallow diamine; cationic; emulsifier, anticorrosive, lubricant, bactericidal, pigment and filler dispersants in natural and syn. rubbers, additives for crude oil extraction, polymerization catalysts in epoxy resins and polyurethane foams; Gardner 5 max. solid; m.w. 340; 85% act.

Lilamin 560. [368] Coconut diamine; cationic; see Lilamin 540; also pigment wetting in coatings; Gardner 6 max. paste; m.w. 265; sp.gr. 0.812 (60 C); visc. 3.85 cps (60 C); m.p. 25 C; flash pt. 125 C (OC); 85% act.

Lilamin 570. [368] Tallow diamine; cationic; see Lilamin 540; Gardner 5 max. paste; m.w. 340; sp.gr. 0.820 (60 C); visc. 4.05 cps (60 C); m.p. 42 C; flash pt. ± 190 C (OC); 85% act.

Lilamin 570 DO. [368] Dioleate of Lilamin 570; cationic; see Lilamin 540; Gardner 12 max. paste; m.w. 885; 95% act.

Lilamin 572. [368] Oleyl diamine; cationic; see Lilamin 540; Gardner 6 max. liq.; m.w. 350; sp.gr. 0.815 (60 C); visc. 3.95 cps (60 C); m.p. 27 C; flash pt. ± 195 C (OC); 85% act.

Lilamin EO. [368] Ethoxylated amine derivs.; cationic; emulsifier, corrosion inhibitor, antistat, used for crude oil extraction and drilling aids, in textile industry, in metallurgy.

Lilaminox. [368] Tert. amine oxides; gasoline additives, shampoo base.

Lilamuls BG. [348] N-tallow propylene diamine; cationic; emulsifier, adhesion agent; paste

Lipal 3 TD. [496] Trideceth-3; nonionic; emulsifier used in emollient creams, lotions, etc., solubilizer, wetting agent, dispersant; Gardner 1 liq.; sol. in IPA, propylene glycol, peanut and min. oil, water disp.; HLB 8.0 ± 1; cloud pt. 40–43 C (1%); 100% act.

Lipal 4 LA. [496] Laureth-4; nonionic; see Lipal 3 TD; Gardner 1 liq.; sol. in IPA, propylene glycol, peanut oil, disp. in water and min. oil; HLB 9.7 ± 1; cloud pt. 52–53 C (1%); 100% act.

Lipal 5 L. [496] PEG-5 laurate; nonionic; emulsifier, solubilizer, wetting agent, dispersant for emollient creams, lotions, shampoos, makeup, etc.; Gardner 2 liq.; sol. in IPA, disp. in water; HLB 9.8 ± 1; sapon. no. 135–150; 100% act.

Lipal 5 OA. [496] Oleth-5; nonionic; see Lipal 3 TD; Gardner 3 liq.; sol. in IPA, propylene glycol, peanut and min. oil; HLB 9.0 ± 1; cloud pt. 57–60 C (1%); 100% act.

Lipal 5 S. [496] PEG-5 stearate; nonionic; see Lipal 5 L; Gardner 4 solid; sol. in IPA, disp. in water; HLB 6.8 ± 1; sapon. no. 105–120; 100% act.

Lipal 6 TD. [496] Trideceth-6; nonionic; see Lipal 3 TD; Gardner 1 liq.; sol. in water, IPA, propylene glycol, min. oil; HLB 11.0 ± 1; cloud pt. 66–69 C (1%); 100% act.

Lipal 9 C. [496] PEG-9 castor oil; nonionic; see Lipal 5 L; Gardner 4 liq.; sol. in IPA, disp. in water; HLB 8.0 ± 1; sapon. no. 120–136; 100% act.

Lipal 9 L. [496] PEG-9 laurate; nonionic; see Lipal 5 L; Gardner 2; sol. in IPA, water; HLB 13.0 ± 1; sapon. no. 89–104; 100% act.

Lipal 9 LA. [496] Laureth-9; nonionic; see Lipal 3 TD; Gardner 1 liq.; sol. in water, IPA, propylene glycol; HLB 13.6 ± 1; cloud pt. 73–76 C (1%); 100% act.

Lipal 9 N. [496] Nonoxynol-9; nonionic; see Lipal 3 TD; Gardner 1 liq.; sol. in water, IPA, propylene glycol; HLB 15.0 ± 1; cloud pt.

52–56 C (1%); 100% act.

Lipal 9 OL. [496] PEG-9 oleate; nonionic; see Lipal 5 L; Gardner 4 liq.; sol. in IPA, disp. in water; HLB 12.0 ± 1; sapon. no. 84–100; 100% act.

Lipal 10 OA. [496] Oleth-10; nonionic; see Lipal 3 TD; Gardner 3, transitional form; sol. in water, IPA; HLB 12.4 ± 1; cloud pt. 48–52 C (1%); 100% act.

Lipal 10 TD. [496] Trideceth-10; nonionic; see Lipal 3 TD; Gardner 1 solid; sol. in water, IPA, propylene glycol; HLB 13.8 ± 1; cloud pt. 75–85 C (1%); 100% act.

Lipal 12 LA. [496] Laureth-12; nonionic; see Lipal 3 TD; Gardner 1 solid; sol. in water, IPA, propylene glycol; HLB 14.8 ± 1; cloud pt. 65–68 C (1%); 100% act.

Lipal 15 CSA. [496] Ceteareth-15; nonionic; see Lipal 3 TD; Gardner 1 solid; sol. in water, IPA, propylene glycol; HLB 14.4 ± 1; cloud pt. 67–70 C (1%); 100% act.

Lipal 15 T. [496] PEG-15 tallate; nonionic; see Lipal 5 L; Gardner 12 liq.; sol. in water, IPA; HLB 10.5 ± 1; sapon. no. 25–40; 100% act.

Lipal 20 OA. [496] Oleth-20; nonionic; see Lipal 3 TD; Gardner 3 solid; sol. in water, IPA; HLB 15.4 ± 1; cloud pt. 67–72 C (1%); 100% act.

Lipal 20 SA. [496] Steareth-20; nonionic; see Lipal 3 TD; Gardner 1 solid; sol. in water, IPA, propylene glycol; HLB 15.3 ± 1; cloud pt. 70–75 C (1%); 100% act.

Lipal 23 LA. [496] Laureth-23; nonionic; see Lipal 3 TD; Gardner 1 solid; sol. in water, IPA, propylene glycol; HLB 16.9 ± 1; cloud pt. 75–78 C (1%); 100% act.

Lipal 25 C. [496] PEG-25 castor oil; nonionic; see Lipal 5 L; Gardner 4 liq.; sol. in water, IPA; HLB 14.5 ± 1; sapon. no. 76–88; 100% act.

Lipal 25 S. [496] PEG-25 stearate; nonionic; see Lipal 5 L; Gardner 4 solid; sol. in water, IPA, propylene glycol HLB 14.0 ± 1; sapon. no. 35–47; 100% act.

Lipal 30 SA. [496] Steareth-30; nonionic; see Lipal 3 TD; Gardner 1 solid; sol. in water, IPA, propylene glycol; HLB 17.0 ± 1; cloud pt. 76–78 C (1%); 100% act.

Lipal 39 S. [496] PEG-40 stearate; nonionic; see Lipal 5 L; Gardner 4 solid; sol. in water, IPA, disp. in propylene glycol, peanut oil; HLB 16.9 ± 1; sapon. no. 23–35; 100% act.

Lipal 50 OA. [496] Oleth-50; nonionic; see Lipal 3 TD; Gardner 3 solid; sol. in water, IPA, propylene glycol; HLB 18.5 ± 1; cloud pt. 75–78 C (1%); 100% act.

Lipal 50 S. [496] PEG-50 stearate; nonionic; see Lipal 5 L; Gardner 2 solid; sol. in water, IPA; HLB 17.9 ± 1; sapon. no. 18–30; 100% act.

Lipal 52 C. [496] PEG-52 castor oil; nonionic; see Lipal 5 L; Gardner 4 solid; sol. in water, IPA, propylene glycol; HLB 16.0 ± 1; sapon. no. 46–58.

Lipal 300 DL. [496] PEG 300 dilaurate; nonionic; emulsifier, solubilizer, wetting agent, dispersant for emollient creams, lotions, shampoos, makeup, etc.; Gardner 1 liq.; sol. in IPA, peanut oil, min. oil, disp. in water; HLB 6.3 ± 1; sapon. no. 148–162; 100% act.

Lipal 300 W. [496] PEG 300 monooleate; nonionic; see Lipal 300 DL; Gardner 5 liq.; sol. in IPA, min. oil, disp. in water; HLB 9.2 ± 1; sapon. no. 102–115; 100% act.

Lipal 400 DL. [496] PEG 400 dilaurate; nonionic; see Lipal 300 DL; Gardner 4 liq.; sol. in IPA, peanut oil, min. oil, disp. in water; HLB 10.4 ± 1; sapon. no. 125–140; 100% act.

Lipal 400 DS. [496] PEG 400 distearate; nonionic; see Lipal 300 DL; Gardner 4 solid; sol. in IPA, peanut and min. oils, disp. in hot water; HLB 7.2 ± 1; sapon. no. 122–132; 100% act.

Lipal 400 DW. [496] PEG 400 dioleate; nonionic; emulsifier for w/o and o/w systems, solubilizer, wetting agent, dispersant, used for cosmetic and pharmaceutical preparations; liq.; 100% act.

Lipal 400 OL. [496] PEG 400 monooleate; nonionic; see Lipal 300 DL; Gardner 13 liq.; sol. in IPA disp. in water; HLB 6.8 ± 1; sapon. no. 83–88; 100% act.

Lipal 400 S. [206; 496] PEG 400 stearate; nonionic; see Lipal 300 DL; Gardner 1 solid; sol. in IPA, disp. in water; HLB 11.3 ± 1; sapon. no. 82–92; 100% act.

Lipal 400 W. [496] PEG 400 oleate; nonionic; see Lipal 400 DW; liq.; 100% act.

Lipal 600 S. [206; 496] PEG 600 stearate; nonionic; see Lipal 400 DW; solid; HLB 13.8; 100% act.

Lipal 600 W. [496] PEG 600 oleate; nonionic; see Lipal 400 DW; liq.; 100% act.

Lipal 610. [496] Trideceth-8; nonionic; emulsifier, solubilizer, wetting agent, dispersant used in emollient creams, lotions, etc.; Gardner 1 liq.; sol. in water, IPA, propylene glycol; HLB 12.5 ± 1; cloud pt. 42–46 C (1%); 85% act.

Lipal CE 38. [496] POE 5 coconut ester; nonionic; emulsifier, solubilizer, wetting agent, coupling agent, dispersant used in emollient creams and lotions, shampoos, etc.; Lovibond 3R, transitional form; sol. in IPA, min. oil, trichloroethylene, alcohol; HLB 7.6 ± 1; sapon. no. 130–150; 100% act.

Lipal CE 55. [496] POE 11 coconut ester; nonionic; see Lipal CE 38; Lovibond 3R, transitional form; sol. in IPA, min. oil, trichloroethylene, alcohol; HLB 11.0 ± 1; sapon. no. 90–115; 100% act.

Lipal CE 64. [496] POE 16 coconut ester; nonionic; see Lipal CE 38; Lovibond 3R, transitional form; sol. in water, IPA, min. oil, trichloroethylene, alcohol; HLB 12.8 ± 1; sapon. no. 70–100; 100% act.

Lipal CE 71. [496] POE 21 coconut ester; nonionic; see Lipal CE 38; Lovibond 3R, transitional form; sol. in water, IPA, min. oil, trichloroethylene, alcohol; HLB 14.2 ± 1; sapon. no. 50–80; 100% act.

Lipal OE 55. [496] POE 12 safflower ester; nonionic; see Lipal CE 38; Lovibond 6R, transitional form; sol. in water, IPA, trichloroethylene, min. oil, alcohol; HLB 11.0 ± 1; sapon. no. 70–90.

Lipal OE 64. [496] POE 17 safflower ester; nonionic; see Lipal CE 38; Lovibond 6R, liq.; sol. in water, IPA, trichloroethylene, min. oil, alcohol; HLB 12.8 ± 1; sapon. no. 55–75.

Lipal OE 70. [496] POE 23 safflower ester; nonionic; see Lipal CE 38; Lovibond 6R, liq.; sol. in water, IPA, trichloroethylene, min. oil, alcohol; HLB 14.0 ± 1; sapon. no. 42–62.

Lipal TE 43. [496] POE 8 tallow ester; nonionic; see Lipal CE 38; Lovibond 6R, liq.; sol. IPA, trichloroethylene, min. oil, alcohol; HLB 8.6 ± 1; sapon. no. 90–115; 100% act.

Lipal TE 55. [496] POE 12 tallow ester; nonionic; see Lipal CE 38; Lovibond 6R, transitional; sol. in water, IPA, trichloroethylene, min. oil, alcohol; HLB 11.0 ± 1; sapon. no. 70–100; 100% act.

Lipal TE 70. [496] POE 22 tallow ester; nonionic; see Lipal CE 38; Lovibond 6R, soft solid; sol. in water, IPA, trichloroethylene, min. oil, alcohol; HLB 14.0 ± 1; sapon. no. 40–65; 100% act.

Lipal TE 76. [496] POE 30 tallow ester; nonionic; see Lipal CE 38; Lovibond 6R, soft solid; sol. in water, IPA, trichloroethylene, min. oil, alcohol; HLB 15.2 ± 1; sapon. no. 30–60; 100% act.

Lipamide MEAA. [371] Acetamide MEA; nonionic; antistat; foam modifier; conditioner for hair and skin; liq.; 75% conc.

Lipamide S. [371] Stearamide DEA; nonionic; emulsifier, thickener, emulsion stabilizer, lubricant, used in skin and hair care products; off-wh. wax; mild, waxy odor; misc. hot with common oil phase ingred. and most org. solv., disp. in hot water; m.p. 41–46 C; pH 9–10.5 (1% aq.); 100% act.

Lipo DGLS. [371] PEG-2 laurate SE; nonionic; spreading agent, emulsifier, dispersant, lubricant used in bath oils, creams, lotions; yel. liq.; HLB 8.3 ± 1; sapon. no. 160–170; 100% act.

Lipo DGS-SE. [371] Diethylene glycol monostearate SE; see Lipo DGLS; wh./off wh. beads, flakes; HLB 4.0±1; sapon. no. 160–180.

Lipo Diglycol Laurate. [371] Emulsifier, defoamer; liq.; 100% conc.

Lipo EGDS. [371] Ethylene glycol distearate; see Lipo DGLS; wh./off wh. beads, flakes; HLB 1.0 ± 1; sapon. no. 190–205.

Lipo EGMS. [371] Glycol stearate; nonionic; see Lipo DGLS; also opacifier, pearlizer for shampoos; wh./off wh. beads, flakes; HLB 2.0 ± 1; sapon. no. 175–190.

Lipo GMS. [371] Glyceryl stearate; nonionic; emulsifier and defoamer for food industry; solid; HLB 3.7; 100% conc.

Lipo GMS 450. [371] Glyceryl stearate; nonionic; general purpose emulsifier, emollient, opacifier and visc. builder in creams and lotions; wh. bead or flake; HLB 3.6 ± 1; sapon. no. 165–182; 100% act.

Lipo GMS 470. [371] Glyceryl stearate SE; nonionic; general purpose emulsifier, emollient, opacifier and visc. builder in creams and lotions; wh. bead or flake; HLB 5.8 ± 1; sapon. no. 138–152; 100% act.

Lipo PGMS. [371] Propylene glycol stearate; nonionic; see Lipo DGLS; wh. solid wax; HLB 3.0 ± 1; sapon. no. 180–192.

Lipocol B. [371] PEG-9 stearate, PEG-9 laurate and PEG-2 laurate SE; nonionic; spreading agent, emulsifier, dispersant, lubricant, used in bath oils, creams, lotions, nonionic; wh. to off-wh., solid wax; HLB 11.0 ± 1; sapon. no. 94–104; 100% act.

Lipocol C-2, -10, -20. [371] Ceteth-2, -10, -20; nonionic; emulsifier, defoamer, wetting agent, solubilizer, conditioning agent for personal care products and pigment disp.;

wh., solid wax; HLB 5.3, 12.9, 15.7 resp.; 100% act.

Lipocol L-1, -4, -12, -23. [371] Laureth-1, -4, –12, -23; nonionic; see Lipocol C-2; colorless liq. to wh. solid wax; HLB 3.6, 9.7, 14.5, 16.9 resp.; 100% act.

Lipocol M-4. [371] Myreth-4; nonionic; see Lipocol C-2; liq.; HLB 8.8; 100% conc.

Lipocol O-2, -10, -20. [371] Oleth-2, -10, -20; nonionic; see Lipocol C-2; yel. liq.; HLB 4.9, 12.4, 15.3 resp.; 100% act.

Lipocol S-2, -10, -20. [371] Steareth-2, -10, -20; nonionic; see Lipocol C-2; wh. solid wax; HLB 4.9, 12.4, 15.3 resp.; 100% act.

Lipocol SC-4, -10, -15, -20. [371] Ceteareth-4, -10, -15, -20; nonionic; see Lipocol C-2; solid; HLB 8.0, 12.5, 14.3, 15.4; 100% conc.

Lipocol TD-3, -6, -12. [371] Trideceth-3, -6, -12; nonionic; emulsifier, wetting and scouring agent, dispersant for essential oils; raw material for sulfation and phosphation; liq., paste (TD-12); HLB 14.6 ± 1 (TD-12); 100% act.

Lipodan. [273] Dist. monoglycerides; nonionic; food emulsifier; flakes; 100% act.

Lipolan. [371] Hydrog. lanolin; nonionic; aux. w/o emulsifier; emollient; conditioner; lubricant; paste; 100% conc.

Lipolan 31. [371] PEG-24 hydrog. lanolin; nonionic; o/w emulsifiers, solubilizer, emollient, conditioner used in cosmetics, toiletries and topical pharmaceuticals; cream waxy solid; bland, char. odor; sol. in water and ethanol; sapon. no. 8; 100% act.

Lipolan 327 F. [370] alpha-Olefin sulfonate; anionic; detergent base; paste; 37% conc.

Lipolan 440. [370] alpha-Olefin sulfonate; anionic; detergent base; emulsifier for cosmetics and emulsion polymerization; liq.; 37% conc.

Lipolan 1400. [370] alpha-Olefin sulfonate; anionic; emulsifier for cosmetics; emulsifier for dispersant for emulsion polymerization; powd.; 100% conc.

Lipolan AO. [370] alpha-Olefin sulfonate; anionic; milling and scouring agent for wool; liq.; 30% conc.

Lipolan AOL. [370] alpha-Olefin sulfonate; anionic; scouring agent for wool and feathers; liq.; 28% conc.

Lipolan G. [370] alpha-Olefin sulfonate; anionic; deinking agent for waste paper; liq.; 30% conc.

Lipolan R. [371] Lanolin oil; emollient, conditioner, plasticizer, and lubricant for personal care products; dispersant for pigments; yel., amber liq.; mild char. odor; sol. in min. and veg. oils, isopropyl esters, and anhyd. IPA; insol. in water; sapon. no. 85–110; cloud pt. 18 C max.

Lipolan TE, TE(P). [370] Alkyl methyl tauride; anionic; scouring agent for textiles; dispersant for pigment; paste, powd. resp.; 25 and 28% conc. resp.

Lipomin COH. [370] Imidazoline; amphoteric; milling and scouring agent for wool; liq.; 20% conc.

Lipomulse 165. [371] Glyceryl stearate and PEG-100 stearate; nonionic; general purpose emulsifier, emollient, opacifier and visc. builder in creams and lotions; wh. bead or flake; HLB 11.0 ± 1; sapon. no. 90–100; 100% act.

Lipon P-106. [370] Alkyl benzene sulfonate; anionic; emulsifier; paste; 60% conc.

Liponic EG-1. [371] Glycereth-26; nonionic; humectant in creams and lotions, lubricant, plasticizer for hair resins, foam stabilizer, pigment dispersant; used in personal care products; colorless, clear to slightly hazy visc. liq.; sol. in water, alcohol, acetone and ethyl acetate; 100% act.

Liponox LCF. [370] POE alkyl ether; nonionic; wetting agent; paste; 100% conc.

Liponox LCR. [370] POE alkyl ether; nonionic; dispersant for pigment; wax; 100% conc.

Liponox N-105. [370] POE alkyl ether; nonionic; penetrant; liq.; 100% conc.

Liponox NC 2Y. [370] Alkylphenol ether, ethoxylated; nonionic; emulsifier for emulsion polymerization; flakes; 100% conc.

Liponox NC 6E, NCG, NCI, NCT. [370] POE alkylphenol ether; nonionic; detergent, penetrant, emulsifier, dispersant used in dyeing application for textiles, min. oil; wax, liq.; 100% conc.

Liponox OCS. [370] POE alkyl ether; nonionic; dye leveling agent; wax; 100% conc.

Lipopeg 2-DL. [371] PEG-4 dilaurate; nonionic; dispersant, emulsifier, spreading agent and lubricant in personal care products, bath oils; yel. liq.; HLB 6.0 ± 1; sapon. no. 170–185.

Lipopeg 4-DL. [371] PEG-8 dilaurate; nonionic; see Lipopeg 2-DL; yel. liq.; HLB 10.0 ± 1; sapon. no. 125–142.

Lipopeg 4-DO. [371] PEG-8 dioleate; see Lipopeg 2-DL; amber liq.; HLB 7.2 ± 1; sapon. no. 113–128.

Lipopeg 4-DS. [371] PEG-8 distearate; see Lipopeg 2-DL; cream soft wax; HLB 8.0± 1; sapon. no. 113–128.

Lipopeg 4-L. [371] PEG-8 laurate; nonionic; spreading agent, emulsifier, dispersant, lubricant for bath oils, creams and lotions; yel. liq.; HLB 13.0± 1; sapon. no. 90–100; 100% act.

Lipopeg 4-O. [371] PEG-8 oleate; nonionic; liq.; HLB 11.0; 100% conc.

Lipopeg 4-S. [371] PEG-8 stearate; nonionic; see Lipopeg 4-L; cream paste; HLB 11.2± 1; sapon. no. 80–90; 100% act.

Lipopeg 6-L. [371] PEG-12 laurate; nonionic; see Lipopeg 4-L; yel. liq.; HLB 14.6 ± 1; sapon. no. 65–76; 100% act.

Lipopeg 10-S. [371] PEG-20 stearate; nonionic; spreading agent, emulsifier, dispersant, and lubricant for personal care products; solid; HLB 15.2; 100% conc.

Lipopeg 15-S. [371] PEG-6-32 stearate; nonionic; see Lipopeg 10-S; solid; HLB 13.8; 100% conc.

Lipopeg 39-S. [371] PEG-40 stearate; nonionic; see Lipopeg 4-L; wh. solid wax; HLB 16.9 ± 1; sapon. no. 23–35; 100% act.

Lipopeg 100-S. [371] PEG-100 stearate; nonionic; see Lipopeg 4-L; tan flake or bead; HLB 18.8 ± 1; sapon. no. 9–20; 100% act.

Lipopeg 6000-DS. [371] PEG-150 distearate; nonionic; see Lipopeg 4-L; off-wh. flake; HLB 18.4 ± 1; sapon. no. 12–20; 100% act.

Lipophos PE9. [371] Nonylphenol ether phosphate ester acid form; anionic; textile scouring, emulsion polymerization, emulsifier and wetting agent; liq.; 100% conc.

Lipophos PL6. [371] Linear alcohol ether phosphate ester acid form; anionic; see Lipophos PE9; liq.; 100% conc.

Lipoproteol LCO, LK. [513] Sodium/TEA-lauroyl hydrolyzed animal protein; anionic; shampoo base; liq.; 22 and 25% conc. resp.

Lipoproteol LCOK. [513] Lipoaminoacid salt; anionic; shampoo base and detergent; liq.; 22–25% conc.

Lipoproteol UCO. [513] Sodium/TEA-undecylenoyl hydrolyzed animal protein; anionic; liq.; 23% conc.

Liposorb L. [371] Sorbitan laurate; nonionic; emulsifier, thickener, lubricant, antistat, all-purpose lipophilic surfactant used with POE Liposorb series; amber liq.; HLB 8.6 ± 1; sapon. no. 158–170; 100% act.

Liposorb L-10. [371] PEG-10 sorbitan laurate; nonionic; o/w emulsifier, lubricant, antistat, all-purpose hydrophilic surfactant used for solubilizing oils and in conjunction with Liposorb esters; yel. liq.; HLB 14.9 ± 1; sapon. no. 66–76; 100% act.

Liposorb L-20. [371] Polysorbate 20; nonionic; see Liposorb L-10; yel. liq.; HLB 16.7 ± 1; sapon. no. 40–50; 100% act.

Liposorb O. [371] Sorbitan oleate; nonionic; see Liposorb L; amber liq.; HLB 4.3 ± 1; sapon. no. 145–160; 100% act.

Liposorb O-5. [371] Polysorbate 81; nonionic; hydrophilic surfactant used for solubilizing oils; emulsifier, lubricant, antistat; liq.; HLB 10.0; 100% conc.

Liposorb O-20. [371] Polysorbate 80; nonionic; see Liposorb L-10; yel. liq.; HLB 15.0 ± 1; sapon. no. 45–55; 100% act.

Liposorb P. [371] Sorbitan palmitate; nonionic; see Liposorb L; tan beads or flakes; HLB 6.7 ± 1; sapon. no. 139–151; 100% act.

Liposorb P-20. [371] Polysorbate 40; nonionic; see Liposorb L-10; yel. liq.; HLB 15.6 ± 1; sapon. no. 40–53; 100% act.

Liposorb S. [371] Sorbitan stearate; nonionic; see Liposorb L; cream beads or flakes; HLB 4.7 ± 1; sapon. no. 147–157; 100% act.

Liposorb S-20. [371] Polysorbate 60; nonionic; see Liposorb L-10; yel. paste; HLB 14.9 ± 1; sapon. no. 45–55; 100% act.

Liposorb SQO. [371] Sorbitan sesquioleate; nonionic; see Liposorb L; amber liq.; HLB 3.7 ± 1; sapon. no. 149–160; 100% act.

Liposorb TO. [371] Sorbitan trioleate; nonionic; see Liposorb L; amber liq.; HLB 1.8 ± 1; sapon. no. 171–185; 100% act.

Liposorb TO-20. [371] Polysorbate 85; nonionic; see Liposorb L-10; yel. liq.; HLB 11.0 ± 1; sapon. no. 82–95; 100% act.

Liposorb TS. [371] Sorbitan tristearate; nonionic; see Liposorb L; cream flakes or beads; HLB 2.1 ± 1; sapon. no. 175–190; 100% act.

Liposorb TS-20. [371] Polysorbate 65; nonionic; see Liposorb L-10; tan solid wax; HLB 10.5 ± 1; sapon. no. 88–98; 100% act.

Lipowax. [371] Cetearyl alcohol, sodium lauryl sulfate, cetyl esters, myristyl alcohol; nonionic; wax used for building visc. in personal care products; solid; 100% conc.

Lipowax D. [371] Cetearyl alcohol, ceteareth-20; nonionic; o/w self-emulsifying wax, used in skin and hair creams and lotions, personal care products; wh. to off-wh. waxy solid; bland, char. odor; sol. in alcohol, misc. warm with most oil phase ingredients, disp. in warm water; m.p. 46–55 C; HLB 11;

sapon. no. 2; 100% act.

Lipowax G. [371] Cetearyl alcohol and ceteareth-20; nonionic; emulsifier for personal care products; solid; water-disp.; 100% conc.

Lipowax NI. [371] Cetearyl alcohol and ceteth-20; nonionic; see Lipowax D; solid; water-disp.; 100% conc.

Lipowax P. [371] Blend of fatty alcohols and ethoxylates; nonionic; o/w self-emulsifying wax for neutral and mildly acidic and alkaline pH systems; creamy, waxy solid; bland, char. odor; sol. in alcohol, misc. warm with most oil phase ingred., disp. in warm water; m.p. 48–52 C; HLB 9; sapon. no. 14; 100% act.

Liptol 40C. [370] Alkyl ethoxylate; nonionic; deinking agent for flotation process; liq.; 93% conc.

Liquid Base. [159] Min. oil and lanolin alcohol; nonionic; surfactant, emollient, penetrant, moisturizer, softener, primary w/o emulsifier; stabilizer for o/w emulsions; used in personal care products; lt. straw liq.; faint pleasant odor; sp.gr. 0.850–0.870; sapon. no. 2.0 max.

Lissapol N. [318] Nonylphenol ethoxylate; nonionic; detergent and wetting agent for textile, metal treatment, dust suppression and general cleaning applications; liq.; HLB 12.8; 27% conc.

Lissapol NX. [318] Nonylphenol ethoxylate; nonionic; general purpose detergent, wetting agent and emulsifier for textile processing; emulsification of med. polarity oils and solvs.; liq.; 100% conc.

Lissapol NXP. [318] Nonylphenol ethoxylate; nonionic; general purpose cleaning and applications; HLB 12.3; 100% conc.

LKE Granules. [45] Lecithin; nonionic; granular; 100% conc.

Lomar BDG. [187] Formaldehyde condensed naphthalene/alkylaryl sulfonate copolymer; sec. or coemulsifier for emulsion polymerization; dark amber liq.; biodegrad.; 33% solids.

Lomar D. [187] Highly polymerized naphthalene sulfonate; anionic; dispersing agent for carbon black pigments in latex systems; clear amber powd.; pH 9.3; 84% act.

Lomar DL. [187] Naphthalene sulfonic acid, sodium salts; liq. version of Lomar D; amber liq.; sp.gr. 1.20; pH 9.0; 30% act.

Lomar LS, PL. [187] Sodium salt of a condensed mono naphthalene sulfonic acid; anionic; dispersant in agric. formulations, emulsifier; powd., amber liq.; 46–95% act.

Lomar PW. [187] Sodium neutralized, condensed naphthalene sulfonic acid; anionic; dispersant, stabilizer, visc. suppressant, used in paint and paper industries; tan powd.; sol. in hard or soft water; pH 9.5 (10% aq.); 87% act.

Lomar PWA. [187] Ammonia neutralized, condensed naphthalene sulfonic acid; anionic; dispersant, visc. depressant, stabilizer; lt. tan powd.; sol. in hard or soft water; pH 7.2 (10% aq.); 92% act.

Lonzaine 12C. [372] Coco betaine; amphoteric; used in personal care products and industrial applications; liq.; 35% conc.

Lonzaine C, CO. [372] Cocoamidopropyl betaine; amphoteric; foaming agent, conditioner, visc. booster, wetting agent, used in cosmetics, toiletries, detergents, metal finishing, textile finishing, etc.; Gardner 2 nonvisc. liq.; char. odor; surf. tens. 33.6 and 33.9 dynes/cm resp. (0.1%); pH 4.5–5.5 and 6–8 resp. (10%); biodeg.; 30% act.

Lonzaine CS. [372] Cocoamidopropyl hydroxysultaine; amphoteric; see Lonzaine 12C; liq.; 50% conc.

Lonzest PEG-4DO. [372] PEG-8 dioleate; nonionic; detergent, lubricant, dispersant, anticorrosive, emulsifier, defoamer, used in textile, petroleum, insecticide industries; yel. liq.; sol. in min. oil, IPA, toluol, disp. in water; sp.gr. 0.98; m.p. 0 C; HLB 9.9; sapon. no. 120–130; 100% conc.

Lonzest PEG 4-L. [372] PEG-8 laurate; nonionic; dispersant, rewetting agent, stabilizer, used in cosmetics, pharmaceuticals, textile, paper, agric. industries; yel. liq.; sol. in water, IPA, toluol; sp.gr. 1.03; m.p. 10 C; HLB 13.1; sapon. no. 90–100; 100% conc.

Lonzest PEG 4-O. [372] PEG-8 oleate; nonionic; detergent, emulsifier, polish remover in pharmaceuticals and cosmetics, rust preventative oils and degreasers; yel. liq.; sol. in IPA, toluol, min. oil, water disp.; sp.gr. 1.01; m.p. 0 C; HLB 11.4; sapon. no. 86–96; 100% conc.

Lonzest SML. [372] Sorbitan laurate; nonionic; emulsifier for o/w systems; fiber lubricant and corrosion inhibitor used in food industry; liq.; 100% conc.

Lonzest SML-20. [372] Polysorbate 20; nonionic; emulsifier used where o/w and w/o emulsions are required; liq.; 100% conc.

Lonzest SMO. [372] Sorbitan monooleate;

nonionic; see Lonzest SML; liq.; 100% conc.

Lonzest SMO-20. [372] Polysorbate 80; nonionic; emulsifier, solubilizer and stabilizer used in foods, personal care products and industrial applications; liq.; 100% conc.

Lonzest SMP. [372] Sorbitan palmitate; nonionic; see Lonzest SML; solid; 100% conc.

Lonzest SMP-20. [372] Polysorbate 40; nonionic; see Lonzest SMO-20; liq.; 100% conc.

Lonzest SMS. [372] Sorbitan stearate; nonionic; see Lonzest SML; flake; 100% conc.

Lonzest SMS-20. [372] Polysorbate 60; nonionic; see Lonzest SMO-20; liq.; 100% conc.

Lonzest STO. [372] Sorbitan trioleate; nonionic; see Lonzest SML; liq.; 100% conc.

Lonzest STO-20. [372] Polysorbate 85; nonionic; see Lonzest SMO-20; liq.; 100% conc.

Lonzest STS. [372] Sorbitan tristearate; nonionic; see Lonzest SML; solid; 100% conc.

Lonzest STS-20. [372] Polysorbate 65; nonionic; see Lonzest SMO-20; solid; 100% conc.

Lonzol LA-300. [372] Ammonium lauryl sulfate; anionic; foaming agent, emulsifier, detergent, used in personal care and household products; visc. liq.; pH 6–7 (10%); 27–30% act.

Lonzol LS-300. [372] Sodium lauryl sulfate; anionic; foaming agent, emulsifier, detergent, used in personal care and household products; clear liq.; pH 7.3–8.5 (10%); 28–30% act.

Lonzol LT-400. [372] TEA-lauryl sulfate; anionic; foaming agent, emulsifier, detergent, used in personal care and household products; clear liq.; pH 7–8 (10%); 38–42% act.

Loraderm 8120. [212] Trigylceride; neutral oil base for cosmetic creams and lotions, solubilizer for act. ingred.; liq.; 100% act.

Loraderm S 1333. [212] Disodium sulfosuccinate of a veg. fatty acid; anionic; detergent used in shampoos, cleaners, personal care products, used for cutting and drilling oils; amber liq.; pH 6.5–7.5 (5%); 40% act.

Loraderm SB/LAN 5. [212] Wool wax alcohol sulfosuccinate; anionic; detergent, moisturizing skin protective agent for creams, lotions, and cleansing formulations; br. liq.; slightly sol. in alcohols, glycols and other solv.; pH 6.5–7.5 (5%); biodeg.; 50% act.

Loraderm SBLAN/5, SBLAN/E. [212] Sodium salt of the sulfosuccinate of wool wax alcohol; anionic; see Loraderm SB/LAN 5; yel. paste; slight typ. odor; sp.gr. 1.0 (SBLAN/5); 50% act.

Loramine AC 28. [212] Groundnut fatty acid polydialkanolamide; nonionic; detergent, emulsifier, used as an anticorrosion additive for metalworking lubricants; br. liq.; dens. 0.98 g/cm³; sp.gr. 1.0; 100% act.

Loramine B 204. [212] Coconut fatty acid alkylamido propyldimethylamine oxide; nonionic; detergent, softener, foam booster in wash-active formulations; pale yel. liq.; pH 5–7 (10%); biodeg.; 35% act.

Loramine C 212. [212] Coconut fatty acid monoethanolamide; nonionic; detergent, emulsion stabilizer; off-wh. flakes; slight amidic odor; m.p. 73 C; 100% act.

Loramine DAS. [212] Special surfactant; disinfectant cleaning agent; liq.; 50% act.

Loramine DC 212. [212] Specialized fatty acid polydiethanolamide; nonionic; lubricating additive, corrosion inhibitor; liq.; 100% act.

Loramine DC 212/S. [212] Coconut fatty acid diethanolamide; nonionic; detergent, foam stabilizer and visc. modifier used for shampoos and detergents; lt. yel. liq.; slight typ. odor; dens. 0.99 g/cm³; 100% act.

Loramine DC 212/SE. [212] Coconut fatty acid diethanolamine; nonionic; detergent, foam stabilizer, detergent builder, used in cosmetic shampoos, bath preparations, personal hygiene toiletries; liq.; 100% act.

Loramine DC 220/SE. [212] Coconut fatty acid diethanolamide; nonionic; detergent, thickening agent, foam stabilizer, superfatting agent; soft paste; 100% act.

Loramine DC 280. [212] Oleic acid polydiethanolamide; nonionic; detergent, used in cosmetic and detergent industry; amber liq.; slight typ. odor; sp.gr. 1.0; 100% act.

Loramine DL 203. [212] Lauric acid polydialkanolamide; nonionic; detergent, used in fixation of perfumes; wh. wax; dens. 0.97 g/cm³; sp.gr. 1.0; 100% act.

Loramine DL 203/S. [212] Lauric acid diethanolamide; nonionic; detergent, foam stabilizer and visc. modifier for shampoos and detergents; off-wh. solid; slight, typ. odor; m.p. 40 C; 100% act.

Loramine DL 240. [212] Coconut fatty acid polydiethanolamide; nonionic; detergent for cosmetic preparations and cleaning agents; yel. liq.; faint, typ. odor; 100% act.

Loramine DLM/SE. [212] Fatty acid diethanolamide; nonionic; detergent, thickening and superfatting raw material for cosmetic preparation; wax; 100% act.

Loramine DO 280/S. [212] Groundnut fatty acid; nonionic; detergent, foam stabilizer and visc. modifier for shampoos and detergents; pale br. liq.; slight, typ. odor; 100% act.

Loramine DO 280/SE. [212] Oleic acid diethanolamide; nonionic; detergent for cosmetic products; honey liq.; dens. 0.96 g/cm³; 100% act.

Loramine DU 185. [212] Undecylenic acid polydialkanolamide; nonionic; detergent, emulsifier, bacteriocide, thickener and foam stabilizer in shampoos; yel. liq.; sol. in water, oil, alcohols, glycols; sp.gr. 1.0; 100% act.

Loramine IM-BT. [212] Imidazoline deriv.; cationic; adhesive additive for bitumen, emulsifier, rust inhibitor in acidic media; green-br., Gardner 13 max. liq.; amine odor; dens. 0.94 g/cm³; 100% act.

Loramine IM-CA. [212] 1-Hydroxyethyl-2-heptadecenyl imidazoline; cationic; corrosion inhibitor, emulsifier, penetrant, wetting agent, additive used in leather and metalworking industries; yel. wax; m.w. 285; sol. in most polar and nonpolar solv., disp. in water; 100% act.

Loramine IM-OA. [212] 1-Hydroxyethyl-2-heptadecenyl imidazoline; cationic; corrosion inhibitor, emulsifier, wetting agent, additive, rust inhibitor, adhesive and coupling agent, detergent, antistat, raw material used in metal, cleaning, and paint industries; amber liq.; m.w. 345; sol. in most polar and nonpolar solv., disp. in water; dens. 0.94 g/cm³; visc. 800 cps; pH 10–11 (1% aq.); 100% act.

Loramine IM-OD. [212] 1-Hydroxyethyl-2-alkyl-imidazoline; corrosion inhibitor, hydrophobing agent; liq.; 100% act.

Loramine IPE 280. [212] Oleic monoisopropanolamide; nonionic; detergent, skin-protecting additive for liq. and paste products; lt. yel. semisolid; slight, typ. odor; 100% act.

Loramine IPL 203. [212] Lauric monoisopropanolamide; nonionic; detergent, additive for solid and paste end-products; wh. flakes; slight, typ. odor; m.p. 55 C; 100% act.

Loramine IPP 240. [212] Coconut fatty acid monoisopropanolamide; nonionic; see Loramine IPL 203; wh. flakes; slight, typ. odor; m.p. 50 C; 100% act.

Loramine L203. [212] Lauric monoethanolamide; nonionic; detergent; wh. flakes; slight, typ. odor; m.p. 84 C; 100% act.

Loramine L 408. [212] Lauryl dimethylamine oxide; nonionic; detergent, foam booster for shampoos and foam baths, hair conditioner; liq.; 30% act.

Loramine MRC. [212] Mixt. of fatty acid alkanolamides; corrosion inhibitor, aluminum cleaners, detergent; liq.; 100% act.

Loramine RE. [212] Nonionic fatty acid blends; anticorrosion additive, emulsifier, used in metalworking industry; dk. br. liq.; dens. 0.98 g/cm³; biodeg.; 100% act.

Loramine S 280. [212] Stearic acid monoethanolamide; nonionic; detergent, emulsion stabilizer; wh. flakes; slight, typ. odor; m.p. 92 C; 100% act.

Loramine SBU 185. [212] Sodium salt of the monosulfosuccinate of undecylenic acid monoethanolamide; anionic; fungicide, detergent, foaming agent, used in foam baths, shampoos, skin cleaners, washing agents; pale yel. liq.; slight, typ. odor; sol. in alcohol, glycol; sp.gr. 1.0; pH 6.5–7.5 (5%); biodeg.; 50% act.

Loramine U 185. [212] Monoethanolamide of undecylenic acid; nonionic; fungicide, detergent, foam booster/stabilizer, superfatting, visc. booster; yel. flakes; strong, typ. odor; sol. in alcohol, glycol, disp. in warm water; m.p. 50 C; 100% act.

Lorapal C 6. [212] Coconut amide polyglycolether; nonionic; dispersant, wetting agent, foaming agent, used as solv. in perfumes in shampoo and foam-bath formulations; amber liq.; cloud pt. 80–90 C (2% aq.); pH 8–10 (1%); biodeg.; 100% act.

Lorapal HV 5, HV 8, HV 9. [212] Nonoxynol-5, -8, -9 resp.; nonionic; emulsifer for emulsions and insecticides, raw material for rinsing, washing, and cleaning agents, wetting agent in paper, textile and cellulose industries, degreasing agent in pickling and alkaline immersion-bath cleaners; Gardner 2 max. liq.; sol. in min. oil, alcohol, ketone, toluene, xylene, chlorinated hydrocarbons; cloud pt. 59–62, 41–45, and 59–62 C resp.; pH 5–7 (1%); 100% act.

Lorapal HV 10, HV 14, HV 25. [212] Nonoxynol-10, -14, and -25 resp.; nonionic; see Lorapal HV 5; Gardner 2 max. liq. wh. soft wax, and wh. wax. resp.; sol. in min. oil, alcohol, ketone, toluene, xylene, chlorinated hydrocarbons; cloud pt. 70–73, 60–63, and 73–76 C resp.; pH 5–7 (1%); 100% act.

Lorapal LA 3. [212] Fatty alcohol polyglycol ether; nonionic; emulsifier, coupling agent, raw material; clear liq.; sol. in min. oil and many org. solv.; pH 5–7 (1%); biodeg.

Lorapal O 8. [212] Oleic acid amide polyglycol ether; nonionic; wetting and dispersing agent, o/w emulsifier, machine detergent; br. liq.; sol. in water, alcohol, ketones, esters, chlorinated hydrocarbons, benzene, fatty oils; cloud pt. 75–85 C; pH 8–10 (1%); biodeg.; 100% act.

Lorapal TA Series. [212] Fatty alcohol polyglycol ether; nonionic; wash-active base; wh. soft wax, wh. powd.; sol. in water and many org. solv.; cloud pt. 69–90 C; pH 5–7 (1%); biodeg.

Loraphat E 1027. [212] Alkylphenol polyglycol ether phosphate; anionic; corrosion inhibitor, emulsifier, dispersant, wetting agent, used in metalworking oils; yel. liq.; 100% act.

Loraphat EAK 8190. [212] Fatty alcohol polyglycol ether phosphate; anionic; see Loraphat E 1027; yel. liq.; water disp.; biodeg.; 100% act.

Lorapon AM-2C. [212] Coconut-based ampholyte; amphoteric; detergent, dispersant, foam booster, wetting agent, used in toiletries; golden-yel. liq.; biodeg.; 49% act.

Lorapon AM-2L. [212] Imidazoline-based ampholyte; amphoteric; detergent, used in mild foam baths, shampoos, personal hygiene toiletries; liq.; 50% act.

Lorapon AM-B 13, 13/T. [212] Alkyl amido betaine; amphoteric; detergent, foam booster, used in shampoos, foam baths, toiletries, cleaning agents; yel. liq.; pH 4–6 (10%); biodeg.; 30–35% act.

Lorapon AM-CA. [212] Coconut-based ampholyte; amphoteric; detergent, foaming agent, used in shampoos and foam baths; cl. yel. liq.; pH 8–8.5 (10%); 30% act.

Lorapon AM-DML. [212] Dimethyl lauryl aminobetaine; amphoteric; detergent used in baby shampoos, cleaners, pickling agents; liq.; 40% act.

Lorapon AM-G 30. [212] Imidazoline-based betaine, modified; detergent used in baby and child care cosmetic formulations; liq.; 30% act.

Lorapon AM-KSF. [212] Imidazoline-based ampholyte, salt free; detergent used in baby and child care cosmetic formulations, personal hygiene toiletries; liq.; 70% act.

Lorapon AM-LL. [212] Linear alkyl-amido amphoteric; detergent with high skin compatibility for cosmetics; liq.; 30% act.

Lorapon AM-V. [212] Short-chain fatty acid based ampholyte; amphoteric; detergent, wetting agent, used in cleaning agents; dk. amber liq.; pH 10–12 (10%); 28% act.

Loraquat B 41. [212] Alkyl triethanol ammonium chloride; cationic; disinfectant; liq.; 50% act.

Loraquat B 50. [212] Lauryl dimethyl benzyl ammonium chloride; disinfectant for all types of cleaners, for dairies and other food manufacturing industries, algicide; liq.; 50% act.

Loraquat CPEM. [212] Coco-pentaethoxy methyl-ammonium methosulfate; softener, emulsifier for polymerization processes, antistatic agent; liq.; 100% act.

Loraquat CR 3099. [212] Di-fatty acid ester dimethyl ammonium methosulfate; softener, dry-cleaning agent, textile and leather aux.; liq.; 100% act.

Loraquat M 5040. [212] 1-Methyl-2 alkyl-3 alkylamidoethyl-imidazolinium-methosulfate; cationic; softener for textiles and paper, detergent, antistatic agent, additive for hair conditioning rinses; amber liq.; disp. in cold water; pH 4–5 (1% solids); 75% act.

Loraquat M 5040 H. [212] Quat. imidazolinium deriv.; softener for textiles; soft paste; 75% act.

Loraquat QA 100. [212] Alkyl dimethylbenzyl ammonium saccharinate; cationic; disinfectant, detergent, used in personal care products, pharmaceuticals, household items; wh. crystals; sol. in glycol ethers, alcohol, glycol, ketones, carbon tetrachloride; m.p. 40–50 C; 94% act.

Lorol CT-M. [15] Syn. fatty alcohol triethoxysulfate and alkanolamide; anionic; pearlescent hair shampoo conc. used in personal care products; wh. pearlescent visc. liq.; dilutable in water; sp.gr. 1.054; visc. 4500 ± 1500 cP; pH 7.0 ± 0.8 (2%); 23.0 ± 1.0% act.

Loropan CME. [610] Coco fatty acid monoethanolamide; nonionic; foam stabilizer and thickening agent for shampoos and detergents; bead; 92–96% conc.

Loropan KD. [610] Coco fatty acid diethanolamide; nonionic; emulsifier, stabilizer, thickener for shampoos and bubble baths; liq., paste; 85–90% conc.

Loropan KM. [610] Coco fatty acid monoethanolamide; nonionic; see Loropan KD; paste; 30% conc.

Loropan LD. [610] Lauric acid diethanolamide; nonionic; foam stabilizer for shampoos, detergents, fortifier for perfumes in soap; solid; 90% conc.

Loropan LM. [610] Lauric acid monoethanolamide; nonionic; foam stabilizer and thickening agent for shampoos and detergents; bead; 92–96% conc.

Loropan LMD. [610] Lauric and myristic acids diethanolamide; nonionic; foam stabilizer; superfatting and thickening agent for shampoos; fortifier for perfume in soaps; paste; 90% conc.

Loropan OD. [610] Oleic acid diethanolamide; nonionic; foam stabilizer, thickener, and superfatting agent for shampoos and bubble bath products; liq.; 90% conc.

Lostat 105D. [136] Lignosulfonate; anionic; dispersant for dyestuffs; powd.; 100% conc

Louryl T-50. [610] Triethanolamine alkylbenzene sulfonate; anionic; detergent, shampoos, and bubble bath products; liq.; 50% conc.

Loving KL. [341] Sucrose ester and food additives; nonionic; cleaning compd. for the food industry; liq.

Loving P. [341] Sucrose ester and food additives; nonionic; peeling agent for food industry; powd.

LSF-54. [653] Anionic; detergent, wetting agent, high flame resistance; foam; sp.gr. 1.032; visc. 5.0 cs; f.p. 5 C; pH 6.5–8.0.

Lubrol 12A-9. [317] Dodecyl alcohol condensate (9.5 moles); nonionic; antistat for polyethylene; paste; HLB 13.8; 100% conc.

Lubrol 17A-10. [317] Oleyl/cetyl alcohol condensate (10 moles); nonionic; antifoaming, emulsifier; used in sea water evaporator; oleines, and paraffin waxes; paste; HLB 12.7; 100% conc.

Lubrol 17A-17. [317] Fatty alcohol oxide condensate, ethoxylated; nonionic; emulsifier used to produce o/w dispersions; flakes; 100% conc.

Lubrol N5. [317] Nonoxynol-6 (5.5 EO); nonionic; emulsifier for paraffins; liq.; HLB 10.5; 100% conc.

Lubrol N13. [317] Nonoxynol-13; nonionic; emulsifier for pest control products; paste; HLB 14.4; 100% conc.

Lumo WW 75. [660] Sodium alkyl benzene sulfonate; anionic; detergents; paste; 75% conc.

Lutensit A-EP. [82] Fatty alcohol alkoxylate acid phosphate ester; anionic; wetting agent, dispersant, solubilizer, and cleaner used in industrial and household cleaning agents; ylsh. liq.; sol. in water, 5% caustic soda, 5% HCl, IPA, wh. spirit, trichlorethylene, xylene; dens. 1.07 g/ml; pH 2 (0.1% aq.); surf. tens. 39 dynes/cm; pH 2 (0.1% aq.); biodeg.; 100% act.

Lutensit A-ES. [82] Alkyl phenol ether sulfate, sodium salt; anionic; wetting and dispersing agents for engineering, industrial, and household cleaners; ylsh. liq.; sol. in water, 5% caustic soda, 5% HCl, IPA; dens. 1.09 g/ml; surf. tens. 31 dynes/cm (1 g/l water); pH 7–9 (0.1% aq.); biodeg.; 40% act.

Lutensit A-FK. [82] Fatty acid condensation product, sodium salt; anionic; see Lutensit A-ES; ylsh. liq.; sol. in water, trichlorethylene, xylene; dens. 1.00 g/ml; pour pt. 5 C; surf. tens. 36 dynes/cm (1 g/l water); pH 8–9.5 (0.1% water); biodeg.; 55% act.

Lutensit A-LBA. [82] Alkylbenzene sulfonate, alkanolamine salt; anionic; see Lutensit A-ES; ylsh. liq.; sol. in water, 5% caustic soda, 5% HCl, IPA, trichlorethylene, xylene; dens. 1.08 g/ml; pour pt. –5 C; surf. tens. 33 dynes/cm (1 g/l water); pH 7–9 (0.1% aq.); biodeg.; 55% act.

Lutensit A-PS. [82] Alkylsulfonate, sodium salt; anionic; wetting, dispersing and cleansing agents for industrial and household uses; yel. liq.; sol. in water, 5% caustic soda, 5% HCl; dens. 1.2 g/ml; surf. tens. 34 dynes/cm (1 g/l water); pH 7–8 (0.1% aq.); biodeg.; 65% act.

Lutensit AN 10, 20. [82] Mixts. of various nonionic alkoxylates and an anionic, acidic, readily neutralizable surfactant; anionic/nonionic; degreasing agent, cold cleaners; ylsh. liq.; dens. 0.96–0.97 g/cm^3; flash pt. 120 C; pour pt. < –15 C; surf. tens. 31 and 36 mN/m resp. (1 g/l sodium salt); pH 4–5 and 3.5–4.5 resp. (0.1% aq.); 100% act.

Lutensit AS 2230, 2270. [82] Sulfated natural alcohol polyglycol ether, sodium salt; anionic; foaming surfactant for chemical and cosmetic industry; liq., paste resp.; 28 and 70% conc. resp.

Lutensit AS 3330. [82] Sodium salt of sulfated oxoalcohol polyglycol ether; anionic; foaming agent, surfactant for the light chemical and cosmetics industry; liq.; 28% act.

Lutensit AS 3334. [82] Sodium salt of sulfated fatty alcohol polyglycol ether; anionic; see Lutensit AS 3330; liq.; 28% act.

Lutensit K-HP. [82] Fatty acid condensation product; cationic; detergent, hydrophobic rinse aid for car washing; liq.; 100% act.

Lutensit K-LC, K-LC 80. [82] Dimethyl $C_{12}C_{14}$ fatty alkyl benzyl ammonium chloride; cationic; biocidal, wetting and dispersing agents for production of disinfectant cleaners for beverages and foodstuffs, trading concerns and household; clear to ylsh. liq.; sol. in water, 5% sodium hydroxide, 5% HCl, 5% sodium chloride, alcohol, chlorinated hydrocarbons, oil, wh. spirit; dens. 0.98 g/cm^3; visc. 80 and 500 mPa•s resp.; cloud pt. –4 and –10 C; surf. tens. 37 mN/m (1 g/l); pH 6 (1% aq.); 80% biodeg.; 50 and 80% act.

Lutensit K-OC. [82] Dimethyl alkyl benzyl ammonium chloride; cationic; see Lutensit K-LC; ylsh. liq.; sol. in water, 5% sodium hydroxide, alcohol, oil, wh. spirit; dens. 0.98 g/cm^3; visc. 100 mPa•s; cloud pt. –3 C; surf. tens. 31 mN/m (1 g/l); pH 6 (1% aq.); 80% biodeg.; 50% act.

Lutensit K-TI. [82] Iodine surfactant complex; disinfectant, biocidal for cleaners; liq.

Lutensol A 7. [82] PEG-7 $C_{16}C_{18}$ fatty alcohol; water-sol. detergent for production of detergents and in chemical processing industry; liq.; 100% act.

Lutensol A 8. [82] PEG-8 $C_{12}C_{14}$ fatty alcohol; nonionic; detergent, wetting, dispersing, and emulsifying agent; clear liq.; sol. in water, 5% caustic soda, 5% HCl, min. oil, gasoline, chlorinated hydrocarbon; cloud pt. 55 C (0.1% aq.); pH 6.0–7.5 (1% aq.); 90% act.

Lutensol A 80. [82] Tallow alcohol ethoxylate; nonionic; detergent, wetting, dispersing, and emulsifying agent; wh. or lt. ylsh., waxy powd.; sol. in water, 5% caustic soda, 5% HCl, alcohol, ketones, benzene, xylene, chlorinated hydrocarbon; cloud pt. > 100 C (1% aq.); pH 6.0–7.5 (1% aq.); 100% act.

Lutensol AO 3, AO 5. [82] PEG-3 and PEG-5 resp. straight chain syn. fatty alcohol; nonionic; detergent, emulsifying and dispersing agent used in household and industrial detergents; clear, dull liq.; sol. in petroleum fractions, alcohols, aromatic hydrocarbons; slightly sol. water; dens. 0.93 and 0.96 g/cm^3; visc. 40 mPa•s; flash pt. 130 and 150 C; pH 7 (1% aq.); > 80% biodeg.; 100% act.

Lutensol AO 7, AO 8. [82] PEG-7 and PEG-8 resp. straight chain syn. fatty alcohol; nonionic; see Lutensol AO 3; clear, dull liq.; sol. water, 5% HCl, petroleum fractions, alcohols, aromatic and chlorinated hydrocarbons; dens. 0.98 and 0.96 g/cm^3; visc. 120 mPa•s, 30 mPa•s (60 C); cloud pt. 43 and 52 C (1%); flash pt. 190 and 200 C; pH 7 (1% aq.); > 80% biodeg.; 100% act.

Lutensol AO 10, AO 11, AO 12. [82] PEG-10, -11, and -12 resp. straight chain syn. fatty alcohol; nonionic; see Lutensol AO 3; wh. soft paste (AO 10, 11), wh. solid paste (AO 12); sol. water, 5% HCl, alcohols, aromatic and chlorinated hydrocarbons; dens. 0.98, 0.99, and 1.00 g/cm^3 (60 C); visc. 40 cps (60 C); cloud pt. 80, 86, and 91 C (1%); flash pt. 200 C; pH 7 (1% aq.); > 80% biodeg.; 100% act.

Lutensol AO 30. [82] PEG-30 straight chain syn. fatty alcohol; nonionic; see Lutensit A-FK; wax; 100% conc.

Lutensol AO 109. [82] PEG-10 straight chain syn. fatty alcohol; nonionic; see Lutensol AO 3; clear, dull liq.; sol. water, 5% HCl, petroleum fractions, alcohols, aromatic and chlorinated hydrocarbons; dens. 1.03 g/cm^3; visc. 200 cps; cloud pt. 80 C (1%); flash pt. 190 C; pH 7 (1% aq.); > 80% biodeg.; 90% act.

Lutensol AP 6. [82] PEG-6 alkylphenol; nonionic; detergent, wetting, emulsifying and dispersing agent, used in cleaners, detergents, leather, fur, paper, paint and dye industries; ylsh. liq.; sol. in min. oil, alcohols, ketones, aromatic and chlorinated hydrocarbons; sp.gr. 1.05; visc. 400 cps; surf. tens. 31 dynes/cm; 100% act.

Lutensol AP 7. [82] PEG-7 alkylphenol; nonionic; see Lutensol AP 6; clear liq.; sol. in min. oil, alcohols, ketones, aromatic and chlorinated hydrocarbons; sp.gr. 1.04; visc. 350 cps; surf. tens. 31 dynes/cm; 100% act.

Lutensol AP 8, AP 9. [82] PEG-8 and PEG-9 resp. alkylphenol; nonionic; see Lutensol AP 6; clear liq.; sol. in water, 5% HCl, alcohols, ketones, aromatic and chlorinated hydrocarbons; sp.gr. 1.05; visc. 350 cps; surf. tens. 31 dynes/cm; 100% act.

Lutensol AP 10. [82] PEG-10 alkylphenol; nonionic; see Lutensol AP 6; clear liq.; sol. in water, 5% HCl, alcohols, ketones, aromatic and chlorinated hydrocarbons; sp.gr. 1.06; visc. 350 cps; surf. tens. 32 dynes/cm; 100% act.

Lutensol AP 14. [82] PEG-14 alkylphenol; nonionic; see Lutensol AP 6; clear liq.; sol. in water, 5% HCl, alcohols, ketones, aromatic and chlorinated hydrocarbons; sp.gr. 1.05 (50 C); visc. 250 cps (30 C); surf. tens. 37 dynes/cm; 100% act.

Lutensol AP 20. [82] PEG-20 alkylphenol;

nonionic; see Lutensol AP 6; wh. soft wax; sol. in water, 5% HCl, alcohols, ketones, aromatic and chlorinated hydrocarbons; sp.gr. 1.06 (50 C); surf. tens. 42 dynes/cm; 100% act.

Lutensol AP 30. [82] PEG-30 alkylphenol; nonionic; see Lutensol AP 6; wax; 100% act.

Lutensol AT 11. [82] PEG-11 sat. $C_{16}C_{18}$ alcohol; nonionic; detergent, wetting, dispersing and emulsifying agent used in chemical industry, for household and industrial detergents; wh. to pale ylsh. paste; sol. in dist. water; visc. 30 cps (60 C); cloud pt. 86 C (0.5% aq.); pH practically neutral (1%); 100% act.

Lutensol AT 18. [82] PEG-18 sat. $C_{16}C_{18}$ alcohol; nonionic; see Lutensol AT 11; paste; 100% act.

Lutensol AT 25. [82] PEG-25 sat. $C_{16}C_{18}$ alcohol; nonionic; see Lutensol AT 11; wh. powd.; sol. in dist. water, benzene, chlorinated hydrocarbons, ethanol, IPA; visc. 70 cps (60 C); cloud pt. > 100 C (0.5% aq.); pH practically neutral (1%); 100% act.

Lutensol AT 50. [82] PEG-50 sat. $C_{16}C_{18}$ alcohol; nonionic; see Lutensol AT 11; wh. powd.; sol. in dist. water, benzene, chlorinated hydrocarbons, ethanol; visc. 125 cps (60 C); cloud pt. > 100 C (0.5% aq.); pH practically neutral (1%); 100% act.

Lutensol AT 80. [82] PEG-80 sat. C16C18 alcohol; nonionic; see Lutensit A-FK; wax; 100% conc.

Lutensol ED 140. [82] EO-PO ethylene diamine compd.; nonionic; antistat, detergent, dispersant, defoamer, wetting agent, gelling agent, solubilizer, emulsifier, demulsifier, lubricant, foam suppressor for household and industrial uses; clear ylsh. liq.; sol. @ 10% in water, 5% HCl, benzene, ethanol, IPA, trichlorethylene; dens. 1.05 g/cm^3; visc. 750 cps; cloud pt. 80 C (1% aq.); pour pt. –20 C; surf. tens. 44 mN/m (1 g/l); pH 9 (5% aq.); 100% act.

Lutensol ED 310. [82] EO-PO ethylene diamine compd.; nonionic; see Lutensol ED 140; clear to dull ylsh. liq.; sol. @ 10% in benzene, ethanol, IPA; dens. 1.00 g/cm^3; visc. 800 cps; cloud pt. 20 C (1% aq.); pour pt. –25 C; surf. tens. 33 mN/m (1 g/l); pH 9 (5% aq.); 100% act.

Lutensol ED 370. [82] EO-PO ethylene diamine compd.; nonionic; see Lutensol ED 140; wh. to ylsh. fine powd.; sol. @ 10% in water, 5% HCl, ethanol, IPA, trichlorethylene; dens. 1.05 g/cm^3; visc. 600 cps; m.p. 50 C; cloud pt. > 100 C (1% aq.); surf. tens. 44 mN/m (1 g/l); pH 8 (5% aq.); 100% act.

Lutensol ED 610. [82] EO-PO ethylene diamine compd.; nonionic; see Lutensol ED 140; colorless to ylsh. dull liq.; sol. @ 10% in benzene, ethanol, IPA; dens. 1.02 g/cm^3; visc. 1300 cps; cloud pt. 15 C (1% aq.); pour pt. –25 C; surf. tens. 33 mN/m (1 g/l); pH 8 (5% aq.); 100% act.

Lutensol FA 12. [82] Oleyl amine ethoxylate; nonionic; emulsifier, dispersant, wetting and degreasing agent used in heavy-duty and other detergents, shampoos; ylsh. liq.; sol. @ 10% in dist. water, 5% caustic soda, 5% HCl, alcohol; dens. 1.02 g/cm^3 (20 C); visc. 200 cps; surf. tens. 35 dynes/cm (1 g/l aq.); pH 8 (1% aq.); > 80% biodeg.; 100% act.

Lutensol FSA 10. [82] Oleic acid amide ethoxylate; nonionic; emulsifier, dispersant, detergent, wetting agent, used in fine detergents, hand cleaners, fur dressing; rdsh. liq. with some sediment; sol. see Lutensol FA 12; dens. 1.01 (60 C); visc. 70 cps (60 C); cloud pt. 80 C (1% aq.); surf. tens. 35 dynes/cm (1 g/l aq.); pH 8 (1% aq.); > 80% biodeg.; 100% act.

Lutensol GD 50, 70. [82] Alkyl glucoside; nonionic; foaming surfactant for formulation of cleaners and degreasing agents; liq.; 50 and 70% conc. resp.

Lutensol LF 220, 221, 223, 224. [82] Alkoxylated straight chain alcohol; nonionic; surfactant for dishwashing powds. and rinse aids; liq.; 95, 95, 98, and 100% conc. resp.

Lutensol LF 400. [82] Alkoxylated straight chain alcohol; nonionic; low-foaming detergent, wetting and dispersing agent, acidic rinse aid, dishwashing, glass cleaners, metal cleaning; colorless clear liq.; sol. @ 10% in water, 5% HCl, wh. spirit, benzene, ethanol, IPA, trichlorethylene; dens. 0.97 g/cm^3; visc. 60 cps; cloud pt. 31 C (1% aq.); surf. tens. 30 dynes/cm (0.1% aq.); pH 7 (5% aq.); 100% act.

Lutensol LF 401. [82] Alkoxylated straight chain alcohol; nonionic; low-foaming detergent, wetting and dispersing agent for heavy-duty detergents, acid pickling, metal cleaning; colorless clear liq.; sol. @ 10% in water, 5% HCl, wh. spirit, benzene, ethanol, IPA, trichlorethylene; dens. 1.05 g/cm^3; visc. 330 cps; cloud pt. 77 C (1% aq.); surf. tens. 40 dynes/cm (0.1% aq.); pH 7 (5% aq.); 100% act.

Lutensol LF 403. [82] Alkoxylated straight chain alcohol; nonionic; surfactant for dishwashing powds.; liq.; 100% conc.

Lutensol LF 404, 405. [82] Alkoxylated straight chain alcohol; nonionic; see Lutensol LF 220; liq.; 100 and 95% conc.

Lutensol LF 431. [82] Alkoxylated straight chain alcohol; nonionic; surfactant for dishwashing powds, rinse aids and bottlewashing; liq.; 100% conc.

Lutensol LF 600. [82] Alkoxylated straight chain alcohol; nonionic; low-foaming detergent, wetting and dispersing agent for car washing, hard surface cleaners; colorless clear liq.; sol. @ 10% in water, 5% HCl, wh. spirit, benzene, ethanol, IPA, trichlorethylene; dens. 0.97 g/cm^3; visc. 110 cps; cloud pt. 52 C (1% aq.); surf. tens. 40 dynes/cm (0.1% aq.); pH 7 (5% aq.); 100% act.

Lutensol LF 700. [82] Alkoxylated straight chain alcohol; nonionic; low-foaming detergent, wetting and dispersing agent for dishwashing, rinse aids, dairy cleaners, household and industrial cleaners; slightly ylsh. clear liq.; sol. @ 10% in 5% HCl, wh. spirit, benzene, ethanol, IPA, trichlorethylene; dens. 0.96 g/cm^3; visc. 110 cps; cloud pt. 77 C (1% aq.); surf. tens. 34 dynes/cm (0.1% aq.); pH 7 (5% aq.); 100% act.

Lutensol LF 711. [82] Alkoxylated straight chain alcohol; nonionic; low-foaming detergent, wetting and dispersing agent for dishwashing machines, rinse aids, metal cleaning; colorless clear liq.; sol. @ 10% in water, 5% HCl, wh. spirit, benzene, ethanol, IPA, trichlorethylene; dens. 1.0 g/cm^3; visc. 110 cps; cloud pt. 36 C (1% aq.); surf. tens. 28 dynes/cm (0.1% aq.); pH 7 (5% aq.); 100% act.

Lutensol LF 1300. [82] Alkoxylated straight chain alcohol; nonionic; see Lutensol LF 700; slightly ylsh. clear liq.; sol. @ 10% in wh. spirit, benzene, ethanol, IPA, trichlorethylene; dens. 0.97 g/cm^3; visc. 140 cps; surf. tens. 35 dynes/cm (0.1% aq.); pH 7 (5% aq.); 100% act.

Lutensol LSV. [82] Alkoxylated straight chain alcohol; nonionic; low-foaming detergent, wetting and dispersing agent for bottle washing, lime and scale dispersant; slightly ylsh. clear liq.; sol. @ 10% in benzene, 5% HCl, ethanol, IPA, trichlorethylene; sol. opaque in water; dens. 1.04 g/cm^3; visc. 500 cps; cloud pt. 30 C (1% aq.); surf. tens. 42 dynes/cm (0.1% aq.); pH 8–9 (5% aq.); 100% act.

Lutensol LT 30. [82] Specially alkoxylated fatty alcohol; nonionic; surfactant for acid and alkaline low-foaming cleaners; liq.; 100% act.

Lutensol ON 30, 50, 60, 70, 80, 110. [82] PEG-3, -5, -6, -7, -8, -11 resp. syn. fatty alcohol; nonionic; see Lutensit A-FK; liq. except ON 110 (paste); 100% conc.

Lutensol TO 3, 5, 8, 12. [82] PEG-3, -5, -8, -12 resp. syn. fatty alcohol; nonionic; surfactant for chemical industry and production of cleaners; liq. except TO 12 (paste); 100% conc.

Lutopon SN. [188] Linear alkylaryl sulfonate and solv.; anionic; detergent used in textile processing; liq.; 100% conc.

Lutostat 171. [192] Betaine; amphoteric; textile antistat; liq.; 30% conc.

Lyogen AFS Liq. [537] Detergent used for scouring, leveling and finishing of nylon hosiery; br. liq.; turbid sol. in water; sp.gr. 1.01; 85–86% act.

Lyogen BE Liq. [537] Dyeing assistant used on nylon; yel. liq.; 55% sol. in water @ 70 F; sp.gr. 1.16; 45% act.

Lyogen DFT Liq. [537] Detergent, wetting agent, emulsifier, dispersant, dyeing assistant for polyester; yel. liq.; misc. with water; sp.gr. 1.05; 30% act.

Lyogen F Liq. [537] Leveling agent, dyeing assistant used on wool; br. liq.; misc. with water; sp.gr. 1.06; 85% act.

Lyogen KF Liq. [537] Detergent, wetting agent, emulsifier, dispersant, used with vat dyes; yel. liq.; sp.gr. 1.05; 17% act.

Lyogen MS Liq. [537] Dyeing and leveling agent for wool and nylon; yel. liq.; misc. with water; sp.gr. 1.08; 33% act.

Lyogen NL Liq., P Liq. [537] Dyeing assistant used on nylon with acid dyes; amber and yel. resp. liq.; misc. with water (NL Liq.); sp.gr. 1.05 and 1.09 resp.; 34 and 35% act. resp.

Lyogen PAA Liq. [537] Antiprecipitant used in dyeing acrylic fibers and blends with wool and nylon where cationic and anionic dyes are used simultaneously; amber liq.; misc. with water; 43% act.

Lyogen SF Liq. [537] Dyeing assistant used in wool, silk, nylon; yel. liq.; sp.gr. 1.03; 25% act.

Lyogen SMK-40 Liq., SMK Paste. [537] Dye leveling agent, antiprecipitating agent for dyes; amber liq., br. gel, paste; misc. with water; sp.gr. 1.03; 20–49.5% act.

Lyogen V (U) Liq. [537] Detergent, wetting

agent, surfactant and leveling agent for scouring and printing of polyester and nylon; amber liq.; water disp.; sp.gr. 1.04.

Lyogen WD Liq. [537] Dyeing assistant, leveling agent; amber liq.; misc. with water; sp.gr. 1.05; 40% act.

M

Mackadet 40K. [401] Potassium coconate; anionic; liq. hand soap; liq.; 40% conc.

Mackam 1C. [401] Cocoamphoacetate; amphoteric; cleanser and shampoos; liq.; 45% conc.

Mackam 1C-SF. [401] Coco imidazoline carboxylate; amphoteric; shampoos, cleaner, dishwash formulation; liq.; 38% conc.

Mackam 1L. [401] Lauramphoacetate; amphoteric; shampoos, cleaners; liq.; 38% conc.

Mackam 1L-30. [401] Lauramphoacetate; amphoteric; see Mackam 1C; liq.; 36% conc.

Mackam 2C. [401] Cocoamphodiacetate; amphoteric; see Mackam lC; liq.; 50% conc.

Mackam 2C-75. [401] Cocoamphodiacetate; amphoteric; see Mackam lC; liq.; 37% conc.

Mackam 2C-SF. [401] Cocoamphodipropionate; amphoteric; see Mackam 1C-SF; liq.; 38% conc.

Mackam 2CT. [401] Caproamphodiacetate and sodium trideceth sulfate; amphoteric; shampoo; liq.; 50% conc.

Mackam 2CY. [401] Caproamphodiacetate; amphoteric; alkaline cleaner; liq.; 50% conc.

Mackam 2CYSF. [401] Caproamphodipropionate; amphoteric; heavy duty, all-purpose and metal cleaners; liq.; 50% conc.

Mackam 2L. [401] Lauric imidazoline dicarboxylate; amphoteric; see Mackam 1L; liq.; 38% conc.

Mackam 2LSF. [401] Lauroamphodipropionate; amphoteric; see Mackam 2 CYSF; liq.; 38% conc.

Mackam 35, 35 HP. [401] Cocoamidopropyl betaine; amphoteric; foamer for shampoos, bubble baths, dishwash; liq.; 35% conc.

Mackam CAP. [401] Cocamidopropyl dimethylaminopropionate; anionic; betaine; liq.; 30% conc.

Mackam CB, CB-35, CB-LS. [401] Coco betaine; amphoteric; foam booster and visc. builder; liq.; 45, 35, and 32% conc. resp.

Mackam CSF. [401] Cocoamphopropionate; amphoteric; see Mackam 2 CYSF; liq.; 39% conc.

Mackam HV. [401] Oleamidopropyl betaine; amphoteric; hair conditioner, foamer, emollient, visc. builder; liq.; 35% conc.

Mackam ISA. [401] Isostearylamidopropyl betaine; amphoteric; conditioning betaine; liq.; 35% conc.

Mackam ISP. [401] Isostearamido dimethylaminopropionate; anionic; betaine; liq.; 30% conc.

Mackam LAP. [401] Lauramidopropyl dimethylamine propionate; anionic; betaine; liq.; 30% conc.

Mackam LMB. [401] Lauramidopropyl betaine; amphoteric; shampoos, bubble baths, dishwash formulations; liq.; 35% conc.

Mackam LMB-LS. [401] Lauramidopropyl betaine; amphoteric; visc. builder and foam booster; liq.; 35% conc.

Mackam LT, MLT. [401] Lauroamphodiacetate and sodium trideceth sulfonate; amphoteric/anionic; shampoos, cleaners, detergents; liq.; 38 and 35% conc. resp.

Mackam OB. [401] Oleyl betaine; amphoteric; hair conditioner foamer, emollient, visc. builder; liq.; 50% conc.

Mackam OB-30. [401] Oleyl betaine; amphoteric; visc. builder, alkaline cleanser; liq.;

30% conc.

Mackamide 100-A. [401] Cocamide DEA; nonionic; foam stabilizer and thickener; liq.; 100% conc.

Mackamide AME-75. [401] Acetamide MEA; nonionic; humectant; liq.; 75% conc.

Mackamide AN55. [401] Alkylolamide DEA; nonionic; foam enhancer, visc. builder; liq.; 100% conc.

Mackamide C. [401] Coco diethanolamide; nonionic; foam stabilizer and thickener for shampoos; liq.; 100% conc.

Mackamide CD. [401] Cocamide DEA and diethanolamine; nonionic; foam enhancer and visc. builder; liq.; 100% conc.

Mackamide CD-8, -25, CDM, CDT. [401] Cocamide, modified; nonionic; degreaser, rust inhibitor, visc. builder; liq.; 100% conc.

Mackamide CDX. [401] Amide, modified; nonionic; see Mackamide CD-8; liq.; 100% conc.

Mackamide CMA. [401] Cocamide MEA; nonionic; foam enhancer and visc. builder; solid; 100% conc.

Mackamide CS. [401] Cocamide DEA; nonionic; see Mackamide CMA; liq.; 100% conc.

Mackamide CSA. [401] Cocamide, modified; nonionic; see Mackamide CD-8; liq.; 100% conc.

Mackamide EC. [401] Cocamide DEA; nonionic; see Mackamide CMA; liq.; 100% conc.

Mackamide ISA. [401] Isotearamide DEA; nonionic; lubricant; liq.; 100% conc.

Mackamide L-95, LLM, LMD. [401] Lauramide DEA; nonionic; foam enhancer and visc. builder; solid and liq. resp.; 100% conc.

Mackamide LMM. [401] Lauramide MEA; nonionic; see Mackamide L-95; solid; 100% conc.

Mackamide MC. [401] Cocamide DEA; nonionic; see Mackamide L-95; liq.; 100% conc.

Mackamide O. [401] Oleamide DEA; nonionic; solv. degreaser, visc. builder; liq.; 100% conc.

Mackamide PK. [401] Palm kernel diethanolamide; nonionic; visc. builder; liq.; 100% conc.

Mackamide R. [401] Ricinoleic diethanolamide; nonionic; emulsifier; softener; liq.; 100% conc.

Mackamide S. [401] Soyamide DEA; nonionic; thickener, hair conditioner; liq.; 100% conc.

Mackamine BAO. [401] Behenamido propylamine oxide; nonionic; conditioner and foam stabilizer; paste; 25% conc.

Mackamine CO. [401] Cocamine oxide; amphoteric; foam booster and stabilizer; liq.; 30% conc.

Mackamine IAO. [401] Isostearamido propylamine oxide; nonionic; conditioner and foam stabilizer; liq.; 30% conc.

Mackamine LAO. [401] Lauramidopropylamine oxide; nonionic; foam booster and visc. builder; liq.; 30% conc.

Mackamine LO. [401] Lauramine oxide; nonionic; see Mackamine LAO; liq.; 30% conc.

Mackamine O2. [401] Oleamine oxide; nonionic; see Mackamine LAO; visc. builder; liq.; 50% conc.

Mackamine OAO. [401] Oleamidopropylamine oxide; nonionic; see Mackamine LAO; liq.; 50% conc.

Mackanate CM, CM-100. [401] Disodium cocamido MEA-sulfosuccinate; anionic; base for rug cleaners, shampoos, bubble baths; 40% liq., 100% powd. resp.

Mackanate CP. [401] Disodium cocamido MIPA-sulfosuccinate; anionic; surfactant; liq; 40% conc.

Mackanate DOS 70M5. [401] Sodium dioctyl sulfosuccinate; anionic; emulsifier for dry cleaning and agric. use; liq.; 70% conc.

Mackanate DOS-75. [401] Sodium dioctyl sulfosuccinic; anionic; wetting agent and penetrant; liq.; 75% conc.

Mackanate EL. [401] Disodium laureth sulfosuccinate; anionic; shampoos and bubble baths; liq.; 40% conc.

Mackanate L-1, -2. [401] Disodium laureth sulfosuccinate; anionic; surfactant for shampoos and skin cleaners; liq.; 40% conc.

Mackanate L-101, -102. [401] Disodium laureth sulfosuccinate; anionic; surfactant for powds.; 100% conc.

Mackanate LA. [401] Diammonium lauryl sulfosuccinate; anionic; surfactant for personal care products; liq.; 40% conc.

Mackanate LM-40. [401] Disodium lauramido MEA-sulfosuccinate; anionic; surfactant; liq.; 40% conc.

Mackanate LO, -100. [401] Disodium lauryl sulfosuccinate; anionic; see Mackanate LA; solid and powd. resp.; 40 and 100% conc. resp.

Mackanate MM. [401] Disodium myristamido MEA-sulfosuccinate; anionic; liq.; see Mackanate LA; liq.; 40% conc.

Mackanate OD. [401] Disodium oleamide PEG-2 sulfosuccinate; anionic; surfactant for baby shampoos; liq.; 30% conc.

Mackanate OD-2. [401] Sulfosuccinate sulfate; anionic; surfactant for shampoos; liq.; 30% conc.

Mackanate ODT. [401] TEA-oleamido PEG-2 sulfosuccinate; anionic; see Mackanate OD-2; liq.; 40% conc.

Mackanate ODT-2. [401] Sulfosuccinate sulfate; anionic; see Mackanate OD-2; liq.; 40% conc.

Mackanate OL. [401] Sulfosuccinate; anionic; foam enhancer; liq.; 35% conc.

Mackanate OM. [401] Disodium oleamido MEA-sulfosuccinate; anionic; foaming surfactant for shampoos; liq.; 35% conc.

Mackanate OP. [401] Disodium oleamido MIPA sulfosuccinate; anionic; surfactant; liq.; 35% conc.

Mackanate RM. [401] Disodium ricinoleamide MEA-sulfosuccinate; anionic; surfactant for personal care products; liq.; 40% conc.

Mackanate TDS. [401] Disodium tridecyl sulfosuccinate; anionic; surfactant; liq.; 40% conc.

Mackanate UD. [401] Undecylenic sulfosuccinate; anionic; surfactant for shampoos; liq.; 45% conc.

Mackanate WGD. [401] Disodium wheat germamido PEG-2 sulfosuccinate; anionic; emollient surfactant; liq.; 35% conc.

Mackazoline C. [401] Cocoyl hydroxyethyl imidazoline; nonionic; emulsifier, corrosion inhibitor; liq.; 100% conc.

Mackazoline L. [401] Lauryl hydroxyethyl imidazoline; nonionic; see Mackazoline C; liq.; 100% conc.

Mackazoline O. [401] Oleyl hydroxyethyl imidazoline; nonionic; see Mackazoline C; liq.; 100% conc.

Mackester Series. [401] Glycol mono/diesters; nonionic; emulsifiers, emollient bases for cosmetics; liq.; 100% conc.

Macol 1, 2, 4, 5, 8, 10. [400] PO-EO block copolymers; nonionic; defoamer, deduster, demulsifier, detergent, dispersant, dye leveler, gelling agent, antistat; liq., paste, flake; HLB 3.0, 7.0, 15.0, 17.0, 29.0, 4.5 resp.; 100% conc.

Macol 15, 18, 19, 20. [400] EO-PO block copolymers; nonionic; defoamer, pulp and paper additive, dispersant, lubricant, leveling aid, wetting agent; liq.; HLB 15.0, 6.0, 8.0, 3.0 resp.; 100% conc.

Macol 21, 24, 25, 30. [400] Oxethylated straight chain alcohol, modified; nonionic; detergent, dispersant, wetting agent, rinse additive; liq.; HLB 7.0, 10.0, 10.0, 9.0 resp.; 100% conc.

Macol 31, 32, 33, 34. [400] EO-PO block copolymers; nonionic; see Macol 15; liq.; HLB 6.3, 4.0, 4.0, 10.0 resp.; 100% conc.

Macol 35, 42, 44, 46, 72, 77, 85, 101, 108, 121. [400] PO-EO block copolymers; nonionic; see Macol 1; liq., flake, paste; HLB 8.0, 12.0, 3.0, 18.5, 2.0, 24.5, 16.0, 1.0, 17.5, 5.0 resp.; 100% conc.

Macol CSA-2. [400] Ceteareth-2; nonionic; detergent, wetting agent, emulsifier, dispersant, solubilizer, and coupling agent for cosmetics, textiles, metalworking lubricants, household products, and industrial applications; wh. solid; sol. in IPA; insol. in water; HLB 5.1; m.p. 39 C.

Macol CSA-4. [400] Ceteareth-4 nonionic; see Macol CSA-2; wh. solid; sol. in IPA; insol. water and min. oil; HLB 7.9; m.p. 38 C.

Macol CSA-10. [400] Ceteareth-10; nonionic; see Macol CSA-2; wh. solid; sol. in IPA; disp. in water; HLB 12.6; m.p. 39 C.

Macol CSA-15. [400] Ceteareth-15; nonionic; see Macol CSA-2; wh. solid; sol. in water, IPA; insol. min. oil; HLB 14.3; m.p. 40 C.

Macol CSA-20. [400] Ceteareth-20; nonionic; see Macol CSA-2; wh. solid; sol. in water, IPA; insol. min. oil; HLB 15.4; m.p. 40 C.

Macol LA-4. [400] Laureth-4; nonionic; detergent, wetting agent, emulsifier, dispersant, solubilizer, coupling agent for cosmetics, textiles, metalworking lubricants, household products, industrial uses; colorless liq.; sol. in IPA, min. oil; HLB 9.7; 100% conc.

Macol LA-9. [400] Laureth-9; nonionic; see Macol LA-4; colorless liq.; sol. in water, IPA; HLB 11.8; 100% conc.

Macol LA-12, -23. [400] Laureth-12 and -23 resp.; nonionic; see Macol LA-4; wh. solid; sol. in water, IPA; HLB 14.5 and 16.9 resp.; m.p. 32 C (LA-23); 100% conc.

Macol LA-790. [400] Laureth-7; nonionic; see Macol CSA-2; colorless liq.; sol. in water, IPA; insol. min. oil; HLB 11.0; pour pt. 5 C.

Macol LF-110, -111, -120. [400] Polyalkoxylated aliphatic ether; nonionic; wetting aid and low surface tension surfactant; used in cleaners and rinse aids; metalworking base fluid; liq.; 100% conc.

Macol NP-4, -6. [400] Nonoxynol-4 and -6

resp.; nonionic; emulsifier, detergent, dispersant, oil surfactant and intermediate; liq.; HLB 8.8 and 10.8; 100% conc.

Macol NP-9.5. [400] Nonoxynol-9 (9.5 EO); nonionic; detergent emulsifier, wetting agent; surfactant used in paints, textile and pesticide formulation; liq.; water-sol.; HLB 13.0; 100% conc.

Macol NP-11, -20. [400] Nonoxynol–11 and –20 resp.; nonionic; wetting agent, emulsifier, and detergent; liq. and waxy solid resp.; water-sol.; HLB 14.2 and 16.0; 100% conc.

Macol NP-30, -30(70). [400] Nonoxynol-30; nonionic; detergent, wetting agent and emulsifier used in pesticide, textile treatment; HLB 17.2; 100 and 70% conc. resp.

Macol NP-70, -100. [400] Nonoxynol–70 and –100 resp.; nonionic; emulsifier; detergent and sanitizer formulations; solid (both), flake (NP-100); HLB 18.4 (NP-70); 100% conc.

Macol OA-2, -4. [400] Oleth-2 and -4 resp.; nonionic; see Macol LA-4; colorless liq.; sol. in IPA, min. oil; HLB 4.9 and 8.0; 100% conc.

Macol OA-5. [400] Oleth-5; nonionic; see Macol CSA-2; colorless liq.; sol. in IPA, min. oil; insol. in water; HLB 8.7.

Macol OA-10, -20. [400] Oleth-10 and -20 resp.; nonionic; see Macol LA-4; cream solid; sol. in water, IPA; disp. min. oil; HLB 12.4 and 15.3; m.p. 16 and 30 C; 100% conc.

Macol SA-2, -5, -10. [400] Steareth-2, -5, and –10 resp.; nonionic; see Macol LA-4; wh. solid; sol. in IPA; HLB 4.9, 8.9, and 12.4; m.p. 43 C (SA-2), 38 C (SA-10); 100% conc.

Macol SA-15, -20, -40. [400] Steareth-15, -20, and -40 resp.; nonionic; see Macol LA-4; wh. solid; sol. in water, IPA; HLB 14.2, 15.3, and 17.4; m.p. 38, 38, and 40 C; 100% conc.

Macol TD-3, -4, -6, -8. [400] Trideceth-3, -4, –6, and -8 resp.; nonionic; see Macol LA-2; liq.; HLB 7.9, 9.3, 11.4, 12.7; 100% conc.

Macol TD-10. [400] Trideceth-10; nonionic; see Macol LA-2; semisolid; HLB 13.7; 100% conc.

Macol TD-100. [400] Trideceth-100; nonionic; emulsifier and wetting agent used in cosmetic, pharmaceutical and industrial applications; solid, flake; 100% conc.

Macol TD-610. [400] Trideceth-8; nonionic; see Macol LA-2; liq.; HLB 12.7; 86% conc.

Mafo 13. [400] Potassium salt of complex *n*-stearyl amino acid; amphoteric; emulsifier, wetting agent, corrosion inhibitor, solubilizer for difficult materials, dairy chain and metal-to-metal lubricant, emollient; clear amber liq.; sol. in water and most popular solvs.; sp.gr. 1.015; visc. 140–200 cps; b.p. 230 F; flash pt. 230 F; pour pt. 10 C; biodeg.; 70% act.

Mafo C-12. [400] Potassium salt of complex *n*-coco amino acid; amphoteric; surfactant used for hard surface cleaners, agric. emulsions, cosmetics, steam cleaners, dishwashing; lt. yel. clear liq.; sol. @ 5% in 18% KOH, 7% NaOH, 15% HCl, 10% trisodium phosphate, sea water @ pH 4 and 8, disp. at pH 9–10; dens. 8.47 lb./gal; sp.gr. 1.02; flash pt. 212 F (COC); pH 10–11; 29–31% solids.

Mafo CAB. [400] Cocamidopropyl betaine; amphoteric; used in personal care, dishwashing, rug and carpet cleaning applications; liq.; 30% conc.

Magathin. [410] Lecithin deriv.; nonionic; natural fat emulsifier for toffees; liq.; 100% conc.

Maisine. [247] Monoglyceride 40%; nonionic; food emulsifier for ice creams, sauces, bread, and rice cakes; liq.; HLB 3.0; 100% conc.

Makon 4. [575] Nonoxynol-4; nonionic; detergent, emulsifier used in chemical specialties, cosmetic, agric., industrial and metal cleaners, textile, paper and petroleum industries; lt. straw liq.; oil-sol.; dens. 8.5 lb/gal; visc. 260 cps; pour pt. –20 C; pH 7.7 (1%); 100% act.

Makon 6. [575] Nonoxynol-6; nonionic; see Makon 4; lt. straw liq.; disp. in water; dens. 8.67 lb/gal; visc. 255 cps; pour pt. –29 C; pH 7.9 (1%); 100% act.

Makon 8, 10, 12. [575] Nonoxynol-8, -10, and -12 resp.; nonionic; see Makon 4; lt. straw liq.; sol. in water; dens. 8.76, 8.85, and 8.9 lb/gal; visc. 205, 235, and 300 cps; cloud pt. 24, 54, and 81 C (1%); pour pt. –5, 2.8, and 12.2 C; pH 7.0, 8.2, and 7.2 (1%); 100% act.

Makon 14. [575] Nonoxynol-14; nonionic; see Makon 4; lt. straw liq.; sol. in water; dens. 8.9 lb/gal; visc. 520 cps; cloud pt. 94 C (1%); pour pt. 18.9 C; pH 7.2 (1%); 100% act.

Makon 20. [575] Nonoxynol-20; nonionic; see Makon 4; off-wh. waxy solid; sol. in water; dens. 9.0 lb/gal; pour pt. 30 C; pH 9.0 (1%); 100% act.

Makon 30. [575] Nonoxynol-30; nonionic; see Makon 4; off-wh. waxy solid; sol. in water; dens. 9.0 lb/gal; pour pt. 40 C; pH 8.5 (1%); 100% act.

Makon NF-5, -12. [575; 576] POE aliphatic base; nonionic; detergent, wetting agent for dishwashing, industrial metal cleaning, bottle and spray washing, etc.; colorless to pale straw liq.; bland odor; sp.gr. 1.021 and 0.995 resp.; pour pt. < 40 F and < 15 F resp.; pH 9; 97+ and 100% act.

Makon NP-40, 40-70. [575; 576] Nonylphenol ethoxylate; nonionic; wetting agent, dispersant, and emulsifier; solid, liq. resp.; 100 and 70% conc. resp.

Makon OP-9. [575; 576] Octoxynol-9 (9–10 EO); nonionic; detergent for hard surface cleaners, textile and paper industries; pale straw clear liq.; m.w. 628 ± 10; water sol.; visc. 200–300 cps; HLB 13.5; cloud pt. 65 ± 2 (1% aq.); pour pt. 45 F; pH 7.5–8.5 (5% in 1:1 water:IPA); 100% act.

Manoxol IB. [383] Sodium diisobutyl sulfosuccinate; anionic; wetting agent used in dispersion, agriculture, battery separators, dry cleaning, laundry; m.w. 332; sol. 76.0 g/100 ml in water; sp.gr. 1.06 (60%); surf. tens. 49.4 dynes/cm (0.1%); 100% solid.

Manoxol MA. [383] Sodium dimethylamyl sulfosuccinate; anionic; see Manoxol IB; sol. 34.3 g/100 ml in water; sp.gr. 1.06 (60%); surf. tens. 49.1 dynes/cm (0.1%); 60% act.

Manoxol N. [383] Sodium dinonyl sulfosuccinate; anionic; see Manoxol IB; m.w. 472; sol. 11.5 g/100 ml in water; surf. tens. 27.3 dynes/cm (0.1%); 100% solid and 60% water/alcohol sol'n.

Manoxol OT. [383] Dioctyl sodium sulfosuccinate; anionic; see Manoxol IB; m.w. 444; sol. in most solvs., oils, org. media, 1.5% sol. in water; sp.gr. 1.1; surf. tens. 30 dynes/cm (0.1%); 100% act.

Manoxol OT 60%. [383] Manoxol OT sol'n.; anionic; see Manoxol IB; water-wh. liq.; water disp.; sp.gr. 1.05; flash pt. 80 F; 60% act. in 15% industrial methylated spirits and 25% water.

Manoxol OT/B. [383] 85% Manoxol OT and 15% sodium benzoate; anionic; see Manoxol IB; also used for making wettable powds.; powd.; 85% act.

Manoxol OT/P. [383] Dioctyl sodium sulfosuccinate; pharmaceutical grade of Manoxol OT.

Manro ALES 27. [384] Ammonium lauryl ether sulfate; anionic; used in shampoos, bubble baths; pale yel. clear liq.; mild odor; sp.gr. 1.03; visc. 3000 cps; biodeg.; 27% act.

Manro ALES 60. [384] Ammonium syn. C_{13}–C_{15} fatty alcohol ether sulfate; anionic; used in liq. detergents, cleaning products, fire fighting foams; amber, slightly hazy, mobile gel; slight alcoholic odor; sp.gr. 1.033; pH 7–8 (10% aq.); biodeg.; 60% act.

Manro ALS 30. [384] Ammonium lauryl sulfate; anionic; used in cosmetics and toiletries, esp. shampoos; pale yel. clear to slightly hazy, visc. liq.; mild odor; sp.gr. 1.02; visc. 5000 cps; pH 6–7 (10% aq.); biodeg.; 28% min. act.

Manro ALS 30S. [384] Ammonium alkyl sulfate; used in cosmetics and toiletries, esp. shampoos; pale yel. clear to slightly hazy, visc. liq.; negligible odor; sp.gr. 1.03; visc. 5000 cps; pH 6–7 (10% aq.); biodeg.; 28% min. act.

Manro BA. [384] Dodecylbenzene sulfonic acid; anionic; used in heavy and lt. duty detergents, emulsion polymerization; dk. br., visc. liq.; char. odor; sp.gr. 1.04; visc. 1500 cps; biodeg.; 95% min. act.

Manro BA 25, 30. [384] Sodium dodecylbenzene sulfonate; anionic; used in liq. detergents, cleaning products, emulsion polymerization systems; pale amber clear liq.; sp.gr. 1.04 and 1.05 resp.; visc. 250 and 1000 cps; pH 7–8 (10% aq.); biodeg.; 25 and 29% min. act.

Manro BEC 28, 70. [384] Alcohol ethoxy sulfate, sodium salt; anionic; foaming and wetting in shampoos and toiletry products; liq., gel resp.; 28 and 70% conc. resp.

Manro BES 27. [384] Sodium alkyl ether sulfate; foaming agent, emulsifier used in cosmetics, toiletries, detergent and industrial products; pale yel., mobile liq.; mild odor; sp.gr. 1.04; visc. 200 cps; pH 6.5–7.5 (10% aq.); biodeg.; 26.5% min. act.

Manro BES 60. [384] Sodium syn. C_{13}–C_{15} fatty alcohol ether sulfate; used in liq. detergents, cleaning products, fire fighting foams; amber, slightly hazy, mobile gel; slight alcoholic odor; sp.gr. 1.090; pH 7–8 (10% aq.); biodeg.; 60% act.

Manro BES 70. [384] Sodium alkyl ether sulfate; anionic; foaming agent, emulsifier for chlorinated solvs., lime soap dispersant, intermediate used in cosmetics, household cleaning products and fire fighting foams; pale yel., mobile gel; mild odor; sp.gr. 1.05; pH 7–9 (10% aq.); biodeg.; 67.5% min. act.

Manro CD, CDS, CDX. [384] Cocamide DEA; foam stabilizer, solubilizer used in liq. detergents, shampoos, bubble baths; clear golden yel. visc. liq.; mild odor; sp.gr. 0.982, 0.990, 1.01 resp.; visc. 950, 800, 950 cps; biodeg.; 90, 85, and 70% act.

Manro CMEA. [384] Cocamide MEA; nonionic; foam booster, foam stabilizer used in liq. and powd. detergent systems; cream waxy flake; mild odor; biodeg.; 94% min. act.

Manro DB 30 S, DB 45. [384] Anionic/nonionic; detergent, foaming agent used in liq. detergents, hard surface cleaners; clear golden yel., visc. liq.; negligible odor; sp.gr. 1.048 and 1.05 resp.; visc. 500 and 1200 cps; cloud pt. 0 and 5 C; pH 6.2–6.8 and 6.5–7.5; 30 and 45% act.

Manro DES. [384] Alcohol ethoxy sulfate, sodium salt; anionic; foamer for wallboard, plasterboard; liq.; 32% conc.

Manro DL 28. [384] Sodium lauryl sulfate, modified; anionic; used in foamable precoats and no-gel foam compds. in latex industry; pale yel. clear liq.; mild char. odor; sp.gr. 1.05; visc. 50 cps; cloud pt. 0 C; pH 7–8 (10% aq.); biodeg.; 28% min. act.

Manro DL 32. [384] Sodium dodecylbenzene sulfonate; anionic; emulsion polymerization; liq.; 32% conc.

Manro DP 60. [384] Detergent used in nonscratch cream cleanser; wh. visc. liq.; negligible odor; sp.gr. 1.47; visc. 3200 cps; pH 10.5 (10% aq.); 60% solids.

Manro D Paste. [384] Sodium cetyl oleyl alcohol sulfate; used in wool scouring and Kier boiling in textile industry; cream paste; char. odor; sp.gr. 1.08 (40 C); pH 7–9.5 (10% aq.); 29% act.

Manro DS 35. [384] Sodium alkyl sulfate; anionic; solubilizer, wetting agent used in industrial cleaners; clear pale yel. liq.; char. odor; sp.gr. 1.07; visc. 50 cps; pH 7–8 (10% aq.); 35% act.

Manro HA. [384] Dodecylbenzene sulfonic acid; anionic; used in heavy and light duty detergents; dk., visc. liq.; char. odor; sp.gr. 1.06; visc. 18,000 cps; 95% min. act.

Manro HCS. [384] Isopropylamine dodecylbenzene sulfonate; anionic; emulsifier in the mfg. of kerosene-based hand cleaning preparations; amber clear visc. liq.; mild odor; sp.gr. 1.03; visc. 3000 cps; pH 5–6 (2% aq.); 89.5% min. act.

Manro KXS 40. [384] Potassium xylene sulfonate; solubilizing agent; pale yel. clear liq.; negligible odor; sp.gr. 1.18 (40 C); pH 7–10 (10% aq.); 39% min. act. in water.

Manro MA 35. [384] Disodium octadecyl sulfosuccinamate; anionic; used in no-gel foam systems based on high solids, noncarboxylated SBR latex or natural rubber latex; lt. cream fluid; mild, char. odor; sp.gr. 1.05; pH 7.5–9.5 (10% aq.); 35% solids.

Manro ML 33. [384] MEA-lauryl sulfate; anionic; used in cosmetics and toiletries; pale yel. clear to slightly hazy visc. liq.; mild odor; sp.gr. 1.03; visc. 7500 cps; pH 6.3–7.3 (10% aq.); biodeg.; 32% act.

Manro ML 33S. [384] MEA alkyl sulfate; used in cosmetics and toiletries; pale yel. clear to slightly hazy visc. liq.; mild odor; sp.gr. 1.03; visc. 7500 cps; pH 6.3–7.3 (10% aq.); biodeg.; 32% min. act.

Manro NA. [384] Dodecylbenzene sulfonic acid; anionic; raw material used as base for liq. heavy and light duty detergents; dk. br. visc. liq.; char. odor; sp.gr. 1.06; visc. 2000 cps; 96% min. act.

Manro NEC, NEC 70. [384] Sodium lauryl ether sulfate; anionic; foaming agent used in shampoos, bubble baths; water wh. clear liq., pale yel. mobile gel; mild odor; sp.gr. 1.05; pH 6–7 (10% aq.); biodeg.; 27 and 68% act.

Manro PC 35N. [384] Anionic/nonionic; pearlized lotion conc. for shampoos, bubble baths; wh., visc. liq.; negligible odor; sp.gr. 1.00; visc. 16,000 cps; pH 6.5–7.0 (10% aq.); 35% act.

Manro SDBS 30. [384] Sodium dodecylbenzene sulfonate; anionic; used in liq. detergents and cleaning products, emulsion polymerization; amber clear visc. liq.; mild odor; sp.gr. 1.05 (30 C); visc. 1000 cps (30 C); pH 7–9 (10% aq.); biodeg.; 29% min. act.

Manro SIOS. [384] Sodium isooctyl sulfate; wetting agent in industrial detergents; pale amber clear liq.; char. odor; sp.gr. 1.12; pH 7–8 (10% aq.); 38% act. in water.

Manro SLS 28, 28S. [384] Sodium lauryl sulfate; anionic; used in cosmetics, toiletries, emulsion polymerization; very pale yel. clear liq.; mild odor; sp.gr. 1.04; visc. 200–7000 cps; pH 7–8 (10% aq.); biodeg.; 28% min. act.

Manro STS 40. [384] Sodium toluene sulfonate; solubilizing agent; pale yel. clear liq.; negligible odor; sp.gr. 1.19; pH 7–10 (10% aq.); 40% act. in water.

Manro SXS 30, 40. [384] Sodium xylene sulfonate; solubilizing agent; pale yel. clear liq.; negligible odor; sp.gr. 1.15–1.17 (30 C); pH

7–10 (10% aq.); 30 and 40% act. in water.

Manro TDBS 60. [384] TEA dodecylbenzene sulfonate; anionic; detergent used in car shampoos; amber clear visc. liq.; slight, alcoholic odor; sp.gr. 1.06; visc. 400 cps; pH 6.5–7.5 (10% aq.); 60% act.

Manro TL 40. [384] TEA lauryl sulfate; anionic; used in shampoos, bubble baths; pale yel. clear liq.; mild odor; sp.gr. 1.025; visc. 100 cps; pH 6.5–7.0 (10% aq.); biodeg.; 40% act.

Manro TL 40S. [384] TEA alkyl sulfate; used in shampoos, bubble baths; pale yel. clear liq.; negligible odor; sp.gr. 1.04; visc. 300 cps; pH 6.5–7.0 (10% aq.); biodeg.; 37% min. act.

Mapeg 200 DL. [400] PEG 200 dilaurate; nonionic; emulsifier, dispersant used in cosmetics, pharmaceuticals, metalworking and fiber lubricants, etc.; yel. clear liq.; sol. in IPA, toluol, soybean and min. oil, water disp.; sp.gr. 0.95; m.p. 10 C; HLB 6.8; sapon. no. 176–192; 100% conc.

Mapeg 200 DO. [400] PEG 200 dioleate; nonionic; see Mapeg 200 DL; yel. clear liq.; sol. in IPA, toluol, soybean and min. oil, water disp.; sp.gr. 0.95; m.p. < –10 C; HLB 6.0; sapon. no. 148–158; 100% conc.

Mapeg 200 DS. [400] PEG 200 distearate; nonionic; see Mapeg 200 DL; wh. solid; sol. in IPA, toluol, soybean and min. oil, hot water disp.; m.p. 34 C; HLB 4.7; sapon. no. 155–165; 100% conc.

Mapeg 200 ML. [400] PEG 200 monolaurate; nonionic; see Mapeg 200 DL; yel. clear liq.; sol. in IPA, toluol, soybean and min. oil, water disp.; sp.gr. 0.99; m.p. 5 C; HLB 9.8; sapon. no. 139–159; 100% conc.

Mapeg 200 MO. [400] PEG 200 monooleate; nonionic; see Mapeg 200 DL; yel. clear liq.; sol. in IPA, toluol, soybean oil, water disp.; sp.gr. 0.98; m.p. < –10 C; HLB 8.0; sapon. no. 115–125; 100% conc.

Mapeg 200 MOT. [400] PEG 200 monotallate; nonionic; see Mapeg 200 DL; liq.; HLB 8.0; 100% conc.

Mapeg 200 MS. [400] PEG 200 monostearate; nonionic; see Mapeg 200 DL; wh. solid; sol. in IPA, toluol, soybean oil, hot water disp.; m.p. 33 C; HLB 8.0; sapon. no. 120–130; 100% conc.

Mapeg 350 MS. [400] PEG 350 monostearate; nonionic; see Mapeg 200 DL; solid; HLB 11.0; 100% conc.

Mapeg 400 DL. [400] PEG 400 dilaurate; nonionic; see Mapeg 200 DL; lt. yel. liq.; sol. in IPA, toluol, soybean oil, water disp.; sp.gr. 0.98; m.p. 18 C; HLB 9.8; sapon. no. 130–140; 100% conc.

Mapeg 400 DO. [400] PEG 400 dioleate; nonionic; see Mapeg 200 DL; yel. liq.; sol. in IPA, toluol, soybean and min. oil, water disp.; sp.gr. 0.98; m.p. < 7 C; HLB 8.5; sapon. no. 114–122; 100% conc.

Mapeg 400 DOT. [400] PEG 400 ditallate; nonionic; see Mapeg 400 DL; liq.; HLB 8.5; 100% conc.

Mapeg 400 DS. [400] PEG 400 distearate; nonionic; see Mapeg 200 DL; wh. solid; sol. in IPA, toluol, soybean and min. oil, hot water disp.; m.p. 36 C; HLB 8.1; sapon. no. 116–125; 100% conc.

Mapeg 400 ML. [400] PEG 400 monolaurate; nonionic; see Mapeg 200 DL; lt. yel., liq.; sol. in IPA, toluol, water; sp.gr. 1.01; m.p. 12 C; HLB 13.1; sapon. no. 89–96; 100% conc.

Mapeg 400 MO. [400] PEG 400 monooleate; nonionic; see Mapeg 200 DL; yel. liq.; sol. in IPA, toluol, soybean oil, water disp.; sp.gr. 1.01; m.p. < 10 C; HLB 11.4; sapon. no. 80–88; 100% conc.

Mapeg 400 MOT. [400] PEG-8 tallate; nonionic; see Mapeg 200 DL; liq.; HLB 11.4; 100% conc.

Mapeg 400 MS. [400] PEG 400 monostearate; nonionic; see Mapeg 200 DL; wh. solid; sol. in IPA, toluol, soybean oil, hot water disp.; m.p. 33 C; HLB 11.5; sapon. no. 84–93; 100% conc.

Mapeg 600 DL. [400] PEG 600 dilaurate; dispersant and emulsifier for metalworking lubricants, fiber lubricants and softeners, solubilizers, defoamers, antistats, cosmetics, pharmaceuticals, and chemical intermediates; lt. yel. semisolid; sol. in IPA, toluol, soybean oil; partly sol. min. oil; disp. water; sp.gr. 0.99; HLB 12.2; m.p. 24 C; sapon. no. 102–112.

Mapeg 600 DO. [400] PEG 600 dioleate; nonionic; see Mapeg 200 DL; yel. liq.; sol. in IPA, toluol, soybean oil, water disp.; sp.gr. 1.00; m.p. 20 C; HLB 10.5; sapon. no. 92–102; 100% conc.

Mapeg 600 DOT. [400] PEG 600 ditallate; see Mapeg 600 DL; amber liq.; sol. IPA, toluol, soybean oil; disp. water; sp.gr. 1.00; HLB 13.2; m.p. 15 C; sapon. no. 85–95.

Mapeg 600 DS. [400] PEG 600 distearate; nonionic; see Mapeg 200 DL; wh. solid or flake; sol. in IPA, toluol, soybean oil, hot water disp.; m.p. 41 C; HLB 10.6; sapon. no. 94–104; 100% conc.

Mapeg 600 ML. [400] PEG-12 laurate; see Mapeg 600 DL; lt. yel. liq.; sol. in water, IPA, toluol; sp.gr. 1.02; HLB 14.9; m.p. 23 C; sapon. no. 64–74.

Mapeg 600 MO. [400] PEG 600 monooleate; nonionic; see Mapeg 200 DL; yel. liq.; sol. in IPA, toluol, soybean oil, water; sp.gr. 1.03; m.p. 25 C; HLB 13.5; sapon. no. 60–70; 100% conc.

Mapeg 600 MOT. [400] PEG 600 monotallate; see Mapeg 600 DL; amber liq.; sol. in water, IPA, toluol; sp.gr. 1.04; HLB 10.3; m.p. 20 C; sapon. no. 60–70.

Mapeg 600 MS. [400] PEG 600 monostearate; nonionic; see Mapeg 200 DL; wh. solid; sol. in IPA, toluol, soybean oil, water, propylene glycol, disp. min. oil; m.p. 36 C; HLB 13.6; sapon. no. 62–70; 100% conc.

Mapeg 1000 MS. [400] PEG 1000 monostearate; nonionic; see Mapeg 200 DL; wh. solid or flake sol. in IPA, toluol, propylene glycol, water; m.p. 42 C; HLB 15.7; sapon. no. 41–49; 100% conc.

Mapeg 1500 MS. [400] PEG 1500 monostearate; nonionic; see Mapeg 200 DL; solid; HLB 16.8; 100% conc.

Mapeg 1540 DS. [400] PEG 1540 distearate; nonionic; see Mapeg 200 DL; wh. solid or flake; sol. in IPA, toluol, soybean oil, propylene glycol, water; m.p. 45 C; HLB 14.8; sapon. no. 49–58; 100% conc.

Mapeg 4000 MS. [400] PEG-75 stearate; nonionic; see Mapeg 200 DL; liq.; HLB 18.7; 100% conc.

Mapeg 6000 DS. [400] PEG 6000 distearate; nonionic; see Mapeg 200 DL; wh. solid or flake; sol. in IPA, toluol, propylene glycol, water; m.p. 55 C; HLB 18.4; sapon. no. 14–20; 100% conc.

Mapeg 6000 MS. [400] PEG-150 stearate; see Mapeg 600 DL; lt. yel. waxy solid, flake; sol. in water, IPA, toluol; HLB 18.8; m.p. 60 C; sapon. no. 7–14.

Mapeg CO-25H. [400] PEG-25 hydrog. castor oil; nonionic; surfactant for formulation of gels; liq.; HLB 10.8; 100% conc.

Mapeg DGLD. [400] Diethylene glycol monolaurate; nonionic; see Mapeg CO-25H; HLB 8.3; 100% conc.

Mapeg EGDS. [400] Ethylene glycol distearate; nonionic; see Mapeg 200 DL; wh. solid or flake; sol. in IPA, toluol, soybean and min. oil, m.p. 63 C; HLB 1.4; sapon. no. 190–199; 100% conc.

Mapeg EGMS. [400] Ethylene glycol monostearate; nonionic; see Mapeg 200 DL; wh. to cream solid or flake; sol. in IPA, toluol, soybean and min. oil; m.p. 56 C; HLB 2.4; sapon. no. 180–188; 100% conc.

Mapeg PGDS. [400] Propylene glycol distearate; see Mapeg 600 DL; lt. yel. solid; sol. in IPA, toluol, propylene glycol, soybean and min. oils; insol. water; HLB 1.2; m.p. 40 C; sapon. no. 183–193.

Mapeg PGMS. [400] Propylene glycol stearate; see Mapeg 600 DL; lt. yel. solid; sol. in IPA, toluol, propylene glycol, soybean and min. oil; insol. in water; HLB 2.3; m.p. 36 C; sapon. no. 172–182.

Mapeg S-40. [400] PEG-40 monostearate; nonionic; see Mapeg 200 DL; wh. solid or flake; sol. in IPA, toluol, propylene glycol, water; m.p. 48 C; HLB 17.4; sapon. no. 25–35; 100% conc.

Mapeg S 100. [400] PEG-100 monostearate; nonionic; emulsifier for cosmetic and pharmaceutical and other chemical specialty formulations; liq.; water-sol.; HLB 18.9; 100% conc.

Mapeg TAO-15. [400] PEG 660 monotallate; nonionic; see Mapeg 200 DL; clear amber liq.; sol. in IPA, toluol, water; m.p. 20 C; HLB 13.8; sapon. no. 55–65; 100% conc.

Maphos 4. [400] Complex org. phosphate acid ester; anionic; wetting agent used in detergent formulations, emulsion polymerization, oil and cutting fluids; liq.; oil-sol.; water-disp.; 100% conc.

Maphos 17. [400] Aromatic phosphate ester; anionic; emulsifier for emulsion polymerization; yel. clear visc. liq.; sol. @ 5% in water @ pH 2.0 and 9.5; sp.gr. 1.11; pour pt. < 0 C; pH 1.5–2.5 (1% aq.); 68.5–71.5% act.

Maphos 60. [400] Aliphatic phosphate ester; anionic; hydrotrope used in hard surface cleaners, dedusting applications, detergent, anticorrosive agent; lt. yel. clear visc. liq.; sol. @ 5% in dist. water @ pH 2.0 and 9.5; sp.gr. 1.08; pour pt. < 0 C; pH 1.5–2.5 (1% aq.); 100% act.

Maphos 60A. [400] Complex org. phosphate acid ester; anionic; textile wetting, hard surface detergent; liq.; water-sol.; 100% conc.

Maphos 66. [400] Potassium salt of phosphate ester; see Maphos 60; colorless clear liq.; sol. in dist. water @ pH 9.5 @ 5%; sp.gr. 1.18; pour pt. < 0 C; pH 9–9.5 (1% aq.); 45–47% act.

Maphos 66H. [400] Phosphate acid ester; anionic; hard surface cleaning hydrotrope; liq.; 50% conc.

Maphos 76. [400] Aromatic phosphate ester; anionic; detergent, antistat, anticorrosive agent, emulsifer used in drycleaning formulations, hard surface cleaners, emulsion polymerization; yel. clear visc. liq.; sol. in dist. water @ pH 9.5 (@ 5%), in perchloroethylene (@ 10%); sp.gr. 1.09; pour pt. < 0 C; pH 1.5–2.5 (1% aq.); 100% act.

Maphos 76 NA. [400] Partial sodium salt of aromatic phosphate ester; anionic; detergent, antistat, anticorrosive agent used in drycleaning formulations; lt. yel. clear liq.; sol. in dist. water @ pH 2.0 and 9.5 (@ 5%), in perchloroethylene (@ 10%); sp.gr. 1.08; pour pt. 8 C; pH 5.0–6.0 (1% aq.); 87–89% act.

Maphos 78. [400] Complex org. phosphate ester; anionic; hydrotrope with detergency properties; liq.; 100% conc.

Maphos 91. [400] Aromatic phosphate ester; anionic; see Maphos 76; also in dedusting, lubricant additives; clear yel. visc. liq.; sol. in dist. water @ pH 2.0 and 9.5 (@ 5%), in perchloroethylene and xylene (@ 10%); sp.gr. 1.10; pour pt. 2 C; pH 1.5–2.5 (1% aq.); 100% act.

Maphos 6600. [400] Complex org. phosphate acid ester; anionic; hydrotrope used in cleaning applications; liq.; 100% conc.

Maphos 8135. [400] Aromatic phosphate ester; anionic; hydrotrope, lubricant additive for greases, syn. cutting oils, drawing compds.; lt. amber visc. liq.; sol. in dist. water @ pH 2.0 and 9.5 (@ 5%); sp.gr. 1.07; pour pt. 2 C; pH 1.5–2.5 (1% aq.); 100% act.

Maphos L-6. [400] Aromatic phosphate ester; anionic; detergent, antistat, anticorrosive agent, lubricant for drycleaning and lubricant formulations; yel. visc. liq.; sol. @ 10% in perchloroethylene, min. oil, and xylene; sp.gr. 1.05; pour pt. 8 C; pH 1.5–2.5 (1% aq.); 100% act.

Maphos L-12. [400] Aliphatic phosphate ester; anionic; detergent, lubricant, emulsifier for hard surface cleaners, lubricant formulations, emulsion polymerization; clear lt. yel. visc. liq.; sol. @ 5% in dist. water @ pH 9.5, @ 10% in min. oil and xylene; sp.gr. 1.01; pour pt. 12 C; pH 1.5–2.5 (1% aq.); 100% act.

Maprofix 563. [456] Sodium lauryl sulfate; anionic; detergent, foamer used as dentifrice, in toothpaste, mouthwashes, food additives, fruit washing compds.; powd. or granule; sp.gr. 0.66; flash pt. > 200 F; 97% min. act.

Maprofix 563-SD. [456] Sodium lauryl sulfate; anionic; dentifrice; used in pharmaceutical and food applications; powd.; sp.gr. 0.40; flash pt. > 200 F; 95% min. act.

Maprofix AEG. [456] Ammonium lauryl sulfate, ammonium laureth sulfate, cocoamido propyl betaine, cocoamido DEA; anionic/amphoteric; surfactant conc. for personal care products; liq.; sp.gr. 1.02; flash pt. 148 F; pH 4.5–6.0; 40–44% solids.

Maprofix DFS. [456] Sodium lauryl sulfate and an amine oxide; anionic; good foaming detergent used in rug and upholstery shampoos, hard surface cleaners; liq.; sp.gr. 0.98; flash pt. 140 F; pH 7.5–9.5; 27.5% min. solids.

Maprofix DLS-35. [456] DEA lauryl sulfate; anionic; surfactant for use in personal care products; sp.gr. 1.03; flash pt. > 200 F; 33–35% act.

Maprofix EP-1. [575] Sodium lauryl sulfate; anionic; emulsion polymerization; liq.; 30% conc.

Maprofix ER. [575] Ammonium lauryl sulfate; anionic; used in personal care products and industrial foams; liq.; 30% conc.

Maprofix ES. [456] Sodium laureth-3 sulfate; anionic; detergent, foamer used in shampoos, bubble baths, cosmetics; liq.; sp.gr. 1.05; flash pt. > 200 F; 27% min. act.

Maprofix ES1. [575] Sodium laureth-1 sulfate; anionic; emulsion polymerization; detergent foamer shampoos, bubble bath; liq.; 25% conc.

Maprofix ES2. [575] Sodium laureth sulfate; shampoos, bubble baths; liq.; 30% conc.

Maprofix ES3. [575] Sodium laureth-3 sulfate; anionic; see Maprofix ES1; liq.; 28% conc.

Maprofix ESY. [456] Sodium laureth-1 sulfate; anionic; detergent, foamer used in shampoos, bubble baths, cosmetics; liq.; sp.gr. 1.04; flash pt. > 200 F; 24.5–25.5% act.

Maprofix LCP. [456] Sodium lauryl sulfate; anionic; foaming and wetting agent used in emulsion polymerization, lt. duty household and industrial formulations; liq.; sp.gr. 1.05; flash pt. > 200 F; 28–30% act.

Maprofix LES-60A. [456] Ammonium laureth-3 sulfate; anionic; detergent used lt. duty liq. detergents, shampoo conc.; liq.;

sp.gr. 1.06; flash pt. 82 F; 57–60% act.

Maprofix LES-60C. [456] Sodium laureth-3 sulfate; anionic; detergent used in lt. duty liq. detergents, in shampoo conc., bubble baths; liq.; sp.gr. 1.06; flash pt. 82 F; 56–59% act.

Maprofix LK (USP). [456] Sodium lauryl sulfate; anionic; foaming agent, dispersant used in powdered detergent formulations for household and personal care products, industrial pigment dispersant; powd.; sp.gr. 0.35; flash pt. > 200 F; 90% act.

Maprofix MG. [456] Magnesium lauryl sulfate; anionic; wetting agent used in rug and upholstery shampoos; liq.; sp.gr. 1.02; flash pt. > 200 F; 27–29% act.

Maprofix MM. [456] Sodium lauryl sulfate; anionic; industrial and cosmetic detergent used in rug and hair shampoos; paste; sp.gr. 1.09; flash pt. > 200 F; 41.5% min. act.

Maprofix NH. [456] Ammonium lauryl sulfate; anionic; detergent, foaming agent used in nonalkaline shampoos, bubble baths, mild cleaners; visc. liq.; sp.gr. 1.01; flash pt. > 200 F; 27–30% act.

Maprofix NHL. [456] Ammonium lauryl sulfate; anionic; foaming agent for industrial use; liq.; sp.gr. 1.02; flash pt. > 200 F; 27–30% act.

Maprofix NHL-22. [456] Ammonium lauryl sulfate; anionic; low visc. version of Maprofix NH; liq.; sp.gr. 1.01; flash pt. > 200 F; 27–28% act.

Maprofix NH-54. [456] Ammonium lauryl sulfate; anionic; industrial foaming agent; liq.; sp.gr. 0.90; flash pt. 82 F; 49–51% act.

Maprofix RS. [456] Sodium lauryl sulfate and sodium lauroyl sarcosinate; anionic; foaming and wetting agent used in rug and upholstery shampoo concs.; liq.; sp.gr. 1.04; flash pt. > 200 F; pH 7.5–9.5; 28% solids.

Maprofix RSN. [456] Ammonium lauryl sulfate and ammonium lauroyl sarcosinate; anionic; foaming and detergent for nonalkaline rug and upholstery shampoo concs.; liq.; sp.gr. 1.01; flash pt. > 200 F; pH 6.5–7.5; 24.5% solids.

Maprofix SP. [456] Anionic/amine oxide blend; high foaming surfactant for acid and alkaline personal care products; liq.; sp.gr. 1.03; flash pt. > 200 F; pH 7.5–9.0; 24.5–27.5% solids.

Maprofix TLS, TLS-65, -106. [456] TEA lauryl sulfate; anionic; detergent for shampoo concs.; paste, sp.gr. 1.05, 1.06, and 1.05 resp.; flash pt. 88, > 200, and 88 F; 79, 64–67, and 79% min. act. resp.

Maprofix TLS-500, -513. [456] TEA lauryl sulfate; anionic; detergent, foaming agent for cosmetic and industrial formulations; liq.; sp.gr. 1.05; cloud pt. 0 and 10 C resp.; flash pt. > 200 F; 40 and 39% min. act. resp.

Maprofix WAC. [456] Sodium lauryl sulfate; anionic; detergent, foamer for shampoos, bubble baths, detergent specialties; very lt. color liq.; sp.gr. 1.04; flash pt. > 200 F; 28–30% act.

Maprofix WAC-LA. [456] Sodium lauryl sulfate; anionic; detergent for personal care and industrial formulations, emulsion polymerization; water-wh. liq.; sp.gr. 1.04; flash pt. > 200 F; 28–30% act.

Maprofix WAM. [456] Sodium lauryl sulfate; anionic; surfactant, foaming agent and detergent for personal care products and detergent specialties; liq.; sp.gr. 1.05; flash pt. > 200 F; 28–30% act.

Maprofix WA Paste. [456] Sodium lauryl sulfate; anionic; detergent used in industrial and cosmetic applications, in rug and hair shampoos; paste; sp.gr. 1.09; flash pt. > 200 F; 41.5% act.

Maprolyte 101. [456] Anionic/amphoteric; detergent, conditioner used in shampoos and bubble baths for conditioning properties; liq.; sp.gr. 1.03; flash pt. > 200 F; pH 7.5–8.5; 33–37% solids.

Maprolyte C. [456] Cocoamidopropyl betaine; nonionic; detergent, wetting and foaming agent, stabilizer used in shampoos and bubble baths; liq.; sp.gr. 1.04; flash pt. > 200 F; 33–37% solids.

Maprolyte LX. [456] DEA-lauryl sulfate, DEA-lauraminopropionate and sodium lauraminopropionate; anionic/amphoteric; foaming agent, conditioner, mild conc. for shampoos, body soaps, bubble baths; liq.; sp.gr. 1.03; flash pt. > 200 F; pH 7.5–8.5; 33–37% solids.

Maprosyl 30. [456] Sodium *n*-lauroyl sarcosinate; anionic; detergent, wetting and foaming agent used in personal care and household detergent products; liq.; sp.gr. 1.03; flash pt. > 200 F; 29.5–30.5% solids.

Maramul SS. [502] Lignosulfonate; anionic; emulsifier for asphalt emulsions; powd., liq.; 45-100% conc.

Maranil A. [289] Linear dodecylbenzene sulfonate; anionic; detergent base for household and industrial cleansers; dishwashing liq.; powd.; 90% conc.

Maranil Paste A 55, A 75, Powder A. [291] Sodium dodecylbenzene sulfonate; anionic; base for mfg. detergents, dishwashing, cleaning and wetting agent; paste, powd.; 55, 75, and 90% conc. resp.

Marasperse C-21. [40] Highly purified calcium lignosulfonate; anionic; dispersant, emulsion stabilizer used in agric. chemical formulations, gypsum board additive, industrial cleaners, in mfg. of brick, tile, refractories, pottery and porcelain ware; br. powd.; 100% water sol.; dens. 35–40 lb/ft^3; surf. tens. 49.4 dynes/cm (1%); pH 7.0–8.2 (3%); 100% conc.

Marasperse CB, N-22. [40] Sodium lignosulfonate; anionic; dispersant, o/w emulsion stabilizer used in textile dyeing, agric. chemical formulations, gypsum board additives, industrial cleaners; black powd., br. powd. resp.; 100% water sol.; dens. 43–47 and 35–40 lb/ft^3; surf. tens. 51.4 and 52.8 dynes/cm (1%); pH 8.5–9.2 and 7.5–8.5 (3%); 100% conc. (N-22).

Marlamid D 1218, DF 1218. [308] Coconut diethanolamide; nonionic; foam stabilizer in wash liquors, superfatting agent; liq.; 100% act.

Marlamid D 1885. [308] Oleic acid diethanolamide; nonionic; foam stabilizer in wash liquors, superfatting agent; liq.; 100% act.

Marlamid KL. [308] Cocamidopropyl lauryl ether; nonionic; pearlescent surfactant for liq. and paste detergents; solid; 100% conc.

Marlamid KLA. [308] Fatty acid alkylolamide and fatty alcohol ether sulfate; nonionic; pearlescent surfactant for personal care products; liq.; 40% conc.

Marlamid M 1218. [308] Coconut monoethanolamide; nonionic; foam stabilizer, thickener in household, personal, industrial detergents, superfatting agent; solid; 100% act.

Marlazin 7265. [308] Cationic surfactant used in the mfg. of products having water-repellent properties; solid; 100% act.

Marlazin 8567. [308] Aromatic amine; cationic surfactant used in floor cleaners and polymer strippers; liq.; 100% act.

Marlazin L 10, L 410. [308] Fatty amine polyglycol ether; nonionic surfactant used in industrial cleaners, in dye and fiber auxiliaries; liq.; 100 and 50% act. resp.

Marlazin OK 1. [308] Alkanolamine fatty condensate; cationic surfactant for the mfg. of products having water-repellent properties; solid; 100% act.

Marlazin OL 20, S 10. [308] Fatty amine polyglycol ether; nonionic; see Marlazin L 10; liq.; 100% act.

Marlazin T 10, T 410. [308] Fatty amine polyglycol ether; nonionic; see Marlazin L 10; liq.; 100% and 50% act. resp.

Marlican. [308] Straight-chain dodecylbenzene; anionic; detergent intermediate, solubilizer; liq.; 100% act.

Marlinat DF 8. [308] Dioctyl sodium sulfosuccinate; anionic; highly act. wetting agent for textile, paint and paper industries used in cleaners, cosmetic preparations; liq.; 70% act.

Marlinat HA 12. [308] Sodium salt of lauric acid monoethanolamide sulfosuccinate; anionic; detergent raw material used in cleaners of sensitive textiles, cosmetic detergents; paste; 45% act.

Marlinat SRN 30. [308] Anionic; see Marlinat HA 12; liq.; 30% act.

Marlipal 34/30, /50, /60, /70, /79, /99, /100, /109, /110, /119, /120, /140, /200, /300, 1039. [308] C_{13}/C_{14}-oxoalcohol polyglycol ethers; nonionic surfactants in powd. and liq. detergents for household and industry, emulsifiers and wetting agents in textile industry; liq. to solid; 100% act.

Marlipal 124, 129, 1218/5, 1618/11, 1618/25, 1618/40, 1850/5, 1850/8, 1850/10, 1850/30, 1850/80. [308] Fatty alcohol polyglycol ethers; nonionic surfactants; sol. in oil and water; liq. to wax; 100% act.

Marlipal BS. [308] Fatty acid polyglycol ester; nonionic surfactant, superfatting agent, thickener in detergents, dishwashing and cosmetic preparations; wax; 100% act.

Marlipal FS. [308] PEG-12 dioleate; nonionic surfactant used for liq. and paste detergents, esp. hair shampoos, cosmetic preparations; liq.; 100% act.

Marlipal KE, KF. [308] Fatty alcohol polyglycol ethers; nonionic surfactant, detergent raw material and additive used in dishwashing agents, in mfg. of tablet soaps; liq. to solid; 100% act.

Marlipal MG. [308] Laureth-7; nonionic; see Marlipal KE; 100% act.

Marlipal ML. [308] Fatty alcohol polyglycol ethers; nonionic; see Marlipal KE; liq. to solid; 100% act.

Marlipal O13/30, /50, /60, /80, /100, /120, /170. [308] C_{13}-oxoalcohol polyglycol ethers; nonionic; surfactant in detergents; emulsifier and aux. agent in textile processing;

finishing and wetting agent; liq., wax; 100% conc.

Marlipal SU. [308] Cetoleth-25; nonionic; see Marlipal KE; liq. to solid; 100% act.

Marlon A 350, 360, 365, 375, 390, 396. [308] Sodium dodecylbenzene sulfonates; anionic; detergent raw materials; liq., powd., paste; biodeg.; 50, 60, 65, 75, 90, 96% act. resp.

Marlon AFM 40, 40 N, 43, 50N. [308] Sodium dodecylbenzene sulfonate, alkyl ether sulfate, and nonionic; anionic/nonionic; base for dishwashing and cleaning agents and industrial cleaners; visc. liq.; 40–50% act.

Marlon AFR. [308] Modified sodium dodecylbenzene sulfonate; anionic; base material for dishwashing agents, cosmetic detergents, car shampoos; liq.; 30% act.

Marlon AM 80. [308] Sodium dodecylbenzene sulfonate/nonionic blend; anionic/nonionic; versatile paste and liq. formulations; liq.; 80% act.

Marlon AMX. [308] Amine dodecylbenzene sulfonate; anionic; base material for production of drycleaning detergents, degreasing agents for metal industry, floor cleaners; liq.; 90% act.

Marlon ARL. [308] Sodium dodecylbenzene sulfonate and sodium toluene sulfonate; anionic; detergent component, foaming, wetting agent; wh.-yel. powd.; dens. 470 g/l; ph 8; 95% act.

Marlon AS_3. [308] Dodecylbenzene sulfonic acid; anionic; intermediate for mfg. of anionic surfactants; liq.; 100% act.

Marlon PS 60. [308] Sec. sodium alkane sulfonate; base for detergents and cleaners; fluid, paste; 60% conc.

Marlophen 83, 84, 85, 86, 86 S, 87, 88, 89, 810, 812, 814, 820, 825, 850. [308] Nonylphenol polyglycol ethers; detergent, wetting and dispersing agent; sol. in oil and water; liq. to wax; 100% act.

Marlophen DNP 16, 18, 30. [308] Dinonylphenol polyglycol ethers; nonionic; nonionic; raw material for textile and paper auxiliaries, dispersant; wax; 100% act.

Marlophen X. [308] Alkylphenol polyglycol ether; nonionic; wetting agent for use in binders for coal dust, pigments, and in concrete mfg.; liq.; 100% act.

Marlophor AS-Acid, CS-Acid, DS-Acid. [308] Partial phosphate ester as acid; anionic surfactant used as base material for acid cleaners; liq. to solid; 100% act.

Marlophor FC-Acid. [308] Partial phosphate ester as acid; anionic surfactant; low-foam wetting agent for alkaline sol'ns; liq.; 100% act.

Marlophor FC-Sodium Salt, N 3-Sodium Salt. [308] Partial phosphate ester as salt; anionic; wetting agent for alkaline sol'ns.; liq.; 100% act.

Marlophor ND. [308] Partial phosphate ester; anionic; wetting agent for textile and paper industries, antistat for natural and syn. fibers used in drycleaning detergents; liq.; water- and oil-sol.; 100% act.

Marlophor ND-Acid. [308] Org. phosphate ester; anionic; detergent; wetting agent; antistat; component for textile aux. agents and drycleaning formulations; antistat; liq.; 100% conc.

Marlophor ND DEA Salt, ND Na-Salt. [308] Org. phosphate ester/alkylphenol polyglycol ether; anionic/nonionic; detergent, wetting agent, antistat; component for textile aux. agents and drycleaning formulations; liq.; 100 and 92% conc. resp.

Marlophor NP 5-Acid, 6-Acid, 7-Acid, 20-Acid. [308] Org. phosphate esters; anionic; see Marlophor AS Acid, also base for drycleaning formulations with antistatic properties; liq., wax; 100% conc.

Marlophor T 10-Acid. [308] Partial phosphate ester as acid; anionic; special liq. formulation for production of drycleaning detergents; liq.; 100% act.

Marlophor T 10-DEA Salt. [308] Diethanolamine alkyl polyglycol ether phosphate; anionic; see Marlophor NP 5-Acid; liq.; 100% conc.

Marlophor T 10-Sodium Salt. [308] Partial phosphate ester as salt; anionic; special liq. formulation for production of drycleaning detergents; liq.; 100% act.

Marlopon ADS 50. [308] DEA dodecylbenzene sulfonate; anionic; raw material for liq. phosphate-containing detergents and cleaners; fluid paste; 50% act.

Marlopon AMS 60. [308] Amine dodecylbenzene sulfonate/nonionic blend; anionic/nonionic; surfactant used for detergents, dishwashing agents and cleaners; fluid paste; 60% act.

Marlopon AT 50. [308] TEA-dodecylbenzene sulfonate; anionic; surfactant used as neutral detergent base; liq.; 50% act.

Marlopon CA. [308] TEA dodecylbenzene sulfonate, modified; anionic; detergent;

superfatting and dishwashing agent; personal care products; liq.; 55% conc.

Marlosol 183, 189, 1820, 1825, OL 7, OL 8, OL 10, OL 15, OL 20. [308] Fatty acid polyglycol esters; nonionic; raw material for finishing agents in the syn. fiber industry; liq. to wax; 100% act.

Marlowet 4508. [308] Org. ammonium salt; anionic; emulsifier for wh. spirit, kerosene, for environmentally compatible cold cleaners; liq.; 100% act.

Marlowet 4603. [308] Carboxylic acid polyglycol ester; nonionic; emulsifer for min. oils, textile finishes; liq.; 100% act.

Marlowet 4702. [308] Carboxylic acid polyglycol ester; nonionic; emulsifer for min. oils, metalworking; liq.; 100% act.

Marlowet 4703. [308] Carboxylic acid polyglycol ester; nonionic; emulsifer for solvs., min. oils, metal cleaning, metalworking; liq.; 100% act.

Marlowet 4800. [308] Alkyl polyglycol ether; nonionic; emulsifier for waxes, polishes, textile lubricants, leather care; wax; 100% act.

Marlowet 4857. [308] Alkyl polyglycol ether; nonionic; emulsifier for paraffins, waxes, silicone oils for textile, wood, paper, mold release, polishes; paste; 100% act.

Marlowet 4862. [308] Alkyl polyglycol ether; nonionic; emulsifier for waxes, paraffins, polishes; flakes; 100% act.

Marlowet 5001. [308] Alkyl polyglycol ether; nonionic; emulsifier for wh. oils, textile aux., cosmetics, textile lubricants and finishes; liq.; 100% act.

Marlowet 5100. [308] Alkylene oxide adduct of aliphatic diamine; nonionic; emulsifier for lubricating aids, for metalworking, foam control; liq.; 100% act.

Marlowet 5300, 5301, 5305, 5306, 5320. [308] Partially esterified phosphoric acid; anionic; emulsifer metalworking, metal cleaning; liq.; 100% act.

Marlowet 5311. [308] Partially esterified phosphoric acid; anionic; emulsifer used in paint removers; liq.; 100% act.

Marlowet 5400. [308] Alkylamine polyglycol ether; nonionic; emulsifier in car wash rinses, furniture polishes; liq.; 100% act.

Marlowet 5401. [308] Alkylamine salt, ethoxylated; cationic; emulsifier for mixts. of waxes and wh. spirits; liq.; 100% conc.

Marlowet 5436, 5439. [308] Carboxylic acid alkylolamine ester; nonionic; emulsifier for min. oils, metalworking, corrosion protection; liq.; 100% act.

Marlowet 5440. [308] Substituted imidazoline; cationic; emulsifier for min. oils, corrosion protection, car wash rinses; liq.; 100% act.

Marlowet 5480. [308] Carboxylic acid alkylolamide; nonionic; emulsifier for min. oils, metalworking, corrosion protection; liq.; 100% act.

Marlowet 5529. [308] Mixt. of alkoxylated amines and carboxylic acids; cationic; emulsifier for wh. spirit, petrol, for fast-separating cold cleaners; liq.; 100% act.

Marlowet 5600. [308] Amine salt of carboxylic acid; anionic; emulsifier for waxes, leather care agents, wood and paper impregnation agents; solid; 100% act.

Marlowet 5606. [308] Amine salt of carboxylic acid; anionic; emulsifier for mfg. of water-repellent chipboard; wax; 100% act.

Marlowet 5609. [308] Amine salt of carboxylic acid; anionic; emulsifier for wh. spirit, trichlorethylene; dewatering agents; liq.; 100% act.

Marlowet 5622. [308] Amine salt of carboxylic acid; anionic; emulsifier for metalworking, corrosion protection, floor care; liq.; 100% act.

Marlowet BIK. [308] Mixt. of carboxylic acid amine salt and alkyl polyglycol ether; anionic/cationic; emulsifier for min. oils and hydrocarbons, bitumen, for metalworking, road construction; liq.; 100% act.

Marlowet BL. [308] Alkyl polyglycol ether; nonionic; emulsifier for spindle oils, textile lubricants; liq.; 100% act.

Marlowet EF. [308] Mixt. of carboxylic acid polyglycol esters; nonionic; emulsifier for textile lubricants, pest control, cosmetic preparations; paste; 100% act.

Marlowet FOX. [308] Alkyl polyglycol ether; nonionic; emulsifier for textile lubricants, car and furniture polishes, leather care; wax; 100% act.

Marlowet GFN. [308] Alkyl polyglycol ether; nonionic; emulsifier for furniture and floor care, emulsion polishes; paste; 100% act.

Marlowet GFW. [308] Alkyl polyglycol ether; nonionic; emulsifier for furniture polishes, emulsion polishes; flake; 100% act.

Marlowet IHF. [308] Mixt. of *n*-alkyl benzene sulfonate, alkyl polyglycol ether, and solv.; nonionic; emulsifier for insecticides, herbicides, pesticides, cleaning agents; liq.; 100% act.

Marlowet ISM. [308] Alkyl polyglycol ether; nonionic; emulsifier for solvs., for pesticides, cold cleaners; liq.; 100% act.

Marlowet LVX. [308] Mixt. of carboxylic acid polyglycol esters; nonionic; emulsifier for animal and vegetable oils, for leather auxiliaries, textile lubricants; liq.; 100% act.

Marlowet MA. [308] Mixt. of carboxylic acid polyglycol esters and alkylolamides; nonionic; emulsifier for min. oils, textile lubricants, solvs.; liq.; 100% act.

Marlowet NF. [308] Mixt. of carboxylic acid polyglycol esters; nonionic; emulsifier for leather, auxiliaries, textile lubricants; liq.; 100% act.

Marlowet OFA. [308] Mixt. of *n*-alkyl benzene sulfonate, carboxylic acid polyglycol esters, and alkyl polyglycol ethers; anionic/nonionic; emulsifier for lubricants, pesticides, metal cleaning; liq.; 100% act.

Marlowet OFW. [308] Mixt. of carboxylic acid polyglycol esters and alkyl polyglycol ethers; nonionic; emulsifier for textile lubricants, metalworking; liq.; 100% act.

Marlowet OTS. [308] Carboxylic acid polyglycol esters; nonionic; emulsifier for metalworking and cleaning, textile lubricants; liq.; 100% act.

Marlowet PA, PW. [308] Alkyl polyglycol ether; nonionic; surfactant, emulsifier for textile and paper impregnation, mold release agents, furniture polishes; paste; 100% act.

Marlowet RA. [308] Mixt. of n-alkylbenzene sulfonate and carboxylic acid polyglycol esters; anionic/nonionic; emulsifier for essential oils, cosmetic preparations; liq.; 100% act.

Marlowet RNP. [308] Mixt. of carboxylic acid polyglycol ether and a glycol deriv.; nonionic; see Marlowet RA; liq.; 100% act.

Marlowet SAF. [308] Mixt. of alkyl polyglycol esters; nonionic; emulsifier for textile lubricants; liq.; 100% act.

Marlowet SLM. [308] Mixt. of carboxylic acid polyglycol esters and alkyloamides; nonionic; emulsifier for textile lubricants; liq.; 100% act.

Marlowet SLS. [308] Carboxylic acid polyglycol ester; nonionic; emulsifier for textile lubricants, cold cleaners, road construction; liq.; 100% act.

Marlowet SW. [308] Mixt. of carboxylic acid polyglycol ether; nonionic; emulsifier for textile lubricants, metalworking; liq.; 100% act.

Marlowet SWN. [308] Mixt. of carboxylic acid polyglycol esters and alkyl polyglycol ethers; nonionic; emulsifier for spindle and wh. oils, solvs., for easily washed-out textile lubricants, metal cleaning; liq.; 100% act.

Marlowet T. [308] Amine salt of carboxylic acid; anionic; emulsifier for min. oils, wh. spirit, bitumen, for metalworking, road construction; paste; 100% act.

Marlowet TM. [308] Alkylaryl polyglycol ether; nonionic; emulsifier for min. oils, for metalworking, cold cleaners; liq.; 100% act.

Marlowet WOE. [308] Alkyl polyglycol ether; nonionic; emulsifier for paraffinic min. oils, textile lubricants; liq.; 100% act.

Marlowet WSD. [308] Mixt. of alkyl and alkylaryl polyglycol ethers; nonionic; emulsifier for min. oils, textile lubricants; liq.; 100% act.

Marlox LM 25/30, LM 75/30, LP 90/20. [308] Polypropylene glycol/EO addition products; nonionic surfactant for liq., paste and powd. detergents and cleaning agents; liq.; 100% act.

Marlox NP 109. [308] Alkylphenol/alkylene oxide addition product; nonionic surfactant for low-foam formulations; liq.; 100% act.

Marlox S 58. [308] Fatty alcohol/alkylene oxide addition products; nonionic; detergent in low-foam formulations and in combination with other detergent raw materials; liq.; 100% act.

Marpol 1 F. [397] Alkylsulfate deriv.; anionic; finishing oil for rayon staple fiber; paste; 50% conc.

Marpol C-180. [397] Alkyl phosphate; anionic; spinning oil for syn. fiber dyeing; paste; 15% conc.

Mars AMLS. [392] Ammonium lauryl sulfate; anionic; foaming agent, detergent, wetting agent, emulsifier used in cosmetics, sanitary supply and industrial cleaners; pale yel. clear liq.; dens. 8.5 lb/gal; sp.gr. 1.02; visc. 600 cps max; pH 7.0–7.5 (10%); 30% act.

Mars Anionic 68. [392] Linear sodium alkylaryl sulfonate and nonionic; anionic; detergent in sanitary supply and industrial cleaners; medium amber clear gel; sol. in water, lower alcohols; dens. 8.8 lb/gal; sp.gr. 1.05; pH 6.8–7.2 (10%); biodeg.; 68% act.

Mars Emulsifier C. [392] Complex difatty quat.; cationic; emulsifier, textile dry cleaning antistat, fabric lubricant and softener used in sanitary supply and industrial chemicals; pale amber clear liq.; mild, fatty odor;

sol. in min. oil, kerosene, xylene, min. spirits, alcohols, disp. in water; dens. 7.8 lb/gal; sp.gr. 0.93; visc. 200 cps; pour pt. 10 C; pH 4.5–5.5 (10%); biodeg.; 74% min. act.

Mars ES-60. [392] Sodium lauryl ether sulfate; anionic; detergent, foaming agent used in cosmetics, cleaners; lt. amber clear liq.; dens. 8.8 lb/gal; sp.gr. 1.04; visc. 64 cs (100 F); pH 8–9 (5%); biodeg.; 58–60% act.

Mars NI-100. [392] PEG ether; nonionic; wetting and rewetting agent, dispersant, foaming agent, emulsifier used in cosmetics, sanitary supply and industrial cleaners, agric. sprays, textile, paper and paint industries; clear pale yel. slightly visc. liq.; sol. in water, alcohols, glycols, certain solvs.; dens. 8.75 lb/gal; sp.gr. 1.06; visc. 30.0 cps; HLB 13.0; cloud pt. 52–56 C (2%); pour pt. 15 C; pH 7–8 (10%); biodeg.; 100% act.

Mars NI-233. [392] Ethoxylated alcohol; nonionic; emulsifier for bath oils, cream and lotion in cosmetics, sanitary supply and industrial cleaners, agric. sprays; colorless clear liq.; sol. in propylene glycol, alcohols, monobutyl ethers, perchloroethylene, heavy aromatic naphtha, kerosene, min. oil, xylene, toluene, water disp.; dens. 7.7 lb/gal; sp.gr. 0.93; visc. 40 cps; HLB 8.0; pH 7–8 (10%); biodeg.; 100% act.

Mars PA-400. [392] Free acid form of a phosphated polyether alcohol; anionic; detergent, coupling agent, corrosion inhibitor used in sanitary supply and industrial cleaners, detergents; yel., visc. liq; sol. in xylene, toluene trichloroethylene, carbon tetrachloride, glycols, alcohols, monobutyl ethers, aq. sol'ns. @ pH 1–14; dens. 9.15 lb/gal; sp.gr. 1.1; visc. 6000 cps; pour pt. < 0 C; pH 1–2 (10%); biodeg.; 100% act.

Mars Pro Conc I. [392] Anionic/nonionic surfactant blend with foam stabilizer and water softening agents; detergent, foaming agent, foam stabilizer, water softener used in sanitary supply and industrial cleaners; clear amber liq.; dens. 9.1 lb/gal; sp.gr. 1.09; visc. 500 cps; pH 8.5–9.5 (10%); 64–66% solids.

Mars Pro Conc II. [392] Anionic/nonionic surfactant blend; detergent for dry foam carpet and upholstery shampoos, animal shampoos, hand soap, car wash, dishwash, cosmetics and shampoos; lt. yel. clear liq.; water sol.; dens. 8.5 lb/gal; sp.gr. 1.02; visc. 1000 cps; cloud pt. 15 C; pH 7–8 (10%); biodeg.; 36–38% solids.

Mars SA-98-S. [392] Linear alkylaryl sulfonic acid; anionic; detergent intermediate, wetting agent used in sanitary supply, institutional, and janitorial cleaners, cosmetics, detergents; reddish amber visc. liq.; misc. in water; dens. 8.8 lb/gal; sp.gr. 1.06; biodeg.; 98% act.

Mars SLS-30, SLS-A95. [392] Sodium lauryl sulfate; anionic; foaming agent, detergent, wetting agent, emulsifer used in cosmetics, sanitary supply and industrial cleaners; yel. clear liq.; dens. 8.6 lb/gal; sp.gr. 1.04 and 1.03 resp.; visc. 550 and 40 cps; cloud pt. 14 and 18 C; pH 7.5–9.0 and 7.0–8.5 (10%); biodeg.; 29 and 30% act. resp.

Mars SXS. [392] Sodium xylene sulfate; solubilizer, coupling agent, hydrotrope for coupling builder with detergents, cloud pt. depressant in cosmetics and sanitary supply industries; clear colorless liq; mild alcohol odor; dens. 9.8 lb/gal; sp.gr. 1.18; visc. < 20 cps; pH 8–9 (10%); 40% act.

Mars TLS-40. [392] TEA lauryl sulfate; anionic; detergent, foaming agent, wetter, emulsifier used in cosmetic industry, sanitary supply and industrial cleaners; pale yel. clear liq.; dens. 8.7 lb/gal; sp.gr. 1.04; visc. 500 cps; cloud pt. 0 C; pH 6.5–7.5 (10%); 40% act.

Marsamid 10. [392] Modified 2:1 coconut diethanolamide with blend of amide, free fatty acid, and free diethanolamine; anionic/nonionic; detergent, coupler, visc. builder, solv. emulsifier in pesticide formulations, sanitary supply, household and industrial cleaners; dk. amber clear liq.; dens. 8.4 lb/gal; sp.gr. 1.01; visc. 1900 cps; pour pt. 10 C min; pH 8.5–9.5 (5%); biodeg.; 100% solids.

Marsamid 40. [392] 2:1 coconut diethanolamide; nonionic; foam stabilizer, visc. builder, wetting agent, detergent used in sanitary supply, general purpose and industrial cleaners; clear amber liq.; sol. in water, polar org. solvs., and many hydrocarbons and chlorinated org. compds.; dens. 8.55 lb/gal; sp.gr. 1.025; visc. 1450 cps; pH 9.5–10.5 (2%); biodeg.; 99+% solids.

Marsamid 50. [392] 1:1 coconut diethanolamide; nonionic; visc. builder, foam builder and stabilizer used in cosmetics, sanitary supply, general purpose and industrial cleaners; pale yel. visc. liq.; disp. in water, sol. in alcohols, glycols, carbon tetrachloride, Cellosolve-type ethers, Stod., pine oil; dens. 8.21 lb/gal; visc. 1500 cps; pH 9.0–9.5 (2%); biodeg.; 100% solids.

Marsonol SF-C2. [392] Disodium salt of a dicarboxylated imidazoline derived from coconut fatty acid and amine condensate; amphoteric; foaming agent and stabilizer, detergent, wetting agent, emulsifier used in industrial cleaners and cosmetics; clear amber liq.; sol. in water; dens. 8.9 lb/gal; sp.gr. 1.07; visc. 50 cps; pH 9.6 (20%, 30 C); biodeg.; 38–40% act.

Marvanol 55% SPO. [391] Sulfated propyl oleate; anionic; detergent, wetting and leveling agent; amber liq.; 55% act.

Marvanol 60% SBO. [391] Sulfated butyl oleate; anionic; detergent, wetting and leveling agent, emulsifier; amber liq.; 60% act.

Marvanol 75% SCO. [391] Sulfated castor oil; detergent, wetting agent, dyeing assistant and lubricant used in finishing operations; amber liq.; 75% act.

Marvanol CO. [391] Ethoxylated phenol; textiles detergent, emulsifier, wetting agent in finishing operations; colorless liq.; 99+% act.

Marvanol DC. [391] Ethoxylated alcohol; textile fiber wetting and dyeing agent; colorless liq.; 35% act.

Marvanol GC. [391] Sulfosuccinamide; anionic; textile fiber wetting and dye leveling agent; amber gel; 65% act.

Marvanol Penetrant 35. [391] Amine neutralized sulfonic acid; detergent, wetting agent, textile dyeing, finishing, leveling and retarding agents for acid dyes; amber liq.; 40% act.

Marvanol RE-1274. [391] 1:1 coconut fatty amide; cationic; emulsifer, visc. builder, dye leveler; amber liq.; sp.gr. 0.950; 97% act.

Marvanol RE-1281. [391] 2:1 coconut fatty amide; cationic; emulsifer, visc. builder, dye leveler; amber liq.; sp.gr. 1.02; 98% act.

Marvanol Solvent Scour 34. [391] Solvs. and detergents; anionic; solv. scour; amber liq.; 100% act.

Marvanscour KW. [391] Solvs. and detergents; nonionic; detergent for prescour and afterscour; straw liq.; 83% act.

Marvantex BS. [391] Alkaline built blend; anionic; detergent, alkaline scour for dye stripping; yel. to wh. powd.; 100% act.

Marvantex GS. [391] Alkaline scour; anionic; detergent used for oil, grease and graphite removal; powd.; 100% act.

Marvelin W-50. [397] Alkylene oxide addition products; nonionic; dye leveling agent of wool; flake; 100% conc.

Masil 2132, 2133, 2134. [400] Organo silicone fluid, modified; surface tension antistat and wetting agent for textile, plastics, lubricants and formulations; solv. based coating, dispersant, and antifoam; liq.; 100% conc.

Matexil AA-NS. [317] Mixed polyglycol fatty acid; anionic; antifoam agent for textile dyeing processes; liq.

Matexil DA-AC. [317] Naphthalene sulfonic acid/formaldehyde condensate, disodium salt; anionic; dispersant for dyeing applications; powd.

Matexil DN-VL500. [317] Fatty alcohol ethoxylate; nonionic; antistat; dye leveling agent; dispersant, emulsifier; flakes

Matexil LC-CWL. [317] EO condensate; nonionic; level dyeing assistant for application of dyeing applications; liq.

Matexil LC-RA. [317] Quat. ammonium compd.; cationic; retarding agent for dyeing applications; liq.

Matexil LN-PT. [317] Alcohol, ethoxylated; nonionic; used in dyeing processes; liq.

Matexil LN-RD. [317] Fatty acid ethoxylate; nonionic; level-dyeing assistant for disperse dyes; liq.

Matexil PN-HT. [317] Fatty acid ethoxylate; nonionic; used in dyeing applications; liq.

Matexil PN-PR. [317] Fatty alcohol ethoxylate; nonionic; emulsifier for o/w emulsion thickenings in textile printing applications; flakes

Mayodan. [272] Stabilizer blends; food emulsifier for mayonnaise and salad dressing.

Maypon 4C. [575] Potassium coco-hydrolyzed animal protein; anionic; detergent used in personal care products, general purpose cleansers; clear to slightly hazy amber liq.; misc. with water; visc. 500 cps max; pH 6.7–7.3; 34–36% solids.

Maypon 4CT. [575] TEA-coco-hydrolyzed animal protein; similar to Maypon 4C with lower cloud pt.; clear to slightly hazy amber liq.; misc. with water; visc. 150 cps max; pH 6.7–7.3; 38–40% solids.

Maypon K. [575] Sodium soya hydrolyzed animal protein; anionic; detergent for industrial cleaners; clear amber liq.; misc. with water @ 50 C; sp.gr. 1.12–1.14; visc. 6000 cps max.; pH 8.5–9.5; 42–44% solids.

Maypon UD. [575] Potassium undecylenoyl hydrolyzed animal protein; anionic; detergent used in personal care products, antifungal properties; dk. amber clear liq.; sol. in alcohol, glycerin, glycols, other polyols, misc. with water; sp.gr. 1.06–1.09; visc. 100

cps max.; pH 6.3–6.8; 34–38% solids.

Mazamide 25. [400] 2:1 coconut sulfonic acid; nonionic/anionic; emulsifier, detergent used in hard surface cleaners, dishwash liq., in metalworking fluids; liq.; sp.gr. 1.053; biodeg.

Mazamide 65, 65CZ, 66. [400] 2:1 mixed fatty acid; nonionic/anionic; emulsifier, detergent used in hard surface cleaners, dishwash liq., in metalworking fluids; liq.; sp.gr. 0.99–1.02; pH 9–10; biodeg.

Mazamide 70. [400] 2:1 Cocamide DEA and diethanolamine; nonionic; emulsifier, detergent used in hard surface cleaners, dishwash liq., in metalworking fluids; liq.; sp.gr. 0.99–1.02; pH 9–10; biodeg.

Mazamide 80. [400] 1:1 Cocamide DEA and diethanolamine; nonionic; thickener, foam stabilizer, emulsifier used in cosmetic and toiletry formulations, hard surface wetters and cleaners; liq.; sp.gr. 0.98–1.00; pH 9–10.5; biodeg.

Mazamide C-5. [400] PEG-6 cocamide; nonionic; emulsifier, lubricant, rust inhibitor, buffing compd.; liq.; sp.gr. 1.0–1.1; pH 9.5–10.5; biodeg.

Mazamide CA-20. [400] 2:1 Cocamide DEA and diethanolamine; nonionic/anionic; emulsifier, detergent used in hard surface cleaners, dishwash liq., in metalworking fluids; liq.; sp.gr. 0.99–1.02; pH 9–10; biodeg.

Mazamide CS-148. [400] 1:1 Cocamide DEA; nonionic; thickener, foam stabilizer, emulsifier used in cosmetic and toiletry formulations; liq.; sp.gr. 0.98–1.00; pH 9.0–10.5; biodeg.

Mazamide L-5. [400] PEG-6 lauramide; nonionic; emulsifier, lubricant, rust inhibitor, buffing compd.; liq.; sp.gr. 1.0–1.1; pH 9.5–10.5; biodeg.

Mazamide L-298. [400] 2:1 lauric alkanolamide; nonionic/anionic; foam builder and stabilizer, emulsifier, dispersant, visc. builder for hard surface cleaners, dishwashing, shampoos, metalworking fluids, automotive specialties, fiber and hair conditioners, dry cleaning, agric. sprays, leather/fur preparations, emulsifiable waxes, rust inhibitors, polishes, paint removers, rug shampoos, fuel oil additives, textile detergents; solid; water sol.; sp.gr. 1.00; pH 9.4 (5% aq.); biodeg.

Mazamide LM-20. [400] 2:1 Lauramide DEA and diethanolamine; nonionic/anionic; emulsifier, detergent used in hard surface cleaners, dishwash liq., in metalworking fluids; liq.; sp.gr. 0.99–1.01; pH 9.2–10.0; biodeg.

Mazamide LM-21. [400] 2:1 lauric myristic alkanolamide; nonionic/anionic; see Mazamide L-298; liq.; water sol.; sp.gr. 1.00; pH 10.0 (5% aq.); biodeg.

Mazamide LS-173, -196. [400] 1:1 Lauramide DEA; nonionic; thickener, foam stabilizer, emulsifier used in cosmetic and toiletry formulations; liq.; sp.gr. 0.98–1.00; pH 9.0–10.5; biodeg.

Mazamide O-10. [400] 1:1 oleic alkanolamide; nonionic; see Mazamide L-298; paste; insol. in aq. systems; sp.gr. 0.94 (40 C).

Mazamide O-20. [400] 2:1 Oleamide DEA and diethanolamine; nonionic; emulsifier, lubricant, rust inhibitor, buffing compd.; liq.; sp.gr. 0.99–1.01; pH 9–10; biodeg.

Mazamide SS-10. [400] 1:1 Soyamide DEA; nonionic; emulsifier, lubricant, rust inhibitor, buffing compd.; liq.; sp.gr. 0.98–1.00; pH 9.0–10.5; biodeg.

Mazamide SS-20. [400] 2:1 soybean oil alkanolamide; nonionic/anionic; see Mazamide L-298; liq.; water sol.; sp.gr. 1.01; pH 9.8 (5% aq.); biodeg.

Mazamide T-10. [400] 1:1 tall oil alkanolamide; nonionic; see Mazamide L-298; liq.; insol. in aq. systems; sp.gr. 0.99; pH 9.5 (5% aq.); biodeg.

Mazamide T-15, -20. [400] 2:1 tall oil alkanolamide; nonionic/anionic and nonionic resp.; see Mazamide L-298; solid and liq. resp.; water sol.; sp.gr. 1.00; pH 9.5 (5% aq.); biodeg.

Mazamide TO-10. [400] 1:1 tall oil alkanolamide; nonionic; emulsifier; liq.; sp.gr. 0.98–1.00; pH 9.0–10.0; biodeg.

Mazamide TO-20. [400] 2:1 tall oil; nonionic; emulsifier, detergent used in hard surface cleaners, in metalworking fluids, syn. coolants; liq.; sp.gr. 0.99–1.01; pH 9–10; biodeg.

Mazawax 163R. [400] Cetearyl alcohol and PEG-12 stearate; nonionic; emulsifier for pharmaceutical and cosmetic applications; flake; 100% conc.

Mazawet 77. [400] Alkyl polyoxyalkylene ether; nonionic; wetting agent, surfactant used in textile processing, spray-dried detergents, emulsion polymerization, drycleaning systems, hard surface cleaners; Gardner 1; m.w. 426 avg.; sol. @ 5% in toluene, perchloroethylene, Stod., propylene glycol, water; sp.gr. 0.965; cloud pt. 45 C (1% aq.); flash pt. 340 F (PMCC); pour pt. 11 C; surf. tens. 28.1 dynes/cm (0.1%); pH 6.5 (3% aq.); 100% act.

Mazawet DF. [400] Alkyl polyoxyalkylene ether; nonionic; wetting surfactant for metalworking fluids, paint, printing inks, polishes, floor waxes, hard surface cleaning, and emulsion polymerization; liq.; 100% conc.

Mazeen 173. [400] Tetrahydroxypropyl ethylenediamine; insecticide and herbicide emulsifier, antistat and rewetting agent, grease additive, textile lubricant; emulsifier for lubricants, inks, and cosmetics; Gardner 2 liq.; m.w. 292; sol. in water, benzene, acetone, IPA; sp.gr. 1.01; surf. tens. 51.4 dynes/cm (0.1%).

Mazeen 174-75. [400] Tetrahydroxypropyl ethylenediamine; see Mazeen 173; Gardner < 1 liq.; m.w. 292; sol. in water, acetone, IPA; sp.gr. 1.00; surf. tens. 54.2 dynes/cm (0.1%).

Mazeen C 2. [400] PEG-2 cocamine; cationic; emulsifier, antistat, rewetting agent, lubricant used in insecticides and herbicides, antistatic agents, grease additives, textile lubricants, water-based inks, cosmetics; Gardner 11 liq..; m.w. 285; sol. in benzene, acetone, IPA, min. oil, forms gel in water; sp.gr. 0.874; surf. tens. 28 dynes/cm (0.1%); 100% conc.

Mazeen C 5. [400] PEG-5 cocamine; cationic; see Mazeen C 2; Gardner 11 liq..; m.w. 425; sol. in water, benzene, acetone, IPA, min. oil; sp.gr. 0.976; surf. tens. 33 dynes/cm (0.1%); 100% conc.

Mazeen C 10. [400] PEG-10 cocamine; cationic; see Mazeen C 2; Gardner 11 liq..; m.w. 645; sol. in water, benzene, acetone, IPA; sp.gr. 1.017; surf. tens. 39 dynes/cm (0.1%); 100% conc.

Mazeen C 15. [400] PEG-15 cocamine; cationic; see Mazeen C 2; Gardner 9 liq..; m.w. 860; sol. in water, benzene, acetone, IPA; sp.gr. 1.042; surf. tens. 41 dynes/cm (0.1%); 100% conc.

Mazeen DBA. [400] Dibutylamino ethanol; see Mazeen 173; Gardner 9 liq.; m.w. 192; sol. in benzene, acetone, IPA, min. oil; insol. in water; sp.gr. 0.860.

Mazeen DBA-1. [400] Polyalkoxylated amine; cationic; surfactant and wetting agent for paints and inks; liq.; 100% conc.

Mazeen S 2. [400] PEG-2 soyamine; cationic; see Mazeen C 2; Gardner 14 liq.; m.w. 350; sol. in benzene, acetone, IPA, min. oil; sp.gr. 0.911; surf. tens. 26 dynes/cm (0.1%); 100% conc.

Mazeen S 5. [400] PEG-5 soyamine; cationic; see Mazeen C 2; Gardner 14 liq.; m.w. 480; sol. in benzene, IPA; partly sol. acetone, min. oil; sp.gr. 0.951; surf. tens. 33 dynes/cm (0.1%); 100% conc.

Mazeen S 10. [400] PEG-10 soyamine; cationic; see Mazeen C 2; Gardner 14 liq.; m.w. 710; sol. in water, benzene, acetone, IPA; sp.gr. 1.020; surf. tens. 40 dynes/cm (0.1%); 100% conc.

Mazeen S 15. [400] PEG-15 soyamine; cationic; see Mazeen C 2; Gardner 18 liq.; m.w. 930; sol. in water, benzene, acetone, IPA; sp.gr. 1.040; surf. tens. 43 dynes/cm (0.1%); 100% conc.

Mazeen T 2. [400] PEG-2 tallow amine; cationic; see Mazeen C 2; Gardner 11 liq.; m.w. 350; sol. in benzene, acetone, IPA, min. oil; sp.gr. 0.916; surf. tens. 29 dynes/cm (0.1%); 100% conc.

Mazeen T 5. [400] PEG-5 tallow amine; cationic; see Mazeen C 2; Gardner 12 liq.; m.w. 480; sol. in benzene, acetone, IPA, min. oil; gels in water; sp.gr. 0.966; surf. tens. 34 dynes/cm (0.1%); 100% conc.

Mazeen T 15. [400] PEG-15 tallow amine; cationic; see Mazeen C 2; Gardner 18 liq.; m.w. 925; sol. in water, benzene, acetone, IPA; sp.gr. 1.028; surf. tens. 41 dynes/cm (0.1%); 100% conc.

Mazol 80 MG. [400] PEG-20 glyceryl stearate; emulsifier for cakes, icings, whipped toppings; yel. liq.; sol. in ethanol, partly sol. in soybean oil, disp. in propylene glycol; m.p. 25–27 C; HLB 11; sapon. no. 65–75.

Mazol 159. [400] PEG-7 glycerol cocoate; emulsifier for food products; used in cosmetics, toiletries, pharmaceuticals, lubricants, mold release compds.; plasticizer in syn. fabrics and plastics; Gardner 1 clear liq.; HLB 13.0; sapon. no. 90.

Mazol 165C. [400] Glyceryl stearate and PEG-100 stearate; see Mazol 159; wh. solid, flake; HLB 11.0; sapon. no. 95.

Mazol 300. [400] Glycerol monooleate; nonionic; antifoam agent for sugar and protein processing, coemulsifier; yel. liq.; sol. in soybean oil, propylene glycol; disp. in ethanol; HLB 3.0; sapon. no. 150–160; 100% conc.

Mazol GDO. [400] Glycerol dioleate; see

Mazol 159; Gardner 4 max. clear liq.; HLB 2.9; sapon. no. 182.

Mazol GMO. [400] Glycerol monooleate; nonionic; see Mazol 300; yel. liq.; sol. in soybean oil; disp. in ethanol, propylene glycol; HLB 2.4; sapon. no. 167–177; 100% conc.

Mazol GMS. [400] Glyceryl stearate; see Mazol 159; wh. solid, flake; HLB 2.8; sapon. no. 172.

Mazol GMS-D. [400] Glycerol stearate SE; see Mazol 159; wh. solid, flake; HLB 6.0; sapon. no. 142.

Mazol GMS-HM. [400] Glycerol monostearate; see Mazol 159; wh. solid, flake; HLB 3.3; sapon. no. 165.

Mazol PGMS. [400] Propylene glycol monostearate; nonionic; coemulsifier for edible oil and shortenings, dispersing aid for nondairy creamers; wh.-cream solid; sol. in propylene glycol, disp. in ethanol; m.p. 39–46 C; HLB 3.4; sapon. no. 170–180; 100% conc.

Mazol PGO-31. [400] Triglycerol monooleate; nonionic; emulsifier for food products, in cosmetics, toiletries, pharmaceuticals, lubricants, mold release compds., plasticizers; Gardner 8 max. liq.; HLB 6.2; sapon. no. 125–150; 100% conc.

Mazol PGO-62. [400] Hexaglycerol dioleate; nonionic; see Mazol PGO-31; Gardner 10 max. liq.; HLB 8.5; sapon. no. 105–125.

Mazol PGO-66. [400] Hexaglycerol hexaoleate; see Mazol PGO-31; Gardner 8 max. liq.; HLB 3.0; sapon. no. 160–180.

Mazol PGO-104. [400] Decaglycerol tetraoleate; nonionic; see Mazol PGO-31; Gardner 8 max. liq.; HLB 6.2; sapon. no. 125–150.

Mazol PGO-108. [400] Decaglycerol octaoleate; nonionic; see Mazol PGO-31; Gardner 8 max. liq.; HLB 3.4; sapon. no. 155–175.

Mazol PGO-1010. [400] Decaglycerol decaoleate; nonionic; see Mazol PGO-31; Gardner 8 max. liq.; HLB 2.5; sapon. no. 165–185.

Mazol PGS-31. [400] Triglycerol monostearate; nonionic; see Mazol PGO-31; Gardner 8 max., solid; HLB 6.2; sapon. no. 120–145.

Mazol PGS-61. [400] Hexaglycerol monostearate; see Mazol 159; Gardner 10 max. solid; HLB 8.87; sapon. no. 100.

Mazol PGS-62. [400] Hexaglycerol distearate; nonionic; see Mazol PGO-31; Gardner 10 max. solid; HLB 8.5; sapon. no. 105–125.

Mazol PGSH-31. [400] Triglycerol mono shortening; nonionic; see Mazol PGO-31; Gardner 10 max. semisolid; HLB 6; sapon. no. 115–140.

Mazol PGSH-62. [400] Hexaglycerol di shortening; nonionic; see Mazol PGO-31; Gardner 10 max., semisolid; HLB 8.3; sapon. no. 100–125.

Mazoline OA. [400] Oleyl imidazoline; cationic; emulsifier for preparing w/o emulsions, used in metalworking; disp. freely in nonpolar solvs., e.g., kerosene and min. oils, lipophilic; 100% conc.

Mazoline T. [400] Tall oil substituted imidazoline; cationic; see Mazoline OA; liq.; 100% conc.

Mazon 21. [400] Alkyl polyoxyalkylene ether; anionic; emulsifier for degreasers and industrial cleaner formulations; liq.; 100% conc.

Mazon 60T. [400] TEA alkylaryl sulfonate; anionic; emulsifier for shampoos, dishwashing, car wash, textile, metal cleaners and other formulations; liq.; 60% conc.

Mazon 86. [400] Polyoxyalkylene ether ester; nonionic; surfactant for metalworking lubricants; liq.; 100% conc.

Mazon 1045A, 1086, 1096. [400] POE fatty acid ester; nonionic; emulsifier for pesticide, herbicide, metalworking, die-cast lubricant formulations, and emulsion polymerization; liq.; 100% conc.

M-C-Thin. [410] Lecithin, standardized; amphoteric; used for food and tech. applications; liq., paste; 100% conc.

M-C-Thin 45. [410] Lecithin; nonionic; release aid based on phopholipids; liq.; 30-40% conc.

M-C-Thin AF-1. [410] Lecithin, natural; amphoteric; emulsifier, wetting agent and stabilizer used in the food industry; liq.; 100% conc.

Medialan Brands. [301] Fatty acid sarcosides; anionic; basic material for cosmetic industry; detergent for textile industry.

Medialan KA Conc. [44] Based on fatty acid sarcoside sodium salt; anionic; detergent used in cream shampoos, foaming agent, mild toiletries, hair shampoos; yel. soft paste; weak odor; m.w. 310 ± 5; misc. with water; biodeg.; 63 ± 2% act.

Medialan KF. [44] Salt of the condensation product of sat. fatty acids and sarcosine; anionic; detergent used in preparation of clear, liq. shampoos and body lotions; clear yel. low-visc. liq.; misc. with water; 40% act.

Melanol LP 1. [551] Isopropanolamine lauryl sulfate; anionic; base for shampoos; gel; 60% conc.

Mercerol GVC-65 Liq. [538] Wetting agent, mercerizing penetrant for efficient wetout in caustic soda saturator; yel. liq.; disp. in water; sp.gr. 1.03; 65% act.

Mercerol NL, QW. [538] Sulfonate, cresol-free, containing solvs.; anionic; wetting agent for causticizing baths; liq.

Mercol 15, 25, 30. [601] Sodium dodecylbenzene sulfonate; anionic; detergent, emulsifier, wetting agent for industrial, household, cosmetic, food uses; wh. beads; neutral odor; water sol.; dens. 0.15, 0.25, and 0.30 g/cc resp.; pH 7.0–8.5; biodeg.; 40% act.

Mercol Special. [601] Sodium dodecylbenzene sulfonate; anionic; see Mercol 15; beads; neutral odor; water sol.; dens. 0.30 g/cc; pH 7.0–8.5; biodeg.; 20% act.

Merpalyt L-15-1. [347] Alkylamide betaine; ampholyte used in cosmetic and shampoo formulations, in alkaline and acid cleaning formulations; yel. liq.; 45% act.

Merpasol DIO 60, 75. [347] Sodium dioctyl sulfosuccinate; anionic; rapid wetting agent for textile and paper industries; lt., visc. liq.; 60 and 75% act. resp.

Merpasol L 29. [347] Disodium fatty alcohol polyglycol ether sulfosuccinate; anionic; surfactant for cosmetic formulations; liq.; 40% conc.

Merpasol L 44. [347] Disodium lauryl sulfosuccinate; anionic; surfactant for carpet cleaners and cosmetic formulations; paste; 40% conc.

Merpasol L 47. [347] Disodium lauramido MEA-sulfosuccinate; anionic; see Merpasol L 44; liq.; 40% conc.

Merpasol L 52. [347] Disodium coco fatty acid ethanolamide sulfosuccinate; anionic; surfactant for carpet cleaners and cosmetic formulations; liq.; 40% conc.

Merpasol L 53. [347] Disodium castor oil fatty acid ethanolamide sulfosuccinate; anionic; surfactant for cosmetic formulations; liq.; 40% conc.

Merpemul 1065, 1070. [347] Calcium alkylbenzene sulfonate, linear; nonionic; used for emulsion concs. of pesticide substances; liq.; 65 and 70% conc. resp.

Merpemul 1088, 1088 L, 1165, 1178. [347] Alkylbenzene sulfonate amine salt, linear; nonionic; emulsifier for petroleum, pesticides, cold cleansers, dry cleaning; liq.; 90, 90, 78, and 65% conc. resp.

Merpemul 3034. [347] Oleyl/cetyl polyglycol ether; nonionic; emulsifier for animal oil; liq.; HLB 7.3; 100% conc.

Merpemul 3050. [347] Oleyl/cetyl polyglycol ether; nonionic; emulsifier for min. oils; liq.; HLB 9.2; 100% conc.

Merpemul 3200. [347] Oleyl/cetyl polyglycol ether; nonionic; emulsifier for pesticides; wax; HLB 15.4; 100% conc.

Merpemul 4065, 4070. [347] Calcium alkylbenzene sulfonate; anionic; see Merpemul 3200; liq.; 65 and 70% conc. resp.

Merpemul 5200, 5400, 5500. [347] Castor oil polyglycol ether; nonionic; see Merpemul 3200; liq., paste, and paste resp.; HLB 9.7, 13.1, and 13.9 resp.; 100% conc.

Merpemul 6060, 6080. [347] Fatty acid polyglycol ether unsat.; nonionic; used in emulsions of min., spindle, wh., silicon oils, and benzins; liq.; HLB 9.7 and 11.2 resp.; 100% conc.

Merpemul 7080, 7100. [347] Tridecyl polyglycol ether; nonionic; used in dispersions of silica, silicic acid, pigments, and solid matter; liq.; HLB 12.7 and 13.7 resp.; 100% conc.

Merpemul 8122, 8610. [347] Alkylaryl polyethoxycarboxylate; nonionic; used in emulsions of crude oil at high electrolytic conc.; liq.; 90 and 70% conc. resp.

Merpemul 9060. [347] C_{18-22} alcohol polyglycol ether; nonionic; used in emulsions of paraffin; wax; HLB 8.8; 100% conc.

Merpinal D 40. [347] Sodium decyl sulfate; anionic; low-foaming wetting agent for alkaline metal baths, degreasing and cleaning of hard surfaces; wh., visc. liq.; biodeg.; 40% act.

Merpinal EH 40. [347] Sodium-2-ethylhexyl sulfate; anionic; wetting agent for alkaline metal baths; degreaser and cleanser for hard surfaces; liq.; 40% conc.

Merpinal LM 40. [347] Lauryl sodium sulfate; anionic; liq. household detergent, also used in shampoo and foam bath preparations; wh. paste; 40% act.

Merpinal LM 50. [347] Sodium lauryl sulfate C_{12}–C_{14}; anionic; high-foaming paste and powd. household and textile detergents; wh. paste; biodeg.; 50% act.

Merpinal LMT 40. [347] Lauryl triethanolamine sulfate; anionic; liq. household detergent, also used in shampoo and foam bath preparations; lt., visc. liq.; 40% act.

Merpinal LS 40. [347] Lauryl sodium sulfate; anionic; paste and powd. household and textile detergents; wh. soft paste; 40% act.

Merpinal LS 50. [347] Sodium lauryl sulfate C_{12}–C_{18}; anionic; high-foaming paste and powd. household and textile detergents; wh. soft paste; biodeg.; 50% act.

Merpinal Q 101. [347] Sodium octadecyl sulfate; anionic; used in paste and powd. detergents; wh. paste; 30% act.

Merpinal Q 147. [347] Sodium 2-ethylhexyl sulfate; anionic; low-foaming wetting agent for alkaline metal baths, used in degreasing and cleaning of hard surfaces; clear yel. liq; biodeg.; 40% act.

Merpinal T 35. [347] Tallow fatty alcohol sodium sulfate; anionic; paste and powd. detergents; yel.-wh. paste; 35% act.

Merpinamid KD 11, 12. [347] Lauramide DEA; nonionic; superfatting agent, foam stabilizer for shampoos, liq. cleaning and washing preparations for household and industry; ylsh. visc. liq.

Merpinamid KM. [347] Lauramide MEA; nonionic; see Merpinamid KD 11; lt. br. soft paste.

Merpinamid LD/E. [347] Lauramide DEA; nonionic; see Merpinamid KD 11; ylsh. visc. liq.

Merpinamid LM. [347] Lauramide MEA; nonionic; see Merpinamid KD 11; ylsh.-wh. flakes.

Merpinamid LMIPA. [347] Lauramide MIPA; nonionic; superfatting agent, foam stabilizer, detergent in shampoo and liq. cleaner used in household and industry; wh. wax.

Merpinamid LSD. [347] Lauryl polydiethanolamide; nonionic; see Merpinamid KD 11; yel. visc. liq.

Merpinamid LSD/E. [347] Lauramide DEA; nonionic; see Merpinamid KD 11; ylsh. visc. liq.

Merpinamid LSM. [347] Lauramide MEA; nonionic; see Merpinamid KD 11; ylsh.-wh. flakes.

Merpinamid OD. [347] Oleamide DEA; nonionic; superfatting agent, foam stabilizer, detergent in shampoo and liq. cleaner used in household and industry; yel. visc. liq.

Merpinol CK 30. [347] Protein coconut fatty acid condensate, potassium salt; protein product for shampoos and skin and body care products; yel.-br. visc. liq.; 30% act.

Merpinol S. [347] Protein oleic acid condensate, sodium salt; nonionic; used as washing, bleaching, and dyeing aux. for textile industry; surfactant in tech. cleansers; liq.; 25% act.

Merpiphos A 30/50. [347] Alkyl ether addition compd., sodium salt; anionic; wetting agent for alkaline metal baths, mercerizing aux.; lt. colored liq.; 100% act.

Merpiphos B 52. [347] C_{16-18} alcohol hydrogen phosphate; anionic; foam controlling surfactant in washing powd. formulations; wax; 90% conc.

Merpiphos B 59. [347] PEG-5 C_{8-10} alcohol hydrogen phosphate; anionic; liq. cleanser with high electrolytic content; liq.; 100% conc.

Merpiphos BP 80. [347] PEG-5 C_{10-12} alcohol phosphate, sodium salt; anionic; wetting agent, emulsifier; used for drilling oils; lt. yel. liq.; 100% act.

Merpiphos EP. [347] Nonoxynol-4 hydrogen phosphate; anionic; emulsifier, antistat for textiles, drycleaning, pesticides; br. liq.; 100% act.

Merpiphos GP. [347] Nonoxynol-9.5 hydrogen phosphate; anionic; emulsifier, detergent for industrial cleaning, drycleaning, pesticides; ylsh. liq.; 100% act.

Merpiphos LP. [347] Nonoxynol-6 hydrogen phosphate; anionic; see Meripiphos EP; lt. br. liq.; 100% act.

Merpiphos MP. [347] PEG-8 C_{16-18} alcohol hydrogen phosphate (unsat.); anionic; emulsifier, detergent, lubricant for textiles; yel. paste; 100% act.

Merpiphos PP. [347] PEG-4 iso-C_{13} alcohol hydrogen phosphate; anionic; hydrogrope, detergent, liq. cleaners; lt. yel. liq.; 100% act.

Merpiphos RP. [347] PEG-4 C_{12-14} alcohol hydrogen phosphate; anionic; see Merpiphos PP; lt. yel. liq.; 100% act.

Merpiphos SP. [347] Org. phosphoric acid ester; anionic; hydrotrope for slurries; act. agent in sprayed washing powds.; lt. br. liq.; 100% act.

Merpiphos TP. [347] PEG-4 C_{16-18} alcohol hydrogen phosphate with hydrotrope; anionic; see Merpiphos SP; lt. br. liq.; 100% act.

Merpiphos UP. [347] PEG-6.5 iso-C_{13} alcohol hydrogen phosphate; anionic; drycleaning products; antistat textile wetting agent; liq.; 100% conc.

Merpiphos VP. [347] PEG-8 C_{16-18} alcohol

hydrogen phosphate; anionic; textile washing agent; lubricant for syn. fibers; emulsifier; pasty substance; 100% conc.

Merpiquat K-8-2. [347] Cocodimethylbenzyl ammonium chloride; cationic; disinfectant and cleanser; liq.; 80% conc.

Merpiquat K-26. [347] PEG-5 coco ammonium ethosulfate; cationic; hair preparations, emulsifier, metal treatment; liq.; 90% conc.

Merpisap A 497. [347] Blend of anionic active and nonionic basic products; anionic; detergent used in liq. washing and cleaning agents, for household and industry; lt.-brnsh. pumpable paste; 97% act.

Merpisap AC 70. [347] Dodecylbenzene calcium sulfonate; anionic; emulsifier used in insecticide formulations; yel. visc. oil; 70% act.

Merpisap AE 50, 60, 70. [347] Dodecylbenzene sodium sulfonate; anionic; detergent for household and industry; wh. paste; 50, 60, and 70% act.

Merpisap AN 90. [347] Blend of anionic active and nonionic basic products; anionic; detergent household and industry; yel. soft paste; 90% act.

Merpisap AP 80 W, 85 W, 90, 90 P. [347] Linear sodium dodecylbenzene sulfonate; anionic; powd. detergents for household and industry, scouring and wetting agents; wh. powd. (90 P contains anticaking agent); biodeg.; 80, 85, 88, and 85% act. resp.

Merpisap AS 98. [347] Linear dodecylbenzene sulfonic acid; anionic; intermediate, basic product for liq., paste, and powd. detergents, wetting and cleansing formulations; dk. br. visc. liq.; biodeg.; 98% act.

Merpisap AT 50, 90. [347] Linear TEA dodecylbenzene sulfonate; anionic; detergent used in liq. washing and cleaning agents for household and industry, textile aux., cosmetic industry; yel. pumpable paste; biodeg., 50 and 90% act. resp.

Merpisap D 397. [347] Blend of anionic active and nonionic basic products; anionic; detergent for household and industry; lt. br. pumpable paste; 97% act.

Merpisap DC 70. [347] Calcium sulfonate; anionic; emulsifier for insecticide formulations; lt. br. visc. oil; 70% act.

Merpisap DE 70. [347] Sodium tetrapropylene benzene sulfonate; anionic; detergent for household and industry, textile aux.; wh. paste; 70% act.

Merpisap DN 90. [347] Blend of anionic active and nonionic basic products; anionic; detergent for household and industry; lt. paste; 90% act.

Merpisap DP 82. [347] Sodium tetrapropylene benzene sulfonate; anionic; powd. detergents for household and industry, scouring and wetting agents, textile aux.; wh. free-flowing powd.; 83% act.

Merpisap DT 90. [347] TEA tetrapropylene benzene sulfonate; anionic; detergent for household and industry, textile aux., cosmetic industry; yel. pumpable paste; 90% act.

Merpisap F-9-3. [347] Blend of anionic and nonionic products; anionic; detergent for semifinished cleaning agents for household and industry; lt. colored, low-visc. opaque liq.; biodeg.; 50% act.

Merpisap KH 30. [347] Liquified linear sodium dodecylbenzene sulfonate; anionic; detergent for household and industry, textile aux.; yel. pumpable paste; biodeg.; 30% act.

Merpisap KO 30. [347] Blend of linear sodium dodecylbenzene sulfonate and nonionic products; anionic; detergent for semifinished cleaning agents for household and industry; yel. pumpable paste; biodeg.; 30% act.

Merpisap TS 98. [347] Tetrapropylene benzene sulfonic acid; anionic; intermediate, basic product for liq., paste, and powd. detergents, wetting and cleansing formulations; dk. br. visc. liq.; 98% act.

Merpol C. [208] Long-chain alcohol sulfate; anionic; wetting agent, detergent, penetrant, emulsifier used in textile and paper industries; lt. amber liq.; misc. with water; dens. 7.5 lb/gal; pH 8 (1%).

Merpol CH-196. [208] Modified EO condensate; nonionic; detergent, wetting agent used in textile industry; clear, colorless liq.; disp. in water; dens. 8.1 lb/gal; sp.gr. 0.973; visc. 24 cps; HLB 13.0; surf. tens. 27.2 dynes/cm (0.1 g/100 ml); pH 6–8.

Merpol DA. [208] EO condensate; nonionic; low-foaming leveling agent and dyeing assistant for textile dyes; lt. yel. clear, slightly visc. liq.; alcoholic odor; sol. in water; > 50% sol. in ethanol, ethylene glycol, cellosolve, acetone, MEK, oleic acid; dens. 8.5 lb/gal; visc. 110 cps; cloud pt. 10 C (lower), > 100 C (upper 1%); clear pt. 12 C; surf. tens. 50 dynes/cm (0.1%); pH 9 (1%); 60% act.

Merpol HCS. [208] EO condensate; nonionic;

wetting agent, penetrant, antistat, leveling agent, detergent, dyeing assistant, emulsifier, stabilizer used in textiles, paints, inks, medicinal ointments, antiperspirants, cosmetics; lt. yel. clear, visc. liq.; mild aromatic odor; 40% sol. in water, sol. in org. solvs. that are misc. with water; dens. 8.63 lb/gal; HLB 15.5; cloud pt. 43 F; flash pt. > 235 F; surf. tens. 42.9 dynes/cm (0.1%); pH 6–8 (10%); 60% act.

Merpol HCW. [208] Higher fatty alcohol EO condensate; nonionic; emulsifier, leveling agent, stabilizer, wetting agent, dispersant, detergent, dyebath additive for all fibers; clear, visc. liq.; mild, alcoholic odor; water sol.; dens. 8.32 lb/gal; f.p. < 3 F; cloud pt. > 212 F; flash pt. > 80 F; pH 6–7 (10%); 30% act.

Merpol LF-H. [208] Polyether; nonionic; detergent, nonfoam wetting, scouring and dyeing assistant for textiles, emulsifier, lubricant; pale yel. clear liq.; mild odor; 40% max. sol. in water below 70 F; dens. 7.96 lb/gal; visc. 28 cps; cloud pt. 78 F (upper 1% aq.); flash pt. > 200 F (TCC); pH 6–8; biodeg.

Merpol OJS. [208] EO condensate; nonionic; wetting and disp. agent for textiles, emulsifier, antistat, detergent; slightly yel. clear liq.; mild fatty odor; sol. in water, sol. or disp. in alcohol, ketones, aliphatic and aromatic hydrocarbons; dens. 8.04 lb/gal; sp.gr. 0.9648; visc. 46 cps; HLB 12.7; cloud pt. 30 F; flash pt. 167 F (TOC); surf. tens. 33.5 dynes/cm (0.1%); pH 6–8; 60% act.

Merpol SE. [208] Long-chain fatty alcohol EO condensate; nonionic; wetting and rewetting agent, emulsifier, dispersant; colorless to pale yel. clear liq.; 0.1% sol. in water, cloudy disp. at higher conc.; dens. 8.1 lb/gal; HLB 10.5; pH 6.2 (0.1%); 95% act.

Merpol SH. [208] Long-chain alcohol-EO condensate; nonionic; detergent, wetting agent used in textile processing; colorless clear liq.; alcoholic odor; misc. with water, sol. or disp. in alcohols, acetone, MEK, ethylene glycol, glycerin, oleic acid; dens. 8.0 lb/gal; visc. 24 cps; HLB 12.9; cloud pt. < 40 F; flash pt. 92 F (TCC); surf. tens. 27.4 dynes/cm (0.1 g/100 ml); pH 5–7 (10% aq.); biodeg.; 50% act.

Merpophal SP. [347] C_{12-14} alcohol hydrogen phosphate; anionic; act. substance in spray-dried washing powds.; paste; 100% conc.

Merpoxal LM 3028. [347] Sodium lauryl ether sulfate; anionic; detergent, emulsifier used in cosmetic, hair shampoo, and foam bath preparations, household and industrial uses; water-clear liq.; biodeg.; 28% act.

Merpoxal LM 3030. [347] Sodium lauryl sulfate; anionic; surfactant used in hair shampoo and foam-bath preparations; water-wh., visc. liq.; 30% act.

Merpoxal LM 3070. [347] Sodium lauryl ether sulfate; anionic; detergent, emulsifier used in cosmetic, hair shampoo, and foam bath preparations, household and industrial uses; water-clear liq.; biodeg.; 70% act.

Merpoxal NOA 4080. [347] Ammonium nonylphenol ether sulfate; anionic; detergent for household and industrial uses, emulsifier; yel. soft paste; 80% act.

Merpoxal TR 4060. [347] Sodium tridecyl ether sulfates; anionic; detergent, emulsifier used in cosmetic, hair shampoo, and foam bath preparations, household and industrial uses; yel. soft paste; 60% act.

Merpoxen DL 52. [347] PEG-5.2 C_{10-12} alcohol; nonionic; wetting agent, basic product for liq. and powd. cleaning and washing preparations; clear, visc. liq.; 100% act.

Merpoxen DNP 160. [347] Dinonyl phenol ethoxylate; nonionic; emulsifier, liq. and powd. detergent, wetting agent; yel., visc. liq.; 100% act.

Merpoxen KM 45, 60. [347] Coconut fatty acid monoethanolamide 4.5 and 6 polyglycol ethers; nonionic; foam stabilizer for shampoos and liq. detergents; yel. liquid-pasty substance and yel. liq. resp.; 100% act.

Merpoxen KM 90. [347] Coconut fatty acid monoethanolamide 9 polyglycol ether; nonionic; foam stabilizer for shampoos and liq. detergents; refattening agent; liq.; 100% conc.

Merpoxen LM 20, 30. [347] PEG-2 and PEG-3 resp. C_{12}–C_{14} alcohol; nonionic intermediate, basic product for ether sulfates; clear liq.; 100% act.

Merpoxen LP 60, 70, 109, 120. [347] PEG-6, -7, -10, and -12 C_{12}–C_{15} alcohol; nonionic; detergent, emulsifier, cleaning agent, textile aux.; clear, opaque liq. (LP 60, 70, 109), wh. soft paste (LP 120); 100% act.

Merpoxen NO 15. [347] Nonoxynol-1 (1.5 EO); nonionic; emulsifier, defoamer; water-wh. visc. liq.; 100% act.

Merpoxen NO 40. [347] Nonoxynol-4; nonionic; emulsifier, intermediate; water-wh. visc. liq.; 100% act.

Merpoxen NO 60, 80, 90. [347] Nonoxynol-6, -8, -9 resp.; nonionic; emulsifier, wetting agent, detergent, textile aux.; water-wh. visc. liq.; 100% act.

Merpoxen NO 95, 100, 110. [347] Nonoxynol-9 (9.5 EO), -10, -11 resp.; nonionic; see Merpoxen NO 60; water-wh., visc. liq.; 100% act.

Merpoxen NO 120, 140. [347] Nonoxynol-12, -14 resp.; nonionic; see Merpoxen NO 60; wh . soft paste; 100% act.

Merpoxen NO 150, 200, 300. [347] Nonoxynol-15, -20, -30 resp.; nonionic; emulsifier, detergent; wh. wax; 100% act.

Merpoxen NO 600. [347] Nonoxynol-60; nonionic; emulsifier; dispersant; wax; 100% conc.

Merpoxen OA 150. [347] Alkyl amine ethoxylate; nonionic; emulsifier, leveling and softening agent used in textile industry; lt. br. soft paste; 100% act.

Merpoxen OFS 60, 80. [347] PEG-6 and -8 oleate; nonionic; emulsifier for textiles and detergents industries; yel. liq.; 100% act.

Merpoxen PEG 200, 300, 400. [347] PEG; nonionic; intermediate, softener, lubricant in cosmetic preparations, industrial use; clear liq.; m.w. 200, 300, 400 resp.; 100% act.

Merpoxen PEG 600, 800, 1000. [347] PEG; nonionic; softener, lubricant for cosmetic preparations, industrial use; clear paste, clear liq., and wh. wax. resp.; m.w. 600, 800, 1000 resp.; 100% act.

Merpoxen PEG 1500 S. [347] PEG; nonionic; see Merpoxen PEG 600; wh. flakes; m.w. 1500; 100% act.

Merpoxen RO 200, 300, 400. [347] PEG-20, -30, -40 castor oil ether; nonionic; emulsifier, textile aux.; yel. visc. liq. (RO 200), wh. wax (RO 300, 400); 100% act.

Merpoxen RO 800. [347] PEG-80 castor oil ether; nonionic; emulsifier for pesticides; textile aux.; wax; 100% conc.

Merpoxen S-5-96/10, /20, /40, /70. [347] Alkylamine polyalkylenoxide ether; nonionic; detergent, cooling lubricant, used in cosmetic industry; lt. yel. opaque liq. (S-5-96/10, /20), lt. yel. soft paste (S-5-96/40, /70); 100% act.

Merpoxen SFS 60, 80, 120. [347] PEG-6, -8, and -12 stearate; nonionic; emulsifier, textile aux.; wh. soft paste (SFS 60, 80), wh. wax-like substance (SFS 120); 100% act.

Merpoxen SP 100, 200, 230. [347] PEG-10, -20, -23 oleyl-cetyl alcohol; nonionic; emulsifier, detergent, intermediate used in textile industry; wh. waxy paste; 100% act.

Merpoxen STA 150, 200, 250. [347] Alkyl amine ethoxylate; nonionic; emulsifier, leveling and softening agent in textile industry; lt. brnsh. soft paste (STA 50), lt. brnsh. wax (STA 200, 250); 100% act.

Merpoxen T-2-3. [347] Fatty alcohol polyalkylene oxide ether; nonionic; surfactant component in washing, cleansing, and polishing products; used in household and industrial products; liq.; 100% conc.

Merpoxen T-4-8. [347] I-tridecyl-alcohol polyglycol ether, modified; nonionic; low-foaming wetting agent; clear, wh. liq.; 100% act.

Merpoxen T-5-5, T-5-7. [347] Alkyl polyglycol ether, modified; nonionic; low-foaming wetting agent; clear whitish. liq.; 100% act.

Merpoxen T-8-3. [347] Alkylaryl polyalkylenoxide ether; nonionic; see Merpoxen T-2-3; liq.; 100% conc.

Merpoxen T 110, 180. [347] Talloweth-11, -18; nonionic; detergent, emulsifier for paraffins, basic product for powd. detergents and cleansers; wh. waxy paste; 100% act.

Merpoxen T 250, 500, 800. [347] Talloweth-25, -50, -80 resp.; nonionic; detergent, basic product for powd. detergents and cleansers; wh. wax, powd. or flake; 100% act.

Merpoxen TA 120. [347] Tall oil fatty acid ethoxylate; nonionic; detergent, emulsifier, washing and lathering agents in textile industry; yel. visc. liq.; 100% act.

Merpoxen TP 185. [347] PEG-10 C_{13-15} alcohol; nonionic; component for detergent powds.; liq.; 85% conc.

Merpoxen TR 40. [347] Trideceth-4; nonionic; intermediate, emulsifier for wettable powders, wetting agent for textile industry, basic product for ether sulfates; water-wh. visc. liq.; 100% act.

Merpoxen TR 70. [347] Trideceth-7; nonionic; see Merpoxen TR 40; whitish. visc. liq.; 100% act.

Merpoxen TR 80, 100. [347] Trideceth-8, -10; nonionic; see Merpoxen TR 40; wh. soft paste; 100% act.

Merpoxen UN 22. [347] PEG-2.2 alkyl C_9–C_{11} alcohol; nonionic; defoamer, intermediate, basic product for the production of ether sulfates; clear, visc. liq.; 100% act.

Merpoxen UN 40, 50. [347] PEG-4 and PEG-5 alkyl C_9–C_{11} alcohol; nonionic; intermediate, basic product cleaning agents; clear, visc. liq.; 100% act.

Merpoxen UN 65. [347] PEG-6.5 alkyl C_9–C_{11} alcohol; nonionic; see Merpoxen UN 40; clear, visc. liq.; 100% act.

Metarin. [410] Lecithin fraction; nonionic; instantizer for powd. foods; powd.; 100% conc.

Metarin F. [410] Lecithin compd.; nonionic; instantizer for foods; liq.; 100% conc.

Metarin P. [410] Lecithin fraction; nonionic; see Metarin F; powd.; 100% conc.

Methocel A4C, A4M, A15-LV. [201] Methylcellulose; food gums used as thickener, stabilizer, emulsifier, adhesive, and gelling agent used in food industry; firm gel structure; water-sol.; visc. 400, 4000, 15 cps. (2% aq.) resp.

Methocel E5, E15-LV, E50-LV, E4M, F50-LV, F4M, K3, K35, K100-LV, K4M, K15M, K100M. [201] Hydroxypropyl methylcellulose or carbohydrate gum; see Methocel A4C; soft, semifirm gel structure; water-sol.; visc. 5, 15, 50, 4000, 50, 4000, 3, 35, 100, 4000, 15,000, and 100,000 cps (2% aq.) resp.

Michelene LS-90. [411] Sodium lauryl sulfate; anionic; detergent; cleaning and wetting agent; extrusion; 90% conc.

Migregal 2N. [442] Alkyl sulfate; anionic; level dyeing of nylon piece goods with acid dyes; liq.

Migregal NC-2. [442] Alkyl sulfate and nonionic surfactant; anionic/nonionic; leveling agent for fabrics; liq.

Minemal 320, 325, 330. [440] POE alkyl phenyl ether; nonionic; degreaser for leather; liq.; 100% conc.

Minemal 350. [440] Nonionic/anionic; emulsifier for machine oil; liq.

Miramine CC. [414] Cocoyl imidazoline; cationic; corrosion inhibitor in cutting oils, lubricant for metal surfaces, used in acid shampoos, with min. spirits to emulsify grease and oil; amber liq.; bland to faintly ammoniacal odor; sol. in many org. solvs., disp. in water; sp.gr. 0.940; biodeg.; 99% act.

Miramine CPC. [414] Caprylic fatty acid substituted imidazoline; cationic; see Miramine CC; liq.; 100% conc.

Miramine DD. [414] Stearamidopropyl dimethylamine; cationic; softener, emulsifier, and conditioner for hair and skin products; solid; 100% conc.

Miramine GS. [414] Stearic fatty acid substituted imidazoline; cationic; emulsifier for asphalt, fuel and cutting oils, metal lubricants; solid; 100% conc.

Miramine GT. [414] Tall oil fatty acid substituted imidazoline; cationic; see Miramine GS; liq.; 100% conc.

Miramine IC. [414] Isostearic acid substituted imidazoline; cationic; corrosion inhibitor, lubricant, fabric softener; liq.; 100% conc.

Miramine OC. [414] Oleyl hydroxyethyl imidazoline; cationic; detergent, thickener of HCl, corrosion inhibitor in metal treating and cutting fluids; amber liq. to pasty solid; ammoniacal odor; oil sol. and water disp.; sp.gr. 0.930; biodeg.; 99% min. act.

Miramine SC. [414] Soya imidazoline; cationic; emulsifier and coemulsifier in solv./water systems, corrosion inhibitor in cutting oils, lubricant for metal surfaces; amber liq.; ammoniacal odor; sol. in many org. solvs., disp. in water; sp.gr. 0.930; biodeg.; 99% act.

Miramine TC. [414] Tallow fatty acid substituted imidazoline; cationic; see Miramine CC; solid; 100% conc.

Miramine TOC. [414] Tall oil hydroxyethyl imidazoline; cationic; see Miramine CC; solid; 100% conc.

Miranate LEC. [414] Sodium laureth-13 carboxylate; anionic; aux. detergent for shampoo systems, lime soap dispersant; clear, hazy gel, pH 8.0 (10%); 70% solids.

Miranate LSS. [414] Disodium lauryl sulfosuccinate; anionic; surfactant for use in shampoo systems; wh. creamy solid; pH 7.0 (10%); 35.0% act.

Miranol 2MCA-ESF. [414] Cocoamphodipropionate and sodium lauryl sulfate; amphoteric; used in shampoos; visc. liq.; 33% conc.

Miranol 2MCA Modified. [414] Cocoamphodiacetate, sodium lauryl sulfate and hexylene glycol; used in formulating nonirritating shampoos; lt. amber clear liq.; water sol.; pH 8.0–8.5 (20% aq.); 46–48% solids.

Miranol 2MCAS Modified. [414] Cocoamphodiacetate, sodium lauryl sulfate, sodium laureth sulfate, propylene glycol; amphoteric surfactant used in formulating nonirritating shampoos; lt. yel. clear liq.; water sol.; pH 7.5–8.0; 45.5–48.0% solids.

Miranol 2MCT. [414] Cocoamphodiacetate, sodium trideceth sulfate and hexylene glycol; amphoteric; see Miranol 2MCA-ESF; liq.; 43% conc.

Miranol 2MCT Modified. [414] Cocoamphodiacetate, sodium trideceth sulfate, hexylene glycol; amphoteric; base detergent for nonirritating make-up removers; lt. amber clear liq.; water sol.; pH 7.5–8.0; 48.5–51.5% solids.

Miranol 2MHT. [414] Lauroamphodiacetate, sodium trideceth sulfate and hexylene glycol; amphoteric; see Miranol 2MCA-ESF; liq.; 42% conc.

Miranol 2MHT Modified. [414] Lauroamphodiacetate, sodium trideceth sulfate, hexylene glycol; amphoteric; detergent base; lt. amber clear liq.; water sol.; pH 8.0–8.5 (20% aq.); 48–50% solids.

Miranol BM Conc. [414] Monocarboxylic lauric/myristic deriv., sodium salt; amphoteric; shampoos, skin cleansers; liq.; 30% conc.

Miranol BT. [414] Lauroamphodiacetate and sodium trideceth sulfate; amphoteric; see Miranol 2MCA-ESF; liq.; 33% conc.

Miranol C2M Anhydrous Acid. [414] Cocoamphodipropionic acid; amphoteric; detergent, wetting agent, high foaming surfactant for disp. on caustic soda and on powd. mixes, etc., leveling agent in tin plating from acid baths; amber stiff gel; sol. in water and alcohol, slightly sol. in nonpolar org. solvs.; pH 4.6 (50%); 100% act.

Miranol C2M Conc. [414] Cocoamphodiacetate; amphoteric; detergent used in nonirritating shampoos, skin cleansers, make-up removers, emulsifier, solubilizer and stabilizer in pharmaceuticals, household and industrial uses; clear, visc. liq.; sol. in water; pH 8.0–8.5 (20% aq.); 38% act.

Miranol C2M-DEA 60%. [414] DEA-cocoamphodipropionate; amphoteric; liq.; 60% conc.

Miranol C2M-SF 70%. [414] Cocoamphodipropionate; amphoteric; detergent used for heavy-duty liq. cleaning compds., steam cleaners, etc.; paste at R.T., pumps and pours at 50 C; pH 9–10 (10% aq.); 70% act.

Miranol C2M-SF Conc. [414] Cocoamphodipropionate; amphoteric; see Miranol C2M-SF 70%; clear, amber liq.; sol. in water and alcohol; pH 9.4–9.8; biodeg.; 39% act.

Miranol CM Conc. [414] Cocoamphoacetate; amphoteric; detergent, wetting and foaming agent, sequestrant, emulsifier, dispersant, germicidal; lt. amber, visc. liq.; pH 9.0–13.0; biodeg.; 37% conc.

Miranol CM-SF Conc. [414] Cocoamphopropionate; amphoteric; see Miranol CM Conc.; lt. amber clear liq.; sol. in water and alcohol; pH 9.5–10.5; biodeg.; 36–38% solids.

Miranol CS 60%. [414] Cocoamphohydroxypropyl sulfonate; amphoteric; shampoos and fabric detergents; liq.; 50% conc.

Miranol CS Conc. [414] Cocoamphohydroxypropylsulfonate; amphoteric; detergent, wetting agent, corrosion inhibitor, emulsifier, sequestrant, foaming agent, for household and industrial uses; liq.; biodeg.; 38% conc.

Miranol DM. [414] Stearamphoacetate; amphoteric; antistat, household softener, lubricant used in textile industry; wh. creamy paste, readily pourable @ 60 C; pH 5.4–6.0 (65 C); biodeg.; 25–27% solids.

Miranol DS. [414] Stearoamphohydroxypropylsulfonate; amphoteric; detergent used in laundry compds., household and dishwash products, shampoos, soaps, metal processing; paste; 21% conc.

Miranol Ester PO-LM4. [414] Oligomeric polyester; nonionic; emulsifier, emollient, conditioner, antistat for personal care products; liq.; 100% conc.

Miranol H2M Conc. [414] Lauroamphodiacetate; amphoteric; similar to Miranol C2M Conc.; lt. amber visc. liq.; water sol.; pH 8.0–8.5 (20% aq.); biodeg.; 49–51% solids.

Miranol H2M-SF 70%. [414] Lauroamphodipropionate; used in place of Miranol C2M-SF Conc. where higher quality is required; for cosmetic formulations; paste at R.T., pours and pumps readily at 50 C; 70% act.

Miranol H2M-SF Conc. [414] Lauroamphodipropionate; see Miranol H2M-SF 70%; lt. amber clear liq.; sol. in water, alcohol; pH 9.2–9.8; biodeg.; 38–40% solids.

Miranol HM Conc. [414] Lauroamphoacetate; amphoteric; detergent, wetting and foaming agent, sequestrant, emulsifier, dispersant, germicidal; lt. amber, visc. liq.; pH 9.0–9.5; biodeg.; 43–45% solids.

Miranol HM-SF Conc. [414] Lauroamphopropionate; amphoteric; emulsion polymerization; liq.; 37% conc.

Miranol HS Conc. [414] Lauroamphohydroxypropylsulfonate; amphoteric; general surfactant, sequestrant for heavy metal, detergent, wetting and foaming agent, emulsifier; liq.; 38% conc.

Miranol ISM. [414] Monocarboxylic isostearic deriv., sodium salt; amphoteric; lubricant for cosmetics and industrial products; visc. booster for shampoos; visc. liq.; 35% conc.

Miranol J2M Anhydrous Acid. [414] Anhydrous acid of Miranol J2M Conc.; amphoteric; wetting agent, penetrant used in bottle washing, wax stripping formulations; stiff amber gel; sol. in water, alcohol, slightly sol. in nonpolar org. solvs.; pH 4.6 (50%); 100% act.

Miranol J2M Conc. [414] Capryloamphocarboxyglycinate; see Miranol JEM Conc., slightly better rinsing properties than JEM Conc., used in emulsion polymerization of syn. rubbers and resins; clear liq.; pH 8.2–8.6; biodeg.; 48–50% solids.

Miranol J2M-SF Conc. [414] Capryloamphocarboxypropionate; salt free version of Miranol J2M Conc.; liq.; pH 8.8–9.3; biodeg.; 38–40% solids.

Miranol JEM Anhydrous Acid. [414] Anhydrous acid of Miranol JEM Conc.; amphoteric; detergent, wetting agent, low foaming surfactant for disp. on caustic soda and on powd. mixes, metal cleaning, etc.; stiff amber gel; sol. in water, alcohol, slightly sol. in nonpolar solvs.; pH 4.6 (50%); 100% act.

Miranol JEM Conc. [414] Sodium salt of a dicarboxylated caprylic and ethylhexoic imidazoline deriv.; amphoteric; see Miranol J2M Anhydrous Acid; clear liq.; surf. tens. 30.5 dynes/cm (0.12% in water); pH 10.0–10.5; biodeg.; 33–35% solids.

Miranol JS Conc. [414] Capryloamphopropylsulfonate; amphoteric; wetting agent, corrosion inhibitor used in pickling acids; liq.; biodeg.; 40% conc.

Miranol L2M-SF Conc. [414] Disodium dicarboxylated imidazoline deriv., salt free; foamer with good oil emulsifying properties; liq.; pH 10.0–10.5; biodeg.; 38–40% conc.

Miranol LM-SF Conc. [414] Tallamphopropionate; similar to L2M-SF Conc. but more hydrophobic, suggested as lubricant for ferrous and nonferrous metals; visc. liq. which sets to a pasty gel on aging at R.T., pourable @ 60 C; pH 9.5–10.5; biodeg.; 36–38% solids.

Miranol MHT. [414] Lauroamphoacetate and sodium trideceth sulfate; amphoteric; base detergent for nonirritating and nonstinging foaming hair and skin cleansing products; lt. amber clear liq.; water sol.; pH 9.4–10.2; 33.5–36.5% solids.

Miranol OM-SF Conc. [414] Oleoamphopropionate; amphoteric; wetting and foaming agent, sequestrant, emulsifier, dispersant, germicidal; sol. in water, alcohol; pH 9.8–10.2; biodeg.; 36–38% solids.

Miranol OS. [414] Oleoamphohydroxypropylsulfonate; amphoteric; general surfactant, sequestrant for heavy metal, detergent, wetting and foaming agent, emulsifier; paste; 21% conc.

Miranol OS-D. [414] Sulfonated oleic deriv., sodium salt; amphoteric; detergent, shampoos; liq.; 21% conc.

Miranol S2M Conc. [414] Caproamphodiacetate; amphoteric; wetting and foaming agent, sequestrant, emulsifier, dispersant, germicidal; clear liq.; pH 8.1–8.3; biodeg.; 50–52% solids.

Miranol S2M-SF Conc. [414] Caproamphodipropionate; see Miranol S2M Conc.; liq.; pH 8.8–9.2; biodeg.; 38–40% solids.

Miranol SM Conc. [414] Caproamphoacetate; amphoteric; wetting agent, foaming agent, detergent used in medicated and germicidal shampoos and hand soaps, rug and upholstery shampoos, in emulsion polymerization; clear liq.; pH 9.0–9.5; biodeg.; 40–42% solids.

Miranol T-60. [414] Dicarboxylic tall oil deriv., disodium salt; amphoteric; detergent emulsifier for wash and wax products and heavy-duty detergents; liq.; 60% conc.

Mirapol A-15. [414] Polyquaternium-2; cationic; softening, conditioning, and surface modifying agent used in cream rinses, conditioning-type shampoos; amber, visc. liq.; m.w. 2260 (avg.); dissolves readily in water; pH 8.5; 64% act.

Mirapol AD-1. [414] Polyquat. ammonium chloride; cationic; conditioner and antistat for personal care products; amber visc. liq.; pH 8.0 (10%); 62.0% act. in water.

Mirapol AZ-1. [414] Polyquaternium-18; cationic; conditioner and antistat for personal care products.

Mirapol WT. [414] Polyquat. ammonium chloride; cationic; antiscaling compd. in water systems; textile softener; corrosion inhibitor; liq.; 63% conc.

Mirapon FM. [414] Alkylamino acid, potassium salt; amphoteric; detergent, emulsifier, wetting agent and lubricant; liq.; 70% conc.

Mirapon JAS 50%. [414] Capryl imidazoline carboxylate deriv.; amphoteric; surfactant for acid and alkaline cleaners, degreasers and wax strippers; 50% conc.

Mirataine BB. [414] Lauramidopropyl betaine; amphoteric; visc. builder and foam booster for shampoos and dishwashing liqs.;

liq. 30% conc.

Mirataine BD. [414] Cocamidopropyl betaine; amphoteric; detergent and visc. builder for shampoos, bubble baths; liq.; 30% conc.

Mirataine CB. [414] Cocamidopropyl betaine; amphoteric; surfactant, mild, high foaming base for shampoo formulations; pale-colored liq.; dens. 8.71 lb/gal; sp.gr. 1.045; pH 8.5; 35% act. in water.

Mirataine CBC, CBR. [414] Cocamidopropyl betaine; amphoteric; visc. builder and foam booster for shampoos and dishwashing liqs.; liq.; 30% conc.

Mirataine CBL. [414] Cocamidopropyl betaine and sodium trideceth sulfate; amphoteric; shampoo base; liq.; 35% conc.

Mirataine CBS. [414] Cocamidopropyl hydroxysultaine; amphoteric; surfactant, mild, high foaming base for shampoo formulations; liq.; dens. 9.1 lb/gal; cloud pt. ≤ −12 C; pH 8.2; 50% act. in water.

Mirataine CDMB. [414] Coco betaine; amphoteric; see Mirataine CB; liq.; 30% conc.

Mirataine CEB. [414] Cocoamidoalkyl betaine; amphoteric; surfactant, mild, high foaming base for shampoo formulations; pale-colored liq.; dens. 8.75 lb/gal; sp.gr. 1.05; cloud pt. clear to freeze; pH 8.5; biodeg.; 40% act. in water.

Mirataine COB. [414] Coco oleamidopropyl betaine; amphoteric; surfactant and conditioner for use in personal care products; yel. visc. liq.; pH 7.0; 34.0% solids.

Mirataine DDW. [414] Cocamidopropyl hydroxysultaine; amphoteric; detergent for drilling foams; liq.; 42% conc.

Mirataine H2C. [414] Disodium lauriminodipropionate; amphoteric; detergent, lubricant, antistat, corrosion inhibitor, and solubilizer used in detergent and industrial formulations, personal care products, leather, and textile fibers; yel. thin liq.; pH 10.5; flash pt. (PM) 132 C; 30.0% act.

Mirataine H2C-HA. [414] Sodium lauriminodipropionate; amphoteric; surfactant for shampoos and skin cleansers and hard surface cleaners; liq.; 30% conc.

Mirataine HC-HA. [414] Sodium lauraminopropionate; amphoteric; surfactant for shampoos and skin cleansers; liq.; 40% conc.

Mirataine HTS. [414] Hydrog. tallow hydroxy sultaine; amphoteric; lime soap dispersant; conditioner for hair and skin preparations; gel; 35% conc.

Mirataine ODMB-35%. [414] Oleyl betaine; amphoteric; visc. builder and conditioner for personal care products; liq.; 30% conc.

Mirataine TABS. [414] Tallowamidopropyl hydroxysultaine; amphoteric; lime soap dispersant in industrial and household cleaners; clear, hazy gel; pH 8.3; 41.0% solids.

Mirataine T2C. [414] Disodium tallowiminodipropionate; amphoteric; detergent, solubilizer, moderate foaming surfactant used in textile, leather, industrial and personal care products; liq.; sol. clear (30% solids in 1% NaOH); surf. tens. 31.6 dynes/cm (1% aq.); pH 11.5; biodeg.; 35% act. in water.

Mirataine TM. [414] Dihydroxyethyl tallow glycinate; amphoteric; conditioner for shampoos; visc. liq.; 35% conc.

Mirataine XL. [414] DEA-lauryl sulfate, DEA-lauraminopropionate, sodium lauraminopropionate, propylene glycol; shampoo base; liq.; 35% conc.

Mirawet B. [414] Sodium butoxyethoxy acetate; wetting agent; liq.; 46% conc.

ML-33-F. [287] Sorbitan monolaurate; nonionic; emulsifier for creams and sun lotions, ointments; coemulsifier for latex stabilization; lubricant; liq.; HLB 8.0; 100% conc.

ML-55-F. [287] Polysorbate 20; nonionic; solubilizer for essential oils, fragrances, vitamins; emulsifier for cosmetics, pharmaceuticals; stabilizer in syrups; liq.; HLB 16.5; 100% conc.

ML-55-F-4. [287] Polysorbate 21; nonionic; coemulsifier for creams and ointments; pigment dispersions, color pastes, fiber preparations, polishes, hydraulic liqs.; liq.; HLB 12.0; 100% conc.

MO-33-F. [287] Sorbitan monooleate; nonionic; emulsifier for w/o creams, lotions, and shampoos; ointments, biocides, paints and varnishes, anti-corrosion oils, polishes, cutting oils; liq.; HLB 4.4; 100% conc.

MO-55-F. [287] Polysorbate 80; nonionic; solubilizer for essential oils; pharmaceuticals and cosmetics; emulsifier for food industry; leather aux. products and other special technical applications; liq.; HLB 15.0; 100% conc.

MO-55-F-5. [287] Polysorbate 81; nonionic; coemulsifier for creams and ointments; pigment dispersions and color pastes; antistat for films, polishes, hydraulic liqs.; liq.; HLB 10.5; 100% conc.

Modicol L. [186; 187] PEG fatty ester; nonionic; coagulant, wetting and dispersing agent, used in latex and resin emulsions; clear, yel. liq.; sol. in ethanol, acetone, ethyl acetate, toluol, veg. oil, water, min. oil; dens. 8.6 lb/gal; pH 7.5 (2%); 99% act.

Mollisan 4NY, 379 Conc. [376] Fatty ester; nonionic; softener, o/w emulsifier; paste; 100% conc.

Mollisan Conc. [376] Fatty condensate; nonionic; lubricant, softener, wetting agent; o/w emulsifier; paste; 100% conc.

Monafax 057. [418] Phosphate ester, aromatic-based; surfactant used as lubricant, corrosion inhibitor in metal-working fluids; lt. yel. clear visc. liq.; sol. @ 10% in water, ethanol, aromatic hydrocarbon, chlorinated paraffin, naphthenic oil; sp.gr. 1.2; dens. 9.96 lb/gal; pH < 2.5 (10% aq.); 100% act.

Monafax 060. [418] Phosphate ester, aliphatic-based; hydrotrope and solubilizer for nonionics in alkaline systems; used in soak tank, alkaline, and steam cleaners; amber clear visc. liq.; sol. @ 10% in water, ethanol, ethylene glycol monobutyl ether; sp.gr. 1.335; dens. 11.3 lb/gal; pH 2.5 (10%); 100% act.

Monafax 785. [418] Mixt. of mono and diphosphate esters; anionic; emulsifier, lubricant, antistat, detergent, corrosion inhibitor for agric., industrial use; clear to slightly hazy visc. liq.; sol. in water, ethanol, perchlorethylene, xylene; dens. 9.2 lb/gal; sp.gr. 1.115; pH < 2.5 (10%); 100% act.

Monafax 786. [418] Nonoxynol-6 phosphate; anionic; see Monafax 785; clear to slightly hazy visc. liq.; sol. @ 10% in ethanol, perchloroethylene, Stod., xylene; disp. water; dens. 9.0 lb/gal; sp.gr. 1.085; pH < 2.5 (10%); 100% act.

Monafax 794. [418] Mixt. of mono and diphosphate esters; anionic; emulsifier for polymerization, corrosion inhibitor; soft waxy paste; sol. @ 10% in water, ethanol, perchloroethylene, xylene; dens. 9.8 lb/gal; sp.gr. 1.175; pH < 2.5 (10%); 100% act.

Monafax 831. [418] Mixt. of mono and diphosphate esters; anionic; emulsifier, lubricant, antistat, detergent, corrosion inhibitor for cleaning and industrial use; clear visc. liq.; sol. @ 10% in water, ethanol, perchloroethylene, min. oil, Stod., xylene; dens. 9.0 lb/gal; sp.gr. 1.08; pH < 2.5 (10%); 100% act.

Monafax 872. [418] Org. phosphate ester, potassium salt; anionic; hydrotrope and solubilizer for detergent systems; lt. straw clear liq.; sol. @ 10% in water, disp. in ethanol, ethylene glycol monobutyl ether, cottonseed oil; dens. 11.24 lb/gal; sp.gr. 1.35; pH 7–9 (10%); 50% act.

Monafax 1214. [418] Phosphate ester; anionic; surfactant; liq.; 100% conc.

Monafax H-15. [418] Mixt. of mono and diphosphate esters; anionic; emulsifier for herbicides and insecticides, lubricant, antistat, detergent, corrosion inhibitor; hazy visc. liq.; sol. in water, ethanol, perchloroethylene, xylene; dens. 8.9 lb/gal; sp.gr. 1.07; pH < 2.5 (10%); 100% act.

Monafax L-10. [418] Mixt. of mono and diphosphate esters; anionic; see Monafax H-15; slightly hazy visc. liq.; sol. @ 10% in ethanol, perchloroethylene, kerosene, Stod., xylene; disp. water and min. oil; dens. 8.8 lb/gal; sp.gr. 1.067; pH < 2.5 (10%); 100% act.

Monamate C-1142. [418] Disodium cocamido MIPA-sulfosuccinate; anionic; surfactant for personal care and household products; yel. clear liq.; sp.gr. 1.102; dens. 9.18 lb/gal; pH 5–7; 40 ± 1% solids.

Monamate CPA-40, CPA-100. [418] Disodium cocamido MIPA-sulfosuccinate; anionic; high foaming, nonirritating surfactant for personal care and household produts; liq., powd. resp.; biodeg.; 40 and 100% act.

Monamate LA-100. [418] Disodium lauryl sulfosuccinate; anionic; surfactant for personal care and household products; wh. powd.; pH 6.0–7.0 (10%); 98.0% min. act.

Monamate LNT-40. [418] Diammonium lauryl sulfosuccinate; anionic; surfactant used in personal care products; lt. liq.; sp.gr. 1.07; dens. 8.8 lb/gal; pH 6–7; 40 ± 1% solids.

Monamate OPA-30. [418] Disodium oleamido PEG-2 sulfosuccinate; anionic; high foaming, nonirritating surfactant for shampoos, bubble baths, etc.; lt. yel. liq.; dens. 8.85 lb/gal; sp.gr. 1.06; biodeg.; 30% solids.

Monamid 7-100. [418] 1:1 Coconut diethanolamide; nonionic; foam booster/stabilizer, emulsifier, detergent, wetting agent, corrosion inhibitor, visc. builder, lubricant, dispersant, used in textiles, cosmetics, metalworking, agric., leather and fur applications; Gvcs-33 6 max. liq.; sol. @ 10% in ethanol, chlorinated and aromatic hydrocarbons, min. spirits, kerosene, wh. min. oil, natural fats and oils, water disp.; dens. 8.00 lb/gal; sp.gr. 0.96; pH 8–9 (10%); biodeg.; 100% act.

Monamid 7-153 CS. [418] Modified coconut diethanolamide; nonionic/anionic; see Monamid 7-100; Gvcs-33 3 max. liq.; sol. @ 10% in ethanol, chlorinated and aromatic hydrocarbons, min. spirits, kerosene, wh. min. oil, disp. in water and natural oils and fats; dens. 8.75 lb/gal; sp.gr. 1.05; pH 8.5–9.5 (10%); biodeg.; 100% act.

Monamid 15-70W. [418] 1:1 Linoleic diethanolamide; nonionic, see Monamid 7-100; Gvcs-33 11 max. liq.; sol. @ 10% in ethanol, chlorinated and aromatic hydrocarbons, min. spirits, kerosene, natural oils and fats, gels in water; dens. 8.00 lb/gal; sp.gr. 0.96; pH 10–11 (10%); biodeg.; 100% act.

Monamid 150-AD, 150-ADD. [418] 1:1 Coconut diethanolamide; nonionic; see Monamid 7-100; Gvcs-33 11 and 4 max. resp., liqs.; sol. @ 10% in water, ethanol, chlorinated and aromatic hydrocarbons, natural oils and fats; dens. 8.25 and 8.30 lb/gal; sp.gr. 0.99 and 1.00; pH 9.8–10.8 and 10–11 (10%); biodeg.; 100% act.

Monamid 150-ADY. [418] 1:1 Linoleamide DEA; nonionic; see Monamid 7-100; Gvcs-33 10 max. liq.; sol. @ 10% in ethanol, chlorinated and aromatic hydrocarbons, min. spirits, kerosene, gels in water; dens. 8.20 lb/gal; sp.gr. 0.97; pH 10–11 (10%); biodeg.; 100% act.

Monamid 150-CW. [418] 1:1 Capric diethanolamide; nonionic; see Monamid 7-100; Gvcs-33 7 max. liq., crystallizes on aging; sol. @ 10% in water, ethanol, chlorinated and aromatic hydrocarbons, natural oils and fats; dens. 8.25 lb/gal; sp.gr. 0.99; pH 10.3–11.3 (10%); biodeg.; 100% act.

Monamid 150-DR. [418] 1:1 Coconut diethanolamide; nonionic; see Monamid 7-100; Gvcs-33 5 max. liq.; sol. @ 10% in ethanol, min. spirits, kerosene, wh. min. oil, natural oils and fats, disp. in water, chlorinated and aromatic hydrocarbons; dens. 8.20 lb/gal; sp.gr. 0.98; pH 8.5–9.5 (10%); biodeg.; 100% act.

Monamid 150-GLT. [418] Modified lauric diethanolamide; nonionic; thickener and foam stabilizer for cosmetic and toiletry products; amber liq.; pH 9.0–10.5 (1%); 100% act.

Monamid 150-IS. [418] 1:1 Isostearic diethanolamide; nonionic; see Monamid 7-100; Gvcs-33 6 max. liq.; sol. 10% in ethanol, chlorinated and aromatic hydrocarbons, min. spirits, kerosene, wh. min. oil, natural oils and fats, water disp.; dens. 8.00 lb/gal; sp.gr. 0.96; pH 8.8–9.8 (10%); biodeg.; 100% act.

Monamid 150-LMW-C. [418] Lauramide DEA; nonionic; see Monamid 7-100; Gvcs-33 3 max. solid; sol. @ 10% in ethanol, chlorinated and aromatic hydrocarbons, min. spirits, gels in water; dens. 8.20 lb/gal; sp.gr. 0.98 (40 C); pH 10.2–11.2 (10%); biodeg.; 100% act.

Monamid 150-LW, 150-LWA. [418] 1:1 Lauric diethanolamide; nonionic; see Monamid 7-100; Gvcs-33 3 max. solid; sol. @ 10% in ethanol, chlorinated and aromatic hydrocarbons, disp. in water; dens. 8.20 lb/gal; sp.gr. 0.98 (40 C); pH 10–11 and 9.5–10.5 (10%); biodeg.; 100% act.

Monamid 150-MW. [418] 1:1 Myristic diethanolamide; nonionic; see Monamid 7-100; Gvcs-33 2 max. solid; sol. @ 10% in ethanol, chlorinated and aromatic hydrocarbons, disp. in water; dens. 8.20 lb/gal; sp.gr. 0.98 (45 C); pH 9.5–10.5 (10%); biodeg.; 100% act.

Monamid 664. [418] Coconut alkanolamide; nonionic; foam booster, stabilizer and thickener for industrial and household detergents; liq.; 100% conc.

Monamid 716. [418] Modified lauric diethanolamide; nonionic; foam booster and stabilizer, visc. builder, solubilizer, coupler, wetting agent, used in shampoos, bubble baths, and liq. detergent formulations, mild; lt. amber, clear liq.; sol. @ 10% in water, ethanol, chlorinated and aromatic hydrocarbons, natural oils and fats; dens. 8.2 lb/gal; sp.gr. 0.98; cloud pt. < 1 C; pH 10.0–11.0 (10%); biodeg.; 100% act.

Monamid 718. [418] Stearic diethanolamide; nonionic; see Monamid 7-100; Gvcs-33 4 max. solid; sol. @ 10% in ethanol, aromatic hydrocarbons, gels in water; dens. 8.00 lb/gal; sp.gr. 0.96 (45 C); pH 9.3–10.3 (10%); biodeg.; 100% act.

Monamid 759. [418] Cocamide DEA; foam booster and stabilizer, visc. builder, and conditioner for personal care products; amber liq.; alkali no. 35 ± 5; pH 10.5 ± 0.5 (10%); cloud pt. 0 C; clear pt. 33 C; 100% act.

Monamid 770. [418] Lauramide DEA; nonionic; see Monamid 7-100; Gvcs-33 4 max. liq.; sol. @ 10% in water, ethanol, chlorinated and aromatic hydrocarbons, natural fats and oils; dens. 8.25 lb/gal; sp.gr. 0.99;

pH 9.2–10.2 (10%); biodeg.; 85% act.

Monamid 853. [418] Alkanolamide, modified; nonionic/anionic; detergent, wetting emulsifier, foam stabilizer used in shampoos, liq. household detergents, dairy cleaners, wool scouring, waterless hand cleaners; liq.; 100% conc.

Monamid 982, 1007. [418] 1:1 Lauramide DEA and linoleamide DEA; foam booster and stabilizer, visc. builder, used in shampoos, bubble baths, skin cleansers, household and industrial liq. detergents; clear, amber liq.; dens. 8.16 lb/gal; 100% act.

Monamid 1034. [418] Lauramide DEA; nonionic; foam booster and visc. builder for shampoos; liq.; 100% conc.

Monamid CMA. [418] 1:1 Coconut monoethanolamide; nonionic; see Monamid 7-100; tan granular; sol. @ 10% in ethanol, disp. in water, chlorinated and aromatic hydrocarbons, min. spirits, kerosene, wh. min. oil, natural oils and fats; pH 9.4–10.8 (10%); biodeg.; 100% act.

Monamid LIPA. [418] 1:1 Lauramide MIPA; nonionic; see Monamid 7-100; tan granular; sol. @ 10% in ethanol, disp. in water, kerosene, wh. min. oil, natural fats and oils; pH 10.3–11.3 (10%); biodeg.; 100% act.

Monamid LMA. [418] 1:1 Lauric monoethanolamide; nonionic; see Monamid 7-100; tan granular; sol. @10 in ethanol, disp. in water, aromatic hydrocarbons, kerosene, wh. min. oil, natural fats and oils; pH 10–11 (10%); biodeg.; 100% act.

Monamid LMIPA. [418] Lauramide MIPA; nonionic; surfactant used in personal care products and detergent systems; off-wh. solid; m.p. 61 ± 2 C; alkali no. 10–25; pH 10.5 ± 0.5 (10% aq.); 100% act.

Monamid LMMA. [418] 1:1 Lauramide MEA; nonionic; see Monamid 7-100; tan granular; sol. @ 10% in ethanol, disp. in water, aromatic hydrocarbons, kerosene, wh. min. oil, natural oils and fats; pH 9.7–10.7 (10%); biodeg.; 100% act.

Monamid R31-42. [418] Lauramide DEA and propylene glycol; nonionic; see Monamid 7-100; 2 max (Gvcs-33) liq.; sol. @ 10% in water, ethanol, chlorinated and aromatic hydrocarbons, disp. in natural oils and fats; dens. 8.25 lb/gal; sp.gr. 0.99; pH 10–11 (10%); biodeg.; 80% act.

Monamid S. [418] 1:1 Stearic monoethanolamide; nonionic; see Monamid 7-100; tan granular; disp. @ 10% in water, chlorinated and aromatic hydrocarbons, min. spirits, kerosene, wh. min. oil, natural oils and fats; pH 9.5–11.0 (10%); biodeg.; 100% act.

Monamine Series. [418] Fatty acid alkanolamide; foam booster and stabilizer, emulsifier, detergent, wetting agent, corrosion inhibitor, visc. builder, lubricant, dispersant for cosmetics, household and industrial cleaners, textiles, agric. sprays, leather and fur preparations.

Monamine 779. [418] DEA-laureth sulfate and cocamide DEA; nonionic/anionic; foaming agent, visc. builder, detergent, wetting and penetrating agent used in shampoos, bubble baths, household and industrial cleaners; lt. yel. clear visc. liq.; sol. @ 10% in water, polar solvs., alcohol, glycols, fatty acid esters, chlorinated and aromatic hydrocarbons, dens. 8.75 lb/gal; sp.gr. 1.05 ± 0.1; cloud pt. 2 C; surf. tens. 29 dynes/cm (0.1%); pH 9.25 ± 0.5 (10%); biodeg.; 100% act.

Monamine 853. [418] Modified alkanolamide; detergent used in gel and lotion type waterless hand cleaners; clear amber liq.; dens. 8.6 lb/gal; 100% act.

Monamine 928. [418] Tallowamide DEA and diethanolamine; nonionic/anionic; emulsifier, lubricant, corrosion inhibitor used in syn. coolants and metalworking fluids; cream paste; dens. 8.0 lb/gal (50 C); 100% act.

Monamine 929. [418] Modified long-chain alkanolamide; nonionic/anionic; emulsifier, lubricant, corrosion inhibitor used in syn. coolants and metalworking fluids; cream paste; dens. 8.3 lb/gal; 100% act.

Monamine AA-100. [418] 1:2 Dist. cocamide DEA and diethanolamine; nonionic/anionic; see Monamine Series; Gvcs-33 7 max. liq.; sol. @ 10% in water, ethanol, chlorinated and aromatic hydrocarbons; dens. 8.30 lb/gal; sp.gr. 1.00; pH 9.5–10.5 (10%); biodeg.; 100% act.

Monamine AC-100. [418] Cocamide DEA and linoleamide DEA; nonionic/anionic; see Monamine Series; Gvcs-33 13 max. liq.; sol. @ 10% in ethanol, chlorinated and aromatic hydrocarbons, gels in water; dens. 8.30 lb/gal; sp.gr. 1.00; pH 9.5–10.5 (10%); biodeg.; 100% act.

Monamine ACO-100. [418] Lauramide DEA and diethanolamine; nonionic/anionic; see Monamine Series; Gvcs-33 5 max. paste; sol. @ 10% in water, ethanol, chlorinated and aromatic hydrocarbons; dens. 8.40 lb/

gal; sp.gr. 1.01; pH 9.5–10.5 (10%); biodeg.; 100% act.

Monamine AD-100, ADD-100. [418] Cocamide DEA and diethanolamine; nonionic; see Monamine Series; Gvcs-33 11 and 3 max. resp., liqs.; sol. @ 10% in water, ethanol, chlorinated hydrocarbons, disp. in aromatic hydrocarbons; dens. 8.50 lb/gal; sp.gr. 1.02; pH 9.5–10.5 (10%); biodeg.; 100% act.

Monamine ADD-100 LE. [418] Cocamide DEA and diethanolamine; nonionic; see Monamine Series; Gvcs-33 3 max. liq.; sol. @ 10% in water, ethanol, chlorinated and aromatic hydrocarbons; dens. 8.50 lb/gal; sp.gr. 1.02; pH 10.5–11.5 (10%); biodeg.; 100% act.

Monamine ADDS-100. [418] 1:2 Mixed fatty acid diethanolamide; nonionic/anionic; see Monamine Series; Gvcs-33 7 max. liq.; sol. @ 10% in water, ethanol, chlorinated and aromatic hydrocarbons; dens. 8.30 lb/gal; sp.gr. 1.00; pH 9.0–10.0 (10%); biodeg.; 100% act.

Monamine ADS-100. [418] Cocamide DEA; nonionic/anionic; see Monamine Series; Gvcs-33 11 max. liq.; sol. @ 10% in water, ethanol, chlorinated and aromatic hydrocarbons; dens. 8.30 lb/gal; sp.gr. 1.00; pH 9.0–10.0 (10%); biodeg.; 100% act.

Monamine ADY-100. [418] Linoleamide DEA and diethanolamine; nonionic/anionic; see Monamine Series; Gvcs-33 9 max. liq.; sol. @ 10% in ethanol, min. spirits, kerosene; disp. in water, chlorinated and aromatic hydrocarbons; dens. 8.20 lb/gal; sp.gr. 0.98; pH 10.5–11.5 (10%); biodeg.; 100% act.

Monamine AF-100. [418] 1:2 Mixed fatty acid diethanolamide; nonionic/anionic; see Monamine Series; Gvcs-33 12 max. liq.; sol. @ 10% in water, ethanol, chlorinated and aromatic hydrocarbons, min. spirits; dens. 8.25 lb/gal; sp.gr. 0.99; pH 9–10 (10%); biodeg.; 100% act.

Monamine ALX-80 SS, ALX-100 S. [418] Cocamide DEA and DEA-dodecylbenzene sulfonate; nonionic/anionic; see Monamine Series; Gvcs-33 6 and 8 max. resp. liq.; sol. @ 10% in water, ethanol; dens. 8.75 and 8.85; sp.gr. 1.05 and 1.06; pH 8.5–9.5 (10%); biodeg.; 80 and 100% act. resp.

Monamine ARA-100. [418] 1:2 Coconut diethanolamide; nonionic/anionic; see Monamine Series; Gvcs-33 7 max. liq.; sol. @ 10% in water, ethanol, chlorinated and aromatic hydrocarbons; dens. 8.30 lb/gal; sp.gr. 1.00; pH 9.5–10.5 (10%); biodeg.; 100% act.

Monamine CD-100. [418] Linoleamide DEA and diethanolamide; lubricant and emulsifier for syn. cutting and drawing fluids; metal corrosion inhibitor; dk. br. visc. liq.; water-disp.; pH 9.5–10.5 (10% disp.); 100% act.

Monamine CF-100 M. [418] Cocamide DEA and diethanolamine; nonionic/anionic; see Monamine Series; Gvcs-33 11 max. liq.; sol. @ 10% in water, ethanol, chlorinated and aromatic hydrocarbons; dens. 8.40 lb/gal; sp.gr. 1.01; pH 8.5–9.5 (10%); biodeg.; 100% act.

Monamine I-76. [418] 1:2 Coconut diethanolamide; nonionic/anionic; see Monamine Series; Gvcs-33 12 max. liq.; sol. @ 10% in water, ethanol, chlorinated and aromatic hydrocarbons; dens. 8.30 lb/gal; sp.gr. 1.00; pH 8.5–9.5 (10%); biodeg.; 100% act.

Monamine LM-100. [418] Lauramide DEA and diethanolamine; nonionic/anionic; see Monamine Series; Gvcs-33 4 max. liq.; sol. @ 10% in water, ethanol, chlorinated and aromatic hydrocarbons; dens. 8.50 lb/gal; sp.gr. 1.02; pH 9.5–10.5 (10%); biodeg.; 100% act.

Monamine R8-26. [418] Linoleamide DEA and diethanolamine; nonionic/anionic; see Monamine Series; Gvcs-33 11 max. liq.; sol. @ 10% in water, ethanol, chlorinated and aromatic hydrocarbons; dens. 8.25 lb./gal; sp.gr. 0.99; pH 9.0–10.0 (10%); biodeg.; 100% act.

Monamine T-100. [418] Tallamide DEA and diethanolamine; nonionic/anionic; see Monamine Series; Gvcs-33 10 max. liq.; sol. @ 10% in ethanol, chlorinated and aromatic hydrocarbons, min. spirits, kerosene, disp. in water; dens. 8.10 lb/gal; sp.gr. 0.97; pH 10.0–11.0 (10%); biodeg.; 100% act.

Monamulse 653-C. [418] Alkanolamide, modified; anionic/nonionic; emulsifier, solubilizer, degreaser for solvs.; liq.; 100% conc.

Monamulse 748. [418] Alkanolamide, modified; emulsifier for and dispersant for crude oil; liq.; 100% conc.

Monamulse 947. [418] Alkylaryl; anionic; emulsifier for aromatic and chlorinated solvs.; liq.; 100% conc.

Monamulse R 10-29M. [418] Imidazoline, modified; nonionic/cationic; emulsifier, solubilizer, corrosion inhibitor for fuel oils; liq.; 100% conc.

Monaprin Emulsifier S. [317] Long chain alcohol EO condensate; emulsifier for natural and syn. oils and waxes; o/w emulsions in textile printing; liq.; 40% conc.

Monaquat AT-1074. [418] Quat. compd.; cationic; thickener with foaming and cleaning properties for HCl sol'ns.; used in acid bowl and pipe line cleaners, rust removers, and oil well sizing applications; Gvcs-33 3 clear to slightly hazy liq.; sol. in HCl; sp.gr. 1.2; dens. 10 lb/gal; solid pt. < 0 C; pH 5.5 (10% aq.); 30% act.

Monoquat ISIES. [418] Isostearyl ethylimidonium ethosulfate; cationic; antistat, lubricant, softener, corrosion inhibitor used in cosmetic industry, in industrial and textile applications; amber liq.; sol. @ 10% in water, ethanol, butyl Cellosolve, aromatic, chlorinated and fluorinated hydrocarbons; sp.gr. 1.03; pH 6.9 (10%); biodeg.; 100% act.

Monaquat P-TC. [418] Cocamidopropyl PEG-dimonium chloride phosphate; cationic; bactericidal, conditioner, antistat, detergent, foamer, emulsifier, solubilizer, dispersant, thickener, and wetting agent used in personal care, household, pharmaceutical, veterinary products, fire fighting foams, petroleum production, photographic processes, agriculture, mining, and textiles; amber clear liq.; sp.gr. 1.10; dens. 9.1 lb/gal; pH 7.0 (10%); surf. tens. 37.7 dynes/cm (1%); 40% act.

Monaquat P-TD. [418] Lauramidopropyl PEG-dimonium chloride phosphate; cationic; see Monaquat P-TC; amber clear liq.; sp.gr. 1.05; dens. 8.7 lb/gal; pH 7.5 (10%); surf. tens. 43.1 dynes/cm (1%); 34% act.

Monaquat P-TL. [418] Laurampho PEG-glycinate phosphate; cationic; see Monaquat P-TC; clear to opaque visc. liq.; sp.gr. 1.10; dens. 9.1 lb/gal; pH 7.5 (10%); surf. tens. 33.7 dynes/cm (1%); 30% act.

Monaquat P-TS. [418] Tris alkylamido tri quat.; cationic; thickener, foamer; paste; 30% conc.

Monaquat P-TZ. [418] Cocohydroxyethyl PEG-imidazolinium chloride phosphate; cationic; see Monaquat P-TC; amber clear liq.; sp.gr. 1.07; dens. 8.9 lb/gal; pH 7.0 (10%); surf. tens. 37.6 dynes/cm (1%); 30% act.

Monaquat TG. [418] Alkyl poly hydroxy-substituted ammonium chloride; cationic; surfactant used in personal care products; antistat and foaming used in fabric laundering and softening products; thickener for acid bowl cleaners, naval gels; Gardner 2 clear to slightly hazy liq.; sp.gr. 1.011; dens. 8.39 lb/gal; solid. pt. –10 to –12 C; cloud pt. 3 C; pH 4 (10%); 30% act.

Monaterge 85. [418] Fatty acid amido complex; nonionic/anionic; detergent, wetting agent, emulsifier for chlorinated solvs., used in cleaners; clear, amber liq.; sol. in water, ethanol, aromatic hydrocarbons, natural oils and fats; dens. 8.5 lb/gal; sp.gr. 1.02; surf. tens. 27.9 dynes/cm (0.05%); pH 8.9 (10%); biodeg.; 85% act.

Monaterge 85 HF. [418] Cocamide DEA, DEA-acrylinoleate, and DEA-dodecylbenzene sulfonate; nonionic/anionic; detergent and wetting agent; liq.; 85% conc.

Monaterge LF-945. [418] Modified imidazoline; amphoteric/nonionic; detergent for heavy duty alkaline and acid cleaners at low temps.; lt. yel., clear to hazy liq.; dens. 8.3 lb/gal; sp.gr. 1.00; pH 10–11 (10%); 100% act.

Monateric 805. [418] Cocoamphodiacetate and disodium cocamido MIPA-sulfosuccinate; amphoteric/anionic; high-foaming, mild surfactant for hair conditioners, skin care products; lt. amber, visc. liq.; dens. 9.16 lb/gal; sp.gr. 1.10; pH 7.8 ± 0.5 (10%); 39–41% act.

Monateric 810-A-50. [418] Caprylic/capric monocarboxylic propionate, imidazoline-derived, salt-free; amphoteric; surfactant used in industrial detergents, cleaners, and cosmetics; hydrotrope, coupling agent, and/or solubilizer; corrosion inhibitor in metalworking systems, oil well flooding, and aerosol pkgs.; br. clear liq.; sp.gr. 1.07; dens. 8.9 lb/gal; pH 4.4 (10%); 50% act.

Monateric 811. [418] Capryloamphodipropionate; amphoteric; corrosion inhibitor, detergent, wetting agent in noncorrosive cleaners and industrial detergents; amber liq.; water sol.; dens. 8.7 lb/gal; sp.gr. 1.04; pH 10.4 (10%); 50% act.

Monateric 951A. [418] Lauroamphodiacetate; amphoteric; surfactant used in personal care products; pale yel. clear liq.; sp.gr. 1.08; dens. 8.96 lb/gal; pH 8.8; 24% act.

Monateric 985A. [418] Lauroamphodiacetate; amphoteric; see Monateric 810-A-50; amber clear liq.; sp.gr. 1.07; dens. 8.9 lb/gal; pH 9.3 (10%); 36% act.

Monateric 1000. [418] Amphoteric based on alkyl imidazoline; corrosion inhibitor, detergent, wetting agent for metal cleaning, cutting fluids, syn. lubricants; amber liq.; water sol.; dens. 8.8 lb/gal; sp.gr. 1.05; pH 11.2 (10%); 50% act.

Monateric 1202. [418] Dihydroxyethyl tallow glycinate; amphoteric; surfactant used as a conditioner, coupling, and visc. control agent in personal care products; clear visc. liq.; sol. in water; pH 5.5; 35.0% act.

Monateric 1203. [418] Sodium hydrog. tallow dimethyl glycinate; amphoteric; surfactant used in hair care products; off-wh. opaque paste; pH 6.5 (5%); 30.0% act.

Monateric ADA. [418] Cocamidopropyl betaine; used in air drilling, foam drilling, foam blanketing, air entraining agent for cement, gypsum, wallboard; clear, amber liq.; pH 7.9; 38% solids.

Monateric ADFA. [418] Cocamidopropyl betaine; amphoteric; surfactant for air and foam drilling; liq.; 33% conc.

Monateric CA-35%. [418] Cocoamphopropionate; amphoteric; detergent, wetting agent, emulsifier, dispersant used in cosmetic, household, and industrial products; amber liq.; m.w. 360; dens. 8.5 lb/gal; sp.gr. 1.02; surf. tens. 29.5 dynes/cm (0.1% conc.); biodeg.; 35% act.

Monateric CAB. [418] Cocamidopropyl betaine; amphoteric; base surfactant for shampoos, bubble baths, hair conditioners; clear, yel. liq.; pH 7.0 ± 0.5; 30% act.

Monateric CAM-40. [418] Cocoamphopropionate; amphoteric; see Monateric 810-A-50; amber clear liq.; sp.gr. 1.05; dens. 8.8 lb/gal; pH 9.3 (10%); 40% act.

Monateric CDL. [418] Cocoamphodiacetate and sodium lauryl sulfate; amphoteric; see Monateric 810-A-50; yel. clear liq.; sp.gr. 1.11; dens. 9.2 lb/gal; pH 8.5 (10%); 31% act.

Monateric CDS. [418] Cocoamphodiacetate and sodium lauryl sulfate; amphoteric; see Monateric 810-A-50; yel. clear liq.; sp.gr. 1.09; dens. 9.1 lb/gal; pH 8.5 (10%); 31% act.

Monateric CDTD. [418] Cocoamphodiacetate and sodium trideceth sulfate; amphoteric; see Monateric 810-A-50; yel. clear liq.; sp.gr. 1.11; dens. 9.2 lb/gal; pH 8.5 (10%); 44% act.

Monateric CDX-38. [418] Cocoamphodiacetate; amphoteric; detergent, foaming agent, mild base surfactant for shampoos and skin cleansers; lt. amber, visc. liq.; dens. 9.8 lb/gal; sp.gr. 1.18; pH 8.4 (10%); biodeg.; 38% act.

Monateric CDX-38 Mod. [418] Cocoamphodiacetate; amphoteric; see Monateric 810-A-50; yel. clear liq.; sp.gr. 1.18; dens. 9.8 lb/gal; pH 8.8 (10%); 39% act.

Monateric CEM-38%. [418] Cocoamphodipropionate; amphoteric; detergent used in liq. detergent systems, wetting agent, emulsifier, dispersant, solubilizer, hydrotrope; amber, clear to hazy liq.; dens. 8.75 lb/gal; surf. tens. 35 dynes/cm (0.1%); pH 8.5 ± 0.5 (10%); biodeg.; 38% act.

Monateric CEM-38CG. [418] Cocoamphopropionate; amphoteric; see Monateric 810-A-50; amber clear to hazy liq.; sp.gr. 1.07; dens. 8.9 lb/gal; pH 9.8 (10%); 38% act.

Monateric CM-36S. [418] Cocoamphoacetate; amphoteric; foaming agent, emulsifier, detergent, wetting agent used in cosmetic and textile industries; clear, yel. liq.; water sol.; surf. tens. 32 dynes/cm (0.015%); pH 12.8; 36% act.

Monateric CNa-40. [418] Coconut monocarboxylic propionate, imidazoline-derived, salt-free; amphoteric; see Monateric 810-A-50; amber clear liq.; sp.gr. 1.09; dens. 9.1 lb/gal; pH 10.9 (10%); 40% act.

Monateric COAB. [418] Cocamidopropyl betaine; amphoteric; see Monateric 810-A-50; yel. clear liq.; sp.gr. 1.04; dens. 8.7 lb/gal; pH 7.9 (10%); 32% act.

Monateric CSH-32. [418] Cocoamphodiacetate; amphoteric; high foaming, nonirritating surfactant for shampoos; clear, yel. liq.; dens. 9.4 lb/gal; sp.gr. 1.13; visc. 2000 cps; f.p. < 0 C; cloud pt. < 0 C; pH 8.4 (10%); biodeg.; 32% act.

Monateric CyA-50, CyMM-40. [418] Caprylic dicarboxylic propionate, imadazoline-derived, salt-free; amphoteric; see Monateric 810-A-50; dk. br. and amber clear liq. resp.; sp.gr. 1.07 and 1.10 resp.; dens. 8.9 and 9.2 lb/gal resp.; pH 5.6 and 9.8 (10%) resp.; 50 and 40% act. resp.

Monateric CyNa-50%. [418] Capryloamphopropionate; amphoteric; detergent, emulsifier, wetting agent, low to moderate foaming surfactant used in conc. electrolyte systems; amber liq.; pH 10.5 ± 0.5 (10%); biodeg.; 50% act.

Monateric ISA-35%. [418] Isostearoamphopropionate; amphoteric; surfactant used in cosmetic and industrial products; amber clear to hazy flowable gel; sol. @ 10% in water, alcohol, glycol ethers and alkanolamines, disp. in min. oil and natural oils and fats; surf. tens. 30 dynes/cm (0.1%); pH 5–6 (10%); biodeg.; 35% act.

Monateric LF-100. [418] 2-Alkyl imidazoline; amphoteric; detergent used in either low or high temperature alkaline and acid cleaners; amber liq.; m.w. 290; readily sol. in aq. acid systems, neutralizes slowly in alkaline systems; dens. 8.7 lb/gal; sp.gr. 1.04; pH 12.0 ± 0.5 (10%); biodeg.; 99% min. act.

Monateric LF Na-50. [418] Sodium salt of Monateric LF-100; similar to Monateric LF-100 with greater sol.; amber liq.; dens. 9.1 lb/gal; sp.gr. 1.09; pH 11.0 ± 0.5 (10%); biodeg.; 50% act.

Monateric LMAB. [418] Lauramidopropyl betaine; amphoteric; see Monateric 810-A-50; lt. yel. clear liq.; sp.gr. 1.04; dens. 8.7 lb/gal; pH 8.3 (10%); 30% act.

Monateric LMM-30. [418] Lauroamphoacetate; amphoteric; see Monateric 810-A-50; amber visc. liq.; sp.gr. 1.09; dens. 9.1 lb/gal; pH 9.2 (10%); 30% act.

Monateric MCB. [418] Cocamidopropyl betaine; amphoteric; see Monateric 810-A-50; lt. yel. clear liq.; sp.gr. 1.02; dens. 8.5 lb/gal; pH 4.8 (10%); 30% act.

Monateric P-1013, P-1023. [418] Alkylamido phosphobetaine (P-1023—modified); amphoteric; foamer, cleanser, wetter, and surf. tens. reducer used in industrial applications; liq.; 30% conc.

Monateric P-BSA. [418] Bis alkylamido phosphobetaine; amphoteric; see Monateric P-1013; paste; 30% conc.

Monateric P-CDL. [418] Alkylamido glyceryl phosphobetaine; amphoteric; see Monateric P-1013; liq.; 30% conc.

Monateric TA-35. [418] Tallamphopropionate; amphoteric; see Monateric 810-A-50; dk. br. gel; sp.gr. 1.02; dens. 8.5 lb/gal; pH 5.2 (10%); 35% act.

Monateric TDB-35. [418] Disodium tallowiminodipropionate; amphoteric; hydrotrope and detergent for high electrolyte systems such as heavy-duty liq. cleaners and wax strippers; yel. clear-hazy liq.; sp.gr. 1.03; dens. 8.6 lb/gal; pH 11.25 (1%); 35% act.

Monawet MB-45. [418] Sodium diisobutyl sulfosuccinate; anionic; wetting, dispersing, emulsifying, penetrating and solubilizing agent used in emulsion polymerization of styrene/butadiene for rug backing; clear, colorless liq.; m.w. 332; sol. in water, fairly sol. in polar and nonpolar solvs.; dens. 9.3 lb/gal; sp.gr. 1.12; cloud pt. 13 C; flash pt. 215 F (PMCC); surf. tens. 54 dynes/cm (0.1%); pH 6 ± 1; 45% act.

Monawet MM-80. [418] Sodium dihexyl sulfosuccinate; anionic; see Monawet MB-45; also used in paint and food industries; clear, colorless liq.; m.w. 388; sol. 33 g/100 g water, in polar and nonpolar solvs.; dens. 9.2 lb/gal; sp.gr. 1.10; cloud pt. < 0 C; flash pt. 110 F (PMCC); surf. tens. 46 dynes/cm (0.1%); pH 6 ± 1; 80% act.

Monawet MO-65-150. [418] Sodium dioctyl sulfosuccinate, anhyd.; anionic; wetting, penetrating and spreading agent, emulsifier used in oil well cleaning, drycleaning detergents, solv. cleaners and strippers; clear, colorless liq.; dens. 8.75 lb/gal; sp.gr. 1.05; flash pt. 42–43 C (COC); pH 5.5 ± 1.0 (10% sol'n.); 65% act.

Monawet MO-70. [418] Sodium dioctyl sulfosuccinate; anionic; wetting, dispersing, emulsifying, penetrating and solubilizing agent used in emulsion and suspension polymerization, textile, fertilizer, cosmetic, food industries; clear, colorless liq.; m.w. 444; sol. in polar and nonpolar solvs.; dens. 9.0 lb/gal; sp.gr. 1.08; cloud pt. < –5 C; flash pt. 325 F (PMCC); surf. tens. 29 dynes/cm (0.1%); pH 6 ± 1; 70% act.

Monawet MO-70-150. [418] Sodium dioctyl sulfosuccinate; anionic; see MO-65-150; liq.; 70% conc.

Monawet MO-70E. [418] Sodium dioctyl sulfosuccinate; anionic; see Monawet MO-70; clear, colorless liq.; m.w. 444; sol. in polar and nonpolar solvs.; dens. 9.0 lb/gal; sp.gr. 1.08; cloud pt. < –5 C; flash pt. 82 F (PMCC); surf. tens. 29 dynes/cm (0.1%); pH 6 ± 1; 70% act.

Monawet MO-70R. [418] Sodium dioctyl sulfosuccinate; anionic; wetting, dispersing, emulsifying, penetrating and solubilizing agent for textile, printing, agric., cosmetic, food industries; clear, colorless liq.; m.w. 444; sol. in polar and nonpolar solvs.; dens. 8.8 lb/gal; sp.gr. 1.06; cloud pt. < –5 C; flash pt. 280 F (PMCC); surf. tens. 29 dynes/cm (0.1%); pH 6 ± 1; 70% act.

Monawet MO-70RP. [418] Sodium dioctyl sulfosuccinate; anionic; wetting for pharmaceuticals; liq.; 70% conc.

Monawet MO-70S. [418] Sodium dioctyl sulfosuccinate; anionic; wetting agent for dry cleaning soaps, spotting compds; wetting agent and emulsifier; liq.; oil and solv. sol.; 70% conc.

Monawet MO-75E. [418] Sodium dioctyl sulfosuccinate; anionic; see Monawet MO-70R; clear, colorless liq.; m.w. 444; sol. in polar and nonpolar solvs.; dens. 9.0 lb/gal; sp.gr. 1.08; cloud pt. < –5 C; flash pt. 80 F (PMCC); surf. tens. 29 dynes/cm (0.1%); pH 6 ± 1; 75% act.

Monawet MO-84R2W. [418] Sodium dioctyl sulfosuccinate; anionic; see Monawet MO-70R; lt. yel. visc. liq.; m.w. 444; sol. in polar and nonpolar solvs.; dens. 9.2 lb/gal; sp.gr. 1.10; cloud pt. < –10 C; flash pt. 223 F (PMCC); surf. tens. 29 dynes/cm (0.1%); pH 5.5 ± 1; 83% act.

Monawet MO-85P. [418] Sodium dioctyl sulfosuccinate; anionic; wetting for wettable powds., pigments; wh. fine powd.; 0.85% sol. in water; surf. tens. 26 dynes/cm (0.1%); pH 6.5; 85% act.

Monawet MT-70. [418] Sodium ditridecyl sulfosuccinate; anionic; wetting, dispersing, emulsifying, penetrating and solubilizing agent used in emulsion and suspension polymerization, indirect food additives; lt. straw clear liq.; m.w. 584; sol. in polar and nonpolar solvs.; dens. 8.5 lb/gal; sp.gr. 1.02; cloud pt. –2 C; flash pt. 230 F (PMCC); surf. tens. 29 dynes/cm (0.1%); pH 6 ± 1; 70% act.

Monawet MT-70E. [418] Sodium ditridecyl sulfosuccinate; anionic; see Monawet MT-70; lt. straw clear liq.; m.w. 584; sol. in polar and nonpolar solvs.; dens. 8.4 lb/gal; sp.gr. 1.01; cloud pt. –15 C; flash pt. 86 F (PMCC); surf. tens. 29 dynes/cm (0.1%); pH 6 ± 1; 70% act.

Monawet MT-80H2W. [418] Sodium ditridecyl sulfosuccinate; anionic; wetting, dispersing, emulsifying, penetrating and solubilizing agent used in printing inks, indirect food additives; lt. yel. clear liq.; m.w. 584; sol. in polar and nonpolar solvs.; dens. 8.5 lb/gal; sp.gr. 1.02; cloud pt. < 0 C; flash pt. 225 F (PMCC); surf. tens. 29 dynes/cm (0.1%); pH 5.5 ± 1; 80% act.

Monawet SNO-35. [418] Tetrasodium dicarboxyethyl stearyl sulfosuccinamate; anionic; wetting agent, solubilizer, emulsifier, dispersant, mild detergent used in textile, cosmetic, agric. products; clear, lt. amber liq.; m.w. 653; sol. in water, high electrolyte salt sol'ns.; dens. 9.5 lb/gal; sp.gr. 1.12–1.16; visc. 16–18 s (#2 Zahn cup); f.p. 45 ± 5 F; surf. tens. 43 dynes/cm (0.1%); pH 7–8; biodeg.; 35% solids.

Monawet TD-30. [418] Disodium deceth-6 sulfosuccinate; anionic; surfactant used in emulsion polymerization, cosmetic and textile industries; lt. yel. liq.; dens. 9.0 lb/gal; sp.gr. 1.08; visc. 45 cps; cloud pt. –3 C; surf. tens. 32.9 dynes/cm (0.1%); pH 5–6; 30% act.

Monazoline Series. [418] 1-Hydroxyethyl-2-alkylimidazolines; cationic; wetting agent, emulsifier for nonpolar liq., detergent, thickener, corrosion inhibitor, antistat, softener, bactericide used in paint and textile industries; liq.; sol. in oil, disp. in water; biodeg.

Monazoline C. [418] Cocoyl hydroxyethyl imidazoline; see Monazoline Series, also in acid dairy cleaners, in oil well acidifying and sec. recovery operations; amber liq., may crystallize on aging; m.w. 282; sol. @ 10% in ethanol, chlorinated hydrocarbons, min. oil, toluene, kerosene, min. spirits, veg. oil, disp. in water; dens. 7.75 lb/gal; sp.gr. 0.93; pH 10.5–12.0 (10% disp.); biodeg.; 90% min. imidazoline.

Monazoline CY. [418] Capryl hydroxyethyl imidazoline; see Monazoline Series; amber liq., may crystallize on aging; m.w. 212; sol. @ 10% in ethanol, chlorinated hydrocarbons, toluene, veg. oil; dens. 8.25 lb/gal; sp.gr. 0.99; pH 10.5–12.0 (10% disp.); biodeg.; 90% min. imidazoline.

Monazoline IS. [418] Isostearyl hydroxyethyl imidazoline; cationic; corrosion inhibitor and lubricant; liq.; 100% conc.

Monazoline O. [418] Oleyl hydroxyethyl imidazoline; see Monazoline Series, also in agric. preparations, acid dairy cleaners; amber liq., may crystallize on aging; m.w. 345; sol. @ 10% in ethanol, chlorinated hydrocarbons, min. oil, toluene, kerosene, min. spirits, veg. oil; dens. 7.66 lb/gal; sp.gr. 0.92; pH 10.0–11.5 (10% disp.); biodeg.; 90% min. imidazoline.

Monazoline T. [418] Tall oil hydroxyethyl imidazoline; see Monazoline Series, also aids gravel-to-asphalt bonding, rinse aid for automatic car washes, printing ink additive, protective metal coatings; amber liq., may crystallize on aging; m.w. 350; sol. @ 10% in ethanol, chlorinated hydrocarbons, min. oil, toluene, kerosene, min. spirits, veg. oil, disp. in water; dens. 7.75 lb/gal; sp.gr. 0.93; pH 10.0–11.5 (10% disp.); biodeg.; 90% min. imidazoline.

Monflor 31. [320] Fluorochemical surfactant; anionic; wetting agent, emulsifier for industrial cleaning, suspension and emulsion polymerization, ink/paint pigments, metal treatment; liq.; dens. 1.15 g/cc; 30% act.

Monflor 32. [320] Fluorochemical surfactant; anionic; leveling and wetting agent, dispersant, emulsifier for ink/paint pigments; liq.; dens. 1.09 g/cc; 30% act.

Monflor 51. [320] Fluorochemical surfactant; nonionic; dispersant, emulsifier, foam stabilizer, leveling agent, wetting agent for mold release, industrial cleaning, plastics, suspension/emulsion polymerization, ink/paint pigments; waxy solid; sol. in alcohols, acetone, MEK, trichloroethylene, water, toluene; dens. 1.25 g/cc; 100% act.

Monflor 52. [320] Fluorochemical surfactant; nonionic; dispersant, emulsifier, foam stabilizer, leveling agent, wetting agent for mold release, industrial cleaning, plastics, ink/paint pigments; visc. liq.; sol. in alcohols, acetone, MEK, trichloroethylene, toluene, water; dens. 1.32 g/cc; 100% act.

Monflor 53. [320] Fluorochemical surfactant; nonionic; wetting agent, emulsifier for photographic film emulsions; oily liq.; sol. in alcohol, acetone, MEK, perchloroethylene, trichloroethylene, toluene; dens. 1.14 g/cc; 100% act.

Monflor 72. [320] Fluorochemical surfactant; cationic; wetting agent, emulsifier for chemical specialties, floor polishes, plastics additives, ink/paint pigments; liq.; dens. 1.18 g/cc; 30% act.

Monflor 91. [320] Fluorochemical surfactant; anionic; wetting agent, emulsifier for mold release, metal treatment; liq.; dens. 1.6 g/cc; 10% act.

Monogen. [172] Sulfated fatty alcohol, sodium salt; anionic; textile scouring and washing agent; dye dispersant and leveling agent; paste; 60% conc.

Monolan 1206/2. [360] EO-PO block polymer; nonionic; antifoam for detergents, wetting agents, syn. lubricants; colorless clear liq., faint odor; water-insol.; sp.gr. 1.003; visc. 445 cs; flash pt. > 400 F (COC); pour pt. < 0 C; pH 6–8 (1% aq.); 100% act.

Monolan 2000 E/12. [360] EO-PO block polymer; nonionic; detergent, wetting agent, antifoam, dispersant, rinse aid; colorless clear liq., faint odor; water-insol.; sp.gr. 1.020; visc. 390 cs; flash pt. > 400 F (COC); pour pt. < 0 C; pH 6–8 (1% aq.); 100% act.

Monolan 2500 E/30. [360] EO-PO block polymer; nonionic; detergent, wetting agent for emulsion polymerization, metal cleaning, resin plasticizers, latex stabilization, textile processing; colorless clear liq., faint odor; water-sol.; sp.gr. 1.043; visc. 650 cs; flash pt. > 400 F (COC); pour pt. < 0 C; pH 6–8 (1% aq.); 100% act.

Monolan 3000 E/50. [360] EO-PO copolymer; nonionic; detergent and rinse aid; solid; 100% conc.

Monolan 3000 E/60, 8000 E/80. [360] EO-PO block polymer; nonionic; detergent, wetting agent, antifoam for industrial and domestic detergents, emulsion polymerization, metal cleaning, resin plasticizers, latex stabilization, textile processing; wh. solid and wh. flake resp., faint odor; water-sol.; sp.gr. 1.052 and 1.070; visc. 355 cs and 1100 cs (60 C) resp.; flash pt. > 400 F (COC); pour pt. 31 and 49 C; pH 6–8 and 7–8 (1% aq.); 100% act.

Monolan 12,000 E/80. [360] EO-PO copolymer; nonionic; latex stabilizer and sec. emulsifier; toilet block component; solid; HLB 16.0; 100% conc.

Monolan M. [360] EO-PO block polymer; nonionic; low-foam emulsifier for herbicides and pesticides; colorless clear liq., faint odor; water-sol.; sp.gr. 1.020; visc. 470 cs; cloud pt. 24 C (1% aq.); flash pt. > 400 F (COC); pour pt. < 0 C; pH 6–8 (1% aq.); 100% act.

Monolan P222. [360] EO-PO block polymer; nonionic; antifoam for detergents, wetting agents, syn. lubricants; colorless clear liq., faint odor; water-insol.; sp.gr. 1.004; visc. 440 cs; flash pt. > 400 F (COC); pour pt. < 0 C; pH 6–8 (1% aq.); 100% act.

Monolan PB, PC. [360] Complex EO–PO copolymer; nonionic; wetting agent, low foam detergent and rinse aid, pigment dispersant; colorless liq.; faint odor; disp. and sol. in water resp.; sp.gr. 1.020 and 1.040; visc. 680 and 800 cs; cloud pt. 19 and 44 C (1% aq.); flash pt. > 400 F (COC); pour pt. < 0 and 8 C; pH 6–8 (1% aq.); 100% act.

Monolan PM7. [360] EO-PO copolymer on an org. substrate; nonionic; low foam detergent and wetting agent for use in bottle washing formulations, rinse aid; clear, colorless liq.; faint odor; sol. in water; sp.gr. 1.050; visc. 710 cs; cloud pt. 41 C (1% aq.); flash pt. > 400 F (COC); pour pt. < 0 C; pH 6–8 (1% aq.); 100% act.

Monolan PPG440, PPG1100, PPG2200. [360] Polypropylene glycol; nonionic; lubricant, antistat, plasticizer, cosolvs. in dyestuff, ink, resin, rubber industries, cosmetic preparations, intermediates in surfactants and plastic production; colorless liq.; faint odor; sol. in water (PPG440), insol. in water (PPG1100, 2200); m.w. 400, 1000, 2000 resp.; sp.gr. 1.010, 1.005, and 1.004; visc. 80, 180, and 450 cs; flash pt. > 450 F (COC); pour pt. < 0 C; 100% act.

Monolan PT. [360] Complex EO–PO copolymer; nonionic; low foam detergent and rinse aid, pigment dispersant; colorless liq.; faint odor; water sol.; sp.gr. 1.037; visc. 750 cs; cloud pt. 28 C (1% aq.); flash pt. > 400 F (COC); pour pt. < 0 C; pH 6–8 (1% aq.); 100% act.

Monomax AH60. [74] Monodiglyceride; nonionic; food emulsifier; beads; HLB 4.0; 98% conc.

Monomax AH90. [74] Monoglyceride, dist.; nonionic; food emulsifier; beads; HLB 4.0; 92% conc.

Monomax VH60. [74] Monodiglyceride; nonionic; food emulsifier; beads; HLB 4.0; 98% conc.

Monomax VH90. [74] Monoglyceride, dist.; nonionic; food emulsifier; beads; HLB 4.0; 92% conc.

Monomer C18. [129] Tall oil fatty dimer acid; mfg. of soaps, surfactants, dispersants, and suspending agents; liq.

Monomuls 60-10. [277] Lard glycerides; nonionic; emulsifier, stabilizer, dispersant, opacifier for cosmetics, foods and drugs; solid; 100% conc.

Monomuls 60-15. [277] Hydrog. lard glycerides; nonionic; see Monomuls 60-10; powd.; 100% conc.

Monomuls 60-20. [277] Tallow glycerides; nonionic; see Monomuls 60-10; solid; 100% conc.

Monomuls 60-25. [277] Hydrog. tallow glycerides; nonionic; see Monomuls 60-10; solid; 100% conc.

Monomuls 60-25/2. [277] Hydrog. tallow glycerides with 2% sodium stearate; nonionic; see Monomuls 60-10; powd.; 100% conc.

Monomuls 60-25/5. [277] Hydrog. tallow glycerides with 5% sodium stearate; nonionic; see Monomuls 60-10; powd.; 100% conc.

Monomuls 60-30. [277] Palm oil glycerides; anionic; see Monomuls 60-10; solid.; 100% conc.

Monomuls 60-35. [277] Hydrog. palm oil glycerides; nonionic; see Monomuls 60-10; powd.; 100% conc.

Monomuls 60-40. [277] Sunflower seed oil glycerides; nonionic; see Monomuls 60-10; solid; 100% conc.

Monomuls 60-45. [277] Hydrog. soybean oil glycerides; nonionic; see Monomuls 60-10; powd.; 100% conc.

Monomuls 90-10. [277] Dist. lard glyceride; nonionic; see Monomuls 60-10; solid; 100% conc.

Monomuls 90-15. [277] Dist. hydrog. lard glyceride; nonionic; see Monomuls 60-10; powd.; 100% conc.

Monomuls 90-20. [277] Dist. tallow glyceride; nonionic; see Monomuls 60-10; solid; 100% conc.

Monomuls 90-25. [277] Dist. hydrog. tallow glyceride; nonionic; see Monomuls 60-10; powd.; 100% conc.

Monomuls 90-25/2, 90-25/5. [277] Dist. hydrog. tallow glyceride with 2% and 5% sodium stearate resp.; anionic; see Monomuls 60-10; powd.; 100% conc.

Monomuls 90-30. [277] Dist. palm oil glyceride; nonionic; see Monomuls 60-10; solid; 100% conc.

Monomuls 90-35. [277] Dist. hydrog. palm oil glyceride; nonionic; see Monomuls 60-10; powd.; 100% conc.

Monomuls 90-40. [277] Dist. sunflower seed oil glyceride; nonionic; see Monomuls 60-10; solid; 100% conc.

Monomuls 90-45. [277] Dist. hydrog. soybean oil glyceride; nonionic; see Monomuls 60-10; powd.; 100% conc.

Monomuls 90-L12. [277] Glyceryl laurate; nonionic; refatting agent and thickener for personal care products; solid; 100% conc.

Monomuls 90-O18. [277] Glyceryl oleate; nonionic; emulsifier, stabilizer for cosmetics, food and drugs; solid; 100% conc.

Monopol Oil 75. [245] Sulfated castor oil; anionic; wetting agent, penetrant, leveling agent, softener/lubricant or softener additive for cellulosic fibers, used in dyeing; lt. yel., clear liq.; sp.gr. 1.05; 75% act.

Monosteol. [248] Propylene glycol stearate; nonionic; stabilizer for ointments, cream lotions; solid; HLB 4.0; 100% conc.

Monosulph. [186] Highly sulfonated castor oil; anionic; penetrant, emulsifier; textile dyeing assistant; fat liquor for suede leather; paper coating evener; plasticizer for starch, glues;

emulsifier for latex; liq.; 68% conc.

Montaline RH. [551] Castor oil sulfonated ester; anionic; wetting agent; liq.; 60% conc.

Montaline SPCV. [551] POE fatty alcohol; nonionic; aux. for production of phosphoric acid and superphosphate; liq.; 50% conc.

Montane 20. [551] Sorbitan monolaurate; nonionic; emulsifier; liq.; HLB 8.6; 100% conc.

Montane 40. [551] Sorbitan monopalmitate; nonionic; emulsifier; solid; HLB 6.7; 100% conc.

Montane 60. [551] Sorbitan monoisostearate; nonionic; emulsifier; solid; HLB 4.7; 100% conc.

Montane 65. [551] Sorbitan tristearate; nonionic; emulsifier; solid; HLB 2.7; 100% conc.

Montane 70. [551] Sorbitan monostearate; nonionic; emulsifier; liq.; 100% conc.

Montane 80. [551] Sorbitan monooleate; nonionic; emulsifier; liq.; HLB 4.3; 100% conc.

Montane 83. [551] Sorbitan sesquioleate; nonionic; emulsifier; liq.; HLB 3.7; 100% conc.

Montane 85. [551] Sorbitan trioleate; nonionic; emulsifier; liq.; HLB 1.8; 100% conc.

Montanox 20. [551] POE (20) sorbitan monolaurate; nonionic; emulsifier; liq.; HLB 16.7; 100% conc.

Montanox 21. [551] POE (5) sorbitan monolaurate; nonionic; emulsifier; liq.; HLB 13.3; 100% conc.

Montanox 40. [551] POE (20) sorbitan monopalmitate; nonionic; emulsifier; gel; HLB 15.6; 100% conc.

Montanox 60. [551] POE (20) sorbitan monostearate; nonionic; emulsifier; gel; HLB 14.9; gel; 100% conc.

Montanox 61. [551] POE (4) sorbitan monostearate; nonionic; emulsifier; solid; HLB 9.6; 100% conc.

Montanox 65. [551] POE (20) sorbitan tristearate; nonionic; emulsifier; solid; HLB 10.5; 100% conc.

Montanox 70. [551] POE (20) sorbitan monoisostearate; nonionic; emulsifier; liq.; 100% conc.

Montanox 80. [551] POE (20) sorbitan monooleate; nonionic; emulsifier; liq.; HLB 15.0; 100% conc.

Montanox 81. [551] POE (5) sorbitan monooleate; nonionic; emulsifier; liq.; HLB 10.0; 100% conc.

Montanox 85. [551] POE (20) sorbitan trioleate; nonionic; emulsifier; liq.; HLB 11.0; 100% conc.

Montapol CST. [551] Oleo/cetyl sulfate; anionic; detergent; paste; 35% conc.

Montegal 150 RG, AP 80, DZ. [551] Alkylpolyethoxy ether; nonionic; textile leveling agent; liq., flakes, wax resp.; 50, 100, and 100% conc. resp.

Montegal OL 50. [551] Castor oil, ethoxylated; nonionic; see Montegal 150 RG; liq.; 100% conc.

Montegal SH 25. [551] Fatty amine, ethoxylated; nonionic; see Montegal 150 RG; liq.; 45% conc.

Monteine LCT. [551] Collagen polypeptide coco fatty acid condensate, triethanolamine salt; anionic; for shampoo and general cosmetic use; liq.; 36% conc.

Montosol IF-21. [494] Sodium laureth-12 sulfate; anionic; used in shampoos; liq.; 35 ± 1% conc.

Montosol IL-13. [494] IPA laureth-2.7 sulfate; anionic; used in personal care products, dishwashing agents, and fire-fighting foam concs.; liq.; 30 ± 1% conc.

Montosol IL-17. [494] Magnesium laureth-2.7 sulfate; anionic; used in personal care products; liq.; 27 ± 1% conc.

Montosol IQ-15. [494] Ammonium alcohol ether (4) sulfate; anionic; foaming agent; liq.; 35 ± 1% conc.

Montosol PB-25. [494] Sodium nonoxynol-25 sulfate; anionic; liq.; emulsifier for emulsion polymerization; 35% conc.

Montosol PF-10. [494] Sodium lauryl ether (2.8) sulfate; anionic; mfg. of personal care products; solubilizer for perfumes; liq.; 27 ± 1% conc.

Montosol PF-14. [494] Sodium lauryl ether (2) sulfate; anionic; see Montosol PF-10; liq.; 27 ± 1% conc.

Montosol PF-16, -18. [494] Ammonium lauryl ether (1.5) sulfate; anionic; wetting agent used in personal care products, detergents and window cleaners; liq.; 27 ± 1% conc.

Montosol PF-26, PG-10. [494] Sodium lauryl ether (2.4) sulfate; anionic; see Montosol PF-10; liq., gel resp.; 27 ± 1 and 70 ± 2% conc. resp.

Montosol PG-12. [494] Ammonium lauryl ether (1) sulfate; anionic; see Montosol PF-16; liq.; 27 ± 1% conc.

Montosol PG-17. [494] MEA lauryl ether (1) sulfate; anionic; dispersant for personal care products; liq.; 27 ± 1% conc.

Montosol PL-14. [494] TEA lauryl ether (2) sulfate; anionic; see Montosol IL-13; liq.; 35 ± 1% conc.

Montosol PL-16. [494] Sodium lauryl ether (2.4) sulfate; anionic; see Montosol PF-10; liq.; 27 ± 1% conc.

Montosol PQ-11, -15. [494] Ammonium laureth-3 sulfate; anionic; see Montosol PF-16; liq.; 60 ± 2 and 27 ± 1% conc. resp.

Montosol PQ-17. [494] Sodium laureth-3 sulfate; anionic; see Montosol PF-10; liq.; 59 ± 2% conc.

Montosol TQ-11. [494] Ammonium laureth-2.4 sulfate; anionic; see Montosol PF-16; liq.; 25 ± 1% conc.

Montovol GF-15. [494] DEA lauryl sulfate; anionic; detergent wetting agent, foamer for personal care products; liq.; 30 ± 1% conc.

Montovol GJ-12. [494] Sodium tallow alcohol sulfate; anionic; detergent, dispersant; paste; 34 ± 1% conc.

Montovol GL-13. [494] Magnesium lauryl sulfate; anionic; detergent, rug shampoos, wetting agent; liq.; 27 ± 1% conc.

Montovol RF-10. [494] Sodium lauryl sulfate; anionic; emulsifier; additive for emulsion polymerization; used in shampoos and specialty cleaners; liq., paste; 29 ± 1% conc.

Montovol RF-11. [494] TEA lauryl sulfate; anionic; hair and rug shampoos; liq.; 40 ± 1% conc.

Montovol RF-12. [494] Ammonium lauryl sulfate; anionic; liq. shampoos and foam bath products; liq.; 27 ± 1% conc.

Montovol RF-13. [494] Isopropanolamine lauryl sulfate; anionic; foamer for personal care products; liq.; 35 ± 1% conc.

Montovol RJ-13. [494] Sodium fatty alcohol sulfate; anionic; emulsifier used in cleaners and detergents; paste; 40 ± 1% conc.

Montovol RL-10. [494] TEA lauryl sulfate; anionic; hair and rug shampoos and specialty cleaners; liq.; 40 ± 1% conc.

Montovol RQ-10. [494] Ammonium fatty alcohol sulfate; anionic; for shampoos and foam bath products; liq.; 27 ± 1% conc.

Morco H-62 Natural, H-70 Natural. [385] Sodium petroleum sulfonate; anionic; dispersant for drilling muds, drycleaning compds., fuel and lube oil additives, metalworking, ore flotation, pigment dispersion; m.w. 520; oil sol.; dens. 8.45 and 8.60 lb/gal resp. (60 F); visc. 300 and 850 cs (210 F); 62 and 70% sulfonate resp.

Morco HM-62 Natural. [385] Sodium petroleum sulfonate; anionic; degreasing compds., demulsifying crude oil, drilling muds, drycleaning compds., fuel and lube oil additives, metalworking, pigment dispersion; m.w. 490; oil sol.; dens. 8.48 lb/gal (60 F); visc. 260 cs (210 F); 62% sulfonate.

Morco L-62 Natural. [385] Sodium petroleum sulfonate; anionic; degreasing compds., demulsifying crude oil, drycleaning compds., fuel oil additives, metalworking and sol. oils; m.w. 430; oil sol.; dens. 8.60 lb/gal (60 F); visc. 150 cs (210 F); 62% sulfonate.

Morco L-70 Natural. [385] Sodium petroleum sulfonate; anionic; emulsifier, wetting agent for degreasing compds., demulsifying crude oil, drycleaning compds., fuel oil additives, metalworking and sol. oils; m.w. 430; oil sol.; dens. 8.80 lb/gal (60 F); visc. 400 cs (210 F); 70% sulfonate.

Morco M-62 Natural. [385] Sodium petroleum sulfonate; anionic; see Morco L-62 Natural; m.w. 465; oil sol.; dens. 8.50 lb/gal (60 F); visc. 225 cs (210 F); 62% sulfonate.

Morco M-70 Natural. [385] Sodium petroleum sulfonate; anionic; emulsifier, dispersant for degreasing compds., demulsifying crude oil, drilling muds, drycleaning compds., fuel and lube oil additives, metalworking, ore flotation; m.w. 465; oil sol.; dens. 8.65 lb/gal (60 F); visc. 500 cs (210 F); 70% sulfonate.

Morwet B. [181] Sodium n-butyl naphthalene sulfonate; anionic; wetting agent for pesticides and dyestuffs; powd.; 75% conc.

Morwet IP. [181] Sodium diisopropyl naphthalene sulfonate; anionic; see Morwet B; powd.; 75% conc.

Morwet M. [181] Sodium mono- and dimethyl naphthalene sulfonate; anionic; wetting agent for pesticide formulations; powd.; 95% conc.

MP-33-F. [287] Sorbitan monopalmitate; nonionic; w/o emulsions for cosmetic products; softener and lubricant for polymers; flakes; HLB 6.5; 100% conc.

MP-55-F. [287] Polysorbate 40; nonionic; solubilizer for essential oils, vitamins and antibiotics; emulsifier for o/w cosmetic, pharmaceutical and tech. applications; liq.; HLB 16.0; 100% conc.

M-Quat 32. [400] Octadecyl diethanol methyl ammonium chloride; cationic; emulsifier, defoamer, and coagulents; liq.; 50% conc.

M-Quat 2475. [400] Dicoco dimethyl ammonium chloride; cationic; see M-Quat 32; liq.; 75% conc.

MS-33-F. [287] Sorbitan monostearate; nonionic; w/o emulsions; used in cosmetic and pharmaceutical applications; emulsifier for food industry; coemulsifier in latex stabilization; flakes; HLB 5.0; 100% conc.

MS-55-F. [287] Polysorbate 60; nonionic; o/w emulsifier for cosmetics, pharmaceuticals, and special tech. applications; latex stabilizer for emulsion polymerization; antistat; liq., paste; HLB 15.0; 100% conc.

MS-55-F-4. [287] Polysorbate 61; nonionic; emulsifier for o/w for suppository bases, coemulsifier with MS-55-F for cosmetics; glass industry; water repellent surface coatings; solid; HLB 9.5; 100% conc.

Mulsifan CB, CPA. [660] Fatty alcohol polyglycol ether; nonionic; emulsifier for creams and lotions; wax and liq. resp.; HLB 10.0 and 9.0; 100% conc.

Mulsifan RT 1, 2. [660] Fatty acid polyglycol ester; nonionic; emulsifier for min. oils; liq.; HLB 10.0; 100% conc.

Mulsifan RT 7. [660] Triglyceride, ethoxylated; nonionic; emulsifier for triglycerides of veg. and animal origin; liq.; HLB 10.0; 100% conc.

Mulsifan RT 11. [660] Fatty alcohol polyglycol ether; nonionic; emulsifier for fatty acids and alcohols, triglycerides, waxes; wax; HLB 16.0; 100% conc.

Mulsifan RT 19. [660] Fatty alcohol polyglycol ether; nonionic; emulsifier for waxes; paste; HLB 13.0; 100% conc.

Mulsifan RT 24. [660] Fatty alcohol polyglycol ether; nonionic; emulsifier for wh. and paraffin oils; liq.; HLB 9.0; 100% conc.

Mulsifan RT 27. [660] Fatty alcohol polyglycol ether; nonionic; see Mulsifan RT 11; wax; HLB 16.0; 100% conc.

Mulsifan RT 37. [660] Alkylphenol polyglycol ether; nonionic; emulsifier for min. oils and solvs.; liq.; HLB 11.0; 100% conc.

Mulsifan RT 69. [660] Triglyceride, ethoxylated; nonionic; emulsifier for fat and oils, solubilizer for perfumes and volatile oils; liq., paste; HLB 13.0; 100% conc.

Mulsifan RT 110. [660] Fatty alcohol polyglycol ether; nonionic; emulsifier for paraffins; wax; HLB 10.0; 100% conc.

Mulsifan RT 113. [660] Fatty acid polyglycol ester; nonionic; emulsifier for min. oils; liq.; HLB 9.0; 100% conc.

Mulsifan RT 141. [660] POE sorbitan monolaurate; nonionic; solubilizer for perfumes and volatile oils; liq.; HLB 16.0; 100% conc.

Mulsifan RT 146. [660] POE sorbitan monooleate; nonionic; see Mulsifan RT 141; liq.; HLB 15.0; 100% conc.

Mulsifan RT 163. [660] Triglyceride, ethoxylated; nonionic; emulsifier for triglycerides; liq.; HLB 6.0; 100% conc.

Mulsifan RT 203/80. [660] Fatty alcohol polyglycol ether; nonionic; see Mulsifan RT 141; liq.; HLB 14.0; 80% conc.

Mulsifan RT 231. [660] Polyglycol ether combination; nonionic; emulsifier for self-brilliant waxes; wax; HLB 12.0; 100% conc.

Mulsifan RT 248. [660] Fatty acid polyglycol ether; nonionic; emulsifier for solvs.; liq.; HLB 12.0; 100% conc.

Mulsifan RT 258. [660] Fatty alcohol polyglycol ether; nonionic; emulsifier for paraffin; wax; HLB 10.0; 100% conc.

Mulsifan RT 282. [660] Alkyl and alkylaryl polyglycol combination; nonionic; emulsifier for wh. oils; liq.; HLB 8.0; 100% conc.

Multinol C. [442] Special sulfonic acid deriv.; anionic; one-bath scouring, bleaching and dyeing agent; paste

Musloid 815D, 815M, 815S. [148] Fatty acid ethoxylate; nonionic; dispersant, stabilizer, leveling agent, low-foaming surfactant in emulsion paint systems; amber visc. liq.; sp.gr. 0.95, 1.01, and 0.98 resp.; 100% act.

Myrj 45. [320] PEG-8 stearate; nonionic; general purpose o/w emulsifiers for cosmetics, pharmaceuticals, etc.; cream-colored soft waxy solid; sol. in alcohol, disp. in water; sp.gr. 1.0; HLB 11.1; pour pt. 28 C; sapon. no. 82–95; 100% conc.

Myrj 49, 51. [319] PEG-20 and -30 stearate; nonionic; o/w emulsifier for cosmetic and pharmaceutical applications; solid; HLB 15.0 and 16.0 resp.; 100% conc.

Myrj 52. [320] PEG-40 stearate; nonionic; see Myrj 45; ivory waxy solid; sol. in water, acetone, ether, alcohol; sp.gr. 1.1; HLB 16.9; pour pt. 38 C; sapon. no. 25–35; 100% conc.

Myrj 52S. [320] PEG-40 stearate; nonionic; see Myrj 45; wh. waxy granular solid; sol. in water, toluol, acetone, ether, Cellosolve, carbon tetrachloride, alcohol; sp.gr. 1.1; HLB 16.9; pour pt. 38 C; sapon. no. 25–35; 100% conc.

Myrj 52C. [319] PEG-40 stearate; nonionic; see Myrj 49; pieces; HLB 16.9; 100% conc.

Myrj 53. [319; 320] PEG-50 stearate; nonionic; see Myrj 45; cream-colored solid; sol. in water, alcohol; HLB 17.9; pour pt. 42 C; sapon. no. 20–28; 100% conc.

Myrj 59. [319] PEG-100 stearate; nonionic; see Myrj 49; solid; HLB 18.8; 100% conc.

Mytab. [297] Myristyl trimethyl ammonium bromide; surfactant used as emulsifier and antistat in hair rinses; wh. powd.; characteric odor; sol. in alcohols, chloroform; m.w. 336.40; pH 5–8 (1% aq.); 99% min. assay.

Myvacet 5-07, 7-00, 7-07K. [216] Dist. acetylated monoglycerides from hydrog. cottonseed oils, hydrog. lard, and hydrog. cottonseed oils resp.; nonionic; emulsifier used in confectionery products, cosmetics, and pharmaceuticals; waxy solid; sol. in most common org. solvs.; m.p. 41–46, 37–40, and 37–40 C resp.; HLB 3.8–4.0; sapon. no. 278–292, 316–331, 316–331 resp.; 100% act.

Myvacet 9-40, 9-45. [216] Dist. acetylated monoglycerides from lard and veg. oil resp.; nonionic; emulsifier for antidusting applications, cosmetics, pharmaceuticals, defoaming, lubricants, release coatings, etc.; clear liq.; sol. in all common org. solvs., veg. and min. oils; m.p. 4–12 C; HLB 3.8–4.0; sapon. no. 375–385; 100% act.

Myvaplex 600. [216] Food-grade conc. glyceryl monostearate; nonionic; emulsifier in macaroni and cereal products; powd.; sp.gr. 0.92 (80 C); 90% min. monoester content.

Myvatem. [216] Diacetyl tartaric acid esters of dist. monoglycerides; dispersant and emulsifier for baked goods; lt. amber sticky, visc. semisolid, faint acetic acid odor; 70% sol. in soybean oil; 50% sol. in water; sp.gr. 1.056 (60 C); HLB 8; m.p. 30 C; flash pt. (TOC) 208 C; sapon. no. 380–425.

Myvatex 3-50. [216] Dist. glyceryl monostearate and dist. propylene glycol monostearate; nonionic; emulsifier in food applications; small beads; 90% max. monoester content.

Myvatex 8-20, 8-20E. [216] Dist. monoglycerides with 20% hydrog. soybean oil; nonionic; emulsifier used in food products; small beads (8-20), powd. or small beads (8-20E); 70% max. monoester content.

Myvatex 25-00, 25-07. [216] Dist. monoglyceride from hydrog. lard and from hydrog. veg. oil resp.; nonionic; emulsifier used in bread and sweet goods shortening; soft solid; 25% act.

Myvatex 40-06S. [216] Myvatex 3-50, lactylic stearate, and water; nonionic; cake emulsifier; soft solid; HLB 3.8–4.0; 25% act.

Myvatex Peanut Butter Stabilizer. [216] Dist. glyceryl monostearate and fully hydrog. palm oil; nonionic; peanut butter emulsifier and stabilizer; small beads; 45% max. monoester content.

Myvatex Super DO. [216] Succinylated monoglycerides and dist. monoglycerides; nonionic; emulsifier for baked goods; powd.; 35–40% monoester content.

Myverol 18-00. [216] Dist. hydrog. animal glyceride; nonionic; emulsifier for baked goods, confectionery products, cosmetics, dehydrated potatoes, etc.; small beads; sp.gr. 0.91 (80 C); 100% act.

Myverol 18-04. [216] Dist. hydrog. palm oil glyceride; nonionic; see Myverol 18-00; small beads; sp.gr. 0.94 (80 C); 100% act.

Myverol 18-06. [216] Dist. hydrog. soy glyceride; nonionic; emulsifier for foods, cosmetics; small beads; sp.gr. 0.92 (80 C); 100% act.

Myverol 18-07. [216] Dist. hydrog. cottonseed glyceride; nonionic; see Myverol 18-06; small beads; sp.gr. 0.92 (80 C); 100% act.

Myverol 18-30. [216] Dist. tallow glyceride; nonionic; emulsifier for baked goods, confectionery products, cosmetics, etc.; plastic; sp.gr. 0.92 (80 C); 100% act.

Myverol 18-35. [216] Dist. palm oil glyceride; nonionic; see Myverol 18-30; plastic; sp.gr. 0.94 (80 C); 100% act.

Myverol 18-40. [216] Dist. lard glyceride; nonionic; see Myverol 18-30; plastic; sp.gr. 0.91 (80 C); 100% act.

Myverol 18-50K. [216] Dist. hydrog. veg. glyceride; nonionic; see Myverol 18-30; plastic; sp.gr. 0.91 (80 C); 100% act.

Myverol 18-85. [216] Dist. cottonseed glyceride; nonionic; see Myverol 18-30; soft plastic; sp.gr. 0.92 (80 C); 100% act.

Myverol 18-98. [216] Dist. safflower glyceride; nonionic; emulsifier for diet margarines, icing and cream-fillings shortenings; soft plastic; sp.gr. 0.90 (80 C); 100% act.

Myverol P-06. [216] Dist. monoester from hydrog. soybean oil and propylene glycol; food emulsifier, stabilizer; small beads; sp.gr. 0.89 (80 C); m.p. 45 C; 90% min. monoester content.

N

Nacconol 35SL. [575] Sodium dodecylbenzenesulfonate; anionic; detergent, wetting and foaming agent, emulsifier, dispersant, penetrant used in textile, cosmetics, paper, leather, food industries; lt. colored clear liq.; faint odor; water sol.; dens. 9.12 lb/gal; visc. 103 cps; cloud pt. 0 C; surf. tens. 41.8 dynes/cm (0.1%); pH 6.0–7.5; 90% min. biodeg.; 35–38% act.

Nacconol 40F. [575; 576] Sodium dodecylbenzenesulfonate; anionic; wetting and foaming agent, dispersant, detergent used in cement, textile, cleaning industries; wh. flakes; faint odor; water sol.; dens. 18.7–22.0 lb/ft^3 (untamped); surf. tens. 31 dynes/cm (0.1%); pH 6.4–7.6 (1% aq.); 90% min. biodeg.; 40% act.

Nacconol 40G. [575] Sodium dodecylbenzene sulfonate; anionic; foamer, dispersant, wetting agent, detergent for agric., cement, dyeing, emulsion polymerization, textile, metal cleaning, metalworking, mining, paper industries; cream gran. powd.; sol. 2 g/100 ml in water; dens. 35 lb/ft^3; pH 6.4–7.6 (1% aq.); biodeg.; 38–42% act.

Nacconol 90F. [575; 576] Sodium dodecylbenzenesulfonate; anionic; wetting agent, emulsifier, dispersant, foaming agent, detergent used in agric., cleaners, shampoos, textile dyeing, metal cleaning, mining, paper; lt. cream colored flakes; water sol.; dens. 22.0 lb/ft^3 (untamped); surf. tens. 35 dynes/cm (0.1%); pH 6.0–7.5 (10% aq.); 90% min. biodeg.; 90% act.

Nacconol 90G. [575] Sodium dodecylbenzene sulfonate; see Nacconol 40G; also emulsifier for latex polymerization; lt. cream gran. powd.; sol. 15 g/100 ml in water; dens. 30 lb/ft^3; pH 6.0–7.5 (1% aq.); biodeg.; 90% act.

Na-Cumene Sulfonate 40, Sulfonate Powd. [308] Sodium cumene sulfonate; anionic; hydrotrope, sol'n. aid for liq. detergents and slurries; 40% act. liq., 96% act. powd. resp.

Nalkylene 500. [152; 634] C_{10-12} linear undecylbenzene sulfonate; surfactant intermediate for production of linear alkylbenzene sulfonate, for formulation of detergent products; Saybolt 30+ liq.; odorless; f.p. < –70 F; 100% act.

Nalkylene 515, 550. [152] Linear alkylbenzene; see Nalkylene 500; water-wh. oily liq. and Saybolt 30+ liq. resp., odorless; sp.gr. 0.861 and 0.862; m.p. < –90 F; 100% act. (550).

Nalkylene 550L. [634] C_{10-14} linear dodecylbenzene sulfonate; see Nalkylene 500; liq.; 100% act.

Nalkylene 600. [152] Linear alkylbenzene; see Nalkylene 500; Saybolt 30+ liq.; odorless; sp.gr. 0.865; visc. 46.9 SSU (100 F); m.p. < –90 F; 100% act.

Nalkylene 600L. [634] C_{12-14} linear tridecylbenzene sulfonate; intermediate to produce surfactants, powd. detergents, industrial and specialty chemicals; liq.; 100% act.

Nansa 1042. [17] Dodecylbenzene sulfonic acid; used in mfg. of detergents, laundry products; dk. br. visc. liq.; sp.gr. 1.05; visc. 1900 cs; 95% act.

Nansa 1042/P. [15; 17] Dodecylbenzene sulfonic acid; intermediate for neutralization;

sodium salts as emulsifiers in emulsion polymerization of plastics and syn. rubbers; dk. br. visc. liq.; sp.gr. 1.05; visc. 1900 cs; 95% act.

Nansa 1106/P, 1169/P. [15; 17] Sodium dodecylbenzene sulfonate; surfactant for emulsion polymerization; golden yel. clear and cloudy resp. liq.; sp.gr. 1.0; visc. 2500 and 5500 cs; cloud pt. 11 and 25 C; pH 8.0 ± 1 (5% aq.); 30 ± 1% act.

Nansa 1192. [17] Foaming agent for gypsum slurries in plaster board production; golden yel. clear liq.; dens. 1.03 g/cc; visc. 290 cs; cloud pt. < 0 C; pH 8 (5%).

Nansa 1909. [15; 17] Dodecylbenzene sulfonic acid; surfactant for high-visc. specialty industrial, institutional cleaners; reddish-br. visc. liq..; sp.gr. 1.04; visc. 1450 cP; > 80% biodeg.; 94% act.

Nansa AS 40. [15; 17] Ammonium dodecylbenzene sulfonate; anionic; formulation of domestic and industrial liq. detergents; golden yel. opaque gel; pH 6.0–6.5 (2% aq.); 40% act.

Nansa BXS. [17] Sodium alkyl naphthalene sulfonate; anionic; wetting and dispersing agent for colloidal systems without detergent properties, used in agric., leather, paper and pulp, textile industries; cream-colored powd.; pH 7.5–9.5 (1%); 38% act.

Nansa EVM 62H, EVM 70. [15] Calcium dodecylbenzene sulfonate; anionic; emulsifier for agric. chemicals; wetting agent for oil-based drilling fluids; liq.; 62 and 67% conc. resp.

Nansa HS40-AU. [15; 17] Sodium dodecylbenzene sulfonate; wetting agent, dispersant in scouring powds., industrial applications, automotive, floor, wall, and hard surface cleaners; cream coarse powd.; dens. 0.55 g/cm³; pH 8.5 ± 1.0 (1%); > 80% biodeg.; 38 + 3% act.

Nansa HS40 Soft. [17] Sodium dodecylbenzene sulfonate; anionic; wetting and dispersing agent, emulsifier used in detergent compd. for domestic and industrial use; cream-colored, free-flowing coarse powd.; dens. 0.55 g/cm³; pH 8.5 ± 1 (1%); > 80% biodeg.; 38 ± 3% act.

Nansa HS80-AU. [15] Sodium alkylbenzene sulfonate; wetting agent, dispersant in laundry, detergent compds., insecticides, metal pickling sol'ns.; cream coarse powd.; dens. 0.55 g/cm³; pH 8.5 ± 1.0 (1%); > 80% biodeg.; 78 + 3% act.

Nansa HS80P. [15; 17] Sodium dodecylbenzene sulfonate; anionic; see Nansa HS40 Soft; cream-colored powd.; dens. 0.52 g/cm³; pH 8.5 ± 1 (1%); > 80% biodeg.; 80 ± 3% act.

Nansa HS80/S. [15] Sodium dodecylbenzene sulfonate; anionic; formulation of detergents, hard surface and bottle cleaners; metal treatment and paper processing; scouring and wetting agent for textile industry; foamer and mortar plasticizer in building industry; cream powd.; sp.gr. 0.65; pH 9–11 (1%); 80% conc.

Nansa HS80SK. [15] Sodium dodecylbenzene sulfonate; built detergent, wetting, dispersing agent for dairy, laundry, heavy-duty, hard surface detergents; off-wh. to cream coarse powd.; dens. 0.55 g/cm³; pH 9.0 ± 0.8 (1%); > 80% biodeg.; 75 + 3% act.

Nansa HS80 Soft. [17] Sodium dodecylbenzene sulfonate; anionic; see Nansa HS40 Soft; cream-colored coarse powd.; dens. 0.55 g/cm³; pH 8.5 ± 1 (1%); > 80% biodeg.; 78 ± 3% act.

Nansa HS85/S. [15; 17] Sodium dodecylbenzene sulfonate; anionic; see Nansa HS80/S; also as rubber/plastics emulsifier; cream flake; dens. 0.5 g/cm³; pH 9–11 (1%); 85 ± 3% act.

Nansa HSA/L. [17] Alkylbenzene sulfonic acid; anionic; basic ingredient in low act. lt. duty liq. detergents; dk. br. visc. liq.; > 80% biodeg.; 86% min. act.

Nansa LES 42. [17] Blended detergent conc.; used in dishwashing and lt. duty clothes washing formulations; golden clear liq.; visc. 250 cs; cloud pt. < 0 C; surf. tens. 30 dynes/cm (1 g/l); pH 6.2–6.8 (5%); 37% act.

Nansa MA 30. [17] Blend of sodium dodecylbenzene sulfonate and ethoxylated nonionic surfactants; anionic/nonionic; base for liq. detergents; golden liq.; sp.gr. 1.046; visc. 500 cs; cloud pt. < 0 C; pH 6.5–7.0 (5% aq.); 30% act.

Nansa MS 45. [15; 17] Magnesium dodecylbenzene sulfonate; anionic; detergent ingred.; liq.; 45% conc.

Nansa S40/S. [17] Sodium alkylbenzene sulfonate; anionic; detergent, wetting agent in insecticides, metal pickling, printing inks, paper processing; wh./pale cream powd.; pH 8–9 (1% aq.); 40% act.

Nansa SB 62. [15] Sodium dodecylbenzene sulfonate; anionic; see Nansa MS 45; paste; 62% conc.

Nansa SBA. [17] Dodecylbenzene sulfonic acid; anionic; detergent, intermediate; dk. br. liq.; visc. 20,000 cs; 96% act.

Nansa SL 30, SS 30. [15; 17] Sodium dodecylbenzene sulfonate; anionic; surfactant used in detergent formulations; pale yel. liq. and golden liq. resp.; visc. 2500–3500 and 250 cs resp.; cloud pt. 15 and 0 C; pH 7.0–7.5 and 6.0–6.8 (2%); 30% act.

Nansa SS 60 Paste. [15; 17] Sodium dodecylbenzene sulfonate; anionic; see Nansa SL 30; cream paste; pH 7.0–9.0 (2%); 60% act.

Nansa SSA. [17] Dodecylbenzene sulfonic acid; anionic; detergent, intermediate; dk. br. liq.; visc. 2000 cs; 96% act.

Nansa SSAL. [15; 17] Alkylbenzene sulfonic acid; anionic; surfactant, basic ingredient in lt.-duty liq. detergents and cleaners used in mfg. of wetting, emulsifying, and foaming agents; dk. br. visc. liq.; > 80% biodeg.; 94% act.

Nansa SSA/P. [17] Dodecylbenzene sulfonic acid; see Nansa 1042/P; dk. br. visc. liq.; sp.gr. 1.05; visc. 1900 cs; 95% act.

Nansa TDB. [17] Tridecylbenzene sulfonic acid; anionic; see Nansa MS-45; liq.; 90% conc.

Nansa TS 60. [17] TEA-dodecylbenzene sulfonate; anionic; detergent, emulsifier, pigment dispersant used in cleaners, desizing agent for syn. fibers; medium br. visc. liq.; sp.gr. 1.09; visc. 9000 cs; cloud pt. < 0 C (50%); pH 6.6–7.0 (2% aq.); 60% act.

Nansa UCA/S, UCP/S. [17] Built powders based on sodium alkylbenzene sulfonate; anionic; domestic and industrial detergent powd.; blue powd.; sol. in hot or cold water; biodeg.; 17 and 15% act. resp.

Nansa YS 94. [15; 17] Isopropylamine dodecylbenzene sulfonate; emulsifier for solv.-based hand cleaners, coupling agent for water in charge detergent systems; amber visc. liq.; visc. 10,000 cs; cloud pt. < 0 C; pH 5–8 (2% aq.); 94% act.

Nansol. [17] Sodium dodecylbenzene sulfonate and sodium alkyl ether sulfate; anionic; detergent for dishwashing and hard surface cleansing operations; golden liq.; sp.gr. 1.05; visc. 400 cs; pH 6.5 ± 0.5; biodeg.; 30% act.

Natrex D 3. [192] Phosphated ester, modified; anionic/nonionic; detergent, wetting agent, emulsifier for processing cotton and syns.; liq.; 60% conc.

Natrex EFB 171. [192] Betaine; amphoteric; foamer; liq.; 30% conc.

Natrex J 3. [192] Phosphated ester, modified; anionic; textile wetting agent; detergent; liq.; 65% conc.

Naturechem EGHS. [115] Ethylene glycol hydroxystearate; aux. emulsifier, emollient, thickener for cosmetics, household products; solid; HLB 2.0; 100% conc.

Naturechem GMHS. [115] Glyceryl hydroxystearate; aux. emulsifier, emollient for cosmetics, household products; solid; HLB 3.4; 100% conc.

Naturechem PGHS [115] Propylene glycol hydroxystearate; aux. emulsifier, dispersant imparting emolliency and stability to cosmetics, household products; solid; HLB 2.6; 100% conc.

Naturechem PGR [115] Propylene glycol ricinoleate; pigment dispersant providing emolliency, gloss, plasticization to cosmetics, household products.

Naturechem THS-200. [115] PEG-200 trihydroxystearin; nonionic; emulsifier imparting emolliency, thickening to cosmetics, household products; wh. wax-like; sol. in water and alcohol; HLB 18; 100% conc.

Naxonate 4AX. [428] Ammonium xylene sulfonate; anionic; hydrotrope, stabilizer, solubilizer used in formulating detergents, inks, electroplating baths, dyestuffs, polymers; Klett 50 max. liq.; pH 6.5–8.5; 40% min. act.

Naxonate 4KT. [428] Potassium toluene sulfonate; anionic; see Naxonate 4AX; Klett 50 max. liq.; pH 7–9; 40% min. act.

Naxonate 4L. [428] Sodium xylene sulfonate; anionic; see Naxonate 4AX; Klett 50 max. liq.; pH 7–9; 40% min. act.

Naxonate 4ST. [428] Sodium toluene sulfonate; anionic; see Naxonate 4AX; Klett 50 max. liq.; pH 7–9; 40% min. act.

Naxonate 5KT. [428] Potassium toluene sulfonate; anionic; see Naxonate 4AX; Klett 50 max. liq.; pH 7–9; 50% min. act.

Naxonate 5L. [428] Sodium xylene sulfonate; anionic; see Naxonate 4AX; Klett 60 max. liq.; pH 7–9; 50% min. act.

Naxonate 6AC. [428] Ammonium cumene sulfonate; anionic; see Naxonate 4AX; Klett 80 max. liq.; pH 7–8; 60% min. act.

Naxonate 45SC. [428] Sodium cumene sulfonate; anionic; see Naxonate 4AX; Klett 75 max. liq.; pH 7–9; 44% min. act.

Naxonate AX. [428] Ammonium xylene sulfonate; anionic; see Naxonate 4AX; cream powd.; 95% min. act.

Naxonate G. [428] Sodium xylene sulfonate; anionic; see Naxonate 4AX; cream powd.; 95% min. act.

Naxonate KT. [428] Potassium toluene sulfonate; anionic; see Naxonate 4AX; cream powd.; 95% min. act.

Naxonate SC. [428] Sodium cumene sulfonate; anionic; see Naxonate 4AX; cream powd.; 95% min. act.

Naxonate ST. [428] Sodium toluene sulfonate; anionic; see Naxonate 4AX; cream powd.; 95% min. act.

Nekal BA-77. [245] Sodium alkyl naphthalene sulfonate; anionic; wetting agent, dispersant, stabilizer without detergent properties, foaming agent used in textiles, paints, pesticides, latex emulsions; tan powd.; m.w. 314; pH 8–10 (5%); 75% act.

Nekal BX-78. [245] Sodium alkyl naphthalene sulfonate; anionic; emulsifier, wetting and dispersing agent, foamer for dyes, pigments, industrial use; cream to tan powd.; m.w. 326; water-sol.; pH 6–8 (5%); 70% act.

Nekal BX Conc. Paste, BX Dry. [82] Sodium alkyl naphthalene sulfonate; anionic; see Nekal BX-78; ylsh. high visc. paste, pale hygroscopic powd. resp.; water sol.; pH 10–12 and 9–11 (5% aq.); 60 and 65% act.

Nekal NF. [245] Alkyl naphthalene sulfonate; anionic; wetting agent, dispersant, penetrant used in textile dyeing; clear, amber liq.; misc. with water; sp.gr. 1.04; pH 6–8.

Nekal WS-21, 25. [245] Sulfonated aliphatic polyester; anionic; wetting, rewetting agent, dispersant for textile processing, latex paints; water-wh. slightly visc. liq.; slightly alcoholic odor; sol. in water; sp.gr. 0.9–1.0 and 1.0–1.1 resp.; pH 6.0–7.5 (10%); 48 and 55–60% conc. resp.

Nekal WT-27. [245] Sodium dioctyl sulfosuccinate; anionic; wetting and rewetting agent used in dyeing, drycleaning detergents, emulsion polymerization, firefighting; water-wh. clear liq.; water sol. up to 1.2% solids and 5.5% @ 70 C; sol. in some polar and nonpolar solvs.; pH 5–6 (1%); 70% act.

Nekanil 907. [82] Low-ethoxylated alkyl phenol; nonionic; detergent, wetting agent used in textile industry; 100% act.

Nekanil 910. [82] Alkyl phenol adduct with 9–10 moles EO; nonionic; detergent, wetting and dispersing agent used in textile industry; 100% act.

Nekanil LN. [82] Low-ethoxylated alkyl phenol; nonionic; detergent, wetting agent used in textile industry; 100% act.

Neobee 18. [496] Hybrid safflower oil; emollient oil for cosmetics and pharmaceuticals, solubilizer, solv.; liq.; bland taste and odor; sol. in alcohol, min. oil, acetone; sp.gr. 0.915; visc. 70 cps; sapon. no. 190–200; surf. tens. 31.6 dynes/cm.

Neobee 62. [496] Tristearin; solubilizer, stabilizer used in food applications; flakes; sapon. no. 189–195.

Neobee 1053. [496] Coconut oil derived triglycerides; diluent vehicle for flavoring, medicinals, colorings, clouding agent in beverages, liq.; odorless and tasteless; sol. in min. oil, acetone, alcohol; sp.gr. 0.930–0.960; sapon.no. 325–355.

Neobee 1054. [496] Propylene glycol diester of coconut fatty acids; cosolv. for cosmetics and pharmaceuticals, carrier for flavors and colors; liq.; sol. in alcohol, min. oil, acetone; sp.gr. 0.910–0.923; sapon. no. 315–335.

Neobee 1062. [496] Coconut oil-derived triglycerides; cosolv., carrier for fat-sol. vitamins, medicinals, and ointments in food, cosmetic and pharmaceutical industries; liq.; sol. in min. oil, acetone; sp.gr. 0.925–0.945; sapon. no. 295–315.

Neobee 1223 (Neobee O). [496] Caprylic/capric triglyceride; see Neobee 1062; also spreading agent, penetrant; liq.; sol. in min. oil, acetone; sp.gr. 0.938; visc. 30 cps; sapon. no. 300–315; surf. tens. 32 dynes/cm.

Neobee 1400 Oil (Neobee M-5). [496] Caprylic/capric triglyceride; diluent vehicle for flavoring, medicinals, colorings used in foods, beverages, cosmetic and pharmaceutical production, emollient; liq.; tasteless and odorless; sol. in alcohol, min. oil, acetone; sp.gr. 0.930–0.960; visc. 23 cps; sapon. no. 335–360; surf. tens. 32.3 dynes/cm.

Neobee M-20. [496] Propylene glycol dicaprylate/dicaprate; emollient oil for cosmetics and pharmaceuticals, carrier for flavors and colors in foods; liq.; sol. in alcohol, min. oil, acetone; sp.gr. 0.920; visc. 9 cps; sapon. no. 315–335; surf. tens. 31.0 dynes/cm.

Neobee O. [496] See Neobee 1223.

Neocation G. [437] Quat. ammonium salt; cationic; surfactant, leveling agent for cationic dyes; liq.; 50% conc.

Neodene 6. [555] Linear C_6 alpha olefin; intermediate for surfactants and specialty industrial chemicals; liq.; 100% conc.

Neodene 8. [555] Linear C_8 alpha olefin; see Neodene 6; liq.; 100% conc.

Neodene 10. [555] Linear C_{10} alpha olefin; see Neodene 6; liq.; 100% conc.
Neodene 12. [555] Linear C_{12} alpha olefin; see Neodene 6; liq.; 100% conc.
Neodene 14. [555] Linear C_{14} alpha olefin; see Neodene 6; liq.; 100% conc.
Neodene 16. [555] Linear C_{16} alpha olefin; see Neodene 6; liq.; 100% conc.
Neodene 18. [555] Linear C_{18} alpha olefin; see Neodene 6; liq.; 100% conc.
Neodene 20. [555] Linear C_{20} alpha olefin; see Neodene 6; wax; 100% conc.
Neodene 1214. [555] C_{12}-C_{14} alpha olefin; see Neodene 6; liq.; 100% conc.
Neodene 1416. [555] C_{14}-C_{16} alpha olefin; see Neodene 6; liq.; 100% conc.
Neodene 1418. [555] C_{14}-C_{18} alpha olefin; see Neodene 6; liq.; 100% conc.
Neodene 1618. [555] C_{16}-C_{18} alpha olefin; see Neodene 6; liq.; 100% conc.
Neodol 23. [555] C_{12}-C_{13} alcohol; detergent intermediate; liq.; 100% conc.
Neodol 23-3. [555] C_{12}–C_{13} pareth-3; detergent intermediate used in preparation of sulfates for high-foaming liq. detergents; APHA 50 max., clear to slightly hazy liq.; m.w. 310–342; dens. 7.7 lb/gal; sp.gr. 0.925; visc. 19 cs (100 F); m.p. 5–6 C; HLB 8.1; flash pt. 300 F (PMCC); pour pt. 4 C; 100% act.
Neodol 23-6.5. [555] C_{12}–C_{13} pareth-7 (6.5 EO); see Neodol 23-3; also wetting agent; practically colorless liq., mild odor; m.w. 484; dens. 8.08 lb/gal (100 F); sp.gr. 0.963; visc. 29.5 cs (100 F); m.p. 11–15 C; HLB 12.0; flash pt. 330 F (PMCC); pour pt. 16 C; 100% act.
Neodol 25. [555] C_{12}-C_{15} alcohol; detergent, emulsifier intermediate; liq.; 100% conc.
Neodol 25-3. [555] C_{12}–C_{15} pareth-3; nonionic/anionic; detergent intermediate used in preparation of sulfates for high-foaming liq. detergents; colorless, clear to slightly hazy liq.; mild odor; m.w. 320–350; dens. 7.70 lb/gal; sp.gr. 0.925; visc. 19 cs (100 F); m.p. 5–6 C; HLB 7.9; flash pt. 315 F (PMCC); pour pt. 0 C; 100% act.
Neodol 25-3A. [555] Ammonium C_{12}–C_{15} pareth sulfate; anionic; detergent used in high-foaming liq. detergents; lt. clear visc. liq.; mild ethanol odor; m.w. 432; dens. 8.5 lb/gal (60 F); sp.gr. 1.02; visc. 45 cs (100 F); flash pt. 74 F (PMCC); pH 7.3; biodeg.; 59% act.
Neodol 25-3S. [555] Sodium C_{12}–C_{15} pareth sulfate; anionic; detergent used in high-foaming liq. detergents; lt., visc. liq.; mild ethanol odor; m.w. 437; dens. 8.76 lb/gal (60 F); sp.gr. 1.02; visc. 45 cs (100 F); flash pt. 73 F (PMCC); pH 7.5–9.0; 59% act.
Neodol 25-7, -9, -12. [555] C_{12}–C_{15} pareth-7, -9, and -12; nonionic; detergent used in formulation of heavy-duty controlled foaming detergents used in textile, paper, and paint processing, metal cleaning; colorless, paste-like; mild odor; m.w. 519, 610, 729 resp.; dens. 8.12, 8.25 8.40 lb/gal (100 F); sp.gr. 0.958, 0.982, 1.003 (50/25 C); visc. 34, 41, 53 cs (100 F); m.p. 12–17, 25–28, 30–33 C; HLB 12.2, 13.3, 14.4; cloud pt. 50, 74, 97 C (1% aq.); flash pt. 350, 370, 380 F (PMCC); pour pt. 21, 24, 27 C; pH 5.5–6.5 (1% aq.); 100% act.
Neodol 45. [555] C_{14}-C_{15} alcohol; see Neodol 23; liq.; 100% conc.
Neodol 45-7. [555] C_{14-15} pareth-7; nonionic; surfactant, detergent; APHA 10 color; m.w. 539; sol. in alcohols, esters, ketones, chlorinated solvs., aromatic and aliphatic hydrocarbons; sp.gr. 0.967; dens. 8.20 lb/gal; visc. 35 cSt (38 C); HLB 12; m.p. 21–25 C; flash pt. 204 C (PMCC); cloud pt. 46 C (1% aq.); surf. tens. 29 dynes/cm (0.1%); 100% act.
Neodol 45-13. [555] C_{14-15} pareth-13; nonionic; surfactant, detergent; APHA 10 color; m.w. 790; sol. in alcohols, esters, ketones, chlorinated solvs., aromatic and aliphatic hydrocarbons; sp.gr. 1.008 (50/25 C); dens. 8.39 lb/gal (49 C); visc. 58 cSt (38 C); HLB 14.4; m.p. 31–36C; flash pt. 241 C (PMCC); cloud pt. 78 C (5% aq. NaCl); surf. tens. 34 dynes/cm (0.1%); 100% act.
Neodol 91. [555] C_9-C_{11} alcohol; nonionic; see Neodol 23; liq.; 100% conc.
Neodol 91-2.5, -6, -8. [555] C_9–C_{11} pareth-3 (2.5 EO), -6, and -8 (8.4 EO) resp.; nonionic; oil-sol. emulsifier and wetting agent, intermediate, dispersant, surfactant; colorless liq.; mild odor; m.w. 270, 424, 529 resp.; sol. in wide range of oxygenated and hydrocarbon solvs., 91-2.5 partly sol. in water, 91-8 readily disp. in water; dens. 7.73, 8.24, 8.42 lb/gal; sp.gr. 0.934, 0.991, 1.002; visc. 12, 23, 30 cs (100 F); HLB 8.1, 12.5, 14.0; flash pt. 280, 334, 349 F (PMCC); pour pt. –12, 7, 16 C; pH 6.0 (1% aq. disp.); biodeg.; 100% act.
Neopelex FS. [341] Alkylbenzene sulfonic acid; anionic; detergent base; liq.; 96% conc.
Neopelex No. 6, No. 25, No. 6F Powder, F-25, F-65. [341] Sodium alkylaryl sulfonate;

anionic; general purpose detergent; paste, liq., beads, liq., paste resp.; 60, 25, 60, 25, and 65% conc. resp.

Neopon LAM. [654] Ammonium lauryl sulfate/sulfonated fatty alcohol; anionic; detergent, base for cosmetics; liq.; 27% conc.

Neopon LOA, LOS. [654] Sulfonated fatty ether/ammonium lauryl ether sulfate; anionic; see Neopon LAM; liq.; 26 and 27% conc. resp.

Neopon LOS 70. [654] Sodium lauryl ether sulfate; anionic; personal care products; paste; 70% conc.

Neopon LOT. [654] TEA lauryl ether sulfate; special shampoos; 27% conc.

Neopon LS. [654] Sodium lauryl sulfate; see Neopon LAM; 29% conc.

Neopon LT. [654] TEA lauryl sulfate/sulfonated fatty alcohol; anionic; see Neopon LOT; liq.; 41% conc.

Neoprotex KM. [437] Methylol amide surfactant; anionic; softener and water-repellent agent for resin treatment; liq.; 40% conc.

Neorate NA-30. [437] Alkyl phosphate; anionic; penetrant for polyester fiber; liq.; 65% conc.

Neoscoa 203C. [606] POE alkyl ether; nonionic; detergent, scouring agent for cotton, wool, and syn. fibers; liq.; 52% act.

Neoscoa 363. [606] Nonionic blend; detergent, scouring agent for raw wool and textiles, metal cleaning; liq.; 100% act.

Neoscoa 500C. [606] Nonionic-anionic blend; detergent, scouring agent for cotton, wool, and syn. fibers; liq.; 50% act.

Neoscoa FS-100. [606] Nonionic blend; deinking agent for paper production by flotation method; liq.

Neoscoa GF 3C. [606] POE alkylaryl ether, POE alkylaryl ether phosphate; nonionic/anionic; detergent, scouring and wetting agent for cotton and chemical syn. fibers; liq.; 80% act.

Neoscoa PRA-60. [606] Nonionic/anionic; detergent, aftersoaping agent esp. for polyester fibers; liq.; 54% act.

Neoscoa SS-10. [606] Nonionic blend; deinking agent for paper production by washing method; liq.

Neoscoa TH-102. [606] Nonionic/anionic; detergent, dyeing machine cleaner for disperse dyeing; liq.; 50% act.

Neospinol 264. [606] Blend of special anionics, nonionics, and min. oil; dispersant for sulfur in rayon production; liq.; 100% conc.

Neospinol 358. [606] Special nonionic complex (nitrogen compd.); dispersant for sulfur in rayon production; liq.; 100% conc.

Neustrene 045, 053. [310; 312] Hydrog. fish glycerides; textile lubricant, pharmaceutical intermediate, emulsifier, mold release agent, buffing compd.; solid; sapon. no. 186–201; 100% conc.

Neustrene 059, 060. [310; 312] Hydrog. tallow glycerides (060—refined); textile lubricant, pharmaceutical intermediate, emulsifier, mold release agent, buffing compd.; Gardner 5 max. (059) solids; sapon. no. 193–205; 100% conc.

Neustrene 064. [310; 312] Hydrog. soybean oil; textile lubricant, pharmaceutical intermediate, emulsifier, mold release agent, buffing compd.; solid; sapon. no. 188–200; 100% conc.

Neutronyx 600, 656. [456] Nonoxynol-9 (9.5 EO) and nonoxynol-11 resp.; nonionic; detergent, dispersant, wetting agent; liq.; sp.gr. 1.05; cloud pt. 54 and 71 C; flash pt. > 200 F; 100% conc.

Neutronyx S-60. [456] Ammonium nonoxynol-4 sulfate; detergent, foaming agent used in disinfectant hand cleaner formulations, lt. duty dishwash, car wash, glass cleaners; liq.; 60% conc.

Newcol 3-80, 3-85. [440] Sorbitan oleate; nonionic; antistat, lubricant, emulsifier, corrosion inhibitor, emulsion solubilizer; liquid; 100% conc.

Newcol 20. [440] Sorbitan laurate; nonionic; antistat, emulsifier, corrosion inhibitor, extreme pressure lubricant, emulsion solubilizer; solid; 100% conc.

Newcol 25. [440] POE sorbitan laurate; nonionic; see Newcol 20; liq.; 100% conc.

Newcol 40. [440] Sorbitan palmitate; nonionic; see Newcol 20; solid; 100% conc.

Newcol 45. [440] POE sorbitan palmitate; nonionic; see Newcol 20; solid; 100% conc.

Newcol 60. [440] Sorbitan stearate; nonionic; see Newcol 20; solid; 100% conc.

Newcol 65. [440] POE sorbitan stearate; nonionic; see Newcol 20; solid; 100% conc.

Newcol 80. [440] Sorbitan oleate; nonionic; see Newcol 20; liq.; 100% conc.

Newcol 85. [440] POE sorbitan oleate; nonionic; see Newcol 20; liq.; 100% conc.

Newcol 150. [440] POE laurate; nonionic; emulsifier, dispersant, lubricant; liq.; 100% conc.

Newcol 170. [440] POE oleate; nonionic; see

Newcol 150; liq.; 100% conc.

Newcol 180. [440] POE stearate; nonionic; see Newcol 150; solid; 100% conc.

Newcol 180T. [440] Nonionic surfactant used in emulsion polymerization; solid; 100% conc.

Newcol 210. [440] Ammonium dodecylbenzene sulfonate; anionic; wetting agent, emulsifier, detergent; liq.; 50% conc.

Newcol 261A, 271A. [440] Sodium alkyl diphenyl ether disulfonate; anionic; detergent, emulsifier used in emulsion polymerization; liq.; 45% conc.

Newcol 290K, 290M, 290P. [440] Sodium alkyl sulfosuccinate; anionic; wetting and dispersing agent, solubilizer, penetrant; liq.; 75, 75, and 30% conc.

Newcol 405, 410, 420. [440] POE lauryl amine; nonionic; corrosion inhibitor, intermediate for textile industry; liq., solid; 100% conc.

Newcol 506. [440] POE nonylphenyl ether; nonionic surfactant used in emulsion polymerization; solid; 100% conc.

Newcol 508. [440] EO-PO block copolymer; nonionic; low-foaming detergent; solid; 100% conc.

Newcol 560. [440] POE nonylphenyl ether; nonionic; wetting agent, penetrant, spreading and dispersing agent, emulsifier, base material for detergents; liq.; 100% conc.

Newcol 560SF. [440] Ammonium POE nonylphenyl ether sulfate; anionic; emulsifier, wetting agent, penetrant, detergent used in emulsion polymerization; liq.; 50% conc.

Newcol 560SN. [440] Sodium POE nonylphenyl ether sulfate; anionic; emulsifier, wetting agent, penetrant, detergent used in emulsion polymerization; liq.; 30% conc.

Newcol 561H, 562, 564, 565. [440] POE nonylphenyl ether; nonionic; antifoaming agent, mold lubricant, wetting agent, penetrant, spreading and dispersing agent, emulsifier, base material for detergents; liq.; 100% conc.

Newcol 565FH. [440] EO-PO block copolymer; nonionic; low-foaming detergent; liq.; 100% conc.

Newcol 566. [440] POE nonylphenyl ether; nonionic; wetting agent, penetrant, spreading and dispersing agent, emulsifier, base material for detergents, agric. chemicals, machine oils; liq.; 100% conc.

Newcol 568. [440] POE nonylphenyl ether; nonionic; surfactant for emulsion polymerization; solid; 100% conc.

Newcol 569E. [440] EO-PO block copolymer; nonionic; low-foaming detergent; paste; 100% conc.

Newcol 600, 700 Series. [440] POE alkylaryl ether; nonionic; emulsifier, solubilizer, detergent used in emulsion polymerization; liqs. and solids; 100% conc.

Newcol 804, 808, 860. [440] POE octylphenyl ether; nonionic; see Newcol 506; liq., solid; 100% conc.

Newcol 861SE. [440] Sodium octyl-phenoxyethoxyethyl sulfonate; anionic; emulsifier, wetting agent, penetrant, detergent used in emulsion polymerization; liq.; 30% conc.

Newcol 862. [440] POE nonylphenyl ether deriv.; nonionic; low-foaming detergent; paste; 100% conc.

Newcol 864. [440] POE octylphenyl ether; nonionic; antifoaming agent, mold lubricant; liq.; 100% conc.

Newcol 865. [440] POE octylphenyl ether; nonionic; wetting, penetrating, spreading, and dispersing agent, emulsifier, base for detergents, agric. chemicals, machine oils; liq.; 100% conc.

Newcol 1010, 1020. [440] POE octyl ether; nonionic; stabilizer of latex; solid; 100% conc.

Newcol 1100. [440] POE lauryl ether; nonionic; stabilizer of latex; solid; 100% conc.

Newcol 1105. [440] POE octyl ether; nonionic; emulsifier of min. oil, lubricant, detergent, dyeing additive agent; liq.; 100% conc.

Newcol 1110. [440] POE lauryl ether; nonionic; stabilizer of latex; solid; 100% conc.

Newcol 1120. [440] POE lauryl ether; nonionic; emulsifier, dyeing aid; solid; 100% conc.

Newcol 1200. [440] POE oleyl ether; nonionic; emulsifier of min. oil, dyeing additive agent; solid; 100% conc.

Newcol 1203. [440] POE lauryl ether; nonionic; emulsifier of min. oil, dyeing additive agent; liq. 100% conc.

Newcol 1204, 1208, 1210. [440] POE oleyl ether; nonionic; emulsifier of min. oil, dyeing additive agent; liq. (1204), solid (1208, 1210); 100% conc.

Newcol 1305. [440] POE tridecyl ether; nonionic; emulsifier for silicone, dyeing additive agent, lubricant; solid; 100% conc.

Newcol 1305SN, 1310SN. [440] Sodium POE tridecyl ether sulfate; anionic; emulsifier, wetting agent, penetrant, detergent used in emulsion polymerization; liq.; 30% conc.

Newcol 1310, 1515, 1525, 1545. [440] POE

tridecyl ether; nonionic; emulsifier for silicone, dyeing additive agent, lubricant; solid (1310, 1545), liq. (1515, 1525); 100% conc.

Newcol 1610, 1620. [440] POE cetyl ether; nonionic; emulsifier of paraffin wax, latex stabilizer; solid; 100% conc.

Newcol 1807, 1820. [440] POE stearyl ether; nonionic; emulsifier of paraffin wax, latex stabilizer; solid; 100% conc.

Newcol B4, B10, B18. [440] POE naphthyl ether; nonionic; dispersant, suspending and diffusion agent; liqs. (B4, B10), solid (B18); 100% conc.

Newkalgen 135R. [590] POE alkylaryl ether, POE alkyl ester and org. sulfonate; anionic/nonionic; emulsifier for spray oil; liq.; 100% conc.

Newkalgen 2360X1. [590] POE alkylaryl ether, POE alkylaryl polymer and org. sulfonate; anionic/nonionic; emulsifier for agric. emulsion concs.; liq.; 80% conc.

Newkalgen 2720X75. [590] POE alkylaryl ether, POE alkyl ether and org. sulfonate; anionic/nonionic; emulsifier for Diazinon emulsion concs.; liq.; 75% conc.

Newkalgen 3000 A & B. [590] POE alkylaryl ether, POE alkyl polymer, POE alkyl ether and org. sulfonate; anionic/nonionic; emulsifier for agric. emulsion concs.; liq.; 80% conc.

Newlon K-1. [590] Polyglycol ether; nonionic; raw wool scouring agent; liq.; 100% conc.

Newpol PE-61, -62, -64, -68, -74, -75, -78, -88. [539] EO/PO block copolymer; nonionic; base material for household and industrial detergents; plasticizer, antistat for phenol resins; emulsifier for agric. pesticides and emulsion polymerization; pigment and pitch dispersant; liq.; solid; HLB 5.8, 6.3, 10.1, 14.9, 10.1, 10.7, 14.8, and 14.6 resp.; 100% conc.

Niaproof 4. [434] Sodium tetradecyl sulfate; anionic; detergent, wetting agent, emulsifier used in adhesives and sealants, coatings, photo chemicals, emulsion polymerization; colorless liq.; misc. with water; dens. 8.58 lb/gal; sp.gr. 1.031; surf. tens. 47 dynes/cm (0.1% aq.); pH 8.5 (0.1% aq.); 27% act.

Niaproof 7. [434] Sodium heptadecyl sulfate; anionic; detergent, wetting agent, emulsifier used in adhesives, metal finishing, leather treating; colorless liq.; misc. with water; dens. 8.82 lb/gal; sp.gr. 1.060; surf. tens. 34 dynes/cm (0.1% aq.); pH 5.3 (0.1% aq.); 26% act.

Niaproof 08. [434] Sodium 2-ethylhexyl sulfate; anionic; detergent, wetting agent, emulsifier used in metal cleaning, electroplating, photo chemicals, adhesives, emulsion polymerization; colorless liq.; misc. with water; dens. 9.23 lb/gal; sp.gr. 1.109; surf. tens. 63 dynes/cm (0.1% aq.); pH 7.3 (0.1% aq.); 40% act.

Nikkol Alaninate LN-30. [437] N-Sodium N-lauroyl-N-methyl alaninate; anionic; shampoo and facial cleanser base, foaming agent, detergent; liq.; 30% conc.

Nikkol Batyl Alcohol 100, EX. [437] Batyl alcohol; nonionic; emulsifier, hydrotrope; cosmetic and pharmaceutical preparations; powd.; 100% conc.

Nikkol BB-5, -10, -20, -30. [437] Beheneth-5, -10, -20, and -30 resp.; nonionic; emulsifier for cosmetics; solid; HLB 7.0, 10.0, 16.5, and 18.0; 100% conc.

Nikkol BC-1SY thru BC-8SY [437] *n*-Hexadecyl ether, ethoxylated; nonionic; used in industrial applications; solid; 100% conc.

Nikkol BC-2. [437] Ceteth-2; nonionic; emulsifier for o/w type creams and lotions; paste; HLB 8.0; 100% conc.

Nikkol BC-5.5, BC-5.5(FF). [437] Ceteth-6; nonionic; emulsifier for cosmetic and pharmaceuticals; paste; HLB 10.5; 100% conc.

Nikkol BC-7. [437] Ceteth-7; nonionic; see Nikkol BC-5.5; paste; HLB 11.5; 100% conc.

Nikkol BC-10TX. [437] Ceteth-10; nonionic; see Nikkol BC-5.5; solid; HLB 13.5; 100% conc.

Nikkol BC-15TX, -15TX(FF). [437] Ceteth-15; nonionic; used in refining techniques; solid; HLB 15.5; 100% conc.

Nikkol BC-20TX, -20TX(FF) [437] Ceteth-20; nonionic; emulsifier for cosmetics, pharmaceuticals; solid; HLB 17.0; 100% conc.

Nikkol BC-23, -23(FF). [437] Ceteth-23; nonionic; see Nikkol BC-5.5; solid; HLB 18.0; 100% conc.

Nikkol BC-25TX, -30TX, -40TX. [437] Ceteth-25, -30, and -40 resp.; nonionic; see Nikkol BC-5.5; solid; HLB 18.5, 19.5, and 20.0 resp.; 100% conc.

Nikkol BD-1SY thru BD-8SY. [437] *n*-Decyl ether, ethoxylated; nonionic; used in industrial applications; liq., solid; 100% conc.

Nikkol BEG-1630. [437] Isoceteth-30; nonionic; emulsifier and solubilizer used for cosmetics and pharmaceuticals; solid; HLB 15.5; 100% conc.

Nikkol BL-1SY. [437] Laureth-1; nonionic; compd. for use in research on ether-type nonionic surfactants; m.w. 230.4; sp.gr. 0.8681 m.p. 19.5; b.p. 135 C (1 mm Hg); 98% pure.

Nikkol BL-2. [437] Laureth-2; nonionic; for emulsifying products required to possess an abundant liquidity; liq.; HLB 9.5; 100% conc.

Nikkol BL-2SY. [437] Laureth-2; nonionic; see Nikkol BL-1SY; m.w. 274.4; sp.gr. 0.8996 m.p. 17.6; b.p. 164 C (1 mm Hg); 98% pure.

Nikkol BL-3SY. [437] Laureth-3; nonionic; see Nikkol BL-1SY; m.w. 318.5; sp.gr. 0.9216 m.p. 16.0; b.p. 194–197 C (1.5 mm Hg); 98% pure.

Nikkol BL-4SY. [437] Laureth-4; nonionic; see Nikkol BL-1SY; m.w. 362.6; sp.gr. 0.9442 m.p. 18.7; b.p. 214–215 C (2 mm Hg); 98% pure.

Nikkol BL-4.2. [437] Laureth-4 (4.2 EO); nonionic; see Nikkol BL-2; liq.; HLB 11.5; 100% conc.

Nikkol BL-5SY. [437] Laureth-5; nonionic; see Nikkol BL-1SY; m.w. 406.6; sp.gr. 0.9586 m.p. 22.8 C; b.p. 225–229 C (2 mmHg); cloud pt. 30.5 C (0.1%); surf. tens. 29.6 dynes/cm; 98% pure.

Nikkol BL-6SY. [437] Laureth-6; nonionic; see Nikkol BL-1SY; m.w. 450.6; sp.gr. 0.9756 m.p. 24.8; cloud pt. 55 C (0.1%); surf. tens. 31.5 dynes/cm; 98% pure.

Nikkol BL-7SY. [437] Laureth-7; nonionic; see Nikkol BL-1SY; m.w. 494.6; sp.gr. 0.9837 m.p. 26.1; cloud pt. 70 C (0.1%); surf. tens. 33.4 dynes/cm; 98% pure.

Nikkol BL-8SY. [437] Laureth-8; nonionic; see Nikkol BL-1SY; m.w. 538.7; sp.gr. 0.9951 m.p. 31.4; cloud pt. 79 C (0.1%); surf. tens. 34.6 dynes/cm; 98% pure.

Nikkol BL-9EX, -9EX(FF). [437] Laureth-9; nonionic; see Nikkol BL-2; liq.; HLB 14.5; 100% conc.

Nikkol BL-21, -25. [437] Laureth-21 and -25 resp.; nonionic; see Nikkol BL-2; solid; HLB 19.0 and 19.5; 100% conc.

Nikkol BM-1SY thru BM-8SY. [437] *n*-Tetradecyl ether; nonionic; used in industrial applications; solid; 100% conc.

Nikkol BO-2. [437] Oleth-2; nonionic; emulsifier for preparations required to possess an abundant liquidity; liq.; HLB 7.5; 100% conc.

Nikkol BO-7, -10TX. [437] Oleth-7 and -10 resp.; nonionic; see Nikkol BO-2; liq.; HLB 10.5 and 14.5; 100% conc.

Nikkol BO-15TX. [437] Oleth-15; nonionic; see Nikkol BO-2; paste; HLB 16.0; 100% conc.

Nikkol BO-20. [437] Oleth-20; nonionic; see Nikkol BO-2; solid; HLB 17.0; 100% conc.

Nikkol BO-50. [437] Oleth-50; nonionic; see Nikkol BO-2; flake; HLB 18.0; 100% conc.

Nikkol BPS-5. [437] POE (5) phytosterol; nonionic; emulsifier for o/w and w/o compds., solubilizer, dispersant, emollient, foam stabilizer used in cosmetics, pharmaceuticals; wh. to pale yel. paste; sol. in propylene glycol, ethanol; partly sol. in water; HLB 9.5; pH 4.8 (5%).

Nikkol BPS-10. [437] POE (10) phytosterol; nonionic; see Nikkol BPS-5; wh. to pale yel. paste or solid; sol. in water, propylene glycol, ethanol; HLB 12.5; pH 5.2 (5%).

Nikkol BPS-15, -20. [437] POE (15) and (20) resp. phytosterol; see Nikkol BPS-5; wh. to pale yel. wax-like solid; sol. in water, propylene glycol, ethanol; HLB 15.0 and 15.5; pH 5.0 and 5.5 (5%).

Nikkol BPS-25, -30. [437] POE (25) and (30) resp. phytosterol; nonionic; see Nikkol BPS-5; wh. to pale yel. wax-like solid; sol. in water, propylene glycol, ethanol; HLB 18.0; pH 5.7 (5%).

Nikkol BWA-5. [437] POE (5) lanolin alcohol; nonionic; emulsifier, solubilizer, and softener for shampoos; paste; HLB 12.5; 100% conc.

Nikkol BWA-20, -40. [437] POE (20) and (40) resp. lanolin alcohol; nonionic; see Nikkol BWA-5; solid; HLB 16.0 and 17.0 resp.; 100% conc.

Nikkol CCK-40. [437] Potassium coco-hydrolyzed animal protein; anionic; detergent used in shampoos, cleansing cream, kitchen cleaning agents; liq.; 30% conc.

Nikkol CCN-40. [437] Sodium coco-hydrolyzed animal protein; anionic; see Nikkol CCK-40; liq.; 30% conc.

Nikkol Chimyl Alcohol 100. [437] Palmityl glyceryl ether; nonionic; emulsifier, hydrotrope used for cosmetic and pharmaceutical products; powd.; 100% conc.

Nikkol CMT-30. [437] Sodium methyl cocoyl taurate; anionic; shampoo base, foamer and detergent; liq.; 30% conc.

Nikkol CO-3. [437] PEG-3 castor oil; nonionic hydrotrope, emulsifier used in cosmetic and pharmaceutical preparations; liq.; HLB 3.0 100% conc.

Nikkol CO-10, -20TX, -40TX. [437] PEG-10, -20, and -40 castor oil; nonionic; see Nikkol CO-3; liq.; HLB 6.5, 10.5, and 12.5; 100% conc.

Nikkol CO-60TX. [437] PEG-60 castor oil; nonionic; see Nikkol CO-3; paste; HLB 14.0; 100% conc.

Nikkol DDP-2. [437] Di-POE (2) alkyl ether phosphate; anionic; emulsifier, stabilizer, dispersant, anticorrosive agent and detergent used in cosmetics, drugs, agric. chemicals and general industrial use; liq.; HLB 6.5; 100% conc.

Nikkol DDP-4. -6. [437] Di-POE (4) and (6) alkyl ether phosphate; anionic; see Nikkol DDP-2; liq.; HLB 9.0; 100% conc.

Nikkol DDP-8, -10. [437] Di-POE (8) and (10) alkyl ether phosphate; anionic; see Nikkol DDP-2; semisolid; HLB 11.5 and 13.5; 100% conc.

Nikkol Decaglyn 1-IS. [437] Decaglycerin monoisostearate; nonionic; emulsifier, solubilizer, dispersant for cosmetics, pharmaceuticals and foods; liq.; HLB 12.5; 100% conc.

Nikkol Decaglyn 1-L. [437] Decaglycerin monolaurate; nonionic; see Nikkol Decaglyn 1-IS; liq.; HLB 17.0; 100% conc.

Nikkol Decaglyn 1-LN. [437] Decaglycerin monolinoleate; nonionic; see Nikkol Decaglyn 1-IS; liq.; HLB 12.0; 100% conc.

Nikkol Decaglyn 1-M. [437] Decaglycerin monomyristate; nonionic; see Nikkol Decaglyn 1-IS; paste; HLB 14.5; 100% conc.

Nikkol Decaglyn 1-O. [437] Decaglycerin monooleate; nonionic; see Nikkol Decaglyn 1-IS; liq.; HLB 13.5; 100% conc.

Nikkol Decaglyn 1-S. [437] Decaglycerin monostearate; nonionic; see Nikkol Decaglyn 1-IS; solid; HLB 12.5; 100% conc.

Nikkol Decaglyn 2-O. [437] Decaglycerin dioleate; nonionic; o/w emulsifier for cosmetics, pharmaceuticals and foods; liq.; HLB 10.0; 100% conc.

Nikkol Decaglyn 2-S. [437] Decaglycerin distearate; nonionic; see Nikkol Decaglyn 2-O; flake; HLB 12.0; 100% conc.

Nikkol Decaglyn 3-O. [437] Decaglycerin trioleate; nonionic; o/w and w/o emulsifier; gelling agent for hydrocarbon; liq.; HLB 6.5; 100% conc.

Nikkol Decaglyn 3-S. [437] Decaglycerin tristearate; nonionic; see Nikkol Decaglyn 3-O; flake; HLB 6.5; 100% conc.

Nikkol Decaglyn 5-IS. [437] Decaglycerin pentaisostearate; nonionic; emulsifier, lubricant, coating agent and anticrystallization agent; liq.; HLB 3.5; 100% conc.

Nikkol Decaglyn 5-O. [437] Decaglycerin pentaoleate; nonionic; see Nikkol Decaglyn 5-IS; liq.; HLB 4.0; 100% conc.

Nikkol Decaglyn 5-S. [437] Decaglycerin pentastearate; nonionic; see Nikkol Decaglyn 5-IS; solid; HLB 3.5; 100% conc.

Nikkol Decaglyn 7-IS. [437] Decaglycerin heptaisostearate; nonionic; see Nikkol Decaglyn 5-IS; liq.; 100% conc.

Nikkol Decaglyn 7-O. [437] Decaglycerin heptaoleate; nonionic; see Nikkol Decaglyn 5-IS; liq.; 100% conc.

Nikkol Decaglyn 7-S. [437] Decaglycerin heptastearate; nonionic; see Nikkol Decaglyn 5-IS; solid; 100% conc.

Nikkol Decaglyn 10-IS. [437] Decaglycerin decaisostearate; nonionic; see Nikkol Decaglyn 5-IS; liq.; 100% conc.

Nikkol Decaglyn 10-O. [437] Decaglycerin decaoleate; nonionic; see Nikkol Decaglyn 5-IS; liq.; 100% conc.

Nikkol Decaglyn 10-S. [437] Decaglycerin decastearate; nonionic; see Nikkol Decaglyn 5-IS; solid; 100% conc.

Nikkol DGDO. [437] Diglyceryl dioleate; nonionic; w/o emulsifier for cosmetics and pharmaceuticals; liq.; HLB 7.0; 100% conc.

Nikkol DGMO-C. [437] Polyglyceryl-2 oleate; nonionic; see Nikkol DGDO; liq.; HLB 5.5; 100% conc.

Nikkol DGMS. [437] Diglyceryl monostearate; nonionic; see Nikkol DGDO; solid; HLB 5.0; 100% conc.

Nikkol DGO-80. [437] Glyceryl dioleate; nonionic; emollient, emulsifier; liq.; 100% conc.

Nikkol DGS-80. [437] Glyceryl distearate; nonionic; see Nikkol DGO-80; powd.; 100% conc.

Nikkol DLP-10. [437] Dilaureth-10 phosphate; anionic; emulsifier, dispersant, hydrotrope; surfactant, solubilizer; paste; 100% conc.

Nikkol DOP-8. [437] Complex org. phosphate, partial sodium salt; anionic; cosmetic and pharmaceutical applications; paste; 100% conc.

Nikkol ECT-3NEX, ECTD-3NEX. [437] Sodium POE (3) alkyl ether carboxylic acid; anionic; shampoo base, foamer, detergent, and emulsifier; liq.; 85% conc.

Nikkol ECTD-6NEX. [437] Sodium POE (6) alkyl ether carboxylic acid; anionic; shampoo base, foamer, detergent and emulsifier; liq.; 85% conc.

Nikkol GBW-25, -125. [437] POE (6) and (20) sorbitol beeswax deriv.; nonionic; emulsifier for cosmetic and pharmaceuticals; solid; HLB 7.5 and 9.5; 100% conc.

Nikkol GM-18IS. [437] Batyl isostearate; nonionic; emulsion stabilizer used in cosmetics; paste; 100% conc.

Nikkol GM-18S. [437] Batyl stearate; nonionic; see Nikkol GM-18IS; solid; 100% conc.

Nikkol GO-430. [437] PEG-30 sorbitan tetraoleate; nonionic; emulsifier, solubilizer, superfatting agent used in drugs and cosmetics, for emulsion polymerization, agric. chemicals, printing inks; pale yel. liq.; sol. in ethanol, ethyl acetate, xylene; partly sol. in water; sp.gr. 1.048; HLB 11.5; 100% conc.

Nikkol GO-440. [437] PEG-40 sorbitan tetraoleate; nonionic; see Nikkol GO-430; pale yel. liq.; sol. in ethanol, ethyl acetate, xylene; partly sol. in water, propylene glycol; sp.gr. 1.054; HLB 12.5; 100% conc.

Nikkol GO-460. [437] PEG-60 sorbitan tetraoleate; nonionic; see Nikkol GO-430; pale yel. liq.; sol. in water, ethanol, ethyl acetate, xylene; partly sol. in propylene glycol; sp.gr. 1.060; HLB 14.0; 100% conc.

Nikkol GS-460. [437] PEG-60 sorbitan tetrastearate; nonionic; see Nikkol GO-430; pale yel. liq.; sol. in water, ethanol, ethyl acetate, xylene, olive oil; m.p. 35 C; HLB 13.0; 100% conc.

Nikkol HCO-5, -7.5, -10, -20. [437] PEG-5, -7.5, -10, and -20 hydrog. castor oil; nonionic; hydrotrope, emulsifier used in cosmetics and pharmaceuticals; liq.; HLB 6.0, 6.0, 6.5, and 10.5; 100% conc.

Nikkol HCO-30, -30(FF). [437] PEG-30 hydrog. castor oil; nonionic; see Nikkol HCO-5; liq.; HLB 11.0; 100% conc.

Nikkol HCO-40, -40(FF). [437] PEG-40 hydrog. castor oil; nonionic; see Nikkol HCO-5; liq.; HLB 12.5; 100% conc.

Nikkol HCO-50, -50(FF). [437] PEG-50 hydrog. castor oil; nonionic; see Nikkol HCO-5; paste; HLB 13.5; 100% conc.

Nikkol HCO-60, -60(FF). [437] PEG-60 hydrog. castor oil; nonionic; see Nikkol HCO-5; semi-solid; HLB 14.5; 100% conc.

Nikkol HCO-80. [437] PEG-80 hydrog. castor oil; nonionic; see Nikkol HCO-5; solid; HLB 15.0; 100% conc.

Nikkol HCO-100, -100(FF). [437] PEG-100 hydrog. castor oil; nonionic; see Nikkol HCO-5; solid; HLB 16.5; 100% conc.

Nikkol Hexaglyn 1-L. [437] Hexaglycerin monolaurate; nonionic; o/w emulsifier, anticrystallizing agent; liq; HLB 13.0; 100% conc.

Nikkol Hexaglyn 1-O. [437] Hexaglycerin monooleate; nonionic; see Nikkol Hexaglyn 1-L; liq.; HLB 9.5; 100% conc.

Nikkol Hexaglyn 1-S. [437] Hexaglyceryl monostearate; nonionic; emulsifier, dispersant, lubricant; solid; HLB 9.5; 100% conc.

Nikkol Hexaglyn 3-S. [437] Hexaglyceryl tristearate; nonionic; see Nikkol Hexaglyn 1-L; flake; HLB 2.5; 100% conc.

Nikkol Hexaglyn 5-O. [437] Hexaglyceryl pentaoleate; nonionic; see Nikkol Hexaglyn 1-L; liq.; HLB 6.0; 100% conc.

Nikkol Hexaglyn 5-S. [437] Hexaglyceryl pentastearate; nonionic; see Nikkol Hexaglyn 1-L; flake; 100% conc.

Nikkol Hexaglyn PR-15. [437] Hexaglyceryl polyricinoleate; nonionic; see Nikkol Hexaglyn 1-L; flake; 100% conc.

Nikkol LMT. [437] Sodium N-lauroyl methyl taurate; anionic; shampoo base, foamer and detergent; solid; 92% conc.

Nikkol LSA. [437] Sodium lauryl sulfoacetate; anionic; acid detergent base, foamer; solid; 92.5% conc.

Nikkol MGO. [437] Glyceryl monooleate; nonionic; food emulsifer; liq.; HLB 3.0; 100% conc.

Nikkol MGS-A, -B, -DEX. [437] Glyceryl monostearate; nonionic; emulsifier, stabilizer used in foods and cosmetics; flake; HLB 4.5, 5.0, and 6.5 resp.; 100% conc.

Nikkol MGS-ASE, -BSE. [437] Glyceryl monostearate SE; nonionic; see Nikkol MGS-A; flake; HLB 6.5 and 5.5 resp.; 100% conc.

Nikkol MGS-F20, -F40, -F50, –F75. [437] Glyceryl monostearate; nonionic; food emulsifier; solid; HLB 7.5, 4.0, 3.5 and 0.5 resp.; 100% conc.

Nikkol MMT. [437] Sodium N-myristoyl methyl taurate; anionic; shampoo base, foaming agent and detergent; solid; 92% conc.

Nikkol MYO-2. [437] Diethylene glycol oleate; nonionic; w/o or o/w emulsifier for creams, lotions and ointments; liq.; HLB 4.5; 100% conc.

Nikkol MYO-6. [437] PEG-6 oleate; nonionic;

see Nikkol MYO-2; liq.; HLB 8.5; 100% conc.

Nikkol MYS-1EX. [437] PEG-1 stearate; nonionic; emulsifier or solubilizer for cosmetics and pharmaceuticals; flake; HLB 2.0; 100% conc.

Nikkol MYS-2. [437] PEG-2 stearate; nonionic; see Nikkol MYS-1EX; flake; HLB 4.0; 100% conc.

Nikkol MYS-4. [437] PEG-4 stearate; nonionic; see Nikkol MYS-1EX; solid; HLB 6.5; 100% conc.

Nikkol MYS-10. [437] PEG-10 stearate; nonionic; see Nikkol MYS-1EX; solid; HLB 11.0; 100% conc.

Nikkol MYS-25, MYS-25(FF). [437] PEG-25 stearate; nonionic; see Nikkol MYS-1EX; solid; HLB 15.0; 100% conc.

Nikkol MYS-40, MYS-40(FF). [437] PEG-40 stearate; nonionic; see Nikkol MYS-1EX; solid; HLB 17.5; 100% conc.

Nikkol MYS-45, -55. [437] PEG-45 and -55 stearate; nonionic; see Nikkol MYS-1EX; flake; HLB 18.0; 100% conc.

Nikkol NES-203. [437] POE alkyl ether sulfate, sodium salt; anionic; shampoo base, foamer, emulsifier and detergent; liq.; 20% conc.

Nikkol NES-303. [437] POE alkyl ether sulfate, triethanolamine salt; anionic; see Nikkol NES-203; liq.; 30% conc.

Nikkol NP-7.5. [437] Nonoxynol-8 (7.5 EO); nonionic; emulsifier, detergent, solubilizer, and wetting agent used in cosmetics, insecticide, and other industrial applications; liq.; HLB 14.0; 100% conc.

Nikkol NP-10, -15. [437] Nonoxynol-10 and -15; nonionic; see Nikkol NP-7.5; liq.; HLB 16.5 and 18.0 resp.; 100% conc.

Nikkol NP-18TX. [437] Nonoxynol-18; nonionic; see Nikkol NP-7.5; paste; HLB 19.0; 100% conc.

Nikkol N-SP. [437] Cetyl palmitate; syn. spermaceti; solid; 100% conc.

Nikkol OP-3. [437] Octoxynol-3; nonionic; emulsifier, detergent, solubilizer, and wetting agent used in cosmetics, insecticide and other industrial applications; liq.; HLB 6.0; 100% conc.

Nikkol OP-10, -30. [437] Octoxynol-10 and -30 resp.; nonionic; see Nikkol OP-3; liq. and paste resp.; HLB 11.5 and 17.0; 100% conc.

Nikkol OTP-100S. [437] Sodium di-2-ethylhexyl sulfosuccinate; anionic; dispersant, wetting agent; sponge-like solid; 98% conc.

Nikkol PBC-31. [437] PPG-4 ceteth-1; nonionic; emulsifier, solubilizer, dispersant used in cosmetic, pharmaceuticals and other industrial applications; liq.; HLB 9.4; 100% conc.

Nikkol PBC-33. [437] PPG-4 ceteth-10; nonionic; see Nikkol PBC-31; paste; HLB 10.5; 100% conc.

Nikkol PBC-34. [437] POE-POP cetyl ether; nonionic; see Nikkol PBC-31; solid; HLB 16.5; 100% conc.

Nikkol PBC-41. [437] PPG-8 ceteth-1; nonionic; see Nikkol PBC-31; liq.; HLB 9.5; 100% conc.

Nikkol PBC-44. [437] PPG-8 ceteth-20; nonionic; see Nikkol PBC-31; HLB 12.5; 100% conc.

Nikkol PBC-44(FF). [437] POE-POP cetyl ether; nonionic; see Nikkol PBC-31; solid; 100% conc.

Nikkol PEN-4612. [437] PPG-6-decyltetradeceth-12; nonionic; solubilizer for cosmetic lotions and toiletry products; solid; HLB 8.5; 100% conc.

Nikkol PEN-4620. [437] PPG-6-decyltetradeceth-20; nonionic; see Nikkol PEN-4612; solid; HLB 11.0; 100% conc.

Nikkol PEN-4630. [437] PPG-6 decyltetradeceth-30; nonionic; see Nikkol PEN-4612; solid; HLB 12.0; 100% conc.

Nikkol PMS-1C. [437] Propylene glycol monostearate; nonionic; emulsifier, dispersant for cosmetics and pharmaceuticals; solid; HLB 3.5; 100% conc.

Nikkol PMS-1CSE. [437] Propylene glycol monostearate SE; nonionic; see Nikkol PMS-1C; solid; HLB 4.0; 100% conc.

Nikkol PMT. [437] Sodium N-palmitoyl methyl taurate; anionic; shampoo base, foamer and detergent; solid; 92% conc.

Nikkol Sarcosinate LN. [437] Lauroyl sarcosinate; anionic; detergent for shampoo and dentifrice; powd.; 100% conc.

Nikkol Sarcosinate MN. [437] Myristoyl sarcosinate; anionic; see Nikkol Sarcosinate LN; powd.; 100% conc.

Nikkol SCS. [437] Sodium cetyl sulfate; anionic; detergent, powd.; 100% conc.

Nikkol SGC-80N. [437] Sodium cocomonoglyceride sulfate; anionic; foamer and detergent for personal care products; powd.; 70% conc.

Nikkol SI-10R, -10T, -15R, -15T. [437] Sorbitan monoisostearate; nonionic; w/o emulsifier for cosmetics and pharmaceuticals; liq.;

HLB 5.0, 5.0, 4.5, and 4.5 resp.; 100% conc.

Nikkol SL-10. [437] Sorbitan monolaurate; nonionic; food emulsifier; liq.; HLB 8.6; 100% conc.

Nikkol SLS. [437] Sodium lauryl sulfate; anionic; detergent; powd.; 97% conc.

Nikkol SMT. [437] Sodium N-stearoyl methyl taurate; anionic; shampoo base, foamer and detergent; solid; 92% conc.

Nikkol SO-10. [437] Sorbitan oleate; nonionic; emulsifier; liq.; HLB 5.0; 100% conc.

Nikkol SO-15. [437] Sorbitan sesquioleate; nonionic; emulsifier; liq.; HLB 4.5; 100% conc.

Nikkol SO-30. [437] Sorbitan trioleate; nonionic; emulsifier; liq.; HLB 4.0; 100% conc.

Nikkol SP-10. [437] Sorbitan monopalmitate; nonionic; food emulsifier; solid; HLB 6.7; 100% conc.

Nikkol SS-10. [437] Sorbitan stearate; nonionic; emulsifier; flake; HLB 4.7; 100% conc.

Nikkol SS-15. [438] Sorbitan sesquistearate; nonionic; food emulsifier; solid; HLB 4.2; 100% conc.

Nikkol SS-30. [437] Sorbitan tristearate; nonionic; emulsifier; flake; HLB 2.1; 100% conc.

Nikkol TDP-2. [437] C_{12-15} pareth-2 phosphate; anionic; emulsifier and solubilizer for cosmetics, drugs, agric. chemicals, dispersant, anticorrosive agent and detergent for general industrial use; liq.; HLB 7.0; 100% conc.

Nikkol TDP-4, -6, -8. [437] Tri-POE (4), (6), and (8) resp. alkyl ether phosphate; anionic; see Nikkol TDP-2; liq.; HLB 7.0, 8.0, and 11.5; 100% conc.

Nikkol TDP-10. [437] Tri-POE (10) alkyl ether phosphate; anionic; see Nikkol TDP-2; semisolid; HLB 14.0; 100% conc.

Nikkol Tetraglyn 1-O. [437] Tetraglyceryl monooleate; nonionic; o/w emulsifier, anticrystallizing agent; liq.; 100% conc.

Nikkol Tetraglyn 1-S. [437] Tetraglyceryl monostearate; nonionic; see Nikkol Tetraglyn 1-O; flake; 100% conc.

Nikkol Tetraglyn 3-S. [437] Tetraglyceryl tristearate; nonionic; see Nikkol Tetraglyn 1-O; flake; 100% conc.

Nikkol Tetraglyn 5-O. [437] Tetraglyceryl pentaoleate; nonionic; see Nikkol Tetraglyn 1-O; liq.; 100% conc.

Nikkol Tetraglyn 5-S. [437] Tetraglyceryl pentastearate; nonionic; see Nikkol Tetraglyn 1-O; flake; 100% conc.

Nikkol TI-10. [437] POE (20) sorbitan monoisostearate; nonionic; emulsifier, solubilizer for cosmetics; liq.; HLB 15.0; 100% conc.

Nikkol TL-10, TL-10(FF). [437] Polysorbate 20; nonionic; emulsifier, hydrophilic; used in cosmetics and pharmaceuticals; liq.; HLB 16.9; 100% conc.

Nikkol TO-10, TO-10(FF). [437] Polysorbate 80; nonionic; see Nikkol TL-10; liq.; HLB 15.0; 100% conc.

Nikkol TO-30. [437] Polysorbate 85; nonionic; see Nikkol TL-10; liq.; HLB 11.0; 100% conc.

Nikkol TO-106. [437] PEG-6 sorbitan oleate; nonionic; see Nikkol TL-10; liq.; HLB 10.0; 100% conc.

Nikkol TS-10, TS-10(FF). [437] Polysorbate 60; emulsifier, solubilizer for o/w products, cosmetics, pharmaceuticals; pale yel. to yel., visc. liq.; sol. in water, ethanol, ethyl acetate, toluene; HLB 14.9; sapon. no. 43–49; pH 5.7–7.7 (5%); 100% conc.

Nikkol TS-30. [437] Polysorbate 65; nonionic; see Nikkol TS-10; liq.; HLB 11.0; 100% conc.

Nikkol TW-10, -20, -30. [437] PEG-10, -20, and -30 lanolin; nonionic; solubilizer or bodying agent for soap, cleanser, shampoos, and rinses; paste; HLB 12.0, 13.0, and 15.0; 100% conc.

Nilo VON. [538] EO sulfonate; nonionic; emulsifier for min. oils; liq; 69% conc.

Nilofoam 60, M, XC. [62] Silicone emulsion; anionic, nonionic, nonionic resp.; defoamer used in pressure machines; beam batching operation for knit goods; wh. visc. emulsion; sol. in water; pH 6.0-6.5 (1%)

Nimco 1780. [217] Lanolin alcohols; nonionic; emulsifier and stabilizer for o/w systems; solid; HLB 8.9; 100% conc.

Nimco 1781 [217] Lanolin acid; nonionic; used in personal care products; solid; 100% conc.

Nimcolan 1747. [217] Petrolatum, lanolin and lanolin alcohol; nonionic; emollient w/o emulsifier and aux. o/w emulsifier; soft solid; 100% conc.

Ninate 401, 401-HF. [575; 576] Calcium alkylbenzene sulfonate; anionic; emulsifier used in pesticide formulations and in self-dispersing liq., 401-HF recommended where higher flash is required; dk. amber liq.; sol. in kerosene, xylene and aromatic naphtha, insol. in water and min. oil; sp.gr. 1.00 and 1.02 resp.; visc. 2500 cps; flash pt. 108 and 123 F resp. (Seta CC); pH 4.5–5.5 (5% in 50% aq. IPA);

65 and 60% act. resp.

Ninate 411. [575; 576] Amine dodecylbenzene sulfonate; anionic; detergent, emulsifier used in emulsifiable kerosene formulations; yel. visc. liq.; m.w. 385; sol. in kerosene, Stod., chlorinated and aromatic solvs., alcohol, pine oil, insol. in min. oil, forms clear sol'n. in water at higher concs.; dens. 8.5 lb/gal; surf. tens. 30 dynes/cm (0.1% aq.); pH 3.6 (20% in 1:1 IPA/water); 95% act.

Ninate ABS-60C. [575] Calcium dodecylbenzene sulfonate; anionic; intermediate for emulsifiers; liq.; 60% conc.

Ninate LAS-931. [575] Dodecylbenzene sulfonic acid, isopropylamine salt; anionic; surfactant for use in dye carrier formulations, agric. and w/o emulsion systems; liq.; oil-sol.; 93% conc.

Ninex 24. [575] Amide sulfonate; detergent, wetting agent for syn. cleaners; lt. amber liq.; sol. in alkali salt sol'ns; lt. amber liq.; 90% conc.

Ninex MT 610, 615. [575] Monotallates; nonionic; detergent and textile scouring agent; liq.; 100% conc.

Ninex TDO 5, 9, 14. [575] Tall oil dioleates; coemulsifier of min. oils, fatty acids; lubricant, sanforizer, fiber processing aid and disperse dye carrier in textile industry; liq.; 100% conc.

Ninex TMO 5, 7, 9, 14. [575] Tall oil monooleate; nonionic; textile emulsifier, lubricant, fiber processing aid; liq.; HLB 8.3, 10.3, 11.5 and 13.4 resp.; 100% conc.

Ninol 25-CM. [575] 1:1 Cocamide DEA; surfactant for personal care products, bubble baths; Gardner 6 liq.

Ninol 41-CO. [575] 1:1 Cocamide DEA; surfactant, visc. modifier for bubble baths, personal care products; Gardner 4 liq.

Ninol 51-LL. [575] Lauramide DEA (1:1); nonionic; surfactant, visc. builder for personal care products; Gardner 6 liq.

Ninol 52-LL. [575] Lauramide DEA (1:1); nonionic; surfactant, foam enhancer, visc. builder for personal care products; Gardner 4 liq.

Ninol 55-LL. [575] Lauramide DEA (1:1); nonionic; see Ninol 52-LL; Gardner 3 liq.

Ninol 71-LL. [575] Lauric/myristic DEA (1:1); nonionic; see Ninol 152-LL; Gardner 2 liq.

Ninol 128 Extra. [575] Cocamide DEA; nonionic; foam booster and stabilizer, detergent, thickener used in shampoos, bubble baths, dishwash; yel. liq.; dens. 8.3 lb/gal; 100% conc.

Ninol 201. [575] Oleamide DEA; nonionic; emulsifier in industrial lubricant systems; liq.; sol. in oils, disp. in water; dens. 8.23 lb/gal; pH 9.5–10.5.

Ninol 1281, 1285. [575] Fatty acid alkylolamide; nonionic; detergent used in floor cleaners and all purpose detergent formulations; amber liq. and straw visc. liq. resp.; dens. 8.5 and 8.35 lb/gal; pH 9.0 (1%); 100% act.

Ninol 1301. [575] Fatty acid alkylolamide derived from coconut oil; nonionic; detergent for syn. floor cleaners, wax strippers, alkali cleaners, sanitizers; lt. tan soft wax; sol. in hard water; congeal pt. 110–120 F; pH 9.5–10.5; 100% act.

Ninol 2012 Extra. [575] Cocamide DEA; nonionic; foam booster and stabilizer, detergent, thickener used in shampoos, bubble baths, dishwash; yel. liq.; dens. 8.2 lb/gal; 100% conc.

Ninol 4821. [575] Lauramide DEA; nonionic; foam booster/stabilizer, visc. builder for personal care products; clear amber liq.; dens. 8.2 lb/gal; 90% act.

Ninol AA-62, AA-62 Extra. [575] Lauramide DEA; nonionic; foam booster and stabilizer, detergent, thickener used in shampoos, bubble baths, dishwash; paste and wh. wax resp.; dens. 8.34 and 8.2 lb/gal; 100% act.

Ninol AC-201. [576] Oleic fatty acid polydiethanolamide; nonionic; anti-corrosion additive for cutting oils; liq.; 100% conc.

Ninol CNR. [576] Coconut monoethanolamide; nonionic; thickener, superfatting agent, foam booster; solid; 100% conc.

Ninol HDA-7. [575] Specialty alkylolamide; nonionic; detergent used in floor cleaners and all purpose detergent formulations; amber liq.; dens. 8.34 lb/gal; pH 9.5 (1%); 93% act.

Ninol LD. [576] Coconut diethanolamide (2:1); nonionic; thickener, detergent builder, foam booster; liq.; 100% conc.

Ninol P-616. [575] Lauramide DEA; see Ninol AA-62 Extra.

Ninol P-621. [575] Lauramide DEA; nonionic; foam booster and stabilizer, detergent, thickener used in shampoos, bubble baths, dishwash; wh. paste; dens. 8.2 lb/gal; 100% conc.

Ninol P-650. [575] Cocamide DEA; see Ninol 2012 Extra.

Ninol SNR. [576] Stearic monoethanolamide; nonionic; thickener, superfatting agent, pearlescent; solid; 100% conc.

Ninox CA. [575] Cocamidopropylamine oxide; nonionic/cationic; surfactant, foamer, conditioner for light-duty detergents, shampoos, etc.; clear straw liq.; 30% act.

Ninox FCA. [576] Cocamidopropyl dimethylamine oxide; nonionic; wetting, foaming, foam stabilizer in personal care products and detergents; liq.; 30% conc.

Ninox L. [575] Lauramine oxide; see Ninox CA; clear liq.; 30% act.

Ninox M. [575] Myristamine oxide; see Ninox CA; clear liq.; 30% act.

Niox EO-12. [494] Linear fatty alcohol, ethoxylated sat.; nonionic; surface act. agent for detergents; dispersant; soap additive; flakes; HLB 16.0; 100% conc.

Niox EO-13. [494] Linear fatty alcohol, ethoxylated sat.; nonionic; powd. detergent mixtures; flakes; HLB 18.6; 100% conc.

Niox EO-14, 23. [494] Linear fatty alcohol ethoxylated sat.; nonionic; foam controller in detergents; dispersant; flakes; HLB 17.8 and 17.4 resp.; 100% conc.

Niox EO-32, -35. [494] Linear fatty alcohol ethoxylated sat.; nonionic; see Niox EO-12; flakes; HLB 16.2 and 15.5 resp.; 100% conc.

Niox KF-12. [494] Linear fatty alcohol ethoxylated sat.; nonionic; detergent, wetting agent, dispersant; emulsifier for waxes, oils, and fats; stabilizer for syn. resins; solid; HLB 14.5; 100% conc.

Niox KF-13. [494] Linear fatty alcohol ethoxylate; nonionic; emulsifier for min. oils, aliphatic solvs.; visc. depressant; intermediate for sulfonation; liq.; HLB 7.6; 100% conc.

Niox KF-17. [494] Linear fatty alcohol ethoxylate; nonionic; emulsifier for min. oils; additive for detergents; intermediate for sulfonation; liq.; oil-sol.; HLB 6.1; 100% conc.

Niox KF-18. [494] Linear fatty alcohol ethoxylate; nonionic; wetting agent and detergent for liq. formulations; wax emulsifier; paste; oil-sol.; HLB 13.3; 100% conc.

Niox KF-22. [494] Linear fatty alcohol ethoxylate; nonionic; emulsifier for silicones and solvs.; assistant in dry cleaning formulations; liq.; HLB 9.4; 100% conc.

Niox KF-25. [494] Linear fatty alcohol ethoxylate; nonionic; detergent and emulsifier for oils, waxes and solvs.; wool scouring and wetting agent; liq., paste; HLB 11.4; 100% conc.

Niox KF-26. [494] Linear fatty alcohol ethoxylate; nonionic; see Niox KF-17; liq.; HLB 6.8; 100% conc.

Niox KG-11. [494] Linear fatty alcohol ethoxylate; nonionic; basic product for sulfates; visc. depressant for plastisols; emulsifier for cosmetics; liq.; HLB 5.0; 100% conc.

Niox KG-14. [494] Linear fatty alcohol ethoxylate; nonionic; wetting agent for detergents, textile liq. formulations

Niox KH Series. [494] Fatty acid, ethoxylated; nonionic; raw material used as emulsifier and coemulsifier for paraffinic waxes and compds., dyeing assistant, lubricant, antistat; emulsifier for the textile industry, min. oils, defoamer, softener, detergent and solubilizer for all industries; 100% conc.

Niox KI Series. [494] Castor oil, ethoxylated; nonionic; surface act. agent used as detergent, emulsifier, dispersant, and hydrosolubilizer for conc. pesticides in the metal, leather, cosmetics, toiletries, pharmaceutical and polymer industries; textile assistant; 100% conc.

Niox KJ-55. [494] Linear fatty alcohol, ethoxylated sat.; nonionic; surface act. agent for detergents; soap additive; dispersant; solid; HLB 13.2; 100% conc.

Niox KJ-56. [494] Linear fatty alcohol ethoxylated sat.; nonionic; emulsifier intermediate raw material for mfg. of ethoxysulfates used in detergent and industrial specialties; solid; HLB 8.1; 100% conc.

Niox KJ-61. [494] Linear fatty alcohol ethoxylated sat.; nonionic; see Niox KJ-55; solid; HLB 14.1; 100% conc.

Niox KJ-66. [494] Linear fatty alcohol ethoxylated sat.; nonionic; see Niox KJ-56; solid; HLB 10.2; 100% conc.

Niox KJ-72. [494] Linear fatty alcohol ethoxylated sat.; nonionic; lipophilic emulsifier used in topical cosmetic applications; solid; oil-sol. and disp.; HLB 4.9; 100% conc.

Niox KL-16. [494] Linear fatty alcohol ethoxylated sat.; nonionic; scouring agent for the textile industry; soaking agent for the leather industry; household and industrial detergent; paste; HLB 12.8; 100% conc.

Niox KL-19. [494] Linear fatty alcohol ethoxylated sat.; nonionic; see Niox KF-17; liq.; HLB 8.0; 100% conc.

Niox KP-62, -63, -67. [494] Isotridecanol, ethoxylated; nonionic; emulsifier for cosmetic and pharmaceutical preparations; wetting agent for industrial applications; liq.; HLB 11.7, 10.4, and 8.0 resp.; 100% conc.

Niox KP-68. [494] Isotridecanol, ethoxylated; nonionic; detergent, foamer, solubilizer; scouring and leveling agent for the textile industry; paste; HLB 14.1; 100% conc.

Niox KP-69. [494] Isotridecanol, ethoxylated; nonionic; detergent, foamer, solubilizer used in foaming detergents; emulsifier and wetting agent for textile fiber finishing; liq.; HLB 12.7; 100% conc.

Niox KQ-20. [494] Fatty oxo-alcohol, ethoxylated; nonionic; aux. for wool scouring; emulsifier for hydrocarbon solv.; wetting and dispersant for detergents; paste; HLB 13.0; 100% conc.

Niox KQ-30. [494] Fatty oxo-alcohol, ethoxylated; nonionic; see Niox KF-17; liq.; HLB 7.8; 100% conc.

Niox KQ-32. [494] Fatty oxo-alcohol, ethoxylated; nonionic; see Niox KQ-20; paste; HLB 12.0; 100% conc.

Niox KQ-33. [494] Fatty oxo-alcohol, ethoxylated; nonionic; emulsifier for silicones and hydrocarbon solvs.; dry cleaning assistant; intermediate raw material for sulfonation; liq.; HLB 9.2; 100% conc.

Niox KQ-34. [494] Oxo-alcohol, ethoxylated; nonionic; wetting agent; emulsifier for fats and oleins; aux. product for emulsion polymerization; solid; HLB 15.4; 100% conc.

Niox KQ-36. [494] Fatty oxo-alcohol, ethoxylated; nonionic; see Niox KQ-20; paste; HLB 11.0; 100% conc.

Niox KQ-51. [494] Fatty oxo-alcohol, ethoxylated; nonionic; w/o emulsifier; coemulsifier of min. oils, waxes and paraffinic hydrocarbons; liq.; HLB 11.6; 100% conc.

Niox KQ-55, -56. [494] Fatty oxo-alcohol, ethoxylated; nonionic; o/w emulsifier for hydrocarbon solvs. and min. oils; detergent; wetting agent and antistat for textile industry; assistant for wool scouring and specialty cleaners; liq., paste resp.; HLB 12.8 and 13.7 resp.; 100% conc.

Niox KQ-70, LQ-13. [494] Fatty oxo-alcohol, ethoxylated; nonionic; see Niox KQ-20; paste, liq. resp.; 100 and 88% conc. resp.

Nipol 2782, 4472, 5595, 5690. [575] EO-PO block polymer; nonionic; emulsifier for pesticides; off-wh. to tan solid; m.w. 2700, 4400, 5500, 5600 resp.; sol. in water, aromatic hydrocarbons, other org. solvs.; sp.gr. 1.04, 1.05, 1.07, 1.08 resp.; m.p. 40–50 C; flash pt. > 200 F; cloud pt. 82, 72, 95, 90 C (1% aq.).

Nissan Amine AB. [441] Primary octadecylamine; intermediate for cationic surfactants; anticorrosive agent, germicide, wetting agent, mold release agent, softener, emulsifier, dispersant, intermediate used in textiles, water treatment, concrete, asphalt, agriculture, ceramics; wh. waxy solid or flake; 98% min. primary amine.

Nissan Amine ABT. [441] Primary hydrog. tallow amine; see Nissan Amine AB; wh. waxy solid or flake; 98% min. primary amine.

Nissan Amine ABT$_2$. [441] Primary tallowalkylamine; see Nissan Amine AB; wh. waxy solid; 98% min. primary amine.

Nissan Amine BB. [441] Primary dodecylamine; see Nissan Amine AB; 120 max APHA, liq. in summer, solid in winter; 98% min. primary amine.

Nissan Amine DT. [441] Tallow-alkyl propylenediamine; asphalt emulsifier, corrosion inhibitor, wetting agent, dispersant used in water treatment, pigment flushing and dispersing, ore flotation, gasoline, grease, fuel oil additive; yel. br. solid.

Nissan Amine DTH. [441] Tallow-hydrog.-alkyl propylenediamine; see Nissan Amine DT; ylsh. br. flake.

Nissan Amine FB. [441] Primary cocoalkylamine; see Nissan Amine AB; 120 max APHA, liq. in summer, solid in winter; 98% min. primary amine.

Nissan Amine MB. [441] Primary tetradecylamine; see Nissan Amine AB; wh. waxy solid; 98% min. primary amine.

Nissan Amine OB. [441] Primary oleylamine; see Nissan Amine AB; 140 max APHA, liq. in summer, solid in winter; 98% min. primary amine.

Nissan Amine PB. [441] Primary hexadecylamine; see Nissan Amine AB; wh. waxy solid; 98% min. primary amine.

Nissan Amine SB. [441] Primary soybean alkylamine; see Nissan Amine OB; liq., solid; 100% conc.

Nissan Amine VB. [441] Behenylamine; see Nissan Amine OB; solid; 100% conc.

Nissan Anon BF. [441] Dimethyl cocoalkyl betaine; amphoteric; plastic antistatic agent, softener, germicide in foods and cleaning industries, asphalt antistripping agent, textile treatment; lt. yel. liq.; 25% min act.

Nissan Anon BL. [441] Dimethyl dodecyl betaine; amphoteric; antistatic agent, softener, germicide in foods and cleaning industries used in textile industry, rinse aid in cosmetics, extraction aid in fermentation; lt. yel. liq.; pH 4.5–6.5; 30–33% min. act.

Nissan Anon LG. [441] Alkyl di-(aminoethyl) glycine; amphoteric; antistatic agent, softener, germicide in foods and cleaning industries used in water treatment, as duckweed killer, extraction aid for fermentation; lt. yel. liq.; pH 8.5–10.5; 30% act.

Nissan Asphasol 10. [441] Alkyl betaine; amphoteric, cationic; emulsifier for asphalt; 100% conc.

Nissan Asphasol 20, 100. [441] Cationic diamine; asphalt emulsifier for penetrating and mixing processes; yel.-br. semisolid, reddish-br. liq.; sol. in aq. sol'n. of org. acid, insol. in water.

Nissan Cation AB. [441] Octadecyl trimethyl ammonium chloride; cationic; germicide in water treatment, petroleum, paper, foods and textile industries, antistat in plastics, pulp and paper industries, rinse agent, pigment coating agent, dispersant, coagulant, softener; lt. yel. liq.; 23% min. act.

Nissan Cation ABT-350, 500. [441] Tallow-hydrog. alkyl trimethyl ammonium chloride; cationic; see Nissan Cation AB; lt. yel. liq.; 33 and 50% act.

Nissan Cation AR-4. [441] Alkyl imidazoline chloride; cationic; see Nissan Cation AB; yel.-br. liq.; 35% min. act.

Nissan Cation BB. [441] Dodecyl trimethyl ammonium chloride; cationic; see Nissan Cation AB; lt. yel. liq.; 30% min. act.

Nissan Cation F_2-10R, -20R, –40E, -50. [441] Coco-alkyl dimethyl benzyl ammonium chloride; cationic; see Nissan Cation AB; colorless to lt. yel. liq.; 9.5, 19–21, 40–44, and 50% act. resp.

Nissan Cation FB, FB-500. [441] Coco-alkyl trimethyl ammonium chloride; cationic; see Nissan Cation AB; lt. yel. liq.; 30 and 50% min act.

Nissan Cation L-207. [441] POE dodecyl monomethyl-ammonium chloride; cationic; see Nissan Cation AB; ylsh.-br., visc. liq.; 90% min. act.

Nissan Cation M_2-100. [441] Tetradecyl dimethyl benzyl ammonium chloride; cationic; see Nissan Cation AB; wh. or lt. yel. cryst. powd.; 85% min. act.

Nissan Cation MA. [441] Tetradecylamine acetate; cationic; corrosion inhibitor and germicide in water treatment and petroleum industries, duckweed killer, pigment dispersant, antistatic agent used in textile industry; wh. to lt. yel. flake.

Nissan Cation PB-40, -300. [441] Hexadecyl trimethyl ammonium chloride; cationic; see Nissan Cation AB; lt. yel. liq.; 40 and 27% act. resp.

Nissan Cation S_2-100. [441] Octadecyl dimethyl benzyl ammonium chloride; cationic; see Nissan Cation AB; wh. or lt. yel. cryst. powd.; 85% min. act.

Nissan Cation SA. [441] Octadecylamine acetate; cationic; see Nissan Cation MA; wh. to lt. yel. flake.

Nissan Diapon T. [441] N-methyl tallow-alkyl taurate, sodium salt; anionic; dyeing aux. for hair dye detergent, emulsifier; powd.; 30% conc.

Nissan Diapon TO. [441] N-methyl oleoyl taurate, sodium salt; anionic; detergent, emulsifier; scouring agent for wool; dyeing aux.; powd.; 30% conc.

Nissan Dispanol 16A. [441] POE alkyl ether; nonionic; water disp. emulsifier for paraffin; APHA 120 max. liq.; sol. in methanol, xylene, kerosene, ethyl ether, disp. in water; HLB 10.7; cloud pt. 0 C max. (1% aq.).

Nissan Dispanol LS-100. [441] Polyoxyalkylene alkylaryl ether; nonionic; low-foam detergent for syn. fibers; APHA 50 max. liq.; sol. in methanol, ethyl ether, water, disp. in xylene; cloud pt. 23–29 C (1% aq.).

Nissan Dispanol N-100. [441] POE alkylaryl ether; nonionic; detergent for animal, cellulosic, and syn. fibers; APHA 120 max. liq.; sol. in water, methanol, ethyl ether, xylene; HLB 12.5; cloud pt. 33–41 C (1% aq.).

Nissan Dispanol TOC. [441] POE alkyl ether; nonionic; detergent, wetting agent, emulsifier used in textile industry; APHA 120 max. liq.; sol. in water, methanol, ethyl ether, disp. in xylene; HLB 12.0; cloud pt. 45–53 C (1% aq.).

Nissan Monogly I. [441] Glycerol monostearate SE; nonionic; emulsifier for creams and lotions of the o/w type; wh. flake; disp. in water; m.p. 55–63 C; HLB 11.0; 37% min. monoglyceride content.

Nissan Monogly M. [441] Glycerol monostearate; nonionic; emulsifier for cosmetics and pharmaceuticals; wh. flake; m.p. 53–61 C; HLB 3.0; 40% min. monoglyceride content.

Nissan Newrex. [441] Alkylbenzene sulfonate; anionic; general detergent; paste, powd.; 52, 30, 60% conc.

Nissan Nonion CP-08R. [441] Sorbitan monocaprylate; nonionic; emulsifier for cosmetic, pharmaceutical and food applications; liq.; oil-sol.; HLB 7.3; 100% conc.

Nissan Nonion DN-202, -203, -209. [441] POE lauryl ether; nonionic; see Nissan Nonion CP-08R; liq.; oil-sol.; HLB 6.2, 7.9, and 13.2 resp.; 100% conc.

Nissan Nonion DS-60HN. [441] PEG distearate; emulsifier, thickener used in cosmetics, textile, industrial uses, nonionic; solid; sol. in water, methanol, warm in diethylene glycol; m.p. 54–62 C; HLB 19; 100% conc.

Nissan Nonion E-205, 215, 230. [441] POE oleyl ether; nonionic; emulsifier, dispersant, detergent, wetting agent used in textile processing, cosmetics, metalworking, agric. preparations, industrial cleaners; APHA 140 max. liq., semisolid, solid resp.; sol. in xylene, kerosene, methanol, disp. in water (E-205); sol. in water, methanol, cloudy in xylene (E-215, -230); HLB 9.0, 14.2 and 16.6; cloud pt. 0, 95 and 100 C resp. (1% aq.); 100% conc.

Nissan Nonion HS-204.5, -206. [441] POE octylphenyl ether; nonionic; detergent, emulsifier for household and industrial cleaners and detergents, textile processing, paints, printing inks, agric., latex polymerization; APHA 120 liq.; sol. in xylene, kerosene, ether, methanol; disp. in water; HLB 9.8 and 11.2; cloud pt. 0 C (1% aq., HS-204.5); 100% conc.

Nissan Nonion HS-208. [441] POE octylphenyl ether; nonionic; see Nissan Nonion HS-204.5; APHA 120 liq.; sol. in xylene, ether, methanol; disp. in water; HLB 12.6; 100% conc.

Nissan Nonion HS-210. [441] POE octylphenyl ether; nonionic; see Nissan Nonion HS-204.5; APHA 120 liq.; sol. in water, xylene, ether, methanol; HLB 13.3; cloud pt. 70–78 C (1% aq.); 100% conc.

Nissan Nonion HS-215. [441] POE octylphenyl ether; nonionic; see Nissan Nonion HS-204.5; APHA 120 liq.; sol. in water, xylene, ether, methanol; HLB 15.0; cloud pt. 91–99 C (1% aq.); 100% conc.

Nissan Nonion HS-220. [441] POE octylphenyl ether; nonionic; see Nissan Nonion HS-204.5; APHA 120 liq.; sol. in water, methanol, cloudy in xylene, ether; HLB 16.2; cloud pt. 100 C min. (1% aq.); 100% conc.

Nissan Nonion HS-240. [441] POE (40) octylphenyl ether; nonionic; see Nissan Nonion HS-204.5; APHA 120 liq.; sol. in water, methanol; HLB 17.9; cloud pt. 100 C min. (1% aq.); 100% conc.

Nissan Nonion HS-270. [441] POE (70) octylphenol ether; nonionic; see Nissan Nonion HS; semisolid; HLB 18.7; 100% conc.

Nissan Nonion K-202, -203. [441] POE lauryl ether; nonionic; emulsifier, dispersant, detergent, wetting agent used in textile processing, cosmetics, metalworking, agric. preparations, industrial cleaners; APHA 200 max. liq.; sol. in xylene, ether, methanol; disp. in kerosene; HLB 6.0 and 8.0 resp.; cloud pt. 0 C max. (1% aq.); 100% conc.

Nissan Nonion K-204. [441] POE lauryl ether; nonionic; see Nissan Nonion K-202; APHA 200 max. semisolid; sol. in xylene, ether, methanol; sol. warm in kerosene; disp. in water; HLB 9.2; cloud pt. 0 C max. (1% aq.); 100% conc.

Nissan Nonion K-207. [441] POE lauryl ether; nonionic; see Nissan Nonion K-202; APHA 50 max. semisolid; sol. in water, xylene, methanol; sol. warm in ether; HLB 12.1; cloud pt. 55–63 C max. (1% aq.); 100% conc.

Nissan Nonion K-211. [441] POE lauryl ether; nonionic; see Nissan Nonion K-202; APHA 50 max. semisolid; sol. in water, xylene, methanol; HLB 14.1; cloud pt. 90–98 C max. (1% aq.); 100% conc.

Nissan Nonion K-215. [441] POE lauryl ether; nonionic; see Nissan Nonion K-202; APHA 140 max. semisolid; sol. in water, xylene, methanol; HLB 15.2; cloud pt. 100 C min. (1% aq.); 100% conc.

Nissan Nonion K-220. [441] POE lauryl ether; nonionic; see Nissan Nonion K-202; APHA 120 max. solid; sol. in water, xylene, methanol; HLB 16.2; cloud pt. 100 C min. (1% aq.); 100% conc.

Nissan Nonion K-230. [441] POE lauryl ether; nonionic; see Nissan Nonion K-202; APHA 120 max. semisolid; sol. in water, xylene, methanol; HLB 17.3; cloud pt. 100 C min. (1% aq.); 100% conc.

Nissan Nonion L-2, -4. [441] PEG monolaurate; nonionic; general emulsifier for textile, leather, industrial uses; Gardner 4 max. liq.; L-2 sol. in methanol, sol. warm in kerosene, disp. in water; L-4 sol. in water, ether, methanol; HLB 9.5 and 13.1 resp.; 100% conc.

Nissan Nonion LP-20•R, LP-20•RS. [441] Sorbitan monolaurate; nonionic; emulsifier for cosmetic, pharmaceutical and food applications, o/w emulsion stabilizer and thickener, fiber lubricant and softener; Gardner 5 max. oily liq.; sol. in methanol, ethanol, acetone, xylene, ethyl ether, kerosene, disp. in water; HLB 8.6; 100% conc.

Nissan Nonion LT-221. [441] Polysorbate 20; nonionic; emulsifier for cosmetic, pharmaceutical and food applications; Gardner 6 max. oily liq.; sol. in water, methanol, ethanol, acetone, xylene, ethyl ether, ethylene glycol, HLB 16.7; 100% conc.

Nissan Nonion MP-30R. [441] Sorbitan monomyristate; nonionic; see Nissan Nonion LP-20R; solid; oil-sol.; HLB 6.6; 100% conc.

Nissan Nonion NS-202. [441] POE nonylphenyl ether; nonionic; detergent, emulsifier for household and industrial cleaners, textile processing, paints, printing inks, agric. preparations, and latex polymerization; APHA 120 max. liq.; sol. in xylene, kerosene, ether, methanol; disp. in water; HLB 5.7; cloud pt. 0 C (1% aq.); 100% conc.

Nissan Nonion NS-204.5. [441] POE nonylphenyl ether; nonionic; see Nissan Nonion NS-202; APHA 120 max. liq.; sol. in xylene, kerosene, ether, methanol; disp. in water; HLB 9.5; cloud pt. 0 C max. (1% aq.); 100% conc.

Nissan Nonion NS-206. [441] POE (6) nonylphenyl ether; nonionic; see Nissan Nonion NS-202; APHA 120 max. liq.; sol. in xylene, ether, methanol; disp. in water; HLB 10.9; cloud pt. 0 C max. (1% aq.); 100% conc.

Nissan Nonion NS-208.5. [441] POE nonylphenyl ether; nonionic; see Nissan Nonion NS-202; APHA 120 max. liq.; sol. in xylene, ether, methanol; disp. in water; HLB 12.6; cloud pt. 33–41 C (1% aq.); 100% conc.

Nissan Nonion NS-210. [441] POE nonylphenyl ether; nonionic; see Nissan Nonion NS-202; APHA 120 max. liq.; sol. in water, xylene, ether, methanol; HLB 13.3; cloud pt. 46–54 C (1% aq.); 100% conc.

Nissan Nonion NS-212. [441] POE (12) nonylphenyl ether; nonionic; see Nissan Nonion NS-202; APHA 120 max. liq.; sol. in water, xylene, ether, methanol; HLB 14.1; cloud pt. 73–84 C (1% aq.); 100% conc.

Nissan Nonion NS-215. [441] POE nonylphenyl ether; nonionic; see Nissan Nonion NS-202; APHA 120 max. liq.; sol. in water, xylene, ether, methanol; HLB 15; cloud pt. 89–97 C (1% aq.); 100% conc.

Nissan Nonion NS-220. [441] POE nonylphenyl ether; nonionic; see Nissan Nonion NS-202; APHA 120 max. semisolid; sol. in water, xylene, ether, methanol; HLB 16; cloud pt. 100 C min. (1% aq.); 100% conc.

Nissan Nonion NS-230. [441] POE nonylphenyl ether; nonionic; see Nissan Nonion NS-202; APHA 120 max. solid; sol. in water, xylene, methanol, cloudy in ether; HLB 17.1; cloud pt. 100 C min. (1% aq.); 100% conc.

Nissan Nonion NS-240, -270. [441] POE nonylphenyl ether; nonionic; see Nissan Nonion NS-202; APHA 120 max. solid; sol. in water, methanol; HLB 17.8 and 18.7 resp.; cloud pt. 100 C min. (1% aq.); 100% conc.

Nissan Nonion O-2, O-3. [441] PEG monooleate; nonionic; general emulsifier for textile, leather, industrial uses; Gardner 10 max. liq.; sol. in kerosene, methanol; disp. in water, xylene, ether; HLB 7.9 and 10.3 resp.; 100% conc.

Nissan Nonion O-4. [441] PEG monooleate; nonionic; general emulsifier for textile, leather, industrial uses; Gardner 9 max. liq.; sol. in xylene, kerosene, methanol; sol. warm in ether; disp. in water; HLB 11.5; 100% conc.

Nissan Nonion O-6. [441] PEG monooleate; nonionic; general emulsifier for textile, leather, industrial uses; Gardner 9 max. liq.; sol. in water, xylene, kerosene, ether, methanol; HLB 13.5; 100% conc.

Nissan Nonion OP-80•R. [441] Sorbitan monooleate; nonionic; see Nissan Nonion LP-20•R; Gardner 9 max. oily liq.; oil-sol.; HLB 4.3; 100% conc.

Nissan Nonion OP-83•RAT. [441] Sorbitan sesquioleate; nonionic; see Nissan Nonionic LP-20•R; Gardner 9 max. oily liq.; sol. in ethanol, acetone, xylene, ethyl ether, kerosene, methanol, warm in water; HLB 3.7; 100% conc.

Nissan Nonion OP-85•R. [441] Sorbitan trioleate; nonionic; see Nissan Nonion LP-20•R; Gardner 9 max. oily liq.; oil-sol.; HLB 1.8; 100% conc.

Nissan Nonion OT-221. [441] POE sorbitan monooleate; nonionic; see Nissan Nonion LT-221; Gardner 6 max. oily liq.; sol. in water, ethanol, acetone, xylene, disp. in methanol; HLB 15.0; 100% conc.

Nissan Nonion P-6. [441] PEG monopalmitate; nonionic; emulsifier and thickener for cosmetic applications, textile and other industrial uses; semisolid; HLB 13.8; 100% conc.

Nissan Nonion P-208. [441] POE cetyl ether; nonionic; emulsifier, dispersant, detergent, wetting agent used in textile processing, cosmetics, metalworking, agric. preparations, industrial cleaners; APHA 120 liq.; sol. in water, xylene, methanol; HLB 11.9; cloud pt. 43–53 C (1% aq.); 100% conc.

Nissan Nonion P-210. [441] POE cetyl ether; nonionic; see Nissan Nonion P-208; APHA 120 max. semisolid; sol. in water, xylene, methanol; HLB 12.9; cloud pt. 64–74 C (1% aq.); 100% conc.

Nissan Nonion P-213. [441] POE cetyl ether; nonionic; see Nissan Nonion P-208; APHA 120 max. semisolid; sol. in water, kerosene, ether, methanol; HLB 14.1; cloud pt. 87–97 C (1% aq.); 100% conc.

Nissan Nonion PP-40•R. [441] Sorbitan monopalmitate; nonionic; see Nissan Nonion LP-20•R; Gardner 7 max. waxy solid; oil-sol.; HLB 6.7; 100% conc.

Nissan Nonion PT-221. [441] POE sorbitan monopalmitate; nonionic; see Nissan Nonion LT-221; Gardner 8 max. oily liq.; sol. in water, methanol, ethanol, acetone, xylene, ethyl ether, ethylene glycol; HLB 15.3; 100% conc.

Nissan Nonion S-2. [441] PEG monostearate; nonionic; emulsifier, thickener used in cosmetics, textile, industrial uses; Gardner 4 max. semisolid; sol. in methanol, warm in xylene, disp. in water, kerosene, ether; m.p. 33–41 C; HLB 8.0; 100% conc.

Nissan Nonion S-4. [441] PEG monostearate; nonionic; see Nissan Nonion S-2; Gardner 4 max. semisolid; sol. in kerosene, ether, methanol, warm in xylene, disp. in water; m.p. 30–40 C; HLB 11.6; 100% conc.

Nissan Nonion S-6. [441] PEG monostearate; nonionic; see Nissan Nonion S-2; Gardner 4 max. semisolid; sol. in water, methanol, warm in xylene, ether, cloudy in kerosene; m.p. 35–41 C; HLB 13.6; 100% conc.

Nissan Nonion S-10. [441] PEG monostearate; nonionic; see Nissan Nonion S-2; Gardner 4 max. semisolid; sol. in water, xylene, ether, methanol; m.p. 38–44 C; HLB 15.2; 100% conc.

Nissan Nonion S-15. [441] PEG monostearate; nonionic; see Nissan Nonion S-2; Gardner 4 max. semisolid; sol. in kerosene, ether, methanol, warm in xylene; disp. in water; m.p. 39–45 C; HLB 12.8; 100% conc.

Nissan Nonion S-15.4. [441] PEG monostearate; nonionic; see Nissan Nonion S-2; Gardner 4 max. semisolid; sol. in water, xylene, ether, methanol; m.p. 42–48 C; HLB 16.7; 100% conc.

Nissan Nonion S-40. [441] PEG monostearate; nonionic; see Nissan Nonion S-2; Gardner 6 max. semisolid; sol. in water, xylene, ether, methanol; m.p. 49–55 C; HLB 18.2; 100% conc.

Nissan Nonion S-207, -215, -220. [441] POE stearyl ether; nonionic; emulsifier, dispersant, detergent, wetting agent for textile processing, cosmetics, metalworking compds., agric. preparations, industrial cleaners; 120 max. APHA, semisolid; sol. in xylene, kerosene, methanol, cloudy in ether; disp. in water (S-207); sol. in water, xylene, kerosene, methanol (S-215, -220); HLB 10.7, 14.2, 15.3; cloud pt. 0 C max. (S-207), 100 C min. (S-215, -220) (1% aq.); 100% conc.

Nissan Nonion SP-60•R. [441] Sorbitan monostearate; nonionic; see Nissan Nonion LP-20•R; Gardner 5 max. waxy solid; sol. in methanol, ethanol, xylene, kerosene, ethyl ether, disp. in warm water; HLB 4.7; 100% conc.

Nissan Nonion ST-221. [441] POE sorbitan monostearate; nonionic; see Nissan Nonion LT-221; Gardner 5 max. oily liq.; sol. in water, methanol, ethanol, acetone, xylene, ethyl ether, kerosene; HLB 14.9; 100% conc.

Nissan Nonion T-15. [441] PEG monotallow acid ester; nonionic; see Nissan Nonion P-6; liq.; HLB 12.8; 100% conc.

Nissan Nymeen DT-203, -208. [441] POE tallow-alkylpropylene diamine; nonionic; pigment wetting and disp. agent, emulsifier and acid cleaner additive for metals, textile antistatic agent and auxiliary; lt. br. liq.; cloud pt. 0 and 100 C resp.

Nissan Nymeen L-201. [441] Oxyethylene dodecylamine; nonionic; see Nissan Nymeen DT-203; colorless to lt. yel. liq.; cloud pt. 0 C max.

Nissan Nymeen L-202, 207. [441] POE dodecylamine; nonionic; see Nissan Nymeen DT-203; lt. yel. and lt. br. liq. resp.; cloud pt. 0 and 80 C.

Nissan Nymeen S-202, S-204. [441] POE octadecylamine; nonionic; see Nissan Nymeen DT-203; ylsh.-br. solid; cloud pt. 0 C max.

Nissan Nymeen S-210. [441] POE octadecylamine; nonionic; see Nissan Nymeen DT-203; lt. br. solid in winter, liq. in summer; cloud pt. 95 C min.

Nissan Nymeen S-215, -220. [441] POE octadecylamine; nonionic; see Nissan Nymeen

DT-203; ylsh.-br. solid; cloud pt. 100 C min.

Nissan Nymeen T_2-206, -210. [441] POE tallow-alkylamine; nonionic; see Nissan Nymeen DT-203; lt. br. liq.; cloud pt. 95 C.

Nissan Nymeen T_2-230, -260. [441] POE tallow-alkylamine; nonionic; see Nissan Nymeen DT-203; lt. br. solid; cloud pt. 100 C.

Nissan Nymide MT-215. [441] POE alkylamide; nonionic; modifier for soaps; solid; 100% conc.

Nissan Ohsen A. [441] Alkylbenzene sulfonate; anionic; general purpose detergent and emulsifier; paste; 60% conc.

Nissan Panacete 810. [441] Triglyceride, medium chain; nonionic; diluent for perfumes; raw material for pharmaceuticals and special foods; liq.; 100% conc.

Nissan Persoft EK. [441] Sulfated fatty alcohol ethoxylate, sodium salt; anionic; emulsifier for vinyl polymerization; textile detergent; dyeing assistant; degreaser; liq.; 30% conc.

Nissan Persoft NK-60, -100. [441] Fatty alcohols, ethoxylated; nonionic; detergent, emulsifier, wetting agent; textile detergent; dyeing assistant; base for liq. detergent; dispersant for pulp pitch; lubricant for bottle cleaning processing; liq.; 90% conc.

Nissan Persoft SK. [441] Sulfated fatty alcohol, sodium salt; anionic; detergent, emulsifier, dispersant, wetting agent; base for shampoo and liq. detergent; liq.; 30% conc.

Nissan Plonon 102, 104. [441] POE-POP ether; nonionic; emulsifier, solubilizer, dispersant, detergent, antifoaming agent, wetting agent used in metal cleaning, emulsion polymerization, fermentation, paper industries; colorless liq.; odorless and tasteless; m.w. 1250 and 1650 resp.; sol. in water, ethanol, acetone, benzol, dioxane; visc. 105 cs (37.8 C); HLB 7 and 15; cloud pt. 22 and 64 C (10% aq.); surf. tens. 41.1 and 43.5 dynes/cm (0.1%); 100% conc.

Nissan Plonon 108. [441] POE-POP ether; nonionic; see Nissan Plonon 102; m.w. 4000; HLB 30.5; cloud pt. 100 C (10% aq.); 100% conc.

Nissan Plonon 171. [441] POE-POP ether; nonionic; see Nissan Plonon 102; m.w. 2000; sol. in ethanol, acetone, benzol, dioxane; visc. 160 cs (37.8 C); HLB 3; cloud pt. 24 C (10% aq.); surf. tens. 39.5 dynes/cm (0.1%); 100% conc.

Nissan Plonon 172. [441] POE-POP ether; nonionic; see Nissan Plonon 102; m.w. 2400; sol. in ethanol, acetone, benzol, dioxane; visc. 210 cs (37.8 C); HLB 7; cloud pt. 28 C (10% aq.); 100% conc.

Nissan Plonon 201. [441] POE-POP ether; nonionic; see Nissan Plonon 102; colorless liq.; odorless and tasteless; m.w. 2200; sol. in ethanol, acetone, benzol, dioxane; visc. 190 cs (37.8 C); HLB 3; cloud pt. 21 C (10% aq.); surf. tens. 38 dynes/cm (0.1%); 100% conc.

Nissan Plonon 204. [441] POE-POP ether; nonionic; see Nissan Plonon 102; colorless paste; odorless and tasteless; m.w. 3400; sol. in water, ethanol; visc. 370 cs (37.8 C); HLB 13.5; cloud pt. 64 C (10% aq.); surf. tens. 39.8 dynes/cm (0.1%); 100% conc.

Nissan Plonon 208. [441] POE-POP ether; nonionic; see Nissan Plonon 102; colorless paste; odorless and tasteless; m.w. 8000; sol. in water; HLB 28; cloud pt. 100 C (10% aq.); surf. tens. 44.1 dynes/cm (0.1%); 100% conc.

Nissan Rapisol B-30, -80, C-70. [441] Sodium dioctyl sulfosuccinate; anionic; wetting agent, polymerization agent for PVC, dyeing aux.; liq.; 30, 80, and 70% conc. resp.

Nissan Soft Osen 550A. [441] Linear alkylbenzene sulfonate; anionic; general purpose detergent, emulsifier; paste; 60% conc.

Nissan Stafoam DF. [441] Coconut diethanolamide; nonionic; detergent and scouring agent for fibers; liq.; 100% conc.

Nissan Stafoam DL. [441] Lauroyl diethanolamide; nonionic; foam stabilizer for liq. detergents; liq.; 100% conc.

Nissan Stafoam DO, DO-S. [441] Oleoyl diethanolamide; nonionic; foam stabilizer for shampoo and liq. detergents; liq.; 100% conc.

Nissan Stafoam L. [441] Lauroyl diethanolamide (1:2); nonionic; foam stabilizer for liq. detergents; liq.; 100% conc.

Nissan Stafoam MF. [441] Coconut monoethanolamide; nonionic; foam stabilizer for paste shampoos; flake; 100% conc.

Nissan Sunalpha T. [441] Alpha-sulfonated fatty acid ester blend, sodium salt; anionic; detergent for household and industrial cleaning; granules; sol. in water; 100% conc.

Nissan Sunamide C-3, CF-3, -10. [441] Fatty alkylolamide ether sulfate; anionic; base for preparation of shampoos and dishwashing detergents; liq.; 30, 30 and 35% conc. resp.

Nissan Sunbase, Powder. [441] Alpha-sulfonated fatty acid ester, sodium salt; anionic; lime soap dispersant; detergent base; builder for household detergent; emulsifier and dispersant; paste, granules; 30 and 40% conc. resp.

Nissan Tertiary Amine AB. [441] Octadecyldimethylamine; intermediate for various surfactants, visc. index improver for lubricating oil, curing catalyst for epoxy resin, corrosion inhibitor, germicide; lt. yel. liq. or half-solid; 95% min. tert. amine.

Nissan Tertiary Amine ABT. [441] Tallowhydrog. alkyl dimethylamine; see Nissan Tert. Amine AB; lt. yel. liq. or half-solid; 95% min. tert. amine.

Nissan Tertiary Amine BB. [441] Dodecyldimethylamine; see Nissan Tert. Amine AB; lt. yel. liq. 95% min. tert. amine.

Nissan Tertiary Amine FB. [441] Coco-alkyl dimethylamine; see Nissan Tert. Amine AB; lt. yel. liq.; 95% min. tert. amine.

Nissan Tertiary Amine MB. [441] Tetradecyl dimethylamine; see Nissan Tert. Amine AB; lt. yel. liq. or half-solid; 95% min. tert. amine.

Nissan Tertiary Amine PB. [441] Hexadecyldimethylamine; see Nissan Tert. Amine AB; lt. yel. liq. or half-solid; 95% min. tert. amine.

Nissan Trax H-45. [441] POE alkylaryl ether sulfate; anionic; penetrant, detergent, wetting agent; scouring agent for wool; dyeing aux. and emulsifier for syn. resins; liq.; 30% conc.

Nissan Trax K-40, -300. [441] POE alkyl ether sulfate; anionic; emulsifier for emulsion polymerization; liq.; 30 and 34% conc. resp.

Nissan Trax N-300. [441] POE alkylaryl ether sulfate; anionic; see Nissan Trax K-40; liq.; 34% conc.

Nissan Unisafe A-LE. [441] Alkyl dihydroxy ethyl amine oxide; nonionic; foam stabilizer, detergent; liq.; 40% conc.

Nissan Unisafe A-LM. [441] Alkyl dimethyl amine oxide; nonionic; see Nissan Unisafe A-LE; liq.; 35% conc.

Nitrene 100 SD. [289] Linoleamide DEA; nonionic; thickener, foamer, corrosion inhibitor; emulsifier for oils/greases used in industrial and household cleaners; liq.; 100% conc.

Nitrene 11120. [290] Modified cocamide DEA (2:1); detergent for truck washes, household, alkaline industrial cleaners; liq.; dens. 8.3 lb/gal 100% conc.

Nitrene 11230. [289; 290] Modified cocamide DEA (2:1); detergent, wetting agent, emulsifier for industrial and specialty cleaners, household hard surface cleaners; liq.; dens. 8.3 lb/gal 100% conc.

Nitrene 13026. [290] Modified cocamide DEA; detergent, wetting agent, emulsifier for industrial and specialty cleaners; liq.; 100% conc.

Nitrene A-309. [289; 290] Modified cocamide DEA (2:1); detergent, dispersant, chain lubricant; liq.; dens. 8.4 lb/gal; 100% conc.

Nitrene A-567. [290] Modified cocamide DEA (2:1); detergent, emulsifier for alkaline industrial and household cleaners; liq.; dens. 8.3 lb/gal; 100% conc.

Nitrene C. [289; 290] Cocamide DEA (1:1); nonionic; emulsifier, dispersant, wetting agent, foam booster, visc. builder for liq. detergents, car shampoos, solv. cleaners, drycleaning; lt. amber liq.; sol. in water, lower alcohols, PEG, acetone, toluene, xylene; dens. 8.41 lb/gal; pH 9–11 (10%); 100% conc.

Nitrene C Extra. [289] Cocamide DEA (1:1); nonionic; emulsifier, dispersant, wetting agent, foam booster, and visc. builder for detergents and cleansers; intermediate for liq.; detergents; foam stabilizer; clarifier for liq. soaps; solv. cleaners, and dry-cleaning products; lt. amber liq.; sol. in water, alcohol, PEG, acetone, toluene, xylene; dens. 8.41 lb/gal; pH 9-11 (10%); 100% conc.

Nitrene L-76. [289; 290] Lauric-myristic diethanolamide (1:1); nonionic; wetting, foaming, dispersing agent, foam stabilizer, thickener for household, institutional, and industrial cleaners; pigment dispersant; soft solid to lt. amber liq.; sol. in lower alcohols, propylene glycol, PEG; disp. water; dens. 8.1 lb/gal; pH 9–11 (10%); biodeg.; 100% conc.

Nitrene L-90. [289; 290] Lauramide DEA (1:1); nonionic; detergent, thickener, foam stabilizer for liq. detergents; solid; dens. 8.3 lb/gal; 100% conc.

Nitrene OE. [289; 290] Polyoxyalkylene DEA amide ester; emulsifier for sol. oils and lubricants; liq.; dens. 8.2 lb/gal; 100% conc.

Noigen EA 33, 73, 83, 143. [172] POE dodecyl phenol ether; nonionic; emulsifier; liq.; oil/water sol.; HLB 4.1, 9.0, 10,0 and 14.0 resp.; 100% conc.

Noigen EA 50, 70, 80, 80E. [172] PEG nonylphenol ether; nonionic; emulsifier; liq.; oil/water sol.; HLB 6.0, 8.0, 10.0 and 9.0 resp.; 100% conc.

Noigen EA92, 102, 110, 112, 120, 120B, 130T, 140, 140L, 142, 150, 152, 160, 160P, 170, 190D. [172] PEG octyl phenol ether; nonionic; emulsifier; liq, paste, solid, flake; oil/water sol.; 9.0, 11.0, 11.0, 12.0, 12.0, 12.0, 13.0, 14.0, 14.0, 14.0, 16.0, 15.0, 15.0, 16.0 and 17.0 resp.; 100% conc.

Noigen ES90, 120, 140, 160. [172] PEG oleic acid ester; nonionic; emulsifier for animal, veg. and min. oil; dispersant for paint and pigments; liq., solid; HLB 9.0, 12.0, 14.0, and 16.0 resp.; 100% conc.

Noigen ET60. [172] PEG oleyl ether; nonionic; emulsifier; liq.; oil-sol.; HLB 6.0; 100% conc.

Noigen ET65, 77. [172] PEG alkyl ether; nonionic; emulsifier; liq.; oil-sol.; HLB 6.0 and 7.0 resp.; 100% conc.

Noigen ET80. [172] PEG oleyl ether; nonionic; emulsifier; liq.; oil-sol.; HLB 8.0; 100% conc.

Noigen ET83. [172] PEG lauryl ether; nonionic; emulsifier; liq.; HLB 8.0; 100% conc.

Noigen ET95, 97. [172] PEG alkyl ether; nonionic; emulsifier; liq.; HLB 9.0; 100% conc.

Noigen ET100. [172] PEG oleyl ether; nonionic; emulsifier; liq.; HLB 10.0; 100% conc.

Noigen ET102. [172] PEG lauryl ether; nonionic; emulsifier; liq.; HLB 10.0; 100% conc.

Noigen ET107, 115. [172] PEG alkyl ether; nonionic; emulsifier; liq.; HLB 10.0 and 11.0 resp.; 100% conc.

Noigen ET120. [172] PEG oleyl ether; nonionic; emulsifier; paste; HLB 12.0; 100% conc.

Noigen ET127, 135. [172] PEG alkyl ether; nonionic; emulsifier; liq.; HLB 12.0 and 13.0 resp.; 100% conc.

Noigen ET140, 150. [172] PEG oleyl ether; nonionic; emulsifier; solid; HLB 14.0 and 15.0 resp.; 100% conc.

Noigen ET143, 160. [172] PEG lauryl ether; nonionic; emulsifier; liq. and solid resp.; HLB 14.0 and 16.0 resp.; 100% conc.

Noigen ET147, 157. [172] PEG alkyl ether; nonionic; emulsifier; liq.; HLB 14.0 and 15.0 resp.; 100% conc.

Noigen ET165, 167. [172] PEG alkyl ether; nonionic; emulsifier; liq., solid resp.; HLB 16.0; 100% conc.

Noigen ET170, 190. [172] PEG lauryl ether; nonionic; emulsifier; solid; HLB 17.0 and 19.0 resp.; 100% conc.

Noigen ET180, 190S. [172] PEG oleyl ether; nonionic; emulsifier; solid; HLB 18.0 and 19.0 resp.; 100% conc.

Noigen ET187. [172] PEG alkyl ether; nonionic; emulsifier; solid; HLB 18.0; 100% conc.

Noigen ET207. [172] PEG alkyl ether; nonionic; emulsifier; solid; HLB 20.0; 100% conc.

Noigen O 100. [172] PEG oleyl ether; nonionic; emulsifier; solid; water-sol.; HLB 20.0; 100% conc.

Noigen YX400, 500. [172] PEG lauryl ether; nonionic; emulsifier; solid; HLB 20.0; 100% conc.

Noiox AK-40. [494] Polyethylene oleic acid ester; nonionic; spreading agent, defoamer, kerosene and min. oil emulsifier; liq.; 100% conc.

Noiox AK-41. [494] Polyethylene oleic acid ester; nonionic; lubricant, solubilizer and emulsifier; wetting agent and pigment dispersant; antistat for plastics and aux. dispersant; liq.; 100% conc.

Noiox AK-43. [494] Polyethylene stearic acid ester; nonionic; lubricant, antistat for textile industry; preparer and after treatment for fibers; paste; 90% conc.

Noiox AK-44. [494] Polyethylene oleic acid ester; nonionic; lubricant, emulsifier for min. and veg. oils; antifoamer for dispersing assistant; liq.; 100% conc.

Noiox AK-45. [494] Polyethylene coconut acid ester; nonionic; preparer and oiler for fibers; liq.; 98% conc.

Noiox AK-46. [494] Polyethylene oleic acid ester; nonionic; see Noiox AK-45; liq.; 90% conc.

Noiox EO-33. [494] Polyethylene stearic acid diester; nonionic; thickener, opacifier, conditioner; used to control visc. in shampoos and lotions; flakes; 100% conc.

Noiox KJ Series. [494] Unsat. fatty alcohols, ethoxylated; nonionic; surface act. agents used in cosmetic and pharmaceutical industries; 100% conc.

Noiox KS-10, -12, -13, -14, -16. [494] PEG; nonionic; lubricant, solv. for dyestuffs and resins; plasticizer for casein, gelatin compositions and printing inks; used in cosmetic and pharmaceutical items; intermediate products; liq., solid, paste; water-sol.; 100% conc.

Nonal 206, 208, 210. [606] Nonylphenol ethoxylate; nonionic; detergent, penetrant, emulsifier, scouring agent; pitch dispersant; liq.; HLB 10.9, 12.3, 13.3; 100% conc.

Nonal 310. [606] Octyl phenol ethoxylate; nonionic; detergent, scouring agent; penetrant; liq.; HLB 13.7; 100% conc.

Nonarox 575, 730. [551] Alkylphenyl polyethoxy ether; nonionic; wetting agent; liq.; 100% conc.

Nonarox 1030, 1230. [551] Alkylphenyl polyethoxy ether; nonionic; detergent; liq.; 100% conc.

Nonasol 3922. [283] Blend of sulfates, sulfonates, and amide; anionic; detergent used in wide variety of compds.; yel. liq.; sp.gr. 1.03; 60% act.

Nonasol N4AS. [283] Sulfated linear alcohol ethoxylate, ammonium salt; anionic; detergent base; yel. liq.; sp.gr. 1.02; biodeg.; 58% act.

Nonasol N4SS. [283] Sulfated linear alcohol ethoxylate, sodium salt; anionic; detergent base; yel. liq.; sp.gr. 1.04; biodeg.; 60% act.

Nonex C5E. [283] PEG 500 monococoate; lubricant for textile applications; liq.; water-sol.; HLB 13.0; 100% conc.

Nonex DL-2. [283] POE 200 dilaurate; nonionic; coemulsifier and lubricant in industrial and textile oils; liq.; HLB 7.4; 100% act.

Nonex DO-4. [283] POE 400 dioleate; nonionic; emulsifier, solubilizer for oils, fats, solvs.; liq.; HLB 7.2; 100% act.

Nonex O4E. [283] POE 400 oleate; nonionic; emulsifier for min. and animal oils; liq.; HLB 11; 100% act.

Nonex S3E. [283] POE 300 stearate; nonionic; lubricant and softener for textiles; liq.; HLB 9.6; 100% act.

Nonion LA4. [15] Syn. fatty alcohol ethoxylate (4 EO); coemulsifier for o/w emulsion systems, e.g., degreasers; off-wh. clear liq.; sol. @5% in ethyl acetate, ethanol, min. oil; disp. veg. oils, water; sp.gr. 0.956; visc. 92 cP; HLB 9.8; m.p. 2 C; pH 7 ± 1 (1%).

Nonionic 1017-R, 1025-R. [300] EO/PO block polymers; detergent, antifoam, rinse aids, visc. control, dispersant, demulsifier; liq.; 100% conc.

Nonionic 1035-L, 1044-L, 1061-L, 1062-L, 1064-L, 1068-L, 1088-L. [300] EO/PO block copolymers; nonionic; surfactant used as a detergent, antifoam, wetting agent, emulsifier, antistat, demulsifier, visc. modifier, deduster, gelation aid, metal working lubricant, dispersant; liq., flake; 100% conc.

Nonionic 2017-R, 2025-R, 4017-R, 4025-R, 5025-R. [300] EO/PO block copolymer; nonionic; see Nonionic 1017-R; liq., paste; 100% conc.

Nonionic E-4, -5, -6, -7, -10, –12, -20, -30. [300] POE alkylaryl ethers; nonionic; detergent and wetting agent used in cosmetics, insecticide and other formulations; liq., solid; HLB 8.6, 10.6, 12.6, 14.0, 17.0, and 17.5 resp.; 100% conc.

Non Ionic Emulsifier 42. [562] Emulsifier for oils and emulsion systems; Gardner 13 color; sol. in paraffinic oils; disp. in water; sp.gr. 0.9756 (60 F); dens. 8.125 lb/gal; visc. 216 SSU (100 F); HLB 13.3; cloud pt. 57 F; sapon. no. 62.6; pH 7.65 (1%); 100% act.

Non Ionic Emulsifier T-9. [562] Emulsifier for oils and emulsion systems; Gardner 6–12 color; sol. in paraffinic oils; disp. in water; sp.gr. 0.967–0.976 (60 F); visc. 195–250 SSU (60 F); HLB 12; cloud pt. 60–70 F; sapon. no. 60–70; pH 6.5–7.0 (1%); 100% act.

Nonionic Emulsifier WGS. [562] Alkyl ether ester; nonionic; emulsifier for oils and emulsion systems; Gardner 10–12 liq.; sol. in paraffinic oils, disp. in water; dens. 8.07 lb/gal; sp.gr. 0.965–0.980; HLB 10; cloud pt. 50–60 F; flash pt. 400 F; Sapon. no. 60–70; pH 6–7 (1% aq.); biodeg.; 100% act.

Nonipol 20, 40, 55. [539] POE nonyl phenyl ether; nonionic; penetrant, wetting agent, spreader-sticker; base material for detergent; emulsifier for agric. pesticides and emulsion polymerization, org. solv., machine oils, liq. paraffins; liq.; HLB 5.7, 8.9, 10.5; 100% conc.

Nonipol 60, 70, 85, 95, 100, 110, 120, 130, 160, 200, 400. [539] POE nonyl phenyl ether; detergent and penetrant; liq.; HLB 10.9, 11.7, 12.6, 13.1, 13.3, 13.8, 14.1, 14.5, 15.2, 16.0, and 17.8 resp.; 100% conc.

Nonipol BX. [539] POE alkylphenyl ether; nonionic; textile scouring agent and penetrant; liq.; 100% conc.

Nonipol D-160. [539] POE dinonyl phenyl ether; nonionic; see Nonipol 20; solid; HLB 13.3

Nonipol Soft SS-50, -70, -90. [539] POE higher alcohol ether; nonionic; penetrant, detergent, emulsifier, dispersant, wetting agent, or base material for household liq. detergent; liq.; HLB 10.5-13.2; 100% conc.

Nonipol T-28. [606] Nonionics and min. oil; nonionic; surfactant used in spinning oil for wool; liq.; 100% act.

Nonisol 100. [137] PEG 400 monolaurate; nonionic; emulsifier, solubilizing and wetting agent, thickener, used in textiles, cosmetics, hand cleaners, spreading agents; yel. liq.; water-sol.; 100% conc.

Nonisol 210. [137] PEG 400 dioleate; nonionic; toxicant emulsifier, dispersant of sludge in fuel oil, textile lubricant; lt. amber liq.; sol. in kerosene; 100% conc.

Nonisol 300. [137] PEG 400 monostearate; nonionic; emulsifier, wetting agent, starch stabilizer, base for ointments, creams, suppositories, hair dressing, liq. makeup; wh. paste; sol. in xylene, disp. in water; m.p. 30 C; 100% conc.

Nopalcol 1-S. [187] Diethylene glycol monostearate; nonionic; emulsifier, plasticizer, lubricant, wetting agent, binding and thickening agent, used in cosmetics, dry cleaning, leather, textile industries; solid; HLB 3.8; 98% conc.

Nopalcol 1-TW. [187] Diethylene glycol monotallowate; nonionic; see Nopalcol 1-S; solid; HLB 4.1; 99% conc.

Nopalcol 2-L. [187] POE (4) monolaurate; nonionic; emulsifier; liq.; HLB 10.5; 100% conc.

Nopalcol 4-C, -CH. [187] POE (8) monococoate; nonionic; emulsifier; liq.; HLB 13.9; 99% conc.

Nopalcol 4-DTW. [187] POE (8) ditallowate; nonionic; emulsifier; solid; HLB 9.1; 99% conc.

Nopalcol 4-L. [187] POE (8) monolaurate; nonionic; emulsifier; liq.; HLB 14.0; 100% conc.

Nopalcol 4-O. [187] POE (8) monooleate; nonionic; emulsifier; dispersant for leather pigments; paper coating defoamer, plasticizer and leveling agent; liq.; HLB 12.0; 100% conc.

Nopalcol 4-S. [187] POE (8) monostearate; nonionic; emulsifier; paste; HLB 12.0; 100% conc.

Nopalcol 6-L. [187] POE (12) monolaurate; nonionic; emulsifier; liq.; HLB 15.5; 100% conc.

Nopalcol 6-O. [187] POE (12) monooleate; nonionic; emulsifier, wetting agent for oil phase systems; liq.; HLB 14.0; 100% conc.

Nopalcol 6-R. [187] PEG 600 monoricinoleate; nonionic; emulsifier; liq.; HLB 13.5; 100% conc.

Nopalcol 6-S. [187] PEG 600 monostearate; nonionic; emulsifier; solid; HLB 14.0; 100% conc.

Nopalcol 10-CO. [187] PEG 1000 castor oil; nonionic; emulsifier; liq.; HLB 10.3; 99% conc.

Nopalcol 19-CO. [187] PEG 1900 castor oil; nonionic; emulsifier; liq.; HLB 14.9; 99% conc.

Nopalcol 30-S. [187] PEG 3000 monostearate; nonionic; emulsifier; solid; HLB 18.5; 100% conc.

Nopalcol 200. [187] PEG 200; nonionic; emulsifier; liq.; 99% conc.

Nopco 1179. [187] Fatty amido condensate; nonionic; wool and worsted fulling and scouring; general purpose detergent; liq.; HLB 14.0; 100% conc.

Nopco 1186A. [187] Sulfonated alkyl ester; anionic; wetting agent for fibrous materials; emulsifier in latex mfg.; liq.; 65% conc.

Nopco 1419-A. [187] Defoamer for SBR, PVAc, acrylic paint and adhesive systems; amber liq.; forms milky emulsion in water; sp.gr. 0.85; dens. 7.1 lb/gal; pH 6.5 (2%); flash pt. 47 C; 99% act.

Nopco 1471, 1471-M. [187] Veg. oil, high sulfated; anionic; emulsifier for min. and veg. oils; liq.; 68% conc.

Nopco 2272-R. [187; 192] Fatty ester, highly sulfated; anionic; wetting in paper towels, latex impregnation and dipping, in metal treating and textile dyeing; liq.; 65% conc.

Nopco Color-Sperse 188-A. [192] Fatty acid, ethoxylated; nonionic; pigment wetting and dispersant; liq.; 100% conc.

Nopco JMY. [187] Defoamer for PVAc, acrylic paint and adhesive systems; amber liq.; forms milky emulsion in water; sp.gr. 0.88; dens. 7.3 lb/gal; pH 6.5 (2%); flash pt. 116 C; 99% act.

Nopco NDW. [187] Defoamer for SBR, PVAc, acrylic, water-sol. resins paint and adhesive systems; amber slightly hazy liq.; disp. in all latex and adhesive systems; sp.gr. 0.90; dens. 7.5 lb/gal; pH 6.5 (2%); flash pt. 79 C; 99.9% act.

Nopco NXZ. [187] Defoamer for syn. latex emulsions, paint and adhesives from SBR, PVAc, acrylic, water-sol. resins; amber hazy liq.; forms milky emulsion in water; sp.gr. 0.91; dens. 7.6 lb/gal; pH 7 (2%); flash pt. 171 C; 99% act.

Nopco PD#1-D. [187] Defoamer for adhesives, paints, joint compds., plaster; off-wh. powd.; water-wettable; dens. 22–26 lb/ft^3; pH 8.7 (1%); 65% act.

Nopco RDY. [187] Nonoxynol-50; nonionic; agric. surfactant; solid; cloud pt. 100 C; pour pt. 46 C.

Nopcocastor, L. [187] Castor oil, sulfated; anionic; emulsifier; superfatting agent for cosmetics; ironing aid; liq.; 75 and 50% conc. resp.

Nopcogen 14-L. [187; 192] Lauric alkylolamine condensate; detergent, wetting, emulsifier; liq.; 100% conc.

Nopcogen 14-LT. [192] Coconut fatty acid alkylamide; nonionic; detergent; cleaning and sterilizer of surgical instruments; liq.; 99+% conc.

Nopcogen 14-S. [192] Stearamide DEA; textile softener; solid; m.p. 48 C; solid; 100% conc.

Nopcogen 16-L. [192] Lauric polyamine condensate; emulsifier and textile finishing agent; m.p. 60 C; solid; 100% conc.

Nopcogen 16-O. [192] Oleic polyamine condensate; emulsifier; w/o emulsions with min. and veg. oils; solid; m.p. 44 C; 100% conc.

Nopcogen 22-O. [187] Oleyl imidazoline; cationic; emulsifier for min. oil, kerosene, wetting agent, corrosion inhibitor, used in textile, asphalt, paper industries; amber, visc. liq.; sol. in water, acetic acid, hydrochloric acid; misc. with pine oil and min. oil; cloud pt. 0 C; 90% act.

Nopcolan SHR3. [192] Thiosulfate, org.; anionic; shrink resister for wool; liq.; 42% conc.

Nopcosant. [187] Sodium salt of condensed naphthalene sulfonate; dispersant for NR, SR latexes; tan powd.; forms clear sol'ns. in water; sp.gr. 1.02; pH 7.5–8.0 (2%).

Nopcosant K. [187] Anionic polymer; dispersant for latex systems; clear liq.; misc. with water in all proportions; dens. 10.9 lb/gal; pH 12 (5%); 34% act.

Nopcosant L. [187] Sulfated naphthalene; anionic; NR, SR latex dispersant for use in aq. systems, paints, paper coatings, textiles, and leather; lt. yel. liq., odorless; water sol.; sp.gr. 1.18; dens. 9.8 lb/gal; pH 9.6 (5%); 25% solids.

Nopcosperse 44. [187] Sodium neutralized polycarboxylate; anionic; dispersant for paint systems; clear lt. amber liq.; sol. in water and glycols; dens. 10.3 lb/gal; pH 7.8; 35% act.

Nopcosulf CA-60, -70. [187; 192] Castor oil, sulfated; anionic; softener in finishing starches, gums; plasticizer for starches; furniture base polish; liq.; 68 and 63% conc. resp.

Nopcosulf TA-30, -45V. [187; 192] Sulfated tallow; anionic; softener for cotton goods; paste; 44 and 78% conc. resp.

Noram 2 C. [117] N-dicoco amine; cationic; synthesis intermediate, anticaking agent, flotation, antistripping for road making, soil stabilization; auxs. for fuel additives, rust inhibition, paint; chemical intermediate for quats., betaines, amine oxides; solid.

Noram 2 SH. [117] NN-dihydrog. tallow amine; cationic; anti-caking agent; solid.

Noram C. [117] N-coco amine; cationic; see Noram 2 C; pasty.

Noram DMC. [117] N-coco dimethylamine; cationic; see Noram 2 C; liq.

Noram DMS. [117] N-tallow dimethylamine; see Noram 2 C; liq.

Noram DMSH. [117] N-hydrog. tallow dimethylamine; cationic; see Noram 2 C; pasty.

Noram M2C. [117] N-dicoco methylamine; cationic; see Noram 2 C; pasty.

Noram M2 SH. [117] N-dihydrog. tallow methylamine; cationic; see Noram 2 C; solid.

Noram O. [117] Oleyl amine; synthesis intermediate, lubricant and textile industries; solid.

Noram S. [117] N-tallow amine; cationic; synthesis intermediate; oil industry; solid.

Noram SH. [117] N-hydrog. tallow amine; cationic; anti-caking agent; solid.

Noramac C. [188] N-coco amine acetate; cationic; flotation, bactericide, emulsifier, anticaking, soil stabilization, flocculation, corrosion inhibitor; pasty/solid.

Noramac O. [188] Oleyl amine acetate; cationic; see Noramac C; pasty/solid.

Noramac S. [188] N-tallow amine acetate; see Noramac C; pasty/solid.

Noramium MC 50. [188] Coco trimethyl ammonium chloride; cationic; additive for antibiotics mfg.; liq.; 50% conc.

Noramium MO 50. [188] Oleyl trimethyl ammonium chloride; cationic; see Noramium MC 50; liq.

Noramium MSH 50. [188] Hydrog. tallow trimethyl ammonium chloride; cationic; see Noramium MC 50; liq.; 75% conc.

Noramium S 75. [188] Tallow dimethyl benzyl ammonium chloride; cationic; see Noramium MC 50.

Noramox C2 to 15. [188] N-ethoxylated coco

amine; cationic; emulsifier, drying assistant, viscose additive, metal treatment, rust inhibitor antistat for ABS, polystyrene; liq.

Noramox S2 to 11. [188] N-ethoxylated tallow amine; cationic; see Noramox C2; pasty.

Norfox 40. [451] Sodium dodecylbenzene sulfonate; anionic; base for industrial and agric. wetting agents and detergents; dry blending applications; beads; 40% conc.

Norfox 85. [451] Sodium dodecylbenzene sulfonate; anionic; base for industrial and agric. wetting agents and detergents; flakes, granules; 85% conc.

Norfox 90. [451] Sodium dodecylbenzene sulfonate; anionic; see Norfox 85; also used in paper industry; flakes, granules; 90% conc.

Norfox 916. [451] Linear primary alcohol, ethoxylated; nonionic; wetting agent; wet processing aid for paper, textiles and leather; liq.; HLB 12.5; 100% conc.

Norfox 1101. [451] Potassium cocoate; anionic; shampoo base; emulsifier; liq.; 40% conc.

Norfox 1115. [451] Potassium veg. oil soap; anionic; soap base; emulsifier; hand soap conc.; liq.; 37% conc.

Norfox ALES-60. [451] Ammonium pareth sulfate; anionic; window cleaner surfactant; liq.; 60% conc.

Norfox ALPHA XL. [451] Sodium alpha olefin sulfonate; anionic; base for shampoo and lt. duty liqs.; liq.; 40% conc.

Norfox ALS. [451] Ammonium lauryl sulfate; anionic; base for woolen carpet shampoo; liq.; 30% conc.

Norfox B. [451] Sodium stearate; anionic; gelling agent; stabilizer in cosmetics; lubricant; granules; 96% conc.

Norfox Coco Betaine. [451] Coco amidoalkyl betaine; amphoteric; base for shampoos; liq.; 30% conc.

Norfox Coco Powder. [451] Sodium cocoate; anionic; soap base; stabilizer; gelling agent; granules; 92% conc.

Norfox DC. [451] Coconut oil DEA condensate; nonionic; intermediate for detergent mfg., cosmetics, suds stabilizer and dedusting agent for dry products; liq.; 100% conc.

Norfox DCO. [451] Oleic acid modified coconut oil DEA condensate; anionic/nonionic; visc. builder; metal weaving, drawing, stamping or cold rolling lubricant; burnishing, heavy duty cleaner for soft water areas; liq.; 100% conc.

Norfox DCS. [451] Alkylaryl sulfonate modified coconut oil DEA condensate; anionic/nonionic; wetting agent and detergent base for textiles; base for specialty formulations for auto and home care products; liq.; 100% conc.

Norfox DESA. [451] Cocamide DEA; nonionic; visc. and foam enhancer; liq.; 100% conc.

Norfox DLSA. [451] Lauramide DEA; nonionic; specialty foam booster and visc. builder for shampoo and bubble baths; waxy solid; 100% conc.

Norfox F-221. [451] Complex fatty amido ester; nonionic; emulsifier; surfactant in oils; liq.; 100% conc.

Norfox GMS-FG. [451] Glycerol monostearate; nonionic; cosmetic opacifier, food grade emulsifier; flake; HLB 3.9; 100% conc.

Norfox IM-38. [451] N-oleyl imidazolinium hydrochloride; cationic; wetting agent and emulsifier; base for automotive rinse-wax formulations; liq.; 52% act.; 93% org.

Norfox KD. [451] Cocamide DEA and diethanolamine; nonionic; visc. enhancer and foam booster; liq.; 100% conc.

Norfox KO. [451] Potassium oleate; anionic; liq. soap for hand cleaners, tire mounting lubricant; emulsifier and corrosion control in paint strippers; liq.; HLB 20.0; 80% conc.

Norfox NP-1. [451] Nonoxynol-1; nonionic; emulsion stabilizer and defoamer; liq.; HLB 4.5; oil-sol.; 100% conc.

Norfox NP-4. [451] Nonoxynol-4; nonionic; emulsifier, detergent and dispersant for petroleum based lubricants; intermediate for sulfanic acid sulfonation to produce foaming agent; liq.; oil-sol.; HLB 9.0; 100% conc.

Norfox NP-6. [451] Nonoxynol-6; nonionic; emulsifier; wetting agent for oil-based systems; liq.; HLB 11.0; 100% conc.

Norfox NP-7. [451] Nonoxynol-7 (7.5 EO); nonionic; detergent for cleaners; liq.; HLB 12-13; 100% conc.

Norfox NP-9. [451] Nonoxynol-9; nonionic; formulation of sanitizers; dedusting oil in detergent; liq.; HLB 13.0; 100% conc.

Norfox NP-11. [451] Nonoxynol-11; nonionic; see Norfox NP-9; liq.; HLB 13.0-14.0; 100% conc.

Norfox PEA-N. [451] Complex fatty amido phosphate; anionic; detergent; textile wetting agent; specialty institutional and metal cleaners; liq.; 100% conc.

Norfox PEW. [451] Mono and dialkyl acid phosphate; anionic; detergent; emulsifier; coupling agent; corrosion inhibitor; liq.; 100% conc.

Norfox 1 Polyol. [451] Hydrophobic block polymer; nonionic; defoamer base for surfactant mixtures; liq.; HLB 3.0; 100% conc.

Norfox 2 Polyol. [451] Mid-range block polymer; nonionic; detergent base for controlled foam; liq.; HLB 7.0; 100% conc.

Norfox 2LF Polyol. [451] Low foam block polymer; nonionic; detergent base for low foam; liq.; HLB 7.0

Norfox 4 Polyol. [451] Hydrophilic block polymer; nonionic; control foam solubilizer; liq.; HLB 11.0; 100% conc.

Norfox SEHS. [451] Sodium 2-ethylhexyl sulfate; anionic; foamer with wetting and detergency; coupling agent; liq.; 40% conc.

Norfox SLES-02. [451] Sodium laureth sulfate; anionic; foaming and wetting agent for household and cosmetic specialties; liq.; 28% conc.

Norfox SLES-03. [451] Sodium laureth sulfate; anionic; foaming agent; liq.; 28% conc.

Norfox SLES-30. [451] Sodium pareth sulfate; anionic; foaming detergent base; ingred. in window cleaners, car wash base; liq.; 30% conc.

Norfox SLES-60. [451] Sodium pareth sulfate; anionic; base for shampoos and lt. duty liq. formulators; liq.; 60% conc.

Norfox SLS. [451] Sodium lauryl sulfate; anionic; foamer and wetting agent for household and cosmetic specialties; paste; HLB 40.0; 29% conc.

Norfox Sorbo S-60FG. [451] Sorbitan monostearate; nonionic; food emulsifiier in veg. and dairy products; flake; HLB 4.7; 100% conc.

Norfox Sorbo-S-80. [451] Sorbitan monooleate; nonionic; hydrophobic emulsion; liq.; HLB 4.3; 100% conc.

Norfox T-60. [451] N,N,N-triethanolamine dodecylbenzene sulfonate; anionic; detergent base; specialty wetting agent and foamer for agric. and industrial use; liq.; 60% conc.

Norfox TB Granules. [451] Sodium tallow/coco soap; anionic; soap base for mfg. specialty bar soap; ingred. in glycerine bar soap; granules; 92% conc.

Norfox Vertex Flakes. [451] Sodium oleate; anionic; textile scouring, fulling, dye leveler; flake; HLB 20.0; 90% conc.

Norfox X. [451] Cocamide DEA; modified; cationic/nonionic; used with acidic builders; wash-wax formulations, glass and appliance cleaners; liq.; 93% conc.

Norlig. [40] Calcium and sodium lignosulfonate; anionic; binder, dispersant, used to manufacture clay and brick products, linoleum mastics, soil stabilizer; liq. or powd.; 50–60% conc. (liq.), 100% conc. (powd.).

Novel 1412-70 Ethoxylate. [634] Linear alcohol, ethoxylated (70% EO); nonionic; surfactant, wetting agent, emulsifier, detergent formulation; wh. solid; HLB 11.0; 100% conc.

Novol. [159] Super refined oleyl alcohol; emollient, stabilizer, superfatting agent, suspending aid, used in cosmetics, personal care products; liq.; mild odor; misc. with fat, oil and wax mixts.

Noxamine CA 30, 02/30. [117] Tert. amine oxide; cationic; foamer; liq.; 30% conc.

Noxamium. [117] Quat., ethoxylated; cationic; emulsifier, corrosion; liq.

Noxamium C2-15. [117] Cocoamine, ethoxylated, ammonium quat. deriv.; cationic; antistat, bactericidal emulsifier; liq.; 50% conc.

Noxamium S2-11. [117] Tallow amines, ethoxylated, ammonium quat. deriv.; cationic; see Noxamium C2-15; liq.; 50% conc.

NP-55-40. [287] Nonoxynol-4; nonionic; wetting agent for pigments; foam suppressing additive to wetting agents; detergent base material; emulsifier for insecticides, resin dispersions; liq.; HLB 9.0; 100% conc.

NP-55-50. [287] Nonoxynol-5; nonionic; wetting agent, component of detergents; emulsifier for insecticides, resin dispersions; liq.; HLB 10.0; 100% conc.

NP-55-60. [287] Nonoxynol-6; nonionic; wetting agent for agric. chemicals, color pigments; detergent component; emulsifier for veg. and castor oils; liq.; HLB 11.0; 100% conc.

NP-55-80. [287] Nonoxynol-8; nonionic; wetting agent for textile industry, coloring material pigments; detergent base material for industrial and household detergents; emulsifier; liq.; HLB 12.0; 100% conc.

NP-55-85. [287] Nonoxynol-8 (8.5 EO); nonionic; wetting agent for textile industry; detergent base material; emulsifier for technical purposes; liq.; HLB 12.5; 100% conc.

NP-55-90. [287] Nonoxynol-9; nonionic; wetting agent for textile, paper, leather and other industries; detergent base material; emulsifier for tech. purposes; liq.; HLB 13.0; 100% conc.

NP-55-95. [287] Nonoxynol-9 (9.5 EO); nonionic; wetting agent, detergent base material, emulsifier for technical purposes; liq.; HLB 13.0; 100% conc.

NP-55-120. [287] Nonoxynol-12; nonionic; see NP-55-90; liq.; HLB 14.0; 100% conc.

NP-55-150. [287] Nonoxynol-15; nonionic; wetting agent for the textile industry, galvanic and other industries; detergent base material; emulsifier for technical purposes; liq.; HLB 15.0; 100% conc.

NP-55-300. [287] Nonoxynol-17; nonionic; wetting agent, detergent base material, corrosion inhibitor galvanic industry; solid; HLB 17.0; 100% conc.

Nutrapon AL 1. [145] Ammonium lauryl ether (1) sulfate; anionic; for personal care products; liq.; 27-29% conc.

Nutrapon AL 30. [145] Ammonium lauryl ether (3) sulfate; anionic; personal care products; dishwashing and lt. duty detergents; liq.; 26-28% conc.

Nutrapon AL 60. [145] Ammonium lauryl ether sulfate; anionic; see Nutrapon AL 30; liq.; 57-60% conc.

Nutrapon B. [145] Sodium lauryl sulfate and ethylene glycol monostearate; anionic/nonionic; shampoo and detergent blend with pearlizing agent for pearlescent formulations; visc. liq., paste; 32-34% conc.

Nutrapon DE 3796. [145] Diethanolamine lauryl sulfate; anionic; bubble bath and shampoo conc.; liq.; 36% min. conc.

Nutrapon ES 2. [145] Sodium lauryl ether (2) sulfate; anionic; used in personal care products; liq.; 25-26% conc.

Nutrapon ES 60. [145] Sodium lauryl ether (3) sulfate; anionic; personal care products; liq.; 56-59% conc.

Nutrapon ESY. [145] Sodium lauryl ether (1) sulfate; anionic; see Nutrapon ES 60; liq.; 24.5-25.5% conc.

Nutrapon HA 3841. [145] Ammonium lauryl sulfate; anionic; personal care products and detergent cleaners; liq.; 27.0-28.5% conc.

Nutrapon KF 3846. [145] Ammonium lauryl ether sulfate; anionic; see Nutrapon ES 60; liq.; 27-29% conc.

Nutrapon RS 1147. [145] Sodium lauryl sulfate/sodium sarcosinate; anionic; surfactant for rug and upholstery shampoo; liq.; 28-31% conc.

Nutrapon TLS-500. [145] Triethanolamine lauryl sulfate; anionic; ingred. in cosmetic and industrial formulations; liq.; 39% min. conc.

Nutrapon W, WAC, WAQ. [145] Sodium lauryl sulfate; anionic; personal care and industrial formulations; visc. liq.; 28-30% conc.

Nutrol 100. [145] POE (9) octyl phenol ether; nonionic; wetting agent and dispersant; used in metal and acid cleaners and pesticide formulations; liq.; HLB 13.0; 99% min. conc.

Nutrol 600. [145] POE (9) nonylphenol ether; nonionic; dispersant, emulsifier; wetting agent, detergent; liq.; HLB 13.0; 99% min. conc.

Nutrol 611. [145] POE (8) nonylphenol ether; nonionic; see Nutrol 600; liq.; HLB 12.2; 99% min. conc.

Nutrol 622. [145] POE (4) nonylphenol ether; nonionic; emulsifier, lt. duty detergents, liq.; HLB 8.8; 99% min. conc.

Nutrol 640. [145] POE (15) nonylphenol ether; nonionic; see Nutrol 622; liq.; HLB 15.0; 99% min. conc.

Nutrol 656. [145] POE (11) nonylphenol ether; nonionic; see Nutrol 622; liq.; HLB 13.6; 99% min. conc.

Nutrol E-710. [145] Linear primary alcohol ethoxylate (10.6 EO); nonionic; see Nutrol 622; soft max.; HLB 14.2; 99% min. conc.

Nutrol S 60. [145] Ammonium nonylphenoxy polyethoxy alcohols (70% EO); anionic; emulsifier for use in polymers; detergent; wetting agent; liq.; 58 ± 1% conc.

Nylomine Assistant DN. [320] Blend of surfactants; anionic; dyeing assistant, leveling agent for dyeing fibers; yel. liq.; sol. in water; sp.gr. 1.0; f.p. 0 C; b.p. 100 C; pH 6.0–7.5 (1%); 55% act.

O

Obazoline 662Y. [606] Imidazoline deriv.; amphoteric; antistat and softener for syn. fibers; base material for shampoos and hair rinse; liq.; 35% conc.

Obazoline LB-40. [606] n-Alkyl betaine; amphoteric; antistat for hair products; liq.; 40% conc.

Ocetox 5525. [654] Fatty alcohol, ethoxylated; nonionic; emulsifier for oleine, veg. oils; detergent; paste.

Octapol 60, 100, 300, 400. [539] POE octylphenyl ether; nonionic; base material for detergents; liq., solid; HLB 11.3, 13.6, 17.3 and 17.9 resp.; 100% conc.

Octaron PS 80. [551] POE nonylphenol ether sulfated, sodium salt; anionic; detergent; paste; 80% conc.

Ogtac-85. [131] Glycidyl trimethyl ammonium chloride; cationic; starch modifier, intermediate for paper, textile and cosmetic industry; liq.; 70% conc.

Ogtac-85 V. [131] Glycidyl trimethyl ammonium chloride; cationic; intermediate for surfactants, starch modifier; liq.; 80% conc.

OHlan. [37; 38] Hydroxylated lanolin; nonionic; primary w/o emulsifier, aux. emulsifier and stabilizer, pigment wetting and dispersing agent, emollient and conditioner in personal care products; yel.-amber to lt. tan waxy solid; misc. with common oil phase ingredients, sol. at low levels in castor oil; m.p. 39–46 C; HLB 4; sapon. no. 95–110; 100% conc.

OL-55-F-2. [287] POE (2) oleyl ether; nonionic; emulsifier for cosmetic and pharmaceutical fluid emulsions; for immiscible oils; liq.; HLB 5.0; 100% conc.

OL-55-F-10. [287] POE (10) oleyl ether; nonionic; emulsifier for cosmetic and pharmaceutical fluid emulsions; solubilizer for essential oils; leveling and stripping agent for textiles; wetting agent; for lubricating oil emulsions; liq., pasty; HLB 13.5; 100% conc.

OL-55-F-20. [287] POE (20) oleyl ether; nonionic; o/w emulsifier for cosmetic and pharmaceutical fluid emulsions; wetting agent for leveling and stripping agents in the textile industry; various tech. applications; soft wax; HLB 15.5; 100% conc.

Olicine. [248] Monoglyceride 40%; nonionic; food emulsifier; liq.; HLB 3.0; 100% conc.

Onamer M. [456] Polyquaternium-1; cationic; personal care compd.; liq.; sp.gr. 1.06; flash pt. > 200 F; 30% solids

Onamine 12. [456] *n*-Dodecyl dimethylamine; cationic; intermediate used in synthesis of surfactants, antioxidants, oil and grease additives; liq.; sp.gr. 0.78; flash pt. > 200 F; 100% conc.

Onamine 14. [456] *n*-Tetradecyl dimethylamine; cationic; see Onamine 12; liq.; sp.gr. 0.80; flash pt. > 200 F; 100% conc.

Onamine 16. [456] *n*-Hexadecyl dimethylamine; cationic; see Onamine 12; liq.; sp.gr. 0.80; flash pt. > 200 F; 100% conc.

Onamine 18. [456] *n*-Stearyl dimethylamine; cationic; see Onamine 12; liq.; sp.gr. 0.80; flash pt. > 200 F; 100% conc.

Onamine 65, 835, 1214, 1416. [456] *n*-Alkyl dimethylamine; cationic; see Onamine 12; liq.; sp.gr. 0.79; flash pt. > 200 F; 100% conc.

Onyxide 75. [456] *n*-Alkenyl dimethyl ethyl ammonium bromides; cationic; algicide used in recirculating water systems, swimming pools, humidifiers; paste; sp.gr. 0.95; flash pt. 68 F; 75% min. act.

Onyxide 172. [456] Quaternium-8; cationic; org. nonvolatile antifungal agent and preservative for latex emulsions, paints, adhesives, coated fabrics, cutting oils; liq.; sp.gr. 1.02; flash pt. > 200 F; 77–81% act.

Onyxide 200. [456] Hexahydro-1.3.5-tris (2-hydroxethyl)-s-triazine; preservative for sol. cutting fluids and coolants; liq.; sp.gr. 1.15; flash pt. > 200 F; 78.5 min. act.

Onyxide 3300. [456] Myristalkonium saccharinate; germicide for toiletry and pharmaceutical formulations; powd.; sp.gr. 0.38; flash pt. > 200 F; 95% min. act.

Onyxol 42. [456] Stearamide DEA; softener, dye carrier, opacifying and thickening agent, used in textile and cosmetic industries; wax; sp.gr. 0.97; flash pt. > 200 F.

Onyxol 336. [456] Lauramide DEA; nonionic; foaming, wetting, and dispersing agent, used in personal care products; liq.; sp.gr. 1.01; flash pt. > 200 F; 97% conc.

Onyxol 345. [456; 575] Lauramide DEA; nonionic; foam stabilizer, wetting agent, dispersant, thickener, detergent used in cosmetic formulations, household detergents; liq.; sp.gr. 1.00; flash pt. > 200 F; 97% conc.

Onyxol 2062. [456] Mixed fatty acid/tall oil fatty acid modified 2:1 diethanolamide; nonionic; surfactant for hard surface cleaners; liq.; sp.gr. 0.99; flash pt. > 200 F.

Onyxol SD. [456; 575] Lauramide DEA: nonionic; foam stabilizer and thickener for personal care products and lt.-duty detergents; liq.; 100% conc.

Onyxol WW. [456] Mixed fatty acid diethanolamide; used in detergents; liq.; sp.gr. 0.99; flash pt. > 200 F.

OP-55. [287] POE (9) nonylphenyl ether; nonionic; wetting agent and detergent conc.; liq.; 80% conc.

OP-2000. [86] Oleic acid-PO adduct; cosmetic emollient, lubricant, defoamer, visc. control agent, dispersant, spreading agent for personal care products; pale yel. liq., mild, pleasant odor; m.w. 2000; sol. in IPM, IPA; insol. water; sp.gr. 0.986; dens. 8.3 lb/gal; visc. 267 cps; pour pt. –20 F; sapon no. 23; 100% act.

Oramide DL 200. [551] Cocamide DEA; nonionic; cosmetic additive; wax; 100% conc.

Oramix L30. [551] Lauryl sarcosinate; anionic; personal care products; liq.; 30% conc.

Orapol HC. [551] Fatty diethanolamide; nonionic; detergent; liq.; 100% conc.

Orgozon 3157. [145] Phosphate ester, neutralized; anionic; detergent; liq.; 47-48% conc.

Orgozon CC 1118. [145] Fatty phosphate ester, potassium salt; anionic; detergent and emulsifier; liq.; 35-36% conc.

Orzan A. [165] Ammonium lignin sulfonate; dispersant used to reduce slurry visc. for fine mineral products, used for skins, leather, metallurgy, etc.; br. powd.; 55% sol. in water; sp.gr. 1.21 (50% aq.); visc. 300 cps (50% aq.); 95% conc.

Orzan AL-50. [332] Ammonium lignosulfonate aq. sol'n.; binder for linoleum paste; dispersant for gypsum boards; also for briquetting, adhesives, resins; liq.; sp.gr. 1.230–1.240; visc. 500–800 cps; pH 5; biodeg.; 65% lignosulfonate.

Orzan LS. [332] Sodium lignosulfonate; dispersant, suspending agent in wettable pesticides; sequestrant in water treatment; emulsifier, stabilizer in oil and wax emulsions; industrial cleaners; ore flotation; yel. powd.; rapidly sol. in water; bulk dens. 30 lb/ft^3; pH 6.5 (25%); biodeg.; 59% lignosulfonate.

Orzan LS-50. [332] Sodium lignosulfonate aq. sol'n.; see Orzan LS; liq.; sp.gr. 1.255–1.265; visc. 100 cps (20 C); pH 7.0 (25%); biodeg.; 47% sol'n., 59% lignosulfonate.

Orzan S. [165] Sodium lignin sulfonate; dispersant used to reduce slurry visc. for fine mineral products, used for skins, leather, metallurgy; br. powd.; 60% sol. in water; sp.gr. 1.264 (50% aq.); visc. 250 cps (50% aq.); 95% conc.

Orzan SL-50. [332] Sodium lignosulfonate aq. sol'n.; binder for carbon black pellets, plywood, particle board; dispersant for portland cements, gypsum board, limestone slurry; also for briquetting, adhesives, resins; liq.; sp.gr. 1.235–1.245; visc. 200 cps (20 C); pH 6.5 (25%); biodeg.; 56% lignosulfonate.

Osimol Grunau 109. [277] Alkylaryl sulfonate; anionic; leveling agent for dyeing polyamide; liq.; 60% conc.

Osimol Grunau 110. [277] Alkylaryl sulfonate and ethylene oxide deriv.; anionic; see Osimol Grunau 109; liq.; 50% conc.

Osimol Grunau DP. [277] Sulfonic acid; anionic; dispersant for dyeing polyester; liq.; 40% conc.

Osimol Grunau EFA. [277] Amide amine deriv.; cationic; leveling agent for dyeing acrylic; liq.; 50% conc.

Osimol Grunau MA. [277] Amide amine deriv.; cationic; migrating agent for dyeing acrylic; liq.; 55% conc.

Osimol Grunau PHT. [277] EO deriv.; cationic; dispersant and leveling agent for polyester dyeing; liq.; 100% conc.

Osimol Grunau RAC. [277] Amide amine deriv.; cationic; see Osimol Grunau EFA; liq.; 50% conc.

Osimol Grunau SF. [277] Sulfonic acid; anionic; dispersant and protective colloid for dyeing applications; liq.; 45% conc.

Ospin Salt ON, TAN. [607] Quat. ammonium salt; cationic; migrator, retarder used for dyeing fibers; liq.; 50% conc.

Oxamin LO. [316] Lauryl dimethyl amine oxide; nonionic; detergent, food foamer and foam stabilizer for personal care products and detergent formulations; liq.; 30% conc.

Oxetal C 110, D 104, O 108/112, S 125, T 103, T 106/110, 111/118. [660] Fatty alcohol polyglycol ether; nonionic; detergent, dispersant, emulsifier, wetting agent used in detergents and cleaners for household and industry; aux. agent for textile, paper and leather industry; paste, liq.; wax; 100% conc.

Oxypon 2145. [660] PEG-15 glyceryl isostearate; nonionic; superfatting agent; liq.; 100% conc.

P

P-10 Acid. [115] Ricinoleic acid; chemical intermediate; liq.

P & G Amide No. 27. [489] Coconut fatty acid monoethanolamide; nonionic; foam stabilizer, thickener; foam booster and visc. builder for detergents and shampoos; flakes; 100% conc.

Panalite 40 HVK. [460] Mono- and diglycerides; emulsifier in food industry in w/o emulsions; flakes; HLB 2.8.

Panalite 40 SVK, 50 SA, 50 SVK. [460] Mono- and diglycerides; see Panalite 40 HVK; plastic; HLB 2.6, 3.5, and 3.5 resp.

Panalite 50 HVK. [460] Mono- and diglycerides; see Panalite 40 HVK; beads; HLB 3.5.

Panalite 100 K. [460] Mono- and diglycerides and polysorbate 80; see Panalite 40 HVK; beads; HLB 5.2.

Panalite EOM-K. [460] Mono- and diglycerides, ethoxylated; see Panalite 40 HVK; plastic; HLB 13.1.

Panalite MP Hydrate. [460] Hydrated mono- and diglycerides, polysorbate 60; hydrated emulsifier system; plastic; HLB 6.3; 7% alpha mono.

Panalite MPH, MPS. [460] Hard and soft mono- and diglycerides and polysorbate 60; see Panalite 40 HVK; flakes and plastic resp.; HLB 6.0 and 8.1.

Panatex, Panatex HM. [460] Dist. monoglycerides; see Panalite MP Hydrate; HLB 3.5; 20 and 35% min. alpha mono resp.

Panodan 235, FDP-K, SD, SD-K. [273] Monoglyceride diacetyl tartaric acid ester; anionic; food emulsifier; wh. fine powd., ivory fine powd., amber semiliq., and amber semiliq. resp.; m.p. 45 C; sapon. no. 395–420.

Panodan FDP-K. [273] Monoglyceride diacetyl tartaric acid ester; see Panodan 235; ivory fine powd.; m.p. 45 C; sapon. no. 395–420.

Panodan SD, SD-K. [273] Monoglyceride diacetyl tartaric acid ester; food emulsifier; amber semi-liq.; m.p. 45 C; sapon. no. 395–420.

Pantex [600] Alkylaryl sulfonate; institutional hand washing; powd.; 10% conc.

Paracol OP. [440] Alkyl phosphate; anionic; anticorrosion agent, emulsion breaker; liq.; 100% conc.

Paracol SV. [440] Alkyl phosphate; anionic/nonionic; emulsifier for org. phosphate insecticides; liq.; 100% conc.

Paraffin Wax Emulsifier CB0674, CB0680. [159] Polyethoxylated fatty alcohols; nonionic; paraffin wax emulsifier; solid; HLB 10.6 and 10.5 resp.; 100% conc.

Paralan. [159] Tech. lanolin; nonionic; emulsifier for liq. emulsions; solid.

Patco 3. [462] Sodium stearoyl lactylate and calcium stearoyl-2-lactylate; conditoner/softener for food industry; lt. tan powd., mild caramel odor; acid no. 55–83; 100% act.

Pationic 122A. [462; 519] Sodium capryl lactylate; anionic; surfactant with microbial inhibitor properties; amber clear visc. liq.; sol. in propylene glycol, IPM, IPA, water; visc. 24 cps; HLB 11.3; sapon. no. 235–265; surf. tens. 24.72 dynes/cm; pH 5.25 (2% aq.); 100% conc.

Pationic 138C. [462; 519] Sodium lauroyl lactylate; anionic; detergent, conditioner in shampoos, visc. builder, lipophilic emulsifier; beige waxy solid; sol. in water; visc. 19 cps (2%); m.p. 55–59 C; HLB 14.4; sapon. no. 175–205; surf. tens. 24.30 dynes/cm; pH 6.90 (2% aq.); 100% conc.

Pationic 145A. [462] Sodium stearoyl-1-lactylate; anionic; emulsifier for cosmetics; wh. flaked solid; disp. in water; visc. 185 cps (2%); m.p. 57–60 C; HLB 8.3; sapon. no. 169–190; surf. tens. 28.87 dynes/cm; pH 5.70 (2% aq.); 100% conc.

Pationic CSL. [462] Calcium stearoyl-2-lactylate; anionic; emulsifier and protein complexer for cosmetics; lt. tan powd.; visc. 85 cps; m.p. 40–45 C; HLB 5.0; sapon. no. 195–230; surf. tens. 37.00 dynes/¢m; pH 4.90 (2% aq.); 100% conc.

Pationic ISL. [462] Sodium isostearoyl-2-lactylate; anionic; surfactant for cosmetics; straw, honey clear visc. liq.; sol. in min. oil, propylene glycol, IPM, IPA; disp. in water; HLB 5.9; sapon. no. 205–225; surf. tens. 26.28 dynes/cm; pH 6.30 (2% aq.); 100% act.

Pationic NMF. [462] Aliphatic esters and alcohol; nonionic; emollient and moisturizer for cosmetics; clear liq.; bland odor; oil-sol.; dens. 7.3 lb/gal; sp.gr. 0.888; visc. 40 cps; sapon. no. 170–190; ref. index 1.4456; 100% conc.

Pationic SSL. [462; 519] Sodium stearoyl 2-lactylate; anionic; emulsifier, visc. builder and protein complexer for cosmetics; lt. tan powd.; disp. in water; visc. 175 cps (2%); m.p. 41–46 C; HLB 6.5; sapon. no. 210–235; surf. tens. 32.34 dynes/cm; pH 5.95 (2% aq.); 100% conc.

Patlac IL. [462] Isostearyl lactate; nonionic; surfactant for cosmetics; pale yel. visc. liq.; sol. in ethanol, propylene glycol, min. oil, IPM; sp.gr. 0.92; HLB 10.3 ± 1.0; pH 4.5 (2% aq.); 100% conc.

Pave 100. [138] Complex amide amine; asphalt additive, wetting and bonding aid; antistripper used in asphalt cement, cutbacks, and emulsions for paving; dk. brn. visc. liq.; dens. 8.2 lb/gal (60 F); sp.gr. 0.985 (60 F); visc. 6600 SUS (100 F); flash pt. 450 F (COC); pour pt. 40 F; 100% act.

Pave 192. [138] Complex amide amine; see Pave 100; dk. brn. liq.; dens. 7.8 lb/gal (60 F); sp.gr. 0.934 (60 F); visc. 1104 SUS (100 F); flash pt. 420 F (COC); pour pt. 10 F; 100% act.

Pave Bond LP. [138] Complex amide amine; coating additive; used in asphalt cement and cutbacks, stockpile mixes; water stripping; dk. brn. liq.; dens. 7.68 lb/gal; sp.gr. 0.921 (77/60 F); visc. 466 cs; flash pt. (COC) 315 F; pour pt. 15 F; 100% act.

Pave Bond Special. [138] Com.p.lex amide amine; see Pave Bond LP; dk. brn liq.; dens. 8.04 lb/gal (100 F); sp.gr. 0.963 (100/60 F); visc. 92.3 cs (100 F); flash pt. 310 F (COC); pour pt. 23 F; 100% act.

PB-92. [503] Calcium lignosulfate, modified; anionic; wetting agent, emulsifier, dispersant; oil well mud thinner; yel. solid; water-sol.; 95% conc.

PC-35. [460] Mono- and diglyceride hydrate; emulsifier used in food industry; plastic; HLB 3.2.

Pegafac CA 600. [97] Phosphate ester; anionic; detergent, dispersant, and wetting agent in textile wet processing; APHA 200 max. clear liq.; sol. in xylene, perchloroethylene; sp.gr. 1.07; pour pt. < 0 C; pH < 2.5 (5%); flash pt. > 93 C; 100% conc.

Pegafac CE 410, 610. [97] Phosphate ester; anionic; see Pegafac CA 600; also antistat, lubricant, and emulsifier for fiber and metal lubricants; APHA 200 max. clear liq.; sol. see Pegafac CA 600; sp.gr. 1.10 and 1.11 resp.; pour pt. 18 and < 0 C resp.; pH < 2.5 (5%); flash pt. > 93 C; 100% act.

Pegafac CP 710, CR 719. [97] Complex org. phosphate ester; anionic; see Pegafac CA 600; liq.; 100 and 90% conc. resp.

Pegafac CS 410, 610. [97] Phosphate ester; anionic; see Pegafac CE 410; APHA 200 max. hazy to cloudy liq.; sol. see Pegafac CA 600; sp.gr. 1.02 and 1.04 resp.; pour pt. < 0 C; pH < 2.5 (5%); flash pt. > 93 C; 100% act.

Pegafac CS 710. [97] Phosphate ester; anionic; see Pegafac CE 410; APHA 200 max. cloudy liq.; sol. in xylene, perchloroethylene; sp.gr. 1.05; pour pt. 18 C; pH < 2.5 (5%); flash pt. > 93 C; 100% act.

Pegafac PDA Free Acid. [97; 245] Phosphate ester; anionic; antistat, lubricant, and emulsifier in fiber and metal lubricants; APHA 100 max. hazy to cloudy liq.; sol. in perchloroethylene; sp.gr. 1.02; pour pt. < 0 C; pH < 2.5 (5%); flash pt. > 93 C; 100% act.

Pegafac PDA (Neutralized). [97] Phosphate ester; anionic; see Pegafac PDA Free Acid; APHA 100 max. clear liq.; sol. in xylene, perchloroethylene; sp.gr. 1.00; pour pt. < 0 C; pH 7.5–8.5 (5%); 90% act.

Pegafac PEH. [97; 245] Complex org. phosphate ester; nonionic; see Pegafac CA 600; liq.; 100% conc.

Pegafac PNP 9. [97; 245] Phosphate ester; see Pegafac CA 600; APHA 200 max. clear liq.; sol. in xylene, perchloroethylene; sp.gr. 1.10; pour pt. < 0 C; pH < 2.5 (5%); flash pt. > 93 C; 100% act.

Pegameen 02. [245] Fatty amine, ethoxylated; cationic/nonionic; wetting agent, emulsifier, corrosion inhibitor, antistat, dispersant and detergent; liq.; HLB 4.9; 100% conc.

Pegameen 030, 30 80%. [245] Fatty amine, ethoxylated; cationic/nonionic; see Pegameen 02; solid and liq. resp.; HLB 16.6; 100 and 80% conc.

Pegameen C2, C5, C10, C25. [245] Fatty amine, ethoxylated; cationic/nonionic; see Pegameen 02; liq.; HLB 6.0, 10.4, 13.7, 16.6 resp.; 100% conc.

Pegameen S2, S5, S20. [245] Fatty amine, ethoxylated; cationic/nonionic; see Pegameen 02; liq.; HLB 4.9, 9.0, 15.3 resp.; 100% conc.

Pegameen T2, T5, T15, T25. [245] Fatty amine, ethoxylated; cationic/nonionic; see Pegameen 02; liq.; HLB 5.0, 9.0, 14.2, 16.1 resp.; 100% conc.

Peganate CO 5, 16, 25, 30. [245] Castor oil ethoxylate; nonionic; lubricant, softener, antistat, emulsifier and detergent; liq., solid; HLB 3.9, 8.8, 10.8, 11.7 resp.; 100% conc.

Peganate CO 36, 40, 200, 200 50%. [245] Castor oil ethoxylate; nonionic; see Peganate CO 5; liq.; HLB 12.5, 12.9, 18.1, 18.1 resp.; 100% conc.

Peganate COH 25. [245] Hydrog. castor oil, ethoxylated; nonionic; see Peganate CO 5; liq.; HLB 11.1; 100% conc.

Peganate COH 200, COH 200 50%. [245] Hydrog. castor oil, ethoxylated; nonionic; see Peganate CO 5; solid and liq. resp.; HLB 18.1; 100 and 50% conc.

Peganate L 5, 20. [245] Sorbitan ester ethoxylate; nonionic; emulsifier, solubilizer, lubricant and detergent; liq.; HLB 13.3 and 16.7 resp.; 100% conc.

Peganate MA 8. [245] Fatty acids, ethoxylated; nonionic; see Peganate CO 5; liq.; HLB 10.0; 100% conc.

Peganate MA 15, 15 80%. [245] Fatty acids, ethoxylated; nonionic; see Peganate CO 5; solid and liq. resp.; HLB 14.0; 100 and 80% conc. resp.

Peganate ML 5, 9, 14. [245] Fatty acids, ethoxylated; nonionic; see Peganate CO 5; liq.; HLB 9.8, 13.1, 14.9 resp.; 100% conc.

Peganate MO 6, 9, 14. [245] Fatty acids, ethoxylated; nonionic; see Peganate CO 5; liq.; HLB 9.8, 11.7, 13.6 resp.; 100% conc.

Peganate MS 8, 14, 23. [245] Fatty acids, ethoxylated; nonionic; see Peganate CO 5; solid; HLB 11.3, 13.8, 15.6 resp.; 100% conc.

Peganate O 5, 20. [245] Fatty acids, ethoxylated; nonionic; emulsifier, solubilizer, lubricant and detergent; solid; HLB 10.0 and 9.6 resp.; 100% conc.

Peganate S 5, 20. [245] Fatty acids, ethoxylated; nonionic; see Peganate O 5; solid; HLB 9.6 and 14.9 resp.; 100% conc.

Peganate TO 9, 16. [245] Fatty acids, ethoxylated; nonionic; see Peganate CO 5; liq.; HLB 11.9 and 13.5 resp.; 100% conc.

Peganol A 24, 26. [245] Alcohol, ethoxylated; nonionic; wetting agent, emulsifier, dispersant and detergent; solid and liq. resp.; HLB 7.3 and 13.2; 100% conc.

Peganol B 26. [245] Alcohol, alkoxylated; nonionic; see Peganol A 24; liq.; HLB 14.0; 100% conc.

Peganol CL 214, 225, 240. [245] Alcohol alkoxylate; see Peganol A 24; liq.; 100% conc.

Peganol CSS 10, 25. [245] Alcohol, ethoxylated; nonionic; see Peganol A 24; solid; HLB 12.5 and 16.1 resp.; 100% conc.

Peganol D 25. [245] Alcohol, alkoxylated; nonionic; see Peganol A 24; liq.; HLB 10.0; 100% conc.

Peganol DA 4, 6. [245] Alcohol, ethoxylated; nonionic; see Peganol A 24; liq.; HLB 10.6 and 12.5 resp.; 100% conc.

Peganol DA 211, 214. [245] Alcohol, alkoxylated; nonionic; see Peganol A 24; liq.; 100% conc.

Peganol DDP 6, 10. [245] Dodecyl phenol ethoxylate; nonionic; see Peganol A 24; liq.; HLB 10.1 and 12.5 resp.; 100% conc.

Peganol DNP 8, 24, 150. [245] Dinonyl phenol ethoxylate; nonionic; see Peganol A 24; liq., solid, solid resp.; HLB 10.0, 15.0, 19.0; 100% conc.

Peganol LA 4, 9. [245] Alcohol, ethoxylated; nonionic; see Peganol A 24; liq.; HLB 8.0 and 13.2 resp.; 100% conc.

Peganol NP 1.5. [97] Nonoxynol-1 (1.5 EO); nonionic; surfactant and chemical intermediate for esters; foam stabilizer for detergents; coemulsifier for surfactant blends;

detergent defoamer; emulsion stabilizer; detergent and dispersant for lubricants; Gardner 2 max. clear visc. liq.; sol. in oil, xylene, perchloroethylene; insol. in water; sp.gr. 0.98; visc. 8 cs (100 C); HLB 4.6; pour pt. –19 C; pH 6.0–7.5 (5%); flash pt. > 200 C.

Peganol NP 4. [97] Nonoxynol-4; nonionic; detergent and dispersant for lubricants; co-emulsifier; plasticizer and antistat; stabilizer for latex emulsions; chemical intermediate for sulfate esters; corrosion inhibitor for engine oils; APHA 100 max. clear visc. liq.; sol. in oil, xylene, perchloroethylene; insol. in water; sp.gr. 1.023; visc. 9 cs (100 C); HLB 8.8; pour pt. –25 C; pH 6.0–7.5 (5%); flash pt. > 200 C.

Peganol NP 5. [97] Nonoxynol-5; nonionic; dispersant for lubricants; surfactant; chemical intermediate; emulsifier and coupling agent for surfactant blends; rust inhibitor in storage tanks; deicing fluid for gasoline and jet aircraft fuels; APHA 100 max. clear visc. liq.; sol. see Peganol NP 4; sp.gr. 1.031; visc. 10 cs (100 C); HLB 10.0; pour pt. –30 C; pH 6.0–7.5 (5%); flash pt. > 200 C.

Peganol NP 6. [97] Nonoxynol-6; nonionic; detergent and dispersant for lubricants; emulsifier for silicones, agric. chemicals, and min. oil; APHA 100 max. clear visc. liq.; sol. see Peganol NP 4; sp.gr. 1.038; visc. 10.5 cs (100 C); HLB 10.8; pour pt. –32 C; pH 6.0–7.5 (5%); flash pt. > 200 C.

Peganol NP 9. [97] Nonoxynol-9; nonionic; detergent, wetting agent, penetrant, corrosion inhibitor, emulsifier in metal processing, textile/leather processing, paper industry, household, institutional, and industrial cleaners, paints, emulsion cleaning; APHA 100 max. clear visc. liq.; sol. in water, xylene, perchloroethylene; sp.gr. 1.055; visc. 11 cs (100 C); HLB 13.0; pour pt. –1 C; pH 6.0–7.5 (5%); cloud pt. 52–56 C (1% aq.).

Peganol NP 10. [97] Nonoxynol-10; nonionic; see Peganol NP 9; APHA 100 max. clear liq.; sol. see Peganol NP 9; sp.gr. 1.057; visc. 12 cs (100 C); HLB 13.2; pour pt. 8 C; pH 6.0–7.5 (5%); flash pt. > 200 C; cloud pt. 61–65 C (1% aq.).

Peganol NP 12. [97] Nonoxynol-12; nonionic; see Peganol NP 9; also detergent in scouring raw wool; detergent formulations; APHA 100 max. clear liq.; sol. see Peganol NP 9; sp.gr. 1.064; visc. 14.5 cs (100 C); HLB 14.2; pour pt. 17 C; pH 6.0–7.5 (5%); flash pt. > 200 C; cloud pt. 80–83 C.

Peganol NP 15. [97] Nonoxynol-15; nonionic; detergent and dispersant; emulsifier for fats, oils, and waxes; wetting agent and penetrant in caustic formulations; APHA 100 max. clear liq.; sol. see Peganol NP 9; sp.gr. 1.074; visc. 17 cs; (100 C); HLB 15.0; pour pt. 23 C; pH 6.0–7.5 (5%); flash pt. > 200 C; cloud pt. 95–100 C (1% aq.).

Peganol NP 20. [97] Nonoxynol-20; nonionic; detergent, wetting agent, and dispersant; emulsifier for fats, oils, waxes, solvs. and polyester resins; demulsifier for crude petroleum oil emulsions; syn. latice stabilizer; component of silicone emulsions used in glass mold release agents; APHA 100 max. (50 C) soft wax; sol. see Peganol NP 9; sp.gr. 1.052; visc. 20 cs (100 C); HLB 16.0; pour pt. 33 C; pH 6.0–7.5 (5%); flash pt. > 200 C; cloud pt. > 100 C.

Peganol NP 30, 30 70%. [97] Nonoxynol-30; nonionic; wetting agent, emulsifier, detergent, dispersant; solid; HLB 17.2; 100 and 70% conc. resp.

Peganol NP 40. [97] Nonoxynol-40; nonionic; wetting agent in conc. electrolyte sol'ns.; syn. latex stabilizer; polymerization emulsifier for vinyl acetate and acrylic emulsions; floor wax and polish emulsifier and stabilizer; dyeing assistant; APHA 100 max. (50 C) wax; sol. in water; sp.gr. 1.087 (50 C); visc. 44 cs (100 C); HLB 17.8; pour pt. 45 C; pH 6.0–7.5 (5%); flash pt. > 200 C; cloud pt. > 100 C.

Peganol NP 40 70%. [245] Nonoxynol-40; nonionic; see Peganol NP 30; liq.; HLB 17.7; 70% conc.

Peganol NP 50. [97] Nonoxynol-50; nonionic; see Peganol NP 40; APHA 100 max. (50 C); wax; sol. in water; sp.gr. 1.095 (50 C); visc. 55 cs (100 C); HLB 18.2; pour pt. 46 C; pH 6.0–7.5 (5% aq.); flash pt. > 200 C; cloud pt. > 100 C.

Peganol NP 100. [97] Nonoxynol-100; nonionic; see Peganol NP 40; APHA 100 max. (50 C) wax; sol. in water; sp.gr. 1.120 (50 C); visc. 120 cs (100 C); HLB 19.0; pour pt. 50 C; pH 6.0–7.5 (5%); flash pt. > 200 C; cloud pt. > 100 C (1% aq.).

Peganol NP 100 70%. [245] Nonoxynol-100; nonionic; see Peganol NP 30; liq.; HLB 19.0; 70% conc.

Peganol OAL 20, 23. [97] Alcohol, ethoxylated; nonionic; wetting agent, emulsifier, detergent, dispersant; solid; HLB 15.3 and 15.8 resp.; 100% conc.

Peganol OP 3. [245] Octoxynol-3; nonionic; wetting agent, emulsifier, detergent, dispersant; liq.; HLB 7.8; 100% conc.

Peganol OP 6. [245] Octoxynol-6; nonionic; see Peganol OP 3; liq.; HLB 10.0; 100% conc.

Peganol OP 8. [245] Octoxynol-8; nonionic; see Peganol OP 3; liq.; HLB 12.8; 100% conc.

Peganol OP 10. [245] Octoxynol-10; nonionic; see Peganol OP 3; liq.; HLB 13.5; 100% conc.

Peganol OP 20. [245] Octoxynol-20; nonionic; see Peganol OP 3; solid; HLB 16.2; 100% conc.

Peganol OP 30, 30 70%. [245] Octoxynol-30; nonionic; see Peganol OP 3; solid; HLB 17.3; 100 and 70% conc. resp.

Peganol OP 40, 40 70%. [245] Octoxynol-40; nonionic; see Peganol OP 3; solid and liq. resp.; HLB 17.9; 100 and 70% conc. resp.

Peganol OP 70, 70 70%. [245] Octoxynol-70; nonionic; see Peganol OP 3; solid and liq. resp.; HLB 18.7; 100 and 70% conc. resp.

Peganol SFE. [245] Octyl phenol ethoxylate; nonionic; see Peganol OP 3; solid; HLB 16.0; 100% conc.

Peganol TDA 6, 8. [245] Octyl phenol ethoxylate; nonionic; see Peganol OP 3; liq.; HLB 11.4 and 12.7 resp.; 100% conc.

Pegeste SML, SMO, SMP, SMS. [245] Sorbitan ester; nonionic; emulsifier, stabilizer, lubricant, corrosion inhibitor and antistat; liq. (SML, SMO), solid (SMP, SMS); HLB 8.6, 4.3, 6.7, and 4.7 resp.; 100% conc.

Pegnol C-14, -18, -20. [606] POE cetyl ether; nonionic; emulsifier, dispersant; solid; HLB 14.4, 15.3, and 15.7 resp.; 100% conc.

Pegnol HA-120. [606] POE alkyl amine; nonionic; emulsifier; liq.; 100% act.

Pegnol L-6. [606] POE lauryl ether; nonionic; emulsifier, dispersant; paste; HLB 11.7; 100% conc.

Pegnol L-8, -10, -12, -15, -20. [606] POE lauryl ether; nonionic; see Pegnol L-6; solid; HLB 13.1, 14.1, 14.8, 15.6, and 16.5 resp.; 100% conc.

Pegnol O-6, -16. [606] POE oleyl ether; nonionic; see Pegnol L-6; solid; HLB 9.6 and 14.5 resp.; 100% conc.

Pegnol OA-400. [606] POE alkyl amine; nonionic; antistat for syn. fibers, emulsifier, dispersant, leveling agent for polyester and polyamide fabrics; solid; HLB 17.4; 100% conc.

Pegol 17 R1, 17 R2, 25 R1, 25 R 2, 31 R 1. [245] Alkoxylated glycol; nonionic; defoamer, dispersant, wetting agent, emulsifier, demulsifier, leveling agent and detergent; liq.; HLB 6.0, 8.0, 4.0, 6.0, and 4.0 resp.; 100% conc.

Pegol E 200, 400, 600, 1000, 4000. [245] PEG; nonionic; surfactant intermediate; liq. (E 200, 400, 600), solid (E 1000, 4000); 100% conc.

Pegol F 68, 88. [245] Alkoxylated glycol; nonionic; see Pegol 17 R1; solid; HLB 29.0 and 18.0 resp.; 100% conc.

Pegol L 31, 35, 43, 61. [245] Alkoxylated glycol (block copolymers); nonionic; see Pegol 17 R1; liq.; HLB 4.5, 18.5, 16.0, 3.0 resp.; 100% conc.

Pegol L 62, 62 LF. [245] Alkoxylated glycol (block copolymers); nonionic; see Pegol 17 R1; liq.; HLB 7.0 and 6.6 resp.; 100% conc.

Pegol L 64. [245] Alkoxylated glycol (block copolymers); nonionic; see Pegol 17 R1; liq.; HLB 15.0; 100% conc.

Pegol L 81, 101, 121. [245] Alkoxylated glycol; nonionic; see Pegol 17 R1; liq.; HLB 2.0, 1.0, 5.0 resp.; 100% conc.

Pegol P 85. [245] Alkoxylated glycol; nonionic; see Pegol 17 R1; paste; HLB 16.0; 100% conc.

Pegol P 400, 700, 1000, 2000. [245] POP; nonionic; see Pegol E 200; liq.; 100% conc.

Pegosperse 50 DS. [256] Glycol distearate; nonionic; emulsifier, stabilizer for suspensions and dispersions; emollient, lubricant and pigment dispersant in pharmaceuticals and cosmetics; thickener, wetting agent and plasticizer in hair products; wh. flakes; sol. in ethanol, min. and veg. oil; insol. in water; HLB 1 ± 1; m.p. 58–63 C; sapon. no. 185–200

Pegosperse 50 MS. [256] Glycol stearate; nonionic; dispersant, emulsifier for o/w emulsions; wh. flakes; sol. in methanol, ethanol, acetone, ethyl acetate, toluol, naphtha, min. and veg. oils; sp.gr 0.96; m.p. 55–60 C; HLB 0.9 ± 0.5; sapon. no. 180–187; pH 4.0–6.0 (3% aq.); 100% conc.

Pegosperse 100 L. [256] PEG-2 laurate; see Pegosperse 50 MS; straw liq.; sol. in ethanol, toluol, naphtha, min. oil; disp. in water; sp.gr. 0.97; HLB 7.4 ± 0.5; solid. pt. < 13 C; sapon. no. 160–170; pH 8–10 (5% aq.); 100% conc.

Pegosperse 100 ML. [256] PEG-2 laurate; see Pegosperse 50 MS; straw liq.; sol. in methanol, ethanol, acetone, ethyl acetate; sp.gr. 0.96; HLB 6.0 ± 0.5; solid. pt. 16–19 C; sapon. no. 180–190; pH 3.0–6.0 (5% aq.); 100% conc.

Pegosperse 100 MR. [256] PEG-2 ricinoleate; nonionic; see Pegosperse 50 MS; amber liq.; sol. see Pegosperse 100 ML; sp.gr. 0.98; HLB 4.8 ± 0.5; solid. pt. < 0 C; sapon. no. 140–150; pH 3–5 (5% aq.); 100% conc.

Pegosperse 100 O. [256] PEG-2 oleate; nonionic; see Pegosperse 50 MS; amber liq.; sol. in methanol, ethanol, toluol, naphtha; disp. in water; sp.gr. 0.93; HLB 3.5 ± 0.5; solid. pt. < 0 C; sapon. no. 160–175; pH 8–9 (5% aq.); 100% conc.

Pegosperse 100 S. [256] PEG-2 stearate; nonionic; see Pegosperse 50 MS; wh. beads; sol. in methanol, ethanol, toluol, naphtha, min. and veg. oils; disp. in water; sp.gr. 0.96; m.p. 46–54 C; HLB 3.8 ± 0.5; sapon. no. 165–175; pH 6.8–7.5 (3% aq.); 100% conc.

Pegosperse 200 DL. [256] PEG-4 dilaurate; nonionic; see Pegosperse 50 DS; lt. yel. liq.; sol. in ethanol, min. and veg. oil; disp. in water; sp.gr. 0.96; HLB 7 ± 1; solid. pt. 3 C; sapon. no. 170–185

Pegosperse 200 ML. [256] PEG-4 laurate; nonionic; see Pegosperse 50 MS; yel. liq.; sol. in methanol, ethanol, acetone, ethyl acetate, toluol; misc. with water, sp.gr. 0.99; HLB 8.6 ± 0.5; solid pt. < 5; sapon. no. 149–159; pH 4.0–6.5 (5% aq.); 100% conc.

Pegosperse 400 DL. [256] PEG-8 dilaurate; nonionic; see Pegosperse 50 MS; yel. liq.; sol. in methanol, ethanol, acetone, ethyl acetate, toluol, naphtha, min. and veg. oils; sp.gr. 0.99; HLB 10.0 ± 0.5; solid. pt. 5–12 C; sapon. no. 127–140; pH 3–5 (5% aq.); 100% conc.

Pegosperse 400 DO. [256] PEG-8 dioleate; nonionic; see Pegosperse 50 MS; amber liq.; sol. in methanol, ethanol, ethyl acetate, toluol, naphtha, min. and veg. oils; disp. in water; sp.gr. 0.97; HLB 7.2 ± 0.5; solid. pt. < 0 C; sapon. no. 115–125; pH 4.5–6.5 (5% aq.); 100% conc.

Pegosperse 400 DS. [256] PEG-8 distearate; nonionic; see Pegosperse 50 MS; cream soft solid; sol. in ethanol, ethyl acetate, toluol, naphtha, min. and veg. oil, methanol; disp. in water; sp.gr. 0.98; m.p. 29–37 C; HLB 7.8 ± 0.5; sapon. no. 115–125; pH 4.0–6.5 (5% aq.); 100% conc.

Pegosperse 400 DTR. [256] PEG-8 ditriricinoleate; nonionic; see Pegosperse 50 MS; amber liq.; sol. in acetone, ethyl acetate, toluol, naphtha, min. and veg. oils; disp. in water; sp.gr. 0.95–0.97; HLB 1.8 ± 0.5; solid. pt. < 10 C; sapon. no. 156–166 pH 3.5–5.0 (5% aq.); 100% conc.

Pegosperse 400 MC. [256] PEG-8 cocoate; nonionic; see Pegosperse 50 MS; yel. liq.; sol. in methanol, ethanol, acetone, ethyl acetate, toluol, veg. and min. oil; misc. with water; sp.gr. 1.03; HLB 13.9 ± 0.5; solid. pt. < 6 C; sapon. no. 88–100; pH 4–6 (5% aq.); 100% conc.

Pegosperse 400 ML. [256] PEG-8 laurate; nonionic; see Pegosperse 50 MS; straw liq.; sol. see Pegosperse 400 MC; sp.gr. 1.03; HLB 13.9 ± 0.5; solid. pt. < 7 C; sapon. no. 90–100; pH 4.0–6.0 (5% aq.); 100% conc.

Pegosperse 400 MO. [256] PEG-8 oleate; nonionic; see Pegosperse 50 MS; amber liq.; sol. in methanol, ethanol, acetone, ethyl acetate, toluol, naphtha; disp. in water; sp.gr. 1.01; HLB 11.0 ± 0.5; solid. pt. < 0 C; sapon. no. 80–88; pH 4–6 (5% aq.); 100% conc.

Pegosperse 400 MOT. [256] PEG-8 tallate; nonionic; see Pegosperse 50 MS; yel. liq.; sol. in methanol, ethanol, acetone, ethyl acetate, toluol, naphtha, min. and veg. oils; disp. in water; sp.gr. 1.02; HLB 11.0 ± 0.5; solid. pt. < 0 C; sapon. no. 83–89; pH 3.5–5.0 (5% aq.); 100% conc.

Pegosperse 400 MS. [256] PEG-8 stearate; nonionic; see Pegosperse 50 MS; wh. soft solid; sol. see Pegosperse 400 MOT; sp.gr. 1.0; m.p. 30 C min.; HLB 11.2 ± 0.5; sapon. no. 83–94; pH 3.5–6.0 (5% aq.); 100% conc.

Pegosperse 600 ML. [256] PEG-12 laurate; nonionic; see Pegosperse 50 MS; yel. liq.; sol. in water, methanol, ethanol, acetone, ethyl acetate, naphtha; sp.gr. 1.01; HLB 14.6 ± 0.5; solid. pt. 16–21 C; sapon. no. 65–75; pH 6–8 (5% aq.); 100% conc.

Pegosperse 600 MS. [256] PEG-12 stearate; nonionic; see Pegosperse 50 MS; wh. soft solid; sol. in methanol, ethanol, acetone, ethyl acetate, toluol, veg. oil; disp. in water; sp.gr. 1.01; m.p. 27–32 C; HLB 13.2 ± 0.5; sapon. no. 72–78; pH 3–5 (5% aq.); 100% conc.

Pegosperse 700 TO. [256] PEG-14 oleate; nonionic; see Pegosperse 50 MS; amber liq.; sol. in water, methanol, ethanol, acetone, ethyl acetate, toluol, veg. oil; sp.gr. 1.10; HLB 13.5 ± 0.5; solid pt. < 15 C; sapon. value 35–45; pH 6–8 (5% aq.); 100% conc.

Pegosperse 1000 MS. [256] PEG-20 stearate;

nonionic; see Pegosperse 50 MS; cream solid; sol. in methanol, ethanol, acetone, ethyl acetate, toluol, naphtha, veg. oil; disp. in water; sp.gr. 1.02; m.p. 37–43 C; HLB 15.2 ± 0.5; sapon. no. 45–55; pH 3–5 (5% aq.); 100% conc.

Pegosperse 1500 DO. [256] PEG-6-32 dioleate; nonionic; see Pegosperse 50 MS; amber soft solid; sol. see Pegosperse 600 MS; sp.gr. 1.05; m.p. 30–38 C; HLB 7.8 ± 0.5; sapon. no. 104–112; pH 3.5–5.0 (5% aq.); 100% conc.

Pegosperse 1500 MS. [256] PEG-6-32 stearate; nonionic; see Pegosperse MS; cream waxy solid; sol. in ethanol, ethyl acetate, toluol, veg. oil, methanol, naphtha; misc. with water; sp.gr. 1.05; m.p. 27–31 C; HLB 13.8 ± 0.5; sapon. no. 57–67; pH 3–5 (5% aq.); 100% conc.

Pegosperse 1750 MS. [256] PEG-40 stearate; nonionic; see Pegosperse 50 DS; wh. flakes; sol. in water, ethanol; HLB 18 ± 1; m.p. 46 C; sapon. no. 25–35

Pegosperse 4000 MS. [256] PEG-75 stearate; nonionic; see Pegosperse 50 MS; cream waxy solid; sol. see Pegosperse 700 TO; sp.gr. 1.10; m.p. 54–61; HLB 18.0 ± 0.5; sapon. no. 17–22; pH 3–5 (5% aq.); 100% conc.

Pegosperse 6000 DS. [256] PEG-150 distearate; nonionic; see Pegosperse 50 DS; sol. in water, ethanol; HLB 18 ± 1; m.p. 55 C; sapon. no. 17

Pegosperse 9000 CO. [256] PEG 9000 castor oil; nonionic; see Pegosperse 50 MS; tan solid; sol. in water, methanol, ethanol, acetone; sp.gr. 1.08; m.p. 41–44 C; HLB 18.1 ± 0.5; sapon. no. 14–20; pH 6.0–7.0 (3% aq.); 100% conc.

Pegosperse EGMS-70. [256] Glycol stearate; see Pegosperse 50 DS; wh. flakes; sol: see Pegosperse 50 DS; HLB 2 ± 1; m.p. 52–56 C; sapon. no. 180–188

Pelex NBL, NB Paste. [341] Sodium alkyl naphthalene sulfonate; anionic; wetting agent and penetrant; liq. and paste resp.; 35 and 50% conc. resp.

Pelex OT-P. [341] Sodium dialkyl sulfosuccinate; anionic; wetting agent for textile processing; emulsifier for emulsion polymerization; liq.; 70% conc.

Pelex SS-H. [341] Sodium dodecyl diphenyl ether disulfonate; anionic; emulsifier for syn. latex and emulsion polymerization; paste; 50% conc.

Pelex SS-L. [341] Sodium dodecyl diphenyloxide disulfonate; anionic; emulsifier for syn. latex and emulsion polymerization; paste; 50% conc.

Pelex TA. [341] Sodium dialkyl sulfosuccinate; anionic; foam controller for latex; paste; 35% conc.

Peltex. [503] Ferro-chromium lignosulfate, modified; anionic; wetting agent, emulsifier, dispersant; oil well drilling mud thinner; blk. powd.; sol. in water; dens. 0.5; 95% act.

Penestrol N-160. [607] Sec. alcohol ethoxylate; nonionic; penetrant and wetting agent; bleaching assistant; liq.; 60% conc.

Penetron Conc. [284] Sodium dioctyl sulfosuccinate; detergent, wetting agent, emulsifier for textile wetting and rewetting, bleaching, wallpaper removal, skins, leather, laundering, dyeing, papermaking, printing, preparation of emulsified paints; colorless liq.; alcoholic odor; sol. in water and polar org. solvs.; sp.gr. 1.05; 70% act.

Pentex 40. [148] Alkylaryl sulfonate; anionic; wetting agent, penetrant, emulsifier; textile and industrial processing; improves dye uniformity; aids rewetting of leather, paper, and paper-mill felts; red-br. liq.; sp.gr. 1.15; pour pt. 0 C; pH 9.5 (5%)

Pentex 99. [148] Dioctyl sodium sulfosuccinate and propylene glycol; anionic; detergent; textile scouring and dispersant for dyes; paper rewetting and felt washing surfactant; wetting agent in cosmetics; detergent additive in dry cleaning fluids; dishwashing compds.; wallpaper removers; emulsion polymerization; water-based paint formulations; clear liq.; sp.gr. 1.08–1.13 (70 F); visc. 500–1000 cps (70 F); pour pt. 40 F; 75% conc.

Pentine 1185. [145] Isopropylamine alkylbenzene sulfonate; anionic; emulsifier for emulsion degreasers, dry cleaning soaps, agric. emulsifier; visc. liq.; 95–97% conc.

Pepol A-0638. [606] Oxyethylated fatty alcohol, modified; nonionic; detergent, wetting agent; liq.; 100% conc.

Pepol A-0858. [606] Oxyethylated fatty alcohol, modified; nonionic; detergent, penetrant; paste; 100% conc.

Pepol AX-1. [606] Oxyethylated fatty alcohol, modified; nonionic; see Pepol A-0638; liq.; 100% conc.

Peregal O. [245] POE fatty alcohol; nonionic; wetting agent, penetrant, dye assistant; leveling agent and stripping assistant for dyes;

clear yel.-brn. liq.; misc. with water; sp.gr. 1.02.

Peregal OK. [245] Methyl polyethanol quat. amine; cationic; wetting agent, solubilizer, textile aux. used for leveling, retarding, anti-stripping agents, and improved penetration for dyes; cloudy pale amber liq.; misc. with water; sp.gr. 1.03; pH 7.6.

Perlankrol ACM2. [360] Fatty alkylolamide ether sulfate; base; preparation of personal care products; 30% act.

Perlankrol ADP3. [360] Sodium fatty alcohol ether sulfate; anionic; ingred.; base for preparation of personal care products; clear water-wh. liq.; mild odor; water sol.; sp.gr. 1.048; visc. 28 cs.; flash pt. > 200 F (COC); pour pt. < O C; pH 7.0–8.5 (1% aq.); 27% act.

Perlankrol ASC2. [360] Fatty alcohol ether sulfate, sodium salt; anionic; see Perlankrol ADP3; water-wh. clear liq.; mild odor; water-sol.; sp.gr. 1.047; visc. 1000 cs; flash pt. > 200 F; pour pt. < 0 C; pH 6.5–7.5 (1% aq.); 27.5% act.

Perlankrol ASC38. [360] Fatty alcohol ethoxylate; anionic; toiletry intermediate; liq.; 28% conc.

Perlankrol ATL40. [360] TEA lauryl sulfate; see Perlankrol ADP3; pale straw liq.; mild odor; water-sol.; sp.gr. 1.048; visc. 60 cs.; flash pt. > 200 F (COC); pour pt. < 0 C; pH 7–8 (1% aq.); 40% act.

Perlankrol DAF25. [360] Syn. primary alcohol sulfate, ammonium salt; anionic; foaming agent, base; formulation of toiletries and carpet shampoos; pale straw liq.; 25% act.

Perlankrol DGS. [360] Sodium primary alcohol sulfate; anionic; solubilizer, textile aux.; detergent base in household and industrial cleaners; foaming agent in syn. latex industry; primary emulsifier in emulsion polymerization; wh. slurry; faint fatty alcohol odor; water-sol.; sp.gr. 1.041; visc. 220 cs (40 C); flash pt. > 200 F (COC); pour pt. 12 C; pH 7.0–8.5 (1% aq.); 28% act.

Perlankrol DSA. [360] Syn. primary alcohol sulfate, sodium salt; see Perlankrol DGS; wh. slurry; faint fatty alcohol odor; water-sol.; sp.gr. 1.046; visc. 300 cs (40 C); flash pt. > 200 F (COC); pour pt. 10 C; pH 7.0–8.5 (1% aq.); 28% act.

Perlankrol EAD-60. [360] Syn. primary alcohol ether sulfate, ammonium salt; anionic; stabilizer, detergent, industrial and household cleaners; foaming agent for fire fighting foam compds.; dispersant and emulsifier for chlorinated solvs.; clear mobile straw gel; faint alcoholic odor; water-sol.; sp.gr. 1.033; visc. 240 cs; flash pt. 85 F (Abel CC); pour pt. < 0 C; pH 7–9 (1% aq.); 60% act. in ethanol.

Perlankrol ESD. [360] Syn. primary alcohol ether sulfate, sodium salt; see Perlankrol EAD-60; water-wh. clear liq.; faint odor; water-sol.; sp.gr. 1.053; visc. 130 cs; flash pt. > 200 F (COC); pour pt. < 0 C; pH 7–9 (1% aq.); 27% act.

Perlankrol ESD-60. [360] Syn. primary alcohol ether sulfate, sodium salt; see Perlankrol EAD-60; water-wh. clear to hazy mobile gel; faint alcoholic odor; water-sol.; sp.gr. 1.090; visc. 210 cs; flash pt. 90 F (Abel CC); pour pt. < 0 C; pH 7–9 (1% aq.); 60% act.

Perlankrol ESK-29. [360] Syn. primary alcohol ether sulfate, sodium salt; anionic; solubilizer for disinfectants; foam additive for industrial detergents; concrete foaming agent; liq.; 29% conc.

Perlankrol ESS-25. [360] Syn. primary alcohol ether sulfate, sodium salt; see Perlankrol EAD-60; water-wh. clear liq.; faint odor; water-sol.; sp.gr. 1.040; visc. 65 cs; flash pt. > 200 F (COC); pour pt. < 0 C; pH 7–9 (1% aq.); 25% act.

Perlankrol FB-25. [360] Alkyl phenol ether sulfate, modified, sodium salt; anionic; primary emulsifier for emulsion polymerization; clear straw liq.; faint alcoholic odor; water-sol.; sp.gr. 1.038; visc. 48 cs; flash pt. 86 F (Abel CC); pour pt. –3 C; pH 6.0–8.5 (1% aq.); 25% act.

Perlankrol FD-63. [360] Alkyl phenol ether sulfate, ammonium salt; see Perlankrol FB-25; clear straw liq.; faint alcoholic odor; water-sol.; sp.gr. 1.056; visc. 106 cs; flash pt. 80 F (Abel CC); pour pt. < 0 C; pH 6.0–8.5 (1% aq.); 63% act.

Perlankrol FF. [360] Alkyl phenol ether sulfate, ammonium salt; anionic; foam booster, stabilizer for industrial detergents; amber hazy visc. liq.; faint ammoniacal odor; water-sol.; sp.gr. 1.115; visc. 4200 cs; flash pt. 84 F (Abel CC); pour pt. 5 C; pH 6.0–8.5 (1% aq.); 90% act.

Perlankrol FN-65. [360] Alkyl phenol ether sulfate, sodium salt; anionic; foam stabilizer used in specialty detergent formulations; primary emulsifier in emulsion polymerization; clear straw liq.; faint alcoholic odor; sp.gr. 1.110; visc. 200 cs; flash pt. 85 F (Abel CC); pour pt. < 0 C; pH 6.0–8.5 (1% aq.); 65% act.

Perlankrol N Range. [360] Alkylaryl phenol ether sulfates, sodium salt; primary emulsifier for emulsion polymerization; pigment dispersant; agrochemical adjuvants; liq.

Perlankrol O. [360] Sodium alkyl sulfate; anionic; clarifying and wetting agent, stabilizer, visc. modifier used in detergent and cleaning formulations; clear lt. amber liq.; char. odor; water-sol.; sp.gr. 1.120; visc. 44 cs; flash pt. > 200 F (COC); pour pt. < 0 C; pH 7.0–8.5 (1% aq.); 40% act.

Perlankrol PA Conc. [360] Alkyl phenol ether sulfate, ammonium salt; anionic; foam stabilizer, foaming agent used in liq. detergent formulations; emulsifier for cresylic acid; hazy visc. amber liq.; faint ammoniacal odor; water-sol.; sp.gr. 1.110; visc. 6000 cs; flash pt. 86 F (Abel CC); pour pt. 7 C; pH 6.0–8.5 (1% aq.); 90% act.

Perlankrol RN-75. [360] Alkyl phenol ether sulfate, sodium salt; anionic; speciality foaming and wetting agent; primary emulsifier in emulsion polymerization; straw clear liq.; faint alcoholic odor; water-sol. sp.gr. 1.120; visc. 470 cs; flash pt. 88 F (Abel CC); pour pt. < 0 C; pH 6.0–8.5 (1% aq.); 75% act.

Perlankrol SN. [360] Alkyl phenol ether sulfate, sodium salt; anionic; foam booster and stabilizer, wetting agent, emulsifier in emulsion polymerization; industrial detergent formulations; water-wh. clear liq.; faint odor; water-sol.; sp.gr. 1.054; visc. 48 cs; flash pt. > 200 F (COC); pour pt. < 0 C; pH 6.0–8.5 (1% aq.); 30% act.

Perlankrol S Range. [360] Triarylphenol ether sulfates, sodium salts; anionic; see Perlankrol N Range; liq.

Perlankrol TM1. [360] Fatty amide ether sulfate, sodium salt; see Perlankrol ADP3; clear pale yel. liq.; mild odor; water-sol.; sp.gr. 1.045; visc. 715 cs; flash pt. > 200 F (COC); pour pt. < 0 C; pH 7–8 (1% aq.); 30% act.

Permalene A-100. [376] Amine-coconut oil condensation product; nonionic; detergent, dye leveler and lubricant for rayon, acetate dyeing; liq.; 100% conc.

Permax AW-2. [658] Quat. ammonium salt polymer; cationic; antistat for syn. fibers; liq.; 25% conc.

Permax PM 420. [658] Quat. ammonium salt polymer; cationic; antistat for polyester textiles; liq.; 20% conc.

Permax PM 503. [658] Alkyl phosphate, ethoxylated; anionic; antistat for syn. fibers; liq.; 50% conc.

Peroxide Stabilizer H. [301] Polycarboxylic acid, sodium salt; anionic; stabilizer for hydrogen peroxide bleaching; liq.; 25% conc.

Pestilizer A. [575] Phosphate ester, acid form; emulsifier for pesticide formulations; dispersant for liq. and powd. formulations; sol. in water, xylene; sp.gr. 1.104; dens. 9.21 lb/gal; visc. 4500 cps; pour pt. –9 C; flash pt. (Seta CC) > 200 F; pH < 2.0.

Pestilizer B. [575] Sulfosuccinate, isopropylamine salt; see Pestilizer A; sol. in water, xylene; sp.gr. 1.088; dens. 9.07 lb/gal; visc. 43,400 cps; pour pt. 9 C; flash pt. (Seta CC); > 200 F; pH 5.5–6.5.

Pestilizer J. [575] Phosphate ester, acid form; see Pestilizer A; sol. in xylene; sp.gr. 1.091; dens. 9.10 lb/gal; visc. 10,750 cps; pour pt. –3 C; flash pt. (Seta CC) > 200 F; pH < 2.0.

Pestilizer M. [575] Phosphate ester, sodium salt; see Pestilizer A; sol. in water, xylene; sp.gr. 1.122; dens. 9.36 lb/gal; visc. 31,250 cps; pour pt. –3 C; flash pt. (Seta CC) > 200 F; pH 6.5–7.0.

Pestilizer N. [575] Phosphate ester, sodium salt; see Pestilizer A; sol. in water, xylene; sp.gr. 1.120; dens. 9.34 lb/gal; visc. 22,250 cps; pour pt. 3 C; flash pt. > 200 F (Seta CC); pH 6.5–7.0.

Petrac 250. [469] Stearic acid, rubber grade; activator, dispersant, plasticizer and lubricant in rubber compd. processing; thickener and gelling agent for greases; emulsifier in polishing/buffing compds.; FAC 5 color; sapon. no. 199.

Petrac 270. [469] Tech. stearic acid; ext. lubricant in flexible PVC processing; chemical intermediate for metallic stearates, esters, etc.; dispersant, plasticizer, activator, lubricant in rubber compounding; thicknener for greases; acid no. 205.

Petrac CP-11, -22G. [181] Calcium stearate; lubricant, emulsifier; powd. and granules resp.; 100% conc.

Petro 11. [181] Alkyl naphthalene sulfonate; germicidal agent, liquid hand soaps, wetting agent in plating and electrolytic baths; 50% act. liq., 95% act. powd.

Petro 22. [181] Alkyl naphthalene sulfonate; surfactant for carpet steam cleaners, tank soak compds., high alkaline and acid cleaners; 50% act. liq., 95% act. powd.

Petro AA. [181] Alkyl naphthalene sulfonate; hydrotrope, detergent; stable in strong acids, alkalies, salts, detergent builders; 50% act.

liq., 95% act. powd.

Petro BA. [181] Alkyl naphthalene sulfonate; high-solids liquid laundry detergents, electrolytic baths, metal and dairy cleaners; 50% act. liq., 95% act. powd.

Petro BAF. [181] Alkyl naphthalene sulfonate; surfactant for rug and upholstery shampoos, car wash, window cleaners; 50% act. liq., 95% act. powd.

Petro Dispersant 98. [181] Alkyl naphthalene sulfonate, sodium salt; dispersant for dyestuffs.

Petro Dispersant 425. [181] Sulfonated naphthalene-formaldehyde condensate, sodium salt; dispersant for dyestuffs, pigments.

Petro LBA. [181] Alkyl naphthalene sodium sulfonate; lighter color form of Petro BA; 50% act. liq., 95% act. powd.

Petro LBAF. [181] Alkyl naphthalene sulfonate; lighter color form of Petro BAF; liq.; 50% act.

Petro P. [181] Alkyl naphthalene sodium sulfonate; high-foaming detergents, high-alkaline and acid cleaners; 50% act. liq., 95% act. powd.

Petro S. [181] Alkyl naphthalene sodium sulfonate; anionic; soil conditioner; 50% act. liq.; 95% act. powd.

Petro ULF. [181] Alkyl naphthalene sodium sulfonate; anionic; wetting agent for jet-beck dyeing, bottle washing; liq.; 50% act.

Petro WP. [181] Alkyl naphthalene sodium sulfonate, modified; anionic; wetting agent used in pesticides and industrial cleaners; powd.; 95% conc.

Petromixes. [650] Petroleum sulfonate, modified; anionic; emulsifier; base for preparation of cutting oils, solv. degreasers; liq.

Petronate 25C, 25H. [650] Calcium petroleum sulfonate; anionic; detergent and rust inhibitor component in lube oil additives; emulsifier for w/o systems; ASTM 3.5 liq.; dens. 8.2 lb/gal; visc. 170 SSU (100 C); flash pt. 360 F; 43–46% calcium sulfonate.

Petronate Basic. [650] Barium petroleum sulfonate; anionic; oil additive for lube, industrial oils, fuels, greases; ASTM 7.0 liq.; visc. 250 SSU (100 C); 40–45% barium sulfonate.

Petronate CR. [650] Sodium petroleum sulfonate; anionic; wetting agent, emulsifier, dispersant, rust preventative; dk. brn. visc. liq.; petroleum odor; sol. in org. solvs.; dens. 8.5 lb/gal; sp.gr. 1.02 (60 F); visc. 175 s (Furol, 100 C); flash pt. 380 F; 61–63% act.

Petronate HL. [650] Sodium petroleum sulfonate; anionic; emulsifier, dispersant, wetting agent; dk. brn. visc. liq.; petroleum odor; sol. in oil., org. solvs.; dens. 8.5 lb/gal; sp.gr. 1.1 (60 F); visc. 135 s (Furol, 100 C); flash pt. 380 F; 61–63% act.

Petronate HMW. [650] Calcium petroleum sulfonate; detergent, rust inhibitor; lubricating oil additives, dens. 8.2 lb/gal; visc. 200 SSU (100 C); flash pt. 350 F; 45–46% calcium sulfonate.

Petronate K. [650] Sodium petroleum sulfonate; anionic; dispersant, emulsifier and wetting agent in dry cleaning soaps, textile processing; lt. br. visc. liq.; oil-sol.; dens. 8.3 lb/gal; sp.gr. 1.1 (60 F); visc. 85 s (Furol, 100 C); flash pt. 380 F; 61–63% act.

Petronate L. [650] Sodium petroleum sulfonate; anionic; dispersant, wetting agent and emulsifier in oil and solv. systems; dr. brn. visc. liq.; petroleum odor; sol. in oil and org. solvs.; dens. 8.5 lb/gal; sp.gr. 1.1 (60 F); visc. 125 s (Furol, 100 C); flash pt. 380 F; 61–63% act.

Petronate Neutral (50-S). [650] Barium petroleum sulfonate; anionic; oil additive for lube, industrial and fuel oils, undercoating, and greases; ASTM 3.5 liq.; visc. 100 SSU (100 C); 50–52% barium sulfonate

Petronate S. [650] Sodium didodecylbenzene sulfonate; anionic; see Petronate CR; liq.; 62% conc.

Petrosan 102. [506] Methyl palmitate oleate; nonionic; enhances lubricity and wettability of petroleum oils for metals; liq.; 100% conc.

Petroscale 101. [181] Org. polyphosphonate; industrial water treatment.

Petrostep 420, 465. [575] Petroleum sulfonate; anionic; used in enhanced oil recovery; liq.; 50% conc.

Petrostep A-70. [575] Pentadecylbenzene sulfonic acid, branched; salts of the acid are emulsifiers for waxes, oils, and org. liqs.; dk. brn. visc. liq.; sol. in water and org. solvs.; dens. 8.5 lb/gal; visc. 16,000 cps; 94–96% act.

Petrostep HMW, LMC, MMW. [575] Petroleum sulfonate; anionic; see Petrostep 420; liq.; 50, 28, and 50% conc. resp.

Petrosul 545. [467] Sodium petroleum sulfonate; anionic; emulsifier, demulsifier and flotation agent; fuel oil and grease additive; leather, textile and metalworking oils; metal cleaners; ASTM-D-1500 3.5 max. visc. liq.; m.w. 460–475; oil-sol.; dens. 8.3 lb/gal; 50% act.

Petrosul 550. [467] Sodium petroleum sulfonate; anionic; solubilizer, rust preventative, dispersant; demulsifier, fuel oil and grease additive; lubricating oil additive mfg.; metal cleaners; ASTM-D-1500 3.5 max. visc. liq.; m.w. 505–520; oil-sol.; dens. 8.3 lb/gal; 50% act.

Petrosul 742. [467] Sodium petroleum sulfonate; anionic; emulsifier, solubilizer; textile and dry cleaning formulations; fat splitting; demulsifier, flotation agent; fuel oil additive; leather oils; pigment grinding agent; ASTM-D-1500 3.5 max. visc. liq.; m.w. 415–430; oil-sol.; dens. 8.4 lb/gal; 62% act.

Petrosul 744 CL. [467] Sodium petroleum sulfonate; anionic; emulsifier, solubilizer, demulsifier, dry cleaning detergent, fat splitting; flotation agent, fuel oil and grease additive; leather, textile and metalworking oils; metal cleaners; pigment grinding agent; ASTM-D-1500 5 max. visc. liq.; m.w. 440–450; oil-sol.; dens. 8.6 lb/gal; 62% act.

Petrosul 745. [467] Sodium petroleum sulfonate; anionic; emulsifier; demulsifier, dry cleaning detergent, fat splitting; flotation agent, fuel oil additive; leather and textile oils; pigment grinding aid; ASTM-D-1500 3.5 max. visc. liq.; m.w. 460–475; oil-sol.; dens. 8.6 lb/gal; 62% act.

Petrosul 750. [467] Sodium petroleum sulfonate; anionic; dispersant, rust preventative; demulsifier, fuel oil and grease additive; lubricating oil additive mfg.; metalworking oils; metal cleaners; ASTM-D-1500 3.5 max. visc. liq.; m.w. 505–520; oil-sol.; dens. 8.6 lb/gal; 62% act.

Petrosul Neutral Barium. [467] Barium petroleum sulfonate; anionic; dispersant, fuel and grease additives; rust preventatives; ASTM-D-1500 5 max. visc. liq.; m.w. 1100–1140; oil-sol.; dens. 8.2 lb/gal; 40% act.

Petrosul Neutral Calcium. [467] Calcium petroleum sulfonate; anionic; dispersant, rust preventative; fuel oil and grease additive; ASTM-D-1500 5 max. visc. liq.; m.w. 1000–1040; oil-sol.; dens. 8.0 lb/gal; 40% act.

Petrowet R. [208] Sodium alkyl sulfonate; anionic; wetting agent, detergent, dispersant, penetrant; petroleum, metalworking, textile, and paper industries; industrial formulations and cleaning; clear yel. to amber liq.; alcoholic odor; water-sol.; dens. 9 lb/gal; visc. 15 cps (27 C); flash pt. 196 F (OC); surf. tens. 47.1 dynes/cm (0.1%); pH 4.0–5.5 (1% aq.); 22% act.

PGE-300. [287] PEG 300; nonionic; pharmaceutical sol'ns.; antistat; various applications in the plastic industry, printing and stamping ink; liq.; 100% conc.

PGE-400. [287] PEG 400; nonionic; applications in pharmaceuticals and cosmetics; plastic industry, pigment colors, printing and stamping inks; liq.; 100% conc.

PGE-400-DO. [287] POE (8) dioleate; nonionic; w/o and o/w emulsifier for cosmetics and pharmaceuticals; insecticides, herbicides, leather industry; liq.; HLB 8.0; 100% conc.

PGE-400-DS. [287] POE (8) distearate; nonionic; pharmaceuticals and cosmetics; plasticizer for plastics; solid; HLB 8.0; 100% conc.

PGE-400-ML. [287] POE (8) monolaurate; nonionic; emulsifier o/w; pharmaceutical and cosmetic applications; textile industry; antistat for plastic industry; insecticides; liq.; HLB 12.0; 100% conc.

PGE-400-MO. [287] POE (8) oleate; nonionic; o/w and w/o emulsifier for cosmetics, perfume oils, insecticides, fat liquors; liq.; HLB 11.5; 100% conc.

PGE-400-MS. [287] POE (8) monostearate; nonionic; emulsifier o/w for cosmetics and pharmaceuticals; paper industry; deinking; solid; HLB 11.5; 100% conc.

PGE-550. [287] PEG oxide blend; nonionic; pharmaceutical ointments; pasty; 100% conc.

PGE-600. [287] PEG 600; nonionic; various applications in pharmaceutical and cosmetics; pigment colors, alkyd resin, antistat; paper finish, plastics; pasty; 100% conc.

PGE-600-DS. [287] POE (12) distearate; nonionic; emulsifier o/w for cosmetics and pharmaceuticals; plasticizer for plastics; various technical applications; flakes; HLB 10.5; 100% conc.

PGE-600-ML. [287] POE (12) monolaurate; nonionic; solubilizer for o/w in pharmaceuticals and cosmetics; liq.; HLB 15.0; 100% conc.

PGE-600-MO. [287] POE (12) monooleate; nonionic; solubilizer for o/w pharmaceutical and cosmetic applications; emulsifier, antistat in textile industry; emulsifier for insecticides, herbicides, tech. applications; liq.; HLB 13.5; 100% conc.

PGE-1000. [287] PEG 1000; nonionic; used in the pharmaceutical and cosmetic industries;

solid; 100% conc.

PGE-1000-MS. [287] POE (20) monostearate; nonionic; emulsifier for cosmetic creams and pharmaceuticals; aq. coloring material disps.; solid; HLB 15.5; 100% conc.

PGE-1500. [287] PEG 1500; nonionic; used in cosmetic and pharmaceutical applications; flakes; 100% conc.

Phenyl Sulphonate CA, CAL. [301] Calcium alkylaryl sulfonate; anionic; basic material for mfg. of biocide emulsifiers; liq.; 70% conc.

Phosfac 1004, 1006, 1044, 1044FA, 1066, 1066FA, 1068FA. [376] Phosphate ester; anionic; emulsifier, penetrant, antistat, wetting agent, detergent, lubricant, corrosion inhibitor and dispersant used in liq. detergent formulations; liq.; 99, 99, 99, 90, 90, 99, and 99% conc. resp.

Phosfac 8000, 8608, 9604, 9609. [376] Phosphate ester; anionic; see Phosfac 1004; liq.; 99% conc.

Phosfetal 201, 204, 600, 601, 602, 603. [660] Phosphoric acid ester; anionic; cleansing agents; liq.; 100% conc.

Phospholan AD-1. [187] Complex phosphate ester, free acid; anionic; surfactant for liq. detergents; amber clear liq.; dens. 1.08 g/ml; pour pt. < 0 C; pH 2.9 (10%)

Phospholan ALF-5. [360] Complex phosphate ester from fatty alcohol base; anionic; foam limiter for use with anionic surfactants; off-wh. solid; mild odor; insol. in water; visc. 120 cs (80 C); flash pt. > 300 F (COC); pour pt. 65 C; pH 2.0–3.0 (1% aq.); 100% act.

Phospholan BH-14. [187] Complex phosphate ester, free acid; anionic; surfactant for cleaners; coupling agent; lt. yel. clear visc. liq.; dens. 1.27 g/ml; pour pt. < 0 C; pH 2.6 (10%)

Phospholan KPE-4 [360] Complex phosphate ester, potassium salt; anionic; hydrotrope; solubilization of low foaming surfactants into highly built liquids; industrial and domestic detergents and cleaners; pale straw liq.; mild odor; water-sol.; sp.gr. 1.294; visc. 180 cs; flash pt. 200 F (COC); pour pt. < 0 C; pH 6.0–8.0 (1% aq.); 65% act.

Phospholan PDB-3. [360] Fatty alcohol ethoxylate complex phosphate ester; anionic; wetting agent, detergent component, textile and drycleaning applications; electrolyte metal cleaning; emulsifier for phenols and in emulsion polymerization; pesticides and herbicides; metalworking lubricants with anticorrosive properties; production of iodophors; pale straw liq.; mild odor; disp. in water; sp.gr. 1.000; visc. 2400 cs; flash pt. > 300 F (COC); pour pt. 10 C; pH 2.0–3.0 (1% aq.); 100% act.

Phospholan PHB-14. [360] Phosphate ester; anionic; hydrotrope and compatibility agent for agrochemical formulations; liq.; 100% conc.

Phospholan PNP-9. [360] Nonylphenol ethoxylate complex phosphate ester, acid form; see Phospholan PDB-3; dk. amber visc. liq.; mild odor; water-sol.; sp.gr. 1.200; visc. 9400 cs (40 C); flash pt. > 300 F (COC); pour pt. 15 C; pH 2.0–3.0 (1% aq.); 100% act.

Pilot SXS-40. [478] Sodium xylene sulfonate; coupling agt., hydrotrope, solubilizer, solv.; used in liq. cleaners, org. polymers and dyestuffs, petroleum industry, pulping, animal glues; lt. yel. clear liq.; dens. 9.9 lb/gal; pH 7.5; 42% total solids.

Pilot SXS-96. [478] Sodium xylene sulfonate; hydrotrope, visc. reducer, dipsersant, extracting aid for detergent formualtions; stabilizes styrene latex opacifier systems for light-duty dishwashing liquids; white fine powd.; bulk dens. 220 cc/100 g; pH 7.5 (1%); 96% act.

Pinamine A. [153] Pine oil, solv. and surfactant; anionic; scouring compd. for cotton and syn. fibers; liq.; biodeg.

Pinamine K. [153] Pine oil and surfactant; nonionic; scour and detergent for soaping off dyed cotton yarn and knit goods; gel; biodeg.

Pluracol E-200. [86] PEG-4; intermediate for preparation of nonionic surfactants; base, coating, stabilizer, solv., vehicle, extender, and coupling agent for pharmaceutical, cosmetic, and toiletries; lubricant for metal applications, rubber industry; wood treatment; textile conditioning, antistat, and sizing agent, softener; colorless clear liq.; sol. in water and org. solvs. except aliphatic hydrocarbons; dens. 9.4 lb/gal; sp.gr. 1.12; visc. 4.36 cs (210 F); flash pt. 360 F; surf. tens. 57.2 dynes/cm (1%); pH 6.5 (5% aq.)

Pluracol E-300. [86] PEG-6; see Pluracol E-200; also dispersant in food tablets and preparations; plasticizer; colorless clear liq.; sol. see Pluracol E-200; dens. 9.4 lb/gal; sp.gr. 1.12; visc. 5.75 cs (210 F); flash pt. 410 F; surf. tens. 62.9 dynes/cm (1%); pH 5.7 (5% aq.).

Pluracol E-400. [86] PEG-8; see Pluracol E-300; colorless clear liq.; sol. see Pluracol E-200; dens. 9.4 lb/gal; sp.gr. 1.12; visc. 7.39

cs (210 F); flash pt. 460 F; surf. tens. 66.6 dynes/cm (1%); pH 6.2 (5% aq.).

Pluracol E-400 NF. [86] PEG-8; chemical intermediate, base, coupling agent, thickener, lubricant, mold release agent, defoamer, softener, conditioner, antistat, sizing agent, dispersant for pharmaceutical, cosmetic, and oral care preparations, in metal polishing and cleaning formulations, rubber products, paper and wood products, textile processing, ink formulations; liquid; m.w. 400; visc. 7.4 cs (99 C); pour pt. 5 C; flash pt. 182 C (COC).

Pluracol E-600. [86] PEG-12; see Pluracol E-300; colorless clear liq.; sol. see Pluracol E-200; dens. 9.4 lb/gal; sp.gr. 1.12; visc. 10.83 cs (210 F); flash pt. 480 F; surf. tens. 65.2 dynes/cm (1%); pH 5.3 (5% aq.).

Pluracol E-600 NF. [86] PEG-12; see Pluracol E-400 NF; liquid; m.w. 600; visc. 10.8 cs (99 C); pour pt. 20 C; flash pt. 249 C (COC).

Pluracol E-1000. [86] PEG-20; see Pluracol E-400 NF; solid; m.w. 1000; visc. 17.5 cs (99 C); m.p. 38 C; flash pt. 255 C (COC).

Pluracol E-1450. [86] PEG; see Pluracol E-400 NF; solid; m.w. 1450; visc. 28.5 cs (99 C); m.p. 45 C; flash pt. 255 C (COC).

Pluracol E-1450 NF. [86] PEG, NF grade; see Pluracol E-400 NF; solid; m.w. 600; visc. 28.5 cs (99 C); m.p. 45 C; flash pt. 255 C (COC).

Pluracol E-1500. [86] PEG-6-32; see Pluracol E-300; wh. waxy solid; sol. see Pluracol E-200; dens. 10.0 lb/gal; sp.gr. 1.20; m.p. 46.0–47.5 C; flash pt. > 490 F; surf. tens. 62.8 dynes/cm (1%); pH 6.7 (5% aq.).

Pluracol E-2000. [86] PEG-40; see Pluracol E-400 NF; solid; m.w. 2000; visc. 43.5 cs (99 C); m.p. 52 C; flash pt. > 260 C (COC).

Pluracol E-4000. [86] PEG-75; see Pluracol E-300; wh. waxy solid; sol. see Pluracol E-200; dens. 10.0 lb/gal; sp.gr. 1.20; m.p. 59.5 C; flash pt. > 490 F; surf. tens. 61.9 dynes/cm (1%); pH 6.7 (5% aq.).

Pluracol E-4000 NF. [86] PEG-75; see Pluracol E-400 NF; solid; m.w. 4000; visc. 134 cs (99 C); m.p. 59 C; flash pt. > 260 C (COC).

Pluracol E-4500. [86] PEG; see Pluracol E-400 NF; solid; m.w. 4500; visc. 170 cs (99 C); m.p. 60 C; flash pt. > 260 C (COC).

Pluracol E-6000. [86] PEG-150; see Pluracol E-200; wh. waxy solid; sol. see Pluracol E-200; dens. 10.0 lb/gal; sp.gr. 1.21; m.p. 61.0 C; flash pt. > 490 F; surf. tens. 62.1 dynes/cm (1%); pH 6.7 (5% aq.).

Pluracol E-8000. [86] PEG; see Pluracol E-400 NF; solid; m.w. 8000; visc. 750 cs (99 C); m.p. 61 C; flash pt. > 260 C (COC).

Pluracol E-8000 NF. [86] PEG, NF grade; see Pluracol E-400 NF; solid; m.w. 8000; visc. 750 cs (99 C); m.p. 61 C; flash pt > 260 C (COC).

Pluracol P-410. [86] PPG-9; chemical intermediate, antifoam agent in fermentation and in paint formulations, antiblooming agent for pentachlorophenol-treated wood; binder and lubricant for ceramics; plasticizer of resin-treated papers; preparation of polyurethane foams; hydraulic and grinding fluids; ore flotation; water-wh. liq.; slight ether-like odor; water-sol.; m.w. 425; dens. 8.36 lb/gal; sp.gr. 1.005; visc. 35 cps (100 F); flash pt. > 400 F (PMCC); pour pt. –35 F; pH 6–7

Pluracol P-710. [86] PPG-12; see Pluracol P-410; water-wh. liq.; slight ether-like odor; water-sol.; m.w. 775; dens. 8.35 lb/gal; sp.gr. 1.004; visc. 65 cps (100 F); flash pt. > 400 F (PMCC); pour pt. 35 F; pH 6–7

Pluracol P-1010. [86] PPG-17; see Pluracol P-410; water-wh. liq.; slight ether-like odor; insol. in water; m.w. 1050; dens. 8.38 lb/gal; sp.gr. 1.007; visc. 80 cps (100 F); flash pt. > 400 F (PMCC); pour pt. –35 F; pH 6–7

Pluracol P-2010. [86] PPG-26; see Pluracol P-410; water-wh. liq.; slight ether-like odor; insol. in water; m.w. 2000; dens. 8.34 lb/gal; sp.gr. 1.002; visc. 175 cps (100 F); flash pt. > 400 F (PMCC); pour pt. –35 F; pH 6–7

Pluracol P-3010. [86] PPG; see Pluracol P-410; water-wh. liq.; slight ether-like odor; insol. in water; m.w. 3000; dens. 8.32 lb/gal; sp.gr. 1.00; visc. 295 cps (100 F); flash pt. > 400 (PMCC); pour pt. –20 F; pH 6–7

Pluracol P-4010. [86] PPG-30; see Pluracol P-410; water-wh. liq.; slight ether-like odor; insol. in water; m.w. 4000; dens. 8.33 lb/gal; sp.gr. 1.00; visc. 550 cps (100 F); flash pt. > 400 (PMCC); pour pt. –20 F; pH 6–7

Pluracol W-170. [86] PPG-5-buteth-7; component in demulsifying and wetting formulations; brake and metalworking fluids; rubber and fiber lubricant; textile applications; defoamer for hot and cold applications, food and chemical processing; cosmetic formulations; APHA 50 max. visc. liq.; sol. in water, alcohols, ketones, esters, benzene, toluene, glycol ethers, chlorinated solvs.; dens. 8.58 lb/gal; sp.gr. 1.03; visc. 160–180 SUS (100 F); cloud pt. 73 C (1%); flash pt. 360 F (OC); pour pt. –45 F; pH 5.5–7.0 (10% aq.)

Pluracol W-352ON. [86] PPG; see Pluracol W-170; APHA 40 max. visc. liq.; sol. see Pluracol W-170; dens. 8.83 lb/gal; sp.gr. 1.06; visc. 3520 SUS (100 F); cloud pt. 57 C (1%); flash pt. 437 F (OC); pour pt. –20 F; pH 6.0–7.5 (10% aq.)

Pluracol W-352ON-RL. [86] PPG; see Pluracol W-170; APHA 200 max. visc. liq.; sol. see Pluracol W-170; dens. 8.87 lb/gal; sp.gr. 1.065; visc. 1752–2500 SUS (100 F); cloud pt. 55.5 C (1%); flash pt. 440 F (OC); pour pt. 28.4 F; pH 6.0–7.5 (10% aq.)

Pluracol W-660. [86] PPG-12-buteth-16; see Pluracol W-170; APHA 50 max. visc. liq.; sol. see Pluracol W-170; dens. 8.79 lb/gal; sp.gr. 1.055; visc. 660 SUS (100 F); cloud pt. 60.5 C (1%); flash pt. 440 F (OC); pour pt. –34 F; pH 5.5–7.5 (10% aq.)

Pluracol W-2000. [86] PPG-20-buteth-30; see Pluracol W-170; APHA 50 max. visc. liq.; sol. see Pluracol W-170; dens. 8.83 lb/gal; sp.gr. 1.06; visc. 2000 SUS (100 F); cloud pt. 57.0 C (1%); flash pt. 440 F (OC); pour pt. –25 F; pH 5.5–7.5 (10% aq.)

Pluracol W-5100, W-5100N. [86] PPG-33-buteth-45; see Pluracol W-170; APHA 40 max. visc. liq.; sol. see Pluracol W-170; dens. 8.83 lb/gal; sp.gr. 1.06; visc. 5100 SUS (100 F); cloud pt. 55 C (1%); flash pt. 437 F (OC); pour pt. –20 F; pH 5.5–7.0 (10% aq.)

Pluradot HA-410. [86] Trifunctional polyoxyalkylene glycol; nonionic; emulsifier, solubilizer, deduster, wetting agent, demulsifier, detergent, dispersant; surfactant for hard surface cleaning, machine dishwashing; liq.; m.w. 3200; sp.gr. 1.03; visc. 450 cps; cloud pt. 25 C (1%); pour pt. –15 F; surf. tens. 36 dynes/cm (0.1%); 100% conc.

Pluradot HA-420. [86] Trifunctional polyoxyalkylene glycol; see Pluradot HA-410; liq.; m.w. 3600; sp.gr. 1.04; visc. 565 cps; cloud pt. 28 C (1%); pour pt. –10 F; surf. tens. 37 dynes/cm (0.1%); 100% conc.

Pluradot HA-430. [86] Trifunctional polyoxyalkylene glycol; see Pluradot HA-410; liq.; m.w. 3900; sp.gr. 1.05; visc. 700 cps; cloud pt. 43 C (1%); pour pt. 20 C; surf. tens. 38 dynes/cm (0.1%); 100% conc.

Pluradot HA-433. [86] Trifunctional polyoxyalkylene glycol; see Pluradot HA-410; liq.; m.w. 3900; sp.gr. 1.05; visc. 700 cps; pour pt. 20 F; 100% conc.

Pluradot HA-440. [86] Trifunctional polyoxyalkylene glycol; see Pluradot HA-410; liq.; m.w. 4450; sp.gr. 1.05; visc. 750 cps; cloud pt. 54 C (1%); pour pt. 25 C; surf. tens. 39 dynes/cm (0.1%); 100% conc.

Pluradot HA-450. [86] Trifunctional polyoxyalkylene glycol; see Pluradot HA-410; paste; m.w. 5700; sp.gr. 1.06; cloud pt. 74 C (1%); pour pt. 60 F; surf. tens. 41 dynes/cm (0.1%)

Pluradot HA-510. [86] Trifunctional polyoxyalkylene glycol; see Pluradot HA-410; liq.; m.w. 4600; sp.gr. 1.03; visc. 710 cps; cloud pt. 24 C (1%); pour pt. –15 F; surf. tens. 35 dynes/cm; 100% conc.

Pluradot HA-520. [86] Trifunctional polyoxyalkylene glycol; see Pluradot HA-410; liq.; m.w. 5000; sp.gr. 1.03; visc. 760 cps; cloud pt. 27 C (1%); pour pt. 5 F; surf. tens. 36 dynes/cm (0.1%); 100% conc.

Pluradot HA-530. [86] Trifunctional polyoxyalkylene glycol; see Pluradot HA-410; liq.; m.w. 5300; sp.gr. 1.04; visc. 850 cps; cloud pt. 42 C (1%); pour pt. 30 F; surf. tens. 37 dynes/cm (0.1%); 100% conc.

Pluradot HA-540. [86] Trifunctional polyoxyalkylene glycol; see Pluradot HA-410; liq.; m.w. 6000; sp.gr. 1.05; visc. 1240 cps; cloud pt. 54 C (1%); pour pt. 40 F; surf. tens. 38 dynes/cm (0.1%); 100% conc.

Pluradot HA-550. [86] Trifunctional polyoxyalkylene glycol; see Pluradot HA-410; paste; m.w. 7500; sp.gr. 1.06; cloud pt. 77 C (1%); pour pt. 75 F; surf. tens. 39 dynes/cm (0.1%); 100% conc.

Plurafac A-24. [86] Straight chain, primary aliphatic oxyalkylated alcohol; nonionic; detergent for commercial and home cleaning formulations; colorless liq.; ref. index 1.4501; sol. in alcohols, acetone, MEK, butyl Cellosolve, propylene glycol, chloroform, perchloroethylene, toluene, xylene, kerosene, water; dens. 7.7 lb/gal; sp.gr. 0.927; visc. 27 cps; HLB 6; cloud pt. 0 C (1%); flash pt. 400 F; pour pt. 56 F; surf. tens. 29.4 dynes/cm; pH 6–7 (1%); 100% conc.

Plurafac A-27. [86] Straight chain, primary aliphatic oxyalkylated alcohol; nonionic; see Plurafac A-24; wh. paste; sol. in water, methanol, ethanol, acetone, MEK, butyl Cellosolve, propylene glycol, chloroform, toluene, xylene; dens. 8.4 lb/gal; HLB 14.5; cloud pt. 100 C (1%); flash pt. 460 F; pour pt. 75 F; surf. tens. 31.0 dynes/cm (0.1%); pH 6–7 (1%); 100% conc.

Plurafac A-38. [86] Ceteareth-27; see Plurafac A-24; wh. solid; sol. in water, methyl, ethyl, IPA, acetone, MEK, butyl Cellosolve, chloroform, perchloroethylene, toluene, xylene; m.p. 114 F; HLB 19; cloud pt. > 100 C (1%); flash pt. 480 F; surf. tens. 43.6 dynes/cm (0.1%); pH 6–7; 100% conc.

Plurafac A-39. [86] Ceteareth-55; see Plurafac A-24; wh. solid; sol. in water, chloroform; m.p. 132 F; HLB 20; cloud pt. > 100 C (1%); flash pt. 480 F; surf. tens. 53.2 dynes/cm (0.1%); pH 6–7; 100% conc.

Plurafac A-46. [86] Straight chain primary aliphatic oxyalkylated alcohol; see Plurafac A-24; wh. paste; sol. see Plurafac A-38; dens. 8.2 lb/gal; HLB 13; cloud pt. 72 C (1%); flash pt. 475 F; pour pt. 75 F; surf. tens. 32.5 dynes/cm (0.1%); pH 6–7 (1%); 100% conc.

Plurafac B-25-5. [86] Straight chain primary aliphatic oxyalkylated alcohol; see Plurafac A-24; colorless clear liq.; ref. index 1.4570; sol. see Plurafac A-38; dens. 8.2 lb/gal; sp.gr. 1.002; visc. 100 cps; HLB 12; cloud pt. 65 C (1%); pour pt. 9 F; surf. tens. 36.4 dynes/cm (0.1%); pH 6–7 (1%); 100% conc.

Plurafac B-26. [86] Straight chain primary aliphatic oxyalkylated alcohol; see Plurafac A-24; colorless clear liq.; ref. index 1.4587; sol. see Plurafac A-38; dens. 8.5 lb/gal; sp.gr. 1.020; visc. 114 cps; HLB 14; cloud pt. 73 C (1%); flash pt. 475 F; pour pt. 35 F; surf. tens. 36.3 dynes/cm (0.1%); pH 6–7 (1%); 100% conc.

Plurafac C-17. [86] Straight chain primary aliphatic oxyalkylated alcohol; see Plurafac A-24; wh. opaque liq.; ref. index 1.4526; sol. see Plurafac A-38; dens. 8.6 lb/gal; sp.gr. 1.030; visc. 100 cps; HLB 16; cloud pt. 81 C (1%); flash pt. 520 F; pour pt. 56 F; surf. tens. 34.4 dynes/cm (0.1%); pH 6–7 (1%); 100% conc.

Plurafac D-25. [86] PPG-6 C_{12-18} pareth-11; see Plurafac A-24; slightly cloudy liq.; ref. index 1.4560; sol. see Plurafac A-38; dens. 8.4 lb/gal; sp.gr. 1.007; visc. 95 cps; HLB 11; cloud pt. 59 C (1%); flash pt. 465 F; pour pt. 0 F; surf. tens. 34.3 dynes/cm (0.1%); pH 6–7 (1%); 100% conc.

Plurafac RA-20. [86] Straight chain primary aliphatic oxyalkylated alcohol; nonionic; detergent, dispersant, wetting agent, emulsifier, deduster used in rinse aids and dishwashing products; colorless clear liq.; ref. index 1.4530; sol. see Plurafac A-24; dens. 8.3 lb/gal; sp.gr. 0.992; visc. 83 cps; HLB 10; cloud pt. 45 C (1%); flash pt. 475 F; pour pt. 45 F; surf. tens. 32.8 dynes/cm (0.1%); pH 6–7 (1%); 100% conc.

Plurafac RA-30. [86] Straight chain primary aliphatic oxyalkylated alcohol; colorless clear liq.; ref. index 1.4534; sol. see Plurafac RA-20; dens. 8.1 lb/gal; sp.gr. 0.973; visc. 56 cps; HLB 9; cloud pt. 35 C (1%); flash pt. 455 F; pour pt. 50 F; surf. tens. 30.6 dynes/cm (0.1%); pH 6–7 (1%); 100% conc.

Plurafac RA-40. [86] Straight chain primary aliphatic oxyalkylated alcohol; see Plurafac RA-20; colorless clear liq.; ref. index 1.4510; sol. see Plurafac A-24; dens. 8.1 lb/gal; sp.gr. 0.970; visc. 91 cps; HLB 7; cloud pt. 25 C (1%); flash pt. 437 F; pour pt. –15 F; surf. tens. 30.5 dynes/cm (0.1%); pH 6–7 (1%); 100% conc.

Plurafac RA-43. [86] Straight chain primary aliphatic oxyalkylated alcohol; see Plurafac RA-20; wh. opaque liq.; dens. 7.9 lb/gal; sp.gr. 0.950; visc. 120 cps; HLB 7; cloud pt. 24 C (1%); flash pt. 437 F; pour pt. 21 F; surf. tens. 30.5 dynes/cm (0.1%); pH 3–4 (1%); 100% conc.

Plurafac RA-50. [86] Straight chain, primary oxyalkylated alcohol; nonionic; surfactant, dispersant, wetting agent, emulsifier, demulsifying agent; for industrial and household detergents, colorless liq.; sol. in benzene, xylene, acetone, carbon tetrachloride, perchlorethylene, alcohols, kerosene, butyl Cellosolve, chloroform, MEK, propylene glycol; sol. < 10% in water; visc. 90 cps; sp.gr. 0.992; HLB 9; cloud pt. 40 C (1% aq.); flash pt 451 F; pour pt 13 C; surf. tens. 31.5 dynes/cm (0.1%); pH 6–7 (1% aq.); biodeg.; 100% act.

Plurafac T-55. [86] Oxyalkylated primary aliphatic alcohol; nonionic; agric. spray adjuvant, emulsifier; laundry and car wash detergent, general purpose cleaner, degreaser; pale yel. liq.; water-sol.; sp.gr. 1.012; visc. 164 cps; cloud pt. 57 C (1%); pour pt. 34 F; surf. tens. 36.2 dynes/cm (0.1%)

Pluraflo E4A. [86] Nonionic dispersant formulated for use in liquid form; EPA cleared for agric. use; liquid; visc. 820 cps; pour pt. 7 C; flash pt. 69 F (CC); surf. tens. 42.8 dynes/cm (0.1%)

Pluraflo E4B. [86] See Pluraflo E4A; liquid; visc. 740 cps; pour pt. 11 C; flash pt. 116 F (CC); surf. tens. 42.3 dynes/cm (0.1%)

Pluraflo E5A. [86] See Pluraflo E4A; liquid; visc. 430 cps; pour pt. 5 C; flash pt. 61 F (CC); surf. tens. 42.5 dynes/cm (0.1%)

Pluraflo E5B. [86] See Pluraflo E4A; liquid;

visc. 520 cps; pour pt. 3 C; flash pt. 110 F (CC); surf. tens. 42.7 dynes/cm (0.1%)

Pluraflo E5BG. [86] See Pluraflo E4A; liquid; visc. 250 cps; pour pt. –19 C; flash pt. 138 F (CC); surf. tens. 43.8 dynes/cm (0.1%)

Pluraflo E5G. [86] See Pluraflo E4A; liquid; visc. 260 cps; pour pt. –22 C; flash pt. 126 F (CC); surf. tens. 43.9 dynes/cm (0.1%)

Pluraflo N5G. [86] See Pluraflo E4A; liquid; visc. 410 cps; pour pt. –21 C; flash pt. 126 F (CC); surf. tens. 46.0 dynes/cm (0.1%)

Plurest DS 150. [86] PEG 6000 distearate, nonionic; high m.w. surfactant; wh. to off wh. waxy flakes; HLB 18.4; m.p. 55 C; sapon. no. 14–20

Pluriol E 200. [82] PEG; nonionic; solubilizer, impregnating agent, humectant, mold release agent; flow improver, thermal and hydraulic fluid, org. intermediate; detergent and cleaner; dye and pigment dispersant; inks; textile and coatings industry; coloring ceramics; softener in paper industry; plasticizer in adhesives industry and in production of cellulose film; ceramics and metalworking lubricant; clear colorless liq.; m.w. 200; water-sol.; dens. 1.12 g/cc; visc. 55–65 cs.; m.p. < –40 C; flash pt. > 150 C; pH 6.0–7.5 (1% aq.); 100% conc.

Pluriol E 300. [82] PEG; nonionic; see Pluriol E 200; colorless clear liq.; m.w. 300; water-sol.; dens. 1.13 g/cc; visc. 75–95 cs; m.p. –20 to –10 C; flash pt. > 150 C; pH 6.0–7.5 (1% aq.); 100% conc.

Pluriol E 400. [82] PEG; nonionic; see Pluriol E 200; colorless clear liq.; m.w. 400; water-sol.; dens. 1.13 g/cc; visc. 105–120 cs; m.p. –10 to 10 C; flash pt. > 150 C; pH 6.0–7.5 (1% aq.); 100% conc.

Pluriol E 600. [82] PEG; nonionic; see Pluriol E 200; semisolid; m.w. 600; water-sol.; dens. 1.15 g/cc; visc. 8–11 cs (99 C); m.p. 10–20 C; flash pt. > 220 C; pH 6.0–7.5 (1% aq.); 100% conc.

Pluriol E 1500. [82] PEG; nonionic; solubilizer, humectant; binder and hardener in personal care products; dispersing dyes and pigments; inks; textile and coating industry; coloring ceramics; paper industry softener; plasticizer in adhesives industry and production of cellulose film; ceramics and metalworking lubricants; aux. for copper and nickel electroplating baths; electrolytic polishing of steel; wh. fine powd.; m.w. 1500; water-sol.; bulk dens. 0.6 kg/l; visc. 25–35 cs (99 C); m.p. 45 C; flash pt. > 230 C; pH 6.0–7.5 (1% aq.); 100% conc.

Pluriol E 4000. [82] PEG; nonionic; see Pluriol E 1500; wh. fine powd.; m.w. 4000; water-sol.; bulk dens. 0.6 kg/l; visc. 80–130 cs (99 C); m.p. 55 C; flash pt. > 240 C; pH 6.0–7.5 (1% aq.); 100% conc.

Pluriol E 6000. [82] PEG; nonionic; see Pluriol E 1500; wh. fine powd.; m.w. 6000; water-sol.; bulk dens. 0.6 kg/l; visc. 300 cs (99 C); m.p. 60 C; flash pt. > 240 C; pH 6.0–7.5 (1% aq.); 100% conc.

Pluriol E 9000. [82] PEG; nonionic; see Pluriol E 1500; also used in rubber processing; mold release agent in production of latex and rubber; wh. fine powd.; m.w. 9000; water-sol.; bulk dens. 0.6 kg/l; visc. 1900 cs (99 C); m.p. 65 C; flash pt. > 240 C; pH 6.0–7.5 (1% aq.); 100% conc.

Pluriol P 600. [82] PPG; nonionic; mold release agent, additive for oils and fluids; lubricant and antifoam for rubber; consistency improver and solubilizer; intermediate in industrial applications; colorless clear liq.; misc. with water; m.w. 600; dens. 1.0 g/cc; visc. 130 cs; flash pt. 216 C; pour pt. –43 C; pH 6.5–7.5 (1% aq.); 100% conc.

Pluriol P 900. [82] PPG; nonionic; see Pluriol P 600; also in solv.-type cleaners; colorless clear liq.; m.w. 900; dens. 1.0 g/cc; visc. 180 cs; cloud pt. 33 C (1% aq.); flash pt. 220 C; pour pt. –38 C; pH 6.5–7.5 (1% aq.); 100% conc.

Pluriol P 2000. [82] PPG; nonionic; see Pluriol P 600; colorless clear liq.; m.w. 2000; dens. 1.0 g/cc; visc. 440 cs; flash pt. 222 C; pour pt. –35 C; pH 6.5–7.5 (1% aq.); 100% conc.

Pluriol PE 3100. [82] PO/EO block polymer; nonionic; surfactant; dispersant for colorants; defoamer in various applications; clear colorless liq.; m.w. 1100; ref. index 1.45; sol. in ethanol, 2-propanol, toluene, tetrachloroethylene, min. spirit; dens. 1.02 g/cc; visc. 175 cps: cloud pt. 40 C (1% aq.); pour pt. –20 C; surf. tens. 45 dynes/cm; pH 7 (5% aq.); 100% conc.

Pluriol PE 4300. [82] PO/EO block polymer; nonionic; surfactant; dishwashing detergent and rinse aid; colorless clear liq.; m.w. 1700; ref. index 1.45; sol. in water, 10% HCl, ethanol, 2-propanol, tetrachloroethylene; dens. 1.04 g/cc; visc. 300 cps; cloud pt. 42 C (1% aq.); pour pt. –20 C; surf. tens. 43 dynes/cm; pH 7 (5% aq.); 100% conc.

Pluriol PE 6100. [82] PO/EO block polymer; nonionic; defoamer in household dishwashing machines, bottlewashing plants, metal cleaners, boiler feedwater, acid dyebaths; antifreeze; colorless clear liq.; m.w. 2000; ref. index 1.45; sol. in ethanol, 2-propanol, toluene, tetrachloroethylene, min. spirit; dens. 1.02 g/cc; visc. 350 cps; cloud pt. 24 C (1% aq.); pour pt. –20 C; surf. tens. 44 dynes/cm; pH 7 (5% aq.); 100% conc.

Pluriol PE 6101. [82] PO/EO block polymer; nonionic; defoamer; rinse aid; clear, slightly dull liq.; m.w. 2000; ref. index 1.45; sol. in 10% HCl, ethanol, 2-propanol, toluene, tetrachloroethylene, min. spirit; dens. 1.02 g/cc; visc. 350 cps; cloud pt. 28 C (1% aq.); pour pt. –10 C; surf. tens. 44 dynes/cm; pH 7 (5% aq.); 100% conc.

Pluriol PE 6200. [82] PO/EO block polymer; nonionic; wetting agent, emulsifier, demulsifier used in mechanical cleaning processes and splitting crude oil emulsions; household rinse aids; phosphatizing baths; emulsifier in polymerization processes; colorless clear liq.; m.w. 2500; ref. index 1.45; sol. in water, 10% HCl, ethanol, 2-propanol, toluene and tetrachloroethylene; dens. 1.04 g/cc; visc. 500 cps; cloud pt. 32 C (1% aq.); pour pt. –14 C; surf. tens. 44 dynes/cm; pH 7 (5% aq.); 100% conc.

Pluriol PE 6400. [82] PO/EO block polymer; nonionic; detergent for dishwashing machines, dairy cleaners; coolant and lubricant for grinding, drilling and cutting; dispersant, emulsifier; emulsion polymerization; slightly dull liq.; m.w. 3000; ref. index 1.45; sol. in Pluriol PE 6200; dens. 1.05 g/cc; visc. 850 cps; cloud pt. 59 C (1% aq.); pour pt. 16 C; surf. tens. 44 dynes/cm; pH 7 (5% aq.); 100% conc.

Pluriol PE 6800. [82] PO/EO block polymer; nonionic; powd. detergent for household and industry; dust binder, dispersant, solubilizer for cosmetics, emulsifier, production of emulsion polymers; wh. fine powd.; m.w. 8500; ref. index 1.45 (70 C); sol. in water, 10% HCl, ethanol, toluene; dens. 1.06 g/cc (70 C); cloud pt. > 100 (1% aq.); surf. tens. 48 dynes/cm; pH 7 (5% aq.); 100% conc.

Pluriol PE 8100. [82] PO/EO polymer; nonionic; defoamer; surfactant for cleaning processes; breweries and dairies; colorless to clear-slightly dull liq.; m.w. 2600; ref. index 1.45 (70 C); sol. in 2-propanol, toluene, tetrachloroethylene; dens. 1.04 g/cc; visc. 600 cps; cloud pt. 18 C (1% aq.); pour pt. –30 C; surf. tens. 35 dynes/cm; pH 7 (5% aq.); 100% conc.

Pluriol PE 9200. [82] PO/EO polymer; nonionic; wetting agent for cleaning processes; surfactant in dishwashing machine detergents for household and canteens; rinse aid; clear to slightly dull liq.; m.w. 3650; ref. index 1.45 (70 C); sol. see Pluriol PE 6200; dens. 1.03 g/cc; visc. 700 cps; cloud pt. 22 C (1% aq.); pour pt. –3 C; surf. tens. 35 dynes/cm; pH 7 (5% aq.); 100% conc.

Pluriol PE 9400. [82] PO/EO polymer; nonionic; emulsifier in oil industry and production of aq. disps., dispersant; dishwashing machines, bottle washing lines and dairy cleaners; wh. waxy; m.w. 4600; ref. index 1.45 (70 C); sol. see Pluriol PE 6200; surf. tens. 40 dynes/cm; pH 7 (5% aq.); 100% conc.

Pluriol PE 10100. [82] PO/EO block polymer; nonionic; wetting and antifoam agent; cleaners and rinse aids for dishwashing machines; clear to slightly dull liq.; m.w. 3550; ref. index 1.45; sol. in 10% HCl, ethanol, 2-propanol, toluene, tetrachloroethylene, min. spirit; dens. 1.02 g/cc; visc. 1000 cps; cloud pt. 15 C (1% aq.); pour pt. –30 C; surf. tens. 35 dynes/cm; pH 7 (5% aq.); 100% conc.

Pluriol PE 10500. [82] PO/EO block polymer; nonionic; emulsifier for agrochemicals; household and industrial cleaners; emulsion polymerization; wh. waxy; m.w. 6500; ref. index 1.45 (70 C); sol. in water, 10% HCl, ethanol, 2-propanol; dens. 1.03 g/cc; cloud pt. > 100 C (1% aq.); surf. tens. 36 dynes/cm; pH 7 (5% aq.); 100% conc.

Pluriol RPE 2540. [82] PEG-PPG; nonionic surfactant; defoaming and foam controlled detergents; liq.; dens. 1.0 g/cc; visc. 1500 cps; surf. tens. 37 dynes/cm; pH 9 (5% aq.); 100% conc.

Pluriol RPE 3110. [82] PEG-PPG; nonionic; defoaming and foam controlled detergents; liq.; 100% conc.

Plurol ISO. [248] Polyglycerol isostearate; nonionic; emulsifier; liq.; HLB 10.0; 100% conc.

Plurol Oleique. [248] Polyglycerol oleate; nonionic; emulsifier; liq.; 100% conc.

Plurol Stearique. [248] Polyglycerol stearate; nonionic; emulsifier; solid; 100% conc.

Pluronic 10R5. [86] Meroxapol 105; nonionic; emulsifier, wetting agent, binder, stabilizer, plasticizer, lubricant, solubilizer, dispersant, visc. control agent, defoamer, intermediate for hard surface detergents, rinse aids, automatic dishwashing, textile processing; cosmetics; pharmaceuticals, pulp, paper, and petroleum industries, agric. products, in iodophors, water treating systems, fermentation, cutting and grinding fluids; liq.; m.w. 1970; ref. index. 1.4587; sol. in water, propylene glycol, xylene, IPA, ethyl acetate, perchloroethylene, IPM; sp.gr. 1.058; visc. 400 cps; HLB 21.0; cloud pt. 69 C (1% aq.); flash pt. > 450 F (COC); pour pt. 15 C; surf. tens. 50.9 dynes/cm; 100% act.

Pluronic 10R8. [86] Meroxapol 108; nonionic; see Pluronic 10R5; flakable solid; m.w. 5000; sol. in water, xylene, ethyl acetate; sp.gr. 1.062 (77 C); m.p. 46 C; HLB 33.0; cloud pt. 99 C (1% aq.); flash pt. > 450 F (COC); surf. tens. 54.1 dynes/cm (0.1%); 100% act.

Pluronic 12R3. [86] PO/EO block copolymer; nonionic surfactant for use as defoamer, emulsifier, demulsifier, solubilizer, detergent, dispersant, binder, stabilizer, gelling agent, wetting agent, rinse aid, chemical intermediate in cosmetic, drug, textile, paper, petroleum, paint, detergent, and metal cleaning industries; liquid; m.w. 1800; visc. 340 cps; HLB 5.0; pour pt. –20 C; cloud pt. 53 C (1% aq.); surf. tens. 43 dynes/cm (0.1%); 100% act.

Pluronic 17R1. [86] Meroxapol 171; nonionic; see Pluronic 10R5; also for foam control in paper sizing operations; liq.; m.w. 1950; ref. index 1.4516; sol. in xylene, IPA, ethyl acetate, min. oil, perchloroethylene, IPM, trichlorotrifluoroethane; sp.gr. 1.018; visc. 300 cps; HLB 4.0; cloud pt. 32 C (1% aq.); flash pt. > 450 F (COC); pour pt. –27 C; surf. tens. 33.0 dynes/cm (0.1%); 100% act.

Pluronic 17R2. [86] Meroxapol 172; nonionic; see Pluronic 10R5; liq.; m.w. 2100; ref. index 1.4535; sol. in water, xylene, IPA; ethyl acetate, perchloroethylene, IPM, trichlorotrifluoroethane; sp.gr. 1.030; visc. 350 cps; HLB 8.0; cloud pt. 39 C (1% aq.); flash pt. > 450 F (COC); pour pt. –25 C; surf. tens. 41.9 dynes/cm (0.1%); 100% act.

Pluronic 17R4. [86] Meroxapol 174; nonionic; see Pluronic 10R5; liq.; m.w. 2700; ref. index. 1.4572; sol. in water, propylene glycol, xylene, IPA, ethyl acetate, perchloroethylene, IPM, trichlorotrifluoroethane; sp.gr. 1.048; visc. 560 cps; HLB 16.0; cloud pt. 47 C (1% aq.); flash pt. > 450 F (COC); pour pt. 18 C; surf. tens. 44.1 dynes/cm (0.1%); 100% act.

Pluronic 17R8. [86] Meroxapol 178; see Pluronic 10R5; also dry toilet bowl cleaners, dye levelers, solubilizer of drugs, stick type cosmetics; soap bars, dispersant in deinking operations; flakable solid; m.w. 7500; sol. in water, ethyl acetate, IPM; sp.gr. 1.064 (77 C); m.p. 53 C; HLB 32.0; cloud pt. 81 C (1% aq.); flash pt. > 450 F (COC); surf. tens. 47.3 dynes/cm (0.1%); 100% act.

Pluronic 22R4. [86] PO/EO block copolymer; see Pluronic 12R3; liq.; m.w. 3350; visc. 950 cps; HLB 6.3; pour pt. 24 C; cloud pt. 40 C (1% aq.); surf. tens. 43 dynes/cm (0.1%); 100% act.

Pluronic 25R1. [86] Meroxapol 251; nonionic; see Pluronic 10R5; also foam control in paper sizing operations and antifreeze; liq.; m.w. 2800; ref. index 1.4521; sol. in xylene, IPA, ethyl acetate, perchloroethylene, IPM, trichlorotrifluoroethane; sp.gr. 1.017; visc. 460 cps; HLB 2.3; cloud pt. 28 C (1% aq.); flash pt. > 450 F (COC); pour pt. –27 C; surf. tens. 36.3 dynes/cm (0.1%); 100% act.

Pluronic 25R2. [86] Meroxapol 252; nonionic; see Pluronic 10R5; also wetting and rinse aid; lubricant and leveling agent for paper coating; liq.; m.w. 3120; ref. index 1.4541; sol. see Pluronic 25R1; sp.gr. 1.039; visc. 680 cps; HLB 6.3; cloud pt. 33 C (1% aq.); flash pt. > 450 F (COC); pour pt. –5 C; surf. tens. 37.5 dynes/cm (0.1%); 100% act.

Pluronic 25R4. [86] Meroxapol 254; nonionic; see Pluronic 10R5; liq.; m.w. 3800; ref. index 1.4574; sol. in water, propylene glycol, ethyl acetate, perchloroethylene, IPM, trichlorotrifluoroethane; sp.gr. 1.046; visc. 1110 cps; HLB 14.3; cloud pt. 40 C (1% aq.); flash pt. > 450 F (COC); pour pt. 25 C; surf. tens. 40.9 dynes/cm (0.1%); 100% act.

Pluronic 25R5. [86] Meroxapol 255; nonionic; see Pluronic 10R5; also thickener for cosmetic pastes and creams; paste; m.w. 4500; sol. see Pluronic 25R4; sp.gr. 1.036 (60 C); m.p. 33 C; HLB 18.3; cloud pt. 44 C (1% aq.); flash pt. > 450 F (COC); surf. tens. 43.5 dynes/cm (0.1%); 100% act.

Pluronic 25R8. [86] Meroxapol 258; nonionic; see Pluronic 10R5; also dry toilet bowl cleaners, dye levelers for fabrics; solubilizer for drugs; thickener for cosmetics; deinking operations; felt washing operations; flakable solid; m.w. 9000; sol. in water; sp.gr. 1.062; m.p. 56 C; HLB 30.3; cloud pt. 80 C (1% aq.); flash pt. > 450 F (COC); surf. tens. 46.1 dynes/cm (0.1%); 100% act.

Pluronic 31R1. [86] Meroxapol 311; nonionic; see Pluronic 10R5; also floating bath oils; foam control in antifreeze; liq.; m.w. 3200; ref. index 1.4522; sol. Pluronic 25R1; sp.gr. 1.018; visc. 578 cps; HLB 1.7; cloud pt. 25 C (1% aq.); flash pt. > 450 F (COC); pour pt. –25 C; surf. tens. 34.1 dynes/cm (0.1%); 100% act.

Pluronic 31R2. [86] Meroxapol 312; nonionic; see Pluronic 10R5; paper coating color additive; deinking and felt washing operations; liq.; m.w. 3400; ref. index 1.4542; sol. see Pluronic 25R4; sp.gr. 1.030; visc. 818 cps; HLB 5.7; cloud pt. 30 C (1% aq.); flash pt. > 450 F (COC); pour pt. 9 C; surf. tens. 38.9 dynes/cm (0.1%); 100% act.

Pluronic 31R4. [86] Meroxapol 314; nonionic; see Pluronic 10R5; paste; m.w. 4300; sol. see Pluronic 25R1; sp.gr. 1.028 (60 C); m.p. 26 C; HLB 13.7; cloud pt. 31 C (1% aq.); flash pt. > 450 F (COC); pour pt. 26 C; surf. tens. 41.2 dynes/cm (0.1%); 100% act.

Pluronic F38. [86] Poloxamer 108; nonionic; wetting agent, emulsifier, demulsifier, foam and visc. control agent; dispersant, antistat and gelling agent; agric. chemicals; cosmetics and personal care products; pharmaceuticals; metal cleaning; pulp and paper industry; rinse aids, textile processing; dyeing assistant and leveler; lubricant; scouring aid; water treatment; prilled; m.w. 5000; sol. in ethanol, propylene glycol, water, toluene, dens. 8.9 lb/gal (77 C); sp.gr. 1.07 (77 C); visc. 260 cps (77 C); m.p. 48 C; HLB 30.5; cloud pt. > 100 C (1% aq.); flash pt. 505 F (COC); surf. tens. 52.2 dynes/cm (0.1% conc.); 100% act.

Pluronic F68. [86] Poloxamer 188; nonionic; see Pluronic F38; prilled; m.w. 8350; sol. in ethanol, water; dens. 8.8 lb/gal (77 C); sp.gr. 1.06 (77 C); visc. 1000 cps (77 C); m.p. 52 C; HLB 29.0; cloud pt. > 100 C (1% aq.); flash pt. 500 F (COC); surf. tens. 50.3 dynes/cm (0.1%); 100% act.

Pluronic F77. [86] Poloxamer 217; nonionic; see Pluronic F38; prilled; m.w. 6600; sol. in ethanol, water, toluene; dens. 8.7 lb/gal (77 C); sp.gr. 1.04 (77 C); m.p. 48 C; HLB 24.5; cloud pt. > 100 C (1% aq.); flash pt. 485 F (COC); surf. tens. 47.0 dynes/cm (0.1%); 100% act.

Pluronic F87. [86] Poloxamer 237; nonionic; see Pluronic F38; prilled; m.w. 7700; sol. see Pluronic F77; dens. 8.7 lb/gal (77 C); sp.gr. 1.04 (77 C); visc. 700 cps. (77 C); m.p. 49 C; HLB 24.0; cloud pt. > 100 C (1% aq.); flash pt. 472 F (COC); surf. tens. 44.0 dynes/cm (0.1%); 100% act.

Pluronic F88. [86] Poloxamer 238; nonionic; see Pluronic F38; prilled; m.w. 10,800; sol. in ethanol and water; dens. 8.8 lb/gal (77 C); sp.gr. 1.06 (77 C); visc. 2300 cps (77 C); m.p. 54 C; HLB 28.0; cloud pt. > 100 C (1% aq.); surf. tens. 48.5 dynes/cm (0.1%); 100% act.

Pluronic F98. [86] Poloxamer 288; nonionic; see Pluronic F38; prilled; m.w. 13,000; sol. in ethanol, water, perchoroethylene; dens. 8.8 lb/gal (77 C); sp.gr. 1.06 (77 C); visc. 2700 cps (77 C); m.p. 55 C; HLB 27.5; cloud pt. > 100 C (1% aq.); flash pt. 491 F (COC); surf. tens. 43.0 dynes/cm (0.1%); 100% act.

Pluronic F108. [86] Poloxamer 338; nonionic; see Pluronic F38; prilled; m.w. 14,000; sol. in ethanol, water; dens. 8.8 lb/gal (77 C); sp.gr. 1.06 (77 C); visc. 8000 cps (77 C); m.p. 57 C; HLB 27.0; cloud pt. > 100 C (1% aq.); flash pt. 495 F (COC); surf. tens. 41.2 dynes/cm (0.1%); 100% act.

Pluronic F127. [86] Poloxamer 407; nonionic; see Pluronic F38; prilled; m.w. 12,500; sol. in ethanol, water, toluene, perchloroethylene; dens. 8.8 lb/gal (77 C); sp.gr. 1.05 (77 C); visc. 3100 cps (77 C); m.p. 56 C; HLB 22.0; cloud pt. > 100 C (1% aq.); surf. tens. 40.6 dynes/cm (0.1%); 100% act.

Pluronic L10. [86] PO/EO block copolymer; nonionic; see Pluronic 12R3; liquid; m.w. 3200; visc. 660 cps; HLB 14.0; pour pt. –5 C; cloud pt. 32 C (1% aq.); surf. tens. 41 dynes/cm (0.1%); 100% act.

Pluronic L31. [86] Poloxamer 101; nonionic; see Pluronic F38; liq.; m.w. 1100; ref. index 1.4515; sol. in ethanol, propylene glycol, water, toluene, xylene, perchloroethylene; dens. 8.5 lb/gal; sp.gr. 1.02; visc. 165 cps; HLB 4.5; cloud pt. 37 C (1% aq.); flash pt. 439 F (COC); pour pt. –32 C; surf. tens. 46.9 dynes/cm (0.1%); 100% act.

Pluronic L35. [86] Poloxamer 105; nonionic; see Pluronic F38; liq.; m.w. 1900; sol. see Pluronic L31; dens. 8.8 lb/gal; sp.gr. 1.06; visc. 340 cps; HLB 18.5; cloud pt. 77 C (1% aq.); pour pt. 7 C; surf. tens. 48.8 dynes/cm (0.1%); 100% act.

Pluronic L42. [86] Poloxamer 122; nonionic; see Pluronic F38; liq.; m.w. 1630; ref. index 1.4541; sol. see Pluronic L31; dens. 8.6 lb/gal; sp.gr. 1.03; visc. 250 cps; HLB 8.0; cloud pt. 37 C (1% aq.); flash pt. 450 F (COC); pour pt. –26 C; surf. tens. 46.5 dynes/

cm (0.1%); 100% act.

Pluronic L43. [86] Poloxamer 123; nonionic; see Pluronic F38; liq.; m.w. 1850; ref. index 1.4563; sol. see Pluronic L31; dens. 8.7 lb/gal; sp.gr. 1.04; visc. 310 cps; HLB 12.0; cloud pt. 42 C (1% aq.); pour pt. –1 C; surf. tens. 47.3 dynes/cm (0.1%); 100% act.

Pluronic L44. [86] Poloxamer 124; nonionic; see Pluronic F38; liq.; m.w. 2200; ref. index 1.4580; sol. see Pluronic F31; dens. 8.8 lb/gal; sp.gr. 1.05; visc. 440 cps; HLB 16.0; cloud pt. 65 C (1% aq.); flash pt. 464 F (COC); pour pt. 16 C; surf. tens. 45.3 dynes/cm (0.1%); 100% act.

Pluronic L61. [86] Poloxamer 181; nonionic; see Pluronic L38; liq.; m.w. 2000; ref. index 1.4520; sol. in ethanol, toluene, xylene, perchloroethylene; dens. 8.4 lb/gal; sp.gr. 1.01; visc. 285 cps; HLB 3.0; cloud pt. 24 C (1% aq.); flash pt. 455 F (COC); pour pt. –29 C; 100% act.

Pluronic L62. [86] Poloxamer 182; nonionic; see Pluronic F38; liq.; m.w. 2500; sol. see Pluronic F31; dens. 8.6 lb/gal; sp.gr. 1.03; visc. 400 cps; HLB 7.0; cloud pt. 32 C (1% aq.); flash pt. 466 F (COC); pour pt. –4 C; surf. tens. 42.8 dynes/cm (0.1%); 100% act.

Pluronic L63. [86] Poloxamer 183; nonionic; see Pluronic F38; liq.; m.w. 2650; ref. index 1.4562; sol. see Pluronic F31; dens. 8.7 lb/gal; sp.gr. 1.04; visc. 490 cps; HLB 11.0; cloud pt. 34 C (1% aq.); pour pt. 10 C; surf. tens. 43.3 dynes/cm (0.1%); 100% act.

Pluronic L64. [86] Poloxamer 184; nonionic; see Pluronic F38; liq.; m.w. 2900; ref. index 1.4575; sol. see Pluronic F31; dens. 8.8 lb/gal; sp.gr. 1.05; visc. 550 cps; HLB 15.0; cloud pt. 61 C (1% aq.); flash pt. 485 F (COC); pour pt. 16 C; surf. tens. 43.2 dynes/cm (0.1%); 100% act.

Pluronic L72. [86] Poloxamer 212; nonionic; see Pluronic F38; liq.; m.w. 2750; ref. index 1.4542; sol. in ethanol, water, toluene, xylene, perchloroethylene; dens. 8.6 lb/gal; sp.gr. 1.03; visc. 510 cps; HLB 6.5; cloud pt. 25 C (1% aq.); flash pt. 442 F (COC); surf. tens. 39.0 dynes/cm (0.1%); 100% act.

Pluronic L81. [86] Poloxamer 231; nonionic; see Pluronic F38; liq.; m.w. 2750; ref. index 1.4526; sol. in ethanol, toluene, xylene, perchloroethylene; dens. 8.5 lb/gal; sp.gr. 1.02; visc. 475 cps; HLB 2.0; cloud pt. 20 C (1% aq.); pour pt. –37 C; 100% act.

Pluronic L92. [86] Poloxamer 282; nonionic; see Pluronic F38; liq.; m.w. 3650; ref. index 1.4547; sol. see Pluronic L81; dens. 8.6 lb/gal; sp.gr. 1.03; visc. 700 cps; HLB 5.5; cloud pt. 26 C (1% aq.); flash pt. 445 F (COC); pour pt. 7 C; surf. tens. 35.9 dynes/cm (0.1%); 100% act.

Pluronic L101. [86] Poloxamer 331; nonionic; see Pluronic F38; liq.; m.w. 3800; ref. index 1.4524; sol. see Pluronic L81; dens. 8.5 lb/gal; sp.gr. 1.02; visc. 800 cps; HLB 1.0; cloud pt. 15 C (1% aq.); pour pt. –23 C; 100% act.

Pluronic L121. [86] Poloxamer 401; nonionic; see Pluronic F38; liq.; m.w. 4400; ref. index 1.4527; sol. see Pluronic L81; dens. 8.4 lb/gal; sp.gr. 1.01; visc. 1200 cps; HLB 5.0; cloud pt. 14 C (1% aq.); pour pt. 5 C; surf. tens. 33.0 dynes/cm (0.1%); 100% act.

Pluronic L122. [86] Poloxamer 402; nonionic; see Pluronic F38; liq.; m.w. 5000; ref. index 1.4558; sol. see Pluronic L81; dens. 8.6 lb/gal; sp.gr. 1.03; visc. 1750 cps; HLB 4.0; cloud pt. 19 C (1% aq.); flash pt. 490 F (COC); pour pt. 20 C; surf. tens. 33.0 dynes/cm (0.1%); 100% act.

Pluronic P65. [86] Poloxamer 185; nonionic; see Pluronic F38; paste; sol. see Pluronic F31; m.w. 3400; dens. 8.8 lb/gal (60 C); sp.gr. 1.06 (60 C); visc. 180 cps (60 C); m.p. 30 C; HLB 17.0; cloud pt. 82 C (1% aq.); pour pt. 27 C; surf. tens. 46.3 dynes/cm (0.1%); 100% act.

Pluronic P75. [86] Poloxamer 215; nonionic; see Pluronic F38; paste; m.w. 4150; sol. see Pluronic F31; dens. 8.8 lb/gal (60 C); sp.gr. 1.06 (60 C); visc. 250 cps (60 C); m.p. 34 C; HLB 16.5; cloud pt. 82 C (1% aq.); pour pt. 27 C; surf. tens. 42.8 dynes/cm (0.1%); 100% act.

Pluronic P84. [86] Poloxamer 234; nonionic; see Pluronic F38; paste; m.w. 4200; sol. see Pluronic F31; dens. 8.6 lb/gal (60 C); sp.gr. 1.03 (60 C); visc. 265 cps (60 C); m.p. 34 C; HLB 14.0; cloud pt. 74 C (1% aq.); flash pt. 442 F (COC); pour pt. 18 C; surf. tens. 42.0 dynes/cm (0.1%); 100% act.

Pluronic P85. [86] Poloxamer 235; nonionic; see Pluronic F38; paste; m.w. 4600; sol. see Pluronic F31; dens. 8.7 lb/gal (60 C); sp.gr. 1.04 (60 C); visc. 310 cps (60 C); m.p. 40 C; HLB 16.0; cloud pt. 85 C (1% aq.); pour pt. 29 C; surf. tens. 42.5 dynes/cm (0.1%); 100% act.

Pluronic P94. [86] Poloxamer 284; nonionic; see Pluronic F38; paste; m.w. 5100; sol. see Pluronic F31; dens. 8.7 lb/gal (60 C); sp.gr.

1.04 (60 C); visc. 110 cps (60 C); m.p. 37 C; HLB 13.5; cloud pt. 76 C (1% aq.); pour pt. 33 C; surf. tens. 39.6 dynes/cm (0.1%); 100% act.

Pluronic P103. [86] Poloxamer 333; nonionic; see Pluronic F38; paste; m.w. 4950; sol. see Pluronic L81; dens. 8.7 lb/gal (60 C); sp.gr. 1.04 (60 C); visc. 285 cps (60 C); m.p. 30 C; HLB 9.0; cloud pt. 86 C (1% aq.); pour pt. 21 C; surf. tens. 34.4 dynes/cm (0.1%); 100% act.

Pluronic P104. [86] Poloxamer 334; nonionic; see Pluronic F38; paste; m.w. 5850; sol. see Pluronic L81; dens. 8.7 lb/gal (60 C); sp.gr. 1.04 (60 C); visc. 550 cps (60 C); m.p. 37.5; HLB 13.0; cloud pt. 81 C (1% aq.); flash pt. 448 F (COC); pour pt. 32 C; surf. tens. 33.1 dynes/cm (0.1%); 100% act.

Pluronic P105. [86] Poloxamer 335; nonionic; see Pluronic F38; paste; m.w. 6500; sol. see Pluronic L81; dens. 8.8 lb/gal (60 C); sp.gr. 1.05 (60 C); visc. 800 cps (60 C); m.p. 42 C; HLB 15.0; cloud pt. 91 C (1% aq.); pour pt. 35 C; surf. tens. 39.1 dynes/cm (0.1%); 100% act.

Pluronic P123. [86] Poloxamer 403; nonionic; see Pluronic F38; paste; m.w. 5750; see Pluronic L81; dens. 8.5 lb/gal (60 C); sp.gr. 1.02 (60 C); visc. 350 cps (60 C); m.p. 31 C; HLB 8.0; cloud pt. 90 C (1% aq.); pour pt. 31 C; surf. tens. 34.1 dynes/cm (0.1%); 100% act.

PMS-33. [287] Propylene glycol monostearate; nonionic; emulsifier and whipping agent for food; consistency regulator and coloring materials for cosmetic applications; solid; HLB 3.5; 100% conc.

Pogol 200. [283] PEG; nonionic; solubilizer, pharmaceuticals, cosmetics; clear liq.; dens. 1.14; 100% act.

Pogol 300. [283] PEG; nonionic; solubilizer, dispersant, emulsifier; herbicides and pesticides; clear liq.; dens. 1.13; 100% act.

Pogol 400. [283] PEG; nonionic; emulsifier, solubilizer, dispersant; see Pogol 300; clear liq.; dens. 1.12; visc. 92 cps; 100% act.

Pogol 400 USP. [283] PEG, USP grade; nonionic; solubilizer, lubricant; mold release agent and lubricant for both natural and syn. products; clear liq.; dens. 1.12; visc; 92 cps; 100% act.

Pogol 600. [283] PEG; nonionic; solubilizer, lubricant; used in paper coating mixes; antisticking agents and oil resistors; clear paste; ref. index 1.4699; dens. 1.13; visc. 130 cps; 100% act.

Pogol 1540. [283] PEG; nonionic; solubilizer; base for mfg. of fatty acid esters, emulsifiers, and dispersants; wh. solid; 100% act.

Polawax A31. [159] Emulsifying wax; nonionic; used in quick-breaking foams; solid; sol. in alcohol and aerosol propellant; solid; 97% conc.

Polawax GP200. [159] Self-emulsifying wax; nonionic; emulsifier for w/o emulsions; used in cosmetics and pharmaceuticals; solid; 97% conc.

Polawax NF. [159] Self-emulsifying wax; nonionic; solid; 97% conc.

Polefine 510N. [590] Alkyl naphthalene sulfonate/formaldehyde condensates; anionic; reducing agent for concrete; liq.

Polyaldo DGDO. [256] Polyglyceryl-10 decaoleate; nonionic; w/o and o/w emulsifier, emollient, and lubricant for cosmetics, toiletries, pharmaceuticals, and household speciality products; amber clear liq.; sol. in ethanol, min. and veg. oils, insol. in water; HLB 3 ± 1; sapon. no. 155–185.

Polyaldo DGHO. [256] Polyglyceryl-10 hexaoleate, nonionic; see Polyaldo DGDO; amber clear liq.; sol. see Polyaldo DGDO; HLB 5 ± 1; sapon. no. 130–160.

Polyaldo HGDS. [256] Polyglyceryl-6 distearate; nonionic; see Polyaldo DGDO; cream beads; sol. in ethanol; disp. in water; HLB 7 ± 1; m.p. 53–57; sapon. no. 120–140.

Polyaldo TGMS. [256] Polyglyceryl-3 stearate; nonionic; see Polyaldo DGDO; cream beads; sol. see Polyaldo DGDO; HLB 7 ± 1; m.p. 54–58 C; sapon. no. 120–140.

Polychol 5. [159] Laneth-5; nonionic; o/w emulsifier for personal care products; Gardner 13 max.; disp. in water; pH 4.5–6.0 (10% aq.).

Polychol 10. [159] Laneth-10; nonionic; see Polychol 5; Gardner 12 max.; sol. see Polychol 5; pH 4.5–6.0 (10% aq.).

Polychol 15. [159; 160] Laneth-15; nononic; emulsifier, dispersant, solubilizer, emollient and gelling agent for cosmetics and pharmaceuticals; solid; HLB 12.7; 100% conc.

Polychol 20. [159] Laneth-20; nonionic; see Polychol 5; Gardner 8–10 max.; water-sol.; pH 3.5–5.5 (10% aq.).

Polychol 40. [159] Laneth-40; nonionic; see Polychol 5; Gardner 8–10 max.; water-sol.; pH 3.0–6.5 (10% aq.).

Polyfac 8110. [642] Syn. alkylbenzene, branched side chain, calcium salt; anionic;

surfactant used in emulsifier blends; dk. liq.; sp.gr. 1.04; dens. 8.67 lb/gal; visc. 5500–6500 cps; pour pt. 45 F; flash pt. (PMCC) 133 F; pH 5–7; 60% act.

Polyfac 8170. [642] Alkylphenoxy polyethoxy ethanol, phosphate ester; compatibility agent in liq. fertilizer systems, emulsifier blend formulations; pale yel. clear liq.; sp.gr. 1.12; dens. 9.34 lb/gal; visc. 3600 cps; pour pt. 28 F; flash pt. (PMCC) > 300 F; pH 2.0 (1% aq.); 99+% act.

Polyfac 8200. [642] Nonylphenol POE adduct; nonionic; see Polyfac 8110; also adjuvant; clear liq.; sp.gr. 1.05; dens. 8.77 lb/gal; visc. 2850 cps; HLB 13; cloud pt. 128 F (1% aq.); pH 5–7 (1% aq.); surf. tens. 31 dynes/cm (1%).

Polyfac 8240. [642] POE castor oil; nonionic; see Polyfac 8110; wh. semisolid; sp.gr. 1.06; dens. 8.84 lb/gal; visc. 600 cps; pour pt. 50 F; flash pt. (PMCC) > 200 F; cloud pt. 168 F (1% aq.); pH 5–7 (1% aq.).

Polyfac 8320. [642] Butyl polyalkylene oxide block copolymer; agric. emulsifier formulations; lt. amber liq.; sp.gr. 1.05; dens. 8.77 lb/gal; visc. 2000 cps; pour pt. 69 F; flash pt. (PMCC) > 200 F; cloud pt. 160 F (1% aq.); pH 5–7 (1% aq.).

Polyfac 8321. [642] Polyalkylene oxide block copolymer; nonionic; see Polyfac 8110; amber liq.; sp.gr. 1.03; dens. 8.57 lb/gal; visc. 1000 cps; HLB 5.5; pour pt. < 30 F; pH 5.7. 100% act.

Polyfac 8322. [642] Polyalkylene oxide block copolymer; nonionic; see Polyfac 8110; sp.gr. 1.02 (150 F); dens. 8.51 lb/gal (150 F); visc. 700 cps (150 F); HLB 14; pour pt. 74 F; pH 5–7; 100% act.

Polyfac ABS-40S. [642] Sodium branched alkylbenzene sulfonate; emulsifier for the polymerization of styrene, vinyl chloride, SBR, and other monomer systems; slurry; sp.gr. 1.00; visc. 3650 cps; pH 7.6; 40% act.

Polyfac ABS-60C. [642] Calcium dodecylbenzene sulfonate; anionic; see Polyfac 8110; dk. liq.; sp.gr. 1.04; visc. 6000 cps; pour pt. 45 F; flash pt. (PMCC) 133 F; pH 6.0; 60% act.

Polyfac ABS-95. [642] Branched alkylbenzene sulfonic acid; mfg. of emulsifiers for agriculture, oil field, and textile applications; rdsh.-amber liq.; m.w. 326; pH 1.5 (1% aq.); 96% sulfonic acid.

Polyfac ALS-30. [642] Ammonium lauryl sulfate; surfactant for household, personal, and pet care products; Klett 70 liq.; dens. 8.5 lb/gal; visc. 900 cps; pH 6.7 (10%) 28.5% act.

Polyfac LAS-40S. [642] Sodium linear alkylbenzene sulfate; slurry; surfactant offering good detergency, high foam; Klett 50.0 liq.; dens. 8.4 lb/gal; visc. 5000 cps; pH 8.0 (5%); biodeg.; 39.5% act.

Polyfac LAS-50S. [642] Sodium linear alkylbenzene sulfate; see Polyfac LAS-40S; Klett 50.0 liq.; dens. 8.4 lb/gal; visc. 6000 cps; pH 7.5 (5%); 50.0% act.

Polyfac LAS-60T. [642] TEA alkylbenzene sulfonate; surfactant used in personal care products; yel. clear liq.; dens. 9.09 lb/gal; visc. 1650 cps; pour pt. < 32 F; pH 7.2 (10%); 59% act.

Polyfac LAS-97. [642] Linear dodecylbenzene sulfonic acid; surfactant for formulation of detergents; rdsh.-amber liq.; m.w. 318; pH 1.6 (1% aq.); 97% sulfonic acid.

Polyfac LAS-93I. [642] Isopropylamine alkylbenzene sulfonate, nonionic; lt. br. clear liq.; oil-sol.; sp.gr. 1.05; visc. 1500 cps; flash pt. (PMCC) 105 F; pH 3.5 (10%); 93% act.

Polyfac LES-60S. [642] Sodium lauryl ether sulfate; surfactant used in detergents and personal care products; Klett 50.0 liq.; dens. 8.5 lb/gal; pH 7.5; 59% act.

Polyfac MT-610. [642] POE (10) monotallate; coemulsifier and emulsifier-solubilizer used in fluids, oils and greases, fiber lubricant, textile applications, preparation of cutting fluids; amber liq.; sol. in IPA, xylene; disp. in water; dens. 8.7 lb/gal; visc. 300 cps; HLB 11.8; pour pt. 41 F; cloud pt. 77 F; sapon. no. 70–80.

Polyfac MT-615. [642] POE (15) monotallate; see Polyfac MT-610; also emulsifier and coemulsifier in softeners, dye carriers, and detergents; scouring agent and dye bath leveling agent; amber liq. to paste; sol. in water, IPA, xylene; dens. 8.8 lb/gal; visc. 350 cps; HLB 13.3; pour pt. 55 F; cloud pt. 127 F; sapon. no. 55–65.

Polyfac NP-40-70. [642] Nonylphenol ethoxylate; emulsifier for emulsion polymerization of syn. latices, paints, and textile finishes; clear liq.; dens. 9.1 lb/gal; visc. 890 cps; cloud pt. 88; pH 7.0 (1% aq.); 70% solids.

Polyfac PCW-92. [642] Sodium alkyl ether sulfate; foaming agent for air drilling of oil and gas wells; yel. clear liq.; sp.gr. 1.09; dens. 9.08 lb/gal; visc. 150 cps; pour pt. 16 F; flash pt. 97 F (PMCC); pH 8.5 (10%); 59.5% act.

Polyfac PCW-160. [642] Sodium and triethanolamine lauryl sulfates, coconut monoethanolamide conc.; liq. detergent offering good foam; yel. clear liq.; sp.gr. 1.02; dens. 8.5 lb/gal; visc. 700 cps; pour pt. 40 F; flash pt. > 200 F (PMCC); pH 7.7; 31.3% solids.

Polyfac PCW-180. [642] Phosphate ester; surfactant used as lubricant additive for oil-based metalworking and hydraulic fluids; amber visc. liq.; sol. in min. oil; insol. in water; dens. 8.8 lb/gal; visc. 3500 cps; pH 2.1.

Polyfac PCW-181, -182. [642] Complex carboxylated fatty amide salt; additive used as corrosion inhibitor and lubricant in metalworking fluids; amber visc. liq.; visc. 2600 and 3100 cps resp.; pH 8.6 and 8.5 (2% aq.) resp.

Polyfac PCW-198. [642] Ammonium alcohol ether sulfate; foaming agent for air and foam drilling for oil and gas wells; used in liq. detergents, cements and wallboard mfg.; yel. clear liq.; sp.gr. 1.065; dens. 8.88 lb/gal; flash pt. 99 F (PMCC); pH 7.3; 60.0% act.

Polyfac PN-209. [642] Nonylphenol phosphate, ethoxylated; anionic; surfactant for liq. detergents, textile wet processing, lubricant and corrosion inhibitor; yel. liq.; dens. 9.34 lb/gal; visc. 3600 cps; flash pt. > 300 F (PMCC); pH 2 (1% aq.); 99% act.

Polyfac SA-60S. [642] Sodium alcohol ether sulfate; anionic; foaming agent used in detergent concs., carwash formulations, and cleaners; yel. clear liq.; sp.gr. 1.09; dens. 9.08 lb/gal; visc. 150 cps; pour pt. 16 F; flash pt. 97 F (PMCC); pH 8.5 (10%); 59.5% act.

Polyfac SD-60A. [642] Linear alcohol ether sulfate, ammonium salt; surfactant for use in detergent and personal care products; Klett 50; dens. 8.5 lb/gal; pH 7.5; 59% act.

Polyfac SD-60S. [642] Sodium lauryl ether sulfate; surfactant used in liq. detergents and personal care products; Klett 50; dens. 8.5 lb/gal; pH 7.5; 59.0% act.

Polyfac SEE-340. [642] POE (20) sorbitan tritallate; surfactant used with other surfactants in an emulsifier blend; amber liq.; disp. in water; sp.gr. 1.04; dens. 8.67 lb/gal; visc. 440 cps; HLB 11; pour pt. < 10 F; cloud pt. < 25 C (1% aq.).

Polyfac SG-60A. [642] Ammonium alcohol ether sulfate, anionic; foaming agent for use in detergent concs., carwash formulations, and cleaners; yel. clear liq.; sp.gr. 1.065; dens. 8.88 lb/gal; flash pt. 99 F (PMCC); pH 7.3; 60.0% act.

Polyfac SJ-30A. [642] Ammonium alkyl ethoxy sulfate; anionic; surfactant for use in emulsion polymerization; pale yel. liq.; sp.gr. 1.03; visc. 33 cps; cloud pt. –3 C; pH 6.9; 30.5% solids.

Polyfac SLS-30. [642] Sodium lauryl sulfate; anionic; surfactant for personal care products; pale yel. liq.; cloud pt. 65 F; pH 8.0 (10%); 29.0% act.

Polyfac SLS-30 Special. [642] Sodium lauryl sulfate; anionic; see Polyfac SLS-30; pale yel. liq.; dens. 8.7 lb/gal; cloud pt. 50 F; pH 8.0 (10%); 29.5% act.

Polyfac SN-60A. [642] Ammonium alkylphenol polyethoxy sulfate; anionic; surfactant for use in dishwash, rug shampoos, and car wash formulations; emulsifier for emulsion polymerization; yel. clear liq.; sp.gr. 1.065; dens. 8.9 lb/gal; visc. 100 cps; pour pt. 5 C; pH 7.0 (10%); 59% act.

Polyfac SP-20S. [642] Sodium alkylaryl polyether sulfate; wetting agent; emulsifier for polymerization of styrene/butadiene and other latices; wh. pourable paste; sp.gr. 1.03; visc. 3500 cps. pour pt. 28 F; pH 8.3; 20.0% solids.

Polyfac TA-20. [642] PEG-20 dist. tallow amine; surfactant for textile auxs.; amber liq.; sp.gr. 1.05; dens. 8.74 lb/gal; visc. 200 cps; flash pt. (PMCC) > 300 F; pH 10.0 (1% aq.).

Polyfac TDO-5. [642] PEG 200 tall oil dioleate; lipophilic surfactant used as a coemulsifier in industrial oils and mold release agent; amber liq.; sol. in IPA, xylene, min. oil; disp. in water; dens. 8.0 lb/gal; visc. 60 cps; HLB 5.4; pour pt. 10 F; cloud pt. < 77 F; sapon. no. 138–148.

Polyfac TDO-9. [642] PEG 400 tall oil dioleate; emulsifier and solubilizer for solvs., fats, and min. oil; emulsifier for kerosene in pesticide formulations, metalworking, and industrial lubricants; amber liq.; sol. see Polyfac TDO-5; dens. 8.2 lb/gal; visc. 130 cps; HLB 8.5; pour pt. 36 F; cloud pt. < 77 F; sapon. no. 110–120.

Polyfac TDO-14. [642] PEG 600 tall oil dioleate; component of defoamers, softeners, and lubricants; amber liq. to paste; disp. in water; sol. in water, IPA, xylene; dens. 8.3 lb/gal; visc. 155 cps; HLB 10.5; pour pt. 64 F; cloud pt. < 77; sapon. no. 90–100.

Polyfac TLS-40. [642] TEA lauryl sulfate; surfactant for use in liq. detergents, personal and pet care products; Klett 60 liq.; dens. 8.75 lb/gal; visc. 56 cps; pour pt. < 32 F; cloud pt. < 32 F; pH 7.3 (10%); 40.0% act.

Polyfac TMO-5. [642] POE (5) tall oil monooleate; emulsifier and solubilizer for solvs., min. and fatty oils; emulsifier for textile lubricants, metalworking fluids, and pesticide formulations; amber liq.; sol. in IPA, xylene, min. oil; disp. in water; dens. 8.3 lb/gal; visc. 95 cps; HLB 8.3; pour pt. 23 F; cloud pt. < 77 F; sapon. no. 110–120.

Polyfac TMO-7. [642] POE (7) tall oil monooleate; see Polyfac TMO-5; amber liq.; sol. in IPA, xylene; disp. in water; dens. 8.4 lb/gal; visc. 95 cps; HLB 10.3; pour pt. 32 F; cloud pt. < 77 F; sapon. no. 90–110.

Polyfac TMO-9. [642] POE (9) tall oil monooleate; emulsifier in industrial lubricants and degreasers; component of textile softeners and lubricants; amber liq.; sol. see Polyfac TMO-7; dens. 8.5 lb/gal; visc. 125 cps; HLB 11.5; pour pt. 45 F; cloud pt. < 77 F; sapon. no. 85–95.

Polyfac TMO-14. [642] POE (14) tall oil monooleate; emulsifier and detergent; textile dye leveling and lubricant formulations; amber liq.; sol. in water, IPA, xylene; dens. 8.7 lb/gal; visc. 195 cps; HLB 13.4; pour pt. 61 F; cloud pt. 126 F; sapon. no. 60–70.

Polyfon F. [642] Sodium lignosulfonate; dispersant industrial applications from agric. to ceramics; br. powd.; dens. 26 lb/ft^3; pH 11.0 (2% aq.); surf. tens. 50.4 dynes/cm (1% aq.).

Polyfon H. [642] Sodium lignosulfonate; dispersant in agric. chemicals and dyestuff formulations; brn. powd.; sol. see Polyfon F; dens. 22–24 lb/ft^3; pH 10.0 (2% aq.); surf. tens. 55 dynes/cm (1% aq.).

Polyfon O. [642] Sodium lignosulfonate; see Polyfon F; brn. powd.; sol. see Polyfon F; dens. 24 lb/ft^3; pH 10.6 (2% aq.).

Polyfon OD. [642] Kraft lignin polymer, sodium salt; see Polyfon F; brn. powd.; sol. see Polyfon F; dens. 24 lb/ft^3; pH 8.0 (2% aq.); surf. tens. 40.0 dynes/cm (1% aq.).

Polyfon T. [642] Sodium lignosulfonate; see Polyfon F; brn. powd.; sol. see Polyfon F; dens. 26 lb/ft^3; pH 10.5 (2% aq.).

Poly-G 200. [455] PEG; chemical intermediate for production of surfactants for cleaners, textiles, paper, cosmetics; carrier for pharmaceuticals; also in cosmetics and personal care products, textiles, rubber mold releases, printing inks and dyes, metalworking fluids, foods, paints, paper, wood products, adhesives, agric. products, ceramics, electrical equipment, petroleum products, photographic products, resins; APHA 25 max. liq.; m.w. 200; sol. in water, acetone, ethanol, ethyl acetate, toluene; sp.gr. 1.125; dens. 9.38 lb/gal; visc. 4.3 cs (99 C); flash pt. 171 C (COC).

Poly-G 300. [455] PEG; see Poly-G 200; APHA 25 max. liq.; m.w. 300; sol. in water, acetone, ethanol, ethyl acetate, toluene; sp.gr. 1.125; dens. 9.38 lb/gal; visc. 5.8 cs (99 C); f.p. –15 to –8 C; flash pt. 196 C (COC).

Poly-G 400. [455] PEG; see Poly-G 200; APHA 25 max. liq.; m.w. 400; sol. in water, acetone, ethanol, ethyl acetate, toluene; sp.gr. 1.127; dens. 9.4 lb/gal; visc. 7.3 cs (99 C); f.p. 4–10 C; flash pt. 224 C (COC); pour pt. 3–10 C.

Poly-G 600. [455] PEG; see Poly-G 200; APHA 25 max. liq.; m.w. 600; sol. in water, acetone, ethanol, ethyl acetate, toluene; sp.gr. 1.127; dens. 9.4 lb/gal; visc. 10.5 cs (99 C); f.p. 20–25 C; flash pt. 246 C (COC); pour pt. 19–24 C.

Poly-G 1000. [455] PEG; see Poly-G 200; wh. waxy solid; m.w. 1000; somewhat less sol. in water than liq. glycols; sp.gr. 1.104 (50/20 C); dens. 9.20 lb/gal (50/20 C); visc. 17.4 cs (99 C); f.p. 38–41 C; flash pt. 260 C (COC); pour pt. 40 C.

Poly-G 1500. [455] PEG; see Poly-G 200; wh. waxy solid; m.w. 1500; somewhat less sol. in water than liq. glycols; sp.gr. 1.104 (50/20 C); dens. 9.20 lb/gal (50/20 C); visc. 28 cs (99 C); f.p. 43–46 C; flash pt. 266 C (COC); pour pt. 45 C.

Poly-G 2000. [455] PEG; see Poly-G 200; APHA 25 max. liq.; m.w. 2000; somewhat less sol. in water than lower m.w. glycols; sp.gr. 1.113; dens. 9.26 lb/gal; visc. 11.7 cs (99 C); f.p. –20 C; 60% aq. sol'n.

Poly-G B1530. [455] PEG; see Poly-G 200; wh. waxy solid; m.w. 900; somewhat less sol. in water than liq. glycols; sp.gr. 1.104 (50/20 C); dens. 9.20 lb/gal (50/20 C); visc. 15 cs (99 C); f.p. 38–41 C; flash pt. 254 C (COC); pour pt. 38 C.

Polylan. [38] Oleyl linoleate and lanolin linoleate; nonionic; dispersant, lubricant, cosolvent, dye solubilizer, hydrophobic dispersant for cosmetics and pharmaceuticals; liq.; HLB 8.0; 100% conc.

Polylev #745. [480] Hydroxyethylated compds.; dispersant, penetrant, and leveling agent for dyestuffs; misc. with hot water.

Polylube GK. [283] Polyoxyalkylene condensate; nonionic; syn. fiber lubricant, antistat; liq.; water-sol.; ref. index 1.467; sp.gr. 1.09; visc. 330 cps; 100% act.

Polylube NPP. [283] Polyoxyalkylene and esters; nonionic; lubricant for processing of syn. fiber blends; yel. liq.; sp.gr. 0.99; visc. 230 cps; 87% act.

Polylube SC. [283] Fatty acid condensate; nonionic; fiber lubricant for polyester; liq.; water-sol.; ref. index 1.467; sp.gr. 1.02; visc. 110 cps; 87% act.

Polylube WS. [283] Polyalkylene glycol ether; nonionic; lubricant for spinning nylon and wool yarns; clear liq.; water-sol.; ref. index 1.4642; sp.gr. 1.086; visc. 500 cps; 100% act.

Polymulse 6. [350] PEG-6 palmitate; nonionic; emulsifier, dispersant, suspending agent, thickener used in personal care products; paste; 100% conc.

Polyol 20. [86] Sorbitol ethoxylate (20 EO); surfactant; Gardner 2 max. liq.; m.w. 1062 (theoret.); sp.gr. 1.09; visc. 450 cps; pH 5.5–7 (5% aq.); 100% act.

Polypeptide 37. [575] Hydrolyzed animal protein; anionic; surfactant for personal care products; dk. amber clear liq.; misc. with water; sp.gr. 1.20–1.23; visc. 50 cps max.; pH 7.0–7.4; 48–50% solids.

Polypeptide AAS 20%. [575] SD alcohol and hydrolyzed animal protein ethyl ester; cationic; see Polypeptide 37; amber clear liq.; alcoholic odor; sol. in anhyd. ethanol, IPA, propylene glycol, hexylene glycol, water; sp.gr. 0.87–0.88; pH 5.5–6.5 (10% aq.); 20–94% min. solids.

Polypeptide LSN. [575] Hydrolyzed animal protein; anionic; see Polypeptide 37; amber clear liq.; misc. with water; visc. 100 cps max.; pH 6.6–7.4; 40–42% solids.

Polypeptide LSN Anhydrous. [575] Protein hydrolysate, ammonium salt; anionic; see Polypeptide 37; wh. powd.; misc. with water; dens. 20–21 lb/ft^3; 94% min. solids.

Polypeptide SF. [575] Protein hydrolysate, ammonium salt; anionic; see Polypeptide 37; amber clear liq.; misc. with water; pH 6.5–7.0; 56–58% solids.

Polyquart H. [290] PEG-15 tallow polyamine; surfactant used in personal care products; amber clear liq.; visc. 1500–4000 cps; pH 5.0–6.0 (1%); 48–51.0% moisture.

Polyquart H-7102. [290] PEG-15 cocopolyamine and stearalkonium chloride; see Polyquart H; Gardner 6.0 max. liq.; visc. 1500–4000 cps; pH 5.5–6.5 (1%); 49–52% solids.

Polyram S. [117] N-tallow polypropylene polyamine; cationic; emulsifier for bitumen; antistripping agent for road making; paste.

Polystate C. [247] PEG 300 stearate; nonionic; base for cosmetic lotions; solid; 100% conc.

Polystep A-2. [575] Sodium xylene sulfonate; hydrotrope; pale to colorless liq.; 40% act.

Polystep A-4, -6, -7. [575] Sodium linear alkyl (C_{12}) benzene sulfonate; anionic; emulsifier for emulsion polymerization; yel. turbid liq., yel. visc. slurry, yel. slurry resp.; 50, 58, 39% act. resp.

Polystep A-9. [575] Triethanolamine linear alkyl (C_{12}) benzene sulfonate; see Polystep A-4; yel. clear liq.; 60% act.

Polystep A-10. [575] Calcium alkylbenzene sulfonate, branched; see Polystep A-4; dk. visc. liq.; oil-sol.; 65% act.

Polystep A-11 [575] Isopropylamine alkylbenzene sulfonate, branched; see Polystep A-4; pale clear visc. liq.; oil-sol.; 88% act.

Polystep A-12. [575] Potassium alkyl (C_{15}) benzene sulfonate, branched; see Polystep A-4; pale clear visc. liq.; 47% act.

Polystep A-13. [575] Linear alkyl (C_{12}) benzene sulfonic acid; see Polystep A-4; visc. liq.; 97% act.

Polystep A-15, A-16. [575; 576] Sodium linear alkyl (C_{12}) benzene sulfonate; anionic; emulsifier for emulsion and styrene polymerization; clear to hazy liq., pale turbid liq. resp.; 22 and 30% act. resp.

Polystep A-17. [575] Dodecylbenzene sulfonic acid, branched; intermediate for the production of sodium salts; neutralized acid as emulsifier; salts as emulsifiers for SBR latex; emulsion polymerization; visc. liq.; water-sol.; dens. 9.1 lb/gal; sp.gr. 1.09; visc. 9000 cps (30 C); 97.4% act.

Polystep A-18. [575] Sodium alpha olefin (C_{14}, C_{16}) sulfonate; for vinyl and vinylidene chloride, SBR polymerization; pale clear liq.; 40% act.

Polystep B-1. [575; 576] Ammonium alkyl phenol ether sulfate (4 EO); anionic; see Polystep A-4; yel. visc. liq.; 59% act.

Polystep B-3. [575] Sodium lauryl sulfate; surfactant for emulsion polymerization; wh. powd.; 97.5% act.

Polystep B-5. [575; 576] Sodium lauryl sulfate; anionic; emulsifier; pale clear liq.; 29% act.

Polystep B-6. [575] Triethanolamine lauryl sulfate; see Polystep A-4; pale clear liq.;

40% act.

Polystep B-7. [575] Ammonium lauryl sulfate; see Polystep A-4; also water resistance in coatings; pale clear liq.; 28% act.

Polystep B-8. [575] Diethanolamine lauryl sulfate; see Polystep A-4; pale clear liq.; 33% act.

Polystep B-9. [575] Magnesium lauryl sulfate; see Polystep A-4; also improves stability of latex; pale clear liq.; 28% act.

Polystep B-10. [575] Sodium lauryl ethoxylate sulfate (4 EO); see Polystep A-4; pale clear liq.; 28% act.

Polystep B-11. [575; 576] Ammonium lauryl ether sulfate (4 EO); anionic; see Polystep A-4; clear liq.; 59% act.

Polystep B-12. [575; 576] Sodium lauryl ether sulfate (4 EO); anionic; see Polystep A-4; pale clear liq.; 60% act.

Polystep B-13. [575] Ammonium alkyl polyether ethoxylate sulfate; see Polystep A-4; pale clear liq.; 60% act.

Polystep B-14. [575] Sodium alkyl polyether ethoxylate sulfate; see Polystep A-4; pale clear liq.; 60% act.

Polystep B-16. [575; 576] Sodium lauryl ethoxylate sulfate; see Polystep A-4; pale clear liq.; 28% act.

Polystep B-19. [575; 576] Sodium lauryl ether sulfate (30 EO); anionic; emulsifier for floor polish, finish polymer and latexes; good water resistance; pale clear liq.; 26% act.

Polystep B-20. [575; 576] Ammonium lauryl ether sulfate (30 EO); anionic; emulsifier for floor finish latexes; good water resistance; pale clear liq.; 26% act.

Polystep B-21. [575] Sodium laureth-12 sulfate; for high carboxyl latex; pale clear liq.; 28% act.

Polystep B-22. [575; 576] Ammonium lauryl ether sulfate (12 EO); anionic; emulsifier for acrylic copolymers; clear liq.; 28% act.

Polystep B-23. [575; 576] Sodium laureth-12 sulfate; high active version of Polystep B-21; oil-sol.; hazy amber liq.; 60% act.

Polystep B-24. [575] Sodium lauryl sulfate; see Polystep A-4; low cloud pt.; pale liq.; 28% act.

Polystep B-25. [575] Sodium decyl sulfate; see Polystep A-4; high surf. tens.; hydrophilic; pale clear liq.; 38% act.

Polystep B-26. [575] Sodium cetyl-stearyl sulfate; see Polystep A-4; pale paste; 25% act.

Polystep B-27. [575] Sodium alkyl phenol ethoxylate sulfate (4 EO); emulsifer for acrylics and SBR; pale clear liq.; 28% act.

Polystep B-28. [575] Sodium alkyl polyether sulfate; see Polystep A-4; low visc.; high surf. tens.; clear amber liq.; 40% act.

Polystep B-29. [576] Sodium octyl sulfate; low-foaming emulsifier; liq.; 33% conc.

Polystep F-1. [575] Nonoxynol-4; nonionic; pigment dispersant; pale clear liq.; oil-sol.; 100% act.

Polystep F-2. [575] Nonoxynol-6; nonionic; pigment dispersant; pale clear liq.; oil-sol.; 100% act.

Polystep F-3. [575] Nonoxynol-8; nonionic; pale clear liq.; water-sol.; 100% act.

Polystep F-4. [575; 576] Nonoxynol-10; nonionic; see Polystep A-4; pale clear liq.; 100% act.

Polystep F-5. [575] Nonoxynol-12; nonionic; emulsifier and stabilizer for emulsion polymerization; clear to turbid liq.; 100% act.

Polystep F-6. [575; 576] Nonoxynol-14; nonionic; see Polystep A-4; clear to turbid liq.; 100% act.

Polystep F-7. [575] Nonoxynol-15; see Polystep A-4; turbid visc. liq.; 100% act.

Polystep F-8. [575] Nonoxynol-20; used in straight nonionic systems; clear to hazy liq. or gel; 70% act.

Polystep F-9. [575; 576] Nonoxynol-30; nonionic; emulsifier for acrylic systems; clear to hazy liq. or gel; 70% act.

Polystep F-10. [575] Nonoxynol-40; nonionic; see Polystep F-9; clear to hazy liq. or gel; 70% act.

Polystep F-12. [575; 576] Laureth-12; nonionic; for preemulsification and post stabilization in latex systems; wh. paste; 100% act.

Polystep F-13. [575; 576] Laureth-30; nonionic; surfactant; wh. paste; 100% act.

Polystep F-95-B. [575] Alkyl phenol ethoxylate; nonionic; emulsifier for acrylics and vinyl acetate; liq.; 70% conc.

Polystep RA-90. [575] Sulfonic acid; see Polystep A-4; rdsh.-br. visc. liq.; 90–95% act.

Polytard. [642] Kraft lignin sulfonic acid deriv.; specialty cement additive; powd.

Poly-Tergent B-150. [455] Nonylphenoxy polyether alcohol; nonionic; surfactant for a wide variety of applications; pale yel. liq.; mild odor; sol. in carbon tetrachloride, ethanol, kerosene, min. spirits, xylene; disp. in water; sp.gr. 1.015–1.025; dens. 8.51 lb/gal; HLB 8.8; pour pt. –20 C; flash pt. 243 C (COC); cloud pt. < 0 C (1% aq.); surf. tens. 27 dynes/cm (0.1%); 100% functional.

Poly-Tergent B-200. [455] Nonylphenoxy polyether alcohol; nonionic; see Poly-Tergent B-150; pale yel. liq.; mild odor; sol. see Poly-Tergent B-150; sp.gr. 1.035–1.045; dens. 8.64 lb/gal; HLB 11.7; pour pt. –30 C; flash pt. 260 C (COC); cloud pt. < O (1% aq.); surf. tens. 28 dynes/cm (0.1%); 100% functional.

Poly-Tergent B-300. [455] Nonylphenoxy polyether alcohol; nonionic; see Poly-Tergent B-150; pale yel. liq.; mild odor; sol. in carbon tetrachloride, ethanol, ethylene glycol, water, xylene; sp.gr. 1.050–1.065; dens. 8.75 lb/gal; HLB 12.5; pour pt. 3 C; flash pt. 288 C (COC); cloud pt. 58 C (1% aq.); surf. tens. 29 dynes/cm (0.1%); 100% functional.

Poly-Tergent B-315. [575] Nonylphenoxy polyether alcohol; nonionic; see Poly-Tergent B-150; APHA 100 max. clear liq.; mild odor; sol. in water, alcohols, glycol ethers, aromatic hydrocarbons, chlorinated solvs.; sp.gr. 1.06; dens. 8.8 lb/gal; HLB 13.1; pour pt. 5 C; flash pt. 282 C (COC); cloud pt. 59 C (1% aq.); pH 5.0–7.0 (1% aq.); surf. tens. 31 dynes/cm (0.1%); 100% functional.

Poly-Tergent B-350. [575] Nonylphenoxy polyether alcohol; nonionic; see Poly-Tergent B-150; pale yel. liq.; mild odor; sol. see Poly-Tergent B-300; sp.gr. 1.060–1.070; dens. 8.83 lb/gal; HLB 13; pour pt. 10 C; flash pt. 288 C (COC); cloud pt. 73 C (1% aq.); surf. tens. 30 dynes/cm (0.1%); 100% act.

Poly-Tergent B-500. [575] Nonylphenoxy polyether alcohol; nonionic; pale yel. liq.; mild odor; sol. see Poly-Tergent B-300; sp.gr. 1.065–1.075; dens. 8.89 lb/gal; HLB 15; pour pt. 22 C; flash pt. 288 C (COC); cloud pt. 100 C (1% aq.); surf. tens. 36 dynes/cm (0.1%); 100% act.

Poly-Tergent J-200. [455] Branched chain alcohols, ethoxylated; nonionic; wetting agent for textile and paper industries; APHA 100 max. clear liq.; mild odor; sol. in alcohols, glycol ethers, aromatic hydrocarbons, chlorinated solvs.; disp. in water; sp.gr. 1.01; dens. 8.4 lb/gal; HLB 11.9; flash pt. 166 C (COC); cloud pt. < 0 C (1% aq.); pH 5.0–7.0 (1% aq.); surf. tens. 28 (0.1%); 100% act.

Poly-Tergent J-300. [455] Branched chain alcohols, ethoxylated; nonionic; see Poly-Tergent J-200; APHA 100 max. clear liq.; mild odor; sol. in water, alcohols, glycol ethers, aromatic hydrocarbons, chlorinated solvs.; sp.gr. 1.03; dens. 8.6 lb/gal; HLB 13.3; flash pt. 185 C (COC); cloud pt. 56 C (1% aq.); pH 6.0–7.0 (1% aq.); surf. tens. 27 dynes/cm (0.1%); 100% act.

Poly-Tergent P-17A. [455] EO/PO block polymer; nonionic; defoamer, deduster, emulsifier used in dishwashing applications and laundry detergents; APHA 50 max. clear liq.; mild odor; sol. in water, alcohols, glycol ethers, aromatic and aliphatic hydrocarbons, chlorinated solvs.; sp.gr. 1.01; dens. 8.4 lb/gal; flash pt. 241 C (COC); cloud pt. 28 C (1% aq.); pH 5.0–7.5 (1% aq.); surf. tens. 37 dynes/cm (0.1%); 100% act.

Poly-Tergent P-17B. [455] EO/PO block polymer; nonionic; see Poly-Tergent P-17A; APHA 50 max. clear liq.; mild odor; sol. in water, alcohols, glycol ethers, aromatic hydrocarbons, chlorinated solvs.; sp.gr. 1.04; dens. 8.6 lb/gal; flash pt. 246 C (COC); cloud pt. 31 (1% aq.); pH 5.0–7.5 (1% aq.); surf. tens. 40 dynes/cm (0.1%); 100% act.

Poly-Tergent P-17D. [455] EO/PO block polymer; nonionic; demulsifier, dedusting dispersant used in rinse aids, dishwashing, metal cleaning and papermaking applications; liq.; 100% conc.

Poly-Tergent P-22A. [455] EO/PO block polymer; nonionic; see Poly-Tergent P-17D; liq.; 100% conc.

Poly-Tergent P-32A. [455] EO/PO block polymer; nonionic; see Poly-Tergent P-17A; APHA 50 max. clear liq.; mild odor; sol. in alcohols, glycol ethers, aromatic hydrocarbons, chlorinated solvs.; insol. in water; sp.gr. 1.02; dens. 8.5 lb/gal; flash pt. 229 C (COC); cloud pt. 15 C (1% aq.); pH 5.0–7.5 (1% aq.); surf. tens. 43 dynes/cm (0.1%); 100% act.

Poly-Tergent S-205LF. [455] Alkoxylated linear alcohol; nonionic; surfactant for dishwashing applications, metal cleaning baths; foam depressant in hard surface cleaners; cleaners for dairy industry; rewetting agents for paper toweling; APHA 100 max. clear liq.; mild odor; sol. in water, alcohols, glycol ethers, aromatic hydrocarbons, chlorinated solvs.; sp.gr. 1.00; visc. 192 cs; flash pt. 229 C (COC); cloud pt. 10 C (1% aq.); pH 5.5–7.0 (1% aq.); surf. tens. 32 dynes/cm (0.1%); 100% act.

Poly-Tergent S-305LF. [455] Alkoxylated linear alcohol; nonionic; see Poly-Tergent S-205LF; APHA max. clear liq.; mild odor; sol. in alcohols, glycol ethers, aromatic hydrocarbons, chlorinated solvs.; disp. in water; sp.gr. 1.00; dens. 8.3 lb/gal; visc. 171 cs; flash pt. 224 C (COC); cloud pt. 19 C (1% aq.); pH 5.5–7.0 (1% aq.); surf. tens. 31 dynes/cm (0.1%); biodeg.; 100% act.

Poly-Tergent S-405LF. [455] Alkoxylated linear alcohol; nonionic; see Poly-Tergent S-205LF; APHA 100 max. clear liq.; mild odor; sol. see Poly-Tergent S-205LF; sp.gr. 1.01; dens. 8.4 lb/gal; visc. 189 cs; flash pt. 227 C (COC); cloud pt. 28 C (1% aq.); pH 5.5–7.0 (1% aq.); surf. tens. 32 dynes/cm (0.1%); biodeg.; 100% act.

Poly-Tergent S-505LF. [455] Alkoxylated linear alcohol; nonionic; see Poly-Tergent S-205LF; APHA 100 max. clear liq.; mild odor; sol. see Poly-Tergent S-205LF; sp.gr. 1.02; dens. 8.5 lb/gal; visc. 125 cs; flash pt. 232 C (COC); cloud pt. 47 C (1% aq.); pH 5.5–7.0 (1% aq.); surf. tens. 33 dynes/cm (0.1%); biodeg.; 100% act.

Poly-Tergent SL-42. [455] Alkoxylated linear alcohol; nonionic; surfactant used in industrial applications; APHA 100 max. clear liq.; mild odor; sol. see Poly-Tergent S-205LF; sp.gr. 1.01; dens. 8.4 lb/gal; HLB 13.0; flash pt. 182 C (COC); cloud pt. 42 C (1% aq.); pH 6.0–8.0 (1% aq.); surf. tens. 28 dynes/cm (0.1%); biodeg.; 100% act.

Poly-Tergent SL-62. [455] Alkoxylated linear alcohol; nonionic; see Poly-Tergent SL-42; APHA 100 max. clear visc. liq.; mild odor; sol. see Poly-Tergent S-205LF; sp.gr. 1.02; dens. 8.5 lb/gal; HLB 14.0; flash pt. 204 C (COC); cloud pt. 62 C (1% aq.); pH 6.0–8.0 (1% aq.); surf. tens. 29 dynes/cm (0.1%); biodeg.; 100% act.

Poly-Tergent SL-92. [455] Alkoxylated linear alcohol; nonionic; see Poly-Tergent SL-42; APHA 100 max. slush to solid; mild odor; sol. see Poly-Tergent S-205LF; sp.gr. 1.05; dens. 8.7 lb/gal; HLB 15.0; flash pt. 274 C (COC); cloud pt. 92 C (1% aq.); pH 6.0–8.0 (1% aq.); surf. tens. 31 dynes/cm (0.1%); biodeg.; 100% act.

Poly-Tergent SLF-18. [455] Alkoxylated linear alcohol; nonionic; surfactant used in removing protein-type soils; rinsing aid in machine dishwashing; APHA 100 max. clear liq.; mild odor; sol. in water, chlorinated, aromatic and polar solvs.; sp.gr. 1.06; dens. 8.8 lb/gal; flash pt. 229 C (COC); cloud pt. 17 C (1% aq.); pH 6.0–8.0 (1% aq.); surf. tens. 33 dynes/cm (0.1%); 100% act.

Polytex 10. [350] Stearamide DIBA-stearate; nonionic; emulsifier, thickener and pearling agent for cosmetic preparations; wax, flakes; 100% conc.

Polywet AX-7. [618] Functionalized oligomer, ammonium salt; anionic; emulsifier for emulsion polymerization of vinyl acetate/acrylate polymers and other acidic systems; liq.; 45% conc.

Polywet KX-3. [618] Functionalized oligomer, potassium salt; anionic; emulsifier; mfg. of styrene-butadiene latices; amber liq.; mild thiol odor; water-sol.; sp.gr. 1.19; visc. 380 cps; HLB 16; pour pt. 5 F; surf. tens. 47 dynes/cm (0.1%) pH 7.0–8.0; 46% act.

Polywet KX-4. [618] Functionalized oligomer, potassium salt; mfg. of acrylate latices; amber liq.; mild thiol odor; water-sol.; sp.gr. 1.177; visc. 5000 cps (15 C); HLB 17; pour pt. 15 F; surf. tens. 47 dynes/cm (0.1%); pH 7–8; 40% act.

Polywet KX-5. [618] Functionalized oligomer, potassium salt; yel. to amber liq.; water-sol.; sp.gr. 1.19; visc. 1200 cps; HLB 26; pour pt. 15 F; surf. tens. 38 dynes/cm (0.1%); pH 6.5–8.0; 45 ± 1% solids.

Polywet ND-1. [618] Polyfunctional oligomer, sodium salt; anionic; dispersant for minerals, pigments and fillers in paints, paper, latex compds., and water treatment applications; slight yel. liq.; sp.gr. 1.13; visc. 35 cps; pH 6.8–7.8; 25% act.

Polywet ND-2. [618] Functionalized oligomer, sodium salt; anionic; dispersant for pigments, minerals, extenders, fillers in aq. systems; latex paints and enamels; coatings, adhesives, paper and paperboard; boiler water; colorless to slight yel. liq.; sp.gr. 1.16; visc. 32 cps; pH 6.8–7.8; 25% act.

Prapagen WK brands. [301] Quat. ammonium salts; textile softener with antistat.

Precirol ATO. [247; 248] Glycerol di/tripalmito stearate; nonionic; additive for tablets, binder, lubricant, sustained release; solid; 100% conc.

Precirol WL 2155. [247; 248] Glycerol ditristearate; nonionic; additive for tablets mfg.; solid; HLB 2.0; 100% conc.

Primarol 1006. [289] 2-Hexyl decanol-1; extender and solv. for dyes and fragrance oils; intermediate for production of comps. for applications in lubricants, emulsifiers, metal processing, and textiles; water-wh. clear liq.; sol. in solvs.; flash pt. 154 C (COC); cloud pt. < –20 C; sapon. no. < 5.

Primarol 1107. [289] C_{16}-C_{20} 2-alkyl alcohols; see Primarol 1006; water-wh. clear liq.;

cloud pt. < 0 C.

Primarol 1208. [289] 2-Octyl-dodecanol-1; carrier for oil-sol. products; pigment dispersant and solubilizer for waxes and fatty alcohols and esters; intermediate for production of comps. for applications in lubricants, emulsifiers, metal processing and textiles; water-wh. clear liq.; cloud pt. < –20 C; sapon. no. 2–6.

Primasol FP. [82] Aliphatic sulfo acids, sodium salts; anionic; wetting agent, textile aux.; low visc. yel.-brn. liq.; sol. in water, alkali; 75% conc.

Primasol NF. [86] Phosphated alcohol; anionic; wetting and deaerating agent, pad bath application; liq.; 50% conc.

Procetyl 10. [159] PPG-10 cetyl ether; nonionic; cosolvent, plasticizer, superfatting, wetting and spreading agent, penetrant; lubricant in cosmetics and personal care products; alcoholic and aq. alcoholic compositions; emollient; APHA 150 max. clear liq.; faint, char. sweet odor; pH 6.0–7.5 (3% disp.); 100% act.

Procetyl 20. [159] PPG-20 cetyl ether; see Procetyl 10; APHA 150 max. clear liq.; faint, char. sweet odor; pH 6.0–7.5 (3% disp.); 100% act.

Procetyl 30. [159] PPG-30 cetyl ether; see Procetyl 10; APHA 150 max. clear liq.; faint, char. sweet odor; pH 6.0–7.5 (3% disp.); 100% act.

Procetyl 50. [159] PPG-50 cetyl ether; see Procetyl 10; APHA 150 max. clear liq.; faint, char. sweet odor; pH 6–7.5 (3% disp); 100% act.

Procetyl AWS. [159] PPG-5 ceteth-20; emulsifier, plasticizer, nonionic; emollient and solubilizer in aq. and aq. alcoholic systems; water-wh. turbid oily liq.; char. sweet odor; water-sol.; 100% act.

Prochols. [159] Propoxylated dist. lanolin alcohols; cosolvent; emollient for cosmetics and skin care products; amber liq.; alcohol-sol.

Product BCO [208] Cetyl betaine; amphoteric wetting agent, detergent, emulsifier, dispersant, surfactant; dyeing applications; softener for textiles; leveling and rewetting agent in the paper industry; dyeing assistant and degreaser in the leather industry; antistat on plastic films; brn. clear liq.; mild odor; misc. in water; dens. 8.4 lb/gal; visc. free-flowing; surf. tens. 27.8 dynes/cm (32 C, 0.1%); 33% solids.

Product LT 10-8-1. [144] Oleamide DEA and diethanolamine; amber visc. liq.; sol. in alcohols, aliphatic and aromatic hydrocarbons, ketones, esters, glycols; insol. in water; sp.gr. 0.963; dens. 8.0 lb/gal; visc. (Saybolt) 61.5 s (210 F); flash pt. (COC) 280 F.

Product LT 14-69. [144] Cocamide DEA (2:1); nonionic; foamer for hard-surface cleaners, emulsifier, corrosion inhibitor in aerosols; amber visc. liq.; sol. in alcohols, water, glycols, ether alcohols, esters, ketones; sp.gr. 0.999; 100% conc.

Product LT 18-48. [144] Lauramide DEA; foam stabilizer and thickener for use in personal care products and dishwashing detergents; yel. liq.; sol. in alcohols, glycols, esters, ketones, aromatic hydrocarbons, chlorinated solvs., veg. oil; disp. in water; sp.gr. 0.99; pH 9–10.5 (1% disp.); 90% min. amide content.

Product LT 19-14. [144] Tallow diethanolamide; emulsifier, lubricant, and thickener for aq. sol'ns.; rust inhibition; amber liq. to off-wh. semisolid; sol. in alcohol, glycol ethers, ketones, aromatic and aliphatic hydrocarbons; disp. in water; sp.gr. 0.998; cloud pt. 27 C; pH 9–10.5 (1% disp.).

Product LT 19-44. [144] Lauric/linoleic diethanolamide; foam stabilizer and visc. builder for aq. systems; amber visc. liq.; sol. in ethanol, isopropanol, esters, ketones, glycols, aromatic and aliphatic hydrocarbons; disp. in water; sp.gr. 0.97; m.p. –8 to –11 C; cloud pt. 13 C.

Product WRS 1-66. [144] Oleamide DEA; surfactant, emulsifier and solubilizer of aliphatic hydrocarbons for laundry products; rust inhibition char.s; dk. amber visc. liq.; sol. in alcohols, glycols, glycol esters, ketones, esters, benzene, chlorinated hydrocarbons, aliphatic hydrocarbons; disp. in water; sp.gr. 0.99; pH 9–10 (1% disp.).

Produkt B 2045. [159] Alkenyl succinic anhydride; corrosion inhibitor additive for nonaq. systems; hardener for cosmetics and pharmaceuticals; forms liq. emulsions; colorless solid; odorless; m.p. 55–63 C; sapon. no. 2 max.; 100% conc.

Produkt B 3010. [159] Disodium alkenyl succinate; corrosion inhibitor for aq. systems; visc. liq.; 40% act.

Produkt B 3032. [159] Trialkanolammonium alkenyl succinate; corrosion inhibitor for aq. systems; visc. liq.; 40% act.

Produkt RT 63. [660] Alkyl polyglycolamine; nonionic; detergent for cleaning agents; liq.;

100% conc.

Produkt RT 237. [660] Alkylaryl polyglycol ether; nonionic; spreading aux. for wax; liq.; 100% conc.

Produkt RT 275. [660] Abietic acid polyglycol ester; nonionic; shampoo additive; paste; 100% conc.

Profan 24 Extra, 128 Extra. [539] 1:1 Coconut oil fatty acid diethanolamide; nonionic; foam stabilizer and thickener for shampoo; liq.; 100% conc.

Profan 2012E. [539] 1:2 Coconut oil fatty acid diethanolamide; nonionic; see Profan 24 Extra; liq.; 100% conc.

Profan AA62. [539] 1:1 Lauric acid diethanolamide; nonionic; see Profan 24 Extra; solid; 100% conc.

Profan AB20. [539] Coconut oil fatty acid monoethanolamide; nonionic; see Profan 24 Extra; solid; 100% conc.

Profan AD31. [539] Lauric acid isopropanolamide; nonionic; see Profan 24 Extra; solid; 100% conc.

Profan ME-20. [539] POE coconut fatty acid amide; nonionic; used in shampoos, dishwashing detergents and acid cleansers; liq.; 97% conc.

Progalan X-13. [376] Ammonium alkylaryl ether sulfate; anionic; foaming and scouring agent for textile applications and detergent formulations; liq.; 60% conc.

Progasol 40. [376] Alkylaryl sulfonate; anionic; wetting, leveling, and scouring agent; emulsifier used in textiles; liq.; 40% conc.

Progasol 230. [376] Fatty alkanolamide; nonionic; textile scouring agent for fibers and formulation of cleaners and detergents; liq.; 100% conc.

Progasol 443, 457. [376] Fatty alkanolamide; nonionic; see Progasol 230; liq.; 100% conc.

Promodan SP, SPV. [273] Fatty acids propylene glycol esters; food emulsifier; m.p. 38–45 C; 90% min. monoester.

Promulgen D. [522] Cetearyl alcohol and ceteareth-20; nonionic; gelling agent; o/w emulsifier, emollient, and stabilizer for cosmetics and pharmaceuticals; colorless solid; odorless; m.p. 47–55 C; sapon. no. 2 max.; 100% conc.

Promulgen G. [522] Stearyl alcohol and ceteareth-20; nonionic; see Promulgen D; yel. liq.; sp.gr. 0.848; visc. 35 cps; 100% act.

Propetal 99. [660] Fatty alcohol polyalkylene; nonionic; detergent, wetting agent, intermediate for detergents and cleaners; emulsifier, antifoaming additive; liq.; 100% conc.

Propetal 103. [660] Fatty alcohol polyalkylene; nonionic; see Propetal 99; liq.; 100% conc.

Propetal 241. [660] Fatty alcohol polyalkylene; nonionic; see Propetal 99; liq.; 100% conc.

Propetal 254. [660] Fatty alcohol polyalkylene; nonionic; emulsifier, antifoaming additive; liq.; 100% conc.

Propetal 281. [660] Fatty alcohol polyalkylene; nonionic; detergent and wetting agent; liq.; 100% conc.

Propoxyl 1695. [217] Propoxylated lanolin; nonionic; cosmetic additive; solid; 100% conc.

Prostearyl 15. [159] PPG-15 stearyl ether; emollient for cosmetics.

Protamine-45. [427] Amide-amine fatty derivative; cationic; emulsifier, softener and lubricant base; solid; 100% conc.

Protegin. [258; 259] Min. oil, petrolatum, ozokerite, glyceryl oleate, lanolin alcohol; nonionic; soft solid; 100% conc.

Protegin W, WX. [258; 259] Petrolatum, ozokerite, hydrog. castor oil, glyceryl oleate, polyglyceryl-4 isostearate; nonionic; emollient, emulsifier; paste; HLB 3.5; 100% conc.

Protegin X. [258; 259] Min. oil, petrolatum, ozokerite, glyceryl oleate, lanolin alcohol; nonionic; paste; HLB 3.5; 100% conc.

Protodye QNF. [490] Naphthalene sulfonate, sodium salt; wetting agent and dispersant for fabrics, dyes, pigments, latices; pasting agent for dyestuffs; wood carbonizing; lt. yel. liq.; water-sol.; 16% act.

Protopan. [490] EO condensate; anionic; detergent, dyeing assistant, wetting agent, emulsifier; textile industry applications; straw clear liq.; water-sol.; pH 7.5 (1%); 70% conc.

Protosan EP. [490] Alkylbenzene sulfonate; anionic; wetting and rewetting agent, dyeing assistant; nonfoaming surfactant for textile applications; liq.; 42% conc.

Protowax 90. [427] Glyceryl monostearate; nonionic; emulsifier, softener base; solid; 100% conc.

Protowax ES-100. [427] Fatty acid, ethoxylated; nonionic; emulsifier, softener, dyeing assistant; solid; 100% conc.

Protowet 5 EROH. [427] Alcohol ethoxylated; nonionic; wool scouring agent; liq.; 40% conc.

Protowet 35, 60. [490] Complex alcohol phosphate; anionic; detergent; cotton preparation; scouring of syns. in jig, pressure machines, or dye beck: straw clear liq.; water-sol.; pH 7 (1%); 60% act.

Protowet 100. [427] EO condensate; nonionic; wetting, rewetting and scouring agent used in textile industry; liq.; 100% conc.

Protowet 3072. [490] Alkyl ester, phosphated, free acid; chemical intermediate; antistat in syn. fibers; lt. brn. liq.; insol. in water; pH 6.5 (1%); 100% act.

Protowet 3098. [427] Phosphate alcohol free acid; anionic; base for scouring agent, dyeing assistant, emulsifier; liq.; 100% conc.

Protowet 5218. [427] Surfactant blend; anionic/nonionic; liq.; scouring agent/ 45% conc.

Protowet 5799. [427] Surfactant blend; detergent, scouring agent with antiredeposition properties; liq.; 80% conc.

Protowet C. [427; 490] Fatty acid ester, sulfonated; anionic; emulsifier, textile wet processing, dyeing and other applications; preparation of wettable insecticide powds.; clear amber oil; dissolves in water; sol. in org. solvs.; 60% conc. .

Protowet C-5. [427] Fatty ester sulfates; anionic; bleach stabilizer containing surfactants and lubricants; liq.; 38% conc.

Protowet D-75, D-75 S. [427; 490] Sodium di(2-ethylhexyl) sulfosuccinate; anionic; textile wetting and rewetting and dyeing applications; emulsifier for insol. fatty products; metal and glass cleaning, pigment disp., dry cleaning comps.; lt. visc. liq.; 75% act.

Protowet E-4. [427; 490] EO condensate; nonionic; wetting, emulsifying, scouring and foaming agent used in textiles; liq.; 25% conc.

Protowet FHL. [490] Sulfated glycerides, sodium salt; anionic; agent; emulsifier and lubricant in spin finish formulations; emulsifies min. oil, butyl stearate, coconut oil; dye bath lubricant; dyeing assistant; med. amber liq.; pH 5–6 (1%); biodeg.; 75% act.

Protowet MB. [427] Ethoxylated alcohol sulfates; anionic; wetting, leveling and dyeing assistant; 68% conc.

Protowet NBF. [427] Surfactant blend; anionic; emulsifier, scouring agent; liq.; 50% conc.

Protowet NRW. [490] Phosphated alcohol; anionic; penetrant, emulsifier; dyeing assistant in textile industry applications; lt. yel. clear liq.; pH 7 (1%); 30% act.

Protowet RW. [490] Sulfonated fatty acid ester; anionic; rewetting; emulsifier for org. solvs.; amber clear oil; dissolves in water; sol. in org. solvs.; 62% act.

Protowet TJ. [490] Phosphated alcohol; anionic; emulsifier, dispersant, wetting agent used in various applications of the textile industry; pale yel. liq.; water-sol.; sp.gr. 1.10–1.20; 85% act.

Protowet XL. [427; 490] Sulfonated fatty acid ester; see Protowet C; amber clear oil; dissolves in water; sol. in org. solvs.; 62% act.

Provol 10. [159] PPG-10 oleyl ether; emollient; pigment dispersant and coupling agent used in personal care products; Gardner 10 max. liq.; sol. in min. oil, IPM, Fluilan lanolin oil; insol. in water; pH 5.5–7.5 (3% aq.).

Provol 30. [159] PPG-30 oleyl ether; see Provol 10; Gardner 10 max. liq.; sol. in acetone IPM, Fluilan lanolin oil; insol. in water; pH 5.5–7.5 (3% aq.).

Provol 50. [159] PPG-50 oleyl ether; see Provol 10; Gardner 10 max. liq.; sol. see Provol 30; pH 5.5–7.5 (3% aq.).

Purton CFD. [660] Coconut diethanolamide; nonionic; foam stabilizer, thickener and superfatting agent for cosmetics, cleaners; liq.; 100% conc.

Purton SFD. [660] Unsat. fatty acids diethanolamide; nonionic; foam stabilizer, thickener and superfatting agent for cosmetics, cleaners; liq. to paste; 100% conc.

Puxol CB-22. [494] TEA dodecylbenzene sulfonate; anionic; used for formulation of liq. detergents, personal care products, car washing shampoos; emulsifier for emulsion polymerization reactions; pigment dispersant; liq.; 50 ± 1% conc.

Puxol FB-11. [494] Sodium dodecylbenzene sulfonate; anionic; scouring and wetting agent in the textile industry and emulsifier for emulsion polymerization in the rubber and plastic industry; paste; 65 ± 2% conc.

Puxol XB-10. [494] Dodecylbenzene sulfonic acid; anionic; intermediate for mfg. of detergents; liq.; 95% conc.

Pycal 94. [320] POE aryl ether; plasticizer; pale yel. liq.; ref. index 1.502; sol. in water, ether, alcohol, ketone, lower aliphatic ester, and aromatic hydrocarbon solv.; sp.gr. 1.1; visc. approx 50 cps; flash pt. > 300 F; pH 4–6 (50% aq.).

Pyronate 40. [650] Sodium petroleum sulfonate; anionic; wetting agent, dispersant; oil and froth flotation; dk. liq.; water-sol.; dens.

9.6 lb/gal; sp.gr. 1.18 (60 F); 42% act.

Pyroter CPI-40. [9] POE (40) hydrog. castor oil monopyroglutamic monoisostearic diester; nonionic; emulsifier, solubilizer and thickener used in personal care products; liq.; water-sol.; HLB 9.4; 100% conc.

Pyroter GPI-25. [9] POE (25) glycerin monopyroglutamic monoisostearic diester; nonionic; moisturizer, emulsifier, solubilizer, dispersant and emollient; liq.; water-sol.; HLB 13.4.

Q

QO-33-F. [287] Sorbitan sesquioleate; nonionic; emulsifier for cosmetics in w/o emulsions; liq.; HLB 4.0; 100% conc.

Quad Opacifier #35. [113] Polymer emulsion; opacifier used in household ammonia, laundry and dishwashing detergents; acid bowl cleaner, fabric softener, hair conditioner; dens. 8.5 lb/gal; pH 2–3; 34–36% solids.

Quadrilan AT. [360] Quat. fatty amine ethoxylate; cationic; antistat used in PVC and other polymers; amber clear liq.; char. odor; water-sol.; sp.gr. 1.081; visc. 400 cs; flash pt. > 300 F (COC); pour pt. < O C; pH 6–9 (1% aq.).

Quadrilan BC. [360] Alkyl benzyl dimethyl ammonium chloride sol'n., BP grade; cationic; germicide in disinfectants and sanitizers, emulsifier in the dyeing industry; pale yel. liq.; mild char. odor; sol. in water; sp.gr. 0.988; visc. 120 cs; flash pt. > 200 F (COC); pour pt. < O C; pH 7–8 (1% aq.); 50% act.

Quadrilan OD-75. [360] Quat. ammonium compd.; cationic; softener, antistat; used in textiles; wh. stiff paste; IPA odor; disp. in water; sp.gr. 0.870 (40 C); visc. 55 cs (50 C); flash pt. 60 F (COC); pour pt. 35–40 C; pH 6.0–9.0 (1% aq.); 75% act.

Quadrilan SK. [360] Quat. fatty amine ethoxylate; cationic; antistat used in polymers; liq.; 100% conc.

Quartamin CPR. [561] Coco trimethyl ammonium chloride; cationic; emulsifier; liq.; biodeg.; 50%.

Quartamin DCP. [341; 561] Dicoco dimethyl ammonium chloride; cationic; emulsifier and corrosion inhibitor; liq.; 75% conc.

Quartamin HTPR. [561] Hydrog. tallow trimethyl ammonium chloride; cationic; emulsifier; liq.; biodeg.; 50%.

Quartamin TPR. [561] Tallow trimethyl ammonium chloride; cationic; emulsifier; liq.; biodeg.; 50%.

Quaternary O. [136] Quat. oleyl imidazoline; cationic; detergent, wetting agent, penetrant used in acids, solvs., germicides, fungistats, acid cleaners and corrosion inhibitor formulations, ore flotation, asphalt wetting; amber liq.; 100% conc.

Quatrene 7670. [290] Fatty amidoamine quat. compd.; cationic; wetting agent, demulsifier, and corrosion inhibitor for petroleum production; liq.; 60% act.

Quatrene C-5-6. [290] Coco benzyl imidazolinium chloride; cationic; corrosion inhibitor for continuous treatment in oil and gas production, pipelines; red liq.; water-sol.; 60% conc.

Quatrene CE. [290] Coco dialkyl benzyl ammonium chloride; cationic; see Quatrene 7670; liq., paste; 75% act.

Querton 14 BCL. [368] Myristyl dimethyl benzyl quat. ammonium chloride; cationic; surfactant; solid; 100% conc.

Querton 16 BCL. [368] Palmityl dimethyl benzyl quat. ammonium chloride; cationic; surfactant; solid; 100% conc.

Querton 16 CL 29. [348] Cetyl trimethyl ammonium chloride; cationic; softener; detergent; liq.; 29% conc.

Querton 24 CL 35. [317] Cetyl trimethyl ammonium chloride; cationic.

Querton 28 CL. [368] Dioctyl dimethyl quat.

ammonium chloride; cationic; surfactant; Gardner 3 max. liq.; m.w. 320; 50% conc.

Querton 210 CL. [368] Didecyl dimethyl quat. ammonium chloride; cationic; surfactant; Gardner 3 max. liq.; m.w. 360; 45% conc.

Querton 280. [348] Trimethyl alkyl ammonium chloride; cationic; surfactant; liq.; 35% conc.

Querton 441-BC. [348] Hydrog. tallow dimethylbenzyl ammonium chloride; cationic; agent for organophilic bentonite production; liq.; 76%.

Querton BGCL50. [348] Trimethyl tallow ammonium chloride; cationic; emulsifier, dispersant; liq.; 50% conc.

Querton HBG BCL. [368] Hydrog. tallow dimethyl benzyl quat. ammonium chloride; cationic; surfactant; Gardner 3 max. solid, liq.; pH 7 (5% aq.); 100% conc.

Querton IL 49. [368] Trialkyl benzyl quat. ammonium chloride; catonic; surfactant; Gardner 3 max. liq.; m.w. 420; pH 7 (5% aq.); 50% conc.

Querton KK BCL. [368] Coconut dimethyl benzyl quat. ammonium chloride; cationic; surfactant; Gardner 3 max. liq.; pH 7 (5% aq.); 50% conc.

Quimipol 9106B. [500] Syn. primary alcohol ethoxylate; nonionic; coemulsifier for min. oil and waxes; emulsifier for alkyl resins; liq.; HLB 12.6; 100% conc.

Quimipol 9108. [500] Syn. primary alcohol ethoxylate; nonionic; component for dishwashing liqs. and surfactants; liq.; HLB 13.8; 100% conc.

Quimipol DEA OC. [500] Coconut diethanolamide; nonionic; foamer booster, emulsifier, thickener, foam stabilizer; liq.; 85% conc.

Quimipol EA 2503. [500] Syn. primary alcohol ethoxylate; nonionic; emulsifier for min. oils and aliphatic solvs.; visc. depressant for PVC plastisols; liq.; HLB 7.8; 100% conc.

Quimipol EA 2504. [500] Syn. primary alcohol ethoxylate; nonionic; emulsifier for silicones and hydrocarbon solvs.; drycleaning assistant; liq.; HLB 9.8; 97% conc.

Quimipol EA 2506. [500] Syn. primary alcohol ethoxylate; nonionic; wool scouring and hydrocarbon solv. emulsifier; wetting agent; liq.; HLB 11.4; 100% conc.

Quimipol EA 2507. [500] Syn. primary alcohol ethoxylate; nonionic; wool scouring, antistat, detergent, wetting agent, emulsifier; liq.; HLB 12.2; 100% conc.

Quimipol EA 2508. [500] Syn. primary alcohol ethoxylate; nonionic; emulsifier for hydrocarbon solvs.; paste; HLB 12.6; 100% conc.

Quimipol EA 2509. [500] Syn. primary alcohol ethoxylate; nonionic; wetting agent and detergent in liq. formulations; wax emulsifier; solid; HLB 13.2; 100% conc.

Quimipol EA 2512. [500] Syn. primary alcohol ethoxylate; nonionic; wetting agent; cosmetic emulsifier; solid; HLB 14.2; 100% conc.

Quimipol EA 4505, 4508. [500] Syn. primary alcohol ethoxylate; nonionic; clothes washing powds.; liq. and solid resp.; HLB 10.0 and 12.3 resp.; 100% conc.

Quimipol EA 6801, 6802. [500] Syn. primary alcohol ethoxylate; nonionic; see Quimipol EA 4505; solid; HLB 2.8 and 5.0 resp.; 100% conc.

Quimipol EA 6803, 6804. [500] Syn. primary alcohol ethoxylate; nonionic; emulsifier and textile lubricant; solid; HLB 6.7 and 8.0 resp.; 100% conc.

Quimipol EA 6806, 6807, 6808. [500] Syn. primary alcohol ethoxylate; nonionic; see Quimipol EA 6803; solid; HLB 10.0, 11.0 and 11.5 resp.; 100% conc.

Quimipol EA 6810. [500] Syn. primary alcohol ethoxylate; nonionic; solubilizer, emulsifier, pigment dispersant, and detergent; solid; HLB 12.5; 100% conc.

Quimipol EA 6812, 6814, 6818, 6820, 6823, 6825, 6850. [500] Syn. primary alcohol ethoxylate; nonionic; see Quimipol EA 6810; solid; HLB 13.5, 14.0, 15.0, 15.5, 15.8, 16.2, 16.7 resp.; 100% conc.

Quimipol EA 9105. [500] Syn. primary alcohol ethoxylate; nonionic; wetting agent and emulsifier; liq.; HLB 11.6; 100% conc.

Quimipol EA 9106. [500] Syn. primary alcohol ethoxylate; nonionic; detergent, wetting agent in textile processing; degreaser in metal cleaning; liq.; HLB 12.5; 100% conc.

Quimipol ED 2021, 2022. [500] Polypropylene glycol ethoxylate; nonionic; antistat, deduster, defoamer, demulsifier, detergent, dispersant; liq.; HLB 3.0 and 7.0 resp.; 100% conc.

Quimipol ENF 15, 20, 30. [500] Nonylphenol ethoxylate; nonionic; emulsifier and degreaser; liq.; HLB 4.5, 5.8, 7.6 resp.; 100% conc.

Quimipol ENF 40. [500] Nonylphenol ethoxylate; nonionic; emulsifier for min. oils, alkyl resins and paraffin hydrocarbons; liq.; HLB 9.0; 100% conc.

Quimipol ENF 55. [500] Nonylphenol ethoxylate; nonionic; emulsifier for min. oil and aliphatic solvs.; coupling agent; liq.; HLB 10.5; 100% conc.

Quimipol ENF 65. [500] Nonylphenol ethoxylate; nonionc; acid wool scouring, drycleaning and emulsifier for oils, hydrocarbon solvs. and waxes; liq.; 100% conc.

Quimipol ENF 80. [500] POE (8) nonyl phenyl ether; nonionic; wool scouring, oil emulsification and drycleaning detergents; liq.; HLB 12.2; 100% conc.

Quimipol ENF 90. [500] Nonylphenol ethoxylate; nonionic; wool scouring, oil emulsification, antistat, detergent, and wetting agent; liq.; HLB 12.9; 100% conc.

Quimipol ENF 95, 100, 110. [500] Nonylphenol ethoxylate; nonionic; textile scouring agent, emulsifier, detergent and wetting agent; liq.; HLB 13.0, 13.4, 13.8 resp.; 100% conc.

Quimipol ENF 120. [500] Nonylphenol ethoxylate; nonionic; foam stabilizer and suspending agent for liq. detergents; liq.; HLB 14.0; 100% conc.

Quimipol ENF 140, 170. [500] Nonylphenol ethoxylate; nonionic; solubilizer for alkyd aryl sulfonates and essential oils, emulsifier for pesticides and herbicides; paste; HLB 14.5 and 15.4 resp.; 100% conc.

Quimipol ENF 200, 230, 300. [500] Nonylphenol ethoxylate; nonionic; emulsifier in emulsion polymers, wetting agent and detergent; solid; HLB 16.0, 16.4, 17.1 resp.; 100% conc.

R

R3124 Ester. [506] PEG 600 ditallate; nonionic; multipurpose emulsifier; liq.; 99% conc.

Radia 7040. [454] *n*-Butyl oleate; chemical intermediate; used in antibiotic culture broths; lubricant in min. oils; formulation of oils, rust inhibitor; textile and leather industry; m.w. 327; ref. index 1.4518; sol. in solvs., min. and veg. oil; sp.gr. 0.854 (37.8 C); visc. 5.30 cps (37.8 C); cloud pt. –15 C; flash pt. 195 C (COC); surf. tens. 31.5 dynes/cm.

Radia 7051. [454] *n*-Butyl stearate; chemical intermediate and synthesis; plasticizer; lubricant in oils; rust inhibitor; textile and leather industry; m.w. 325; ref. index 1.4432; sol. see Radia 7040; sp.gr. 0.844 (37.8 C); visc. 6.10 cps (37.8 C); cloud pt. 22.5 C; flash pt. 200 C (COC); surf. tens. 32.00 dynes/cm.

Radia 7060. [454] Methyl oleate; chemical intermediate; lubricant in min. oil; formulation of oils, rust inhibitor; textile and leather industry; m.w. 286; ref. index 1.4513; sol. see Radia 7040; sp.gr. 0.862 (37.8 C); visc. 4.10 cps (37.8 C); cloud pt. –11 C; flash pt. 178 C (COC); surf. tens. 31.50 dynes/cm.

Radia 7070. [454] *n*-Butyl myristate; chemical intermediate; lubricant in oils; rust inhibitor; textile and leather industry; m.w. 282; ref. index 1.4388; sol. see Radia 7040; sp.gr. 0.845 (37.8 C); visc. 4.10 cps (37.8 C); cloud pt. 0 C; flash pt. 172 C (COC); surf. tens. 30.00 dynes/cm.

Radia 7104. [454] Glycerol capromyristate; lubricant, chemical intermediate; formulation of oils, corrosion inhibitor; m.w. 603; sol. in trichlorethylene, hexane, benzene, IPA, min. oil; sp.gr. 0.881 (98.9 C); visc. 6.25 cps (98.9 C); m.p. 36 C; flash pt. 216 C (COC).

Radia 7110. [454] Methyl stearate; see Radia 7070; m.w. 285; sol. see Radia 7040; m.p. 31.0 C; flash pt. 166 C (COC).

Radia 7118. [454] Methyl laurate; see Radia 7070; m.w. 214; ref. index 1.4322; sol. see Radia 7040; sp.gr. 0.855 (37.8 C); visc. 2.30 cps (37.8 C); cloud pt. 1.5 C; flash pt. 132 C (COC).

Radia 7120. [454] Methyl palmitate; see Radia 7070; m.w. 267; sol. see Radia 7040; sp.gr. 0.855 (37.8 C); visc. 4.30 cps (37.8 C); m.p. 27.5 C; flash pt. 149 C (COC).

Radia 7127. [454] Ethylhexyl laurate; see Radia 7070; m.w. 330; ref. index 1.4422; sol. see Radia 7040; sp.gr. 0.846 (37.8 C); visc. 4.50 cps (37.8 C); cloud pt. –25 C; flash pt. 171 C (COC); surf. tens. 30.50 dynes/cm.

Radia 7129. [454] Ethylhexyl palmitate; see Radia 7070; m.w. 357; ref. index 1.4467; sol. see Radia 7040; sp.gr. 0.847 (37.8 C); visc. 7.60 cps (37.8 C); cloud pt. –1 C; flash pt. 206 C (COC); surf. tens. 31.00 dynes/cm.

Radia 7130. [454] Isooctyl stearate; see Radia 7070, also as plasticizer in paper industry; m.w. 377; ref. index 1.4479; sol. see Radia 7040; sp.gr. 0.846 (37.8 C); visc. 9.00 cps (37.8 C); cloud pt. 12.5 C; flash pt. 220 C (COC).

Radia 7161. [454] Glycerol trioleate; see Radia 7104; also used in antibiotic culture broths; m.w. 880; ref. index 1.4685; sol. in hexane, benzene, IPA, trichlorethylene, min. and veg. oils; sp.gr. 0.908 (37.8 C); visc. 49.20 cps (37.8 C); cloud pt. –9 C; flash pt. 274 C (COC).

Radia 7163. [454] Glycerol trioleate; see Radia 7161; m.w. 880; ref. index 1.4688; sol. see Radia 7161; sp.gr. 0.906 (37.8 C); visc. 42.30 (37.8 C); cloud pt. –12 C; flash pt. 283 C (COC).

Radia 7171. [454] Pentaerythritol tetraoleate; see Radia 7104; m.w. 1160; ref. index 1.4733; sol. in trichlorethylene, min. oil, hexane, benzene, IPA, veg. oil; sp.gr. 0.913 (37.8 C); visc. 70.80 cps (37.8 C); cloud pt. –9.5 C; flash pt. 302 C (COC).

Radia 7176. [454] Pentaerythritol tetrastearate; see Radia 7104; m.w. 1140; sol. in trichlorethylene; sp.gr. 0.868 (98.9 C); visc. 14.50 cps (98.9 C); m.p. 64.5 C; flash pt. 290 C.

Radia 7190. [454] Isopropyl myristate; see Radia 7070, also as emollient, plasticizer, solubilizer of act. components in cosmetics and pharmaceuticals; m.w. 271; ref. index 1.4345; sol. see Radia 7040; sp.gr. 0.840 (37.8 C); visc. 3.50 cps (37.8 C); cloud pt. –3.5 C; flash pt. 157 C (COC); surf. tens. 27.50 dynes/cm.

Radia 7194. [454] Diisopropyl adipate; see Radia 7190; m.w. 230; ref. index 1.4237; sol. see Radia 7040; sp.gr. 0.938 (37.8 C); visc. 2.80 cps (37.8 C); cloud pt. –14.5 C; flash pt. 131 C (COC); surf. tens. 30.00 dynes/cm.

Radia 7195. [454] Isopropyl stearate; see Radia 7190; m.w. 315; ref. index 1.4382; sol. see Radia 7040; sp.gr. 0.842 (37.8 C); visc. 5.10 cps (37.8 C); cloud pt. 18.5 C; flash pt. 175 C (COC); surf. tens. 30.50 dynes/cm.

Radia 7197. [454] Diisobutyl adipate; see Radia 7070; m.w. 258; ref. index 1.4300; sol. see Radia 7040; sp.gr. 0.939 (37.8 C); visc. 3.90 cps (37.8 C); cloud pt. –30 C; flash pt. 153 C (COC); surf. tens. 30.70 dynes/cm.

Radia 7200. [454] Isopropyl palmitate; see Radia 7190; m.w. 300; ref. index 1.4373; sol. see Radia 7040; sp.gr. 0.843 (37.8 C); visc. 5.05 cps (37.8 C); cloud pt. 11.5 C; flash pt. 167 C (COC); surf. tens. 30.50 dynes/cm.

Radia 7230. [454] Isobutyl oleate; see Radia 7040; m.w. 327; ref. index 1.4499; sol. see Radia 7040; sp.gr. 0.854 (37.8 C); visc. 5.60 cps (37.8 C); cloud pt. –17 C; flash pt. 191 C (COC); surf. tens. 31.50 dynes/cm.

Radia 7231. [454] Isopropyl oleate; see Radia 7060; m.w. 313; ref. index 1.4479; sol. see Radia 7040; sp.gr. 0.853 (37.8 C); visc. 5.05 cps (37.8 C); cloud pt. –17 C; flash pt. 169 C (COC); surf. tens. 31.00 dynes/cm.

Radia 7240. [454] Isobutyl stearate; see Radia 7070, also as gloss promoter in lacquers; plasticizer in paper industry; m.w. 325; ref. index 1.4421; sol. see Radia 7040; sp.gr. 0.844 (37.8 C); visc. 6.20 cps (37.8 C); cloud pt. 17 C; flash pt. 181 C (COC); surf. tens. 30.20 dynes/cm.

Radia 7303. [454] Glycerol trioleate; see Radia 7161; m.w. 880; ref. index 1.4709; sol. see Radia 7161; sp.gr. 0.910 (37.8 C); visc. 45.20 cps (37.8 C); cloud pt. 5 C; flash pt. 275 C (COC).

Radia 7330. [454] Isooctyl oleate; see Radia 7060; m.w. 373; ref. index 1.4542; sol. see Radia 7040; sp.gr. 0.857 (37.8 C); visc. 7.50 cps (37.8 C); cloud pt. –20 C; flash pt. 208 C (COC); surf. tens. 31.50 dynes/cm.

Radia 7331. [454] Ethylhexyl oleate; see Radia 7060; m.w. 373; ref. index 1.4538; sol. see Radia 7040; sp.gr. 0.857 (37.8 C); visc. 7.50 cps (37.8 C); cloud pt. –23 C; flash pt. 193 C (COC); surf. tens. 32.50 dynes/cm.

Radia 7363. [454] Glycerol trioleate; see Radia 7161; m.w. 880; ref. index 1.4691; sol. see Radia 7161; sp.gr. 0.906 (37.8 C); visc. 41.20 cps (37.8 C); cloud pt. –13 C; flash pt. 289 C (COC).

Radia 7370. [454] Trimethylpropane trioleate; see Radia 7104; m.w. 900; ref. index 1.4712; sol. see Radia 7161; sp.gr. 0.904 (37.8 C); visc. 45.90 cps (37.8 C); cloud pt. –24 C; flash pt. 307 C (COC).

Radia 7500. [454] Cetyl palmitate; see Radia 7070, also used for wax formulation; m.w. 484; sol. in benzene and trichlorethylene; sp.gr. 0.805 (98.9 C); visc. 4.30 (98.9 C); m.p. 49 C; flash pt. 214 C (COC).

Radia 7501. [454] Cetyl stearate; see Radia 7500; m.w. 534; sol. in trichlorethylene; sp.gr. 0.807 (98.9 C); visc. 4.60 cps. (98.9 C); m.p. 56 C; flash pt. 216 C (COC).

Radia 7510. [454] Isononyl stearate; see Radia 7070; m.w. 380; ref. index 1.4493; sol. see Radia 7040; sp.gr. 0.851 (37.8 C); visc. 11.30 cps (37.8 C); cloud pt. 11 C; flash pt. 201 C (COC); surf. tens. 31.50 dynes/cm.

Radia 7514. [454] Pentaerythritol tetrabehenate; see Radia 7104; m.w. 1300; sol. in trichlorethylene, min. oil; sp.gr. 0.860 (98.9 C); visc. 17.20 cps (98.9 C); m.p. 73.5 C; flash pt. 289 C (COC).

Radiamac 6148, 6149. [454] Hydrog. tallow amine acetate; cationic; flotation reagent, anticaking aid for fertilizer, corrosion inhibitor, emulsifier; solid; HLB 11; 100% conc.

Radiamac 6169. [454] Coconut oil amine acetate; cationic; see Radiamac 6148; solid; HLB 11; 100% conc.

Radiamac 6179. [454] Tallow amine acetate; cationic; see Radiamac 6148; solid; HLB 11; 100% conc.

Radiamine 6140, 6141. [454] Hydrog. tallow amine (6141—dist.); cationic; mineral flotation, corrosion inhibitor, pigment dispersant; cosmetics, lubrication and mold release of hard rubber, textile chemical, chemical synthesis; solid; 100% conc.

Radiamine 6160, 6161. [454] Coconut oil amine (6161—dist.); cationic; see Radiamine 6140; liq.; 100% conc.

Radiamine 6163, 6164. [454] Laurylamine (6164—dist.); cationic; see Radiamine 6140; liq., solid; 100% conc.

Radiamine 6170, 6171. [454] Tallow amine (6171—dist.); cationic; see Radiamine 6140; paste; 100% conc.

Radiamine 6172, 6173. [454] Oleyl amine (6173—dist.); cationic; see Radiamine 6140; liq., paste; 100% conc.

Radiamine 6540. [454] Hydrog. tallow diamine; cationic; corrosion inhibitor, dispersant, emulsifier, intermediate for chemical synthesis; solid; 100% conc.

Radiamine 6560. [454] Coconut oil diamine; cationic; see Radiamine 6540; paste; 100% conc.

Radiamine 6570. [454] Tallow diamine; cationic; see Radiamine 6540; paste; 100% conc.

Radiamine 6572. [454] Oleyl diamine; cationic; see Radiamine 6540; liq.; 100% conc.

Radiamine AA 23, 27, 57. [454] *n*-Alkyl propylene diamine; cationic; bitumen emulsifier in road construction; paste; 100% conc.

Radiamine AA 60. [454] *n*-Alkyl propylene diamine; cationic; see Radiamine AA 23; liq.; 100% conc.

Radiamine AC 14, 20, 30, 81, 82, 505, 6144. [454] Fatty amine compd.; cationic; anticaking and dust inhibitor for fertilizers; paste; 100% conc.

Radiamuls 125. [454] Sorbitan monolaurate; nonionic; food emulsifier, dispersant, antistaling agent, spray drying aid, solubilizer, dryness improver, whipping aid; liq.; HLB 6.8.

Radiamuls 135. [454] Sorbitan monopalmitate; see Radiamuls 125; powd., flake; m.p. 50 C; HLB 5.3.

Radiamuls 137. [454] Sorbitan POE (20) monolaurate; see Radiamuls 125; liq.; HLB 16.

Radiamuls 141. [454] Glycerol monostearate SE; nonionic; food emulsifier; used for binding of water; powd., flakes; m.w. 509; sol. in IPA, trichlorethylene, min. oil, hexane and veg. oil; sp.gr. 0.902 (98.9 C); visc. 20.10 cps (98.9 C); m.p. 58 C; HLB 3.5; flash pt. 198 C (COC); sapon. no. 145–160.

Radiamuls 142. [454] Glycerol monostearate; see Radiamuls 141; powd. or flakes; m.w. 515; sol. in benzene, IPA, trichlorethylene, min. and veg. oil; sp.gr. 0.885 (98.9 C); visc. 12.00 cps (98.9 C); m.p. 58 C; HLB 3.0; flash pt. 233 C (COC) sapon. no. 169–176.

Radiamuls 143. [454] Veg. glycol monostearate; see Radiamuls 141; powd., flakes; m.p. 56–60 C; HLB 3.0; sapon. no. 165–180.

Radiamuls 145. [454] Sorbitan monostearate; see Radiamuls 125; powd., flakes; m.p. 50–60 C; HLB 5.

Radiamuls 147. [454] POE (20) sorbitan monostearate; see Radiamuls 125; liq.; HLB 15.

Radiamuls 152. [454] Glycerol monooleate; see Radiamuls 141; liq.; HLB 2.8; cloud pt. 10 C max.; sapon. no. 165–178.

Radiamuls 155. [454] Sorbitan monooleate; see Radiamuls 125; liq.; HLB 4.7.

Radiamuls 157. [454] Sorbitan POE (20) monooleate; see Radiamuls 125; liq.; HLB 15.7.

Radiamuls 341. [454] Glycerol monostearate SE; see Radiamuls 141; powd., flakes; m.p. 56–60 C; HLB 5; sapon. no. 164–174.

Radiamuls 345. [454] Sorbitan tristearate; see Radiamuls 125; powd., flakes; m.p. 50–60 C; HLB 2.7.

Radiamuls 600. [454] Glycerol monostearate; see Radiamuls 141; powd., flakes; m.w. 460; sol. see Radia 7161; sp.gr. 0.893 (98.9 C); visc. 13.40 cps (98.9 C); m.p. 59 C; HLB 3.3; flash pt. 223 C (COC); sapon. no. 165–175.

Radiamuls 601. [454] Glycerol monostearate; see Radiamuls 141; powd., flakes; m.w. 460; sol. see Radia 7161; sp.gr. 0.900 (98.9 C); visc. 10.70 cps (98.9 C); m.p. 52 C; HLB 3.3; flash pt. 217 C (COC); sapon. no. 164–174.

Radiamuls 602. [454] Glycerol monosoyate; see Radiamuls 141; liq.; m.w. 464; ref. index 1.475; sol. in hexane, benzene, IPA, min. and veg. oil; sp.gr. 0.951 (37.8 C); HLB 3.1; cloud pt. 17.5 C; flash pt. 212 C (COC); sapon. no. 165–175.

Radiamuls 900. [454] Glycerol monostearate; see Radiamuls 141; powd., flake; m.p. 70 C; HLB 4.6; sapon. no. 155–170.

Radiamuls AMG 2021, 2134. [454] Monoglycerides, acetylated; nonionic; food emulsifier; liq., paste; m.p. 35 and 5 C resp.

Radiamuls CMG 2930. [454] Citrylated monoglycerides; anionic/nonionic; food emulsifier and antispattering agent; powd., flake; m.p. 56 C.

Radiamuls CMG 2931. [454] Citrylated monoglycerides; anionic/nonionic; food emulsifier in meats; powd., flake; m.p. 56 C.

Radiamuls CSL 2980. [454] Calcium stearoyl lactylate; anionic; food emulsifier; flake, powd.; m.p. 45–55 C.

Radiamuls DATA 2001. [454] Diacetyl tartaric monoglycerides; food emulsifier; flake.

Radiamuls DATA 2003, 2004. [454] Diacetyl tartaric monoglycerides; food emulsifier; flake and paste resp..

Radiamuls DATA 2006, 2007. [454] Diacetyl tartaric monoglycerides; food emulsifier; powd. and liq. resp..

Radiamuls DATA 2008, 2009. [454] Diacetyl tartaric monoglycerides; food emulsifier; powd.

Radiamuls LMG 2950, 2951. [454] Lactylated monoglycerides; nonionic; food emulsifier; m.p. 45 C.

Radiamuls MG 2141 SE. [454] Glyceryl monostearate SE; food emulsifier; powd., flake; m.p. 56–60 C; 40% conc.

Radiamuls MG 2142. [454] Glyceryl monostearate; food emulsifier; powd., flake, liq.; m.p. 56–60 C; HLB 3.0; 40% conc.

Radiamuls MG 2143. [454] Veg. glycerol monostearate; food emulsifier; powd., flake, liq.; m.p. 56–60 C; HLB 3.0; 40% conc.

Radiamuls MG 2151. [454] Unsat. monoglycerides; food emulsifier; liq., paste; m.p. 45 C; 60% conc.

Radiamuls MG 2152. [454] Glycerol monooleate; food emulsifier; liq., paste; m.p. 10 C max.; HLB 2.8; 40% conc.

Radiamuls MG 2341 SE. [454] Saturated monoglycerides SE; food emulsifier; powd., flake; m.p. 56–60 C; 40% conc.

Radiamuls MG 2600. [454] Glyceryl monostearate; nonionic; food emulsifier; powd., flake, liq.; m.p. 58–60 C; HLB 3.3; 60% conc.

Radiamuls MG 2602. [454] Glycerol monosoyate; food emulsifier; liq., paste; m.p. 20 C; HLB 3.1; 60% conc.

Radiamuls MG 2603. [454] Saturated monoglycerides; food emulsifier; powd., flake, liq.; m.p. 60–65 C; 60% conc.

Radiamuls MG 2605 SE. [454] Saturated monoglycerides SE; food emulsifier; powd., flake; m.p. 58–60 C; 60% conc.

Radiamuls 2606. [454] Glycerol monoester; food emulsifier; liq., paste; m.p. 8–13 C; HLB 3.2; 60% conc.

Radiamuls MG 2900. [454] Glyceryl monostearate; food emulsifier; powd., flake, liq.; m.p. 60–70 C; 90% conc.

Radiamuls PEG. [454] Fatty acid PEG esters; nonionic; emulsifier and fat dispersant for milk powd. for animal feed; liq., paste, flakes, powd.; 100% conc.

Radiamuls PG 2201, 2206. [454] Propylene glycol esters; food emulsifier; paste, liq.; m.p. 40 C and –10 C resp.

Radiamuls POLY. [454] Fatty acids polyglycerol esters; nonionic; aerator and stabilizer for food products; visc. liq., flakes, powd.; 100% conc.

Radiamuls SORB 2125. [454] Sorbitan monolaurate; food emulsifier; liq.; HLB 7.6; 100% conc.

Radiamuls SORB 2135. [454] Sorbitan monopalmitate; food emulsifier; powd., flake; HLB 6.3; m.p. 50 C; 100% conc.

Radiamuls SORB 2137. [454] POE (20) sorbitan monolaurate; food emulsifier; liq.; HLB 16.4; 100% conc.

Radiamuls SORB 2145. [454] Sorbitan monostearate; food emulsifier; HLB 5; m.p. 50–60 C; 100% conc.

Radiamuls SORB 2147. [454] POE (20) sorbitan monostearate; food emulsifier; liq.; HLB 15.1; 100% conc..

Radiamuls SORB 2155. [454] Sorbitan monooleate; food emulsifier; liq.; HLB 4.7; 100% conc.

Radiamuls SORB 2157. [454] POE (20) sorbitan oleate; food emulsifier; liq.; HLB 14.9; 100% conc.

Radiamuls SORB 2345. [454] Sorbitan tristearate; food emulsifier; powd., flake; HLB 2.3; m.p. 50–60 C; 100% conc.

Radiamuls SORB 2355. [454] Sorbitan trioleate; nonionic; food emulsifier; liq.; HLB 2.2; 100% conc.

Radiamuls SSL 2990. [454] Sodium stearoyl lactylate; anionic; food emulsifier; flake, powd.; m.p. 45–55 C; 100% conc.

Radianol 7106. [454] Glycerol tricaprate/caprylate; emollient, plasticizer, solubilizer used in the food and feed industry, cosmetics and pharmaceuticals; low visc. oil; m.w. 518; ref. index 1.4402; sol. in hexane, benzene, IPA, min. and veg. oils, trichlorethylene; sp.gr. 0.933 (37.8 C); visc. 15.90 cps. (37.8 C); cloud pt. < –30 C: flash pt. 178 C (COC).

Radianol 7376. [454] Glycerol triheptanoate; see Radianol 7106, also used as butter tracers; low visc. oil; m.w. 432; ref. index 1.4444; sol. see Radianol 7106; sp.gr. 0.952 (37.8 C); visc. 10.10 cps (37.8 C); cloud pt. < –30 C; flash pt. 207 C (COC).

Radiaquat 6442. [454] Dihydrog. tallow dimethyl ammonium chloride; cationic; softener; paste; 74% conc.

Radiaquat 6443. [454] Dialkyl dimethyl ammonium chloride; cationic; softener; liq.; 75% conc.

Radiaquat 6444. [454] Palmityl trimethyl ammonium chloride; cationic; softener; liq.; 50 or 29% conc.

Radiaquat 6462. [454] Dicoconut oil dimethyl ammonium chloride; cationic; softener; paste; 75% conc.

Radiaquat 6470. [454] Dialkyl dimethyl ammonium chloride; cationic; softener; liq., paste; 75% conc.

Radiaquat 6471. [454] Tallow trimethyl ammonium chloride; cationic; softener; liq.; 50% conc.

Radiasurf 7000. [454] POE (20) glycerol monostearate; nonionic; chemical intermediate and synthesis, emulsifier, detergent, wetting agent; o/w or w/o cosmetic and pharmaceutical emulsions; detergency and cleaning products; formulation of oils, rust inhibitor; pigment wetting, paints, lacquers, and printing inks; lubricant and mold release in plastics; liq.; m.w. 1395; ref. index 1.4672; sol. in trichlorethylene, hexane, benzene, IPA, min. and veg. oil; disp. in water; sp.gr. 1.032 (37.8 C); visc. 113.30 cps (37.8 C); HLB 13.0; cloud pt. 27.5 C; flash pt. 244 C (COC).

Radiasurf 7125. [454] Sorbitan monolaurate; nonionic; emulsifier, descouring aid, antistat, anticorrosive agent, superfatting, bodying and antifog aid; emulsion of solvs.; cutting oils; textile lubricant additive; leather aux.; cosmetics and pharmaceuticals; plastics; pesticides; dry cleaning formulations; paper additive; pigment dispersant; detergent additive in concrete; amber liq.; m.w. 450; ref. index 1.4718; sol. see Radiasurf 7000; sp.gr. 1.025 (37.8 C); visc. 1193.0 cps (37.8 C); HLB 6.8; cloud pt. 23 C; flash pt. 198 C (COC); sapon. no. 155–175.

Radiasurf 7135. [454] Sorbitan monopalmitate; see Radiasurf 7125; wh. solid; m.w. 500; sol. in IPA, trichlorethylene, min. oil; disp. in water; sp.gr. 0.943 (98.9 C); visc. 51.70 cps (98.9 C); m.p. 54.5 C; HLB 5.3; flash pt. 214 C (COC); sapon. no. 140–160.

Radiasurf 7136. [454] Sorbitan-POE-monopalmitate; nonionic; emulsifier, wetting agent, defoamer, rust inhibitor, pigment grinding, antistat; liq.; HLB 15.5.

Radiasurf 7137. [454] POE (20) sorbitan monolaurate; see Radiasurf 7125; amber liq.; m.w. 1340; ref. index 1.4712; sol. in water, benzene, IPA, trichlorethylene, min. and veg. oils; sp.gr. 1.095 (37.8 C); visc. 172.3 cps (37.8 C); HLB 16; cloud pt. –10 C; flash pt. 248 C (COC); sapon. no. 40–55.

Radiasurf 7140. [454] Glycerol monostearate; nonionic; see Radiasurf 7136; powd., flakes; HLB 3.0; 40% conc.

Radiasurf 7141. [454] Glycerol monostearate SE; nonionic; see Radiasurf 7136; powd., flakes.

Radiasurf 7145. [454] Sorbitan monostearate; see Radiasurf 7125; wh. solid; m.w. 510; sol. in trichlorethylene; disp. in water; sp.gr. 0.943 (98.9 C); visc. 48.20 cps. (98.9 C); m.p. 58 C; HLB 5; flash pt. 232 C (COC); sapon. no. 146–158.

Radiasurf 7146. [454] Glycerol mono-12-hydroxystearate; chemical synthesis; personal care products and pharmaceuticals; detergency and cleaning products; oils and rust inhibitor; pigment wetting, paints, lacquers, printing inks; paper industry; lubricant and mold release in plastics; textile and leather processing; solid; m.w. 550; sol. in benzene, IPA, trichlorethylene; sp.gr. 0.925 (98.9 C); visc. 25.80 cps. (98.9 C); m.p. 70 C; HLB 1.9; flash pt. 260 C (COC).

Radiasurf 7147. [454] POE (20) sorbitan monostearate; see Radiasurf 7125; amber liq.; m.w. 1400; ref. index 1.4694; sol. in trichlorethylene, min. and veg. oil, water, benzene, IPA; sp.gr. 1.068 (37.8 C); visc. 202.6 cps (37.8 C); HLB 15; cloud pt. 21.5 C; flash pt. 260 C (COC); sapon. no. 44–56.

Radiasurf 7151. [454] Glycerol monooleate SE; nonionic; chemical intermediate and synthesis, emulsifier, detergent, wetting agent; cosmetics and pharmaceuticals; detergency and cleaning products; oil and rust inhibitor; pigment wetting; paints, lacquers, and printing inks; paper industry; lubricant and mold release in plastics; textile and leather processing; liq.; m.w. 515; ref. index 1.4696; sol. in IPA, min. and veg. oil, hexane, benzene, trichlorethylene; sp.gr. 0.947 (37.8 C); visc. 137.70 cps (37.8 C); HLB 3.3; cloud pt. 12.0 C; flash pt. 208 C (COC).

Radiasurf 7152. [454] Glycerol monooleate; see Radiasurf 7151; liq.; m.w. 521; ref. index 1.4697; sol. see Radiasurf 7151; sp.gr. 0.934 (37.8 C); visc. 77.10 cps (37.8 C); HLB 2.8; cloud pt. 7.5 C; flash pt. 232 C (COC).

Radiasurf 7153. [454] Glycerol monoricinoleate; see Radiasurf 7151, also antistat and antiflogging additive in plastics; liq.; m.w. 554; ref. index 1.4774; sol. in benzene, IPA, trichlorethylene, min. and veg. oil; sp.gr. 0.982 (37.8 C); visc. 356.20 cps (37.8 C); HLB 2.5; cloud pt. –16.0 C; flash pt. 229 C (COC).

Radiasurf 7155. [454] Sorbitan monooleate; see Radiasurf 7125; amber liq.; m.w. 510; ref. index 1.4778; sol. in hexane, IPA, trichlorethylene, min. and veg. oil; sp.gr. 0.987 (37.8 C) visc. 395.2 cps (37.8 C); HLB 4.7; cloud pt. 7 C; flash pt. 249 C (COC); sapon. no. 145–161.

Radiasurf 7157. [454] Sorbitan POE (20) monooleate; see Radiasurf 7125; amber liq.; m.w. 1400; ref. index 1.4726; sol. in water, benzene, trichlorethylene, IPA, min. and veg. oil; sp.gr. 1.071 (37.8 C); visc. 240.4 cps (37.8 C); HLB 15.7; cloud pt. 5.5 C; flash pt. 287 C (COC); sapon. no. 44–56.

Radiasurf 7175. [454] Pentaerythritol monostearate; see Radiasurf 7146; solid; m.w. 675; sol. in trichlorethylene; sp.gr. 0.886 (98.9 C); visc. 17.60 cps (98.9 C); m.p. 53.0 C; HLB 2.3; flash pt. 259 C (COC).

Radiasurf 7196. [454] Propylene glycol monomyristate; nonionic; wetting agent, lubricant, opacifier, antistat, dispersant, w/o emulgent, scouring and detergent aid, defoamer, plasticizer, rust inhibitor; cosmetics and pharmaceuticals, oils, textile and leather aid, pigment grinding in paints and printing inks, plastics, waxes and maintenance products, insecticides; wh.-amber liq.; m.w. 378; ref. index 1.4379; sol. see Radiasurf 7000; sp.gr. 0.902 (37.8 C); visc. 17.30 cps (37.8 C); HLB 3.9; cloud pt. 16 C; flash pt. 145 C (COC); sapon. no 190–205.

Radiasurf 7201. [454] Propylene glycol monostearate; see Radiasurf 7196; wh.-cream solid; m.w. 457; sol. in hexane, trichlorethylene, min. oil; sp.gr. 0.849 (98.9 C); visc. 4.50 cps (98.9 C); m.p. 39 C; HLB 1.9; flash pt. 206 C (COC); sapon. no. 177.

Radiasurf 7206. [454] Propylene glycol monooleate; see Radiasurf 7196; amber liq.; m.w. 460; ref. index 1.4606; sol. in benzene, IPA, trichlorethylene, veg. and min. oil, hexane; sp.gr. 0.900 (37.8 C); visc. 18.80 cps (37.8 C); HLB 2.3; cloud pt. –9 C; flash pt. 157 C (COC); sapon. no. 172–180.

Radiasurf 7269. [454] Ethylene glycol distearate; see Radiasurf 7196; wh. solid; sol. in trichlorethylene, hexane, benzene, IPA, min. and veg. oil; m.p. 65 C; HLB 1.5; sapon. no. 190–200.

Radiasurf 7270. [454] Ethylene glycol monostearate; see Radiasurf 7196; wh. solid; m.w. 445; sol. in benzene, IPA, trichlorethylene, veg. oil; sp.gr. 0.851 (98.9 C); visc. 4.80 cps (98.9 C); m.p. 61 C; HLB 1.8; flash pt. 214 C (COC); sapon. no. 180–190.

Radiasurf 7345. [454] Sorbitan tristearate; see Radiasurf 7125; solid; m.w. 900; sol. in trichlorethylene, min. and veg. oils; sp.gr. 0.893 (98.9 C); visc. 19.50 cps (98.9 C); m.p. 60 C; HLB 2.7; flash pt. 253 C (COC).

Radiasurf 7372. [454] Trimethylolpropane monooleate; nonionic; see Radiasurf 7151; liq.; m.w. 514; ref. index 1.4705; sol. in hexane, benzene, IPA, trichlorethylene, min. and veg. oil; sp.gr. 0.935 (37.8 C); visc. 90 cps (37.8 C); HLB 4.4; cloud pt. –18.5 C; flash pt. 224 C (COC).

Radiasurf 7400. [454] Diethylene glycol monooleate; see Radiasurf 7196; amber liq.; m.w. 478; ref. index 1.4642; sol. see Radiasurf 7000; sp.gr. 0.929 (37.8 C); visc. 58.50 cps (37.8 C); HLB 3.8; cloud pt. –6.5 C; flash pt. 180 C (COC); sapon. no. 155–168.

Radiasurf 7402. [454] PEG 200 monooleate; nonionic; wetting agent, lubricant, opacifier, antistat, dispersant, o/w emulgent, scouring and detergent aid, defoamer, plasticizer, rust inhibitor, visc. modifier, antifog aid; cosmetics and pharmaceuticals; lubricating and cutting oils; textile and leather aid; pigment grinding aid in paints and printing inks; plastics; waxes and maintenance products; glass fiber; insecticides; silicones; amber liq.; m.w. 512; ref. index 1.4645; sol. in benzene, isopropyl, trichlorethylene, hexane, min. and veg. oil; disp. in water; sp.gr. 0.962 (37.8 C); visc. 34.00 cps (37.8 C);

HLB 8.4; cloud pt. –10 C; flash pt. 218 C (COC); sapon. no. 105–125; surf. tens. 32.00 dynes/cm (0.1% aq.)

Radiasurf 7403. [454] PEG 400 monooleate; see Radiasurf 7402; amber liq.; m.w. 728; ref. index 1.4672; sol. see Radiasurf 7000; sp.gr. 1.007 (37.8 C); visc. 49.50 cps (37.8 C); HLB 11.7; cloud pt. –1 C: flash pt. 261 C (COC); sapon. no. 75–90; surf. tens. 34.00 dynes/cm (0.1% aq.).

Radiasurf 7404. [454] POE (12) monooleate; see Radiasurf 7402; amber liq.; m.w. 920; ref. index 1.4669; sol. in water, benzene, IPA, trichlorethylene, hexane, min. and veg. oil; sp.gr. 1.030 (37.8 C); visc. 66.30 cps (37.8 C); HLB 13.2; cloud pt. 19 C; flash pt. 254 C (COC); sapon. no. 60–75; surf. tens. 37.00 dynes/cm (0.1% aq.).

Radiasurf 7410. [454] Diethylene glycol monostearate SE; see Radiasurf 7196; wh. solid; m.w. 475; sol. in hexane, benzene, IPA, trichlorethylene, min. and veg. oil; sp.gr. 0.873 (98.9 C); visc. 5.95 cps (98.9 C); m.p. 51.5 C: HLB 3.5; flash pt. 191 C (COC); sapon. no. 160–175.

Radiasurf 7411. [454] Diethylene glycol monostearate SE; see Radiasurf 7196; wh. solid; m.w. 470; sol. in hexane, benzene, IPA, min. and veg. oil, trichlorethylene; sp.gr. 0.878 (98.9 C); visc. 5.90 cps (98.9 C); m.p. 52 C: HLB 4.1; flash pt. 176 C (COC); sapon. no. 160–170.

Radiasurf 7412. [454] POE (4) monostearate; see Radiasurf 7402; wh. paste; m.w. 522; sol. in benzene, IPA, hexane, min. oil, trichlorethylene; sp.gr. 0.913 (98.9 C); visc. 6.70 cps (98.9 C); m.p. 36 C; HLB 7.5; flash pt. 186 C (COC); sapon. no. 120–135.

Radiasurf 7413. [454] PEG 400 monostearate; see Radiasurf 7402; wh. paste; m.w. 722; sol. in benzene, trichlorethylene, min. oil, hexane; sp.gr. 0.951 (98.9 C); visc. 9.45 cps (98.9 C); m.p. 34 C; HLB 11.9; flash pt. 248 C (COC); sapon. no. 75–90.

Radiasurf 7414. [454] POE (12) monostearate; see Radiasurf 7402; wh. paste; m.w. 906; sol. in benzene IPA, hexane, min. and veg. oil, trichlorethylene; sp.gr. 0.981 (98.9 C); visc. 12.60 cps (98.9 C); m.p. 38 C; HLB 13.3; flash pt. 241 C (COC); sapon. no. 60–75.

Radiasurf 7417. [454] PEG 1500 monostearate; see Radiasurf 7402; wh. solid; m.w. 1812; sol. in benzene, trichlorethylene; sp.gr. 1.023 (98.9 C); visc. 28.40 cps (98.9 C); m.p. 47 C; HLB 16.5; flash pt. 248 C (COC); sapon. no. 30–40; surf. tens. 46.50 dynes/cn (0.1% aq.).

Radiasurf 7420. [454] Diethylene glycol monolaurate; see Radiasurf 7196; wh. paste; m.w. 352; sol. in IPA, trichlorethylene, hexane, min. and veg. oil; sp.gr. 0.942 (37.8 C); visc. 15.20 cps (37.8 C); HLB 5.7; cloud pt. 28.5 C; flash pt. 164 C (COC); sapon. no. 195–205.

Radiasurf 7421. [454] Diethylene glycol monolaurate SE; see Radiasurf 7196; wh. liq.; m.w. 346; sol. in trichlorethylene, hexane, benzene, IPA, min. and veg. oil; sp.gr. 0.901 (98.9 C); visc. 3.60 cps (98.9 C); HLB 6.3; cloud pt. 29.5 C; flash pt. 173 C (COC); sapon. no. 183–193.

Radiasurf 7422. [454] POE (4) monolaurate; see Radiasurf 7402; wh. liq.; m.w. 379; ref. index 1.4542; sol. see Radiasurf 7000; sp.gr. 0.983 (37.8 C); visc. 23.40 cps (37.8 C); HLB 9.6; cloud pt. –3 C: flash pt. 188 C (COC); sapon. no. 140–150; surf. tens. 30.00 dynes/cm (0.1% aq.).

Radiasurf 7423. [454] POE (8) monolaurate; see Radiasurf 7402; wh. liq.; m.w. 610; ref. index 1.4596; sol. in benzene, IPA, trichlorethylene, water, hexane, min. and veg. oil; sp.gr. 1.023 (37.8 C): visc. 41.70 cps (37.8 C); HLB 13.2; cloud pt. 4 C; flash pt. 238 C (COC); sapon. no. 90–100; surf. tens. 31.00 dynes/cm (0.1% aq.).

Radiasurf 7431. [454] PEG 300 monooleate; see Radiasurf 7402; amber liq.; sol. see Radiasurf 7000; HLB 9.5; cloud pt. –8 C; sapon. no. 95–115.

Radiasurf 7432. [454] PEG 300 monostearate; see Radiasurf 7402; wh. paste; sol. in benzene, IPA, trichlorethylene, min. oil, hexane; m.p. 28–35 C; HLB 9.7; sapon. no. 95–115.

Radiasurf 7443. [454] POE (8) dioleate; see Radiasurf 7402; amber liq.; m.w. 911; ref. index 1.4655; sol. see Radiasurf 7000; sp.gr. 0.962 (37.8 C); visc. 47.00 cps (37.8 C); HLB 7.4; cloud pt. –3 C: flash pt. 248 C (COC); sapon. no. 120–130.

Radiasurf 7453. [454] POE (8) distearate; see Radiasurf 7402; wh. paste; m.w. 901; sol. in benzene, IPA, trichlorethylene; sp.gr. 0.920 (98.9 C); visc. 9.85 cps (98.9 C); m.p. 39 C; HLB 7.7; flash pt. 242 C (COC); sapon. no. 120–130.

Radiasurf 7454. [454] POE (12) distearate; see Radiasurf 7402; wh. solid; m.w. 1101; sol in benzene, trichlorethylene, hexane, IPA, min. oil; sp.gr. 0.940 (98.9 C); visc. 12.20 cps (98.9 C); m.p. 40 C; HLB 9.7; flash pt. 249 C; sapon. no. 100–110.

Radiasurf 7600. [454] Glycerol monostearate; nonionic; see Radiasurf 7125; powd., flakes; HLB 3.3; 60% conc.

Radiasyn 7174. [454] Pentaerythritol tetra-C_5/C_9; syn. ester lubricant; liq.; m.w. 580; ref. index 1.4543; sol. in hexane, benzene, IPA, trichlorethylene, min. and veg. oil; sp.gr. 0.971 (37.8 C); visc. 29.50 cps (37.8 C); cloud pt. –35 C; flash pt. 275 C (COC).

Radiasyn 7177. [454] Pentaerythritol tetra-C_7; syn. ester lubricant; liq.; m.w. 580; ref. index 1.4526; sol. see Radiasyn 7174; sp.gr. 0.971 (37.8 C); visc. 24.20 cps (37.8 C); cloud pt. –33 C; flash pt. 251 C (COC).

Radiasyn 7178. [454] Pentaerythritol tetra-C_8/C_{10}; syn. ester lubricant; liq.; m.w. 690; ref. index 1.4546; sol. see Radiasyn 7174; sp.gr. 0.945 (37.8 C); visc. 31.50 cps (37.8 C); cloud pt. –2 C; flash pt. 282 C (COC).

Radiasyn 7364. [454] Trimethylolpropane tri-C_5/C_9; syn. ester lubricant; liq.; m.w. 465; ref. index 1.4511; sol. see Radiasyn 7174; sp.gr. 0.951 (37.8 C); visc. 15.40 cps (37.8 C); cloud pt. < -35 C; flash pt. 255 C (COC).

Radiasyn 7367. [454] Trimethylolpropane tri-C_7; syn. ester lubricant; liq.; m.w. 465; ref. index 1.4514; sol. see Radiasyn 7174; sp.gr. 0.950 (37.8 C); visc. 14.70 cps (37.8 C); cloud pt. < -35 C; flash pt. 223 C (COC).

Radiasyn 7368. [454] Trimethylolpropane tri-C_8/C_{10}; syn. ester lubricant; liq.; m.w. 550; ref. index 1.4527; sol. see Radiasyn 7174; sp.gr. 0.932 (37.8 C); visc. 22.30 cps (37.8 C); cloud pt. < -35 C; flash pt. 248 C (COC).

Raylig. [332] Sodium lignosulfonate; anionic; dispersant, deflocculation and visc. control in water dispersions of solids; brn. liq.; vanilla odor; water-sol.; sp.gr. 1.23; visc. 500 cps. b.p. 100 C; pH 5.5; 50% act.

Reax 15B. [642] Sodium lignosulfonate; anionic; treatment of industrial water supplies; dyestuff dispersant for org. pesticides; powd.; pH 10.5; 100% conc.

Reax 45A. [642] Sodium lignosulfonate, modified; wetting agent and dispersant for pesticides; brn. powd.; water sol.; pH 10 (2% aq.).

Reax 77. [642] Amine acetate and sulfonated lignin; anionic; grinding aid; flow promoter; production of portland cements; liq.; sp.gr. 1.200; visc. 40 cps; 70% act.

Reax 77XF. [642] Amine acetate and polyhydroxy alcohol; anionic; liq.; sp.gr. 1.220; visc. 75 cps.

Reax 80A. [642] Sodium lignosulfonate; dispersant for dyestuffs to lead acid storage batteries; brn. powd.; dens. 29 lb/ft^3; pH 11.2 (2% aq.); surf. tens. 54.0 dynes/cm (1% aq.).

Reax 80C. [642] Sodium lignosulfonate, modified; see Reax 80A; brn. powd.; dens. 29 lb/ft^3; pH 9.8 (2% aq.); surf. tens. 54.0 dynes/cm (1% aq.); 100% conc.

Reax 81A. [642] Sodium lignosulfonate, modified; dispersant in dye applications and liq. fertilizer sol'ns.; brn. powd.; dens. 30 lb/ft^3; pH 10.5 (2% aq.); surf. tens. 57.0 dynes/cm (1% aq.); 100% conc.

Reax 82. [642] Sodium lignosulfonate, modified; dispersant; brn. powd.; dens. 26 lb/ft^3; pH 11.0 (2% aq.); surf. tens. 45 dynes/cm (1% aq.); 99% conc.

Reax 83A. [642] Sodium lignosulfonate, modified; dispersant in dye applications, liq. fertilizer and agricultural chemical formulations; brn. powd.; dens. 31 lb/ft^3; pH 10.2 (2% aq.); surf. tens. 61.5 dynes/cm (1% aq.); 100% conc.

Reax 85A. [642] Sodium lignosulfonate, modified; see Reax 81A; brn. powd.; dens. 28 lb/ft^3; pH 10.9 (2% aq.); surf. tens. 59.3 dynes/cm (1% aq.); 100% conc.

Reax 88B. [642] Sodium lignosulfonate, modified; anionic; used in aq. systems; dispersant, visc. reducer used in pesticides and plant foods, fertilizers; visc. reducer in liq. dye formulations; powd.; visc. 95 cps; pH 11.5 (aq.); 100% act.

Reax 90P. [642] Sodium lignosulfonate; dispersant in dye systems; brn. powd.; dens. 24 lb/ft^3; pH 10.3 (2% aq.); surf. tens. 42 dynes/cm (1% aq.).

Reax 92. [642] Sulfonated kraft lignin; dye dispersant; brn. powd.; dens. 33 lb/ft^3; pH 9.8 (2% aq.); surf. tens. 50.2 dynes/cm (1% aq.).

Reax 95A. [642] Sodium lignosulfonate; brn. powd.; dens. 30 lb/ft^3; pH 9.5 (2% aq.); surf. tens. 53.0 dynes/cm (1% aq.).

Reax 100M. [642] Kraft lignosulfonate, modified; dispersant for inorg. materials in aq. media; complexing agent for micronutrient formulations; brn. powd.; dens. 24 lb/ft^3; visc. 65 cps; pH 8.9 (2% aq.); surf. tens. 48.9 dynes/cm (1% aq.).

Reax 100M (Liquid). [642] Kraft lignosulfonate, modified; see Reax 100M; brn. liq.; visc. 20 cps; pH 8.9 (2% aq.); surf. tens. 48.9 dynes/cm (1% aq.).

Reax SR-1. [642] Sodium lignosulfonate, modified; anionic; dyestuff dispersant; brn. powd.; dens. 24 lb/ft^3; surf. tens. 55.8 dynes (1% aq.); pH 9.9 (2% aq.); 100% conc.

Reax SR-7. [642] Sodium lignosulfonate, modified; anionic; see Reax SR-1; brn. powd.; surf. tens. 58.8 dynes (1% aq.); pH 6.7 (2% aq.); 100% conc.

Regitant S-78. [607] Phenol deriv.; anionic; resisting agent for textile products; liq.; 40% conc.

Remanol 1300. [347] Collagen protein hydrolysate; component for shampoos; brn. visc. liq.; m.w. 1300; 50% act.

Remanol 1300/35. [347] Collagen protein hydrolysate; used in personal care products and building industry; liq.; m.w. 1300; 35% conc.

Renex 20. [319; 320] PEG-16 tallate; nonionic; detergent for home laundry, textile scouring, spotting agent, rug shampoos; clear amber liq.; sol. in water, lower alcohols, toluene, acetone, diethyl ether, dioxane, Cellosolve, ethyl acetate, aniline; sp.gr. 1.1; visc. 350 cps; HLB 13.8; cloud pt. 126 F (1% aq.); flash pt. > 300 F; pour pt. 55 F; surf. tens. 41 dynes/cm (0.01%); 100% act.

Renex 22. [320] Mixed fatty and resin acids POE ester (12 EO); see Renex 20; also for mechanical dishwashing, metal cleaning; amber liq.; sol. in acetone, ethyl acetate, dioxane, Cellosolve, aniline, ethyl alcohol; disp. in water; sp.gr. 1.1; visc. 300 cps; HLB 12.2; cloud pt. 32 F (1% aq.); flash pt. > 300 F; pour pt. 35 F; surf. tens. 39 dynes/cm (0.01%).

Renex 25. [320] Renex 20 and urea; nonionic; detergent; formulates cleaning products; used in home laundry, rug shampoos; tan powd.; sol. in water, lower alcohols, Cellosolve; HLB 13.5; cloud pt. 124 F (1%); flash pt. > 300 F; surf. tens. 39 dynes/cm (0.01%); pH 7 (1% aq.); 50% act.

Renex 26. [320] Fatty acid ester, ethoxylated; nonionic; detergent; amber liq.; sol. in water, petroleum solv. and cleaners' naphtha; visc. 600 cps; flash pt. > 300 F.

Renex 30. [319; 320] POE (12) tridecyl ether; nonionic; detergent and wetting agent; used for raw wool scouring and carbonizing, metal cleaning, bottle washing, home laundry; clear-hazy colorless liq.; sol. in water, lower alcohols, xylene, butyl Cellosolve, carbon tetrachloride, ethylene glycol, propylene glycol; sp.gr. 1.0; visc. 60 cps; HLB 14.5; cloud pt. 183 F (1% aq.); flash pt. > 300 F; pour pt. 55 F; surf. tens. 28 dynes/cm (0.01%); pH 6 (1%); 100% conc.

Renex 31. [320] POE (15) tridecyl ether; nonionic; detergent used for liq. hand dishwashing formulations and metal cleaning; wh. soft paste; sol. in water, propylene glycol, lower alcohols, butyl Cellosolve, carbon tetrachloride, xylene; sp.gr. 1.0; visc. 130 cps; HLB 15.4; cloud pt. 210 F (1% aq.); flash pt. > 300 F; pour pt. 61 F; surf. tens. 34 dynes/cm (0.01%); 100% conc.

Renex 35. [320] POE (12) tridecyl ether and urea complex; see Renex 30; formulating cleaning prods.; ivory wh. powd.; sol. in water, ethylene glycol, lower alcohols; m.p. 257 F; HLB 14.5; cloud pt. 180 F (1% aq.); flash pt. > 300 F; surf. tens. 33 dynes/cm (0.01%); pH 7 (1% aq.); 50% conc.

Renex 36. [319; 320] POE (6) tridecyl ether; nonionic; detergent, wetting agent used in alkaline cleaners; dispersant for solids and paint pigments; dishwashing; cloudy-clear colorless thin oily liq.; sol. in acetone, carbon tetrachloride, ethyl acetate, Cellosolve, ethanol, toluene, aniline; disp. in water; sp.gr. 1.0; visc. 80 cps.; HLB 11.4; cloud pt. < 32 F (1% aq.); flash pt. > 300 F; surf. tens. 27 dynes/cm (0.01%); pH 6 (1%); 100% conc.

Renex 647. [71; 319] POE (4) nonylphenol ether; nonionic; emulsifier, detergent; liq.; 100% conc.

Renex 648. [319; 320] POE (5) nonylphenol ether; nonionic; detergent, solv. and emulsion cleaning applications; metal cleaning; clear liq.; sol. in propylene glycol, perchlorethylene, IPA, cottonseed oil; sp.gr. 1.03; visc. 223 cps; HLB 10; flash pt. > 200 F; pour pt. –20 F; 100% act.

Renex 649. [71; 319] POE (20) nonylphenol ether; nonionic; emulsifier, detergent; solid; 100% conc.

Renex 650. [319; 320] POE (30) nonylphenol ether; nonionic; detergent, emulsifier, stabilizer; surfactant for detergent use; emulsification, and latex stabilizer; wh. solid; sol. in water, propylene glycol, IPA glycol, IPA, xylene; sp.gr. 1.15; HLB 17.1; cloud pt. 212 F; flash pt. > 300 F; pour pt. 91 F; surf. tens. 42 dynes/cm (0.01%); 100% act.

Renex 678. [319; 320] POE (15) nonylphenol ether; anionic; detergent, wetting agent used in dishwashing; clear liq.; sol. in water, propylene glycol, perchloroethylene, IPA, xylene; sp.gr 1.08; visc. 319 cps; HLB 15; cloud pt. 211 F (1%); flash pt. > 300 F; pour pt. 69 F; surf. tens. 33 dynes/cm (0.01%); 100% act.

Renex 679. [71; 319] POE (13) nonylphenol ether; nonionic; emulsifier, detergent; liq.; HLB 14.4; 100% conc.

Renex 682. [319] POE (12) nonylphenol ether; nonionic; emulsifier and detergent; liq.; HLB 13.9; 100% conc.

Renex 688. [319; 320] POE (8) nonylphenol ether; nonionic; detergent, wetting agent; surfactant for dishwashing, textile scouring; liq.; sol. in propylene glycol, perchloroethylene, IPA, xylene, water and cottonseed oil; sp.gr. 1.05; visc. 240 cps; HLB 12.3; cloud pt. 87 F (1%); flash pt. > 300 F; pour pt. 12 F; surf. tens. 30 dynes/cm (0.01%); 100% act.

Renex 690. [319; 320] POE (10) nonylphenol ether; nonionic; wetting agent, detergent for home and commercial laundry products, textile scouring, dyeing and desizing, acid metal cleaner, sanitizer, and hard surface cleaning; liq.; sol. see Renex 688; sp.gr. 1.06; visc. 260 cps; HLB 13.3; cloud pt. 150 F (1%); flash pt. > 300 F; pour pt. 33 F; surf. tens. 31 dynes/cm (0.01%); 100% act.

Renex 697. [319; 320] POE (6) nonylphenol ether; see Renex 648; clear liq.; sol. in propylene glycol, perchloroethylene, IPA, xylene, cottonseed oil; sp.gr. 1.04; visc. 250 cps; HLB 10.9; cloud pt. < 32 F (1%); flash pt. > 300 F; pour pt. –21 F; surf. tens. 30 dynes/cm (0.01%); 100% act.

Renex 698. [319; 320] POE (9–9.5) nonylphenol ether; see Renex 690; clear liq.; sol. see Renex 688; sp.gr. 1.06; visc. 245 cps; HLB 13; cloud pt. 129 F (1%); flash pt. > 300 F; pour pt. 28 F; surf. tens. 30 dynes/cm (0.01%); 100% act.

Renex 702. [71; 319] POE (2) syn. primary C_{13-15} alcohol; nonionic; general purpose detergent, emulsifier; liq.; HLB 5.9; biodeg.; 100% conc.

Renex 703. [71; 319] POE (3) syn. primary C_{13-15} alcohol; nonionic; general purpose detergent, emulsifier; liq.; HLB 7.8; biodeg.; 100% conc.

Renex 704. [71] POE alkyl ether; nonionic; general purpose detergent, emulsifier; liq.; biodeg.; 100% conc.

Renex 707. [71; 319] POE (7) syn. primary C_{13-15} alcohol; nonionic; see Renex 702; liq.; HLB 12.2; biodeg.; 100% conc.

Renex 711. [71; 319] POE (11) syn. primary C_{13-15} alcohol; nonionic; see Renex 702; liq.; HLB 13.9; biodeg.; 100% conc.

Renex 714. [71] POE alkyl ether; nonionic; see Renex 702; liq.; biodeg.; 100% conc.

Renex 720. [71; 319] POE (20) syn. primary C_{13-15} alcohol; nonionic; see Renex 702; liq.; HLB 16.2; biodeg.; 100% conc.

Renex 750. [319] POE (10) octylphenol; emulsifier, deteregent; liq.; HLB 13.6; biodeg.; 100% conc.

Renex 751. [71] POE octyl phenol ether; nonionic; lubricant, emulsifier, textile processing; liq.; 100% conc.

Reserve Salt Flake. [137] Sodium *m*-nitrobenzene sulfonate; anionic; stabilizer for dyeing of fibers; assistant in discharge printing; flake.

Resogen DM. [163] Aminoplast precondensate; cationic; fixing agent for dyes on cellulosics; liq.

Retarder A. [163] Quat. ammonium deriv.; cationic; retarder for level dyeing on acrylics.

Retarder CA. [283] N-alkyl dimethylbenzyl ammonium chloride; cationic; retarder for acrylic fiber dyeing with cationic dyestuff; clear liq.; sp.gr. 0.98; 40% act.

Retarder N, N-85. [284] Lauryl dimethyl benzyl ammonium chloride; retarder in dyeing, antistatic agent; liq.; 40 and 85% conc. resp.

Rewocid DC 212. [509] Fatty acid polydiethanolamide; nonionic; lubricating additive, corrosion inhibitor; med. visc. liq.; 100% conc.

Rewocid DU 185. [212] Undecylenamide DEA and diethanolamine; nonionic; detergent, emulsifier used as bacteriocide, thickener, foam stabilizer in shampoos; yel. visc. liq.; sol. in water, oil, alcohols, glycols; sp.gr. 1.0; high visc.; 100% act.

Rewocid SBU 185. [212] Disodium undecylenamido MEA-sulfosuccinate; anionic; fungicide, detergent, foamer used in personal care products; pale yel. powd.; slight odor; sol. in alcohol, glycol; sp.gr. 1.0; pH 6.5–7.5 (5% solids); 50% act.

Rewocid SBU 185/P. [509] Undecylenic acid alkanolamide sulfosuccinate; anionic; antidandruff shampoo surfactant; powd., 95% conc.

Rewocid U 185. [212] Undecylenamide MEA; nonionic; fungicide, detergent; compatible with all surface act. agents; yel. flakes; strong odor; sol. in alcohol, glycol, disp. in water; m.p. 50 C; 100% act.

Rewocid UTM 185. [509] Undecylenic acid propylamido trimethyl ammonium methosulfate; cationic; bactericide, fungicide for toiletries; conditioner for shampoos, antistat; low visc. liq.; 45% conc.

Rewocor AC 28. [509] Fatty acid alkylolamide; nonionic; corrosion inhibitor for metal working fluids; med. visc. liq.; 100% conc.

Rewocor B 2045. [509] Alkenyl sulfosuccinic acid anhydride; nonionic; corrosion inhibitor; visc. liq.; 100% conc.

Rewocor B 3010. [509] Alkenyl succinate, sodium salt; anionic; anticorrosion agent; liq.; 40% conc.

Rewocor B 3032. [509] Alkenyl succinate, triethanolamine salt; anionic; see Rewocor B 3010; liq.; 40% conc.

Rewocor BAC. [509] Butyl ammonium caprylate; anionic; see Rewocor B 3010; liq.; water-sol.; 95% conc.

Rewocor RA 178. [509] Aliphatic/aromatic carboxylic acid, alkanolammonium salt; anionic; anticorrosion agent for aq. systems; med. visc. liq.; 80% conc.

Rewocor RA 280. [509] Oleic acid dibutylamide; nonionic; corrosion inhibitor for oils; visc. liq.; 100% conc.

Rewocor RA-B 90. [509] Boric acid ester, modified; anionic; see Rewocor RA 178; visc. liq.; 85% conc.

Rewocor RA-SI. [509] Alkenyl succinimide; nonionic; corrosion inhibitor for lubricants and hydraulic fluids; visc. liq.; 100% conc.

Rewocor TPAC 100. [509] Tallow propylene diammonium caprylate; anionic; anticorrosion agent for oils and boiler feed water systems; visc. liq.; 100% conc.

Rewoderm ES 90. [509] Fatty acid monoglyceride polyglycol ether; nonionic; emulsifier for cosmetics, superfatting agent; visc. liq.; 100% conc.

Rewoderm LI 48. [509] Tallow fatty acid monoglyceride polyglycol ether; nonionic; surfactant for personal care products; waxy solid; 100% conc.

Rewoderm LI 63, 67, 67-75. [509] Coconut fatty acid monoglyceride polyglycol ether; nonionic; see Rewoderm LI 48; soft paste, waxy solid, and med.-visc. liq. resp.; 100, 100, and 75% conc.

Rewoderm LI 420. [509] Tallow fatty acid monoglyceride polyglycol ether; nonionic; see Rewoderm LI 48; waxy solid; 100% conc.

Rewoderm S 1333. [212; 509] Veg. fatty acid disodium sulfosuccinate; anionic; detergent; used for skin protection, washing up liqs., personal care products; amber liq.; pH 6.5–7.5 (5% solids); 40% act.

Rewoderm S 1333/P. [509] Ricinoleic acid alkanolamide sulfosuccinate; anionic; surfactant for syndet soaps, powd. detergents; powd.; 100% conc.

Rewoderm SB/LAN 5. [212] Wool wax alcohol sulfosuccinate; anionic; detergent; moisturizer for personal care products; brn. liq.; sol. in alcohol, glycol, and solvs.; pH 6.5–7.5 (5% solids); 50% act.

Rewolan 5. [212; 509] Wool wax alcohol sulfosuccinate; anionic; see Rewoderm SB/LAN 5; pH 6.5–7.5 (5% solids); 50% act.

Rewolan AWS. [509] Lanolin, ethoxylated; nonionic; superfatting agent for personal care products; visc. liq.; 100% conc.

Rewolan E. [509] Wool wax alcohol sulfosuccinate; anionic; moisturizing skin-protective agent for creams, lotions, and cleansing formulations; liq., paste; 50% conc.

Rewolan E 50, 100. [509] Lanolin polyglycol ether; nonionic; superfatting agent for personal care products; visc. liq. and hard wax resp.; 50 and 100% conc. resp.

Rewolan SPS. [509] Beta-sitosterol polyglycol ether sulfosuccinate; anionic; surfactant for personal care products; med. visc. liq.; 35% conc.

Rewolub KSM 14. [509] Dicarboxylic acid ethoxylate; nonionic; surfactant for syn. cooling oils, lubricant; visc. liq.; 100% conc.

Rewolub KSM 80. [509] Dicarboxylic acid diamide modified; nonionic; surfactant for syn. cooling oils; visc. liq.; 83% conc.

Rewomat B 2003. [212] Tetrasodium (1,2-dicarboxyethyl)-N-alkyl sulfosuccinamide; anionic; detergent, foaming agent and stabilizer; emulsion polymerization additive; industrial, metal, household and all-purpose cleaner; solubilizer for detergent raw materials; soldering aid; pigment dispersant; emulsifier for wax and oil; cosmetics; amber clear liq.; sol. in water, alkali and electrolytes; pH 7–9 (10% solids); 35% solids.

Rewomat TMS. [212] Disodium alkyl sulfosuccinamide; anionic; foaming agent, stabilizer, emulsifier, foam additive for natural and syn. lattices; emulsion polymerization; wetting agent for latex impregnation; soft creamy paste or liq.; pH 9–10 (10% solids); 35% solids.

Rewomid 203/S. [212] Lauric acid diethanolamide; nonionic; detergent, foam stabilizer and visc. modifier for shampoos and detergents; off-wh. solid; slight odor; m.p. 40 C; 100% act.

Rewomid AC 28. [212] Groundnut fatty acid polydialkanolamide; nonionic; detergent, emulsifer used as an anticorrosion additive for metal-working lubricants and oils; brn. liq.; dens. 0.98 g/cm^3; sp.gr. 1.0; visc. medium; 100% act.

Rewomid C 212. [212] Coconut fatty acid monoethanolamide; nonionic; detergent used in detergent products; stabilizer of emulsions; off-wh. flakes; amidic slight odor; m.p. 73 C; 100% act.

Rewomid C 220SE. [509] Coconut diethanolamide; nonionic; detergent, foam stabilizer, detergent builder; liq.; 100% conc.

Rewomid CD. [507] Coconut monoethanolamide; nonionic; detergent, foam stabilizer, detergent builder; liq.; 100% conc.

Rewomid DC 212/S. [212] Coconut fatty acid diethanolamide; nonionic; detergent, foam stabilizer and visc. modifier used for shampoos and detergents; lt. yel. visc. liq.; slight odor; dens. 0.99 g/cm^3; 100% act.

Rewomid DC 212/SE. [509] Coconut fatty acid diethanolamide; nonionic; foam booster, thickener for toiletries; med. visc. liq.; 100% conc.

Rewomid DC 220/LS. [509] Coconut fatty acid diethanolamide modified; nonionic; thickener and superfatting agent; med. visc. liq.; 100% conc.

Rewomid DC 220/SE. [212] Coconut fatty acid diethanolamide; nonionic; detergent, thickener, foam stabilizer and superfatting agent; soft paste; 100% act.

Rewomid DL 203. [212] Lauric acid polydialkanolamide; nonionic; detergent; increases visc.; good superfatting; stabilizer of emulsions; fixation of perfumes; wh. wax; dens. 0.97 g/cm^3; sp.gr. 1.0; 100% act.

Rewomid DL 203/S. [509] Lauric acid diethanolamide; nonionic; foam booster and thickener for personal care products and general purpose cleaners; wax; 100% conc.

Rewomid DL 240. [212] Coconut fatty acid polydiethanolamide (2:1); nonionic; detergent added to cosmetic preparations and cleaning and washing agents; stabilizer of emulsions; superfatting; yel. liq.; faint odor; visc. med.; 100% act.

Rewomid DLMS. [509] Lauric/myristic diethanolamide; nonionic; foam stabilizer and thickener in detergent systems; wax; 100% conc.

Rewomid DLM/SE. [509] Myristic diethanolamide; nonionic; thickener and superfatting agent for cosmetics; wax; 100% conc.

Rewomid DO 280. [212; 509] Oleic acid polydiethanolamide; nonionic; detergent; products in the cosmetic, cleaning, and detergent industries; amber liq.; slight odor; sp.gr. 1.0; 100% act.

Rewomid DO 280/S. [509] Mixed fatty acid diethanolamide; nonionic; aerosol anticorrosive agent; hair cosmetic specialty; cream bath additive; emulsifier; superfatter; dye and perfume solubilizer; paste; 100% conc.

Rewomid DO 280/SE. [212; 509] Oleic acid diethanolamide; nonionic; detergent for cosmetic products; honey liq.; dens. 0.96 g/cm^3; 100% act.

Rewomid F. [509] Linoleic diethanolamide; nonionic; superfatting agent and thickener for shampoos; med. visc. liq.; 100% conc.

Rewomid IPE 280. [212; 509] Oleic monoisopropanolamide; nonionic; detergent; additive for skin protecting products; yel. semisolid; slight odor; 100% act.

Rewomid IPL 203. [212; 509] Lauric monoisopropanolamide; nonionic; detergent; additive for solid and paste end products; improved washing power; stabilizer of emulsions; wh. flakes; slight odor; m.p. 55 C; 100% act.

Rewomid IPP 240. [212; 509] Coconut fatty acid monoisopropanolamide; nonionic; see Rewomid IPL 203; wh. flakes; slight odor; m.p. 50 C; 100% act.

Rewomid L203. [212; 509] Lauric monoethanolamide; nonionic; detergent for detergent preparations; fixation of perfumes; stabilizer of emulsions; wh. flakes; slight odor; m.p. 84 C; 100% act.

Rewomid OM 101/G, 101/IG. [507] Mixed fatty acid ethanolamide; nonionic; emulsifier, detergent intermediate; solv.; specialty cosmetic product; wax; 100% conc.

Rewomid OM 101/IG/ER. [507] Mixed fatty acid ethanolamide distillate; nonionic; solv.; specialty cosmetic product; wax; 100% conc.

Rewomid R 280. [509] Ricinoleic acid monoethanolamide; nonionic; surfactant for soaps; foam stabilizer; wax; 100% conc.

Rewomid S 280. [212; 509] Stearic acid monoethanolamide; nonionic; detergent;

superfatting agent; stabilizer of emulsions; wh. flakes; slight odor; m.p. 92 C; 100% act.

Rewomid SD. [507] Specialty alkylolamide; nonionic; amphoteric foam booster; liq.; 100% conc.

Rewomid U 185. [509] Undecylenic acid monoethanolamide; nonionic; fungicide; antimycotic; flake; 100% conc.

Rewomine IM-BT. [212] Imidazoline deriv.; cationic; adhesive additive/agent and emulsifier in acidic bitumen emulsions; rust inhibitor in acidic media; Gardner 13 max. gr.-brn. liq.; amine odor; dens. 0.94 g/cm^3; 100% act.

Rewomine IM-CA. [212] 1-Hydroxyethyl-2-heptadecenyl imidazoline; cationic; leather and metalworking industry; wetting agent for carbonization baths; yel. wax; m.w. 285; sol. in nonpolar solvs.; disp. in water; 100% act.

Rewomine IM-OA. [212] 1-Hydroxyethyl-2-heptadecenyl imidazoline; cationic; coupling agent for rust prevention; emulsifier for the emulsification of oils and bitumens; antistat; raw material for quat. reactions; paint industry; amber liq.; m.w. 345; sol. in polar and nonpolar solvs.; disp. in water; dens. 0.94 g/cm^3; visc. 800 cps; pH 10–11 (1% aq.); 100% act.

Rewominox B 204. [509] Alkyl amidopropyl dimethylamine oxide; nonionic; foam booster; antistat; hair conditioner; liq.; 35% conc.

Rewominox L 408. [509] Lauryl dimethyl amine oxide; nonionic; foam booster for personal care products; liq.; 30% conc.

Rewominox S 300. [509] Stearyl dimethylamine oxide; nonionic; foam booster, antistat, conditioner; liq.; 25% conc.

Rewominoxid B 204. [212] Cocamidopropylamine oxide; nonionic; emulsifier, softener; calcium soap dispersant used in personal care products; pale clear liq.; almost odorless; pH 5–7 (10% solids); 35% min. solids.

Rewomul MG. [509] Glyceryl stearate; nonionic; emulsifier for cosmetics; flake; 100% conc.

Rewomul MG SE. [509] Glyceryl stearate SE; nonionic; emulsifier; wax; HLB 5.5; 100% conc.

Rewopal C 6. [212; 509] PEG-6 cocamide; nonionic; dispersant, emulsifier, wetting agent for calcium soap, personal care products; amber liq.; cloud pt. 80–90 C (2%); pH 8–10 (1% solids); 100% act.

Rewopal CSF 11. [509] Cetostearyl alcohol polyglycol ether (11 EO); nonionic; solubilizer; emulsifier for cosmetic and pharmaceutical preparations; flake; 100% conc.

Rewopal CSF 20. [509] Cetostearyl alcohol polyglycol ether (20 EO); nonionic; flakes; 100% conc.

Rewopal EO 70. [509] Oleic acid polyglycol ester; nonionic; emulsifier for min. oils; med. visc. liq.; 100% conc.

Rewopal HV 4. [509] POE (4) nonyl phenyl ether; nonionic; emulsifier for min. oils, petroleum, aliphatic hydrocarbons; med. visc. liq.; 100% conc.

Rewopal HV 5. [212; 509] POE (5) nonyl phenyl ether; nonionic; emulsifier used in emulsions and insecticides, raw material for rinsing, washing, and cleaning agents; wetting agent in the paper and cellulose industry, degreaser used in pickling and alkaline immersion-bath cleaners; Gardner 2 max. liq.; sol. in min. oil, alcohol, ketone, toluene, xylene, chlorinated hydrocarbons; cloud pt. 59–62 C; pH 5–7 (1% solids); 100% act.

Rewopal HV 6. [509] POE (6) nonyl phenyl polyglycol ether; nonionic; emulsifier for min. oils, petroleum, aliphatic hydrocarbons in degreasers; med. visc. liq.; 100% conc.

Rewopal HV 8. [212; 509] POE (8) nonyl phenyl ether; nonionic; see Rewopal HV 5; Gardner 2 max. liq.; sol. see Rewopal HV 5; cloud pt. 41–45 C; pH 5–7 (1% solids); 100% act.

Rewopal HV 9. [212] POE (9) nonyl phenyl ether; nonionic; see Rewopal HV 5; Gardner 2 max. liq.; sol. see Rewopal HV 5; cloud pt. 59–62 C (2%); pH 5–7 (1% solids).

Rewopal HV 10. [212] POE (10) nonyl phenyl ether; nonionic; see Rewopal HV 5; Gardner 2 max. liq.; sol. see Rewopal HV 5; cloud pt. 70–73 C (2%); pH 5–7 (1% solids); 100% act.

Rewopal HV 14. [212; 509] POE (14) nonyl phenyl ether; nonionic; see Rewopal HV 5; wh. soft wax; sol. see Rewopal HV 5; cloud pt. 60–63 C (2%); pH 5–7 (1% solids); 100% act.

Rewopal HV 25. [212; 509] POE (25) nonyl phenyl ether; nonionic; see Rewopal HV 5; wh. wax; sol. see Rewopal HV 5; cloud pt. 73–76 C (2%); pH 5–7 (1% solids); 100% act.

Rewopal LA 3. [212] Fatty alcohol polyglycol ether; raw material for the production of ether sulfates; shampoos; clear liq.; sol. in min. and org. solvs.; pH 5–7 (1% solids).

Rewopal M 365. [509] Fatty acid polyglycol ester; nonionic; additive for metal working fluids and oils; med. visc. liq; 100% conc.

Rewopal MPG 10. [509] Monophenyl glycol; nonionic; solv., solubilizer for preservatives; liq.; 100% conc.

Rewopal MPG 12, 40. [509] Phenol polyglycol ether; nonionic; solv.; liq.; 100% conc.

Rewopal MT 65. [509] Fatty alcohol PEG methyl ether; nonionic; surfactant for textile aux.; visc. liq.; 100% conc.

Rewopal MT 2455. [509] Fatty alcohol EO/PO methylether; nonionic; fiber lubricant, detergent; liq.; 100% conc.

Rewopal MT 2540. [509] Fatty alcohol polyalkylene oxide methyl ether; nonionic; surfactant for textile aux.; visc. liq; 100% conc.

Rewopal MT 5722. [509] Fatty alcohol EO/PO methylether; nonionic; fiber lubricant, detergent; liq.; 100% conc.

Rewopal O 8. [212] Oleic acid amide polyglycol ether; nonionic; detergent; wetting agent and dispersant for calcium soap; o/w emulsifier; suitable for machine washing formulations; brn. liq.; sol. in water, alcohol, ketone, ester, chlorinated hydrocarbons, benzene, fatty oils; cloud pt. 75–85 C; pH 8–10 (1%); 100% act.

Rewopal PEG 6000 DS. [509] PEG-150 distearate; nonionic; thickener for toiletries; flake; 100% conc.

Rewopal PG 280. [509] Glycol distearate; anionic; pearlizing agent; flake; 100% conc.

Rewopal PG 340. [509] Ethylene glycol dibehenate; nonionic; pearlizing agent for nonionics; flake; 100% conc.

Rewopal PO. [509] EO/PO block polymers; nonionic; wetting agent for nonfoaming cleaning agents; med. visc. liq.; 100% conc.

Rewopal TA 11. [212; 509] Tallow alcohol polyglycol ether; nonionic; wash-act. base, wetting agent, dispersant, emulsifier, and cleaning power; wh. soft wax; sol. in water and org. solvs.; cloud pt. 69–73 C; pH 5–7 (1% solids).

Rewopal TA 25. [212; 509] Tallow alcohol polyglycol ether; nonionic; see Rewopal TA 11; wh. powd.; sol. see Rewopal TA 11; cloud pt. 75–90 C (2%); pH 5–7 (1% solids).

Rewopal TA 25/S. [509] Tallow fatty alcohol polyglycol ether (25 EO); nonionic; detergents surfactant, dispersant; flakes; 100% conc.

Rewopal TA 50. [212] Fatty alcohol polyglycol ether; wh. powd.; cloud pt. 72–76 C (2%); pH 5–7 (1% solids).

Rewopal TPD 30. [509] Fatty amine ethoxylate; nonionic; emulsifier for min. oils; visc. liq.; 100% conc.

Rewophat E 1027. [212; 509] Alkylphenol polyglycol ether phosphate; anionic; corrosion inhibitor, emulsifier, dispersant, wetting agent; used in pure oil for metals; antistat; yel. liq.; 100% act.

Rewophat EAK 8190. [212; 509] Fatty alcohol polyglycol ether phosphate; see Rewophat E 1027; yel. liq.; disp. in water; 100% act.

Rewophat NP 90. [509] Nonylphenol polyglycol ether phosphate; anionic; emulsifier, antistat, raw material for industrial cleaners; liq.; 100% conc.

Rewophat OP 80. [509] Fatty alcohol polyglycol ether phosphate; anionic; additive for synthetic fiber; min. oil emulsifier, antistat; paste; 100% conc.

Rewophat TD 40. [509] Fatty alcohol polyglycol ether phosphate; anionic; hydrotrope; solubilizer for acid cleaners; liq.; 100% conc.

Rewophat TD 70. [509] Fatty alcohol polyglycol ether phosphate; anionic; raw material for chemical and acid cleaners; antisoiling finish, flame retarder; liq.; 100% conc.

Rewopol 15/L. [509] Sodium lauryl sulfate; anionic; surfactant for emulsion polymerization; liq.; 15% conc.

Rewopol AL 3, 60%. [509] Ammonium lauryl ether sulfate; anionic; raw material for dishwashing and cleansing agents; visc. liq.; 60% conc.

Rewopol B 1003. [509] Alkyl sulfosuccinate; anionic; foaming and antigelling agent for latex foam backings and coatings; paste; 35% conc.

Rewopol B 2003. [509] Tetrasodium dicarboxyethyl stearyl sulfosuccinamate; anionic; emulsion polymerization; flotation agent; low visc. liq.; 35% conc.

Rewopol CHT 12. [509] Tricarboxymethyl diamino alkyl amide; anionic; sequestering agent, complex building surfactant for detergents; visc. liq.; 40% conc.

Rewopol CT. [509] Trideceth-7 carboxylic acid; anionic; surfactant for personal care products; textile aux.; opaque liq.; 90% conc.

Rewopol CTN. [509] Sodium trideceth-7 carboxylate; anionic; surfactant for textile auxs.; liq.; 60% conc.

Rewopol DBC. [509] Dodecyl benzyl chloride; nonionic; alkylating agent; visc. liq.; 100% conc.

Rewopol-Emulsifier BW. [212] Fatty acid deriv., modified; nonionic; forms cleansing liqs.; hand cleaners; oily or greasy surface cleaner; clear yel. visc. liq.; oil sol.; 100% act.

Rewopol-Emulsifier BWA. [212] Fatty acid deriv., modified; nonionic; emulsifier for metal cleaning, solv. degreaser, paraffin wax and cold cleaners to dewax cars; clear yel. visc. liq.; 100% act.

Rewopol MLS 30. [509] MEA-lauryl sulfate; anionic; surfactant for personal care products; visc. liq.; 30% conc.

Rewopol MLS 35. [507] MEA-lauryl sulfate; anionic; raw material for liq. detergents, foam baths and shampoos; liq.; 35% conc.

Rewopol NEHS 40. [509] Sodium dioctyl sulfosuccinate; anionic; wetting agent for alkaline cleaners, mercerizing, electroplating; hydrotrope; liq.; 40% conc.

Rewopol NI 56. [509] Sodium 2-hydroxyethyl sulfate; anionic; surfactant for syndet soaps, textile aux.; low visc. liq.; 56% conc.

Rewopol NL 3 28%. [212; 509] Sodium laureth sulfate; anionic; detergent, foamer; cleaning formulations; personal care products; colorless clear liq.; low odor; visc. 50–150 cps; pH 7–8 (5% solids); 28% act.

Rewopol NL 3 S-70. [509] Sodium lauryl ether sulfate; anionic; see Rewopol NL 3 28%; liq.; 70% conc.

Rewopol NLS 28. [212; 509] Sodium lauryl sulfate; anionic; detergent, emulsifier in emulsion polymerization; raw material for washing and cleaning agents; cosmetics; lt. yel. liq.; m.w. 300; pH 7–8 (1% solids); 28% act.

Rewopol NLS 30/L. [509] Sodium lauryl sulfate; anionic; see Rewopol 15/L; liq.; 30% conc.

Rewopol NLS 90 Powder. [212] Sodium lauryl sulfate; see Rewopol NLS 28; wh.-lt. yel. powd.; m.w. 300; pH 7–9 (1% solids); 85% min. act.

Rewopol NOS 5, 8, 10, 25. [509] Nonyl phenol polyglycol ether sulfate; anionic; emulsion polymerization surfactant; liq.; 20–35% conc.

Rewopol SBC 212. [509] Disodium cocamido MEA sulfosuccinate; anionic; surfactant for cleaners, detergents, and personal care products; low visc. liq.; 40% conc.

Rewopol SBC 212/G. [509] Disodium cocamido MEA sulfosuccinate; anionic; see Rewopol SBC 212; granules; 95% conc.

Rewopol SB-DO 70. [212; 509] Dioctyl sodium sulfosuccinate; anionic; wetting agent, solubilizer; cosmetic and personal care preparations, household and metal cleaners; used in paper, textile, paint, and dye industries; dry cleaning; colorless liq.; surf. tens. < 30 dynes/cm (0.1%); pH 6.5–7.5 (5% solids); 64% min. act.

Rewopol SBF 12. [212; 509] Disodium lauryl alcohol sulfosuccinate; anionic; detergent raw material for personal care products; carpet and upholstery shampoos; wh. paste; pH 6.5–7.5 (5% solids); 39% solids.

Rewopol SBF 18. [509] Disodium stearyl sulfosuccinamate; anionic; raw material for detergents; additive to syndet soap; flakes; 95% conc.

Rewopol SBFA 30, 40%. [212; 509] Disodium laureth sulfosuccinate; anionic; detergent raw material for personal care products; colorless clear liq.; visc. 100–299 cps; pH 6.5–7.5 (5% solids); 39% min. solids.

Rewopol SBFA 50. [509] Fatty alcohol polyglycol ether sulfosuccinate; anionic; emulsion polymerization surfactant; low visc. liq.; 30% conc.

Rewopol SBL 203. [212; 509] Disodium lauramido MEA sulfosuccinate; anionic; detergent for cosmetics; aids spray-drying; carpet and upholstery shampoos; soaps; yel. turbid liq.; pH 6.5–7.5 (5% solids); 39% min. solids.

Rewopol SBL 203/G, 203/P. [509] Fatty acid alkylolamide sulfosuccinate; anionic; raw material for carpet shampoos, soaps, and household cleaners; granules and powd. resp.; 95% conc.

Rewopol SBLC, SBLC/G, SBLC/P. [509] Fatty acid alkylolamide sulfosuccinate; anionic; surfactant for cleaners, detergents, shampoos; paste, granules, powd. resp.; 40, 95, and 95% conc.

Rewopol SB MB 80. [509] Diisohexyl sulfosuccinate; anionic; emulsion polymerization surfactant; visc. liq.; 80% conc.

Rewopol SBR 12-Powder. [212] Disodium lauryl alcohol sulfosuccinate; anionic; detergent for soaps, shampoos, moisture sensitive tech. products; wh., lt. yel. powd.; water-sol.; dens. 250 g/l; pH 6.5–7.5 (5% solids); 95% solids.

Rewopol SBV. [212] [509] Sulfosuccinate,

modified; surfactant for personal care products and detergents; lt. yel. clear liq.; visc. 2500 cps; pH 6.5–7.5 (1% solids); 28% min. solids.

Rewopol SBZ. [509] Fatty acid polyglycol ether sulfosuccinate; anionic; see Rewopol SBV; low visc. liq.; 50% conc.

Rewopol SMS 35. [509] Alkyl disodium sulfosuccinamate; anionic; emulsifier for emulsion polymerization; foaming agent for latex emulsion; antigelling and cleaning agent for paper mill felts; paste; 35% conc.

Rewopol TLS 40. [509] Triethanolamine lauryl sulfate; anionic; surfactant for personal care products; low visc. liq.; 40% conc.

Rewopol TLS 45/B. [509] Triethanolammonium lauryl sulfate; anionic; surfactant for fire fighting foams; textile aux.; liq.; 45% conc.

Rewopol TLS 90/L. [509] Triamine lauryl sulfate; anionic; personal care product concs.; med. visc. liq.; 90% conc.

Rewopol TMS/F. [509] Alkyl sulfosuccinamide; anionic; emulsifier for emulsions; foaming agent for latex emulsion; antigelling agent; liq.; 31% conc.

Rewopon AM-2C. [212] Coconut-based ampholyte; amphoteric; detergent, foam booster, wetting agent, foam stabilizer used in toiletries; golden-yel. liq.; 49% act.

Rewopon AM-2L. [212] Imidazoline-based ampholyte; amphoteric; detergent for personal care products; liq.; 50% act.

Rewopon AM-B 13. [212] Alkyl amido betaine; amphoteric; detergent, foam booster, personal care products; yel. liq.; pH 4–7 (10% solids); 30% act.

Rewopon AM-CA. [212] Coconut-based ampholyte; amphoteric; see Rewopon AM-B 13; clear yel. liq.; pH 8–8.5 (10% solids); 30% act.

Rewopon AM-V. [212] Fatty acid-based ampholyte, short-chain; amphoteric; detergent; preparation of cleaning agents; wetting agent for acidic and alkaline cleaners; dk. amber liq.; pH 10–12 (10% solids); 28% act.

Rewopon IM-AN. [509] Imidazoline deriv.; cationic; water-repellent for car rinses; inhibitor for acid pickling baths; visc. liq., paste; 100% conc.

Rewopon IM-BT. [509] Imidazoline deriv.; cationic; improves adhesive properties for coatings; visc. liq., paste; 100% conc.

Rewopon IM-CA. [509] 1-Hydroxyethyl-2-alkyl-imidazoline; cationic; corrosion inhibitor, antistat, lubricant adhesive, raw material for quarternization; firm paste; 100% conc.

Rewopon IM-OA. [509] 1-Hydroxyethyl-2-alkyl-imidazoline; cationic; corrosion inhibitor, emulsifier, antistat; med. visc. liq.; 100% conc.

Rewopon JMBT. [509] Long-chain imidazoline deriv.; cationic; adhesion agent and emulsifier for acid asphalt; liq.; 100% conc.

Rewopon JMCA. [509] Coconut imidazoline deriv.; cationic; emulsifier, anticorrosive agent; firm paste; 100% conc.

Rewopon JMOA. [509] Oleic imidazoline deriv.; cationic; rust inhibitor, emulsifier for antistatic preparations; liq.; 100% conc.

Rewoquat B 50. [509] Benzalkonium chloride; cationic; disinfectant for cleaners, dairy and food industries; algicide; textile dyeing aux.; low visc. liq.; 50% conc.

Rewoquat CPEM. [509] Coco pentaethoxy methyl ammonium methosulfate; cationic; conditioner for shampoos, emulsifier in emulsion polymerization, antistat; visc. liq.; 100% conc.

Rewoquat CR 3099. [509] Difatty acid isopropyl ester dimethyl ammonium methylsulfate; cationic; fabric softener; visc. liq.; 95% conc.

Rewoquat DQ 35. [509] Tallow propylene diamine PEG ammonium methosulfate; cationic; antistat and wetting agent; liq.; 35% conc.

Rewoquat QA 100. [509] Myristalkonium saccharinate; cationic; personal care products; granulate; 95% conc.

Rewoquat RTM 50. [509] Ricinoleic acid propylamido trimethyl ammonium methosulfate; cationic; conditioner for personal care products; antistat; low visc. liq.; 50% conc.

Rewoquat W 222 LM. [509] Ditallow amidoammonium methosulfate; fabric softener; liq.; 90% conc.

Rewoquat W 3690, W 3690/PG. [509] Dioleyl imidazoline methosulfate; cationic; fabric softener; visc. liq.; 75% conc.

Rewoquat W 7500. [509] Quaternium-27; cationic; fabric softener; visc. liq.; 75% conc.

Rewoquat W 7500/H. [509] Quat. imidazoline deriv.; cationic; hair softener; soft paste; 75% conc.

Rewoquat W 7500/PG. [509] Dialkyl imidazoline methosulfate, IPA free; cationic; fabric softener; visc. liq.; 75% conc.

Rewoquat W 9000, 9000/PG. [509] Quat.

imidazoline deriv.; cationic; fabric softener; soft paste; 90% conc.

Rewoquat WKH 80. [509] Quat. imidazoline deriv.; cationic; textile softener; paste; 80% conc.

Reworyl ACS 60. [509] Ammonium cumene sulfonate; anionic; hydrotrope for detergent and cleaning systems; liq.; 60% conc.

Reworyl B 70. [509] Benzene sulfonic acid; anionic; catalyst for esterification, polymerization, and polycondensation in foundry resins; straw clear liq.; 70% conc.

Reworyl C 65. [509] Cumene sulfonic acid; anionic; see Reworyl B 70; liq.; 65% conc.

Reworyl K. [509] Dodecylbenzene sulfonic acid; anionic; raw material for anionic detergent systems; med. visc. liq.; 97% conc.

Reworyl NCS 40. [509] Sodium cumene sulfonate; anionic; see Reworyl ACS 60; liq.; 40% conc.

Reworyl NKS 50, 100. [509] Sodium dodecylbenzene sulfonate; anionic; detergents and cleaners; textile industry; paste and powd. resp.; 50 and 98% conc. resp..

Reworyl NTS 40. [509] Sodium toluene sulfonate; anionic; see Reworyl NCS 40; liq.; 40% conc.

Reworyl NXS 40. [509] Sodium xylene sulfonate; anionic; see Reworyl NCS 40; liq.; 40% conc.

Reworyl-Sulfonic Acid K. [212] Dodecylbenzene sulfonic acid; anionic; detergent base for dishwashing, washing, and cleaning formulations; br. visc. liq.; 97% min. act.

Reworyl T 65. [212; 509] *p*-Toluene sulfonic acid; anionic; industrial applications; intermediate; leather tanning agent; toxic crop-dusting agents; production of syn. fibers; pale clear liq.; char. odor; 65% act.

Reworyl TKS 90/F. [212; 509] Triammonium dodecylbenzene sulfonate; anionic; detergent raw material for cleansing and defatting agents, car shampoos; chain lubricant, dk. amber visc. liq.; pH 7.0–7.5 (1% solids); 85% min. act.

Reworyl TKS 90/L. [509] Triamine dodecylbenzene sulfonate; anionic; raw material for liq. detergents; visc. liq.; 90% conc.

Reworyl X 65. [509] Xylene sulfonic acid; anionic; catalyst for esterification, polymerization, and polycondensation in foundry resins; liq.; 65% conc.

Rewoteric AM-2 C. [509] Cocoamphodiacetate; amphoteric; shampoos and foam baths; visc. liq.; 50% conc.

Rewoteric AM-2 C/NM. [509] Cocoamphoacetate; amphoteric; personal care products; visc. liq.; 50% conc.

Rewoteric AM-2L. [509] Lauroamphoacetate; amphoteric; see Rewoteric AM-2 C/NM; low visc. liq.; 50% conc.

Rewoteric AM-B 13. [509] Cocamidopropyl betaine; amphoteric; foam booster for personal care products; low visc. liq.; 35% conc.

Rewoteric AM-CA. [509] Lauroamphoacetate and sodium laureth sulfate; amphoteric; see Rewoteric AM-2 C/NM; med. visc. liq.; 30% conc.

Rewoteric AM-CAS. [509] Sulfo betaine; amphoteric; used in personal care products and acid and alkaline cleaners; low visc. liq.; 50% conc.

Rewoteric AM-DML. [509] Lauryl betaine; amphoteric; used in personal care products, cleaners, and pickling baths; liq.; 40% conc.

Rewoteric AM-G 30. [509] Lauroamphoacetate, sodium lauryl sulfate and hexylene glycol; see Rewoteric AM-2 C/NM; low visc. liq.; 38% conc.

Rewoteric AM-KSF. [509] Cocoamphopropionate; amphoteric; used in personal care products and industrial cleaners; visc. liq.; 40% conc.

Rewoteric AM-R 40. [509] Ricinoleamidopropyl betaine; amphoteric; cosmetic and medicated cleaning agents; med. visc. liq.; 40% conc.

Rewoteric AM-TEG. [509] Betaine, ethoxylated; cationic; surfactant for acidic cleaning products; thickener; liq.; 40% conc.

Rewoteric AM-V. [509] Capryloamphoglycinate; amphoteric; wetting agent for pickling baths; used in acid and alkaline cleaners; low visc. liq.; 35% conc.

Rewowax CL 1263. [509] Amide wax; nonionic; gloss emulsion for floors, varnishes, plastics; flake; 100% conc.

Rexobase PW. [218] Naphthenic deriv.; cationic; emulsifier for min. oil and paraffin; paste

Rexobase XX. [218] Amide condensate; anionic; emulsifier; liq.

Rexol 25/1. [283] Nonoxynol-1; nonionic; defoamer for detergent systems; coemulsifier for surfactant and solv. blends; liq.; HLB 4.6; 100% act.

Rexol 25/4. [283] Nonoxynol-4; nonionic; emulsifier, intermediate for anionic sulfonates; clear liq.; ref. index 1.5000; oil-sol.; sp.gr. 1.02; visc. 200 cps; HLB 8.7;

100% act.

Rexol 25/6. [283] Nonoxynol-6; nonionic; dispersant in petroleum systems; intermediate for anionic sulfates; antistat plasticizer for plastics; clear liq.; ref. index 1.4909; oil-sol.; sp.gr. 1.04; visc. 250 cps; HLB 10.8; 100% act.

Rexol 25/7. [283] Nonoxynol-7; nonionic; emulsifier; clear liq.; HLB 11.6; 100% act.

Rexol 25/8. [283] Nonoxynol-8; nonionic; detergent, wetting agent, emulsifier; base surfactant for household and industrial detergents; used for leather, paint, textile, pesticides, pulp and paper industries; clear liq.; water-sol.; 100% act.

Rexol 25/9. [283] Nonoxynol-9; nonionic; detergent, dispersant, emulsifier and wetting agent; used in leather, paint, textile, pulp and paper industries; base for industrial and household detergents; liq.; water-sol.; HLB 13.0; 100% conc.

Rexol 25/10. [283] Nonoxynol-10; nonionic; see Rexol 25/9; liq.; water-sol.; HLB 13.4; 100% conc.

Rexol 25/11. [283] Nonoxynol-11; nonionic; detergent, wetting agent; scouring agent for natural and syn. textile fibers; clear liq.; ref. index 1.4893; sp.gr. 1.06; visc. 314 cps; HLB 13.8; 100% act.

Rexol 25/14. [283] Nonoxynol-14; nonionic; detergent and wetting agent used for metal cleaning; ref. index 1.4848; sp.gr. 1.055; visc. 250 cps; 100% act.

Rexol 25/15. [283] Nonoxynol-15; nonionic; wetting agent and detergent for hard surfaces; clear paste; ref. index 1.4870; sp.gr. 1.06; visc. 400 cps (35 C); HLB 14.8; 100% act.

Rexol 25/20. [283] Nonoxynol-20; nonionic; particle dispersant in aq. systems; solid; HLB 16.0; cloud pt. 72–74 C (1%); 100% act.

Rexol 25/30. [283] Nonoxynol-30; nonionic; solubilizer for essential oils and pesticides; solid; HLB 17.2; cloud pt. 74–76 C (1%); 100% act.

Rexol 25/40. [283] Nonoxynol-40; nonionic; emulsifier used in polymer emulsification for paints and coatings; HLB 17.8; cloud pt. 74–76 C (1%); 100% act.

Rexol 25/50. [283] Nonoxynol-50; nonionic; emulsifier and stabilizer used in floor finishes; solid; HLB 18.2; cloud pt. 74–76 C (1%); 100% act.

Rexol 25/100-70%. [283] Nonoxynol-100; nonionic; demulsifier for petroleum oils; pressured textile scouring; emulsifier/stabilizer for latices, asphalt, and floor finishes; liq.; HLB 19.0; cloud pt. 74–76 C (1%); 70% act.

Rexol 25/307. [283] Nonoxynol-30; nonionic; see Rexol 25/30; liq.; HLB 17.2; cloud pt. 74–76 C (1%); 70% act.

Rexol 25/407. [283] Nonoxynol-40; nonionic; detergent, wetting agent; emulsifier for vinyl acetate and acrylate polymerization; demulsifier of petroleum oils; textile scouring; stabilizer for syn. latices; yel. liq.; water-sol.; sp.gr. 1.10; 70% act.

Rexol 25/507. [283] Nonoxynol-50; nonionic; see Rexol 25/50; liq.; HLB 18.2; cloud pt. 74–76 C (1%); 70% act.

Rexol 25J. [283] Alkyl phenol ethoxylate; nonionic; detergent and wetting agent for textile, pulp and paper, leather, paint, and pesticide industries; emulsifier; clear liq.; ref. index 1.4823; sp.gr. 1.06; visc. 250 cps; HLB 13.0; 100% act.

Rexol 25JM1. [283] Alkyl phenol ethoxylate; nonionic; detergent, wetting agent, emulsifier used in Kraft and sulfite digesters, screen rooms, bleach plants and paper machines; yel. liq.; sp.gr. 1.05; visc. 290 cps; 100% act.

Rexol 25JWC. [283] Alkyl phenol ethoxylate; nonionic; detergent, wetting agent, emulsifier; scouring of raw wool; foamer for Kuster dyeing; clear liq.; HLB 3.4; 100% act.

Rexol 35/3. [283] Alcohol ethoxylate; nonionic; emulsifier for o/w; intermediate for sulfation and phosphation; liq.; HLB 8.0; cloud pt. 49–51 C (1%); 100% act.

Rexol 35/5. [283] Alcohol ethoxylate; nonionic; detergent in textile and leather applications; liq.; HLB 10.6; cloud pt. 64–65 C (1%); 100% act.

Rexol 35/8. [283] Alcohol ethoxylate; nonionic; detergent intermediate; wetting agent used in textile applications; liq.; HLB 12.8; cloud pt. 79–83 C (1%); 85% act.

Rexol 35/11. [283] Alcohol ethoxylate; nonionic; hard surface cleaner used in bottle washing and metal cleaning; paste; HLB 14.2; cloud pt. 72–76 C; 100% act.

Rexol 35/100. [283] Fatty alcohol, ethoxylated; nonionic; detergent, wetting agent used in scouring of paper machine felts; preparation of industrial detergents; emulsifier, dispersant; clear paste; sp.gr. 1.003; visc. 100 cps; 100% act.

Rexol 45/1. [283] POE (1) octyl phenyl ether;

nonionic; emulsifier, detergent, dispersant; coemulsifier for surfactant and solv. preparation blends; liq.; oil-sol.; HLB 3.6; 100% act.

Rexol 45/3. [283] POE (3) octyl phenyl ether; nonionic; emulsifier, detergent, dispersant, coemulsifier for petroleum oils and solvs.; liq.; HLB 7.8; 100% act.

Rexol 45/5. [283] POE (5) octyl phenyl ether; nonionic; emulsifier for fats and oils, detergent, dispersant; liq.; HLB 10.4; cloud pt. 63–66 C (10%); 100% act.

Rexol 45/7. [283] POE (7) octyl phenyl ether; nonionic; hard surface detergent; emulsifier, dispersant; liq.; HLB 12.4; cloud pt. 21–23 C (1%); 100% act.

Rexol 45/10. [283] POE (10) octyl phenyl ether; nonionic; detergent and wetting agent used in textile, paint, plastic, pulp and paper industries; base surfactant for household and industrial detergents; agricultural formulations; liq.; HLB 13.5; cloud pt. 64–67 C (1%); 100% act.

Rexol 45/12. [283] POE (12) octyl phenyl ether; nonionic; emulsifier, detergent, dispersant used in metal cleaning, industrial and household liq. detergents and cleaners; liq.; HLB 14.5; cloud pt. 86–90 C (1%); 100% act.

Rexol 45/16. [283] POE (16) octyl phenyl ether; nonionic; detergent and wetting agent used in metal cleaning and bottle washing; emulsifier; dispersant; paste; HLB 15.8; cloud pt. 65–69 C (1%); 100% act.

Rexol 45/307. [283] POE (30) octyl phenyl ether; nonionic; emulsifier for vinyl acetate and acrylate emulsion polymerization; detergent, dispersant; liq.; HLB 17.4; cloud pt. 73–77 C (1%); 70% act.

Rexol 45/407. [283] POE (40) octyl phenyl ether; nonionic; see Rexol 45/307; liq.; HLB 18.0; cloud pt. 73–77 C (1%); 70% act.

Rexol 130. [283] Alcohol ethoxylate; nonionic; emulsifier, dyeing assistant; clear liq.; ref. index 1.3850; sp.gr. 1.02; visc. 93 cps; 35% act.

Rexol 2000 HWM. [283] Amide ethoxylate, modified; nonionic; antistat emulsifier for mfg. of textile lubricants; amber liq.; ref. index 1.4618; sp.gr. 0.982 visc. 300 cps; 77% act.

Rexol 4736. [283] Alcohol ethoxylate, branch chained; nonionic; surfactant for greases; wetting agent for carpet dyeing; clear liq.; water-sol.; 85% act.

Rexol AE-1, AE-2. [283] Alcohol ethoxylate; nonionic; intermediate for shampoo and detergent mfg.; liq.; HLB 4.2 and 6.1 resp.; cloud pt. 36–39 and 50–52 C (1%) resp.; 100% act.

Rexol AE-3. [283] Linear alcohol ethoxylate; nonionic; detergent, mfg. of ether sulfates; emulsifier for hydrocarbons; clear liq.; ref. index 1.4494; sp.gr. 0.901; visc. 24 cps; HLB 8.2; 100% act.

Rexol AE-23. [283] Alcohol ethoxylate; nonionic; dispersant, emulsifier in cosmetic applications; solid; HLB 17.0; cloud pt. 90–95 C (1%); 100% act.

Rexol C14. [283] Castor oil ethoxylate; nonionic; emulsifier, antistat, demulsifier, leveling agent for dyeing of syn. fibers; solid; 90% conc.

Rexol CCN. [283] Castor oil ethoxylate; nonionic; emulsifier, lubricant; syn. fiber mfg.; clear liq.; sp.gr. 0.98; visc. 610 cps; 90% act.

Rexol CTX. [283] Polyether deriv.; nonionic; scouring agent for wool oil, fats, and greases; clear liq.; 100% act.

Rexonic 6X. [283] Linear alcohol ethoxylate; nonionic; emulsifier, wetting agent and detergent for cleaning of paper machine wet felts; clear liq.; ref. index 1.3695; sp.gr. 1.01; visc. 400 cps; 30% act.

Rexonic 1006. [283] Linear alcohol ethoxylate; nonionic; dispersant and leveling assistant for dyeing leather; prescouring aid; clear liq.; 75% act.

Rexonic 1012-6. [283] Linear alcohol ethoxylate; nonionic; emulsifier; detergent and wetting agent for scouring of natural and syn. textile fibers; clear liq.; ref. index 1.4532; sp.gr. 0.986; visc. 34 cps; HLB 12; 100% act.

Rexonic 1218-6. [283] Linear alcohol ethoxylate; nonionic; detergent and wetting agent for textile applications; clear liq.; ref. index 1.4492; sp.gr. 1.002; visc. 140 cps; 90% act.

Rexonic N-4. [283] Linear alcohol ethoxylate; nonionic; intermediate; preparation of liq. detergents; used for wetting, back wetting, sizing, emulsification and dispersant of petroleum oils; clear liq.; oil-sol.; sp.gr. 0.92; visc. 50 cps; HLB 8.0; 100% act.

Rexonic N6.6. [283] Linear alcohol ethoxylate; nonionic; detergent, wetting agent, emulsifier; formulation of liq. detergents; foam stabilizer; clear paste; sp.gr. 0.98; HLB 11.9; 100% act.

Rexonic N7. [283] Linear alcohol ethoxylate; nonionic; detergent; clear solid; sp.gr. 1.001;

HLB 14.0; 100% act.

Rexonic N7.2. [283] Linear alcohol ethoxylate; nonionic; wetting agent, emulsifier, hard surface detergent; clear solid; sp.gr. 1.0; 100% act.

Rexonic N7.6. [283] Linear alcohol ethoxylate; nonionic; detergent, wetting agent, emulsifier; base surfactant for household and industrial detergents; clear solid; 100% act.

Rexonic N23-3. [283] Linear alcohol ethoxylate; nonionic; detergent, surfactant and chemical intermediate used in solv. emulsion cleaners and dry cleaning detergents; clear liq.; oil-sol.; sp.gr. 0.93; visc. 19 cps; 100% act.

Rexonic N23-6.5. [283] Linear alcohol ethoxylate; nonionic; detergent, wetting agent, emulsifier; industrial maintenance, metal cleaning, paper, paint, and leather processing; clear liq.; water-sol.; sp.gr. 0.96; visc. 30 cps; HLB 11.2; 100% act.

Rexonic N25-3. [283] Alcohol ethoxylate; nonionic; emulsifier for oil; backwetting agent, intermediate for detergents; liq.; HLB 7.8; cloud pt. 59–61 C (1%); 100% act.

Rexonic N25-7. [283] Linear alcohol ethoxylate; nonionic; emulsifier, wetting agent and formulation of foaming detergents; clear paste; sp.gr. 0.958; 100% act.

Rexonic N25-9, N25-9 85%. [283] Alcohol ethoxylate; nonionic; detergent and wetting agent for textile, pulp and paper, paint and pesticide industries; HLB 13.1; cloud pt. 69–78 C (1%); 100 and 85% act. resp.

Rexonic N25-9 85% [283] Alcohol ethoxylate; nonionic; see Rexonic N25-9; liq.; HLB 13.1; cloud pt. 69–78 C (1%); 85% act.

Rexonic N25-12. [283] Alcohol ethoxylate; nonionic; detergent; solid; HLB 14.4; cloud pt. 90–95 C (1%); 100% act.

Rexonic N25-14. [283] Alcohol ethoxylate; nonionic; detergent for metal cleaning and bottle washing; solid; HLB 15.2; cloud pt. 83–86 C (1%); 100% act.

Rexonic N25-14 85%. [283] Linear alcohol ethoxylate; nonionic; detergent for scouring textile fibers; base for mfg. of ether sulfates; dispersant, emulsifier, and wetting agent; used in leather, paint, textile, pulp and paper industries; base surfactant for household and industrial cleaners; liq.; water-sol.; 85% conc.

Rexonic N91-1.6. [283] Alcohol ethoxylate; nonionic; intermediate for mfg. of ethoxysulfates; emulsifier, detergent, dispersant; liq.; HLB 6.1; cloud pt. 39–41 C (1%); 100% act.

Rexonic N91-2.5. [283] Alcohol ethoxylate; nonionic; emulsifier for oils and solvs.; intermediate for ethoxysulfates; detergent, dispersant; liq.; HLB 8.1; cloud pt. 50–54 C (1%); 100% act.

Rexonic N91-6, N91-8. [283] Alcohol ethoxylate; nonionic; detergent for industrial use; emulsifier; shampoo intermediate, dispersant; liq.; water-sol.; HLB 12.5 and 14.1 resp.; cloud pt. 51–53 and 80–82 C resp. (1%); 100% act.

Rexonic P-1. [283] Linear alcohol, propoxylated, ethoxylated; nonionic; detergent, wetter, emulsifier, dispersant for natural and syn. fibers; clear liq.; water-sol.; ref. index 1.455; sp.gr. 0.97; visc. 90 cps; 100% act.

Rexonic P-3. [283] Polyoxyalkylene alcohol; nonionic; detergent, emulsifier; liq.; HLB 9.0; cloud pt. 36 C (1%); 100% act.

Rexonic P-4. [283] Polyoxyalkylene alcohol; nonionic; detergent used in detergent tablets, sanitizers, and metal cleaners; HLB 10.0; cloud pt. 57 C (1%); 100% act.

Rexonic P-5. [283] Polyoxyalkylene alcohol; nonionic; detergent used in machine dishwashing and rinse aids; liq.; HLB 10.0; cloud pt. 45 C (1%); 100% act.

Rexonic P-6. [283] Polyoxyalkylene alcohol; nonionic; detergent for mechanical dishwashing; liq.; cloud pt. 35 C (10%); 100% act.

Rexonic P-7. [283] Polyoxyalkylene alcohol; nonionic; detergent used in rinse aid formulations; liq.; cloud pt. 56 C (10%); 100% act.

Rexonic RL. [283] Alcohol ethoxylate; nonionic; detergent; clear liq.; 25% act.

Rexopal 3928. [283] Linear alcohol ethoxylate, modified; nonionic; detergent; scouring and wetting agent for natural and syn. fibers; yel. liq.; water-sol.; ref. index 1.4465; sp.gr. 1.0; visc. 45 cps; 85% act.

Rexopal 3960. [283] Linear alcohol ethoxylate, modified; nonionic; detergent for prescouring of fabrics; yel. liq.; water-sol.; sp.gr. 0.977; 85% act.

Rexopal 4356. [283] Linear alcohol ethoxylate, modified; nonionic; detergent and wetting and scouring agent; clear liq.; 78% act.

Rexopal C52. [283] Sulfonates and linear alcohol ethoxylate; anionic/nonionic; detergent; prescouring woven fabrics; yel. liq; ref. index 1.4058; sp.gr. 1.0; visc. 32.5 cps; 40% act.

Rexopal EP1. [283] Linear alcohol polyoxyalkylene; nonionic; detergent, prescouring of polyester and blends; clear liq.; water-sol.; sp.gr. 0.990; visc. 112 cps; 50% act.

Rexopal NB7. [283] Linear alcohol ethoxylate, modified; nonionic; detergent; scouring and dyeing of polyester knits; clear liq.; sp.gr. 0.978; 40% act.

Rexopene. [218] Sodium alkylaryl sulfonate; anionic; wetting, leveling, penetrant, scouring assistant; liq.

Rexophos 25/67. [283] Alkylaryl phosphate ester; anionic; emulsifier for solvs.; liq.; 100% act.

Rexophos 25/97. [283] Phosphated alkyl phenol ethoxylate; anionic; prescouring aid for cotton fabrics; pale yel. liq.; 100% act.

Rexophos 4668. [283] Linear alcohol phosphate ester; anionic; detergent; liq.; 80% act.

Rexophos N25-38. [283] Linear alcohol phosphate ester; anionic; emulsifier for solvs. and oils; liq.; 100% act.

Rexopon E. [218] Amide amine condensate, modified; nonionic; detergent; liq.

Rexopon V. [218] Alkylaryl sulfonate sulfonated amide condensate; anionic; detergent; scouring; paste

Rexowet 77. [218] Heptadecyl sodium sulfate; anionic; wetting agent for textiles; liq.; 30% conc.

Rexowet CR. [218] Sulfonated isopropyl oleate and cresylic acid; anionic; wetting agent and penetrant for cotton fabrics; liq.

Rexowet GR. [218] Sulfated diester; anionic; penetrant, wetting agent; liq.

Rexowet MS. [218] Sodium alkyl naphthalene sulfonate; anionic; penetrant for textiles; liq.

Rexowet NF, NFX. [218] Alcohol sulfate salt; anionic; penetrant, wetting agent; textile dyeing assistant; liq.

Rexowet RW. [218] Aliphatic mono and diester, sulfonated; anionic; wetting, rewetting and leveling agent, penetrant; enzyme activator; gel

Rheodol 430, 450. [341] POE polyol fatty acid ester; nonionic; emulsifier for cosmetics; liq. and paste resp.; HLB 10.5 and 13.8; 100% conc.

Rheodol 440. [341] PEG-40 sorbitan tetraoleate; nonionic; see Rheodol 430; liq.; HLB 11.8; 100% conc.

Rheodol AO-10. [341] Sorbitan monooleate; nonionic; emulsifier for pharmaceuticals and cosmetics; liq.; HLB 4.3; 100% conc.

Rheodol AO-15. [341] Sorbitan sesquioleate; nonionic; see Rheodol AO-10; liq.; HLB 3.7; 100% conc.

Rheodol AS-10. [341] Sorbitan monostearate; nonionic; see Rheodol AO-10; solid; HLB 4.7; 100% conc.

Rheodol MO-60. [341] Glycerol monooleate; nonionic; see Rheodol AO-10; liq.; HLB 2.8; 100% conc.

Rheodol MS-50, MS-60, SEM. [341] Glycerol monostearate; nonionic; see Rheodol AO-10; powd.; HLB 2.8, 3.5, and 11.0 resp.; 100% conc.

Rheodol SP-010. [341] Sorbitan monooleate; nonionic; emulsifier and dispersant for printing inks, pastes, and paints; liq.; HLB 4.3; 100% conc.

Rheodol SP-030. [341] Sorbitan trioleate; nonionic; see Rheodol SP-010; liq.; HLB 1.8; 100% conc.

Rheodol SP-S30. [341] Sorbitan tristearate; nonionic; see Rheodol SP-010; beads; HLB 2.1; 100% conc.

Rheodol SP-L10. [341] Sorbitan monolaurate; nonionic; emulsifier for cosmetics; liq.; HLB 8.6; 100% conc.

Rheodol SP-P10. [341] Sorbitan monopalmitate; nonionic; emulsifier for cosmetics; beads; HLB 6.7; 100% conc.

Rheodol SP-S10. [341] Sorbitan monostearate; nonionic; see Rheodol SP-010; beads; HLB 4.7; 100% conc.

Rheodol TW-L106, -L120. [341] POE sorbitan monolaurate; nonionic; emulsifier for cosmetics and pharmaceuticals; liq.; HLB 13.3 and 16.7 resp.; 100% conc.

Rheodol TW-O106, O120. [341] POE sorbitan monooleate; nonionic; stabilizer in emulsion polymerization; liq.; HLB 10.0 and 15.5 resp.; 100% conc.

Rheodol TW-O320. [341] POE sorbitan trioleate; nonionic; see Rheodol TW-0106; liq.; HLB 11.0; 100% conc.

Rheodol TW-P120. [341] POE (20) sorbitan monopalmitate; nonionic; emulsifier for pharmaceuticals, cosmetics; solubilizer for colorants; stabilizer for emulsion polymerization; liq.; HLB 15.6; 100% conc.

Rheodol TW-S106, -S120. [341] POE sorbitan monostearate; nonionic; stabilizer in emulsion polymerizaton; solid; HLB 9.6 and 14.9 resp.; 100% conc.

Rheodol TW-S320. [341] POE sorbitan tristearate; nonionic; see Rheodol TW-S106; solid; HLB 10.5; 100% conc.

Rheotol. [631] Polymerized alkyl phosphate;

dispersant and wetting agent for pigmented coatings; pale liq.; dens. 0.97 mg/m³.

Richamate 3780. [515] Stearamidopropyl dimethylamine lactate; cationic; surfactant for use as solubilizer; emulsifier in personal care products; Gardner 3 slurry; dens. 8.3 lb/gal; pH 4–5; 25% solids.

Richamate 6613. [515] Isostearamidopropyl dimethylamine lactate; cationic; see Richamate 3780; Gardner 12 liq.; dens. 7.8 lb/gal; pH 6–7; 95% solids.

Richamide 5085. [515] Oleamide DEA; alkanolamide used as a corrosion inhibitor, detergent aid, foam booster and stabilizer; emulsifier, solubilizer and thickener for shampoos, dishwashing comps., and hard surface cleaners; liq.; dens. 8.3 lb/gal.

Richamide 6310. [515] Lauramide DEA; see Richamide 5085; paste; dens. 8.1 lb/gal.

Richamide 6404. [515] Cocamide DEA; see Richamide 5085; liq.; dens. 8.3 lb/gal.

Richamide 6445. [515] 2:1 Cocamide DEA, modified; see Richamide 5085; liq.; dens. 8.3 lb/gal.

Richamide 6625. [515] 1:1 Cocamide DEA, modified; see Richamide 5085; liq.; dens. 8.2 lb/gal.

Richamide 6737, 6754. [515] 2:1 Cocamide DEA, modified; see Richamide 5085; liq.; dens. 8.4 lb/gal.

Richamide CD. [515] 2:1 Cocamide DEA; see Richamide 5085; liq.; dens. 8.4 lb/gal.

Richamide CDA. [515] 2:1 Cocamide DEA, modified; see Richamide 5085; liq.; dens. 8.3 lb/gal.

Richamide M-3. [515] Cocamide DEA; see Richamide 5085; liq.; dens. 8.4 lb/gal.

Richamide MX. [515] 1:1 Cocamide MEA, modified; see Richamide 5085; paste; dens. 9.0 lb/gal.

Richamide S-771, S-780. [515] 2:1 Cocamide DEA, modified; see Richamide 5085; liq.; dens. 8.3 lb/gal.

Richamide STD-HP. [515] 1:1 Lauramide DEA; see Richamide 5085; liq., wax; 8.1 lb/gal.

Richonate 40B, 45B. [515] Sodium dodecylbenzene sulfonate; solubilizer, emulsifier, detergent aid used in dishwashing compds., industrial, household and dry cleaning formulations; Gardner 3 solid; dens. 9.0 and 9.2 lb/gal resp.; pH 6.5–7.5; 36 and 40% act.

Richonate 45BX, 50BD. [515] Sodium dodecylbenzene sulfonate; see Richonate 40B; Gardner 3 solid; dens. 9.1 lb/gal; pH 7–8 and 8–9 resp.; 42 and 37% act. resp.

Richonate 60B, 1850, 1850U. [515] Sodium dodecylbenzene sulfonate; see Richonate 40B; Gardner 3 solid; dens. 9.2, 8.9, and 8.9 lb/gal resp.; pH 7.5–8.5, 7.5–8.5, and 8–9 resp.; 57, 37, and 37% act. resp.

Richonate C-50H. [515] Sodium dodecylbenzene sulfonate; see Richonate 40B; Gardner 3 solid; dens. 8.5 lb/gal; pH 7.0–8.5; 44% act.

Richonate CS-6021H. [515] Calcium dodecylbenzene sulfonate; see Richonate 40B; Gardner 3 solid; dens. 8.7 lb/gal; pH 6.0–7.0; 60% act.

Richonate S-1280, TAB. [515] TEA-dodecylbenzene sulfonate; see Richonate 40B; Gardner 5; dens. 9.2 and 9.0 lb/gal resp.; pH 7.0–8.0; 75 and 52% act.

Richonate SXS. [515] Sodium xylene sulfonate; see Richonate 40B, also used as coupling agent; Gardner 1; dens. 9.8 lb/gal; pH 8.0–9.5; 39% act.

Richonate TDB. [515] Sodium tridecylbenzene sulfonate; see Richonate 40B; Gardner 3 solid; dens. 8.7 lb/gal; pH 7.0–8.0; 44% act.

Richonate YLA. [515] Isopropylamine dodecylbenzene sulfonate; see Richonate 40B; Gardner 5; dens. 8.6 lb/gal; pH 5.0–7.0; 95% act.

Richonic Acid B. [515] Dodecylbenzene sulfonic acid; solubilizer, emulsifier, detergency aid in household and industrial cleaning compds; Klett 300; dens. 8.8 lb/gal; 97% act.

Richonic 94H, 97H. [515] Sulfonic acid; see Richonic Acid B; Klett 325 and 300 resp.; dens. 8.8 lb/gal; 94 and 97% act. resp.

Richonol 2310. [515] Sodium lauryl sulfate; solubilizer, emulsifier used in cleaning and personal care products; Gardner 2 liq.; dens. 9.1 lb/gal; pH 6.5–7.0; cloud pt. 40 F; 49% act.

Richonol 3715. [515] Amine alkyl sulfate; see Richonol 2310; Gardner 5 paste; dens. 9.3 lb/gal; pH 6.5–7.5; 85% act.

Richonol 6522. [515] Sodium lauryl sulfate; see Richonol 2310; Gardner 2 liq.; dens. 8.7 lb/gal; pH 7.5–8.5; cloud pt. 50 F; 29% act.

Richonol 7031. [515] Sodium alkyl sulfate; see Richonol 2310; Gardner 2 liq.; dens. 9.2 lb/gal; pH 7.0–8.5; cloud pt. 32 F; 39% act.

Richonol A, A Powder. [515] Sodium lauryl sulfate; see Richonol 2310; Gardner 2 liq. and wh. powd. resp.; dens. 8.7 and 1.6 lb/gal;

pH 7.5–8.5 and 8.5–10.0; cloud pt. 60 F; 28.5 and 93% act. resp.

Richonol AM. [515] Ammonium lauryl sulfate; see Richonol 2310; Gardner 2 liq.; dens. 8.6 lb/gal; pH 6.0–7.0; cloud pt. 65 F; 28.5% act.

Richonol C. [515] Sodium lauryl sulfate; see Richonol 2310; wh. paste; 9.2 lb/gal; pH 7.0–8.0; 28.5% act.

Richonol Mg. [515] Magnesium lauryl sulfate; see Richonol 2310; Gardner 3 liq.; dens. 8.7 lb/gal; pH 6.0–7.0; cloud pt. 50 F; 28.5% act.

Richonol OME. [515] Sodium laureth sulfate; see Richonol 2310; Gardner 1 liq.; dens. 8.7 lb/gal; pH 7.5–8.5; cloud pt. 41 F; 25% act.

Richonol S-1285C. [515] Sodium laureth sulfate; see Richonol 2310; Gardner 3 liq.; dens. 9.1 lb/gal; pH 8.0–9.0; cloud pt. 65 F; 52% act.

Richonol S-1300C. [515] Ammonium laureth sulfate; see Richonol 2310; Gardner 3 liq.; dens. 8.6 lb/gal; pH 6.5–7.5; cloud pt. 70 F; 57% act.

Richonol S-5260. [515] Sodium laureth sulfate; see Richonol 2310; Gardner 2 liq.; dens. 8.8 lb/gal; pH 7.5–9.5; cloud pt. 40 F; 27% act.

Richonol T. [515] TEA-lauryl sulfate; see Richonol 2310; Gardner 2 liq.; dens. 8.8 lb/gal; pH 7.0–7.5; cloud pt. 40 F; 40% act.

Ritacetyl. [519] Acetylated lanolin; nonionic; water resistant films; solid; 100% conc.

Ritachol. [519] Min. oil and lanolin alcohol; nonionic; used in cosmetics and toiletries; stabilizer for emulsions, dispersions, and suspensions; primary or aux. emulsifier; epidermal moisturizer, lubricant, and emollient; pale straw liq.; faint odor; sp.gr. 0.84–0.86; sapon. no. 2 max.; 100% act.

Ritachol 1000. [519] Cetearyl alcohol, polysorbate 60, PEG-150 stearate, steareth-20; nonionic; emulsifer for personal care products, pharmaceuticals, and household specialties; wh. waxy, flakes; low odor; m.p. 48–52 C; sapon. no. 9–14.

Ritachol 2000. [519] Cetearyl alcohol and polysorbate 60; nonionic; emulsifier for cosmetics and pharmaceuticals; used in powd. suspension formulas; wh. waxy, flakes; m.p. 49–53 C; sapon. no. 9–14.

Ritaderm. [519] Petrolatum, lanolin, sodium PCA, polysorbate 85; nonionic; emollient, moisturizer, and lubricant used in cosmetics; yel. cream; m.p. 42–48 C; 100% act.

Ritahydrox. [519] Hydroxylated lanolin; nonionic; emulsifier, hypoallergenic emollient; paste; HLB 4.0; 100% conc.

Ritalafa. [519] Lanolin acid; nonionic; rewetting of makeup preparations; solid; 100% conc.

Ritalan. [519] Lanolin USP, dewaxed fraction; nonionic; moisturizer, plasticizer, penetrant, emollient; skin lubricant; liq.; 100% conc.

Ritalan AWS. [519] Alkoxylated lanolin oil; nonionic; aux. emulsifier, moisturizer, emollient; liq.; sol. in water and alcohol; HLB 18.0; 100% conc.

Ritalan C. [519] Isopropyl palmitate and lanolin oil; nonionic; blending agent, emollient, and solubilizer for waxes and other oil-sol. or disp. materials; personal care products; ASTM 2 max. liq.; bland odor; sol. in min. oil; 100% act.

Ritapro 100. [519] Cetearyl alcohol, steareth-20, and steareth-10; nonionic; o/w emulsifier; solid; 100% conc.

Ritapro 200, 300. [519] POE fatty alcohol ether; nonionic; o/w emulsifier for personal care products; flake; 100% conc.

Ritasol. [519] Isopropyl lanolate; nonionic; film former for lip preparations; solid; 100% conc.

Ritawax. [519] Lanolin alcohol; nonionic; emulsifier emollient for skin preparations; solid; 100% conc.

Ritawax 5. [519] Laneth-5; nonionic; emulsifier, solubilizer, emollient for creams and lotions; solid; 100% conc.

Ritawax 15. [519] Laneth-15; nonionic; see Ritawax 5; solid; 100% conc.

Ritawax 40. [519] Laneth-40; nonionic; see Ritawax 5; solid; 100% conc.

Ritawax AEO. [519] Laneth-10 acetate; nonionic; emollient, lubricant, moisturizer, penetrant; liq.; 100% conc.

Ritawax ALA. [519] Cetyl acetate and acetylated lanolin alcohol; nonionic; emollient, lubricant, moisturizer, penetrant; solid; 100% conc.

Ritoleth 2. [519] Oleth-2; nonionic; o/w emulsifier, solubilizer; liq.; HLB 4.9; 100% conc.

Ritoleth 5. [519] Oleth-5; nonionic; see Ritoleth 2; liq.; HLB 9.6; 100% conc.

Ritoleth 10. [519] Oleth-10; nonionic; see Ritoleth 2; semisolid; HLB 12.4; 100% conc.

Ritoleth 20. [519] Oleth-20; nonionic; see Ritoleth 2; semisolid; HLB 15.1; 100% conc.

Roccal 50% Technical. [299] Alkyl dimethyl benzyl ammonium chloride; cationic; disinfectant and sanitizer used in food or beverage processing plants, food service establishments, industrial plants, and farm use (not for use in hospitals); lt. yel. liq.; mild odor; sol. in water, lower alcohols, ketone, and glycols; dens. 8.2 lb/gal; sp.gr. 0.98; pH 6.8–8.0 (10%); 50% act.

Roccal II 50%. [299] Benzalkonium chloride USP; cationic; antimicrobial agent, disinfectant, sanitizer; see Roccal 50% Technical; antistat for fabrics; lt. yel. liq. mild pleasant odor; m.w. 359; sol. in water, lower alcohols, ketone, and glycols; dens. 7.8 lb/gal; sp.gr. 0.94; pH 6.8–8.0 (10%); 50% act.

Roccal MC-14. [299] Tetradecyl benzyl dimethyl ammonium chloride dihydrate; cationic; sanitizer, disinfectant, detergent, germicidal laundry products, swimming pool algicides and slimicides, medicated cosmetics and toiletries; wh. powd.; m.w. 404; sol. in water, ketone, alcohols and glycols; m.p. 60 C min.; pH 7.5 ± 0.5 (10%); 90% act.

Rodea. [144] Ricinoleamide DEA (2:1); emulsifier for aliphatic hydrocarbons; rust inhibitor; formulation of conveyor chain lubricants; dk. amber visc. liq.; sol. in alcohol, glycol, glycol-ether, esters, ketone, aliphatic and aromatic hydrocarbons; disp. in water; sp.gr. 1.016; pH 9.0–10.5 (1%); 4% max. free fatty acid.

Rotolan. [519] Proprietary lanolin deriv.; aids in pigment dispersion for use in inks; paste; 100% conc.

RS-55-9. [287] POE (9) stearate; nonionic; emulsifier o/w for cosmetics, pharmaceuticals; textile industry; spinning oils; wetting agent for deinking; coemulsifier for emulsion polymerization; solid; HLB 11.0; 100% conc.

RS-55-40. [287] POE (40) monostearate; nonionic; emulsifier o/w for cosmetics and pharmaceuticals; glass surface finishing agent; solid; HLB 17.0; 100% conc.

RS-55-100. [287] POE (100) monostearate; nonionic; o/w emulsifier for cosmetics and pharmaceuticals; binder for additives; solid; HLB 18.5; 100% conc.

Rueterg 97-G. [237] Alkylaryl sulfonate, amine salt; detergent base for laundry, dishwashing, car washes; liq.; 95% conc.

Rueterg 97-S. [237] Alkylaryl sulfonate, amine salt; anionic; detergent base for laundry, dishwashing and car washes; liq.; 95% conc.

Rueterg 97-T. [237] TEA-dodecylbenzene sulfonate; anionic; household and industrial detergent base; liq.; 95% conc.

Rueterg IPA-HP. [237] Alkylaryl sulfonate, amine salt; anionic; detergent and car wash, dry cleaning; liq.; 96% conc.

Rueterg SA. [237] Alkylaryl sulfonic acid; anionic; intermediate for detergents and emulsifiers; liq.; 97% conc.

Rychem 17. [533] Polyglycol ester; nonionic; emulsifier for fat and min. oils; lubricant, softener, and rewetting agent; liq.; 90–95% conc.

Rychem 21, 33. [182; 533] Polyglycol ester; nonionic; oil emulsifier used in textile, leather and other industrial uses; liq.; oil-sol. (21); 95% conc.

Rycofax 618. [182; 533] Fatty amido-amine salt; cationic; softener/debonder; release agent for tissue mfg.; liq.; 98% conc.

Rycofax 3115. [533] Amine deriv.; cationic; softener and antistat for fabrics; liq.; 99% conc.

Rycofax O. [533] PEG ester; nonionic; wetting and rewetting in pulp/paper and wet process industries such as textile, metal; emulsifier for oils and hydrocarbons; liq.; 100% conc.

Rycomid 2120. [533] Fatty alkanolamide; nonionic; detergent used for floor, textile, and metal cleaners; liq.; 100% conc.

Ryoto Sugar Ester LWA-1540. [534] Sucrose monolaurate; nonionic; o/w and w/o emulsifier, softener, conditioner, and aerating agent in foods; paste; HLB 15.0; 40% conc.

Ryoto Sugar Ester OWA-1570. [534] Sucrose monooleate; nonionic; see Ryoto Sugar Ester LWA-1540; paste; HLB 15.0; 40% conc.

Ryoto Sugar Ester P-1570, -1670. [534] Sucrose monopalmitate; nonionic; see Ryoto Sugar Ester LWA-1540; powd.; HLB 15.0 and 16.0 resp.; 100% conc.

Ryoto Sugar Ester S-170. [534] Sucrose polystearate; nonionic; see Ryoto Sugar Ester LWA-1540; powd.; HLB 1.0; 100% conc.

Ryoto Sugar Ester S-270, -370. [534] Sucrose ditristearate; nonionic; see Ryoto Sugar Ester LWA-1540; powd.; HLB 2.0 and 2.3 resp.; 100% conc.

Ryoto Sugar Ester S-570, -770. [534] Sucrose distearate; nonionic; see Ryoto Sugar Ester LWA-1540; powd.; HLB 5.0 and 7.0 resp.; 100% conc.

Ryoto Sugar Ester S-970, -1170. [534] Sucrose monodistearate; nonionic; see Ryoto Sugar Ester LWA-1540; powd.; HLB 9.0 and 11.0 resp.; 100% conc.

Ryoto Sugar Ester S-1570, -1670. [534] Sucrose monostearate; nonionic; see Ryoto Sugar Ester LWA-1540; powd.; HLB 15.0 and 16.0 resp.; 100% conc.

S

Sactol 2 OS 2. [367] Sodium laureth sulfate; anionic; base for detergents and shampoos; liq.; 50% conc.

Sactol 2 OS 28. [367] Sodium laureth sulfate; anionic; see Sactol 2 OS 2; liq.; 28% conc.

Sactol 2 OT. [367] TEA laureth sulfate; anionic; see Sactol 2 OS 2; liq.; 35% conc.

Sactol 2 S 3. [367] Sodium lauryl sulfate; anionic; see Sactol 2 OS 2; paste; 49% conc.

Sactol 2 T. [367] TEA lauryl sulfate; anionic; see Sactol 2 OS 2; liq.; 35% conc.

Sanac C. [66] 2-(2-carboxyethoxy) ethyl 2-(2-(2-hydroxy) ethoxylethyl methyldodecyl ammonium methyl sulfate, potassium salt; amphoteric; detergent, wetting agent, foamer, solubilizer, heavy duty surface cleaner, metal treatment; straw to amber semivisc. liq.; sol. in water and polar solvs.; dens. 8.7 lb/gal; sp.gr. 1.05; pH 7.5–8.5; 35% act.

Sanac S. [66] [2-(2-carboxyethoxy) ethyl] [2-[2-(hydroxy) ethoxyethyl] methyloctadecenyl ammonium methyl sulfate, potassium salt; see Sanac C; straw to amber semivisc. liq.; mild typ. odor; water-sol.; dens. 8.7 lb/gal; sp.gr. 1.05; pH 7.5–8.5; 35% act.

Sanac T. [111] Complex carboxylated tallow quat.; amphoteric; detergent, wetting agent and foamer, solubilizer; liq.; 60% conc.

Sandet 60. [539] Sodium dodecylbenzene sulfonate; anionic; detergent base and emulsifier for emulsion polymerization of vinyl resin; liq.

Sandet ALH. [539] Alkyl diphenyl ether sodium disulfate; anionic; wetting agent, penetrant and base material for making liq. bleaching detergents; liq.; 30% conc.

Sandet EN. [539] Sodium alkyl ether sulfate; anionic; shampoo base; liq.; 31.5% conc.

Sandet END, ENM. [539] Sodium alkyl ether sulfate; anionic; see Sandet EN; liq.; 26.5% conc.

Sandobet SC. [536] Cocoamido sulfo betaine; amphoteric; cosmetic and toiletry applications; liq.; 50% conc.

Sandopan 2N. [538] Polyglycol ether, modified; anionic/nonionic; scouring agent and emulsifier for min. oils; liq., solid.

Sandopan CBH Paste. [537] Sulfonate EO adduct, modified; anionic; detergent, emulsifier, wetting agent used for applications in the fabric industry; yel. paste; mild odor; misc. with water; 51% act.

Sandopan DTC. [537] Sodium trideceth-7 carboxylate; anionic; detergent, wetting agent; cosmetics, household, specialty, industrial and institutional cleaner; dishwashing detergent; clear gel; water-sol.; sp.gr. 1.06; visc. 5000 cps; b.p. 220 F; 68% act.

Sandopan DTC Linear P Acid. [536] C_{12-15} pareth-7 carboxylic acid; anionic; surfactant for industrial, personal care, and household use; liq.; 90% conc.

Sandopan KD Conc. [538] Sodium alkyl sulfate, modified; anionic; scouring, milling, wetting agent and dispersant used in textiles; paste.

Sandopan KD (U) Liquid. [537] Sodium alkyl sulfate, modified; anionic; detergent; scouring of fibers; fulling of wool; clear liq.; mild odor; misc. with water; sp.gr. 1.02; 21% act.

Sandopan LF Liquid. [537] EO adduct; nonionic; detergent, emulsifier for scouring/dyeing and prescouring of polyester knits; clear liq.; misc. with water; mild odor; sp.gr. 1.02; pH 6.0; 25% act.

Sandopan LFW. [538] EO adduct; nonionic; detergent and wetting agent for polyester and cellulosic fibers; liq.; 100% conc.

Sandopan TFL Conc. [536; 537] Oleoamphohydroxy propylsulfonate; amphoteric; detergent, cosmetic and toiletries products; leather industry; amber gel; mild fatty odor; water-sol.; sp.gr. 1.14; b.p. 220 F; 48% conc.

Sandopur DK. [538] Fatty alkyl polyglycol ether, modified; amphoteric; washing-off agent for textiles; liq., paste.

Sandoxylate 206. [536] Alkoxylated alcohol; nonionic; wetting agent, dispersant used in cleaning and treatment products and processes; liq.; HLB 12–14; 100% conc.

Sandoxylate 224. [536] Alkoxylated alcohol; nonionic; see Sandoxylate 206; semisolid; HLB 16–17; 100% conc.

Sandoxylate 408. [536] Alkoxylated alcohol; nonionic; see Sandoxylate 206; liq.; HLB 10–12; 100% conc.

Sandoxylate 412. [536] Alkoxylated alcohol; nonionic; see Sandoxylate 206; liq.; HLB 12–14; 100% conc.

Sandoxylate 418. [536] Alkoxylated alcohol; nonionic; see Sandoxylate 206; liq.; HLB 16.0; 100% conc.

Sandoxylate 424. [536] Alkoxylated alcohol; nonionic; see Sandoxylate 206; semisolid; HLB 16.0; 100% conc.

Sandoxylate AC-9. [536] Alcohol ethoxylate; nonionic; see Sandoxylate 206; solid; HLB 12.0; 100% conc.

Sandoxylate AC-24. [536] Alcohol ethoxylate; nonionic; see Sandoxylate 206; solid, flakes; HLB 15.8; 100% conc.

Sandoxylate AD-4. [536] Alcohol ethoxylate; nonionic; see Sandoxylate 206; liq.; 100% conc.

Sandoxylate AD-6. [536] Alcohol ethoxylate; nonionic; see Sandoxylate 206; liq.; 100% conc.

Sandoxylate AD-9. [536] Alcohol ethoxylate; nonionic; see Sandoxylate 206; liq.; 100% conc.

Sandoxylate AL-4. [536] Alcohol ethoxylate; nonionic; see Sandoxylate 206; liq.; 100% conc.

Sandoxylate AO-12. [536] Alcohol ethoxylate; nonionic; see Sandoxylate 206; liq.; HLB 12.7; 100% conc.

Sandoxylate AO-20. [536] Alcohol ethoxylate; nonionic; see Sandoxylate 206; semisolid; HLB 15.4; 100% conc.

Sandoxylate AO-60. [536] Alcohol ethoxylate; nonionic; see Sandoxylate 206; solid; HLB 18.9; 100% conc.

Sandoxylate AT-6.5. [536] Alcohol ethoxylate; nonionic; see Sandoxylate 206; liq.; HLB 11.5; 100% conc.

Sandoxylate AT-12. [536] Alcohol ethoxylate; nonionic; see Sandoxylate 206; liq.; HLB 14.5; 100% conc.

Sandoxylate C-10. [536] Castor oil ethoxylate; nonionic; see Sandoxylate 206; liq.; 100% conc.

Sandoxylate C-15. [536] Castor oil ethoxylate; nonionic; see Sandoxylate 206; liq.; HLB 13.5; 100% conc.

Sandoxylate C-32. [536] Castor oil ethoxylate; nonionic; see Sandoxylate 206; liq.; HLB 15.0; 100% conc.

Sandoxylate FO-9. [536] Fatty acid ethoxylate; nonionic; surfactant, wetting agent, dispersant used in cleaning and treatment products and processes; liq.; HLB 11.9; 70% conc.

Sandoxylate FO-30/70. [536] Fatty acid ethoxylate; nonionic; see Sandoxylate FO-9; liq.; HLB 13.9; 100% conc.

Sandoxylate FS-9. [536] Fatty acid ethoxylate; nonionic; see Sandoxylate FO-9; solid; HLB 11.5; 100% conc.

Sandoxylate FS-35. [536] Fatty acid ethoxylate; nonionic; see Sandoxylate FO-9; solid; HLB 16.9; 100% conc.

Sandoxylate NC-5. [536] Amine ethoxylate; nonionic; see Sandoxylate FO-9; liq.; 100% conc.

Sandoxylate NC-15. [536] Amine ethoxylate; nonionic; see Sandoxylate FO-9; liq.; 100% conc.

Sandoxylate NSO-30. [536] Amine ethoxylate; nonionic; see Sandoxylate FO-9; liq.; 100% conc.

Sandoxylate NT-5. [536] Amine ethoxylate; nonionic; see Sandoxylate FO-9; liq., paste; 100% conc.

Sandoxylate NT-15. [536] Amine ethoxylate; nonionic; see Sandoxylate FO-9; liq., paste; 100% conc.

Sandoxylate PDN-7. [536] Amine ethoxylate; nonionic; see Sandoxylate FO-9; liq.; HLB 9.9; 100% conc.

Sandoxylate PN-6. [536] Amine ethoxylate; nonionic; see Sandoxylate FO-9; liq.; 100% conc.

Sandoxylate PN-9. [536] Alkyl phenol ethoxylate; nonionic; see Sandoxylate FO-9; liq.; HLB 12.9; 100% conc.

Sandoxylate PN-10.5. [536] Alkyl phenol ethoxylate; nonionic; see Sandoxylate FO-9; liq.; 100% conc.

Sandoxylate PO-5. [536] Alkyl phenol ethoxylate; nonionic; see Sandoxylate FO-9; liq.; HLB 10.4; 100% conc.

Sandoxylate SX-208. [537] PO/EO alkoxylate; nonionic; wetting agent for household and industrial applications, textile wet processing; intermediate for anionic surfactants; Gardner < 1 liq.; water-sol.; sp.gr. 0.9625; HLB 11; cloud pt. 22 ± 2 C (1% aq.); pH 6.5 ± 1 (1%); surf. tens. 28.0 dynes/cm; 98 ± 2% act.

Sandoxylate SX-224. [537] PO/EO alkoxylate; nonionic; see Sandoxylate SX-208; Gardner < 1 liq.; water-sol.; sp.gr. 1.0262; HLB 15; cloud pt. 93 ± 2 C (1% aq.); pH 6.5 ± 1 (1%); surf. tens. 30.2 dynes/cm; 98 ± 2% act.

Sandoxylate SX-408. [537] PO/EO alkoxylate; nonionic; see Sandoxylate SX-208; Gardner < 1 liq.; water-sol.; sp.gr. 0.9650; HLB 11; cloud pt. 22 ± 2 C (1% aq.); pH 6.5 ± 1 (1%); surf. tens. 28.8 dynes/cm; 98 ± 2% act.

Sandoxylate SX-412. [537] PO/EO alkoxylate; nonionic; see Sandoxylate SX-208; Gardner < 1 liq.; water-sol.; sp.gr. 0.9822; HLB 13; cloud pt. 38 ± 2 C (1% aq.); pH 6.5 ± 1 (1%); surf. tens. 28.6 dynes/cm; 98 ± 2% act.

Sandoxylate SX-418. [537] PO/EO alkoxylate; nonionic; see Sandoxylate SX-208; Gardner < 1 liq.; water-sol.; sp.gr. 1.007; HLB 15; cloud pt. 65 ± 2 C (1% aq.); pH 6.5 ± 1 (1%); surf. tens. 29.4 dynes/cm; 98 ± 2% act.

Sandoxylate SX-424. [537] PO/EO alkoxylate; nonionic; see Sandoxylate SX-208; Gardner < 1 solid; water-sol.; sp.gr. 1.0267; HLB 15; cloud pt. 89 ± 2 C (1% aq.); pH 6.5 ± 1 (1%); surf. tens. 30.3 dynes/cm; 98 ± 2% act.

Sandoz Amide CO. [537] Coconut alkanolamide (2:1); detergent, foamer, wetting and rewetting agent, visc. modifier used in metal cleaning, heavy duty detergents, hard surface cleaning, textile wet processing; amber liq.; 97% act.

Sandoz Amide NP. [537] Coconut alkanolamide (2:1); see Sandoz Amide CO; amber liq.; 97% act.

Sandoz Amide NT. [536] Alkanolamide; nonionic; surfactant used as a detergent, emulsifier and visc. aid for household care and cosmetic applications; liq.; 100% conc.

Sandoz Amide PE. [537] Lauryl alkanolamide (1:1); detergent, foamer, wetting and rewetting agent, visc. modifier used for liq. dishwashing products and personal care products; wh. solid; 90–95% act.

Sandoz Amide PL. [537] Lauryl alkanolamide (2:1); see Sandoz Amide PE; amber liq.; 100% act.

Sandoz Amide PN. [537] Coconut alkanolamide (2:1); detergent, foamer, wetting and rewetting agent, visc. modifier used for heavy duty cleaning formulations, oil emulsification; amber liq.; 97% act.

Sandoz Amide PO. [537] Coconut alkanolamide (1:1); detergent, foamer, wetting and rewetting agent, visc. modifier used for personal care products and cleaning formulations; lt. yel. visc. liq.; 85% act.

Sandoz Amide PS. [537] Coconut alkanolamide (2:1); detergent, foamer, wetting and rewetting agent, visc. modifier used in hard surface cleaners, detergents, dishwashing compds.; textile scouring; amber liq.; 97% act.

Sandoz Amine Oxide XA-C. [537] Cetyl amine oxide; nonionic; surfactant, foam booster, visc. improver for personal care and liq. dishwashing formulations; antistat; textile wet processing; softener and emollient; amber liq.; sol. in water, lower alcohols, glycols; biodeg.; 20% act.

Sandoz Amine Oxide XA-L. [537] Lauryl amine oxide; nonionic; surfactant, foam booster, visc. improver for personal care and liq. dishwashing formulations; antistat, textile wet processing; amber liq.; sol. in water, lower alcohols, glycols; biodeg.; 30% act.

Sandoz Amine Oxide XA-M. [537] Myristyl amine oxide; nonionic; surfactant; see Sandoz Amine Oxide XA-L; amber liq.; sol. in water, lower alcohols, glycols; biodeg.; 30% act.

Sandozin AM. [538] Alkylaryl sulfonate; anionic; wetting agent, textile products; paste.

Sandozin NE. [538] Fatty acid deriv., sulfonated; anionic; wetting agent for bleaching in textile applications; liq.

Sandozin NIT. [538] EO adduct; nonionic; wetting agent and detergent for textiles; liq.

Sandozol KB. [538] Sulfonated oil; anionic; finishing and sanforizing oil; liq.

Sandoz Phosphorester 510. [537] Complex phosphate ester; anionic; detergent, emulsifier in waterless hand cleaners, dry cleaning detergents; antistat, corrosion inhibitor in textile fiber lubricants; primary emulsifier in emulsion polymerization of PVAc, acrylic; clear visc. liq.; water-disp.; sol. in conc. electrolyte systems; pH 1.5–2.5 (10%); 99% solids.

Sandoz Phosphorester 600. [537] Complex phosphate ester; hydrotrope; solubilizer for nonionic surfactants; wetting agent used in acid and solv. cleaners; emulsifier for solvs.; clear visc. liq.; sol. in conc. electrolyte systems; pH 1.5–2.5 (10%); 99% solids.

Sandoz Phosphorester 610. [537] Complex phosphate ester; emulsifier for aromatic and chlorinated solvs.; solubilizer for nonionic surfactants; emulsion polymerization surfactant used in vinyl acetate and acrylic systems; clear visc. liq.; sol. in conc. electrolyte systems; pH 1.5–2.5 (10%); 99% solids.

Sandoz Phosphorester 690. [537] Complex phosphate ester; emulsifier for emulsion polymerization; household and industrial cleaners; soft waxy paste; pH 1.5–2.5; 95% solids.

Sandoz Sulfate 216. [537] Ammonium laureth sulfate (3 EO); foamer and visc. modifier for personal care products; basic ingred. in shampoos and liq. dishwashing products; textile scouring; emulsifier and suspending agent in fabrics; pale yel. liq.; visc. 1000 cps max.; cloud pt. 15 C; pH 6.5–7.5 (10%); 58.5–60% act.

Sandoz Sulfate 219. [537] Sodium laureth sulfate (3 EO); foamer and visc. modifier in personal care products; textile scouring; pale yel. liq.; visc. 1000 cps max.; cloud pt. 15 C; pH 7.5–8.5 (15%); 58.5–60% act.

Sandoz Sulfate 219 Special. [536] Laureth sulfate, sodium salt; anionic; surfactant used in personal and cosmetic products; liq.; 30% conc.

Sandoz Sulfate 830. [536] Sodium N-octyl sulfate; anionic; surfactant for household and industrial cleaning formulations; liq.; 30% conc.

Sandoz Sulfate 1030. [536] Sodium decyl sulfate; anionic; see Sandoz Sulfate 830; liq.; 30% conc.

Sandoz Sulfate A. [537] Lauryl sulfate, ammonium salt; foamer and visc. modifier in personal care products; basic ingred. in shampoos and liq. dishwashing products; lt. yel. liq.; visc. 6500 cps; cloud pt. 5 C; pH 6.3–7.0 (10%); 28–30% act.

Sandoz Sulfate EP. [537] DEA-lauryl sulfate; foamer and visc. modifier for personal care products; lt. yel. liq.; visc. 225 cps; cloud pt. 6 C; pH 7.5–8.5 (10%); 35–37.5% act.

Sandoz Sulfate ES-3. [537] Sodium laureth sulfate (3 EO); foamer and visc. modifier for personal care products; textile scouring; wetting agent; emulsifier and suspending agent in fabrics; pale yel. liq.; visc. 1000 cps max.; cloud pt. 15 C; pH 7.5–8.5 (10%); 28–30% act.

Sandoz Sulfate K. [537] Mixed alkyl sulfate, ammonium salt; see Sandoz Sulfate A; brn. liq.; visc. 1000 cps max.; cloud pt. 15 C; pH 6.5–7.0 (10%); 45% act.

Sandoz Sulfate TL. [537] TEA-lauryl sulfate; see Sandoz Sulfate EP; lt. yel. liq.; visc. 225 cps; cloud pt. 2.5 C; pH 6.4–7.7 (10%); 38.5–42.5% act.

Sandoz Sulfate W2-30. [536] Sodium 2-ethylhexyl sulfate; anionic; see Sandoz Sulfate 830; liq.; 30% conc.

Sandoz Sulfate WA-9. [536] Sodium lauryl sulfate; anionic; used in oral products; powd.; 97% conc.

Sandoz Sulfate WA Dry. [537] sodium lauryl sulfate; foamer and visc. modifier for personal care, dental, and household products; wh. powd.; pH 6.5–7.5 (10%); 90–96% act.

Sandoz Sulfate WAG. [537] Sodium lauryl sulfate; see Sandoz Sulfate WA Dry; wh. needles; pH 6.5–7.5 (10%); 90–96% act.

Sandoz Sulfate WAS. [537] Sodium lauryl sulfate; see Sulfate WA Special; lt. yel. liq.; visc. 1000–3000 cps; cloud pt. 17–23 C; pH 7.4–8.6 (10%); 27.5–30.5% act.

Sandoz Sulfate WA Special. [537] Sodium lauryl sulfate; foamer, visc. modifier for personal care, household products; wh. paste; visc. 1000–5000 cps; cloud pt. 17–21 C; pH 7.4–8.6 (10%); 27.5–30.5% act.

Sandoz Sulfate WE. [537] Sodium laureth sulfate (3.5 EO); foamer and visc. modifier for personal care products; textile scouring; colorless liq.; visc. 750 cps; cloud pt. 13 C; pH 7.8–8.5 (10%); 27–29% act.

Sandoz Sulfonate 2A1. [537] Dodecyl diphenyl ether sulfonate, sodium salt; anionic; solubilizer; household detergents, industrial, disinfectant, and metal cleaners, emulsifier in emulsion polymerization; textile aux.; also avail in acid form; yel./brn. liq.; sol. in electrolyte sol'ns., caustic, HCl, TKPP; sp.gr. 1.16; visc. 145 cps; 45% act.

Sandoz Sulfonate 3B2. [537] Decyl diphenyl ether sulfonate, sodium salt; anionic; wetting agent; agric. formulations; spreader and penetrant for farm usage; also avail. in acid form; amber liq.; sol. see Sandoz Sulfonate 2A1; sp.gr. 1.16; visc. 120 cps; 45% act.

Sandoz Sulfonate AAS 35S. [537] Linear alkyl (C_{12}) benzene sulfonate, sodium salt; anionic; wetting agent and foamer for household and industrial detergent systems; hard surface cleaners; liq.; biodeg.; 31.5% act.

Sandoz Sulfonate AAS 40FS. [537] Linear alkyl (C_{12}) benzene sulfonate, sodium salt; anionic; see Sandoz Sulfonate AAS 35S; powd.; 40% act.

Sandoz Sulfonate AAS 45S. [537] Linear alkyl (C_{12}) benzene sulfonate, sodium salt; anionic; see Sandoz Sulfonate AAS 35S; also for textile wet processing; liq.; 41-43% act.

Sandoz Sulfonate AAS 50MS. [537] Linear alkyl (C_{12}) benzene sulfonate, ammonium salt; anionic; see Sandoz Sulfonate AAS 35S; also used in dishwashing liq.; metal cleaning; liq.; 40% act.

Sandoz Sulfonate AAS 60S. [537] Linear alkyl (C_{12}) benzene sulfonate, triethanolamine salt; anionic; see Sandoz Sulfonate AAS 35S; also liq. detergents, personal and animal care products; textile wet processing; liq.; 60% act.

Sandoz Sulfonate AAS 70S. [537] Linear alkyl (C_{12}) benzene sulfonate, sodium salt; anionic; see Sandoz Sulfonate AAS 35S; also for liq. detergents, metal and hard surface cleaners; liq.; 60% act.

Sandoz Sulfonate AAS 75S. [537] Linear alkyl (C_{12}) benzene sulfonate, calcium salt; anionic; see Sandoz Sulfonate AAS 35S; also for agric. products; liq.; oil-sol.; 60% act.

Sandoz Sulfonate AAS 90. [537] Linear alkyl (C_{12}) benzene sulfonate, sodium salt; anionic; see Sandoz Sulfonate AAS 35S; powd.; 85% act.

Sandoz Sulfonate AAS 98S. [537] DDBSA; anionic; conc. which is neutralized to any of a variety of salts as required; liq.; 96–98% act.

Sandoz Sulfonate AAS-Special 3. [537] Linear alkyl (C_{12}) benzene sulfonate, isopropylamine salt; anionic; surfactant with good detergency; metal cleaning, dry cleaning systems; liq.; oil-sol.; 97% min. act.

Sanmorin 11. [539] POE alkyl ether; nonionic; penetrant and wetting agent for textile processing; liq.

Sanmorin OT 70. [539] Sodium dioctyl sulfosuccinate; anionic; wetting agent and penetrant for industrial uses; liq.; 70% conc.

Sanstat 2012-A. [539] Sodium dioctyl sulfosuccinate; cationic; antistat for plastics; liq.

Santelle-EOM K. [210] Mono- and diglycerides, ethoxylated; nonionic; food emulsifier; emulsion stabilizer; dispersant aid; paste; m.p. 80–85 F; HLB 13.1; 100% act.

Santone 3-1-S. [210] Triglycerol monostearate; emulsifier; aeration of lipid systems; solubilizer; color/flavor dispersant in water; cream bead; m.p. 128–130 F; HLB 7.2; sapon. no. 122–139; 27–34% total mono- and diglycerides.

Santone 3-1-SH. [210] Polyglyceryl-3 oleate; emulsifier and aerating agent used in food industry, textile and plastic lubricant; color dispersant; HLB 7.2; sapon no. 115–135.

Santone 8-1-O. [210] Polyglyceryl-8 oleate; emulsifier for personal care products; visc. reducer; emulsion stabilizer; beverage clouding agent; HLB 13.0; sapon. no. 75–85.

Santone 8-1-S. [210] Polyglyceryl-8 stearate; emulsifier for personal care products; whipping agent; emulsion stabilizer; HLB 13.0; sapon. no. 110–119.

Santone 10-10-0. [210] Polyglycerol-10-decaoleate; food emulsifier; solubilizer; emulsion stabilizer; dispersant aid in flavors; liq.; HLB < 1.0; sapon. no. 165–180.

Sapogenat T Brands. [301] Tributyl phenol polyglycol ether (4–50 EO); surface act. basic material; 100% conc.

Sarkosyl L. [136] Lauroyl sarcosine; anionic; detergent, corrosion inhibitor, foam booster and stabilizer, wetting agent, lubricant, emulsifier used dentifrices, personal care, and household cleaning products, pharmaceuticals, metal processing and finishing, metalworking and cutting oils; powd.; m.w. 264–285; sol. in org. solvs.; insol. in water; sp.gr. 0.969; m.p. 35–37 C; 94% min. purity.

Sarkosyl LC. [136] Cocoyl sarcosine; see Sarkosyl L; powd.; m.w. 285–300; sol. in org. solvs.; insol. in water; sp.gr. 0.969; m.p. 23–28 C; 94% min. purity.

Sarkosyl NL-30. [136] Sodium lauroyl sarcosinate; see Sarkosyl L; dentifrices, shampoos, rug and window cleaners and fine fabric detergents; colorless liq.; pH 7.5–8.5 (1%); 30% act.

Sarkosyl O. [136] Oleoyl sarcosine; see Sarkosyl L; powd.; m.w. 340–360; sol. in org. solvs.; insol. in water; sp.gr. 0.948; 94% min. purity.

Satexlan. [159] PEG-20 hydrog. lanolin; nonionic; o/w emulsifier, emollient; perfume solubilizer; Gardner 4 max.; sol. in water and aq./alcoholic sol'ns.; m.p. 42–50 C; sapon no. 7 max.

Satulan. [160] Hydrog. lanolin; nonionic; w/o emulsifier, emollient; solid; HLB 1.0; 100% conc.

Schercamox C-AA. [541] Cocoamido propylamine oxide; nonionic; detergent, wetting agent; foamer; skin conditioner in personal care products; antistat; foam stabilizer in dishwashing liqs.; water-wh. clear to hazy liq.; m.w. 320; sol. in water and hydrophilic solvs.; sp.gr. 0.986; pH 7.0 ± 0.5; 30% act.

Schercamox CMA. [541] Dihydroxyethyl cocoamine oxide; nonionic; detergent, foam stabilizer, conditioner, emollient, lubricant, and visc. builder in personal care products; water-wh. clear liq.; mild odor; m.w. 301; sol. see Schercamox C-AA; sp. gr. 0.99 ± 0.05; pH 7.0 ± 0.5 (1%); 40% act.

Schercamox DMA. [541] Myristamine oxide; nonionic; detergent, wetting agent; surfactant, foam booster in personal care products and liq. dishwashing products; water-wh. clear liq.; mild odor; m.w. 256; sol. see Schercamox C-AA; pH 7.0 ± 1.0; 30% act.

Schercamox DML. [541] Lauramine oxide; nonionic; wetting agent, foamer and emulsifier used in personal care products; water-wh. clear liq.; mild odor; m.w. 235; sol. in water and hydrophilic solvs.; sp.gr. 0.99 ± 0.01; pH 7.0 ± 1.0 (1%); 30% act.

Schercamox DMM. [541] Myristamine oxide; nonionic; detergent, wetting agent and foamer; emulsifier, antistat, stabilizer, visc. builder used in personal care products and liq. dishwashing products; water-wh. clear liq.; mild odor; m.w. 263; sol. in water and hydrophilic solvs.; sp.gr. 0.98 ± 0.05; pH 7.0 ± 1.0; 30% act.

Schercamox DMS. [541] Stearamine oxide; nonionic; detergent, softener, visc. builder, foam stabilizer, emollient, and conditioner used in personal care products; off-wh. to wh. paste; typ. odor; m.w. 326; sol. in alcohols, glycols, triols, polyols, glycol ethers; sp.gr. 0.90; pH 7.0 ± 1.0 (1%); 25% act.

Schercamox T-12. [541] Dihydroxyethyl tallowamine oxide; wetting agent, softener, and foam stabilizer in dishwashing compds., shampoos and hair rinses; clear liq.; 49% min. amine oxide.

Schercemol DEGMS. [541] PEG-2 stearate; primary emulsifier in creams and lotions; wh. to cream solid; m.p. 48 C; sapon. no. 165–175; 100% conc.

Schercemol DEIS. [541] Decyl isostearate; lubricant, dispersant, penetrant, coupler, and moisturizer for personal care products; yel. clear liq.; char. odor; m.w. 451; sol. in hydrophobic solvs.; insol. in water; sp.gr. 0.87; dens. 7.25 lb/gal; sapon. no. 120–135; ref. index 1.4580.

Schercemol DID. [541] Diisopropyl dimerate; ingred. and emollient in personal care products; dispersant in pigmented products; yel. hazy liq.; sapon. no. 155–185.

Schercemol DIS. [541] Diisopropyl sebacate; emollient, solubilizer, and coupler in personal care products; colorless clear liq.; sapon. no. 380–400.

Schercemol EGMS. [541] Glycol stearate; nonionic; emulsifier, opacifier, pearling agent, thickener, and visc. controller for personal care products; wh. flakes; m.w. 312; sol. in org. solvs.; insol. in water; m.p. 55–60 C; sapon. no. 170–190; 100% conc.

Schercemol GMIS. [541] Glycerol isostearate; nonionic; emulsifier and emollient for creams and lotions; lt. amber clear liq., soft solid; typ. odor; m.w. 385; sol. in org. solvs.; insol. in water; sp.gr. 0.960; dens. 8.0 lb/gal; sapon. no. 150–170; ref. index 1.4715; flash pt. (OC) > 170 C; 100% conc.

Schercemol GMS. [541] Glycerol stearate; nonionic; emulsifier, opacifier, stabilizer, and thickener in personal care products; wh. to cream flakes; m.w. 342; sol. in org. solvs.; insol. in water; m.p. 55–60 C; sapon. no. 160–176; 100% conc.

Schercemol PGML. [541] Propylene glycol laurate; cosmetic base, emulsion stabilizer, solvent, solubilizer, coupling agent for perfumes, flavors, liq. soaps, shampoos, pesticides; defoamer, plasticizer/stabilizer for vinyls; Gardner 5 max. clear liq.; mild odor; m.w. 258; sol. in most org. solv.; sp.gr. 0.905 ± 0.01; dens. 7.45 lb/gal; f.p. 5 C max.; acid no. 5 max.; sapon. no. 225-240; ref. index 1.4425; pH 7 (10% disp.); flash pt. 160 C min. (OC).

Schercemol PGMS. [541] Propylene glycol stearate; nonionic; emulsifier, opacifier, thickener, visc. controller, and stabilizer for personal care products; wh. to cream solid; typ. odor; m.w. 326; sol. see Schercemol GMS; m.p. 35 C; sapon. no. 180 ± 5; flash pt. (OC) > 170 C; 100% conc.

Schercodine B. [541] Behenamidopropyl dimethylamine; cationic; emulsifier for skin and hair preparations; tan hard wax; m.w. 394; m.p. 63–67 C; 98% min. amide.

Schercodine C. [541] Cocamidopropyl dimethylamine; cationic; surfactant for hair and bath preparations; tan soft solid; m.w. 304; 98% min. amide.

Schercodine L. [541] Lauramidopropyl dimethylamine; cationic; surfactant, intermediate for betaine amphoterics; lt. tan solid; m.w. 284; m.p. 48–52 C; 98% min. amide.

Schercodine M. [541] Myristamidopropyl dimethylamine; cationic; o/w emulsifier, conditioner, and visc. builder; lt. tan wax; m.w. 312; m.p. 52–56 C; 98% min. amide.

Schercodine O. [541] Oleamidopropyl dimethylamine; emollient, conditioner for hair and skin preparations; amber liq.; m.w. 366; 98% min. amide.

Schercodine P. [541] Palmitamidopropyl dimethylamine; cationic; conditioner and emulsifier in creams, lotions, and rinses; solid; 100% conc.

Schercodine S. [541] Stearamidopropyl dimethylamine; softener, emulsifier, and conditioner for hair and skin preparations; tan hard wax; m.w. 366; m.p. 65–70; 98% min. amide.

Schercodine T. [541] Tallamidopropyl dimethylamine; conditioner for cationic emulsions; substantive thickener; amber liq.; 98% min. amide.

Schercomid 1-102. [541] Cocamide DEA; nonionic; thickener, foam booster, and emulsifier for cosmetic and detergent applications; yel. clear visc. liq.; mild odor; sol. in water and org. solvs.; sp.gr. 0.99; flash pt. > 180 C (OC); 100% act.

Schercomid 304. [541] Coco amide, modified; nonionic; dry-cleaning detergent; liq.; 98% act.

Schercomid 1214. [541] Lauramide DEA; nonionic/anionic; foam booster, stabilizer for detergent compositions; thickener, visc. builder for personal care products, hard surface cleaners; yel. clear liq.; typ. odor; sol. in water and org. solvs.; sp.gr. 1.01; dens. 8.4 lb/gal; flash pt. > 170 C (OC); 100% act.

Schercomid AME. [541] Acetamide MEA; nonionic; solubilizer, humectant, skin and hair conditioner, intermediate, coupling agent, pigment dispersant; Gardner 2.0 max. clear liq.; mild organoleptic odor; sol. in alcohols, glycols, diols, polyols, glycol ethers, ketones, water; sp.gr. 1.120; dens. 9.3 lb/gal; ref. index. 1.4700; pH 6.0–8.5 (50% aq.); flash pt. > 180 C (OC); 95% min. act.

Schercomid AME-70. [541] Acetamide MEA; nonionic; see Schercomid AME; Gardner 2.0 max. clear liq.; mild organoleptic odor; sol. in alcohols, glycols, diols, triols, polyols, glycol ethers and water; sp.gr. 1.100 ± 0.01; dens. 9.2 lb/gal; ref. index. 1.4395; pH 6.0–8.5; flash pt. > 180 C (OC); 70% min. act.

Schercomid CCD. [541] Cocamide DEA; nonionic/anionic; soil dispersant and suspending agent, detergent base used in floor cleaners, hand dishwashing compds., car waxes, personal care products, industrial, household and hard surface cleaners, agric. sprays, lubricating oils; dk. amber liq.; mild odor; sol. in water and org. solvs.; sp.gr. 0.99 ± 0.01; dens. 8.3 lb/gal; pH 9.9 ± 0.5 (10%); flash pt. > 170 C (OC); 100% act.

Schercomid CDA. [541] Cocamide DEA; nonionic/anionic; see Schercomid CCD; dk. amber visc. liq.; sol. see Schercomid CCD; sp.gr. 1.01 ± 0.01; dens. 8.35 lb/gal; visc. 1000 cps min.; pH 9.6 ± 0.5 (10%); flash pt. > 170 C (OC); 100% act.

Schercomid CDA-H. [541] Cocamide DEA; nonionic/anionic; thickener, foam booster and stabilizer, detergent, lime soap dispersant used in cosmetics; emulsifier for hydrocarbons; solv. cleaner for grease and tar; lubricant for wool oil; yel. clear liq.; slight odor; sol. see Schercomid CCD; sp.gr. 1.01 ± 0.01; dens. 8.4 lb/gal; visc. 1100 cps min.; pH 10 ± 0.5 (10% aq.).

Schercomid CDO-Extra. [541] Cocamide DEA and diethanolamine; nonionic/anionic; detergent, wetting agent, foam stabilizer, visc. builder; personal care products, household and industrial cleaners, liq. dishwashing compds., rug and floor cleaners; yel. visc. liq.; slight odor; sol in water, alcohols, diols, triols, glycol ethers, polyols, aromatic and chlorinated hydrocarbons; sp.gr. 1.00 ± 0.01; visc. 950 cps. min.; flash pt. > 170 C (OC); pH 10 ± 0.5 (10%); 100% act.

Schercomid CME. [541] Cocamide MEA; nonionic; emulsifier, foam booster and visc. builder for personal care products and detergent compositions; tan wax; ammoniacal odor; sol. in org. solvs.; disp. in water; m.p. 63 ± 3 C; flash pt. > 180 C (OC); 100% act.

Schercomid CMI. [541] Cocamide MIPA; nonionic; foam booster and stabilizer, antidefatting agent for use in detergents; lt. tan wax; 90% min. amide.

Schercomid EAC. [541] Fatty amide, modified; nonionic; wetting agent, emulsifier; detergent for scouring textile fibers; amber liq.; mild ammoniacal odor; water-sol.; dens. 8.16 lb/gal (50 C); sp.gr. 0.98; pH 9.8 ± 1.0 (1%); 100% act.

Schercomid EAC-S. [541] Coco amide, solubilized; nonionic; scouring and fulling agent for wool; liq.; 30% act.

Schercomid HT-60. [541] PEG-50 tallow amide; nonionic; thickener, detergent, emulsifier, dispersant; Gardner 7 max. hard waxy solid; ammoniacal odor; sol. in hydrophilic solvs.; sp.gr. 1.064 (60 C); dens. 9.6 lb/gal; m.p. 50–55 C; pH 9.0–10.0 (10% aq.); flash pt. (OC) > 180 C; 100% act.

Schercomid ID. [541] Isostearamide DEA; nonionic; w/o emulsifier, thickener, and conditioner used in personal care products; lt. amber liq.; 70% min. amide.

Schercomide IMI. [541] Isostearamide MIPA; nonionic; visc. builder, emulsifier, conditioner, emollient, and lubricant in personal care products; lt. yel. soft wax; mild ammoniacal odor; sol. in org. solvs.; disp. in water; sp.gr. 0.885 ± 0.01 (50 C); dens. 7.4 lb/gal (50 C); pH 10.0 ± 0.7 (10% disp.); flash pt. > 180 C; 100% act.

Schercomid LD. [541] Lauramide DEA and diethanolamine; nonionic/anionic; detergent, emulsifier, foam stabilizer and booster, visc. builder, detergent formulations and personal care products; yel. visc. liq.; typ. odor; sol. in water and org. solvs.; dens. 8.4 lb/gal; sp.gr. 1.01 ± 0.01; visc. 1000 cps. min.; flash pt. > 170 C (OC); pH 9.7 ± 0.5 (10%); 100% act.

Schercomid M. [541] Isostearamide DEA; nonionic; visc. builder, conditioner, and emulsifier in personal care products; clear amber liq., soft solid; 85% min. amide.

Schercomid MD-Extra. [541] Myristamide DEA; nonionic; foam stabilzier, visc. controller for shampoos, creams and lotions; emulsifier; cream solid; slight, typ. odor; m.w. 420; sol. in most alcohols, glycols, glycol ethers; partly sol. in hydrocarbons; disp. in water; m.p. 47–52 C; acid no. 3 max.; pH 10.5 ± 1 (10% disp.); flash pt. > 170 C (OC); 100% act.

Schercomid MME. [541] Myristamide MEA; nonionic; emulsifier, pearling agent, thickener, opacifier, visc. builder, and foam booster in personal care products, pharmaceuticals, and detergents; pale wax; ammoniacal odor; m.w. 271; sol. in org. solvs.; disp. in water; m.p. 88 ± 4.0 C; flash pt. (OC) > 170 C; 100% act.

Schercomid ODA. [541] Oleamide DEA and diethanolamine; nonionic/anionic; visc. builder, foam booster and stabilizer; detergent in industrial cleaners; w/o emulsifier for hydrocarbon solvs.; dispersant for mineral clays or pigments; personal care products; amber liq.; char. odor; sol. in alcohols, glycols, glycol ethers, aliphatic and chlorinated hydrocarbons; disp. in water; dens. 7.9 lb/gal; sp.gr. 0.950 ± 0.01; visc. 1200 cps min.; flash pt. > 170 C (OC); pH 9.9 ± 0.5 (10% disp.); 100% act.

Schercomid OMI. [541] Oleamide MIPA; nonionic; emulsifier, thickener, lubricant, emollient, softener, and hair conditioner for personal care products; amber liq. to soft solid; mild ammoniacal odor; sol. in org. solvs.; disp. in water; sp.gr. 0.90 ± 0.01; dens. 7.5 lb/gal; flash pt. (OC) 180 C; 100% act.

Schercomid SCE. [541] Cocamide DEA; nonionic; visc. builder, thickener, softener, and foam stabilizer for personal care products; liq. dishwashing detergent and rug shampoos; lt. amber clear visc. liq.; mild odor; sol. in org. solvs.; sp.gr. 0.97; dens. 8.1 lb/gal; flash pt. (OC) > 180 C; 100% act.

Schercomid SCO-Extra. [541] Cocamide DEA; nonionic; emulsifier, foam stabilizer, and wetting agent used in household and industrial products; lt. amber clear visc. liq.; 80% act.

Schercomid SD-DS. [541] Stearamide DEA distearate; nonionic; emulsifier, pearling agent, thickener, and opacifier for personal care products; lt. tan wax; 40% min. amide.

Schercomid SI. [541] Isostearamide DEA; nonionic; detergent, emulsifier, corrosion inhibitor, thickener; lt. amber clear liq.; 87% min. amide.

Schercomid SI-M. [541] Isostearamide DEA; nonionic; visc. builder in shampoos; conditioner, emulsifier, low-irritation surfactant; clear amber liq. to soft solid; acid no. 2 max.; 85% min. amide.

Schercomid SLE. [541] Linoleamide DEA; nonionic; visc. builder, conditioner for personal care products; w/o emulsifier; emulsion stabilizer; amber clear liq.; mild fruity odor; sol. in org. solvs.; disp. in water; dens. 8.0 lb/gal; sp.gr. 0.960; flash pt. > 180 C

(OC); pH 10.2 ± 0.5 (10% disp.); 100% act.

Schercomid SL-Extra. [541] Lauramide DEA; nonionic; detergent, wetting agent, foam stabilizer and booster, thickener used in personal care products; dishwashing, all-purpose, and laundry detergents; emulsifier; wh. cryst. solid; mild fruity odor; sol. in alcohols, glycols, glycol ethers, polyols, aliphatic, aromatic, and chlorinated hydrocarbons, and natural fats; disp. in water; dens. 8.1 lb/gal (45 C); sp.gr. 0.97 ± 0.01 (45 C); visc. 1800 cps min. (10% aq.); m.p. 42 ± 3 C; pH 10.2 ± 0.5 (10% aq.); 100% act.

Schercomid SLM. [541] Lauramide DEA; nonionic; visc. builder and foamer for personal care products; o/w emulsifier; lt. yel. liq.; mild odor; sol. see Schercomid SL-Extra; sp.gr. 0.97 ± 0.01 (45 C); dens. 8.1 lb/gal (45 C); visc. 1800 cps min. (10% disp.); pH 10.2 ± 0.5 (10% disp.); flash pt. (OC) > 170 C; 100% act.

Schercomid SLM-C. [541] Lauramide DEA; nonionic; thickener and foam stabilizer used in personal care products; amber clear liq.; 85% min. amide.

Schercomid SL-ML. [541] Lauramide DEA; nonionic; visc. builder and emulsifier used in personal care products and detergent formulations; amber clear liq.; mild odor; sol. in org. solvs.; sp.gr. 0.97; dens. 8.0 lb/gal; flash pt. (OC) > 180 C; 100% act.

Schercomid SLM-LC. [541] Lauramide DEA; nonionic; wetting agent, thickener and foam stabilizer; clear amber liq.; 85% amide.

Schercomid SLM-S. [541] Lauramide DEA; nonionic; thickener, foam stabilizer, wetting agent, and detergent used in personal care products; wh. cryst. solid; 87% min. amide.

Schercomid SLS. [541] Linoleamide DEA; nonionic; w/o emulsifier and emulsion stabilizer for o/w emulsions; thickener; amber clear liq.; 82% min. amide.

Schercomid SME, SME-A. [541] Stearamide MEA; nonionic and nonionic/anionic resp.; detergent, emulsifier, pearlescing agent, and thickener for stick deodorant; lt. tan wax; 90 and 85% min. amide resp.

Schercomid SME-S. [541] Stearamide MEA stearate; nonionic; emulsifier, pearling agent, thickener, opacifier in cosmetic and pharmaceutical applications; pale yel. wax; typ. odor; m.w. 565; sol. in hydrophobic org. solvs.; disp. in water; m.p. 77–82 C; flash pt. (OC) > 180 C; 100% act.

Schercomid SO-A. [541] Oleamide DEA; nonionic/anionic; w/o emulsifier; lt. amber clear liq.; typ. oleic odor; sol. in org. solvs.; disp. in water; dens. 7.9 lb/gal; sp.gr. 0.95 ± 0.01; visc. 450 cps. min.; pH 9.8 ± 0.5 (10% disp.); 100% act.

Schercomid SO-T. [541] Tallowamide DEA; nonionic/anionic; w/o emulsifier; amber clear liq.; 85% min. amide.

Schercomid TO-2. [541] Tallamide DEA; nonionic/anionic; w/o emulsifier, thickener, lubricant, corrosion inhibitor; amber clear liq.; 65% min. amide.

Schercomul H. [541] Fatty acid diethanolamide; nonionic; w/o emulsifier for kerosene; liq.; 100% conc.

Schercomul K. [541] Fatty acid diethanolamide; anionic; emulsifier for perchloroethylene; liq.; 100% conc.

Schercophos L. [541] Complex phosphate; anionic; detergent builder and water conditioner for the textile industry; liq.; 60% act.

Schercopol CMIS-Na. [541] Cocamido MIPA sulfosuccinate; nonionic; detergent, foamer, ingred. and detergent for personal care products, rug and upholstery shampoos, and lt.-duty detergents; yel. clear liq.; mild char. odor; m.w. 491; water-sol.; dens. 9.4 lb/gal; sp.gr. 1.12 ± 0.01; visc. 400 cps max.; cloud pt. 1.0 C max.; pH 5–7; 40% act.

Schercopol CMS-Na. [541] Disodium cocamido MEA sulfosuccinate; anionic; detergent, foamer, softener; primary surfactant in personal care products; yel. clear liq.; mild char. odor; m.w. 471; water-sol.; dens. 9.3 lb/gal; sp.gr. 1.12 ± 0.01; visc. 100 cps max.; cloud pt. 5.0 C; pH 5–7; 30% act.

Schercopol DOS-70. [541] Dioctyl sodium sulfosuccinate; wetting agent, surf. tens. depressant; visc. liq.; pH 6–8; 70% min. dry solids.

Schercopol DOS-PG-85. [541] Dioctyl sodium sulfosuccinate; see Schercopol DOS-70; visc. liq.; 85% min. dry solids.

Schercopol LPS. [541] Disodium laureth sulfosuccinate; surfactant; visc. enhancer; yel. clear liq.; pH 5–7; 39% min. dry solids.

Schercopol OMES-A. [541] Diammonium oleamido PEG-2 sulfosuccinate; anionic; surfactant for personal care products; yel. clear liq.; mild char. odor; sp.gr. 1.05; dens. 8.75 lb/gal; visc. 500 cps max.; cloud pt. 5.0 C max.; pH 6–8; 34% min. solids.

Schercopol OMES-Na. [541] Disodium oleamido PEG-2 sulfosuccinate; see Schercopol OMES-A; lt amber liq.; pH 5–7;

34% min. dry solids.

Schercopol OMIS-Na. [541] Oleamido MIPA; anionic; detergent, foamer; surfactant used in personal care products; yel. clear liq.; mild char. odor; water-sol.; dens. 9.16 lb/gal; sp.gr. 1.1 ± 0.05; visc. 500 cps max.; cloud pt. 5.0 C max.; pH 6.0 ± 1.0; 35% act.

Schercopol OMS-Na. [541] Disodium oleamido MEA sulfosuccinate; anionic; detergent, foamer, conditioner, and emollient in personal care products and dishwashing detergents; yel. clear liq.; mild char. odor; water-sol.; dens. 9.16 lb/gal; sp.gr. 1.10 ± 0.05; visc. 500 cps max.; cloud pt. 5.0 C max.; pH 6.0 ± 1.0; 35% act.

Schercopol RMS-Na. [541] Disodium ricinoleamide MEA sulfosuccinate; anionic; surfactant; yel. clear liq.; mild char. odor; m.w. 565; sp.gr. 1.120; dens. 9.3 lb/gal; visc. 500 cps. max.; cloud pt. 5.0 C max.; pH 5–7; 40% min. solids.

Schercopol UMS-Na. [541] Disodium undecylenamido MEA sulfosuccinate; anionic; surfactant; amber pearlescent liq.; mild char. odor; m.w. 453; sp.gr. 1.16; dens. 9.7 lb/gal; cloud pt. 40 C max.; pH 5–7; 43% min. solids.

Schercopon 2WD. [541] Sulfosuccinate, ethoxylated; anionic; dry-cleaning detergent; water-wh. liq.; 90% act.

Schercoquat DAB. [541] Quaternium-63; cationic; quat. used in personal care products; amber visc. liq.; m.w. 994; 90% min. solids.

Schercoquat DAS. [541] Quaternium-61; conditioner used in personal care products; amber visc. liq.; m.w. 1050; 90% min. dry solids.

Schercoquat FOAS. [541] N-safflower-(3-amidopropyl)-N-N-dimethyl-N-ethyl ammonium ethyl sulfate; quat. for hair conditioners; visc. amber liq.; 90% min. dry solids.

Schercoquat IAS-LC. [541] Isostearamidopropyl ethyl dimonium ethosulfate; see Schercoquat DAB; amber visc. liq.; m.w. 550; 90% min. dry solids.

Schercoquat IB. [541] Isostearamodopropyl alkonium chloride; see Schercoquat DAB; amber visc. liq.; m.w. 520; 90% min. dry solids.

Schercoquat IEP. [541] Quaternium-62; see Schercoquat DAB; amber visc. liq.; good water sol.; m.w. 487; 80% min. dry solids.

Schercoquat IIB. [541] Isostearyl benzyl imidonium chloride; see Schercoquat DAB; dk. amber visc. liq.; m.w. 504; 90% min. dry solids.

Schercoquat IIS. [541] Isostearyl ethyl imidonium ethosulfate; see Schercoquat DAB; dk. amber visc. liq.; m.w. 532; 99% min. dry solids.

Schercoquat IIS-RD. [541] 2-Isoalkyl (C_{14}-C_{20}), 1-hydroxyethyl, 1-ethyl imidazolinium ethyl sulfate; see Schercoquat DAB; dk. amber visc. liq.; m.w. 532; 99% min. dry solids.

Schercoquat ROAB. [541] Rapeseedamidopropyl benzyldimonium chloride; see Schercoquat DAB; dk. amber visc. liq.; m.w. 532; 90% min. dry solids.

Schercoquat ROAS. [541] Rapeseedamidopropyl ethyldimonium ethosulfate; see Schercoquat DAB; amber visc. liq.; m.w. 560; 90% min. dry solids.

Schercoquat ROEP. [541] N-rapeseed-(3-amidopropyl)-N-N dimethyl-N-(2,3 epoxypropyl)-ammonium chloride; see Schercoquat DAB; dk. amber visc. liq.; m.w. 498; 80% min. dry solids.

Schercoquat SAB. [541] N-stearyl-(3-amidopropyl)-N,N-dimethyl-N-benzyl ammonium chloride; cationic; see Schercoquat DAB; yel. hazy liq.; mild amine odor; m.w. 480; sp.gr. 0.945; dens. 8.0 lb/gal; pH 4–7 (5% aq.); 80% min. nonvolatiles.

Schercoquat SAS. [541] N-stearyl-(3-amidopropyl)-N,N dimethyl-N-ethyl ammonium ethyl sulfate; cationic; see Schercoquat DAB; yel. hazy liq.; mild amine odor; m.w. 508; sp.gr. 0.965; dens. 8.0 lb/gal; pH 4–7 (5% aq.); 80% min. nonvolatiles.

Schercoquat SOAB. [541] Soyamidopropyl benzyldimonium chloride; see Schercoquat DAB; dk. amber visc. liq.; m.w. 488; 90% min. dry solids.

Schercoquat SOAS. [541] Soyamidopropyl ethyldimonium ethosulfate; see Schercoquat DAB; amber visc. liq.; m.w. 516; 90% min. dry solids.

Schercotaine CAB. [541] Cocamidopropyl betaine; amphoteric; detergent, wetting agent, foamer, antistat, and softener used in personal care products; pale clear liq.; char. odor; m.w. 386; water-sol.; dens. 8.75 lb/gal; sp.gr. 1.05 ± 0.01; visc. 100 cps max.; cloud pt. –2.0 C max.; pH 5.5 ± 1.0; 30% act.

Schercotaine CAB-A. [541] N-[3-(cocoamido)-propyl]-N,N-dimethyl betaine, ammonium salt; amphoteric; surfactant; yel. clear liq.; typ. odor; m.w. 380; sp.gr. 1.04;

dens. 8.7 lb/gal; pH 5.0–7.0; 40% min. act.

Schercotaine CAB-G. [541] Cocoamidopropyl betaine; amphoteric; detergent, wetting agent, foamer, and cloud pt. depressant used in personal care products; lt. yel. clear liq.; pH 5–7; 45% min. dry solids.

Schercotaine CAB-KG. [541] N-[3-(cocoamido)-propyl]-N,N-dimethyl betaine, potassium salt; amphoteric; surfactant; yel. clear liq.; typ. odor; m.w. 401; sp.gr. 1.06; dens. 8.8 lb/gal; pH 5.0–7.0; 40% min. act.

Schercotaine CAB-Z. [541] Zinc-bis (cocoamidopropyl betaine); amphoteric; surfactant; yel. clear liq.; char. odor; m.w. 789; sp.gr. 1.08; dens. 9.00 lb/gal; pH 4–7; 35.0% min. nonvolatiles.

Schercotaine IAB. [541] Isostearamidopropyl betaine; amphoteric; conditioner and detergent for personal care products; amber soft gel; pH 5–7; 35% min. dry solids.

Schercotaine MAB. [541] Myristamidopropyl betaine; amphoteric; detergent, wetting agent and thickener for personal care products; lt. yel. clear liq.; pH 5–7; 30% min. solids.

Schercotaine OAB. [541] Oleylamidopropyl betaine; amphoteric; visc. control agent, textile softener and conditioner; liq.; 35% conc.

Schercotaine PAB. [541] Palmitamidopropyl betaine; amphoteric; thickener used in personal care products; lt. yel. soft gel; pH 5–7; 35% min. dry solids.

Schercotaine SCAB. [541] Cocamidopropyl hydroxy sultaine; amphoteric; detergent, wetting agent and foamer, cloud pt. depressant used in personal care products; lt. lemon clear liq.; typ. odor; m.w. 500; water-sol.; dens. 8.9 lb/gal; sp.gr. 1.07 ± 0.01; visc. 100 cps max.; cloud pt. –7 C; pH 6.0 ± 1.0; 35% act.

Schercotaine SCAB-A. [541] N-[3-(Cocoamido)-propyl]-N-(2-hydroxy-3-sulfopropyl)-N-N-dimethyl betaine, ammonium salt; amphoteric; surfactant; liq.; 45% conc.

Schercotaine SCAB-G. [541] Cocoamidopropyl hydroxy sultaine; amphoteric; detergent, wetting agent and foamer; lt. amber clear liq.; pH 5–7; 50% min. dry solids.

Schercotaine SCAB-KG. [541] N-[3-(cocoamido)-propyl]-N-(2-hydroxy-3-sulfopropyl)-N,N-dimethyl betaine, potassium salt; amphoteric; surfactant used in personal care products; yel. clear liq.; typ. odor; m.w. 480; sp.gr. 1.10; dens. 9.1 lb/gal; pH 5.0–7.0; 42% min. act.

Schercotaine UAB. [541] Bis(undecylenic amido-propyl dimethyl glycinate; amphoteric; surfactant with germicidal/bactericidal properties for shampoos; liq.; 30% conc.

Schercotaine UAB-Z. [541] Zinc-bis (undecylenic amidopropyl dimethyl glycinate); amphoteric; surfactant; amber clear liq.; char. odor; m.w. 717; sp.gr. 1.09; dens. 9.1 lb/gal; pH 4–7; 35.0% max. chloride.

Schercoterge 140. [541] Fatty amide, ethoxylated, modified; nonionic; detergent, wetting and textile scouring agent, emulsifier, wool fulling; dyeing assistant; post scouring agent; dye bath stabilizer; lemon liq.; mild ammoniacal odor; water-sol.; dens. 8.3 lb/gal; sp.gr. 1.0; pH 9.5 ± 0.5 (1%); 98% act.

Schercoteric CY-2. [541] Capryloamphocarboxyglycinate; amphoteric; surfactant for household and industrial cleaning products; amber clear liq.; 50% min. dry solids.

Schercoteric CY-SF-2. [541] Capryloamphocarboxypropionate; amphoteric; see Schercoteric CY-2; amber soft paste; 75% min. dry solids.

Schercoteric I-AA. [541] Isostearoamphopropionate; amphoteric; surfactant for cosmetic and industrial cleaners; amber visc. liq.; 34% min. dry solids.

Schercoteric IS-SF-2. [541] Isostearoamphocarboxypropionate; amphoteric; surfactant used in personal care products; amber clear liq.; 38% min. dry solids.

Schercoteric LS. [541] Lauroamphoacetate; amphoteric; detergent for shampoos, dishwashing compds., textiles; yel. clear visc. liq.; 45% min. dry solids.

Schercoteric LS-2. [541] Lauroamphodiacetate; amphoteric; see Schercoteric LS; yel. clear visc. liq.; 50% min. dry solids.

Schercoteric LS-2ES MOD. [541] Lauroamphodiacetate, sodium laureth sulfate, SD alcohol 40; amphoteric; see Schercoteric IS-SF-2; lt. yel. clear liq.; 40% min. dry solids.

Schercoteric LS-2TE. [541] Lauroamphocarboxyglycinate, sodium trideceth sulfate; amphoteric; see Schercoteric IS-SF-2; lt.yel. clear liq.; 33% min. dry solids.

Schercoteric LS-2TE MOD. [541] Lauroamphocarboxyglycinate, sodium trideceth sulfate, hexylene glycol; amphoteric; see Schercoteric IS-SF-2; lt. yel. clear liq.; 33% min. dry solids.

Schercoteric LS-EP. [541] Lauroamphohydroxypropylsulfonate; amphoteric; detergent for shampoos, dishwashing compds., textiles; yel. clear liq.; 45% min. dry solids.

Schercoteric LS-SF, LS-SF Conc. [541] Lauroamphopropionate; amphoteric; see Schercoteric IS-SF-2; yel. clear liq. and lt. amber soft paste resp.; 38 and 75% min. dry solids resp.

Schercoteric LS-SF-2. [541] Lauroamphodipropionate; amphoteric; detergent; surfactant for personal care products; liq.; 38% conc.

Schercoteric LS-SF-2 Conc. [541] Lauric imidazolinium deriv. dicarboxylate; see Schercoteric LS-SF-2; liq.; 38% conc.

Schercoteric LS-SF-2 Super Conc. [541] Lauric imidazolinium deriv. dicarboxylate; amphoteric; surfactant for personal care products; paste; 75% conc.

Schercoteric MS. [541] Cocoamphoacetate; amphoteric; detergent used in personal care products and industrial cleaners; amber clear visc. liq.; 50% min. dry solids.

Schercoteric MS-2. [541] Cocoamphodiacetate; amphoteric; detergent used in personal care products and industrial cleaners; amber clear visc. liq.; 50% min. dry solids.

Schercoteric MS-2 MOD. [541] Cocoamphodiacetate, sodium lauryl sulfate, hexylene glycol; amphoteric; base for personal care products and cleaning preparations; foamer; yel. clear liq.; 37 min. dry solids.

Schercoteric MS-2ES MOD. [541] Cocoamphodiacetate, sodium laureth sulfate; amphoteric; see Schercoteric MS-2 MOD; yel. clear liq.; 40% min. dry solids.

Schercoteric MS-2TE MOD. [541] Cocoamphodiacetate, sodium trideceth sulfate; amphoteric; see Schercoteric MS-2 MOD; yel. clear liq.; 32% min. dry solids.

Schercoteric MS-EP. [541] Cocoamphohydroxypropylsulfonate; amphoteric; see Schercoteric IS-SF-2; amber clear liq.; 45% min. dry solids.

Schercoteric MS-SF (38%). [541] Coco imidazolinium deriv. monocarboxylate; amphoteric; surfactant for shampoos, industrial cleaners, dishwashing compds.; yel. clear liq.; mild, typ. odor; m.w. 406; dens. 8.7 lb/gal; sp.gr. 1.05; pH 8–10; 38% act.

Schercoteric MS-SF (70%). [541] Coco imidazolinium deriv. monocarboxylate; amphoteric; see Schercoteric MS-SF (38%); lt. amber soft gel; mild, typ. odor; sol. in water; m.w. 406; dens. 9.1 lb/gal; sp.gr. 1.1; pH 8–10; 70% act.

Schercoteric MS-SF-2 (38%). [541] Cocoamphodipropionate; amphoteric; detergent, foaming agent, base for shampoos; yel. clear liq.; typ. odor; sp.gr. 1.07; cloud pt. –8 C; pH 9.5 ± 0.5 (50%); 38% act.

Schercoteric MS-SF-2 (70%). [541] Cocoamphodipropionate; see Schercoteric MS-SF-2 (38%); lt. pourable gel; typ. odor; water-sol.; sp.gr. 1.10; pH 9.5 ± 0.5 (50%); 70% act.

Schercoteric MS-SF-2 Super Conc. [541] Coco imidazolinium deriv. dicarboxylate; amphoteric; see Schercoteric LS-SF-2 Super Conc.; paste; 75% conc.

Schercoteric OS-SF. [541] Oleoamphopropionate; amphoteric; surfactant for cosmetic and industrial cleaners; amber soft paste; 38% min. dry solids.

Schercoteric STS. [541] Stearoamphodiacetate; amphoteric; surfactant and conditioner for personal care products; pearly cream paste; 25% min. dry solids.

Schercozoline B. [541] Behenyl imidazoline; cationic; antistat, dispersant, wetting agent, emulsifier, microbicide used in acid and emulsion cleaning, cleaners, polishes, surf. treatment, textile and leather processing, agriculture and cosmetic intermediate; tan solid; m.w. 383; m.p. 45–49 C; 90% min. imidazoline.

Schercozoline C. [541] Cocoyl imidazoline; cationic; see Schercozoline B; also detergent and used in primer paints; tan semisolid; m.w. 278; 90% min. imidazoline.

Schercozoline I. [541] Isostearyl imidazoline; surfactant, softener, antistat; amber clear liq.; m.w. 378; 90% min. imidazoline.

Schercozoline L. [541] Lauryl imidazoline; see Schercozoline C; cream solid; m.w. 268; m.p. 38–42 C; 90% min. imidazoline.

Schercozoline O. [541] Oleic imidazoline; cationic; see Schercozoline B; also as w/o emulsifier, corrosion inhibitor, and intermediate for quat. ammonium compds.; dk. amber liq.; m.w. 350; 90% min. imidazoline.

Schercozoline S. [541] Stearyl imidazoline; see Schercozoline I; cream solid; m.w. 336; m.p. 44–48 C; 90% min. imidazoline.

SCMS. [299] Sodium coconut monoglyceride sulfonate; anionic; wetting agent and foamer, dispersant, emulsifier, detergent for water-base systems, personal care products; wh. powd.; mild char. odor; m.w. 366; low water-sol.; HLB 40; pH 6–8; 93% min. act.

Scourol 95. [341] POE alkylaryl ether with org. sulfonates; anionic; bleach assistant for textiles; liq.; 40% conc.

Scourol 700. [341] POE alkyl ether; nonionic; textile processing assistant; scouring and desizing agent; liq.

Scourtex 1063. [283] Fatty alkanolamide; anionic; scouring and fulling agent for fabrics; amber liq.; ref. index 1.4794; disp. in water; sp.gr. 1.012; visc. 700 cps; 92% act.

Scripset 500. [419] Styrene maleic anhydride, sodium salt; anionic; emulsifier, dispersant, thickener; powd.; 100% conc.

Scripset 520. [419] Styrene maleic anhydride copolymer; emulsifier, binder, sizing agent, visc. modifier, stabilizer, starch modifier, pigment dispersant, protective colloid, sizing, coating, water-paint calsomines, adhesives, printing, preparation of emulsified paints; off-wh. powd.; faint aromatic odor; m.w. 50,000; sol. in alkaline aq. systems.

Scripset 540. [419] Styrene/maleic anhydride copolymer; see Scripset 520; off-wh. powd.; faint aromatic odor; m.w. 20,000; sol. in alkaline aq. systems.

Scripset 550. [419] Styrene copolymer, maleic anhydride; off-wh. powd.; faint aromatic odor; m.w. 10,000; sol. alkaline aq. systems.

Scripset 700. [419] Styrene/maleic anhydride copolymer; see Scripset 520; m.w. 50,000; water-sol.; dens. 9.5 lb/gal; 30% act. in water.

Scripset 720. [419] Styrene/maleic anhydride copolymer; see Scripset 520; m.w. 50,000; water-sol.; 25% act. in water.

Scripset 808. [419] Styrene maleic anhydride amide/NH_4OH acid salt; anionic; see Scripset 500; powd.; 100% conc.

Seabomul B. [546] Alcohol ethoxylate; nonionic; detergent used for scouring, degreasing and heavy duty dispersant; liq.; HLB 10.0; 33% conc.

Seabomul CFXB. [546] Alcohol ethoxylate; nonionic; surfactant used as detergent and dispersant; liq.; water-sol.; 70% conc.

Seabomul Z. [546] Alcohol ethoxylate; nonionic; detergent and emulsifier; liq.; HLB 12.0; 50% conc.

Seachem 105. [546] Sulfated oil; anionic; min. oil emulsifier, softener and penetrant; liq.; 75% conc.

Seachem SC. [546] Sulfated castor oil; anionic; softener, emulsifier, lubricant and dye leveling assistant in textile industry; liq.; 75% conc.

Seachem SF. [546] Sulfated fatty acid; anionic; emulsifier, softener, dispersant, leveling agent; liq.; 75% conc.

Seachem SS. [546] Esterified oil; anionic; min. oil emulsifier, wetting agent and dispersant for pigments and dyes; paste; 80% conc.

Seachem V900. [546] Sulfated veg. oil; anionic; textile softener, emulsifier for min. and veg. oils; liq.; 75% conc.

Secolat. [576] Alkyl disodium sulfosuccinamate; anionic; foamer for latex emulsions; emulsifier for emulsion polymerization; liq.; 35% conc.

Secosol AL/959. [576] Sodium mono lauryl sulfosuccinate; anionic; mild foamer for shampoos and foam baths; paste; 25% conc.

Secosol ALL40. [576] Sodium mono lauryl ether sulfosuccinate; anionic; see Secosol AL/959; liq.; 30% conc.

Secosol AS. [576] Sodium cetyl-stearyl sulfosuccinate; anionic; additive to toilet and syndet soaps; cleaning agent; paste; 45% conc.

Secosol DOS/70. [576] Sodium dioctyl sulfosuccinate; anionic; dispersant and wetting/rewetting agent; textile additive; liq.; 70% conc.

Secosol EA/40. [576] Sodium mono alkylethanolamide sulfosuccinate; anionic; rug and hair shampoos, foam baths; paste; 40% conc.

Secoster CL 10. [576] PEG-20 sorbitan laurate; nonionic; emulsifier, dispersant, solubilizer; liq.; 100% conc.

Secoster CP 10. [576] PEG-20 sorbitan palmitate; nonionic; see Secoster CL 10; liq.; 100% conc.

Secoster CS 10. [576] PEG-20 sorbitan stearate; nonionic; see Secoster CL 10; liq.; 100% conc.

Secoster DMS. [576] Glycol distearate; nonionic; emollient, pearlescent, emulsifier for creams and cleansing milks; solid; 100% conc.

Secoster DO 600. [576] PEG 600 dioleate; nonionic; additive for cutting oils; emulsifier for creams, cleansing milks and pesticides; liq.; 100% conc.

Secoster EMS. [576] Glycol stearate; nonionic; see Secoster DMS; solid; 100% conc.

Secoster KL 10. [576] Sorbitan laurate; nonionic; emulsifier, dispersant, solubilizer; liq.; 100% conc.

Secoster KP 10. [576] Sorbitan palmitate; nonionic; emulsifier, dispersant, solubilizer; solid; 100% conc.

Secoster KS 10. [576] Sorbitan stearate; nonionic; emulsifier, dispersant, solubilizer; solid; 100% conc.

Secoster MA 300. [576] PEG 300 abietate; nonionic; see Secoster DO 600; liq.; 100% conc.

Secoster ML 300. [576] PEG 300 laurate; nonionic; see Secoster DO 600; liq.; 100% conc.

Secoster ML 4000. [576] PEG 4000 laurate; nonionic; emulsifier, dispersant; solid; 100% conc.

Secoster MO 100. [576] Diethylene glycol oleate; nonionic; emulsifier, dispersant; liq.; 100% conc.

Secoster MO 400. [576] PEG 400 oleate; nonionic; see Secoster DO 600; liq.; 100% conc.

Secosyl. [576] Sodium N-lauroyl sarcosinate; anionic; rug shampoos, anticorrosion additive; base for personal care products; liq.; 30% conc.

Sellogen 641. [187] Alkyloxyalkyl sodium sulfate; anionic; textile detergent and emulsifier; pickling; yel. clear liq.; cloud pt. 5 C; flash pt. 43 C (PMCC); pH 8.5 (2%) 30% act.

Sellogen DFL. [186] Sodium alkyl naphthalene sulfonate; anionic; wetting agent for pesticide formulations; powd.; 68% conc.

Sellogen HR. [187] Sodium dialkyl naphthalene sulfonate; anionic; wetting agent, dispersant; heavy duty household and industrial cleaners; fire control, paint strippers, agric. chemical formulations, emulsion and suspension polymerization aids, latex paints; off-wh. powd.; water-sol.; dens. 0.40 g/cc; surf. tens. 44.1 dynes/cm (0.25%); pH 9.0; 75.5% act.

Sellogen HR-90 Liquid. [186] Sodium alkyl naphthalene sulfonate; anionic; wetting agent for insecticides, acid and alkaline mediums; liq.; 35.5% conc.

Sellogen NS-50. [186] Sodium alkyl aromatic sulfonate; anionic; anticaking agent in fertilizer and clay; used in phosphate rock acidulation; slurry; 50% conc.

Sellogen W. [187] Sodium dialkyl naphthalene sulfonate; anionic; see Sellogen HR; off-wh. powd.; water-sol.; dens. 0.38 g/cc; surf. tens. 42.1 dynes/cm (0.25%); pH 9.0 (5%); 66.5% act.

Sellogen WL. [187] Sodium alkyl naphthalene sulfonate; anionic; see Sellogen HR; dk. amber liq.; water-sol.; dens. 9.5 lb/gal; sp.gr. 1.140; surf. tens. 48.9 dynes/cm (0.25%); pH 9.5 (10%); 34% act.

Separol AF 27. [82] EO/PO adduct; nonionic; demulsifier used in crude oil emulsions and dehydrating equipment; yel. clear liq.; sol. in alcohols; dens. 0.96 g/cc; visc. 30 cs; flash pt. > 23 C; pour pt. –30 C.

Separol WF 22. [82] EO/PO adduct; nonionic; see Separol AF 27; yel. clear liq.; sol. in water and alcohols; dens. 1.02 g/cc; visc. 200 cs; flash pt. > 23 C; pour pt. –50 C.

Separol WF 34. [82] EO/PO adduct; nonionic; see Separol AF 27; yel. clear liq.; sol. in alcohols; dens. approx 0.95 g/cc; visc. 200 cs; flash pt. 16 C; pour pt. –50 C.

Separol WF 41. [82] EO/PO adduct; nonionic; see Separol AF 27; yel. clear liq.; sol. in alcohols; dens. 0.95 g/cc; visc. 50 cs; flash pt. 17 C; pour pt. < –50 C.

Separol WF 221. [82] EO/PO adduct; nonionic; see Separol AF 27; yel. clear liq.; sol. in alcohols and water; dens. 1.02 g/cc; visc. 200 cs; flash pt. > 23 C; pour pt. –50 C.

Separol WK 25. [82] EO/PO adduct; nonionic; see Separol AF 27; yel. clear liq.; sol. in alcohols and water; dens. 1 g/cc; visc. 300 cs; flash pt. 24 C; pour pt. –50 C.

Serdet DCK 3/70. [552] Sodium laureth sulfate (2 EO); anionic; used in personal care products and dishwashing formulations; liq.; biodeg.; 70% conc.

Serdet DCK 30. [552] Sodium laureth sulfate (3 EO); anionic; see Serdet DCK 3/70; liq.; biodeg.; 27% conc.

Serdet DCN 30. [552] Ammonium C_{12}–C_{14} alkyl ether sulfate (3 EO); anionic; see Serdet DCK 3/70; liq.; biodeg.; 30% conc.

Serdet DFK 40. [552] Sodium lauryl sulfate; anionic; personal care and dishwashing formulations; emulsifier in emulsion polymerization; paste; biodeg.; 41% conc.

Serdet DFL 40. [552] TEA-lauryl sulfate; anionic; foamer for personal care products; liq.; 40% conc.

Serdet DFM 33. [552] MEA-lauryl sulfate; anionic; see Serdet DFL 40; liq.; 33% conc.

Serdet DFN 30. [552] Ammonium lauryl sulfate; anionic; see Serdet DFL 40; liq.; 27% conc.

Serdet DM. [552] Dodecylbenzene sulfonic acid; anionic; used in scouring powds. and liq. detergents; emulsifier for emulsion polymerization; liq.; 97% conc.

Serdet DMK 75. [552] Sodium dodecylbenzene sulfonate; anionic; see Serdet DFN 30; paste; 75% conc.

Serdet DML 45. [552] Triethanolammonium

dodecylbenzene sulfonate; anionic; personal care products; liq.; 45% conc.

Serdet DNK 30. [552] Sodium nonoxynol-4 sulfate; anionic; detergent base; emulsifier for emulsion polymerization; liq.; 31% conc.

Serdet DPK 3/70. [552] Sodium C_{12-15} alkyl ether sulfate (3 EO); anionic; personal care products and dishwashing formulations; paste; 70% conc.

Serdet DPK 30. [552] Sodium C_{12-15} pareth sulfate; anionic; see Serdet DCK 30; liq.; 27% conc.

Serdet DSK 40. [552] Sodium 2-ethylhexyl sulfate; anionic; wetting agent; latex stabilizer; liq.; 40% conc.

Serdet NZ 60. [552] Lauryl ether sulfate and alkylolamide; anionic; base for personal care products; liq.; 60% conc.

Serdet Perle Conc. [552] Lauryl ether sulfate and alkylolamide; anionic; see Serdet DSK 40; liq.; 34% conc.

Serdolamide POF 61. [552] Oleic acid polydiethanolamide; nonionic; foam stabilizer, refatting agent, visc. modifier in detergents and personal care products; liq.; 100% conc.

Serdolamide POF 61 C. [552] Oleic acid polydiethanolamide; nonionic; component in emulsifiers for cutting oils; corrosion inhibitor; liq.; 100% conc.

Serdolamide PYF 77. [552] Fatty acid diethanolamide; nonionic; foam stabilizer, refatting agent, visc. improver in personal care products; liq.; 100% conc.

Serdox NBS 4. [552] C_{9-11} alkyl polyglycol ether (4 EO); nonionic; scouring agent for textiles; soaking assistant in leather industry; liq.; HLB 10.5; 100% conc.

Serdox NBS 5.5. [552] C_{9-11} alkyl polyglycol ether (5.5 EO); nonionic; see Serdox NBS 4; liq.; HLB 12.0; 100% conc.

Serdox NBS 6. [552] C_{9-11} alkyl polyglycol ether (6 EO); nonionic; see Serdox NBS 4; liq.; HLB 12.5; 100% conc.

Serdox NBS 6.6. [552] C_{9-11} alkyl polyglycol ether (6.6 EO); nonionic; see Serdox NBS 4; liq.; HLB 13.0; 100% conc.

Serdox NBS 8.5. [552] C_{9-11} alkyl polyglycol ether (8.5 EO); nonionic; detergent and wetting agent used in electrolyte concentrations; paste; HLB 14.0; 100% conc.

Serdox NBSQ 5/5. [552] Alkyl polyglycol ether; nonionic; foamer and wetting agent; liq.; biodeg.; 100% conc.

Serdox NDI 100. [552] Dinonylphenol polyglycol ether (100 EO); nonionic; intermediate for textile auxs.; HLB 18.5; 100% conc.

Serdox NEL 3. [552] C_{12-14} alkyl polyglycol ether (3 EO); nonionic; raw material for sulfation and phosphation; emulsifier, detergent; liq.; HLB 8; biodeg.; 100% conc.

Serdox NEL 12/80. [552] C_{12-14} alkyl polyglycol ether (12 EO); nonionic; emulsifier for fats, oils, and waxes; stabilizer for syn. latices; liq.; HLB 14.5; biodeg.; 100% conc.

Serdox NES 3. [552] C_{12-15} alkyl polyglycol ether (3 EO); nonionic; liq.; HLB 8.0; biodeg.; 100% conc.

Serdox NES 8/85. [552] C_{12-15} alkyl polyglycol ether (8 EO); nonionic; scouring agent for textiles; soaking agent in leather industry; detergent for household and industrial purposes; solid; HLB 12.5; biodeg.; 85% conc.

Serdox NES 12. [552] C_{12-15} alkyl polyglycol ether (12 EO); nonionic; paste; HLB 14.5; biodeg.; 100% conc.

Serdox NJAD 15. [552] Tallow amine polyglycol ether, dist. (15 EO); nonionic; emulsifier and dispersant in acid and neutral formulations, oil and wax emulsifier; wetting agent in metal cleaning compds.; antistat; solid; 100% conc.

Serdox NJAD 20. [552] Tallow amine polyglycol ether, dist. (20 EO); nonionic; see Serdox NJAD 15; solid; 100% conc.

Serdox NJAD 30. [552] Tallow amine polyglycol ether, dist. (30 EO); nonionic; see Serdox NJAD 15; solid; 100% conc.

Serdox NNP 4. [552] POE (4) nonyl phenyl ether; nonionic; detergent; dispersant; emulsifier for insecticides and herbicides; liq.; oil-sol.; HLB 9.0; biodeg.; 100% conc.

Serdox NNP 5. [552] POE (5) nonyl phenyl ether; nonionic; detergent; emulsifier for insecticides and herbicides; liq.; HLB 10.0; biodeg.; 100% conc.

Serdox NNP 6. [552] POE (6) nonyl phenyl ether; nonionic; see Serdox NNP 5; liq.; oil-sol.; HLB 11.0; biodeg.; 100% conc.

Serdox NNP 7. [552] Nonylphenol polyglycol ether (7 EO); nonionic; see Serdox NNP 5; liq.; oil-sol.; HLB 12.0; biodeg.; 100% conc.

Serdox NNP 8.5. [552] POE (8.5) nonyl phenyl ether; nonionic; scouring agent for textiles; soaking assistant in the leather industry; detergent for household and industrial purposes; emulsifier for insecticides and herbicides; plasticizer for mortar and concrete; liq.; HLB 12.5; 100% conc.

Serdox NNP 10. [552] POE (10) nonyl phenyl

ether; nonionic; used in textiles; rinsing agent for washing paper machine felts; rewetting agent for paper towels and tissues; liq.; HLB 13.5; biodeg.; 100% conc.

Serdox NNP 11. [552] POE (11) nonyl phenyl ether; nonionic; see Serdox NNP 10; liq.; HLB 14.0; biodeg.; 100% conc.

Serdox NNP 12. [552] POE (12) nonyl phenyl ether; nonionic; scouring agent for textiles; soaking assistant in leather industry; detergent for household and industrial purposes; emulsifier for insecticides and herbicides; liq.; HLB 14.0; 100% conc.

Serdox NNP 13. [552] POE (13) nonyl phenyl ether; nonionic; see Serdox NNP 12; liq.; HLB 14.5; 100% conc.

Serdox NNP 14. [552] POE (14) nonyl phenyl ether; nonionic; detergent and wetting agent for electrolyte concentrations; emulsifier for fatty acids and waxes; liq.; HLB 14.5; biodeg.; 100% conc.

Serdox NNP 15. [552] POE (15) nonyl phenyl ether; nonionic; see Serdox NNP 14; liq.; HLB 15.0; biodeg.; 100% conc.

Serdox NNP 20/70. [552] POE (20) nonyl phenyl ether; nonionic; stabilizer for syn. latices; solid; HLB 16.0; 70% conc.

Serdox NNP 25. [552] POE (25) nonyl phenyl ether; nonionic; emulsifier for emulsion polymerization; solid; HLB 16.5; 100% conc.

Serdox NNP 30/70. [552] POE (30) nonyl phenyl ether; nonionic; dyeing assistant; lime soap dispersant; emulsifier and stabilizer for emulsion polymerization; liq.; HLB 17.0; 70% conc.

Serdox NNPQ 7/11. [552] Alkylphenol polyglycol ether; nonionic; wetting agent; liq.; 100% conc.

Serdox NOG 200, 400. [552] Fatty acid, ethoxylated; nonionic; emulsifier for fats and oils; water-sol.; 100% conc.

Serdox NOG 440. [552] Oleyl polyglycol ester (10 EO); nonionic; emulsifier for crude and veg. oils; liq.; biodeg.; 100% conc.

Serdox NOL 2. [552] POE (2) oleyl ether; nonionic; co-emulsifier for min. oils and non-polar solvs.; liq.; HLB 5.0; biodeg.; 100% conc.

Serdox NOL 8. [552] POE (8) oleyl ether; nonionic; see Serdox NOL 2; liq.; HLB 11.5; biodeg.; 100% conc.

Serdox NOL 15. [552] POE (15) oleyl ether; nonionic; see Serdox NOL 2; paste; HLB 14.5; biodeg.; 100% conc.

Serdox NOP 9. [552] POE (9) octyl phenyl ether; nonionic; see Serdox NNP 10; liq.; HLB 13.0; 100% conc.

Serdox NOP 30/70. [552] POE (30) octyl phenyl ether; nonionic; emulsifier and stabilizer for emulsion polymerization; liq.; HLB 17.5; 70% conc.

Serdox NOP 40/70. [552] POE (40) octyl phenyl ether; nonionic; see Serdox NOP 30/70; liq.; HLB 18.0; 70% conc.

Serdox NSG 400. [552] Stearyl polyglycol ester (9 EO); nonionic; textile processing and cosmetic emulsions; antistat for plastics; solid; biodeg.; 100% conc.

Sermul EA 54. [552] Sodium nonoxynol-4 sulfate; anionic; emulsifier for emulsion polymerization; liq.; 31% conc.

Sermul EA 88. [552] Calcium dodecylbenzene sulfonate; anionic; emulsifier for pesticide formulations; liq.; biodeg.; 65% conc.

Sermul EA 129. [552] Ammonium lauryl sulfate; anionic; see Sermul EA 54; liq.; biodeg.; 27% conc.

Sermul EA 136. [552] Acid nonylphenol polyglycol ether (15 EO) phosphate; anionic; see Sermul EA 129; liq.; 100% conc.

Sermul EA 139. [552] Acid 2-ethylhexyl polyglycol ether (3 EO) phosphate; anionic; see Sermul EA 129; liq.; biodeg.; 100% conc.

Sermul EA 146. [552] Sodium nonylphenol polyglycol ether (15 EO) sulfate; anionic; see Sermul EA 129; liq.; 35% conc.

Sermul EA 150. [552] Sodium lauryl sulfate; anionic; see Sermul EA 129; paste; biodeg.; 41% conc.

Sermul EA 151. [552] Sodium nonylphenol polyglycol ether (10 EO) sulfate; anionic; see Sermul EA 129; liq.; 35% conc.

Sermul EA 176. [552] Sodium mononony1phenol polyglycol ether (10 EO) sulfosuccinate; anionic; see Sermul EA 129; liq.; biodeg.; 35% conc.

Sermul EA 188. [552] Acid nonylphenol polyglycol ether (10 EO) phosphate; anionic; see Sermul EA 129; liq.; 100% conc.

Sermul EA 205. [552] Acid nonylphenol polyglycol ether (50 EO) phosphate; anionic; see Sermul EA 129; liq.; 100% conc.

Sermul EA 211. [552] Nonylphenol polyglycol ether (6 EO) phosphoric acid; anionic; see Sermul EA 129; liq.; 100% conc.

Sermul EA 221. [552] Sodium C_{12-15} alcohol polyglycol ether (3 EO); anionic; emulsifier for emulsion polymerization; liq.; 40% conc.

Sermul EN 20/70. [552] POE (20) nonyl phenyl ether; nonionic; post stabilizer for emulsion polymerization; solid; 70% conc.

Sermul EN 30/70. [552] Nonylphenol polyglycol ether (30 EO); nonionic; see Sermul 20/70; solid; 70% conc.

Sermul EN 145. [552] Nonylphenol polyglycol ether (30 EO); nonionic; see Sermul EN 20/70; liq.; 70% conc.

Servamine KEO 260. [552] N (oleyl amido ethyl) N (ethanol amine); cationic; adhesion agent in bitumen and corrosion inhibitor; liq.; 100% conc.

Servamine KEP 4527. [552] N (palmityl amido propyl) N,N,N trimethyl ammonium chloride; cationic; emulsifier with bactericide properties; liq.; biodeg.; 50% conc.

Servamine KET 350. [552] N (tall oil amido propyl) N,N dimethyl amine; cationic; see Servamine KEO 260; liq.; 100% conc.

Servamine KET 4542. [552] N (alkyl amido propyl) N ethyl N,N dimethyl ammonium etho sulfate; cationic; emulsifier; liq.; biodeg.; 50% conc.

Servamine KOO 330 B. [552] Oleyl amido ethyl oleyl imidazoline; cationic; basic material in mfg. of quat. imidazoline; liq.; biodeg.; 100% conc.

Servirox OEG 45. [552] Castor oil polyglycol ester (17 EO); nonionic; used in textile, phytopharmaceutical, leather, paper and cosmetic industries; liq.; 100% conc.

Servirox OEG 55. [552] Castor oil polyglycol ester (26 EO); nonionic; see Servirox OEG 45; liq.; 100% conc.

Servirox OEG 65. [552] Castor oil polyglycol ester (32 EO); nonionic; see Servirox OEG 45; liq.; 100% conc.

Servirox OEG 90/50. [552] Castor oil polyglycol ester (180 EO); nonionic; see Servirox OEG 45; liq.; 50% conc.

Servo Amfolyt JA 110. [552] Imidazoline, modified; amphoteric; shampoo formulations; liq.; biodeg.; 40% conc.

Servo Amfolyt JA 140. [552] Imidazoline, modified; amphoteric; hair shampoos; liq.; biodeg.; 40% conc.

Servo Brillant Olie B AZ 75. [552] Sodium castor oil; anionic; softener and finishing oil; pasting oil for dyestuffs; liq.; 66% conc.

Servo XRK 60. [552] 3-Chlor-2-hydroxy propyl trimethyl ammonium chloride; agent for preparation of cationic starch; water-sol.

Servoxyl VLA 2170. [552] Sodium di-2-ethylhexyl sulfosuccinate; anionic; wetting and rewetting agent; liq.; biodeg.; 70% conc.

Servoxyl VLB 1123. [552] Sodium monoalkyl polyglycol ether sulfosuccinate; anionic; raw material for personal care products; cleaning agent; liq.; biodeg.; 40% conc.

Servoxyl VPGZ 7/100. [552] Cetyl oleyl ether phosphate, acid form (7 EO); anionic; detergent in dry-cleaning, emulsifier in formulation of metal cleaners in emulsion polymerization; pesticide and cosmetic preparations; liq.; 100% conc.

Servoxyl VPIZ 100. [552] Acid butyl phosphate; see Servoxyl VPGZ 7/100; liq.; 100% conc.

Servoxyl VPNZ 10/100. [552] Nonylphenol ether phosphate, acid form (10 EO); anionic; see Servoxyl VPGZ 7/100; liq.; 100% conc.

Servoxyl VPQZ 9/100. [552] Dinonylphenol polyglycol ether phosphate, acid form (9 EO); anionic; see Servoxyl VPGZ 7/100; liq.; 100% conc.

Servoxyl VPTZ 3/100. [552] 2-Ethylhexyl polyglycol ether phosphate, acid form; anionic; see Servoxyl VPGZ 7/100; liq.; 100% conc.

Servoxyl VPTZ 100. [552] 2-Ethylhexyl phosphate, acid form; anionic; see Servoxyl VPGZ 7/100; liq.; 100% conc.

Servoxyl VPUZ. [552] Acid methyl phosphate; see Servoxyl VPGZ 7/100; liq.; 100% conc.

Servoxyl VPYZ 500. [552] Acid phosphate ester, modified; anionic; see Servoxyl VPGZ 7/100; liq.; 100% conc.

Serwet WH 170. [552] Sodium di-2-ethylhexyl sulfosuccinate; anionic; wetting agent; sol. in water and alcohol; 65% conc.

Setacin 103 Spezial. [660] Lauryl alcohol polyglycol ether sulfosuccinate, sodium salt; anionic; detergent for personal care products; cleaning agent; liq.; 40% conc.

Setacin F Special Paste. [660] Coconut alcohol sulfosuccinate, sodium salt; anionic; detergent, cosmetics, washing and cleaning agent; paste; 40% conc.

Setacin M. [660] Sulfosuccinate, alkanolamine salt; anionic; detergent, personal care products and cleaning agents; liq.; 40–42% conc.

Simulsol 52. [551] Ceteth-2; nonionic; emulsifier; solid; HLB 5.3; 100% conc.

Simulsol 56. [551] Ceteth-10; nonionic; emulsifier; solid; HLB 12.9; 100% conc.

Simulsol 58. [551] Ceteth-20; nonionic; emulsifier; solid; HLB 15.7; 100% conc.

Simulsol 72. [551] Steareth-2; nonionic; emulsifier; solid; HLB 4.9; 100% conc.

Simulsol 76. [551] Steareth-10; nonionic; emulsifier; solid; HLB 12.4; 100% conc.

Simulsol 78. [551] Steareth-20; nonionic; emulsifier; solid; HLB 15.3; 100% conc.

Simulsol 92. [551] Oleth-2; nonionic; emulsifier; liq.; HLB 4.9; 100% conc.

Simulsol 96. [551] Oleth-10; nonionic; emulsifier; liq.; HLB 12.4; 100% conc.

Simulsol 98. [551] Oleth-20; nonionic; emulsifier; solid; HLB 15.3; 100% conc.

Simulsol A. [551] Alkyl polyethoxy ester; nonionic; emulsifier; liq.; 100% conc.

Simulsol A 686. [551] Alkyl polyethoxy ester; nonionic; emulsifier; powd.; 100% conc.

Simulsol M 45. [551] PEG-8 stearate; nonionic; emulsifier; solid; HLB 11.1; 100% conc.

Simulsol M 49. [551] PEG-20 stearate; nonionic; emulsifier; solid; HLB 15.0; 100% conc.

Simulsol M 51. [551] PEG-30 stearate; nonionic; emulsifier; solid; HLB 16.0; 100% conc.

Simulsol M 52. [551] PEG-40 stearate; nonionic; emulsifier; solid; HLB 16.9; 100% conc.

Simulsol M 53. [551] PEG-50 stearate; nonionic; emulsifier; solid; HLB 17.9; 100% conc.

Simulsol M 59. [551] PEG-100 stearate; nonionic; emulsifier; solid; HLB 18.8; 100% conc.

Simulsol NP 575. [551] Alkylphenyl polyethoxy ether; nonionic; emulsifier; liq.; 100% conc.

Simulsol O. [551] Alkyl polyethoxy ether; nonionic; emulsifier; paste; 100% conc.

Simulsol P4. [551] Laureth-4; nonionic; emulsifier; liq.; 100% conc.

Sipenol IC2. [20] Coco amine, ethoxylated; cationic; coemulsifier and intermediate, dyeing assistant, antistat, softener and lubricant treatment; liq.; 100% conc.

Sipenol IC10. [20] Coco amine, ethoxylated; nonionic; see Sipenol IC2; liq.; 100% conc.

Sipenol IO12. [20] Oleyl amine, ethoxylated; nonionic; see Sipenol IC2; liq.; 100% conc.

Sipenol IS05. [20] Fatty amine, ethoxylated; nonionic; see Sipenol IC2; liq.; 100% conc.

Sipenol IT11. [20] Tallow amine, ethoxylated; nonionic; see Sipenol IC2; liq.; 100% conc.

Sipenol IT50. [20] Tallow amine, ethoxylated; nonionic; see Sipenol IC2; wax; 100% conc.

Sipex 206C-SP. [20] Alkyl phenol ethoxylate, modified; anionic; surfactant for rug and fabric shampoo concs.; clear liq.; visc. 80 cps; cloud pt. < 10 C; pH 7; 29% act.

Sipex ADS. [20] Alkyl sulfate; anionic; surfactant and wetting agent for textile, industrial and agric. applications; liq.; 85% conc.

Sipex BOS. [20] Sodium octyl sulfate; anionic; wetting agent, emulsifier, detergent, foamer, rinse aid; post-stabilizer in latex paints; metal treatment; textile and plywood mfg.; visc. control in plywood mfg.; fruit and veg. washing; hard surface cleaners; clear liq.; visc. 50 cps; HLB 42; cloud pt. < 10 C; pH 9.5–10.5 (10%); 39–40% act.

Sipex CAV. [20] Sodium isodecyl sulfate; see Sipex BOS, also aid for emulsion polymerization of vinyl, acrylic, SBR, and PVC systems; clear liq.; dens. 8.95 lb/gal; visc. 150 cps; cloud pt. < 10 C; pH 9.5–10.5 (10%); 39–40% act.

Sipex EC-111. [20] Sodium cetyl sulfate; emulsifier, detergent; collector in ore flotation; softener/lubricant for textiles; cosmetics and toiletries; wh. thick paste; pH 8 (10%); 25% act.

Sipex EST-30. [249] Sodium trideceth sulfate; anionic; wetting agent, foamer, detergent base; sole emulsifier and post-stabilizer in emulsion polymerization of styrene systems; household and industrial cleaners; scouring of fibers; liq. dishwashing compds.; Gardner 2 max. liq.; cloud pt. 14 C max.; pH 7.5–8.5 (10%); 29–30% act.

Sipex EST-60. [249] Sodium trideceth sulfate; wetting agent, detergent; household cleaners; industrial metal cleaning processes; clear liq.; pH 9 (10%); 57% act.

Sipex EST-75. [249] Sodium trideceth sulfate; anionic; emulsifier, wetting agent; personal care and cleaning products; liq.; pH 9.0–10.5 (10%); 70% act.

Sipex FA30. [20] Alkyl sulfate, modified; anionic; industrial foamer; latex foam, carpet backing; liq.; 28–30% conc.

Sipex LCP. [20] Sodium lauryl sulfate; surfactant for emulsion polymers; visc. 50 cps; cloud pt. 1 C; pH 8.0 (10%); 30% act.

Sipex OK. [20] Alcohol sulfate; anionic; specialty surfactant for use as textile lubricant and spin-finish ingred.; liq.; 65% act.

Sipex OLS. [20] Sodium octyl sulfate; anionic; wetting agent; rinse aid; mercerizing agent for cotton goods; surfactant in electrolyte baths for metal cleaning; hard surface cleaning; neoprene dispersant; clear liq.; visc. 100 cps; HLB 42; pH 8 (10%); 33% act.

Sipex OS. [20] Sodium oleyl sulfate; anionic;

emulsifier, emulsion polymerization; hair products and cosmetics; food pkg.; clear liq.; HLB 42; pH 8 (10%); 26% act.

Sipex SB. [20] Sodium lauryl sulfate; anionic; emulsifier, emulsion polymerization; food washing and pkg. applications; metal cleaning, textile scouring; commercial car washing; clear liq.; visc. 150 cps; cloud pt. 20 C; pH 8 (10%); 29% act.

Sipex SD. [20] Sodium lauryl sulfate; anionic; surfactant used in metal processing, electroplating, textile mfg., leather treatment, rug and upholstery shampoos; wh. powd.; HLB 40; pH 7.5 (3%); 95% act.

Sipex TDS. [20] Sodium tridecyl sulfate; anionic; emulsifier, wetting agent, emulsion polymerization of vinyl chloride, styrene, and styrene/acrylic monomers; detergent formulations; colorless liq.; visc. 100 cps; cloud pt. 10 C; pH 8.5 (10%); 25% act.

Sipex TEA. [20] Triethanolamine lauryl sulfate; detergent, emulsifier; pigment dispersant and stabilizer; wool scouring; ore flotation; rug shampoos; latex paint formulations; dishwashing detergent; clear liq.; visc. 30 cps; cloud pt. 2 C; pH 7 (10%); 40% act.

Sipex UB. [20] Sodium lauryl sulfate; anionic; emulsifier, surfactant for vinyl chloride, vinylidene chloride, styrene, butadiene, and acrylic ester emulsion polymers; liq.; visc. 50 cps; HLB 40; cloud pt. 18 C; pH 8 (10%); 30% act.

Sipomide 843. [20] Lauramide DEA; visc. and foam booster for personal care, household and industrial products used in cleaning; solid; 100% act.

Sipomide 1500. [20] Cocamide DEA; see Sipomide 843; liq.; 100% act.

Sipon 201-10, ES. [20] Sodium laureth (3) sulfate; surfactant; personal care products; liq.; 58 and 27% act. respectively.

Sipon 201-20. [20] Ammonium laureth (3) sulfate; surfactant used in shampoos, bubble baths, lt. duty household products; liq.; 58% act.

Sipon EA. [20] Ammonium laureth sulfate; anionic; detergent, wetting agent, foamer, solubilizer, penetrant, personal care products and pet shampoos; liq. detergent for fabrics; liq.; mild, pleasant odor; visc. 300 cps; cloud pt. –5 C; pH 6.7 (10%); 27% act.

Sipon EA2. [20] Ammonium laureth sulfate; anionic; detergent used in personal care products; liq.; 26% conc.

Sipon EAY. [20] Ammonium laureth sulfate; anionic; see Sipon EA; thixotropic product; cloud pt. 0 C; pH 6.7 (10%); 26% act.

Sipon ES. [20] Sodium lauryl ether sulfate; anionic; detergent used in personal care products; polymerization; liq.; 25–27% conc.

Sipon ES-2. [20] Sodium laureth (2) sulfate; see Sipon 201-10; liq.; 26% act.

Sipon ES-7. [20] Sodium laureth (7) sulfate; see Sipon 201-10; liq.; 28% act.

Sipon ES-12. [20] Sodium laureth (12) sulfate; see Sipon 201-10; liq.; 60% act.

Sipon ESY. [20] Sodium laureth (1) sulfate; see Sipon 201-10; liq. 25% act.

Sipon L-22. [20] Ammonium lauryl sulfate; anionic; detergent, foamer; personal care products; pet shampoos; wool scouring agent; assistant for pigment dispersion; emulsion polymerization aid; fire fighting foams; clear liq.; visc. 1000 cps; HLB 31; cloud pt. 14 C; pH 6.8 (10%); 28% act.

Sipon L-22HV. [20] Ammonium lauryl sulfate; surfactant for personal care products and SBR latex froth applications; liq.; 28% act.

Sipon LD. [20] DEA-lauryl sulfate; surfactant for shampoos and bubble baths; liq.; 40% act.

Sipon LM. [20] Magnesium lauryl sulfate; anionic; detergent, surfactant for shampoos, bubble baths, rug and unholstery shampoos; food pkg. applications; clear liq.; visc. 50 cps; cloud pt. 2 C; pH 6.5 (10%); 27% act.

Sipon LS. [20] Sodium lauryl sulfate; anionic; detergent, surfactant for personal care products; visc. liq.; visc. 25,000 cps; HLB 40; cloud pt. 20 C; pH 8 (10%); 28% act.

Sipon LSB. [20] Sodium lauryl sulfate; anionic; detergent for personal care products; pale clear liq.; visc. 150 cps; HLB 40; cloud pt. 20 C; pH 8 (10%); 29% act.

Sipon LT-6. [20] TEA-lauryl sulfate; anionic; shampoo ingred.; pale yel. clear liq.; visc. 50 cps; HLB 34.0; cloud pt. 2 C; pH 7 (10%); 40% act.

Sipon MLS. [20] Magnesium lauryl sulfate, pharmaceutical grade; lubricant in drug tableting operations; spray-dried; pH 6.5 (3%); 88% act.

Sipon WD. [20] Sodium lauryl sulfate, high purity; anionic; foaming agent; food processing emulsifier, film former on fresh citrus; pharmaceuticals; household formulations; toiletries; industrial applications; crystal; dens. 5 lb/gal; HLB 40; pH 8.6 (1%); 98% act.

Siponate 301-10 F. [20] Sodium C_{14-16} olefin sulfonate; detergent for personal care products, soaps, and fabrics, flake; 90% act.

Siponate 330. [20] Alkylaryl sulfonate; anionic; emulsifier, solubilizer, solv., penetrant; emulsifier for agric. pesticides; degreaser and wax removers; drycleaning; latex paints, pigment dispersant, metal cleaning; amber liq.; sp.gr. 1.03; HLB 11.7; visc. 6500 cps; pH 3–6 (5%); 90% conc.

Siponate A-167. [20] Sodium C_{16-18} olefin sulfonate; detergent, liq.; 30% act.

Siponate A-168. [20] Sodium alpha-olefin C_{15-18} sulfonate; anionic; detergent and foamer; textile and industrial applications; personal care products; liq.; 30% conc.

Siponate A-246 L. [20] Sodium C_{14-16} olefin sulfonate; detergent and emulsifier; personal care products; dishwashing detergents; car washes; pet shampoos, general purpose cleaners; liq.; 40% act.

Siponate A 246 LX. [20] Sodium C_{14-16} olefin sulfonate; detergent, visc. builder; liq.; 40% act.

Siponate DS-4. [20] Sodium dodecylbenzene sulfonate; anionic; emulsifier for polymers; food pkg. applications; surfactant for washing fruits and vegs.; dk. brn. liq.; 23% act.

Siponate DS-10. [20] Sodium dodecylbenzene sulfonate; anionic; emulsifier and surfactant for emulsion polymerization; heavy duty household and institutional cleaning products; oil recovery operations; washing fruits and vegs.; food pkg. materials; flake; pH 7.5 (10%); 98% act.

Siponate SA. [20] Linear alkylate sulfonic acid; anionic; detergent, emulsifier used in detergent formulations; liq.; 97% conc.

Siponate X200. [20] Sodium octoxynol-2 ethane sulfonate; anionic; detergent and emulsifier for polymerization; used in personal care and pharmaceutical applications; household and industrial cleaners; slurry; 28% conc.

Siponic 218. [20] PEG-10 isolauryl thioether; nonionic; detergent, emulsifier, wetting agent for metal cleaning, steel processing, textile treatment, cosmetics, agric. formulations, wood pulp and paper industries; Gardner 6 max. liq.; sol. in water and alkaline detergent builders; dens. 8.6 lb/gal; sp.gr. 1.05; cloud pt. 52 C (1%); surf. tens. 28 dynes/cm (0.05%); 95% act.

Siponic 260. [20] PEG-6 isolauryl thioether; nonionic; see Siponic 218; Gardner 6 max. liq.; sol. in org. solvs.; dens. 8.4 lb/gal; sp.gr. 1.01; surf. tens. 29 dynes/cm (0.05%); 98% act.

Siponic 1173. [20] C_{22-24} pareth-33; nonionic; paraffin wax and heavy oil emulsifier; solid; HLB 15.0; 100% conc.

Siponic C-20. [20] Cetearyl alcohol, ceteareth-20; nonionic; emulsifier for creams and lotions; HLB 5.3; 100% act.

Siponic E-2. [20] Ceteareth-4; nonionic; emulsifier, lubricant, emollient, and conditioner for personal care products; dyeing assistant, leveling agent, emulsion polymerization; HLB 8.0; 100% act.

Siponic E-3. [20] Ceteareth-6; nonionic; see Siponic E-2; HLB 10.1; 100% act.

Siponic E-5. [20] Ceteareth-10; nonionic; rubber emulsifier, lubricant, dye and fulling assistant; conditioner and emollient for personal care products; emulsion polymerization; floor waxes, paper finishes, pharmaceuticals; wax; HLB 12.4; cloud pt. 73 C (1%); pH 6.5 (1%); 100% act.

Siponic E-7. [20] Ceteareth-14; nonionic; see Siponic E-5; also textile lubricant; HLB 14.1; cloud pt. 93 C (1%); pH 6.5 (1%); 100% act.

Siponic E-10. [20] Ceteareth-20; nonionic; see Siponic E-2; HLB 15.3; 100% act.

Siponic E-15. [20] Ceteareth-30; nonionic; see Siponic E-5; wax; HLB 16.7; cloud pt. > 95 C (1%); pH 6.5 (1%); 100% act.

Siponic F-160. [20] Octoxynol-16; nonionic; primary surfactant in emulsion polymerization of latices; textile scouring and dye coupling applications; cosmetic products, household and industrial cleaners; HLB 15.8; 100% act.

Siponic F-300. [20] Octoxynol-30; nonionic; see Siponic F-160; HLB 17.3; 70% act.

Siponic F-400. [20] Octoxynol-40; nonionic; see Siponic F-160; HLB 17.9; 70% act.

Siponic F-707. [20] Octoxynol-70; nonionic; see Siponic F-160; HLB 18.7; 70% act.

Siponic L1. [20] Laureth-1; nonionic; thickener in shampoos and bubble baths; stabilizer for emulsion polymers; emulsifier for monomer systems, floor waxes, paper finishes, rubber; emollient for pharmaceuticals; liq.; HLB 3.7 (1%); pH 6.5 (1%); 100% act.

Siponic L4. [20] Laureth-4; nonionic; see Siponic L1; also emulsifier for personal care products; liq.; HLB 9.4; pH 6.5 (1%); 100% act.

Siponic L7. [20] Laureth-7; nonionic; emulsifier for personal care products, rubber and monomer systems; stabilizer for emulsion polymers; floor waxes, paper finishes, emollient for pharmaceuticals; wax; HLB 12.5; cloud pt. 56 C (1%); pH 6.5 (1%); 100% act.

Siponic L-7-90. [20] Laureth-7; nonionic; thickener, emulsifier, visc. control, lubricant, solubilizer for personal care products; emulsion polymerization surfactant; textile spin finishes; HLB 12.1; 90% act.

Siponic L12. [20] Laureth-12; see Siponic L7; wax; HLB 14.6; cloud pt. 89 C (1%); pH 6.5 (1%); 100% act.

Siponic L16. [20] Laureth-16; see Siponic L7; wax; HLB 15.8; cloud pt. > 95 C (1%); pH 6.5 (1%); 100% act.

Siponic L25. [20] Laureth-23; see Siponic L7; wax; HLB 16.9; cloud pt. > 95 C (1%); pH 6.5 (1%); 100% act.

Siponic NP4. [20] Nonoxynol-4; nonionic; emulsifier; detergent, wetting agent, coupler, dispersant; liq.; oil-sol.; HLB 8.1; 100% conc.

Siponic NP6. [20] Nonoxynol-6; nonionic; coemulsifier; intermediate for the synthesis of anionic surfactants; liq.; HLB 10.7; 100% conc.

Siponic NP9. [20] Nonoxynol-9; nonionic; emulsifier, detergent, wetting agent; used in polymerization, paints, textiles, leather, paper, cleaning compds.; oil-sol.; HLB 12.7; 100% conc.

Siponic NP10.5. [20] Nonoxynol-10; nonionic; see Siponic NP9; liq.; HLB 13.4; 100% conc.

Siponic NP15. [20] Nonoxynol-15; nonionic; see Siponic NP9; liq.; HLB 14.9; 100% conc.

Siponic NP40. [20] Nonoxynol-40; nonionic; detergent, wetting agent; coemulsifier and vinyl/acrylic latex stabilizer; emulsifier for floor waxes and polishes; liq.; water-sol.; HLB 17.7; 70% conc.

Siponic NP75. [20] Nonoxynol-75; nonionic; emulsifier for emulsions; wetting agent and dispersant; liq.; HLB 17.7; 70% conc.

Siponic OP1.5. [20] POE (1) octyl phenyl ether (1.5 EO); nonionic; emulsifier, detergent, wetting agent, coupler; liq.; oil-sol.; HLB 4.7; 100% conc.

Siponic OP7. [20] POE (7) octyl phenyl ether; nonionic; emulsifier and wetting agent; liq.; HLB 11.5; 100% conc.

Siponic OP10. [20] POE (10) octyl phenyl ether; nonionic; emulsifier for vinyl and acrylic polymerization; detergent and wetting agent; liq.; water-sol.; HLB 14.4; 100% conc.

Siponic OP30. [20] POE (30) octyl phenyl ether; nonionic; detergent, wetting agent; coemulsifier and vinyl/acrylic latex stabilizer; emulsifier for fats and waxes; liq.; HLB 17.3; 70% conc.

Siponic OP40. [20] POE (40) octyl phenyl ether; nonionic; see Siponic OP30; liq.; HLB 17.9; 70% conc.

Siponic SK. [20] PEG-8 isolauryl thioether; nonionic; wetting agent, metal cleaning; heavy duty detergent; inhibitor in steel processing; scouring of textiles; agric. formulations; antiskinning agent for paints; hair products; wood and paper industry; emulsifier for petroleum oils, chlorinated solvs., silicones, metallic soaps, and emulsion polymers; Gardner 6 max. liq.; sol. in water; dens. 8.5 lb/gal; sp.gr. 1.03; cloud pt. 28 C (1%); surf. tens. 31 dynes/cm (0.05%); 98% act.

Siponic TD-3. [20] Trideceth-3; nonionic; emulsifier used in textile scouring, household and metal cleaners and industrial applications; surfactant in cosmetic emulsions; HLB 7.9; 100% act.

Siponic TD-6. [20] Trideceth-6; nonionic; emulsifier, solubilizer used in dry cleaning solvs., agric. sprays; degreaser; metal cleaning; clear oily liq.; oil-sol.; dens. 8.2 lb/gal; sp.gr. 0.98; visc. 80 cps; HLB 11.4; flash pt. 355 F; pH 6 (1%); 99% act.

Siponic TD-12. [20] Trideceth-12; nonionic; see Siponic TD-3; HLB 14.5; 100% act.

Siponic TD990. [20] Trideceth-9; nonionic; emulsifier, emollient and lubricant for skin creams, lotions; liq.; HLB 12.9; 90% conc.

Siponic Y500-70. [20] Oleth-25; nonionic; wetting agent, emulsifier for polymerization; stabilizer for latices; dispersant in polyurethane foam; textile industry; foamer for latex; floor waxes; pharmaceutical emulsions; metal cleaning; leather industry; food pkg. applications; liq.; mild odor; visc. 1650 cps; HLB 16.2; cloud pt. 6 C (1%); pH 7.5 (1%); 70% act.

Siponic Y501. [20] Oleth-20; nonionic; see Siponic Y500-70; solid; mild odor; sol. in water; m.p. 35 C; HLB 15.4; cloud pt. 65 C (1%); pH 7.5 (1%); 100% act.

Sipoteric 1112. [20] DEA-C_{12-15} alcohols sulfate, cocoaminopropionic acid; amphoteric; surfactant for shampoo and bubble baths; liq.; 35% act.

Sipoteric 1127. [20] Lauroamphoacetate; amphoteric; surfactant for nonirritating shampoos; liq.; 40% act.

Sipoteric 1304. [20] Sodium trideceth sulfate, cocoamphodiacetate; amphoteric; baby shampoo base; liq.; 37% act.

Sipoteric 1398. [20] Cocoamphodiacetate; amphoteric; surfactant for nonsting and nonirritating shampoos; liq.; 40% act.

Sipoteric 1442. [20] Sodium laureth sulfate and stearoamphodiacetate; amphoteric/anionic; liq. pearl conc. useful in making "no heat" and "cold mix" formulas; liq.; 21% act.

Sipoteric 1444. [20] Ammonium lauryl sulfate and stearoamphodiacetate; amphoteric/anionic; see Sipoteric 1442; liq.; 21% act.

Sipoteric 1450. [20] Sodium C_{14-16} olefin sulfonate and stearoamphodiacetate; amphoteric/anionic; see Sipoteric 1442; liq.; 35% act.

Sipoteric CAPA. [20] Cocaminopropionic acid; amphoteric; surfactant used as emulsifier, hard surface cleaner, general purpose applications; liq.; 40% act.

Sipoteric CB, COB. [20] Cocamidopropyl betaine; amphoteric; visc. and foam builder for personal care products and toiletries; liq.; 35% act.

Sipoteric GPA. [20] Cocoamphodiacetate and stearoamphodiacetate; amphoteric; surfactant for personal care products; paste; 45% act.

Sipoteric GSC. [20] Sodium laureth sulfate and cocoamphodiacetate; anionic/amphoteric; surfactant used in pH formulation; liq.; 30% act.

Sipoteric SLIP. [20] Sodium laurimino dipropionate; amphoteric; see Sipoteric CAPA; liq.; 30% act.

SK Flot 1. [561] Amine base, acetate salt; cationic; min. flotation reagent; solid, paste; 100% conc.

SK Flot 2. [561] Amine base, acetate salt; cationic; see SK Flot 1; paste; 100% conc.

SK Flot 3. [561] Amine base, acetate salt; cationic; see SK Flot 1; solid; 100% conc.

SK Flot 4. [561] Amine base, acetate salt; cationic; see SK Flot 1; solid, flakes; 100% conc.

Skinotan S 10. [660] Polysiloxane polyglycol ether; nonionic; shampoo additive; liq.; 100% conc.

Skliro. [159] Lanolin acid; industrial uses; paste.

Skyllex. [2] Dialkyl dimethyl ammonium chloride; fabric softener; Gardner 3; m.p. 50 C; 75% act.

Slik No. 5. [447] Alkanolamide and oxyethylated alkylphenol; nonionic; degreaser; solv. cleaner or detergent for metal and industrial cleaning; liq.; water-sol.; dens. 8.38 lb/gal; sp.gr. 1.004 (68 F); visc. 25 SUS (70 F); pour pt. 32 F; 18% conc.

Slik No. 52. [446] Alkanolamide and oxyethylated alkylphenol; nonionic; degreaser and multi-purpose cleaner; liq.; 100% conc.

S-Maz 20. [400] Sorbitan laurate; nonionic; surfactant for w/o emulsions; antistat, textile softener, lubricant; process defoamer, opacifier, coemulsifier; cosmetics, food formulations, industrial oils and household products; amber liq.; sol. in ethanol, naphtha; disp. in water; sp.gr. 1.0; visc. 4500 cps; HLB 8.6; sapon. no. 158–170; 100% conc.

S-Maz 40. [400] Sorbitan palmitate; nonionic; see S-Maz 20; tan flakes; m.p. 45–46 C; HLB 6.7; sapon. no. 140–150; 100% conc.

S-Maz 60. [400] Sorbitan stearate; nonionic; see S-Maz 20; cream flakes; m.p. 50–53 C; HLB 4.7; sapon. no. 147–157; 100% conc.

S-Maz 65. [400] Sorbitan tristearate; nonionic; see S-Maz 20; cream flakes; sp.gr. 1.0; m.p. 53–55 C; HLB 2.1; sapon. no. 176–188; 100% conc.

S-Maz 67. [400] Sorbitan monoisostearate; nonionic; see S-Maz 20; lt. yel. liq.; sol. in min. and veg. oil; disp. in water; sp.gr. 1.0; visc. 1000 cps; HLB 4.7; sapon. no. 146–157; 100% conc.

S-Maz 80. [400] Sorbitan oleate; nonionic; see S-Maz 20; amber liq.; sol. in min. and veg. oil; disp. in water; sp.gr. 1.0; visc. 1000 cps; HLB 4.3; sapon. no. 149–160; 100% conc.

S-Maz 83R. [400] Sorbitan sesquioleate; nonionic; solubilizer, emulsifier and dispersant; liq.; sol. in oils and solvs.; HLB 3.7; 100% conc.

S-Maz 85. [400] Sorbitan trioleate; nonionic; see S-Maz 20; amber liq.; sol. in toluol, naphtha, min. and veg. oil; disp. in water; sp.gr. 1.0; visc. 200 cps; HLB 1.8; sapon. no. 172–186; 100% conc.

S-Maz 90. [400] Sorbitan monotallate; nonionic; see S-Maz 20; amber liq.; sol. in min. and veg. oil; disp. in water; sp.gr. 1.0; visc. 800 cps; HLB 4.1; sapon. no. 145–160; 100% conc.

S-Maz 95. [400] Sorbitan tritallate; nonionic; lipophilic surfactant used for preparing w/o emulsions, antistat, textile softener, lubricant, process defoamer, opacifier, and coemulsifier; cosmetics, food formulations, industrial oils, and household products; amber liq.; sol. in toluol, naphtha, min. and veg. oil; disp. in water; sp.gr. 0.9; visc. 200 cps; HLB 1.7; sapon. no. 168–182; 1.0% max. water.

Sochamine 2662. [654] Quat. ammonium compd.; catonic; textile softener; liq.; 75% conc.

Sochamine A 755. [654] Monocarboxylate coconut deriv., sodium salt; amphoteric; detergent, wetting agent; base for shampoos; liq.; 40% conc.

Sochamine A 7525. [654] Dicarboxylate coconut deriv., sodium salt; amphoteric; see Sochamine A 755; liq.; 40% conc.

Sochamine A 7527. [654] Dicarboxylate coconut deriv., sodium salt, salt free; amphoteric; detergent; wetting agent; base for shampoos; liq.; 40% conc.

Sochamine A 8955. [654] Dicarboxylic caprylic deriv.; amphoteric; foamer and wetting agent in alkaline formulations; liq.; 38% conc.

Sochamine OX 30. [654] Amine oxide; cationic; lt. duty detergent; liq.; 30% conc.

Sochamine SG 8, 100. [654] Fatty acid, ethoxylated; nonionic; textile softener; antistat; paste.

Sofbon C-1-S. [590] Quat. ammonium compd.; cationic; softener for polyester jersey; paste.

Sofnon 105G. [590] Imidazoline deriv.; cationic; softener for textiles; wax; 80% conc.

Sofnon GF-2, -5. [606] Alkyl amide imidazoline compd. acetate; cationic; softener for glass fibers; liq.

Softal 300. [607] Polyamine; cationic; softener and lubricant for syn. fibers; liq.; 15% conc.

Softal A-3. [607] Aliphatic amine deriv.; cationic; softener for acrylic fibers; paste; 15% conc.

Softal MR-30. [607] Special polyamide deriv.; cationic/nonionic; softener for syn. fibers and blends; liq.

Soft Detergent 60. [370] Alkyl benzene sulfonate; anionic; general use detergent; powd.; 60% conc.

Soft Detergent 95. [370] alpha-Olefin sulfonate; anionic; emulsifier and dispersant for emulsion polymerization; powd.; 95% conc.

Softenol 3100. [213] C_{12-18} fatty acid triglyceride; lubricant for machinery used in mfg. of foodstuffs; additive to textile and glass fiber finishing, sizing agent; block pallets; neutral odor and taste; sol. in MEK, MIBK, ethyl acetate; dens. 0.876 g/cm^3 (80 C); visc. 32.9 mm^2/s (40 C); m.p. 33–35 C; sapon. no. 235–245; biodeg.

Softenol 3107. [213] C_7 fatty acid triglyceride; lubricant for machinery for confectionary and food industries; mold release agent in processing of plastics; additive to cutting oils, lacquer and varnish systems; antiblocking agent for artificial skins; liq.; neutral odor and taste; dens. 0.930 g/cm^3; visc. 20.6 mm^2/s; sapon. no. 385–395; biodeg.

Softenol 3108. [213] C_{8-10} fatty acid triglyceride; lubricant for machinery used in food and confectionary industries and medical/technical equipment; antiblocking agent for artificial skins; oil in mfg. of aluminum foils; mold release agent for plastics; additive to cutting oils and lubricants; hydrophobing agent against moisture; mfg. of textile and glass-fiber sizings; liq.; neutral odor and taste; dens. 0.950 g/cm^3; visc. 31 mm^2/s; sapon. no. 325–345; biodeg.

Softenol 3114. [213] C_{14} fatty acid triglyceride; compressing aid in tableting technical substances; hydrophobing agent against moisture; lubricant for machinery, lacquer and varnish formulations; additive to liq. lubricating systems; powd.; neutral odor and taste; dens. 8.73 g/cm^3 (80 C); visc. 11.2 mm^2/s (80 C); m.p. 52–55 C; sapon. no. 227–235; biodeg.

Softenol 3118. [213] C_{18} fatty acid triglyceride; see Softenol 3114; microfine powd.; neutral odor and taste; dens. 0.863 g/cm^3 (80 C); visc. 17.1 mm^2/s (80 C); m.p. 69–71 C; sapon. no. 188–190; biodeg.

Softenol 3178. [213] C_{8-18} fatty acid triglyceride; lubricant; paste; neutral odor and taste; dens. 0.878 g/cm^3 (80 C); visc. 31.5 mm^2/s (40 C); m.p. 37–40 C; sapon. no. 240–255.

Softenol 3408. [213] C_{8-10} fatty acid-1,2-propanediol ester; additive in mfg. of textile and glass fibers; lubricant for cutting devices; mold release agent in processing of plastics; liq.; neutral odor and taste; dens. 0.920 g/cm^3; visc. 10.9 mm^2/s; flash pt. 193 C; sapon. no. 320–340; biodeg.

Softenol 3819. [213] C_{6-10} fatty acid triglyceride; compressing aid in mfg. of dry substances; carrier; lubricant in cutting devices; raw material in preparation of lubricants; liq.; neutral odor and taste; dens. 0.964 g/cm^3; visc. 22.6 mm^2/s; flash pt. 219 C; sapon. no. 350–450; biodeg.

Softenol 3829. [213] C_{8-10} fatty acids triglyceride, modified; oil in mfg. of aluminum foils; lubricant for machinery, medical/technical equipment, instruments; additive to cutting oils and lubricants; liq.; neutral odor and taste; dens. 1.012 g/cm³; visc. 230 mm²/s; flash pt. 265 C; sapon. no. 400–450; biodeg.

Softenol 3900. [213] Glycerol monostearate; nonionic; emulsifier, lubricant, antistat, demolding and antiblocking agent, greaser, plasticizer, transparency and pigment dispersant used in mfg. and processing of plastics, textiles, leather, oils and polishing materials; powd.; neutral odor and taste; dens. 0.893 g/cm³ (80 C); visc. 22.9 mm²/s (80 C); m.p. 58–65 C; sapon. no. 155–175; biodeg.

Softenol 3991. [213] Glycerol monostearate; nonionic; see Softenol 3900; liq.; neutral odor and taste; dens. 0.914 g/cm³ (80 C); visc. 34.5 mm²/s (80 C); m.p. 65 C; sapon. no. 155–165; biodeg.; 90% 1-monoester.

Softex K-304. [341] Polyamide deriv., long chain; cationic; textile softener; liq.

Softex K-363. [341] Polyamide deriv., long chain; cationic; see Softex K-304; liq.

Softigen 701. [213; 344] Glyceryl ricinoleate; nonionic; refatting and skin protecting agent; emulsifier; personal care products and pharmaceuticals; yel.-wh. clear oily liq.; char. odor; sol. in ether, benzene, toluene, xylene, chlorofoam, and dichlorethylene; dens. 0.979–0.981; visc. 640–660 cps; HLB 4.5; sapon. no. 155–170; 100% conc.

Softigen 767. [213; 344] PEG-6 caprylic/capric glycerides; nonionic; refatting and wetting agent, solubilizer, emollient used in cosmetics and pharmaceuticals; detergents; yel. visc. oily liq.; char. odor; sol. in water, acetone, ethyl and butyl acetate; dens. 1.068 g/ml; visc. 110–130 cps; HLB 19; sapon. no. 90–100; surf. tens. 29.7 dynes/cm (1%); 100% conc.

Softisan 100. [213] Trilaurin; nonionic; ointment base for personal care products and pharmaceutical industry; wh. solid; neutral odor and taste; ref. index 1.4490–1.4510 (40 C); sol. in benzene, ether, xylene, toluene, chloroform, carbon tetrachloride, dioxane; sp.gr. 0.950–0.980; m.p. 33–36 C; visc. 30 cps (40 C); solid. pt. 29–33 C; sapon. no. 230–240.

Softisan 133. [213] Hydrog. coco-glycerides; hard wax used in cosmetic and pharmaceutical preparations; wh. blocks; neutral odor and taste; sol. in benzene, toluene, acetone, chloroform, petroleum spirit; insol. in water; m.p. 32–34 C; solid. pt. 29–32 C; sapon. no. 240–255; ref. index 1.445–1.449 (50 C).

Softisan 134. [213] Hydrog. coco-glycerides; see Softisan 133; wh. blocks; neutral odor and taste; sol. see Softisan 133; m.p. 33–36 C; solid. pt. 27–32 C; sapon. no. 225–235; ref. index 1.445–1.449 (50 C).

Softisan 138. [213] Hydrog. coco-glycerides; see Softisan 133; wh. blocks; neutral odor and taste; sol. see Softisan 133; m.p. 37–39 C; solid. pt. 32–36 C; sapon. no. 220–230; ref. index 1.445–1.449 (50 C).

Softisan 142. [213] Hydrog. coco-glycerides; see Softisan 133; wh. blocks; neutral odor and taste; sol. see Softisan 133; m.p. 42–44 C; solid. pt. 37–42 C; sapon. no. 220–230; ref. index 1.445–1.449 (50 C).

Softisan 154. [213] Hydrog. coco-glycerides; see Softisan 133; wh. flakes; neutral odor and taste; sol. see Softisan 133; m.p. 53–58 C; solid. pt. 48–53 C; sapon. no. 200–215; ref. index 1.445–1.460 (50 C).

Softisan 378. [213] Caprylic/capric/stearic triglyceride; nonionic; stabilizer; ointment base for personal care products; sapon. no. 240–255; 100% act.

Softisan 601. [213] Glyceryl cocoate, hydrog. coconut oil, and ceteareth-25; nonionic; cosmetic and pharmaceutical base for o/w products; wh.-yel. solid; sp.gr. 0.97 (60 C); sapon. no. 120–140; 100% conc.

Softisan Hard Fats. [344] Glycerol trilaurate/stearate; nonionic; stabilizer for emulsions, sticks; solid, pastille; m.p. 30–55 C; HLB 1.0; 100% conc.

Soft Touch. [460] Hydrated mono- and diglycerides, stearoyl lactylate, and polysorbate 60; hydrated emulsifier system; HLB 4.8; 13.5% min. alpha mono.

S-Oil. [347] Oleic acid chloride; anionic; intermediate for condensation reactions; liq.

Sol Mul 440. [565] Sodium petroleum sulfonates; anionic; additive for metal working industries; thick liq.; 60% conc.

Soitem 5 C/70. [565] Sodium dodecylbenzene sulfonate; anionic; surfactant for pesticides; liq.; HLB 4.0; 70% conc.

Soitem 8 FL/N. [565] Phosphate derivate, amminic salt; anionic; wetting agent and dispersant for pesticide formulations; thick liq.; 100% conc.

Soitem 13 FL/N. [565] Phosphate derivate,

amminic salt; anionic; see Soitem 8 FL/N; wax; 100% conc.

Soitem 101. [565] Ethoxylated phenolic compd. and org. sulfonate; anionic/nonionic; emulsifier for pesticides; thick liq.; HLB 11.5; 90% conc.

Soitem 251. [565] Ethoxylated fatty acid and org. sulfonate; anionic/nonionic; see Soitem 101; thick liq.; HLB 11.0; 87% conc.

Soitem 258. [565] Ethoxylated fatty acid and org. sulfonate; anionic/nonionic; see Soitem 101; liq.; 80% conc.

Soitem 480. [565] Ethoxylated, modified fatty acid and org. sulfate; anionic/nonionic; stabilizer and emulsifier for pesticides; thick liq.; 95% conc.

Soitem B. [565] Sodium alkylnaphthalene sulfonate, modified; anionic; wetting agent for pesticides; powd.; 50% conc.

Soitem SFL/1. [565] Polyglycol ethers; nonionic; wetting agent and dispersant for pesticide formulations; liq.; 100% conc.

Solamine BO5. [551] Imidazoline, modified; cationic; softener; paste; 90% conc.

Solan. [159] PEG-60 lanolin; nonionic; emulsifier for cosmetic formulation; wax; HLB 16.2; 100% conc.

Solan E. [159] POE lanolin; nonionic; emollient and emulsifier; solid; water-sol.; HLB 16.2; 100% conc.

Solangel 401. [159] PEG-75 lanolin; nonionic; emulsifier, humectant for soap; nonionic; liq.; HLB 16.0; 50% conc.

Solangel 501. [159] POE (50) lanolin; nonionic; see Solangel 401; liq.; HLB 15.8; 50% conc.

Solangel 851. [159] POE (85) lanolin; nonionic; see Solangel 401; liq.; HLB 16.4; 50% conc.

Solar 25. [585] Coconut oil fatty acid amine condensate and amine sulfonate; nonionic/anionic; wetting agent, detergent, air entraining agent in home and industrial products; straw visc. liq.; sp.gr. 1.05; pour pt. 0 C; surf. tens. 34.5 dynes/cm (0.1% aq.); pH 8.5 (1%); 100% act.

Solar 40 BD Flakes and Granules. [585] Alkylaryl sulfonate; anionic; wetting agent, detergent base for built compds.; used in metal and dairy cleaning, dishwashing compds., building maintenance, agric. chemicals, automotive specialities; cream wh. flakes, granules; slight odor; dens. 37.4 lb/ft^3; surf. tens. 35.8 dynes/cm (0.1%); pH 9.05 (1%); 40% act.

Solar 80 Flakes & Granules. [585] Alkylaryl sulfonate; anionic; detergent, wetting agent; mfg. of detergents, bubble baths, dishwashing products, industrial cleaners, auto wash products; rewetting agent; paint and cement industries; cream wh. flakes, granules; slight pleasant odor; dens. 21.8–33.5 lb/ft^3; surf. tens. 35.8 dynes/cm (0.1%); pH 7.7 (1%); 85% act.

Solar Air. [585] Coconut fatty acid amine condensate; nonionic; detergent, air-entraining agent used in building industry; yel. liq.; ref. index 1.3427; sp.gr. 1.0015; visc. 5 cps; 25% act.

Solar Air-2. [585] Coconut fatty acid amine condensate; nonionic; see Solar Air; lt. liq.; ref. index 1.3520; sp.gr. 1.0040; visc. 10 cps.

Solar Air-4. [585] Coconut fatty acid amine condensate; nonionic; see Solar Air; lt. liq.; ref. index 1.3695; sp.gr. 1.0090; visc. 22 cps.

Solar CI-387. [585] Imidazoline hydrochloride; cationic; car wax compd. and cleaner, corrosion inhibitor; amber liq.; mild odor; sp.gr. 0.951; visc. 40 cps; pH 5–7; 52% act.

Solar CO. [585] Coconut oil fatty acid amine condensate; nonionic; cosmetic and detergent formulations; intermediate; dedusting and suds stabilizer in detergents; lt. straw visc. liq.; slight odor; sp.gr. 1.02; flash pt. 166 C; pour pt. 23 C; pH 9.5 (1%); 100% act.

Solar F221. [585] Amide, modified; nonionic; emulsifier, w/o emulsions; making of cutting and sol. oils; dk. amber liq.; odorless; dens. 8.5 lb/gal; pour pt. 10 F; 100% act.

Solar F342. [585] Amide, modified; nonionic; see Solar F221; also for laundering; dk. amber liq.; odorless; dens. 8 lb/gal; pour pt. –6 F; 100% act.

Solar NP. [585] Nonyl phenoxy POE ethanol; nonionic; detergent, emulsifier, wetting agent, penetrant, base for cleaning compds.; cosmetics; leather and textile processes; paint, plastics, and rubber dispersant; agric. wetting and emulsification; visc. clear liq.; odorless; sol. in water, ethylene glycol, carbon tetrachloride, xylene; sp.gr. 1.06; cloud pt. 125–133 F (1%); flash pt. 535–555 F; pour pt. 37 F; solid. pt. 26 F; surf. tens. 30 dynes/cm (0.01%); pH 7.0 (1%); 100% act.

Solar Regular. [585] Coconut oil fatty acid amine condensate and diethanolamine oleate; nonionic; detergent, thickener, wetting agent, corrosion inhibitor, degreaser, base for cleaning compds. used in textile, household and industrial industries; making of cutting and sol. oils; amber visc. liq.; sweet odor; water-sol.; sp.gr. 1.07; flash pt.

179 C; pour pt. –4 C; pH 9.5 (1%); 100% act.

Solar TE. [585] Coconut oil amide, modified; nonionic; detergent, lubricant; hard surface cleaners; textile fulling and scouring; coupler for alkalis; amber liq.; flash pt. 360 F; pour pt. 20 F; surf. tens. 32.0 dynes/cm (0.1%); pH 9.15 (1%).

Solegal W Conc. [301] Alkyl phenyl polyglycol ester; nonionic; emulsifier for textile printing; liq.

Sole-Onic CDS. [300] Diethylene glycol monolaurate; anionic; defoamer, emulsifier for emulsion paints, industrial oil emulsions; liq.; 100% conc.

Sole-Terge 8. [300] Oleic acid isopropanolamide sulfosuccinate; nonionic; detergent, emulsifier for personal care products; ore flotation and emulsion polymerization; liq.; 100% conc.

Sole-Terge TS-2-S. [300] Sodium 2-ethylhexyl sulfate; anionic; wetting agent, penetrant, industrial penetrant for textiles; liq.; 35% conc.

Solocod G. [506] Fish oil, sulfated; anionic; emulsifier oil for processing leather; liq.; 75% conc.

Solricin 135. [446] Potassium ricinoleate; anionic; emulsifier, lubricant, and stabilizer in rubber processes; germicide; making of cutting and sol. oils; household and cosmetic products; Gardner 2 liq.; dens. 8.6 lb/gal; sp.gr. 1.034; visc. 0.9 stokes; 32% act.

Solulan 5. [37] Laneth-5, ceteth-5, oleth-5 and steareth-5; nonionic; emulsifier, wetting agent, dispersant, conditioner, plasticizer, emollient used in personal care and dermatology products; foam stabilizer for detergent systems; amber semisolid; faint pleasant odor; sol. @ 5% in anhyd. ethanol; disp. in water; sapon. no. 14 max.; 100% conc.

Solulan 16. [37] Laneth-16, ceteth-16, oleth-16 and steareth-16; nonionic; see Solulan 5; lt. tan waxy solid; sol. @ 5% in water, 30% ethanol; HLB 15; cloud pt. 64–70 C; sapon. no. 8 max.; pH 4.5–7.5 (10% aq.); 100% conc.

Solulan 25. [37] Laneth-25, ceteth-25, oleth-25 and steareth-25; nonionic; yel. waxy solid; faint pleasant odor; sol. see Solulan 16; HLB 16; cloud pt. 85–92 C; sapon. no. 8 max.; pH 4.5–7.5 (10% aq.); 100% conc.

Solulan 75. [37] PEG-75 lanolin; nonionic; see Solulan 5; lt. yel.-amber waxy solid; faint pleasant odor; sol. see Solulan 16; HLB 15; cloud pt. 80–87 C; sapon. no. 10–20; pH 4.5–7.5 (10% aq.); 100% conc.

Solulan 97. [37] Polysorbate 80 acetate, cetyl acetate and acetylated lanolin alcohol; nonionic; see Solulan 5; lt. amber visc. liq.; faint pleasant odor; sol. @ 5% in water, ethanol; HLB 15; sapon. no. 110–130; pH 4.5–7.5 (10% aq.); 100% conc.

Solulan 98. [37] Polysorbate 80, acetylated lanolin alcohol, and cetyl acetate; nonionic; see Solulan 5, also aids pearling of stearic acid emulsions; lt. amber visc. liq.; faint pleasant odor; sol. @ 5% in water, ethanol; HLB 13; sapon. no. 65–75; pH 4.5–7.5 (10% aq.); 100% conc.

Solulan C-24. [37] Choleth-24 and ceteth-24; nonionic; see Solulan 5; off-wh. to pale yel. waxy solid; faint pleasant odor; sol. @ 5% in water; HLB 14; cloud pt. 88–95 C; sapon. no. 3 max.; 100% conc.

Solulan L-575. [37] PEG-75 lanolin; nonionic; see Solulan 5; lt. yel.-amber visc. liq.; faint pleasant odor; sol. @ 5% in water, 30% ethanol; HLB 15; cloud pt. 80–87 C; sapon. no. 10 max.; 50% conc.

Solulan PB-2. [37; 38] PPG-2 lanolin ether; nonionic; spreading agent, dispersant, plasticizer; emollient and conditioner for personal care products, detergents, pharmaceuticals, waxes, polishes, leather treatment; amber semisolid, liq.; sol. in castor oil, IPM, IPP, anhyd. ethanol; HLB 8.0; sapon. no. 7 max.; 100% conc.

Solulan PB-5. [37; 38] PPG-5 lanolin ether; nonionic; see Solulan PB-2; lt. amber clear visc. liq.; sol. see Solulan PB-2; HLB 10.0; sapon. no. 6 max.; 100% conc.

Solulan PB-10. [37; 38] PPG-10 lanolin ether; nonionic; see Solulan PB-2; straw clear heavy-visc. liq.; sol. see Solulan PB-2; HLB 12.0; sapon. no.a 4 max.; 100% conc.

Solulan PB-20. [37; 38] PPG-20 lanolin ether; nonionic; see Solulan PB-2; lt. straw clear med-visc. liq.; sol. see Solulan PB-2; HLB 14.0; sapon. no. 3 max.; 100% conc.

Solusol 75%. [43] Dioctyl sodium sulfosuccinate; anionic; wettng agent, cosmetic formulations; liq.; sp.gr. 1.09; 75% act.

Solusol 84%. [43] Dioctyl sodium sulfosuccinate, propylene glycol; anionic; see Solusol 75%; liq.; sp.gr. 1.10; b.p. 350 F; 84% act.

Solusol 85%. [43] Dioctyl sodium sulfosuccinate, sodium benzoate; anionic; see Solusol 75%; wh. solid; sp.gr. 1.1; m.p. < 300 C; 85% act.

Solusol 100%. [43] Dioctyl sodium sulfosuccinate; anionic; see Solusol 75%; wh. solid; sp.gr. 1.1; m.p. 153–157 C; surf. tens. 28.7 dynes/cm (0.1%); 100% act.

Solvent Scour 2527. [283] Alkanolamides and alkyl phenol ethoxylate; nonionic; scouring agent; yel. liq.; dens. 0.899; 100% act.

Solwax C-24. [630] Polyethoxylated cholesterol; nonionic; o/w emulsifier; stabilizer for w/o emulsions; used in personal care products; solubilizer for steroids in aq. systems; lt. yel. waxy solid; faint pleasant odor; sol. in water and alcohol; m.p. 45–48 C; cloud pt. 85–95 C; sapon. no. 5.0 max.; 100% act.

Solwax L 20. [630] Alkoxylated lanolin deriv., pharmaceutical; nonionic; foam stabilizer; o/w emulsifier for personal care products; lt. yel. waxy solid; faint pleasant odor; dissolves in water; softening pt. 45–55 C; cloud pt. 75–80 C; sapon. no. 10–30; 100% act.

Solwax LG 35. [630] Alkoxylated lanolin alcohols deriv.; nonionic; emulsifier, emollient, solubilizer, conditioner; personal care products; yel.-br. waxy solid; faint pleasant odor; sol. in water and alcohol; softening pt. 40–50 C; cloud pt. 75–85; sapon. no. 5–15; 100% act.

Sopralub ACR 265. [192] Polyoxyalkylene fatty glycerides; nonionic; textile softener, plasticizer; liq.; 30% conc.

Sopralub ACR 275. [192] Polyoxyalkylene fatty glycerides; nonionic; softener, plasticizer for textile finishes and coatings; fluid cream; 30% conc.

Soprofor BC/Series. [253] Octylphenol, ethoxylated; nonionic; emulsifier, detergent, wetting agent; 100% conc.

Soprofor NP/20, /30, /100. [253] Nonoxynol-20, -30, -100; nonionic; finishing auxs. used as antifoaming agents, detergents, dispersants, emulsifiers, softeners, and dyeing assistants; solid, solid, powd. resp.; 100% conc.

Soprofor NPF/10. [253] Acid nonylphenyl ether phosphate; anionic; detergent in industrial cleaners; household detergents; paint stripping agent; dedusting; emulsifier for acrylic, polyvinyl acetate emulsion polymerization; liq.; 100% conc.

Soprofor PE/220. [253] PEG 200 monolaurate; nonionic; visc. depressant for PVC emulsions; liq.; 100% conc.

Soprofor PE/400. [253] PEG 400 monolaurate; nonionic; see Soprofor PE/220; liq.; 100% conc.

Soprofor PL/Series. [253] POE/POP block polymer; nonionic; detergent, emulsifier, wetting and antistat, defoamer, deduster; 100% conc.

Soprofor RHS. [253] Alkyl polyglycolic ether; nonionic; emulsifier for wh. oils; liq.; 100% conc.

Soprofor S/20. [253] Sorbitan monolaurate; nonionic; emulsifier, lubricant and softener for the textile industry; liq.; 100% conc.

Soprofor S/60. [253] Sorbitan monostearate; nonionic; see Soprofor S/20; solid; 100% conc.

Soprofor S/65. [253] Sorbitan tristearate; nonionic; see Soprofor S/20; solid; 100% conc.

Soprofor S/80. [253] Sorbitan monooleate; nonionic; see Soprofor S/20; liq.; 100% conc.

Soprofor S/85. [253] Sorbitan trioleate; nonionic; see Soprofor S/20; liq.; 100% conc.

Soprofor T/20. [253] PEG-20 sorbitan monolaurate; nonionic; emulsifier, solubilizer, antistat and lubricant for textile industry; liq.; 100% conc.

Soprofor T/60. [253] PEG-20 sorbitan monostearate; nonionic; see Soprofor T/20; paste; 100% conc.

Soprofor T/65. [253] Sorbitan tristearate, ethoxylated; nonionic; see Soprofor T/20; solid; 100% conc.

Soprofor T/80. [253] PEG-20 sorbitan monooleate; nonionic; see Soprofor T/20; liq.; 100% conc.

Soprofor T/85. [253] PEG-20 sorbitan trioleate; nonionic; see Soprofor T/20; liq.; 100% conc.

Soprofor TR/Series. [253] Tridecyl alcohol, ethoxylated; nonionic; emulsifier for Vaseline and min. oil; liq.; 100% conc.

Soprofor VR/50. [253] POE stearic mono diglycerides; nonionic; thickener for shampoo; visc. control; liq.; 50% conc.

Sorban AL. [654] Sorbitan ester; nonionic; antifoaming; wax; 100% conc.

Sorban AO. [654] Sorbitan ester; nonionic; emulsifier; liq.; 100% conc.

Sorban AO 1. [654] Sorbitan ester; nonionic; oil additive, corrosion inhibitor; liq.; 100% conc.

Sorban AST. [654] Sorbitan ester; nonionic; emulsifier for oils and greases; antifoamer, corrosion inhibitor; cosmetic applications; flakes; 100% conc.

Sorban CO. [654] Sorbitan ester; nonionic; antifoamer; corrosion inhibitor; liq.; 100% conc.

Sorbanox AL. [654] Sorbitan ester, ethoxylated; nonionic; detergent used in veg. cleaning, antistat; liq.; 100% conc.
Sorbanox AOM. [654] Sorbitan ester, ethoxylated; nonionic; emulsifier; solubilizer of vitamin; liq.; 100% conc.
Sorbanox AP. [654] Sorbitan ester, ethoxylated; nonionic; emulsifier; liq.; 100% conc.
Sorbanox AST. [654] Sorbitan ester, ethoxylated; nonionic; emulsifier for paraffin and min. oil; insecticides, cosmetic applications; liq.; 100% conc.
Sorbanox CO. [654] Sorbitan ester, ethoxylated; nonionic; emulsifier for min. oil; liq.; 100% conc.
Sorbax HO-50. [122] POE sorbitol ester; o/w emulsifier for solvs., veg. and petroleum oils used in the textile and metal lubricant industries; liq.; HLB 11.4.
Sorbax MO-40. [122] POE sorbitol ester; see Sorbax HO-50; liq.; HLB 10.2.
Sorbax PML-20. [122] Polysorbate 20; nonionic; o/w emulsifier, solubilizer for perfumes and flavors; liq.; water-sol.; HLB 16.7; 100% conc.
Sorbax PMO-5. [122] Polysorbate 81; nonionic; see Sorbax PML-20; liq.; water-sol.; HLB 10.0; 100% conc.
Sorbax PMO-20. [122] Polysorbate 80; nonionic; see Sorbax PML-20; liq.; water-sol.; HLB 15.0; 100% conc.
Sorbax PMP-20. [122] Polysorbate 40; nonionic; see Sorbax PML-20; liq.; water-sol.; HLB 15.6; 100% conc.
Sorbax PMS-20. [122] Polysorbate 60; nonionic; see Sorbax PML-20; liq.; water-sol.; HLB 14.9; 100% conc.
Sorbax PTO-20. [122] Polysorbate 85; nonionic; see Sorbax PML-20; liq., gel; water-sol.; HLB 11.0; 100% conc.
Sorbax PTS-20. [122] Polysorbate 65; nonionic; see Sorbax PML-20; solid; water-sol.; HLB 10.5; 100% conc.
Sorbax SML. [122] Sorbitan monolaurate; nonionic; emulsifier, surfactant for o/w emulsion stabilizers and thickeners used in cosmetic and textile industries; liq.; sol. in oil and org. water systems; HLB 8.6; 100% conc.
Sorbax SMO. [122] Sorbitan monooleate; nonionic; see Sorbax SML; liq.; sol. see Sorbax SML; HLB 4.3; 100% conc.
Sorbax SMP. [122] Sorbitan monopalmitate; nonionic; see Sorbax SML; solid; sol. see Sorbax SML; HLB 6.7; 100% conc.
Sorbax SMS. [122] Sorbitan monostearate; nonionic; see Sorbax SML; solid; sol. see Sorbax SML; HLB 4.7; 100% conc.
Sorbax STO. [122] Sorbitan trioleate; nonionic; see Sorbax SML; liq.; sol. see Sorbax SML; HLB 1.8; 100% conc.
Sorbax STS. [122] Sorbitan tristearate; nonionic; see Sorbax SML; solid; sol. see Sorbax SML; HLB 2.1; 100% conc.
Sorbeth 40. [159] POE (4) sorbitol hexaoleate; nonionic; o/w emulsifier; liq.; 100% conc.
Sorbeth 40HO. [159] Sorbitol POE (40) hexaoleate; nonionic; emulsifier/dispersant for agric. chemicals and the oil industry; liq.; HLB 10.5; 100% conc.
Sorbeth 55. [159] POE (55) sorbitol; nonionic; o/w emulsifier; liq.; 100% conc.
Sorbeth 55HO. [159] Sorbitol POE (55) hexaoleate; nonionic; emulsifier/dispersant for agric. chemicals and oil industry; liq.; HLB 11.5; 100% conc.
Sorbit P. [136] Alkyl naphthalene sulfonate; anionic; wetting agent and foamer; emulsifier, detergent; germicides, textile leveling; powd.; 65% conc.
Sorbitol. [498] Sorbitol; water-wh. liq.; odorless; ref. index 1.458; sp.gr. 1.30; visc. 110 cps; pH 7; 70% act.
Sorbon S-20. [606] Sorbitan monolaurate; nonionic; emulsifier, dispersant for w/o emulsion; liq.; 100% conc.
Sorbon S-40. [606] Sorbitan monopalmitate; nonionic; used with Sorbon T series; solid; 100% conc.
Sorbon S-60. [606] Sorbitan monostearate; nonionic; used with Sorbon T series; solid; 100% conc.
Sorbon S-80. [606] Sorbitan monooleate; nonionic; used with Sorbon T series; liq.; 100% conc.
Sorbon T-20. [606] POE (20) sorbitan monolaurate; nonionic; emulsifier, dispersant, solubilizer for o/w emulsion; liq.; HLB 16.7; 100% conc.
Sorbon T-40. [606] POE sorbitan monopalmitate; nonionic; used with Sorbon S series; liq.; HLB 15.7; 100% conc.
Sorbon T-60. [606] POE sorbitan monostearate; nonionic; used with Sorbon S series; liq.; HLB 15.3; 100% conc.
Sorbon T-80. [606] POE sorbitan monooleate; nonionic; used with Sorbon S series; liq. 100% conc.
Sorbon TR 814. [606] POE sorbitol oleate; nonionic; detergent, emulsifier, cosmetics;

liq.; HLB 17.5; 100% conc.

Sorbon TR 843. [606] POE sorbitol oleate; nonionic; see Sorbon TR 814; liq.; HLB 11.2; 100% conc.

Sorgen 30. [518] Sorbitan sesquioleate; nonionic; antifoamer and emulsifier for foods, cosmetics, and pharmaceutical products; liq.; HLB 3.7; 100% conc.

Sorgen 40. [518] Sorbitan monooleate; nonionic; see Sorgen 30; liq.; HLB 4.3; 100% conc.

Sorgen 50. [518] Sorbitan monostearate; nonionic; see Sorgen 30; flake; HLB 4.7; 100% conc.

Sorgen 90. [518] Sorbitan monolaurate; nonionic; see Sorgen 30; liq.; HLB 8.6; 100% conc.

Sorgen S-30-H. [518] Sorbitan sesquioleate; nonionic; see Sorgen 30; liq.; HLB 3.7; 100% conc.

Sorgen S-40-H. [518] Sorbitan monooleate; nonionic; see Sorgen 30; liq.; HLB 4.3; 100% conc.

Sorgen TW20. [518] POE-20 sorbitan monolaurate; nonionic; emulsifier; liq.; HLB 16.7; water-sol.; 100% conc.

Sorgen TW60. [518] POE sorbitan monostearate; nonionic; emulsifier; paste; water-sol.; HLB 14.9; 100% conc.

Sorgen TW80. [518] POE sorbitan monooleate; nonionic; emulsifier; liq.; water-sol.; HLB 15.0; 100% conc.

Sorpol 320, 355. [606] POE alkylaryl ether, POE sorbitan alkylate, and sulfonate; anionic/nonionic; emulsifier for Malathion E.C. and other pesticides; liq.; 100% conc.

Sorpol 560. [606] POE alkylaryl ether, POE sorbitan alkylate, and sulfonate; anionic/nonionic; emulsifier for Malathion E.C.; liq.; 100% conc.

Sorpol 900A, 1200, 2020K, 3005X. [606] POE alkylaryl ether, POE sorbitan alkylate, and sulfonate; anionic/nonionic; emulsifier for pesticides; liq.; 100% conc.

Sorpol 3044. [606] POE alkylaryl ether, POE fatty acid ether, and sulfonate; anionic/nonionic; see Sorpol 1200; liq.; 100% conc.

Sorpol H-770, L-550. [606] POE alkylaryl ether, POE sorbitan alkylate, and sulfonate; anionic/nonionic; see Sorpol 1200; liq.; 100% conc.

Sotex 3CW. [422] Fatty acid ester, long chain; nonionic; dispersant for paints and coatings, ink vehicle; wetting agent for dyes; visc. liq.; sol. in aliphatic and aromatic hydrocarbons; sp.gr. 0.930–0.960; b.p. > 200 C; 100% act.

Sotex COS #2. [422] Fatty acid ester, long chain; cationic; dispersant for paints and coatings; surface treating pigments; liq.; oil-sol.; sp.gr. 0.900–0.940; b.p. > 200 C; 100% act.

Sotex CW. [422] Fatty acid ester, long chain; nonionic; visc. depressant and stabilizer in vinyl plastisols; dispersant for paints and coatings; pigments in plasticizers; cosolvent; visc. liq.; sol. in hydrocarbons, ketones; sp.gr. 0.970–1.010; b.p. > 200 C; 100% act.

Sotex N. [422] Fatty acid ester, long chain; nonionic; dispersant for polar solvs., and inks; visc. liq.; sol. in polar solvs.; sp.gr. 1.02–1.06; b.p. > 200 C; 100% act.

Sovatex DS/C5. [571] Sodium dodecylbenzene sulfonate and fatty alcohol ethoxylate; anionic/nonionic; conc. detergent, wetting and dye control agent for textile processes; pale amber clear liq.; water-sol.; sp.gr. 1.04; pH 7–8.5; 50% solids.

Span 20. [320] Sorbitan laurate; nonionic; emulsifier, stabilizer, thickener, lubricant, softener, antistat, foods, pharmaceuticals, cosmetics, cleaning compds., textiles; amber liq.; sol. in propylene glycol; visc. 4250 cs; HLB 8.6; 100% act.

Span 40. [320] Sorbitan palmitate; nonionic; see Span 20; tan solid; sol. in perchloroethylene; HLB 6.7; pour pt. 48 C; 100% act.

Span 60. [320] Sorbitan stearate; nonionic; see Span 20; tan solid; sol. in perchloroethylene, xylene; HLB 4.7; pour pt. 53 C; 100% act.

Span 65. [320] Sorbitan tristearate; nonionic; see Span 20; cream solid; HLB 2.1; pour pt. 53 C; 100% act.

Span 80. [320] Sorbitan oleate; nonionic; see Span 20; amber liq.; visc. 1000 cs; HLB 4.3; 100% act.

Span 85. [320] Sorbitan trioleate; nonionic; see Span 20; amber liq.; visc. 210 cs; HLB 1.8; 100% act.

Spermwax. [521] Syn. spermaceti cetyl esters wax; solid.

Sponto 102. [650] Metal sulfonate and POE ether; anionic/nonionic; insecticides; liq.; oil-sol.

Sponto 150 TH. [652] Sulfonate and POE ether; anionic/nonionic; agric. emulsifier for pesticides; liq.; oil-sol.; 100% conc.

Sponto 150 TL. [652] Sulfonate and POE ether; anionic/nonionic; see Sponto 150 TH; liq.; 100% conc.

Sponto 168-D. [650] Phosphoric acid and alkylphenoxy polyethoxyethanol mono and diesters; anionic; emulsifier and compatibility agent for pesticides and fertilizers; liq.; 100% conc.

Sponto 200. [650] POE ether; nonionic; emulsifier for citrus and crop oils; liq.

Sponto 217. [652] Sulfonate and POE ether; anionic/nonionic; see Sponto 150 TH; liq.; 100% conc.

Sponto 221. [652] Sulfonate and POE ether; anionic/nonionic; see Sponto 150 TH; liq.; 100% conc.

Sponto 232T. [650] Sulfonate and POE ether; anionic/nonionic; pesticides; liq.; oil-sol.

Sponto 234T. [650] Sulfonate and POE ether; anionic/nonionic; see Sponto 232T; liq.; oil-sol.

Sponto 710T. [650] Sulfonate and POE ether; anionic/nonionic; emulsifier for herbicides; liq.; oil-sol.

Sponto 712T. [650] Sulfonate and POE ether; anionic/nonionic; see Sponto 710T; liq.; oil-sol.

Sponto 714T. [650] Sulfonate and POE ether; anionic/nonionic; see Sponto 710T; liq.; oil-sol.

Sponto 2174 H. [652] Sulfonate and POE ether; anionic/nonionic; see Sponto 150 TH; liq.; 100% conc.

Sponto 2224T. [650] Sulfonate and POE ether; anionic/nonionic; emulsifier for chlorinated insecticides; liq.

Sponto AC60-02D. [650] POE ether; nonionic; see Sponto 200; liq.

Sponto AD4-10N. [650] Sulfonate and POE ether; anionic/nonionic; emulsifier for aromatic solvs. in weed control; liq.; oil-sol.

Sponto AD6-39A. [650] Sulfonate and POE ether; anionic/nonionic; aux. for emulsifier; liq.; oil-sol.

Sponto AG3-55N. [650] Sulfonate and POE ether; anionic/nonionic; see Sponto 710T; liq.; oil-sol.

Sponto AK30-23. [650] Phosphoric acid mono- and diester and alkylphenoxy; anionic; emulsifier for Vapona; liq.

Sponto AL69-49. [650] POE ether; nonionic; emulsifier for Dimethoate; liq.

Sponto AM2-07. [650] Sulfonate, phosphate ester and POE ether; anionic/nonionic; agric. emulsifier; liq.; oil-sol.

Sponto H-44C. [650] Polyglycol fatty acid ester; nonionic; agric. emulsifier for w/o systems; liq.

Sponto N-139B. [650] Sulfonate and POE ether; anionic/nonionic; emulsifier for insecticides; liq.; oil-sol.

Spreading Agent ET0672. [160] Complex alkyl polyglycol ether; o/w emulsifier; bath oil spreading agent and dispersant; pale straw clear liq.

Squalane. [160] Squalane; emollient, penetrant used in personal care products; water-wh. clear liq.; odorless.

ST-55-2. [287] POE (2) stearyl ether; nonionic; w/o emulsifier for cosmetics and pharmaceuticals; textile industry; additive for antistats; component for leveling and stripping agents; additive for latex stabilizing; emulsion polymerization; solid; HLB 5.0; 100% conc.

ST-55-20. [287] POE (20) stearyl ether; nonionic; o/w emulsifier for cosmetics and pharmaceuticals; co-emulsifier; textile industry; wetting agent for leveling and stripping agents; dispersant for dyestuffs; emulsifier for silicone oil; silicone antifoam agent, polish; flakes; HLB 15.0; 100% conc.

Stabilisol S. [301] Polysulfide; anionic; stabilizer for hydrogen peroxide bleaching baths; liq.

Stabilizer T. [277] Protein fatty acid condensate; anionic; stabilizer for bleaching; paste; 50% conc.

Stafoam. [441] Fatty acid diethanolamide; rust inhibitor; detergent for laundry and drycleaning use; textile softener and dyeing aux.; metal cleaning agent; lt. yel. liq.; water-sol.; sp.gr. 0.996; visc. 810 cps; surf. tens. 8 dynes/cm (0.1%); pH 9.5–10.5 (1% aq.).

Stafoam DF-1. [441] Coconut oil fatty acid diethanolamide (1:1); foam booster; used in shampoos, dishwashing detergents; drycleaning, textile softening aux., antistat, metal cleaning; lt. yel. liq.; sp.gr. 0.981; visc. 625 cps; surf. tens. 14 dynes/cm (0.1%); pH 9.6–10.6 (1% aq.).

Stafoam DF-4. [441] Coconut oil fatty acid diethanolamide (1:1); see Stafoam DF-1; lt. yel. liq.; sp.gr. 0.974; visc. 675 cps; m.p. 15 C max.; surf. tens. 16 dynes/cm (0.1%); pH 9.6–10.6 (1% aq.).

Stafoam DL. [441] Lauric acid diethanolamide (1:1); detergent, foam stabilizer and booster, visc. modifier; dishwashing and laundry detergent, drycleaning, personal care products; textile softener and dyeing aux.; anticlouding agent; wh. solid; sp.gr. 0.969 (50 C); visc. 179 cps (50 C); m.p. 42–47 C; surf.

tens. 5 dynes/cm (0.1%); pH 9.6–10.6 (1% aq.).

Stafoam F. [441] Coconut oil fatty acid diethanolamide (1:2); rust inhibitor, foamer, penetrant, laundry and drycleaning detergent, metal cleaning, textile softener and dyeing aux.; lt. yel. liq.; sol. in water; sp.gr. 0.995; visc. 755 cps; surf. tens. 7 dynes/cm (0.1%); pH 9.5–10.5 (1% aq.).

Stamid HT. [145] Coco fatty acid diethanolamide; nonionic; foam stabilizer, emulsifier and thickener used in a variety of household, industrial and cosmetic formulations; liq.; 85% conc.

Standamid 100. [288] Coconut fatty acid monoethanolamide; foam stabilizer; superfatting agent and thickener for personal care products and detergents; solid; 90% amide.

Standamid CD. [289; 291] Capramide DEA and diethanolamide; nonionic; foam stabilizer and booster, wetting agent, superfatting, formulating cosmetic, household, and industrial detergents and other chemical specialities; amber visc. liq.; pH 9.0–10.0 (10%); 96–98% act.

Standamid CM. [291] Cocamide MEA; nonionic; foamer, stabilizer used in personal care products; granules; m.p. 140–148 F; pH 9.5–10.5 (10%); 100% conc.

Standamid CMG. [291] Cocamide MEA, super amide (1:1); foam booster, stabilizer, thickener for personal care products and dishwashing detergents; solid; 100% conc.

Standamid ID. [290] Isostearamide DEA; nonionic; conditioner, lubricant, foamer, emulsifier used in the processing of paper and personal care products; mold-release agent; amber liq.; sol. in lower alcohols; disp. in water; sp.gr. 0.964; dens. 8.04 lb/gal; pH 8–10 (1.0%); 100% conc.

Standamid KD. [289; 291] Cocamide DEA; nonionic; thickener and superfatting agent, foam stabilizer used in personal care products; emulsifier and dissolving intermediate; amber paste; m.p. 100 F; pH 9.0–9.5 (10%); 85–90% amide.

Standamid KDM. [291] Coconut fatty acid alkanolamine condensate; nonionic; foam builder and stabilizer, thickener, wetting agent, dispersant used in personal care products; household detergents; amber visc. liq.; sol. in water, alcohol, acetone and glycols; pH 9.5–10.0 (1%); 100% conc.

Standamid KDO. [289; 290] Cocamide DEA (1:1); nonionic; visc. builder for anionic systems; personal care products; amber liq.; sol. in lower alcohols, propylene alcohol, polyethylene glycols; insol. in water; sp.gr. 0.982; pH 9–11 (1.0%); 100% conc.

Standamid KDS. [289; 290] Lauramide DEA (1:1); nonionic; enhances foam density, lubricity, stability; lt. amber liq.; 80% conc.

Standamid KM. [290] Cocamide MEA; see Standamid KDO; flakes/beads; sol. in lower alcohols; insol. in water; sp.gr. 0.884; m.p. 80 C; pH 9–11 (1.0%).

Standamid LD. [289; 290] Lauramide DEA; nonionic; foam stabilizer, superfatting and thickener used in personal care products; paste; m.p. 105 F; pH 9.5–10.5 (10%); > 90% amide.

Standamid LD 80/20. [289; 290] Lauric/myristic acid diethanolamide, modified; nonionic; foam booster and stabilizer, emulsifier, thickener, conditioner used in personal care products and dishwashing detergents; amber clear liq.; cloud pt. 0 C; pH 9.5–10.5 (10%); 80% conc.

Standamid LDM. [290] Lauric diethanolamide, modified; nonionic; foam stabilizer, visc. builder used in personal care products; clear visc. liq.; cloud pt. 0 C; 100% conc.

Standamid LDO. [289; 290] Lauramide DEA (1:1); nonionic; see Standamid KDO; amber liq.; sol. see Standamid KDO; sp.gr. 0.985; pH 9–11 (1.0%); 100% conc.

Standamid LDS. [289; 290] Lauramide DEA; nonionic; solubilizer for liqs. and oils; used in personal care products, dishwashing detergents; lt. amber liq.; sol. in lower alcohols, propylene glycol, polyethylene glycol; disp. in water; sp.gr. 0.98; dens. 8.2 lb/gal; pH 9–11 (1.0%); 100% conc.

Standamid LM. [290] Lauramide MEA; see Standamid KM; flakes/beads; sol. in lower alcohols; insol. in water; sp.gr. 0.915; m.p. 80 C; pH 9–11 (1.0%).

Standamid LP. [288] Lauric acid isopropanolamide; nonionic; foam stabilizer, superfatting agent, thickener; shampoos and other detergents; fortifier for perfumes in soaps; solid; > 90% amide.

Standamid PD. [289; 290] Cocamide DEA (2:1) and diethanolamine; nonionic; foam builder, stabilizer, thickener used in personal care products; amber visc. liq.; 100% conc.

Standamid SD. [289; 290] Cocamide DEA; nonionic; foam stabilizer used in personal care products; dishwashing and laundry detergents; liq.; m.p. 25 F; pH 9.5–10.5 (10%); 100% conc.

Standamid SDO. [289; 290] Cocamide DEA (1:1); nonionic; visc. builder, foam enhancer and conditioner used in personal care products; amber liq.; sol. see Standamid KDO; sp.gr. 0.991; pH 9–11 (10%); 100% conc.

Standamid SM. [289; 290] Cocamide MEA (1:1); see Standamid KM; flakes/beads; sol. in lower alcohols; insol. in water; sp.gr. 0.934; m.p. 62 C; pH 9–11 (1.0%).

Standamid SOD. [290] Linoleic diethanolamide (1:1); nonionic; visc. enhancer, thickener in industrial, institutional and household cleaners; corrosion inhibitor and emulsifier for oils and lubricants; dk. amber liq.; sol. in lower alcohols, glycols, acetone, toluene, xylene; disp. in water; dens. 8.20 lb/gal; pH 9–11 (1%); 100% conc.

Standamid SOMD. [289; 290] Linoleamide DEA; visc. builder for personal care products; paste; sol. see Standamid SM; sp.gr. 0.935; pH 9–11.

Standamox CAW. [289; 290] Cocamidopropylamine oxide; foam builder, stabilizer, thickener, emollient for personal care products; clear liq.; 29.5% conc.

Standamox C 30. [288] Coco amido amine oxide; wetting agent, foam stabilizer, visc. builder for personal care products and dishwashing detergents; liq.; pH 6–7; 28–30% act.

Standamox LMW-30. [289] Lauryl/myristyl dimethylamine oxide; cationic; foamer and conditioner in household and industrial cleaners; liq.; 30% conc.

Standamox O1. [289; 290] Oleamine oxide; nonionic; lubricant, softener used in industrial applications; amber clear liq.; 55% conc.

Standamul 302. [290] Propylene glycol dicaprylate/dicaprate; base, emollient used in pharmaceutical and personal care products; clear visc. oily liq.; char. odor; sol. in min. and castor oil, IPM, oleyl alcohol, anhyd. ethanol, and silicone fluid; sp.gr. 0.921; HLB 9; sapon. no. 315–335; 100% conc.

Standamul 318. [290] Caprylic/capric triglyceride; base; carrier used in pharmaceutical, personal care and food products; clear visc. oily liq.; odorless; sol. in min. and castor oil, IPM, oleyl alcohol and anhyd. ethanol; sp.gr. 0.950; visc. 25 cps; HLB 12; cloud pt. –5 C max.; 100% conc.

Standamul 1414-E. [290] Myreth-3 myristate; carrier, self-emulsifying base, emulsifier, coupler, emollient and moisturizer used in pharmaceutical and personal care products; clear visc. oily liq.; sol. in min. and castor oil, IPM, oleyl alcohol, anhyd. ethanol; HLB 12; cloud pt. 25 C max.; sapon. no. 90–100; 100% conc.

Standamul 1616. [290] Cetyl palmitate; spermaceti replacement, consistency giving agent used in pharmaceutical and personal care products; wh. waxy solid; faint char. odor; sol. in min. and castor oil, IPM, oleyl alcohol; m.p. 45–55 C; HLB 9; sapon. no. 115–130; 100% conc.

Standamul 7061. [290] Isocetyl stearate; base, emollient used in pharmaceutical and personal care products; clear visc. oil; faint char. odor; sol. see Standamul 1616; sp.gr. 0.865; visc. 25 cps; HLB 8; cloud pt. 0 C max.; sapon. no. 105–115; 100% conc.

Standamul 7063. [290] Octyl dodecyl stearate; see Standamul 7061; clear visc. oil; faint char. odor; sol. see Standamul 1616; sp.gr. 0.865; visc. 100 cps; HLB 8; cloud pt. 0 C max.; sapon. no. 90–105; 100% conc.

Standamul 7105. [290] Caprylic/capric triglyceride; base used in pharmaceutical and personal care products; clear visc. oil; faint char. odor; sol. see Standamul 318; sp.gr. 1.064; visc. 100 cps max.; HLB 12; cloud pt. –5 C max.; sapon. no. 340–350; 100% conc.

Standamul B-1. [290] Ceteareth-12; nonionic; emulsifier and solubilizer for personal care and household products, essential oils; waxy solid; sp.gr. 0.97 (70 C); HLB 12; solid. pt. 35 C; 99.5–100% solids.

Standamul B-2. [290] Ceteareth-20; nonionic; emulsifier and solubilizer for personal care products; additive in the paper and textile industry; household cleaning and leather products; waxy solid; sp.gr. 1.0 (70 C); HLB 14; solid. pt. 40 C; 99.7–100% solids.

Standamul B-3. [290] Ceteareth-30; nonionic; see Standamul B-2; waxy solid; sp.gr. 1.023 (70 C); HLB 15; solid. pt. 43–46 C; 99.5–100% solids.

Standamul Conc. 1002. [290] Cetearyl alcohol, PEG-40 hydrog. castor oil, stearalkonium chloride; conditioner and softener used in hair care preparations; off-wh. waxy flakes; m.p. 58 C; pH 5.0 (1%); 99.0% min. solids.

Standamul CTA. [290] Hexyl laurate; spreading agent, base, dispersant used in pharmaceutical and personal care products; clear visc. oil; odorless; sol. see Standamul 302; sp.gr. 0.85–0.87; HLB 9; cloud pt. 0 C max.; sapon. no. 190–205; 100% conc.

Standamul CTV. [290] Decyl oleate; see Standamul CTA; clear visc. liq.; faint char. odor; sol. see Standamul 318; sp.gr. 0.86–0.87; HLB 9; cloud pt. 10 C max.; sapon. no. 130–150; 100% conc.

Standamul G. [290] Octyl dodecanol; carrier for oil-sol. act. ingred.; dissolves dyes; pigment dispersant; coupler for waxes and other materials in personal care products; water-wh. clear liq.; odorless; cloud pt. –50 C; sapon. no. 5.0 max.; 100% conc.

Standamul G-16. [290] Isocetyl alcohol; emollient; extender and solv. for pigments and personal care products; binder in cosmetics; water-wh. clear liq.; visc. 100 cps; cloud pt. –50 C; sapon. no. 5.0 max.; 100% conc.

Standamul G-3236 Stearate. [290] Myristyl eicosyl stearate; stabilizer, base, lubricant, opacifier, emollient used in pharmaceutical and personal care products; lt. soft wax; faint char. odor; sol. in min. and castor oil, IPM, oleyl alcohol; m.p. 33–40 C; HLB 7; sapon. no. 65–75.

Standamul GTO-26. [290] Undecylpentadecanol; skin lubricant and pigment wetting agent in cosmetic emulsion products; used in personal care products; water-wh. clear liq.; insol. in water; visc. 300 cps; cloud pt. –40 C.

Standamul HE. [290] PEG-7 glyceryl cocoate; nonionic; emollient, refatting agent, solubilizer, coupler, emulsifier used in pharmaceutical and personal care products; clear low visc. oil; faint char. odor; sol. in oleyl alcohol, castor oil, anhyd. ethanol, water; sp.gr. 1.050; visc. 200 cps; HLB 16; cloud pt. 0 C; sapon. no. 90–100; 100% conc.

Standamul M-8. [290] Oleyl and cetyl alcohol PEG ether; nonionic; o/w emulsifier used in personal care products; paste; sp.gr. 1.000; HLB 11; 79–81% solids.

Standamul O-5. [290] Oleth-5; nonionic; emulsifier, emollient, o/w and w/o emulsions; used in personal care products; visc. liq.; sol. in alcohols, hydrocarbons, and org. solvs.; 100% conc.

Standamul O-10. [290] Oleth-10; see Standamul O-5; soft waxy solid; sol. see Standamul O-5; 100% conc.

Standamul O-20. [290] Oleth-20; see Standamul O-5; waxy solid; sol. see Standamul O-5; 100% conc.

Standamul OXL. [290] PPG-10 ceteareth-20; nonionic; emollient, superfatting agent for personal care products; water-wh. liq.; sol. in water; 98% conc.

Standamul STC-25. [290] Stearalkonium chloride; cationic; antistat, surfactant for personal care products and fabrics; wh. slurry; 25% conc.

Standapol 125-E Conc. [290] Sodium laureth-12 sulfate; detergent, foamer used in personal care products; straw yel. clear liq.; sol. in water and hydrophilic solvs.; cloud pt. 10 C max.; pH 7.5–9.0 (10% aq.); 58–60% act.

Standapol 130-E. [290] Sodium laureth-12 sulfate; see Standapol 125-E Conc.; pale yel. clear liq.; sol. see Standapol 125-E Conc.; pH 7.5–9.0 (10% aq.); 28–30% act.

Standapol 230-E Conc. [290] Ammonium laureth-12 sulfate; see Standapol 125-E Conc.; straw-yel. clear liq.; sol. see Standapol 125-E Conc.; cloud pt. 0 C; pH 6.5–7.0 (10% aq.); 28–30% act.

Standapol A. [289; 290] Ammonium lauryl sulfate; anionic; see Standapol 125-E Conc.; water-wh. visc. liq.; visc. 1500–3000 cps; pH 6.5–7.0 (10% aq.); 27.5–28.5% act.

Standapol AA-1. [290] DEA-lauryl sulfate, DEA-lauraminopropionate, and sodium lauraminopropionate; anionic-amphoteric; surfactant used in personal care products; amber liq.; cloud pt. 12 C; pH 7.5–8.5; 34.5–38.5% solids.

Standapol AB-45. [290] Coco betaine; amphoteric; foam and visc. booster, conditioner, antistat, and gelling agent used in personal care products; pale yel. liq.; sol. in alcohols, glycols, alkanolamides and water; cloud pt. < 0 C; pH 6.5–7.5; 36.5% min. act.

Standapol A-HV. [289; 290] Ammonium lauryl sulfate; anionic; surfactant used in personal care products; water-wh. to pale yel. liq.; water-sol.; dens. 1.025 g/ml; visc. < 10,000 cP; pH 6.0–7.0; 27.5–28.5% act.

Standapol AL-60. [290] Ammonium laureth sulfate; surfactant for personal care, household and industrial cleaning products; yel. clear liq.; water-sol.; dens. 1.025 g/ml; visc. 100–200 cP; cloud pt. 10 C max.; pH 6.0–7.0; 58–60% act.

Standapol AMS-100. [290] Ammonium lauryl sulfate; see Standapol 125-E Conc.; water-wh. visc. liq.; visc. 1500–3000 cps; cloud pt. < 10 C; pH 6.5–7.5 (10% aq.); 27.5–28.5% act.

Standapol AP-60. [290] Ammonium C_{12-15} pareth sulfate; see Standapol AL-60; yel. clear visc. liq.; water-sol.; dens. 1.025 g/ml; visc. 100–200 cps; cloud pt. 10 C max.; pH 6.0–7.0 (10%); 58–60% act.

Standapol BAW. [290] Cocamidopropyl betaine; amphoteric; conditioner and foamer, antistat; surfactant used in personal care products; lt. amber liq.; 34% conc.

Standapol BC-35. [290] Cocamidopropyl betaine; amphoteric; conditioner, emollient, visc. builder, gelling agent used in personal care products; lt. amber liq.; 35% conc.

Standapol Blend EA-K. [290] Ammonium myreth sulfate, cocamide DEA; used in personal care products; amber clear liq.; misc. in R.T. water; visc. 200 cps; pH 6.3 ± 0.3; 61.0 ± 1.5% solids.

Standapol CIM-40. [290] Cocoamphoacetate; amphoteric; surfactant used in personal care products; lt. clear liq.; pleasant mild odor; cloud pt. 0 C; pH 8.0–9.0 (10%); 40–42% act.

Standapol Conc. 7021. [290] DEA-lauryl sulfate and TEA-lauryl sulfate; anionic; see Standapol 125-E Conc.; anhyd. slurry; 100% conc.

Standapol Conc. 7023. [290] Cocamide DEA; anionic; detergent, foamer used in personal care products; solubilizer for fatty alcohols, fatty acid esters, and min. oil; straw-yel. clear visc. liq.; visc. 250 cps max. (50 C); cloud pt. 10 C max.; pH 8.5–9.5 (1% aq.); 100% conc.

Standapol CS-50. [290] Sodium lauryl sulfate, sodium cetyl sulfate, and laureth-3; see Standapol 125-E Conc.; pearly wh. visc. liq.; mild char. odor; pH 6.5–7.5 (10% aq.); 28.5–30% act.

Standapol CS Paste. [290] Sodium lauryl sulfate, sodium cetyl sulfate, and laureth-3; see Standapol 125-E Conc.; pearly creamy paste; mild char. odor; pH 6.5–7.5 (10% aq.); 57–60% act.

Standapol DEA. [289; 290] DEA-lauryl sulfate; anionic; base, detergent, foamer used in personal care products; lt. yel. clear liq.; 36% conc.

Standapol EA-1. [289; 290] Ammonium laureth sulfate; anionic; see Standapol DEA; water-wh. visc. liq.; visc. 2500 cps; cloud pt. 5 C max.; pH 6.5–7.0 (10% aq.); 25.5–26.5% act.

Standapol EA-2. [289; 290] Ammonium laureth sulfate; anionic; see Standapol DEA; water-wh. visc. liq.; visc. 2500 cps max.; cloud pt. 0 C max.; pH 6.5–7.0 (10% aq.); 24.5–25.5% act.

Standapol EA-3. [289; 290] Ammonium laureth sulfate; solubilizer used in personal care products; water-wh. visc. liq.; water-sol.; dens. 1.020 g/ml; visc. 2500 cps; cloud pt. 5 C; pH 6.0–7.0 (10%); 26–28% act.

Standapol EA-40 Conc. [289; 290] Ammonium myreth sulfate; anionic; see Standapol 125-E Conc.; pale-yel. clear liq.; cloud pt. 10 C max.; pH 6.0–7.0 (10% aq.); 58–60% act.

Standapol ES-1. [289; 290] Sodium laureth sulfate; anionic; see Standapol DEA; water-wh. visc. liq.; visc. 500–1200 cps; cloud pt. 5 C max.; pH 7.5–8.5 (10% aq.); 24.5–25.5% act.

Standapol ES-2. [289; 290] Sodium laureth sulfate; anionic; see Standapol DEA; water-wh. visc. liq.; visc. 4000–5000 cps; cloud pt. 0 C max.; pH 6.5–7.5 (10% aq.); 25–26% act.

Standapol ES-3. [289; 290] Sodium laureth sulfate; anionic; see Standapol DEA; water-wh. visc. liq.; visc. 500 cps max.; cloud pt. 0 C; pH 6.5–8.0 (10% aq.); 28–29% act.

Standapol ES-40 Conc. [290] Sodium myreth sulfate; anionic; see Standapol 125-E Conc.; pale-yel. clear liq.; cloud pt. 10 C max.; pH 7.5–9.0 (10% aq.); 58–60% act.

Standapol ES-7099. [290] Sodium laureth sulfate; anionic; see Standapol DEA; water-wh. visc. liq.; visc. 4000–5000 cps; cloud pt. 0 C max.; pH 6.5–7.5 (10% aq.); 25–26% act.

Standapol LF. [289; 290] Sodium octyl sulfate; anionic; wetting agent for metal degreasers, hard surface cleaners, food equipment cleaners, dust control; solubilizer, hydrotrope; liq.; 33% conc.

Standapol MG. [290] Magnesium lauryl sulfate; anionic; see Standapol 125-E Conc.; straw yel. liq.; cloud pt. 3 C max.; pH 6–7 (10% aq.); 28–30% act.

Standapol MLS. [290] MEA-lauryl sulfate; anionic; see Standapol EA-40; lt. yel. clear visc. liq.; visc. 3000–5000 cps; cloud pt. 0 C; pH 7–8 (10% aq.); 32–34% act.

Standapol OLB-30. [290] Oleyl betaine; amphoteric; detergent, conditioner and foamer; visc. builder used in personal care products; amber visc. liq.; 30% conc.

Standapol OLB-50. [290] Oleyl betaine; amphoteric; see Standapol OLB-30; lt. amber gel; 50% conc.

Standapol Pearl Conc. 7130. [290] Glycol

distearate, sodium laureth sulfate, propylene glycol, cocamide MEA, laureth-9; used in personal care products; pearly liq.; misc. in R.T. water; visc. 5000 cps max.; pH 6.5 ± 1.0% (10%); 41.0 ± 3.0% solids.

Standapol S. [290] Sodium lauryl sulfate, sodium laureth sulfate, lauramide MIPA, cocamide MEA, glycol stearate, and coceth-8; anionic; see Standapol 125-E Conc.; pearly-wh. visc. liq.; visc. 1200–1700 cps; pH 6.6–7.4 (10% aq.); 36.8–37.3% act.

Standapol SCO. [290] Sulfated castor oil; emulsifier and dispersant, wetting agent, foam depressant, emollient surfactant and solubilizer used in personal care products; amber visc. liq.

Standapol SH-100. [289; 290] Disodium monooleamido PEG-2 sulfosuccinate; anionic; detergent used in personal care products; lt. amber liq.; mild pleasant odor; cloud pt. < 0 C; pH 6.0–7.0 (5%); 29–31% act.

Standapol SH-135 Special. [289; 290] Disodium monooleamido PEG-2 sulfosuccinate; anionic; see Standapol SH-100; lt. amber liq. mild pleasant odor; cloud pt. 10 C max.; pH 6–7 (5%); 34–36% act.

Standapol SH-200. [290] Ammonium monooleamido PEG-2 sulfosuccinate; see Standapol SH-100; lt. amber liq.; mild pleasant odor; cloud pt. < 0 C; pH 6–7 (5%); 29–31% act.

Standapol SH-300. [290] TEA-monooleamido PEG-2 sulfosuccinate; see Standapol SH-100; lt. amber liq.; char. odor; cloud pt. < 0 C; pH 6–7 (5%); 39–41% act.

Standapol SHC-101. [290] Disodium monooleamido PEG-2 sulfosuccinate and sodium lauryl sulfate; anionic; see Standapol EA-40; lt. amber liq.; mild pleasant odor; visc. 1000 cps max.; cloud pt. 0 C max.; pH 6.5–7.5 (5%); 30% min. solids.

Standapol SHC-301. [290] TEA monooleamido PEG-2 sulfosuccinate and TEA lauryl sulfate; see Standapol SHC-101; lt. amber liq.; mild pleasant odor; visc. 1200 cps max.; cloud pt. 0 C max.; pH 6–7 (5%); 39–41% solids.

Standapol SL-60. [290] Sodium laureth sulfate; surfactant for personal care, household, and industrial cleansing products; lt. yel. clear liq.; water-sol.; dens. 1.050 g/ml; visc. 100–200 cps; cloud pt. 10 C max.; pH 7.0–9.0 (10%); 58–60% act.

Standapol SP-60. [290] Sodium C_{12-15} pareth sulfate; see Standapol SL-60; yel. clear liq.; water-sol.; dens. 1.050 g/ml; visc. 100–200 cps; cloud pt. 10 C max.; pH 7.0–9.0 (10%); 58–60% act.

Standapol T. [289; 290] TEA-lauryl sulfate; anionic; see Standapol EA-40; water-wh. liq.; 39% conc.

Standapol TS-100. [290] TEA-lauryl sulfate; anionic; see Standapol 125-E Conc.; water-wh. liq.; 39% conc.

Standapol WA-AC. [289; 290] Sodium lauryl sulfate; anionic; see Standapol 125-E Conc.; pearly-wh. paste; visc. 20,000 cps min.; cloud pt. 25 C max.; pH 7.5–8.5 (10% aq.); 28–30% conc.

Standapol WAQ-115. [290] Sodium C_{12-15} alcohols sulfate; see Standapol 125-E Conc.; water-wh. visc. liq.; visc. 1000 cps max.; cloud pt. 12 C max.; pH 7.5–8.5 (10% aq.); 29–31% conc.

Standapol WAQ-LC. [289; 290] Sodium lauryl sulfate; see Standapol EA-40; water-wh. liq.; visc. 100 cps max.; cloud pt. 8 C max.; pH 7–9 (10% aq.); 28.5–31.5% conc.

Standapol WAQ-LCX. [289] Sodium lauryl sulfate; anionic; used in rug shampoos; liq.; 28–30% conc.

Standapol WAQ Special. [289; 290] Sodium lauryl sulfate; see Standapol EA-40; water-wh. visc. liq.; visc. 1500 cps max.; cloud pt. 15 C max.; pH 7.5–8.5 (10% aq.); 28–30% conc.

Standapol WAS-100. [290] Sodium lauryl sulfate; see Standapol EA-40; water-wh. visc. liq.; visc. 1500 cps max.; cloud pt. 15 C max.; pH 7.5–8.5 (10% aq.); 28–31% conc.

Standapon 95, 100. [289] Alkylaryl polyglycol ether; nonionic; detergent, rewetting, wetting and textile processing; liq.; 95% conc.

Stantex 322. [289] Ester sulfate; anionic; wetting and rewetting agent for dyeing and finishing; liq.; 64% conc.

Starfol D. [69] Glyceryl monostearate, ethoxylated; nonionic; emulsifier used in bakery products; APHA 250 max. solid; m.p. 26 C; HLB 13.6; sapon. no. 65–75; 100% conc.

Starfol IPM. [69] Isopropyl myristate; nonionic; emollient; APHA 30 max. liq.; sapon. no. 202–212; 100% conc.

Starfol IPP. [69] Isopropyl palmitate; nonionic; emollient; APHA 30 max. liq.; sapon. no. 182–192; 100% conc.

Starfol Wax CG. [69] Difatty ester; nonionic; syn. spermaceti wax; APHA 100 max. solid; m.p. 46–49 C; sapon. no. 109–117; 100% conc.

Stave WA Special. [573A] Lauryl sulfate; foam stabilizer, dispersant for latex foams; polymer additive in SBR, vinyl and ABS-nitrile copolymers; liq. to paste; pH 7.4–8.6 (10%); 33% act.

Stearal. [37] Stearyl alcohol; emollient, aux. emulsifier, texturizer; wh. waxy solid; mild char. odor; m.p. 55–60 C; sapon. no. 2.0 max.

Stedbac. [297] Stearyl dimethyl benzyl ammonium chloride; cationic; hair conditioner, emulsifier and antistat; powd.; 100% conc.

Steinamat B 2003. [212] Tetrasodium (1,2-dicarboxyethyl)-N-alkyl sulfosuccinamide; anionic; detergent, foamer, emulsifier, and stabilizer; emulsion polymerization additive; household, industrial, and metal cleaners; solubilizer for detergent raw materials; soldering aid; pigment dispersant; used in personal care products; amber clear liq.; sol. in water, alkali, and electrolytes; pH 7–9 (10%); 35% solids.

Steinamat TMS. [212] Disodium alkyl sulfosuccinamide; anionic; stabilizer, emulsifier, foam additive for lattices; emulsion polymerization; wetting agent for latex impregnation; soft creamy paste or liq.; pH 9–10 (10%); 35% solids.

Steinamid 203/S. [212] Lauric acid diethanolamide; nonionic; detergent, foam stabilizer, and visc. modifier for shampoos and detergents; off-wh. solid; slight odor; m.p. 40 C; 100% act.

Steinamid AC 28. [212] Groundnut fatty acid polydialkanolamide; nonionic; detergent, emulsifier; anticorrosion additive for min. oil metalworking lubricants; brn. liq.; dens. 0.98 g/cm^3; sp.gr. 1.0; 100% act.

Steinamid C 212. [212] Coconut fatty acid monoethanolamide; nonionic; detergent, superfatting agent; stabilizer of emulsions; off-wh. flakes; amidic odor; m.p. 73 C; 100% act.

Steinamid DC 212/S. [212] Coconut fatty acid diethanolamide; nonionic; detergent; foam stabilizer, visc. modifier, and superfatting agent for shampoos and detergents; lt. yel. liq.; typ. odor; dens. 0.99 g/cm^3; 100% act.

Steinamid DC 220/SE. [212] Coconut fatty acid diethanolamide; nonionic; detergent, thickener, foam stabilizer, superfatting agent; soft paste; 100% act.

Steinamid DL 240. [212] Coconut fatty acid polydiethanolamide; nonionic; detergent used in cosmetics, cleaning agents; stabilization of emulsions; superfatting; yel. liq.; typ. odor; 100% act.

Steinamid DO 280. [212] Oleic acid polydiethanolamide; nonionic; detergent used in cosmetic, cleaning and detergent industries; amber liq.; typ. odor; sp.gr. 1.0; 100% act.

Steinamid DO 280/SE. [212] Oleic acid diethanolamide; nonionic; detergent used in cosmetic products; honey liq.; dens. 0.96 g/cm^3; 100% act.

Steinamid IPE 280. [212] Oleic monoisopropanolamide; nonionic; detergent; skin-protecting additive; lt. yel. semisolid; typ. odor; 100% act.

Steinamid IPL 203. [212] Lauric monoisopropanolamide; nonionic; detergent; additive for solid and paste end-products; emulsion stabilizer; wh. flakes; typ. odor; m.p. 55 C; 100% act.

Steinamid IPP 240. [212] Coconut fatty acid monoisopropanolamide; nonionic; see Steinamid IPL 203; wh. flakes; m.p. 50 C; 100% act.

Steinamid L203. [212] Lauric monoethanolamide; nonionic; used in detergent preparations; personal care products; emulsion stabilizer; wh. flakes; typ. odor; m.p. 84 C; 100% act.

Steinamid S 280. [212] Stearic acid monoethanolamide; nonionic; detergent; superfatting agent; emulsion stabilizer; wh. flakes; typ. odor; m.p. 92 C; 100% act.

Steinamin IM-BT. [212] Imidazoline deriv.; cationic; adhesive agent and emulsifier in acidic bitumen emulsions; rust inhibitor in acidic media; grn.-brn. liq.; amine odor; dens. 0.94 g/cm^3; 100% act.

Steinamin IM-CA. [212] 1-Hydroxy- thyl-2-heptadecenyl imidazoline; cationic; corrosion inhibitor in leather and metal industry, emulsifier, penetrant industries; wetting agent for carbonization baths; additive for preparation of glass fiber; yel. wax; m.w. 285; sol. in polar and nonpolar solvs.; disp. in water; 100% act.

Steinamin IM-OA. [212] 1-Hydroxy-ethyl-2-heptadecenyl imidazoline; cationic; rust inhibitor, adhesive agent, coupler, emulsifier, detergent, antistat; used in industrial and household applications; paint industry; amber liq.; sol. see Steinamin IM-CA; m.w. 345; dens. 0.94 g/cm^3; visc. 800 cps; pH 10–11 (1% aq.); 100% act.

Steinaminoxid B 204. [212] Alkylamido-propyldimethylamine oxide; nonionic; emulsifier, softener; foam booster in wash-act. formulations; calcium soap dispersant; used in shampoos; pale yel. clear liq.; odorless; pH 5–7 (10%); 35% min. solids.

Steinapal C 6. [212] Coconut amide polyglycol ether; nonionic; dispersant, wetting agent, foamer; solv. used in personal care products; amber liq.; cloud pt. 80–90 C (2% aq.); pH 8–10 (1%); 100% act.

Steinapal HV 5. [212] Nonylphenol polyglycol ether; nonionic; emulsifier for emulsions and insecticides; raw material for rinsing, cleaning agents, textile treatment; wetting agent in the paper and cellulose industry; degreaser in pickling and alkaline immersion-bath cleaners; Gardner 2 max. liq.; sol. in min. oil, alcohol, ketone, toluene, xylene, chlorinated hydrocarbons; cloud pt. 59–62 C; pH 5–7 (1%); 100% act.

Steinapal HV 8. [212] Nonylphenol polyglycol ether; nonionic; see Steinapal HV 5; Gardner 2 max. liq.; sol. see Steinapal HV 5; cloud pt. 41–45 C; pH 5–7; 100% act.

Steinapal HV 9. [212] Nonylphenol polyglycol ether; nonionic; see Steinapal HV 5; Gardner 2 max. liq.; sol. see Steinapal HV 5; cloud pt. 59–62 C; pH 5–7; 100% act.

Steinapal HV 10. [212] Nonylphenol polyglycol ether; nonionic; see Steinapal HV 5; Gardner 2 max. liq.; sol. see Steinapal HV 5; cloud pt. 70–73 C; pH 5–7; 100% act.

Steinapal HV 14. [212] Nonylphenol polyglycol ether; nonionic; see Steinapal HV 5; wh. soft wax; sol. see Steinapal HV 5; cloud pt. 60–63 C; pH 5–7; 100% act.

Steinapal HV 25. [212] Nonylphenol polyglycol ether; nonionic; see Steinapal HV 5; wh. wax; sol. see Steinapal HV 5; cloud pt. 73–76 C; pH 5–7; 100% act.

Steinapal LA 3. [212] Fatty alcohol polyglycol ether; nonionic; emulsifier, coupler; raw material for production of ether sulfates; used in shampoos; clear liq.; sol. in min. oil and org. solvs.; pH 5–7.

Steinapal O 8. [212] Oleic acid amide polyglycol ether; nonionic; detergent, wetting and calcium soap dispersant; o/w emulsifier; machine washing formulations; brn. liq.; water-sol.; cloud pt. 75–85 C; pH 8–10 (1% solids); 100% act.

Steinapal TA 11. [212] Fatty alcohol polyglycol ether; nonionic; wash-act. base; wetting agent, dispersant, emulsifier; wh. soft wax; sol. in water and org. solvs.; cloud pt. 69–73 C (2%); pH 5–7 (1%).

Steinapal TA 25. [212] Fatty alcohol polyglycol ether; nonionic; see Steinapal TA 11; wh. powd.; sol. see Steinapal TA 11; cloud pt. 75–90 C; pH 5–7.

Steinapal TA 50. [212] Fatty alcohol polyglycol ether; nonionic; see Steinapal TA 11; wh. powd.; cloud pt. 72–76 C; pH 5–7.

Steinaphat E 1027. [212] Alkylphenol polyglycol ether phosphate; anionic; corrosion inhibitor, emulsifier, dispersant, wetting agent used in the metal industry; antistat; yel. liq.; 100% act.

Steinphat EAK 8190. [212] Fatty alcohol polyglycol ether phosphate; see Steinaphat E 1027; yel. liq.; disp. in water; 100% act.

Steinapol-Emulsifier BW. [212] Fatty acid deriv., modified; nonionic; emulsifier, cleansing liq. for hand cleaners and oily and greasy surfaces; yel. clear visc. liq.; oil-sol.; 100% act.

Steinapol-Emulsifier BWA. [212] Fatty acid deriv., modified; nonionic; emulsifier for metal cleaning and paraffin waxes; yel. clear visc. liq.; 100% act.

Steinapol NL, 3, 28%. [212] Sodium lauryl ether sulfate; anionic; detergent, foamer, cleaning formulations; personal care products; clear liq.; visc. 50–150 cps; pH 7–8 (5% solids); 28% act.

Steinapol NLS 28. [212] Sodium lauryl sulfate; anionic; detergent; raw material for washing and cleaning agents; personal care products; emulsifier in emulsion polymerization; lt. yel. liq.; m.w. 300; pH 7–8 (1% solids); 28% act.

Steinapol NLS 90-Powder. [212] Sodium lauryl sulfate; anionic; see Steinapol NLS 28; wh.-lt. yel. powd.; m.w. 300; pH 7–9 (1% solids); 85% min. act.

Steinapol SB-DO 70. [212] Sodium diisooctyl sulfosuccinate; anionic; wetting agent, solubilizer, personal care products; household and metal cleaners; used in paper, paint, dye and textile industries; dry cleaning; colorless liq.; surf. tens. < 30 dynes/cm (0.1%); pH 6.5–7.5 (5% solids); 64% min. act.

Steinapol SBF 12. [212] Disodium lauryl alcohol sulfosuccinate; anionic; foamer; detergent raw material for personal care products, carpet and upholstery shampoos; wh. paste; pH 6.5–7.5 (5% solids); 39% solids.

Steinapol SBFA 30, 40%. [212] Disodium lauryl alcohol polyglycol ether sulfosuccinate; anionic; detergent raw material for personal care products; colorless clear liq.;

visc. 100–200 cps; pH 6.5–7.5 (5% solids); 39% min. solids.

Steinapol SBL 203. [212] Disodium lauric acid ethanolamide sulfosuccinate; anionic; detergent used in personal care products, carpet and upholstery shampoos; aids spray-drying; yel. turbid liq.; pH 6.5–7.5 (5% solids); 39% min. solids.

Steinapol SBR 12-Powder. [212] Disodium lauryl alcohol sulfosuccinate; anionic; detergent for cleansing, tech., and personal care products; improves spray-drying properties; wh., lt. yel. powd.; water-sol.; dens. 250 g/l; pH 6.5–7.5 (5% solids); 95% solids.

Steinapol SBV. [212] Sulfosuccinate, modified; nonionic; detergent used in personal care products; lt. yel. clear liq.; visc. 2500 cps; pH 6.5–7.5 (1% solids); 28% min. solids.

Steinapon AM-2C. [212] Coconut-based ampholyte; amphoteric; detergent, foam booster, wetting agent; calcium soap dispersant; used in personal care products; golden-yel. liq.; 49% act.

Steinapon AM-2L. [212] Imidazoline-based ampholyte; amphoteric; detergent used in personal care products; liq.; 50% act.

Steinapon AM-B 13. [212] Alkyl amido betaine; amphoteric; detergent, foam booster used in personal care products; yel. liq.; pH 4–6 (10% solids); 30% act.

Steinapon AM-CA. [212] Coconut-based ampholyte; amphoteric; detergent, foamer used in personal care products; yel. clear liq.; pH 8–8.5 (10% solids); 30% act.

Steinaryl-Sulfonic Acid K. [212] Dodecylbenzene sulfonic acid; anionic; detergent base for dishwashing, washing, and cleaning formulations; brn. visc. liq.; 97% min. act.

Steinaryl T 65. [212] *p*-Toluene sulfonic acid; anionic; catalyst for preparation of esters, acid hydrolysis, polymerization, dehydration, and alkylation reactions; electrolytic processes; intermediate for preparation of cresol; leather tanning agent; crop-dusting agent; production of syn. fibers; vulcanization of buna-rubber; processing of phenol resins for foundry industry; pale clear liq.; char. odor; 65% act.

Steinaryl TKS 90/F. [212] Triammonium dodecylbenzenesulfonate; anionic; detergent, raw material for cleansing agents; dk. amber visc. liq.; pH 7.0–7.5 (1% solids); 85% min. act.

Steinazid DU 185. [212] Undecylenic acid polyethanolamide; nonionic; detergent, emulsifier, bactericide, thickener, foam stabilizer in shampoos; yel. visc. liq.; sol. in water, oil, alcohols, glycols; sp.gr. 1.0; visc. high; 100% act.

Steinazid SBU 185. [212] Sodium undecylenic acid monoethanolamide monosulfosuccinate; anionic; fungicide, detergent, foamer used in personal care products; pale yel. powd.; typ. odor; sol. in alcohol, glycol; sp.gr. 1.0; pH 6.5–7.5 (5% solids); 50% act.

Steinazid U 185. [212] Undecylenic acid monoethanolamide; nonionic; fungicide, detergent; yel. flakes; typ. odor; sol. in alcohol, glycol, disp. in water; m.p. 50 C; 100% act.

Steol 4N. [575] Sodium laureth sulfate; anionic; detergent, emulsifier, foamer used in personal care products; cleaning products; textile mill applications; emulsion polymerization; water-wh. liq.; water-sol.; sp.gr. 1.045; visc. 31 cps; cloud pt. –4 C; pH 8.0; 28% act.

Steol 7N. [575] Linear fatty alcohol ether sulfate, sodium salt; anionic; detergent, wetting agent, foamer used in personal care products; pale yel. liq.; water-sol.; sp.gr. 1.054; visc. 100 cps; cloud pt. –2 C; surf. tens. 37.7 dynes/cm (0.1%); pH 8.0 (10%); 28% act.

Steol 7T. [575] Linear fatty alcohol ether sulfate, triethanolamine salt; see Steol 7N; pale yel. liq.; water-sol.; sp.gr. 1.046; visc. 100 cps; cloud pt. –1 C; surf. tens. 38.4 dynes/cm (0.1%); pH 7.3 (10%); 28% act.

Steol CA-460. [575] Ammonium laureth sulfate; anionic; detergent, emulsifier, foamer; personal care and cleaning products; textile mill applications; emulsion polymerization; pale yel. liq.; water-sol.; sp.gr. 1.016; visc. 67 cps; cloud pt. 19 C; pH 7.0; 59% act.

Steol CS-125, -225. [575] Sodium laureth sulfate; anionic; foamer and detergent for personal care products; water-wh. liq.; 24–26% act.

Steol CS-130. [575] Sodium laureth sulfate; surfactant for personal care applications; pH 7.5–8.5 (10% aq.); 24.5–26% act.

Steol CS-230. [575] Sodium lauryl ether sulfate; see Steol CS-130; pH 6.5–8.5 (10% aq.); 24.5–26.5% act.

Steol CS-330. [575] Sodium lauryl ether sulfate; see Steol CS-130; pH 6.5–8.5 (10% aq.); 27.5–29.5% act.

Steol CS-460. [575] Sodium laureth sulfate; see Steol CA-460; pale yel. liq.; water-sol.; sp.gr. 1.037; visc. 69 cps; cloud pt. 17 C; pH

8.0 (1% aq.); 59.5% act.

Steol CS-760. [575] Sodium laureth-7 sulfate; see Steol 7N; pale yel. liq.; water-sol.; sp.gr. 1.059; visc. 100 cps; cloud pt. 1 C; surf. tens. 46.2 dynes/cm (0.1%); pH 8.0 (10%); 59% act.

Steol FA-50. [575] Fatty ether sulfate; foamer for gypsum board production; Klett 50 max. clear liq.; water-sol.; dens. 8.42 lb/gal; flash pt. 80–100 F; 49–51.0% act.

Steol FA-60. [575] Fatty ether sulfate; see Steol FA-50; Klett 50 max. clear liq.; water-sol.; dens. 8.40 lb/gal; flash pt. 100–115 F; 58.5–61% act.

Steol KA-460. [575] Linear fatty alcohol ether sulfate, ammonium sulfate; detergent, foamer and wetting agent used in personal care products; straw liq.; water-sol.; dens. 8.65 lb/gal; cloud pt. 0 C; pH 7.0 (10% aq.); 60% act.

Steol KS-460. [575] Linear fatty alcohol ether sulfate, sodium salt; see Steol KA-460; straw liq.; water-sol.; dens. 8.94 lb/gal; cloud pt. 0 C; pH 8.0 (10% aq.); 60% act.

Stepan C25. [575] Methyl caprylate-caprate; intermediate for mfg. of detergents, emulsifiers, wetting agents, stabilizers, lubricants, plasticizers, resins, and textile specialties; water-wh.; m.p. –30 C; sapon. no. 329–335.

Stepan C40, C41. [575] Methyl laurate; see Stepan C25; water-wh.; m.p. 5 and 1 C resp.; sapon. no. 260–264 and 258–264 resp.

Stepan C42, C43. [575] Methyl laurate; see Stepan C25; water-wh.; sapon. no. 254–258 and 252–256 resp.

Stepan C49. [575] Whole coco methyl esters; see Stepan C25; Gardner 1 max.; sapon. no. 250–256.

Stepan C50. [575] Methyl myristate; see Stepan C25; Gardner 1 max.; m.p. 15 C; sapon. no. 230–234.

Stepan C60. [575] Methyl palmitate; see Stepan C25; Gardner 1 max.; m.p. 28 C; sapon. no. 203–210.

Stepan C65. [575] Methyl palmitate-oleate; see Stepan C25; Gardner 1 max.; m.p. 14 C; sapon. no. 197–203.

Stepan C66. [575] Methyl stearate; see Stepan C25; Gardner 1 max.; m.p. 30 C; sapon. no. 195–203.

Stepan C68. [575] Methyl oleate; see Stepan C25; Gardner 2 max.; m.p. 16 C; sapon. no. 191–197.

Stepan Cationic 10. [575] Dialkyl quat. ammonium methosulfate; cationic; surfactant for household and industrial fabric softeners; antistat; metal cleaners; waterproofing, leather conditioner; textile lubricant; penetrating oils; buffing/polishing compds.; corrosion inhibitor; organoclay thickener; cream fluid; sol. in aromatic solvs.; dens. 8.0 lb/gal; pour pt. 23 C; flash pt. 80 F (CC); pH 3.0 ± 0.5 (5% aq.); surf. tens. 42 dynes/cm (0.1% aq.); 83 ± 3% solids.

Stepan D-50. [575] Isopropyl myristate; emollient used in personal care products; water-wh. clear oily liq.; bland fatty odor; ref. index. 1.4325; oil-sol.; 100% act.

Stepan D-70. [575] Isopropyl palmitate; see Stepan D-50; water-wh. clear liq.; odorless; ref. index 1.437; oil-sol.; sapon. no. 185–195; 100% act.

Stepan HDA-7. [575] Fatty acid alkylolamide, sodium cation; anionic/nonionic; detergent; foam base for all-purpose and floor cleaners; amber liq.; water-sol.; 90% act.

Stepan P-621. [575] Lauric-myristic alkylolamide; nonionic; liq. detergent foam stabilizer; lt. solid; water-disp.

Stepanate AM. [575] Ammonium xylene sulfonate; anionic; hydrotrope, solubilizer, coupler in detergent field; heavy duty cleaners, wax strippers, dishwashing detergents; solv. or fluidizer; freeze-thaw stabilizer; additive for high electrolyte or brine systems; colorless liq.; water-sol.; dens. 9.05 lb/gal; pH 6.5–7.2 (1% aq.); 40% act.

Stepanate C-S. [575] Sodium cumene sulfonate; coupler or solubilizer for liq. cleaners; detergent slurries; Klett 70 max.; dens. 9.65 lb/gal; 44–46% act.

Stepanate T. [575] Sodium toluene sulfonate; see Stepanate AM; dens. 10.1 lb/gal; pH 7–9 (1% aq.); 40% act.

Stepanate X. [575] Sodium xylene sulfonate; see Stepanate AM; also as cloud pt. depressant; colorless liq.; water-sol.; dens. 9.83 lb/gal; pH 7–9 (1% aq.); 40% act.

Stepanol ALS-30. [575] Ammonium lauryl sulfate; anionic; detergent, cosmetic use, industrial foam; liq.; 30% conc.

Stepanol AM. [575] Ammonium lauryl sulfate; anionic; detergent, foamer used in personal care products; rug and upholstery shampoos; household, metal, and industrial cleaners; fruit washing; insecticides; textile and leather processing; pharmaceuticals; pale yel. visc. liq.; water-sol.; pH 6–7 (10%); 28–30% act.

Stepanol AM-V. [575] Ammonium lauryl sulfate; anionic; see Stepanol AM; visc. liq., lt. gel; water-sol.; pH 6–7 (10%); 27–29% act.

Stepanol DEA. [575] DEA-lauryl sulfate; see Stepanol AM; pale yel. clear liq.; water-sol.; pH 7.3–7.7 (10%); 33–35% act.

Stepanol ME Dry. [575] Sodium lauryl sulfate; see Stepanol AM; wh. powd.; water-sol.; pH 7.5–1.0 (10%); 93% min. act.

Stepanol MG. [575] Magnesium lauryl sulfate; see Stepanol AM; water-sol.; pH 6.5–7.5 (10%); 28–30% act.

Stepanol SLS-30. [575] Sodium lauryl sulfate; anionic; wetting and emulsion polymerization agent; detergent for cosmetics; liq.; 30% conc.

Stepanol TLS-40. [575] Triethanolamine lauryl sulfate; anionic; foaming surfactant for use in liq. detergents and in personal and pet care products; liq.; 40% conc.

Stepanol WA-100. [575] Sodium lauryl sulfate; anionic; detergent, foamer, wetting and suspending agent; dentrifrice formulations; pharmaceuticals; emulsion polymerization; clay dispersions; wh. powd.; water-sol.; pH 7.5–10 (10%); 98.5% min. act.

Stepanol WAC. [575] Sodium lauryl sulfate; anionic; see Stepanol AM; pale yel. clear liq.; water-sol.; pH 7.5–8.5 (10%); 28–30% act.

Stepanol WA Extra. [575] Sodium lauryl sulfate; anionic; see Stepanol AM; clear liq.; water-sol.; pH 7.5–8.5 (10%); 28–30% act.

Stepanol WA Paste. [575] Sodium lauryl sulfate; anionic; see Stepanol AM; water-wh. clear paste; water-sol.; pH 7.5–8.5 (10%); 28–30% act.

Stepanol WAQ. [575] Sodium lauryl sulfate; anionic; see Stepanol AM; water-wh. clear visc. liq.; water-sol.; pH 7.5–8.5 (10%); 28–30% act.

Stepanol WA Special. [575] Sodium lauryl sulfate; anionic; see Stepanol AM; water-wh. clear liq.; water-sol.; pH 7.5–8.5 (10%); 28–30% act.

Stepanol WAT. [575] TEA-lauryl sulfate; see Stepanol AM; water-wh. clear liq.; water-sol.; pH 7.0–8.5 (10%); 39–41.5% act.

Stepantan 29N. [575] Sodium linear olefin sulfonate; surfactant for oil field applications; Klett 400 color; sp.gr. 1.05; pH 9; 20% act.

Stepantan 39N. [575] Sodium linear olefin sulfonate; see Stepantan 29N; Klett 400; sp.gr. 1.05; visc. 300 cps; pH 9; 40% act.

Stepantan A. [575] Naphthalene-formaldehyde condensation sodium sulfonate; anionic; tanning agent; pitch control agent in papermaking and dye mordant; dispersant for pigments, acid or alkaline media; cream powd.; water-sol.; dens. 0.37–0.42 g/cc; pH 7.4 (2% aq.).

Stepantan H-100. [575] Dodecylbenzene sulfonic acid; emulsifier for oils, solvs., waxes and oil field applications; air entraining agent and foamer for cellular concrete; wetting agent; water-sol.; sp.gr. 1.09; dens. 9.1 lb/gal; visc. 9000 cps.

Stepantex DA-52. [575] Alkylamine alkylaryl sulfonate; water coupler with solvs. for cleaning products; emulsion degreaser; emulsifier for dye carriers; dispersant for dyes and pigments; drycleaning detergent; demulsifier in oil recovery; sol. in kerosene, perchlorethylene, Stod., xylene, IPA; disp. in water.

Stepantex Q-90B. [575] Fatty quat. sulfate; cationic; softener for home and commercial use; pale amber clear liq.; disp. in water; dens. 8.03 lb/gal; sp.gr. 0.96; pour pt. 77 F; 90% act.

Stepfac PN 209. [575] Complexed phosphate esters, acid form; compatibilizer, deresinator, hydrotrope for cleaners, derusters; liq.; 100% conc.

Stepfac SLS-30 Special. [575] Sodium lauryl sulfate; anionic; detergent, wetting and emulsion polymerization agent; liq.; 30% conc.

Steposol CA-207. [575] Alkyl ether sulfate, ammonium salt; anionic; used for gypsum board, cellular concrete, air drilling, foam cleaners; liq.; 60% conc.

Steramine CD. [192] Polyoxyalkylene fatty deriv.; nonionic; softener assistant in textile products; fluid paste; 26% conc.

Steramine FPA 197. [192] Alkyl amido amine; antistat, lubricant, softener for acrylics and blends; fluid paste; 21% conc.

Steramine PNA 75. [192] Alkyl amido amine; antistat, lubricant for acrylics and blends; fluid paste; 20% conc.

Steramine S2. [192] Fatty alcohol and acid EO adduct; nonionic; finishing agent for textile products; wax; 100% conc.

Steramine TV. [192] Polyoxyalkylene fatty deriv.; nonionic; softener assistant for textile products; fluid paste; 32% conc.

Sterling 2XS. [110] Sodium xylene sulfonate; hydrotrope, coupler, cloud pt. depressant; used for liq. detergents and general-purpose applications; liq.; 40% act.

Sterling 66/33. [110] Cocamide DEA; nonionic; foam stabilizer, aux. detergent, visc. builder; liq., paste; 100% act.

Sterling AB-40. [110] Sodium linear dodecylbenzene sulfonate; anionic; detergent and general cleaners; powd.; 40% act.

Sterling AB-80. [110] Sodium linear dodecylbenzene sulfonate; see Sterling AB-40; flake; 80% act.

Sterling AM. [110] Ammonium lauryl sulfate; anionic; detergent used in personal care products; visc. liq.; 27.7% act.

Sterling AM-HV. [110] Ammonium lauryl sulfate; see Sterling AM; visc. liq.; 28.0% act.

Sterling Amide 374. [110] Cocamide DEA; nonionic; foamer, visc. control agent; base for detergent formulations; oil solubilizer; liq.; 100% act.

Sterling Amide 460. [110] Lauramide/myristamide DEA; see Sterling 66/33; liq.; 76% act.

Sterling AOS. [110] Sodium C_{14-16} olefin sulfonate; anionic; detergent, foamer and stabilizer used in personal care products; general-purpose cleaners; liq. detergents; liq.; 39% act.

Sterling CAB. [110] Cocamidopropyl betaine; amphoteric; detergent; foamer, visc. builder used in personal care products; liq.; 32% conc.

Sterling DEA. [110] Cocamide DEA; nonionic; see Sterling 66/33; liq.; 100% act.

Sterling Emulsifier #3. [110] C_{12-15} pareth-3; nonionic; emulsifier; detergent additive; liq.; oil-sol.; 100% act.

Sterling Emulsifier #15. [110] Pareth-25-15; nonionic; emulsifier and detergent additive; solubilizer; paste; water-sol.; 100% act.

Sterling ES 600. [110] Ammonium laureth sulfate; anionic; used in personal care products, detergents, degreasers, and cleaners; liq.; 57.5% act.

Sterling Granulated Wax. [110] Cocamide MEA; detergent, visc. builder, foam stabilizer; flake; 100% act.

Sterling LA Acid-C, Reg. [110] Linear dodecylbenzene sulfonic acid; anionic; intermediate for detergent preparations and industrial cleaners; liq.; 96.3% act.

Sterling LA Oil. [110] TEA-dodecylbenzene sulfonate; anionic; liq. detergents and general cleaners; amber liq.; 57% act.

Sterling LA Paste 55%, 60%. [110] Sodium linear dodecylbenzene sulfonate; anionic; intermediate for detergent preparations, general purpose and industrial cleaning formulations; paste; 55 and 57% act. resp.

Sterling LDEA-90. [110] Lauramide DEA; nonionic; see Sterling 66/33; also used in cosmetics; solid; 100% act.

Sterling LM-70. [110] Lauramide/myristamide DEA; see Sterling LDEA-90; solid; 100% act.

Sterling NPX. [110] Nonoxynol-9; nonionic; detergent intermediate; household and industrial cleaner formulations; emulsifier; liq.; 100% act.

Sterling Snow White 750. [110] Ditallow imidazolinium quat.; softener for home and industrial applications; liq.; 75% act.

Sterling Super 1E. [110] Sodium laureth sulfate; anionic; detergent for personal care products; liq.; 25% act.

Sterling Super 2E-25%, -60%. [110] Sodium laureth-2 sulfate; see Sterling Super 1E; liq.; 25 and 58% act. resp..

Sterling Super 3E. [110] Sodium lauryl ether sulfate; anionic; personal care products and cleansing formulations; liq.; 28% conc.

Sterling Super 3.5E. [110] Sodium laureth-3.5 sulfate; detergent; personal care products; polymerization emulsifier; liq.; 28% act.

Sterling Super A. [110] Sodium laureth sulfate; anionic; detergent used in personal care products, degreasers, and cleaners; liq.; 57.5% act.

Sterling WADE. [110] DEA-lauryl sulfate; anionic; detergent used in personal care products; and conc. cleansing preparations; liq.; 35% act.

Sterling WA Paste. [110] Sodium lauryl sulfate; anionic; see Sterling AM; visc. liq.; 28.2% act.

Sterling WA Powder. [110] Sodium lauryl sulfate; anionic; detergent, shampoo and cosmetic base material; pharmaceutical applications and polymerization emulsifier; powd.; 99% conc.

Sterling WAQ-CH. [110] Sodium lauryl sulfate; anionic; detergent for personal care products; polymerization emulsifier; liq.; 29.3% act.

Sterling WAQ-Cosmetic. [110] Sodium lauryl sulfate; anionic; see Sterling AM; visc. liq.; 29.5% act.

Sterling WAQ-LO. [110] Sodium lauryl sulfate; see Sterling WAQ-CH; visc. liq.; 29.5% act.

Sterling WAQ-P. [110] Sodium lauryl sulfate; see Sterling AM; visc. liq.; 29% act.

Sterling WAT. [110] TEA-lauryl sulfate; anionic; see Sterling AM; liq.; 40% act.

Sterling XE. [110] Pareth-25-10; nonionic; detergent and wetting agent used in industrial and household formulations; textile scouring; emulsifier; liq.; 85% act.

Sterox AJ. [419] Tridecyl alcohol ethoxylate (8.5 EO); nonionic; detergent, wetting agent; rewetting; deinking; agric. emulsifier; industrial cleaner; textile processing; laundering; transport industry; yel. clear liq.; mild odor; water-sol.; ref. index 1.4469; sp.gr. 1.002; HLB 9.7; cloud pt. 39–45 C; pour pt. –12 F; flash pt. 380 F; surf. tens. 26.2 dynes/cm; 85% act.

Sterox AJ-100. [419] Tridecyl alcohol ethoxylate (8.5 EO); nonionic; see Sterox AJ; wh. paste; mild odor; water-sol.; ref. index 1.4600; sp.gr. 1.003; HLB 9.7; cloud pt. 39–45 C; flash pt. 385 F; pour pt. 82 F; surf. tens. 26.4 dynes/cm; 99% act.

Sterox CD. [419] Tall oil ethoxylate (10 EO); nonionic; see Sterox AJ; also used in dedusting, clear brn. liq.; mild odor; misc. in water; ref. index 1.4830; sp.gr. 1.06–1.07; HLB 12.3; flash pt. 555 F; pour pt. 46 F; surf. tens. 38.6 dynes/cm; 97.8% act.

Sterox DF. [419] Dodecyl phenol ethoxylate (6 EO); nonionic; wetting agent, emulsifier, paper rewetting, intermediate for metal cleaning compds., o/w agric. emulsifier, dedusting agent, felt washing, latex emulsion stabilizer, paints, cutting and sol. oils; clear yel. visc. liq.; mild odor; misc. with water; sol. in nonpolar solvs. and oils; m.w. 535; ref. index 1.4922; sp.gr. 1.0277; visc. 343 cps; HLB 10.0; flash pt. 495 F; pour pt. 0 F; surf. tens. 29.9 dynes/cm; pH 6.0–7.5 (1%); 99.5% act.

Sterox DJ. [419] Dodecyl phenol ethoxylate (10.5 EO); nonionic; detergent, wetting agent, emulsifier; household, metal and industrial cleaners; textiles, laundering, transport, paper and leather processing; toxicant concs.; latex formulations; clear yel. visc. liq.; mild odor; water-sol.; m.w. 715; ref. index 1.4868; sp.gr. 1.0469; visc. 1310 cps; HLB 12.6; cloud pt. 38–42 C; flash pt. 555 F; pour pt. 40 F; surf. tens. 29.7 dynes/cm; pH 6.0–7.5 (1%); 99.1% act.

Sterox ND. [419] Nonylphenol ethoxylate (4 EO); nonionic; detergent, wetting agent, emulsifier; intermediate for surfactant mfg.; solv. cleaners; additive in petroleum products; dedusting; making of cutting and sol. oils; polyvinyl acetate plasticizer; latex emulsion stabilizer; agric. emulsions; animal skin cleaners; clear yel. visc. liq.; mild odor; misc. with water; sol. in nonpolar solvs.; m.w. 396; ref. index 1.4962; sp.gr. 1.0216; visc. 252 cps; HLB 8.9; flash pt. 435 F; pour pt. –24 F; surf. tens. 29.7 dynes/cm; pH 6.0–7.5 (1%); 99.5% act.

Sterox NE. [419] Nonylphenol ethoxylate (5 EO); nonionic; detergent, wetting agent, emulsifier; intermediate for surfactant mfg.; solv. cleaners, additive in petroleum products; making of cutting and sol. oils; clear yel. liq.; mild odor; sol. see Sterox ND; ref. index 1.4955; sp.gr. 1.0322; visc. 236 cps; HLB 10.0; flash pt. 445 F; pour pt. –30 F; surf. tens. 28.7 dynes/cm; 99.5% act.

Sterox NF. [419] Nonylphenol ethoxylate (6 EO); nonionic; see Sterox NE; also used in emulsified paints; clear yel. liq.; mild odor; sol. see Sterox ND; ref. index 1.4940; sp.gr. 1.0385; visc. 207 cps; flash pt. 480 F; pour pt. –20 F; surf. tens. 28.8 dynes/cm; 99.5% act.

Sterox NG. [419] Nonylphenol ethoxylate (6.5 EO); nonionic; see Sterox NE; also additive in solv.-based paints; clear yel. liq.; mild; sol. see Sterox ND; ref. index 1.4933; sp.gr. 1.0426; visc. 229 cps; HLB 11.3; flash pt. 485 F; pour pt. 0 F; surf. tens. 29.1 dynes/cm; 99.5% act.

Sterox NJ. [419] Nonylphenol ethoxylate (9.5 EO); nonionic; surfactant used in detergent products; textile applications; dairy, metal and hard surf. cleaning; paper processing; latex paint wetting; agric. emulsifier; veg. washing; laundering; clear yel. visc. liq.; mild odor; water-sol.; m.w. 625; ref. index 1.4884; sp.gr. 1.0538; visc. 241 cps; HLB 13.0; cloud pt. 51–56 C; flash pt. 540 F; pour pt. 26 F; surf. tens. 29.8 dynes/cm; pH 6.0–7.5 (1%); 99.5% act.

Sterox NK. [419] Nonylphenol ethoxylate (10.5 EO); nonionic; detergent, emulsifier; household, dairy and industrial cleaner; paper and textile processing; metal cleaning; latex paint wetting; agric. emulsions; veg. washing; laundering; clear yel. visc. liq.; mild odor; water-sol.; m.w. 673; ref. index 1.4879; sp.gr. 1.062; visc. 304 cps; HLB 13.5; cloud pt. 66–70 C; flash pt. 560 F; pour pt. 36 F; surf. tens. 30.9 dynes/cm; pH 6.0–7.5 (1%); 99.5% act.

Sterox NL. [419] Nonylphenol ethoxylate (11

EO); nonionic; detergent, wetting agent, emulsifier; household and industrial cleaners; textiles; laundering; transport; clear yel. liq.; mild odor; water-sol.; ref. index 1.4867; sp.gr. 1.065; visc. 283 cps; cloud pt. 74–79 C; flash pt. 535 F; pour pt. 34 F; surf. tens. 32.3 dynes/cm; 99.5% act.

Sterox NM. [419] Nonylphenol ethoxylate (12.5 EO); nonionic; detergent, wetting agent, emulsifier; metal, dairy, household and industrial cleaners; paper processing; latex paint wetting; agric. emulsions; veg. washing; transport; clear yel. visc. liq.; mild odor; water-sol.; m.w. 770; ref. index 1.4860; sp.gr. 1.0674; visc. 275 cps; HLB 14.3; cloud pt. 88–94 C; flash pt. 535 F; pour pt. 52 F; surf. tens. 33.1 dynes/cm; pH 6.0–7.5 (1%); 99.5% act.

Strodex MO-100. [183] 2-Ethylhexyl polyphosphoric ester acid anhydride; anionic; emulsifier; pigment dispersant used in oil-based paints and leather coating specialties; liq.; sol. in polar and nonpolar solvs.; insol. in water; 100% conc.

Strodex MOK-70. [183] Strodex MO-100, potassium salt; anionic; dispersant in paints; wetting and rewetting agent; paste; 70% conc.

Strodex MR-100. [183] Polyphosphoric ester acid anhydride; anionic; emulsifier, dispersant for org. pigments in polar and nonpolar solvs.; liq.; 100% conc.

Strodex MRK-98. [183] Strodex MR-100, potassium salt; anionic; emulsifier for grinding and dispersing org. pigments; liq.; 98% conc.

Strodex P-100. [183] Polyphosphoric ester acid anhydride; anionic; dispersant for pigments in polar and nonpolar solvs.; coupler for emulsifiers in aq. and nonaq. systems; used in metal and tile cleaners; pigment grinding aid; liq.; 100% conc.

Strodex PK-90. [183] Strodex P-100, potassium salt; anionic; emulsifier, dispersant for extender pigments in latex paints, barium sulfate, and iron oxides; wetting agent; oxidation-corrosion inhibitor; used in heavy-duty alkaline cleaners; liq.; 90% conc.

Strodex PSK-28. [183] Phosphate esters, potassium salt; anionic; emulsifier, wetting agent, pigment dispersant; liq.; water-sol.; 58% conc.

Strodex SE-100. [183] Polyphosphoric ester acid anhydride; anionic; dispersant, emulsifier and coemulsifier in aq. and nonaq. systems; liq.; oil-sol.; 100% conc.

Strodex SEK-50. [183] Strodex SE-100, potassium salt; anionic; dispersant, emulsifier, stabilizer for latex emulsions, paints, titanium dioxide and extender pigments; liq.; 50% conc.

Strodex Super V-8. [183] Phosphate ester, potassium salt; anionic; emulsifier, detergent, dispersant, wetting agent used in textile processing applications; clear liq.; sol. in water, aliphatic, aromatic, and chlorinated solvs.; sp.gr. 1.09–1.11; pH 7–8 (1%); 75–80% act.

STXS. [154] Sodium toluene xylene sulfonate; hydrotrope used in detergent formulations; colorless clear liq.; mild odor; dens. 9.94; sp.gr. 1.18; m.p. 12 F; b.p. 212 F; 42% act.

Sucro Ester W.E. 7. [247] Saccharose distearate; nonionic; food emulsifier; pulver; HLB 7.0; 100% conc.

Sucro Ester W.E. 11. [247] Saccharose mono/distearate; nonionic; food emulsifier; pulver; HLB 11.0; 100% conc.

Sucro Ester W.E. 15. [247] Saccharose monopalmitate; nonionic; food emulsifier; pulver; HLB 14.0; 100% conc.

Sufatol LS/3. [571] Sulfated fatty alcohol and fatty alcohol ethoxylate; anionic/nonionic; scouring, wetting and washing-off agent for textiles; liq.; 25% conc.

Sufatol LX/B. [571] Sulfated fatty alcohol; anionic; scouring agent, softener, and dyebath aux. for textiles; paste; 25% conc.

Sufatol LX/C. [571] Sulfated fatty alcohol; detergent; textile scouring; paste.

Sulfamin. [90] Triethanolamine alkylaryl sulfonate; anionic; detergent, foamer; washing and cleaning agent; textile wetting agent; penetrant; Gardner 5 clear liq.; water-sol.; dens. 1.080 g/cc; visc. 300 cps; clear pt. < 5 C; flash pt. > 100 C; surf. tens. 34 dynes/cm; pH 6.5–7.5 (1% aq.); 50% act.

Sulfatol 33 MO. [2] MEA lauryl sulfate; detergent, foamer; used in personal care products, household cleaning and industrial applications; pale yel. clear liq.; char. odor; visc. 2000–3000 cP; cloud pt. 2 C; pH 6–7 (2%); 29–31% act.

Sulfatol 33 Paste. [2] Sodium lauryl sulfate; raw material for personal care products; emulsifier in mfg. of cosmetic and industrial products; carpet shampoos and other chemical specialities; wh. paste; pH 7.5–9.5 (2%); 39–41% act.

Sulfatol 33 TR. [2] Triethanolamine lauryl

sulfate; surfactant and emulsifier for personal care products; pale amber clear visc. liq.; char. odor; visc. 300–500 cP; cloud pt. –3 C; pH 6–7 (2%); 39–41% act.

Sulfatol 88.22. [2] Sodium lauryl ether sulfosuccinate; surfactant used in personal care products; colorless clear visc. liq; char. odor; sp.gr. 1.10; visc. 50–100 cP; cloud pt. –2 C; pH 6.0–6.5 (5%); 35% act.

Sulfatol B6. [2] Ammonium/triethanolamine lauryl sulfate; surfactant used in personal care products; golden clear visc. liq.; visc. 3000–5000 cP; cloud pt. 6 C; pH 6–7 (2%); 29–31% act.

Sulfatol B7. [2] Sodium lauryl ether sulfate; see Sulfatol 33 MO; liq.; 30% act.

Sulfatol B10. [2] Sodium lauryl ether sulfate; raw material for personal care products, auto shampoos, chemical specialties; wh. emulsion; visc. 1000–2000 cP; pH 8–9 (2%); 34–36% act.

Sulfatol E3. [2] Sodium lauryl ether sulfate; surfactant used in personal care products, auto shampoos, chemical specialties; raw material for dishwashing detergents; colorless visc. liq.; visc. 200–300 cP; cloud pt. –3 C; pH 5.5–7.5 (5%); 27–28% act.

Sulfatol KE3. [2] Sodium lauryl ether sulfate; see Sulfatol E3; gelled paste; pH 6–8 (2%); 60% min. act.

Sulfetal 4069. [660] Sodium alkylsulfate; anionic; detergent and wetting agent; liq.; 45% conc.

Sulfetal 4105. [660] Sodium isoalkylsulfate; anionic; detergent, wetting agent in industrial and metal cleaners; liq.; 45% conc.

Sulfetal C 38. [660] Sodium alkyl sulfate; anionic; detergent and washing agents; paste; 38% conc.

Sulfetal C 90. [660] Sodium coco fatty alcohol sulfate; anionic; detergent, washing and cleaning agent; powd.; 90% conc.

Sulfetal CA 30. [660] Amonium lauryl sulfate; anionic; detergent for cosmetics; liq.; 30% conc.

Sulfetal C JOT 38. [660] Lauryl alcohol sulfate, alkanolamine salt; anionic; shampoos, cosmetics, detergents; liq.; 38% conc.

Sulfetal FA 40. [660] Sodium alkyl sulfate, modified; anionic; detergent and wetting agent; used in industrial cleaners, metal and tank cleaning; liq.; 40% conc.

Sulfetal KT 400. [660] TEA lauryl sulfate; anionic; detergent in personal care products; liq.; 40% conc.

Sulfetal MG 30. [660] Magnesium lauryl sulfate; anionic; detergent for cosmetics; liq.; 30% conc.

Sulfetal TC 50. [660] Tallow coco fatty alcohol sulfate, sodium salt; anionic; detergent; washing and cleaning agents; hand cleaning pastes; paste; 50% conc.

Sul-fon-ate AA-9. [140] Sodium dodecylbenzene sulfonate; anionic; wetting agent; air pollution control; cement, food, commercial laundry and industrial industries; cosmetics; fertilizers, insecticides; leather, paper, petroleum, and rubber processing; metal cleaning, electroplating, etching, pickling; mining; wh. crisp flake; odorless; surf. tens. 30.6 dynes/cm (86 F); pH 7–8 (1%); 90% act.

Sul-fon-ate AA-10. [140] Sodium dodecylbenzene sulfonate; anionic; see Sul-fon-ate AA-9; wh. crisp flake; pH 7–8 (1%); 96% act.

Sul-fon-ate LA-10. [140] Sodium dodecylbenzene sulfonate; surfactant; wh. crisp flake; water-sol.; pH 6.

Sul-fon-ate OA-5. [140] Sulfonated oleic acid, sodium salt; anionic; wetting agent; foamer; lt. brn. liq.; 41% act.

Sul-fon-ate OA-5R. [140] Sulfonated oleic acid, sodium salt; surfactant; lt. brn. liq.; visc. 200 cps; cloud pt. 50–56 F; pH 5.5–6.5; 42% act.

Sul-fon-ate OE-500. [140] Sulfonated oleic acid, sodium salt, amyl-ester; solubilizer, emulsifier for dyes and additives used in polyesters.

Sulfonated Castor Oil 50%. [427] Fatty glyceride sulfate; anionic; emulsifier, lubricant, dyeing assistant and dispersant; liq.; biodeg.; 50% conc.

Sulfonated Castor Oil 75%. [427] Fatty glyceride sulfate; anionic; see Sulfonated Castor Oil 50%; liq.; 70% conc.

Sulfonated Castor Oil GTO. [427] Fatty glycerine sulfate; anionic; emulsifier, lubricant, dyeing assistant, dye dispersant and leveler; liq.; biodeg.; 77% conc.

Sulfonated Neatsfoot 3089. [145] Sulfated neatsfoot oil; anionic; emulsifier, fat liquor for leather industry; liq.; 49–51% conc.

Sulfonated Syn. Sperm Oil 2091. [145] Sulfated syn. sperm oil; anionic; emulsifier, lubricant, fat liquor for leather industry; liq.; 59–61% conc.

Sulfonic 800. [595] Sulfonated oleic acid; intermediate used in textile and industrial cleaning; dk. brn. visc. liq.; water-sol.; sp.gr. 1.09; m.p. < 16 F; b.p. 248–257 F; flash pt.> 230 F

(PMCC); pH 0.6; 50–55% act.

Sulfonic 864. [595] Sulfonic acids; anionic; detergent and penetrant; liq.; biodeg.

Sulfonic Acid LS. [283] Linear alkylbenzene sulfonic acid; anionic; intermediate for detergent formulations; amber liq.; dens. 1.055; visc. 1000 cps; 96% act.

Sulfopon 101, 101 Special. [291] Sodium lauryl sulfate; anionic; personal care products and lt.-duty detergents; paste; 30% conc.

Sulfopon 102. [291] Sodium lauryl sulfate C_{10}-C_{16}; anionic; detergent, emulsifier for emulsion polymerization of syn. rubber, PVC, and other polymers; liq.; 30% conc.

Sulfopon 103. [291] Sodium lauryl sulfate; anionic; detergent and wetting agent for personal care products; paste; 30% conc.

Sulfopon K35. [291] Sodium lauryl sulfate; anionic; base for detergents, washing and cleansing pastes; paste; 35% conc.

Sulfopon KT 115. [291] Coconut/tallow fatty alcohol sulfate; anionic; base for detergents and hand cleaning pastes; paste; 40–42% conc.

Sulfopon O. [291] Oleyl/cetyl sulfate; anionic; detergent; paste; 25% conc.

Sulfopon T 55. [291] Tallow fatty alcohol sulfate; anionic; detergent; paste; 55% conc.

Sulfopon WA 1. [289] Sodium alkyl sulfate; anionic; detergent, wetting, emulsifier for SLS; dispersant; solubilizer for emulsion polymers; base for rug and upholstery shampoos; liq.; 30% conc.

Sulfopon WA-3. [289] Sodium lauryl sulfate; surfactant used in personal care products; wh. paste; 28–30% act.

Sulfopon WA 30. [289] Sodium lauryl sulfate; raw material for personal care products; paste; pH 6.5–8.0 (5%); 29–31% act.

Sulfopon WAQ LCX. [289] Sodium lauryl sulfate; anionic; detergent, foamer; base for shampoos and bubble baths; rug shampoos; 30% conc.

Sulfopon WAQ Special. [289] Sodium lauryl sulfate; surfactant for personal care products and rug shampoos; water-wh. liq.; 28–30% act.

Sulfosil P-491. [650] Sodium silicate/sodium sulfate-based builder; detergent; booster for household and industrial cleaners; powd.; 85% conc.

Sulfosoft. [90] Linear alkylbenzene sulfonic acid; anionic; intermediate used in production of detergents; household, industrial, and washing compds.; emulsifier for solvs. and oils; Hazen 150 visc. liq.; water-sol.; m.w. 320; dens. 1.04 g/cc; visc. 3500 cps; pour pt. 6 C; surf. tens. 33 dynes/cm; pH 2.2 (1% aq.); 97 ± 1.0% act.

Sulfotex 110. [290] Sodium *n*-decyl sulfate; anionic; wetting agent and emulsifier used in cleaning formulations, rug shampoos; used in sealing food containers; pale amber liq.; dens. 8.8 lb/gal; pH 9–10 (10%); biodeg.; 30–32% act.

Sulfotex 130. [289] Sulfated castor oil; anionic; wetting agent; liq.; 75% conc.

Sulfotex DOS. [289] Disodium monooleamido PEG sulfosuccinate; anionic; detergent, emulsifier, lathering agent for skin cleansers; foamer for carpet backing and fire fighting; emulsifier for polymerization; liq.; 28% conc.

Sulfotex LCX. [290] Sodium lauryl sulfate; anionic; wetting agent and emulsifier; primary surfactant for rug and upholstery shampoos, hard surface cleaner; water-wh. liq.; dens. 8.73 lb/gal; cloud pt. 2 C; pH 7.0–9.0 (10%); biodeg.; 28.5–30.5% act.

Sulfotex LMS-E. [290] Sodium lauryl ether sulfate; anionic; surfactant used in cleansing products; pale yel. clear liq.; dens. 8.75 lb/gal; cloud pt. 10 C; pH 7.0–9.0 (10%); biodeg.; 58–60% act.

Sulfotex OA. [290] Sodium octyl sulfate; anionic; wetting agent for animal glue on paper and paperboard, high electrolyte concs., and food processing; detergent and dispersant in industrial cleaners; mercerizing agent in dyeing of fibers; lt. amber; dens. 9.2 lb/gal; pH 8.0–10.0 (10%); biodeg.; 38.5–40.5% act.

Sulfotex OT. [290] Ammonium lauryl ether sulfate; anionic; surfactant used in household and industrial detergents; pale yel. visc. liq.; dens. 8.54 lb/gal; cloud pt. 10 C; pH 6.0–7.0 (10%); biodeg.; 58–60% act.

Sulfotex PAI. [290] Ammonium alkyl ether sulfate; anionic; foamer and dispersant used in cleaning formulations; pale yel. liq.; dens. 8.31 lb/gal; pH 7.0–7.5; biodeg.; 46–50% act.

Sulfotex PAI-S. [290] Sodium alkyl ether sulfate; anionic; see Sulfotex PAI; pale yel. liq.; dens. 8.69 lb/gal; pH 7.5–8.5; biodeg.; 45–49% act.

Sulfotex RIF. [290] Sulfated linear alcohol ethoxylate, ammonium salt; anionic; foamer for petroleum industry applications, household and industrial cleaners; liq.; biodeg.; 47% act.

Sulfotex SAL, SAT. [290] Sulfated linear alcohol ethoxylate, ammonium salt; anionic; emulsifier, wetting agent, and foamer for liq. detergents, car shampoos, household and industrial cleaners, petroleum industry applications; liq.; biodeg.; 58% act.

Sulfotex WA. [289] Sodium lauryl sulfate; anionic; detergent base for personal care products and liq. cleaners; liq.; 28% conc.

Sulfotex WAQ-LCX. [290] Sodium lauryl sulfate, sodium decyl sulfate, anionic; surfactant for rug and upholstery shampoos, hard surface cleaners, spot removers; water-wh. to slightly yel. liq.; minimum odor; water-sol.; dens. 1.046 g/ml; visc. 100 cps; cloud pt. 5 C; pH 7.0–9.0 (10%); biodeg.; 29.5 ± 1.0% act.

Sulfotex WAT. [290] TEA lauryl sulfate, anionic; detergent base, foamer and wetting agent for soaps, shampoos, and detergents; liq.; biodeg.; 40% act.

Sulframin 40. [650] Linear alkylaryl sodium sulfonate; anionic; detergent base and wetting agent for household and industrial specialty compds., personal care products, metal cleaning; lubricant, emulsifier, dye dispersant, penetrant, scouring agent, antistat in textile industry; defoamer for petroleum industry; flakes, granules; 40% conc.

Sulframin 40DA. [650] Linear alkylaryl sodium sulfonate; anionic; detergent for car washing, household and industrial cleaning; wetting agent, lubricant, emulsifier, antistat; metal cleaning; textile surfactant; beads; 40% conc.

Sulframin 40RA. [650] Linear alkylaryl sodium sulfonate; anionic; see Sulframin 40DA; also used as base in personal care products; beads; 40% conc.

Sulframin 40T. [650] Linear alkylaryl triethanolamine sulfonate; anionic; see Sulframin 40DA; liq.; 40% conc.

Sulframin 45. [650] Linear alkylaryl sodium sulfonate; anionic; detergent, wetting agent, base for liq. detergent, household, and industrial specialty compds.; metal cleaning; textile surfactant; liq.; cloud pt. 14 C; 45% conc.

Sulframin 45LX. [650] Linear alkylaryl sodium sulfonate; anionic; see Sulframin 45; liq.; 44% conc.

Sulframin 60T. [650] Linear alkylaryl triethanolamine sulfonate; see Sulframin 40T; also for personal care products; liq.; 60% conc.

Sulframin 85. [650] Linear alkylaryl sodium sulfonate; anionic; detergent, lubricant, emulsifier, antistat, wetting agent for household and industrial specialty compds., metal cleaning, personal care products, textiles; flakes, powd.; 85% conc.

Sulframin 90. [650] Linear alkylaryl sodium sulfonate; anionic; see Sulframin 85; also surfactant for emulsion polymerization and latex stabilization; flakes; 91% conc.

Sulframin 1240, 1245. [650] Linear alkylaryl sodium sulfonate; anionic; see Sulframin 85; slurry; 40 and 45% conc. resp.

Sulframin 1250, 1260. [650] Linear alkylaryl sodium sulfonate; anionic; see Sulframin 60T; slurry; 51 and 60% conc. resp.

Sulframin 1288. [650] Linear alkylaryl sulfonic acid; anionic; penetrant, lubricant, dispersant, detergent, antistat, intermediate for wetting agents, emulsifiers, and detergents for household and industrial specialties, personal care products; textile surfactant; liq.; 88% conc.

Sulframin 1298. [650] Linear alkylaryl sulfonic acid; anionic; see Sulframin 1288; also used in personal care products; surfactant for emulsion polymerization; liq.; 98% conc.

Sulframin 1388. [650] Linear alkylaryl sulfonic acid; anionic; see Sulframin 1288; liq.; 88% conc.

Sulframin 4010D, 4010R. [650] Linear alkylaryl sodium sulfonate; detergent for car washing, household and industrial cleaning; beads.

Sulframin AOS, AOS 90. [650] Sodium alpha olefin sulfonate; detergent base for personal care products; surfactant for emulsion polymerization and latex stabilization; liq. and flakes resp..

Sulframin CSA. [650] Cumene sulfonic acid; hydrotrope, coupler, solubilizer, catalyst; used in liq. detergents.

Sulframin HD, Low-Foam HD. [650] Linear alkylaryl sodium sulfonate; laundry and all-purpose detergent; beads.

Sulframin LX. [650] Linear alkylaryl sodium sulfonate, modified; anionic; detergent base and wetting agent for household and industrial specialty products; metal cleaning; textile surfactant; flakes; 85% conc.

Sulframin Phos-Free HD. [650] Linear alkylaryl sodium sulfonate; see Sulframin HD; beads.

Sulframin TX. [650] Toluene sulfonic acid, modified; hydrotrope; coupler, catalyst;

solubilizer for liq. detergent.

Sulphonated 'Lorol' Liquid MA, MR. [527] Monoethanolamine lauryl sulfate; anionic; base for shampoos; liq.; 32% conc.

Sulphonated 'Lorol' Liquid NH. [527] Ammonium lauryl sulfate; anionic; preparation of hair and carpet shampoos; liq.; 30% conc.

Sulphonated 'Lorol' Liquid TA, TN, TNR. [527] Triethanolamine lauryl sulfate; base for shampoos; liq.; 42, 36, and 32% conc. resp.

Sulphonated 'Lorol' Paste. [527] Sodium lauryl sulfate; anionic; preparation of cream shampoos; paste; 40% conc.

Sulphonic Acid LS. [283] Alkylbenzene sulfonic acid, anionic; base for detergent compds.; liq.; 96% act.

Sunaptol CA 25, 120, 350, 400, 750. [317] Castor oil, ethoxylated; nonionic; defoamer, emulsifier, dispersant, softener, leveling agent and antistat for textiles; liq.; 100% conc.

Sunaptol CFA 90, 110. [317] Fatty acid, ethoxylated; nonionic; emulsifier and antistat; liq.; 100% conc.

Sunaptol DB. [317] Linear alcohol, ethoxylate; nonionic; detergent, wetting and scouring agent for textile products; liq.; 90% conc.

Sunaptol DL Conc. [317] Alkyl phenol, ethoxylated; nonionic; wetting agent and detergent for scouring in textiles; antistat; emulsifier for waxes and greases; liq.; 85% conc.

Sunaptol HB4. [317] Alkyl phenol, ethoxylated; nonionic; emulsifier, solubilizer and wetting agent used for iodine in Iodophor preparations; liq.; 100% conc.

Sunaptol OA. [317] Oleic acid, ethoxylated; nonionic; emulsifier, oil slick dispersant, antistat for PVC, ABS; liq., paste.

Sunaptol OA 70, 90, 100, 140. [317] Oleic acid, ethoxylated; nonionic; see Sunaptol OA; liq., paste.

Sunaptol OP Conc. [317] Alkyl phenol, ethoxylated; nonionic; wetting agent and detergent; liq.; 100% conc.

Sunaptol P Extra Liquid. [317] Fatty alcohol, ethoxylated; nonionic; dispersant and leveling agent for dyes; dyeing assistant in textiles; emulsifier; liq.; 40% conc.

Sunaptol PP75B. [317] Coco MEA, ethoxylated; nonionic; foam stabilizer for liq. detergent formulations and personal care products; paste; 100% conc.

Sunaptol S 60, 80, 230. [317] Stearic acid, ethoxylated; nonionic; emulsifier and softener; solid; 100% conc.

Sunaptol TO 28, 85. [317] Tall oil, ethoxylated; nonionic; defoamer and emulsifier; liq.; 100% conc.

Sunnol 605 S, 710 H. [370] POE alkyl ether sulfate; anionic; detergent and personal care products; liq.; 27% conc.

Sunnol DOS. [370] POE alkylphenyl ether sulfate; anionic; emulsion polymerization agent; liq.; 30% conc.

Sunnol LBN. [370] POE lauryl ether sulfate; anionic; shampoos; liq.; 24.5–25.5% conc.

Sunnol LDF-110. [370] Alcohol sulfate; anionic; foamer for personal care products; powd.; 100% conc.

Sunnol LL-103. [370] Alcohol sulfate; anionic; see Sunnol LDF-110; liq.; 30% conc.

Sunnol LST. [370] TEA lauryl sulfate; anionic; detergents and personal care products; liq.; 39–41% conc.

Sunnol NES. [370] POE alkylphenyl ether sulfate; anionic; dyeing assistant in textiles; liq.; 30% conc.

Sunny Safe. [172] Sucrose fatty acid ester compd.; nonionic; cleansing agent for foods; liq.

Sunsoflon CK. [437] Alkyl polyamide; cationic; surfactant, softener for textiles; liq.

Sunsoflon K-2. [437] Alkyl polyamide; cationic; see Sunsoflon CK; liq.; 18% conc.

Sunsoflon MT-100. [437] Alkyl polyamide; cationic; see Sunsoflon CK; liq.

Sunsoflon PXP-705. [437] Imidazoline; cationic; softener for textiles; liq.

Sunsoflon PXP-2000. [437] Imidazoline; cationic; softener and lubricant for textiles; soft paste.

Sunsolt RZ-2, -6. [437] Alkylaryl ester; anionic; dispersant and leveling agent for textiles; liq.; 26 and 40% conc. resp.

Super Amide GR. [456] Cocoamide DEA; nonionic; foam stabilizer, emulsifier, thickener, household, industrial, and cosmetic formulations; sp.gr. 0.99; flash pt. 146 F; 80% conc.

Super Amide L-9. [456] Lauramide DEA; nonionic; thickener, foam stabilizer; cosmetic, industrial and household formulations; paste; sp.gr. 0.98; flash pt. 178 F; 90% conc.

Super-Amide L-9A. [456] Lauramide DEA; nonionic; foam stabilizer, emulsifier, thickener used for personal care products; paste; 97% conc.

Super Amide L-9C. [456] Lauramide DEA; nonionic; see Super Amide L9; paste; sp.gr. 0.96; flash pt. 148 F; 97% conc.

Super Amide LL. [456] Lauramide DEA, modified; nonionic; surfactant used as foam booster, stabilizer, and visc. builder; liq.; sp.gr. 0.98; flash pt. 178 F; 97% conc.

Super Amide LM. [456] Lauramide DEA; see Super Amide L9; paste; sp.gr. 0.98; flash pt. 178 F.

Supercol U. [290] Guar gum; visc. builder, texture and slip modifier used in creams and lotions; wh. powd.; 100% conc.

Super Corona. [159] Anhyd. lanolin, refined; superfatting emollient, emulsifier, and plasticizer in cosmetics and pharmaceuticals; Gardner 8.5 soft solid; odorless; m.p. 36–42 C.

Super Hartolan. [159; 160] Lanolin alcohol; nonionic; w/o emulsifier, dispersant for pigments; emollient, conditioner and moisturizer in superfatting in cosmetics and pharmaceuticals; solid; HLB 1.0; 100% conc.

Supermontaline SLT65. [551] Sodium dioctyl sulfosuccinate; anionic; wetting and emulsifier; liq.; 65% conc.

Super Pro #5A. [575] TEA-coco-hydrolyzed animal protein; anionic; detergent, conditioner, emulsifier, moisturizer, foamer used in personal care products; Gardner 10 max. liq.; sp.gr. 1.15 min.; visc. 400 cps min.; pH 7.0–7.5; 68% act.

Super Pro 100. [575] Hydrolyzed animal protein, TEA-coco-hydrolyzed animal protein, and sorbitol; used in personal care products; amber clear liq.; water-sol.; sp.gr. 1.20; visc. 400 cps; pH 6.75; 64% solids.

Super-Sat. [519] Hydrog. lanolin; nonionic; plasticizer, emollient, cosmetic and pharmaceutical preparations; wh. solid; trace odor; sp.gr. 0.85–0.86; m.p. 48–52 C; 100% act.

Super-Sat AWS-4. [519] PEG-20 hydrog. lanolin; nonionic; emollient, emulsifier, plasticizer for emulsion systems; solid; 100% conc.

Super Wet. [519] Ethoxylated fatty acid and ether; nonionic; wetting agent and emulsifier for pesticides; soil penetrant; liq.; 100% conc.

Supratol VF. [283] Sulfonated castor oil; anionic; penetrant, lubricant and conditioner in leather processing; leveler and dispersant for acid and disperse dyestuffs; amber liq.; ref. index 1.4022; dens. 1.112; visc. 90 cps; 30% act.

Surco 40-SX. [456] Sodium alkylaryl sulfonate; anionic; detergent base for liq. cleaning systems; hydrotrope coupler; liq.; sp.gr. 1.09; flash pt. > 200 F; pH 7.5–8.5; 37–40% act.

Surco 55S. [456] Sodium alkylaryl sulfonate; anionic; detergent for liq. cleaning systems; slurry; sp.gr. 1.02; flash pt. > 200 F; pH 7–8; 49–52% act.

Surco 60S. [456] Sodium alkylaryl sulfonate; anionic; see Surco 55S; slurry; sp.gr. 1.02; flash pt. > 200 F; pH 7–8; 58–61% act.

Surco 60T. [456] TEA-dodecylbenzene sulfonate; anionic; detergent; liq.; sp.gr. 1.08; flash pt. > 200 F; pH 6–7; 56–60% act.

Surco 128T. [456] Cocamide DEA (1:1); nonionic; foam stabilizer, thickener, emulsifier; household, industrial, and cosmetic formulations; liq.; sp.gr. 0.99; flash pt. 146 F; 100% conc.

Surco 233. [456] Sodium alkylaryl sulfonate; anionic; base for detergents; liq.; sp.gr. 1.09; flash pt. > 200 F; pH 7.0–9.0; 37–40% act.

Surco 5024. [456] 2:1 Mixed fatty acid diethanolamide, modified; nonionic; self-coupling surfactant used for soils; sp.gr. 1.00; flash pt. > 200 F; 100% conc.

Surco A 10 MM, AA 10 MM. [456] Mixed fatty acid diethanolamide, modified; nonionic; detergent; hard surface cleaner, self-coupling; liq.; sp.gr. 0.98; flash pt. > 200 F; 100% conc.

Surco AOS. [456] Sodium alpha olefin sulfonate; anionic; surfactant for detergents and in personal care products; sp.gr. 1.05; flash pt. > 200 F; pH 8.0–9.0; 38–40% act.

Surco CMEA. [456] Cocamide MEA; nonionic; foam stabilizer, thickener, opacifier; industrial, household, and cosmetic formulations; wax or flake; sp.gr. 0.92; flash pt. 146 F; 100% conc.

Surco Coco Betaine. [456] Cocoamidopropyl betaine; amphoteric; surfactant in household, industrial, and dishwashing formulations; sp.gr. 1.04; flash pt. > 200 F; 40.0–44.0% solids.

Surco Coco Condensate. [456] Coconut 2:1 diethanolamide; nonionic; surfactant for detergents; liq.; sp.gr. 1.01; flash pt. > 200 F.

Surco DDBSA. [456] Linear dodecylbenzene sulfonic acid; anionic; detergent intermediate; liq.; sp.gr. 1.05; flash pt. > 200 F; 96% min. act.

Surco EA58M. [456] Ammonium lauryl ether sulfate/alkanolamide/urea blend; anionic;

detergent, foamer used in personal care products; liq.; sp.gr. 1.03; pH 6–7; 58% min. act.

Surco HDL Base. [456] Coconut fatty acid/ mixed fatty acid DDBSA; anionic; detergent, foamer used in hard surface cleaner systems; liq.; sp.gr. 1.02; flash pt. > 200 F; 98% conc.

Surco MA. [456] Coconut fatty acid and mixed fatty acid DDBSA; nonionic; detergent, foamer, visc. builder and degreaser used in hard surface cleaners; liq.; sp.gr. 1.00; flash pt. > 200 F.

Surco MG-LS. [456] Magnesium lauryl sulfate; used in rug and upholstery shampoos; liq.; sp.gr. 1.02; flash pt. > 200 F.

Surco MGS-35. [456] Magnesium lauryl sulfate and coco monoethanolamide; anionic; detergent, foamer used in rug and upholstery shampoos; liq.; sp.gr. 1.02; flash pt. > 200 F; pH 6–7; 33% min. act.

Surco SF42B. [456] Sodium alkylaryl sulfonate; anionic; detergent; liq.; sp.gr. 1.02; flash pt. 81 F; pH 6.5–7.5; 40–44% act.

Surco SR-100. [456] Coconut/oleic fatty acid diethanolamide, modified; nonionic; detergent, emulsifier, wetting agent, hard surface cleaner; liq.; sp.gr. 0.98; flash pt. > 200 F; 100% conc.

Surco SR-200. [456] Coconut fatty acid/ DDBSA diethanolamide; anionic; detergent, foamer, visc. builder; hard surface cleaning systems; liq.; sp.gr. 1.03; flash pt. > 200 F; 98% conc.

Surco SXS. [456] Sodium xylene sulfonate; anionic; hydrotrope, coupler, and solubilizer; liq.; sp.gr. 1.17; flash pt. > 200 F; pH 7–10; 39–42% act.

Surco TY. [456] Coconut/mixed fatty acid diethanolamide; nonionic; component in conveyor lubricants for hard surface cleaners; liq.; sp.gr. 0.99; flash pt. > 200 F.

Surco WC Conc. [456] Cocamide DEA; nonionic; detergent, foam booster and stabilizer, thickener used in personal care products and dishwash formulations; liq.; sp.gr. 0.99; flash pt. 178 F; 100% conc.

Surco WHC. [456] Coconut/oleic fatty acid diethanolamide; nonionic; emulsifier used in waterless hand cleaners; liq.; sp.gr. 1.02; flash pt. > 200 F; 100% conc.

Surfac 18-EHP. [392] Primary amine phosphate salt; cationic; corrosion inhibitor used in metal processing; amber visc. liq.; oil-sol.; dens. 7.9 lb/gal; sp.gr. 0.95; 100% act.

Surfac P14B. [392] Primary ether amine acetate; cationic; surfactant; m.w. 268; disp. in water; dens. 7.60 lb/gal; 100% conc.

Surfac P24M. [392] Primary ether amine acetate; cationic; surfactant; m.w. 261; water-disp.; dens. 7.61 lb/gal; 100% conc.

Surfactant AR 150. [295] Rosin oxyethylene glycol ester; nonionic; detergent, emulsifier for detergents and food related areas; liq., wax; 100% conc.

Surfactol 13. [446] Castor oil, modified; nonionic; wetting agent, wax plasticizer, mold release agent, antifoamer for textiles, leather, paints, household, cosmetics, dyeing, tanning, finishing, sizing, making of cutting and sol. oils, dispersing waxes, pigments, resins, dry powds.; Gardner 4 liq.; faint odor; water-sol.; dens. 8.36 lb/gal; sp.gr. 1.005; visc. 17 stokes; 100% conc.

Surfactol 318. [446] PEG-5 castor oil; nonionic; emulsifier for oils, waxes; solubilizer for fragrances; emollient; used in textiles, paints, household, cosmetics, dyeing, tanning, finishing, sizing, insecticides, herbicides, fungicides, kier boiling, making of cutting and sol. oils, dispersing waxes, pigments, resins, rewetting dried skins; Gardner 3 liq.; faint odor; sol. in toluene, butyl acetate, MEK; sp.gr. 0.984; visc. 690 cps; HLB 3.6; 100% conc.

Surfactol 365. [446] PEG-40 castor oil; see Surfactol 318; Gardner 3 liq.; sol. in toluene, butyl acetate, MEK, ethanol, water; sp.gr. 1.054; visc. 500 cps; HLB 13; 100% conc.

Surfactol 575. [446] PEG-66 trihydroxystearin; see Surfactol 318; wh. solid; water-sol.; alcohol; sp.gr. 1.08; m.p. 39 C; HLB 15.

Surfactol 590. [446] PEG-200 trihydroxystearin; see Surfactol 318; wh. solid; sol. in water, alcohol; sp.gr. 1.18; m.p. 53 C; HLB 18.

Surfageen S 30. [131] Fatty alcohol ether sulfosuccinate; anionic; raw material for personal care products; liq.; 40% conc.

Surfageen S 30 40%. [131] Alkyl polyglycol ether sulfosuccinate, sodium salt; anionic; detergent for shampoo and foambath; liq.; 90% conc.

Surfam D89DA. [392] Alkyl ether diamine; corrosion inhibitor; flotation agent for phosphate and iron ores; lt. liq.; mild ammonia odor; dens. 7.3 lb/gal; pour pt. 13 C.

Surfam P5. [392] Ether amine; cationic; intermediate for ore flotation, corrosion inhibitor, textile aux., lube and fuel additive, water treatment chemical; liq.; 100% conc.

Surfam P5 Dist. [392] Methoxypropylamine; intermediate; emulsifier in wax emulsions for floor care products; textile finishing, paints, and metal working oils; insecticide emulsions; Gardner 2 max. clear liq.; m.w. 90; sol. in water and org. solvs.; dens. 7.29 lb/gal; b.p. 115–123 C; flash pt. 102 F; 98% min. act.

Surfam P5 Tech. [392] Methoxypropylamine; see Surfam P5 Dist.; Gardner 7 max. clear liq.; sol. see Surfam P5 Dist.; m.w. 91; dens. 7.30 lb/gal; 90% min. act.

Surfam P10. [392] Primary ether amine; cationic; corrosion inhibitor; emulsifier and lubricant in fiber and fabric processing; raw material in industrial applications; intermediate in formulation of anticaking agents, specialty surfactant; gasoline and fuel additive; Gardner 4 max. clear liq.; m.w. 169; dens. 7.09 lb/gal; 100% conc.

Surfam P12B. [392] Primary ether amine; cationic; see Surfam P10; Gardner 4 max. clear liq.; m.w. 200; dens. 7.11 lb/gal; 100% conc.

Surfam P14B. [392] Primary ether amine; cationic; see Surfam P10; Gardner 4 max. clear liq.; m.w. 229; dens. 7.11 lb/gal; 100% conc.

Surfam P17B. [392] Primary ether amine; cationic; see Surfam P10; Gardner 4 max. clear liq.; m.w. 279; dens. 7.30 lb/gal; 100% conc.

Surfam P24M. [392] Primary ether amine; cationic; see Surfam P10; Gardner 4 max. clear liq.; m.w. 212; dens. 7.10 lb/gal; 100% conc.

Surfam P86M. [392] Primary ether amine; cationic; see Surfam P10; Gardner 4 max. clear liq.; m.w. 280; dens. 7.05 lb/gal; 100% conc.

Surfam P89MB. [392] Primary ether amine; cationic; see Surfam P10; Gardner 4 max. clear liq.; m.w. 305; dens. 7.10 lb/gal; 100% conc.

Surfam P-MEPA. [392] Methoxyethoxypropylamine; specialty chemical; used in textile dye chemicals, emulsifier intermediate in floor waxes and agric. products; Gardner 1 liq.; dens. 7.82 lb/gal; sp.gr. 0.93–0.94; b.p. 210–230 C; flash pt. < 212 F (COC); 97% min. act.

Surfax WO. [305] Sulfated ester; anionic; penetrant; softener, wetting and rewetting agent; liq.; 64% conc.

Surfine AZI-A. [237] Carboxylate; nonionic; detergent, wetter, and dispersant; liq.; 90% conc.

Surfine T-A. [237] Trideceyl ether carboxylic acid; nonionic; see Surfine WNT A; liq.; 90% conc.

Surfine WCT-Gel. [237] Sodium C_{11-15} pareth-7 carboxylate; nonionic; for cosmetic formulations; gel; 60% conc.

Surfine WLL. [237] Carboxylate; solubilizer and coupler, dispersant; gel; 70% conc.

Surfine WNG-A. [237] Carboxylate; nonionic; detergent, wetting agent, emulsifier, solubilizer and coupler used in detergents; liq.; 90% conc.

Surfine WNT A. [237] Carboxylate; nonionic; used in cosmetic formulations; liq.; 90% conc.

Surfine WNT Conc. [237] Carboxylate; detergent, wetting agent; dispersant for household and cosmetic formulations; liq.; 85% conc.

Surfine WNT Gel. [237] Sodium C_{12-15} pareth-7 carboxylate; see Surfine WNT Conc.; gel; 60% conc.

Surfine WNT LC. [237] Sodium C_{12-15} pareth-7 carboxylate; see Surfine WNT Conc.; liq.; 50% conc.

Surfine WNT LS. [237] Sodium C_{12-15} pareth-7 carboxylate; see Surfine WNT Conc.; liq.; 95% conc.

Surflo-CW3. [446] Alcohol ethoxylate, sulfated, ammonium salt; anionic; surfactant used in drilling operations; liq.; 40% conc.

Surflo-OW-1. [447] Fatty acid glycerol ester, oxyethylated; nonionic; o/w emulsifier for elimination of oil slicks; liq.; sol. in IPA and water; dens. 8.56 lb/gal; sp.gr. 1.028 (65 F); visc. 160 SUS; flash pt. > 175 F (TOC); pour pt. 5 F; pH 7.5; 58% conc.

Surflo-S24. [447] Alkylphenol and alcohol, oxyethylated; nonionic; dispersant, detergent, wetting and surf. and interfacial tens. reducing agent; used in producing wells and flotation units; liq.; water-sol.; dens. 8.30 lb/gal; sp.gr. 0.996 (68 F); visc. 125 SUS; flash pt. 82 F (PMCC); pour pt. < –10 F; 30% conc.

Surflo-S30. [447] Sulfated alcohol ethoxylate, ammonium salt; anionic; detergent, foamer, surfactant and foamer in wells, production equipment, and air drilling applications; liq.; water-sol.; dens. 8.16 lb/gal; sp.gr. 0.979 (68 F); visc. 52 SUS; flash pt. 81 F (TOC); pour pt. –3 F; biodeg.; 40% conc.

Surflo-S32. [447] Sodium alkylaryl sulfonate; anionic; detergent, dispersant, emulsifier; surfactant in oil industry; liq.; water-sol.; dens. 8.16 lb/gal; sp.gr. 0.980 (68 F); visc. 44

SUS; flash pt. 80 F (TOC); pour pt. < –10 F; 20% conc.

Surflo-S40. [447] Polyglycol and alcohol ethoxylate sulfated, ammonium salt; anionic; demulsifier, antifoulant, surfactant in oil industry; liq.; water-sol.; dens. 8.22 lb/gal; sp.gr. 0.9873 (68 F); visc. 43 SUS; flash pt. 90 F (TCC); pour pt. –10 F; pH 7.3 (1%); 27% conc.

Surflo-S41. [447] Polyglycol; see Surflo-S40; liq.; water-sol.; dens. 8.04 lb/gal; sp.gr. 0.9662 (68 F); visc. 47 SUS; flash pt. 120 F (TOC); pour pt. –5 F; pH 5.5 (1%); 20% conc.

Surflo-S242. [446] Oxyethylated alkylphenol and oxyethylated alcohol; nonionic; surfactant in oil production operations; liq.; 40% conc.

Surflo-S302. [446] Sulfated alcohol ethoxylate, ammonium salt; anionic; surfactant in oil production operations; liq.; 94% conc.

Surflo-S362. [446; 447] Sulfated alcohol ethoxylate, ammonium salt; anionic; foamer used in air drilling operations; liq.; water-sol.; dens. 8.65 lb/gal; sp.gr. 1.038 (68 F); visc. 95 SUS; flash pt. 79 F (TCC); pour pt. –10 F; pH 8.8 (5%); 100% conc.

Surflo-S375. [447] Sulfated alcohol ethoxylate, ammonium salt; anionic; see Surflo-S362; liq.; dens. 8.50 lb/gal; sp.gr. 1.020 (68 F); visc. 192 SUS; flash pt. 74 F (PMCC); pour pt. –10 F; pH 6.4 (1%); 53% conc.

Surflo-390. [447] Alcohol ethoxylate sulfated, ammonium salt; anionic; see Surflo-S362; liq.; dens. 8.46 lb/gal; sp.gr. 1.016; visc. 195 SUS; flash pt. 82 F (PMCC); pour pt. –10 F; pH 6.4 (1%); 38% conc.

Surflo-S398. [446] Alpha olefin sulfonate; anionic; foamer in air drilling operations; liq.; 40% conc.

Surflo-S412. [446] Polyglycol; anionic; surfactant in oil production operations; liq.; 40% conc.

Surflo-S1720. [446] Sulfated alcohol ethoxylate, ammonium salt; anionic; see Surflo-S398; liq.; 41% conc.

Surflo S-3902. [446] Sulfated alcohol ethoxylate, ammonium salt; anionic; see Surflo-S398; liq.; 51% conc.

Surfonic HDL. [599] Surfonic N-85 (85%), triethanolamine (15%); heavy-duty detergent formulations; Pt-Co 100 max. liq.; dens. 1.0608 g/cm^3; visc. 301 cSt; pour pt. 20 F; flash pt. (PMCC) 425 F; ref. index 1.4885.

Surfonic J-4. [335] Alkyl polyoxyalkalene ether; nonionic; detergent; industrial cleaning compds.; wetting agent for clays in brick and ceramics mfg.; concrete additive; dust control compositions; textile and paper wetting agent; wallpaper removal; clear liq.; water-sol.; dens. 8.4 lb/gal; sp.gr. 1.0072; visc. 267 SUS (68 F); cloud pt. 61 C (1% aq.); flash pt. 375 F (COC); pour pt. 23 F; surf. tens. 29.24 dynes/cm (0.1%); pH 5–7 (1% aq.); 100% act.

Surfonic LF-6. [335] Aliphatic alcohol-EO adduct, modified; nonionic; surfactant, mechanical dishwashing, laundry detergent, pulp and paper, metal cleaning; pale yel. liq.; ref. index 1.4559; sol. in water, acetone, benzene, toluene, perchloroethylene, carbon tetrachloride, kerosene, petroleum ether, and naphtha; sp.gr. 1.0023; visc. 60 cs (100 F); cloud pt. 35.5 C (1% aq.); surf. tens. 29.6 dynes/cm (0.1%); pH 6–7 (1% aq.); 100% conc.

Surfonic LF-7. [335] Aliphatic alcohol-EO adduct, modified; see Surfonic LF-6; pale yel. clear liq.; sol. see Surfonic LF-6; ref. index 1.4565; sp.gr. 1.0068; visc. 67 cs (100 F); HLB 12.0; cloud pt. 30.5 (1% aq.); surf. tens. 30.0 dynes/cm (0.1%); pH 6–7 (1% aq.); 100% conc.

Surfonic LF-17. [335] Primary alcohol-EO adduct, modified; nonionic; surfactant; also see Surfonic LF-6; clear liq.; HLB 12.2; cloud pt. 34.8 C (1% aq.); surf. tens. 33.0 dynes/cm (0.1%).

Surfonic N-10. [335] Nonoxynol-1; nonionic; emulsifier, wetting agent, dry cleaning detergent, penetrant, solubilizer, lime soap dispersant, antifoamer used in agric. chemicals, cosmetics, industrial cleaners, ceramics and concrete, dust control, wallpaper removal, photographic film developing, fire fighting, emulsion polymerization, indirect food additives, cutting oil emulsifiers; stabilizer for rubber latex and drilling-mud additives; degreaser for leather industry; liq.; sol. in acetone, methanol, xylene, carbon tetrachloride, min. oil; m.w. 264; ref. index 1.5090; dens. 8.1 lb/gal; sp.gr. 0.98; visc. 675 SUS (100 F); HLB 3.4; flash pt. 355 F (OC); 100% act.

Surfonic N-31.5. [335] Nonoxynol-3; nonionic; see Surfonic N-10; liq.; sol. see Surfonic N-10; m.w. 358; ref. index 1.4950; dens. 8.4 lb/gal; sp.gr. 1.01; visc. 480 SUS (100 F); HLB 7.7; flash pt. 410 F (OC); 100% act.

Surfonic N-40. [335] Nonoxynol-4; nonionic;

see Surfonic N-10; liq.; sol. in acetone, methanol, xylene, carbon tetrachloride, Stoddard; m.w. 396; ref. index 1.4979; dens. 8.5 lb/gal; sp.gr. 1.026; visc. 445 SUS (100 F); HLB 8.9; flash pt. 435 F; surf. tens. 27.5 dynes/cm (0.1%); 100% act.

Surfonic N-60. [335] Nonoxynol-6; nonionic; see Surfonic N-10; liq.; sol. in acetone, methanol, xylene, carbon tetrachloride, Stoddard solv.; disp. in water; m.w. 484; ref. index 1.4938; dens. 8.7 lb/gal; sp.gr. 1.041; visc. 440 SUS (100 F); HLB 10.9; flash pt. 475 F (OC); surf. tens. 28.7 dynes/cm (0.1%); 100% act.

Surfonic N-85. [335] Nonoxynol-8.5; nonionic; see Surfonic N-10; liq.; sol. in acetone, methanol, xylene, carbon tetrachloride, water; m.w. 596; ref. index 1.4923; dens. 8.8 lb/gal; sp.gr. 1.056; visc. 485 SUS (100 F); HLB 12.6; cloud pt. 44 C (1% aq.); flash pt. 500 F (OC); surf. tens. 30.5 dynes/cm (0.1%); 100% act.

Surfonic N-95. [335] Nonoxynol-10; nonionic; see Surfonic N-10; liq.; sol. see Surfonic N-85; m.w. 632; ref. index 1.4893; dens. 8.8 lb/gal; sp.gr. 1.061; visc. 510 SUS (100 F); HLB 12.9; cloud pt. 54.2 (1% aq.); flash pt. 500 F (OC); surf. tens. 30.8 dynes/cm (0.1%); 100% act.

Surfonic N-100. [335] Nonoxynol-10; nonionic; see Surfonic N-10; liq.; sol. see Surfonic N-85; m.w. 660; ref. index 1.4888; dens. 8.8 lb/gal; sp.gr. 1.064; visc. 69 SUS (210 F); HLB 13.2; cloud pt. 65 C (1% aq.); flash pt. 510 F (OC); surf. tens. 31.0 dynes/cm (0.1%); 100% act.

Surfonic N-102. [335] Nonoxynol-10; nonionic; see Surfonic N-10; liq.; sol. see Surfonic N-85; m.w. 668; ref. index 1.4884; dens. 8.8 lb/gal; sp.gr. 1.065; HLB 13.4; cloud pt. 81 C (1% aq.); flash pt. 510 F (OC); surf. tens. 31.2 dynes/cm (0.1%); 100% act.

Surfonic N-120. [335] Nonoxynol-12; nonionic; see Surfonic N-10; liq.; sol. see Surfonic N-85; m.w. 748; ref. index 1.4869; dens. 8.9 lb/gal; sp.gr. 1.070; visc. 560 SUS (100 F); HLB 14.1; cloud pt. 81 C (1% aq.); flash pt. 525 F (OC); surf. tens. 32.3 dynes/cm (0.1%); 100% act.

Surfonic N-150. [335] Nonoxynol-15; nonionic; see Surfonic N-10; wh. semisolid; sol. see Surfonic N-85; m.w. 880; ref. index 1.4815 (30 C); dens. 8.95 lb/gal; sp.gr. 1.065 (30/4 C); visc. 89 SUS (210 F); HLB 15.0; cloud pt. 94 C (1% aq.); flash pt. > 500 F (OC); surf. tens. 34.2 dynes/cm (0.1%); 100% act.

Surfonic N-200. [335] Nonoxynol-20; nonionic; see Surfonic N-10; wh. semisolid; sol. see Surfonic N-85; m.w. 1100; ref. index 1.4720 (50 C); dens. 9.0 lb/gal; visc. 105 SUS (210 F); HLB 15.8; cloud pt. > 100 C (1% aq.); flash pt. > 500 F (OC); 100% act.

Surfonic N-300. [335] Nonoxynol-30; nonionic; see Surfonic N-10; wh. waxy solid; sol. see Surfonic N-85; m.w. 1540; ref. index 1.4690 (50 C); dens. 9.1 lb/gal; visc. 160 SUS (210 F); HLB 17.1; cloud pt. > 100 C (1% aq.); flash pt. > 500 F (OC); 100% act.

Surfonic N-400. [335] Nonoxynol-40; nonionic; see Surfonic N-10; wh. waxy solid; sol. see Surfonic N-85; m.w. 1980; dens. 9.1 lb/gal; visc. 210 SUS (210 F); HLB 7.8; cloud pt. > 100 C (1% aq.); flash pt. > 500 F (OC); 100% act.

Surfonic NB-5. [335] Nonoxynol-30; nonionic; see Surfonic N-10; clear liq.; 70% act.

Surfonic NB-14. [335] Nonoxynol-40; nonionic; see Surfonic N-10; clear liq.; 70% act.

Surfynol 61. [8] Dimethyl hexynol; nonionic; wetting agent used for paper coatings, floor polishes, and glass cleaning formulations; colorless clear liq.; sp.gr. 0.855; b.p. 151 C; surf. tens. 50.2 (0.1% aq.); 100% act.

Surfynol 82. [8] Dimethyl octynediol; nonionic; defoamer, wetting agent used in pesticide concs.; cosmetic ingred.; developer compds.; electroplating baths; wh. cryst. powd.; sp.gr. 0.933; m.p. 51 C; b.p. 222 C; surf. tens. 55.3 (0.1% aq.); 100% conc.

Surfynol 104. [8] Tetramethyl decynediol; nonionic; defoamer and dye dispersant in paint and ink formulations; surfactant in rinse aids; substrate pigment wetting agent for industrial coatings and adhesives; defoamer in dyestuff formulations; wetting agent for industrial cleaners; visc. reducer for vinyl dispersions; wh. waxy solid; sp.gr. 0.893; m.p. 37 C; b.p. 260 C; surf. tens. 31.6 (0.1% aq.); 100% conc.

Surfynol 104A. [8] Tetramethyl decynediol and 2-ethyl hexanol; nonionic; see Surfynol 104; lt. yel. liq.; sp.gr. 0.869; surf. tens. 33.0 (0.1% aq.); 50% conc.

Surfynol 104E. [8] Tetramethyl decynediol and ethylene glycol; nonionic; see Surfynol 104; lt. yel. clear liq.; sp.gr. 1.001; surf. tens. 36.2 (0.1% aq.); 50% conc.

Surfynol 104H. [8] Tetramethyl decynediol and ethylene glycol; nonionic; see Surfynol

104; lt. yel. clear liq.; sp.gr. 0.951; surf. tens. 33.8 (0.1% aq.); 75% conc.

Surfynol 440. [8] Tetramethyl decynediol, ethoxylated (3.5 moles); nonionic; water-based industrial finishes; defoamer, rewetting, and leveling agent for paperboard coatings; metal cleaning and plating bath additive; straw clear liq.; sp.gr. 0.982; surf. tens. 31.0 (0.1% aq.); 100% conc.

Surfynol 465. [8] Tetramethyl decynediol, ethoxylated (10 moles); nonionic; surfactant for emulsion polymerization; electroplating additive; straw clear liq.; sp.gr. 1.038; surf. tens. 33.2 (0.1% aq.); 100% conc.

Surfynol 485. [8] Tetramethyl decynediol, ethoxylated (30 moles); nonionic; see Surfynol 465; straw clear liq.; sp.gr. 1.080 (30 C); surf. tens. 40.1 (0.1% aq.); 100% conc.

Surfynol GA. [8] Acetylenic diol; nonionic; pigment wetting agent in coatings and other pigmented systems.

Surfynol PC. [8] Acetylenic glycol; nonionic; defoamer used in paper coating latices; anti-shock agent for paper coatings; lt. yel. clear liq.; sp.gr. 0.9585; 100% conc.

Surfynol SE. [8] Acetylenic diol; nonionic; wetting agent, foam control agent in pressure-sensitive adhesives, aq. lubricants, water based paints, inks, dye processing, and paper coatings; liq.

Surfynol TG. [8] Acetylenic glycol and ethylene glycol; nonionic; pigment and substrate wetting agent used in latex and water-reducible paints, adhesives, paper coatings, and pigmented aq. systems; lt. yel. clear liq.; sp.gr. 1.000; surf. tens. 27.6 (0.1% aq.); 83% conc.

Surfynol TG-E. [8] Actylenic glycol and propylene glycol; nonionic; wetting agent used in pesticide formulations; liq.; 83% conc.

Swanic 6L, 7L. [584] Nonylphenol ethoxylate; nonionic; wetting agent, emulsifier and detergent used in wool scouring, paint formulation and textile processing; clear liq.; pH 6–7 (1% in dist. water); 100 and 25% act. resp.

Swanic 51. [584] Coconut monoethanolamide (oil-based); foam booster/stabilizer in detergent powds.; wh. solid paste.

Swanic 52L. [584] Coconut alkylolamide; anionic; foam booster, visc. controller used in shampoos and liq. detergent formulations; yel. liq.; 100% conc.

Swanic 110. [584] Alcohol ethoxylate; nonionic; retarder and dye leveling agent used in textiles; liq.; 30% conc.

Swanic C-30. [584] Alcohol ethoxylate; nonionic; detergent; liq.; 100% conc.

Swanic CD-48. [584] Alkyl phenol ethoxylate; nonionic; emulsifier; liq.; 100% conc.

Swanic CD-72. [584] Alkyl phenol ethoxylate; nonionic; emulsifier; paste; 100% conc.

Swanic CM. [584] Alkyl phenol POE ether; nonionic; air entraining agent; concrete additive; liq.; 25% conc.

Swanic CWT. [584] Sulfonated ester; anionic; wetting and rewetting agent used in textiles; liq.; 30% conc.

Swanic D-16, -24, -52, -60, -80, –120. [584] Alkyl phenol POE ether; nonionic; wetting agent, detergent and emulsifier ingred. in formulation of pesticides; liq., paste, solid.

Swanic SD-36. [584] Alkyl phenol POE ether; nonionic; surfactant for pigment paste preparation; liq.

Swanic T-51. [584] Alcohol sulfate; anionic; wetting agent for textiles; liq.; 30% conc.

Swanic X-100. [584] Alkyl phenol POE ether; nonionic; wetting agent; liq.; 100% conc.

Swanol AM-301. [437] Lauryl betaine; amphoteric; detergent, wetting, dispersant; liq.; 35% conc.

Swascol 1P. [584] Sodium lauryl sulfate; anionic; detergent for toothpaste mfg.; emulsifier in emulsion polymerization; powd.; 95% conc.

Swascol 1PC. [584] Sodium lauryl sulfate; anionic; wetting agent and foamer used in personal care products, liq. detergents; powd.; 96% act.

Swascol 3L. [584] Sodium lauryl sulfate; anionic; wetting agent, foamer and detergent used in personal care products and liq. detergents; liq., paste; 30% act.

Swascol 4L. [584] TEA lauryl sulfate; anionic; foamer, detergent, base material for shampoos and liq. detergents; liq.; 40% act.

Swascol L-327. [584] Sodium laureth sulfate; anionic; foamer, detergent, base material in formulation of shampoos; liq.; cloud pt. 0 C; 27% act.

Swastik Detergent Powder. [584] Sodium alkylaryl sulfonate with alkaline builders, STPP, etc.; detergent powd. for heavily soiled clothes; blue powd.

Sway. [584] Sodium alkylaryl sulfonate with STPP, alkaline builders, etc.; heavy duty detergent for cottons; blue powd.

Swedex 100, 100V. [74] Mono/diglyceride

diacetyl tartaric acid ester of animal and veg. origin resp.; anionic; food emulsifier; powd.; HLB 10.0; 90% conc.

Swedex AR, AR41, AR58P. [74] Mono/diglyceride diacetyl tartaric acid ester; anionic; food emulsifier; paste, paste, and powd. resp.; HLB 10.0; 90% conc.

Swedex E 16. [74] Polyglycol polyricinoleate; nonionic; food emulsifier; liq.; HLB 2.0; 92% conc.

Swedex SSL. [74] Sodium stearoyl-2-lactylate; anionic; food emulsifier; powd.; HLB 15.0; 95% conc.

SXS. [154] Sodium xylene sulfonate; hydrotrope used in detergent formulations; colorless liq.; mild odor; dens. 9.80; sp.gr. 1.17; visc. 6.0; m.p. 31 F; b.p. 212 F; 41% act.

Sykanol DKM 45, 80. [291] Sulfated castor oil; anionic; wetting agent for mfg of bath preparations; liq.; 45 and 80% conc. resp.

Syncrowax AW1C. [159] C_{18-36} acid; emulsifier; HLB 2.0; 100% conc.

Syncrowax BB4, BB5. [160] Syn. beeswax; used in creams and sticks; pale cream solid.

Syncrowax ERL-C. [160] C_{18-36} acid glycol ester; reducer of bleeding and sweating; gloss improver; sticks, creams; lt. tan solid.

Syncrowax HGL-C. [160] C_{18-36} acid triglyceride; strength improver, stabilizer, gloss improver; used in cosmetic makeup; lt. tan solid.

Syncrowax HR-C. [160] Glyceryl tribehenate; gloss improver used in personal care products; off-wh. solid.

Syncrowax PRLC. [160] Fatty wax acid propylene glycol ester; reducer of bleeding and sweating; gloss improver; lt. tan solid.

Syn Fac 334. [413] Aryl POE ether; nonionic; emulsifier, dispersant for pigments, insecticides, solvs., cleaning compds. and latex paint; liq. and semiliq. resp.; HLB 11.6 and 12 resp.; 87 and 100% conc. resp.

Syn Fac 905. [413] POE (9.5) nonylphenol; nonionic; wetting agent and dispersant, emulsifier; liq.; 100% conc.

Syn Fac 8216. [413] Aryl POE ether; nonionic; pigment dispersant for paints, coatings, and textile print systems; solid; HLB 15.0; 100% conc.

Syn Fac 8337. [413] Phosphated alkoxylated aryl phenol, potassium salt; anionic; dispersant for org. pigments; lt. yel. visc. liq.; dens. 9.0 lb/gal; visc. 1000 cps; HLB 20; flash pt. 310 C; pH 7.0 (5%).

Syn Fac TDA-92. [413] POE (8) triethanolamine; nonionic; emulsifier and wetting agent used in fiber applications; liq.; HLB 12.5; 100% conc.

Syn Fac TEA-97. [413] Ethoxylated amine; nonionic; solv. and dispersant in the dye industry; liq.; 100% conc.

Syn Lube 106 (60%). [413] POE (200) glyceride ester; nonionic; emulsifier, lubricant; spin finish and viscose additive, tufting aid; liq.; 60% conc.

Syn Lube 107. [413] POE (25) glyceride ester; nonionic; see Syn Lube 106 (60%); liq.; HLB 15.5; 100% conc.

Synotol 119 N. [496] Cocamide DEA; nonionic; detergent, wetting agent, thickener, foam booster and stabilizer, lime soap dispersant; used in personal care products; liq.; 50% amide.

Synotol CN 60. [496] Cocamide DEA; nonionic; see Synotol 119 N; liq.; 50% amide.

Synotol CN 80. [496] Cocamide DEA; nonionic; see Synotol 119 N; liq.; 75% amide.

Synotol CN 90. [496] Cocamide DEA; foam booster and stabilizer, wetting agent, thickener, superfatting agent; household and cosmetic applications; liq.; 85% conc.

Synotol Detergent E. [496] Cocamide DEA; see Synotol 119 N; liq.; 50% amide.

Synotol L 60. [496] Lauramide DEA; see Synotol 119 N; liq.; 50% amide.

Synotol L 90. [206] Lauramide DEA; see Synotol 119 N; solid; 88% conc.

Synotol LM 60. [496] Lauramide DEA; see Synotol 119 N; liq.; 50% amide.

Synotol LM 90. [206] Lauric/myristic acid diethanolamide; see Synotol 119 N; paste; 88% conc.

Synotol ME-90. [206] Coconut fatty acid monoethanolamide; nonionic; foam booster, wetting agent, thickener, foam stabilizer, and superfatting agent for cosmetic and household preparations; flakes; 85% conc.

Synperonic 3S25. [318] Syn. primary alcohol ethoxylate sulfate, sodium salt; used in cosmetic formulations; colorless to pale yel. mobile liq.; dens. 1.040 g/ml; surf. tens. 40.2 dynes/cm (0.1%); pH 7.0–8.5 (5% aq.); 24.5–25.5% act.

Synperonic 3S27S. [318] Syn. primary alcohol ethylene oxide condensate, sodium salt; liq. detergent blend used in personal care products; liq.; 27% conc.

Synperonic 3S60A. [318] Syn. primary alcohol ethoxylate sulfate, ammonium salt; anionic;

detergent, foamer; used in household formulations; yel. clear to hazy mobile liq.; dens. 1.020 g/ml; surf. tens. 39.4 dynes/cm (0.1%); pH 6.5–8.0 (5% aq.); 59–61% act.

Synperonic 3S60S. [318] Syn. primary alcohol ethoxylate sulfate, sodium salt; see Synperonic 3S60A; colorless to pale yel. mobile liq.; dens. 1.046 g/ml; surf. tens. 40.5 dynes/cm (0.1%); pH 7.0–8.5 (5% aq.); 59–61% act.

Synperonic 3S70. [318] Syn. primary alcohol ethoxylate; anionic; dishwashing liqs. and personal care products; visc. liq.; 70% conc.

Synperonic 91/2.5. [318] Syn. primary C_{9-11} alcohol ethoxylate; nonionic; detergent and emulsifier; intermediate for sulfation; liq.; HLB 8.2; 100% conc.

Synperonic 91/5. [318] Syn. primary C_{9-11} alcohol ethoxylate; nonionic; wetting and scouring agent for wool textile processing, degreasing hard surface cleaners; coupler; o/w emulsifier; liq.; HLB 11.5; 100% conc.

Synperonic 91/6. [318] Syn. primary C_{9-11} alcohol ethoxylate; nonionic; see Synperonic 91/5; also detergent cleaners; liq.; HLB 12.5; 100% conc.

Synperonic 91/8. [318] Syn. primary C_{9-11} alcohol ethoxylate; nonionic; see Synperonic 91/6; HLB 14; 100% conc.

Synperonic 91/10. [318] Syn. primary C_{9-11} alcohol ethoxylate; nonionic; liq.; HLB 14.6; 100% conc.

Synperonic A2. [318] Syn. primary alcohol ethoxylate; nonionic; detergent, emulsifier, coemulsifier; base for emulsifiers, textile antistats, scouring and wetting agents, and specialty cleaners; intermediate for sulfation used in personal care products; Hazen 50 max. liq.; sol. in alcohol, glycol ethers, kerosene, and min. oil; dens. 0.897 g/ml; visc. 25 cps; HLB 5.9; pour pt. 2 C; pH 6–8 (1% aq.); 99% min. act.

Synperonic A3. [318] Syn. primary alcohol ethoxylate; see Synperonic A2; Hazen 50 max. liq.; sol. see Synperonic A2; dens. 0.918 g/ml; visc. 29 cps; HLB 7.8; pour pt. 5 C; pH 6–8 (1% aq.); 99% min. act.

Synperonic A4. [318] Syn. primary alcohol ethoxylate; see Synperonic A2; Hazen 50 max. liq.; sol. in alcohol, glycol ethers, kerosene; sol./disp. in min. oil; dens. 0.937 g/ml; visc. 34 cps; HLB 9.1; pour pt. 9 C; pH 6–8 (1% aq.); 99% min. act.

Synperonic A6. [318] Syn. primary alcohol ethoxylate; see Synperonic A2; Hazen 50 max. liq.; sol. in alcohol, glycol ethers; dens. 0.966 g/ml; visc. 46 cps; HLB 11.2; pour pt. 17 C; surf. tens. 29.8 dynes/cm (0.1%); pH 6–8 (1% aq.); 99% min. act.

Synperonic A7. [318] Syn. primary alcohol ethoxylate; nonionic; wetting and detergent leveling agent; emulsifier and coemulsifier for oils, solvs., and waxes; Hazen 50 max. visc. liq.; sol. in water, glycol ethers and alcohol; dens. 0.977 g/ml; visc. 60 cps; HLB 12.2; cloud pt. 44–48 C (1% aq.); pour pt. 21 C; surf. tens. 30.2 dynes/cm (0.1%); pH 6–8 (1% aq.); 99% min. act.

Synperonic A9. [318] Syn. primary alcohol ethoxylate; see Synperonic A7; wh. paste; sol. see Synperonic A7; dens. 0.984 g/ml (40 C); visc. 35 cps (40 C); HLB 12.5; cloud pt. 62–68 C (1% aq.); pour pt. 24 C; surf. tens. 31.4 dynes/cm (0.1%); pH 6–8 (1% aq.); 99% min. act.

Synperonic A11. [318] Syn. primary alcohol ethoxylate; see Synperonic A7; wh. paste; sol. see Synperonic A7; dens. 0.998 g/ml (40 C); visc. 42 cps (40 C); HLB 13.9; cloud pt. 85–89 C (1% aq.); pour pt. 27 C; surf. tens. 33.2 dynes/cm (0.1%); pH 6–8 (1% aq.); 99% min. act.

Synperonic A14. [318] Syn. primary alcohol ethoxylate; nonionic; detergent, solubilizer, dispersant, stabilizer; wh. wax; sol. see Synperonic A7; dens. 1.014 g/ml (40 C); visc. 55 cps (40 C); HLB 14.9; cloud pt. > 100 C (1% aq.); pour pt. 30 C; surf. tens. 36.5 dynes/cm (0.1%); pH 6–8 (1% aq.); 99% min. act.

Synperonic A20. [318] Syn. primary alcohol ethoxylate; nonionic; see Synperonic A14; wh. hard wax; sol. see Synperonic A7; dens. 1.028 g/ml (50 C); visc. 58 cps (50 C); HLB 16.2; cloud pt. > 100 C (1% aq.); pour pt. 37 C; surf. tens. 40.1 dynes/cm (0.1%); pH 6–8 (1% aq.); 99% min. act.

Synperonic BD100. [318] Syn. primary alcohol ethoxylate with PO; nonionic; detergent, emulsifier; Hazen 150 max. liq.; sol. in water, alcohol, glycol ethers, kerosene; dens. 0.985 g/ml; visc. 92 cps; HLB 12.0; pour pt. 5 C; surf. tens. 29.7 dynes/cm (0.1%); pH 6–8 (1% aq.); 95% min. act.

Synperonic FS. [318] Syn. quat. ammonium compd.; cationic; textile products; liq.; 80% conc.

Synperonic K87. [318] Syn. primary alcohol ethoxylate with PO; nonionic; detergent for textile and raw wool scouring; Hazen 150 max. liq.; sol. see Synperonic A7; dens. 0.980 g/ml; visc. 72 cps; HLB 12.0; cloud pt. 44–49 C; pour pt. 11 C; surf. tens. 29.5 dynes/cm (0.1%); pH 6–8 (1% aq.); 99% min. act.

Synperonic KB. [318] Alcohol ethoxylate; nonionic; wetting agent; wool scouring; liq.; 94% conc.

Synperonic N. [318] Nonylphenol ethoxylate; nonionic; wetting agent and detergent for textile, metal treatment, dust suppression, and general cleaning applications; liq.; 27% conc.

Synperonic NP4. [318] Nonoxynol-4; nonionic; detergent, emulsifier, coemulsifier; intermediate for sulfation in mfg. lubricants and antistats; Hazen 150 max. liq.; sol. see Synperonic A3; dens. 1.022 g/ml; visc. 400 cps; HLB 8.9; pour pt. < 0 C; pH 6–8 (1% aq.); 99% min. act.

Synperonic NP5. [318] Nonoxynol-5; nonionic; see Synperonic NP4; Hazen 150 max. liq.; sol. in alcohol, glycol ethers, kerosene, min. oil; insol./disp. in water; dens. 1.035 g/ml; visc. 350 cps; HLB 10.5; pour pt. < 0 C; pH 6–8 (1% aq.); 99% min. act.

Synperonic NP6. [318] Nonoxynol-6; nonionic; see Synperonic NP4; Hazen 150 max. liq.; sol. in alcohol, glycol ethers, min. oil; disp. in water; dens. 1.041 g/ml; visc. 355 cps; HLB 10.9; pour pt. < 0 C; pH 6–8 (1% aq.); 99% min. act.

Synperonic NP8. [318] Nonoxynol-8; nonionic; detergent for textile scouring; wetting agent, emulsifier; Hazen 150 max. liq.; sol. in water, alcohol, glycol ethers; dens. 1.053 g/ml; visc. 355 cps; HLB 12.3; cloud pt. 30–34 C (1% aq.); pour pt. < 0 C; surf. tens. 29.4 dynes/cm (0.1%); pH 6–8 (1% aq.); 99% min. act.

Synperonic NP9. [318] Nonoxynol-9; nonionic; see Synperonic NP8; Hazen 150 max. liq.; sol. in water, alcohol, glycol ethers; dens. 1.058 g/ml; visc. 340 cps; HLB 12.8; cloud pt. 51–56 C (1% aq.); pour pt. 0 C; surf. tens. 30.6 dynes/cm; pH 6–8 (1% aq.); 99% min. act.

Synperonic NP10. [318] Nonoxynol-10; nonionic; detergent, solubilizer, dispersant, stabilizer; Hazen 150 max. liq.; sol. see Synperonic NP9; dens. 1.061 g/ml; visc. 360 cps; HLB 13.3; cloud pt. 62–67 C (1% aq.); pour pt. 5 C; surf. tens. 30.6 dynes/cm; pH 6–8 (1% aq.); 99% min. act.

Synperonic NP12. [318] Nonoxynol-12; nonionic; see Synperonic NP10; Hazen 150 max. liq.; sol. Synperonic NP9; dens. 1.062 g/ml; visc. 265 cps; HLB 13.9; cloud pt. 79–84 C (1% aq.); pour pt. 14 C; surf. tens. 35.2 dynes/cm; pH 6–8 (1% aq.); 99% min. act.

Synperonic NP13. [318] Nonoxynol-13; nonionic; emulsifier for solvs. and agrochemical pesticides and herbicides; Hazen 150 max. paste; sol. see Synperonic NP9; dens. 1.068 g/ml; visc. 280 cps; HLB 14.4; cloud pt. 87–92 C (1% aq.); pour pt. 17 C; surf. tens. 34.8 dynes/cm; pH 6–8 (1% aq.); 99% min. act.

Synperonic NP15. [318] Nonoxynol-15; nonionic; see Synperonic NP13; Hazen 150 max. paste; sol. see Synperonic NP9; dens. 1.058 g/ml (40 C); visc. 125 cps (40 C); HLB 15.0; cloud pt. 95–99 C (1% aq.); pour pt. 21 C; surf. tens. 33.4 dynes/cm; pH 6–8 (1% aq.); 99% min. act.

Synperonic NP20. [318] Nonoxynol-20; nonionic; solubilizer and emulsifier for polar substrates; Hazen 150 max. solid; sol. see Synperonic NP9; dens. 1.073 g/ml (40 C); visc. 168 cps (40 C); HLB 16.0; cloud pt. > 100 C (1% aq.); pour pt. 30 C; surf. tens. 41.7 dynes/cm; pH 6–8 (1% aq.); 99% min. act.

Synperonic NP30. [318] Nonoxynol-30; nonionic; see Synperonic NP20; Hazen 150 max. solid; sol. see Synperonic NP9; dens. 1.074 g/ml (50 C); visc. 150 cps (50 C); HLB 7.1; cloud pt. > 100 C (1% aq.); pour pt. 40 C; surf. tens. 42.8 dynes/cm; pH 6–8 (1% aq.); 99% min. act.

Synperonic NP35. [318] Nonoxynol-35; nonionic; solubilizer and emulsifier for high polarity materials; solid; HLB 17.4%; 100% conc.

Synperonic NPE1800. [318] Nonylphenol ethoxylate; nonionic; used in agricultural industry; solid; 100% conc.

Synperonic NX. [318] Nonylphenol ethoxylate; nonionic; detergent, wetting agent, emulsifier for textile processing; emulsification of oils and solvs.; liq.; HLB 12.3; 100% conc.

Synperonic NXP. [318] Nonylphenol ethoxylate; nonionic; general purpose cleaning and applications; liq.; HLB 12.8; 100% conc.

Synperonic OP10. [318] Octyl phenol ethoxylate; nonionic; detergent, wetting agent, emulsifier; Hazen 150 max. liq.; sol. see Synperonic NP9; dens. 1.062 g/ml; visc. 393 cps; HLB 13.6; cloud pt. 63–67 C (1% aq.);

pour pt. 7 C; surf. tens. 31.8 dynes/cm; pH 6–8 (1% aq.); 99% min. act.

Synperonic OP11. [318] Octyl phenol ethoxylate; nonionic; see Synperonic OP10; Hazen 150 max. liq.; sol. see Synperonic NP9; dens. 1.067 g/ml; visc. 371 cps; HLB 14.0; cloud pt. 79–82 C (1% aq.); pour pt. 8 C; surf. tens. 31.0 dynes/cm; pH 6–8 (1% aq.); 99% min. act.

Synperonic OP16. [318] Octyl phenol ethoxylate; nonionic; emulsion polymerization; liq.; HLB 15.6; 70% conc.

Synperonic OP40. [318] Octyl phenol ethoxylate; nonionic; see Synperonic OP16; liq.; HLB 17.9; 70% conc.

Synperonic PE30/10. [318] PO/EO block copolymer; nonionic; defoamer, demulsifier, wetting agent, dispersant, rinse aid for machine dishwashing; liq.; m.w. 1900; sp.gr. 1.017; visc. 325 cs; m.p. < 0 C; HLB 3; cloud pt. 24 C (1% aq.); pH 6–8 (2.5%); 99% min. act.

Synperonic PE30/20. [318] PO/EO block copolymer; nonionic; see Synperonic PE30/10; liq.; m.w. 2500; sp.gr. 1.032; visc. 420 cs; m.p. < 0 C; HLB 7; cloud pt. 32 C (1% aq.); surf. tens. 42 dynes/cm; pH 6–8 (2.5%); 99% min. act.

Synperonic PE30/40. [318] PO/EO block copolymer; nonionic; see Synperonic PE30/10; visc. liq.; m.w. 2900; sp.gr. 1.05; visc. 600 cs; m.p. < 0 C; HLB 15; cloud pt. 58 C (1% aq.); surf. tens. 41.1 dynes/cm; pH 6–8 (2.5%); 99% act.

Synperonic PE30/80. [318] PO/EO block copolymer; nonionic; emulsifier, dispersant, stabilizer; emulsion polymerization; used in pigments in emulsion paints, inks; flake; m.w. 8500; m.p. 50 C; HLB 29; cloud pt. > 100 C (1% aq.); surf. tens. 48 dynes/cm; pH 6–8 (2.5%); 99% min. act.

Synperonic PE39/70. [318] PO/EO block copolymer; nonionic; emulsifier, dispersant; wh. flake; m.w. 7500; m.p. 49 C; HLB 24; cloud pt. > 100 C (1% aq.); surf. tens. 45.2 dynes/cm; pH 6–8 (2.5%); 99% min. act.

Synperonic TO5. [318] Acid ethoxylate; nonionic; general purpose emulsification; liq.; 100% conc.

Synprol Alcohol. [318] Syn. primary alcohol, fully sat.; intermediate for production of ethoxylates, sulfates and ether sulfates; ref. index. 1.4457; sp.gr. 0.837; visc. 29.1 cps; flash pt. 290 F (PMCC); 99% act.

Synprolam 35. [318] C_{13-15} alkyl primary amine; anticaking agent for fertilizer; flotation agent, corrosion inhibitor; metal cleaning formulations, pigment dispersions, rubber processing auxs., bitumen emulsifier; intermediate for production of amine salts, ethoxylates and sulfosuccinates; Hazen 50 clear liq.; sp.gr. 0.8; visc. 4.5 cps; flash pt. 127 C (PMCC); 99% primary amine.

Synprolam 35A. [318] Syn. C_{13-15} alkyl primary amine; cationic; emulsifier, fertilizer anticaking agent; min. flotation; solid; 100% conc.

Synprolam 35BQC (50). [318] Syn. C_{13-15} dimethyl tert. amine benzyl ammonium chloride; cationic; emulsifier, sanitizer, biocide, corrosion inhibitor, textile drying aux., timber preservative; liq.; 50% conc.

Synprolam 35BQC (80). [318] Syn. C_{13-15} dimethyl tert. amine benzyl ammonium chloride; cationic; see Synprolam 35BQC (50); liq.; 80% conc.

Synprolam 35 DM. [318] Syn. tert. fatty amine; chemical intermediate for production of quat. salts and amine oxides used in cosmetics; catalyst for polyurethane foam; corrosion inhibitor; Hazen 50 clear liq.; sp.gr. 0.79; visc. 3.9 cps; flash pt. 143 C (PMCC); 97% tert. amine.

Synprolam 35DMA. [318] Syn. C_{13-15} dimethyl tert. amine, acetic acid salt; cationic; emulsifier, biocide, timber preservative; liq.; 100% conc.

Synprolam 35DMO. [318] C_{13-15} alkyl dimethyl amine oxide; mildly cationic; foam booster/stabilizer, thickener in personal care products; Hazen 50 clear liq.; sp.gr. 0.97; surf. tens. 29 dynes/cm (0.1% aq.); 30% act.

Synprolam 35M. [318] Syn. (C_{13-15}) alkyl methyl sec. amine; cationic; surfactant intermediate; liq.; 100% conc.

Synprolam 35N3. [318] N-(C_{13-15}) alkyl-1,3-propane diamine; cationic; corrosion inhibitor, bitumen adhesion agent emulsifier; liq.; 100% conc.

Synprolam 35N3X3, 35N3X5, 35N3X10, 35N3X15, 35N3X25 [318] N-(C_{13-15}) alkyl-1,3-propane diamine ethoxylate; cationic; emulsifier, corrosion inhibitor, textile processing aids, wetting agent and antistat; liq.; 100% conc.

Synprolam 35TMQC. [318] (C_{13-15}) alkyl trimethyl ammonium chloride; cationic; biocide, emulsifier, wetting agent and antistat; liq.; 100% conc.

Synprolam 35X2, 35X5, 35X10, 35X15,

35X20, 35X25, 35X35, 35X50. [318] POE C_{13-15} alkyl amine (2, 5, 10, 15, 20, 25, 35, and 50 EO resp.); cationic; emulsifier, polyurethane catalyst, textile and plastic processing aid, textile dyeing auxs., antistat; liq.; 100% conc.

Synprolam 35X2/0. [318] C_{13-15} alkyl bis (2-hydroxyethyl) amine oxide; nonionic; foam stabilizer for personal care products and dishwashing detergents; liq.; 30% conc.

Synprolam FS. [318] Syn. quat. ammonium compd.; cationic; textile treatment; liq.; 80% conc.

Synprolam VC. [318] Syn. quat. ammonium compd.; cationic; vehicle cleaning formulation component; liq.; 100% conc.

Syntens KMA 40. [287] POE (4) alkyl ether; nonionic; wetting agent for agric. chemicals, pigments; component of detergent; emulsifier for oils; produces resin dispersions; liq.; HLB 10.5; 100% conc.

Syntens KMA 55. [287] POE (5.5) alkyl ether; nonionic; wetting agent for textile, paper, and leather industry; detergent base for industrial and household detergents; emulsifier and tech. purposes; liq.; HLB 12.5; 100% conc.

Syntens KMA 70-W-02, 70-W-07, 70-W-12, 70-W-17, 70-W-22, 70-W-32. [287] Alcohol PO/EO polymer; nonionic; surfactant for household and industrial cleansing agents and detergents; textile applications; liq.; biodeg.; 100% conc.

Syntens KMA 80. [287] POE (8) alkyl ether; nonionic; wetting agent for textile, leather, and other industries; detergent base for industrial and household detergents; emulsifier for agric. chemicals, bitumen emulsions; liq.; HLB 14.0; 100% conc.

Syntens KMA 100. [287] POE (10) alkyl ether; nonionic; wetting agent for the textile, galvanic, and other industries; detergent base for industrial and household detergents; emulsifier for tech. purposes; pigment carrier, resin and bitumen emulsions; liq.; HLB 15.0; 80–100% conc.

Syntens KMA 100-Y-01, 100-Y-11. [287] Alcohol PO/EO polymer; nonionic; see Syntens KMA 70-W-02; liq.; 100% conc.

Syntens KMA 200. [287] POE (20) alkyl ether; nonionic; base with cleansing effect; binding material; detergent base for industrial and household detergents, personal care products; consistency builder and thickener; solid; HLB 17.0; 100% conc.

Syntens QSA 110. [287] POE (11) alkyl ether; nonionic; wetting agent for textile industry; detergent base for industrial detergents; solid; HLB 14.0; 100% conc.

Syntens QSA 250. [287] POE (25) alkyl ether; nonionic; base material with cleaning effect; binder for polishing pastes, abrasives; personal care products; consistency builder and thickener; solid; HLB 19.0; 100% conc.

Syntens QSA 1000. [287] POE (100) alkyl ether; nonionic; see Syntens QSA 250; solid; HLB 19.0; 100% conc.

Syntergent K. [186] Fatty amido condensate; nonionic; detergent for textile, leather, paper, and metalworking; liq.; 30% conc.

Syntergent TER-1. [187] Phosphate ester; anionic; wetting agent, textile surfactant, calcium sulfate dispersant; liq.; 82% conc.

Synthionic D 7520, 7525, 9500 [654] Block polymer; nonionic; wetting agent, detergent, rinse aid, defoamer; liq.

Synthionic P 6040, 8020. [654] Block polymer; nonionic; see Synthionic D 7520; liq.

Synthrapol KB. [318] Aliphatic alcohol ethoxylate; nonionic; wetting and rewetting agent, detergent; used in textile processing; colorless clear liq.; alcohol odor; water-sol.; sp.gr. 1.0; cloud pt. 40 C; pH 5.5–7.0 (1%); 96 ± 2% act.

Synthrapol N. [320] Nonyl phenol ethoxylate; nonionic; detergent, wetting agent, emulsifier, penetrant/dyeing assistant used in textile processing; colorless liq.; mild pleasant odor; sp.gr. 1.0; b.p. 100 C; pH 5.5–7.0; 26% conc.

Syntofor A03, A04, AB03, AB04, AL3, AL4, B03, B04. [654] Polyglycol fatty ester; nonionic; emulsifier for textiles and cosmetics; auxs. for textile, leather, metal industries, insecticides; liq.

Syntol N77. [2] Sodium lauryl ether sulfate; surfactant for personal care products, lt. duty and dishwashing detergents, car shampoos, and chemical specialties; pale yel. clear low-visc. liq.; visc. 300–500 cP; cloud pt. 0 C; pH 5.5–7.5 (5%); biodeg.; 26–27% act.

Syntophos O6N. [654] Phosphate ester; anionic; wetting agent and detergent for textiles; liq.

Syntopon 8 A. [654] Octyl phenol ethoxylate; anionic; wetting agent for textile and leather; detergent; emulsifier; liq.; 100% conc.

Syntopon 8 B. [654] Octyl phenol ethoxylate; anionic; detergent for metal cleaning; liq.; 100% conc.

Syntopon 8 C. [654] Octyl phenol ethoxylate; anionic; detergent for textiles; liq.; 100% conc.

Syntopon 8 D. [654] Octyl phenol ethoxylate; anionic; detergent for degreasing; wetting agent; emulsifier for insecticides; liq.; 100% conc.

Syntopon A. [654] Nonylphenol ethoxylate; nonionic; detergent; emulsifier for solvs. and min. oils; wetting agent; liq.; 100% conc.

Syntopon A 100. [654] Nonylphenol ethoxylate; nonionic; detergent, emulsifier for latex; wetting agent; liq.; 100% conc.

Syntopon B. [654] Nonylphenol ethoxylate; nonionic; wetting agent, detergent for textiles; emulsifier; liq.; 100% conc.

Syntopon B 300. [654] Nonylphenol ethoxylate; nonionic; see Syntopon 8 B; liq.; 100% conc.

Syntopon C. [654] Nonylphenol ethoxylate; nonionic; wetting agent and detergent for textiles; liq.; 100% conc.

Syntopon D. [654] Nonylphenol ethoxylate; nonionic; detergent, emulsifier for perfume; liq.; 100% conc.

Syntopon E. [654] Nonylphenol ethoxylate; nonionic; detergent, emulsifier for metal cleaning; dispersant; wax; 100% conc.

Syntopon F. [654] Nonylphenol ethoxylate; nonionic; detergent, emulsifier, cleaning agent; wax; 100% conc.

Syntopon G. [654] Nonylphenol ethoxylate; nonionic; see Syntopon F; wax; 100% conc.

Syntopon H. [654] Nonylphenol ethoxylate; nonionic; crude oil demulsifier; wax; 100% conc.

Syntopon S 493, 630, 1030. [654] Alkylaryl ether sulfate; anionic; household detergent; paste, liq., liq. resp.

T

T-9. [603] Ethoxylated linear alcohol tallate; nonionic; liq.; disp. in water; liq.; 100% conc.

Tagat I, I 2. [258; 259] POE glycerol monoisostearate; nonionic; preparation of o/w emulsions; solubilizer for flavors, perfumes, vitamin oils; dispersant and antistat; liq.; HLB 15.6 and 14.2 resp.; 100% conc.

Tagat L. [258; 259] PEG-30 glyceryl laurate; nonionic; see Tagat I; liq.; HLB 17.0; 100% conc.

Tagat L-2. [258; 259] PEG-20 glyceryl laurate; nonionic; see Tagat I; liq.; HLB 15.7; 100% conc.

Tagat O. [258; 259] PEG-30 glyceryl oleate; nonionic; see Tagat I; liq.; HLB 16.4; 100% conc.

Tagat O 2. [258; 259] PEG-20 glyceryl oleate; nonionic; see Tagat I; liq.; HLB 15.0; 100% conc.

Tagat R 1. [258; 259] PEG-15 glyceryl ricinoleate; nonionic; see Tagat I; liq.; HLB 14.0; 100% conc.

Tagat S. [258; 259] PEG-30 glyceryl stearate; nonionic; see Tagat I; liq.; HLB 16.4; 100% conc.

Tagat S 2. [258; 259] PEG-20 glyceryl stearate; nonionic; see Tagat I; liq.; HLB 15.0; 100% conc.

Tagat TO. [258; 259] PEG-25 glyceryl trioleate; nonionic; see Tagat I; liq.; HLB 11.3; 100% conc.

Tairygent CA-1. [242] Linear dodecylbenzene sulfonic acid; br. paste; 96% min. act.

Tairygent CA-2. [242] Dodecylbenzene sulfonic acid, branched; br. paste; 96% min. act.

Tairygent CB-1. [242] Linear sodium dodecylbenzene sulfonate; wh. paste; pH 7 ± 1 (1%); 60 ± 2% act.

Tairygent CB-2. [242] Sodium dodecylbenzene sulfonate, branched; wh. paste; pH 7 ± 1 (1%); 60 ± 2% act.

Tallates, K. [424] K salt of tall oil fatty acid; anionic; detergent, emulsifier for petroleum and agric. products.

Tally 100. [210] Mono- and diglycerides and polyglycerate 60; emulsifier, dough strengthener and crumb softener used in breads; m.p. 118–125 F.

Tally 100 Plus. [210] Glyceryl stearate, PEG-20 glyceryl stearate, hydrog. soybean oil; see Tally 100; also as textile lubricant and fabric softener; HLB 6.1; m.p. 55–57 C.

Talofloc. [559] Quat. ammonium salt; cationic; sugar for clarification of sugar syrup; Gardner 5 max. paste; disp. in water; pH 6–9 (5%).

Tanapal ME Acid. [587] Phosphate ester; anionic; leveling agent for disperse dyes; liq.; 100% conc.

Tanapon NF-200. [587] Phosphated ester; anionic; wetting agent for alkaline cleaning compds., pigments, adhesive; liq.; 60% conc.

Tanaquad T-85. [587] Quat. ammonium compd.; cationic; wetting agent, bactericide; sanitizer, dye retarder, antistat; liq.; 85% conc.

Tanaterge PE-67. [587] Phosphated deriv.; anionic; detergent for industrial cleaners; liq.; 67% conc.

Tandem 5K, 8. [320] Mono and diglycerides,

polysorbate 60 with BHA, citric acid; nonionic; food emulsifier, conditioner/softener for yeast-raised baked goods; cream-wh. plastic solid; sol. in veg. oils; disp. in water; m.p. 128 F; HLB 8.1; flash pt. > 300 F; 100% conc.

Tandem 9. [320] Mono- and diglycerides and polysorbate 60 (25%); see Tandem 5K; cream-wh. flakes; bland odor; disp. in water; m.p. 135 F; HLB 6.0; flash pt. > 300 F; 100% conc.

Tandem 11H. [320] Hydrated mono- and diglycerides, polysorbate 60 with sodium propionate and lactic acid; see Tandem 5K; cream-wh. soft plastic; disp. in water; HLB 6.3; flash pt. > 300 F; pH 5; 50% conc.

Tandem 552. [320] Mono- and diglycerides, polysorbate 60 with BHA, citric acid; see Tandem 5K; golden clear liq.; sol. in veg. oils; disp. in water; m.p. 45 F; HLB 8.1; flash pt. > 300 F; 100% conc.

Tauranol I-78, I-78 Flakes, I-78-3. [237] Sodium cocoyl isethionate; anionic; detergent; foamer and dispersant; conditioner; used in personal care products; powd., flakes, and paste resp.; 83, 83, and 27% conc.

Tauranol M. [237] Sodium methyl oleoyl taurate; anionic; personal care products; powd.; 75% conc.

Tauranol M-35. [237] Sodium N-methyl-N oleoyl taurate; anionic; detergent for textile processes; liq.; 23% conc.

Tauranol ML. [237] Sodium methyl oleoyl taurate; anionic; see Tauranol M; liq.; 33% conc.

Tauranol MS. [237] Sodium methyl oleoyl taurate; anionic; wetting agent and detergent scour for personal care products; paste; 33% conc.

Tauranol T-Gel. [237] Sodium N-methyl-N oleoyl taurate; anionic; wetting and detergent scour for textiles; gel; 14% conc.

Tauranol WS, WS Conc. [237] Sodium methyl cocoyl taurate; anionic; detergent used in cosmetic preparations; slurry; 24 and 31% conc. resp.

Tauranol WS H.P. [237] Sodium methyl cocoyl taurate; anionic; personal care and pharmaceutical preparations; powd.; 95% conc.

Tauranol WSP. [237] Sodium methyl cocoyl taurate; anionic; see Tauranol WS: powd.; 70% conc.

T-Det A-026. [603] EO and straight-chain primary alcohols (C_{10}-C_{12}); nonionic; wetting agent, detergent; textile processing; surfactant for formulated detergents for laundry use and cleaners; APHA 100 max. clear liq.; sol. in water, ethanol, glycol ether, HAN, isopropanol, xylene; dens. 8.2 lb/gal (60/60 F); sp.gr. 0.98 (60/60 F); visc. 21 cps (100 F); HLB 13.5; cloud pt. 36 C (1%); flash pt. 250 F (OC); pH 6–8 (1%); biodeg.; 100% conc.

T-Det BP-1. [603] PO/EO block copolymer; nonionic; emulsifier, detergent, dispersant and polymerization agent; pesticide formulations; solid; HLB 15.7; 100% conc.

T-Det C-40. [603] PEG-40 castor oil; nonionic; emulsifier, solubilizer, degreaser, lubricant, dispersant, penetrant used in leather, textile and metal processing; paint for dispersing pigment slurries, leveling agent, defoamer, stabilizer; paper mfg.; wax and polish preparations; rubber lubricant; yel. liq.; mild, oily odor; sol. in acetone, butyl Cellosolve, carbon tetrachloride, ethanol, ether, ethylene glycol, methanol, veg. oil, water, xylene; dens. 8.7 lb/gal; sp.gr. 1.05; HLB 14.2; flash pt. > 200 F (TOC); pour pt. 60 F; pH 6.0 (1%); 99.5% min. act.

T-Det D-70. [603] PEG-70 dinonyl phenyl ether; nonionic; emulsifier, detergent, surfactant; solid; HLB 18.0; 100% conc.

T-Det D-150. [603] PEG-150 dinonyl phenyl ether; nonionic; used in hard surface cleaners; solid; HLB 19.0; 100% conc.

T-Det DD-5. [603] Dodoxynol-5; nonionic; emulsifier, surfactant for aromatic and aliphatic hydrocarbon solvs.; dry cleaning soap additive; pale yel. liq.; mild aromatic odor; oil-sol.; dens. 8.42 lb/gal; sp.gr. 1.01; HLB 9.6; flash pt. > 300 F; pour pt. < 0 F; pH 7 (1% aq.); 99.5% min. act.

T-Det DD-7. [603] Dodoxynol-7; nonionic; emulsifier; surfactant for solvs.; solv. and emulsion cleaning; pale yel. liq.; mild aromatic odor; disp. in water; dens. 8.66 lb/gal; sp.gr. 1.04; HLB 11.1; flash pt. > 300 F; pour pt. < 0 F; pH 7 (1%); 99.5% min. act.

T-Det DD-9. [603] Dodoxynol-9; nonionic; detergent, emulsifier, wetting agent; pale yel. liq.; mild aromatic odor; disp. in water; sol. in aromatic hydrocarbons; dens. 8.7 lb/gal; sp.gr. 1.05; HLB 12.0; flash pt. > 300 F; pour pt. 42 F; pH 7 (1%); 99.5% min. act.

T-Det DD-10. [603] Dodoxynol-10; nonionic; emulsifier; surfactant for solvs.; pale yel. liq.; mild aromatic odor; disp. in water; dens. 8.75 lb/gal; sp.gr. 1.05; cloud pt. 112 F (1%); flash pt. > 300 F; pour pt. 48 F; pH 7 (1%); 99.5% min. act.

T-Det DD-11. [603] Dodoxynol-11; nonionic; surfactant, detergent, emulsifier; pale yel. liq.; water-sol.; sp.gr. 1.06 (68 F); dens. 8.8 lb/gal (68 F); HLB 13.5; pour pt. 53 F; flash pt. (PMCC) > 300 F; cloud pt. 134 F (1%); pH 5–7 (1% aq.); 99.5% min. act.

T-Det DD-14. [603] Dodoxynol-14; nonionic; surfactant, detergent, emulsifier; pale yel. semisolid; water-sol.; sp.gr. 1.05 (68 F); dens. 8.7 lb/gal (68 F); HLB 14.5; pour pt. 75 F; flash pt. (PMCC) > 400 F; cloud pt. 190 F (1%); pH 5–7; 99.5% min. act.

T-Det EPO-61. [603] EO/PO; anionic; defoamer; crude oil demulsifier; foam control in paper mfg.; water treating compds.; emulsifier and wetting agent in agric. toxicant formulations; clear liq.; sol. in ethanol, toluene, xylene, perchlorethylene; dens. 8.34 lb/gal; sp. gr. 1.01; visc. 271 cps; cloud pt. 23 C (1%); flash pt. > 200 C; pour pt. –28 C; 99.5% min. act.

T-Det N 1.5. [603] Nonoxynol-1 (1.5 EO); nonionic; surfactant for petroleum oils applications; stabilizer; coemulsifier; pale yel. liq.; oil-sol.; sp.gr. 0.99 (68 F); dens. 8.2 lb/gal (68 F); HLB 1.5; pour pt. < 0 F; flash pt. (PMCC) 340 F; pH 5–7 (1% aq.); 99.5% min. act.

T-Det N-4. [603] Nonoxynol-4; nonionic; detergent, wetting agent; agric. toxicant formulations; intermediate for household detergents; dry cleaning soap additive, sludge dispersant additive in fuel oils; gasoline additive; solv. cleaner for metal processing; color development aid, stabilizer and emulsifier for latex paints; pale yel. oily liq.; mild aromatic odor; sol. in butyl Cellosolve, carbon tetrachloride, min. and corn oil, ethanol, kerosene, min. spirits, perchloroethylene, Stod., toluene; dens. 8.5 lb/gal; sp.gr. 1.02; visc. 52 SUS (210 F); HLB 8.9; flash pt. > 400 F (OC); pour pt. < O F; pH 7; 99.5% min. act.

T-Det N-6. [603] Nonoxynol-6; nonionic; emulsifier for silicones, household waxes and polishes; wetting agent; detergent and dispersant for petroleum and fuel oils; agric. toxicant formulations; dry cleaning soap additive; leather and metal processing; solv. cleaner; deinking agent for paper; pale yel. liq.; mild aromatic odor; sol. in butyl Cellosolve, carbon tetrachloride, corn oil, diesel fuel, ethanol, kerosene, Stod., xylene; dens. 8.7 lb/gal; sp.gr. 1.04; visc. 58 SUS (210 F); HLB 10.9; flash pt. > 400 F (OC); pour pt. < 0 F; pH 7; 99.5% min. act.

T-Det N-8. [603] Nonoxynol-8; nonionic; detergent, wetting agent, emulsifier, solubilizer, degreaser; textile, leather and paint processing; agric. toxicant, household and industrial cleaning formulations; pigment dispersant; rewetting, deinking, felt washing, leveling in paper mfg.; oil recovery operations; clear visc. liq.; mild aromatic odor; dens. 8.75 lb/gal; sp.gr. 1.05; visc. 63 SUS (210 F); HLB 12.4; cloud pt. 78 F (1%); flash pt. 500 F (OC); pH 6 (1%); 99.5% min. act.

T-Det N-9.5. [603] Nonoxynol-9; nonionic; see T-Det N-8; also used as a foamer; clear visc. liq.; mild aromatic odor; sol. in butyl Cellosolve, carbon tetrachloride, ethanol, ethylene glycol, HAN, toluene, water, xylene; dens. 8.8 lb/gal; sp.gr. 1.06; visc. 68 SUS (210 F); HLB 13.1; cloud pt. 135 F (1%); flash pt. 500 F (OC); pH 7 (1%); 99.5% min. act.

T-Det N-10.5. [603] Nonoxynol-10; nonionic; detergent, wetting agent, dispersant, emulsifier, solubilizer; textile processing; household and industrial cleaning products; clear visc. liq.; mild aromatic odor; sol. in butyl Cellosolve, ethanol, ethylene glycol, HAN, toluene, water, xylene; dens. 8.8 lb/gal; sp.gr. 1.06; HLB 10.4; cloud pt. 155 F; flash pt. 500 F (OC); pH 7 (1%); 99.5% min. act.

T-Det N-12. [603] Nonoxynol-12; nonionic; detergent; base for detergent and sanitizer formulations; wetting agent used in metal cleaning; opaque liq.; mild aromatic odor; dens. 8.9 lb/gal; sp.gr. 1.07; HLB 12; cloud pt. 180 F (1%); flash pt. > 400 F (OC); pour pt. 57 F; pH 7 (1%); 99.5% min. act.

T-Det N-14. [603] Nonoxynol-14; nonionic; detergent; emulsifier for oils, fats and waxes; wetting agent for metal cleaners; penetrant; corrosion inhibitor; off-wh. semisolid; mild aromatic odor; sol. in butyl Cellosolve, carbon tetrachloride, corn oil, ethanol, ethylene glycol, water, xylene; dens. 8.9 lb/gal (68 F); sp.gr. 1.07 (68 F); visc. 89 SUS (210 F); HLB 14.7; cloud pt. 203 F (1%); flash pt. > 400 F (OC); pour pt. 67 F; pH 6 (1%); 99.5% min. act.

T-Det N-20. [603] Nonoxynol-20; nonionic; surfactant for detergents, emulsion polymerization, petroleum processing; emulsifier; opaque solid; water-sol.; sp.gr. 1.08 (105 F); dens. 9.0 lb/gal (105 F); HLB 16.0; pour pt. 90 F; flash pt. (PMCC) > 500 F;

cloud pt. > 212 F (1%); pH 5–7 (1% aq.); 99.5% min. act.

T-Det N-30. [603] Nonoxynol-30; nonionic; detergent, pressure scouring textile operations; emulsifier for oils, fats and waxes; wetting agent in electrolytes; pale yel. waxy solid; mild aromatic odor; sol. in butyl Cellosolve, ethanol, ethylene glycol, water, xylene; sp.gr. 1.07 (132 F); HLB 17.0; cloud pt. 212 F; flash pt. > 500 F; pour pt. 110 F; pH 6.5–8.0 (1%); 99.5% min. act.

T-Det N-40. [603] Nonoxynol-40; nonionic; see T-Det N-30; pale yel. waxy solid; mild aromatic odor; sp.gr. 1.07 (132 F); HLB 17.7; cloud pt. 212 F (1%); flash pt. > 500 F; pour pt. 120 F; pH 6.5–8.0 (1%); 99.5% min. act.

T-Det N-70. [603] Nonoxynol-70; nonionic; detergent, emulsifier for oils, fats and waxes; stabilizer in latex and alkyd emulsions; emulsion polymerization; scouring textile operations; pale yel. waxy solid; mild aromatic odor; dens. 9.0 lb/gal (135 F); sp.gr. 1.08 (135 F); cloud pt. 212 F (1%); flash pt. > 500 F; pour pt. 125 F; pH 6.0–7.0 (1%); 99.5% min. act.

T-Det N-100. [603] Nonoxynol-100; nonionic; see T-Det N-30; also used as emulsifier for asphalt; pale yel. waxy solid; mild aromatic solid; sol. in ethylene dichloride, ethanol and water; sp.gr. 1.08 (135 F); cloud pt. 212 F (1%); flash pt. > 500 F; pour pt. 127 F; pH 6–7 (1%); 99.5% min. act.

T-Det N-307. [603] Nonoxynol-30; nonionic; see T-Det N-30; APHA 100 liq.; dens. 9.1 lb/gal; visc. 806 cps; cloud pt. > 100 C (1%); flash pt. > 212 F (TOC); pour pt. 30 F; 70 ± 0.5% act.

T-Det N-407. [603] Nonoxynol-40; nonionic; see T-Det N-30; also as emulsifier in paint systems; APHA 100 liq.; dens. 9.0 lb/gal; sp.gr. 1.08; visc. 900 cps; cloud pt. > 212 F (1%); flash pt. > 212 F (TOC); pour pt. 30 F; 70 ± 0.5% act.

T-Det N-507. [603] Nonoxynol-50; see T-Det N-30; also as emulsifier in various paint systems; APHA 100 liq.; dens. 9.1 lb/gal; sp.gr. 1.09; visc. 760 cps; cloud pt. > 100 C (1%); flash pt. > 212 F (TOC); pour pt. 40 F; 70 ± 0.5% act.

T-Det N-705. [603] Nonoxynol-70; nonionic; see T-Det N-70; clear to hazy liq.; mild aromatic odor; dens. 9.0 lb/gal; sp.gr. 1.08; cloud pt. 212 F (1%); flash pt. > 500 F; pour pt. 40 F; pH 5.0–6.5 (1%); 50% min. act.

T-Det N-1007. [603] Nonoxynol-100; nonionic; see T-Det N-100; clear-opaque visc. liq.; mild aromatic odor; sp.gr. 1.1; cloud pt. 212 F (1%); flash pt. > 500 F; pour pt. 62 F; pH 6–7 (1%); 70 ± 0.5% act.

T-Det O-4. [603] Octoxynol-4 (3–4 EO); nonionic; surfactant for agric., leather and metal processing, paint, drycleaning detergents, degreasing applications; clear liq.; water-disp.; sp.gr. 1.03 (68 F); dens. 8.6 lb/gal (68 F); HLB 8.5; pour pt. < 0 F; flash pt. (PMCC) > 500 F; pH 6–8 (1%); 99.5% min. act.

T-Det O-6. [603] Octoxynol-6 (6–7 EO); nonionic; surfactant for detergents, textile and leather processing, agric. formulations; dispersant; clear liq.; water-disp.; sp.gr. 1.05 (68 F); dens. 8.7 lb/gal (68 F); HLB 11.6; pour pt. < 0 F; flash pt. (PMCC) > 500 F; pH 6–8 (1%); 99.5% min. act.

T-Det O-8. [603] Octoxynol-8 (7–8 EO); nonionic; surfactant for detergents, agric. formulations, oil well drilling, leather processing, paints; clear liq.; water-sol.; sp.gr. 1.04 (68 F); dens. 8.7 lb/gal (68 F); HLB 12.4; pour pt. 15 F; flash pt. (PMCC) > 500 F; cloud pt. 70 F (1%); pH 5–7 (1%); 99.5% min. act.

T-Det O-9. [603] Octoxynol-9; nonionic; surfactant for general purpose applications, detergents, solv. emulsifier; clear liq.; water-sol.; sp.gr. 1.06 (68 F); dens. 8.8 lb/gal (68 F); HLB 13.5; pour pt. 45 F; flash pt. (PMCC) > 500 F; cloud pt. 150 F (1%); pH 5–7 (1% aq.); 99.5% min. act.

T-Det O-40. [603] Octoxynol-40; nonionic; surfactant for agric. formulations; detergents, textiles, emulsion polymerization; emulsifier, stabilizer; wh. solid; water-sol.; sp.gr. 1.06 (150 F); dens. 8.8 lb/gal (150 F); HLB 17.9; pour pt. 120 F; flash pt. (PMCC) > 500 F; cloud pt. > 212 F (1%); 99.5% min. act.

T-Det O-407. [603] Octoxynol-40; nonionic; detergent, emulsifier, stabilizer for various applications; APHA 100 max. liq.; water-sol.; dens. 9.1 lb/gal; visc. 600 cps; cloud pt. > 100 C (1%); flash pt. > 100 C (TOC); pour pt. 30 F; pH 7 (1%); 70 ± 0.5% act.

T-Det TDA-60. [603] Tridecyl alcohol 60% EO adduct; nonionic; detergent, dispersant, wetting agent; liq.; water-sol.; HLB 12.0; 100% conc.

T-Det TDA-65. [603] Tridecyl alcohol 65% EO adduct; nonionic; emulsifier, solubilizer; textile processing; hard surface detergent;

disperses and suspends solids; wetting and rewetting agent for dust settling, corrosion inhibitor, metal cleaning and pickling; hazy liq.; mild aliphatic odor; water-sol.; dens. 8.20 lb/gal; cloud pt. 130 F (1%); flash pt. > 290 F (PMCC); pour pt. 62 F; pH 5–7; 99.5% min. act.

T-Det TDA-70. [603] Tridecyl alcohol 70% EO adduct; nonionic; see T-Det TDA-60; paste; water-sol.; HLB 14.0; 100% conc.

T-Det TDA-150. [603] Trideceth-150; nonionic; used in block type hard surface cleaners; solid; HLB 19.4; 100% conc.

Teepol CM 44. [555] Alkylbenzene sulfonate, sodium salt; anionic; detergent, emulsifier; used in dishwashing and general cleaning liqs.; pale amber liq.; misc. with water; dens. 1.05 kg/l; visc. 100–500 cs; clear pt. > 16 C; pH 8.0–9.5 (5% aq.); biodeg.; > 25% act.

Teepol GC 56P. [555] Alkylbenzene sulfonate and primary alcohol ethoxysulfate; detergent; used in dishwashing and general cleaning applications; clear liq.; misc. with water; dens. 1.06 kg/l; visc. 1900 cs; cloud pt. 1 C; pH 6.5–8.0 (5%); biodeg.; > 33% act.

Teepol GD7P. [555] Alkylbenzene sulfonate and primary alcohol ethoxysulfate; anionic/nonionic; see Teepol GC 56P; clear liq.; misc. with water; dens. 1.07 kg/1; visc. 620 cs; cloud pt. 2 C; pH 6.5–8.0 (5%); > 40% act.

Teepol GD53. [555] Alkylbenzene sulfonate and primary alcohol ethoxysulfate; anionic/nonionic; see Teepol GC 56P; clear liq.; misc. with water; dens. 1.07 kg/l; visc. 24–300 cs; cloud pt. 1 C; pH 6.5–8.0 (5%); > 30% act.

Teepol HB 6. [555] Primary alcohol, sulfated, sodium salt; solubilizer, emulsifier; component of hard surface and germicidal cleaners; pale amber liq.; water-sol.; dens. 1.06 kg/1; visc. 20 cs; clear pt. 10 C; pH 8.0 (5% aq.); biodeg.; 34.2% act.

Teepol HB 7. [556] C_9-C_{13} primary alcohol sulfate, sodium salt; anionic; solubilizer and emulsifier component of hard surface and germicidal cleaners; liq.; 40% conc.

Teepol PB. [555] Primary alcohol ethoxysulfate (sodium cation); foamer; used in plasterboard mfg.; clear liq.; misc. with water; dens. 1.06 kg/1; visc. 50 cs; clear pt. 0 C; pH 6.5–8.0 (5%); biodeg.; > 32% act.

Tefose 63. [247; 248] PEG-6-32 stearate and glycol stearate; nonionic; o/w pharmaceutical ointments; waxy solid; 100% conc.

Tefose 1500. [247; 248] PEG-6-32 stearate; nonionic; base for o/w lotions; solid; 100% conc.

Tefose 2000. [247; 248] PEG-6 stearate and ceteth-20; nonionic; base for o/w cosmetic or pharmaceutic lotions; solid; HLB 10.2; 100% conc.

Tefose 2561. [247; 248] PEG-6 stearate and glyceryl stearate and ceteth-20; nonionic; see Tefose 2000; waxy solid; HLB 10–11; 100% conc.

Tegacid Regular V.A, Special V.A. [259] Glycerol monostearate SE; flake; HLB 4.2 and 4.9 resp.; 100% conc.

Tegin. [258; 259] Glyceryl monostearate SE; anionic; emulsifier for cosmetic and pharmaceutical o/w emulsions; powd.; HLB 5.5; 100% conc.

Tegin 90. [258; 259] Glyceryl stearate; nonionic; flake; HLB 4.5; 100% conc.

Tegin 90 NSE. [259] Glyceryl monostearate; nonionic; stabilizer for o/w emulsions for cosmetic and pharmaceuticals; powd.; HLB 4.5; 100% conc.

Tegin 200 L. [259] PEG 200 laurate; nonionic; see Tegin 90; liq.; HLB 7.1; 100% conc.

Tegin 515. [258; 259] Glyceryl stearate; nonionic; emulsifier for creams and lotions; powd.; HLB 3.8; 100% conc.

Tegin 515 V.A. [259] Glyceryl monostearate; nonionic; see Tegin 515; flake; HLB 3.8; 100% conc.

Tegin 4011. [258; 259] Glyceryl monostearate SE; emulsifier for pharmaceuticals; powd.; HLB 3.8; 100% conc.

Tegin 4100. [258; 259] Glyceryl monostearate; nonionic; stabilizer for cosmetics and pharmaceuticals; powd.; HLB 3.8; 100% conc.

Tegin 4433. [258; 259] Glyceryl monostearate SE; anionic; emulsifier for pharmaceuticals; powd.; HLB 12.0; 100% conc.

Tegin 4480. [258; 259] Glyceryl laurate; nonionic; emulsifier and stabilizer for cosmetics and pharmaceuticals; semisolid; HLB 3.8; 100% conc.

Tegin 4600. [258; 259] Glyceryl caprylate/caprate; nonionic; antistat and lubricant for syn. materials; emulsifier for cosmetic and pharmaceutical products; liq.; HLB 4.0; 100% conc.

Tegin 4600 NSE. [259] Glyceryl caprylate; nonionic; antistat for syn. materials; emulsifier for cosmetics and pharmaceuticals; liq.; HLB 3.8; 100% conc.

Tegin A-412. [258; 259] Propylene glycol

monostearate; emulsifier for pharmaceuticals; waxy; HLB 2.8; 100% conc.

Tegin A-422. [258; 259] Diethylene glycol monostearate; emulsifier for pharmaceuticals; waxy; HLB 2.8; 100% conc.

Tegin C-1R. [259] Glyceryl stearate SE and hydrog. tallow glyceride citrate; anionic; solubilizer and co-emulsifier for cosmetics; liq.; HLB 17.0; 100% conc.

Tegin C-61 NSE. [259] Hydrog. tallow glyceride citrate; emulsifier for cosmetic and foodstuffs; powd.; 100% conc.

Tegin C 62. [258; 259] Hydrog. tallow glyceride citrate; anionic; hydrophilic emulsifier for cosmetics and foodstuffs; powd.; HLB 10.0; 100% conc.

Tegin C-62 SE. [259] Hydrog. tallow glyceride citrate; see Tegin C-61 NSE; powd.; HLB 10.0; 100% conc.

Tegin C 63. [258; 259] Monodiglyceride citric acid ester; anionic; hydrophilic emulsifier; powd.; HLB 10.0; 100% conc.

Tegin C-63 SE. [259] Hydrog. tallow glyceride citrate; see Tegin C-61 NSE; powd.; HLB 10.0; 100% conc.

Tegin C-611 SE. [259] Glyceryl stearate SE and hydrog. tallow glyceride citrate; anionic; o/w for acidic emulsions; powd.; HLB 11.0; 100% conc.

Tegin D 6100. [258; 259] Diethylene glycol monostearate; nonionic; lipophilic coemulsifier for o/w emulsions; waxy; HLB 2.8; 100% conc.

Tegin DGS. [259] Diglycol stearate; anionic; coemulsifier for o/w emulsions; solid; HLB 2.8; 100% conc.

Tegin E 41. [258; 259] Acetylated hydrog. tallow glyceride; nonionic; emulsifier for cosmetic and technical products; emollient, plasticizer, food additive; solid; HLB 2.0–3.0; 100% conc.

Tegin E-41 NSE. [259] Acetylated hydrog. tallow glycerides; nonionic; emulsifier for cosmetic and technical products; emollient/plasticizer food additives; semisolid; HLB 2.0–3.0; 100% conc.

Tegin E 61. [258; 259] Acetylated hydrog. lard glyceride; nonionic; see Tegin E 41; solid; HLB 2.0–3.0; 100% conc.

Tegin E-61 NSE. [259] Acetylated hydrog. lard glyceride; nonionic; see Tegin E-41 NSE; semisolid; HLB 2.0–3.0; 100% conc.

Tegin E 66. [258; 259] Acetylated lard glyceride; nonionic; see Tegin E 41; liq.; HLB 2.0–3.0; 100% conc.

Tegin E-66 NSE. [259] Acetylated lard glyceride; nonionic; see Tegin E-61 NSE; liq.; HLB 2.0–3.0; 100% conc.

Tegin G. [258; 259] Glycol stearate SE; anionic; for cosmetics and pharmaceuticals; o/w emulsions; powd.; HLB 12.0; 100% conc.

Tegin G 6100. [258; 259] Ethylene glycol monostearate; nonionic; lipophilic coemulsifier for o/w emulsions; powd.; HLB 3.2; 100% conc.

Tegin GO. [259] Ethylene glycol monooleate; anionic; coemulsifier and stabilizer; liq.; HLB 2.8; 100% conc.

Tegin GRB. [259] Glyceryl stearate; nonionic; emulsifier for o/w systems; powd.; 100% conc.

Tegin GRB NSE. [259] Glycerol monostearate; nonionic; stabilizer for cosmetic and pharmaceuticals; powd.; HLB 3.8; 100% conc.

Tegin ISO. [258; 259] Glyceryl stearate; anionic; w/o emulsifier for creams; paste; HLB 3.4; 100% conc.

Tegin L 61, L-61 NSE, L 62, L-62 NSE. [258; 259] Hydrog. tallow glyceride lactate; nonionic; emulsifier, emollient, plasticizer for cosmetics; food additive; powd.; HLB 5.0, 5.0, 6.0, and 6.0 resp.; 100% conc.

Tegin M. [258; 259] Glyceryl stearate; anionic; emulsifier for o/w systems; solid; HLB 3.8; 100% conc.

Tegin MAV NSE. [259] Glyceryl monostearate; anionic; see Tegin M NSE; powd.; HLB 3.8; 100% conc.

Tegin M NSE. [259] Glycerol monostearate; nonionic; stabilizer for cosmetics and pharmaceuticals; powd.; HLB 3.8; 100% conc.

Tegin O. [258; 259] Glyceryl oleate; nonionic; emulsifier for w/o emulsions; paste; HLB 3.3; 100% conc.

Tegin O NSE. [259] Glyceryl oleate; anionic; w/o emulsions; semisolid; HLB 3.3; 100% conc.

Tegin O Special. [258; 259] Glyceryl monooleate; anionic; emulsifier for o/w systems; paste; HLB 3.3; 100% conc.

Tegin O Special NSE. [259] Glyceryl oleate; anionic; w/o emulsions; semisolid; HLB 3.3; 100% conc.

Tegin P. [258; 259] Propylene glycol stearate SE; anionic; emulsifier for o/w lotions and creams; solubilizer for dyestuffs; solid; HLB 4.4; 100% conc.

Tegin P 411. [258; 259] Propylene glycol monodistearate; nonionic; lipophilic coemulsifier for o/w emulsions; waxy; HLB 2.8; 100% conc.

Tegin P-411 SE. [259] Propylene glycol stearate SE; o/w creams and lotions; solubilizer for dyestuffs; waxy; HLB 2.8; 100% conc.

Tegin PL. [259] Propylene glycol laurate; nonionic; see Tegin P-411 SE; liq.; 100% conc.

Tegin P-SE. [259] Propylene glycol stearate, SE; anionic; o/w cosmetic and pharmaceutical formulations; powd.; HLB 12.0; 100% conc.

Tegin RZ. [258; 259] Glyceryl ricinoleate; nonionic; emulsifier, solubilizer, emollient, plasticizer; paste; HLB 3.0; 100% conc.

Tegin RZ NSE. [259] Glyceryl ricinoleate; nonionic; o/w emulsifier, solubilizer, emollient, plasticizer; semisolid; HLB 3.0; 100% conc.

Tegin SE. [259] Glyceryl stearate, SE; anionic; cosmetic and pharmaceutical emulsions of o/w type; powd.; HLB 12.0; 100% conc.

Tegin Special. [259] Glyceryl monostearate; anionic; emulsifier for cosmetic and pharmaceutical o/w emulsions; powd.; 100% conc.

Tegin Special SE. [259] Glyceryl monostearate SE; anionic; cosmetic and pharmaceutical emulsions of o/w type; powd.; HLB 12.0; 100% conc.

Tegin VA. [259] Glycerol monostearate SE; anionic; emulsifier for cosmetic creams and lotions; flake; HLB 5.5; 100% conc.

Tegin VA 55G. [259] Glycerol monostearate; nonionic; emollient for creams and lotions; flake; HLB 3.8; 100% conc.

Tegin VA 515. [259] Glycerol monostearate; nonionic; emulsifier for creams; flake; HLB 3.8; 100% conc.

Teginacid. [258; 259] Glyceryl stearate and ceteareth-20; nonionic; emulsifier for o/w emulsions; powd.; HLB 12.0; 100% conc.

Teginacid H. [258; 259] Glycerol stearate and ceteth-20; nonionic; see Teginacid; powd.; HLB 11.2; 100% conc.

Teginacid H-SE. [259] Glycerol monodistearate and other nonionics; nonionic; o/w emulsions; powd.; HLB 11.2; 100% conc.

Teginacid ML. [258; 259] Glyceryl stearate and PEG-40 stearate; emulsifier for finely dispersed emulsions.

Teginacid ML-SE. [259] Glycerol monodistearate and PEG-40 stearate; cationic; o/w emulsions; powd.; HLB 10.6; 100% conc.

Teginacid R. [258; 259] Glyceryl stearate and cocamidopropyl dimethylamine; cationic; emulsifier for personal care products; powd.; HLB 12.0; 100% conc.

Teginacid R-SE. [259] Glycerol monodistearate; cationic; emulsifier for personal care products; powd.; HLB 12.0; 100% conc.

Teginacid SE. [259] Glycerol monodistearate; nonionic; o/w emulsions; powd.; HLB 12.0; 100% conc.

Teginacid Special. [258; 259] Glyceryl monostearate; anionic; emulsifier for cosmetic creams and lotions; solid; HLB 4.8; 100% conc.

Teginacid Special SE. [259] Glycerol monostearate; anionic; o/w emulsions; powd.; HLB 12.0; 100% conc.

Teginacid X. [258; 259] Glyceryl stearate and ceteareth-20; nonionic; emulsifier for o/w emulsions; solid; HLB 12.0; 100% conc.

Teginacid X-SE. [259] Glycerol monodistearate; nonionic; o/w emulsions; powd.; HLB 12.0; 100% conc.

Tego-Amid D 5040. [258] Cocamidopropyl dimethylamine; cationic; emulsifier for creams and lotions; conditioner for hair care products; waxy.

Tego-Amid O 18. [258] Oleamidopropyl dimethylamine; cationic; see Tego-Amid D 5040; liq.

Tego-Amid S 18. [258] Stearamidopropyl dimethylamine; cationic; see Tego-Amid D 5040; flakes.

Tego-Betaine C. [259] Cocamidopropyl betaine; amphoteric; surfactant used as foam stabilizer and visc. builder in personal care products, dishwash, liq. soap; liq.; 30% conc.

Tego-Betaine HS. [258; 259] Fatty acid amido alkyl betaine; amphoteric; cosmetic products; liq.; 30% conc.

Tego-Betaine L-7. [258; 259] Cocamidopropyl betaine; amphoteric; surfactant, foam stabilizer, visc. builder in personal care products; liq.; 30% conc.

Tego-Betaine L-10. [258; 259] Cocamidopropyl betaine; amphoteric; for personal care and pharmaceutical products; liq.; 30% conc.

Tego-Betaine L-90. [259] Lauramido propyl betaine; amphoteric; see Tego-Betaine C; liq.; 30% conc.

Tego-Betaine S. [259] Cocamidopropyl betaine; amphoteric; see Tego-Betaine C; liq.; 30% conc.

Tego-Betaine T. [258; 259] Cocamidopropyl betaine; amphoteric; household cleansers, floor cleaner, car wash; liq.; 25% conc.

Tegomuls 90, 90 S. [258] Glyceryl monostearate palmitate; nonionic; food emulsifier; powd.; HLB 4.5 and 4.8 resp.; 100% conc.

Tegomuls 4070. [258] Glyceryl monodistearate; nonionic; food emulsifier; powd.; 100% conc.

Tegomuls 4100. [258] Glyceryl stearate; nonionic; food emulsifier; powd.; HLB 3.8; 100% conc.

Tegomuls 4101. [258] Glyceryl stearate; nonionic; food emulsifier; powd.; HLB 3.8; 100% conc.

Tegomuls 6070, 9050. [258] Glyceryl monodistearate; nonionic; food emulsifier; powd.; 100% conc.

Tegomuls 9102. [258] Glyceryl monostearate; nonionic; food emulsifier; powd.; HLB 4.5; 80% conc.

Tegomuls 9102 S. [258] Glyceryl monostearate; anionic; food emulsifier; powd.; HLB 4.8; 80% conc.

Tegomuls B. [258] Glyceryl monodistearate; anionic; food emulsifier; powd.; HLB 4.5; 100% conc.

Tegomuls M. [258] Glyceryl monodistearate; nonionic; food emulsifier; HLB 3.2; 100% conc.

Tegomuls O. [258] Glyceryl monodioleate; nonionic; food emulsifier; semisolid; HLB 3.3; 100% conc.

Tegomuls O Spezial. [258] Glyceryl monodioleate; nonionic; defoamer in food industry; semisolid; HLB 3.3; 100% conc.

Tegomuls P 411. [258] 1,2-Propylene glycol monodistearate; nonionic; food emulsifier; solid; HLB 2.8; 100% conc.

Tephal Grunau AB conc. [130] Protein fatty acid condensate; anionic; washing, milling and wetting agent for textile products; paste; 65% conc.

Tephal Grunau FL conc. [130] Protein fatty acid condensate; anionic; washing, milling and wetting agent; liq.; 65% conc.

Tequat RO. [598] Acyl amido-amine; cationic; car washing; paste; 100% conc.

Terg-A-Zyme. [22] Alkylaryl sulfonate, lauryl alcohol sulfate, phosphate and carbonate and protease enzyme; anionic; wetting agent, sequestering and synergistic agents, anionic; used in hospitals, laboratories, dairies; cleaning agent in dairy and pollution processing; powd.; pH 9.0–9.5.

Tergelan 1790. [217] Mixed isopropanolamines myristic acid and lanolin acid; anionic; detergent and foam builder in personal care products; gel; 50% conc.

Tergenol G. [284] Sodium *n*-methyl-*n*-oleoyl taurate; detergent, wetting agent, emulsifier; used in textiles, skins, leather, laundering, dyeing, household and industrial cleaners, cosmetic products, dispersing waxes, pigments, resins, dry powd. and printing; corrosion inhibitor in chlorite bleaching; lt. gel; water-sol.; 15% act.

Tergenol S Liquid. [284] Sodium-*n*-methyl-*n*-oleoyltaurate; anionic; see Tergenol G; amber liq.; 38% act.

Tergenol Slurry. [284] Sodium-*n*-methyl-*n*-oleoyltaurate; anionic; see Tergenol G; wh. paste; water-sol.; 60% act.

Tergitol 15-S-3. [616] C_{11-15} pareth-3; nonionic; detergent, emulsifier, wetting agent used in textile processing applications; textile lubricant; Pt.Co 75 max.; m.w. 332; dens. 7.68 lb/gal; sp.gr. 0.924; visc. 37 cs; HLB 8.0; cloud pt. < 0 (1% aq.); flash pt. 370 F (COC); pour pt. –32 C; biodeg.; 100% conc.

Tergitol 15-S-5. [616] C_{11-15} pareth-5; nonionic; see Tergitol 15-S-3; Pt.Co 75 max. liq.; m.w. 420; dens. 8.03 lb/gal; sp.gr. 0.965; visc. 51 cs; HLB 10.5; cloud pt. < 0 C (1% aq.); flash pt. 410 F (COC); pour pt. –15 C; pH 6–8 (1%); biodeg.; 100% conc.

Tergitol 15-S-7. [616] C_{11-15} pareth-7; nonionic; see Tergitol 15-S-3; Pt.Co 75 max. liq.; m.w. 508; dens. 8.26 lb/gal; sp.gr. 0.994; visc. 67 cs; HLB 12.1; cloud pt. 37 C (1% aq.); flash pt. 440 F (COC); pour pt. 7 C; surf. tens. 29.1 dynes/cm (0.1%); pH 6–8 (1%); biodeg.; 100% conc.

Tergitol 15-S-9. [616] C_{11-15} pareth-9; nonionic; see Tergitol 15-S-3; Pt.Co 50 max. liq.; m.w. 596; dens. 8.37 lb/gal; sp.gr. 1.006; visc. 78 cs; HLB 13.3; cloud pt. 60 C (1% aq.); flash pt. 470 F (COC); pour pt. 13 C; surf. tens. 30.3 dynes/cm (0.1%); pH 6–8 (1%); biodeg.; 100% conc.

Tergitol 15-S-12. [616] C_{11-15} pareth-12; nonionic; see Tergitol 15-S-3; Pt.Co 100 max. semisolid; m.w. 728; dens. 8.49 lb/gal; sp.gr. 1.020; visc. 120 cs; HLB 14.5; cloud pt. 90 C (1% aq.); flash pt. 460 F (COC); pour pt. 17 C; surf. tens. 31.5 dynes/cm (0.1%); pH 6–8 (1%); biodeg.; 100% conc.

Tergitol 15-S-15. [616] C_{11-15} pareth-15; Pt.Co 100 max. solid; m.w. 860; dens. 8.40 (55 C);

sp.gr. 1.025; visc. 63 cs (40 C); HLB 15.4; cloud pt. > 100 C (1% aq.); flash pt. 490 F (COC); pour pt. 23 C; pH 6–8 (1% aq.); biodeg.; 100% conc.

Tergitol 15-S-20. [616] C_{11-15} pareth-20; nonionic; see Tergitol 15-S-3; Pt.Co 100 max. solid; m.w. 1080; dens. 8.66 lb/gal (40 C); sp.gr. 1.040; visc. 102 cs (40 C); HLB 16.3; cloud pt. > 100 C (1% aq.); flash pt. 480 F (COC); pour pt. 34 C; pH 6–8 (1%); biodeg.; 100% conc.

Tergitol 15-S-30. [616] C_{11-15} pareth-30; nonionic; see Tergitol 15-S-3; Pt.Co 100 max. solid; m.w. 1520; dens. 8.78 lb/gal (55 C); sp.gr. 1.063; visc. 102 cs (40 C); HLB 17.4; cloud pt. > 100 C (1% aq.); flash pt. 505 F (COC); pour pt. 40 C; pH 6–8 (1%); biodeg.; 100% conc.

Tergitol 15-S-40. [616] C_{11-15} pareth-40; nonionic; see Tergitol 15-S-3; Pt.Co 100 max. solid; m.w. 1960; dens. 8.83 lb/gal (55 C); sp.gr. 1.066; HLB 18.0; cloud pt. > 100 C (1% aq.); flash pt. 515 F (COC); pour pt. 45 C; pH 6–8 (1%); biodeg.; 100% conc.

Tergitol 24-L-3. [616] Primary alcohol PEG ether; nonionic; detergent, emulsifier, surfactant, defoamer, intermediate; liq.; oil-sol.; HLB 7.7; biodeg.; 100% conc.

Tergitol 24-L-50. [616] Primary alcohol PEG ether; nonionic; detergent, wetting agent, emulsifier, surfactant used in household and industrial detergents; liq.; HLB 12.4; biodeg.; 100% conc.

Tergitol 24-L-60. [616] Primary alcohol PEG ether; nonionic; see Terigtol 24-L-50; paste; HLB 12.2; 100% conc.

Tergitol 24-L-92. [616] Primary alcohol PEG ether; nonionic; see Tergitol 24-L-50; solid; HLB 14.0; 100% conc.

Tergitol 25-L-3. [616] C_{12-15} pareth-3; nonionic; detergent, wetting agent, emulsifier, textile surfactant and lubricant; Pt-Co 50 max. liq.; m.w. 341; dens. 7.73 lb/gal; sp.gr. 0.929; visc. 36 cs; HLB 7.7; cloud pt. < 0 C (1% aq.); flash pt. 260 F (COC); pour pt. 8 C; pH 5–7; biodeg.; 100% conc.

Tergitol 25-L-5. [616] C_{12-15} pareth-5; nonionic; see Tergitol 25-L-3; Pt.Co 50 max. liq.; m.w. 440; dens. 7.94 lb/gal (30 C); sp.gr. 0.955; visc. 95 cs; HLB 10.4; cloud pt. < 0 C (1% aq.); flash pt. 325 F (COC); pour pt. 16 C; pH 5–7 (1%); biodeg.; 100% conc.

Tergitol 25-L-7. [616] C_{12-15} pareth-7; nonionic; see Tergitol 25-L-3; also in desizing; Pt.Co 50 max. liq.; m.w. 550; dens. 8.19 lb/gal (30 C); sp.gr. 0.985; visc. 51 cps (38 C); HLB 12.4; cloud pt. 50 C (1% aq.); flash pt. 340 F (COC); pour pt. 23 C; surf. tens. 28.7 dynes/cm (0.1%); pH 5–7 (1%); biodeg.; 100% conc.

Tergitol 25-L-9. [616] C_{12-15} pareth-9; nonionic; see Tergitol 25-L-7; Pt.Co 50 max. semisolid; m.w. 585; dens. 8.18 lb/gal (40 C); sp.gr. 0.991; visc. 33 cps (40 C); HLB 12.8; cloud pt. 60 C (1% aq.); flash pt. 285 F (COC); pour pt. 24 C; surf. tens. 29.0 dynes/cm (0.1%); pH 5–7 (1%); biodeg.; 100% conc.

Tergitol 25-L-12. [616] C_{12-15} pareth-12; nonionic; see Tergitol 25-L-7; Pt.Co 100 max. solid; m.w. 730; dens. 8.37 lb/gal (40 C); sp.gr. 1.013; visc. 51 cs (38 C); HLB 14.2; cloud pt. 90 C (1% aq.); flash pt. 410 F (COC); pour pt. 30 C; surf. tens. 32.0 dynes/cm (0.1%); pH 5–7 (1%); biodeg.; 100% conc.

Tergitol NP-4. [616] Nonoxynol-4; emulsifier used in dry-cleaning detergents and industrial cleaners; colorless clear liq.; char. odor; oil-sol.; m.w. 396; sol. in kerosene, aliphatic hydrocarbons, butyl acetate, butyl Cellosolve, corn and wh. min. oil, diesel fuel, anhyd. IPA, Stod., toluene; sp.gr. 1.031; dens. 8.57 lb/gal; visc. 445 cs; HLB 8.9; b.p. > 300 C; flash pt. 480 F (COC); cloud pt. 0 C; 100% act.

Tergitol NP-6. [616] Nonoxynol-6; nonionic; emulsifier used in cleaning applications; colorless clear liq.; char. odor; m.w. 484; sol. see Tergitol NP-4; sp.gr. 1.055; dens. 8.67 lb/gal; visc. 373 cs; HLB 10.9; b.p. > 300 C; flash pt. 360 F (PMCC); cloud pt. 0 C (0.5% aq.); pH 5–7.5; 100% act.

Tergitol NP-7. [616] Nonoxynol-7; nonionic; emulsifier, wetting agent, dispersant used in agric. toxicant formulations, cleaners and santizers; cleaning booster; Pt.Co 75 max. clear liq.; mild char. odor; m.w. 528; sol. in butyl acetate, butyl Cellosolve, corn oil, anhyd. IPA, toluene, water; sp.gr. 1.055; dens. 8.71 lb/gal; visc. 338 cs; HLB 11.7; solid. pt. –6 C; b.p. > 250 C; flash pt. 525 F (COC); cloud pt. 20 C; pH 5–8 (10% aq.); 100% act.

Tergitol NP-8. [616] Nonoxynol-8; nonionic; removal or emulsification of grease or oils; wetting agent, dispersant, emulsifier; Pt.Co 50 max. clear liq.; mild char. odor; m.w. 572; sol. in butyl acetate, butyl Cellosolve, ethylene glycol, anhyd. IPA, toluene, water;

sp.gr. 1.056; visc. 325; HLB 12.3; solid. pt. –3 C; b.p. 250 C; flash pt. 400 F (PMCC); cloud pt. 43 C; pH 5.0–8.0 (10% aq.); surf. tens. 30 dynes/cm (0.1% aq.); 100% act.

Tergitol NP-9. [616] Nonoxynol-9; nonionic; detergent, wetting agent, dispersant, emulsifier; yel. clear liq.; mild char. odor; m.w. 616; sol. see Tergitol NP-8; sp.gr. 1.057; dens. 8.80 lb/gal; visc. 318 cs; HLB 12.9; solid. pt. 0 C; b.p. > 250 C; flash pt. 540 F (COC); cloud pt. 54 C; pH 5–8 (10% aq.); surf. tens. 30 dynes/cm (0.1% aq.); 100% act.

Tergitol NP-10. [616] Nonoxynol-10; nonionic; see Tergitol NP-9; yel. clear liq.; mild char. odor; m.w. 682; sol. see Tergitol NP-8; sp.gr. 1.062; dens. 8.84 lb/gal; visc. 327 cs; HLB 13.6; solid. pt. 7 C; b.p. > 250 C; flash pt. 500 F (COC); cloud pt. 63 C (0.5% aq.); 100% act.

Tergitol NP-13. [616] Nonoxynol-13; nonionic; cleaning and industrial applications; yel. clear liq.; mild char. odor; m.w. 792; sol. see Tergitol NP-8; sp.gr. 1.071; dens. 8.90 lb/gal; visc. 410 cs; HLB 14.4; solid. pt. 16 C; b.p. > 250 C; cloud pt. 83 C (0.5% aq.); pH 5.0–8.0 (10% aq.); surf. tens. 34 dynes/cm (0.1% aq.); 100% act.

Tergitol NP-14. [616] Nonoxynol-4; nonionic; detergent, wetting agent, emulsifier, intermediate; used in textile applications; Pt.Co 100 max. liq.; m.w. 396; dens. 8.57 lb/gal; sp.gr. 1.031; visc. 448 cs; HLB 8.9; cloud pt. < 0 C (1% aq.); flash pt. 480 F (COC); solid. pt. –40 C; pH 5–8 (10%); 100% conc.

Tergitol NP-27. [616] Nonoxynol-7; nonionic; see Tergitol NP-14; Pt.Co 75 max. liq.; m.w. 528; dens. 8.71 lb/gal; sp.gr. 1.038; visc. 338 cs; HLB 11.7; cloud pt. 22 C (1% aq.); flash pt. 525 F (COC); solid. pt. –6 C; surf. tens. 30.4 dynes/cm (0.1%); pH 5–8 (10%); 100% conc.

Tergitol NP-40. [616] Nonoxynol-40; nonionic; detergent, wetting agent, asphalt emulsifier; wh. solid; mild char. odor; m.w. 1980; water-sol.; sp.gr. 1.080; dens. 8.97 lb/gal (55 C); HLB 17.8; m.p. 48 C; b.p. > 250 C; flash pt. 525 F(COC); cloud pt. 100 (0.5%); pH 4.0–8.0 (10% aq.); surf. tens. 45 dynes/cm (0.1% aq.); 100% act.

Tergitol NP-40 (70% Aqueous). [616] Nonoxynol-40; nonionic; clear liq.; mild odor; water-sol.; sp.gr. 1.104; dens. 9.19 lb/gal; visc. 533 cs; solid. pt. –2 C; 70% act.

Tergitol NP-44. [616] Nonoxynol-40; nonionic; see Tergitol NP-14; Gardner 5 max. solid; m.w. 1980; water-sol.; dens. 8.97 lb/gal (55 C); sp.gr. 1.082; HLB 17.8; cloud pt. > 100 C (1% aq.); flash pt. 525 F (COC); solid. pt. 48 C; pH 4–8 (10%); 100% conc.

Tergitol NPX. [616] Nonoxynol-10 (10.5 EO); see Tergitol NP-14; Pt.Co. 50 max. liq.; m.w. 682; dens. 8.84 lb/gal; sp.gr. 1.062; visc. 318 cs; HLB 13.6; cloud pt. 63 C (0.5% aq.); flash pt. 500 F (COC); solid. pt. 7 C; pH 5–8 (10%); 100% conc.

Tergitol TMN-3. [616] Isolaureth-3; nonionic; wetting agent, textile processing; liq.; m.w. 318; dens. 7.74 lb/gal; sp.gr. 0.936; HLB 8.3; cloud pt. < 0 C (1% aq.); flash pt. 320 F (COC); solid. pt. –40 C; pH 5–8 (10%); 100% conc.

Tergitol TMN-6. [616] Isolaureth-6; nonionic; see Tergitol TMN-3; Pt.Co 150 max. liq.; m.w. 450; dens. 8.40 lb/gal; sp.gr. 1.024; visc. 105 cs; HLB 11.7; cloud pt. 36 C (0.5% aq.); solid. pt. –22 C; pH 5–8 (10%); 90% conc.

Tergitol TMN-10. [616] Isolaureth-10; nonionic; see Tergitol TMN-3; Pt.Co 200 max. liq.; m.w. 626; dens. 8.69 lb/gal; sp.gr. 1.044; visc. 153 cs; HLB 14.1; cloud pt. 76 C (1% aq.); solid. pt. –7.5 C; pH 5–8 (10%); 90% conc.

Tergitol TP-9. [616] Nonoxynol-9; see Tergitol NP-4; Pt-Co 50 max. liq.; m.w. 616; dens. 8.80 lb/gal; sp.gr. 1.057; visc. 318 cs; HLB 12.9; cloud pt. 54 C (0.5% aq.); flash pt. 540 F (COC); solid. pt. 0 C; surf. tens. 31.2 dynes/cm (0.1%); pH 5–8 (10%); 100% act.

Tergitol XD. [616] PPG-24-buteth-27; nonionic; dispersant, emulsifier used in dye pigments in textile baths and latex systems; Pt.Co 100 solid; m.w. 2500; dens. 8.66 lb/gal (40 C); sp.gr. 1.041; visc. 428 cs (40 C); cloud pt. 75 C; flash pt. 480 (COC); solid. pt. 33 C; pH 4–7; 100% conc.

Tergitol XH. [616] Polyalkylene glycol ether; see Tergitol XD; Pt.Co 100 solid; m.w. 3500; dens. 8.72 lb/gal (50 C); sp.gr. 1.048; visc. 325 cs (40 C); cloud pt. 95 C (0.5% aq.); solid. pt. 41 C; pH 4–7 (10%); 100% conc.

Tergitol XJ. [616] Polyalkylene glycol ether; see Tergitol XD; Gardner 2 solid; m.w. 2300; dens. 8.51 lb/gal (45 C); sp.gr. 1.027; cloud pt. 50 C (1% aq.); flash pt. 495 F (COC); solid pt. 26 C; pH 5.5–8.0 (1%); 100% conc.

Teric 12A3. [316] C_{12-15} pareth-3 (2.6 EO); nonionic; wetting agent and dispersant, coemulsifier, detergent, intermediate used in oil and solv.-based systems; horticultural applications; w/o emulsions; household and industrial cleaners; mfg. of sulfate and phosphate surfactants; Hazen 100 liq.; sol. in benzene, ethyl acetate, ethyl Icinol, perchlorethylene, ethanol, kerosene, min. and veg. oil, olein; sp.gr. 0.912; visc. 28 cps; m.p. < 0 C; HLB 7.1; pH 6–8 (1% aq.); biodeg.; 100% act.

Teric 12A6. [316] C_{12-15} pareth-6; nonionic; wetting agent, dispersant, detergent, chemical intermediate; used in textile processing, hard surface abrasives, metal descaling, detergent for household and industrial use in laundries; metal cleaning, sanitizer; Hazen 100 liq. to paste; sol. in benzene, ethyl acetate, ethyl Icinol, perchlorethylene, ethanol, kerosene, min., veg. and paraffin oil, olein; disp. in water; sp.gr. 0.955 (50 C); visc. 24 cps (50 C); m.p. 10 ± 2 C; HLB 11.2; cloud pt. 40 ± 2 C; surf. tens. 27.8 dynes/cm; pH 6–8 (1% aq.); biodeg.; 100% act.

Teric 12A9. [316] C_{12-15} pareth-9; nonionic; see Teric 12A6; Hazen 100 paste; sol. in water, benzene, ethyl acetate, ethyl Icinol, perchlorethylene, ethanol, olein; sp.gr. 0.984 (50 C); visc. 32 cps (50 C); m.p. 20 ± 2 C; HLB 13.2; cloud pt. 82 ± 2 C (1%); surf. tens. 30.2 dynes; pH 6–8 (1% aq.); biodeg.; 100% act.

Teric 12A12. [316] C_{12-15} pareth-12 (11.5 EO); see Teric 12A6; Hazen 100 solid; sol. in benzene, ethyl acetate, ethyl Icinol, perchlorethylene, ethanol, olein; sp.gr. 0.993 (50 C); visc. 37 cps (50 C); m.p. 24 ± 2 C; HLB 14.4; cloud pt. 92 ± 2 C (1%); surf. tens. 35.6 dynes/cm; pH 6–8 (1% aq.); biodeg.; 100% act.

Teric 12A16. [316] C_{12-15} pareth-16; nonionic; wetting agent, dispersant, emulsifier, detergent; used in degreasing compds., industrial and household hard surface cleaners; sanitizers; Hazen 100 solid; sol. in water, benzene, ethyl acetate, ethyl Icinol, perchlorethylene, ethanol, olein; sp.gr. 1.009 (50 C); visc. 38 cps (50 C); m.p. 27 ± 2 C; HLB 15.3; cloud pt. 98 ± 2 C; surf. tens. 33.7 dynes/cm; pH 6–8 (1% aq.); biodeg.; 100% act.

Teric 12A23. [316] C_{12-15} pareth-23; nonionic; see Teric 12A16; Hazen 100 solid; sol. see Teric 12A16; sp.gr. 1.032 (50 C); visc. 75 cps (50 C); m.p. 35 ± 2 C; HLB 16.6; cloud pt. > 100 C; surf. tens. 35.7 dynes/cm; pH 6–8 (1% aq.); biodeg.; 100% act.

Teric 12M2. [316] PEG-2 cocamine; nonionic; dispersant, emulsifier, stabilizer; wetting agent of hydrophobic surfaces; used in metal, stone, paper and textile processing; formulation of lubricants and dye bath auxs.; emulsion stabilization; fat liquoring compds.; liq.; sol. in benzene, ethyl acetate, ethyl Icinol, perchlorethylene, ethanol, kerosene, min. and veg. oil; disp. in water; sp.gr. 0.913; visc. 166 cps; m.p. –1 ± 2 C; HLB 11.4; surf. tens. 28.2 dynes/cm; pH 8–10 (1% aq.); 100% act.

Teric 12M5. [316] PEG-5 cocamine; nonionic; see Teric 12M2; liq.; sol. in benzene, ethyl acetate, ethyl Icinol, perchlorethylene, ethanol, min., veg. and paraffin oil, olein; disp. in water; sp.gr. 0.971; visc. 311 cps; m.p. –9 ± 2 C; HLB 12.4; surf. tens. 29.2 dynes/cm; pH 8–10 (1% aq.); 100% act.

Teric 12M15. [316] PEG-15 cocamine; nonionic; see Teric 12M2; liq.; sol. in water, benzene, ethyl acetate, ethyl Icinol, perchlorethylene, ethanol, veg. oil, olein; sp.gr. 1.038; visc. 343 cps; m.p. –11 ± 2 C; HLB 15.7; cloud pt. > 100 C; surf. tens. 39.6 dynes/cm; pH 8–10 (1% aq.); 100% act.

Teric 13A9. [316] Trideceth-9; nonionic; wetting agent and dispersant, emulsifier, detergent used in pigment disps. and agric. and horticultural applications; solv. emulsion cleaners; Hazen 150 liq.; sol. in water, ethyl acetate, ethyl Icinol, ethanol, olein; sp.gr. 1.018; visc. 142 cps; m.p. < 0 C; HLB 13.3; cloud pt. 63 ± 2 C; surf. tens. 27.9 dynes/cm; pH 6–8 (1% aq.); 90% act.

Teric 16A16. [316] Ceteareth-16; nonionic; wetting agent, emulsifier, detergent, solubilizer, dispersant in hydrophobic conditions; dye and pigment carriers; textile applications; mfg. of wax emulsions and polishes for household and industrial use; antistat; Hazen 100 flakes; sol. in water, benzene, ethyl acetate, ethyl Icinol, perchlorethylene, ethanol, olein; sp.gr. 1.01 (50 C); visc. 65 cps (50 C); m.p. 38 ± 2 C; HLB 14.9; cloud pt. > 100 C; surf. tens. 37.1 dynes/cm; pH 6–8 (1% aq.); biodeg.; 100% act.

Teric 16A22. [316] Ceteareth-22; nonionic; see Teric 16A16; Hazen 100 flakes; sol. see Teric 16A16; sp.gr. 1.01 (50 C); visc. 92 cps (50 C); m.p. 40 ± 2 C; HLB 15.8; cloud pt. > 100 C; surf. tens. 41.3 dynes/cm; pH 6–8 (1% aq.); biodeg.; 100% act.

Teric 16A29. [316] Ceteareth-29; nonionic; see Teric 16A16; Hazen 100 flakes; sol. see

Teric 16A16; sp.gr. 1.05 (50 C); visc. 105 cps (50 C); m.p. 45 ± 2 C; HLB 17.7; cloud pt. > 100 C; surf. tens. 46.0 dynes/cm; pH 6–8 (1% aq.); biodeg.; 100% act.

Teric 16M2. [316] PEG-2 soya amine; nonionic; wetting agent, dispersant, emulsifier for waxes and fats; formulation of leather dressing and metal cleaning compds., fiber lubricant applications in the bldg. industry; liq.; sol. in benzene, ethyl acetate, ethyl Icinol, perchlorethylene, ethanol, kerosene, min., paraffin and veg. oil; olein; sp.gr. 0.906; visc. 191 cps; m.p. –15 ± 2 C; HLB 9.5; surf. tens. 26.6 dynes/cm; pH 8–10 (1% aq.); 100% act.

Teric 16M5. [316] PEG-5 soya amine; nonionic; wetting agent and dispersant; used in metal, cleaning, leather dressing and dye leveling compds., agricultural sprays; emulsifier for fats and waxes; fiber lubricant; wax emulsion for the bldg. industry; liq.; sol. see Teric 16M2; sp.gr. 0.954; visc. 182 cps; m.p. –16 ± 2 C; HLB 10.5; surf. tens. 29.7 dynes/cm; pH 8–10 (1% aq.); 100% act.

Teric 16M10. [316] PEG-10 soya amine; nonionic; wetting agent, dispersant, emulsifier for fats and waxes; used in leather dressing compds., fiber lubricant; wax emulsions for fiber board pkg. and particle board for bldg. industry; liq.; sol. in water, benzene, ethyl acetate, ethyl Icinol, perchlorethylene, ethanol, kerosene, min., veg. and paraffin oil, olein; sp.gr. 0.994; visc. 219 cps; m.p. –22 ± 2 C; HLB 12.1; cloud pt. 89 ± 2 C; surf. tens. 36.1 dynes/cm; pH 8–10 (1% aq.); 100% act.

Teric 16M15. [316] PEG-15 soya amine; nonionic; see Teric 16M5; liq.; sol. in water, benzene, ethyl acetate, ethyl Icinol, perchlorethylene, ethanol; sp.gr. 1.021; visc. 266 cps; m.p. –18 ± 2 C; HLB 13.8; cloud pt. > 100 C; surf. tens. 37.4 dynes/cm; pH 8–10 (1% aq.); 100% act.

Teric 17A2. [316] Cetoleth-2 nonionic; emulsifier, mfg. of textile lubricants, solv. and waterless hand cleaners; coemulsifier; detergent additive in petroleum oils; intermediate for anionic surfactants; Hazen 150 liq.; sol. in benzene, ethyl acetate, ethyl Icinol, perchlorethylene, ethanol, kerosene, veg., paraffin and min. oils, olein; disp. in water; sp.gr. 0.902; visc. 37 cps; m.p. 11 ± 2 C; HLB 6.0; pH 6–8 (1% aq.); biodeg.; 100% act.

Teric 17A3. [316] Cetoleth-3; nonionic; see Teric 17A2; Hazen 150 liq.; sol. see Teric 17A2; sp.gr. 0.918; visc. 41 cps; m.p. 13 ± 2 C; HLB 7.5; pH 6–8 (1% aq.); biodeg.; 100% act.

Teric 17A6. [316] Cetoleth-6; nonionic; see Teric 17A2; Hazen 150 liq.; sol. in benzene, ethyl acetate, ethyl Icinol, perchlorethylene, ethanol, veg. oil, olein; disp. in water; sp.gr. 0.957; visc. 60 cps; m.p. 15 ± 2 C; HLB 10.2; pH 6–8 (1% aq.); biodeg.; 100% act.

Teric 17A8. [316] Cetoleth-8; nonionic; wetting agent, dispersant, emulsifier; used in processing of yarns and fabrics; textile dyeing; horticultural sprays; removal of oil slicks; Hazen 150 paste; sol. in water, benzene, ethyl acetate, ethyl Icinol, perchlorethylene, ethanol, olein; sp.gr. 0.963 (50 C); visc. 35 cps (50 C); m.p. 17 ± 2 C; HLB 11.6; cloud pt. 41 ± 2 C; surf. tens. 30.8 dynes/cm; pH 6–8 (1% aq.); biodeg.; 100% act.

Teric 17A10. [316] Cetoleth-10; nonionic; see Teric 17A8; also as solubilizer for sanitation chemicals; Hazen 150 solid; sol. see Teric 17A8; sp.gr. 0.973 (50 C); visc. 43 cps (50 C); m.p. 24 ± 2 C; HLB 12.7; cloud pt. 72 ± 2 C; surf. tens. 32.9 dynes/cm; pH 6–8 (1% aq.); biodeg.; 100% act.

Teric 17A13. [316] Cetoleth-13; nonionic; see Teric 17A10; Hazen 150 solid; sol. see Teric 17A8; sp.gr. 0.986 (50 C); visc. 45 cps (50 C); m.p. 25 ± 2 C; HLB 13.8; cloud pt. 84 ± 2 C; surf. tens. 33.0 dynes/cm; pH 6–8 (1% aq.); biodeg.

Teric 17A25. [316] Cetoleth-25; nonionic; see Teric 17A10; Hazen 150 solid; sol. in water, benzene, ethyl acetate, ethyl Icinol, perchlorethylene, ethanol; sp.gr. 1.030 (50 C); visc. 92 cps (50 C); m.p. 37 ± 2 C; HLB 16.2; cloud pt. > 100 C; surf. tens. 42.0 dynes/cm; biodeg.; 100% act.

Teric 17M2. [316] PEG-2 tallow amine; nonionic; wetting agent and dispersant; dewatering agent in electrical components and road aggregates; emulsifier for fats, waxes, min. oils and metal working lubricants; corrosion inhibitor; paste; sol. in benzene, ethyl acetate, ethyl Icinol, perchlorethylene, ethanol, kerosene, min. paraffin and veg. oil, olein; sp.gr. 0.896; visc. 56 cps (40 C); m.p. 25 ± 2 C; HLB 9; surf. tens. 26.5 dynes/cm; pH 8–10 (1% aq.); 100% act.

Teric 17M5. [316] PEG-5 tallow amine; nonionic; see Teric 17M2; semi-liq.; sol. see Teric 17M2; sp.gr. 0.953; visc. 177 cps; m.p. 11 ± 2 C; HLB 10; surf. tens. 31.5 dynes/cm; pH 8–10 (1% aq.); 100% act.

Teric 17M15. [316] PEG-15 tallow amine;

nonionic; see Teric 17M2; liq.; sol. in water, benzene, ethyl acetate, ethyl Icinol, perchlorethylene, ethanol, veg. oil., olein; sp.gr. 1.028; visc. 428 cps; m.p. –19 ± 2 C; HLB 13.3; cloud pt. > 100 C; surf. tens. 40.0 dynes/cm; pH 8–10 (1% aq.); 100% act.

Teric 18M2. [316] PEG-2 stearamine; nonionic; wetting agent, dispersant; emulsion stabilizer; emulsifier used in agricultural toxicants and processing of textiles, paper, leather and bldg. board; corrosion inhibitor in lubricants and greases; solid; sol. in benzene, ethyl acetate, ethyl Icinol, perchlorethylene, ethanol, kerosene, min., paraffin and veg. oil, olein; sp.gr. 0.887 (50 C); visc. 75 cps (50 C); m.p. 50 ± 2 C; HLB 8.5; pH 8–10 (1% aq.); 100% act.

Teric 18M5. [316] PEG-5 stearamine; nonionic; see Teric 18M2; also softener and antistat in solv. cleaning of textiles and compd. of plastics; solid; sol. see Teric 18M2; also disp. in water; sp.gr. 0.935 (50 C); visc. 46 cps (50 C); m.p. 35 ± 2 C; HLB 9.5; pH 8–10 (1% aq.); 100% act.

Teric 18M10. [316] PEG-10 stearamine; nonionic; see Teric 18M5; liq.; sol. in water, benzene, ethyl acetate, ethyl Icinol, perchlorethylene, ethanol, kerosene, veg. oil, olein; sp.gr. 0.997; visc. 255 cps; m.p. 18 ± 2 C; HLB 11.2; cloud pt. > 100 C; surf. tens. 34.0 dynes/cm; pH 8–10 (1% aq.); 100% act.

Teric 18M20. [316] PEG-20 stearamine; nonionic; see Teric 18M2; paste; sol. in water, benzene, ethyl acetate, ethyl Icinol, perchlorethylene, ethanol, olein; sp.gr. 1.025 (50 C); visc. 66 cps (50 C); m.p. 28 ± 2 C; HLB 14.5; cloud pt. > 100 C; surf. tens. 40.3 dynes/cm; pH 8–10 (1% aq.); 100% act.

Teric 18M30. [316] PEG-30 stearamine; nonionic; see Teric 18M2; solid; sol. in water, benzene, ethyl acetate, ethyl Icinol, perchlorethylene, ethanol; sp.gr. 1.038 (50 C); visc. 116 cps (50 C); m.p. 28 ± 2 C; HLB 17.8; cloud pt. > 100 C; surf. tens. 43.0 dynes/cm; pH 8–10 (1% aq.); 100% act.

Teric 121. [316] Fatty acid EO condensate; nonionic; antifoam agent, emulsifier; coemulsifier of veg. and min oils, paraffinic waxes and solvs.; lubricant for textile processing; foam control agent in paper mfg., industrial fermentation and distillation, textile dyeing, adhesives and waste water treatment; Hazen > 250 liq.; sol. in benzene, ethyl acetate, ethyl Icinol, perchlorethylene, ethanol, veg. oil, olein; sp.gr. 0.987; visc. 353 cps; m.p. < 0 C; HLB 7.4; pH 8 (1% aq.); biodeg.; 100% act.

Teric 124. [316] Fatty acid EO condensate; nonionic; see Teric 121; Hazen > 250 liq.; sol. see Teric 121; also sol. in min. oil; sp.gr. 0.966; visc. 118 cps; m.p. < 0 C; HLB 4.5; pH 8 (1% aq.); biodeg.; 100% act.

Teric 127. [316] Fatty acid EO condensate; nonionic; see Teric 121; Hazen > 250 liq.; sol. see Teric 124; sp.gr. 0.978; visc. 87 cps; m.p. < 0 C; HLB 7.7; pH 8 (1% aq.); 100% act.

Teric 151. [316] Fatty acid EO condensate; nonionic; emulsifier, antistat, lubricant; textile processing aid for fibers; o/w emulsions; Hazen > 250 solid; sol. in benzene, ethyl acetate, ethyl Icinol, perchlorethylene, ethanol, kerosene, veg. oil, olein; disp. in water; sp.gr. 0.994 (50 C); visc. 45 cps (50 C); m.p. 30 ± 2 C; HLB 11.6; pH 7–9 (1% aq.); biodeg.; 100% act.

Teric 154. [316] Fatty acid EO condensate; nonionic; antistat and emulsifier; formulation of textile processing aids; veg. and lt. petroleum oils; Hazen > 250 liq.; sol. see Teric 18M10; sp.gr. 1.031; visc. 89 cps; m.p. 7 ± 2 C; HLB 13.3; surf. tens. 30.0 dynes/cm; pH 6–8 (1% aq.); biodeg.; 100% act.

Teric 158. [316] Alkyl phenol ethoxylate; nonionic; formulation of powds. for agricultural, surf. coatings and ceramic applications; Hazen > 250 fine powd.; HLB 11.6; pH 6–8 (1% aq.).

Teric 160. [316] Alkylaryl polyoxyalkylene ether; nonionic; wetting agent and dispersant; mfg. of surf. coatings, detergent and sanitizer; Hazen 250 liq.; sol. in water, benzene, ethyl acetate, ethyl Icinol, perchlorethylene, ethanol, veg. oil, olein; sp.gr. 1.032; visc. 312 cps; m.p. < 0 C; HLB 11.3; cloud pt. 29 ± 2 C; surf. tens. 30.4 dynes/cm; pH 6–8 (1% aq.); 100% act.

Teric 161. [316] Alkylaryl polyoxyalkylene ether; nonionic; see Teric 160; Hazen 250 liq.; sol. see Teric 160; sp.gr. 1.034; visc. 340 cps; m.p. < 0 C; HLB 12.3; cloud pt. 35 ± 2 C; surf. tens. 33.4 dynes/cm; pH 6–8 (1% aq.); 100% act.

Teric 163. [316] Alkylaryl polyoxyalkylene ether; nonionic; demulsifier, detergent for machine dishwashing, in-situ cleaning of milking plants, detergent sanitizer and iodophor mfg; Hazen 250 liq.; sol. see Teric 160; sp.gr. 1.110; visc. 1240 cps; m.p. < 0 C; HLB 14.9; cloud pt. 97 ± 2 C; surf. tens. 50.4

dynes/cm; pH 4–6 (1% aq.); 100% act.

Teric 164. [316] Alcohol ethyoxylate, modified; see Teric 163; Hazen 250 liq.; sol. see Teric 18M10; sp.gr. 0.937; visc. 65 cps; m.p. 2 ± 2 C; HLB 10.8; cloud pt. 40 ± 2 C; surf. tens. 30.5 dynes/cm; pH 6–8; biodeg.; 100% min. act.

Teric 165. [316] Alcohol ethoxylate, modified; nonionic; wetting and dedusting agent, additive for cleaning applications; mfg. of detergents; dk. br. liq.; sol. in water, ethyl Icinol, ethanol, olein; sp.gr. 1.035; visc. 68 cps; HLB 12.5; cloud pt. 30–35 C (1% aq.); pour pt. 2–3 C; surf. tens. 28.2 dynes/cm (0.1%); pH 5; biodeg.; 89% min. act.

Teric 200. [316] Alkylene oxide derivative; nonionic; emulsifier; mfg. of concs. of insecticides, pesticides and herbicides in solvs.; Hazen 150 paste; sol. see Teric 18M20; sp.gr. 1.034 (50 C); visc. 123 cps (50 C); m.p. 30 ± 2 C; HLB 16.1; cloud pt. 90 ± 2 C; surf. tens. 36.0 dynes/cm; pH 6–8 (1% aq.); 100% act.

Teric 203. [316] Alkylene oxide derivative; nonionic; agricultural emulsifier; dispersant for conc. formulations; paste; 100% conc.

Teric 313. [316] Alkylaryl POE; nonionic; solubilizer, emulsifier; liq.; 100% conc.

Teric 350. [316] Fatty acid condensate; nonionic; emulsifier and coemulsifier for solvs. and min. oils; formulation of bases for solv. cleaners, degreasing and decarbonizing preparations; lubricant and corrosion inhibitor in lubricant systems used in metal working industry; Hazen > 250 liq.; sol. in benzene, ethyl Icinol, perchlorethylene, ethanol, olein; disp. in water; sp.gr. 1.010; visc. 600 cps; HLB 16.6; pH 8–10 (1% aq.); 100% conc.

Teric 351B. [316] Fatty acid condensate; nonionic; see Teric 350; Hazen > 250 liq.; sol. in perchlorethylene, olein; disp. in water; sp.gr. 1.017; visc. 750 cps; HLB 8.2; pH 8–10 (1% aq.).

Teric BL8. [316] Syn. alcohol polyalkylene oxide deriv.; detergent; wetting agent and dispersant in printing inks, textiles, wool scouring, dust suppression, household and industrial surf. cleaning; emulsifier of solvs., greases and oils; Hazen 250 liq.; sol. see Teric 18M10; sp.gr. 1.003; visc. 63 cps; m.p. < 0 C; HLB 13.7; cloud pt. 59 ± 2 C; surf. tens. 30.3 dynes/cm; pH 6–8 (1% aq.); biodeg.; 100% act.

Teric BL9. [316] Syn. alcohol polyalkylene oxide deriv.; see Teric BL8; Hazen 250 liq.; sol. see Teric 160; sp.gr. 1.018; visc. 100 cps; m.p. < 0 C; HLB 14.5; cloud pt. 67 ± 2 C; surf. tens. 30.0 dynes/cm; pH 6–8 (1% aq.); biodeg.; 98% act.

Teric C12. [316] PEG-12 castor oil; nonionic; detergent, emulsifier and coemulsifier in formulation of sol. oils, solv. cleaners and temp. protective coatings; corrosion resistant, used in industrial and institutional cleaning preparations; die lubricant in metal forming operations; mold release agent in plastics; fiber lubricant in textile processing; Hazen > 250 liq.; sol. in water, benzene, ethyl acetate, ethyl Icinol, ethanol, veg. oil, olein; sp.gr. 1.048; visc. 5000 cps.; m.p. 3 ± 2 C; HLB 12.7; cloud pt. 67–69 C; surf. tens. 42.0 dynes/cm; pH 6–8 (1% aq.); 100% act.

Teric CDE. [316] Cocamide DEA; nonionic; emulsifier of perchlorethylene, power kerosene; foam booster, thickener and detergent in domestic dishwashing formulations; Hazen > 250 liq.; sol. see Teric 350; sp.gr. 1.010; visc. 600 cps; m.p. < 0 C; HLB 16.6; pH 8–10; biodeg.; 100% act.

Teric CME3. [316] PEG-3 cocamide MEA; nonionic; dispersant, detergent; wetting agent for textile applications; coemulsifier for various oils; emulsion stabilizer; foam booster in detergent blends; Hazen > 250 liq., paste; low odor; sol. in benzene, ethyl acetate, ethyl Icinol, perchlorethylene, ethanol, olein; disp. in water; sp.gr. 0.972 (50 C); visc. 56 cps (50 C); m.p. 18 ± 2 C; HLB 13.8; pH 8–10 (1% aq.); biodeg.; 100% act.

Teric CME7. [316] PEG-7 cocamide MEA; nonionic; wetting agent for textile applications; dispersant, detergent, foam stabilizer; cleaning compds.; Hazen > 250 liq.; low odor; sol. see Teric 18M30; sp.gr. 1.048; visc. 264 cps; m.p. 5 ± 2 C; HLB 15.1; surf. tens. 30.0 dynes/cm; pH 8–10; biodeg.; 100% act.

Teric DD5. [316] Dodoxynol-5; nonionic; emulsifier, coemulsifier in solv. emulsion cleaners; formulation of concs. for metal lubricants, cutting, milling and grinding aids, textile processing aid and min. drilling lubricants; Hazen 150 liq.; sol. in benzene, ethyl acetate, ethyl Icinol, perchlorethylene, ethanol, kerosene, min. and veg. oil, olein; sp.gr. 1.014; visc. 550 cps; m.p. < 0 C; HLB 8.6; surf. tens. 29.5 dynes/cm; pH 6–8 (1% aq.); biodeg.; 100% act.

Teric DD9. [316] Dodoxynol-9; nonionic; see

Teric DD5; Hazen 150 liq.; sol. in benzene, ethyl Icinol, ethanol, kerosene, olein; sp.gr. 1.040; visc. 453 cps; m.p. < 0 C; HLB 12.1; surf. tens. 28.2 dynes/cm; pH 6–8 (1% aq.); 100% act.

Teric G9A5. [316] C_{9-11} pareth-5; nonionic; emulsifier for solvs.; formulation of hard surf. cleaners, degreasers and dispersants in industrial and domestic applications; Hazen 250 liq.; sol. in water, veg. oil, olein; sp.gr. 0.976; visc. 52 cps; m.p. < 0 C; HLB 11.6; cloud pt. 36 ± 2 C; surf. tens. 20.1 dynes/cm; pH 5–7 (1% aq.); 100% act.

Teric G9A6. [316] C_{9-11} pareth-6; see Teric G9A5; Hazen 250 liq.; sol. in water, benzene, ethyl acetate, ethyl Icinol, perchlorethylene, ethanol, veg. oil, olein; sp.gr. 0.987; visc. 38 cps; m.p. < 0 C; HLB 12.4; cloud pt. 53 ± 2 C; surf. tens. 24.0 dynes/cm; pH 5–7 (1% aq.); 100% act.

Teric G9A8. [316] C_{9-11} pareth-8; nonionic; emulsifier, dispersant, wetting agent; used in detergent formulations for domestic use; degreaser in textile applications; Hazen 250 paste; sol. see Teric G9A6; sp.gr. 0.988 (50 C); visc. 21 cps (50 C); m.p. 13 ± 2 C; HLB 13.7; cloud pt. 78 ± 2 C; surf. tens. 27.0 dynes/cm; pH 5–7 (1% aq.); 100% act.

Teric G9A12. [316] C_{9-11} pareth-12; nonionic; see Teric G9A8; Hazen 250 solid; sol. see Teric G9A6; sp.gr. 1.023; visc. 40 cps (50 C); m.p. 24 ± 2 C; HLB 15.4; cloud pt. > 100 C; surf. tens. 34.5 dynes/cm; pH 5–7 (1% aq.); 100% act.

Teric G12A4. [316] C_{12-15} pareth-4 (4.5 EO); nonionic; wetting agent and dispersant, detergent; dust suppression; aq. pigment dispersion; coemulsifier for w/o-type emulsion concs.; liq. and powd. detergents for industrial and domestic laundry, dishwashing, metal cleaning, sanitizing, textile processing aids, abrasive cleaners; intermediate in mfg. of sulfate and phosphate surfactants; Hazen 250 liq.; sol. see Teric 18M2; also disp. in water; sp.gr. 0.950; visc. 40 cps; m.p. 3 ± 2 C; HLB 9.8; pH 6–8 (1% aq.); 100% act.

Teric G12A6. [316] C_{12-15} pareth-6; nonionic; wetting agent and dispersant, intermediate, textile processing, hard surf. abrasives, metal descaling, domestic and industrial use in laundries, intermediate for metal cleaning, sanitizing, dairy detergents; Hazen 250 liq. to paste; sp.gr. 0.955; visc. 24 cps (50 C); m.p. 10 ± 2 C; HLB 11.2; cloud pt. 40 ± 2 C; surf. tens. 27.8 dynes/cm; pH 6–8 (1% aq.); biodeg.; 100% act.

Teric G12A8. [316] C_{12-15} pareth-8 (7.8 EO); nonionic; see Teric G12A6; Hazen 250 paste; sol. see Teric G12A4; sp.gr. 0.971 (50 C); visc. 31 cps (50 C); m.p. 16 ± 2 C; HLB 12.6; cloud pt. 58 ± 2 C; surf. tens. 27.2 dynes/cm; pH 6–8 (1% aq.); biodeg.; 100% act.

Teric G12A12. [316] C_{12-15} pareth-12 (11.5 EO); nonionic; see Teric 12A6; Hazen 250 solid; sp.gr. 0.993; visc. 37 cps; m.p. 24 ± 2 C; HLB 14.4; cloud pt. 92 ± 2 C; surf. tens. 35.6 dynes/cm; pH 6–8 (1% aq.); biodeg.; 100% act.

Teric LA4. [316] C_{12-15} pareth-4 (4.5 EO); nonionic; see Teric G12A4; liq.; sol. in water, ethyl acetate, ethyl Icinol, ethanol, kerosene, min. oil, olein; sp.gr. 0.955; visc. 45 cps; m.p. 0 ± 2 C; HLB 9.8; pH 6–8 (1% aq.); biodeg.; 88% act.

Teric LA8. [316] C_{12-15} pareth-8 (7.8 EO); nonionic; see Teric G12A8; lt. amber liq.; sol. see Teric 18M30; sp.gr. 1.002; visc. 100 cps; m.p. 10 ± 2 C; HLB 12.6; cloud pt. 60 ± 2 C; surf. tens. 27.2 dynes/cm; pH 6–8 (1% aq.); biodeg.; 88% act.

Teric LAN70. [316] PEG-70 lanolin; nonionic; emulsifier, dispersant, textile lubricant; yel.-amber soft waxy solid; sol. in water, benzene, ethyl Icinol, perchlorethylene, ethanol, olein; sp.gr. 1.13; visc. 246 cps (50 C); m.p. 42 ± 2 C; HLB 16.3; cloud pt. > 100 C; surf. tens. 48.1 dynes/cm; pH 4.5–7.0 (1% aq.); 100% act.

Teric N2. [316] Nonoxynol-2; nonionic; defoamer, wetting agent and dispersant in solv. and oil-based systems; coemulsifier for o/w emulsions; emulsifier for w/o emulsions; used in agricultural toxicant products, industrial solv.-cleaning systems, surf. coating preparations; intermediate for production of anionics; Hazen 100 liq.; sol. see Teric 18M2; also disp. in water; sp.gr. 1.001; visc. 620 cps; m.p. < 0 C; HLB 5.7; pH 6–8 (1% aq.); 100% act.

Teric N3. [316] Nonoxynol-3 (3.5 EO); nonionic; see Teric N2; Hazen 100 liq.; sol. see Teric N2; sp.gr. 1.017; visc. 395 cps; m.p. < 0 C; HLB 8.2; pH 6–8 (1% aq.); 100% act.

Teric N4. [316] Nonoxynol-4; nonionic; see Teric N2; Hazen 100 liq.; sol. see Teric N2; sp.gr. 1.023; visc. 370 cps; m.p. < 0 C; HLB 8.9; pH 6–8 (1% aq.); 100% act.

Teric N5. [316] Nonoxynol-5 (5.5 EO); nonionic; see Teric N2; Hazen 100 liq.; sol. in benzene, ethyl acetate, ethyl Icinol, perchlorethylene, ethanol, kerosene, veg. oil, olein; disp. in water; sp.gr. 1.035; visc. 355 cps; m.p. < 0 C; HLB 10.5; pH 6–8 (1% aq.); 100% act.

Teric N8. [316] Nonoxynol-8 (8.5 EO); nonionic; wetting agent and dispersant, emulsifier, solubilizer, detergent; used in concrete mfg., agricultural sprays, solv. cleaners, paints; detergents; Hazen 100 liq.; sol. see Teric 160; sp.gr. 1.056; visc. 350 cps; m.p. 0 ± 2 C; HLB 12.3; cloud pt. 32 ± 2 C; surf. tens. 29.4 dynes/cm; pH 6–8 (1% aq.); 100% act.

Teric N9. [316] Nonoxynol-9; nonionic; see Teric N8; Hazen 100 liq.; sol. in water, benzene, ethyl acetate, ethyl Icinol, perchorethylene, veg. oil, olein; sp.gr. 1.060; visc. 330 cps; m.p. 0 ± 2 C; HLB 12.8; cloud pt. 52 ± 2 C; surf. tens. 30.6 dynes/cm; pH 6–8 (1% aq.); 100% act.

Teric N10. [316] Nonoxynol-10; nonionic; see Teric N8; Hazen 100 liq.; sol. see Teric N9; sp.gr. 1.063; visc. 360 cps; m.p. 5 ± 2 C; HLB 13.3; cloud pt. 67 ± 2 C; surf. tens. 30.6 dynes/cm; pH 6–8 (1% aq.); 100% act.

Teric N11. [316] Nonoxynol-11; nonionic; see Teric N8; Hazen 100 liq.; sol. in water, benzene, ethyl acetate, ethyl Icinol, perchlorethylene, ethanol, olein; sp.gr. 1.069; visc. 410 cps; m.p. 7 ± 2 C; HLB 13.7; cloud pt. 74 ± 2 C; surf. tens. 34.8 dynes/cm; pH 6–8 (1% aq.); 100% act.

Teric N12. [316] Nonoxynol-12; nonionic; see Teric N8; Hazen 100 liq.; sol. see Teric N11; sp.gr. 1.045 (50 C); visc. 67 cps (50 C); m.p. 11 ± 2 C; HLB 13.9; cloud pt. 82 ± 2 C; surf. tens. 35.2 dyens/cm; pH 6–8 (1% aq.); 100% act.

Teric N13. [316] Nonoxynol-13; nonionic; see Teric N8; Hazen 100 liq.; sol. see Teric N11; sp.gr. 1.049 (50 C); visc. 75 cps (50 C); m.p. 14+2 C; HLB 14.4; cloud pt. 89 ± 2 C; surf. tens. 34.9 dynes/cm; pH 6–8 (1% aq.); 100% act.

Teric N15. [316] Nonoxynol-15; nonionic; wetting agent, coemulsifier; used in paints, perfumes, detergents and sanitizers; Hazen 100 paste; sol. see Teric N11; sp.gr. 1.051 (50 C); visc. 82 cps (50 C); m.p. 21 ± 2 C; HLB 15.0; cloud pt. 95 ± 2 C; surf. tens. 33.4 dynes/cm; pH 6–8 (1% aq.); 100% act.

Teric N20. [316] Nonoxynol-20; nonionic; see Teric N15; Hazen 100 solid; sol. see Teric N11; sp.gr. 1.061 (50 C); visc. 100 cps (50 C); m.p. 30 ± 2 C; HLB 16.0; cloud pt. > 100 C; surf. tens. 41.7 dynes/cm; pH 6–8 (1% aq.); 100% act.

Teric N30. [316] Nonoxynol-30; nonionic; see Teric N15; Hazen 100 flakes; sol. see Teric 18M30; sp.gr. 1.066 (50 C); visc. 150 cps (50 C); m.p. 40 ± 2 C; HLB 17.2; cloud pt. > 100 C; surf. tens. 42.8 dynes/cm; pH 6–8 (1% aq.); 100% act.

Teric N40. [316] Nonoxynol-40; nonionic; see Teric N15; Hazen 100 flakes; sol. see Teric 18M30; sp.gr. 1.080 (50 C); visc. 250 cps (50 C); m.p. 45 ± 2 C; HLB 18.0; cloud pt. > 100 C; surf. tens. 41.0 dynes/cm; pH 6–8 (1% aq.); 100% act.

Teric N100. [316] Nonoxynol-100; nonionic; see Teric N15; Hazen 100 flakes; sol. see Teric 18M30; sp.gr. 1.113 (60 C); m.p. 52 ± 2 C; HLB 19.1; cloud pt. > 100 C; surf. tens. 46.0 dynes/cm; pH 6–8 (1% aq.); 100% act.

Teric OF4. [316] PEG-4 oleate; nonionic; emulsifier used in o/w emulsions of veg. and min. oils, fats, waxes and solvs.; formulation of cutting oils; metal lubricant in metal cleaning formulations and in solv. degreasers; textile and paper finishing applications; antistat in textile processing and mfg. of syn. fibers; Hazen > 250 liq.; sol. see Teric 151; sp.gr. 0.978; visc. 87 cps; m.p. < 0 C; HLB 7.7; pH 8 (1% aq.); biodeg.; 100% act.

Teric OF6. [316] PEG-6 oleate; nonionic; see Teric OF4; Hazen > 250 liq.; sol. see Teric 151; sp.gr. 0.992; visc. 82 cps; m.p. < 0 C; HLB 9.7; surf. tens. 33.9 dynes/cm; pH 7–9 (1% aq.); biodeg.; 100% act.

Teric OF8. [316] PEG-8 oleate; nonionic; emulsifier, antistat; aux. product and antistat in textile and paper finishing applications; Hazen > 250 liq.; sol. see Teric 151; sp.gr. 1.012; visc. 120 cps; m.p. < 0 C; HLB 11.1; pH 8–10 (1% aq.); biodeg.; 100% act.

Teric PE61. [316] POP + 5.7 EO; nonionic; demulsifier, intermediate, emulsifier, detergent, wetting agent and dispersant; pigments in latex paints; detergent sanitizer and alkaline cleaner; dewatering aid for treatment of crude oil emulsions; defoamer in paper pulp liquors, starch sizing, glues, and boiler water systems; Hazen 100 liq.; sol. in benzene, ethyl acetate, ethyl Icinol, perchlorethylene, veg. oil, olein; sp.gr. 1.017; visc. 388 cps; m.p. < 0 C; HLB 3; cloud pt. 24 ± 2 C; surf. tens. 39.0 dynes/cm; pH 6 (1% aq.); 100% act.

Teric PE62. [316] POP + 17 EO; nonionic; see

Teric PE61; also for bloat control in cattle; Hazen 100 liq.; sol. see Teric 160; sp.gr. 1.032; visc. 440 cps; m.p. < 0 C: HLB 7; cloud pt. 32 ± 2 C; surf. tens. 42.1 dynes/cm; pH 6 (1% aq.); 100% act.

Teric PE64. [316] POP + 25.5 EO; nonionic; wetting agent, dispersant, emulsifier, detergent, intermediate, demulsifier; used for pigments in latex paints, diary cleaners and sanitizers; bloat control in cattle; oil well applications; Hazen 100 liq.; sol. see Teric N11; sp.gr. 1.051; visc. 1332 cps; m.p. 8 ± 2 C; HLB 15; cloud pt. 58 ± 2 C; surf. tens. 43.2 dynes/cm; pH 6 (1% aq.); 100% act.

Teric PE68. [316] POP + 150 EO; nonionic; wetting agent, dispersant, emulsifier, detergent, intermediate; stabilizer in mfg. of latex paints, paper coatings, diary cleaners and santizers; plasticizer for resins; Hazen 100 flakes; sol. see Teric N11; m.p. 50 ± 2 C; HLB 29; cloud pt. > 100 C; surf. tens. 49.0 dynes/cm; pH 6 (1% aq.); 100% act.

Teric SF9. [316] PEG-9 stearate; nonionic; coemulsifier, preparation of self-emulsifying waxes for waterproofing of fiber and chip boards, bldg. sheets, and cardboard pkg.; component of floor wax emulsions and wax strippers; Hazen 250 solid; sol. see Teric 164; sp.gr. 0.994 (50 C); visc. 45 cps (50 C); m.p. 30 ± 2 C; HLB 10.8; pH 8 (1% aq.); biodeg.; 100% act.

Teric SF15. [316] PEG-15 stearate; nonionic; coemulsifier; textile and leather dressing oils and a dye bath additive; Hazen 250 solid; sol. in water, benzene, ethyl acetate, ethyl Icinol, perchlorethylene; sp.gr. 1.014 (50C); visc. 55 cps (50 C); m.p. 32 ± 2 C; HLB 14.0; surf. tens. 40.7 dynes/cm; pH 8 (1% aq.); biodeg.; 100% act.

Teric T2. [316] PEG-2 tallate; nonionic; wetting agent and dispersant for oil-based systems; emulsifier, detergent, grinding aid for wet milling of pigments and resins; mfg. of lubricants for metal and textile industries, solv. cleaners, and powd. detergent formulations; Hazen > 250 liq.; sol. in benzene, ethyl acetate, ethyl Icinol, perchlorethylene, ethanol, min. and veg. oil, olein; disp. in water; sp.gr. 0.966; visc. 118 cps; m.p. < 0 C; HLB 4.5; pH 6–8 (1% aq.); biodeg.; 100% act.

Teric T5. [316] PEG-5 tallate; nonionic; see Teric T2; Hazen > 250 liq.; sol. see Teric T2; sp.gr. 1.018; visc. 196 cps; m.p. < 0 C; HLB 8.2; pH 8–10 (1% aq.); biodeg.; 100% act.

Teric T7. [316] PEG-7 tallate; nonionic; see Teric T2; Hazen > 250 liq.; sol. see Teric 121; sp.gr. 1.031; visc. 151 cps; m.p. < 0 C; HLB 10.0; pH 8–10 (1% aq.); biodeg.; 100% act.

Teric T10. [316] PEG-10 tallate; nonionic; see Teric T2; Hazen > 250 liq.; sol. see Teric 121; also disp. in water; sp.gr. 1.042; visc. 253 cps; m.p. 5 ± 2 C; HLB 11.7; pH 8–10 (1% aq.); biodeg.; 100% act.

Teric X5. [316] Oxtoxynol-5; nonionic; wetting agent and dispersant, detergent used in metal cleaning compds., alkaline and metal cleaners, agricultural powds., emulsions, pigment and wax dispersions; textile auxs.; emulsifier in aromatic and aliphatic solvs., emulsions, cleaners and paints; Hazen 100 liq.; sol. see Teric 18M2; sp.gr. 1.046; visc. 465 cps; m.p. < 0 ± 2 C; HLB 10.4; pH 6–8 (1% aq.); 100% act.

Teric X7. [316] Octoxynol-7 (7.5 EO); nonionic; see Teric X5; Hazen 100 liq.; sol. see Teric 160; sp.gr. 1.059; visc. 420 cps; m.p. < 0 ± 2 C; HLB 12.3; cloud pt. 28 ± 2 C; surf. tens. 29.5 dynes/cm; pH 6–8 (1% aq.); 100% act.

Teric X8. [316] Octoxynol-8 (8.5 EO); nonionic; see Teric X5; Hazen 100 liq.; sol. see Teric 18M20; sp.gr. 1.060; visc. 402 cps; m.p. < 0 ± 2 C; HLB 12.6; cloud pt. 47 ± 2 C; surf. tens. 30.2 dynes/cm; pH 6–8 (1% aq.); 100% act.

Teric X10. [316] Octoxynol-10; nonionic; see Teric X5; Hazen 100 liq.; sol. see Teric 18M20; sp.gr. 1.062; visc. 393 cps; m.p. 7 ± 2 C; HLB 13.6; cloud pt. 65 ± 2 C; surf. tens. 31.8 dynes/cm; pH 6–8 (1% aq.); 100% act.

Teric X11. [316] Octoxynol-11; nonionic; see Teric X5; Hazen 100 liq.; sol. see Teric 18M20; sp.gr. 1.067; visc. 371 cps; m.p. 8 ± 2 C; HLB 14.0; cloud pt. 80 ± 2 C; surf. tens. 31.0 dynes/cm; pH 6–8 (1% aq.); 100% act.

Teric X13. [316] Octoxynol-13; nonionic; see Teric X5; Hazen 100 liq.; sol. see Teric 18M20; sp.gr. 1.052; visc. 70 cps; m.p. 15 ± 2 C; HLB 14.7; cloud pt. 90 ± 2 C; surf. tens. 32.0 dynes/cm; pH 6–8 (1% aq.); 100% act.

Teric X16. [316] Octoxynol-16; nonionic; see Teric X5; Hazen 100 semisolid; sol. see Teric 18M2; sp.gr. 1.059 (50 C); visc. 95 cps (50 C); m.p. 24 ± 2 C; HLB 15.2; cloud pt. > 100 C; surf. tens. 35.6 dynes/cm; pH 6–8 (1% aq.); 100% act.

Teric X40. [316] Octoxynol-40; nonionic; see Teric X5; Hazen 100 solid; sol. see Teric 18M30; sp.gr. 1.082 (50 C); visc. 245 cps (50

C); m.p. 44 ± 2 C; HLB 17.6; cloud pt. > 100 C; surf. tens. 36.4 dynes/cm; pH 6–8 (1% aq.); 100% act.

Tetradecene-1. [227] C_{14} alpha olefins; intermediate for surfactants and industrial chemicals; liq.; 100% conc.

Tetralex CFP. [128] Ethoxylated fatty ester; nonionic; detergent, wetting agent and emulsifier for textiles; liq.; HLB 6.5; 100% conc.

Tetranol. [537] Sulfated fatty ester; nonionic; wetting and leveling agent for dyeing processes; assistant for vat dyeing; emulsifier and detergent in scouring processes; water-sol.; pH 5.8–6.0.

Tetronic 50R1. [86] EO/PO ethylene diamine block copolymer; defoamer, penetrant, plasticizer; demulsifier of crude and industrial waste oil; cutting and grinding fluid; machine dishwashing; solid rinse aid; wetting agent and dispersant for cements and dyes; wetting agent for inks; pigment dispersant and visc. regulator for paints; tablet coating; antistat for plastics; rubber activator; liq.; sp.gr. 1.02; visc. 672 cps; cloud pt. 29 C (1%); pour pt. –18 C; surf. tens. 40.4 dynes/cm (0.1%); 100% act.

Tetronic 50R4. [86] EO/PO ethylene diamine block copolymer; see Tetronic 50R1; liq.; sp.gr. 1.04; visc. 1080 cps; cloud pt. 53 C (1%); pour pt. –4 C; surf. tens. 46.3 dynes/cm (0.1%); 100% act.

Tetronic 50R8. [86] EO/PO ethylene diamine block copolymer; see Tetronic 50R1; solid; m.p. 38 C; cloud pt. 88 C (1%); surf. tens. 52.7 dynes/cm (0.1%); 100% act.

Tetronic 70R1. [86] EO/PO ethylene diamine block copolymer; see Tetronic 50R1; liq.; sp.gr. 1.02; visc. 800 cps; cloud pt. 25 C (1%); pour pt. –17 C; surf. tens. 38.7 dynes/cm (0.1%); 100% act.

Tetronic 70R2. [86] EO/PO ethylene diamine block copolymer; see Tetronic 50R1; liq.; sp.gr. 1.03; visc. 884 cps; cloud pt. 31 C (1%); pour pt. –22 C; surf. tens. 39.8 dynes/cm (0.1%); 100% act.

Tetronic 70R4. [86] EO/PO ethylene diamine block copolymer; see Tetronic 50R1; liq.; sp.gr. 1.05; visc. 2160 cps; cloud pt. 42 C (1%); pour pt. 6 C; surf. tens. 42.9 dynes/cm (0.1%); 100% act.

Tetronic 90R1. [86] EO/PO ethylene diamine block copolymer; see Tetronic 50R1; liq.; sp.gr. 1.02; visc. 945 cps; cloud pt. 21 C (1%); pour pt. –17 C; 100% act.

Tetronic 90R4. [86] EO/PO ethylene diamine block copolymer; see Tetronic 50R1; liq.; sp.gr. 1.05; visc. 3867 cps; cloud pt. 38 C (1%); pour pt. 12 C; surf. tens. 42.7 dynes/cm (0.1%); 100% act.

Tetronic 90R8. [86] EO/PO ethylene diamine block copolymer; see Tetronic 50R1; solid; m.p. 47 C; cloud pt. 81 C (1%); surf. tens. 50.4 dynes/cm (0.1%); 100% act.

Tetronic 110R1. [86] EO/PO ethylene diamine block copolymer; see Tetronic 50R1; liq.; sp.gr. 1.02; visc. 1000 cps; cloud pt. 21 C (1%); pour pt. –18 C; 100% act.

Tetronic 110R2. [86] EO/PO ethylene diamine block copolymer; see Tetronic 50R1; liq.; sp.gr. 1.03; visc. 1320 cps; cloud pt. 27 C (1%); pour pt. –20 C; surf. tens. 37.5 dynes/cm (0.1%); 100% act.

Tetronic 110R7. [86] EO/PO ethylene diamine block copolymer; see Tetronic 50R1; solid; m.p. 47 C; cloud pt. 52 C (1%); surf. tens. 46.5 dynes/cm (0.1%); 100% act.

Tetronic 130R1. [86] EO/PO ethylene diamine block copolymer; see Tetronic 50R1; liq.; sp.gr. 1.02; visc. 1240 cps; cloud pt. 20 C (1%); pour pt. –18 C; 100% act.

Tetronic 130R2. [86] EO/PO ethylene diamine block copolymer; see Tetronic 50R1; liq.; sp.gr. 1.03; visc. 1880 cps; cloud pt. 25 C (1%); pour pt. –3 C; surf. tens. 35.0 dynes/cm (0.1%); 100% act.

Tetronic 150R1. [86] EO/PO ethylene diamine block copolymer; see Tetronic 50R1; liq.; sp.gr. 1.01; visc. 1840 cps; cloud pt. 20 C (1%); pour pt. –17 C; 100% act.

Tetronic 150R4. [86] EO/PO ethylene diamine block copolymer; see Tetronic 50R1; paste; m.p. 34 C; cloud pt. 25 C (1%); surf. tens. 38.9 dynes/cm (0.1%); 100% act.

Tetronic 150R8. [86] EO/PO ethylene diamine block copolymer; see Tetronic 50R1; solid; m.p. 53 C; cloud pt. 38 C (1%); surf. tens. 44.4 dynes/cm (0.1%); 100% act.

Tetronic 304. [86] Poloxamine 304; nonionic; emulsifier, thickener, wetting agent, dispersant, solubilizer, stabilizer, cosmetics and pharmaceuticals; demulsifier in petroleum industry; detergent ingred.; antistat for polyethylene and resin molding powds.; metal treatment; emulsion polymerization; used in latex-based paints, aq.-based syn. cutting fluids and vulcanization of rubber; colorless liq.; m.w. 1650; ref. index 1.4649; water-sol.; sp.gr. 1.06; visc. 450 cps; HLB 16; cloud pt. 94 C (1%); pour pt. –11 C; surf.

tens. 53.0 dynes/cm (0.1%); 100% act.

Tetronic 504. [86] Poloxamine 504; nonionic; see Tetronic 304; colorless liq.; m.w. 3400; ref. index 1.4612; water-sol.; sp.gr. 1.04; visc. 800 cps; HLB 15.5; cloud pt. 68 C (1%); pour pt. 7 C; surf. tens. 44.2 dynes/cm (0.1%); 100% act.

Tetronic 701. [86] Poloxamine 701; nonionic; see Tetronic 304; colorless liq.; m.w. 3400; ref. index. 1.4553; sp.gr. 1.02; visc. 575 cps; HLB 3; cloud pt. 22 C (1%); pour pt. –21 C; surf. tens. 36.1 dynes/cm; 100% act.

Tetronic 702. [86] Poloxamine 702; nonionic; see Tetronic 304; colorless liq.; m.w. 4000; ref. index 1.4572; water-sol.; sp.gr. 1.03; visc. 770 cps; HLB 7; cloud pt. 27 C (1%); pour pt. –7 C; surf. tens. 36.8 dynes/cm (0.1%); 100% act.

Tetronic 704. [86] Poloxamine 704; nonionic; see Tetronic 304; colorless liq.; m.w. 5500; ref. index 1.4613; water-sol.; sp.gr. 1.04; visc. 850 cps; HLB 15; cloud pt. 65 C (1%); pour pt. 18 C; surf. tens. 40.3 dynes/cm (0.1%); 100% act.

Tetronic 707. [86] Poloxamine 707; nonionic; see Tetronic 304; flakes; m.w. 12,000; water-sol.; m.p. 49 C; HLB 27; cloud pt. > 100 C (1%); surf. tens. 47.6 dynes/cm (0.1%); 100% act.

Tetronic 901. [86] Poloxamine 901; nonionic; see Tetronic 304; colorless liq.; m.w. 4750; ref. index 1.4545; sp.gr. 1.02; visc. 700 cps; HLB 2.5; cloud pt. 20 C (1%); pour pt. –23 C; 100% act.

Tetronic 904. [86] Poloxamine 904; nonionic; see Tetronic 304; colorless liq.; m.w. 7500; ref. index 1.4604; water-sol.; sp.gr. 1.04; visc. 6000 cps; HLB 14.5; cloud pt. 64 C (1%); pour pt. 29 C; surf. tens. 35.4 dynes; 100% act.

Tetronic 908. [86] Poloxamine 908; nonionic; see Tetronic 304; flakes; m.w. 27,000; water-sol.; m.p. 58 C; HLB 30.5; cloud pt. > 100 C (1%); surf. tens. 45.7 dynes/cm (0.1%); 100% act.

Tetronic 1101. [86] Poloxamine 1101; nonionic; see Tetronic 304; colorless liq.; m.w. 5600; ref. index 1.4540; sp.gr. 1.02; visc. 700 cps; HLB 2; cloud pt. 17 C (1%); pour pt. –15 C; surf. tens. 34.0 dynes/cm; 100% act.

Tetronic 1102. [86] Poloxamine 1102; nonionic; see Tetronic 304; colorless liq.; m.w. 6300; ref. index 1.4557; water-sol.; sp.gr. 1.03; visc. 820 cps; HLB 6; cloud pt. 31 C (1%); pour pt. 7 C; surf. tens. 33.7 dynes/cm (0.1%); 100% act.

Tetronic 1104. [86] Poloxamine 1104; nonionic; see Tetronic 304; paste; m.w. 8300; water-sol.; m.p. 34 C; HLB 14; cloud pt. 72 C (1%); surf. tens. 36.3 dynes/cm (0.1%); 100% act.

Tetronic 1107. [86] Poloxamine 1107; nonionic; see Tetronic 304; solid; m.w. 14,500; water-sol.; m.p. 51 C; HLB 24; cloud pt. > 100 C; surf. tens. 42.9 dynes/cm (0.1%); 100% act.

Tetronic 1301. [86] Poloxamine 1301; nonionic; see Tetronic 304; colorless liq.; m.w. 6800; ref. index 1.4545; sp.gr. 1.02; visc. 1000 cps; HLB 1.5; cloud pt. 16 C (1%); pour pt. –9 C; surf. tens. 33.4 dynes/cm; 100% act.

Tetronic 1302. [86] Poloxamine 1302; nonionic; see Tetronic 304; colorless liq.; m.w. 7800; ref. index 1.4562; water-sol.; sp.gr. 1.03; visc. 1300 cps; HLB 5.5; cloud pt. 20 C (1%) pour pt. 13 C; surf. tens. 34.1 dynes/cm (0.1%); 100% act.

Tetronic 1304. [86] Poloxamine 1304; nonionic; see Tetronic 304; paste; m.w. 10,500; water-sol.; m.p. 36 C; HLB 13.5; cloud pt. 78 C (1%); surf. tens. 35.5 dynes/cm (0.1%); 100% act.

Tetronic 1307. [86] Poloxamine 1307; nonionic; see Tetronic 304; solid; m.w. 18,600; water-sol.; m.p. 54 C; HLB 23.5; cloud pt. > 100 C (1%); surf. tens. 43.8 dynes/cm (0.1%); 100% act.

Tetronic 1501. [86] Poloxamine 1501; nonionic; see Tetronic 304; colorless liq.; m.w. 7900; ref. index 1.4537; sp.gr. 1.02; visc. 1170 cps; HLB 1.0; cloud pt. 15 C (1%); pour pt. –4 C; surf. tens. 33.9 dynes/cm; 100% act.

Tetronic 1502. [86] Poloxamine 1502; nonionic; see Tetronic 304; colorless liq.; m.w. 9000; ref. index 1.4560; water-sol.; sp.gr. 1.03; visc. 1570 cps; HLB 5.0; cloud pt. 70 C (1%); pour pt. 18 C; surf. tens. 34.1 dynes/cm (0.1%); 100% act.

Tetronic 1504. [86] Poloxamine 1504; nonionic; see Tetronic 304; paste; m.w. 12,500; water-sol.; m.p. 41 C; HLB 13.0; cloud pt. 90 C (1%); surf. tens. 37.3 dynes/cm (0.1%); 100% act.

Tetronic 1508. [86] Poloxamine 1508; nonionic; see Tetronic 304; solid; m.w. 27,000; water-sol.; m.p. 60 C; HLB 27; cloud pt. > 100 C (1%); surf. tens. 43.8 dynes/cm (0.1%); 100% act.

Tewax TC 1, TC 2. [598] Cetyl/stearyl alcohol with nonionic emulsifiers; nonionic; cosmetics, hair dye preparations; flakes; 100% conc.

Texamine 84(L). [659] Alkanolamide and ethanolamine alkyl benzene sulfonate; anionic/nonionic; raw material for mfg. of liq. detergents; paste; 84% conc.

Texapon 842, 890. [291] Sodium octyl sulfate; anionic; wetting agent for hydrophobic powds.; liq. and powd. resp.; 40 and 100% conc.

Texapon 1030, 1090. [291] Sodium decyl sulfate; anionic; wetting agent and foamer for fire fighting foams; liq. and powd. resp.; 30 and 90% conc.

Texapon A. [289; 290; 291] Ammonium lauryl sulfate; personal care products surfactant; water-wh. liq.; 27.5–28.5% act.

Texapon A 400. [291] Ammonium lauryl sulfate; anionic; detergent and shampoo base; liq.; 33–35% conc.

Texapon ASV. [289; 290; 291] Sodium laureth sulfate, magnesium laureth sulfate, sodium laureth-8-sulfate, sodium oleth sulfate, magnesium oleth sulfate; surfactant for personal care products; lt. yel. clear liq.; 28–30% act.

Texapon CS Paste. [289; 291] Fatty alcohol sulfate mixture; anionic; conc. for personal care products; paste; 60% conc.

Texapon DEA. [289; 290] DEA-lauryl sulfate; personal care products surfactant; lt. yel. clear liq.; 36–38% act.

Texapon EA-1, -2, -3. [289; 290] Ammonium laureth sulfate; surfactant, visc. builder, solubilizer in personal care products; water-wh. visc. liq.; 25–27, 24–26, and 26–28% act. resp.

Texapon EA-40. [289; 290] Ammonium myristyl ether sulfate; surfactant for personal care products; lt. yel. liq.; 58–60% act.

Texapon ES-1, -2, -3. [289; 290] Sodium laureth sulfate; surfactant and solubilizer for personal care products; water-wh. visc. liq.; 24–26, 25–26, and 27.5–29.5% act. resp.

Texapon ES-40. [289; 290] Sodium laureth sulfate; surfactant for personal care products; lt. yel. liq.; 57.5–60.0% act.

Texapon EVR. [289; 291] Sodium lauryl sulfate, sodium laureth sulfate, lauramide MIPA, cocamide MEA, glycol stearate and coceth-8; anionic; detergent, foamer; base for shampoos; liq.; 40% conc.

Texapon F. [291] Fatty alcohol sulfate and alkylaryl sulfate compd.; anionic; detergent, dishwashing agents; powd.; 45% conc.

Texapon F35. [291] Fatty alcohol sulfate and alkylaryl sulfate compd.; anionic; see Texapon F; powd.; 35% conc.

Texapon GYP 2. [291] Fatty alcohol polyglycol ether; anionic; foamer for production of plaster boards; liq.; 40% conc.

Texapon IES. [291] Cocamide DEA; anionic; conc. for personal care products; liq.; 60% conc.

Texapon K-12. [289] Sodium lauryl sulfate; anionic; wetting agent, dispersant and foamer used in personal care products, dentifrice mfg.; wh. fine powd.; pH 7.5–9.0 (1% aq.); 90% min. act.

Texapon K-12-94. [291] Sodium lauryl sulfate (C_{12}); anionic; foamer, wetting agent and emulsifier for personal care products; powd.; 94% conc.

Texapon K-12-96. [290] Sodium lauryl sulfate; cleansing agent, foamer for dentifrices; wetting agent and disintegrant in personal care products; wh. powd.; dens. 2.08 lb/gal; pH 6–9 (1%); 96% act.

Texapon K 12 Granules. [291] Sodium lauryl sulfate (C_{12}); anionic; foamer for pharmaceuticals; cleaning agent; granules; 85% conc.

Texapon K 12 Needles. [291] Sodium lauryl sulfate; anionic; foamer for toothpaste; needles; 85% conc.

Texapon K 14 S Spec. [289; 291] Sodium lauryl/myristyl ether sulfate; anionic; detergent for personal care products; liq.; 28% conc.

Texapon L-100. [290] Sodium lauryl sulfate; wetting and cleansing agent, dispersant used in dentifrice mfg.; wh. powd.; pH 7.0 (1% aq.); 99% act.

Texapon LLS. [291] Lithium lauryl sulfate; anionic; wetting agent for carpet shampoos and cleaners; liq.; 30% conc.

Texapon LS Highly Conc. Needles. [291] Sodium lauryl sulfate (C_{12}-C_{14}); anionic; detergent, foamer used in personal care products and detergents; needles; 90% conc.

Texapon MG. [291] Magnesium laureth sulfate; anionic; liq. shampoo and bath preparation base; liq.; 29–31% conc.

Texapon MLS. [291] MEA-lauryl sulfate; anionic; detergent and shampoos; liq.; 28% conc.

Texapon N 25. [291] Sodium laureth sulfate; anionic; base for personal care products; dishwashing agent; fire-fighting foam concs.; liq.; 27–28% conc.

Texapon N 40. [288] Sodium laureth sulfate; anionic; raw material; mfg. of shampoos and bubble baths; liq.; pH 6–7.5 (5%); 25–26% act.

Texapon N 42. [288] TEA-lauryl sulfate; see Texapon N 40; liq.; pH 6–7 (5%); 39–41% act.

Texapon N 70. [291] Sodium laureth sulfate; anionic; detergent in personal care products; paste; 70% conc.

Texapon N 70-88. [291] Fatty alcohol ether sulfate; anionic; see Texapon N Conc.; paste; 70% conc.

Texapon N 70 LS. [291] Sodium laureth sulfate; anionic; see Texapon N 70; paste; 70% conc.

Texapon N 70 N. [289; 291] Sodium laureth sulfate; anionic; used in personal care products; water-wh. liq.; 70% conc.

Texapon N 103. [291] Sodium laureth sulfate (3 EO); anionic; liq. shampoo and bubble bath base; dishwashing agent and fire-fighting foam concs.; liq.; 27–28% conc.

Texapon NA. [288; 291] Ammonium lauryl ether sulfate; see Texapon N 40; liq.; pH 6.5–7.0 (5); 23–24% act.

Texapon N Conc. [291] Sodium lauryl ether sulfate; anionic; detergent for personal care products; paste; 57% conc.

Texapon NSO. [291] Sodium lauryl ether sulfate; anionic; see Texapon N 70; liq.; 27–28% conc.

Texapon NT. [291] TEA laureth sulfate; anionic; see Texapon N 70; liq.; 35% conc.

Texapon OT Highly Conc. Needles. [291] Sodium lauryl sulfate C_{12}-C_{18}; anionic; see Texapon N 70; needles; 91% conc.

Texapon SBN. [288; 291] Sodium laureth sulfate and disodium laureth sulfosuccinate; anionic; detergent base for personal care products; liq.; 30% act.

Texapon SCO. [289] Sulfated/sulfonated castor oil; anionic; wetting agent; liq.; 75% conc.

Texapon SG. [291] Sodium laureth sulfate, PEG-8, cocamide MEA, and glycol distearate; anionic; see Texapon N 70; paste, liq.; 22% conc.

Texapon SH 100. [289] Disodium mono oleamide PEG-2 sulfosuccinate; anionic; detergent; base for shampoos and skin cleansers; liq.; 30% conc.

Texapon SH-135 Special. [289] Disodium mono oleamide PEG-2 sulfosuccinate; anionic; surfactant for personal care products; liq.; 35% conc.

Texapon Special. [291] Ammonium lauryl sulfate; anionic; see Texapon N 70; liq.; 44–45% conc.

Texapon T. [289; 290] TEA-lauryl sulfate; surfactant for personal care products; water-wh. liq.; 39–41% act.

Texapon T 35, T 42, TH. [291] TEA-lauryl sulfate; anionic; see Texapon N 70; liq.; 35, 42, and 47–49% conc. resp.

Texapon V Highly Conc. Needles. [290] Sodium lauryl sulfate C_{12}-C_{16}; anionic; wetting agent and foamer; detergent and cosmetic base; emulsifier for personal care products; wh. fine needles; pH 7–8 (1% aq.); 90% act.

Texapon WW 99. [291] Cocamide DEA; anionic; conc. detergent base and emulsifier for oily foam bath preparations; liq.; 90% conc.

Texapon Z. [291] Sodium lauryl sulfate C_{12}-C_{14}; anionic; detergent, wetting agent, personal care, pharmaceutical preparations, pesticides, paint; needles, powd.; 58–62% conc.

Texapon Z Granules. [291] Sodium lauryl sulfate C_{12}-C_{14}; anionic; detergent, wetting agent, dispersant for personal care products; granules; 50% conc.

Texapon Z Highly Conc. Needles. [290] Sodium lauryl sulfate C_{12}-C_{18}; detergent, wetting agent and foamer; detergent and cosmetic base; wh. fine needles; pH 6.8–7.5 (1% aq.); 85–89% act.

Texapon Z Highly Conc. Powder. [290] Sodium lauryl sulfate C_{12}-C_{18}; see Texapon V Highly Conc. Needles; wh. fine powd.; pH 7.5–8.5 (0.25% aq.); 90% act.

Texin 128 Powder. [291] Disodium sulfosuccinic acid monolauryl ester; anionic; surfactant for hand cleaning preparations and syndet bars; powd.; 90% conc.

Texnol IL. [440] Alkyl imidazoline laurate; nonionic; antistat; solid; 100% act.

Texnol R 5. [440] Alkyl benzyl ammonium salt; anionic; emulsifier, fungicide, softener; liq.; 100% act.

Texo 227. [600] Alkylaryl sulfonate; anionic; used in general purpose cleaners; liq.; HLB 11.0; biodeg.

Texo 284. [600] Alkylaryl sulfonate; nonionic; wetting agent; liq.; HLB 10.0; biodeg.

Texo 1060. [600] Alkylaryl sulfonate; anionic; cleaner; liq.; HLB 11.0; biodeg.

Texo LP 528A. [600] Ethoxylated alkylaryl; nonionic; mold release agent; liq.; 40% conc.

Textamine 05. [289] Fatty imidazoline; cationic; oil-emulsifying agent, corrosion inhibitor and ink dispersant; liq.; 100% act.

Textamine 1839. [289] Fatty-amido tert. amine; cationic; foamer, thickener, ore flotation agent and starting material for quats., betaines, biocides and amine oxides; liq.; 100% act.

Textamine A-3417. [289] Fatty amido diamine; cationic; corrosion inhibitor, asphalt emulsifier; intermediate for spray wax emulsifier formulations for car washes; visc. liq.; dens. 7.8 lb/gal; 100% conc.

Textamine Carbon Detergent K. [289] Fatty amino complex; anionic; emulsifier for R.T. solv. degreasing formulations; aid in carbon removal; soft paste; dens. 7.8 lb/gal; 80% conc.

Textamine O-1. [289] 1-Hydroxyethyl-2-oleyl imidazoline; nonionic; emulsifier for hydrocarbons and nonpolar solvs.; fuel additive for sludge dispersions; used in fungicide, pesticide, and herbicide formulations; thickener and corrosion inhibitor for paints; used in industrial applications; amber liq.; sol. in alcohol, chlorinated hydrocarbons, oils; disp. in water; dens. 7.66 lb/gal; pH 10.5 (10); 90% imidazoline.

Textamine O-5. [289] Oleo-fatty imidazoline; cationic; corrosion inhibitor for gasoline and diesel fuels; emulsifier, textile softener and leather working agent; liq.; 100% conc.

Textamine Oxide CA. [289] Cocoyl amidoalkylamine; nonionic/cationic; wetting agent and foamer in brines and conc. acid and alkaline sol'ns.; liq.; 50% act.

Textamine Oxide CAW. [289] Cocoyl amidoalkylamine oxide; nonionic/cationic; see Textamine Oxide CA; liq.; 30% act.

Textamine Oxide LMW. [289] Lauryl dimethylamine oxide; nonionic/cationic; see Textamine Oxide CA; liq.; 30% act.

Textamine T-1. [289] 1-Hydroxyethyl-2-tall oil imidazoline; nonionic; see Textamine O-1; amber liq.; sol. in alcohol, chlorinated hydrocarbons, aliphatic, and aromatic hydrocarbons, oils; disp. in water; dens. 7.75 lb/gal; pH 11 (10%); 90% min. imidazoline.

Textamine T-5-D. [289; 290] Fatty imidazoline; cationic; emulsifier, oil-sol. corrosion inhibitor, dispersant; liq.; 100% act.

Textol 80 (L). [659] Ethanolamine alkyl benzene sulfonate; anionic; raw material for mfg. of liq. detergents; paste; 80% conc.

Tex-Wet 1001. [330] Sodium dioctyl sulfosuccinate; anionic; wetting and rewetting agent used in textile wet processing; clear liq.; dens. 8.8 lb/gal; pH 6 ± 0.5 (1%); biodeg.; 60 ± solids.

Tex-Wet 1002. [330] Phosphated alcohol; anionic; dispersant, wetting agent and penetrant for textile wet processing; dyeing assistant for woolen goods; lt. amber liq.; water-sol.; dens. 8.75 lb/gal; pH 7.0 ± 0.5; 55% conc.

Tex-Wet 1004. [330] Alkylaryl sulfonate and sulfated nonionic; anionic; detergent, wetting and rewetting agent; wet processing, preparation and/or desizing assistant; dyeing assistant; dens. 8.2 lb/gal; pH 7–8; biodeg.; 35 ± 0.5% solids.

Tex-Wet 1010. [330] Long chain phosphated alcohol, modified; anionic; wetting agent, dyeing assistant, detergent, penetrant, boil-off assistant for textiles; liq.; 50% conc.

Tex-Wet 1034. [330] Phosphate ester; anionic; wetting agent, detergent, emulsifier, antistat; liq.; 70% conc.

Tex-Wet 1048. [330] POE linear alcohol; nonionic; wetting agent, penetrant in textile products; liq.; HLB 12.5; 20% conc.

Tex-Wet 1070. [330] Phosphate ester; anionic; wetting agent in kier boiling; liq.; 68% conc.

Tex-Wet 1103. [330] Phosphated alcohol; anionic; intermediate for preparing penetrants; wetting agent for textile processing; liq.; 100% conc.

Tex-Wet 1131. [330] Coconut alkanolamide; nonionic; detergent, emulsifier, foam stabilizer; liq.; 100% conc.

Tex-Wet 1155. [330] Ethoxylated alkyl phenol; nonionic; detergent, wetting, emulsifier for insecticides and herbicides; household and industrial cleaners, textile processing; liq.; HLB 13.5; 100% conc.

Tex-Wet 1158. [330] Sodium alcohol ether sulfate; anionic; wetting agent, dispersant, emulsifier, detergent, dyeing assistant; fabric scouring agent; cotton processing; stabilizer for dye bath; dyeing of syns.; pale amber liq.; dens. 8.7 lb/gal; 57% min. act.

Tex-Wet 1197. [330] Dodecylbenzene sulfonic acid; anionic; detergent intermediate; emulsifier, coemulsifier used in textile dyeing and scouring; dish detergent and cleaning formulations; red-br. liq.; dens. 8.8 lb/gal; 97% act.

Thorowet G-40, -75. [145] Sodium dioctyl sulfosuccinate; anionic; wetting and rewetting agent; liq.; 42–45 and 73–76% conc.

Tinegal KD. [137] EO condensate, modified; amphoteric; dyeing and printing aux. for nylon; liq.

Tinegal MR-50. [137] Quat. ammonium salt; cationic; leveling agent for fibers; liq.

Tinegal NA. [137] Polyoxethylated fatty alcohol; nonionic; dyebath stabilizer, surfactant, emulsifier; liq.

Tinegal PM. [137] Alkylamine polyglycol ether; cationic; leveling agent for nylon; liq.

Titadine TA. [605] Alkyl naphthalene sulfonic acid ester and sulfated alcohol mixt.; wetting agent; textile, leather, paper, household detergents; liq.

Titadine TCP. [605] Sulfonated alcohol deriv.; wetting agent, emulsifier for textiles, leather, paper, and household detergents; liq.

Titan Castor No. 75. [605] Sulfonated castor oil; anionic; emulsifier for textile processing; liq.; 62% conc.

Titan Decitrene. [605] Alkylated aromatic sulfonate; see Titadine TA.

Titanole RMA. [605] Alkylated aryl sodium sulfonate; see Titadine TA; liq.

Titanterge CAC. [605] Coconut amine condensate; nonionic; detergent for textile processing; liq.; 100% conc.

Titanterge DBS Conc. [605] Dodecylbenzene sulfonate; anionic; wetting agent, detergent for texile processing; liq.; 92% conc.

Titapene Conc. [605] Sodium sulfosuccinate, modified; anionic; wetting agent in textile processing; liq.

Titazole HTX. [605] Fatty acid amino sulfonate; anionic; detergent for textile processing; liq.

Titazole SA. [605] Sodium alkyl naphthalene sulfonate; see Titadine TA; liq.

T-Maz 20. [400] Polysorbate 20; nonionic; emulsifier, solubilizer, wetting agent, antistat, stabilizer, dispersant, visc. modifier used in the food, cosmetic, drug, textile and metalworking industries; yel. liq.; sol. in water, ethanol, acetone; sp.gr. 1.1; visc. 400 cps; HLB 16.7; sapon. no. 40–50; 97% act.

T-Maz 28. [400] PEG-80 sorbitan monolaurate; emulsifier and solubilizer in essential oils, wetting agent, visc. modifier, antistat, stabilizer and dispersant used in food, cosmetic, drug, textile and metalworking industries; yel. liq.; sol. see T-Maz 20; sp.gr. 1.0; visc. 1100 cps; sapon. no. 5–15; 30% max. water.

T-Maz 40. [400] Polysorbate 40; see T-Maz 20; yel. liq.; sol. see T-Maz 20; sp.gr. 1.0; visc. 600 cps; HLB 15.6; sapon. no. 43–49; 97% min. act.

T-Maz 60. [400] Polysorbate 60; see T-Maz 20; yel. liq.; sol. in water, toluol; sp.gr. 1.1; visc. 550 cps; HLB 14.9; pour pt. 23–25 C; sapon. no. 45–55; 97 min. act.

T-Maz 60K. [400] Polysorbate 60, kosher; nonionic; emulsifier for foods; liq.; HLB 14.9; 100% conc.

T-Maz 61. [400] Polysorbate 61; nonionic; see T-Maz 20; solid; water-sol.; HLB 9.6; 100% conc.

T-Maz 65. [400] Polysorbate 65; see T-Maz 20; sol. in ethanol, acetone, naphtha; disp. in water; sp.gr. 1.1; m.p. 30–32 C; HLB 10.5; sapon. no. 88–98; 97% min. act.

T-Maz 65K. [400] Polysorbate 65, kosher; nonionic; food emulsifier; solid; HLB 10.5; 100% conc.

T-Maz 67. [400] PEG-20 sorbitan isostearate; see T-Maz 20; yel. liq.; sol. see T-Maz 60; sp.gr. 1.0; visc. 400 cps; HLB 14.8; sapon. no. 45–55; 97% min. act.

T-Maz 80. [400] Polysorbate 80; see T-Maz 20; yel. liq.; sol. in water, ethanol, veg. oil, toluol; sp.gr. 1.0; visc. 400 cps; HLB 15.0; sapon. no. 45–55; 97% min. act.

T-Maz 80K. [400] Polysorbate 80, kosher; nonionic; solubilizer and emulsifier for foods; liq.; HLB 15.0; 100% conc.

T-Maz 81. [400] Polysorbate 81; nonionic; see T-Maz 20; liq.; water-sol.; HLB 10.0; 100% conc.

T-Maz 85. [400] Polysorbate 85; see T-Maz 20; yel. liq.; sol. in ethanol; disp. in water; sp.gr. 1.0; visc. 300 cps; HLB 11.0; sapon. no. 83–93; 95% min. act.

T-Maz 90. [400] PEG-20 sorbitan monotallate; see T-Maz 20; amber liq.; sol. in water, ethanol, toluol, veg. oil; sp.gr. 1.0; visc. 500 cps; HLB 15.0; sapon. no. 45–55; 97% min. act.

T-Maz 95. [400] PEG-20 sorbitan tritallate; see T-Maz 28; amber liq.; sol. see T-Maz 85; sp.gr. 1.0; visc. 350 cps; HLB 10.8; sapon. no. 45–55; 3.0% max. water.

T-Mulz 540. [603] Calcium alkylaryl sulfonate and POE ether; anionic/nonionic; emulsifier for herbicide formulations; liq.

T-Mulz 734-2. [603] Phosphate ester; anionic; emulsifier for agric. formulations, solvs., detergents, emulsion polymerization, alkaline cleaners; hydrotrope; amber clear liq.; sp.gr. 1.18 (68 F); dens. 9.8 lb/gal (68 F); pour pt. 50 F; flash pt. > 350 F (PMCC); pH 2–3 (1% aq.).

T-Mulz 808 A. [603] POE ether; nonionic; emulsifier, agric. crop oil formulations; straw clear liq.; dens. 8.33 lb/gal (68 F); sp.gr. 1.00 (68 F); flash pt. > 200 F (TCC); pour pt. < 10 F; pH 4–6 (1% aq. disp.).

T-Mulz AS-1151, -1152, -1153. [603] Amine alkylaryl sulfonate with polyethylene ether; anionic/nonionic; emulsifier for herbicide formulations; liq; HLB 11.5, 12.5, and 12.6 resp.

TO-33-F. [287] Sorbate 85 sorbitan trioleate; nonionic; emulsifier w/o for cosmetics and pharmaceuticals; antifoamer solubilizer for various technical applications; liq.; HLB 2.5; 100% conc.

TO-55-A. [287] POE (2) sorbitan trioleate; nonionic; w/o emulsifier, antifoam agent for cosmetics, pharmaceuticals and technical applications; liq.; HLB 4.5; 100% conc.

TO-55-E. [287] POE (18) sorbitan trioleate; nonionic; emulsifier for emulsions of min., veg. and trail oils, waxes; cattle feed, textile industry, biocide, paint varnishes, plastics, leather and fur industry; cosmetics and pharmaceuticals; liq.; HLB 10.5; 100% conc.

TO-55-EL. [287] POE (17) sorbitan trioleate; nonionic; see TO-55-E; liq.; HLB 10.0; 100% conc.

TO-55-F. [287] Polysorbate 85; nonionic; emulsifier for cosmetics and pharmaceuticals; technical applications; emulsions for wood preservation, pigment stabilizer, furniture polishes; liq.; HLB 11.0; 100% conc.

Tohol N-220. [606] Coconut diethanolamide; nonionic; foam stabilizer and thickener for personal care products; liq.; 100% conc.

Tohol N-220X. [606] Coconut diethanoamide; nonionic; see Tohol N-220; solid; 100% conc.

Tohol N-230. [606] Lauryl diethanolamide; nonionic; see Tohol N-220; liq.; 100% conc.

Tohol N-230X. [606] Lauryl diethanolamide; nonionic; see Tohol N-220; solid; 100% conc.

Tomah AO-14-2, AO-17-2. [608] Amine oxide; cationic; shampoo formulations; liq.; 50% conc.

Tomah DA 62. [608] Aliphatic diamine; cationic; corrosion inhibitor; oilfield chemical; asphalt emulsifier; liq.; 80% conc.

Tomah E-14-2. [608] Ethoxylated fatty amines; emulsifier, corrosion inhibitor, lubricant used in min. acid inhibition, textile processing; amber liq.; sp.gr. 0.94; HLB 8.3; 95–100% act.

Tomah E-14-5. [608] Ethoxylated fatty amines; see Tomah E-14-2; sp.gr. 0.99; 95–100% act.

Tomah E-14-15. [608] Ethoxylated fatty amines; see Tomah E-14-2; sp.gr. 1.05; 95–100% act.

Tomah E-17-2. [608] Ethoxylated fatty amines; see Tomah E-14-2; amber liq.; sp.gr. 0.93; HLB 5.6; 95–100% act.

Tomah E-18-2. [608] Ethoxylated fatty amines; see Tomah E-14-2; wax; sp.gr. 0.95; HLB 3.0; 95–100% act.

Tomah E-18-5. [608] Ethoxylated fatty amines; see Tomah E-14-2; paste; sp.gr. 0.97; HLB 11.0; 95–100% act.

Tomah E-18-10. [608] Ethoxylated fatty amines; see Tomah E-14-2; amber liq.; sp.gr. 1.02; 95–100% act.

Tomah E-18-15. [608] Ethoxylated fatty amines; see Tomah E-14-2; amber liq.; sp.gr. 1.03; HLB 16.0; 95–100% act.

Tomah E-C-2. [608] Ethoxylated amine; cationic; emulsification; corrosion inhibitor; min. acid inhibition, textile processing and lubricants; liq.; HLB 2.0; 100% conc.

Tomah E-C-5. [608] Ethoxylated amine; cationic; see Tomah E-C-2; liq.; HLB 5.0; 100% conc.

Tomah E-C-15. [608] Ethoxylated amine; cationic; see Tomah E-C-2; liq.; HLB 15.0; 100% conc.

Tomah E-S-2. [608] Ethoxylated fatty soya amines; see Tomah E-14-2; amber liq.; sp.gr. 0.91; HLB 2.0; 95–100% act.

Tomah E-S-5. [608] Ethoxylated fatty soya amines; see Tomah E-14-2; amber liq.; sp.gr. 0.95; HLB 5.0; 95–100% act.

Tomah E-S-10. [608] Ethoxylated fatty soya amines; see Tomah E-14-2; amber liq.; sp.gr. 1.02; 95–100% act.

Tomah E-S-15. [608] Ethoxylated fatty soya amines; see Tomah E-14-2; amber liq.; sp.gr. 1.03; HLB 15.0; 95–100% act.

Tomah PA-14. [608] Primary amine; cationic; emulsification; liq.; 100% conc.

Tomah PA-17. [608] Tridecyloxypropylamine; corrosion inhibitor, cationic emulsification and replacement for oleyl and soya amines; colorless to lt. amber liq.; m.w. 274; sp.gr. 0.85; pour pt. –30 F; 95% min. amine.

Tomah PA-62. [608] Primary amine; cationic; emulsifier, corrosion inhibitor; liq.; 75% conc.

Tomah Q-17-2. [608] Tridecyloxypropyl dihydroxyethyl methyl ammonium chloride; see

Tomah Q-14-2; amber liq.; sp.gr. 0.97; pH 6–9 (5% aq.); insol. in water; dens. 8.5 lb/gal; visc. 472 cSt (100 F); HLB 8.9; pour pt. –10 C; flash pt. 430 F; cloud pt. < 25 C.

Tomah Q-18-2. [608] Octadecyl dihydroxyethyl methyl ammonium chloride; cationic; see Tomah Q-14-2; paste; sp.gr. 0.92; pH 6–9 (5% aq.).

Tomah Q-18-15. [608] Octadecyl POE (15) methyl ammonium chloride; cationic; see Tomah Q-14-2; amber liq.; sp.gr. 1.06; pH 6–9 (5% aq.); 100% act.

Tomah Q-311. [608] Quat. ammonium chloride; cationic; acid inhibitor, emulsifier; liq.; HLB 16.3–18.0; 75% conc.

Tomah Q-511. [608] Quat. ammonium chloride; cationic; acid cleaner, wetting agent; cosmetics; liq.; 75% conc.

Tomah Q-D-T. [608] Tallow dimethyl trimethyl propylene diammonium chloride; cationic; see Tomah Q-14-2; amber liq.; sp.gr. 0.95; pH 6–9 (5% aq.).

Tomah Q-S. [608] Soya trimethyl ammonium chloride; cationic; see Q-14-2; amber liq.; sp.gr. 0.91; pH 6–9 (5% aq.).

Tomah Tallow Amine. [608] Tallow primary amine, tech.; cationic; ore flotation agent, emulsifier, corrosion inhibitor; solid.; 100% conc.

Tomah Tallow Diamine. [608] Tallow diamine; cationic; emulsifier, corrosion inhibitor; solid; 100% conc.

Tomah Tallow Triamine. [608] Tallow triamine; cationic; emulsifier; solid; 100% conc.

Topcithin. [410] Soya lecithin, highly purified fluid; nonionic; improvement of flowing properties and physical appearance; extends shelf-life; liq.; 100% conc.

Toximul 80. [575] Alkylaryl polyethoxy ethanol, isopropanol; nonionic; wetting and spreading agent, penetrant, surfactant adjuvant used as spray additive in herbicide formulations; liq.; water-sol.

Toximul 351. [575] Ammonium alkylaryl ether sulfate, water and alcohol cosolvs.; anionic; surfactant ingred. in aq. conc. formulation of herbicides; emulsifier in solv.-based formulations; straw clear liq.; sol. in water and PCP emulsifiable formulations; sp.gr. 1.055; dens. 8.8 lb/gal; flash pt. 79 F (CC); pH 7.0 max. (10% aq.).

Tridecanol 6.5. [318] Alcohol ethoxylate; nonionic; wetting agent and detergent with oil emulsification properties; liq.; HLB 11.8; 100% conc.

Trinoram C. [117] N-coco dipropylene triamine; cationic; emulsifier for bitume; antistripping agent for roadmaking.

Trinoram S. [117] N-tallow dipropylene triamine; cationic; see Trinoram C.

Triodan. [273] Fatty acid polyglycerol ester; nonionic; food emulsifier; beads, liq.; HLB 5.5–7.4; 100% conc.

Triodan 20. [273] Fatty acid polyglycol ester; food emulsifier; visc. liq.; sapon. no. 140–155.

Triodan 55. [273] Fatty acid polyglycol ester; food emulsifier; powd.; sapon. no. 120–135.

Triodan 55/S6. [273] Fatty acid polyglycerol ester; food emulsifier; block; sapon. no. 110–130.

Triton 770 Conc. [524] Sodium alkylaryl polyether sulfate; anionic; wetting agent, penetrant used in detergents, cleaners, textile and leather processing; amber liq.; dens. 8.3 lb/gal; visc. 15 cps; flash pt. 78 F (PMCC); pour pt. –20 F; surf. tens. 28 dynes/cm (1%); 30% act.

Triton AG-98. [524] Octyl phenoxy polyethoxy ethanol; nonionic; spray adjuvant for herbicidal activity; agricultural adjuvant; liq.; 80% conc.

Triton AG-120. [524] Octylphenoxy polyethoxy ethanol; surfactant, emulsifier; wh. powd.; water-sol.; sp.gr. 0.33; pH 5–7 (5%).

Triton B. [525] Alkylpolyether alcohols acrylic emulsion; nonionic; agricultural spreader sticker; liq.; 50% conc.

Triton B-1956. [524] Phthalic glycerol alkyd resin, modified; nonionic; emulsifier, detergent, insecticide formulation; fungicide spray for fruits and vegetables; amber liq.; dens. 8.8 lb/gal; visc. 100 cps; flash pt. 74 F (Setaflash CC); pour pt. 25 F; surf. tens. 32 dynes/cm (0.01%); 77% act.

Triton CF-10. [524] Alkylaryl polyether; nonionic; detergent for mechanical dishwashing, rinse aids, laundering, metal and dairy cleaning; nylon dyeing assistant; VCS 4 liq.; dens. 8.9 lb/gal; visc. 250 cps; HLB 14.0; cloud pt. 28 C (1%); flash pt. > 500 F (TOC); pour pt. 60 F; surf. tens. 32 dynes/cm (1%); 100% act.

Triton CF-21. [524] Alkylaryl polyether; nonionic; detergent for mechanical dishwashing, final rinse additive, laundering, metal and dairy cleaning; APHA 125 liq.; dens. 8.7 lb/gal; visc. 250 cps; cloud pt. 40 C (1%); flash pt. > 480 F (TOC); pour pt. –30 F; surf.

tens. 31 dynes/cm (1%); 100% act.

Triton CF-32. [524] Amine polyglycol condensate; nonionic; wetting and antifoamer; detergent for mechanical dishwashing, metal and dairy cleaning; VCS 5 liq.; dens. 8.6 lb/gal; visc. 550 cps; cloud pt. 25 C (1%); flash pt. > 300 F (TOC); pour pt. 15 F; surf. tens. 32 dynes/cm (1%); 95% act.

Triton CF-54. [524] Polyethoxy adduct, modified; nonionic; see Triton CF-10; VCS 2 liq.; dens. 8.7 lb/gal; visc. 180 cps; HLB 13.6; cloud pt. 38 C (1%); flash pt. > 200 F (PMCC); pour pt. 35 F; surf. tens. 29 dynes/cm (0.01%); 100% act.

Triton CF-76. [524] Polyethoxy adduct, modified; nonionic; detergent; dishwashing formulations; lt. yel. liq.; dens. 8.7 lb/gal; visc. 295 cps; cloud pt. 31 C (1%); flash pt. > 300 F (TOC); pour pt. 40 F; surf. tens. 29 dynes/cm (0.01%); 100% act.

Triton CF-87. [524] Alkylaryl ether, modified; nonionic; surfactant; rinse aid in mechanical dishwashing; straw liq.; dens. 9.0 lb/gal; visc. 240 cps; cloud pt. 32 C (1%); flash pt. > 215 F (TOC); pour pt. 30 F; surf. tens. 32 dynes/cm (0.01%); 90% act.

Triton CG-110. [525] Caprylyl/capryl glucoside; nonionic; foamer for personal care products; liq.; 60% conc.

Triton CG-400, -500. [524] Stearalkonium chloride; cationic; surfactant; wh. liq.; dens. 8.0 and 7.7 lb/gal resp. (50 C); flash pt. 111 and 90 F (Seta CC); pour pt. 110 F; 25 and 50% act.

Triton CS-7. [524] Alkylaryl polyether alcohol and org. sulfonate; spreading-binding agent; used in agricultural concs. with insecticides, fungicides, and miticides; yel. liq.; dens. 8.5 lb/gal; flash pt. 98 F (Setaflash CC); pour pt. –35 F; 60% act.

Triton DF-12. [524] Polyethoxylated alcohol, modified; nonionic; see Triton CF-10; lt. straw liq.; dens. 8.6 lb/gal; visc. 60 cps; cloud pt. 17 C (1%); flash pt. 430 F (COC); pour pt. 30 F; surf. tens. 33 dynes/cm (0.01%); 100% act.

Triton DF-16. [524] Ethoxylated linear alcohol; nonionic; detergent suited for rinse additives; APHA < 200 liq.; dens. 8.2 lb/gal; visc. 35 cps; cloud pt. 36 C (1%); flash pt. > 200 F (PMCC); pour pt. 20 F; surf. tens. 28 dynes/cm (1%); 100% act.

Triton DF-18. [524] Polyethoxylated alcohol, modified; nonionic; detergent for dishwashing, cleaner for food processing equipment, spray metal cleaner; yel. liq.; dens. 8.5 lb/gal; visc. 140 cps; cloud pt. < 0 C (1%); flash pt. > 220 F (TOC); pour pt. 65 F; surf. tens. 32 dynes/cm (0.01%); biodeg.; 90% act.

Triton DN-14. [524] Alkylpolyether alcohol; nonionic; detergent, dye leveling, scouring agent and emulsifier; yel. liq.; dens. 8.8 lb/gal; visc. 400 cps; cloud pt. 90 C (1%); flash pt. > 300 F (TOC); pour pt. 55 F; surf. tens. 39 dynes/cm (1%); biodeg.; 90% act.

Triton GR-5M. [524] Dioctyl sodium sulfosuccinate; anionic, wetting and rewetting agent, emulsifier; APHA 200 liq.; dens. 8.4 lb/gal; visc. 40 cps; flash pt. 75 F (Setaflash CC); pour pt. –60 F; surf. tens. 23 dynes/cm (1%); 60% act.

Triton GR-7M. [524] Dioctyl sodium sulfosuccinate; anionic; emulsifier used in drycleaning detergents, agricultural emulsions; VCS 2 liq.; dens. 8.4 lb/gal; visc. 110 cps; flash pt. 130 F (TOC); pour pt. –70 F; 64% act.

Triton H-55. [524] Phosphate ester, potassium salt; anionic; hydrotrope used in surfactants for liq. concs.; VCS 3 liq.; dens. 11.2 lb/gal; visc. 40 cps; flash pt. 154 F (PMCC); 50% act.

Triton H-66. [524] Phosphate ester, potassium salt; anionic; see Triton H-55; amber liq.; dens. 10.5 lb/gal; visc. 120 cps; flash pt. > 200 F (PMCC); pour pt. –5 F; 50% act.

Triton N-40. [525] Nonylphenol polyethoxy ethanol; nonionic; antifogging in plasticized PVC, corrosion inhibitor; dispersant of petroleum oils; liq.

Triton N-42. [524; 525] Nonoxynol-4; nonionic; surfactant; colorless-yel. clear visc. liq.; m.w. 405; water-disp.; sp.gr. 1.068; dens. 8.5 lb/gal; visc. 250 cps; HLB 9.1; pour pt. –26 C; flash pt. 149 C (TOC); pH 7.5; surf. tens. 29 dynes/cm (0.01% aq.); 100% act.

Triton N-57. [524; 525] Nonoxynol-5; nonionic; detergent, emulsifier, intermediate; APHA 100 liq.; m.w. 440; oil-sol.; water-disp; dens. 8.5 lb/gal; visc. 240 cps; HLB 10.0; cloud pt. < 0 C (1%); flash pt. > 515 F (COC); pour pt. –25 F; surf. tens. 29 dynes/cm (1%); 100% conc.

Triton N-60. [524] Nonoxynol-6; nonionic; see Triton N-57; APHA 125 liq.; m.w. 484; dens. 8.7 lb/gal; visc. 300 cps; HLB 10.9; cloud pt. 53 C (1%); pour pt. –25 F; surf. tens. 28 dynes/cm (1%); 100% act.

Triton N-87. [524; 525] Nonoxynol-8; nonionic; detergent for soils; surfactant in laundry detergents and hard surface cleaners; emulsifier for o/w systems; pale straw clear liq.; mild odor; m.w. 594; sol. in water, aromatic and chlorinated hydrocarbons and org. solvs.; sp.gr. 1.055; dens. 8.8 lb/gal; visc. 260 cps; HLB 12.6; solid. pt. 21 F; pour pt. 1.7 C; b.p. 249 C; flash pt. > 218 C (PMCC); cloud pt. 42 C (1% aq.); pH 5–7 (5% aq.); surf. tens. 30 dynes/cm (0.1% aq.); biodeg.; 100% act.

Triton N-101. [524; 525] Nonoxynol-10; nonionic; wetting agent, detergent, dispersant, emulsifier for pesticides; household and industrial cleaners, textile processing; APHA 100 liq.; m.w. 642; dens. 8.7 lb/gal; visc. 240 cps; HLB 13.4; cloud pt. 54 C (1%); pour pt. 40 C; surf. tens. 29 dynes/cm (1%); 100% act.

Triton N-111. [524; 525] Nonoxynol-11; nonionic; wetting agent, detergent, emulsifier; textile processing; APHA 100 liq.; m.w. 704; water-sol.; dens. 8.8 lb/gal; visc. 310 cps; HLB 13.8; cloud pt. 72 C (1%); flash pt. > 300 F (TOC); pour pt. 55 F; 100% act.

Triton N-150. [524; 525] Nonoxynol-15; nonionic; detergent, emulsifier; APHA 160 liq.; m.w. 880; dens. 9.0 lb/gal; visc. 4350 cps; HLB 15.0; cloud pt. 95 C (1%); pour pt. 65 F; surf. tens. 33 dynes/cm (1%); 100% act.

Triton N-302. [524; 525] Nonylphenoxy polyethoxy ethanol; nonionic; emulsifier used in mfg. of emulsion polymers and agricultural emulsion concs.; base surfactant in household and industrial chemical specialties; APHA 250 clear visc. liq.; mild odor; m.w. 1543; sol. in water, alcohol, dichloroethane; sp.gr. 1.092; dens. 9.1 lb/gal; visc. 649 cps; HLB 17.1; flash pt. > 93 C (SFCC); cloud pt. > 100 C (1% aq.); pH 5–7 (5% aq.); biodeg.; 69–71% act.

Triton N-401. [524; 525] Nonoxoynol-40; nonionic; emulsifier; emulsion polymerization; APHA 250 liq.; m.w. 1980; dens. 9.2 lb/gal; visc. 850 cps; HLB 17.8; cloud pt. > 100 (1%); flash pt. > 230 F (Setaflash CC); pour pt. 18 F; 70% act.

Triton N-998. [524; 525] Nonoxynol-100; nonionic; emulsifier for asphalt, hydrocarbon oils and waxes; surfactant for detergents; textile scouring and processing; pale beige waxy solid; m.w. 4630; sol. in water, alcohol, dichloroethane; sp.gr. 1.07 (65.6 C); dens. 8.91 lb/gal; HLB 19.0; pour pt. 46 C; flash pt. 110 C (PMCC); cloud pt. > 100 C (1% aq.); pH 6.0–7.0 (5% aq.); biodeg.; > 99.5% act.

Triton N-998-70%. [524] Nonoxynol-100; see Triton N-998; yel. clear visc. liq.; mild odor; m.w. 4630; sol. see Triton N-998; HLB 19.0; sp.gr. 1.10; dens. 9.16 lb/gal; pour pt. 18 C; flash pt. > 500 F (TOC); pH 6.0–7.0 (5% aq.); 69–71% act.

Triton RW-10. [524] Polyethoxy amine; nonionic/cationic; surfactant and detergent used in industrial applications; emulsifier for metalworking for coolants, fiber lubricant; degreaser in annealing nylon fibers; lt. amber clear liq.; amine odor; insol. in water; sp.gr. 0.89; visc. 400 cps; pour pt. –9 C; flash pt. 129 C (SFCC); pH 10.0 (5% aq.).

Triton RW-20. [524] Polyethoxy amine; nonionic/cationic; see Triton RW-10; lt. amber clear liq.; amine odor; insol. in water; sp.gr. 0.913; dens. 7.61 lb/gal; visc. 240 cps; HLB 6–8; b.p. 73 C; flash pt. > 167 C (SFCC); pH 9.5–11.0 (5% aq.).

Triton RW-50. [524] Polyethoxy amine; nonionic/cationic; surfactant; lt. amber clear liq.; amine odor; sp.gr. 0.968; dens. 8.06 lb/gal; HLB 12–14; pour pt. –38 F; flash pt. 149 C (CC); cloud pt. < 0 C (1% aq.); pH 10.0–11.5 (5% aq.); surf. tens. 32 dynes/cm (0.01% aq.); pH 10.

Triton RW-75. [524] Polyethoxy amine; nonionic/cationic; see Triton RW 10; lt. amber clear liq.; amine odor; sp.gr. 0.981; dens. 8.17 lb/gal; HLB 14–16; flash pt. > 149 C (CC); cloud pt. 32 C (1% aq.); pH 9.5–11.0 (5% aq.).

Triton RW-100. [524] Polyethoxy amine (10 EO); nonionic/cationic; lt. amber clear liq.; amine odor; sp.gr. 1.004; dens. 8.36 lb/gal; visc. 180 cps; HLB 16; pour pt. –25 F; flash pt. 149 C (CC); cloud pt. 67 C (1% aq.); pH 9.5–11.0 (5% aq.); surf. tens. 39 dynes/cm (0.01% aq.).

Triton RW-150. [524] Polyethoxy amine (15 EO); nonionic/cationic; see Triton RW-10; lt. amber clear liq.; amine odor; sp.gr. 1.024; dens. 8.58 lb/gal; visc. 480 cps; HLB > 16; pour pt. –8 F; flash pt. > 149 C (CC); cloud pt. 96 C (1% aq.); pH 9.5–11.0 (5% aq.); surf. tens. 46 dynes/cm (0.01% aq.)

Triton W-30 Conc. [524] Sodium alkylaryl ether sulfate; anionic; penetrant, wetter for textiles; emulsifier in emulsion polymerization; amber liq.; dens. 8.2 lb/gal; visc. 9 cps; flash pt. 74 F (PMCC); pour pt. –15 F; surf. tens. 29 dynes/cm (1%); 27% act.

Triton X-15. [524] Octoxynol-1; nonionic; coupling agent, emulsifier; APHA 250 liq.; m.w. 250; dens. 8.2 lb/gal; visc. 790 cps; HLB 3.6; flash pt. > 300 F (TOC); pour pt. 15 F; 100% act.

Triton X-35. [524] Octoxynol-3; nonionic; see Triton X-15; APHA 125 liq.; m.w. 338; dens. 8.5 lb/gal; visc. 370 cps; HLB 7.8; flash pt. > 300 F (TOC); pour pt. –10 F; surf. tens. 29 dynes/cm (0.01%); 100% act.

Triton X-45. [524] Octoxynol-5; nonionic; drycleaning detergent, dispersant, insecticide emulsifier, wetting agent; APHA 100 liq.; m.w. 426; dens. 8.7 lb/gal; visc. 290 cps; HLB 10.4; cloud pt. < 0 C (1%); flash pt. > 300 F (TOC); pour pt. –15 F; surf. tens. 28 dynes/cm (1%); 100% act.

Triton X-67. [524] Alkylpolyether alcohol; nonionic; surfactant; wh. solid; HLB 16.8; cloud pt. > 100 C (1%); surf. tens. 43 dynes/cm (1%); 100% act.

Triton X-100. [524] Octoxynol-9; nonionic; wetting agent, dispersant, detergent, household and industrial cleaners; textile processing, wool scouring; emulsifier for insecticides and herbicides; APHA 100 liq.; m.w. 628; dens. 8.9 lb/gal; visc. 240 cps; HLB 13.5; cloud pt. 65 C (1%); flash pt. > 300 F (TOC); pour pt. 45 F; surf. tens. 30 dynes/cm (1%); 100% act.

Triton X-102. [524] Octoxynol-13; nonionic; detergent, wetting agent; metal cleaning, industrial, household liq. detergents and cleaners; APHA 100 liq.; m.w. 756; dens. 8.9 lb/gal; visc. 330 cps; HLB 14.6; cloud pt. 88 C (1%); flash pt. > 300 F (TOC); pour pt. 60 F; surf. tens. 32 dynes/cm (1%); 100% act.

Triton X-114. [524] Octoxynol-8; nonionic; detergent, wetting agent used in household and industrial laundry uses; APHA 100 liq.; m.w. 536; dens. 8.8 lb/gal; visc. 260 cps; HLB 12.4; cloud pt. 22 C (1%); flash pt. > 300 F (TOC); pour pt. 15 F; surf. tens. 29 dynes/cm (1%); 100% act.

Triton X-114SB. [524] Octoxynol-8; nonionic; see Triton X-114; APHA 100 liq.; m.w. 536; dens. 8.2 lb/gal; HLB 12.4; flash pt. 61 F (Setaflash CC); pour pt. –5 F; 80% act.

Triton X-120. [524] Octoxynol-9 (9–10 EO); nonionic; wetting agent and dispersant; agricultural wettable powds.; wh. solid; m.w. 628; 40% act.

Triton X-151. [524] Alkylaryl polyether alcohol and org. sulfonate; anionic; surfactant, emulsifier used in insecticides; liq.; dens. 8.7 lb/gal; visc. 80 cps; flash pt. 56 F (Setaflash CC); pour pt. 10 F; surf. tens. 37 dynes/cm (0.01%); 100% act.

Triton X-152. [524] Alkylaryl polyether alcohol and org. sulfonate; anionic; surfactant; VCS 11 liq.; dens. 8.5 lb/gal; visc. 155 cps; flash pt. 56 F (Setaflash CC); pour pt. 35 F; 100% act.

Triton X-155. [524] Methylenebisdiamyl phenoxy polyethoxy ethanol; nonionic; surfactant in emulsification of aromatic hydrocarbons; wetting agent and detergent; antistat and lubricant for textile fiber processing; amber clear visc. liq.; mild odor; sol. in water and aromatic hydrocarbons; sp.gr. 1.052; dens. 8.76 lb/gal; visc. 790 cps; HLB 12.5; pour pt. 0 F; flash pt. (TOC) 430 F; cloud pt. < 10 C (1% aq.); pH 7.0–9.0 (1%); surf. tens. 30 dynes/cm (0.1%); 100% act.

Triton X-155-90%. [524] Methylenebisdiamyl phenoxy polyethoxy ethanol; nonionic; see Triton X-155; yel.-amber clear liq.; mild odor; sol. see Triton X-155; sp.gr. 1.02; dens. 8.50 lb/gal; visc. 250 cps; HLB 12.5; pour pt. –20 F; flash pt. (SFCC); 53 C; cloud pt. < 10 C (1% aq.); pH 8.50 (5%); surf. tens. 32 dynes/cm (0.01% aq.); 90% act.

Triton X-161. [524] Alkylaryl polyether alcohol and org. sulfonate; anionic; surfactant; VCS 8–10 liq.; dens. 8.7 lb/gal; visc. 175 cps; flash pt. 185 F (TOC); pour pt. –45 F; 100% act.

Triton X-165. [524] Octoxynol-16; nonionic; detergent, emulsifier, wetting agent; surfactant; APHA 125 liq.; m.w. 910; dens. 9.0 lb/gal; visc. 540 cps; HLB 15.8; cloud pt. > 100 C (1%); flash pt. > 300 F (TOC); pour pt. 55 F; surf. tens. 34 dynes/cm (1%); 70% act.

Triton X-171. [524] Alkylaryl polyether alcohol and org. sulfonate; see Triton X-151; liq.; dens. 8.7 lb/gal; visc. 270 cps; flash pt. 44 (Setaflash CC); pour pt. –20 F; 100% act.

Triton X-172. [524] Alkylaryl polyether alcohol and org. sulfonate; anionic; surfactant; VCS 14 liq.; dens. 8.7 lb/gal; visc. 6640 cps; flash pt. 71 F (Setaflash CC); pour pt. 5 F; 100% act.

Triton X-180. [524] Alkylaryl polyether alcohol and org. sulfonate; anionic; emulsifier for pesticide formulations; hazy amber liq.; dens. 8.6 lb/gal; visc. 130 cps; flash pt. 54 F (Setaflash CC); pour pt. 40 F; 100% act.

Triton X-185. [524] Alkylaryl polyether alcohol and org. sulfonate; anionic; see Triton X-

180; clear amber liq.; dens. 8.7 lb/gal; visc. 465 cps; flash pt. 34 F (Setaflash CC); pour pt. 40 F; 100% act.

Triton X-185A. [524] Alkylaryl polyether alcohol and org. sulfonate; anionic; surfactant; lt. br. solid; dens. 8.8 lb/gal; flash pt. 88 F (Setaflash CC); pour pt. 45 F; 100% act.

Triton X-190. [524] Alkylaryl polyether alcohol and org. sulfonate; anionic; surfactant; clear amber liq.; dens. 8.8 lb/gal; visc. 715 cps; flash pt. 46 F (Setaflash CC); pour pt. –15 F; 100% act.

Triton X-193. [524] Alkylaryl polyether alcohol and org. sulfonate; anionic; emulsifier for pesticide formulations; hazy amber liq.; dens. 8.7 lb/gal; visc. 740 cps; flash pt. 74 F (Setaflash CC); pour pt. 60 F; 100% act.

Triton X-200. [524] Sodium octoxynol-2 ethane sulfonate; anionic; detergent, metal cleaning, pickling and plating baths; household cleaners; polymerization emulsifier and latex post-stabilizer; wetting agent in strong alkaline baths; dispersant for fulling in lime soap; dye leveling agent for acid dyestuffs; wh. liq.; dens. 8.9 lb/gal; visc. 7000 cps; flash pt. > 300 F (TOC); pour pt. 25 F; surf. tens. 29 dynes/cm (1%); 28% act.

Triton X-202. [524] Sodium alkylaryl polyether sulfonate; anionic; detergent, wetting agent, emulsifier; metal cleaning, pickling and plating baths; household cleaners; latex post stabilizer; wh. liq.; dens. 8.9 lb/gal; visc. 4800 cps; flash pt. > 212 F (TOC); pour pt. 45 F; surf. tens. 30 dynes/cm (1%); 30% act.

Triton X-207. [524] Alkylaryl polyether alcohol; nonionic; emulsifier for agricultural sprays and industrial oils; VCS 3 liq.; dens. 8.2 lb/gal; visc. 600 cps; HLB 10.7; flash pt. > 200 F (PMCC); pour pt. –10 F; 100% act.

Triton X-300. [524] Alkylaryl polyether alcohol and org. sulfonate; anionic; emulsifier/adjuvant for use in oil-surfactant concs.; pale yel. clear liq.; sol. in paraffin-based petroleum oils; sp.gr. 0.982 (75 F); dens. 8.18 lb/gal; visc. 110 cps; pour pt. 25 F; flash pt. (PMCC) 162 F; 80% act.

Triton X-301. [524] Sodium alkylaryl polyether sulfate; anionic; wetting agent, penetrant; detergent for household and industrial cleaners; cream liq.; dens. 8.8 lb/gal; visc. 4200 cps; flash pt. > 300 F (TOC); pour pt. 30 F; surf. tens. 28 dynes/cm (1%); 20% act.

Triton X-305. [524] Octoxynol-30; nonionic; surfactant; APHA 150 liq.; m.w. 1526; dens. 9.1 lb/gal; visc. 470 cps; HLB 17.3; cloud pt. > 100 C (1%); flash pt. > 300 F (TOC); pour pt. 35 F; surf. tens. 41 dynes/cm (1%); 70% act.

Triton X-363M. [524] Alkylaryl polyether alcohol; nonionic; emulsifier used in agricultural and industrial oil sprays; food contact uses; lt. straw liq.; dens. 8.2 lb/gal; visc. 100 cps; HLB 9.1; flash pt. > 198 F (Setaflash CC); pour pt. –25 F; 100% act.

Triton X-400. [524] Stearyl dimethyl benzyl ammonium chloride; cationic; surfactant; cream-wh. liq.; dens. 8.0 lb/gal; flash pt. > 212 F (TOC); surf. tens. 34 dynes/cm (1%); 25% act.

Triton X-405. [524] Octoxynol-40; nonionic; surfactant; APHA 250 liq.; m.w. 1966; dens. 9.2 lb/gal; visc. 490 cps; HLB 17.9; cloud pt. > 100 C (1%); flash pt. > 212 F (TOC); pour pt. 25 F; surf. tens. 44 dynes/cm (1%); 70% act.

Triton X-700. [524] Alkylaryl polyether alcohol and org. sulfonate; anionic; surfactant; liq.; VCS 8–10 liq.; dens. 8.5 lb/gal; visc. 175 cps; flash pt. 58 F (Setaflash CC); pour pt. 45 F; 100% act.

Triton X-705. [524] Octoxynol-70; nonionic; surfactant; wh. solid; m.w. 3286; dens. 9.4 lb/gal; HLB 18.7; cloud pt. > 100 C (1%); flash pt. > 500 (COC); surf. tens. 41 dynes/cm (1%); 100% act.

Triton X-705-50%. [524] Octoxynol-70; nonionic; surfactant; APHA 200 liq.; m.w. 3286; dens. 9.0 lb/gal; visc. 625 cps; HLB 18.7; cloud pt. > 100 C (1%); flash pt. > 212 F (TOC); pour pt. 5 F; surf. tens. 41 dynes/cm (1%); 50% act.

Triton X-800. [524] Alkylaryl polyether alcohol and org. sulfonate; anionic; surfactant; VCS 8–11 liq.; dens. 8.8 lb/gal; visc. 1010 cps; flash pt. 56 F (Setaflash CC); pour pt. –20 F; 100% act.

Triumphnetzer ZSN. [660] Diisooctyl sulfosuccinate; anionic; wetting agent for textile and chem-tech products; liq.; 75% conc.

Troykyd Lecithin W.D. [611] Lecithin; pigment dispersant for aq. systems; visc. stabilizer; used in paint processing; amber visc. liq.; water-disp.; sp.gr. 1.02–1.05 (80 F); dens. 8.5–8.8 lb/gal (80 F); Gardner visc. Z1–Z4; pH 7.3–8.2 (5% disp.); 95% solids.

Trycol 5878. [217] Ethoxylated fatty alcohol ethers; nonionic; emulsifier, lubricant for textiles and detergent shampoo base; solubilizer; personal care products; liq.; HLB 9.5; 100% conc.

Trycol 5940, 5941, 5942. [217] Ethoxylated tridecyl ether; nonionic; foam builder and detergent for detergent systems; liq.; HLB 11.4, 13.0, and 13.8 resp.; 100% conc.

Trycol 5943, 5946. [217] Ethoxylated tridecyl ether; nonionic; emulsifier, detergent, wetting agent used in paper and textile industries; leveling agent, solubilizer; liq.; HLB 14.5 and 16.0 resp.; 100% conc.

Trycol 5950. [217] Ethoxylated aliphatic alcohol; nonionic; wetting agent and penetrant for textile applications; liq.; water disp.; HLB 10.5; 100% conc.

Trycol 5951. [217] Alkyl POE glycol; nonionic; penetrant, wetting agent and emulsifier for heavy duty cleaners; Gardner 1 liq.; dens. 8.2 lb/gal; visc. 23 cs; HLB 11.6; cloud < 25 C; 100% act.

Trycol 5952. [217] Ethoxylated aliphatic alcohol; nonionic; textile wetting agent; liq.; water-sol.; HLB 12.5; 97% conc.

Trycol 5964. [217] Ethoxylated fatty alcohol ethers; nonionic; emulsifier, lubricant used in textiles and detergent shampoo bases; solubilizer in personal care products; solid; HLB 14.0; 100% conc.

Trycol 5967. [217] Ethoxylated fatty alcohol ethers; nonionic; see Trycol 5964; solid; HLB 16.9; 100% conc.

Trycol 5971. [217] Ethoxylated fatty alcohol; nonionic; see Trycol 5964; liq.; HLB 15.6; 60% conc.

Trycol 5972. [217] Ethoxylated oleyl alcohol; nonionic; emulsifier, lubricant used in textiles, cosmetics and processing of animal fibers; solid; HLB 15.9; 100% conc.

Trycol 6720 (LF-1). [217] Alkyl polyether; nonionic; detergent and wetting agent; liq.; 100% conc.

Trycol 6961. [217] Ethoxylated nonylphenol; nonionic; detergent and emulsifier; chemical intermediate; liq.; oil-sol.; HLB 8.9; 100% conc.

Trycol 6964. [217] Ethoxylated nonylphenol; nonionic; household and industrial cleaning formulations; liq.; HLB 13.0; 100% conc.

Trycol 6965. [217] Ethoxylated nonylphenol; nonionic; for high temps.; tolerant of builders; liq.; HLB 13.5; 100% conc.

Trycol 6979. [217] Alkylaryl POE glycol; nonionic; emulsifier, detergent; surfactant; gel; oil-sol.; water-disp.; HLB 9.8; 100% conc.

Trycol CE. [217] Ethoxylated alcohol; emulsifier for min. oils and dye carriers; Gardner 1 liq.; disp. in water; dens. 7.8 lb/gal; visc. 23 cSt (100 F); HLB 8.7; pour pt. –7 C; flash pt. 340 F; cloud pt. < 25 C.

Trycol DA-4. [217] POE (4) decyl alcohol; nonionic; wetting agent, penetrant for yarn and fabrics in dyeing systems and resin padbath applications; intermediate in the production of anionics; Gardner < 1 liq.; sol. in glycerol trioleate; disp. in water; dens. 7.9 lb/gal; visc. 30 cs; HLB 10.5; cloud pt. < 25 C; 100% act.

Trycol DA-6. [217] POE (6) decyl alcohol; nonionic; wetting agent in clay soils, fire fighting; penetrant and emulsifier in repellent applications; Gardner 1 liq.; dens. 8.3 lb/gal; visc. 15 cs; HLB 12.5; cloud pt. 40 C; 100% act.

Trycol DA-69. [217] POE (6) decyl alcohol; nonionic; see Trycol DA-6; Gardner 1 liq.; dens. 8.3 lb/gal; visc. 28 cSt (100 F); HLB 12.4; pour pt. –15 C; flash pt. 315 F; cloud pt. 42 C; 90% act.

Trycol DNP-8. [217] POE (8) dinonyl phenol; foam control agent; emulsifier for polar and nonpolar solvs. and textile jet dye carrier applications; spreading agent in pigment printing; post-stabilizer in emulsion polymerization; intermediate in mfg. of anionics; surfactant component in acidic cleaners, aerosols, insecticides and wax emulsions; Gardner 2 liq.; disp. in water; dens. 8.4 lb/gal; visc. 410 cs; HLB 10.4; cloud pt. < 25 C; 100% act.

Trycol DNP-150. [217] POE (150) dinonyl phenyl ether; dispersant and wetting agent for pesticides; emulsifier, detergent and leveling agent in textile industry; controlled-subs component in detergents and cleaners; Gardner 1 solid, flakes; m.p. 55 C; HLB 19.0; cloud pt. > 100 C; 100% act.

Trycol DNP-150/50. [217] POE (150) dinonyl phenyl ether; see Trycol DNP-150; Gardner < 1 liq.; dens. 9.0 lb/gal; HLB 19.0; cloud pt. > 100 C; 50% act.

Trycol LAL-4. [217] POE (4) lauryl alcohol; nonionic; intermediate; emulsifier for oils and fats in industrial lubricants, textile coning oils; coemulsifier for silicone in cleaner polishes and mold release agents; Gardner 1 liq.; disp. in water; dens. 7.7 lb/gal; visc. 35 cs; HLB 9.2; cloud pt. < 25 C; biodeg.; 100% act.

Trycol LAL-8. [217] POE (8) lauryl alcohol; nonionic; detergent, emulsifier, wetting agent; Gardner 1 liq.; dens. 8.2 lb/gal; HLB

12.6; cloud pt. 62 C; biodeg.; 100% act.

Trycol LAL-12. [217] POE (12) lauryl alcohol; nonionic; see Trycol LAL 8; Gardner 1 solid; m.p. 32 C; HLB 14.0; cloud pt. 90 C; biodeg.; 100% act.

Trycol LAL-23. [217] POE (23) lauryl ether; nonionic; solubilizer; coemulsifier for silicone in cleaner polishes and mold release agents; solv. emulsifier for textile dye carriers; Gardner < 1 solid; m.p. 46 C; HLB 16.9; cloud pt. > 100 C; 100% act.

Trycol NP-1. [217] POE (1) nonyl phenyl ether; nonionic; defoamer; surfactant and coemulsifier; Gardner 1 liq.; dens. 8.3 lb/gal; visc. 460 cs; HLB 4.6; cloud pt. < 25 C; 100% act.

Trycol NP-4. [217] POE (4) nonyl phenyl ether; nonionic; coemulsifier often used as a corrosion inhibitor in two cycle engine oils; Gardner 1 liq.; dens. 8.5 lb/gal; visc. 300 cs; HLB 8.9; cloud pt. < 25 C; 100% act.

Trycol NP-6. [217] POE (6) nonyl phenyl ether; dispersant and wetting agent; coemulsifier in acid cleaning sol'ns, solv. emulsions and detergents; Gardner 2 liq.; disp. in water; dens. 8.5 lb/gal; visc. 100 cSt (100 F); HLB 10.9; pour pt. –10 C; flash pt. 515 F; cloud pt. < 25 C.

Trycol NP-7. [217] POE (7) nonyl phenol; nonionic; wetting agent in solv. emulsions, cleaners and detergents; Gardner 1 liq.; disp. in water; dens. 8.6 lb/gal; visc. 325 cs; HLB 12.3; cloud pt. 20 C; 100% act.

Trycol NP-9. [217] POE (9) nonyl phenyl ether; nonionic; detergent, wetting agent, emulsifier; Gardner 1 liq.; dens. 8.8 lb/gal; visc. 325 cs; HLB 12.9; cloud pt. 55 C; 100% act.

Trycol NP-11. [217] POE (11) nonyl phenyl ether; surfactant used in formulation for detergents and emulsification; Gardner 2 liq.; dens. 8.7 lb/gal; visc. 116 cSt (100 F); HLB 13.5; pour pt. 5 C; flash pt. 480 F; cloud pt. 71 C.

Trycol NP-20. [217] POE (20) nonyl phenyl ether; surfactant used as detergent, wetting agent and emulsifier for emulsion polymerization; Gardner 1 solid; HLB 16.0; m.p. 34 C; flash pt. 515 F; cloud pt. 88 C.

Trycol NP-30. [217] POE (30) nonyl phenol; nonionic; detergent, wetting agent; emulsifier for fats, oils, and waxes and emulsion polymerization; Gardner 1 solid; m.p. 40 C; HLB 17.1; cloud pt. > 100 C; 100% act.

Trycol NP-40. [217] POE (40) nonyl phenol; detergent; wetting agent in sol'ns. of electrolytes; stabilizer for syn. fabrics; coemulsifier for fats, waxes and oils; Gardner 1 solid; HLB 17.8; m.p. 40 C; flash pt. 560 F; cloud pt. 90 C.

Trycol NP-50. [217] POE (50) nonyl phenol; see Trycol NP-40; Gardner 1 solid; HLB 18.2; m.p. 54 C; flash pt. 520 F; cloud pt. 76 C.

Trycol NP-307. [217] POE (30) nonyl phenol; detergent and wetting agent; emulsifier in emulsion polymerization and for fats, oils and waxes; Gardner 1 liq.; dens. 9.0 lb/gal; visc. 260 cSt (100 F); HLB 17.1; pour pt. 5 C; cloud pt. 92 C.

Trycol NP-407. [217] POE (40) nonyl phenol; see Trycol NP-40; Gardner 1 liq.; dens. 9.2 lb/gal; visc. 385 cSt (100 F); HLB 17.8; pour pt. 7 C; cloud pt. 90 C.

Trycol NP-507. [217] POE (50) nonyl phenol; see Trycol NP-50; Gardner 1 liq.; dens. 9.0 lb/gal; visc. 440 cSt (100 F); HLB 18.2; pour pt. –4 C; cloud pt. 76 C.

Trycol NP-1007. [217] POE (100) nonyl phenol; surfactant used as wetting agent in high electrolyte sol'ns.; stabilizer in syn. latices; Gardner 1 liq.; dens. 9.2 lb/gal; HLB 19.0; pour pt. 18 C; cloud pt. 72 C; 70% act.

Trycol OAL-23. [217] POE (23) oleyl alcohol; nonionic; dispersant, solubilizer, detergent, stabilizer and anticoagulant for latices and dye pastes; dyeing assistant for wool/acrylic blends; detergent and lubricant for fiber/fabric scouring; emulsifier for waxes used in coating citrus fruits; Gardner 2 solid; m.p. 40 C; HLB 15.9; cloud pt. > 100 C; 100% act.

Trycol OP-407. [217] POE (40) octyl phenyl ether; hydrophilic dispersant and wetting agent; primary emulsifier in emulsion of polymerization of acrylic and vinyl monomers; Gardner 1 liq.; dens. 9.0 lb/gal; visc. 220 cSt (100 F); HLB 17.9; pour pt. 13 C; cloud pt. 74 C.

Trycol SAL-20. [217] POE (20) stearyl ether; nonionic; emulsifier and solubilizer used in dyeing assistants and textile dye carriers; stabilizer in latices; wax emulsifier in coatings for citrus fruits; Gardner 1 solid; HLB 15.3; m.p. 40 C; flash pt. 560 F; cloud pt. 91 C.

Trycol TDA-3. [217] POE (3) tridecyl ether; nonionic; emulsifier and antifoam in textile formulations; intermediate for production of anionics by sulfation and phosphation; Gardner < 1 liq.; disp. in water; dens. 7.8 lb/gal; visc. 30 cs; HLB 8.0; cloud pt. < 25 C; 100% act.

Trycol TDA-6. [217] POE (6) tridecyl ether; nonionic; dispersant and wetting agent, emulsifier and detergent in degreasers and cutting oils; wool scouring; intermediate for production of surfactants by sulfation or phosphation; Gardner < 1 liq.; disp. in water; dens. 8.2 lb/gal; visc. 50 cs; HLB 11.4; cloud pt. < 25 C; 100% act.

Trycol TDA-8. [217] POE (8) tridecyl alcohol; nonionic; coemulsifier, foam builder and solubilizer for alkylaryl sulfonates, essential oils, aromatic solvs., waxes and fats; Gardner < 1 liq.; dens. 8.3 lb/gal; visc. 65 cs; HLB 12.5; cloud pt. 44 C; 100% act.

Trycol TDA-9. [217] POE (9) tridecyl ether; nonionic; wetting agent, surfactant used with alkylaryl sulfonates in cleaners; rewetter for paper towels; raw wool detergent; coemulsifier for aromatic and aliphatic solvs.; Gardner < 1 liq.; dens. 8.3 lb/gal; visc. 75 cs; HLB 13.2; cloud pt. 55 C; 100% act.

Trycol TDA-11. [217] POE (11) tridecyl ether; nonionic; detergent; lt. and heavy duty cleaning formulations; Gardner < 1 liq.; dens. 8.4 lb/gal; HLB 13.7; cloud pt. 74 C; 100% act.

Trycol TDA-12. [217] POE (12) tridecyl alcohol; nonionic; hydrophilic surfactant for lt. and heavy duty cleaning formulations; Gardner 1 semisolid; dens. 8.4 lb/gal; HLB 14.5; m.p. 16 C; flash pt. 440 F; cloud pt. 70 C.

Trycol TDA-18. [217] POE (18) tridecyl alcohol; nonionic; dispersant, solubilizer, emulsifier in cleaners; intermediate in mfg. of anionic surfactants; textile acid dye leveling agent and wet processing detergent; foam stabilizer; Gardner < 1 solid; m.p. 39 C; HLB 16.0; cloud pt. > 100 C; 100% act.

Trycol TP-2. [217] POE (2) tridecyl phenol; nonionic; coemulsifier, foam reducing agent; Gardner 6 liq.; dens. 8.1 lb/gal; visc. 600 cs; HLB 4.6; cloud pt. < 25 C; 100% act.

Trycol TP-6. [217] POE (6) tridecyl phenol; nonionic; dispersant, detergent, emulsifier in metal cleaning, cutting oil and drycleaning formulations, emulsifiable solv. cleaners; Gardner 6 liq.; disp. in water; dens. 8.5 lb/gal; visc. 300 cs; HLB 9.9; cloud pt. < 25 C; 100% act.

Trydet 19. [217] Ethoxylated mixed rosin and fatty acids, anhyd.; nonionic; see Trydet 20; Gardner 12 liq.; dens. 8.9 lb/gal; visc. 200 cSt (100 F); HLB 13.4; pour pt. 15 C; flash pt. 570 F; cloud pt. 50 C.

Trydet 20. [217] Ethoxylated mixed esters of rosin and fatty acids; nonionic; detergent and emulsifier; textile leveling agent in dyeing applications; coemulsifier for xylene, kerosene, trichlorobenzene, o-dichlorobenzene; dens. 9.0 lb/gal; visc. 480 cs; HLB 13.4; cloud pt. > 100 C; 96% act.

Trydet 22. [217] Ethoxylated tall oil fatty acids, nonionic; surfactant used in formulating controlled foaming detergents for commercial laundries, textile scouring, metal and hard surface cleaners and degreasers; Gardner 12 liq.; dens. 9.0 lb/gal; visc. 196 cSt (100 F); HLB 12.1; pour pt. 20 C; flash pt. 530 F; cloud pt. 35 C.

Trydet DO-9. [217] POE (9) dioleate; nonionic; lipophilic emulsifier and solubilizer for min. oils, fats and solvs.; emulsifier for kerosene in agric. and pesticides sprays; emulsification of latex paints, metalworking fluids, solvs., specialty and industrial lubricants; Gardner 3 liq.; disp. in water; dens. 8.1 lb/gal; visc. 45 cSt (100 F); HLB 8.8; pour pt. –6 C; flash pt. 515 F; cloud pt. < 25 C.

Trydet ISA-4. [217] POE (4) isostearic acid; nonionic; lubricant used in textile fiber processing, conc. fabric softeners; Gardner 2 liq.; dens. 8.2 lb/gal; visc. 100 cs; HLB 7.7; cloud pt. < 25 C; 100% act.

Trydet ISA-9. [217] POE (9) isostearic acid; nonionic; see Trydet ISA-4: Gardner 1 liq.; dens. 8.5 lb/gal; visc. 125 cs; HLB 11.7; cloud pt. < 25 C; 100% act.

Trydet ISA-14. [217] POE (14) isostearic acid; nonionic; surfactant used in fiber lubricants and processing aids; emulsifier in finish formulations; Gardner 5 liq.; disp. in water; dens. 8.6 lb/gal; visc. 92 cSt (100 F); HLB 13.0; pour pt. 10 C; flash pt. 490 F; cloud pt. < 25 C

Trydet LA-5. [217] POE (5) laurate; nonionic; emulsifier and coupling agent; defoamer in water base coatings; visc. depressant in vinyl plastic sols.; control additive in hair rinse formulations; paper softener; Gardner liq.; dens. 8.2 lb/gal; visc. 123 cSt (100 F); HLB 9.3; pour pt. 9 C; flash pt. 445 F; cloud pt. < 25 C.

Trydet LA-7. [217] POE (7) lauric acid; emulsifier, nonionic; lubricant component and scrooping agent for textile fibers and yarns; visc. control agent for plastisols; Gardner 2 liq.; dens. 8.4 lb/gal; visc. 55 cs; HLB 11.8; cloud pt. < 25 C; 100% act.

Trydet MP-9. [217] POE (9) pelargonic acid;

nonionic; surfactant as a base lubricant for syn. fiber finishes; coemulsifier and coupling agent; Gardner 1 liq.; dens. 8.7 lb/gal; visc. 34 cSt (100 F); HLB 14.3; pour pt. 5 C; flash pt. 440 F; cloud pt. 37 C.

Trydet OA-5. [217] POE (5) oleic acid; nonionic; emulsifier for min., cutting, fatty oils and solvs.; solv. emulsifier in metal cleaners and degreasers; w/o emulsifier for pesticide control; textile processing; softener and lubricant leather during tanning; Gardner 3 liq.; disp. in water; dens. 8.1 lb/gal; visc. 55 cs; HLB 8.2; cloud pt. < 25 C; 100% act.

Trydet OA-7. [217] POE (7) oleic acid; nonionic; see Trydet OA-5; Gardner 3 liq.; dens. 8.3 lb/gal; visc. 160 cs; HLB 10.2; cloud pt. < 25 C; 100% act.

Trydet OA-10. [217] POE (10) oleic acid; nonionic; lubricant, emulsifier for solvs. in pesticide carriers, metal cleaners and neatsfoot oil in leather fat liquoring; textile specialty detergent and dyeing assistant; rewetting agent for paper; Gardner 2 liq.; dens. 8.5 lb/gal; visc. 110 cs; HLB 12.1; cloud pt < 25 C; 100% act.

Trydet SA-5. [217] POE (5) stearic acid; nonionic; emulsifier for min. oils and fats; lubricant in leather industry; napping softener in textile industry; min. oil and lard oil emulsions for polish and metal buffing compds.; Gardner 1 solid; m.p. 35 C; HLB 8.7; cloud pt. < 25 C; 100% act.

Trydet SA-7. [217] POE (7) stearic acid; nonionic; waxy emulsifier for oils and fats in industrial lubricants; softener and lubricant to textiles and leather; Gardner < 1 solid; m.p. 30 C; HLB 11.1; cloud pt. < 25 C; 100% act.

Trydet SA-8. [217] POE (8) stearic acid; nonionic; lubricant, softener and scrooping agent for textile and industrial applications; stabilizer in starch sol'ns.; Gardner < 1 solid; m.p. 30 C; HLB 11.1; cloud pt. < 25 C; 100% act.

Trydet SA-23. [217] POE (23) stearic acid; nonionic; emulsifier for textile lubricants and softeners; emulsifier and thickener in personal care products and pharmaceuticals; antigellant in starch sol'ns.; Gardner 1 solid; m.p. 45 C; HLB 15.6; cloud pt. > 100 C; 100% act.

Trydet SA-40. [217] POE (40) stearic acid; nonionic; emulsifier for glycerol monostearate and waxy esters in production of conc. textile lubricants and softeners; stabilizer and antigellant for starch sol'ns.; emulsifier for cosmetics and pharmaceuticals; Gardner < 1 solid, flakes; m.p. 50 C; HLB 17.3; cloud pt. < 25 C; 100% act.

Trydet SA-50/30. [217] POE (50) stearic acid; nonionic; hydrophilic emulsifier for preparing solubilized oils; visc. modifier, softener or plasticizer in acrylic or vinyl resin emulsions; Gardner 1 liq.; dens. 8.5 lb/gal; visc. 671 cSt (100 F); HLB 17.8; pour pt. 0 C; flash pt. 540 F; cloud pt. 81 C.

Trydet SO-9. [217] POE (9) sesquioleate; nonionic; oil and fat emulsifier in industrial and textile lubricants; Gardner 3 liq.; dens. 8.2 lb/gal; visc. 50 cSt (100 F); HLB 9.4; pour pt. –6 C; flash pt. 550 F; cloud pt. < 25 C.

Tryfac 525-K. [217] Phosphate ester; detergent and wetting agent for built liq. detergents; antistat in fiber finishes; Gardner 2 liq.; dens. 8.8 lb/gal; visc. 560 cSt (100 F); pour pt. 18 C; flash pt. 435 F; cloud pt. 76 C.

Tryfac 610-K. [217] Phosphate ester potassium salt; anionic; wetting agent, dispersant, emulsifier for chlorinated hydrocarbons in cleaners, detergents and dye carriers; component in metalworking compds.; textile emulsifier and detergent in solv. scouring; scouring agent for cotton goods, antistat in processing oils for fibers; corrosion inhibitor; Gardner < 1 liq.; dens. 9.2 lb/gal; visc. 700 cs; 88% act.

Tryfac 910-K. [217] Phosphate ester potassium salt; anionic; detergent and wetting agent in textile desizing; polymer emulsification stabilizer; Gardner 1 liq.; dens. 9.4 lb/gal; visc. 480 cs; 90% act.

Tryfac 5365, 5553, 5554. [217] Phosphate ester; anionic; detergent, wetting and antistat, emulsifier; liq.; 90–100% conc.

Tryfac 5571. [217] Ethoxylated alcohol phosphate ester potassium salt; detergent, dispersant and wetting agent used in fabric preparation, hard surface cleaning and antistat protection of fibers; Gardner 1 liq.; dens. 9.2 lb/gal; visc. 650 cSt (100 F); pour pt. < –15 C; cloud pt. 45 C.

Tryfac 5576. [217] Ethoxylated alcohol phosphate ester potassium salt; hydrophilic surfactant used in electrolytes and caustic sol'ns.; antistat, anticorrosive and dispersant; Gardner 1 liq.; dens. 9.5 lb/gal; visc. 190 cSt (100 F); pour pt. < –10 C; cloud pt. 49–54 C.

Trylon 6751. [217] Fatty ester ethoxylate; coemulsifier for solvs., oils and esters; fiber-to-metal lubricant; Gardner 3 liq.; disp. in water; dens. 8.4 lb/gal; visc. 79 cSt (100 F); pour pt. 10 C; flash pt. 530 F; cloud pt. < 25 C.

Trylox 1086. [217] Ethoxylated sorbitol hexaoleate; nonionic; o/w emulsifier for petroleum and veg. oils, and solvs.; used in metal lubricants, textile processing aids; emulsifier component for aliphatic and aromatic solvs. in textile dye carriers; Gardner 6 liq.; disp. in water; dens. 8.4 lb/gal; visc. 235 cs; HLB 10.2; cloud pt. < 25 C; 100% act.

Trylox 1186. [217] Ethoxylated sorbitol hexatallate; see Trylox 1086; Gardner 9 liq.; disp. in water; dens. 8.5 lb/gal; visc. 330 cs; HLB 10.9; cloud pt. < 25 C; 100% act.

Trylox 5900. [217] Ethoxylated castor oil; nonionic; emulsifier, lubricant; liq. to solid; HLB 4.9; 100% conc.

Trylox 6746, 6753. [217] Ethoxylated sorbitol and hexaoleate esters of sorbitol; nonionic; emulsifier for textile and cosmetic use; liq.; HLB 10.2; 100% conc.

Trylox 6747. [217] Ethoxylated sorbitol hexaoleate; nonionic; emulsifier for org. solvs.; used for paraffin oils in industrial lubricants; Gardner 4 liq.; disp. in water; dens. 8.4 lb/gal; visc. 300 cs; HLB 11.4; cloud pt. < 25 C; 100% act.

Trylox CO-5. [217] POE (5) castor oil; emulsifier, nonionic; coemulsifier for chlorinated and aromatic solvs.; dispersant for pigment slurries in water-based paint and clay; carrier for paper coatings; textile foam control agent and emulsifier in dye carriers; Gardner 4 liq.; disp. in water; dens. 8.2 lb/gal; visc. 800 cs; HLB 3.8; cloud pt. < 25 C; 100% act.

Trylox CO-10. [217] POE (10) castor oil; nonionic; emulsifier for w/o emulsions; Gardner 4 liq.; oil-sol.; dens. 8.3 lb/gal; visc. 332 cSt (100 F); HLB 6.4; pour pt. –5 C; flash pt. 560 F.

Trylox CO-16. [217] POE (16) castor oil; nonionic; emulsifier, lubricant, coemulsifier for metalworking oils, hydraulic fluids, rayon delustrants and fiber lubricants; component in textile leveling and dispersant for vat and naphthol dyes; Gardner 3 liq.; dens. 8.5 lb/gal; visc. 600 cs; HLB 8.6; cloud pt. < 25 C; 100% act.

Trylox CO-25. [217] POE (25) castor oil; nonionic; emulsifier and lubricant in formulation of sol. oils, cutting fluids and fiber finishes; Gardner 4 liq.; dens. 8.6 lb/gal; visc. 396 cSt (100 F); HLB 10.8; pour pt. –5 C; flash pt. 565 F; cloud pt. 66 C.

Trylox CO-30. [217] POE (30) castor oil; nonionic; emulsifier, degreaser, dispersant, visc. stabilizer, lubricant for oils, solvs., waxes, syn. fiber lubricants, pigments, fat liquoring, paints, urethane foams and polyester resins; coemulsifier in fabric softener and dye carrier systems; dyeing assistant; Gardner 2 liq.; dens. 8.8 lb/gal; visc. 720 cs; HLB 11.7; cloud pt. > 100 C; 100% act.

Trylox CO-36. [217] POE (36) castor oil; nonionic; emulsifier for solvs. and oils; lubricant and softener for textiles and leather; coemulsifier with anionic emulsifier in formulating textile dye carriers; Gardner 2 liq.; dens. 8.8 lb/gal; visc. 700 cs; HLB 12.6; cloud pt > 100 C; 100% act.

Trylox CO-40. [217] POE (40) castor oil; see Trylox CO-36; Gardner 2 liq.; dens. 8.8 lb/gal; visc. 650 cs; HLB 12.9; cloud pt. > 100 C; 100% act.

Trylox CO-80. [217] POE (80) castor oil; nonionic; o/w emulsifier and solubilizer; used in fiber finish and textile lubricant formulations; Gardner 2 liq.; dens. 8.9 lb/gal; visc. 322 cSt (100 F); HLB 15.9; pour pt. 20 C; cloud pt. 77 C.

Trylox CO-200, -200/50. [217] POE (200) castor oil; nonionic; lubricant, textile fiber processing as antistat humectant and scrooping agent; emulsifier in blended lubricants for fibers and yarns; Gardner 1 solid and Gardner < 1 liq. resp.; dens. 8.6 lb/gal; m.p. 40 C; HLB 18.1; cloud pt. > 100 C; 100 and 50% act.

Trylox HCO-5. [217] POE (5) hydrog. castor oil; nonionic; emulsifier, coemulsifier for syn. esters used in the textile industry; fiber lubricant and softener; Gardner 1 liq.; disp. in water; dens. 7.9 lb/gal; visc. 1200 cs; HLB 3.8; cloud pt. < 25 C; 100% act.

Trylox HCO-16. [217] POE (16) hydrog. castor oil; nonionic; emulsifier for castor oil; lubricant and softener in fabric softeners and aerosol fabric sprays; Gardner 1 liq.; low odor; dens. 8.4 lb/gal; visc. 2100 cs; HLB 8.2; cloud pt. < 25 C; 100% act.

Trylox HCO-25. [217] POE (25) hydrog. castor oil; nonionic; emulsifier for textile resin finishing and softener lubricant systems; coemulsifier for lanolin and dye carrier systems; Gardner 1 liq.; disp. in water, dens. 8.6 lb/gal; visc. 1100 cs; HLB 10.8; cloud pt. 25 C; 100% act.

Trylox HCO-200/50. [217] POE (200) hydrog.

castor oil; nonionic; emulsifier, lubricant; Gardner < 1 clear liq.; dens. 8.8 lb/gal; HLB 18.1; cloud pt. > 100 C; 50% act.

Trylox SS-20. [217] POE (20) sorbitol; nonionic; humectant, plasticizer; intermediate in the synthesis of fatty acid esters; Gardner 1 liq.; dens. 9.7 lb/gal; visc. 400 cs; HLB 16.1; cloud pt. > 100 C; 100% act.

Trymeen 6601, 6602. [217] Coco amine POE deriv.; cationic; textile antistat, emulsifier, lubricant, dispersant, softener; substantive to metals, fibers, and clays; liq.; HLB 13.6 and 15.2 resp.; 100% conc.

Trymeen 6606. [217] Ethoxylated tallow amine; cationic; emulsifier; liq.; HLB 14.3; 100% conc.

Trymeen 6607, 6610, 6617. [217] Ethoxylated tallow amine; cationic; emulsifier; liq.; HLB 15.4, 17.4, and 17.8 resp.; 100% conc.

Trymeen 6640. [217] POE (15) N-tallow 1,3 propylene diamine; cationic; dispersant in acidic sol'ns.; retarder of acidic, disperse and premetalized dyestuffs; Gardner 18 liq.; dens. 8.4 lb/gal; visc. 120 cSt (100 F); HLB 13.1; pour pt. < –10 C; flash pt. 440 F; cloud pt. 86 C.

Trymeen CAM-10. [217] POE (10) coconut amine; cationic; coemulsifier and antistat for textile processing; dispersant for inorg. salts in viscose spinning; emulsifier for fats and oils in industrial lubricants; Gardner 8 liq.; dens. 8.4 lb/gal; visc. 260 cs; HLB 13.8; cloud pt. > 100 C; 100% act.

Trymeen CAM-15. [217] POE (15) coconut amine; cationic; emulsifier, antistat, softener; substantive to clay, metals, minerals and textiles; additive in the mfg. of rayon fiber; prevents sludge build-up in spin baths; textile processing; industrial lubricant; Gardner 6 liq.; dens. 8.6 lb/gal; visc. 120 cs; HLB 15.4; cloud pt. > 100 C; 100% act.

Trymeen OAM-30/80. [217] POE (30) oleyl amine; cationic; emulsifier; antiprecipitant in cross dyeing; stripping agent and dye leveler for acid dyes; Gardner 6 liq.; dens. 8.9 lb/gal; HLB 15.3; cloud pt. > 100 C; 80% act.

Trymeen SAM-50. [217] POE (50) stearyl amine; cationic; antistat, emulsifier in metal buffing compds.; lubricant for fiberglass; leveling agent for dye applications; Gardner 4 solid; m.p. 50 C; HLB 17.8; cloud pt. > 100 C; 100% act.

Trymeen TAM-8. [217] POE (8) tallow amine; cationic; emulsifier for min. and veg. oils, waxes, solvs., polyethylene textile softeners, fats and oils in industrial lubricants; solv. emulsifier for dye carriers; Gardner 8 liq.; dens. 8.2 lb/gal; visc. 200 cs; HLB 11.5; cloud pt. 80 C; 100% act.

Trymeen TAM-15. [217] POE (15) tallow amine; cationic; emulsifier; antiprecipitant for mixed dye baths; leveling agent for acid dyes; migrating agent for dispersed dyes; intermediate for quat. ammonium compds.; antistat for processing syn. fibers; Gardner 8 liq.; dens. 8.5 lb/gal; visc. 175 cs; HLB 14.2; cloud pt. > 100 C; 100% act.

Trymeen TAM-20. [217] POE (20) tallow amine; cationic; textile antistat, coemulsifier and lubricant for wool and syn. fiber processing; antiprecipitant, leveling and migrating agent in dyeing; Gardner 6 liq.; dens. 8.7 lb/gal; visc. 240 cs; HLB 15.4; cloud pt. > 100 C; 100% act.

Trymeen TAM-25. [217] POE (25) tallow amine; see Trymeen TAM-20; Gardner 6 liq.; dens. 8.8 lb/gal; visc. 170 cs; HLB 16.1; cloud pt. > 100 C; 100% act.

Trymeen TAM-40. [217] POE (40) tallow amine; cationic; emulsifier, antistat additive for commercial carpet maintenance; dyeing assistant and stabilizer for latices; Gardner 6 solid; m.p. 40 C; HLB 17.4; cloud pt. > 100 C; 100% act.

TS-33-F. [287] Sorbitan tristearate (sorbate 65); nonionic; emulsifier for w/o cosmetics and pharmaceuticals; stabilizing textile printing inks; silicone antifoamer, polish mfg., wax working; flakes; HLB 2.0; 100% conc.

TS-55-F. [287] POE (20) sorbitan tristearate (polysorbate 65); nonionic; emulsifier o/w for cosmetics and pharmaceuticals, paint industry, silicone antifoam and paraffin emulsion, polishes; solid; HLB 10.5; 100% conc.

T.S.S. [283] Trisodium sulfosuccinate; anionic; anticaking agent in powd. detergents; compatibilizer for detergent blends; liq. 42% act.

Tween 20. [71] Polysorbate 20; anionic; solubilizer; o/w emulsifier; antistat and fiber lubricant used in textile industry; pale yel. liq.; sol. in water, methanol, ethanol, IPA, propylene glycol, ethylene glycol, cottonseed oil; sp.gr. 1.1; visc. 400 cps; HLB 16.7; flash pt. > 300 F; sapon. no. 40–50; 100% act.

Tween 20 SD. [71] POE sorbitan monolaurate; anionic; see Tween 20; liq.; 100% act.

Tween 21. [320] Polysorbate 21; anionic; see Tween 20; yel. oily liq.; sol. in corn oil, dioxane, Cellosolve, carbon tetrachloride, methanol, ethanol, aniline; disp. in water; sp.gr. 1.1; visc. 500 cps; HLB 13.3; flash pt. > 300 F; sapon. no. 100–115; 100% act.

Tween 40. [320] Polysorbate 40; anionic; see Tween 20; pale yel. liq.; sol. in water, methanol, ethanol, IPA, ethylene glycol, cottonseed oil; sp.gr. 1.08; HLB 15.6; flash pt. > 300 F; sapon. no. 41–52; 100% act.

Tween 60. [320] Polysorbate 60; anionic; see Tween 20; pale yel. liq.; sol. in water, IPA, ethyl alcohol; sp.gr. 1.1; visc. 600 cps; HLB 14.9; flash pt. > 300 F; sapon. no. 45–55; 100% act.

Tween 61. [320] Polysorbate 61; anionic; tan waxy solid; sol. in methanol, ethanol; disp. in water; sp.gr. 1.06; HLB 9.6; flash pt. > 300 F; pour pt. 100 F; sapon. no. 95–115; 100% act.

Tween 65. [320] Polysorbate 65; anionic; see Tween 20; tan waxy solid; sol. in IPA, ethyl alcohol, veg. and min. oil; sp.gr. 1.05; HLB 10.5; flash pt. > 300 F; pour pt. 92 F; sapon. no. 88–98; 100% act.

Tween 80. [320] Polysorbate 80; anionic; see Tween 20; pale yel. liq.; sol. in water, IPA, ethanol; sp.gr. 1.08; visc. 425 cps; HLB 15.0; flash pt. > 300 F; sapon. no. 45–55; 100% act.

Tween 81. [320] Polysorbate 81; anionic; see Tween 20; amber oily liq.; sol. in min. and corn oil, dioxane, Cellosolve, methanol, ethanol, ethyl acetate, aniline; disp. in water; sp.gr. 1; visc. 450 cps; HLB 10.0 flash pt. > 300 F; sapon. no. 96–104; 100% act.

Tween 85. [320] Polysorbate 85; anionic; see Tween 20; amber liq.; sol. in veg. oil, Cellosolve, lower alcohols, aromatic solvs., ethyl acetate, min. oils and spirits, acetone, dioxane, carbon tetrachloride and ethylene glycol; disp. in water; sp.gr. 1.0; visc. 300 cps; HLB 11.0; flash pt. > 300 F; sapon. no. 80–95; 100% act.

Twitchell 6805 Oil. [217] Sulfonated min. oil; anionic; rewetting agent, softener and textile fiber lubricant, deduster; liq.; 100% conc.

Twitchell 6808 Oil. [217] Sulfated fatty min. deriv.; anionic; lubricant, wetting agent, and dyestuff deduster; liq.; 100% conc.

Twitchell 6809 Oil. [217] Mineral fatty derivatives; nonionic; wetting, lubricant, dyestuff deduster; liq.; 100% conc.

Twix 20TO. [86] Sorbitol trioleate ethoxylate; emulsifier; Gardner 12 liq.; disp. in water; sp.gr. 1.02; visc. 320 cps; cloud pt. < 25 C (1% aq.); 100% act.

Tylorol 2S, 2S/25. [610] Sodium lauryl ether (2) sulfate; anionic; detergent base for personal care products; liq.; 28 and 25% conc. resp.

Tylorol 3S. [610] Sodium lauryl ether (3) sulfate; anionic; detergent base for liq. shampoos and foam baths; liq.; 28% conc.

Tylorol A. [610] Ammonium lauryl ether (2) sulfate; anionic; detergent base for personal care products; liq.; 24% conc.

Tylorol BS. [610] Sodium lauryl ether (2) sulfate; anionic; detergent base for shampoos; liq.; 22% conc.

Tylorol LM. [610] Monoethanolamine lauryl ether (1) sulfate; anionic; see Tylorol BS; liq.; 30% conc.

Tylorol LT50. [610] Triethanolamine lauryl ether (1) sulfate; anionic; see Tylorol BS; liq.; 42% conc.

Tylorol MG. [610] Magnesium lauryl ether (3) sulfate; anionic; detergent base for the cosmetic industry; liq.; 30% conc.

Tylorol S. [610] Sodium lauryl ether (1) sulfate; anionic; see Tylorol BS; liq.; 25% conc.

Tylose C, CB Series. [301] Sodium carboxymethyl cellulose; anionic; binder in pencil leads; thickener in batteries, rubber industry, cosmetics, foodstuffs, pharmaceuticals, tobacco and textile industry; dispersant, emulsifier for insecticidal, fungicidal and herbicidal products; plasticizer in ceramics; surface sizing in paper industry; press aid and lubricant in welding electrodes; granules, powd.

U

Ucar Super Wetter FP. [616] Org. silicon; nonionic; wetting agent, penetrant, coemulsifier, lubricant, dispersant; liq.; sol./disp. in water; 100% conc.

Udet 950. [181] Linear alkylaryl sulfonate; anionic; used for detergent powds.; powd.; 95% conc.

Ufanon K-80. [613] 2:1 Coconut diethanolamide; nonionic; visc. modifier, foam booster; liq.; 48–56% conc.

Ufanon KD-S. [613] 1:1 Coconut diethanolamide; nonionic; see Ufanon K-80; liq.; 90% conc.

Ufaryl DB80. [613] Sodium dodecylbenzene sulfonate, branched; anionic; detergent powd. emulsifier; powd.; 79–82% conc.

Ufaryl DL85. [613] Sodium alkylbenzene sulfonate linear; anionic; see Ufaryl DB80; powd.; 83–87% conc.

Ufasan 35. [613] Linear sodium alkylbenzene sulfonate; anionic; liq. detergents; liq.; 34–36% conc.

Ufasan 65. [613] Linear sodium alkylbenzene sulfonate; anionic; liq.; detergents; paste; 64–66% conc.

Ufasan Tea. [613] Triethanolamine alkylbenzene sulfonate, linear; anionic; liq. detergents; liq.; 49–51% conc.

Ultra Blend 60, 100. [650] Linear alkylaryl sulfonate, alkanolamide and ethoxylated adduct; anionic; detergent and car washing conc.; liq.; biodeg.; 60 and 100% conc. resp.

Ultra Blend P-280. [650] Linear alkylaryl sulfonate, alcohol ether sulfate and ethoxylated adduct; anionic; detergent, high-foaming conc.; biodeg.; liq.

Ultra NCS Liquid. [650] Ammonium cumene sulfonate; anionic; hydrotrope, coupling agent, solubilizer; cloud point depressant for liq. detergents; antiblocking agent for powd. detergents; liq.; 40% conc.

Ultra NXS Liquid. [650] Ammonium xylene sulfonate; see Ultra NCS Liquid; liq.; 45% conc.

Ultra SCS Liquid. [650] Sodium cumene sulfonate; see Ultra NCS Liquid; liq.; 40% conc.

Ultra SXS Liquid, Powder. [650] Sodium xylene sulfonate; see Ultra NCS Liquid; 40% liq.; 90% powd.

Ultraphos 11, 12, 21A, 41. [654] Ethoxylated fatty alcohol phosphoric ester; anionic; metal working additive; detergent; liq.; 100% conc.

Ultra Sulfate AE-3. [650] Ammonium alcohol ether sulfate; anionic; wetting agent, penetrant, lubricant, emulsifier, dye dispersant, scouring aid, antistat; base for personal care products; detergent for specialty household and industrial cleaners; textile surfactant; liq.; 58% conc.

Ultra Sulfate SE-5. [650] Sodium alcohol ether sulfate; see Ultra Sulfate AE-3; liq.; 58% conc.

Ultra Sulfate SL-1. [650] Sodium lauryl sulfate; see Ultra Sulfate AE-3; also used as surfactant for emulsion polymerization and latex stabilization; liq.; 30% conc.

Ultratex ESB, WK. [137] Polysiloxane, modified; cationic; used in textiles; liq.

Unamide C-2. [372] PEG-3 cocamide; nonionic; foam booster, stabilizer, visc. builder, emulsifier; used in personal care, household and industrial products; wh. solid; sapon. no. 15 max.; pH 9–10 (5%); 100% act.

Unamide C-5. [372] PEG-6 cocamide; nonionic; see Unamide C-2; lt. yel. liq.; sapon. no. 12 max.; pH 9–10 (5%); 100% act.

Unamide C-72-3. [372] Coco diethanolamide; nonionic; visc. builder, foam stabilizer; used for lt. duty liqs., industrial, household hard surface cleaners, surfactant, emulsifier, corrosion inhibitor, lubricant and personal care products; amber liq.; pH 9–10.5 (5%).

Unamide C-7649. [372] Coco diethanolamide, modified; anionic/nonionic; detergent base for floor cleaners; foam stabilizer, emulsifier; liq.; 100% conc.

Unamide C-7944. [372] Coco diethanolamide, modified; anionic/nonionic; industrial cleaners, floor strippers and degreasers; liq.; 100% conc.

Unamide CDX. [372] Cocamide DEA; nonionic; foam booster, stabilizer, visc. builder used in personal care products; straw liq.; water-sol.; dens. 8.2 lb/gal; sp.gr. 0.99; 100% act.

Unamide CMX. [372] 1:1 Coco monoethanolamide; nonionic; see Unamide CDX; lt. solid; pH 9–10.5 (5%); 100% act.

Unamide D-10, GC-75. [372] 2:1 Coco diethanolamide, modified; nonionic; see Unamide C-72-3; dk. amber and yel. resp. liqs.; pH 9–10.5 (5%); 100% act.

Unamide J-56. [372] Lauramide DEA; nonionic; see Unamide C-72-3; amber liq.; pH 9–10.5 (5%); 100% act.

Unamide JJ-35. [372] Cocamide DEA; nonionic; see Unamide CDX; lt. straw liq.; pH 9–10.5 (5%); 100% act.

Unamide L-2. [372] PEG-3 lauramide; nonionic; see Unamide C-2; wh. solid; sapon. no. 15 max.; pH 9–10 (5%); 100% act.

Unamide L-5. [372] PEG-6 lauramide; nonionic; see Unamide C-2; wh. soft wax, pH 9–10 (5%); 100% act.

Unamide LDL. [372] Cocamide DEA; nonionic; see Unamide C-72-3; lt. straw liq.; dens. 8.2 lb/gal; sp.gr. 0.98; pH 9–10.5 (5%); biodeg.; 100% act.

Unamide LDX. [372] Lauramide DEA; nonionic; see Unamide CDX; also used in lt. duty detergents; wh. solid; water-sol.; dens. 8.2 lb/gal; sp.gr. 0.99; pH 9–10.5 (5%); biodeg.; 100% act.

Unamide LMDX. [372] Lauramide DEA; nonionic; see Unamide LDX; wh. solid; water-sol.; dens. 8.2 lb/gal; sp.gr. 0.99; pH 9–10.5 (5%); 100% act.

Unamide N-72-3. [372] Cocamide DEA; nonionic; see Unamide C-72-3; yel. liq.; pH 9–10.5 (5%); 100% act.

Unamide S. [372] Stearamide DEA; nonionic; thickener and opacifier in personal care products; visc. builder, gelling agent, lubricant, industrial and household products; textile lubricant and finishing agent; wh. waxy solid; 100% act.

Unamide SI. [372] 1:1 Stearic (ethyl amino hydroxyethyl) amide; nonionic; visc. builder, gelling agent, opacifier used in household and industrial products; textile lubricant and finishing agent; wh. waxy solid.

Unamide W. [372] 1:1 Stearic diethanolamide; see Unamide SI; nonionic; amber liq.

Unamine-T. [372] Tall oil hydroxyethyl imidazoline; cationic; surfactant; w/o emulsifier, corrosion inhibitor, filming agent, de-watering compd.; flushing agent for pigments; liq.; 92% conc.

Ungerol AM3-60. [613] Ammonium lauryl ether sulfate (3 EO); anionic; detergent, shampoos; paste; 58–62% conc.

Ungerol CG27. [613] Sodium lauryl ether sulfate; anionic; liq. detergents and shampoos; anionic; liq.; 26–28% conc.

Ungerol LES 2-70. [613] Sodium lauryl ether sulfate; anionic; shampoos; visc. liq.; 68–70% conc.

Ungerol LES 3-28. [613] Sodium lauryl ether sulfate (3 EO); anionic; liq. detergents, shampoos; liq.; 26–28% conc.

Ungerol LES 3-54. [613] Sodium lauryl ether sulfate; anionic; liq. detergents, shampoos; paste; 53–55% conc.

Ungerol LES 3-70. [613] Sodium lauryl ether sulfate (3 EO); anionic; liq. detergents, shampoos; paste; 70% conc.

Ungerol LS, LSN. [613] Sodium lauryl sulfate; anionic; carpet shampoos, furniture cleaners, polymerization emulsifier; paste; 29–41% conc.

Union Carbide L-720, -721. [616] Organosilicone; nonionic; stabilizer, lubricant, antifoamer, mold release and wetting agent, penetrant; stabilizer, lubricant, antifoamer, mold release and wetting agent; penetrant; solid; water-sol.; 100% conc.

Union Carbide L-722, -727. [616] Organosilicone; nonionic; see Union Carbide L-720; liq.; sol. in alcohol and hydrocarbon; 100% conc.

Union Carbide L-7001, -7002. [616] Organosilicone; nonionic; wetting agent, penetrant, coemulsifier, lubricant, dispersant; liq.; water sol. and disp.; 100% conc.

Union Carbide L-7500. [616] Organosilicone; nonionic; penetrant and dispersant for pigments; liq.; oil-sol.; 100% conc.

Union Carbide L-7600, -7602, –7604. [616] Organosilicone; nonionic; see Union Carbide L-7001; liq.; water sol. and disp.; 100% conc.

Union Carbide L-7605. [616] Organosilicone; nonionic; see Union Carbide L-7001; solid; water sol. and disp.; 100% conc.

Union Carbide L-7607. [616] Organosilicone; nonionic; see Union Carbide L-7001; liq.; water sol. and disp. 100% conc.

Uniperol AC. [82] Oxyethylated fatty amine; cationic; dyeing aux., leveling agent; used in the textile industry; lt. br. clear visc. liq.; water-sol.

Uniperol AC Highly Conc. [82] Oxyethylated fatty amine; cationic; see Uniperol AC; br. visc. oil; water-sol.

Uniperol O. [82] Fatty alcohol-polyglycol ether; nonionic; emulsifier, dyeing aux., leveling and wetting agent, dispersant, detergent used in textile processing; colorless liq.; water-sol.; 20% act.

Uniperol O Micropearl. [82] Fatty alcohol-polyglycol ether; nonionic; see Uniperol O; off-wh. powd.

Unisol UA 20, 80, 150, 300, 400. [317] Ethoxylated tallow amine, nonionic; emulsifier, bitumen dispersant, antistat, dyebath leveling aid; paste, solid; 100% conc.

Usolan. [159] Ethoxylated lanolin; nonionic; solid

V

Valdet 561. [627] Nonylphenol ethoxylate; nonionic; detergent; desizing cotton and rayon; wetting and dispersant in dyeing; scouring fabrics; resin finish penetrant; emulsifier for wool lubricants; APHA 150 clear to slightly hazy liq.; aromatic odor; sol. in water; dens. 8.7 lb/gal; sp.gr. 1.055–1.065; cloud pt. 158–165 (1%); flash pt. 427–445 F; pH 7.0 ± 1.0 (10%); 99% min. solids.

Valdet 4016. [627] Alcohol ethoxylate, straight chain; nonionic; wetting agent; nonrewetting surfactant and emulsion stabilizer used in fluorocarbon finishes; solv. emulsion scouring operations; emulsifier; APHA 60 max. clear to slightly hazy liq.; water-sol.; sp.gr. 1.011; cloud pt. 39–45 C (1%); pH 7.0 ± 0.5 (10%); 90% act.

Valdet AC-40. [627] Amine condensate; anionic; detergent for textile wet processing; cleaning agent for machinery, floors and fixtures; amber clear liq.; misc. with water; dens. 8.3 lb/gal; sp.gr. 1.010–1.030; pH 11.0 ± 1; 40% act.

Valdet CC. [627] Amine condensate; anionic; see Valdet AC-40; dk. amber clear liq.; coconut oil odor; misc. with water; dens. 8.3 lb/gal; sp.gr. 0.990–1.010; 100% act.

Valscour SWR-2. [627] Phosphate ester; anionic; emulsifier used in textiles; liq.; 23% conc.

Value-1209C. [393] POE mono coco-fatty acid ester; nonionic; emulsifier for general use; solid, liq.; water-sol.; HLB 13.4; 100% conc.

Value 1407. [393] POE monooleate; nonionic; emulsifier for general use; liq.; HLB 10.3; water-disp.; 100% conc.

Value 1414. [393] POE monooleate; nonionic; emulsifier for general use; solid, liq.; HLB 13.6; 100% conc.

Value 3608. [393] POE octyl phenyl ether; nonionic; emulsifier for jute batching emulsion; liq.; HLB 12.6; 100% conc.

Value-3706. [393] POE nonyl phenyl ether; nonionic; dispersant, emulsifier, wetting agent, detergent; liq.; water-disp.; HLB 10.9; 100% conc.

Value-3710. [393] POE nonyl phenyl ether; nonionic; emulsifier, textile scouring and wetting agent; liq.; water-sol.; HLB 13.3; 100% conc.

Value-MS-1. [393] Syn. sperm oil sodium sulfate; anionic; softener, emulsifier for veg., animal oil and wax; paste, liq.; 75% conc.

Valwet 092. [627] Arylalkyl sulfonate; anionic; wetting agent, penetrant; liq.; 35% conc.

Vanade. [462] Sorbitan monostearate and polysorbate 60; emulsifier blend used in the food industry; lt. tan soft, plastic solid; slight propionate odor; 32% sorbitan monostearate.

Vanall. [462] Sorbitan monostearate, mono and diglycerides, polysorbate 60; emulsifier used in the food industry; wh. creamy, soft plastic paste; sapon. no. 45–55; pH 4.5–5.0; 70% max. volatiles.

Vanease. [462] Polysorbate 80, hydrog. soybean oil, glyceryl lacto palmitate, and sodium carboxymethyl cellulose; emulsifier used in food industry; wh. creamy soft plastic; 38–43% solids.

Van Gel B. [631] Magnesium aluminum silicate; thickener and visc. stabilizer for dispersions; off-wh. small flakes; dens. 2.60±0.03 mg/m³; visc. 300–900 cps (4%).

Vantoc CL. [316] Quat. ammonium; cationic surfactant and biocide.

Varamide A-2. [559] Coconut diethanolamide, refined; nonionic; detergent; 100% act.

Varamide A-7. [559] Oleic diethanolamide; nonionic; detergent; 100% act.

Varamide A-10. [559] Coco diethanolamide, modified; nonionic; detergent used in visc. floor cleaners; liq.; 100% act.

Varamide A-12, -83. [559] Coco diethanolamide, modified; nonionic; detergent for cleaners; liq.; 100% act.

Varamide DO 280. [559] Modified diethanolamide; anticorrosive lubricant; liq.; 100% conc.

Varamide L-1. [559] Lauramide DEA and diethanolamine; nonionic; detergent; thickener and foam booster, stabilizer for shampoos; paste; 100% act.

Varamide MA-1. [559] Coco superamide; nonionic; detergent, foam stabilizer and booster, thickener; used in dishwashing compds. and shampoos; Gardner 3+ liq.; readily disp. in wqter; dens. 8.2 lb/gal; pH 9.8 (1%); 100% act.

Varamide MA-4. [559] Cocamide DEA; nonionic; detergent, foam booster and stabilizer, thickener used in dishwashing detergents and shampoos; liq.; 100% conc.

Varamide ML-1. [559] Lauric superamide; see Varamide MA-1; wh. wax; water-disp.; dens. 8.1 lb/gal; pH 10.2 (1%); 100% act.

Varamide ML-4. [559] Lauramide DEA; nonionic; see Varamide MA-1; wh. wax; dens. 8.2 lb/gal; pH 10.2 (1%); 100% act.

Varamide SL-9. [559] Lauramide DEA and propylene glycol; nonionic; detergent, foam stabilizer and booster used in shampoos, detergents, laundry products and hard surface cleaners; Gardner 2 clear liq.; dens. 8.3 lb/gal; solid. pt. 84 F; pH 12.5; surf. tens. 33.0 dynes/cm.

Varamide SL-910. [559] Lauric-palmitic superamide; see Varamide SL-9; liq. 100% act.

Varamide T-55. [559] Ethoxylated monoethanolamide; surfactant for floor and hard-surface cleaners; liq.; 100% conc.

Varine C. [559] Coco hydroxyethyl imidazoline; cationic; emulsifier; shampoo base, penetrating oils, antistats, corrosion inhibitors, paints, printing inks, textiles; Gardner 9 liq.; m.w. 274; disp. in water; sol. in polar solvs. and hydrocarbons; m.p. 31 C; 100% act.

Varine O. [559] Oleyl imidazoline; cationic; see Varine C; Gardner 8 liq.; m.w. 352; sol. see Varine C; m.p. –10 C; 100% act.

Varine T. [559] Tall oil hydroxyethyl imidazoline; see Varine C; Gardner 8; m.w. 350; sol. in polar solvs. and hydrocarbons; disp. in water; m.p. –10 C; 85% tert. amine.

Varion 2C. [559] N-B-hydroxyethyl-N-B carboxy methyl coconut fatty acid amido-ethylamine; amphoteric; surfactant; Gardner 5 liq.; pH 8.0; 50% solids.

Varion 2L. [559] N-B hydroxyethyl-N-B carboxy methyl fatty acid amido ethylamine; amphoteric; surfactant; Gardner 4 liq.; pH 8.0; 50% solids.

Varion AMK-SF. [559] Salt-free amphoteric; surfactant for heavy-duty cleaners, steam cleaners, metal cleaners; liq.; 40% conc.

Varion AM-R40. [559] Ricinoleamidopropyl betaine; amphoteric; baby and child cosmetic formulations; liq.; 40% conc.

Varion AM-V. [559] Amphoteric glycine deriv.; low-foam wetting agent in alkaline systems; liq.; 35% conc.

Varion CADG. [559] Cocamidopropyl betaine; amphoteric; detergent, foamer, conditioner, solublizer used in shampoo products; liq.; 32% act.

Varion CADG-HS, CADG-LS. [559] Cocamidopropyl dimethyl acetyl betaine; amphoteric; surfactant; detergent; foam booster and visc. modifier; personal care products and heavy-duty detergents; yel. clear liq.; pH 6.5–8.5 and 5.0–7.0 resp.; 35–37% solids.

Varion CAS. [559] Cocamido 2-hydroxypropyl dimethyl sulfobetaine; amphoteric; emulsifier and foamer for drilling operations, personal care products, detergents, car care formulations; Gardner 4 max. clear liq.; water-sol.; flash pt. > 200 F (PMCC); pH 6–8; surf. tens. 30.0 dynes/cm (0.1%); 38% min. act.

Varion CDG. [559] Lauryl betaine; amphoteric; foamer, detergent, solubilizer, conditioner used in shampoos; liq.; 31% act.

Varion CDS. [559] Coco sulfobetaine; foamer for alkaline floods and saturated brine sol'ns.; surfactant for oil recovery; liq.; flash pt. > 200 F (CC); 35% conc.

Varion SDG. [559] Stearyl betaine; amphoteric; detergent, softener for fabric and hair applications; liq.; 50% conc.

Varion TEG. [559] Tallow dihydroxyethyl betaine; amphoteric; foamer in personal care products; thickener in household and industrial detergents; Gardner 7 clear liq.; pH 4.1–5.3 (5% aq.); 41–44% act.

Variquat 50MC. [559] Alkyl dimethylbenzyl ammonium chloride; cationic; germicide, algicide, disinfectant, deodorant; used in pesticides and mfg. of sanitizers; food processing, dairy, restaurant, industrial and household products; liq.; 50% act.

Variquat 50ME. [559] Dimethyl alkyl (C_{12}–C_{16}) benzyl ammonium chloride; specialty quat., germicide; liq.; Gardner 2 max. liq.; m.w. 358; sp.gr. 0.96; dens. 8.0 lb/gal; flash pt. 120 F (P-M); pH 5–8; 51–52% quat.

Variquat 60LC. [559] Alkyl dimethylbenzyl ammonium chloride; cationic; algicide used in swimming pools and water treatment; liq.; 50% act.

Variquat 80MC. [559] Alkyl dimethylbenzyl ammonium chloride; cationic; germicide, disinfectant, deodorant used in pesticides and sanitizers; liq.; 80% act.

Variquat 80ME. [559] Dimethyl alkyl (C_{12}–C_{16}) benzyl ammonium chloride; specialty quat.; liq.; Gardner 2 max.; m.w. 358; flash pt. 100 F (P-M); 81.5–82.5% quat.

Variquat 190. [559] Distearyl dimethyl ammonium methyl sulfate; specialty quat.; paste; Gardner 4 max.; flash pt. 120 F (P-M); 74–76% solids.

Variquat 473. [559] Oleyl imidazoline methyl sulfate; emulsifier for oil-based drilling fluids; liq.; flash pt. 66 F (CC); 75% conc.

Variquat 638. [559] Methyl bis (2-hydroxyethyl) coco ammonium chloride; cationic; specialty quat. for degreasers, antistats; liq.; Gardner 9 max.; m.w. 353; flash pt. 56 F (PM); 74–75% quat.

Variquat A200. [559] Alkyl trimethyl ammonium chloride; cationic; emulsifier; liq.; 50% act.

Variquat B200. [559] Benzyltrimethyl ammonium chloride; cationic; dispersant, dye leveler and retarder, emulsifier used in textile industry; APHA 50 max. liq.; pH 6.5–8.0; 60% act.

Variquat B343. [559] Dihydrog. tallow methyl benzyl ammonium chloride; specialty quat.; Gardner 2 max. paste; m.w. 656; flash pt. 84 F (PM); 74–76% quat.

Variquat B345. [559] Dimethyl alkyl (C_{14}–C_{18}) benzyl ammonium chloride; specialty quat.; liq.; Gardner 3 max.; m.w. 417; flash pt. 81 F (PM); 75–77% solids.

Variquat E228. [559] Cetrimonium chloride; cationic; softener, conditioner used in hair applications; liq.

Variquat E290. [559] Cetyl trimethyl ammonium chloride; cationic; quat. for personal care products; Gardner 1 max. clear liq.; 28.5–30.0% in water.

Variquat K300. [559] Dimethyl dicoco ammonium chloride; cationic; emulsifier, dispersant; Gardner 5 max. liq.; m.w. 439; flash pt. 50 F (PM); 75% conc.

Variquat K375. [559] Dicoco dimethyl ammonium chloride in hexylene glycol; specialty quat.; liq.; Gardner 5 max.; m.w. 439; flash pt. > 200 F (P-M); 74–77% quat.

Variquat K-1215. [559] Methyl bis (polyethoxy ethanol) alkyl ammonium chloride; specialty quat.; Gardner 11 max. liq.; m.w. 910; flash pt. 258 F (PM); 95% quat. min.

Variquat LC80. [559] Alkyl dimethylbenzyl ammonium chloride; cationic; algicide for use in swimming pools and industrial water treatment; liq. 80% act.

Varisoft 110. [559] Complex quat. sulfate; cationic; fabric softener conc. for home and commercial laundries and textile processing; paste; 75% solids.

Varisoft 135. [559] Sulfated quat. blend; softener, antistat for fabric treatment in dryers; Gardner 4 max. clear liq.; dens. 7.4 lb/gal (140 F); m.p. 40–50 C; flash pt. 76 F (PMCC); pH 6.5; 91–93% solids.

Varisoft 137. [559] Dimethyl distearyl ammonium methyl sulfate; cationic; quat. for home and commercial laundry fabric and tissue softeners, debonding agent and antistat; emulsifier; Gardner 5 solid paste; m.w. 649; disp. in water; dens. 7.4 lb/gal; flash pt. 78 F (PMCC); pH 5–8; 89–91% solids.

Varisoft 190-100P. [559] Dimethyl distearyl ammonium methyl sulfate; see Varisoft 110; powd.; 97% solids.

Varisoft 222, 238. [559] Complex difatty quat. sulfate; cationic; see Varisoft 110; paste; 75–90% solids.

Varisoft 222-90%, 222-HV. [559] Methyl bis (tallowamidoethyl) 2-hydroxyethyl ammonium methyl sulfate; specialty quat.; Gardner 6 max. paste; flash pt. 85 F (PM); 89–91% solids.

Varisoft 222-LT. [559] Methyl bis (oleylamidoethyl) 2-hydroxyethyl ammonium methyl sulfate; specialty quat.; Gardner 7 max. liq.;

flash pt. 80 F (PM); 89–91% solids.

Varisoft 222LM-90. [559] Methyl bis (tallowamidoethyl) 2-hydroxyethyl ammonium methyl sulfate; specialty quat.; Gardner 6 max. liq.; flash pt. 85 F (PM); 89–91% solids.

Varisoft 238-90%. [559] Methyl bis (tallowamidoethyl) 2-hydroxypropyl ammonium methyl sulfate; specialty quat.; Gardner 6 max. liq.; flash pt. 85 F (P-M); 89–91% solids.

Varisoft 445. [559] Methyl (1) hydrog. tallow amidoethyl (2) hydrog. tallow imidazolinium methosulfate; see Varisoft 110; liq.; 77% solids.

Varisoft 475. [559] Quaternium-27; see Varisoft 110; liq.; 77% solids.

Varisoft 3262. [559] Quaternium-18, IPA, stearyl alcohol, and glycol stearate; cationic; cream rinse conc.; solid; 75% conc.

Varisoft 3690. [559] Methyl (1) oleylamidoethyl (2) oleyl imidazolinium methosulfate; see Varisoft 110; liq.; 77% solids.

Varisoft 6112. [559] Stearyl dimethylbenzyl ammonium chloride; cationic; softener used in hair care products; paste; 25% conc.

Varisoft DHT. [559] Dimethyl dihydrog. tallow ammonium chloride; cationic; hair conditioner and antistat in cream rinse concs.; Gardner 3 max. soft solid; m.w. 569; disp. in water; dens. 7.12 lb/gal; flash pt. 68 F (PMCC); pH 6–9; 74–77% act. in aq. IPA.

Varisoft E228. [559] Alkyl trimethyl ammonium chloride; cationic; conditioner, softener; liq.; 28% conc.

Varisoft SDC. [559] Stearalkonium chloride; cationic; softener in hair care products; paste; 25% conc.

Varisoft SDC-65%. [559] Stearalkonium chloride; cationic; used in hair care products; wh. paste; pH 3–5; 65–70% act.

Varisoft SDC-W. [559] Stearalkonium chloride; cationic; antistat used in hair care products; wh. paste; flash pt. > 200 F (PM); pH 3.5–4.0; 24–26% solids in water.

Varonic 320. [559] POE oleyl alcohol; nonionic; emulsifier, lubricant.

Varonic 400DL. [559] PEG 400 dilaurate; nonionic; emulsifier for cosmetics; liq.; 100% act.

Varonic 400MS. [559] PEG 400 monostearate; nonionic; emulsifier, thickener used in emollients, creams and lotions; solid; 100% act.

Varonic 1000MS. [559] PEG 1000 monostearate; nonionic; solubilizer, humectant, surfactant; solid.

Varonic 1800MS. [559] PEG 1800 monostearate; nonionic; see Varonic 1000MS; solid.

Varonic K 202, 205, 205LC, 210, 210LC, 215, 215LC. [559] Ethoxylated coco amines; nonionic; antistat, emulsifier, dispersant for oils; liq.; sol. in water, IPA, benzene, carbon tetrachloride (K210, K215, K215LC); sp.gr. 0.995 (K210), 1.04 (K215, K215LC); 100% act.

Varonic L 202. [559] Ethoxylated soya amine; nonionic; antistat, dispersant; emulsifier for latex, dyes, oils; liq.; 100% act.

Varonic L 205. [559] PEG-5 soyamine; nonionic; see Varonic L 202; liq.; 100% act.

Varonic L 205LC. [559] Ethoxylated soya amine; nonionic; see Varonic L 202; liq.; 100% act.

Varonic L 230, L 230 (60%), L 230 (80%). [559] Ethoxylated soya amine; nonionic; see Varonic L 202; liq.; 100, 60, and 80% act. resp.

Varonic LI-42. [559] PEG-20 glyceryl monotallowate; nonionic; low-irritation detergent, emulsifier, solubilizer for personal care products; Gardner 2 paste; HLB 13; m.p. 27 C; surf. tens. 44 dynes/cm (1%); 100% solids.

Varonic LI-48. [559] PEG-80 glyceryl tallowate; nonionic; emulsifier, solubilizer, antiirritant surfactant used in personal care products; Gardner 2 hard solid; sol. in water, methanol, ethanol, IPA, acetone, and butyl Cellosolve; HLB 18 ± 1; m.p. 41 C; pH 7.0; surf. tens. 49.5 dynes/cm.

Varonic LI-63. [559] PEG-30 glyceryl monococoate; nonionic; low-irritation detergent, emulsifier, solubilizer for personal care products; Gardner 2 paste; HLB 15; m.p. 27 C; surf. tens. 40 dynes/cm (1%); 100% solids.

Varonic LI-67. [559] PEG-80 glyceryl cocoate; nonionic; see Varonic LI-63; Gardner 2 solid; HLB 18; m.p. 42 C; surf. tens. 47 dynes/cm (1%); 100% solids.

Varonic LI-420. [559] PEG-200 glyceryl tallowate; nonionic; see Varonic LI-48; Gardner 2 hard solid; sol. see Varonic LI-48; HLB 19 ± 1; m.p. 53 C; pH 7.0; surf. tens. 54.0 dynes/cm.

Varonic MT 65. [559] Fatty alcohol polyglycol methyl ether; nonionic; surfactant used in detergent; lubricant in tech. oils; textile aids; Gardner 2 clear liq.; sol. in polar and nonpolar solvs.; pH 3.5 (10%); surf. tens. 32 dynes/cm.

Varonic T 105. [559] Ethoxylated alcohol; surfactant for drilling aids and emulsifiers; thickener in hydraulic fracturing and defoamer in cementing; soft paste; flash pt. > 200 F (CC); 100% conc.

Varonic T 202, 205, 205LC, 210, 215, 215LC. [559] Ethoxylated tallow amine; nonionic; emulsifier for latex, dyes and oils; antistat, dispersant; liq.; 100% act.

Varonic U 105. [559] Ethoxylated alcohol; see Varonic T 105; hard paste; flash pt. > 200 F (CC); 100% conc.

Varonic U 205, 205LC. [559] Ethoxylated hydrog. tallow amine; nonionic; antistat, emulsifier, dispersant, dye intermediate; semiliq., solid; 100% act.

Varonic U215. [559] Ethoxylated cetyl stearyl amine; wetting aid and emulsifier in tanning industry; Gardner 7 solid; sol. in water, IPA, benzene, carbon tetrachloride, MEK; sp.gr. 1.025; 97% act.

Varonic U 250. [559] PEG-50 tallow amine; nonionic; see Varonic U 205; semiliq., solid; 100% act.

Varox 185E. [559] Dihydroxyethyl C_{12-15} alkoxypropylamine oxide; foam stabilizer for anionic surfactants used in liq. dishwashing, shampoos and fabric detergents; Gardner 1 liq.; pH 4.5–5.5; surf. tens. 31.5 dynes/cm; 39–43% amine oxide.

Varox 185PH. [559] Hydroxyethyl hydroxypropyl C_{12-15} alkoxypropylamine; see Varox 185E; Gardner 1 max. liq.; pH 4.5–5.5; surf. tens. 30.9 dynes/cm.

Varox 188E. [559] Dihydroxyethyl C_{8-10} alkoxypropylamine oxide; see Varox 185E; pH 4.5–5.5; surf. tens. 34.8 dynes/cm.

Varox 191E. [559] Dihydroxyethyl C_{9-11} alkoxypropylamine oxide; see Varox 185E; Gardner 1 max. liq.; pH 4.5–5.5; surf. tens. 34.0 dynes/cm.

Varox 365. [559] Lauryldimethyl amine oxide; foam stabilizer for anionic surfactants for shampoo and detergent systems; Gardner 1 liq.; 30% amine oxide.

Varox 743. [559] Coco dialkoxy amine oxide; nonionic; detergent, foam booster and stabilizer; liq.; 50% conc.

Varox 1770. [559] Cocamidopropyl amine oxide; nonionic; foam booster; detergent; liq.; 50% act.

Varstat 55. [559] Quat.; cationic; antistat used in treating fibers and textiles; hydrophilic.

Varstat 66. [559] Ethyl bis (poly hydroxyethyl) alkyl ammonium ethyl sulfate; cationic; antistat used in treating fibers and textiles; Gardner 8 max. liq.; m.w. 1110; flash pt. > 200 F (PM); 68–76% conc.

Varstat K22, T22. [559] Ethoxylated fatty amine; cationic; antistat for internal use in polymeric resins; liq.; 100% conc.

Varsulf 5. [559] Lanolin sulfosuccinate; surfactant for hand cleaners; paste; 50% conc.

Varsulf 60L. [559] Triethanolamine alkylaryl sulfonate; anionic; detergent, foamer; liq.; 60% conc.

Varsulf NOS 25. [559] Sodium alkyl phenol polyglycol ether sulfate; anionic; emulsifier for emulsion polymerization; lt. yel. clear liq.; pH 7.0 (1%); 35% solids.

Varsulf S-1333. [559] Disodium ricinoleic acid monoethanolamide sulfosuccinate aq. disp.; anionic; detergent used in dishwash and personal care products; pale yel. liq.; dens. 9.4 lb/gal; visc. 300 cps; pH 6.5–7.5; surf. tens. 28.8 dynes/cm; 39.% min. solids.

Varsulf SBDO-70. [559] Dioctyl sulfosuccinate; anionic; wetting agent and solubilizer for textiles, emulsion polymerization, agric. chemicals; liq.; 70% conc.

Varsulf SBFA 30. [559] Disodium laureth-3 sulfosuccinate aq. disp.; anionic; see Varsulf S-1333; wh.-pale yel. liq.; dens. 9.2 lb/gal; visc. 200 cps; pH 6.5–7.5; surf. tens. 21.7 dynes/cm; 39.0% min. solids.

Varsulf SBL 203. [559] Disodium lauric acid monoethanolamide sulfosuccinate aq. disp.; anionic; see Varsulf S-1333; also used in rug and upholstery shampoos; Gardner 3 max. liq.; dens. 9.5 lb/gal; visc. 100–500 cps; pH 6.5–7.5; surf. tens. 24.3 dynes/cm; 39.0% min. solids.

Varsulf SBU-185. [559] Disodium undecylenamido MEA sulfosuccinate; anionic; detergent; Gardner 4 liq., powd.; pH 7.0; 50% solids.

Veegum. [631] Magnesium aluminum silicate; thickener for emulsions, suspensions and sol'ns used in personal care products, pharmaceuticals, paints, textile finishes, and industrial applications; emulsion stabilizer; suspending agent; wh. flakes; sol. in water and alcohol; dens. 2.6 mg/m³; visc. 250 cps ± 25% (5% aq. disp.); pH 9 (5% aq. disp.).

Veegum HS. [631] Magnesium aluminum silicate; thickener and suspending agent with electrolyte compatibility; wh. flakes; sol. in water and alcohol; visc. 150 cps ± 50%

(5.5% aq. disp.); pH 9 (5% aq. disp.); 92% min. conc.

Veegum HV. [631] Magnesium aluminum silicate; see Veegum; wh. flakes; visc. 250 cps ± 25% (4% disp.).

Veegum T. [631] Magnesium aluminum silicate; see Veegum; wh. flakes; dens. 2.8 mg/m^3; visc. 200 cps ± 50% (4% disp.).

Velvamine HSB. [376] Alkyl imidazoline deriv.; cationic; softener in hair rinses, for leather; paste; 45% conc.

Velvetex 610L. [289] Sodium lauriminodipropionate; amphoteric; surfactant used in personal care products; sec. emulsifier with nonionics; amber clear liq.; sol. in water, acid or alkali systems, polar solvs.; sp.gr. 1.038; visc. < 100 cps; pH 7.0–8.0 (1%); 29.0–31.0% solids.

Velvetex 710L. [289] Lauraminopropionic acid; amphoteric; surfactant used in personal care products; corrosion inhibition for aerosol pkg.; amber clear liq.; sol. in water, acid and alkali systems and alcohol; sp.gr. 0.994; visc. 2500 cps; pH 5.5–6.5 (1%); 48.0–52.0% solids.

Velvetex AB-45. [289] Coco betaine; amphoteric; surfactant used in cleansing emulsions and personal care products; Gardner 2 clear liq.; water-sol.; sp.gr. 1.03; visc. < 100 cps; cloud pt. < 0 C; pH 6.5–8.5 (10%); 43–45% solids.

Velvetex BA. [289] Cocamidopropyl betaine; amphoteric; surfactant used in foam drilling and blanketing; air-inhibiting agent for cement gypsum board and cleaners for oily surfaces; used in personal care products; liq.; dens. 8.2 lb/gal; 60% conc. in alcohol.

Velvetex BA-35. [289] Cocamidopropyl betaine; amphoteric; see Velvetex BA; clear yel. liq.; water-sol.; sp.gr. 1.05; visc. < 100 cps; cloud pt. < 0 C; 34–37% solids.

Velvetex BC. [289] Coco betaine, amphoteric; see Velvetex AB-45; liq.; dens. 7.7 lb/gal; 59% conc.

Velvetex BC-35. [289] Cocamidopropyl betaine; amphoteric; surfactant, antistat, foamer, and conditioner for personal care products; lt. amber liq.; 35% solids.

Velvetex BK-35. [289] Cocamidopropyl betaine; surfactant for hair care products; visc. builder; yel. clear liq.; water-sol.; sp.gr. 1.039; visc. < 100 cps; pH 4.5–5.5; 34.0–36.0% solids.

Velvetex OLB-50. [289] Oleyl betaine; amphoteric; surfactant used in personal care products, mfg. of fibers and cutting oils; amber translucent gel; water-sol.; sp.gr. 0.953 (60 C); pH 6.0–8.0 (10%); 48.0–52.0% solids.

Versilan MX 112. [192] Alcohol ethoxylate, modified; nonionic/anionic; solv. used in degreasers; liq.

Versilan MX 123. [192] Alcohol ethoxylate, modified; nonionic; surfactant used in highly built liqs.; liq.

Versilan MX 134. [192] EO/PO copolymer, modified; see Versilan MX 123; liq.

Versilan MX 145. [192] Quat. ammonium compd., modified; cationic; surfactant; conc. for vehicle cleaning formulations; liq.

Versilan MX 167. [192] Alcohol ether sulfate; nonionic; surfactant used in personal care products; liq.

Versilan MX 233. [360] Alcohol ethoxylate, modified; nonionic; detergency booster for household and domestic laundry powd. formulations; liq.; 100% conc.

Verv. [462] Calcium stearoyl lactylate; anionic; softener and complexing agent used in food industry; cream powd.; mild caramel odor; slightly tart taste; acid no. 50–86; 100% conc.

Victawet 12. [573] Ethylene oxide, 2-ethyl hexanol and P_2O_5; nonionic; wetting agent, penetrant, dispersant, stabilizer; for pkg. dyeing of nylon, acid-type cleaners, emulsion polymerization, starch coatings; APHA 500 max. liq.; sol. in alcohol, acetone, toluene; disp. in water; dens. 9.34 lb/gal; sp.gr. 1.121; visc. 91.2 cs (100 F); b.p. 325 F; flash pt. 250 F (COC); pour pt. –48 F; surf. tens. 29.4 dynes/cm^2 (0.2%); pH 7.0–7.4 (0.5%); 100% act.

Victawet 35B. [573] 2-Ethyl hexanol P_2O_5, NaOH; anionic; wetting agent, penetrant, dispersant, stabilizer; surfactant; lt. tan to wh. soft paste; sol. in water, ethylene glycol; dens. 9.7 lb/gal; sp.gr. 1.17; visc. > 100,000 cps; b.p. 200 F; flash pt. 200 F (COC); surf. tens. 26.2 dynes/cm^2; pH 6.9–7.4 (0.5%); 68–72% act.

Victawet 58B. [573] Capryl alcohol, P_2O_5, NaOH; anionic; wetting agent, penetrant, dispersant, stabilizer, surfactant; lt.tan to wh. soft paste; sol. in water, ethylene glycol; dens. 9.77 lb/gal; sp.gr. 1.121; visc. > 100,000 cps; flash pt. 180 F (COC); surf. tens. 29.0 dynes/cm^2; pH 6.8–7.3 (5%); 68–72% act.

Vifcoll CCN-40, CCN-40 Powder. [437] N-cocoyl collagen peptide, sodium salt; detergent, emulsifier used in personal care products, pharmaceuticals, food industry, and household cleaning products; pale yel. liq. and wh. or pale yel. powd. resp.; m.w. 600; pH 5.5–7.5 (both, 3% aq. for powd.); 30% act. in water (CCN-40).

Vigilan [232] Lanolin oil; nonionic; emulsifier, emollient; liq.; 100% conc.

Vigilan AWS. [232] PEG-75 lanolin oil; o/w emulsifier; plasticizer, emollient, solubilizer, and wetting agent in personal care products; sol. in water and alcohol; sapon. no. 10–25.

Vikosperse KDS. [633] Sulfonated naphthalene formaldehyde condensate; anionic; dispersant for dyes; 95% conc.

Vista C-550, -560. [634] Dodecylbenzene sodium sulfonate, straight chain; anionic; detergent, foamer used in liq. detergent compds.; liq.; 50 and 60% conc. resp.

Volpo 3. [159] Oleth-3; nonionic; emollient, lubricant, emulsifier, solubilizer used in personal care products; hazy liq.; sol. in alcohol, glycol, ketone, chlorinated and aromatic solvs.; HLB 6.6; pH 5–7 (3% aq.).

Volpo 5. [159] Oleth-5; nonionic; see Volpo 3; HLB 8.8; pH 5–7 (3% aq.).

Volpo 10. [159] Oleth-10; nonionic; see Volpo 3; wh. semisolid; sol. see Volpo 3; HLB 12.4; pH 5–7 (3% aq.).

Volpo 20. [159] Oleth-20; nonionic; nonionic; see Volpo 3; wh. solid; sol. see Volpo 3; HLB 15.4; pH 5–7 (3% aq.).

Volpo 25D3. [159] POE (3) C_{12-15} ether; nonionic; wetting, scouring, leveling agent, dispersant, solubilizer, emulsifier, antistat, plasticizer in personal care products, pharmaceuticals, hard-surface cleaners, textile, agric., paper, and metal industries; water-wh. liq.; sol. in ethanol, min. oil, kerosene, trichloroethylene, oleic acid, oleyl alcohol; pH 6.0–7.5 (3%).

Volpo 25D5. [159] POE (5) C_{12-15} ether; see Volpo 25D3; off-wh. opaque liq.; sol. in ethanol, trichloroethylene, oleic acid; HLB 10.3; surf. tens. 29.5 dynes/cm (0.1%); pH 6.0–7.5 (3%).

Volpo 25D10. [159] POE (10) C_{12-15} ether; see Volpo 25D3; off-wh. soft paste; sol. in water, ethanol, trichloroethylene; HLB 13.6; cloud pt. 83 C (1% aq.); surf. tens. 33.5 dynes/cm (0.1% aq.); pH 6.0–7.5 (3%).

Volpo 25D15. [159] POE (15) C_{12-15} ether; see Volpo 25D3; wh. soft solid; sol. see Volpo 25D10; HLB 15.3; cloud pt. 73 C; surf. 36.5 dynes/cm (0.1% aq.); pH 6.0–7.5 (3%).

Volpo 25D20. [159] POE (25) C_{12-15} ether; see Volpo 25D3; wh. soft solid; sol. see Volpo 25D10; HLB 16.1; cloud pt. 80 C; surf. tens. 38.5 dynes/cm (0.1% aq.); pH 6.0–7.5 (3%).

Volpo CS3. [159] POE (3) ceto stearyl ether; see Volpo 25D3; sol. in ethanol, kerosene, trichloroethylene, oleic acid, oleyl alcohol; HLB 7.3; pH 6.0–7.5 (3%).

Volpo CS5. [159] POE (5) ceto stearyl ether; see Volpo 25D3; off-wh. soft waxy solid; sol. see Volpo 25D5; HLB 9.5; surf. tens. 33.0 dynes/cm (0.1% aq.); pH 6.0–7.5 (3%).

Volpo CS10. [159] POE (10) ceto stearyl ether; see Volpo 25D3; off-wh. soft waxy solid; sol. see Volpo 25D5; HLB 12.9; cloud pt. 68 C (1% aq.); surf. tens. 34.5 dynes/cm (0.1% aq.); pH 6.0–7.5 (3%).

Volpo CS15. [159] POE (15) ceto stearyl ether; see Volpo 25D3; off-wh. waxy solid; sol. see Volpo 25D5; HLB 14.6; cloud pt. 94 C (1% aq.); surf. tens. 35.5 dynes/cm (0.1% aq.); pH 6.0–7.5 (3%).

Volpo CS20. [159] POE (20) ceto stearyl ether; see Volpo 25D3; off-wh. hard waxy solid; sol. in water, ethanol, oleic acid, trichloroethylene; HLB 15.6; cloud pt. 78 C; surf. tens. 41.5 dynes/cm (0.1% aq.); pH 6.0–7.5 (3%).

Volpo L4. [159] POE (4) lauryl ether; nonionic; emulsifier and dispersant; liq.; 97% conc.

Volpo L23. [159] POE (23) lauryl ether; nonionic; o/w emulsifer and solubilizer for cosmetics, pharmaceuticals and household products; solid; 97% conc.

Volpo N3. [159] POE (3) oleyl ether, dist.; nonionic; emulsifier, dispersant, wetting, scouring and gelling agent, solubilizer; liq.; 97% conc.

Volpo N5. [159] POE (5) oleyl ether, dist.; nonionic; see Volpo N3; liq.; 97% conc.

Volpo N10. [159] POE (10) oleyl ether, dist.; nonionic; see Volpo N3; paste; 97% conc.

Volpo N15. [159] POE (15) oleyl ether, dist.; nonionic; see Volpo N3; paste; 97% conc.

Volpo N20. [159] POE (20) oleyl ether, dist.; nonionic; see Volpo N3; solid; 97% conc.

Volpo O3. [159] POE (3) oleyl ether; see Volpo 25D3; pale straw liq.; sol. see Volpo D5D3; HLB 6.7; pH 6.0–7.5 (3%).

Volpo O5. [159] POE (5) oleyl ether; see Volpo 25D3; pale straw liq.; sol. see Volpo 25D5; HLB 9.0; surf. tens. 31.0 dynes/cm (0.1% aq.); pH 6.0–7.5 (3%).

Volpo O10. [159] POE (10) oleyl ether; see

Volpo 25D3; pale straw paste; sol. see Volpo 25D10; HLB 12.4; cloud pt. 57 C (1% aq.); surf. tens. 33 dynes/cm (0.1% aq.); pH 6.0–7.5 (3%).

Volpo O15. [159] POE (15) oleyl ether; see Volpo 25D3; pale straw paste; sol. see Volpo 25D10; HLB 14.2; cloud pt. 91 C (1% aq.); surf. tens. 34.5 dynes/cm (0.1% aq.); pH 6.0–7.5 (3%).

Volpo O20. [159] POE (20) oleyl ether; see Volpo 25D3; pale straw soft solid; sol. see Volpo 25D10; HLB 15.5; cloud pt. 78 C; surf. tens. 41.0 dynes/cm (0.1% aq.); pH 6.0–7.5 (3%).

Volpo S2. [159] Steareth-2; nonionic; emulsifier for cosmetics; solid; HLB 5.6; 100% conc.

Volpo S10. [159] Steareth-10; nonionic; see Volpo S2; solid; HLB 12.4; 100% conc.

Volpo S20. [159] Steareth-20; nonionic; see Volpo S2; solid; HLB 15.5; 100% conc.

Volpo T3. [159] POE (3) tridecyl ether; see Volpo 25D3; water-wh. liq.; sol. see Volpo 25D3; HLB 8.0; pH 6.0–7.5 (3%).

Volpo T5. [159] POE (5) tridecyl ether; see Volpo 25D3; water-wh. opaque liq.; sol. in ethanol, kerosene, trichloroethylene, oleic acid; HLB 10.4; surf. tens. 29.0 dynes/cm (0.1% aq.); pH 6.0–7.5 (3%).

Volpo T10. [159] POE (10) tridecyl ether; see Volpo 25D3; off-wh. soft paste; sol. see Volpo 25D10; HLB 13.7; cloud pt. 61 C (1% aq.); surf. tens. 29.5 dynes/cm (0.1% aq.); pH 6.0–7.5 (3%).

Volpo T15. [159] POE (15) tridecyl ether; see Volpo 25D3; off-wh. soft paste; sol. see Volpo 25D10; HLB 15.5; cloud pt. 94 C (1% aq.); surf. tens. 32.0 dynes/cm (0.1% aq.); pH 6.0–7.5 (3%).

Volpo T20. [159] POE (20) tridecyl ether; see Volpo 25D3; off-wh. soft solid; sol. see Volpo 25D10; HLB 16.3; cloud pt. 75C; surf. tens. 36.5 dynes/cm (0.1% aq.); pH 6.0–7.5 (3%).

Vykacet L, T. [159] Acetylated monoglyceride; nonionic; mold release agent in food industry; liq. and solid resp.; 100% conc.

Vykamol N/E, S/E. [159] Glyceryl monostearate; nonionic; emulsifier, dispersant, wetting, scouring and gelling agent, solubilizer; solid.

Vykasoid. [159] Monoglyceride and veg. oil; nonionic; w/o emulsifier for dairy spreads; solid; 100% conc.

W

Wayhib. [455] Triethanolamine phosphate ester; anionic; detergent builder, sequestration agent, scale inhibitor; liq.; 70% conc.

Wayplex NTP-S. [455] Polyphosphonate, sodium salt; anionic; detergent builder, scale inhibitor; liq.; 40% conc.

Wecobee. [496] Hydrog. palm kernel oil; nonionic; cocoa butter substitute; liq.; 40% conc.

Wecobee FS, FW. [496] Lauric triglyceride; used in personal care products, pharmaceuticals, food industry; Lovibond 30/3.0 solid; bland odor; visc. 13.5 and 14.0 cps resp. (150 F); m.p. 107–109 and 101–103 F; sapon. value 230–250.

Wecobee M, R. [496] Hydrog. veg. oil; see Wecobee FS; Lovibond 30/3.0 solid; bland odor; visc. 13.5 and 14.0 cps resp. (150 F); m.p. 96–98 and 101–103 F; sapon. value 238–250 and 230–250.

Wecobee S, SS. [496] Hydrog. veg. oil; see Wecobee FS; Lovibond 30/3.0 solid; bland odor; visc. 14.0 and 13.5 cps resp. (150 F); m.p. 113–115 and 116–118 F; sapon. value 230–250.

Wecobee W. [496] Hydrog. veg. oil; see Wecobee FS; Lovibond 30/3.0 solid; bland odor; visc. 13.5 cps (150 F); m.p. 94–96 F; sapon. value 240–254.

Wecotop 1013. [496] Soya and palm oil veg. fat; food emulsifier; Lovibond 20/2.0 solid; m.p. 120–125 F.

Westvaco Diacid 1550. [642] Dicarboxylic org. acid (C_{21}) 6-carboxy-4-hexyl-2-cyclohexeneoctanoic acid; surfactant; liq.; biodeg.; 100% conc.

Wetting Agent 611, CG 16. [130] Org. sulfonic acid; anionic; wetting and washing agent; liq. and paste resp.; 35 and 70% conc. resp.

Wettol EM 1. [82] Calcium alkylaryl sulfonate; anionic; liq.; 68–70% conc.

Wettol EM 2. [82] Alkylphenol oxalkylation product; nonionic; emulsifier for crop protection concs.; liq.; 82.5% conc.

Wettol EM 3. [82] Oxalkylated veg. oil; anionic; liq.; 85% conc.

Wettol NT 1. [82] Sodium alkyl naphthalene sulfonate; anionic; wetting agent for crop protection; powd.; 65% conc.

White Swan. [159] Lanolin BP, anhyd.; conditioner, moisturizer; w/o emulsifier, and emollient for personal care products; superfatting agent in soap; aids pigment dispersion; yel. unctuous mass.

Wibarco. [131] Linear alkylbenzene; anionic; liq.; biodeg.

Wickenol 535. [646] Wheat germ glyceride; emollient, emulsifier; liq.; 100% conc.

Wickenol 545. [646] Glucose glutamate; humectant used in personal care products; sol. in water.

Witbreak DGE Series. [650] Glycol ester; demulsifier for petroleum industry and waste disposal of processing oils in metals.

Witbreak DPG Series, DTG-62. [650] Polyoxyalkylene glycol; see Witbreak DGE-60.

Witbreak DRA Series, DRB Series, DRC Series. [650] Oxyalkylated phenolic resin; see Witbreak DGE-60.

Witbreak GBG Series. [650] Polyoxyalkylated modified resins; see Witbreak DGE-60.

Witbreak RTC-323, -1315, -1316. [650] Polymeric amine; anionic; o/w demulsifier used in petroleum industry and for waste disposal of processing oils in metals; liq.; water-sol.

Witbreak RTC-6010, -6020, –6030, -6040. [650] Polymeric amine; demulsifier in petroleum industry.

Witcamide 61. [650] Oleamide MIPA and isopropanolamine; nonionic; conditioner, emulsifier, lubricant; cosmetics and toiletries.

Witcamide 70. [650] Stearamide MEA; nonionic; opacifier, conditioner, lubricant, gelling agent, mold release agent, binder; cosmetics and toiletries; wax.

Witcamide 82. [650] Cocamide DEA and diethanolamine; nonionic; foam stabilizer, visc. modifier, lubricant, conditioner, emulsifier, wetting agent, penetrant, dye dispersant, scouring aid, antistat; cosmetics and toiletries, base for scrub soap; thickener; industrial and textile surfactant; metal processing; liq.; 100% conc.

Witcamide 272. [650] Alkanolamide, modified; nonionic; detergent, emulsifier, lubricant, wetting agent, penetrant, dye dispersant, scouring aid, antistat for industrial, textile, drilling fluids, metal processing; liq.; sol. in water, high concs. of electrolytes; 90% conc.

Witcamide 511. [650] Alkanolamide; nonionic; lubricant, wetting agent, penetrant, dye dispersant, scouring aid, antistat, w/o emulsifier for aerosol and industrial applications, drilling muds, metal processing, textiles; liq., paste.

Witcamide 823/10. [650] Isostearic alkanolamide; nonionic; conditioner, lubricant used in personal care products; lt. amber liq.; disp. in water; cloud pt. –5 C; pH 9.5 ± 0.5 (10% disp.); 100% act.

Witcamide 1017. [650] Cocamide DEA; nonionic; drilling mud surfactant; liq.; 100% conc.

Witcamide 5130. [650] Cocamide DEA; anionic/nonionic; emulsifier, visc. modifier, lubricant, dispersant, suspending and wetting agent, penetrant, dye dispersant, scouring aid, antistat, foam stabilizer for aerosol formulations, textiles, industrial use, metal processing; water-sol.; liq.; 98% conc.

Witcamide 5133. [650] Diethanolamine; nonionic; emulsifier, foam stabilizer, visc. modifier, lubricant, conditioner, wetting agent, penetrant, dye dispersant, scouring aid, antistat for textiles, industrial use, personal care products, metal processing; liq.; 100% conc.

Witcamide 5138. [650] Alkanolamide; nonionic; emulsifier, detergent, lubricant, wetting agent, penetrant, dye dispersant, scouring aid, antistat for industrial use, textiles, metal processing, dry cleaning; liq.; oil-sol.; disp. in water.

Witcamide 5140. [650] Alkanolamide, modified; nonionic; detergent, wetting agent, dispersant; industrial cleaning agent; foam stabilizer in drilling mud surfactants and aq. media; air-entraining agent; liq.; 100% conc.

Witcamide 5145M. [650] Alkanolamide, modified; anionic/nonionic; detergent for built liqs., dry cleaning; liq.; 100% conc.

Witcamide 5168. [650] Diethanolamido oleamide DEA; cationic; surfactant for cosmetics, textiles, detergents, and industrial uses; liq.; 100% conc.

Witcamide 5195. [650] Lauramide DEA; nonionic; visc. modifier, lubricant, conditioner, lubricant, emulsifier, wetting agent, thickener, penetrant, dye dispersant, scouring aid, antistat; cosmetics and toiletries; industrial foamer and stabilizer; metal processing; textile surfactant; aerosol formulations.

Witcamide 6310. [650] Lauramide DEA; nonionic; conditioner, foam stabilizer, gelling agent, lubricant, and visc. modifier for cosmetics and toiletries; industrial detergent; paste; water-sol.

Witcamide 6445. [650] Alkanolamide, modified; anionic/nonionic; industrial detergent, o/w emulsifier, lubricant, visc. modifier, foamer and stabilizer; liq.; water-sol.

Witcamide AL69-58. [650] Alkanolamide; cationic/nonionic; lubricant, wetting agent, penetrant, dye dispersant, scouring aid, antistat; industrial w/o emulsifier; metal processing; textile surfactant; liq.; 100% conc.

Witcamide CD. [650] Alkanolamine condensate; detergent, thickener; liq.; 100% conc.

Witcamide CDA. [650] Alkanolamide, modified; nonionic; see Witcamide 6445; liq.; water-sol.

Witcamide CDS. [650] Alkylolamide; nonionic; additive for cutting oils, corrosion inhibitor; liq.; 100% conc.

Witcamide LDTS. [650] Alkylolamide; nonionic; detergent, foam stabilizer, softener, visc. control for cosmetics; liq.; 100% conc.

Witcamide LM, LMT. [650] Alkylolamide; nonionic; thickener for personal care products; flakes; 100% conc.

Witcamide M-3. [650] Cocamide DEA; nonionic; cleansing agent, conditioner, lubricant, and visc. modifier for cosmetics and toiletries; detergent, foamer and stabilizer for industrial detergents; liq.; water-sol.

Witcamide MAS. [650] Stearamide MEA-stearate; nonionic; opacifier, conditioner, lubricant, gelling and mold release agent, binder; cosmetics; wax.

Witcamide MM. [650] Myristamide MEA; nonionic; see Witcamide MAS; wax.

Witcamide N. [650] Alkylolamide; nonionic; metal cleaning, insecticides; liq.; 100% conc.

Witcamide NA, NA1. [650] Alkylolamide; nonionic; foam stabilizer, softener; liq.; 100% conc.

Witcamide S-771. [650] Alkanolamide, modified; anionic/nonionic; industrial foamer and stabilizer, lubricant; liq.; water-sol.

Witcamide S-780. [650] Alkanolamide; anionic/nonionic; see Witcamide S-771; liq.; water-sol.

Witcamide STD-HP. [650] Lauramide DEA; nonionic; conditioner, foam stabilizer, lubricant for cosmetics and toiletries; solid; water-sol.

Witcamine 209. [650] 1-(2-Aminoethyl)-2-*n*-alkyl-2-imidazoline; cationic; emulsifier, lubricity and wetting agent; penetrant, dye dispersant, scouring aid, antistat, corrosion inhibitor intermediate in petroleum industry, metal processing, textiles; liq.

Witcamine 210. [650] Alkyl amidoamine; cationic; emulsifier, lubricant; corrosion inhibitor intermediate in petroleum industry and metal processing.

Witcamine 211. [650] Mixed 1-(2-aminoethyl)-2-imidazoline; see Witcamine 209; liq.

Witcamine 235. [650] 1-Polyaminoethyl-2-*n*-alkyl-2-imidazoline; see Witcamine 209.

Witcamine 3164. [650] Alkylamidoamine; see Witcamine 210.

Witcamine AL42-12. [650] Fatty imidazoline; cationic; industrial antistat, corrosion inhibitor, dispersant, o/w emulsifier, beading agent; textile defoamer; liq.; sol. in oil.

Witcamine PA-60B. [650] Fatty imidazoline salt; cationic; industrial antistat, corrosion inhibitor, dispersant and wetting agent; liq.; sol. in water.

Witcamine PA-78B. [650] Fatty imidazoline salt; cationic; dispersant for inorg. media; corrosion inhibitor for solvs.; wetting agent in oil-based systems; dk. amber clear liq.; sol. in kerosene, xylene, IPA; sp.gr. 0.925; flash pt. > 93 C (PMCC); pH 6.5 (3% in 1:1 IPA/water); 99.5% act.

Witcamine RAD-0500, 0515, –1100, -1110. [650] POE amine; cationic/nonionic; corrosion inhibitor intermediate, scouring agent, softener, dye assistant, emulsifier for acid-stable emulsions; liq.; 100% conc.

Witco 97H Acid. [650] Alkylaryl sulfonic acid; anionic; industrial detergent, dispersant, o/w emulsifier, wetting agent; liq.; sol. in water and oil.

Witco 912. [650] POE carboxylic acid esters and sulfonates; pigment dispersant for aq. systems; solubilizer for methyl cellulose; paint surfactant.

Witco 915. [650] Imidazoline; cationic; pigment dispersant and suspending agent, wetting and grinding aid for oil-based paint systems; emulsifier for latex paints.

Witco 916. [650] Fatty acid ester; spreading agent for aq. paint systems.

Witco 918. [650] Alkylaryl sulfonate; pigment wetting and grinding aid for paints; pigment dispersant; oil-sol.

Witco 934. [650] Alkanolamide; pigment dispersant and suspending agent, wetting and grinding aid for oil-based paint systems; surfactant for latex paints.

Witco 936. [650] Alkylphenol/EO adduct; pigment wetting and grinding aid in oil-based paint systems.

Witco 942. [650] Glycerol monooleate; emulsifier for latex paints.

Witco 960. [650] Alkylphenol/EO adduct; pigment wetting and grinding aid for aq. paint systems; emulsifier for polymerization.

Witco 1298 Hard Acid. [650] Dodecylbenzene sulfonic acid; anionic; see Witco 97H Acid; liq.; sol. in oil and water.

Witco 1298 Soft Acid. [650] Dodecylbenzene sulfonic acid; anionic; detergent intermediate, o/w emulsifier, solubilizer, wetting agent, and detergent for metal cleaning; emulsion polymerization surfactant for latex stabilization and pigment dispersion; liq.; sol. in oil and water.

Witco Acid B. [650] Dodecylbenzene sulfonic acid; anionic; detergent intermediate; detergent, dispersant, o/w emulsifier, wetting agent for industrial and metal cleaning; solubilizer, emulsion polymerization surfactant for latex stabilization and pigment dispersion; liq.; sol. in oil and water.

Witco Acid TPB. [650] Alkylaryl sulfonic acid; anionic; base for detergent formulations; liq.; 98% conc.

Witco CHB Acid. [650] Alkylaryl sulfonic aicd; anionic; industrial detergent and o/w emulsifier; liq.; sol. in oil.

Witco D51-29. [650] Alkylaryl sulfonic acid; anionic; industrial detergent, dispersant, o/w emulsifier; liq.; water-sol.

Witco DTA-150, -180. [650] Fatty acid polymers; corrosion inhibitor intermediate; petroleum industry.

Witco DTA-350. [650] Fatty acid polymer; anionic; corrosion inhibitor for petroleum industry; liq.; sol. in IPA, kerosene, xylene.

Witcolate 58. [650] Ammonium alcohol ether sulfate; wetting agent, penetrant, lubricant, emulsifier, dye dispersant, scouring aid, antistat; detergent base; detergent for metal cleaning; textile surfactant.

Witcolate 1050. [650] Sodium pareth-25 sulfate; anionic; detergent, detergent base, emulsifier, foamer for the detergent industry; liq.; water-sol.

Witcolate 1247H. [650] Alcohol ether sulfate; anionic; industrial detergent, foamer; petroleum industry intermediate; electrolyte tolerant; sol. in IPA and water.

Witcolate 1259. [650] Alcohol ether sulfate; anionic; industrial coupler, detergent, foamer, solubilizer, o/w emulsifier, wetting agent for metal cleaning, dry cleaning, petroleum industry; electrolyte tolerant; liq.; sol. in water, naphthenic oils, IPA, perchloroethylene.

Witcolate 1276. [650] Alcohol ether sulfate; anionic; wetting agent, penetrant, lubricant, emulsifier, dye dispersant, scouring aid, antistat; foamer for wallboard mfg. and petroleum industry; detergent for metal cleaning; textile surfactant; emulsion polymerization and latex stabilization; liq.

Witcolate 1390. [650] Alcohol ether sulfate; anionic; foamer for petroleum industry; liq.; sol. in IPA and water.

Witcolate 7031. [650] Alcohol ether sulfate; anionic; industrial detergent, dispersant, foamer, wetting agent, penetrant; electrolyte tolerance; liq.; water-sol.

Witcolate 7093. [650] Sodium deceth sulfate; antionic; industrial detergent, foamer and wetting agent; electrolyte tolerance; liq.; water-sol.

Witcolate 7103. [650] Alcohol ether sulfate; anionic; see Witcolate 7093; liq.; water-sol.

Witcolate A. [650] Sodium lauryl sulfate; anionic; cleansing agent, detergent base, foamer, emulsifier, solubilizer for cosmetics and toiletries, industrial detergents; liq.; water-sol.

Witcolate A Powder. [650] Sodium lauryl sulfate; anionic; detergent and foamer used in personal care products; polymerization emulsifier; powd.; 90% conc.

Witcolate AE-3. [650] Ammonium pareth-25 sulfate; anionic; see Witcolate 1050; liq.; water-sol.

Witcolate AE3S. [650] Sodium alkyl ether sulfate; anionic; base for liq. detergents; liq.; 27% conc.

Witcolate AM. [650] Ammonium lauryl sulfate; anionic; see Witcolate AE-3; liq.; water-sol.

Witcolate D51-51. [650] Sodium alkylaryl polyether sulfate; anionic; antistat for syn. fibers and polymer products; surfactant, wetting agent and dispersant; clear liq.; sol. in water, ethanol; sp.gr. 1.06; visc. 2500; pH 8.0 (5% aq.); flash pt. (PMCC) > 93.3 C; surf. tens. 29.6 dynes/cm (at CMC); 29% act.

Witcolate D51-52. [650] Sodium alkylaryl polyether sulfate; anionic; emulsifier for polymerization; liq.

Witcolate D-510. [650] Sodium 2-ethylhexyl sulfate; wetting agent and penetrant for industrial use and polymerization reactions; liq.; water-sol.

Witcolate OME. [650] Sodium laureth sulfate; anionic; used in detergents and personal care products; liq.; 24% conc.

Witcolate S1285C. [650] Sodium laureth sulfate; anionic; cleansing agent, detergent, wetting agent, detergent base, and foamer for cosmetics, toiletries, industrial detergents, textiles; liq.; water-sol.

Witcolate S1300C. [650] Ammonium laureth sulfate; anionic; see Witcolate S1285C; liq.; water-sol.

Witcolate SE-5. [650] Sodium laureth sulfate; anionic; detergent, detergent base, foamer, o/w emulsifier, wetting agent for industrial detergents, textiles; liq.; water-sol.

Witcolate T. [650] TEA-lauryl sulfate; anionic; cleansing agent, and foamer, emulsifier, solubilizer for cosmetics and toiletries; detergent base; wetting agent for industrial detergents; liq.; water-sol.

Witcomul 78. [650] Sorbitan monotallate; w/o emulsifier, thickener, emulsion stabilizer for industrial applications; drilling fluid additive.

Witcomul 1557. [650] POE sorbitan monotallate; drilling fluid additive in petroleum industry.

Witcomul 3126. [650] Amine ester; drilling fluid additive for petroleum industry.

Witcomul 4016. [650] Complex alkylate; see Witcomul 3126.

Witconate 45 Liquid. [650] Sodium dodecylbenzene sulfate; anionic; detergent base, emulsifier, foamer, and wetting agent for detergent industry and industrial surfactants; electrolyte tolerant; water-sol.

Witconate 45BX. [650] Sodium dodecylbenzene sulfonate; anionic; detergent base; liq.; 42% conc.

Witconate 45LX Liquid. [650] Linear alkylaryl sodium sulfonate and sodium xylene sulfonate; anionic; detergent, wetting agent, base for household and industrial specialties; liq.; 45% conc.

Witconate 60B. [650] Sodium dodecylbenzene sulfonate; anionic; cleansing agent, foamer, solubilizer for cosmetics and toiletries; detergent base, foamer and wetting agent, emulsifier for industrial detergents; liq.; water-sol.

Witconate 60T. [650] TEA-dodecylbenzene sulfonate; anionic; see Witconate 60B; liq.; water-sol.

Witconate 79S. [650] TEA-dodecylbenzene sulfonate; emulsifier, wetting agent; industrial surfactant; water-sol.

Witconate 90 Flakes. [650] Dodecylbenzene sulfate; anionic; detergent base, emulsifier, foamer, and wetting agent for detergent industry; o/w emulsifier; wetting agent for metal cleaning; textile latex frothing agent; water-sol.

Witconate 93S. [650] Amine alkylaryl sulfonate; anionic; detergent, detergent base, emulsifier, foamer, wetting agent, solubilizer for the detergent industry; liq.; oil-sol.

Witconate 605A. [650] Calcium alkylaryl sulfonate; anionic; industrial detergent, dispersant, o/w and w/o emulsifier, lubricant, wetting agent; liq.; oil-sol.

Witconate 605T. [650] Alkylaryl sulfonate; penetrant, lubricant, scouring aid, antistat, o/w emulsifier, dispersant, oil wetting agent for industrial uses; additive for lube oils; textile surfactant; oil-sol.

Witconate 702. [650] Alkylaryl sulfonic acid, ammonium salt; demulsifier for the petroleum industry.

Witconate 1075X. [650] Amine alkylaryl sulfonate; anionic; industrial detergent, dispersant, o/w emulsifier, foamer, and wetting agent; liq.; water-sol.

Witconate 1240 Slurry. [650] Sodium alkylaryl sulfonate; anionic; detergent base, emulsifier, foamer, and wetting agent for the detergent industry; liq.; water-sol.

Witconate 1250. [650] Sulfonated alkylaryl; anionic; base for liq. detergents, emulsion polymerization; liq.

Witconate 1250 Slurry. [650] Sodium dodecylbenzene sulfonate; anionic; see Witconate 60B; liq.; water-sol.

Witconate 1260 Slurry. [650] Sodium dodecylbenzene sulfonate, sodium xylene sulfonate; anionic; see Witconate 60B; also industrial and textile surfactant; liq.; water-sol.

Witconate 1840X. [650] Fatty acid sulfonate; detergent, coupler, wetting agent, emulsifier, penetrant, lubricant, dye dispersant, scouring aid, antistat; industrial surfactant for alkaline systems; metal processing; textile surfactant; dk. liq.; sol. in water, IPA; sp.gr. 1.10; flash pt. > 93 C; pH 6–8 (5% aq.).

Witconate 3009-15. [650] Sodium alkylaryl ether sulfate; latex emulsifier and stabilizer for polymerizations.

Witconate AOS. [650] Sodium C_{14-16} olefin sulfonate; anionic; see Witconate 60B; also textile and emulsion polymerization surfactant; liq.; water-sol.

Witconate C50H. [650] Sodium dodecylbenzene sulfonate; anionic; industrial detergent, foamer and wetting agent; electrolyte tolerant; liq.; water-sol.

Witconate CHB. [650] Alkyaryl sulfonic acid; demulsifier used in the petroleum industry.

Witconate D51-51. [650] Sodium alkylaryl polyether sulfonate; anionic; emulsion polymerization surfactant; liq.; water-sol.

Witconate LX Flakes. [650] Sodium dodecylbenzene sulfonate; anionic; detergent base, emulsifier, foamer, and wetting agent for detergent industry and industrial surfactants; water-sol.

Witconate NCS. [650] Ammonium cumene sulfonate; anionic; antiblocking agent, coupler, solubilizer, cloud pt. depressant, and hydrotrope for detergent industry; liq.; water-sol.

Witconate NXS. [650] Ammonium xylene sulfonate; hydrotrope, solubilizer, coupler and processing aid in detergent mfg. and industrial processes; antiblocking and anticaking agent in powd. products; formulates shampoos, aerosols, cutting oils, glue; textile finishing; Klett 30 liq.; dens. 1.1 g/cc; pH 8.0; 40% act.

Witconate P-1052N. [652] Dodecylbenzene sulfonic acid, amine salt; anionic; o/w emulsifier, fuel oil additive, solubilizer, degreaser for emulsions; liq.

Witconate P-1059. [650] Dodecylbenzene sulfonic acid, amine salt; anionic; emulsifier, solubilizer, detergent, and wetting agent for oil-based systems; dispersant in oil and water-based systems; used in dry-cleaning surfactants; amber clear liq.; sol. in water, kerosene, xylene, IPA; sp.gr. 1.02; flash pt. > 93 C; pH 4.8 (20% in 25% IPA).

Witconate P-1073F. [650] Alkylaryl sulfonic acid, amine salt; drilling fluid additive used in the petroleum industry.

Witconate SCS. [650] Sodium cumene sulfonate; see Witconate NXS; Klett 50 liq., powd.; dens. 0.32–1.2 g/cc; pH 7.5–9.0; 45% act.

Witconate STS. [650] Sodium toluene sulfonate; see Witconate NXS; Klett 150 liq., powd.; dens. 0.32–1.2 g/cc; pH 9.0–10.0; 40% act.

Witconate SXS. [650] Sodium xylene sulfonate; see Witconate NXS; Klett 40 liq., powd.; dens. 0.32–1.2 g/cc; pH 8.6–9.6; 40% act.

Witconate TAB. [650] TEA-dodecylbenzene sulfonate; anionic; see Witconate 60B; detergent, foamer, and wetting agent in industrial surfactants, paints; electrolyte tolerant; liq.; water-sol.

Witconate TDB. [650] Tridecylbenzene sulfonate; anionic; detergent base, emulsifier, foamer, and wetting agent for the detergent industry; used in industrial surfactants; liq.; water-sol.

Witconate YLA. [650] Amine alkylaryl sulfonate; anionic; detergent base, emulsifier, solubilizer, foaming and wetting agent for the detergent, metal cleaning, and textile industries; liq.; oil-sol.

Witconol 14. [650] Polyglyceryl-4 oleate; nonionic; w/o emulsifier, lubricant, and antifoamer for industrial use; liq; 100% conc.

Witconol 14F. [650] Polyglycerol fatty acid ester; nonionic; w/o emulsifier for food processing; liq.; oil-sol.

Witconol 18F. [650] Polyglyceryl-4 stearate; nonionic; see Witconol 14F; solid; oil-sol.

Witconol 18L. [650] Polyglyceryl-4 isostearate; nonionic; see Witconol 14; also used as solubilizer; liq.; 100% conc.

Witconol 171, 172. [650] Polyalkylene glycol ether; defoamer for soap and detergent foam control; surfactant.

Witconol 1206. [650] Alkyl POE glycol ether; wetting agent for industrial uses; visc. and flow control agent for polymerization reactions.

Witconol 1207. [650] Alkyl POE glycol ether; wetting agent for industrial uses.

Witconol APEM. [650] PPG-3-myreth-3; nonionic; lubricant, emulsifier, wetting agent, penetrant, dye dispersant, scouring aid, antistat; syn. oils for personal care products; metal processing; textile surfactant; liq.

Witconol APM. [650] PPG-3 myristyl ether; nonionic; see Witconol APEM; liq.

Witconol APS. [650] PPG-11 stearyl ether; nonionic; see Witconol APEM; liq.

Witconol CA. [650] Glyceryl stearate SE; nonionic; lubricant, bodying agent, emulsifier, opacifier used in industrial, cosmetic and aerosol formulations; wax; 100% conc.

Witconol CAD. [650] Diethylene glycol monostearate; anionic; bodying agent, lubricant, and opacifier for cosmetics and toiletries; wax; 100% conc.

Witconol CD-17. [650] PPG-34; nonionic; antistat, emulsifier for personal care products; syn. lubricant oil and foam modifier for aerosol formulations.

Witconol CD-18. [650] Polypropoxylated polyol; nonionic; syn. oil for personal care products, cosmetic emulsifier, lubricant, antistat.

Witconol DOS. [650] Diethylene glycol monooleate; nonionic; o/w emulsifier for cosmetic and industrial formulations; liq.; 100% conc.

Witconol F26-46. [650] PPG-36 oleate; nonionic; cosmetic oil, spreading agent and coupler for oil systems; dispersant for industrial use; spreading and anticaking agent for aerosol products; visc. and flow control agent; liq.

Witconol GMOP. [650] Mono- and diglycerides; nonionic; o/w emulsifier for food processing; liq.; sol. in oil.

Witconol GOT. [650] Glycerol mono- and dioleate; nonionic; surfactant and emulsifier for industrial use; liq.; sol. in oil.; 100% conc.

Witconol H-31. [650] PEG-8 tallate; nonionic; industrial defoamer, dispersant and o/w emulsifier; metal processing emulsifiable oil; liq.; sol. in oil.

Witconol H-31A. [650] PEG-8 oleate; nonionic; emulsifier and spreading agent for personal care products and industrial formulations; liq.; 100% conc.

Witconol H-33. [650] PEG-8 ditallate; nonionic; industrial detergent; liq.; oil-sol.

Witconol H-35A. [650] PEG-8 stearate; nonionic; o/w emulsifier for cosmetic and industrial uses; wax.

Witconol MST. [650] Glycerol mono- and distearate; nonionic; cosmetic, pharmaceutical, aerosol formulations; internal lubricant, plasticizer, and emulsifier in industrial applications; flow control agent for polymerization reactions; solid; 100% conc.

Witconol NP-40. [650] Alkylaryl polyether alcohol; nonionic; detergent, o/w emulsifier, solubilizer for oils in metal processing; liq.; sol. in naphthenic and paraffinic oil.

Witconol NP-100. [650] Alkyl polyether; pigment dispersant, latex stabilizer, and leveling agent for polymerization reactions.

Witconol NP-300. [650] Alkylaryl polyether alcohol; nonionic; paint industry emulsifier for emulsion polymerization; pigment wetting and grinding agent for aq. systems; spreading agent; solid, liq.; water-sol.

Witconol NS500K. [650] Polyalkoxylated butyl ether; nonionic; industrial antistat, coupler, detergent, dispersant, o/w emulsifier, lubricant, solubilizer, spreading agent; paste; water-sol.

Witconol O. [650] Glycerol mono- and dioleate; industrial surfactant and emulsifier; oil-sol.

Witconol RDC-D. [650] Diglycol coconate; nonionic; surfactant, emulsifier; cosmetic and industrial use; liq.; oil-sol.

Witconol RHP. [650] Propylene glycol monostearate SE; nonionic; emollient, conditioner, emulsifier, lubricant, opacifier and visc. modifier for cosmetics and toiletries; flake; sol. in oil.

Witconol RHT. [650] Glyceryl stearate SE; nonionic; lubricant, plasticizer, emulsifier used in industrial applications; paste; 100% conc.

Witcor 3192. [650] Complex amine phosphate; cationic; corrosion inhibitor for petroleum industry; liq.; sol. in IPA, kerosene, xylene; disp. in water.

Witcor 3194, 3195. [650] Fatty amide; cationic/nonionic; corrosion inhibitor for petroleum industry; liq.; sol. in IPA, kerosene, xylene.

Witcor SI-3065. [650] Complex phosphate ester; scale inhibitor; anionic used in petroleum industry; liq.

Witcosperse 201. [650] Alkylaryl polyethoxyethanol; nonionic; wetting agent for pesticide formulations; powd.; 50% conc.

Witcosperse 205. [650] Polyoxyalkylated glycol; see Witcosperse 201; powd.; 50% conc.

Witcosperse 206. [650] Ethoxylated sorbitan ester; see Witcosperse 201; powd.; 50% conc.

Witcosperse 207. [650] Alkyl polyoxyalkylene ether; see Witcosperse 201; powd.; 50% conc.

YZ

Yeoman. [159] Anhyd. lanolin BP; used in personal care products; superfatting agent for soap; aids dispersion of pigments into anhyd. systems; yel. unctuous mass; lipophilic.

Zeebase 1, 2, 3. [177] Carboxylated alcohol; nonionic; base for detergents and emulsifier; liq.; 85% conc.

Zeeterge 444. [177] Carboxylated alcohol; nonionic; specialty cleaner; liq.; 70% conc.

Zelec NE. [208] Fatty alcohol phosphate; anionic; antistat for textiles, plastics, and films; yel. visc. paste; mild fatty alcohol odor; water-sol.; dens. 9.3 lb/gal; sp.gr. 1.10; flash pt. > 210 F (PMCC); pH 7.0–7.5 (40%); 100% act.

Zelec NK. [159] Fatty alcohol phosphate; see Zelec NE; lt. yel. visc. paste; mild fatty alcohol odor; disp. in water; dens. 8.7 lb/gal; sp.gr. 1.05; flash pt. > 210 F (PMCC); pH 7.0–7.5 (10% disp.); 100% act.

Zelec TY. [208] Alcohol phosphate; anionic; antistat used on polyolefin fibers, plastics, and films; lt. amber liq.; mild odor; water-sol.; dens. 9.8 lb/gal; pH 6.7–7.3 (10%).

Zelec UN. [208] Unneutralized fatty alcohol phosphate; anionic; antistat; additive for oils to improve lubricity and provide corrosion protection; formulation of textile lubricants and finishes; release agent in plastic molding operations; pale yel. liq.; sol. in perchloroethylene, kerosene, Stod., toluene, glycerin; disp. in water; dens. 8.22 lb/gal; sp.gr. 0.99; pH 2 (5% disp.); 100% act.

Zetesol 100. [660] Fatty alcohol ether sulfate with nonionics; anionic; detergent and emulsifier for personal care products; liq.; 99% conc.

Zetesol 856, D. [660] Lauryl alcohol ether sulfate, alkanolamine salt; anionic; detergent for personal care products; liq.; 56% conc.

Zetesol 856 DT, T. [660] Lauryl alcohol ether sulfate with betaine alkanolamine salt; anionic; detergent for personal care products; liq.; 56% conc.

Zetesol 2056. [660] Fatty alcohol ether sulfate; anionic; detergent for personal care products, household and industrial cleaners; liq.; 56% conc.

Zetesol 2210. [660] Lauryl alcohol ether sulfate, alkanolamine salt; anionic; detergent for personal care products; liq.; 25% conc.

Zetesol AP. [660] Fatty alcohol ether sulfate, ammonium salt; anionic; detergent for personal care products; liq.; 60% conc.

Zetesol MS. [660] Fatty alcohol alcohol ether sulfate; anionic; detergent for personal care products; liq.; 28% conc.

Zetesol NL. [660] Lauryl alcohol ether sulfate, sodium salt; anionic; detergent for personal care, household and industrial cleaners; liq.; 28% conc.

Zetesol SE 35. [660] Fatty alcohol ether sulfate with pearlescent; anionic; detergent, cosmetic, emulsion and personal care products; 35% conc.

Zohar 60 SD. [659] Sodium alkylbenzene sulfonate and builders; anionic; conc. for powd. mfg.; 60% conc.

Zohar 70 SD. [659] Sodium alkylbenzene sulfonate and builders; anionic; conc. for powd. mfg.; 70% conc.

Zohar Automat SD. [659] Sodium alkylbenzene sulfonate and builders; powd. for machine washing; powd.; 60% conc.

Zohar Export SD. [659] Sodium alkylbenzene sulfonate and builders; spray built powd.; powd.

Zohar KAL SD. [659] Sodium alkylbenzene sulfonate and builders; spray built powd. for machine washing; powd.

Zoharex A10, B10. [659] Fatty acid polyglycol ester; nonionic; detergent for laundry liqs. and powds.; liq.; 100% conc.

Zoharlab. [659] Linear alkylbenzene sulfonate; anionic; raw material for mfg. of detergents; paste; 60% conc.

Zoharpon DT-80. [659] Alkanolamine lauryl sulfate; anionic; raw material for shampoo and foam baths; liq.; 80% conc.

Zoharpon ETA 27. [659] Sodium lauryl ether sulfate; anionic; raw material for personal care products; liq.; 27% conc.

Zoharpon ETA 35. [659] Sodium lauryl ether sulfate and nonionic compds.; anionic/nonionic; raw material for mfg. of emulsion, pearlescent shampoos; paste; 35% conc.

Zoharpon ETA 70, 703. [659] Sodium lauryl ether sulfate (2 and 3 EO resp.); anionic; raw material for liq. shampoos and cosmetic preparations; lt.-duty detergents; paste; 70% conc.

Zoharpon K. [659] Sulfosuccinate and alkyl sulfate; anionic; conc. for carpet and upholstery shampoos; liq.; 25% conc.

Zoharpon LAA. [659] Ammonium lauryl sulfate; anionic; raw material for personal care products; liq.; 29% conc.

Zoharpon LAM. [659] Monoethanolamine lauryl sulfate; anionic; raw material for shampoos and foam baths; liq.; 29% conc.

Zoharpon LAS. [659] Sodium lauryl sulfate; anionic; raw material for personal care and pharmaceutical products; liq.; 30% conc.

Zoharpon LAS Special. [659] Sodium lauryl sulfate; anionic; raw material for shampoos; liq., paste; 33% conc.

Zoharpon LAS Spray Dried. [659] Sodium lauryl sulfate; anionic; raw material for mfg. of shampoo, toothpaste, pharmaceutical preparations; powd.; 92% conc.

Zoharpon LAT. [659] Triethanolamine lauryl sulfate; anionic; raw material for personal care preparations; liq.; 40% conc.

Zoharpon SA. [659] Alcohol sulfosuccinate; anionic; see Zoharpon LAT; paste; 50% conc.

Zoharpon SE. [659] Alcohol ether sulfosuccinate; anionic; see Zoharpon LAT; liq.; 40% conc.

Zoharpon SM. [659] Alkanolamide sulfosuccinate; anionic; raw material for shampoos; paste; 40% conc.

Zoharpon ST. [659] Sulfosuccinate and ether sulfate; anionic; base for mild shampoos; liq; 28% conc.

Zoharsoft, 90. [277] Quat. imidazoline deriv.; cationic; fabric softener base; liq., paste; 75 and 90% conc. resp.

Zoharsoft DAS. [277] Fatty acid deriv.; cationic; softener base for textiles; flakes; 100% conc.

Zoharteric D. [659] Cocoamphodiacetate; amphoteric; component for personal care products; detergent for cleaners; liq. to gel; liq., gel; 50% conc.

Zoharteric D-2, D-3, D-3 50%. [659] Cocoamphodiacetate and trideceth sulfate; amphoteric; component of personal care products; liq.; 38, 38, and 50% conc. resp.

Zoharteric M. [659] Cocoamphoacetate; amphoteric; component for personal care products; detergent for cleaners; liq.; 44% conc.

Zoharteric M-2, M-3, M-3 50%. [659] Cocoamphoacetate and sodium trideceth sulfate; amphoteric; personal care products; liq.; 35, 35, and 50% conc. resp.

Zontes PD-101. [397] Alkyl polyamide deriv.; cationic; softener for textiles; flake; 100% conc.

Zontes TA-460-15. [397] Alkyl polyamide deriv.; cationic; see Zontes PD-101; paste; 15% conc.

Zonyl A. [208] EO-ester condensate; nonionic; emulsifier, lubricant; textile wetting agent for fibers and fabrics during dyeing; leveling agent and antifoam in pigment coatings in paper industry; used in metal processing; clear-pale yel. liq.; mild odor; sol. in water, polar and nonpolar solvs.; sp.gr. 1.07; HLB 6.7; surf. tens. 26 dynes/cm (0.1% aq.); 100% act.

Zonyl FSA. [208] Fluorochemical; anionic; wetting agent, emulsifier, dispersant, corrosion inhibitor for adhesives, agriculture, polymerization, cleaners, films and coatings, paper industry, petroleum, plastics, textiles, etc.; liq.; sol. > 2% in water, methanol; dens. 9.8 lb/gal; flash pt. 69 F (PMCC); surf. tens. 18 dynes/cm (0.1%); 50% solids in IPA/water.

Zonyl FSB. [208] Fluorochemical; amphoteric; see Zonyl FSA; also foaming agent for elastomers, fire fighting; liq.; sol. > 2% in water, IPA, methanol; dens. 8.8 lb/gal; flash pt. 67 F (PMCC); surf. tens. 17 dynes/cm (0.1%); 40% solids in IPA/water.

Zonyl FSC. [208] Fluorochemical; cationic; see Zonyl FSB; liq.; sol. > 2% in water, IPA, methanol, acetone; dens. 9.7 lb/gal; flash pt. 70 F (PMCC); surf. tens. 18 dynes/cm (0.1%); 50% solids in IPA/water.

Zonyl FSJ. [208] Fluorochemical; anionic; wetting agent for adhesives, agric., cleaners, fire fighting, petroleum, resin molds, textiles; liq.; sol. > 2% in water, methanol; dens. 9.3 lb/gal; flash pt. 83 F (PMCC); surf. tens. 20 dynes/cm (0.1%); 40% solids in IPA/water.

Zonyl FSN. [208] Fluorochemical; nonionic; see Zonyl FSA; liq.; sol. > 2% in water, IPA, methanol, acetone, ethyl acetate; dens. 8.8 lb/gal; flash pt. 72 F (PMCC); surf. tens. 23 dynes/cm (0.1%); 40% solids in IPA/water.

Zonyl FSP. [208] Fluorochemical; see Zonyl FSA; liq.; sol. > 2% in water, methanol; dens. 9.6 lb/gal; flash pt. 75 F (PMCC); surf. tens. 19 dynes/cm (0.1%); 35% solids in IPA/water.

Zonyl RP. [208] Fluorochemical; anionic; wetting agent, paper fluoridizer; amber liq.; sol. in water; dens. 9.5 lb/gal; visc. 10 cps; flash pt. > 28 C; pH 7–8; 34% conc.

Zusolat 1005/85, 1008/85, 1010/85, 1012/85. [660] Alkyl polyglycol ether; nonionic; dispersant, emulsifier, wetting agent; liq.; 85% conc.

Zusomin C 108. [660] Fatty amine ethoxylate; nonionic; basic material and component for dyeing and textile auxs.; intermediate for quaternization; metal cleaners; liq.; 100% conc.

Zusomin O 102, O 105, O 112. [660] Fatty amine ethoxylate; nonionic/cationc; basic material for textile and dyeing auxs., intermediate for quaternization, metal cleaners; liq.; 100% conc.

Zusomin S 110, S 125. [660] Fatty amine ethoxylate; nonionic; see Zusomin O 102; liq., paste; 100% conc.

Chemical Component-to-Tradename Cross Reference

Acetamide MEA [CAS #142-26-7]
Synonyms: N-Acetyl ethanolamine
Acetamide MEA, MEA-70, MEA-100; Incromectant AMEA-70; Lipamide MEAA; Mackamide AME-75; Schercomid AME, AME-70

Acetylated hydrogenated cottonseed glyceride [RD #977055-83-6]
Synonyms: Glycerides, cottonseed-oil, mono-, hydrogenated, acetates
Myvacet 5-07, 7-07K

Acetylated hydrogenated lard glyceride [CAS #8029-91-2]
Synonyms: Glycerides, lard mono-, hydrogenated, acetates
Myvacet 7-00; Tegin E 61, E-61 NSE

Acetylated hydrogenated tallow glyceride [CAS #68990-58-9]
Synonyms: Glycerides, tallow mono-, hydrogenated, acetates
Lamegon EE; Tegin E 41

Acetylated hydrogenated tallow glycerides [RD #977063-59-4]
Synonyms: Glycerides, tallow mono-, di- and tri-, hydrogenated, acetates
Tegin E-41 NSE

Acetylated lanolin [CAS #61788-48-5]
Lanacet 1705; Ritacetyl

Acetylated lanolin alcohol [CAS #61788-49-6]
Crodalan LA; Ritawax ALA; Solulan 97, 98

Acetylated lard glyceride [CAS #8029-92-3]
Synonyms: Glycerides, lard mono-, acetates
Myvacet 9-40; Tegin E 66, E-66 NSE

Acetylated sucrose distearate
Synonyms: Sucrose distearate, acetates
Crodesta A-10, A-20

Acetylated tartrated vegetable monoglyceride
Drewlate 30

Acetylated vegetable oil glyceride
Myvacet 9-45

Acrylamide/β-methacrylyloxyethyl trimethyl ammonium methosulfate copolymer. See *Polyquaternium-5*

Adipic acid/dimethylaminohydroxypropyl diethylenetriamine copolymer [CAS #61840-27-5]
Cartaretin F-4, F-8

Algin [CAS #9005-38-3]
Synonyms: Sodium alginate
Kelco-Pac: Kelcosol; Kelgin F, HV, LV, MV, QH, QL, QM, RL, XL; Keltex, P, S; Keltone; Kelvis

Alginic acid [CAS #9005-32-7]
Kelacid

Alkyl dimethyl benzyl ammonium chloride. See *Benzalkonium chloride*

Alkyl dimethyl benzyl ammonium saccharinate
Loraquat QA 100

Alkyl dimethyl dichlorobenzyl ammonium chloride
Bio-Quat 50-MAC, T-502

Alkyl dimethyl ethyl ammonium chloride
Bio-Quat 50-36

Alkyl dimethyl ethylbenzyl ammonium chloride
Bio-Quat 50-35, 80-35, 80-36; BTC 1326, 2125 80%, 2125 M 80%, 2125 MP-40, 2125 P-40

Alkyl dimethyl ethylbenzyl ammonium cyclohexyl sulfamate. See *Quaternium-8*

Alkyl trimethyl ammonium bromide
Catinal HTB

Alkyl trimethyl ammonium chloride
Variquat A200; Varisoft E228

Aluminum distearate [CAS #300-92-5]
Aluminum Stearate EA

Aluminum stearate [CAS #7047-84-9]
Dehymuls E, K; Hostacerin WO

Amine dodecylbenzene sulfonate
Amsul 70; Hetsulf 60T; Marlon AMX; Marlopon AMS 60; Ninate 411; Witconate P-1059, P-1052N

2-Amino-2-ethyl-1,3-propanediol
Synonyms: Aminoamylene glycol; AEPD
AEPD

Aminomethyl propanol [CAS #124-68-5]
Synonyms: Isobutanolamine; AMP
AMP, AMP-95

Ammonium alginate [CAS #9005-34-9]
Keltose

Ammonium C12-15 pareth sulfate [RD #977064-04-2]
Synonyms: Ammonium pareth-25 sulfate; POE (1–4) C12-15 alcohol ether sulfate, ammonium salt
Dobanol 25-3A/27 EU, 25-3A/60; Neodol 25-3A; Standapol AP-60; Witcolate AE-3

Ammonium cocamidopropyl dimethyl betaine
Schercotaine CAB-A

Ammonium cocamidopropyl hydroxy sulfopropyl dimethyl betaine
Schercotaine SCAB-A

Ammonium cumenesulfonate [CAS #37475-88-0]
Eltesol ACS 60; Naxonate 6AC; Reworyl ACS 60; Witconate NCS; Ultra NCS Liquid

Ammonium dodecylbenzene sulfonate [CAS #1331-61-9]
Conco AAS-50M; Hetsulf 50A; Nansa AS 40; Newcol 210; Sandoz Sulfonate AAS 50MS

Ammonium laureth sulfate [CAS #32612-48-9 (generic)]
Synonyms: Ammonium lauryl ether sulfate
Alkasurf EA-60; Carsonol SES-A; Cedepal SA-406; Conco Sulfate 216, WM; Cycloryl LDC, MA 330, MA 360; DeSonol AE, AES; Drewpon EKZ; Empicol EAA70, EAB; Geropon AB/20; Lakeway 201-20, 201-29; Manro ALES 27; Maprofix AEG, LES-60A; Montosol PF-16, PF-18, PG-12, PQ-11, PQ-15, TQ-11; Neopon LOA, LOS; Nutrapon AL 1, Al 30, AL 60, KF 3846; Polystep B-11; Rewopol AL 3 60%; Richonol S-1300C; Sandoz Sulfate 216; Sipon 201-20, EA, EA2, EAY; Standapol AL-60, EA-1, EA-2, EA-3; Steol CA-460; Sterling ES 600; Sulfotex OT; Surco EA58M; Texapon EA-1, -2, -3; Tylorol A; Ungerol AM3-60; Witcolate S1300C

Ammonium laureth-12 sulfate [CAS #32612-48-9 (generic)]
Equex AEM; Polystep B-22; Standapol 230-E Conc.

Ammonium laureth-30 sulfate
Polystep B-20

Ammonium lauroyl sarcosinate [CAS #68003-46-3]
Synonyms: N-methyl-N-(1-oxododecyl) glycine, ammonium salt
Maprofix RSN

Ammonium lauryl sulfate [CAS #2235-54-3]
Akyposal ALS 33; Alkasurf ALS; Avirol 200; Carsonol ALS, ALS Special; Cedepon LA-30, LS-30, LS-30P; Conco Sulfate A; Cycloryl MA; DeSonol A; Drewpon NH, NHAC; Emal AD-25; Emersal 6430; Empicol AL 30, AL-30T, AL-70, TC30/T, TC 34, TCR; Incronol ALS; Jordanol AL 300; Lakeway 101-20; Lonzol LA-300; Manro ALS 30; Mapeg AEG; Maprofix ER, NH, NHL, NHL-22, NH-54, RSN; Mars AMLS; Montovol RF-12; Neopon LAM; Norfox ALS; Nutrapon HA 3841; Polyfac ALS-30; Polystep B-7; Richonol AM; Sandoz Sulfate A; Serdet DFN 30; Sermul EA 129; Sipon L-22, L-22HV; Sipoteric 1444; Standapol A, A-HV, AMS-100; Stepanol ALS-30, AM, AM-V; Sterling AM, AM-HV; Sulfetal CA 30; Sulphonated Lorol Liquid NH; Texapon A, A 400, Special; Witcolate AM; Zoharpon LAA

Ammonium lignosulfonate
Lignosite 17; Lignosol TS, TSD; Orzan A, AL-50

Ammonium myreth sulfate [CAS #27731-61-9]
Synonyms: Ammonium myristyl ether sulfate
Standapol Blend EA-K, EA-40 Conc.; Texapon EA-40

Ammonium naphthalene sulfonate
Lomar PWA

Ammonium nonoxynol-4 sulfate [CAS #31691-97-1 (generic); 63351-73-5; RD #977060-75-5]
Alipal CO-436; Neutronyx S-60

Ammonium oleamido PEG-2 sulfosuccinate
Standapol SH-200

Ammonium pareth-25 sulfate. See *Ammonium C12-15 pareth sulfate*

Ammonium stearate [CAS #1002-89-7]
Ammonium Stearate 33% Liquid

Ammonium/TEA lauryl sulfate
Sulfatol B6

Ammonium xylenesulfonate [CAS #26447-10-9]
AXS; Eltesol AX 40; Hartotrope AXS 40; Naxonate 4AX, AX; Stepanate AM; Ultra NXS Liquid; Witconate NXS

Apricot kernel oil PEG-6 complex
Labrafil M 1944 CS

Arachidic acid [CAS #506-30-9]
Synonyms: Eicosanoic acid
Hystrene 7022

Arachidic/behenic fatty acid
Industrene 4022, 5022

Arachidyl-behenyl amine
Kemamine P-150, P-150D, P-190, P-190D; Lilamin 101 D

Arachidyl-behenyl 1,3-propylene diamine
Kemamine D-150, D-190

Batyl alcohol [CAS #544-62-7]
Synonyms: Octadecyl ether of glycerin; Stearyl glyceryl ether; 3-(Octadecyloxy)-1,2-propanediol)
Nikkol Batyl Alcohol 100, EX

Batyl isostearate
Nikkol GM-18IS

Batyl stearate [CAS #1323-26-3]
Nikkol GM-18S

Beeswax [CAS #8006-40-4 (white), 8012-89-3 (yellow)]
Alcolan 40; Cyclochem 326A; Dehymuls E, K; Hostacerin WO

Behenalkonium chloride [CAS #16841-14-8]
Synonyms: Behenyl dimethyl benzyl ammonium chloride
Incroquat Behenyl BDQ/P; Kemamine BQ-2802C

Behenamide [CAS #3061-75-4]
Synonyms: Behenic acid amide; Docosanamide
Kemamide B

Behenamidopropylamine oxide
Mackamine BAO

Behenamidopropyl dimethylamine [RD #977063-18-5]
Incromine BB; Lexamine B-13; Schercodine B

Behenamine
Synonyms: Behenyl amine
Nissan Amine VB

Behenamine oxide
Incromine Oxide B-30P

Beheneth-5
Synonyms: PEG-5 behenyl ether
Nikkol BB-5

Beheneth-10
Nikkol BB-10

Beheneth-20
Nikkol BB-20

Beheneth-30
Nikkol BB-30

Behenic acid [CAS #112-85-6]
Synonyms: Docosanoic acid
Hystrene 7022, 9022

Behentrimonium chloride [CAS #17301-53-0]
Synonyms: Behenyl trimethyl ammonium chloride
Incroquat Behenyl TMC/P

Behentrimonium methosulfate
Synonyms: Behenyl trimethyl ammonium methyl sulfate
Incroquat Behenyl TMS

Behenyl alcohol [CAS #661-19-8]
Synonyms: 1-Docosanol
Lanette 22

Behenyl betaine
Incronam B-40

Behenyl hydroxyethyl imidazoline
Synonyms: Behenyl imidazoline
Schercozoline B

Benzalkonium bromide
Synonyms: Alkyl dimethyl benzyl ammonium bromide
Bio-Quat 50-MAB

Benzalkonium chloride [CAS #8001-54-5]
Synonyms: Alkyl dimethyl benzyl ammonium chloride)
Alacsan 7LUF; Alkaquat DMB-451; Arquad B-50 USP, B-90 USP, B-1000, DMCB, DMMCB-50, DMMCB-75; Barquat 1552; Bio-Dac 205; Bio-Quat 50-24, 50-25, 50-28, 50-30, 50-35, 50-36, 50-40, 50-42, 50-60, 50-65, 80-24, 80-28, 80-35, 80-36, 80-40, 80-42, T-501; BTC 50, 50 USP, 65, 65 USP, 100, 776, 835, 885, 1326, 2125 80%, 2125 M 80%, 2125 MP-40; BTC 2125 P-40, 2565, 8248, 8249, E-8358; Carsoquat 621, 621 80%; Crapol AU-20, AU-21, AU-24; Empigen BAC, BCB50, BCF 80; Gardiquat 12H, 1450, 1480, SV 480; Hyamine 3500; Lutensit K-LC, K-LC 80, K-OC; Quadrilan BC; Retarder CA; Rewoquat B 50; Roccal 50% Tech., II 50%; Variquat 50MC, 50ME, 60LC, 80MC, 80ME, B345, LC80

Benzethonium chloride [CAS #121-54-0]
Synonyms: Diisobutyl phenoxy ethoxy ethyl dimethyl benzyl ammonium chloride
Hyamine 1622

Benzyl trimonium chloride
Synonyms: Benzyl trimethyl ammonium chloride
Variquat B200

BHA [CAS #25013-16-5]
Synonyms: Butylated hydroxyanisole
Tandem 5K, 8, 552

BHT [CAS #128-37-0]
Synonyms: Butylated hydroxytoluene
Hydagen DEO

Bis(undecylenic amidopropyl dimethyl glycinate)
Schercotaine UAB

Butyl alcohol [CAS #71-36-3]
Synonyms: *n*-Butyl alcohol, 1-Butanol; *n*-Butanol; Propyl carbinol
Alfol 4

Butyl ammonium caprylate
Rewocor BAC

Butyl myristate [CAS #110-36-1]
Synonyms: Butyl *n*-tetradecanoate)
Radia 7070

Butyl naphthalene sodium sulfonate
Emkal BNS, BNX Powder

Butyl naphthalene sulfonic acid
Emkal BNS Acid

Butyl oleate [CAS #142-77-8]
Synonyms: Butyl 9-octadecenoate
Grocor 4000 ; Iberwet-BO; Kemester 4000; Kessco Butyl Oleate; Radia 7040

Butyl stearate [CAS #123-95-5]
Synonyms: Butyl octadecanoate
Crodamol BS; Emerest 2325, 2326; Grocor 5410, 5510; Kessco Butyl Stearate Cosmetic, Distilled; Lexate TL; Radia 7051

C6 alcohol
Epal 6

C6-10 acid triglyceride
Softenol 3819

C6-10 alcohol
Alfol 610; Epal 108, 610

C7 acid triglyceride
Softenol 3107

C8 alcohol
Epal 8

C8-10 acid-1,2-propanediol ester
Softenol 3408

C8-10 acid triglyceride
Softenol 3108, 3829

C8-10 alcohols
Alfol 810; Epal 810

C8-10 pareth-5 hydrogen phosphate
Synonyms: PEG-5 C8-10 alcohol hydrogen phosphate
Merpiphos B 59

C8-10 trimethyl ammonium chloride
Adogen 464

C8-18 acid triglyceride
Softenol 3178

C9-11 alcohols [CAS #68551-08-6]
Dobanol 91; Neodol 91

C9-11 pareth-2
Synonyms: Pareth-91-2; PEG-2 C9-11 alcohol ether
Merpoxen UN 22

C9-11 pareth-3 [CAS #68439-46-3 (generic)]
Synonyms: Pareth-91-3; PEG-3 C9-11 alcohol ether
Dobanol 91-2.5; Lenetol 9130 AH; Neodol 91-2.5

C9-11 pareth-4
Synonyms: Pareth-91-4; PEG-4 alkyl C9-11 alcohol
Merpoxen UN 40; Serdox NBS 4

C9-11 pareth-5
Synonyms: Pareth-91-5; PEG-5 alkyl C9-11 alcohol
Dobanol 91-5; Merpoxen UN 50; Serdox NBS 5.5 (5.5 EO); Teric G9A5

C9-11 pareth-6 [CAS #68439-46-3 (generic)]
Synonyms: Pareth-91-6; PEG-6 C9-11 alcohol ether
Dobanol 91-6; Merpoxen UN 65 (6.5 EO); Neodol 91-6; Serdox NBS 6; Serdox NBS 6.6 (6.6 EO); Teric G9A6

C9-11 pareth-8 [CAS #68439-46-3 (generic)]
Synonyms: Pareth-91-8
Dobanol 91-8; Neodol 91-8; Serdox NBS 8.5 (8.5 EO); Teric G9A8

C9-11 pareth-12
Teric G9A12

C10-12 alcohols
Alfol 1012; Epal 1012

C10-12 pareth-5
Synonyms: PEG-5 C10-12 alcohol
Merpoxen DL 52 (5.2 EO)

C10-12 undecylbenzene sulfonate
Nalkylene 500

C10-14 dodecylbenzene sulfonate
Nalkylene 550L

C11-15 pareth-3
Synonyms: Pareth-15-3
Tergitol 15-S-3

C11-15 pareth-5
Synonyms: Pareth-15-5
Tergitol 15-S-5

C11-15 pareth-7
Synonyms: Pareth-15-7
Tergitol 15-S-7

C11-15 pareth-9
Synonyms: Pareth-15-9
Tergitol 15-S-9

C11-15 pareth-12
Synonyms: Pareth-15-12
Tergitol 15-S-12

C11-15 pareth-15
Synonyms: Pareth-15-15
Tergitol 15-S-15

C11-15 pareth-20
Synonyms: Pareth-15-20
Tergitol 15-S-20

C11-15 pareth-30
Synonyms: Pareth-15-30
Tergitol 15-S-30

C11-15 pareth-40
Synonyms: Pareth-15-40
Tergitol 15-S-40

C12-13 alcohols [RD #977063-82-3]
Dobanol 23; Neodol 23

C12-13 pareth-2
Synonyms: Pareth-23-2
Dobanol 23-2

C12-13 pareth-3 [CAS #66455-14-9 (generic)]
Synonyms: Pareth-23-3; PEG-3 C12-13 fatty alcohol ether
Dobanol 23-3; Neodol 23-3

C12-13 pareth-7 [CAS #66455-14-9 (generic)]
Synonyms: Pareth-23-7; PEG-7 C12–13 fatty alcohol ether
Dobanol 23-6.5; Neodol 23-6.5

C12-14 alcohols
Alfol 1214; Epal 12/70, 12/85

C12-14 alcohols hydrogen phosphate
Merpophal SP

C12-14 pareth-2
Synonyms: PEG-2 C12-14 alcohol
Merpoxen LM 20

C12-14 pareth-3
Synonyms: PEG-3 C12-14 alcohol
Merpoxen LM 30; Serdox NEL 3

C12-14 pareth-4 hydrogen phosphate
Synonyms: PEG-4 C12-14 alcohol hydrogen phosphate
Merpiphos PP

C12-14 pareth-8
Synonyms: PEG-8 C12-14 fatty alcohol
Lutensol A 8

C12-14 pareth-12
Synonyms: PEG-12 C12-14 alkyl ether
Serdox NEL 12/80

C12-14 tridecylbenzene sulfonate
Nalkylene 600L

C12-15 alcohols [RD #977063-83-4]
Dobanol 25; Neodol 25

C12-15 alcohols benzoate
Finsolv TN

C12-15 alcohols octanoate
Hetester FAO

C12-15 pareth-2 phosphate
Synonyms: Pareth-25-2 phosphate; PEG-2-C12-15 alcohols phosphate
Crodafos 25 D2 Acid; Nikkol TDP-2

C12-15 pareth-3 [CAS #68131-39-5 (generic); RD #977061-52-1]
Synonyms: Pareth-25-3; PEG-3 C12–15 fatty alcohol ether
Alkasurf LA-3; ; Dobanol 25-3; Neodol 25-3; Serdox NES 3; Sterling Emulsifier #3; Tergitol 25-L-3; Teric 12A3 (2.6 EO); Volpo 25D3

C12-15 pareth-4
Teric G12A4 (4.5 EO), LA4 (4.5 EO)

C12-15 pareth-5
Synonyms: PEG-5 C12-15 alkyl ether
Tergitol 25-L-5; Volpo 25D5

C12-15 pareth-5 phosphate
Crodafos 25 D5 Acid

C12-15 pareth-6
Synonyms: PEG-6 C12-15 alcohol
Merpoxen LP 60; Teric 12A6, G12A6

C12-15 pareth-7 [CAS #68131-39-5 (generic); RD #977054-91-3]
Synonyms: Pareth-25-7; PEG-7 C12–15 fatty alcohol ether
Alkasurf LA-7; Dobanol 25-7; Merpoxen LP 70; Neodol 25-7; Tergitol 25-L-7

C12-15 pareth-7 carboxylic acid
Synonyms: Pareth-25-7 carboxylic acid
Sandopan DTC Linear P Acid

C12-15 pareth-8
Synonyms: PEG-8 C12-15 alkyl ether
Serdox NES 8/85; Teric G12A8.(7.8 EO), LA8 (7.8 EO)

C12-15 pareth-9 [CAS #68131-39-5 (generic)]
Synonyms: Pareth-25-9
Alkasurf LA-9, LA-9 85%; Dobanol 25-9; Neodol 25-9; Tergitol 25-L-9; Teric 12A9

C12-15 pareth-10
Merpoxen LP 109; Sterling XE; Volpo 25D10

C12-15 pareth-10 phosphate
Crodafos 25 D10 acid

C12-15 pareth-12
Synonyms: Pareth 25-12
Alkasurf LA-12; Merpoxen LP 120; Neodol 25-12; Serdox NES 12; Tergitol 25-L-12; Teric 12A12 (11.5 EO), G12A12 (11.5 EO)

C12-15 pareth-15
Alkasurf LA-15; Sterling Emulsifier #15; Volpo 25D15

C12-15 pareth-16
Alkasurf LA-16; Teric 12A16

C12-15 pareth-23
Teric 12A23

C12-15 pareth-25
Synonyms: PEG-25 C12-15 alkyl ether
Volpo 25D20

C12-16 alcohols [CAS #68855-56-1]
Alfol 1216; Epal 1214

C12-18 acid triglyceride
Softenol 3100

C12-18 alcohols
Alfol 1218; Epal 1218

C13-15 alkyl bis (2-hydroxyethyl) amine oxide
Synprolam 35X2/0

C13-15 alkyl-1,3-propane diamine
Synprolam 35N3

C13-15 alkyl trimonium chloride
Synprolam 35TMQC

C13-15 amine
Synprolam 35, 35A

C13-15 dimethyl amine benzyl ammonium chloride
Synprolam 35BQC (50), 35BQC (80)

C13-15 dimethyl amine oxide
Synprolam 35DMO

C13-15 methyl amine:
Synprolam 35M

C13-15 pareth-2
Renex 702

C13-15 pareth-3
Renex 703

C13-15 pareth-7
Renex 707

C13-15 pareth-10
Merpoxen TP 185

C13-15 pareth-11
Renex 711

C13-15 pareth-20
Renex 720

C14 acid triglyceride
Softenol 3114

C14-15 alcohols
Dobanol 45; Neodol 45

C14-15 pareth-7
Synonyms: Pareth 45-7; PEG-7 C14-15 alcohol ether
Dobanol 45-7; Neodol 45-7

C14-15 pareth-11
Synonyms: Pareth-45-11
Dobanol 45-11

C14-15 pareth-13
Synonyms: Pareth-45-13; PEG-13 C14-15 alcohol ether
Neodol 45-13

C14-15 pareth-18
Synonyms: Pareth-45-18
Dobanol 45-18

C14-16 alcohols
Epal 1416

C14-18 alcohols
Epal 1418

C16-18 alcohols
Alfol 1618; Epal 618, 1618

C16-18 alcohol hydrogen phosphate
Merpiphos B 52

C16-18 pareth-4 hydrogen phosphate
Synonyms: PEG-4 C16-18 alcohol hydrogen phosphate
Merpiphos TP

C16-18 pareth-7
Synonyms: PEG-7 C16-18 fatty alcohol
Lutensol A 7

C16-18 pareth-8 hydrogen phosphate
Synonyms: PEG-8 C16-18 alcohol hydrogen phosphate
Merpiphos MP, VP

C16-18 pareth-11
Synonyms: PEG-11 sat. C16-18 alcohol
Lutensol AT 11

C16-18 pareth-18
Lutensol AT 18

C16-18 pareth-25
Lutensol AT 25

C16-18 pareth-50
Lutensol AT 50

C16-18 pareth-80
Lutensol AT 80

C16-20 alcohols
Alfol 1620

C18 acid triglyceride
Softenol 3118

C18 alcohol
Epal 18NF

C18-22 alcohols
Lanette 18-22

C18-36 acid
Syncrowax AW1C

C18-36 acid glycol ester
Syncrowax ERL-C

C18-36 acid triglyceride
Syncrowax HGL-C

C22-24 pareth-33
Siponic 1173

Calcium alginate [CAS #9005-35-0]
Keltose

Calcium dodecylbenzene sulfonate
Agrilan X98; Alkasurf CA; Arylan CA; Casul 70 HF; Chemannate C-60B, C-70B; Chimipon CA; Conco AAS-75S; Drynol E/20, E/30, E/40; Geronol AZ82, V/87, V/497; Merpisap AC 70; Nansa EVM 62H, EVM 70; Ninate ABS-60C; Polyfac ABS-60C; Richonate CS-6021H; Sandoz Sulfonate AAS 75S; Sermul EA 88

Calcium lignosulfonate [CAS #8061-52-7]
Lignosite, 401, 704, 1840; Lignosol B, BD, SF, SFL; Marasperse C-21; Norlig; PB-92

Calcium sodium lignosulfonate
Lignosol R30

Calcium stearate [CAS # 1592-23-0]
Calcium Stearate Regular; Petrac CP-11, -22G

Calcium stearoyl lactylate [CAS #5793-94-2]
Admul CSL 2007, CSL-2008; Grindtek FAL 2; Lamegin CSL; Patco 3, CSL; Radiamuls CSL 2980; Verv

Calcium sulfate [CAS #7778-18-9]
Synonyms: Anhydrite; Calcium sulfonate; Gypsum; Plaster of paris
Emcol D 24-25; Hybase C-300; Merpisap DC 70

Capramide DEA [CAS #136-26-5]
Synonyms: Capric acid diethanolamide
Alkamide 1509; Alrosol C; Comperlan CD; Emid 6544; Monamid 150-CW; Standamid CD

Capric acid [CAS #334-48-5]
Synonyms: *n*-Decanoic acid
Hystrene 9010

Capric/caprylic amidopropyl betaine
Cycloteric BET OD-40

Capric triglyceride
Coconad SK

Caproamphoacetate
Synonyms: Caproamphoglycinate
Miranol SM Conc.

Caproamphodiacetate
Synonyms: Caproamphocarboxyglycinate
Mackam 2CT, 2CY; Miranol S2M Conc.

Caproamphodipropionate
Synonyms: Caproamphocarboxypropionate
Mackam 2CYSF; Miranol S2M-SF Conc.

Capryl hydroxyethyl imidazoline [CAS #37478-68-5]
Synonyms: Caprylic imidazoline
Alkateric EHCIB; Crodazoline Cy; Miramine CPC; Monazoline CY

Caprylic acid [CAS # 124-07-2]
Hystrene 9008; Industrene 365

Caprylic alcohol [CAS #111-87-5]
Synonyms: *n*-Octyl alcohol; 1-Octanol; *n*-Octanol
Alfol 8; Kalcohl 08H

Caprylic/capric/stearic triglyceride [RD #977064-06-4]
Softisan 378

Caprylic triglyceride
Coconad MT

Caprylic/capric glycerides
Imwitor 742

Caprylic/capric triglyceride [RD #977059-83-8]
Coconad PL, RK; Neobee 1223, 1400 Oil, M-5, O; Standamul 318, 7105

Capryloamphoacetate
Synonyms: Capryloampglycinate
Rewoteric AM-V

Capryloamphodiacetate [CAS #7702-01-4]
Synonyms: Capryloamphocarboxyglycinate
Miranol J2M Conc.; Schercoteric CY-2

Capryloamphodipropionate
Synonyms: Capryloamphocarboxypropionate
Miranol J2M-SF Conc.; Monateric 811; Schercoteric CY-SF-2

Capryloamphohydroxypropylsulfonate
Synonyms: Capryloamphopropylsulfonate
Miranol JS Conc.

Capryloamphopropionate
Monateric CyNa-50%

Caprylyl/capryl glucoside
Triton CG-110

Carbomer 910
Carbopol 910

Carbomer 934 [CAS #9007-16-3]
Carbopol 934

Carbomer 940 [CAS #9003-01-4; 9007-17-4]
Carbopol 940

Carbomer 941 [CAS #9003-01-4; RD #977056-22-6]
Carbopol 941

Carboxymethylcellulose [CAS #9004-32-4]
Synonyms: Sodium carboxymethylcellulose; CMC
CMTC-T; Tylose C, CB Series; Vanease

Castor oil [CAS #8001-79-4]
Incroquat CR Conc.; Surfactol 13

Ceresin [CAS #8001-75-0]
Dehymuls K

Cetalkonium chloride [CAS #122-18-9]
Synonyms: Cetyl dimethyl benzyl ammonium chloride
Ammonyx T; Querton 16 BCL

Ceteareth-2
Synonyms: PEG-2 cetyl/stearyl ether; POE (2) cetyl/stearyl ether; PEG 100 cetyl/stearyl ether
Macol CSA-2

Ceteareth-2 phosphate
Crodafos CS2 Acid

Ceteareth-3 [RD #977054-58-2]
Hostacerin T-3; Volpo CS3

Ceteareth-4 [RD #977054-59-3]
Hetoxol CS-4; Lipocol SC-4; Macol CSA-4; Siponic E-2

Ceteareth-5 [RD #977054-60-6]
Hetoxol CS-5; Volpo CS5

Ceteareth-5 phosphate
Crodafos CS5 Acid

Ceteareth-6 [RD #977054-61-7]
Siponic E-3

Ceteareth-9
Hetoxol CS-9

Ceteareth-10 [RD #977054-63-9]
Lipocol SC-10; Macol CSA-10; Siponic E-5; Volpo CS10

Ceteareth-10 phosphate
Crodafos CS10 Acid

Ceteareth-11
Cremophor A 11; Rewopal CSF 11

Ceteareth-12 [RD #977054-64-0]
Eumulgin B-1; Incropol CS-12; Standamul B-1

Ceteareth-14
Siponic E-7

Ceteareth-15 [RD #977054-65-1]
Hetoxol 15 CSA, CS-15; Lipal 15 CSA; Lipocol SC-15; Macol CSA-15; Volpo CS15

Ceteareth-16
Teric 16A16

Ceteareth-20 [RD #977054-66-2]
Cetomacrogol 1000 BP; Cosmowax J, K; Cyclogol NI; Empilan KM 20; Emulgade 1000NI; Eumulgin B-2; Hetoxol CS-20; Incrochem SBJ; Incropol CS-20; Lexemul CS-20; Lipocol SC-20; Lipowax D, G; Macol CSA-20; Promulgen D, G; Rewopal CSF 20; Siponic C-20, E-10; Standamul B-2; Teginacid, Teginacid X; Volpo CS20

Ceteareth-22
Teric 16A22

Ceteareth-25
Cremophor A 25; Industrol CSS-25; Softisan 601

Ceteareth-27 [RD #977063-71-0]
Plurafac A-38; Ceteareth-29; Teric 16A29

Ceteareth-30 [RD #977063-72-1]
Eumulgin B-3; Hetoxol CS-30; Incropol CS-30; Siponic E-15; Standamul B-3

Ceteareth-40
Incropol CS-40; Ceteareth-50; Hetoxol CS-50, CS-50 Special; Incropol CS-50

Ceteareth-55
Plurafac A-39; Ceteareth-60; Incropol CS-60

Cetearyl alcohol [CAS #8005-44-5]
Synonyms: Cetostearyl alcohol; Cetyl/stearyl alcohol
Carsoquat 816-C; Cetomacrogrol Wax BP; Cosmowax J; Crodex A, C, N; Cutina CBS; Cyclochem EM 560, POL; Cyclogol Cetyl-Stearyl Alcohol; Cyclogol NI; Dehydag Wax N, O, SX, W; Empiwax SK, SK/BP; Emulgade 1000NI, A, F, F Special; Incrochem SBJ; Incroquat Behenyl TMS; Incroquat CR Conc.; Kalcohl 68, 86; Lanbritol Wax N21; Lanette N, O, SX, W, Wax CAT, Wax SX, Wax SXBP; Lanol CS; Laurex CS, CS/D; Lexemul CS-20; Lipowax; Lipowax D, G, NI; Mazawax 163R; Promulgen D; Ritachol 1000, 2000; Ritapro 100; Siponic C-20; Standamul Conc. 1002; Tewax TC 1, TC 2

Ceteth-2 [CAS #9004-95-9 (generic); RD #977054-67-3]
Synonyms: PEG-2 cetyl ether; POE (2) cetyl ether; PEG 100 cetyl ether
Brij 52; CE-55-2; Ethosperse CA-2; Hetoxol CA-2; Lexate CRC; Lipocol C-2; Nikkol BC-2; Simulsol 52

Ceteth-5 [CAS #9004-95-9 (generic)]
Solulan 5

Ceteth-6 [CAS #9004-95-9 (generic); RD #977054-69-5]
Nikkol BC-5.5, BC-5.5(FF)

Ceteth-7
Nikkol BC-7

Ceteth-10 [CAS #9004-95-9 (generic); RD #977054-70-8]
Brij 56; Hetoxol CA-10; Lipocol C-10; Nikkol BC-10TX; Simulsol 56

Ceteth-12 [CAS #9004-95-9 (generic)]
Lanbritol Wax N21

Ceteth-15
Nikkol BC-15TX, -15TX(FF)

Ceteth-16 [CAS #9004-95-9 (generic); RD #977064-09-7]
Atlas G-3816; Solulan 16

Ceteth-20 [CAS #9004-95-9 (generic)]
Brij 58; CE-55-20; Cetomacrogol Wax BP; Crodex N; Cyclochem POL; Hetoxol CA-20; Lipocol C-20; Lipowax NI; Nikkol BC-20TX, -20TX(FF); Simulsol 58; Tefose 2000, 2561; Teginacid H

Ceteth-23
Nikkol BC-23, -23(FF)

Ceteth-24 [CAS #9004-95-9 (generic)]
Solulan C-24

Ceteth-25 [CAS #9004-95-9 (generic); RD #977069-13-8]
Nikkol BC-25TX; Solulan 25

Ceteth-30 [CAS #9004-95-9 (generic); RD #977054-71-9]
Nikkol BC-30TX

Ceteth-40
Nikkol BC-40TX

Cetethyldimonium bromide [CAS #124-03-8]
Synonyms: Cetyl dimethyl ethyl ammonium bromide; Quaternium-17
Ammonyx DME

Cetethyl morpholinium ethosulfate [CAS #78-21-7]
Synonyms: Cethyl ethyl morpholinium ethosulfate; Quaternium-25
Atlas G-263, G-263 Anhydrous

Cetoleth-2
Synonyms: PEG-2 cetyl/oleyl ether; POE (2) cetyl/oleyl ether
Teric 17A2

Cetoleth-3
Teric 17A3

Cetoleth-6
Teric 17A6

Cetoleth-7 phosphate
Synonyms: PEG-7 cetyl/oleyl ether phosphate
Servoxyl VPGZ 7/100

Cetoleth-8
Teric 17A8

Cetoleth-10
Lubrol 17A-10; Merpoxen SP 100; Teric 17A10

Cetoleth-13
Teric 17A13

Cetoleth-20
Merpoxen SP 200

Cetoleth-23
Merpoxen SP 230

Cetoleth-25
Marlipal SU; Teric 17A25

Cetrimonium bromide [CAS #57-09-0]
Synonyms: Cetyl trimethyl ammonium bromide
Acetoquat CTAB; Bromat; Cycloton CT 100; Lanette Wax CAT

Cetrimonium chloride [CAS #112-02-7]
Synonyms: Cetyl trimethyl ammonium chloride; Palmityl trimethyl ammonium chloride; Hexadecyl trimethyl ammonium chloride
Adogen 444; Ammonyx CETAC, CETAC-30; Arquad 16-29, 16-50; Carsoquat CTM-29, CTM-429; Chemquat 16-50; Dehyquart A; Incroquat CTC-25, CTC-30; Nissan Cation PB-40, -300; Querton 16 CL 29, 24 CL 35; Radiaquat 6444; Variquat E228, E290

Cetrimonium tosylate [CAS #138-32-9]
Synonyms: Cetyl trimethyl ammonium *p*-toluene sulfonate
Cetats

Cetyl acetate [CAS #629-70-9]
Ritawax ALA; Solulan 97, 98

Cetyl alcohol [CAS #36653-82-4]
Synonyms: Palmityl alcohol; C16 linear primary alcohol
Adol 52 NF, 54, 520 NF; Alfol 16; Amerchol 400; Cetal; Cyclogol Cetyl Alcohol NF; Epal 16NF; Kalcohl 60; Lanette 16; Lanol C; Lexemul LPM

Cetyl betaine [CAS #693-33-4]
Product BCO

Cetyl esters [RD #977067-67-6]
Synonyms: Synthetic spermaceti wax
Cetina; Cyclochem EM 560; Kessco Synthetic Spermaceti N.F.; Lipowax; Spermwax

Cetyl ethyl... See *Cetethyl*

Cetyl palmitate [CAS #540-10-3]
Cutina CBS; Kessco 653; Nikkol N-SP; Radia 7500; Standamul 1616

Cetylpyridinium bromide
Acetoquat CPB

Cetylpyridinium chloride [CAS #123-03-5]
Synonyms: Hexadecylpyridinium chloride
Acetoquat CPC

Cetyl stearate [CAS #1190-63-2]
Synonyms: *n*-Hexadecyl stearate
Radia 7501

Cetyl/stearyl. See *Cetearyl*

Chlor-2-hydroxypropyl trimonium chloride
Servo XRK 60

Choleth-24 [CAS #27321-96-6 (generic)]
Synonyms: PEG-24 cholesteryl ether
Solulan C-24

Citric acid [CAS #77-92-9]

Synonyms: 2-Hydroxy-1,2,3-propanetricarboxylic acid

Incrochem SC1, SC4; Tandem 5K, 8, 552

Cocamide [CAS #61789-19-3]

Synonyms: Coconut oil amides; Coconut acid amide

Armid C; Ecconol 638; Schercomid 304, EAC-S

Cocamide DEA [CAS #61791-31-9; 68603-42-9]

Synonyms: Coconut diethanolamide; Cocoyl diethanolamide; N,N-bis (2-hydroxyethyl) coco amides

Accomid C; Alkamide 1002, 1423, 2104, 2106, 2204, CDE, CDE-Extra, CDO; Alrosol B; Aminol COR-2, COR-2C, HCA, KDE; Ardet DB, DC, DCAE, DCX, DMA, DMC, DMF, WO; Bio Soft N-21; Calamide C, CW-100; Calsuds CD-6; Carsamide 7644, C-3, C-7944, CA, LE, SAC; Carsofoam SC; Cedemide DX; Chimipal DCL; Clindrol 101CG, 200CG, 200CGN, 200HC, 200RC, 201CGN, 202CGN, 203CG, 204CGN; Clindrol 206CGN, 207CGN, 209CGN; Clindrol Superamide 100C, 100CG; Comperlan COD, KD, KDO, KM, LS, PD, SD, SDO; Conco Emulsifier K; Condensate CO, NP, PA, PC, PN, PO, PS; Crillon CDY; Cyclomide DC212, DC212M, DC212/S, DC212/SE, KD, WHC; Cycloryl JB-8, NWC; Detergyl S; Emid 6514, 6515, 6531, 6533, 6534, 6538; Empilan CDE, 2502, CDE, CDEY, CDX, FD, FD20, FE; Equex AEM; ESI-Terge 10, B-15, C-5, S-10; Ethylan A15, LD, LDG, LDS; Gafamide CDD-518; Hartamide OD; Hetamide MC, RC; Hymolon CWC, K-90; Iconol 28, COA; Incromide CA, CAL, LEP; Jordamide CCO, CLD, JT-128, WC Conc.; Lakeway 100-CA; Lauramide 11, D, ME; Lauridit KD, KDG; Loramine DC 212/S, DC 212/SE, DC 220/SE, DL 240; Loropan KD; Mackamide 100-A, C, CD, CS, EC, MC; Manro CD, CDS, CDX; Maprofix AEG; Marlamid D 1218, DF 1218; Marsamid 10, 40, 50; Mazamide 70, 80, CA-20, CS-148; Monamid 7-100, 7-153 CS, 150-AD, 150-ADD, 150-DR, 759; Monamine 779, AA-100, AC-100, AD-100, ADD-100, ADD-100 LE, ADS-100, ALX-80 SS, ALX-100 S, ARA-100, CF-100 M, I-76; Monaterge 85 HF; Ninol 25-CM, 41-CO, 128 Extra, 2012 Extra, LD, P-650; Nissan Stafoam DF; Nitrene 11120, 11230, 13026, A-309, A-567, C, C Extra; Norfox DC, DCO, DCS, DESA, KD, X; Oramide DL 200; Product LT 14-69; Profan 24 Extra, 128 Extra, 2012E; Purton CFD; Quimipol DEA OC; Rewomid C 220SE, DC 212/S, DC 212/SE, DC 220/LS, DC 220/SE, DL 240; Richamide 6404, 6445, 6625, 6737, 6754, CD, CDA, M-3, S-771, S-780; Schercomid 1-102, CCD, CDA, CDA-H, CDO-Extra, SCE, SCO-Extra; Sipomide 1500; Stafoam DF-1, DF-4, F; Stamid HT; Standamid KD, KDO, PD, SD, SDO; Standapol Blend EA-K, Conc. 7023; Steinamid DC 212/S, DC 220/SE, DL 240; Sterling 66/33, Amide 374, DEA; Super Amide GR; Surco 128T, Coco Condensate, WC Conc.; Synotol 119 N, CN 60, CN 80, CN 90, Detergent E; Teric CDE; Texapon IES, WW 99; Tohol N-220, N-220X; Ufanon K-80, KD-S; Unamide C-72-3, C-7649, C-7944, CDX, D-10, GC-75, JJ-35, LDL, N-72-3; Varamide A-2, A-10, A-12, A-83, MA-4; Witcamide 82, 1017, 5130, M-3

Cocamide MEA [CAS #68140-00-1]

Synonyms: Coconut monoethanolamide; N-(2-hydroxyethyl) coco fatty acid amide

Alkamide CME, CMO; Aminol CM, CM Flakes, CM-C Flakes, CM-D Flakes; Ardet CEA; Carsamide CMEA; Chimipal MC; Clindrol 100MC, 100MCG; Comperlan 100, SM; Cyclomide C212; Cycloryl M1; Cyclosheen 100; Emid 6500; Empilan CM, CME; Equex STM; Euperlan PK 771, PK 776, PK-810; Genapol PGM Conc.; Incromide CM, CME; Intermediate 325; Jordamide CFAM, CMEA, CMEA Extra; Lauridit KM; Loramine C 212; Loropan CME, KM; Mackamide CMA; Manro CMEA; Marlamid M 1218; Monamid CMA; Ninol CNR; Nissan Stafoam MF; P & G Amide No. 27; Polyfac PCW-160; Profan AB20; Rewomid C 212, CD; Richamide MX; Schercomid CME; Standamid 100, CM, CMG, KM, SM; Standapol Pearl Conc. 7130; Standapol S; Steinamid C 212; Sterling Granulated Wax; Surco CMEA, MGS-35; Swanic 51; Synotol ME-90; Texapon EVR, SG; Unamide CMX

Cocamide MIPA [CAS #68333-82-4]

Synonyms: Coconut monoisopropanolamide

Empilan CIS; Loramine IPP 240; Rewomid IPP 240; Schercomid CMI; Steinamid IPP 240

Cocamido betaine

Alkateric CAB; Amido Betaine C; Deriphat BA, BAW; Norfox Coco Betaine

Cocamidopropylamine oxide [CAS #68155-09-9]

Synonyms: Cocamidopropyl dimethylamine oxide

Alcolec 634; Alkamox CAPO; Aminoxid WS 35; Ammonyx CDO; Barlox C; Cyclomox CO; Drew BRX-354; Incrochem SC1; Incromine Oxide C; Jordamox CAPA; Loramine B 204; Ninox CA, FCA; Rewominoxid B 204; Schercamox C-AA; Standamox C 30, CAW; Varox 1770

Cocamidopropyl betaine [CAS #61789-40-0]

Synonyms: CADG; Cocamidopropyl dimethyl glycine

Aerosol 30; Amonyl 380 BA; Amphosol CA, CG; Aremsol A; Carsonam 3, 32, 33-S, 3147; Chemadene NA-30; Chimin AX; Cycloteric BET C-30; Dehyton K; Emcol DG, NA-30; Emery 6748; Incronam 30; Jortaine C, CAB-35, CFA-35; Lebon 2000; Lexaine C, CG-30, CG-50, CS, IBC-70, IBC-70S; Lexate CL-60; Lonzaine C, CO; Mackam 35, 35 HP; Mafo CAB; Maprofix AEG; Maprolyte C; Mirataine BD, CB, CBC, CBL, CBR; Monateric ADA, ADFA, CAB, COAB, MCB; Rewoteric AM-B 13; Schercotaine CAB, CAB-G; Sipoteric CB, COB; Standapol BAW, BC-35; Sterling CAB; Surco Coco Betaine; Tego-Betaine C, L-7, L-10, S, T; Varion CADG; Velvetex BA, BA-35, BC-35, BK-35

Cocamidopropyl dimethyl acetyl betaine

Varion CADG-HS, CADG-LS

Cocamidopropyl dimethylamine [CAS #68140-01-2]

Synonyms: N-[3-Dimethylamino)propyl]coco amides

Carsamine CB; Cyclomide CODI; Incromine CB; Jordamine DAPL, DMCAPA, SHCFA; Lexamine C-13; Schercodine C; Teginacid R, Tego-Amid D 5040

Cocamidopropyl dimethylamine lactate [CAS #68425-42-3]

Synonyms: N-[3-(Dimethylamino)propyl] cocamide lactate

Incromate CDL

Cocamidopropyl dimethylamine propionate [CAS #68425-43-4]

Synonyms: N-[3-(Dimethylamino)propyl]coco amides, propionates

Emcol 1655; Incromate CDP; Mackam CAP

Cocamidopropyl hydroxysultaine [CAS #68139-30-0]

Synonyms: (3-Cocamidopropyl)(2-hydroxy-3-sulfopropyl)dimethyl quaternary ammonium compounds, hydroxides, inner salt

Cycoteric BET-CS; Jortaine COSB, CSB, CSB-50; Lexaine CSB-35, CSB-50; Lonzaine CS; Mirataine CBS, DDW; Schercotaine SCAB, SCAB-G; Varion CAS

Cocamidopropyl lauryl ether

Marlamid KL

Cocamidopropyl morpholine

Incromine CPM

Cocamidopropyl morpholine lactate

Incromate CPML

Cocamidopropyl PEG-dimonium chloride phosphate
Monaquat P-TC

Cocamine [CAS #61788-46-3]
Synonyms: Coconut amine
Amine KK; Armeen C, CD; Kemamine P-650, P-650D; Lilamin 160, 160 D; Nissan Amine FB; Noram C; Radiamine 6160, 6161

Cocamine acetate
Acetamin C; Armac C; Kemamine A650; Lilamac 160; Noramac C; Radiamac 6169

Cocamine oxide [CAS #61788-90-7]
Synonyms: Coco dimethylamine oxide
Aromox DMC, DMCD, DMC-W; Barlox 12; Conco XA-Y; Cyclomox C; Genaminox CS, KC; Hartox DMCD; Mackamine CO

Cocaminobutyric acid [CAS #68649-05-8]
Armeen Z, Z-9

Cocaminopropionic acid [RD #977056-35-1]
Synonyms: N-coco-2-aminopropionic acid
Amphoram CP1; Deriphat 151C; Sipoteric 1112, CAPA

Coceth-8 [CAS #61791-13-7 (generic); RD #977064-10-0]
Synonyms: PEG-8 coconut alcohol; POE (8) coconut ether; PEG 400 coconut ether
Standapol S; Texapon EVR

Cocoamphoacetate [CAS #68334-21-4; 68390-66-9]
Synonyms: Cocoamphoglycinate
Alkateric CIB; Amphoterge W; Mackam 1C; Miranol CM Conc.; Monateric CM-36S; Rewoteric AM-2 C/NM; Schercoteric MS; Standapol CIM-40; Zoharteric M, M-2, M-3, M-3 50%

Cocoamphodiacetate [CAS #68650-39-5]
Synonyms: Cocoamphocarboxyglycinate
Alkateric 2CIB; Amphoterge W2; Dehyton G; Lexoteric 2; Mackam 2C, 2C-75; Miranol 2MCA Modified, 2MCAS Modified, 2MCT, 2MCT Modified, C2M Conc.; Monateric 805, CDL, CDS, CDTD, CDX-38, CDX-38 Mod., CSH-32; Rewoteric AM-2 C; Schercoteric MS-2, MS-2ES MOD, MS-2 MOD, MS-2TE MOD; Sipoteric 1304, 1398, GPA, GSC; Zoharteric D, D-2, D-3, D-3 50%

Cocoamphodipropionate
Synonyms: Cocoamphocarboxypropionate
Amphoterge K2; Cycloteric DC-SF; Mackam 2C-SF; Miranol 2MCA-ESF, C2M-SF 70%, C2M-SF Conc.; Monateric CEM-38%; Schercoteric MS-SF-2 (38%), MS-SF-2 (70%)

Cocoamphodipropionic acid
Synonyms: Cocoamphocarboxypropionic acid
Miranol C2M Anhydrous Acid

Cocoamphohydroxypropyl sulfonate
Synonyms: Cocoamphopropylsulfonate
Amphoterge SB; Miranol CS 60%, CS Conc.; Schercoteric MS-EP

Cocoamphopropionate
Synonyms: Cocoamphocarboxypropionate
Amphoterge K; Cycloteric MV-SF; Mackam CSF; Miranol CM-SF Conc.; Monateric CA-35%, CAM-40, CEM-38CG; Rewoteric AM-KSF

Coco-betaine [RD #977056-36-2]
Synonyms: Coco dimethyl glycine; Coconut betaine
Accobetaine CL; Alkateric BC; Amonyl 265 BA; Ampho B11-34; Amphoram CB A30; Carsonam BCW; Dehyton AB-30; Emcol CC-37-18; Hartaine CB-40; Jortaine CB-40; Lonzaine 12C; Mackam CB, CB-35, CB-LS; Mirataine CDMB; Standapol AB-45; Velvetex AB-45, BC

Coco dialkyl benzyl ammonium chloride
Quatrene CE

Coco dimethyl amine. See *Dimethyl cocamine*

Coco dimethyl benzyl ammonium chloride
Merpiquat K-8-2; Nissan Cation F2-10R, -20R, -40E, -50; Querton KK BCL

N-Coco dipropylene triamine
Trinoram C

Cocoglycerides
Synonyms: Glycerides, coconut, mono-, di-, and tri-
Cutina CBS

Cocohydroxyethyl PEG-imidazolinium chloride phosphate
Monaquat P-TZ

Coco imidazoline betaine
Carsonam C, C-SF, C-SPCL; Empigen CDR10, CDR30, XDR 112, XDR121, XDR123, XDR302

Coco imidazoline betaine dicarboxylate
Carsonam DC, DC-SF, DC-70%-SF

Coco imidazoline dicarboxylate
Antaron MC-44

Coco/lauric DEA
Alkamide CL63

Coco nitrile
Arneel C

Coconut acid [CAS #61788-47-4]
Industrene 325, 328

Coconut alcohol [CAS #68425-37-6]
Synonyms: Coconut fatty alcohol
Laurex CH

Coconut diamine
Lilamin 560; Radiamine 6560

Coconut diethanolamide. See *Cocamide DEA*

Coconut monoethanolamide. See *Cocamide MEA*

Coconut monoisopropanolamide. See *Cocamide MIPA*

Coconut oil amides. See *Cocamide*

Coconut oil soap
Hipochem LH-Soap

Coco/oleic DEA
Surco SR-100, WHC

Coco/oleamidopropyl betaine
Mirataine COB

Coco-pentaethoxy methyl-ammonium methosulfate
Loraquat CPEM; Rewoquat CPEM

Coco-1,3-propanediamine
Synonyms: Coco-1,3 diaminopropane
Duomeen C, CD, C Special

Coco-1,3-propanediamine diacetate
Synonyms: Coco-1,3-diaminopropane diacetate
Duomac C

Coco 1,3-propylene diamine
Dinoram C; Kemamine D-650

Coco propylene diamine acetate
Dinoramac C

Coco-1,3-propylene diamine diacetate
Kemamine AD 650

Coco sulfobetaine
Varion CDS

Coco/tallow MEA
Intermediate 300

Coco taurine
Amphoram CT 30

Cocotrimonium chloride [CAS #61789-18-2]
Synonyms: Cocoyl trimethyl ammonium chloride
Adogen 461; Arquad C-33, C-33-W, C-50; Jet Quat C-50; Nissan Cation FB, FB-500; Noramium MC 50; Rewopon JMCA; Quartamin CPR

N² cocoyl-L-arginine ethyl ester DL-pyrrolidone carboxylic acid salt
CAE

Cocoyl hydroxyethyl imidazoline [CAS #61791-38-6]
Synonyms: Coconut hydroxyethyl imidazoline; Cocoyl imidazoline
Alkaquat C; Alkazine C; Amphotensid 9 M, CT, GB 2009; Crapol AV-10; Mackazoline C; Miramine CC; Monazoline C; Schercozoline C; Varine C

Cocoyl sarcosine [CAS #68411-97-2]
Synonyms: N-Cocoyl-N-methyl glycine
Hamposyl C; Sarkosyl LC

Corn oil PEG-6 complex
Labrafil M 2125 CS

Cottonseed acid [CAS #68308-51-0]
Industrene 138, 239

Cottonseed glyceride [CAS #8029-44-5]
Synonyms: Cottonseed oil monoglyceride; Glyceryl mono cottonseed oil
Dimodan CP; Drewmulse 15; Myverol 18-85; Cresylic acid; Rexowet CR

Cumene sulfonic acid
Eltesol CA 65, CA 96; Reworyl C 65; Sulframin CSA

DDBSA. See *Dodecylbenzene sulfonic acid*

DEA-acrylinoleate
Synonyms: DEA-C21-dicarboxylate
Monaterge 85 HF

DEA-C12-15 alcohols sulfate
Sipoteric 1112

DEA-cetyl phosphate
Synonyms: Cetyl DEA phosphate
Crodafos CDP

DEA-cocoamphodipropionate
Synonyms: DEA-cocoamphocarboxypropionate
Miranol C2M-DEA 60%

DEA-dodecylbenzene sulfonate [CAS #26545-53-9]
Marlopon ADS 50; Monamine ALX-80 SS, ALX-100 S; Monaterge 85 HF

DEA-lauraminopropionate [CAS #65104-36-1]
Carsonol BDM; Maprolyte LX; Mirataine XL; Standapol AA-1

DEA-laureth sulfate [RD #977061-59-8]
Synonyms: Diethanolamine laureth sulfate
Cedepal SD-409; Monamine 779

DEA-lauryl sulfate [CAS #143-00-0]
Synonyms: Diethanolamine lauryl sulfate
Alkasurf DLS, WADX; Berol 463; Carsofoam SC; Carsonol BDM, DLS; Conco Sulfate EP; Condanol DLS 35; Cycloryl DA; Duponol EP; Empicol 0031/T, DA, DLS; Lexaine IBC-70, IBC-70S; Maprofix DLS-35; Maprolyte LX; Mirataine XL; Montovol GF-15; Nutrapon DE 3796; Polystep B-8; Sandoz Sulfate EP; Sipon LD; Standapol AA-1, Conc. 7021, DEA; Stepanol DEA; Sterling WADE; Texapon DEA

DEA-oleth-3 phosphate [RD #977060-94-8]
Crodafos N3 Neutral

DEA-oleth-10 phosphate [RD #977060-97-1]
Crodafos N10 Neutral

Decaglyceryl. See *Polyglyceryl-10*

Deceth-3
Synonyms: PEG-3 decyl ether; POE (3) decyl ether
Alkasurf DA-3

Deceth-4
Alkasurf DA-4; Chemal DA-4; DeSonic DA-4; Iconol DA-4; Trycol DA-4

Deceth-4 phosphate [CAS #9004-80-2 (generic)]
Synonyms: PEG-4 decyl ether phosphate
Gafac RA-600

Deceth-5
Iconol DA-5.5 (5.5 EO)

Deceth-6
Alkasurf DA-6; Chemal DA-6; DeSonic DA-6; Iconol DA-6, DA-6-90%; Trycol DA-6, DA-69

Deceth-9
Chemal DA-9; Iconol DA-9

Decyl alcohol [CAS #112-30-1]
Synonyms: C10 linear primary alcohol
Alfol 10; Epal 10; Kalcohl 10H

Decyl amine
Amine 10D

Decylamine oxide [CAS #2605-79-0]
Synonyms: Capric dimethyl amine oxide; Decyl dimethyl amine oxide
Barlox 10S

Decyl diphenyloxide disulfonate
Dowfax 3B0

Decyl diphenyloxide disulfonic acid
Conco Sulfate 3B2 Acid

Decyl isostearate
Synonyms: Decyl isooctadecanoate
Schercemol DEIS

Decyl oleate [CAS #3687-46-5]
Dehymuls K; Standamul CTV

Diacetylated tartaric acid ester glyceride
Admul Data; Datagel; Datamuls 42, 43, 47, 4720, 4741, 4820; Datem esters; Imwitor 1330, 2020, 2320; Myvatem; Panodan 235, FDP-K, SD, SD-K; Radiamuls DATA 2001, 2003, 2004, 2006, 2007, 2008, 2009

Diacetylated tartaric acid ester glycerides
Lamegin DWF, DWH, DWP; Swedex 100, 100V, AR, AR41, AR58P

Dialkyl dimethyl ammonium chloride
Radiaquat 6443, 6470; Skyllex

Dialkyl methyl benzyl ammonium chloride
BTC 776

Diammonium cocoyl sulfosuccinate
Empimin MSS

Diammonium lauramido MEA sulfosuccinate
Incrosul LMA

Diammonium lauryl sulfosuccinamate
Mackanate LA

Diammonium lauryl sulfosuccinate
Incrosul LSA; Monamate LNT-40

Diammonium oleamido PEG-2 sulfosuccinate [RD #977055-96-1]
Schercopol OMES-A

Diamyl sodium sulfosuccinate [CAS #922-80-5]
Synonyms: Sodium diamyl sulfosuccinate
Aerosol AY, AY-65, AY-100; Geropon AY

Diarachidyl-behenyl sec. amine
Kemamine S-190

Dibehenyl/diarachidyl dimonium chloride
Kemamine Q-1302C, Q-1902C

Dibehenyldimonium chloride
Kemamine Q-2802C

Dibutylamino ethanol [CAS #102-80-8]
Mazeen DBA

Dicocamine [CAS #61789-76-2]
Synonyms: Dicoco alkyl amine
Armeen 2C; Kemamine S-650; Noram 2 C

Dicocodimonium chloride [CAS #61789-77-3]
Synonyms: Dicoco dimethyl ammonium chloride
Accoquat 2C-75, 2C-75H; Adogen 462, R-6; Arquad 2C-75, S-2C-50, T-2C-50; Jet Quat 2C-75; Kemamine Q-6502C, Q-6503B; M-Quat 2475; Quartamin DCP; Radiaquat 6462; Variquat K 300, K375

Dicoco methylamine
Kemamine T-6501; Noram M2C

Dicyclohexyl sodium sulfosuccinate [CAS #23386-52-9]
Synonyms: Sodium dicyclohexyl sulfosuccinate
Aerosol A-196; Alconate 2CH

Didecyldimonium chloride [CAS #7173-51-5]
Synonyms: Didecyl dimethyl ammonium chloride
Bio-Dac 50-22; BTC 818, 885, 1010; BTCO 1010; Querton 210 CL

Diethanolamidooleamide DEA
Witcamide 5168

Diethanolamine [CAS #111-42-2]
Synonyms: DEA; 2,2´-Iminobisethanol
Aminol COR-2C, N-1918; Calamide CW-100; Clindrol 200CG, 200CGN, 200HC, 200L, 200RC, 200S, 202CGN, 203CG, 206CGN, 207CGN, 209CGN; Condensate CO, NP, PL, PN, PS; Cyclomide DC212, DIN 295/S, DO280; Emid 6531, 6533, 6534, 6541, 6544; Empilan LDX; Mackamide CD; Marsamid 10; Mazamide 70, 80, CA-20, LM-20, O-20; Monamine

928, AA-100, ACO-100, AD-100, ADD-100, ADD-100 LE, ADY-100, CD-100, CF-100 M, LM-100, R8-26, T-100; Norfox KD; Product LT 10-8-1; Rewocid DU 185; Schercomid CDO-Extra, LD, ODA; Standamid CD, PD; Varamide L-1; Witcamide 82, 5133

Diethylaminoethyl stearate [CAS #3179-81-5]
Cerasynt 303

Diethylene glycol. See *PEG-2*

Diglyceryl. See *Polyglyceryl-2*

Diglycol. *See PEG-2*

Dihexyl sodium sulfosuccinate [CAS #3006-15-3]
Synonyms: Sodium dihexyl sulfosuccinate
Aerosol MA-80; Alcopol OS; Astrowet H-80; Alkasurf SS-MA-80; Empimin MA; Lankropol KMA; Monawet MM-80

Dihydrogenated tallow amine
Noram 2SH

Dihydrogenated tallow benzylmonium chloride
Synonyms: Dihydrogenated tallow methyl benzyl ammonium chloride
Variquat B343

Dihydrogenated tallow dimonium chloride. See *Quaternium-18*

Dihydrogenated tallow dimonium methosulfate
Synonyms: Dimethyl dihydrog. tallow ammonium methyl sulfate
Accosoft 748

Dihydrogenated tallow methylamine [CAS #61788-63-4]
Kemamine T-9701; Noram M2 SH

Dihydroxyethyl C8-10 alkoxypropylamine oxide
Varox 188E

Dihydroxyethyl C9-11 alkoxypropylamine oxide
Varox 191E

Dihydroxyethyl C12-15 alkoxypropylamine oxide
Varox 185E

Dihydroxyethyl cocamine oxide [CAS #61791-47-7]
Synonyms: N,N (bis (2-hydroxyethyl) cocamine oxide
Alkamox C2-0; Aromox C/12, C/12-W, CD/12; Schercamox CMA

Dihydroxyethyl stearamine oxide [CAS #14048-77-2]
Synonyms: Bis (2-hydroxyethyl)stearylamine oxide
Aromox 18/12

Dihydroxyethyl tallowamine oxide [CAS #61791-46-6]
Aromox T/12; Schercamox T-12

Dihydroxyethyl tallow glycinate
Synonyms: Tallow dihydroxyethyl betaine
Jortaine TM; Mirataine TM; Monateric 1202

Diisobutyl adipate [CAS #141-04-8]
Synonyms: DIBA
Radia 7197

Diisobutyl sodium sulfosuccinate [CAS #127-39-9]
Synonyms: Sodium diisobutyl sulfosuccinate
Aerosol IB-45; Alcopol OB; Alkasurf SS-IB-45; Astrowet B-45; Geropon CYA/45; Manoxol IB; Monawet MB-45

Diisohexyl sulfosuccinate
Rewopol SB MB 80

Diisooctyl sulfosuccinate
Triumphnetzer ZSN

Diisopropyl adipate
Radia 7194

Diisopropyl dimerate
Synonyms: Bis(1-methylethyl)dimerate; Diester of isopropyl alcohol and dimer acid
Schercemol DID

Diisopropyl sebacate [CAS #7491-02-3]
Schercemol DIS

Dilaureth-10 phosphate
Nikkol DLP-10

Dimethicone [CAS #9006-65-9; 9016-00-6; 63148-62-9]
Synonyms: Dimethylpolysiloxane; Dimethyl silicone; Polydimethylsiloxane
AF-60, AF-66; Antifoam Compound SWS-202, SWS-203; Dow Corning 200 Fluid; Dow Corning F1-1630, FS-1265 Fluid

Dimethicone copolyol [CAS #64365-23-7; RD #977058-72-2]
Synonyms: Dimethylsiloxane-glycol copolymer
Dow Corning 190, 193, 196, 198 Surfactant; Dow Corning 470A Fluid, 1315 Surfactant; Dow Corning 3225C Formulation Aid; Dow Corning 5043, 5098 Surfactant; Dow Corning FF-400

2-Dimethylamino-2-methyl-1-propanol
DMAMP; DMAMP-80

Dimethylaminopropyl ricinoleamide benzyl chloride
ES-1239

Dimethyl arachidyl-behenyl amine
Kemamine T-1902D

Dimethyl behenamine [CAS #215-42-9]
Synonyms: Behenyl dimethyl amine
Crodamine 3.ABD

Dimethyl C10/18 alkyl amine
Armeen DMAD

Dimethyl C12/C14 alkyl amine
Adma 24

Dimethyl C12/C14/C16 alkyl amine
Adma 246-451, 246-621

Dimethyl C14/C16 alkyl amine
Adma 46

Dimethyl caproamide
Hallcomid M-6

Dimethyl caprylamide-capramide
Hallcomid M 8-10

Dimethyl capryl amine
Lilamin 310 D

Dimethyl caprylyl amine
Lilamin 308 D

Dimethyl cocamine [CAS #61788-93-0]
Synonyms: Coco dimethyl amine
Amine 2MKK; Armeen DMCD, DMMCD; Kemamine T-6502D; Lilamin 367 D, 369; Nissan Tertiary Amine FB; Noram DMC

Dimethyl decylamine
Synonyms: Decyl dimethylamine; C10 alkyl dimethylamine
Adma C10; Armeen DM10D

Dimethyl di(hydrogenated tallow) ammonium chloride. See *Quaternium-18*

Dimethyl erucylamine
Crodamine 3.AED

Dimethyl hexynol
Surfynol 61

Dimethyl hydrogenated tallow amine [CAS #61788-95-2]
Synonyms: Hydrogenated tallow dimethyl amine
Amine 2MHBG; Armeen DMHTD; Kemamine T-9702D; Lilamin 343, 345 D; Nissan Tertiary Amine ABT; Noram DMSH

Dimethyl lauramide
Hallcomid M-12

Dimethyl lauramine [CAS #112-18-5]
Synonyms: Lauryl dimethyl amine; Dodecyldimethylamine
Adma 2; Amine 2M12; Armeen DM12D; Empigen AB; Lilamin 312 D; Nissan Tertiary Amine BB; Onamine 12

Dimethyl lauryl aminobetaine
Lorapon AM-DML

Dimethyl myristamine [CAS #112-75-4]
Synonyms: Dimethyl myristylamine; Myristyl dimethyl amine; Tetradecyl dimethylamine
Adma 4; Amine 2M14; Armeen DM14D; Lilamin 314 D; Nissan Tertiary Amine MB; Onamine 14

Dimethyl octylamine
Synonyms: C8 alkyl dimethylamine; Octyl dimethylamine
Adma C8; Armeen DM8D

Dimethyl octynediol [CAS #1321-87-5]
Surfynol 82

Dimethyl oleamide
Hallcomid M 18-OL

Dimethyl oleamine
Synonyms: Oleyl dimethyl amine
Crodamine 3.AOD; Lilamin 372 D

Dimethyl palmitamine [CAS #112-69-6]
Synonyms: Palmityl dimethyl amine; Hexadecyl dimethylamine
Adma 6; Amine 2M16; Armeen DM16D; Crodamine 3.A16D; Lilamin 316 D; Nissan Tertiary Amine PB; Onamine 16

Dimethylsiloxane-glycol copolymer. See *Dimethicone copolyol*

Dimethyl soyamine [CAS #61788-91-8]
Synonyms: Soya dimethyl amine
Kemamine T-9972D

Dimethyl stearamide
Hallcomid M-18

Dimethyl stearamine [CAS #124-28-7]
Synonyms: Stearyl dimethyl amine; Octadecyl dimethyl amine
Adma 8; Amine 2M18; Armeen DM18D; Crodamine 3.A18D; Kemamine T-9902D; Lilamin 342 D; Nissan Tertiary Amine AB; Onamine 18

Dimethyl tallowamine [RD #977056-62-4]
Synonyms: Tallow dimethylamine
Kemamine T-9742D; Noram DMS

Dinonoxynol-9 citrate
Synonyms: Di(PEG-9 nonyl phenyl ether) citrate
Citrest TD

Dinonyl phenol [CAS #1323-65-5]
Berol Dinonylphenol

Dioctyl dimethyl ammonium chloride
Bardac LF, LF80; BTC 818, 885; Querton 28 CL

Dioctyl sodium sulfosuccinate [CAS #577-11-7]
Synonyms: Sodium dioctyl sulfosuccinate; Sodium di-(2-ethylhexyl) sulfosuccinate
Ablusol C-78; Aerosol GPG, OT-70 PG, OT-75, OT-100, OT-B, OT-S; Alcopol O 60%, O 70 PG; Alkasurf SS-O-30, SS-O-40, SS-O-60, SS-O-60PG, SS-O-70PG, SS-O-75; Alrowet D-65; Astrowet O-70-PG, O-75; Atcowet W7; Avirol SO-70P; AZdry 40, 70; Chemax DOSS-75E; Complemix 100; Condanol SBDO60; Crestopen 5X; Cyclopol SBDO; Denwet CM; Drewfax 528; Elfanol 883; Emcol 4500, 4560; Empimin OP45, OP70, OT; Gemtex PA-75, -85P, PAX-60, SC-40, -70, -75, -75E, SC Powd.; Geropon SDS; Hipochem EK-18; Karawet DOSS; Lankropol KO Special, KO 2; Mackanate DOS 70M5, DOS-75; Manoxol OT, OT 60%, OT/B, OT/P; Marlinat DF 8; Merpasol DIO 60, 75; Monawet MO-65-150, MO-70, MO-

70-150, MO-70E, MO-70R, MO-70RP, MO-70S, MO-75E, MO-84R2W, MO-85P; Nekal WT-27; Nikkol OTP-100S; Nissan Rapisol B-30, -80, C-70; Penetron Conc.; Pentex 99; Protowet D-75, D-75 S; Rewopol NEHS 40, SB-DO 70; Sanmorin OT 70; Sanstat 2012-A; Schercopol DOS-70, DOS-PG-85; Secosol DOS/70; Servoxyl VLA 2170; Serwet WH 170; Solusol 75%, 84%, 85%, 100%; Supermontaline SLT65; Tex-Wet 1001; Thorowet G-40, -75; Triton GR-5M, GR-7M

Dioctyl sulfosuccinate
Arylene M40, M60; Atlas WA-100; Varsulf SBDO-70

Dioleyl imidazoline methosulfate
Rewoquat W 3690, W 3690/PG

Dipropylene glycol. See *PPG-2*

Disodium C12-15 pareth sulfosuccinate
Synonyms: Disodium pareth-25 sulfosuccinate
Emcol 4300

Disodium castor oil MEA sulfosuccinate
Merpasol L 53

Disodium cetearyl sulfosuccinamate
Empimin MKK/L

Disodium cetearyl sulfosuccinate [RD #977063-91-4]
Synonyms: Disodium cetyl-stearyl sulfosuccinate
Empimin MKK98

Disodium cocamido MEA-sulfosuccinate [CAS #61791-66-0; 68784-08-7]
Mackanate CM, CM-100; Merpasol L 52; Rewopol SBC 212, SBC 212/G; Schercopol CMS-Na

Disodium cocamido MIPA-sulfosuccinate [CAS #68515-65-1; 68909-21-7]
Cyclopol SBCP; Mackanate CP; Monamate C-1142, CPA-4, CPA-100; Monateric 805; Schercopol CMIS-Na

Disodium cocamido sulfosuccinate
Alconate CPA; Cyclopol SBF 212

Disodium coconut sulfosuccinamate
Condanol CMS

Disodium cocoyl sulfosuccinate
Empimin MHH

Disodium deceth-4 sulfosuccinate
Alkasurf SS-DA4-HE

Disodium deceth-6 sulfosuccinate [CAS #68311-03-5 (generic); 68630-97-7 (generic)]
Aerosol A-102; Alconate D-6; Monawet TD-30

Disodium hydrogenated cottonseed glyceride sulfosuccinate
Emcol 4072

Disodium isodecyl sulfosuccinate [CAS #37294-49-8; 55184-70-8]
Aerosol A-268

Disodium laneth-5 sulfosuccinate [CAS #68890-92-6 (generic)]
Incrosul LAFS

Disodium lauramido MEA-sulfosuccinate [CAS #25882-44-4; 55101-78-5]
Alconate LEA; Alkasurf SS-L9ME; Condanol SBL 203; Cyclopol SBL203; Incrosul LMS, LMS-Special; Mackanate LM-40; Merpasol L 47; Rewopol SBL 203; Steinapol SBL 203; Varsulf SBL 203

Disodium laureth sulfosuccinate [CAS #39354-45-5 (generic); 40754-59-4; 42016-08-0]
Synonyms: Disodium lauryl ether sulfosuccinate
Alkasurf SS-LA-3; Cyclopol SBFA30; Geronol ACR/4; Incrosul LTS; Jordawet DSLES; Mackanate EL, L-1, L-2, L-101, L-102; Rewopol SBFA 30 40%; Schercopol LPS; Steinapol SBFA 30 40%; Texapon SBN; Varsulf SBFA 30

Disodium lauriminodipropionate [CAS #3655-00-3]
Synonyms: Disodium N-lauryl beta-iminodipropionate
Alkateric A2P-LPS 30%; Deriphat 160; Mirataine H2C

Disodium lauryl sulfosuccinate [CAS #13192-12-6; 19040-44-9; 26838-05-1]
Alconate L3; Aflkasurf SS-L-HE; Condanol SBF 12, SBF 12-Powd.; Emcol 4400-1; Incrosul LS; Jordawet LA-40; Mackanate LO-, LO-100; Merpasol L 44; Miranate LSS; Monamate LA-100; Rewopol SBF 12, SBR 12-Powd.; Steinapol SBF 12, SBR 12-Powd.; Texin 128 Powder

Disodium maleic anhydride/diisobutylene copolymer
Empicryl APD, APD/B

Disodium myristamido MEA-sulfosuccinate [CAS #37767-42-3]
Emcol 4100M; Mackanate MM

Disodium naphthalene sulfonic acid/formaldehyde condensate
Matexil DA-AC

Disodium nonoxynol-10 sulfosuccinate [CAS #67999-57-9 (generic); RD #977069-21-8]
Aerosol A-103

Disodium oleamido MEA-sulfosuccinate [CAS #68479-64-1]
Incrochem SC1, SC4; Incrosul OMS; Mackanate OM; Schercopol OMS-Na

Disodium oleamido MIPA-sulfosuccinate [CAS #43154-85-4; 67815-88-7]
Synonyms: Disodium oleoyl isopropanolamide sulfosuccinate
Emcol 4161L; Mackanate OP; Schercopol OMIS-Na

Disodium oleamido PEG-2 sulfosuccinate [CAS #56388-43-3; 68227-80-5]
Synonyms: Disodium oleamido diglycol sulfosuccinate
Alkasurf SS-OA-HE; Anionyx 12S; Cyclopol SBF 280; Jordawet DMDS; Mackanate OD; Monamate OPA-30; Schercopol OMES-Na; Standapol SH-100, SH-135 Special, SHC-101; Texapon SH 100, SH-135 Special

Disodium N-oleyl sulfosuccinamate
Alkasurf SS-OA; Empimin MTT

Disodium oleyl sulfosuccinate
Empimin MTT/A

Disodium ricinoleamido MEA-sulfosuccinate [CAS #40754-60-7; 65277-54-5; 67893-42-9]
Cyclopol SBR-3; Mackanate RM; Schercopol RMS-Na; Varsulf S-1333

Disodium stearyl sulfosuccinamate [CAS #14481-60-8; 26446-37-7]
Synonyms: Disodium octadecyl sulfosuccinamate
Aerosol 18; Alcopol FA; Alkasurf SS-TA; Empimin MKK; Lankropol ODS/LS, ODS/PT; Manro MA 35; Rewopol SBF 18

Disodium tallowiminodipropionate [CAS #61791-56-8]
Synonyms: Disodium N-tallow-beta iminodipropionate
Alkateric A2P-TS 30%; Deriphat 154; Mirataine T2C; Monateric TDB-35

Disodium tridecyl sulfosuccinate
Incrosul TS; Mackanate TDS

Disodium undecylenamido MEA-sulfosuccinate [CAS #26650-05-5; 37311-67-4; 40839-40-5]
Synonyms: Disodium undecylenic monoethanolamide sulfosuccinate
Cyclopol SBU-185; Empicol SEE; Incrosul UMS, UMS-45; Rewocid SBU 185; Schercopol UMS-Na; Varsulf SBU-185

Disodium wheat germamido PEG-2 sulfosuccinate
Mackanate WGD

Disoya dimethyl ammonium chloride
Arquad 2S-75

Distearyldimonium chloride [CAS # 107-64-2]
Synonyms: Distearyl dimethyl ammonium chloride
Arosurf TA-100, TA-101; Comperlan LD 9

Distearyldimonium methosulfate
Variquat 190; Varisoft 137, 190-100P

Ditallow amidoammonium methosulfate
Rewoquat W 222 LM

Ditallow diamido methosulfate
Incrosoft T-75

Ditallow dimonium chloride [CAS #68783-78-8]
Synonyms: Ditallow dimethyl ammonium chloride; Quaternium-48
Adogen 470; Armosoft L

Ditridecyl sodium sulfosuccinate [CAS #2673-22-5]
Synonyms: Sodium bistridecyl sulfosuccinate; Sodium ditridecyl sulfosuccinate
Aerosol TR-70; Alcopol OD; Emcol 4600; Geropon BIS/SODICO; Monawet MT-70, MT-70E, MT-80H2W

Dodecyl... See *Lauryl*

Dodecylbenzene sulfonic acid [CAS #27176-87-0]
Synonyms: DDBSA
Alkasurf LA Acid; Ardet LAS; Arsul DDB, LAS, S Acid; Arylan SBC Acid, SC Acid, SP Acid; Bio Soft JN, LAS-97, S-100; Calsoft LAS-99; Carsosulf UL-100 Acid; Conco AAS-98S; Condasol Sulfonic Acid K; Cycloryl ABSA; DDBSA 99-b; DeSonate SA, SA-H; Elfan WA

Sulfosaure; Emulsifier 99; Hetsulf Acid; Jordamine DDBSA; Jordanol DDBSA; Lakeway SA; Manro BA, HA, NA; Marlon AS3; Merpisap AS 98; Nansa 1042, 1042/P, 1909, SBA, SSA, SSA/P; Polyfac LAS-97; Polystep A-13, -17; Puxol XB-10; Reworyl K; Reworyl-Sulfonic Acid K; Richonic Acid B; Sandoz Sulfonate AAS 98S; Serdet DM; Steinaryl-Sulfonic Acid K; Stepantan H-100; Sterling LA Acid-C, Reg.; Surco DDBSA; Tairygent CA-1, CA-2; Tex-Wet 1197; Witco 1298 Hard Acid, 1298 Soft Acid, Acid B

Dodecylbenzyl trimonium chloride [CAS #1330-85-4; 19014-05-2]
Synonyms: Dodecylbenzyl trimethyl ammonium chloride
Hyamine 2389

Dodecyl dimethyl ethylbenzyl ammonium chloride. See *Quaternium-14*

Dodecyl diphenyloxide disulfonic acid
Conco Sulfate 2A1 Acid; Dowfax 2A0

Dodecylxylylditrimonium chloride [RD #977065-94-3]
Synonyms: Quaternium-29
Hyamine 2389

Dodoxynol-5 [CAS #9002-92-0 (generic); 9014-92-0 (generic); 26401-47-8 (generic); RD #977064-27-9]
Synonyms: PEG-5 dodecyl phenyl ether; POE (5) dodecyl phenyl ether
DeSonic 5D; T-Det DD-5; Teric DD5

Dodoxynol-6 [CAS #9002-92-0 (generic); 9014-92-0 (generic); 26401-47-8 (generic); RD #977064-28-0]
DeSonic 6D; Sterox DF

Dodoxynol-7 [CAS #9002-92-0 (generic); 9014-92-0 (generic); 26401-47-8 (generic); RD #977064-29-1]
T-Det DD-7

Dodoxynol-9 [CAS #9002-92-0 (generic); 9014-92-0 (generic); 26401-47-8 (generic); RD #977064-30-4]
DeSonic 9D; T-Det DD-9; Teric DD9

Dodoxynol-10
DeSonic 10D; Iconol DDP-10; Sterox DJ (10.5 EO); T-Det DD-10

Dodoxynol-11
T-Det DD-11

Dodoxynol-12 [CAS #9002-92-0 (generic); 9014-92-0 (generic); 26401-47-8 (generic); RD #977064-31-5]
DeSonic 12D

Dodoxynol-14
DeSonic 14D; T-Det DD-14

Egg Oil [CAS #8001-17-0]
Emcon E

Erucamide [CAS #112-84-5]
Synonyms: Erucic acid amide; 13-Docosenamide
Kemamide E

Erucyl erucamide
Kemamide E-221

Erucyl stearamide
Kemamide S-221

Ethyl bis (polyhydroxyethyl) alkyl ammonium ethosulfate
Varstat 66

Ethylene distearamide [CAS #110-30-5]
Synonyms: N,N´-Ethylene bisstearamide
Alkamide STEDA

Ethylene glycol. See *Glycol*

Ethyl hexanediol [CAS #94-96-2]
Lanoquat 1756

2-Ethylhexanol
Surfynol 104A

2-Ethylhexyl. See *Octyl, Caprylic*

Ethylhexyl laurate
Radia 7127

Ethylhexyl oleate
Radia 7331

Ethylhexyl palmitate
Radia 7129

Ethyl hydroxymethyl oleyl oxazoline [CAS #68140-98-7]
Alkaterge-E

Ethyl PEG-15 cocamine sulfate [RD #977069-64-9]
Synonyms: Quaternium-54; N-ethyl-N-cocoammonium ethoxylate (15) sulfate
Antaron PC-37

Ferro-chromium lignosulfate
Peltex

Formaldehyde condensed naphthalene/alkylaryl sulfonate copolymer
Lomar BDG

Glucose glutamate [RD #977056-78-2]
Wickenol 545

Glycereth-7
Synonyms: PEG-7 glyceryl ether; POE (7) glyceryl ether
Hetoxide G-7

Glycereth-26 [CAS #31694-55-0 (generic); RD #977054-72-0]
Ethosperse G-26; Hetoxide G-26; Liponic EG-1

Glycerides, acetates. See *Acetylated glycerides*

Glycerides, hydrogenated. See *Hydrogenated glycerides*

Glycerin [CAS #56-81-5]
Synonyms: Glycerol; 1,2,3-Propanetriol
Croderol G7000; Lexaine IBC-70, IBC-70S; Lexamine R-13

Glyceryl arachidate
GMA-33

Glyceryl capromyristate
Radia 7104

Glyceryl caprylate [CAS #26402-26-6]
Tegin 4600 NSE

Glyceryl caprylate-caprate
Drewmulse GMC-8; Labrasol; Tegin 4600

Glyceryl cocoate [CAS #61789-05-7]
Synonyms: Glycerides, coconut oil mono-; Glycerol mono coconut oil; Glyceryl coconate
Aldo MC; Drewmulse 75; Drewmulse CNO; Softisan 601

Glyceryl di/tribehenate
Compritol 888

Glyceryl dilaurate [CAS #27638-00-2]
Synonyms: Dilaurin
Cithrol GDL N/E, GDL S/E

Glyceryl dioleate [CAS #25637-84-7]
Cithrol GDO N/E, GDO S/E; Emerest 2426; Grocor 2500; Kessco Glycerol Dioleate; Mazol GDO; Nikkol DGO-80

Glyceryl distearate [CAS #1323-83-7]
Cithrol GDS N/E, GDS S/E; Nikkol DGS-80

Glyceryl di/tripalmitostearate
Precirol ATO

Glyceryl ditristearate
Precirol WL 2155

Glyceryl hydroxystearate [CAS #1323-42-8]
Naturechem GMHS

Glyceryl 12-hydroxystearate
Radiasurf 7146

Glyceryl isostearate [CAS #32057-14-0; 66085-00-5]
Imwitor 780; Schercemol GMIS

Glyceryl laurate [CAS #142-18-7; 27215-38-9]
Synonyms: Glyceryl monolaurate
Aldo MLD; Alkamuls GML, GML-45; Cithrol GML N/E, GML S/E; Grindtek ML 90; Kessco Glycerol Monolaurate; Lamesoft LMG; Monomuls 90-L12; Tegin 4480

Glyceryl linoleate [CAS #2277-28-3; 26545-74-4]
Synonyms: Monolinolein
Grindtek MOL 90

Glyceryl mono/dilaurate
Aldo ML, MLD FG

Glyceryl mono/dioleate
Aldo HMO FG, MO FG, MOD FG; Caplube 8350; Tegomuls O, O Spezial; Witconol O

Glyceryl mono/distearate
Aldo HMS FG, MS FG, MS LG FG, MSD FG; Teginacid H-SE, ML-SE, R-SE, SE, X-SE; Tegomuls 4070, 6070, 9050, B, M; Witconol MST

Glyceryl mono/distearate citrate
Axol C62, C63

Glyceryl myristate [CAS #589-68-4; 27214-38-6]
Synonyms: Monomyristin
GMM-33; Grindtek MM 90

Glyceryl oleate [CAS #111-03-5; 25496-72-4; SE grade RD #977064-71-3]
Synonyms: Monoolein; Glyceryl monooleate; GMO
Aldo HMO, MO, MO Tech., MOD; Alkamuls GMO, GMO-45, GMO-45LG, GMO-55LG; Arlacel 186; Capmul GMO; Cithrol GMO N/E, GMO S/E; CPH-31-N, -205-NX, -362-N; Cyclochem GMO; Drewmulse 85, 200, GMO; Dur-Em 114, 204, GMO; Emerest 2421; Emrite 6008; GMO-33; Grocor 2000; Kemester 2000; Kessco Glycerol Monooleate; Mazol 300, GMO; Monomuls 90-O18; Nikkol MGO; Protegin, W, WX, X; Radiamuls 152, MG 2152; Radiasurf 7151 (SE), 7152; Rheodol MO-60; Tegin O, O NSE, O Special., O Special NSE; Witco 942

Glyceryl palmitate lactate
Aldo GLP FG; Grindtek LA 15; Vanease

Glyceryl ricinoleate [CAS #141-08-2; 1323-38-2]
Synonyms: Glyceryl monoricinoleate; Monoricinolein
Aldo MR; Alkamuls GMR; Cithrol GMR N/E, GMR S/E; Drewmulse GMRO; Flexricin 13; Radiasurf 7153; Softigen 701; Tegin RZ, RZ NSE

Glyceryl stearate [CAS # 123-94-4; 11099-07-3; SE grade RD #977053-96-5]
Synonyms: Glyceryl monostearate; GMS; Monostearin
Ahcovel Base N-15; Aldo HMS, MS, MS Industrial, MS LG, MSA, MSC, MSD, PME; Aldosperse 40/60, 40/60 FG, AA 40/60, O-20, O-20 FG, TS-20, TS-20 FG, TS-40, TS-40 FG; Alkamuls GMS, GMS-45; Arlacel 165; Atmos 150; Capmul GMS; Cerasynt 945, M, MN, Q, SD, WM; Cithrol GMS Acid Stable, GMS A/E ES 0743, GMS N/E, GMS S/E; CPH-53-N, -250-SE; Cutina CBS, GMS, KD 16 (SE), MD, MD-A; Cyclochem GMS, GMS 21, GMS 165, GMS/SE; Drewmulse 20/200/V, 21, 30/900, 300, 365, 700, 900; Drewmulse 1128, 1129, GMS, GMS-AE, HM-100, TP, V, V-SE (SE); Drewsoft 100; Dur-Em 117, 207-E; Emerest 2400, 2401, 2407; Empilan GMS LSE32, GMS LSE40, GMS LSE80, GMS MSE40, GMS NSE32, GMS NSE40, GMS NSE90; Empilan GMS SE32, GMS SE40, GMS SE70; Emrite 6003; Geleol; Gelot 64; GMS-33; Grindtek MSP 32-6, 40, 40F, 52; Grocor 5500, 6000, 6000 E, 6000 SE; Ice #2; Imwitor 191, 900 K, 960, 960 K; Kemester 5500, 6000, 6000E, 6000SE; Kessco Glycerol Monostearate 860, DH-1, Pure, SE; Lexate TA, TL; Lexemul 55G, 503, 515, 530 (SE), 561, AR, AS; Lexemul T (SE); Lipo GMS, GMS 450, GMS 470 (SE); Lipomulse 165; Mazol 165C, GMS, GMS-D (SE), GMS-HM; Myvaplex 600; Myvatex 3-50, 40-06S, Peanut Butter Stabilizer; Nikkol MGS-A, MGS-ASE (SE), MGS-B, MGS-BSE (SE), MGS-DEX, MGS-F20, MGS-F40, MGS-F50, MGS-F75; Nissan Monogly I (SE), M; Norfox GMS-F6; Protowax 90; Radiamuls 141 (SE), 142, 341 (SE), 600, 601, 900, MG 2141 SE (SE), MG 2412, MG 2143, MG 2600, MG 2900; Radiasurf 7140, 7141 (SE), 7600; Rewomul MG, MG SE (SE); Rheodol MS-50, MS-60, SEM; Schercemol GMS; Softenol 3900, 3991; Tally 100

Plus; Tefose 2561; Tegacid Regular V.A. (SE), Special V.A. (SE); Tegin (SE), 90, 90 NSE, 515, 515 V.A., 4011 (SE), 4100, 4433 (SE), C-1R (SE), C-611 SE (SE), GRB NSE, GRB, ISO, M, M NSE, MAV NSE, SE (SE), Special., Special SE (SE), VA (SE), VA 55G, VA 515; Teginacid, H, ML, R, Special, Special SE, X; Tegomuls 4100, 4101, 9102, 9102 S; Vykamol N/E, S/E; Witconol CA (SE), RHT (SE)

Glyceryl stearate citrate [CAS #39175-72-9]
Imwitor 370

Glyceryl stearate lactate
Grindtek LA 30; GLS

Glyceryl stearate palmitate
GMS-33-SES, -66, -99, -99-SES; Imwitor 940 K, 965 K; Tegomuls 90, 90 S

Glyceryl stearate SE. See *Glyceryl stearate*

Glyceryl triacetate. See *Triacetin*

Glyceryl tribehenate
Syncrowax HR-C

Glyceryl tricaprate/caprylate
Radianol 7106

Glyceryl triheptanoate
Radianol 7376

Glyceryl triisostearate. See *Triisostearin*

Glyceryl trilaurate. See *Trilaurin*

Glyceryl trilaurate/stearate
Softisan Hard Fats

Glyceryl trioleate. See *Triolein*

Glyceryl tristearate. See *Tristearin*

Glycidyl trimethyl ammonium choride
Ogtac-85, -85V

Glycol [CAS #107-21-1]
Synonyms: Ethylene glycol; 1,2-Ethanediol
Surfynol 104E, 104H

Glycol dibehenate [CAS #344-64-8]
Synonyms: 1,2-Ethanediyl docosanoate
Rewopal PG 340

Glycol dilaurate
Synonyms: Ethylene glycol dilaurate
Cithrol EGDL N/E, EGDL S/E

Glycol dioleate
Synonyms: Ethylene glycol dioleate
Cithrol EGDO N/E, EGDO S/E

Glycol distearate [CAS #627-83-8]
Synonyms: Ethylene glycol distearate; EGDS
Cithrol EGDS N/E, EGDS S/E; Cutina AGS; Cyclochem EGDS; Drewmulse EGDS; Elfan L 310; Emerest 2355; Euperlan PK 771, PK 776; Genapol PGM Conc.; Kemester EGDS; Kessco Ethylene Glycol Distearate; Lexemul EGDS; Lipo EGDS; Mapeg EGDS; Pegosperse 50 DS; Radiasurf 7269; Rewopal PG 280; Secoster DMS; Standapol Pearl Conc. 7130; Texapon SG

Glycol hydroxystearate [CAS #33907-46-9]
Synonyms: Ethylene glycol hydroxystearate
Naturechem EGHS

Glycol laurate
Synonyms: Ethylene glycol monolaurate
Cithrol EGML N/E, EGML S/E

Glycol oleate [CAS #4500-01-0]
Synonyms: Ethylene glycol monooleate
Cithrol EGMO N/E, EGMO S/E; Tegin GO

Glycol propargyl ether
Synonyms: Ethylene glycol monopropargyl ether
Hetoxol PA-1

Glycol ricinoleate [CAS #106-17-2]
Synonyms: Ethylene glycol monoricinoleate
Cithrol EGMR N/E, EGMR S/E; Flexricin 15

Glycol stearate [CAS #111-60-4; SE grade RD #977064-73-5]
Synonyms: Ethylene glycol monostearate; EGMS
Alkamuls EGMS; AMS-33; Cerasynt IP; Cithrol EGMS N/E, EGMS S/E; Clindrol SEG; Cyclochem EGMS, EM 324, POL; Cycloryl M1; Cyclosheen 100; Drewmulse EGMS; Emerest 2350; Empilan EGMS; Euperlan PK-810; Incrochem PPI; Incromate PPI; Kemester EGMS; Kessco Ethylene Glycol Monostearate, 70; Lexate CRC; Lexemul EGMS; Lipo EGMS; Mapeg EGMS; Nutrapon B; Pegosperse 50 MS, EGMS-70; Radiasurf 7270; Schercemol EGMS; Secoster EMS; Standapol S; Tefose 63; Tegin G (SE), G 6100; Texapon EVR; Varisoft 3262

Guar hydroxypropyltrimonium chloride [CAS #65497-29-2]
Cationic Guar C-261

2-Heptadecenyl-4,4-bishydroxymethyl oxazoline
Anfomul 01T

Hexadecyl. See *Cetyl, Palmityl*

Hexaethylene glycol. See *PEG-6*

Hexaglyceryl. See *Polyglyceryl-6*

Hexahydro-1.3.5-tris (2-hydroxethyl)-s-triazine
Onyxide 200

Hexyl alcohol [CAS #111-27-3]
Synonyms: Hexanol
Alfol 6

Hexyl decanol
Primarol 1006

2-Hexyldecyl stearate
Grocor 5820 G

Hexylene glycol [CAS #107-41-5]
Miranol 2MCA Modified, 2MCT, 2MCT Modified, 2MHT, 2MHT Modified; Rewoteric AM-G 30; Schercoteric LS-2TE MOD, MS-2 MOD; Variquat K375

Hexyl laurate
Standamul CTA

Hybrid safflower oil [RD #977064-78-0]
Neobee 18

Hydrogenated animal glyceride
Myverol 18-00

Hydrogenated animal glycerides
Dur-Em 187, 195, 287, 287E

Hydrogenated castor oil [CAS #8001-78-3]
Protegin W, WX

Hydrogenated coco-glycerides
Softisan 133, 134, 138, 142, 154

Hydrogenated coconut acid [RD #977058-93-7]
Industrene 333

Hydrogenated coconut oil [RD #977056-87-3]
Softisan 601

Hydrogenated cottonseed glyceride
Myverol 18-07

Hydrogenated ditallow amine [CAS #61789-79-5]
Synonyms: Dihydrogenated tallow amine; Bis(hydrogenated tallow alkyl) amines
Armeen 2HT; Kemamine S-970

Hydrogenated fish acid
Industrene 3022

Hydrogenated fish glycerides
Neustrene 045, 053

Hydrogenated lanolin [CAS #8031-44-5]
Fancol HL; Lipolan; Satulan; Super-Sat

Hydrogenated lard glyceride [CAS #8040-05-9]
Lactomul 925; Monomuls 9015; Myvatex 25-00

Hydrogenated lard glycerides
Synonyms: Hydrogenated lard mono-, di-, and tri- glycerides
Monomuls 60-15

Hydrogenated palm glycerides
Synonyms: Hydrogenated palm mono-, di-, and tri-glycerides
Lexemul LPM; Monomuls 60-35

Hydrogenated palm kernel oil [RD #977056-90-8]
Wecobee

Hydrogenated palm oil [CAS #8033-29-2]
Myvatex Peanut Butter Stabilizer

Hydrogenated palm oil glyceride
Monomuls 90-35; Myverol 18-04

Hydrogenated palm/palm kernel oil PEG-6 complex
Labrafil M 2130 BS, M 2130 CS

Hydrogenated shark liver oil [RD #977062-72-8]
Robeyl

Hydrogenated soybean oil [CAS #8016-70-4]
Neustrene 064; Tally 100 Plus; Vanease

Hydrogenated soybean oil glycerides
Synonyms: Hydrogenated soybean oil mono-, di- and tri- glycerides
Monomuls 60-45

Hydrogenated soy glyceride
Synonyms: Hydrogenated soybean oil monoglyceride
Monomuls 90-45; Myverol 18-06

Hydrogenated tallowalkonium chloride
Synonyms: Hydrogenated tallow dimethyl benzyl ammonium chloride
Arquad DMHTB-75; Empigen BCM75, BCM75/A; Kemamine BQ-9702C; Querton 441-BC, HBG BCL

Hydrogenated tallow amide [CAS #61790-31-6]
Armid HT; Armoslip HT

Hydrogenated tallow amine [CAS #61788-45-2]
Amine HBG; Armeen HT, HTD; Crodamine 1.HT; Kemamine P-970, P-970D; Lilamin 140, 140 D; Nissan Amine ABT; Noram SH; Radiamine 6140, 6141

Hydrogenated tallow amine acetate
Acetamin HT; Armac HT; Kemamine A970; Lilamac 140; Radiamac 6148, 6149

Hydrogenated tallow amine oxide [CAS #61788-94-1]
Synonyms: Hydrogenated tallow dimethyl amine oxide
Aromox DMHT, DMHTD

Hydrogenated tallow diamine
Lilamin 540; Radiamine 6540

Hydrogenated tallow dimethyl amine. See *Dimethyl hydrogenated tallow amine*

Hydrogenated tallow glyceride [CAS #61789-09-1]
Synonyms: Hydrogenated tallow monoglyceride
Monomuls 90-25, 90-25/2, 90-25/5

Hydrogenated tallow glyceride citrate [CAS #68990-59-0]
Grindtek CA-P; Lamegin ZE 30, ZE 60; Tegin C-1R, C-61 NSE, C 62, C-62 SE, C-63 SE, C-611 SE

Hydrogenated tallow glyceride lactate [CAS #68990-06-7]
Lamegin GLP 10, GLP 20; Tegin L 61, L-61 NSE, L 62, L-62 NSE

Hydrogenated tallow glycerides [CAS #68308-54-3]
Synonyms: Hydrogenated tallow mono-, di- and tri- glycerides
Monomuls 60-25, 60-25/2, 60-25/5; Neustrene 059, 060

Hydrogenated tallow hydroxysultaine
Mirataine HTS

Hydrogenated tallow 1,3-propylene diamine
Dinoram SH; Kemamine D-970; Nissan Amine DTH

Hydrogenated tallowtrimonium chloride [CAS #61788-78-1]
Synonyms: Hydrogenated tallow trimethyl ammonium chloride
Adogen 441; Nissan Cation ABT-350, -500; Noramium MSH 50; Quartamin HTPR

Hydrogenated vegetable glyceride
Myvatex 25-07; Myverol 18-50K

Hydrogenated vegetable glycerides
Synonyms: Hydrogenated vegetable oil mono- di and tri-glycerides
Dur-Em 104K, 107K, 114K, 117K, 127K, 136K, 195, 204K; Dur-Em 207K, 227K, 295

Hydrogenated vegetable glycerides phosphate
Emphos F27-85

Hydrogenated vegetable oil [CAS #68334-28-1; RD #977025-26-5]
Wecobee M, R, S, SS, W

Hydrolyzed animal protein [CAS #9015-54-7]
Lanotein AWS 30; Polypeptide 37, LSN; Super Pro 100

Hydroxyethyl carboxymethyl cocamido ethylamine
Varion 2C

1-Hydroxyethyl-2-heptadecenyl imidazoline
Loramine IM-CA, IM-OA; Rewomine IM-CA, IM-OA; Steinamin IM-CA, IM-OA

Hydroxyethyl hydroxypropyl C12-15 alkoxypropylamine
Varox 185PH

Hydroxyethyl stearamide-MIPA [RD #977068-21-5]
Incropearl Liquid, Liquid NI

Hydroxylated lanolin [CAS #68424-66-8]
OHlan; Ritahydrox

Hydroxylated lecithin [CAS #8029-76-3]
Alcolec Z-3

Hydroxyoctacosanyl hydroxystearate
Elfacos C26

Hydroxypropyl bisstearyldimonium chloride
Jordaquat Dimer 18

Hydroxypropylcellulose [CAS #9004-64-2]
Klucel

Hydroxypropyl methylcellulose [CAS #9004-65-3]
Synonyms: Methyl hydroxypropyl cellulose)
Methocel E5, E15-LV, E50-LV, E4M, F50-LV, F4M, K3, K35, K100-LV, K4M, K15M, K100M

2-Isoalkyl (C14-20), 1-hydroxyethyl, 1-ethyl imidazolinium ethyl sulfate
Schercoquat IIS-RD

Isobutylated lanolin oil [RD #977064-89-3]
Argonol 40

Isobutyl oleate
Radia 7230

Isobutyl stearate [CAS #646-13-9]
Radia 7240

Isoceteth-20
Synonyms: PEG-20 isocetyl ether
Arlasolve 200

Isoceteth-30
Nikkol BEG-1630

Isocetyl alcohol [CAS #36311-34-9]
Synonyms: Isohexadecanol; Isohexadecyl alcohol
Standamul G-16

Isocetyl stearate [CAS #25339-09-7]
Standamul 7061

Isodeceth-4
Synonyms: PEG-4 isodecyl ether
Emulphogene DA-530

Isodeceth-6
Emulphogene DA-630

Isolaureth-3
Synonyms: PEG-3 isolauryl ether
Tergitol TMN-3

Isolaureth-6
Tergitol TMN-6

Isolaureth-10
Tergitol TMN-10

Isononyl stearate
Radia 7510

Isooctyl oleate
Radia 7330

Isooctyl stearate
Radia 7130

Isopropanolamine [CAS #78-96-6]
Synonyms: 1-Amino-2-propanol; Monoisopropanolamine; MIPA; 2-Hydroxypropylamine
Witcamide 61

Isopropyl alcohol [CAS #67-63-0]
Synonyms: Isopropanol; IPA
Aromox C/12, DM16, DMCD, DMHT, T/12; Arquad 2C-75, 2HT-75, 12-33, 12-50, 16-50, 18-50, B-100, C-33, C-50, S-50, T-50; Carsosoft V-90; Emcol 150; Ethoquad 18/12, C/12, O/12; Varisoft 3262

Isopropylamine decyl diphenyloxide disulfonate
Dowfax 3B-Monoisopropylamine

Isopropylamine dodecylbenzenesulfonate [CAS #26264-05-1]
Alkasurf IPAM; Calimulse PRS; Cindet GE; Conco AAS 3H, AAS Special 3; Manro HCS; Nansa YS 94; Ninate LAS-931; Richonate YLA; Sandoz Sulfonate AAS-Special 3

Isopropylamine dodecyl diphenyloxide disulfonate
Dowfac 2A-Monoisopropylamine

Isopropyl lanolate [CAS #63393-93-1]
Amerlate P, W; Crodalan IPL; Ritasol

Isopropyl myristate [CAS #110-27-0]
Lexate TA; Radia 7190; Starfol IPM; Stepan D-50

Isopropyl oleate [CAS #112-11-8]
Synonyms: 1-Methylethyl-9-octadecenoate
Radia 7231

Isopropyl palmitate [CAS #142-91-6]
Synonyms: Isopropyl *n*-hexadecanoate
Radia 7200; Ritalan C; Starfol IPP; Stepan D-70

Isopropyl stearate [CAS #112-10-7]
Synonyms: 1-Methylethyl octadecanoate
Radia 7195

Isostearamide DEA [CAS #52794-79-3]
Comperlan ID; Lameform GE 2; Mackamide ISA; Monamid 150-IS; Schercomid ID, M, SI, SI-M; Standamid ID

Isostearamide MIPA
Synonyms: Isostearic acid monoisopropylamide; Isostearoyl monoisopropanolamide
Schercomid IMI

Isostearamidopropalkonium chloride [CAS #67633-59-4]
Schercoquat IB

Isostearamidopropylamine oxide
Incromine Oxide I; Mackamine IAO

Isostearamidopropyl betaine
Cycloteric BET I-30; Incronam I-30; Mackam ISA; Schercotaine IAB

Isostearamidopropyl dimethylamine [CAS #67799-04-6]
Cyclomide IODI; Incromine IB; Jordamine DAPI

Isostearamidopropyl dimethylamine lactate [CAS #55852-15-8]
Incromate IDL; Richamate 6613

Isostearamido dimethylamine propionate
Mackam ISP

Isostearamidopropyl ethyldimonium ethosulfate [CAS #67633-63-0]
Jordaquat 522; Schercoquat IAS-LC

Isostearamidopropyl morpholine
Incromine ISM

Isostearamidopropyl morpholine betaine
Incronam ISM-30

Isostearamidopropyl morpholine lactate
Emcol ISML; Incromate ISML

Isostearamidopropyl morpholine oxide
Incromine Oxide ISMO

Isosteareth-2 [CAS #52292-17-8 (generic); RD #977064-96-2]
Synonyms: PEG-2 isostearyl ether; POE (2) isostearyl ether; PEG 100 isostearyl ether
Arosurf 66-E2

Isosteareth-10 [CAS #52292-17-8 (generic); RD #977059-85-0]
Arosurf 66-E10

Isosteareth-20 [CAS #52292-17-8 (generic); RD #977064-97-3]
Arosurf 66-E20

Isostearoamphodipropionate
Synonyms: Isostearoamphocarboxypropionate
Schercoteric IS-SF-2

Isostearoamphopropionate [CAS #68630-96-6]
Monateric ISA-35%; Schercoteric I-AA

Isostearyl alcohol [CAS #27458-93-1; 41744-75-6]
Adol 66

Isostearyl benzylimidonium chloride
Schercoquat IIB

Isostearyl diglyceryl succinate
Imwitor 780 K

Isostearyl ethylimidonium ethosulfate [RD #977065-96-5]
Synonyms: Quaternium-32
Monaquat ISIES; Schercoquat IIS

Isostearyl hydroxyethyl imidazoline [CAS #68966-38-1]
Synonyms: Isostearyl imidazoline
Miramine IC; Monazoline IS; Schercozoline I

Isostearyl lactate
Patlac IL

Lactic acid [CAS #50-21-5]
Tandem 11H

Lactylic stearate
Myvatcx 40-06S

Laneth-5 [CAS #61791-20-6 (generic); RD #977054-73-1]
Synonyms: PEG-5 lanolin ether; POE (5) lanolin ether
Polychol 5; Ritawax 5; Solulan 5

Laneth-10 [CAS #61791-20-6 (generic); RD #977054-75-3]
Polychol 10

Laneth-10 acetate [RD #977054-76-4]
Synonyms: Acetylated PEG-10 lanolin ether
Ritawax AEO

Laneth-15 [CAS #61791-20-6 (generic); RD #977054-77-5]
Polychol 15; Ritawax 15

Laneth-16 [CAS #61791-20-6 (generic); RD #977054-78-6]
Solulan 16

Laneth-20 [CAS #61791-20-6 (generic); RD #977054-79-7]
Aqualose W20/50; Polychol 20

Laneth-25 [CAS #61791-20-6 (generic); RD #977054-80-0]
Solulan 25

Laneth-40 [CAS #61791-20-6 (generic); RD #977054-81-1]
Polychol 40; Ritawax 40

Lanolin [CAS #8006-54-0 (anhyd.), 8020-84-6 (hyd.)]
Albalan, 36W, 40; Amerchol 400, BL, C, H-9; Clearlan K50; Corona Lanolin; Coronet Lanolin; Dusoran MD, NG; Golden Dawn Grade 1, 2; Golden Dawn Superfine; Golden Fleece DF, P-80, P-95, RA; Lexate PX; Nimcolan 1747; Paralan; Ritaderm; Ritalan; Super Corona; White Swan; Yeoman

Lanolin acid [CAS #68424-43-1]
Synonyms: Lanolic acids; Lanolin fatty acids
Amerlate LFA, WFA; Fancor LFA; Lanolic acid; Lanolin Fatty Acids G, O; Nimco 1781; Ritalafa; Skliro; Tergelan 1790

Lanolin alcohol [CAS #8027-33-6]
Alcolan; Amerchol 400, BL, C, CAB, H-9, L-99, L-101, L-500, RC; Argobase 125; Argowax Distilled, Standard; Ceralan; Fancol LA, LAO; Hartolan, Hartolan Super; Lanolin Alcohols LG, LO, THG, THO; Liquid Base; Nimco 1780; Nimcolan 1747; Protegin, X; Ritachol; Ritawax; Super Hartolan

Lanolin alcohol sulfosuccinate
Rewoderm SB/LAN; Rewolan 5, E; Varsulf 5

Lanolin linoleate [RD #977057-13-8]
Polylan

Lanolin oil [CAS #8038-43-5]
Fluilan; Fluilanol; Lanogene; Lantrol; Lipolan R; Ritalan C; Vigilan

Lanolin wax [CAS #68201-49-0]
Juniorlan; Lanfrax

Lanosterol [CAS #79-63-0]
Lanosterol

Lapyrium chloride [CAS #6272-74-8]
Emcol E-607L

Lard glyceride [CAS #61789-10-4]
Synonyms: Lard monoglyceride
Dimodan P, PM, PM 300, S; Grindtek MOP 90; Monomuls 90-10; Myverol 18-40

Lard glycerides
Synonyms: Lard mono, di- and tri-glycerides
Monomuls 60-10

Lauralkonium chloride [CAS #139-07-1]
Synonyms: Lauryl dimethyl benzyl ammonium chloride
Hartex SAN Q 50; Loraquat B 50; Retarder N, N-85

Lauramide DEA [CAS #120-40-1]
Synonyms: Lauric diethanolamide
Agent 804; Alkamide 1182, 2124, L9DE; Aminol COR-4, COR-4C, LM-30C Special; Aramide LDS; Ardet DLA, LDA, LMC; Bio Soft LD-95; Carsamide SAL-7, SAL-9, SAL-82; Cedemide AX; Chimipal LDA; Clindrol 200L; Clindrol Superamide 100L, 100LM, LL660, LL743, LL931, LL1843, LL2141; Comperlan LD, LDO, LDS, LMD; Condensate PE, PL; Crillon LDE; Cyclomide DL203, DL203/S, DL207/S, DL207/SL, LE; Emid 6510, 6511, 6541; Empilan LDE, LDX; Ethylan MLD; Hartamide LDA 70, 90; Hetamide ML; Incrochem SC1, SC4; Incromide L-90, LCL, LL, LLA, LM-70, LR; Intermediate 512; Jordamide 1214, CLM; Loramine DL 203/S; Loropan LD; Mackamide L-95, LLM, LMD; Mazamide LM-20, LS-173, LS-196; Merpinamid KD 11, 12, LD/E; Merpinamid LSD; Merpinamid LSD/E; Monamid 150-GLT, 150-LMW-C, 150-LW, 150-LWA, 716, 770, 982, 1007, 1034, R31-42; Monamine ACO-100, LM-100; Ninol 51-LL, 52-LL, 55-LL, 4821, AA-62, AA-62 Extra, P-616, P-621; Nissan Stafoam DL, L; Nitrene L-90; Norfox DLSA; Onyxol 336, 345, SD; Product LT 18-48; Profan AA62; Rewomid 203/S, DL 203/S; Richamide 6310, STD-HP; Schercomid 1214, LD, SL-Extra. SLM, SLM-C, SL-ML, SLM-LC, SLM-S; Sipomide 843; Stafoam DL; Standamid KDS, LD, LDM, LDO, LDS; Steinamid 203/S; Sterling LDEA-90; Super Amide L-9, L-9A, L-9C, LL, LM; Synotol L 60, L 90, LM 60; Tohol N-230, N-230X; Unamide J-56, LDX, LMDX; Varamide L-1, ML-4, SL-9; Witcamide 5195, 6310, STD-HP

Lauramide MEA [CAS #142-78-9]
Synonyms: Lauric monoethanolamide
Alkamide L9ME; Ardet LEA; Cedemide MX; Comperlan LM; Crillon LME; Cyclomide L203; Empilan LME; Hartamide LMEA-90; Incromide LMM; Lauridit LM; Loramine L 203; Loropan LM; Mackamide LMM; Merpinamid KM, LM, LSM; Monamid LMA; Monamid LMMA; Rewomid L203; Standamid LM; Steinamid L203

Lauramide MIPA [CAS # 142-54-1]
Synonyms: Lauric monoisopropanolamide
Alkamide LIPA; Ardet LIPA; Clindrol 101LI, 102LI; Comperlan LP; Cyclochem LIPA; Cyclomide LIPA; Cycloryl M1; Cyclosheen 100; Empicol TCR, TLP, TLR; Empilan LIS; Incromide LI; Loramine IPL 203; Merpinamid LMIPA; Monamid LIPA, LMIPA; Profan AD31; Rewomid IPL 203; Standamid LP; Standapol S; Steinamid IPL 203; Texapon EVR

Lauramidopropyl acetamidodimonium chloride [CAS #68259-01-8; RD #194-03-2]
Bina Qat-43

Lauramidopropylamine oxide [CAS #61792-31-2]
Synonyms: Lauramidopropyl dimethylamine oxide
Cyclomox LO; Mackamine LAO

Lauramidopropyl betaine [CAS #4292-10-8]
Alkateric LAB; Jortaine LMAB; Mackam LMB, LMB-LS; Mirataine BB; Monateric LMAB; Tego-Betaine L-90

Lauramidopropyl dimethylamine [CAS #3179-80-4]
Lexamine L-13; Schercodine L

Lauramidopropyl dimethylamine propionate
Mackam LAP

Lauramidopropyl PEG-dimonium chloride phosphate
Monaquat P-TD

Lauramine [CAS #124-22-1]
Synonyms: Lauryl amine; Dodecanamine; Dodecylamine
Amine 12D; Armeen 12D; Kemamine P-690, P-690D; Lilamin 163, 163 D; Nissan Amine BB; Radiamine 6163, 6164

Lauramine oxide [CAS #1643-20-5]
Synonyms: Lauryl dimethyl amine oxide
Alkamox LO; Ammonyx DMCD-40, LO; Aromox DMMC-W; Conco XA-L; Cyclomox L; Emcol L; Empigen OB; Incromine L-40; Incromine Oxide L; Jordamox LDA; Loramine L 408; Mackamine LO; Ninox L; Oxamin LO; Rewominox L 408; Sandoz Amine Oxide XA-L; Schercamox DML; Textamine Oxide LMW; Varox 365

Lauraminopropionic acid [CAS #1462-54-0; 3614-12-8]
Synonyms: N-dodecyl-beta-alanine
Velvetex 710L

Laureth-1 [CAS #4536-30-5; 9002-92-0 (generic); RD #977054-82-2]
Synonyms: PEG-1 lauryl ether; Ethylene glycol monolauryl ether; Ethylene glycol dodecyl ether
Alkasurf LAN-1; Lipocol L-1; Nikkol BL-1SY; Siponic L1

Laureth-2 [CAS #3055-93-4; 9002-92-0 (generic); RD #977062-02-4]
Synonyms: Diethylene glycol dodecyl ether; PEG-2 lauryl ether; POE (2) lauryl ether
Dehydol LS 2; Empilan KB 2; Incropol L-7; Nikkol Bl-2, BL-2SY

Laureth-3 [CAS #3055-94-5; 9002-92-0 (generic); RD #977064-47-3]
Synonyms: Triethylene glycol dodecyl ether
AE-1214/3; Dehydol LS 3; Empilan KB 3, KC 3; Hetoxol L-3N; LA-55-3; Nikkol BL-3SY; Standapol CS-50, CS Paste

Laureth-3 phosphate [CAS #39464-66-9 (generic)]
Cyclophos PL3

Laureth-4 [CAS #5274-68-0; 9002-92-0 (generic); RD #977054-83-3]
Ameroxol LE-4; Brij 30, 30 SP; Chemal LA-4; Dehydol LS 4; Emthox 5882; Ethosperse LA-4; Hetoxol L-4N; LA-55-4; Lipal 4 LA; Lipocol L-4; Macol LA-4; Nikkol BL-4SY, BL-4.2; Simulsol P4; Siponic L4; Trycol LAL-4; Volpo L4

Laureth-5 [CAS #3055-95-6; 9002-92-0 (generic); RD #977068-22-6]
Nikkol BL-5SY

Laureth-5 carboxylic acid
Akypo RLM 45

Laureth-6 [CAS #3055-96-7; 9002-92-0 (generic); RD #977065-03-4]
Nikkol BL-6SY

Laureth-6 citrate
Citrest LSU

Laureth-7 [CAS #3055-97-8; 9002-92-0 (generic); RD #977062-41-1]
Incropol L-7, L-7-90; Macol LA-790; Marlipal MG; Nikkol BL-7SY; Siponic L7, L-7-90

Laureth-8 [CAS #9002-92-0 (generic); RD #977068-23-7]
Nikkol BL-8SY; Trycol LAL-8

Laureth-9 [CAS #9002-92-0 (generic); RD #977007-24-1]
Carsonon L-985; Chemal LA-9; Hetoxol L-9N, LS-9; Lipal 9 LA; Lubrol 12A-9 (9.5 EO); Macol LA-9; Nikkol BL-9EX, -9EX(FF); Standapol Pearl Conc. 7130

Laureth-10 [CAS #9002-92-0 (generic); RD #977065-01-2]
Dehydol 100; Emulgade A

Laureth-10 carboxylic acid
Akypo RLM 100

Laureth-12 [CAS #9002-92-0 (generic); RD #977054-84-4]
Chemal LA-12; Comperlan LS; Cyclogol EL; Emthox 5967; Ethosperse LA-12; Hetoxol L-12; Incropol L-12; Lanette Wax CAT; Lipal 12 LA; Lipocol L-12; Macol LA-12; Polystep F-12; Siponic L12; Trycol LAL-12

Laureth-15 [CAS #9002-92-0 (generic); RD #977069-29-6]
Alkasurf LAN-15

Laureth-16
Siponic L16

Laureth-21
Nikkol Bl-21

Laureth-23 [CAS #9002-92-0 (generic); RD #977054-85-5]
Alkasurf LAN-23; Ameroxol LE-23; Brij 35, 35 SP; Cerasynt 945; Chemal LA-23; Ethosperse LA-23; Hetoxol L-23N; Incropol L-23; LA-55-23; Lipal 23 LA; Lipocol L-23; Macol LA-23; Siponic L25; Trycol LAL-23; Volpo L23

Laureth-25 [CAS #9002-92-0 (generic); RD #977069-30-9]
Nikkol BL-25

Laureth-30 [CAS # 9002-92-0 (generic); RD #977069-31-0]
Incropol L-30; Polystep F-13

Lauric acid [CAS #143-07-7]
Synonyms: Dodecanoic acid
Hetamide LA; Hystrene 9512

Lauric/coconut DEA
Ardet C28

Lauric diethanolamide. See *Lauramide DEA*

Lauric diglycolamide
Incromide LMD

Lauric/linoleic DEA
Product LT 19-44

Lauric monoethanolamide. See *Lauramide MEA*

Lauric monoisopropanolamide. See *Lauramide MIPA*

Lauric/myristic DEA
Alkamide L7DE, L7DE-BT; Ardet LMD; Clindrol Superamide 100LM; Loropan LMD; Ninol 71-LL; Nitrene L-76; Rewomid DLMS; Standamid LD 80/20; Sterling Amide 460; Sterling LM-70; Synotol LM 90

Lauric/myristic MEA
Alkamide L7ME; Hartamide LMEA-70

Lauric/myristic MIPA
Incromide LMI

Lauric triglyceride
Wecobee FS, FW

Lauroamphoacetate
Synonyms: Lauroamphoglycinate
Mackam 1L, 1L-30; Miranol HM Conc., MHT; Monateric LMM-30; Rewoteric AM-2L, AM-CA, AM-G 30; Schercoteric LS; Sipoteric 1127

Lauroamphodiacetate
Synonyms: Lauroamphocarboxyglycinate
Mackam LT, MLT; Miranol 2MHT, 2MHT Modified, BT, H2M Conc.; Monateric 951A, 985A; Schercoteric LS-2, LS-2 ES MOD, LS-2TE, LS-2TE MOD

Lauroamphodipropionate
Synonyms: Lauroamphocarboxypropionate
Mackam 2LSF; Miranol H2M-SF 70%, H2M-SF Conc.; Schercoteric LS-SF-2

Lauroamphohydroxypropylsulfonate
Synonyms: Lauroamphopropylsulfonate
Miranol HS Conc.; Schercoteric LS-EP

Lauroampho PEG-glycinate phosphate
Monaquat P-TL

Lauroamphopropionate
Miranol HM-SF Conc.; Schercoteric LS-SF, LS-SF Conc.

Lauroyl glutamic acid
Amiter LGS-2

Lauroyl glutamic acid, dioxyethylene 2-octyldodecyl ether diester
Amiter LGOD-2

Lauroyl glutamic acid, 2-octyldodecyl diester
Amiter LGOD

Lauroyl-1-lysine
Amihope LL-11

Lauroyl sarcosine [CAS #97-78-9]
Synonyms: N-methyl-N-(1-oxododecyl) glycine
Hamposyl L; Nikkol Sarcosinate LN; Oramix L30; Sarkosyl L

Laurtrimonium chloride [CAS #112-00-5]
Synonyms: Lauryl trimethyl ammonium chloride; Dodecyl trimethyl ammonium chloride
Arquad 12-33, 12-50; Chemquat 12-33, 12-50; Dehyquart LT; Kemamine Q-6903B; Kosti Kat PE; Nissan Cation BB

Lauryl alcohol [CAS #112-53-8]
Synonyms: 1-Dodecanol; C12 linear primary alcohol; Dodecyl alcohol
Alfol 12; Epal 12; Kalcohl 20; Laurex L1, NC

Lauryl amine. See *Lauramine*

Lauryl betaine [CAS #683-10-3]
Synonyms: Lauryl dimethyl glycine
Empigen BB; Rewoteric AM-DML; Swanol AM-301; Varion CDG

Lauryl dimethylamine. See *Dimethyl lauramine*

Lauryl dimethylamine oxide. See *Lauramine oxide*

Lauryl dimethyl benzyl ammonium chloride. See *Lauralkonium chloride*

Lauryl dimethyl betaine
Synonyms: Dimethyl dodecyl betaine
Nissan Anon BL

Lauryl hydroxyethyl imidazoline [CAS #136-99-2]
Synonyms: Lauryl imidazoline
Crapol AV-11; Imidrol LC; Mackazoline L; Schercozoline L

Lauryl imidazoline betaine
Carsonam L; Empigen CDL 10, CDL 30

Lauryl/myristyl alcohol
Kalcohl 5-24, 6-24, 7-24

Lauryl/myristyl beta-aminopropionic acid
Deriphat 170C

Lauryl/myristyl dimethylamine oxide
Bio-Surf PBC-460; Carsamine Oxide LM; Standamox LMW-30

Laurylpyridinium chloride [CAS #104-74-5]
Charlab LPC; Dehyquart C, C Crystals

Lauryl trimethyl ammonium chloride. See *Laurtrimonium chloride*

Lecithin [CAS #8002-43-5]
Actiflo 68, 70; Alcolec 439-C, 495, 619-B, 621, 628G, 634, 638, 650, 658, 662, 4135, DS, Extra A; Alcolec Granules, HS-3, Powd., RCX-1, S; Asol; Centrocap; Centrol Series; Centrolene

Series; Centrolex Series, P; Centromix Series; Centrophase Series, 31, HR; Centrophil Series; Chocothin; Clearate LV, WD, Special Extra, WDF; Dugista; Dulectin, Granular, Powdered; EmCon E-5, E-20, E-L; Emulbesto; Emulfluid; Emulpur, N; Emulthin M-35, M-501; Lecithin W.D.; Lexin K; LKE Granules; Magathin; M-C-Thin, 45, AF-1; Metarin, F, P; Topcithin; Troykyd Lecithin W.D.

Linoleamide/cocamide DEA
Jordamide SCD

Linoleamide DEA [CAS #56863-02-6]
Synonyms: Linoleic diethanolamide
Clindrol LT 15-73-1; Comperlan F, VOD; Cyclomide DIN 295/S, DOTS; Incromide LA, LLA; Jordamide LLD; Monamid 15-70W, 150-ADY, 982, 1007; Monamine AC-100, ADY-100, CD-100, R8-26; Nitrene 100 SD; Rewomid F; Schercomid SLE, SLS; Standamid SOD, SOMD

Linoleamidopropyl dimethylamine
Incromine LSB

Linoleamidopropyl dimethylamine lactate
Incromate LDL

Lithium lauryl sulfate
Texapon LLS

Magnesium aluminum silicate [CAS #1327-43-1]
Van Gel B; Veegum, Veegum HS, HV, T

Magnesium dodecylbenzene sulfonate
Nansa MS 45

Magnesium laureth sulfate [RD #977069-33-2]
Synonyms: Magnesium lauryl ether sulfate
Drewpon ESG; Empicol EGB; Montosol IL-17; Texapon ASV, MG; Tylorol MG

Magnesium lauryl sulfate [CAS #3097-08-3]
Carsonol MLS; Conco Sulfate M; Cycloryl MG; Drewpon MG; Empicol ML 26; Maprofix MG; Montovol GL-13; Polystep B-9; Richonol Mg; Sipon LM, MLS; Standapol MG; Stepanol MG; Sulfetal MG 30; Surco MG-LS, MGS-35

Magnesium oleth sulfate
Texapon ASV

Magnesium stearate [CAS #557-04-0]
Hostacerin WO

Magnesium sulfonate
Hybase M-300, M-400

Mannide oleate
Arlacel A

MEA decyl diphenyloxide disulfonate
Dowfax 3B-Monoethanolamine

MEA dodecyl diphenyloxide disulfonate
Dowfax 2A-Monoethanolamine

MEA-laureth sulfate [RD #977067-77-8]
Synonyms: Monoethanolamine lauryl ether sulfate
Akyposal MLES 35; Drewpon ECM; Montosol PG-17; Tylorol LM

MEA-lauryl sulfate [CAS #4722-98-9]
Synonyms: Monoethanolamine lauryl sulfate
Alkasurf MLS; Berol 484; Condanol MLS 35; Cycloryl SA; Drewpon MEA; Elfan 240M; Empicol EL, LQ33, LQ33/T, LQ 70, XT 45; Manro ML 33; Rewopol MLS 30, MLS 35; Serdet DFM 33; Standapol MLS; Sulfatol 33 MO; Sulphonated Lorol Liquid MA, MR; Texapon MLS; Zoharpon LAM

Meroxapol 105
Pluronic 10R5

Meroxapol 108
Pluronic 10R8

Meroxapol 171
Pluronic 17R1

Meroxapol 172
Pluronic 17R2

Meroxapol 174
Pluronic 17R4

Meroxapol 178
Pluronic 17R8

Meroxapol 251
Pluronic 25R1

Meroxapol 252
Pluronic 25R2

Meroxapol 254
Pluronic 25R4

Meroxapol 255
Pluronic 25R5

Meroxapol 258
Pluronic 25R8

Meroxapol 311
Pluronic 31R1

Meroxapol 312
Pluronic 31R2

Meroxapol 314
Pluronic 31R4

Methoxyethoxypropylamine
Surfam P-MEPA

Methoxy PEG-22/dodecyl glycol copolymer
Elfacos E200

Methoxy PEG-400 monooleate
Emery 6779

Methoxypropylamine
Surfam P5 Dist., 95 Dist.

1-Methyl-2-alkyl-3 alkylamidoethyl-imidazonolinium methosulfate
Loraquat M 5040

Methylbenzethonium chloride [CAS #25155-18-4]
Synonyms: Diisobutyl cresoxy ethoxy ethyl dimethyl benzyl ammonium chloride
Hyamine 10-X

Methyl bis (2-hydroxyethyl) coco ammonium chloride
Variquat 638

Methyl bis (oleylamidoethyl) 2-hydroxyethyl ammonium methosulfate
Varisoft 222-LT

Methyl bis (polyethoxyethanol) alkyl ammonium chloride
Variquat K-1215

Methyl bis (tallowamidoethyl) 2-hydroxyethyl ammonium methosulfate
Varisoft 222-90%, 222-HV, 222LM-90, 238-90%

Methyl caprylate/caprate
Stepan C25

Methylcellulose [CAS #9004-67-5]
Methocel A4C, A4M, A15-LV

Methyl cocoate [CAS #61788-59-8]
Kemester 325

Methylenebisdiamyl phenoxy polyethoxy ethanol
Triton X-155, X-155-90%

Methyl gluceth-10 [RD #977065-07-8]
Synonyms: PEG-10 methyl glucose ether
Glucam E-10

Methyl gluceth-20 [RD #977065-08-9]
Glucam E-20

Methyl gluceth-20 distearate
Glucam E-20 Distearate

Methyl glucose dioleate
Glucate DO

Methyl glucose sesquistearate [CAS #68936-95-8]
Glucate SS

Methyl (1) hydrogenated tallow amidoethyl (2) hydrogenated tallow imidazolinium methosulfate
Varisoft 445

Methyl lardate
Base ML

Methyl laurate [CAS #111-82-0]
Synonyms: Methyl dodecanoate
Radia 7118; Stepan C40, C41, C42, C43

Methyl myristate [CAS #124-10-7]
Synonyms: Methyl tetradecanoate
Stepan C50

Methyl oleate [CAS #112-62-9]
Grocor 8002, 8008; Kemester 104, 105, 205; Radia 7060; Stepan C68

Methyl oleylamidoethyl oleyl imidazolinium methosulfate
Incrosoft CFI, CFI-75; Varisoft 3690

Methyl palmitate [CAS #112-39-0]
Synonyms: Methyl hexadecanoate
Radia 7120; Stepan C60

Methyl palmitate oleate
Petrosan 102; Stepan C65

Methyl soyate
Kemester 226

Methyl stearate [CAS #112-61-8]
Grocor 8058; Radia 7110; Stepan C66

Methyl tallowate
Base MT; Grocor 8041

Mineral oil [CAS #8012-95-1]
Amerchol BL, L-99, L-101, L-500; Dehymuls K; Fancol LAO; Hostacerin WO; Liquid Base; Protegin, X; Ritachol

MIPA-dodecylbenzene sulfonate [CAS #42504-46-1; 54590-52-2]
Synonyms: (Mono)isopropanolamine dodecylbenzene sulfonate
Hetsulf IPA

MIPA-laureth sulfate
Synonyms: (Mono)isopropanolamine lauryl ether sulfate
Montosol IL-13

MIPA-lauryl sulfate [CAS #21142-28-9]
Synonyms: (Mono)isopropanolamine lauryl sulfate
Empicol MIPA; Incronol ILS; Melanol LP 1; Montovol RF-13

Mixed isopropanolamines lauryl sulfate [CAS #68877-25-8]
Carsonol ILS

Mixed isopropanolamines myristate
Lanamine; Tergelan 1790

Monoethanolamine. See *MEA*

Monoisopropanolamine. See *MIPA*

Monoolein. See *Glyceryl oleate*

Monostearin. See *Glyceryl stearate*

Myreth-3 [CAS #27306-79-2 (generic); RD #977054-86-6]
Synonyms: PEG-3 myristyl ether; POE (3) myristyl ether
Hetoxol M-3

Myreth-3 myristate [RD #977054-88-8]
Synonyms: PEG-3 myristyl ether myristate
Standamul 1414-E

Myreth-4 [CAS #27306-79-2 (generic); RD #977054-89-9]
Lipocol M-4

Myristalkonium chloride [CAS #139-08-2]
Synonyms: Myristyl dimethyl benzyl ammonium chloride; Tetradecyl dimethyl benzyl ammonium chloride
Arquad DM14B-90; BTC 824, 824 P-100, 2125, 2125 M; Cyncal 80%; Dibactol; Nissan Cation M2-100; Querton 14 BCL; Roccal MC-14

Myristalkonium saccharinate [CAS #68989-01-5]
Synonyms: Myristyl dimethyl benzyl ammonium saccharinate; Quaternium-3
Onyxide 3300; Rewoquat QA 100

Myristamide DEA [CAS #7545-23-5]
Synonyms: Myristic diethanolamide
Condensate PM; Monamid 150-MW; Rewomid DLM/SE; Schercomid MD-Extra, MME

Myristamide MEA [CAS #142-58-5]
Synonyms: Myristic monoethanolamide; Myristoyl monoethanolamide
Witcamide MM

Myristamidopropyl betaine [CAS #59272-84-3]
Synonyms: Myristamidopropyl dimethyl glycine
Schercotaine MAB

Myristamidopropyl dimethylamine [CAS #45267-19-4]
Synonyms: Dimethylaminopropyl myristamide
Schercodine M

Myristamine
Synonyms: Myristyl amine; Tetradecylamine
Amine 14D; Kemamine P-790, P-790D; Nissan Amine MB

Myristamine acetate
Synonyms: Tetradecylamine acetate
Nissan Cation MA

Myristamine oxide [CAS #3332-27-2]
Synonyms: Myristyl dimethyl amine oxide; Tetradecyl dimethyl amine oxide
Ammonyx MCO, MO; Aromox DM14D-W; Barlox 14; Conco XA-M; Emcol M; Empigen OH; Incromine Oxide M; Jordamox MDA; Ninox M; Sandoz Amine Oxide XA-M; Schercamox DMA, DMM

Myristic acid [CAS #544-63-8]
Hystrene 9014

Myristic diethanolamide. See *Myristamide DEA*

Myristic monoethanolamide. See *Myristamide MEA*

Myristoyl sarcosine [CAS #52558-73-3]
Hamposyl M; Nikkol Sarcosinate MN

Myristyl alcohol [CAS #112-72-1]
Synonyms: C14 linear primary alcohol; 1-Tetradecanol
Alfol 14, 1412; Cyclochem EM 560; Epal 14; Kalcohl 40; Lanette 14; Lipowax

Myristyl amine. See *Myristamine*

Myristyl/cetearyl amine oxide
Incromine Oxide MC

Myristyl/cetyl dimethylamine oxide
Conco XA-MC

Myristyl dimethylamine. See *Dimethyl myristamine*

Myristyl dimethyl benzyl ammonium chloride. See *Myristalkonium chloride*

Myristyl dimethyl-naphthylmethyl ammonium chloride
BTC 1100

Myristyl eicosyl stearate
Standamul G-3236 Stearate

Myrtrimonium bromide [CAS #1119-97-7]
Synonyms: Myristyl trimethyl ammonium bromide
Mytab

Naphthalene-formaldehyde sulfonate
Dyesperse DC

Nonoxynol-1 [CAS #9016-45-9 (generic); 26027-38-3 (generic); 27986-36-3; 37205-87-1 (generic)]
Synonyms: Ethylene glycol nonyl phenyl ether; PEG-1 nonyl phenyl ether; 2-(Nonylphenoxy) ethanol
Alkasurf NP-1; Cedepal CO-210; Chemax NP-1.5 (1.5 EO); DeSonic 1.5N (1.5 EO); Iconol NP 1.5 (1.5 EO); Igepal CO-210 (1.5 EO); Merpoxen NO 15 (1.5 EO); Norfox NP-1; Peganol NP 1.5 (1.5 EO); Rexol 25/1; Surfonic N-10; T-Det N 1.5 (1.5 EO); Trycol NP-1

Nonoxynol-2 [CAS #9016-45-9 (generic); 26027-38-3 (generic); 27176-93-8; 37205-87-1 (generic); RD #977057-30-9]
Synonyms: PEG-2 nonyl phenyl ether; POE (2) nonyl phenyl ether; PEG 100 nonyl phenyl ether
Conco NI-21; Teric N2

Nonoxynol-3
Surfonic N-31.5; Teric N3 (3.5 EO]

Nonoxynol-4 [CAS #7311-27-5; 9016-45-9 (generic); 26027-38-3 (generic); 37205-87-1 (generic)]
Ablunol NP4; Alkasurf NP-4; Arkopal N040; Carsonon N-4; Cedepal CO-430; Chemax NP-4; Chemcol NPE-40; Conco NI-43; DeSonic 4N; Hetoxide NP-4; Hyonic NP-40; Iconol NP-4; Igepal CO-430; Macol NP-4; Makon 4; Merpoxen NO 40; Norfox NP-4; NP-55-40; Nutrol 622; Peganol NP 4; Polystep F-1; Renex 647; Rewopal HV 4; Rexol 25/4; Serdox NNP 4;

Siponic NP4; Sterox ND; Surfonic N-40; Synperonic NP4; T-Det N-4; Tergitol NP-4, NP-14; Teric N4; Triton N-42; Trycol NP-4

Nonoxynol-4 hydrogen phosphate

Merpiphos EP

Nonoxynol-5 [CAS #9016-45-9 (generic); 26027-38-3 (generic); 37205-87-1 (generic); RD #977057-31-0]

Alkasurf NP-5; DeSonic 5N; Iconol NP-5; Igepal CO-520; Lorapal HV 5; NP-55-50; Peganol NP 5; Renex 648; Rewopal HV 5; Serdox NNP 5; Sterox NE; Synperonic NP5; Teric N5 (5.5 EO); Triton N-57

Nonoxynol-6 [CAS #9016-45-9 (generic); 26027-38-3 (generic); 37205-87-1 (generic); RD #977057-32-1]

Ablunol NP6; Alkasurf NP-6; Arkopal N060; Carsonon N-6; Cedepal CO-530; Chemax NP-6; Chemcol NPE-60; Conco NI-60; DeSonic 6N; Hyonic NP-60; Iconol NP-6; Igepal CO-530; Lubrol N5 (5.5 EO); Macol NP-6; Makon 6; Merpoxen NO 60; Nissan Nonion NS-206; Norfox NP-6; NP-55-60; Peganol NP 6; Polystep F-2; Renex 697; Rewopal HV 6; Rexol 25/6; Serdox NNP 6; Siponic NP6; Sterox NF, NG (6.5 EO); Surfonic N-60; Synperonic NP6; T-Det N-6; Tergitol NP-6; Triton N-60; Trycol NP-6

Nonoxynol-6 hydrogen phosphate

Merpiphos LP

Nonoxynol-6 phosphate

Synonyms: PEG-6 nonyl phenyl ether phosphate

Monafax 786

Nonoxynol-6 phosphoric acid

Sermul EA 211

Nonoxynol-7 [CAS #9016-45-9 (generic); 26027-38-3 (generic); 27177-05-5; 37205-87-1 (generic); RD #977057-33-2]

Alkasurf NP-7LM; DeSonic 7N; Iconol NP-7; Norfox NP-7 (7.5 EO); Rexol 25/7; Serdox NNP 7; Tergitol NP-7, NP-27; Trycol NP-7

Nonoxynol-8 [CAS #9016-45-9 (generic); 26027-38-3 (generic); 37205-87-1 (generic); RD #977057-34-3]

Ablunol NP8; Alkasurf NP-8; Arkopal N080; Carsonon N-8; Cedepal CO-610; Igepal CO-610 (7–8 EO); Igepal CO-620; Lorapal HV 8; Makon 8; Merpoxen NO 80; Nikkol NP-7.5; NP-55-80, -55-85 (8.5 EO); Nutrol 611; Polystep F-3; Quimipol ENF 80; Renex 688; Rewopal HV 8; Rexol 25/8; Serdox NNP 8.5 (8.5 EO); Surfonic HDL (8.5 EO), N-85 (8.5 EO); Synperonic NP8; T-Det N-8; Tergitol NP-8; Teric N8 (8.5 EO); Triton N-87

Nonoxynol-9 [CAS #9016-45-9 (generic); 26027-38-3 (generic); 26571-11-9; 37205-87-1 (generic); RD #977004-65-1]

Ablunol NP9; Alkasurf NP-9; Arkopal N090; Carsonon N-9; Cedepal CO-630; Chemax NP-9; Conco NI-90; DeSonic 9N; Empilan NP9; Gradonic N-95 (9.5 EO); Hetoxide NP-9; Hyonic NP-90; Iconol NP-9; Igepal CO-630; Lipal 9 N; Lorapal HV 9; Macol NP-9.5 (9.5 EO); Merpoxen NO 90; Merpoxen NO 95 (9.5 EO); Neutronyx 600 (9.5 EO); Norfox NP-9; NP-55-90, 55-95 (9.5 EO); Nutrol 600; OP-55; Peganol NP 9; Renex 698 (9–9.5 EO); Rewopal HV 9; Rexol 25/9; Siponic NP9; Sterling NPX; Sterox NJ (9.5 EO); Syn Fac 905 (9.5 EO); Synperonic NP9; T-Det N-9.5; Tergitol NP-9, TP-9; Teric N9; Trycol NP-9

Nonoxynol-9 iodine [RD #977059-98-5]
Synonyms: PEG-9 nonyl phenyl ether iodine complex
Biopal VRO-20

Nonoxynol-9 phosphate [CAS #66197-78-2]
Synonyms: PEG-9 nonyl phenyl ether phosphate
Gafac RE-610

Nonoxynol-9.5 hydrogen phosphate
Merpiphos GP

Nonoxynol-10 [CAS #9016-45-9 (generic); 26027-38-3 (generic); 27177-08-8; 37205-87-1 (generic); RD #977057-35-4]

Alkasurf NP-10; Arkopal N100; Carsonon N-10; Cedepal 710; Chemax NP-10; Chemcol NPE-100; Conco NI-100; DeSonic 10N; Hyonic NP-100, PE-100; Iconol NP-10; Igepal CO-660; Igepal CO-710 (10–11 EO); Lorapal HV 10; Makon 10; Merpoxen NO 100; Nikkol NP-10; Peganol NP 10; Polystep F-4; Renex 690; Rewopal HV 10; Rexol 25/10; Serdox NNP 10; Siponic NP10.5; Sterox NK (10.5 EO); Surfonic N-95, N-100, N-102; Synperonic NP10; T-Det N-10.5; Tergitol NP-10, NPX (10.5 EO); Teric N10; Triton N-101

Nonoxynol-10 phosphate [RD #977065-10-3]
Synonyms: PEG-10 nonyl phenyl ether phosphate
Servoxyl VPNZ 10/100

Nonoxynol-11 [CAS #9016-45-9 (generic); 26027-38-3 (generic); 37205-87-1 (generic); RD #977065-12-5]

Alkasurf NP-11; Arkopal N110; Carsonon N-11; DeSonic 11N; Hyonic NP-110; Macol NP-11; Merpoxen NO 110; Neutronyx 656; Norfox NP-11; Nutrol 656; Rexol 25/11; Serdox NNP 11; Sterox NL; Teric N11; Triton N-111; Trycol NP-11

Nonoxynol-12 [CAS #9016-45-9 (generic); 26027-38-3 (generic); 37205-87-1 (generic); RD #977061-17-8]

Alkasurf NP-12; Carsonon N-12; DeSonic 12N; Hyonic NP-120; Iconol NP-12; Igepal CO-720; Makon 12; Merpoxen NO 120; Nissan Nonion NS-212; NP-55-120; Peganol NP 12; Polystep F-5; Renex 682; Serdox NNP 12; Sterox NM (12.5 EO); Surfonic N-120; Synperonic NP12; T-Det N-12; Teric N12

Nonoxynol-12 iodine
Synonyms: PEG-12 nonyl phenyl ether iodine complex
Biopal NR-20

Nonoxynol-13 [CAS #9016-45-9 (generic); 26027-38-3 (generic); 37205-87-1 (generic); RD #977065-13-6]

Arkopal N130; DeSonic 13N; Lubrol N13; Renex 679; Serdox NNP 13; Synperonic NP13; Tergitol NP-13; Teric N13

Nonoxynol-14 [CAS #9016-45-9 (generic); 26027-38-3 (generic); 37205-87-1 (generic); RD #977061-55-4]

Chemcol NPE-140; Lorapal HV 14; Makon 14; Merpoxen NO 140; Polystep F-6; Rewopal HV 14; Rexol 25/14; Serdox NNP 14; T-Det N-14

Nonoxynol-15 [CAS #9016-45-9 (generic); 26027-38-3 (generic); 37205-87-1 (generic); RD #977006-97-5]

Ablunol NP15; Alkasurf NP-15, NP-15 80%; Arkopal N150; Chemax NP-15; Conco NI-150; DeSonic 15N; Iconol NP-15; Igepal CO-730; Merpoxen NO 150; Nikkol NP-15; NP-55-150;

Nutrol 640; Peganol NP 15; Polystep F-7; Renex 678; Rexol 25/15; Serdox NNP 15; Siponic NP15; Surfonic N-150; Synperonic NP15; Teric N15; Triton N-150

Nonoxynol-17
NP-55-300

Nonoxynol-18
Nikkol NP-18TX

Nonoxynol-20 [CAS #9016-45-9 (generic); 26027-38-3 (generic); 37205-87-1 (generic); RD #977057-36-5]
Alkasurf NP-20, NP-20 70%; Chemax NP-20; Chemcol NPE-200; Conco NI-185; DeSonic 20N; Ethylan 20; Iconol NP-20; Igepal CO-850; Macol NP-20; Makon 20; Merpoxen NO 200; Peganol NP 20; Polystep F-8; Renex 649; Rexol 25/20; Serdox NNP 20/70; Sermul EN 20/70; Soprofor NP/20; Surfonic N-200; Synperonic NP20; T-Det N-20; Teric N20; Trycol NP-20

Nonoxynol-23 [CAS #9016-45-9 (generic); 26027-38-3 (generic); 37205-87-1 (generic); RD #977065-15-8]
Arkopal N-230

Nonoxynol-25
Lorapal HV 25; Rewopal HV 25; Serdox NNP 25

Nonoxynol-30 [CAS #9016-45-9 (generic); 26027-38-3 (generic); 37205-87-1 (generic); RD #977006-96-4]
Alkasurf NP-30 70%; Arkopal N-300; Carsonon N-30, N-30 70%; Chemax NP-30, NP-30/70; Conco NI-187; DeSonic 30N, 30N70; Iconol NP-30, NP-30-70%; Igepal CO-880, CO-887; Macol NP-30, NP-30(70); Makon 30; Merpoxen NO 300; Peganol NP 30, 30 70%; Polystep F-9; Renex 650; Rexol 25/30, 25/307; Serdox NNP 30/70; Sermul EN 30/70, EN 145; Soprofor NP/30; Surfonic N-300, NB-5; Synperonic NP30; T-Det N-30, N-307; Teric N30; Trycol NP-30, NP-307

Nonoxynol-35
Alkasurf NP-35 70%; Synperonic NP35

Nonoxynol-40 [CAS #9016-45-9 (generic); 26027-38-3 (generic); 37205-87-1 (generic); RD #977057-37-6]
Alkasurf NP-40, NP-40 70%; Carsonon N-40 70%; Chemax NP-40, NP-40/70; Chemcol NPE-400; Conco NI-190, NI-197; DeSonic 40N, 40N70; Hetoxide NP-40; Hyonic NP-407; Iconol NP-40, NP-40-70%; Igepal CO-890, CO-897; Peganol NP 40, 40 70%; Polystep F-10; Rexol 25/40, 25/407; Siponic NP40; Surfonic N-400, NB-14; T-Det N-40, N-407; Tergitol NP-40, NP-40 (70% Aq.), NP-44; Teric N40; Triton N-401; Trycol NP-40, NP-407

Nonoxynol-50 [CAS #9016-45-9 (generic); 26027-38-3 (generic); 37205-87-1 (generic); RD #977057-38-7]
Alkasurf NP-50 70%; Carsonon N-50 70%; Chemax NP-50, NP-50/70; Conco NI-2000; DeSonic 50N, 50N70; Hyonic NP-500; Iconol NP-50; Igepal CO-970, CO-977; Nopco RDY; Peganol NP 50; Rexol 25/50, 25/507; T-Det N-507; Trycol NP-50, NP-507

Nonoxynol-60
Merpoxen NO 600

Nonoxynol-70
Iconol NP-70, NP-70-70%; Macol NP-70; T-Det N-70, N-705

Nonoxynol-75
Siponic NP75

Nonoxynol-100 [CAS #9016-45-9 (generic); 26027-38-3 (generic); 37205-87-1 (generic); RD #977065-11-4]
Alkasurf NP-100 70%; Carsonon N-100 70%; Chemax NP-100, NP-100/70; DeSonic 100N, 100N70; Iconol NP-100; Igepal CO-990, CO-997; Macol NP-100; Peganol NP 100, 100 70%; Rexol 25/100-70%; Soprofor NP/100; T-Det N-100, N-1007; Teric N100; Triton N-998, N-998-70%; Trycol NP-1007

Nonyl naphthalene sodium sulfonate
Emkal NNS

Nonyl naphthalene sulfonic acid
Emkal NNS Acid

Nonyl nonoxynol-4
Synonyms: PEG-4 dinonyl phenyl ether; POE (4) dinonyl phenyl ether
Hetoxide DNP-4

Nonyl nonoxynol-7 phosphate [CAS #66172-78-9; 66172-83-6]
Synonyms: PEG-7 dinonyl phenyl ether phosphate]
Gafac RM-410

Nonyl nonoxynol-8
Chemax DNP-8; Iconol DNP-8; Trycol DNP-8

Nonyl nonoxynol-9
Hetoxide DNP-9.6 (9.6 EO)

Nonyl nonoxynol-9 phosphate [CAS #66172-82-5; RD #977065-19-2]
Synonyms: PEG-9 dinonyl phenyl ether phosphate
Servoxyl VPQZ 9/100

Nonyl nonoxynol-10 [CAS #9014-93-1 (generic); RD #977061-24-7]
Synonyms: PEG-10 dinonyl phenyl ether
Igepal DM-530

Nonyl nonoxynol-10 phosphate
Gafac RM-510

Nonyl nonoxynol-15
Chemax DNP-15

Nonyl nonoxynol-15 phosphate
Gafac RM-710

Nonyl nonoxynol-18
Chemax DNP-18

Nonyl nonoxynol-24
Iconol DNP-24

Nonyl nonoxynol-49 [CAS #9014-93-1 (generic); RD #977061-25-8]
Igepal DM-880

Nonyl nonoxynol-70
T-Det D-70

Nonyl nonoxynol-100 [CAS #9014-93-1 (generic)]
Serdox NDI 100

Nonyl nonoxynol-150 [CAS #9014-93-1 (generic); RD #977069-35-4]
Chemax DNP-150; Iconol DNP-150; Igepal DM-970; T-Det D-150; Trycol DNP-150, DNP-150/50

Octadecyl. See *Stearyl*

Octoxynol-1 [CAS #9002-93-1 (generic); 9004-87-9 (generic); 9036-19-5 (generic); RD #977054-43-5]
Synonyms: Ethylene glycol octyl phenyl ether; PEG-1 octyl phenyl ether; POE (1) octyl phenyl ether
Alkasurf OP-1; Cedepal CA-210 ; Igepal CA-210 (1.5 EO); Rexol 45/1; Siponic OP1.5 (1.5 EO); Triton X-15

Octoxynol-3 [CAS #9002-93-1 (generic); 9004-87-9 (generic); 9010-43-9; 9036-19-5 (generic); RD #977057-42-3]
Synonyms: PEG-3 octyl phenyl ether; POE (3) octyl phenyl ether
Chemax OP-3; Igepal CA-420; Nikkol OP-3; Peganol OP 3; Rexol 45/3; Triton X-35

Octoxynol-4
DeSonic S-45 (4.5 EO); Hyonic OP-40; T-Det O-4 (3–4 EO)

Octoxynol-5 [CAS #9002-93-1 (generic); 9004-87-9 (generic); 9036-19-5 (generic); RD #977054-44-6]
Synonyms: PEG-5 octyl phenyl ether; POE (5) octyl phenyl ether
Alkasurf OP-5; Cedepal CA-520; Chemax OP-5; Iconol OP-5; Igepal CA-520; Rexol 45/5; Teric X5; Triton X-45

Octoxynol-6
Alkasurf OP-6; Peganol OP 6; T-Det O-6 (6–7 EO)

Octoxynol-7 [CAS #9002-93-1 (generic); 9004-87-9 (generic); 9036-19-5 (generic)]
Alkasurf OP-8 (7–8 EO); Chemax OP-7; DeSonic S-114 (7–8 EO); Hyonic OP-70; Iconol OP-7; Igepal CA-620; Rexol 45/7; Siponic OP7; Teric X7 (7.5 EO)

Octoxynol-8
Peganol OP 8; T-Det O-8 (7–8 EO); Teric X8 (8.5 EO); Triton X-114, X-114SB

Octoxynol-9 [CAS #9002-93-1 (generic); 9004-87-9 (generic); 9010-43-9; 9036-19-5 (generic)]
Cedepal CA-630; Chemax OP-9; Hyonic OP-100, PE-250; Igepal CA-630; Makon OP-9 (9–10 EO); Nutrol 100; Serdox NOP 9; T-Det O-9; Triton X-100, X-120 (9–10 EO)

Octoxynol-10 [CAS #9002-93-1 (generic); 9004-87-9 (generic); 9036-19-5 (generic)]
Alkasurf OP-10; DeSonic S-100 (9–10 EO); Iconol OP-10; Nikkol OP-10; Peganol OP 10; Renex 750; Rexol 45/10; Siponic OP10; Teric X10

Octoxynol-11 [CAS #9004-87-9 (generic); 9036-19-5 (generic)]
Teric X11

Octoxynol-12
Alkasurf OP-12 (12–13 EO); Cedepal CA-720; DeSonic S-102 (12–13 EO); Rexol 45/12

Octoxynol-13 [CAS #9002-93-1 (generic); 9004-87-9 (generic); 9036-19-5 (generic); RD #977057-43-4]
Igepal CA-720 (12.5 EO); Teric X13; Triton X-102

Octoxynol-16 [CAS #9004-87-9 (generic); 9036-19-5 (generic)]
Alkasurf OP-16; Rexol 45/16; Siponic F-160; Teric X16; Triton X-165

Octoxynol-20
Peganol OP 20

Octoxynol-30 [CAS #9004-87-9 (generic); 9036-19-5 (generic)]
Alkasurf OP-30, OP-30 70%; Igepal CA-887; Nikkol OP-30; Peganol OP 30, 30 70%; Rexol 45/307; Serdox NOP 30/70; Siponic F-300; Siponic OP30; Triton X-305

Octoxynol-40 [CAS #9002-93-1 (generic); 9004-87-9 (generic); 9036-19-5 (generic); RD #977066-20-5]
Alkasurf OP-40, OP-40 70%; Cedepal CA-890, CA-897; Chemax OP-40, OP-40/70; DeSonic S-405; Hyonic OP-407; Iconol OP-40, OP-40-70%; Igepal CA-890, CA-897; Nissan Nonion HS-240; Peganol OP 40, 40 70%; Rexol 45/407; Serdox NOP 40/70; Siponic F-400; Siponic OP40; T-Det O-40, O-407; Teric X40; Triton X-405; Trycol OP-407

Octoxynol-70 [CAS #9004-87-9 (generic); 9036-19-5 (generic)]
Alkasurf OP-70 50%; Hyonic OP-705; Nissan Nonion HS-270; Peganol OP 70, 70 70%; Siponic F-707; Triton X-705, X-705-50%

Octyl. See also *Caprylic*

Octyl amine
Amine 8D; Armeen 8D

Octyl decyl dimethyl ammonium chloride. See Quaternium-24

Octyl dodecanol [CAS #5333-42-6]
Synonyms: 2-Octyl dodecanol
Amerchol L-500; Primarol 1208; Standamul G

Octyldodecyl dimethyl ammonium chloride
BTC 812

Octyldodecyl stearate [CAS #22766-82-1]
Standamul 7063

Octyl phosphate. See *Trioctyl phosphate*

Olealkonium chloride [CAS #37139-99-4]
Synonyms: Oleyl dimethyl benzyl ammonium chloride
Ammonyx KP; Empigen BCJ-50; Jordaquat JO-50

Oleamide [CAS #301-02-0]
Synonyms: 9-Octadecenamide; Oleyl amide
Armid O; Armoslip CPM; Kemamide O, U

Oleamide DEA [CAS #93-83-4]
Synonyms: Oleic diethanolamide
Alrosol O; Aminol OF; Calamide O; Chimipal OLD; Clindrol 200-O; Comperlan OD; Crillon ODE; Cyclomide DO280, DO280/S; Emid 6545; Hartamide 9137; Incromide OPD; Jordamide 201; Lauridit OD; Lexate CL-60; Loramine DC 280, DO 280/SE; Loropan OD; Mackamide O; Marlamid D 1885; Mazamide O-20; Merpinamid OD; Ninol 201; Ninol AC-201; Nissan Stafoam DO, S; Product LT 10-8-1, WRS 1-66; Rewomid DO 280, DO 280/SE; Richamide 5085; Schercomid ODA, SO-A; Serdolamide POF 61, POF 61 C; Steinamid DO 280, DO 280/SE; Varamide A-7

Oleamide MEA [CAS #111-58-0]
Synonyms: Oleic monoethanolamide
Incromide OM, OPM

Oleamide MIPA [CAS #111-05-7]
Synonyms: Oleic monoisopropanolamide
Loramine IPE 280; Rewomid IPE 280; Schercomid OMI; Steinamid IPE 280; Witcamide 61

Oleamide MIPA sulfosuccinate
Sole-Terge 8

Oleamidopropylamine oxide [CAS #25159-40-4]
Synonyms: 9-Octadecenamide, N-[3-(dimethylamino) propyl]-, N-oxide
Mackamine OAO

Oleamidopropyl betaine [CAS #25054-76-6]
Synonyms: Oleamidopropyl dimethyl glycine
Cycloteric BET O-30; Incronam OP-30; Lexaine O; Mackam HV; Schercotaine OAB

Oleamidopropyl dimethylamine [CAS #109-28-4]
Synonyms: Dimethylaminopropyl oleamide
Incromine OPM, OPB; Lexamine O-13; Lexate CL-60; Schercodine O; Tego-Amid O 18

Oleamidopropyl dimethylamine lactate
Incromate ODL

Oleamidopropyl dimethylamine oxide
Incromine Oxide O

Oleamine [CAS #112-90-3]
Synonyms: Oleyl amine
Amine OL; Armeen O, OD; Crodamine 1.O, 1.OD; Kemamine P-989, P-989D; Lilamin 172, 172 D; Nissan Amine OB; Noram O; Radiamine 6172, 6173

Oleamine acetate
Noramac O

Oleamine oxide [CAS #14351-50-9; 61792-38-9]
Synonyms: Oleyl dimethyl amine oxide
Alkamox ODM; Incromine Oxide OD-50; Jordamox ODA; Mackamine O2; Standamox O1

Oleic acid [CAS #112-80-1]
Emersol 210, 220, 6321; Industrene 105, 205, 206

Oleic acid chloride
S-Oil

Oleic acid dibutylamide
Rewocor RA 280

Oleic diethanolamide. See *Oleamide DEA*

Oleoamphohydroxypropylsulfonate
Synonyms: Oleoamphopropylsulfonate
Miranol OS; Sandopan TFL Conc.

Oleoamphopropionate [CAS #67892-37-9]
Miranol OM-SF Conc.; Schercoteric OS-SF

Oleo/cetyl sulfate
Montapol CST

Oleoyl sarcosine [CAS #110-25-8]
Hamposyl O; Sarkosyl O

Oleth-2 [CAS #9004-98-2 (generic); 25190-05-0 (generic); RD #977057-49-0]
Synonyms: PEG-2 oleyl ether; POE (2) oleyl ether; PEG 100 oleyl ether
Alkasurf OA-2; Ameroxol OE-2; Brij 92, 93; Ethosperse OA-2; Hetoxol OL-2; Lipocol O-2; Macol OA-2; Nikkol BO-2; OL-55-F-2; Ritoleth 2; Serdox NOL 2; Simulsol 92

Oleth-2 phosphate
Synonyms: PEG-2 oleyl ether phosphate
Crodafos O2 Acid

Oleth-3 [CAS #9004-98-2 (generic); 25190-05-0 (generic); RD #977057-50-3]
Fluilanol; Hetoxol OA-3 Special; Volpo 3, N3, O3

Oleth-3 phosphate [CAS #39464-69-2 (generic); RD #977060-93-7]
Crodafos N3 Acid; Cyclophos PO3

Oleth-4 [CAS #9004-98-2 (generic); 25190-05-0 (generic); RD #977065-24-9]
Chemal OA-4; Hetoxol OL-4; Macol OA-4

Oleth-4 phosphate [CAS #39464-69-2 (generic); RD #977065-25-0]
Chemfac PB-184

Oleth-5 [CAS #9004-98-2 (generic); 25190-05-0 (generic); RD #977057-51-4]
Chemal OA-5; Hetoxol OA-5 Special, OL-5; Hostacerin O-5; Lipal 5 OA; Macol OA-5; Ritoleth 5; Solulan 5; Standamul O-5; Volpo 5, N5, O5

Oleth-5 phosphate
Crodafos N5 Acid, O5 Acid

Oleth-6 [CAS #9004-98-2 (generic); 25190-05-0 (generic)]
Empilan KL 6

Oleth-7 [CAS #9004-98-2 (generic); 25190-05-0 (generic)]
Nikkol BO-7

Oleth-8 [CAS #9004-98-2 (generic); 25190-05-0 (generic)]
Serdox NOL 8

Oleth-9 [CAS #9004-98-2 (generic); 25190-05-0 (generic)]
Chemal OA-9

Oleth-10 [CAS #9004-98-2 (generic); 25190-05-0 (generic); RD #977057-52-5]
Alkasurf OA-10; Ameroxol OE-10; Brij 96, 97; Empilan KL 10; Eumulgin O10; Hetoxol OA-10 Special, OL-10, OL-10H; Lipal 10 OA; Lipocol O-10; Macol OA-10; Nikkol BO-10TX; OL-55-F-10; Ritoleth 10; Simulsol 96; Standamul O-10; Volpo 10, N10, O10

Oleth-10 phosphate [CAS #39464-69-2 (generic); RD #977060-95-9]
Crodafos N10 Acid, O10 Acid

Oleth-12 [CAS #9004-98-2 (generic); 25190-05-0 (generic)]
Lanbritol Wax N21

Oleth-15 [CAS #9004-98-2 (generic); 25190-05-0 (generic); RD #977065-21-6]
Nikkol BO-15TX; Serdox NOL 15; Volpo N15, O15

Oleth-16 [CAS #9004-98-2 (generic); 25190-05-0 (generic)]
Solulan 16

Oleth-20 [CAS #9004-98-2 (generic); 25190-05-0 (generic); RD #977057-53-6]
Ahco 3998; Alkasurf OA-20; Ameroxol OE-20; Brij 98, 99; Chemal OA-20, OA-20/70; Empilan KL 20; Emulphor ON-870; Hetoxol OA-20 Special, OL-20; Industrol LG-70, LG-100, OAL-20; Lipal 20 OA; Lipocol O-20; Macol OA-20; Nikkol BO-20; OL-55-20; Ritoleth 20; Simulsol 98; Siponic Y501; Standamul O-20; Volpo 20, N20, O20

Oleth-23 [CAS #9004-98-2 (generic); 25190-05-0 (generic); RD #977065-23-8]
Hetoxol OL-23; Trycol OAL-23

Oleth-25 [CAS #9004-98-2 (generic); 25190-05-0 (generic); RD #977057-54-7]
Siponic Y500-70; Solulan 25

Oleth-40
Hetoxol OL-40

Oleth-50 [CAS #9004-98-2 (generic); 25190-05-0 (generic); RD #977057-55-8]
Lipal 50 OA; Nikkol BO-50

Oleyl alcohol [CAS #143-28-2]
Synonyms: 9-Octadecen-1-ol
Adol 80, 85 NF, 320, 330, 340; Fancol OA 50, OA 70, OA 80, OA 90, OA 95; Novol

Oleyl amidoethyl ethanolamine
Servamine KEO 260

Oleyl amidoethyl oleyl imidazoline
Servamine KOO 330 B

Oleyl betaine [CAS #871-37-4]
Synonyms: Oleyl dimethyl glycine
Alkateric OB; Mackam OB, OB-30; Mirataine ODMB-35%; Standapol OLB-30, OLB-50; Velvetex OLB-50

Oleyl/cetyl sulfate
Sulfopon O

Oleyl diamine
Lilamin 572; Radiamine 6572

Oleyl dimethyl benzyl ammonium chloride. See *Olealkonium chloride*

Oleyl hydroxyethyl imidazoline [CAS #95-38-5; 21652-27-7; 27136-73-8]
Synonyms: 1-Hydroxyethyl-2-oleyl imidazoline; Oleic acid imidazoline; Oleyl imidazoline
Alkaquat O; Alkazine O; Amine O; Crodazoline O; Imidrol OC; Mackazoline O; Mazoline OA; Miramine OC; Monazoline O; Nopcogen 22-O; Quaternary O; Rewopon JMOA; Schercozoline O; Textamine O-1; Varine O

Oleyl imidazoline methosulfate
Variquat 473

Oleyl imidazolinium hydrochloride
Norfox IM-38

Oleyl linoleate
Polylan

Oleyl nitrile
Arneel OD

Oleyl palmitamide [CAS #16260-09-6]
Kemamide P-181

N-Oleyl-1,3-propanediamine
Synonyms: Oleyl-1,3 diaminopropane
Duomeen C

Oleyl 1,3-propylene diamine
Dinoram O; Kemamine D-989

Oleyl propylene diamine acetate
Dinoramac O

Oleyl trimethyl ammonium chloride
Arquad S-2C-50; Noramium MO 50

Ozokerite [CAS #8021-55-4]
Lexate PX; Protegin, W, WX, X

Palmamidopropyl betaine
Schercotaine PAB

Palmitamidopropyl betaine [CAS #32954-43-1]
Incronam P-30

Palmitamidopropyl dimethylamine [CAS #39669-97-1]
Incromine PB; Lexamine P-13; Schercodine P

Palmitamidopropyl trimonium chloride
Servamine KEP 4527

Palmitamine [CAS #143-27-1]
Synonyms: Cetyl amine; Palmityl amine; Hexadecylamine
Amine 16D; Armeen 16D; Crodamine 1.16D; Kemamine P-880, P-880D; Nissan Amine PB

Palmitamine oxide [CAS #7128-91-8]
Synonyms: Cetamine oxide; Cetyl dimethyl amine oxide; Palmityl dimethylamine oxide; Hexadecyl dimethyl amine oxide
Ammonyx CO; Aromox DM16, DM16D-W; Conco XA-C; Jordamox CDA-40; Sandoz Amine XA-C

Palmitic acid [CAS #57-10-3]
Synonyms: Hexadecanoic acid; Cetylic acid
Hydrofol Acid 1655-CG-NF, 1690; Hystrene 4516, 8016, 9016; Industrene 4516

Palmitic/stearic acid glycerol monodiester
Imwitor 945

Palmityl. See also *Cetyl*

Palmityl dimethyl amine. See *Dimethyl palmitamine*

Palmityl glyceryl ether
Nikkol Chimyl Alcohol 100

Palm kernel alcohol
Laurex PKH

Palm kernelamide DEA [RD #977069-44-5]
Synonyms: Palm kernel oil acid diethanolamide; Diethanolamine palm kernel oil acid amide; N,N-bis (2-hydroxyethyl) palm kernel oil acid amide
Cyclomide DP240/S; Lauridit PPD (Palm kernel polydiethanolamide); Mackamide PK

Palm kernel glycerides
Grindtek PK 60

Palm oil glyceride
Imwitor 965; Monomuls 90-30; Myverol 18-35

Palm oil glycerides
Imwitor 940; Monomuls 60-30

Pareth 15-.... See *C11-15 pareth-...*

Pareth 23-.... See *C12-13 pareth-...*

Pareth 25-.... See *C12-15 pareth-...*

Pareth 45-.... See *C14-15 pareth-...*

Pareth 91-.... See *C9-11 pareth-...*

PEG-4 [CAS #25322-68-3 (generic); RD #977007-39-8]
Synonyms: PEG 200
Alkapol PEG-200; Merpoxen PEG 200; Nopalcol 200; Pluracol E-200

PEG-6 [CAS #25322-68-3 (generic); RD #977007-40-1]
Synonyms: PEG 300
Alkapol PEG-300; Merpoxen PEG 300; PGE-300; Pluracol E-300

PEG-6-32 [CAS #25322-68-3 (generic); RD #977053-63-6]
Synonyms: PEG 1500
Alkapol PEG-1500; Merpoxen PEG 1500 S; PGE-1500; Pluracol E-1500

PEG-8 [CAS #25322-68-3 (generic); RD #977007-41-2]
Synonyms: PEG 400
Merpoxen PEG 400; PGE-400; Pluracol E-400, E-400 NF; Texapon SG

PEG-9 [CAS #25322-68-3 (generic)]
Synonyms: PEG 450
Alkapol PEG-400

PEG-12 [CAS #25322-68-3 (generic); RD #977053-61-4]
Synonyms: PEG 600
Merpoxen PEG 600; PGE-600; Pluracol E-600, E-600 NF

PEG-14 [CAS #25322-68-3 (generic); RD #977059-82-7]
Alkapol PEG-600

PEG-16
Synonyms: PEG 800
Merpoxen PEG 800

PEG-20 [CAS #25322-68-3 (generic); RD #977053-62-5]
Synonyms: PEG 1000
Alkapol PEG-1000; Merpoxen PEG 1000; PGE-1000; Pluracol E-1000

PEG-40 [CAS #25322-68-3 (generic); RD #977065-62-5]
Synonyms: PEG 2000
Incroquat CR Conc.; Pluracol E-2000

PEG-75 [CAS #25322-68-3 (generic); RD #977007-43-4]
Synonyms: PEG 4000
Pluracol E-4000, E-4000 NF

PEG-150 [CAS #25322-68-3 (generic); RD #977053-64-7; PEG 6000
Alkapol PEG-6000; Pluracol E-6000

PEG 100.... See *PEG-2...*

PEG 200... See *PEG-4...*

PEG 300... See *PEG-6...*

PEG 400... See *PEG-8...*

PEG 450... See *PEG-9...*

PEG 500... See *PEG-10...*

PEG 600... See *PEG-12...*

PEG 800... See *PEG-16...*

PEG 1000... See *PEG-20...*

PEG 1500... See *PEG-6-32...*

PEG 1540... See *PEG-32...*

PEG 1800... See *PEG-36...*

PEG 2000... See *PEG-40...*

PEG 3350
Alkapol PEG-3350

PEG 4000... See *PEG-75...*

PEG 6000... See *PEG-150...*

PEG 8000
Alkapol PEG-8000

PEG-6 abietate
Secoster MA 300

PEG-2 butynediol
Hetoxide BY-1.8 (1.8 EO)

PEG-3 butynediol
Hetoxide BY-3

PEG-2 C13-15 alkyl amine
Synprolam 35X2

PEG-5 C13-15 alkyl amine
Synprolam 35X5

PEG-10 C13-15 alkyl amine
Synprolam 35X10

PEG-15 C13-15 alkyl amine
Synprolam 35X15

PEG-20 C13-15 alkyl amine
Synprolam 35X20

PEG-25 C13-15 alkyl amine
Synprolam 35X25

PEG-35 C13-15 alkyl amine
Synprolam 35X35

PEG-50 C13-15 alkyl amine
Synprolam 35X50

PEG-8 caprylate/caprate
Synonyms: POE (8) caprylate/caprate; PEG 400 caprylate/caprate
Labrasol

PEG-6 caprylic/capric glycerides
Softigen 767

PEG-4 caprylic ether
Synonyms: POE (4) octyl ether; POE (4) 2-ethylhexyl ether
Dehydol O4; Hetoxol CD-4

PEG-2 castor oil
Synonyms: POE (2) castor oil; PEG 100 castor oil
Hetoxide C-2

PEG-3 castor oil [CAS #61791-12-6 (generic)]
Nikkol CO-3

PEG-5 castor oil [CAS #61791-12-6 (generic)]
Alkasurf CO-5; Chemax CO-5; Surfactol 318; Trylox CO-5

PEG-6 castor oil
DeSonic 6C

PEG-9 castor oil [CAS #61791-12-6 (generic); RD #977065-86-3]
Hetoxide C-9; Lipal 9 C

PEG-10 castor oil [CAS #61791-12-6 (generic)]
Alkasurf CO-10; Etocas 10; Incrocas 10; Nikkol CO-10; Trylox CO-10

PEG-12 castor oil
Teric C12

PEG-15 castor oil
Alkasurf CO-15; Hetoxide C-15

PEG-16 castor oil
Chemax CO-16; Trylox CO-16

PEG-17 castor oil
Servirox OEG 45

PEG-20 castor oil [CAS #61791-12-6 (generic)]
Alkasurf CO-20; Merpoxen RO 200; Nikkol CO-20TX; Nopalcol 10-CO

PEG-25 castor oil [CAS #61791-12-6 (generic); RD #977063-56-1]
Alkasurf CO-25; Chemax CO-25; Hetoxide C-25; Industrol CO-25; Lipal 25 C; Trylox CO-25

PEG-26 castor oil
Servirox OEG 55

PEG-28 castor oil
Chemax CO-28; Chemcol CO-28

PEG-30 castor oil [CAS #61791-12-6 (generic); RD#977059-28-1]
Alkasurf CO-30; Chemax CO-30; Chemcol CO-30; DeSonic 30C; Emulphor EL-620; Hetoxide C-30; Incrocas 30; Industrol CO-30; Merpoxen RO 300; Trylox CO-30

PEG-32 castor oil
Servirox OEG 65

PEG-35 castor oil
Etocas 35

PEG-36 castor oil [CAS #61791-12-6 (generic); RD #977054-99-1]
Chemax CO-36; Cremophor EL; DeSonic 36C; Industrol CO-36; Trylox CO-36

PEG-40 castor oil [CAS #61791-12-6 (generic); RD #9777055-00-7]
Alkasurf CO-40, CO-40M; Carsoquat 816-C; Chemax CO-40; Chemcol CO-40; DeSonic 40C; Emulgade F, F Special; Emulphor EL-719; Etocas 40; Hetoxide C-40; Incrocas 40; Industrol CO-40; Merpoxen RO 400; Nikkol CO-40TX; Surfactol 365; T-Det C-40; Trylox CO-40

PEG-52 castor oil
Lipal 52 C

PEG-54 castor oil
DeSonic 54C

PEG-60 castor oil [CAS #61791-12-6 (generic); RD #977065-79-4]
Etocas 60; Incrocas 60; Nikkol CO-60TX

PEG-80 castor oil
Chemax CO-80; Industrol CO-80-80%; Merpoxen RO 800; Trylox CO-80

PEG-100 castor oil [CAS #61791-12-6 (generic); RD #977063-24-3]
Etocas 100; Incrocas 100

PEG-180 castor oil
Servirox OEG 90/50

PEG-200 castor oil [CAS #61791-12-6 (generic); RD #977063-54-9]
Alkasurf CO-200; Chemax CO-200/50; Hetoxide C-200, C-200-50%; Industrol CO-200, CO-200-50%; Trylox CO-200, -200/50

PEG 1900 castor oil
Nopalcol 19-CO

PEG 9000 castor oil
Pegosperse 9000 CO

PEG cetyl ethers. See *Ceteth Series*

PEG cetyl/oleyl ethers. See *Cetoleth Series*

PEG cetyl/stearyl ethers. See *Ceteareth Series*

PEG-3 cocamide [CAS #61791-08-0 (generic); RD #977061-46-3]
Synonyms: PEG (5) coconut amide; POE (5) coconut amide
Alkamidox CME-2; Amidox C-2; Unamide C-2

PEG-4 cocamide MEA
Merpoxen KM 45 (4.5 EO)

PEG-5 cocamide [CAS #61791-08-0 (generic)]
Hetoxamide C-4

PEG-6 cocamide [CAS #61791-08-0 (generic); RD #977061-47-4]
Alkamidox CME-5; Amidox C-5; Empilan MAA; Mazamide C-5; Unamide C-5

PEG-6 cocamide MEA
Merpoxen KM 60; Rewopal C 6; Teric CME3

PEG-7 cocamide MEA
Teric CME7

PEG-9 cocamide MEA
Merpoxen KM 90

PEG-2 cocamine [CAS #61791-14-8 (generic); RD #977061-61-2]
Synonyms: PEG 100 coconut amine; POE (2) coconut amine
Accomeen C2; Alkaminox C2; Chemeen C-2; Ethomeen C/12; Hetoxamine C-2; Mazeen C 2; Teric 12M2

PEG-5 cocamine [CAS #61791-14-8 (generic); RD #977066-78-6]
Accomeen C5; Alkaminox C-5; Chemeen C-5; Ethomeen C/15; Hetoxamine C-5; Mazeen C 5; Teric 12M5

PEG-10 cocamine [CAS #61791-14-8 (generic); RD #977063-21-0]
Accomeen C10; Alkaminox C-10; Chemeen C-10; Ethomeen C/20; Mazeen C 10; Trymeen CAM-10

PEG-15 cocamine [CAS #8051-52-3; 61791-14-8 (generic)]
Accomeen C15; Chemeen C-25; Hetoxamine C-15; Mazeen C 15; Teric 12M15; Trymeen CAM-15

PEG-2 cocoate
Synonyms: Diglycol coconate; POE (2) monococoate; PEG 100 cocoate
Witconol RDC-D

PEG-5 cocoate [CAS #61791-29-5 (generic); RD #977065-64-7]
Ethofat C/15; Lipal CE 38

PEG-8 cocoate [CAS #61791-29-5 (generic); RD #977055-01-8]
Alkamuls 400-MC; Emulsan K; Nopalcol 4-C, -CH; Pegosperse 400 MC

PEG-10 cocoate
Synonyms: PEG 500 cocoate
Nonex C5E

PEG-11 cocoate
Lipal CE 55

PEG-15 cocoate [CAS #61791-29-5 (generic); RD #977065-34-1]
Ethofat C/25

PEG-16 cocoate
Lipal CE 64

PEG-21 cocoate
Lipal CE 71

PEG-2 cocomonium chloride [RD #977066-72-0]
Ethoquad C/12

PEG-15 cocomonium chloride [RD #977066-75-3]
Ethoquad C/25

PEG-15 cocopolyamine
Polyquart H-7102

PEG-5 coco ammonium ethosulfate
Merpiquat K-26

PEG decyl ethers. See *Deceth Series*

PEG-2 dilaurate
Synonyms: Diethylene glycol dilaurate; POE (2) dilaurate; PEG 100 dilaurate
Cithrol DGDL N/E, DGDL S/E

PEG-4 dilaurate [CAS #9005-02-1 (generic); RD #977055-02-9]
Alkamuls 200-DL; Cithrol 2DL; Emerest 2622; Ionet DL-200; Kessco PEG 200 Dilaurate; Lexomul PEG 200 DL; Lipopeg 2-DL; Mapeg 200 DL; Nonex DL-2; Pegosperse 200 DL

PEG-6 dilaurate [CAS #9005-02-1 (generic); RD #977055-03-0]
Kessco PEG 300 Dilaurate; Lipal 300 DL

PEG-8 dilaurate [CAS #9005-02-1 (generic); RD #977055-04-1]
Alkamuls 400-DL; Cithrol 4DL; CPH-79-N; Emerest 2652; Industrol DL-9; Kessco PEG 400 Dilaurate; Lexomul PEG 400 DL; Lipal 400 DL; Pipopeg 4-DL; Mapeg 400 DL; Pegosperse 400 DL; Varonic 400DL

PEG-12 dilaurate [CAS #9005-02-1 (generic); RD #977063-27-6]
Alkamuls 600-DL; Cithrol 6DL; Kessco PEG 600 Dilaurate; Mapeg 600 DL

PEG-20 dilaurate [CAS #9005-02-1 (generic); RD #977063-47-0]
Cithrol 10DL; Kessco PEG 1000 Dilaurate

PEG-32 dilaurate [CAS #9005-02-1 (generic); RD #977065-53-4]
Kessco PEG 1540 Dilaurate

PEG-75 dilaurate [CAS #9005-02-1 (generic); RD #977065-82-9]
Kessco PEG 4000 Dilaurate

PEG-150 dilaurate [CAS #9005-02-1 (generic); RD #977055-05-2]
Kessco PEG 6000 Dilaurate

PEG dinonyl phenyl ethers. See *Nonyl nonoxynol Series*

PEG-2 dioleate
Synonyms: Diethylene glycol dioleate; POE (2) dioleate; PEG 100 dioleate
Alkamuls DEG-DO; Cithrol DGDO N/E, DGDO S/E

PEG-4 dioleate [CAS #9005-07-6 (generic); 52688-97-0 (generic); RD #977065-57-8]
Alkamuls 200-DO; Chemax PEG 200 DOf; Cithrol 2DO; Ionet DO-200; Kessco PEG 200 Dioleate; Mapeg 200 DO

PEG-6 dioleate [CAS #9005-07-6 (generic); 52688-97-0 (generic); RD #977055-06-3]
Kessco PEG 300 Dioleate

PEG-6-32 dioleate [CAS #9005-07-6 (generic); 52688-97-0 (generic); RD #977065-78-3]
Pegosperse 1500 DO

PEG-8 dioleate [CAS #9005-07-6 (generic); 52688-97-0 (generic); RD #977051-75-4]
Alkamuls 400-DO; Chemax PEG 400 DO; Cithrol 4DO; CPH-211-N; Emerest 2648; Industrol DO-9; Ionet DO-400; Kessco PEG 400 Dioleate; Lexomul PEG 400 DO; Lipal 400 DW; Lipopeg 4-DO; Lonzest PEG-4DO; Mapeg 400 DO; Nonex DO-4; Nonisol 210; Pegosperse 400 DO; PGE-400-DO; Radiasurf 7443

PEG-9 dioleate
Trydet DO-9

PEG-12 dioleate [CAS #9005-07-6 (generic); 52688-97-0 (generic)]
Alkamuls 600-DO; Chemax PEG 600 DO; Cithrol 6DO; CPH-213-N; Emerest 2665; Industrol DO-13; Ionet DO-600; Kessco PEG 600 Dioleate; Mapeg 600 DO; Marlipal FS; Secoster DO 600

PEG-20 dioleate [CAS #9005-07-6 (generic); 52688-97-0 (generic); RD #977063-48-1]
Cithrol 10DO; Ionet DO-1000; Kessco PEG 1000 Dioleate

PEG-32 dioleate [CAS #9005-07-6 (generic); 52688-97-0 (generic); RD #977065-54-5]
Kessco PEG 1540 Dioleate

PEG-75 dioleate [CAS #9005-07-6 (generic); 52688-97-0 (generic); RD #977065-83-0]
Kessco PEG 4000 Dioleate

PEG-150 dioleate [CAS #9005-07-6 (generic); 52688-97-0 (generic); RD #977063-37-8]
Kessco PEG 6000 Dioleate

PEG-2 distearate [CAS #109-30-8; 9005-08-7 (generic)]
Synonyms: Diethylene glycol distearate; POE (2) distearate; PEG 100 distearate
Cithrol DGDS N/E, DGDS S/E

PEG-4 distearate [CAS #9005-08-7 (generic); RD #977065-58-9]
Alkamuls 200-DS; Cithrol 2DS; Kessco PEG 200 Distearate; Mapeg 200 DS

PEG-6 distearate [CAS #9005-08-7 (generic); RD #977059-16-7]
Cithrol 3DS; Ionet DS-300; Kessco PEG 300 Distearate

PEG-8 distearate [CAS #9005-08-7 (generic); RD #977053-29-4]
Alkamuls 400-DS; Cithrol 4DS; Cyclochem PEG 400 DS; Emerest 2642; Ionet DS-400; Kessco PEG 400 Distearate; Lipal 400 DS; Lipopeg 4-DS; Mapeg 400 DS; Pegosperse 400 DS; PGE-400-DS; Radiasurf 7453

PEG-12 distearate [CAS #9005-08-7 (generic); RD #977055-07-4]
Alkamuls 600-DS; Cithrol 6DS; Kessco PEG 600 Distearate; Mapeg 600 DS; PGE-600-DS; Radiasurf 7454

PEG-20 distearate [CAS #9005-08-7 (generic); RD #977055-08-5]
Cithrol 10DS; Kessco PEG 1000 Distearate

PEG-32 distearate [CAS #9005-08-7 (generic); RD #977055-09-6]
Kessco PEG 1540 Distearate; Mapeg 1540 DS

PEG-75 distearate [CAS #9005-08-7 (generic); RD #977055-10-9]
Kessco PEG 4000 Distearate

PEG-150 distearate [CAS #9005-08-7 (generic); RD #977055-11-0]
Alkamuls 6000-DS; Atlas G-1821; Cycloryl ALC; Emulvis; Kessco PEG 6000 Distearate; Lipopeg 6000-DS; Mapeg 6000 DS; Pegosperse 6000 DS; Plurest DS 150; Rewopal PEG 6000 DS

PEG-2 ditallate
Synonyms: Diethylene glycol ditallate; POE (2) ditallate; PEG 100 ditallate
Alkamuls DEG-DT

PEG-8 ditallate [CAS #61791-01-3 (generic); RD #977066-81-1]
Mapeg 400 DOT; Witconol H-33

PEG-12 ditallate [CAS #61791-01-3 (generic); RD #977063-28-7]
Industrol DT-13; Mapeg 600 DOT; R3124 Ester

PEG-8 ditallowate
Nopalcol 4-DTW

PEG-8 ditriricinoleate
Pegosperse 400 DTR

PEG-22/dodecyl glycol copolymer
Elfacos ST 37

PEG-45/dodecyl glycol copolymer
Elfacos ST 9

PEG dodecyl phenyl ethers. See *Dodoxynol Series*

PEG-25 glycerin pyroglutamic isostearic diester
Pyroter GPI-25

PEG-7 glyceryl cocoate [RD #977064-68-8]
Synonyms: POE (7) glyceryl monococoate; PEG (7) glyceryl monococoate
Mazol 159; Standamul HE

PEG-30 glyceryl cocoate
Varonic LI-63

PEG-80 glyceryl cocoate
Varonic LI-67

PEG glyceryl ethers. See *Glycereth Series*

PEG-15 glyceryl isostearate
Oxypon 2145

PEG-8 glyceryl laurate
Glycerox L8

PEG-12 glyceryl laurate [CAS #59070-56-3 (generic)]
Lamacit GML 12

PEG-15 glyceryl laurate
Glycerox L15

PEG-20 glyceryl laurate [CAS #59070-56-3 (generic); RD #977065-41-0]
Tagat L-2

PEG-23 glyceryl laurate
Aldosperse ML-23, ML-230

PEG-30 glyceryl laurate [CAS #59070-56-3 (generic); RD #977065-48-7]
Tagat L

PEG-40 glyceryl laurate
Glycerox L40

PEG-20 glyceryl oleate [CAS #68889-49-6 (generic); RD #977065-42-1]
Tagat O 2

PEG-25 glyceryl oleate [CAS #68889-49-6 (generic)]
Lamacit GMO 25

PEG-30 glyceryl oleate [CAS #68889-49-6 (generic); RD #977065-49-8]
Tagat O

PEG-15 glyceryl ricinoleate [CAS #51142-51-9 (generic); RD #977065-35-2]
Tagat R 1

PEG-20 glyceryl ricinoleate [CAS #51142-51-9 (generic); RD #97065-43-2]
Lamacit CR

PEG-20 glyceryl stearate [RD #977055-13-2]
Synonyms: PEG 1000 glyceryl stearate; POE (20) glyceryl stearate
Aldo MS-20 FG; Aldosperse 40/60, 40/60 FG, AA 40/60, MS-20, MS-20 FG; Capmul EMG; Cutina E-24; Durfax EOM, EOM K; Mazol 80 MG; Radiasurf 7000; Tagat S 2; Tally 100 Plus

PEG-30 glyceryl stearate [RD #977065-50-1]
Tagat S

PEG-120 glyceryl stearate [RD #977063-30-1]
Drewmulse 1128

PEG-20 glyceryl tallowate
Varonic LI-42

PEG-80 glyceryl tallowate
Varonic LI-48

PEG-200 glyceryl tallowate
Varonic LI-420

PEG-25 glyceryl trioleate
Tagat TO

PEG-5 hydrogenated castor oil [CAS #61788-85-0 (generic); RD #977065-66-9]
Synonyms: POE (5) hydrogenated castor oil; PEG (5) hydrogenated castor oil
Chemax HCO-5; Nikkol HCO-5; Trylox HCO-5

PEG-7 hydrogenated castor oil [CAS #61788-85-0 (generic)]
Arlacel 989; Cremophor WO7; Nikkol HCO-7.5 (7.5 EO)

PEG-10 hydrogenated castor oil
Croduret 10; Nikkol HCO-10

PEG-16 hydrogenated castor oil [CAS #61788-85-0 (generic); RD #977065-36-3]
Chemax HCO-16; Hetoxide HC-16; Trylox HCO-16

PEG-20 hydrogenated castor oil [CAS #61788-85-0 (generic)]
Nikkol HCO-20

PEG-25 hydrogenated castor oil [CAS #61788-85-0 (generic); RD #977055-14-3]
Arlatone G; Chemax HCO-25; Industrol COH-25; Mapeg CO-25H; Trylox HCO-25

PEG-30 hydrogenated castor oil [CAS #61788-85-0 (generic); RD #977069-47-8]
Croduret 30; Nikkol HCO-30, -30(FF)

PEG-40 hydrogenated castor oil [CAS #61788-85-0 (generic); RD #977055-15-4]
Cremophor RH 40, RH 410, RH 455; Croduret 40; Eumulgin RO-40; Hetoxide HC-40; Nikkol HCO-40, HCO-40(FF); Standamul Conc. 1002

PEG-40 hydrogenated castor oil pyroglutamic isostearic diester
Pryoter CPI-40

PEG-50 hydrogenated castor oil [CAS #61788-85-0 (generic)]
Nikkol HCO-50, HCO-50(FF)

PEG-60 hydrogenated castor oil [CAS #61788-85-0 (generic); RD #977060-08-4]
Cremophor RH 60; Croduret 60; Hetoxide HC-60; Nikkol HCO-60, HCO-60(FF)

PEG-80 hydrogenated castor oil [CAS #61788-85-0 (generic)]
Nikkol HCO-80

PEG-100 hydrogenated castor oil [CAS #61788-85-0 (generic); RD #977063-25-4]
Croduret 100; Nikkol HCO-100, HCO-100(FF)

PEG-200 hydrogenated castor oil [CAS #61788-85-0 (generic); RD #977063-55-0]
Chemax HCO-200/50; Industrol COH-200; Trylox HCO-200/50

PEG-20 hydrogenated lanolin [CAS #68648-27-1 (generic); RD #977055-17-6]
Satexlan; Super-Sat AWS-4

PEG-24 hydrogenated lanolin [CAS #68648-27-1 (generic); RD #977055-18-7]
Lipolan 31

PEG-3 hydrogenated tallow propylene diamine
Dinoramox SH 3

PEG-4 iso-C13 alcohol hydrogen phosphate
Merpiphos PP

PEG-6.5 iso-C13 alcohol hydrogen phosphate
Merpiphos UP

PEG isocetyl ethers. See *Isoceteth Series*

PEG isodecyl ethers. See *Isodeceth Series*

PEG-6 isolauryl thioether
Synonyms: POE (6) isolauryl thioether; PEG 300 isolauryl thioether
Siponic 260

PEG-8 isolauryl thioether
Siponic SK

PEG-10 isolauryl thioether
Siponic 218

PEG-4 isostearate
Synonyms: POE (4) monoisostearate; PEG 200 monoisostearate
Emerest 2625; Trydet ISA-4

PEG-8 isostearate
Emerest 2644

PEG-9 isostearate
Industrol MIS-9; Trydet ISA-9

PEG-12 isostearate [CAS #56002-14-3 (generic); RD #977063-29-8]
Emerest 2664

PEG-14 isostearate
Trydet ISA-14

PEG isostearyl ether. See *Isosteareth Series*

PEG-5 lanolate [CAS #68459-50-7 (generic); RD#977061-92-9]
Synonyms: PEG-5 lanolin acids
Lanpol 5

PEG-10 lanolate [CAS #68459-50-7 (generic); RD#977061-91-8]
Lanpol 10

PEG-20 lanolate [CAS #68459-50-7 (generic)]
Lanpol 20

PEG-5 lanolin [CAS #61790-81-6 (generic)]
Synonyms: PEG (5) lanolin derivative; POE (5) lanolin derivative
Nikkol BWA-5 (PEG-5 lanolin alcohol)

PEG-10 lanolin
Nikkol TW-10

PEG-20 lanolin [CAS #61790-81-6 (generic); RD #977058-73-3]
Atlas G-1790; Nikkol BWA-20 (PEG-20 lanolin alcohol); Nikkol TW-20

PEG-27 lanolin
Lanogel 21

PEG-30 lanolin [CAS #61790-81-6 (generic); RD #977065-51-2]
Aqualose L30; Nikkol TW-30

PEG-40 lanolin [CAS #8051-82-9; 61790-81-6 (generic)]
Lan-Aqua-Sol Hydrophilic 50, 100; Laneto 40; Lanogel 31; Nikkol BWA-40 (PEG-40 lanolin alcohol)

PEG-50 lanolin [CAS #61790-81-6 (generic); RD #977061-54-3]
Atlas G-1795; Lan-Aqua-Sol Hydrophilic-Plus 50, 100; Solangel 501

PEG-60 lanolin [CAS #61790-81-6 (generic)]
Laneto 60; Solan

PEG-70 lanolin
Teric LAN70

PEG-75 lanolin [CAS #8039-09-6; 61790-81-6 (generic)]
Aqualose L75, L75/50; Lan-Aqua-Sol xtra-Hydrophilic 50, 100; Laneto 50, 100; Lanogel 41; Solangel 401; Solulan 75, L-575

PEG-85 lanolin [CAS #61790-81-6 (generic); RD #977055-21-2]
Lan-Aqua-Sol Super-Hydrophilic 50, 100; Lanogel 61; Solangel 851

PEG-5 lanolinamide
Lanpolamide 5

PEG lanolin ethers. See *Laneth Series*

PEG-75 lanolin oil [CAS #68648-38-4 (generic); RD #977055-20-1]
Lanotein AWS 30; Vigilan AWS

PEG-75 lanolin wax [RD #977059-05-4]
Lanfrax WS 55

PEG-3 lauramide [CAS #26635-75-6 (generic); RD #977061-49-6]
Synonyms: POE (3) lauryl amide; PEG (3) lauryl amide
Amidox L-2; Unamide L-2

PEG-6 lauramide [CAS #26635-75-6 (generic); RD #977061-48-5]
Amidox L-5; Mazamide L-5; Unamide L-5

PEG-3 lauramine oxide
Synonyms: POE (3) lauryl dimethyl amine oxide; PEG (3) lauryl dimethyl amine oxide
Empigen OY

PEG-2 laurate [CAS #141-20-8; SE grade RD #977063-42-5]
Synonyms: Diethylene glycol laurate; Diglycol laurate; PEG 100 monolaurate; POE (2) monolaurate
Alkamuls DEG-ML; Alkasurf L-2; Cithrol DGML N/E, DGML S/E; Lipo DGLS (SE), Diglycol Laurate; Lipocol B (SE); Mapeg DGLD; Pegosperse 100 L, 100 ML; Radiasurf 7420, 7421 (SE); Sole-Onic CDS

PEG-4 laurate [CAS #9004-81-3 (generic); RD #977055-22-3]
Ablunol 200ML; Alkamuls 200-ML; Chemax E-200 ML; Cithrol 2ML; CPH-27-N; Crodet L4; Emerest 2620; Kessco PEG 200 Monolaurate; Mapeg 200 ML; Nopalcol 2-L; Pegosperse 200 ML; Radiasurf 7422; Soprofor PE/220; Tegin 200 L

PEG-5 laurate
Alkasurf L-5; Hetoxamate LA-5; Industrol ML-5; Lipal 5 L; Trydet LA-5

PEG-6 laurate [CAS #9004-81-3 (generic); RD #977065-75-0]
Emerest 2630; Kessco PEG 300 Monolaurate; Secoster ML 300

PEG-7 laurate
Trydet LA-7

PEG-8 laurate [CAS #9004-81-3 (generic); RD #977055-23-4]
Ablunol 400ML; Alkamuls 400-ML; Chemax E-400 ML; Cithrol 4ML; Collemul L4; CPH-30-N; Crodet L8; Cyclochem PEG 400 ML; Emerest 2650; Gradonic 400-ML; Kessco PEG 400 Monolaurate; Lexomul PEG 400 ML; Lipopeg 4-L; Lonzest PEG 4-L; Mapeg 400 ML; Nonisol 100; Nopalcol 4-L; Pegosperse 400 ML; PGE-400-ML; Radiasurf 7423; Soprofor PE/400

PEG-9 laurate [CAS #106-08-1; 9004-81-3 (generic)]
Alkasurf L-9; Hetoxamate LA-9; Industrol ML-9; Lipal 9 L; Lipocol B

PEG-12 laurate [CAS #9004-81-3 (generic); RD #977055-24-5]
Ablunol 600ML; Alkamuls 600-ML; Cithrol 6ML; CPH-43-N; Crodet L12; Emerest 2661; Kessco PEG 600 Monolaurate; Lipopeg 6-L; Mapeg 600 ML; Nopalcol 6-L; Pegosperse 600 ML; PGE-600-ML

PEG-14 laurate [CAS #9004-81-3 (generic)]
Alkasurf L-14; Chemax E-600 ML; Industrol ML-14

PEG-20 laurate [CAS #9004-81-3 (generic); RD #977063-49-2]
Cithrol 10ML; Ethylan L10; Kessco PEG 1000 Monolaurate

PEG-24 laurate
Crodet L24

PEG-32 laurate [CAS #9004-81-3 (generic); RD #977065-55-6]
Kessco PEG 1540 Monolaurate

PEG-40 laurate
Crodet L40

PEG-75 laurate [CAS #9004-81-3 (generic); RD #977065-84-1]
Kessco PEG 4000 Monolaurate; Secoster ML 4000

PEG-100 laurate
Crodet L100

PEG-150 laurate [CAS #9004-81-3 (generic); RD #977063-38-9]
Cithrol 60ML; Kessco PEG 6000 Monolaurate

PEG lauryl ether. See *Laureth Series*

PEG-20 mannitan laurate
Atlas G-9046T

PEG-120 methyl glucose dioleate
Glucamate DOE-120

PEG-20 methyl glucose sesiquistearate
Glucamate SSE-20

PEG myristyl ethers. See *Myreth Series*

PEG nonyl phenyl ethers. See *Nonoxynol Series*

PEG octyl phenyl ethers. See *Octoxynol Series*

PEG-5 oleamide [RD #977065-67-0]
Synonyms: POE (5) oleyl amide; PEG (5) oleyl amide
Ethomid O/15; Icomid O-5

PEG-2 oleamine [RD #977065-39-6]
Synonyms: PEG-2 oleyl amine; POE (2) oleyl amine; PEG 100 oleyl amine
Alkaminox O-2; Ethomeen O/12; Hetoxamine O-2

PEG-5 oleamine [RD #977065-68-1]
Ethomeen O/15; Hetoxamine O-5

PEG-15 oleamine [RD #977063-33-4]
Ethomeen O/25; Hetoxamine O-15

PEG-30 oleamine
Chemeen O-30, O-30/80; Icomeen O-30, O-30-80%; Katapol OA-860, OA-910; Trymeen OAM-30/80

PEG-2 oleamonium chloride [RD #977066-73-1]
Ethoquad O/12

PEG-15 oleamonium chloride [RD #977066-76-4]
Ethoquad O/25

PEG-2 oleate [CAS #106-12-7; SE grade RD #977063-43-6]
Synonyms: Diethylene glycol monooleate; Diglycol oleate; POE (2) monooleate; PEG 100 monooleate
Alkamuls DEG-MO; Alkasurf O-2; Cithrol DGMO N/E, DGMO S/E; DMO-33; Hetoxamate MO-2; Nikkol MYO-2; Pegosperse 100 O; Radiasurf 7400; Secoster MO 100; Witconol DOS

PEG-4 oleate [CAS #9004-96-0 (generic); RD #977060-09-5]
Ablunol 200MO; Alkamuls 200-MO; Cithrol 2MO; Collemul H2; CPH-39-N; Emerest 2624; Ethylan A2; Ionet MO-200; Kessco PEG 200 Monooleate; Mapeg 200 MO; Radiasurf 7402; Teric OF4

PEG-5 oleate [CAS #9004-96-0 (generic); RD #977065-69-2]
Alkasurf O-5; Chemax E-200 MO; Emulphor VN-430; Ethofat O/15; Hetoxamate MO-5; Industrol MO-5; Trydet OA-5

PEG-6 oleate [CAS #9004-96-0 (generic); RD #977055-25-6]
Acconon 300 MO; Emerest 2632; Ethylan A3; Grocor PEG 300; Industrol MO-6; Kessco PEG 300 Monooleate; Lipal 300 W; Merpoxen OFS 60; Nikkol MYO-6; Radiasurf 7431; Teric OF6

PEG-7 oleate [CAS #9004-96-0 (generic)]
Alkasurf O-7; Industrol MO-7; Trydet OA-7

PEG-8 oleate [CAS #9004-96-0 (generic); RD #977055-26-7]
Ablunol 400MO; Acconon 400 MO; Cithrol 4MO, A; Collemul H4; Cyclochem PEG 400 MO; Durpeg 400MO; Emerest 2646; Emulsan O; Ethylan A4; Ionet MO-400; Kessco PEG 400 Monooleate; Lexomul PEG 400 MO; Lipal 400 OL, 400 W; Lipopeg 4-O; Lonzest PEG 4-O; Mapeg 400 MO; Merpoxen OFS 80; Nonex O4E; Nopalcol 4-O; Pegosperse 400 MO; PGE-400-MO; Radiasurf 7403; Secoster MO 400; Teric OF8; Witconol H-31A

PEG-9 oleate [CAS #9004-96-0 (generic); RD #977065-87-4]
Alkamuls 400-MO; Alkasurf O-9; Chemax E-400 MO; Hetoxamate MO-9; Industrol MO-9; Lipal 9 OL

PEG-10 oleate [CAS #9004-96-0 (generic); RD #977063-22-1]
Atlas G-2143; Ethofat O/20; Serdox NOG 440; Trydet OA-10

PEG-12 oleate [CAS #9004-96-0 (generic); RD #977055-27-8]
Ablunol 600MO; Alcolec HS-3; Alkamuls 600-MO; Cithrol 6MO; Collemul H6; CPH-41-N; Emerest 2660; Ethylan A6; Industrol MO-13; Kessco PEG 600 Monooleate; Lipal 600 W; Mapeg 600 MO; Nopalcol 6-O; PGE-600-MO; Radiasurf 7404

PEG-14 oleate [CAS #9004-96-0 (generic); RD #977063-32-3]
Alkasurf O-14; Chemax E-600 MO; Pegosperse 700 TO

PEG-15 oleate
Hetoxamate MO-15

PEG-20 oleate [CAS #9004-96-0 (generic); RD #977063-50-5]
Ablunol 1000MO; Chemax E-1000 MO; Cithrol 10MO; Kessco PEG 1000 Monooleate

PEG-32 oleate [CAS #9004-96-0 (generic); RD #977065-56-7]
Kessco PEG 1540 Monooleate

PEG-75 oleate [CAS #9004-96-0 (generic); RD #977065-85-2]
Cithrol 40MO; Kessco PEG 4000 Monooleate

PEG-150 oleate [CAS #9004-96-0 (generic); RD #977063-39-0]
Cithrol 60MO; Kessco PEG 6000 Monooleate

PEG oleyl ethers. See *Oleth Series*

PEG-6 palmitate
Synonyms: POE (6) monopalmitate
Polymulse 6

PEG-20 palmitate [CAS #9004-94-8 (generic)]
Atlas G-2079

PEG-6 pelargonate
Synonyms: POE (6) monopelargonate; PEG 300 monopelargonate
Emerest 2634

PEG-8 pelargonate
Emerest 2654

PEG-9 pelargoniate
Trydet MP-9

PEG-5 phytosterol
Nikkol BPS-5

PEG-10 phytosterol
Nikkol BPS-10

PEG-15 phytosterol
Nikkol BPS-15

PEG-20 phytosterol
Nikkol BPS-20

PEG-25 phytosterol
Nikkol BPS-25

PEG-30 phytosterol
Nikkol BPS-30

PEG-6 PPG-2.5 C9-C11 alcohols ether
Hetoxol 916P

PEG-20-PPG-10 glyceryl stearate
Acconon TGH

PEG-25 propylene glycol stearate [RD #977055-29-0]
Atlas G-2160, G-2162

PEG-120 propylene glycol stearate [RD #977063-31-2]
Drewmulse 1128

PEG-2 ricinoleate [CAS #5401-17-2; 9004-97-1 (generic)]
Synonyms: Diethylene glycol monoricinoleate; PEG-2 monoricinoleate; POE (2) ricinoleate
Pegosperse 100 MR

PEG-12 ricinoleate
Nopalcol 6-R

PEG-12 safflower ester
Lipal OE 55

PEG-17 safflower ester
Lipal OE 64

PEG-23 safflower ester
Lipal OE 70

PEG-8 sesquioleate [RD #977061-31-6]
Synonyms: PEG 400 sesquioleate
Emerest 2647

PEG-9 sesquioleate
Trydet SO-9

PEG-12 sesquioleate
Industrol SO-13

PEG-6 sorbitan beeswax [CAS #8051-15-8]
Synonyms: PEG 300 sorbitan beeswax; POE (6) sorbitol beeswax
Atlas G-1702; Nikkol GBW-25

PEG-20 sorbitan beeswax [RD #977055-32-5]
Atlas G-1726; Nikkol GBW-125

PEG-4 sorbitan hexaoleate
Sorbeth 40

PEG-40 sorbitan hexaoleate
Atlas G-1086; Sorbeth 40HO

PEG-50 sorbitan hexaoleate
Atlas G-1096

PEG-55 sorbitan hexaoleate
Sorbeth 55HO

PEG-20 sorbitan isostearate [RD #977055-33-6]
Synonyms: PEG 1000 sorbitan monoisostearate
Crillet 6; Montanox 70; Nikkol TI-10; T-Maz 67

PEG-20 sorbitan lanolate
Synonyms: POE (20) sorbitol lanolin derivative
Atlas G-1425

PEG-40 sorbitan lanolate [CAS #8036-77-9]
Atlas G-1441

PEG-75 sorbitan lanolate [CAS #8051-13-6]
Atlas G-1471

PEG-4 sorbitan laurate. See *Polysorbate 21*

PEG-5 sorbitan laurate
Synonyms: POE (5) sorbitan monolaurate
Montanox 21

PEG-10 sorbitan laurate [CAS #9005-64-5 (generic); RD #977055-35-8]
Atlas G-7596-J; Hetsorb L-10; Liposorb L-10

PEG-20 sorbitan laurate. See *Polysorbate 20*

PEG-80 sorbitan laurate [CAS #9005-64-5 (generic)]
Cycloryl ALC; Hetsorb L-80-72%; T-Maz 28

PEG-5 sorbitan oleate. See *Polysorbate 81*

PEG-6 sorbitan oleate [CAS #9005-65-6 (generic)]
Synonyms: POE (6) sorbitan oleate; PEG 300 sorbitan monooleate)
Nikkol TO-106

PEG-20 sorbitan oleate. See *Polysorbate 80*

PEG-30 sorbitan oleate
Alkamuls PSMO-30

PEG-20 sorbitan palmitate. See *Polysorbate 40*

PEG-40 sorbitan peroleate
Arlatone T

PEG-4 sorbitan stearate. See *Polysorbate 61*

PEG-20 sorbitan stearate. See *Polysorbate 60*

PEG-16 sorbitan tallate
Atlox 8916P

PEG-20 sorbitan tallate
DeSonic SMT-20; DeSotan SMT-20; T-Maz 90

PEG-30 sorbitan tetraoleate
Nikkol GO-430

PEG-40 sorbitan tetraoleate
Nikkol GO-440; Rheodol 440

PEG-60 sorbitan tetraoleate
Nikkol GO-460

PEG-60 sorbitan tetrastearate
Nikkol GS-460

PEG-2 sorbitan trioleate
TO-55-A

PEG-17 sorbitan trioleate
TO-55-EL

PEG-18 sorbitan trioleate
TO-55-E

PEG-20 sorbitan trioleate. See *Polysorbate 85*

PEG-16 sorbitan tristearate
Atlas G-7166P

PEG-20 sorbitan tristearate. See *Polysorbate 65*

PEG-20 sorbitan tritallate
Polyfac SEE-340; T-Maz 95

PEG sorbitol ethers. See *Sorbeth Series*

PEG-2 soyamine [CAS #61791-24-0 (generic); RD #977063-44-7]
Synonyms: POE (2) soya amine; POE (2) soya amine; PEG 100 soya amine
Accomeen S2; Chemeen S-2; Ethomeen S/12; Hetoxamine S-2; Mazeen S 2; Teric 16M2

PEG-5 soyamine [CAS #61791-24-0 (generic); RD #977055-36-9]
Accomeen S5; Alkaminox SO-5; Chemeen S-5; Ethomeen S/15; Hetoxamine S-5; Icomeen S-5; Mazeen S 5; Teric 16M5; Varonic L 205

PEG-10 soyamine [CAS #61791-24-0 (generic); RD #977063-23-2]
Accomeen S10; Ethomeen S/20; Mazeen S 10; Teric 16M10

PEG-15 soyamine [CAS #61791-24-0 (generic); RD #977063-34-5]
Ethomeen S/25; Hetoxamine S-15; Mazeen S 15; Teric 16M15

PEG-30 soyamine
Chemeen S-30, S-30/80

PEG-5 soya sterol [RD #977065-70-5]
Generol 122 E 5

PEG-10 soya sterol [RD #977065-31-8]
Generol 122 E 10

PEG-16 soya sterol [RD #977065-37-4]
Generol 122 E 16

PEG-25 soya sterol [RD #977065-46-5]
Generol 122 E 25

PEG-2 stearamine [RD #977063-45-8]
Synonyms: PEG-2 stearyl amine; PEG 100 stearyl amine; POE (2) stearyl amine
Chemeen 18-2; Ethomeen 18/12; Hetoxamine ST-2; Teric 18M2

PEG-5 stearamine [RD #977065-71-6]
Chemeen 18-5; Ethomeen 18/15; Hetoxamine ST-5; Icomeen 18-5; Teric 18M5

PEG-10 stearamine [RD #977065-32-9]
Ethomeen 18/20; Teric 18M10

PEG-15 stearamine [RD #977063-35-6]
Ethomeen 18/25; Hetoxamine ST-15

PEG-20 stearamine
Teric 18M20

PEG-30 stearamine
Teric 18M30

PEG-50 stearamine [RD #977065-74-9]
Chemeen 18-50; Ethomeen 18/60; Hetoxamine ST-50; Icomeen 18-50; Trymeen SAM-50

PEG-1 stearate
Alkasurf S-1; Nikkol MYS-1EX

PEG-2 stearate [CAS #106-11-6; 9004-99-3 (generic); SE grade RD #977063-17-4]
Synonyms: Diethylene glycol stearate; Diglycol stearate; PEG 100 monostearate; POE (2) monostearate
Aldo MSA; Alkamuls DEG-MS; Alkasurf S-2; Arlacel 165; Cithrol DGMS N/E, DGMS S/E); Clindrol SDG; DMS-33; Drewmulse DGMS; Grocor 5221 SE; Kemester 5221SE; Kessco Diethylene Glycol Monostearate, Diglycol Stearate Neutral, Diglycol Stearate SE; Lexemul 561; Lipo DGS-SE (SE); Nikkol MYS-2; Nopalcol 1-S; Pegosperse 100 S; Radiasurf 7410 (SE), 7411 (SE); Schercemol DEGMS; Tegin A-422, D 6100, DGS; Witconol CAD

PEG-4 stearate [CAS #106-07-0; 9004-99-3 (generic)]
Ablunol 200MS; Alkamuls 200-MS; Cithrol 2MS; Crodet S4; Kessco PEG 200 Monostearate; Mapeg 200 MS; Nikkol MYS-4; Radiasurf 7412

PEG-5 stearate [CAS #9004-99-3 (generic); RD #977065-72-7]
Alkasurf S-5; Chemax E-200 MS; Ethofat 60/15; Hetoxamate SA-5; Industrol MS-5; Lipal 5 S; Trydet SA-5

PEG-6 stearate [CAS #9004-99-3 (generic); RD #977053-32-9]
Cithrol 3MS; Emerest 2636; Kessco PEG 300 Monostearate; Merpoxen SFS 60; Nonex S3E; Polystate C; Radiasurf 7432; Tefose 2000, 2561

PEG-6-32 stearate [CAS #9004-99-3 (generic); RD #977055-38-1]
Synonyms: PEG 1500 monostearate; POE 1500 monostearate)
Cithrol 15MS; Lipopeg 15-S; Mapeg 1500 MS; Pegosperse 1500 MS; Radiasurf 7417; Tefose 63, 1500

PEG-7 stearate [CAS #9004-99-3 (generic); RD #977065-81-8]
Hetoxamate SA-7; Industrol MS-7; Mapeg 350 MS; Trydet SA-7

PEG-8 stearate [CAS #9004-99-3 (generic); RD #977055-39-2]
Ablunol 400MS; Alkamuls 400-MS; Alkasurf S-8; Cithrol 4MS; Crodet S8; Emerest 2640; Eumulgin ST-8; Industrol MS-8; Ionet MS-400; Kessco PEG 400 Monostearate; Lexemul 400 MS; Lipal 400 S; Lipopeg 4-S; Mapeg 400 MS; Merpoxen SFS 80; Myrj 45; Nonisol 300; Nopalcol 4-S; Pegosperse 400 MS; PGE-400-MS; Radiasurf 7413; Simulsol M 45; Trydet SA-8; Varonic 400MS; Witconol H-35A

PEG-9 stearate [CAS #9004-99-3 (generic); RD #977068-31-7]
Chemax E-400 MS; Cremophor S 9; Emulphor VT-650; Hetoxamate SA-9; Industrol MS-9; Lipocol B; RS-55-9; Serdox NSG 400; Teric SF9

PEG-10 stearate [CAS #9004-99-3 (generic); RD #977065-33-0]
Ethofat 60/20; Nikkol MYS-10

PEG-12 stearate [CAS #9004-99-3 (generic); RD #977055-40-5]
Ablunol 600MS; Alkamuls 600-MS; Cithrol 6MS; Crodet S12; Emerest 2662; Hetoxamate SA-13; Ionet MS-600; Kessco PEG 600 Monostearate; Lipal 600 S; Mapeg 600 MS; Mazawax 163R; Merpoxen SFS 120; Nopalcol 6-S; Pegosperse 600 MS; Radiasurf 7414

PEG-14 stearate [CAS #9004-99-3 (generic)]
Alkasurf S-14; Chemax E-600 MS

PEG-15 stearate
Ethofat 60/25; Teric SF15

PEG-20 stearate [CAS #9004-99-3 (generic); RD #977055-41-6]
Ablunol 1000MS; Cerasynt 840; Chemax E-1000 MS; Cithrol 10MS; Hetoxamate SA-23; Ionet MS-1000; Kessco PEG 1000 Monostearate; Lipopeg 10-S; Mapeg 1000 MS; Myrj 49; Pegosperse 1000 MS; PGE-1000-MS; Simulsol M 49; Varonic 1000MS

PEG-23 stearate
Industrol MS-23; Trydet SA-23

PEG-24 stearate
Crodet S24

PEG-25 stearate
Lipal 25 S; Nikkol MYS-25, MYS-25(FF)

PEG-30 stearate [CAS #9004-99-3 (generic); RD #977055-42-7]
Atlas G-2151; Myrj 51; Simulsol M 51

PEG-32 stearate [CAS #9004-99-3 (generic); RD #977055-43-8]
Synonyms: PEG 1540 stearate
Kessco PEG 1540 Monostearate

PEG-35 stearate [CAS #9004-99-3 (generic); RD #977068-25-9]
Hetoxamate SA-35

PEG-36 stearate [CAS #9004-99-3 (generic); RD #977068-26-0]
Varonic 1800MS

PEG-40 stearate [CAS #9004-99-3 (generic); RD #977009-12-3]
AF-72, AF-75; Alkasurf S-40; Atlas G-2198; Crodet S40; Drewmulse HM 100; Hetoxamate SA-40; Industrol MS-40; Lipal 39 S; Lipopeg 39-S; Mapeg S-40; Myrj 52, 52S. 52C; Nikkol MYS-40, MYS-40(FF); Pegosperse 1750 MS; RS-55-40; Simulsol M 52; Teginacid ML, ML-SE; Trydet SA-40

PEG-45 stearate [CAS #9004-99-3 (generic); RD #977065-63-6]
Nikkol MYS-45

PEG-50 stearate [CAS #9004-99-3 (generic); RD #977055-45-0]
Lipal 50 S; Myrj 53; Simulsol M 53; Trydet SA-50/30

PEG-55 stearate
Nikkol MYS-55

PEG-60 stearate
Nopalcol 30-S

PEG-75 stearate [CAS #9004-99-3 (generic); RD #977055-46-1]
Cithrol 40MS; Gelot 64; Kessco PEG 4000 Monostearate; Mapeg 4000 MS; Pegosperse 4000 MS

PEG-90 stearate [CAS #9004-99-3 (generic); RD #977065-88-5]
Drewmulse 1129; Hetoxamate SA-90

PEG-100 stearate [CAS #9004-99-3 (generic); RD #977055-47-2]
Crodet S100; Cyclochem GMS 165; Industrol MS-100; Kessco Glycerol Monostearate SE; Lipomulse 165; Lipopeg 100-S; Mapeg S 100; Mazol 165C; Myrj 59; RS-55-100; Simulsol M 59

PEG-150 stearate [CAS #9004-99-3 (generic); RD #977055-48-3]
Kessco PEG 6000 Monostearate; Mapeg 6000 MS; Ritachol 1000

PEG-2 stearmonium chloride [RD #977066-74-2]
Ethoquad 18/12

PEG-15 stearmonium chloride [RD #977066-77-5]
Ethoquad 18/25

PEG stearyl ethers. See *Steareth Series*

PEG-2 tallate
Synonyms: POE (2) monotallate; PEG 100 monotallate; PEG-2 tall oil acid ester
Teric T2

PEG-4 tallate [CAS #61791-00-2 (generic); RD #977065-59-0]
Hetoxamate FA-5; Mapeg 200 MOT

PEG-5 tallate
Teric T5

PEG-7 tallate
Teric T7

PEG-8 tallate [CAS #61791-00-2 (generic); RD #977055-49-4]
Chemax TO-8; Industrol 400-MOT; Mapeg 400 MOT; Pegosperse 400 MOT; Witconol H-31

PEG-10 tallate [CAS #61791-00-2 (generic)]
Chemax TO-10; Chemcol TO-10; Emulphor TO-530; Industrol TO-10; Polyfac MT-610; Sterox CD; Teric T10

PEG-12 tallate [CAS #61791-00-2 (generic); RD #977068-27-1]
Mapeg 600 MOT

PEG-15 tallate
Chemcol TO-15; Lipal 15 T; Mapeg TAO-15; Polyfac MT-615

PEG-16 tallate [CAS #61791-00-2 (generic); RD #977055-50-7]
Chemax TO-16; Chemcol TO-16; Emulphor TO-639; Industrol TO-16; Renex 20, 25

PEG-20 tallate [CAS #61791-00-2 (generic); RD #977063-51-6]
Hetoxamate FA-20

PEG-4 tall oil dioleate
Polyfac TDO-5

PEG-8 tall oil dioleate
Polyfac TDO-9

PEG-12 tall oil dioleate
Polyfac TDO-14

PEG-5 tall oil oleate
Polyfac TMO-5

PEG-7 tall oil oleate
Polyfac TMO-7

PEG-9 tall oil oleate
Polyfac TMO-9

PEG-14 tall oil oleate
Polyfac TMO-14

PEG-13 tallow amide
Synonyms: PEG-13 hydrogenated tallow amide; POE (13) hydrogenated tallow amide
Ethomid HT/23 (12.5 EO)

PEG-50 tallow amide [CAS #8051-63-6]
Synonyms: POE (50) hydrogenated tallow amide
Ethomid HT/60 ; Icomid HT-50; Schercomid HT-60

PEG-2 tallow amine [CAS #61791-26-2 (generic); RD #977063-46-9]
Synonyms: PEG 100 tallow amine; POE (2) tallow amine; PEG-2 hydrogenated tallow amine
Accomeen T2; Alkaminox T-2; Chemeen HT-2, T-2; DeSomeen TA-2; Ethomeen T/12; Hetoxamine T-2; Icomeen T-2; Mazeen T 2; Teric 17M2

PEG-5 tallow amine [CAS #61791-26-2 (generic)]
Accomeen T5; Alkaminox T-5; Chemeen HT-5, T-5; Ethomeen T/15; Hetoxamine T-5; Icomeen T-5; Katapol PN-430; Mazeen T 5; Teric 17M5

PEG-7 tallow amine
Icomeen T-7

PEG-8 tallow amine
Trymeen TAM-8

PEG-10 tallow amine
Alkaminox T-10; Chemcol T-10; Chemeen T-10

PEG-12 tallow amine
Alkaminox T-12

PEG-15 tallow amine [CAS # 61791-26-2 (generic); RD #977063-36-7]
Accomeen T15; Alkaminox T-15; Chemcol T-15; Chemeen HT-15, T-15; DeSomeen TA-15; Ethomeen T/25; Hetoxamine T-15; Icomeen T-15; Katapol PN-730; Mazeen T 15; Serdox NJAD 15; Teric 17M15; Trymeen TAM-15

PEG-20 tallow amine [CAS # 61791-26-2 (generic); RD #977063-52-7]
Atlas G-3780A; Chemcol T-20; Chemeen T-20; DeSomeen TA-20; Hetoxamine T-20; Icomeen T-20; Katapol PN-810; Polyfac TA-20; Serdox NJAD 20; Trymeen TAM-20

PEG-25 tallow amine
Alkaminox T-25; Chemcol T-25; Icomeen T-25; Trymeen TAM-25

PEG-30 tallow amine
Alkaminox T-30; Serdox NJAD 30

PEG-40 tallow amine [CAS #61791-26-2 (generic); RD #977065-61-4]
Icomeen T-40, T-40-80%; Trymeen TAM-40

PEG-50 tallow amine [CAS #61791-26-2 (generic); RD #977069-79-6]
Alkaminox T-50; Chemeen HT-50; Varonic U 250

PEG-3 tallow aminopropylamine [RD #977066-69-5]
Ethoduomeen T/13

PEG-10 tallow aminopropylamine [RD #977066-70-8]
Ethoduomeen T/20

PEG-15 tallow aminopropylamine [RD #977066-71-9]
Chemeen DT-15; Ethoduomeen T/25

PEG-2 tallowate
Synonyms: Diethylene glycol monotallowate; POE (2) tallowate; PEG-2 tallow acid ester
Nopalcol 1-TW

PEG-8 tallowate
Lipal TE 43

PEG-12 tallowate
Lipal TE 55

PEG-22 tallowate
Lipal TE 70

PEG-30 tallowate
Lipal TE 76

PEG-3 tallow diamine
Chemeen DT-3

PEG-30 tallow diamine
Icodimeen T-30

PEG-30 tallow diamine
Chemeen DT-30

PEG tallow ethers. See *Talloweth Series*

PEG-15 tallow polyamine [RD #977069-49-0]
Polyquart H

PEG-15 N-tallow 1,3-propylene diamine
Trymeen 6640

PEG-3.5 tetramethyl decynediol
Surfynol 440

PEG-10 tetramethyl decynediol
Surfynol 465

PEG-30 tetramethyl decynediol
Surfynol 485

PEG tridecyl ethers. See *Trideceth Series*

PEG-2 tridecyl phenol
Trycol TP-2

PEG-6 tridecyl phenol
Trycol TP-6

PEG-66 trihydroxystearin [RD #977069-50-3]
Synonyms: POE (66) trihydroxystearin
Surfactol 575

PEG-200 trihyroxystearin [RD #977069-51-3]
Naturechem THS-200; Surfactol 590

Pentadecylbenzene sulfonic acid
Petrostep A-70

Pentaerythrityl stearate
Radiasurf 7175

Pentaerythrityl tetrabehenate [CAS #61682-73-3]
Radia 7514

Pentaerythrityl tetra-C5/C9
Radiasyn 7174

Pentaerythrityl tetra-C7
Radiasyn 7177

Pentaerythrityl tetra-C8/C10
Radiasyn 7178

Pentaerythrityl tetraoleate [CAS #19321-40-5]
Synonyms: Pentaerythritol tetraoleate
Radia 7171

Pentaerythrityl tetrapelargonate
Emerest 2485

Pentaerythrityl tetrastearate [CAS #115-83-3]
Synonyms: Pentaerythritol tetrastearate
Radia 7176

Pentamethyl tallow propanediammonium dichloride
Duoquad T-50

Petrolatum [CAS #8009-03-8 (NF); 8027-32-5 (USP)]
Alcolan, 36W, 40; Amerchol 400, C, CAB, H-9, RC; Dehymuls K; Lexate PX; Nimcolan 1747; Protegin, W, WX, X; Ritaderm

Phenol sulfonic acid
Synonyms: Sulfocarbolic acid
Eltesol PSA 65

POE. See under *PEG*

POE POP block copolymer. See *Poloxamer*

Poloxamer 101 [CAS #9003-11-6 (generic); RD #977057-62-7]
Pluronic L31

Poloxamer 105 [CAS #9003-11-6 (generic); RD #977057-63-8]
Pluronic L35

Poloxamer 108 [CAS #9003-11-6 (generic); RD #977066-68-4]
Pluronic F38

Poloxamer 122 [CAS #9003-11-6 (generic); RD #977057-65-0]
Pluronic L42

Poloxamer 123 [CAS #9003-11-6 (generic); RD #977057-66-1]
Pluronic L43

Poloxamer 124 [CAS #9003-11-6 (generic); RD #977057-67-2]
Pluronic L44

Poloxamer 181 [CAS #9003-11-6 (generic); RD #977057-68-3]
Pluronic L61

Poloxamer 182 [CAS #9003-11-6 (generic); RD #977057-69-4]
Pluronic L62

Poloxamer 183 [CAS #9003-11-6 (generic); RD #977057-70-7]
Pluronic L63

Poloxamer 184 [CAS #9003-11-6 (generic); RD #977054-47-9]
Pluronic L64

Poloxamer 185 [CAS #9003-11-6 (generic); RD #977057-71-8]
Pluronic P65

Poloxamer 188 [CAS #9003-11-6 (generic)]
Pluronic F68

Poloxamer 212 [CAS #9003-11-6 (generic); RD #977057-72-9]
Pluronic L72

Poloxamer 215 [CAS #9003-11-6 (generic); RD #977057-73-0]
Pluronic P75

Poloxamer 217 [CAS #9003-11-6 (generic); RD #977057-74-1]
Pluronic F77

Poloxamer 231 [CAS #9003-11-6 (generic); RD #977057-75-2]
Pluronic L81

Poloxamer 234 [CAS #9003-11-6 (generic); RD #977057-76-3]
Pluronic P84

Poloxamer 235 [CAS #9003-11-6 (generic); RD #977057-77-4]
Pluronic P85

Poloxamer 237 [CAS #9003-11-6 (generic); RD #977057-78-5]
Pluronic F87

Poloxamer 238 [CAS #9003-11-6 (generic); RD #977057-79-6]
Pluronic F88

Poloxamer 282 [CAS #9003-11-6 (generic); RD #977057-80-9]
Pluronic L92

Poloxamer 284 [CAS #9003-11-6 (generic); RD #977057-81-0]
Pluronic P94

Poloxamer 288 [CAS #9003-11-6 (generic); RD #977057-82-1]
Pluronic F98

Poloxamer 331 [CAS #9003-11-6 (generic); RD #977057-83-2]
Pluronic L101

Poloxamer 333 [CAS #9003-11-6 (generic); RD #977057-84-3]
Pluronic P103

Poloxamer 334 [CAS #9003-11-6 (generic); RD #977057-85-4]
Pluronic P104

Poloxamer 335 [CAS #9003-11-6 (generic); RD #977057-86-5]
Pluronic P105

Poloxamer 338 [CAS #9003-11-6 (generic); RD #977057-87-6]
Pluronic F108

Poloxamer 401 [CAS #9003-11-6 (generic); RD #977057-88-7]
Pluronic L121

Poloxamer 402 [CAS #9003-11-6 (generic); RD #977057-89-8]
Pluronic L122

Poloxamer 403 [CAS #9003-11-6 (generic); RD #977057-90-1]
Pluronic P123

Poloxamer 407 [CAS #9003-11-6 (generic); RD #977057-91-2]
Pluronic F127

Poloxamine 304
Tetronic 304

Poloxamine 504
Tetronic 504

Poloxamine 701
Tetronic 701

Poloxamine 702
Tetronic 702

Poloxamine 704
Tetronic 704

Poloxamine 707
Tetronic 707

Poloxamine 901
Tetronic 901

Poloxamine 904
Tetronic 904

Poloxamine 908
Tetronic 908

Poloxamine 1101
Tetronic 1101

Poloxamine 1102
Tetronic 1102

Poloxamine 1104
Tetronic 1104

Poloxamine 1107
Tetronic 1107

Poloxamine 1301
Tetronic 1301

Poloxamine 1302
Tetronic 1302

Poloxamine 1304
Tetronic 1304

Poloxamine 1307
Tetronic 1307

Poloxamine 1501
Tetronic 1501

Poloxamine 1502
Tetronic 1502

Poloxamine 1504
Tetronic 1504

Poloxamine 1508
Tetronic 1508

Polydimethylsiloxane. See *Dimethicone*

Polyether glycol. See *PEG*

Polyethylene glycol. See *PEG*

Polyglyceryl-2 borate sesquioleate
Synonyms: Diglyceryl borate sesquioleate
Emulbon S-83

Polyglyceryl-10 decaisostearate
Synonyms: Decaglycerin decaisostearate
Nikkol Decaglyn 10-IS

Polyglyceryl-10 decaoleate [RD #977057-94-5]
Synonyms: Decaglycerol decaoleate
Caplube 8442; Caprol 10G100; Drewmulse 10-10-O; Drewpol 10-10-O; Mazol PGO-1010; Nikkol Decaglyn 10-O; Polyaldo DGDO; Santone 10-10-O

Polyglyceryl-10 decastearate [RD #977067-09-6]
Synonyms: Decaglycerol decastearate
Caplube 8448; Caprol JB; Drewmulse 10-10-S; Drewpol 10-10-S; Nikkol Decaglyn 10-S

Polyglyceryl-3 diisostearate
Synonyms: Diglyceryl diisostearate
Lameform GE 2

Polyglyceryl-2 dioleate
Synonyms: Diglyceryl dioleate
Nikkol DGDO

Polyglyceryl-6 dioleate [RD #977067-14-3]
Synonyms: Hexaglycerol dioleate
Caprol 6G20; Mazol PGO-62

Polyglyceryl-10 dioleate
Synonyms: Decaglycerol dioleate; Decaglycerin dioleate
Caprol 10G20; Nikkol Decaglyn 2-O

Polyglyceryl-6 distearate [RD #977067-15-4]
Synonyms: Hexaglycerol distearate
Caprol 6G2S; Drewmulse 6-2-S; Drewpol 6-2-S; Mazol PGS-62; Polyaldo HGDS

Polyglyceryl-10 distearate
Synonyms: Decaglycerin distearate
Nikkol Decaglyn 2-S

Polyglyceryl-10 heptaisostearate
Synonyms: Decaglycerin heptaisostearate
Nikkol Decaglyn 7-IS

Polyglyceryl-10 heptaoleate
Synonyms: Decaglycerin heptaoleate
Nikkol Decaglyn 7-O

Polyglyceryl-10 heptastearate
Synonyms: Decaglycerin heptastearate
Nikkol Decaglyn 7-S

Polyglyceryl-6 hexaoleate
Mazol PGO-66

Polyglyceryl-10 hexaoleate
Polyaldo DGHO

Polyglyceryl-4 isostearate [RD #977068-33-9]
Synonyms: Tetraglyceryl monoisostearate
Protegin W, WX; Witconol 18L

Polyglyceryl-10 isostearate
Synonyms: Decaglycerin monoisostearate
Nikkol Decaglyn 1-IS

Polyglyceryl-6 laurate
Synonyms: Hexaglycerin monolaurate
Nikkol Hexaglyn 1-L

Polyglyceryl-10 laurate
Synonyms: Decaglycerin monolaurate
Nikkol Decaglyn 1-L

Polyglyceryl-10 linoleate
Synonyms: Decaglycerin monolinoleate
Nikkol Decaglyn 1-LN

Polyglyceryl-10 monodioleate
Synonyms: Decaglycerol monodioleate
Caprol PGE860

Polyglyceryl-10 myristate
Synonyms: Decaglycerin monomyristate
Nikkol Decaglyn 1-M

Polyglyceryl-10 octaoleate
Synonyms: Decaglycerol octaoleate
Drewmulse 10-8-O; Drewpol 10-8-O; Mazol PGO-108

Polyglyceryl-2 oleate
Nikkol DGMO-C

Polyglyceryl-3 oleate [CAS #9007-48-1 (generic); RD #977067-11-0]
Synonyms: Triglyceryl oleate
Caplube 8410; Caprol 3GO; Drewmulse 3-1-O; Drewpol 3-1-O; Grindtek PGE 25; Mazol PGO-31; Santone 3-1-SH

Polyglyceryl-4 oleate [CAS #9007-48-1 (generic); RD #977057-95-6]
Synonyms: Tetraglyceryl monooleate
Nikkol Tetraglyn 1-O; Witconol 14

Polyglyceryl-6 oleate
Synonyms: Hexaglycerin monooleate
Nikkol Hexaglyn 1-O

Polyglyceryl-8 oleate [CAS #9007-48-1 (generic)]
Santone 8-1-O

Polyglyceryl-10 oleate
Synonyms: Decaglycerin monooleate
Nikkol Decaglyn 1-O

Polyglyceryl-2-PEG-4 stearate
Synonyms: Diglyceryl PEG-4 stearate
Hostacerin DGS

Polyglyceryl-10 pentaisostearate
Synonyms: Decaglycerin pentaiosostearate
Nikkol Decaglyn 5-IS

Polyglyceryl-4 pentaoleate
Synonyms: Tetraglyceryl pentaoleate
Nikkol Tetraglyn 5-O

Polyglyceryl-6 pentaoleate
Synonyms: Hexaglyceryl pentaoleate
Nikkol Hexaglyn 5-O

Polyglyceryl-10 pentaoleate
Synonyms: Decaglycerin pentaoleate
Nikkol Decaglyn 5-O

Polyglyceryl-4 pentastearate
Synonyms: Tetraglyceryl pentastearate
Nikkol Tetraglyn 5-S

Polyglyceryl-6 pentastearate
Synonyms: Hexaglyceryl pentastearate
Nikkol Hexaglyn 5-S

Polyglcyeryl-10 pentastearate
Synonyms: Decaglycerin pentastearate
Nikkol Decaglyn 5-S

Polyglyceryl-6 polyricinoleate
Synonyms: Hexaglyceryl polyricinoleate
Nikkol Hexaglyn PR-15

Polyglyceryl-2 sesquiisostearate
Hostacerin WO

Polyglyceryl-2 sesquioleate [RD #977067-16-5]
Synonyms: Diglyceryl sesquioleate
Hostacerin DGO

Polyglyceryl-3 shortening
Drewpol 3-1-SH

Polyglyceryl-2 stearate
Synonyms: Diglyceryl monostearate
Nikkol DGMS

Polyglyceryl-3 stearate [CAS #37349-34-1 (generic); RD #977067-12-1]
Synonyms: Triglyceryl stearate
Caprol 3GS; Drewmulse 3-1-S; Drewpol 3-1-S; Grindtek PGE 55, 55-6; Mazol PGS-31; Polyaldo TGMS; Santone 3-1-S

Polyglyceryl-4 stearate [CAS #37349-34-1 (generic)]
Synonyms: Tetraglyceryl monostearate
Nikkol Tetraglyn 1-S; Witconol 18F

Polyglyceryl-6 stearate
Synonyms: Hexaglycerol stearate
Mazol PGS-61; Nikkol Hexaglyn 1-S

Polyglyceryl-8 stearate [CAS #37349-34-1 (generic)]
Santone 8-1-S

Polyglyceryl-10 stearate
Synonyms: Decaglycerin monostearate
Nikkol Decaglyn 1-S

Polyglyceryl-10 tetracocate
Synonyms: Decaglycerol tetracocate
Caplube 8445

Polyglyceryl-10 tetraoleate [CAS #34424-98-1]
Synonyms: Decaglycerol tetraoleate
Caplube 8440; Caprol 10G40; Drewmulse 10-4-O; Drewpol 10-4-O; Mazol PGO-104

Polyglyceryl-2 tetrastearate [RD#977067-17-6]
Synonyms: Diglycerol tetrastearate
Caprol 2G4S

Polyglyceryl-10 trioleate
Synonyms: Decaglycerin trioleate
Nikkol Decaglyn 3-O

Polyglyceryl-4 tristearate
Synonyms: Tetraglyceryl tristearate
Nikkol Tetraglyn 3-S

Polyglyceryl-6 tristearate
Synonyms: Hexaglycerin tristearate
Nikkol Hexaglyn 3-S

Polyglyceryl-10 tristearate
Synonyms: Decaglycerin tristearate
Nikkol Decaglyn 3-S

Polyglycol. See *PEG*

Polyoxyethylene (POE). See under *PEG*

Polyoxypropylene (POP). See under *PPG*

Polypropylene glycol. See *PPG*

Polyquaternium-1 [CAS #68518-54-7]
Onamer M

Polyquaternium-2 [CAS #63451-27-4]
Mirapol A-15

Polyquaternium-5 [CAS #26006-22-4]
Synonyms: Quaternium-39; Acrylamide/β–methacrylyloxyethyl trimethyl ammonium methosulfate copolymer
Emcol Q

Polyquaternium-18
Mirapol AZ-1

Polysorbate 20 [CAS #9005-64-5 (generic)]
Synonyms: POE (20) sorbitan monolaurate; Sorbimacrogol laurate 300
Accosperse 20; Ahco 7596T; Alcolec 4135; Alkamuls PSML-20; Armotan PML 20; Capmul POE-L; Crillet 1; Drewmulse POE-SML; Durfax 20; Emasol L-120; Emsorb 6915; Ethylan GEL-2; Glycosperse L-20, L-20X; Hetsorb L-20; Industrol L-20-S; Ionet T-20 C; Liposorb L-20; Lonzest SML-20; ML-55-F; Montanox 20; Nikkol TL-10, TL-10(FF); Nissan Nonion LT-221; Radiamuls 137, SORB 2137; Radiasurf 7137; Secoster CL 10; Soprofor T/20; Sorbax PML-20; Sorbon T-20; Sorgen TW20; T-Maz 20; Tween 20

Polysorbate 21 [CAS #9005-64-5 (generic); RD #977053-95-4]
Synonyms: POE (4) sorbitan monolaurate
Ahco 7596D; Alkamuls PSML-4; Crillet 11; Emasol L-106; Emsorb 6916; Ethylan GLE-21; Hetsorb L-4; ML-55-F-4; Tween 21

Polysorbate 40 [CAS #9005-66-7]

Synonyms: POE (20) sorbitan monopalmitate; Sorbimacrogol palmitate 300

Ahco DFP-156; Crillet 2; Emasol P-120; Emsorb 6910; Ethylan GEP-4; Glycosperse P-20; Hetsorb P-20; Liposorb P-20; Lonzest SMP-20; Montanox 40; MP-55-F; Rheodol TW-P120; Secoster CP 10; Sorbax PMP-20; T-Maz 40; Tween 40

Polysorbate 60 [CAS #9005-67-8 (generic)]

Synonyms: POE (20) sorbitan monostearate; Sorbimacrogol stearate 300

Accosperse 60; Ahco DFS-149; Alkamuls PSMS-20; Alpha 500; Atmos 729, 758, 2462; Atmul 700H; Capmul POE-S; Crillet 3; Drewmulse POE-SMS; Drewpone 60; Durfax 60, 60K; Emasol S-120; Emrite 6125; Emsorb 6905; Ethylan GES-6; Glycosperse S-20; Hetsorb S-20; Incrochem SC1, SC4; Incrosorb S-60; Industrol S-20-S; Liposorb S-20; Lonzest SMS-20; Montanox 60; MS-55-F; Nikkol TS-10, TS-10(FF); Radiamuls 147, SORB 2147; Radiasurf 7147; Ritachol 1000, 2000; Secoster CS 10; Soprofor T/60; Sorbax PMS-20; Tandem 5K, 8, 9, 11H, 552; T-Maz 60, 60K; Tween 60; Vanade; Vanall

Polysorbate 61 [CAS #9005-67-8 (generic); RD #977053-94-3]

Synonyms: POE (4) sorbitan monostearate

Ahco DFS-96; Alkamuls PSMS-4; Crillet 31; Emasol S-106; Emsorb 6906; Hetsorb S-4; Montanox 61; MS-55-F-4; T-Maz 61; Tween 61

Polysorbate 65 [CAS #9005-71-4]

Synonyms: POE (20) sorbitan tristearate; Sorbimacrogol tristearate 300

Ahco 7166T; Aldosperse TS-20, TS-20 FG, TS-40, TS-40 FG; Alkamuls PSTS-20; Crillet 35; Drewmulse 365, POE-STS; Drewpone 65; Durfax 65, 65K; Emasol S-320; Emsorb 6907; Glycosperse TS-20; Hetsorb TS-20; Ice #10K, #12, #12K; Industrol STS-20-S; Liposorb TS-20; Lonzest STS-20; Montanox 65; Nikkol TS-30; Soprofor T/65; Sorbax PTS-20; T-Maz 65, 65K; TS-55-F; Tween 65

Polysorbate 80 [CAS #9005-65-6 (generic)]

Synonyms: POE (20) sorbitan monooleate; Sorbimacrogol oleate 300

Accosperse 80; Ahco DFO-150; Aldosperse O-20, O-20 FG; Alkamuls PSMO-20; Armotan PMD 20, PMO 20; Atlas G-4905; Capmul POE-O; Crillet 4; DeSonic SMO-20; DeSotan SMO-20; Drewmulse 700, POE-SMO; Drewpone 80; Durfax 80, 80K; Emasol O-120; Emrite 6120; Emsorb 6900; Ethylan GEO-8; Glycosperse O-20, O-20 Veg., O-20X; Hetsorb O-20; Ice #2, #2FK; Industrol O-20-S; Liposorb O-20; Lonzest SMO-20; MO-55-F; Montanox 80; Nikkol TO-10, TO-10(FF); Radiamuls 157, SORB 2157; Radiasurf 7157; Solulan 98; Soprofor T/80; T-Maz 80, 80K; Tween 80; Vanease

Polysorbate 80 acetate

Synonyms: POE (20) sorbitan monooleate acetate

Solulan 97; Sorbax PMO-20

Polysorbate 81 [CAS #9005-65-5 (generic); RD #977053-93-2]

Synonyms: POE (5) sorbitan monooleate

Ahco DFO-100; Alcolan 40; Alkamuls PSMO-5; Crillet 41; Emasol O-106; Emsorb 6901; Ethylan GOE-21; Glycosperse O-5; Hetsorb O-5; Liposorb O-5; MO-55-F-5; Montanox 81; Sorbax PMO-5; T-Maz 81; Tween 81

Polysorbate 85 [CAS #9005-70-3]

Synonyms: POE (20) sorbitan trioleate; Sorbimacrogol trioleate 300

Ahco DFO-110; Alkamuls PSTO-20; Crillet 45; Emasol O-320; Emsorb 6903; Ethylan GPS-85; Glycosperse TO-20; Hetsorb TO-20; Liposorb TO-20; Lonzest STO-20; Montanox 85; Nikkol TO-30; Ritaderm; Soprofor T/85; Sorbax PTO-20; T-Maz 85; TO-55-F; Tween 85

POP (Polyoxypropylene). See under *PPG*

Potassium alginate [CAS #9005-36-1]
Kelmar, Kelmar Improved

Potassium 2-(2-carboxyethoxy) ethyl 2-(2-(2-hydroxy) ethoxyethyl methyldodecyl ammonium methosulfate
Sanac C

Potassium 2-(2-carboxyethoxy) ethyl 2-(2-(2-hydroxy) ethoxyethyl methyloctadecenyl ammonium methosulfate
Sanac S

Potassium cocamidopropyl dimethyl betaine
Schercotaine CAB-KG

Potassium cocamidopropyl hydroxy sulfopropyl dimethyl betaine
Schercotaine SCAB-KG

Potassium cocoate [CAS #61789-30-8]
Synonyms: Potassium coconate
Mackadet 40K; Norfox 1101

Potassium coco-hydrolyzed animal protein [CAS #68920-65-0]
Lamepon S, S 2; Lexein S620; Maypon 4C; Nikkol CCK-40

Potassium lauryl phosphate
Crafol AP-31, AP-33

Potassium lauryl sulfate [CAS #4706-78-9]
Conco Sulfate P

Potassium oleate [CAS #143-18-0]
Synonyms: Potassium 9-octadecenoate
Norfox KO

Potassium ricinoleate [CAS #7492-30-0]
Solricin 135

Potassium stearate [CAS #593-29-3]
Imwitor 965, 965 K

Potassium tallate
Synonyms: Tall oil rosin potassium soap
Arizona DRS-40, DRS-42; Tallates, K

Potassium talloweth sulfate
Synonyms: Potassium tallow ether sulfate
Elfan NS 682 KS

Potassium toluenesulfonate [CAS #16106-44-8; 30526-22-8]
Eltesol PT 93; Hartotrope KTS 50; Naxonate 4KT, 5KT, KT

Potassium undecylenoyl hydrolyzed animal protein [CAS #68951-92-8]
Lamepon UD; Maypon UD

Potassium xylene sulfonate
Eltesol PX 40, PX 93; Manro KXS 40

PPG-1.... See also under *Propylene glycol...*

PPG-9 [CAS #253322-69-4 (generic); RD #977055-51-8]
Synonyms: Polyoxypropylene (9); Polypropylene glycol (9)
Pluracol P-410

PPG-12 [CAS #25322-69-4 (generic); RD #977066-97-9]
Pluracol P-710

PPG-17 [CAS #25322-69-4 (generic); RD #977066-99-1]
Pluracol P-1010

PPG-26 [CAS #25322-69-4 (generic); RD #977055-52-9]
Pluracol P-2010

PPG-30 [CAS #25322-69-4 (generic); RD #977055-53-0]
Pluracol P-4010

PPG-34 [CAS #25322-69-4 (generic)]
Witconol CD-17

PPG 425
Alkapol PPG-425

PPG 1000
Alkapol PPG 1000

PPG 1200
Alkapol PPG 1200

PPG 2000
Alkapol PPG 2000

PPG 4000
Alkapol PPG 4000

PPG-5-buteth-7 [CAS #9038-95-3 (generic); RD #977066-93-5]
Synonyms: POE (7) POP (5) monobutyl ether
Pluracol W-170

PPG-12-buteth-16 [CAS #9038-95-3 (generic); RD #977055-56-3]
Pluracol W-660

PPG-20-buteth-30 [CAS #9038-95-3 (generic); RD #977055-58-5]
Pluracol W-2000

PPG-24-buteth-27
Tergitol XD

PPG-33-buteth-45 [CAS #9038-95-3 (generic); RD #977055-60-9]
Pluracol W-5100, W-5100N

PPG-12 C12-18 pareth-11
Plurafac D-25

PPG-2-ceteareth-9
Synonyms: POP (2) POE (9) cetyl/stearyl ether; POE (9) POP (2) cetyl/stearyl ether
Eumulgin L

PPG-10-ceteareth-20 [RD #977067-21-2]
Standamul OXL

PPG-1 ceteth-3 acetate
Synonyms: Propylene glycol POE (3) cetyl ether acetate
Hetester PCA

PPG-4 ceteth-1 [CAS #9087-53-0 (generic)]
Synonyms: POE (1) POP (4) cetyl ether; POP (4) POE (1) cetyl ether
Nikkol PBC-31

PPG-4-ceteth-10 [CAS #9087-53-0 (generic)]
Nikkol PBC-33

PPG-5 ceteth-10 phosphate [RD #977061-68-9]
Synonyms: POP (5) POE (10) cetyl ether phosphate
Crodafos SG

PPG-5-ceteth-20 [CAS #9087-53-0 (generic); RD #977061-66-7]
Incropearl Liquid NI; Procetyl AWS

PPG-8-ceteth-1 [CAS #9087-53-0 (generic)]
Nikkol PBC-41

PPG-8-ceteth-20 [CAS #9087-53-0 (generic)]
Nikkol PBC-44

PPG-10 cetyl ether [CAS #9035-85-2 (generic); RD #977055-66-5]
Synonyms: POP (10) cetyl ether; PPG (10) cetyl ether
Procetyl 10

PPG-10 cetyl ether phosphate [RD #977060-96-0]
Synonyms: POP (10) cetyl ether phosphate
Crodafos CAP

PPG-20 cetyl ether
Procetyl 20

PPG-30 cetyl ether
Procetyl 30

PPG-50 cetyl ether
Procetyl 50

PPG-6-decyltetradeceth-12
Synonyms: POE (12) POP (6) decyl tetradecyl ether
Nikkol PEN-4612

PPG-6-decyltetradeceth-20
Nikkol PEN-4620

PPG-6-decyltetradeceth-30
Nikkol PEN-4630

PPG-9 diethylmonium chloride [CAS #9042-76-6]
Synonyms: POP (9) methyl diethyl ammonium chloride; Quaternium-6
Emcol CC-9

PPG-25 diethylmonium chloride [RD #977062-01-3]
Synonyms: POP (25) methyl diethyl ammonium chloride; Quaternium-20
Emcol CC-36

PPG-40 diethylmonium chloride [CAS #9076-43-1]
Synonyms: POP (40) methyl diethyl ammonium chloride; Quaternium-21
Emcol CC-42

PPG-1 isoceteth-3 acetate
Synonyms: Propylene glycol POE (3) isocetyl ether acetate
Hetester PHA

PPG-3-isosteareth-9 [RD #977068-36-2]
Synonyms: POE (9) POP (3) isostearyl ether; POP (3) POE (9) isostearyl ether
Arosurf 66-PE12

PPG-2 lanolin ether [CAS #68439-53-2 (generic); RD #977067-06-3]
Synonyms: POP (2) lanolin ether; PPG (2) lanolin ether
Solulan PB-2

PPG-5 lanolin ether [CAS #68439-53-2 (generic); RD #977067-04-1]
Solulan PB-5

PPG-10 lanolin ether [CAS #68439-53-2 (generic); RD #977061-53-2]
Solulan PB-10

PPG-20 lanolin ether [CAS #68439-53-2 (generic); RD #977064-51-9]
Solulan PB-20

PPG-2 laurate
Synonyms: Dipropylene glycol monolaurate; POP (2) laurate; PPG (2) laurate
Cithrol DPGMO N/E, DPGML S/E

PPG-3-laureth-9
Synonyms: POE (9) POP (3) lauryl ether; POP (3) POE (9) lauryl ether
Acconon 1300

PPG-10 methyl glucose ether [RD #977067-01-8]
Synonyms: POP (10) methyl glucose ether; PPG (10) methyl glucose ether
Glucam P-10

PPG-20 methyl glucose ether [RD #977067-02-9]
Glucam P-20

PPG-3-myreth-3 [CAS #37311-04-9 (generic); RD #977069-61-6]
Synonyms: POE (3) POP (3) myristyl ether; POP (3) POE (3) myristyl ether
Witconol APEM

PPG-3 myristyl ether [CAS #63793-60-2 (generic); 78436-05-3 (generic); RD #977068-35-1]
Synonyms: Tripropylene glycol myristyl ether; POP (3) myristyl ether; PPG (3) myristyl ether
Witconol APM

PPG-2 oleate
Synonyms: Dipropylene glycol monooleate; POP (2) monooleate; PPG (2) monooleate
Cithrol DPGMO N/E, DPGMO S/E

PPG-36 oleate [CAS #31394-71-5 (generic); RD #977055-75-6]
Witconol F26-46

PPG-10 oleyl ether [CAS #52581-71-2 (generic); RD #977055-76-7]
Provol 10

PPG-30 oleyl ether [CAS #52581-71-2 (generic); RD #977055-77-8]
Provol 30

PPG-50 oleyl ether [CAS #52581-71-2 (generic); RD #977055-78-9]
Provol 50

PPG-12-PEG-50 lanolin [CAS #684858-58-8 (generic); RD #977062-71-7]
Lanexol AWS

PPG-40-PEG-60 lanolin oil [RD #977065-80-7]
Aqualose LL100

PPG-2 stearate
Synonyms: Dipropylene glycol monostearate; POP (2) monostearate; PPG (2) monostearate
Cithrol DPGMS N/E, DPGMS S/E

PPG-11 stearyl ether [CAS #25231-21-4 (generic); RD #977069-62-7]
Synonyms: POP (11) stearyl ether; PPG (11) stearyl ether
Witconol APS

PPG-15 stearyl ether [CAS #25231-21-4 (generic); RD #977061-23-6]
Arlamol E; Prostearyl 15

PPG-3 trideceth-6
Synonyms: POE (6) POP (3) tridecyl ether; POP (3) POE (6) tridecyl ether; PEG-6 PPG-3 trideceyl ether
Hetoxol TDEP-63

PPG-15 trideceth-10
Hetoxol TDEP-15

PPG 1000 ditallate
Industrol 1025-DT

Propylene glycol [CAS #57-55-6]
Synonyms: 1,2-Propanediol. See also under *PPG-1*
Alcolec HS-3; Arlacel 186; Atmos 2462; Cremophor RH455; Incrochem PPI, SBJ, SC1, SC4; Incromate PPI; Lexaine IBC-70, IBC-70S, O; Lexate CL-60; Miranol 2MCAS Modified; Mirataine XL; Monamid R31-42; Pentex 99; Solusol 84%; Standapol Pearl Conc. 7130; Varamide SL-9

Propylene glycol alginate [CAS #9005-37-2]
Kelcoloid D, DH, DO, DSF, HVF, LVF, O, S

Propylene glycol dicaprylate/dicaprate [RD #977060-55-1]
Aldo DC; Edenol 302; Neobee M-20; Standamul 302

Propylene glycol distearate [CAS #6182-11-2]
Mapeg PGDS

Propylene glycol hydroxystearate [CAS #33907-47-0]
Naturechem PGHS

Propylene glycol laurate [CAS #142-55-2; 27194-74-7]
Alcolec 4135; Cithrol PGML N/E, PGML S/E; Drewmulse PGML; Schercemol PGML; Tegin PL

Propylene glycol monodistearate
Tegin P 411; Tegomuls P 411

Propylene glycol myristate [CAS #29059-24-3]
Radiasurf 7196

Propylene glycol oleate [CAS # 1330-80-9]
Atlas G-2185; Cithrol PGMO N/E, PGMO S/E; Radiasurf 7206

Propylene glycol palmitate
Atlas G-2183

Propylene glycol ricinoleate [CAS #26402-31-3]
Cithrol PGMR N/E, PGMR S/E; Flexricin 9; Naturechem PGR

Propylene glycol stearate [CAS #1323-39-3]
Synonyms: Propylene glycol monostearate
Alcolec 4135; Aldo PGHMS, PME, PMS; Cerasynt PA; Cithrol PGMS N/E, PGMS S/E; Cyclochem PGMS; Drewlene 10; Drewmulse PGMS; Durpro 107-55; Grindtek PGMS 90; Homotex PS-90; Kessco Propylene Glycol Monostearate Pure; Lipo PGMS; Mapeg PGMS; Mazol PGMS; Monosteol; Myvatex 3-50, 40-06S; Nikkol PMS-1C; PMS-33; Radiasurf 7201; Schercemol PGMS; Tegin A-412; Witconol RHP (SE)

Propylene glycol stearate SE [RD #977067-33-6]
Emerest 2381; Lexemul P; Nikkol PMS-1CSE; Tegin P, P-411 SE, P-SE

Propyl oleate
Grocor 3000

PVM/MA copolymer [CAS #9011-16-9]
Gantrez AN-119, -139, -149, -169, -179

DL-Pyrrolidone carboxylic acid monoolein ester
Amifat P-30

DL-Pyrrolidone carboxylic acid monostearin ester
Amifat P-21

Quaternium-2. See *Soyaethyl morpholinium ethosulfate*

Quaternium-3. See *Myristalkonium saccharinate*

Quaternium-6. See *PPG-9 diethylmonium chloride*

Quaternium-7. See *Steapyrium chloride*

Quaternium-8 [RD #977066-07-1]
Synonyms: Alkyl dimethyl ethylbenzyl ammonium cyclohexyl sulfamate
Onyxide 172

Quaternium-9. See *Soytrimonium chloride*

Quaternium-14 [CAS #27479-28-3]
Synonyms: Dodecyl dimethyl ethylbenzyl ammonium chloride
BTC 2125, 2125 M

Quaternium-17. See *Cetethyldimonium bromide*

Quaternium-18 [CAS #61789-80-8]

Synonyms: Dimethyl di (hydrogenated tallow) ammonium chloride; Dihydrogenated tallow dimethyl ammonium chloride

Accosoft 707; Adogen 442, 442 H; Arquad 2HT-75; Jet Quat 2HT-75; Kemamine Q-9702C; Radiaquat 6442; Varisoft 3262, DHT

Quaterium-18 methosulfate [CAS #61789-81-9]

Synonyms: Quaternary ammonium compounds, bis (hydrogenated tallow alkyl) dimethyl, methyl sulfates

Carsosoft V-90, V-100

Quaternium-21. See *PPG-40 diethylmonium chloride*

Quaternium-24 [CAS #32426-11-2]

Synonyms: Decyl dimethyl octyl ammonium chloride; Octyl decyl dimethyl ammonium chloride

BTC 818, 885

Quaternium-25. See *Cetethyl morpholinium ethosulfate*

Quaternium-27 [RD #977065-92-1]

Carsosoft S-75, S-90; Rewoquat W 7500; Varisoft 475

Quaternium-32. See *Isostearyl ethylimidonium ethosulfate*

Quaternium-33

Lanoquat 1756

Quaternium-39. See *Polyquaternium-5*

Quaternium-52 [CAS #58069-11-7]

Dehyquart SP

Quaternium-61

Schercoquat DAS

Quaternium-62

Schercoquat IEP

Quaternium-63

Schercoquat DAB

Rapeseedamidopropyl benzyldimonium chloride

Schercoquat ROAB

Rapeseedamidopropyl dimethyl epoxypropyl ammonium chloride

Schercoquat ROEP

Rapeseedamidopropyl ethyldimonium ethosulfate

Schercoquat ROAS

Ricineolamide DEA [CAS #40716-42-5]

Synonyms: Ricinoleic diethanolamide

Alkamide RDO; Aminol CA-2; Lamacit ER; Mackamide R; Rodea

Ricinoleamide MEA [CAS #106-16-1]

Synonyms: Ricinoleoyl monoethanolamide

Rewomid R 280

Ricinoleamidopropyl betaine
Synonyms: Ricinoleamidopropyl dimethyl glycine
Rewoteric AM-R 40; Varion AM-R40

Ricinoleamidopropyl dimethylamine [CAS #20457-75-4]
Lexamine R-13

Ricinoleamidopropyl trimonium methosulfate
Rewoquat RTM 50

Ricinoleic acid [CAS #141-22-0]
Synonyms: 12-Hydroxy-9-octadecenoic acid
P-10 Acid

Saccharose. See *Sucrose*

Safflowreramidopropyl dimethylethyl ammonium ethyl sulfate
Schercoquat FOAS

Safflower glyceride [RD #977058-10-8]
Myverol 18-98

SD Alcohol 40 [RD #977021-81-0]
Schercoteric LS-2ES MOD

SD Alcohol 40-B [RD #977021-83-2]
Equex AEM

Silica [CAS #7631-86-9]
AF-72, AF-75

Simethicone [CAS #8050-81-5]
Dow Corning Antifoam A, AF, C

Sodium alginate. See *Algin*

Sodium alpha olefin sulfonate
Agent 804; Bio Soft LD-70, LD-80; Conco AOS-40, AOS-90F; DeSonate AOS; Hostapur OS Brands; Lakeway 301-10, 301-10F; Norfox ALPHA XL

Sodium-amine dodecylbenzenesulfonate
Ardet A-25

Sodium benzoate [CAS #532-32-1]
Aerosol OT-B; Manoxol OT/B; Solusol 85%

Sodium butoxyethoxy acetate [CAS #67990-17-4]
Mirawet B

Sodium butyl naphthalene sulfonate
Morwet B

Sodium C9-13 alcohols sulfate
Teepol HB 7

Sodium C10-12 pareth-5 phosphate
Synonyms: PEG-5 C10-12 alcohol phosphate, sodium salt
Merpiphos BP 80

Sodium C11-15 pareth-7 carboxylate
Surfine WCT-Gel

Sodium C12-13 pareth sulfate
Synonyms: Sodium pareth-23 sulfate
Akyposal DS 56

Sodium C12-15 alcohols sulfate [RD #977067-40-5]
Lanette Wax SX, SXBP; Standapol WAQ-115

Sodium C14-16 olefin sulfonate [CAS #68439-57-6]
Bio Terge AS-40; DeSonate AUS; Elfan OS 46; Polystep A-18; Siponate 301-10 F, A-246 L, A-246 LX; Sipoteric 1450; Sterling AOS; Witconate AOS

Sodium C15-18 olefin sulfonate
Siponate A-168

Sodium C16-18 olefin sulfonate [RD #977067-81-4]
Siponate A-167

Sodium C12-15 pareth sulfate
Synonyms: Sodium pareth-25 sulfate; POE (1–4) C12–15 fatty alcohol ether sulfated, sodium salt
Dobanol 25-3S/27, 25-3S/60, 25-3S/70; Neodol 25-3S; Serdet DPK 3/70, 30; Standapol SP-60; Witcolate 1050

Sodium C12-15 pareth-3 sulfosuccinate
Sermul EA 221

Sodium C12-15 pareth-7 carboxylate
Synonyms: Sodium pareth-25-7 carboxylate
Surfine WNT Gel, WNT LC, WNT LS

Sodium C13-15 pareth sulfate
Manro BES 60

Sodium C14-17 sec alcohol sulfonate [CAS #68037-49-0]
Synonyms: Sodium C14-17 alkyl sec sulfonate
Hostacerin CG; Hostapur SAS 30, SAS 60, SAS 93

Sodium capryl lactylate [RD #977067-37-0]
Pationic 122A

Sodium carboxymethylcellulose. See *Carboxymethylcellulose*

Sodium castorate [CAS #8013-06-7]
Synonyms: Castor oil, sodium salt
Servo Brillaint Olie B AZ 75

Sodium cetearyl sulfate [CAS #59186-41-3]
Synonyms: Sodium cetostearyl sulfate; Sodium cetyl/stearyl sulfate
Dehydag Wax E, N; Emulgade F; Lanette E, N; Polystep B-26

Sodium cetearyl sulfosuccinate
Secosol AS

Sodium cetyl diphenyl ether disulfonate
Synonyms: Sodium hexadecyl diphenyl ether disulfonate
Conco Sulfate 4C3

Sodium cetyl/oleyl sulfate
Empicol CHC 30; Gardinol CX, CXM; Manro D Paste

Sodium cetyl sulfate [CAS #1120-01-0]
Berol 472; Nikkol SCS; Sipex EC-111; Standapol CS-50, CS Paste

Sodium chloride [CAS #7647-14-5]
Incrochem SC1

Sodium cocamido DEA sulfosuccinate
Alkasurf SS-CDE

Sodium cocaminobutyrate
Synonyms: Sodium N-coco-beta-amino butyrate
Armeen SZ

Sodium cocaminopropionate [CAS #8033-69-0; 12676-37-8; 68608-68-4]
Synonyms: Sodium-N-coco-beta-aminopropionate
Deriphat 151

Sodium cocoate [CAS #61789-31-9]
Synonyms: Coconut oil fatty acid, sodium salt
Norfox Coco Powder

Sodium coco-hydrolyzed animal protein [CAS #68188-38-5]
Nikkol CCN-40

Sodium cocomonoglyceride sulfate [CAS #61789-04-6]
Synonyms: Sodium coconut monoglyceride sulfate
Nikkol SGC-80N

Sodium cocomonoglyceride sulfonate
SCMS

Sodium coconut sulfate
Elfan 280; Sulfetal C 90

Sodium coconut sulfosuccinate
Setacin F Special Paste

Sodium coco-tallow sulfate
Elfan KT 550

Sodium cocoyl collagen peptide
Vifcoll CCN-40, CCN-40 Powd.

Sodium cocoyl glutamate [CAS #68187-32-6]
Synonyms: Monosodium N-cocoyl-L-glutamate
Acylglutamate CS-11, GS-11; Amisoft GS-11

Sodium cocoyl isethionate [CAS #61789-32-0]
Igepon AC-78, AC-941; Jordapon CI, CI Paste; Tauranol I-78, I-78 Flakes, I-78-3

Sodium cocoyl sarcosinate [CAS #61791-59-1]
Closyl 30; Hamposyl C-30

Sodium cumenesulfonate [CAS #32073-22-6]
Na-Cumene Sulfonate 40, Powd.; Naxonate 45SC, SC; Reworyl NCS 40; Stepanate C-S; Ultra SCS Liquid; Witconate SCS

Sodium cyclohexyl palmitoyl taurate
Igepon CN42

Sodium deceth sulfate
Synonyms: Sodium decyl ether sulfate
Witcolate 7093

Sodium deceth-3 sulfosuccinate
Alkasurf SS-DA-3

Sodium deceth-6 sulfosuccinate
Alkasurf SS-DA-6

Sodium decyl diphenyl ether sulfonate
Sandoz Sulfonate 3B2

Sodium decyl diphenyloxide disulfonate
Conco Sulfate 3B2; Dowfax 3B2

Sodium decyl sulfate [CAS #142-87-0]
Avirol SA-4110; Empimin SDS; Merpinal D40; Polystep B-25; Sandoz Sulfate 1030; Sulfotex 110, WAQ-LCX; Texapon 1030, 1090

Sodium diamyl sulfosuccinate. See *Diamyl sodium sulfosuccinate*

Sodium dibutyl naphthalene sulfonate
Geropon NK

Sodium dicyclohexyl sulfosuccinate. See *Dicyclohexyl sodium sulfosuccinate*

Sodium didodecylbenzene sulfonate
Petronate S

Sodium dihexyl sulfosuccinate. See *Dihyexyl sodium sulfosuccinate*

Sodium diisobutyl sulfosuccinate. See *Diisobutyl sodium sulfosuccinate*

Sodium diisooctyl sulfosuccinate
Condanol SBDO70; Geropon CYA/DEP; Steinapol SB-DO 70

Sodium diisopropyl naphthalene sulfonate
Geropon IN; Morwet IP

Sodium dimethylamyl sulfosuccinate
Manoxol MA

Sodium dinaphthyl methane sulfonate
Geropon RM/77

Sodium dinonyl sulfosuccinate
Manoxol N

Sodium dioctyl sulfosuccinate. See *Dioctyl sodium sulfosuccinate*

Sodium diphenyl methane sulfonate
Geropon FMS

Sodium ditridecyl sulfosuccinate. See *Ditridecyl sodium sulfosuccinate*

Sodium dodecylbenzenesulfonate [CAS #25155-30-0]

Synonyms: Sodium lauryl benzene sulfonate

Arylan SBC 25, SC 30, SX Flake; Bio Soft D-35X, D-40, D-60, D-62, LD-95; Calsoft F-90, L-40, L-60; Conco AAS-35, AAS-40F, AAS-40G, AAS-40S, AAS-45S, AAS-65S; Conco AAS-90S; Cycloryl DDB-40; DeSonate 50-S, 60-S; Elfan WA, WA Powder; Emulsifier 30; Hetsulf 40, 40X, 60S; Kadif 50 Flakes; Manro BA 25, 30, DL 32, SDBS 30; Maranil Paste A 55, A 75, Powder A; Marlon A 350, 360, 365, 375, 390, 396, ARM 40, 40N, 43, 50N, AFR, AM 80, ARL; Mercol 15, 25, 30, Special; Merpisap AE 50, 60, 70, AP 80W, 85W, 90, 90P, KH 30, KO 30; Nacconol 35SL, 40F, 40G, 90F, 90G; Nansa 1106/P, 1169/P, HS40-AU, HS40 Soft, HS80P, HS80/S, HS80SK, HS80 Soft, HS85/S, MA 30, SB 62, SL 30, SS 30, SS 60 Paste; Nansol; Norfox 40, 85, 90; Polystep A-4, -6, -7, -15, -16; Puxol FB-11; Reworyl NKS 50, 100; Richonate 40B, 45B, 45BX, 50BD, 60B, 1850, 1850U, C-50H; Sandet 60; Sandoz Sulfonate AAS 35S, AAS-40FS, AAS 45S, AAS 70S, AAS 90; Serdet DMK 75; Siponate DS-10; Soitem 5 C/70; Sovatex DS/C5; Sterling AB-40, AB-80; Sterling LA Paste 55%, 60%; Sul-fon-ate AA-9, AA-10, LA-10; Tairygent CB-1, CB-2; Ufaryl DB80; Vista C-550, C-560; Witconate 45 Liq., 45BX, 60B, 1250 Slurry, 1260 Slurry, C50H, LX Flakes

Sodium dodecyl diphenyl ether disulfonate

Eleminol MON-7; Pelex SS-H

Sodium dodecyl diphenyl ether sulfonate

Sandoz Sulfonate 2A1

Sodium dodecyl diphenyloxide disulfonate

Conco Sulfate 2A1; Dowfax 2A1, 2EP; Dowfax XDS 30599; Pelex SS-L

Sodium glyceryl oleate phosphate [RD #977067-44-9]

Emphos D70-30C

Sodium heptadecyl sulfate

Niaproof 7; Rexowet 77

Sodium hexadecyl diphenyloxide disulfonate

Dowfax XDS 8390.00

Sodium hexyl diphenyloxide disulfonate

Dowfax XDS 8292.00

Sodium hydrogenated tallow dimethyl glycinate

Monateric 1203

Sodium hydrogenated tallow glutamate [RD #977067-45-0]

Synonyms: Sodium hydrogenated tallowyl glutamate

Acylglutamate GS-11, HS-11; Amisoft HS-11

Sodium 2-hydroxyethyl sulfate

Rewopol NI 56

Sodium isethionate [CAS #1562-00-1]

Synonyms: Sodium 2-hydroxyethanesulfonic acid

Emery 5440

Sodium isodecyl sulfate

Sipex CAV

Sodium isooctyl sulfate

Manro SIOS

Sodium isopropyl naphthalene sulfonate
Aerosol OS

Sodium isostearoyl lactylate [CAS #66988-04-3]
Pationic ISL

Sodium laneth-5 sulfosuccinate
Cyclopol SB-5

Sodium lauramido DEA-sulfosuccinate
Alkasurf SS-L7DE, SS-L9DE

Sodium lauramido MEA-sulfosuccinate
Condanol SBL; Marlinat HA 12

Sodium lauraminopropionate [CAS #3546-96-1]
Synonyms: N-dodecyl-β-alanine, monosodium salt
Carsonol BDM; Maprolyte LX; Mirataine HC-HA, XL; Standapol AA-1

Sodium laureth-13 carboxylate
Miranate LEC

Sodium laureth MIPA-sulfosuccinate
Condanol SBIL/3, SBZ

Sodium laureth/myreth sulfate
Synonyms: Sodium lauryl/myristyl ether sulfate
Texapon K 14 S Spec.

Sodium laureth phosphate
Forlanit P; Forlanon

Sodium laureth sulfate [CAS #1335-72-4; 3088-31-1; 9004-82-4 (generic); 13150-00-0; 15826-16-1]
Synonyms: Sodium lauryl ether sulfate (*n*=1–4)
Agent 804; Akyposal 23 ST70, DS 28, EO 20; Alkasurf ES-60; Alscoap TAP-30; Berol 452; Bio Soft LD-95; Calfoam ES-30, SEL-60; Carsofoam SC; Carsonol SES-S, SLES, SLES-2; Cedepal SS-203, SS-306, SS-403, SS-406; Conco Sulfate 219, 219 Special, WE; Condanol NL2, NL3, NL3 28%; Cycloryl M1, N 70, NA, NA-2, NA 61, NA 61 CG; DeSonol SE, SE-2; Drewfax 218; Drewpon EBO, EG, ELB, ETH, EVE; Elfan NS 242, NS 243 S, NS 243 S Conc., NS 252 S, NS 252 S Conc., NS 423 SH, NS 423 SH Conc.; Emersal 6453; Empicol ESB, ESB3, ESB3GA, ESB 3/S, ESB30, ESB50, ESB70, ESB70-AU, ESC3, ESC70, ESC70-AU ; Empicol ESC-AU; Empigen XDR121, XDR123, XDR302; Empimin KSN27, KSN60, KSN70; Euperlan PK 771, PK 776, PK-810 ; Genapol LRO Liq., Paste; Genapol PGM Conc., ZRO Brands; Incrochem SC1, SC4; Incropearl Liquid; Lakeway 201-10, 201-11; Manro NEC, NEC 70; Maprofix ES, ES1, ES2, ES3, ESY, LES-60C; Mars ES-60; Merpoxal LM 3028.LM 3070; Miranol 2MCAS Modified; Montosol PF-10, PF-14, PF-26, PG-10, PL-16, PQ-17; Neopon LOS 70; Norfox SLES-02, SLES-03; Nutrapon ES 2, ES 60, ESY; Polyfac LES-60S, SD-60S; Polystep B-10, -12, -16; Rewopol NL 3 28%, NL 3 S-70; Rewoteric AM-CA; Richonol OME, S-1285C, S-5260; Sactol 2 OS 2, 2 OS 28; Sandoz Sulfate 219, 219 Special, ES-3, WE; Schercoteric LS-2ES MOD, MS-2ES MOD; Serdet DCK 3/70, DCK 30; Sipon 201-10, ES, ES-2, ESY; Sipoteric 1442, GSC; Standapol ES-1, ES-2, ES-3, ES-7099, Pearl Conc. 7130, S, Sl-60; Steinapol NL 3 28%; Steol 4N, CS-125, CS-130, CS-225, CS-230, CS-330, CS-460; Sterling Super 1E, 2E-25%, 2E-60%, 3E, 3.5E, A; Sulfatol B7, B10, E3, KE3; Sulfotex LMS-E; Swascol L-327; Syntol N77; Texapon ASV, ES-1, -2, -3, -40, EVR, N 25, N 40, N 70, N 70 LS, N 70 N, N 103, NA, N Conc., NSO, SBN, SG; Tylorol 2S, 2S/25, 3S,

BS, S; Ungerol CG27, LES 2-70, LES 3-28, LES 3-54, LES 3-70; Witcolate OME, S1285C, SE-5; Zetesol NL; Zoharpon ETA 27, 35, 70, 703

Sodium laureth-7 sulfate [CAS #9004-82-4 (generic); 66197-75-9; RD #977067-48-3]
Synonyms: Sodium POE (7) lauryl ether sulfate
Sipon ES-7; Steol CS-760

Sodium laureth-8 sulfate
Texapon ASV

Sodium laureth-12 sulfate [CAS #9004-82-4 (generic); 66161-57-7; RD #977069-66-1]
Montosol IF-21; Polystep B-21, -23; Sipon ES-12; Standapol 125-E Conc., 130-E

Sodium laureth-30 sulfate
Polystep B-19

Sodium laureth sulfosuccinate
Condanol SBA/1, SBFA/3, SBFA 30 40%; Secosol ALL40; Setacin 103 Spezial; Sulfatol 88.22

Sodium lauriminodipropionate [CAS #14960-06-6; 26256-79-1]
Synonyms: Sodium N-lauryl-β-iminodipropionate
Deriphat 160C; Mirataine H2C-HA; Sipoteric SLIP; Velvetex 610L

Sodium lauroyl glutamate [CAS #29923-31-7 (L-form); 29923-34-0 (DL-form); 42926-22-7 (L-form)
Acylglutamate LS-11; Amisoft LS11

Sodium lauroyl lactylate [CAS #13557-75-0; RD #977067-51-8]
Pationic 138C

Sodium lauroyl sarcosinate [CAS #137-16-6]
Synonyms: N-methyl-N-(1-oxod-decyl) glycine, sodium salt
Crodasinic LS30, LS35; Hamposyl L-30, L-95; Maprofix RS; Maprosyl 30; Sarkosyl NL-30; Secosyl

Sodium lauroyl taurate [CAS #70609-66-4]
Synonyms: Sodium 2-[(1-oxododecyl)amino] ethanesulfonate
Hostapon KTW New

Sodium lauryl/oleyl sulfate
Duponol D Paste

Sodium lauryl sulfate [CAS #151-21-3]
Akyposal 100 DAL, NLS; Alcolec 638; Alkasurf WAQ; Alscoap LN-40, LN-90; Arsul WAQ; Avirol SL-2010; Berol 474; Calfoam SLS-30; Carsonol SLS, SLS Paste B, SLS Special; Cerasynt WM; Conco Sulfate WA, WAG, WAS, WA Special, WB-45, WBS-45, WR, WX; Condanol NLS 30, NLS 90; Cosmopon ESC, LE 50; Crodex A; Cyclochem EM 560; Cycloryl 21, 21 LS, 21 SP, TS30; Dehydag Wax SX, W; DeSonol S; Drewpon 100, AS, CN; Duponol C, ME Dry, QC, WA Dry, WA Paste, WAQ, WAQE; Elfan 200, 240, 260 S; Emal 0, 10; Emersal 6400; Empicol 0045, 0185, 0919, LM, LM45, LM/T, LMV, LMV/T, LS30, LS30B, LS30P, LX, LX28; Empicol LXS95, LXV, LY28/S, LZ, LZ34, LZ/D, LZ/E, LZG 30, LZGV, LZGV/C, LZP, LZV, LZV/E; Empicol WA, WAK; Empimin LR28; Empiwax SK, SK/BP; Equex S, SP, STM, SW; Gardinol WA Paste; Hartenol LAS-30, LES-60; Incronol SLS; Jordanol SL 300; Lakeway 101-10, 101-10P,N, 101-11, 101-19; Lanette SX, W; Lexemul AS; Lipowax; Lonzol LS-300; Manro DL 28, SLS 38, 28S; Maprofix 563, 563-SD, DFS, EP-1, LCP, LK (USP), MM, RS, WAC, WAC-LA, WAM, WA Paste; Mars SLS-30, SLS-A95;

Merpinal LM 40, LM 50, LS 40, LS 50; Merpoxal LM 3030; Michelene LS-90; Miranol 2MCA-ESF, 2MCA Modified, 2MCAS Modified; Monateric CDL, CDS; Montovol RF-10; Neopon LS; Nikkol SLS; Norfox SLS; Nutrapon B, RD 1147, W, WAC, WAQ; Polyfac PCW-160, SLS-30, SLS-30 Special; Polystep B-3, -5, -24; Rewopol 15/L, NLS 28, NLS 30/L, NLS 90 Powd.; Rewoteric AM-G 30; Richonol 2310, 6522, A, A Powd., C; Sactol 2 S 3; Sandoz Sulfate WA-9, WA Dry, WAG, WAS, WA Special; Schercoteric MS-2 MOD; Serdet DFK 40; Sermul EA 150; Sipex LCP, SB, SD, UB; Sipon LS, LSB, WD; Standapol CS-50, CS Paste, S, SHC-101, WA-AC, WAQ-LC, WAQ-LCX, WAQ Special, WAS-100; Steinapol NLS 28, NLS 90-Powder; Stepanol ME Dry, SLS-30, WA-100, WAC, WA Extra, WA Paste, WAQ, WA Special; Stepfac SLS-30 Special; Sterling WA Paste, WA Powder, WAQ-CH, WAQ-Cosmetic, WAQ-LC, WAQ-P; Sulfatol 33 Paste; Sulfopon 101, 101 Special, 102, 103, K35, WA-3, WA 30, WAQ LCX, WAQ Special; Sulfotex LCX, WA, WAQ-LCX; Sulphonated Lorol Paste; Swascol 1P, 1PC, 3L; Texapon EVR, K-12, K-12-94, K-12-96, K 12 Granules, K 12 Needles, L-100, LS Highly Conc. Needles, OT Highly Conc. Needles, V Highly Conc. Needles, Z, Z Granules, Z Highly Conc. Needles, Z Highly Conc. Powder; Ultra Sulfate SL-1; Ungerol LS, LSN; Witcolate A, A Powd.; Zoharpon LAS, LAS Special, LAS Spray Dried

Sodium lauryl sulfoacetate [CAS #1847-58-1]

Lathanol LAL Flake, Powd.; Nikkol LSA

Sodium lauryl sulfosuccinate

Secosol AL/959

Sodium lignosulfonate [CAS #8061-51-6]

Synonyms: Sodium polignate

Dyqex; Irgasol N; Kelig 100; Lignosite 431, 458, 823, 889, GPM-38; Lignosol AXD, D-10, D-30, DXD, FTA, HCX, NSX 120, SFX, X, XD, XD-65; Marasperse CB, N-22; Norlig; Orzan LS, LS-50, S, SL-50; Polyfon F, H, O, T; Raylig; Reax 15B, 45A, 80A, 80C, 81A, 82, 83A, 85A, 88B, 90P, 95A, SR-1, SR-7

Sodium magnesium laureth sulfonate

Synonyms: Sodium magnesium lauryl ethoxy sulfonate

Empicol BSD

Sodium maleic anhydride/diisobutylene copolymer

Empicryl APD90, APD B90

Sodium methallyl sulfonate

Geropon MLS/A

Sodium methyl cocoyl taurate [CAS #12765-39-8; 61791-42-2]

Synonyms: Sodium cocoyl methyl taurate

Adinol CT95; Hostapon CT Paste; Igepon TC-42; Nikkol CMT-30; Tauranol WS, WS Conc., WS H.P., WSP

Sodium methyl lauroyl alaninate

Nikkol Alaninate LN-30

Sodium methyl lauroyl taurate

Synonyms: Sodium lauroyl methyl taurate

Nikkol LMT

Sodium methyl myristoyl taurate

Synonyms: Sodium myristoyl methyl taurate

Nikkol MMT

Sodium methyl oleoyl taurate [CAS #137-20-2]
Synonyms: Sodium N-oleoyl-N-methyl taurate; Oleyl methyl tauride
Adinol T, T35; Consamine K-Gel; Hostapon T; Igepon T-33, T-43, T-51, T-77, TM-43; Nissan Diapon TO; Tauranol M, M-35, ML, MS, T-Gel; Tergenol G, S Liquid, Slurry

Sodium methyl palmitoyl taurate
Synonyms: Sodium palmitoyl methyl taurate
Igepon TN-74; Nikkol PMT

Sodium methyl stearoyl taurate
Synonyms: Sodium stearoyl methyl taurate
Nikkol SMT

Sodium methyl tall oil acid taurate
Igepon TK-32

Sodium methyl tallow taurate
Nissan Diapon T

Sodium myreth sulfate [CAS #25446-80-4; RD #977059-34-9]
Synonyms: Sodium myristyl ether sulfate
Cycloryl ME 60, ME 361; Standapol ES-40 Conc.

Sodium myristoyl glutamate [CAS #38517-37-2; 71368-20-2]
Acylglutamate MS-11

Sodium myristoyl sarcosinate [CAS #30364-51-3]
Hamposyl M-30

Sodium myristyl sulfate [CAS #1191-50-0]
Synonyms: Sodium tetradecyl sulfate
Cycloryl MS; Niaproof 4

Sodium naphthalene-formaldehyde sulfonate
Ablusol NL; Anchoid; Arylan SNS; Blancol, N; Petro Dispersant 425; Stepantan A

Sodium naphthalene sulfonate
Lomar DL, LS, PL, PW; Nopcosant; Protodye QNF

Sodium *m*-nitrobenzenesulfonate [CAS #127-68-4]
Reserve Salt Flake

Sodium nonoxynol-6 phosphate [CAS #12068-19-8]
Synonyms: Sodium POE (6) nonyl phenyl ether phosphate
Gafac LO-529

Sodium nonoxynol-9 phosphate [RD #977058-21-1]
Emphos CS-1361

Sodium nonoxynol-4 sulfate [CAS #9014-90-8 (generic); RD #977069-67-2]
Alipal CO-433; Serdet DNK 30; Sermul EA 54

Sodium nonoxynol-10 sulfate
Sermul EA 151

Sodium nonoxynol-15 sulfate
Sermul EA 146

Sodium nonoxynol-25 sulfate
Montosol PB-25

Sodium nonoxynol-10 sulfosuccinate
Sermul EA 176

Sodium nonyl benzene sulfonate
Emkal NOBS

Sodium octoxynol-2 ethane sulfonate [CAS #2917-94-4]
Siponate X200; Triton X-200

Sodium N-octyl-beta-iminodipropionic acid
Deriphat 130-C

Sodium octyl/decyl sulfate
Duponol WN

Sodium octylphenoxyethoxyethyl sulfonate
Newcol 861SE

Sodium octyl sulfate [CAS #126-92-1]
Synonyms: Sodium 2-ethylhexyl sulfate
Avirol SA-4106; Carsonol SHS; Chemsulf 2EH-Na; Chimin DOS; Duponol 80; Emersal 6465; Merpinal EH 40, Q 147; Niaproof 08; Norfox SEHS; Polystep B-29; Sandoz Sulfate 830, W2-30; Serdet DSK 40; Sipex BOS, OLS; Sole-Terge TS-2-S; Standapol LF; Sulfotex OA; Texapon 842, 890; Witcolate D-510

Sodium oleate [CAS #143-19-1]
Synonyms: Sodium 9-octadecenoate
Norfox Vertex Flakes

Sodium oleic acid sulfate
Dyasulf 2031

Sodium oleic acid sulfonate
Sul-fon-ate OA-5, OA-5R

Sodium oleic acid sulfonate, amyl ester
Sul-fon-ate OE-500

Sodium N-oleoyl sarcosinate
Crodasinic OS35

Sodium oleth sulfate
Texapon ASV

Sodium oleyl/cetyl sulfate
Elfan 680

Sodium oleyl sulfate
Duponol LS Paste; Sipex OS

Sodium oleyl sulfosuccinamate
Cosmopon BN

Sodium PCA [CAS #28874-51-3; 54571-67-4]
Synonyms: PCA Soda; 5-oxo-DL-proline, sodium salt; Sodium pyroglutamate; Sodium DL-2-pyrrolidone-5-carboxylate
Ritaderm

Sodium poly-L-glutamate
Ajicoat SPG

Sodium propionate [CAS #137-40-6]
Tandem 11H

Sodium sarcosinate [CAS #4316-73-8]
Synonyms: Sodium N-methylglycine
Nutrapon RS 1147

Sodium soya hydrolyzed animal protein [CAS #68188-31-8]
Maypon K

Sodium stearate [CAS #822-16-2]
Monomuls 60-25/2, 60-25/5, 90-25/2, 90-25/5; Norfox B

Sodium stearoyl lactylate [CAS #25383-99-7]
Synonyms: Sodium stearyl-2-lactylate
Admul SSL-2003, SSL-2004; Artodan ; CW-6; Emplex; Grindtek FAL 1; Lamegin NSL; Patco 3, 145A, SSL; Radiamuls SSL 2990; Swedex SSL

Sodium stearyl sulfate
Merpinal Q 101

Sodium stearyl sulfosuccinamate
Synonyms: Sodium octadecyl sulfosuccinamate
Alkasurf SS-TA 125; Astromid 18, 18 LV

Sodium styrene maleic anhydride
Scripset 500

Sodium 2-sulfoethyl myristate
Igepon AM-78

Sodium tallate
Synonyms: Tall oil rosin sodium salt
Arizona DRS-43, DRS-44; Dresinate TX

Sodium tallowamide MEA-sulfosuccinate
Elfanol 510

Sodium tallow coco sulfate
Sulfetal TC 50

Sodium tallow sulfate [CAS #8052-50-4]
Empicol TAS30, TAS80V, TAS90; Merpinal T 35; Montovol GJ-12

Sodium/TEA-lauroyl hydrolyzed animal protein
Lipoproteol LCO, LK

Sodium/TEA-undecylenoyl hydrolyzed animal protein
Lipoproteol UCO

Sodium tetraborate decahydrate
Borax

Sodium tetrahydronaphthalene sulfonate
Alkanol S

Sodium tetrapropylene benzene sulfonate
Merpisap DE 70, DP 82

Sodium toluenesulfonate [CAS #12068-03-0]
Synonyms: Methylbenzenesulfonic acid, sodium salt
Eltesol ST 34, ST 40, ST 90, ST Pellets; Hartotrope STS 40, Powder; Manro STS 40; Marlon ARL; Naxonate 4ST, ST; Reworyl NTS 40; Stepanate T; Witconate STS

Sodium toluene xylene sulfonate
STXS

Sodium trideceth-7 carboxylate [RD #977067-36-9]
Rewopol CTN; Sandopan DTC

Sodium trideceth sulfate [CAS #25446-78-0 (*n*=3); 66161-58-8 (*n*=4)]
Synonyms: Sodium tridecyl ether sulfate; Sodium POE tridecyl sulfate
Cedepal TD-404, TD-407; Cycloryl ALC; Mackam 2CT, LT, MLT; Merpoxal TR 4060; Miranol 2MCT, 2MCT Modified, 2MHT, 2MHT Modified, BT, MHT; Mirataine CBL; Monateric CDTD; Schercoteric LS-2TE, LS-2TE MOD, MS-2TE MOD; Sipex EST-30, EST-60, EST-75; Sipoteric 1304; Zoharteric M-2, M-3, M-3 50%

Sodium tridecylbenzene sulfonate [CAS #26248-24-8]
Richonate TDB

Sodium tridecyl sulfate [CAS #3026-63-9]
Avirol SA-4113; Sipex TDS

Sodium undecylenamide-MEA sulfosuccinate
Loramine SBU 185; Rewocid SBU 185; Steinazid SBU 185

Sodium xylenesulfonate [CAS #1300-72-7]
Alkatrope SX-40; Carsosulf SXS-Liq.; Conco SXS; Eltesol SX 30, SX 93, SX Pellets; Hartotrope SXS 40; Lakeway SXS; Lankrosol SXS-30; Manro SXS 30, 40; Mars SXS; Naxonate 4L, 5L, G; Pilot SXS-40, SXS-96; Polystep A-2; Reworyl NXS 40; Richonate SXS; Stepanate X; Sterling 2XS; Surco SXS; SXS; Ultra SXS Liquid, Powder; Witconate 45LX Liq., 1260 Slurry, SXS

Sorbeth-20 [RD #977058-26-6]
Synonyms: PEG-20 sorbitol ether; POE (20) sorbitol ether; PEG 1000 sorbitol ether
Ethosperse SL-20; Polyol 20; Trylox SS-20

Sorbeth-55
Synonyms: PEG-55 sorbitol ether
Sorbeth 55

Sorbitan caprylate
Nissan Nonion CP-08R

Sorbitan dioleate [CAS #29116-98-1]
Synonyms: Sorbide dioleate; Sorbitan, di-9-octadecenoate; Anhydrosorbitol dioleate
Atlas G-950; DO-33-F

Sorbitan distearate

Emasol S-20

Sorbitan isostearate [CAS #1338-39-2; 5959-89-7]

Synonyms: Anhydrosorbitol monoisostearate

Anfomul S6; Arlacel 987; Crill 6; Montane 60; Nikkol SI-10R, SI-10T, SI-15R, SI-15T; S-Maz 67

Sorbitan laurate [CAS #1338-39-2; 5959-89-7]

Synonyms: Sorbitan monolaurate; Anhydrosorbitol monolaurate; Sorbitan monododecanoate

Ablunol S-20; Ahco 759; Alcolec 4135; Alkamuls SML; Arlacel 20; Armotan ML; Crill 1; Dehymuls SML; Drewmulse SML; Durtan 20; Emasol L-10; Emsorb 2515; Ethylan GL-20; Glycomul L, LC; Ionet S-20; Kuplur SML; Liposorb L; Lonzest SML; ML-33F; Montane 20; Newcol 20; Nikkol SL-10; Nissan Nonion LP-20•R, LP-20•RS; Radiamuls 125, SORB 2125; Radiasurf 7125; Rheodol SP-L10; Secoster KL 10; S-Maz 20; Soprofor S/20; Sorbax SML; Sorbon S-20; Sorgen 90; Span 20

Sorbitan myristate

Nissan Nonion MP-30R

Sorbitan oleate [CAS #1338-43-8; 5938-38-5]

Synonyms: Sorbitan monooleate; SMO; Anhydrosorbitol monooleate; Sorbitan mono-9-octadecenoate

Ablunol S-80; Ahco 832, 944; Alkamuls SMO; Anfomul S4, S50; Arlacel 80; Armotan MO; Atlas G-4884; Atpet 80; Capmul O; Crill 4, 50; Dehymuls SMO; DeSonic SMO; DeSotan SMO; Drewmulse SMO; Durtan 80; Emasol O-10; Emsorb 2500; Ethylan GO-80; Glycomul O; Ionet S-80; Kuplur SMO; Liposorb O; Lonzest SMO; MO-33-F; Montane 80; Newcol 3-80, 3-85, 80; Nikkol SO-10; Nissan Nonion OP-80•R; Norfox Sorbo-S-80; Radiamuls 155, SORB 2155; Radiasurf 7155; Rheodol AO-10, SP-010; S-Maz 80; Soprofor S/80; Sorbax SMO; Sorbon S-80; Sorgen 40, S-40-H; Span 80

Sorbitan palmitate [CAS #26266-57-9]

Synonyms: Sorbitan monopalmitate

Ahco FP-67; Arlacel 40; Armotan MP; Crill 2; Emasol P-10; Emsorb 2510; Ethylan GP-40; Glycomul P; Kuplur SMP; Liposorb P; Lonzest SMP; Montane 40; MP-33-F; Newcol 40; Nikkol SP-10; Nissan Nonion PP-40•R; Radiamuls 135, SORB 2135; Radiasurf 7135; Rheodol SP-P10; Secoster KP 10; S-Maz 40; Sorbax SMP; Sorbon S-40; Span 40

Sorbitan sesquioleate [CAS #8007-43-0]

Synonyms: Anhydrosorbitol sesquioleate; Anhydrohexitol sesquioleate

Alcolan, 36W, 40; Anfomul S43; Arlacel 83, C; Crill 43; Dehymuls E, K, SSO; Emasol O-15; Emsorb 2502; Glycomul SOC; Liposorb SQO; Montane 83; Nikkol SO-15; Nissan Nonion OP-83•RAT; QO-33-F; Rheodol AO-15; S-Maz 83R; Sorgen 30, S-30-H

Sorbitan sesquistearate [RD #977067-60-9]

Synonyms: Anhydrosorbitol sesquistearate

Nikkol SS-15

Sorbitan stearate [CAS #1338-41-6]

Synonyms: Sorbitan monostearate; SMS; Anhydrosorbitol monostearate; Sorbitan monooctadecanoate

Ablunol S-60; AF-72, AF-75; Ahco 909; Alkamuls SMS; Arlacel 60; Armotan MS; Atmos 729, 758, 2462; Capmul S; Crill 3; Dehymuls SMS; Drewmulse SMS; Drewsorb 60; Durtan 60, 60K; Emasol S-10; Emrite 6105; Emsorb 2505; Ethylan GS-60; Glycomul S; Grindtek SMS; Ionet S-60 C; Kuplur SMS; Liposorb S; Lonzest SMS; Montane 70; MS-33-F; Newcol 60; Nikkol SS-10; Nissan Nonion SP-60•R; Norfox Sorbo S-60FG; Radiamuls 145, SORB

2145; Radiasurf 7145; Rheodol AS-10, SP-S10; Secoster KS 10; S-Maz 60; Soprofor S/60; Sorbax SMS; Sorbon S-60; Sorgen 50; Span 60; Vanade; Vanall

Sorbitan tallate
DeSonic SMT; DeSotan SMT; S-Maz 90; Witcomul 78

Sorbitan trioleate [CAS #26266-58-0]
Synonyms: Anhydrosorbitol trioleate; Sorbitan tri-9-octadecenoate
Ahco FO-18; Alkamuls STO; Arlacel 85; Atlas G-4885; Atlox 4885; Crill 45; Emasol O-30; Emsorb 2503; Ethylan GT-85; Glycomul TO; Ionet S-85; Liposorb TO; Lonzest STO; Montane 85; Nikkol SO-30; Nissan Nonion OP-85•R; Radiamuls SORB 2355; Rheodol SP-030; S-Maz 85; Soprofor S/85; Sorbax STO; Span 85; TO-33-F

Sorbitan tristearate [CAS #26658-19-5]
Synonyms: Anhydrosorbitol tristearate; Sorbitan trioctadecanoate
Ahco FS-21; Alkamuls STS; Crill 35; Drewmulse STS; Emasol S-30; Emsorb 2507; Glycomul TS; Grindtek STS; Kuplur STS; Liposorb TS; Lonzest STS; Montane 65; Nikkol SS-30; Radiamuls 345, SORB 2345; Radiasurf 7345; Rheodol SP-S30; S-Maz 65; Soprofor S/65; Sorbax STS; Span 65; TS-33-F

Sorbitan tritallate
S-Maz 95

Sorbitol [CAS #50-70-4]
Synonyms: D-Glucitol
Lexein S620S; Sorbitol; Super Pro 100

Soy acid [CAS #68308-53-2]
Industrene 225, 225 FG, 226, 226 FG

Soyaethyldimonium ethosulfate
Synonyms: Soya dimethyl ethyl ammonium ethosulfate
Jordaquat 1033; Larostat 264-A, 264-A Anhyd.

Soyaethyl morpholinium ethosulfate [CAS #61791-34-2]
Synonyms: N-Soya-N-ethyl morpholinium ethosulfate; Quaterium-2
Atlas G-271

Soya hydroxyethyl imidazoline
Synonyms: Soya imidazoline
Alkaquat S; Miramine SC

Soyamide DEA [CAS #68425-47-8]
Synonyms: Soya diethanolamide; N,N-bis(hydroxyethyl) soya amides
Empigen 2125-AU; Empilan 2125-AU; Mackamide S; Mazamide SS-10

Soyamidopropyl benzyldimonium chloride
Schercoquat SOAB

Soyamidopropyl ethyldimonium ethosulfate
Schercoquat SOAS

Soyamine [CAS #61790-18-9]
Synonyms: Soya primary amine
Armeen SD; Kemamine P-997, P-997D; Lilamin 115, 115 D; Nissan Amine SB

Soya oil glyceride
Synonyms: Glyceryl mono soya oil; Glyceryl monosoyate
Drewmulse 10; Radiamuls 602, MG 2602

Soybean oil glyceride
Dimodan PV, PV 300

Soy sterol [RD #977066-08-2]
Generol 122

Soytrimonium chloride [CAS #61790-41-8]
Synonyms: Soya trimethyl ammonium chloride; Quaternium-9
Adogen 415; Arquad S-50; Jet Quat S-50; Tomah Q-S

Spermaceti, synthetic. See *Cetyl esters*

Steapyrium chloride [CAS #1341-08-8; 14492-68-3; 42566-92-7]
Synonyms: Quaternium-7; N-stearoyl colamino formyl methyl pyridinium chloride
Emcol E-607S

Stearalkonium chloride [CAS #122-19-0]
Synonyms: Stearyl dimethyl benzyl ammonium chloride; Octadecyl dimethyl benzyl ammonium chloride
Ammonyx 4, 4B, 485, 490, 4002; Arquad DM18B-90; Carsoquat 816-C, SDQ-25, SDQ-85; Crapol AU-23; Dehyquart STC-25; Incroquat CR Conc., S-85, SDQ-25; Jordaquat JS-25; Nissan Cation S2-100; Polyquart H-7102; Standamul Conc. 1002, STC-25; Stedbac; Triton CG-400, -500, X-400; Varisoft 6112, SDC, SDC-65%, SDC-W

Stearamide [CAS #124-26-5]
Synonyms: Octadecanamide; Stearic acid amide
Armid 18; Armoslip 18; Kemamide S

Stearamide DEA [CAS #93-82-3]
Synonyms: Stearic acid diethanolamide; Stearoyl diethanolamide; N,N-bis(2-hydroxyethyl) octadecanamide
Alkamide HTDE; Aminol N-1918; Cetina; Clindrol 200S, Superamide 100S; Lipamide S; Monamid 718; Nopcogen 14-S; Onyxol 42; Unamide S, W

Stearamide DEA-distearate
Synonyms: Stearic diethanolamide distearate
Schercomid SD-DS

Stearamide DIBA-stearate [RD #977058-30-2]
Polytex 10

Stearamide MEA [CAS #111-57-9]
Synonyms: Stearic acid monoethanolamide; Stearoyl monoethanolamide; N-(2-hydroxyethyl) octadecanamide; Monoethanolamine stearic acid amide
Alkamide HTME; Ardet SEA; Clindrol 200MS, LT 14-8; Comperlan HS; Cyclomide S280; Incromide SM; Loramine S 280; Monamid S; Ninol SNR; Rewomid S 280; Schercomid SME, SME-A; Steinamid S 280; Witcamide 70

Stearamide MEA-stearate [CAS #14351-40-7]
Cerasynt D; Schercomid SME-S; Witcamide MAS

Stearamidoethyl diethylamine [CAS #16889-14-8]
Synonyms: Diethylaminoethyl stearamide
Chemical Base 6532; Lexamine 22; Lexemul AR

Stearamidopropalkonium chloride [CAS #65694-10-2]
Synonyms: N,N-dimethyl-N-[3-[(1-oxooctadecyl)amino] propyl]benzenemethanaminium chloride; Stearamidopropyl dimethyl benzyl ammonium chloride; Stearamidopropyl benzyldimonium chloride
Schercoquat SAB

Stearamidopropylamine oxide
Cyclomox SO

Stearamidopropyl diethylamine
Cyclomide SODI

Stearamidopropyl dimethylamine [CAS #7651-02-7]
Synonyms: Dimethylaminopropyl stearamide
Incrochem SBJ; Incromine SB; Jordamine DAPSA, S-13; Lexamine S-13; Lexate CRC; Miramine DD; Schercodine S; Tego-Amid S 18

Stearamidopropyl dimethylamine lactate [CAS #55819-53-9]
Emcol 3780; Incrochem PPI; Incromate PPI, SDL; Jordamine S-13 Lactate; Lexamine S-13 Lactate; Richamate 3780

Stearamidopropyl ethyldimonium ethosulfate
Synonyms: Stearamidopropyl dimethyl ethyl ammonium ethyl sulfate
Schercoquat SAS

Stearamine [CAS #124-30-1]
Synonyms: Stearyl amine; Octadecylamine; 1-Octadecanamine
Amine 18D; Armeen 18, 18D; Crodamine 1.18D; Kemamine P-990, P-990D; Lilamin 142, 142 D; Nissan Amine AB

Stearamine acetate
Synonyms: Stearyl amine acetate; Octadecylamine acetate
Armac 18D; Kemamine A990; Nissan Cation SA

Stearamine oxide [CAS #2571-88-2]
Synonyms: Stearyl dimethylamine oxide; Octadecyl dimethylamine oxide
Ammonyx SO; Aromox DM18D-W; Conco XA-S; Incromine Oxide S; Jordamox SDA; Rewominox S 300; Schercamox DMS;

Steareth-2 [CAS #9005-00-9 (generic); RD #977055-79-0]
Synonyms: PEG-2 stearyl ether; PEG 100 stearyl ether; POE (2) stearyl ether
Alkasurf SA-2; Brij 72; Hetoxol STA-2; Lipocol S-2; Macol SA-2; Simulsol 72; ST-55-2; Volpo S2

Steareth-5
Macol SA-5; Solulan 5

Steareth-10 [CAS #9005-00-9 (generic); RD #977055-80-3]
Alkasurf SA-10; Brij 76; Cosmowax; Hetoxol STA-10; Lipocol S-10; Macol SA-10; Ritapro 100; Simulsol 76; Volpo S10

Steareth-11 [CAS #9005-00-9 (generic)]
Empilan KM 11

Steareth-15 [CAS #9005-00-9 (generic)]
Macol SA-15

Steareth-16 [CAS #9005-00-9 (generic)]
Solulan 16

Steareth-20 [CAS #9005-00-9 (generic); RD #977053-41-0]
Alkasurf SA-20; Brij 78; Cosmowax; Hetoxol STA-20; Lipal 20 SA; Lipocol S-20; Macol SA-20; Ritachol 1000; Ritapro 100; Simulsol 78; ST-55-20; Trycol SAL-20; Volpo S20

Steareth-25 [CAS #9005-00-9 (generic)]
Solulan 25

Steareth-30 [CAS #9005-00-9 (generic); RD #977055-81-4]
Hetoxol STA-30; Lipal 30 SA

Steareth-40 [CAS #9005-900-9 (generic)]
Macol SA-40

Steareth-50 [CAS #9005-00-9 (generic)]
Empilan KM 50

Steareth-100 [CAS #9005-99-9 (generic)]
Brij 700

Stearic acid [CAS #57-11-4]
Synonyms: *n*-Octadecanoic acid
Hydrofol Acid 1655-CG-NF, 1855, 1895; Hystrene 5016, 7018, 8018, 9018, 9718; Industrene 5016, 5016 FG, 7018, 7018 FG, 8518, 8718 FG, 9018; Kessco Diglycol Stearate SE; Petrac 250, 270

Stearic diethanolamide. See *Stearamide DEA*

Stearic (ethyl amino hydroxyethyl) amide
Unamide SI

Stearic monoethanolamide. See *Stearamide MEA*

Stearic-oleic DEA
Emid 6543

Stearoamphoacetate
Synonyms: Stearoamphoglycinate
Amphoterge S; Miranol DM

Stearoamphodiacetate
Synonyms: Stearoamphocarboxyglycinate
Schercoteric STS; Sipoteric 1442, 1444, 1450, GPA

Stearoamphohydroxypropylsulfonate
Synonyms: Stearamphopropylsulfonate
Miranol DS

Stearone [CAS #504-53-0]
Synonyms: Diheptadecyl ketone
Amerchol 400, RC

Stearoyl lactylic acid
Grindtek FAL 3

Stearoyl sarcosine [CAS #142-48-3]
Synonyms: Stearoyl N-methylglycine; Stearoyl N-methylaminoacetic acid
Hamposyl S

Steartrimonium chloride [CAS #112-03-8]
Synonyms: Stearyl trimethyl ammonium chloride; Octadecyl trimethyl ammonium chloride
Arquad 18-50; Nissan Cation AB

Stearyl alcohol [CAS #112-92-5]
Synonyms: *n*-Octadecanol
Adol 61 NF, 62 NF, 63, 64; Alfol 18; Amerchol RC; Cerasynt WM; Cosmowax, K; Cremophor A 6; Cyclogol Stearyl Alcohol; Cycloton CT 100; Emcol 150; Kalcohl 80; Lanette 18; Lanol S; Promulgen G; Stearal; Varisoft 3262

Stearyl amine. See *Stearamine*

Stearyl betaine [CAS #820-66-6]
Synonyms: Stearyl dimethyl glycine
Alkateric STB; Chimin BX; Varion SDG

Stearyl diethanol methyl ammonium chloride
M-Quat 32

Stearyl dihydroxyethyl methyl ammonium chloride
Tomah Q-18-2

Stearyl dimethyl amine. See *Dimethyl stearamine*

Stearyl dimethyl benzyl ammonium chloride. See *Stearalkonium chloride*

Stearyl erucamide [CAS #10094-45-8]
Synonyms: N-Octadecyl-13-docosenamide
Kemamide E-180

Stearyl hydroxyethyl imidazoline [CAS #95-19-2]
Synonyms: Stearyl imidazoline
Alkazine ST; Amine S; Atlas Softener AD; Crodazoline S; Imidrol SC; Miramine GS; Schercozoline S

Stearyl nitrile
Arneel 18 D

Stearyl PEG-15 methyl ammonium chloride
Tomah Q-18-15

Stearyl stearamide
Kemamide S-180

Stearyl stearate [CAS #2778-96-3]
Lexate TA, TL

Styrene maleic anhydride amide/ammonium hydroxide acid salt
Scripset 808

Styrene/maleic anhydride copolymer [CAS #9011-13-6]
Scripset 520, 540, 550, 700, 720

Sucrose cocoate
Crodesta SL-40; Grilloten LSE87K

Sucrose distearate [CAS #27195-16-0]
Crodesta F-10, F-50, F-110; Ryoto Sugar Ester S-570, -770; Sucro Ester W.E. 7

Sucrose di/tristearate
Ryoto Sugar Ester S-170, -270, -370

Sucrose laurate [CAS #25339-99-5]
Grilloten LSE87; Ryoto Sugar Ester LWA-1540

Sucrose mono/distearate
Ryoto Sugar Ester S-970, -1170; Sucro Ester W.E. 11

Sucrose oleate
Ryoto Sugar Ester OWA-1570

Sucrose palmitate
Ryoto Sugar Ester P-1570, -1670; Sucro Ester W.E. 15

Sucrose ricinoleate
Grilloten ZT12, ZT40, ZT80

Sucrose stearate [CAS #25168-73-4]
Crodesta F-110, F-160; Grilloten PSE141G; Ryoto Sugar Ester S-1570, -1670

Sulfated butyl oleate
Chemax SBO; Hipochem Dispersol SB; Maravanol 60% SBO

Sulfated castor oil [CAS #8002-33-3]
Synonyms: Castor oil sulfated; Sulfonated castor oil; Turkey-red oil
Actrasol 167A, C50, C75, C85, PSR; Ahco AJ-110; Alcolec HS-3; Castor Oil Sulfated 50%, 68%; Chemax SCO; Consos Castor Oil; Cordon NU 890/75; Densol 6920; Eureka 102, 102-WK; Haroil SCO-65, SCO-7525; Hartex V63, V64; Hipochem Dispersol SCO; Marvanol 75% SCO; Monopol Oil 75; Monosulph; Nopcocastor, L; Nopcosulf CA-60, -70; Seachem SC; Standapol SCO; Sulfonated Castor Oil 50%, 75%, GTO; Sulfotex 130; Supratol VF; Sykanol DKM 45, 80; Texapon SCO; Titan Castor No. 75

Sulfated glyceryl trioleate
Actrasol EO; Hipochem Dispersol GTO

Sulfated neatsfoot oil
Atlas Sul L-2; Crestoil; Sulfonated Neatsfoot 3089

Sulfated oleic acid
Actrasol SR75, SRK 75; Densol P-82; Sulfonic 800

Sulfated propyl oleate
Cinwet BOS; Hipochem Dispersol SP; Marvanol 55% SPO

Sulfated soyabean oil
Actrasol CS75, MY75, OY75

Sulfated sperm oil
Densol 6930; Dymsol S; Eureka 400; Sulfonated Syn. Sperm Oil 2091

Sulfated tall oil
Actrasol SP175K; Cordon PB-870

Sulfated tallow
Densulf TA-75; Nopcosulf TA-30, -45V

Sulfated TOFA
Actrasol SP

Sulfonated isopropyl oleate
Rexowet CR

Sulfonated oleyl alcohol
Hipochem LCA

Sunflower seed oil glyceride
Dimodan LS; Monomuls 90-40

Sunflower seed oil glycerides
Synonyms: Sunflower seed mono-, di- and tri-glycerides
Monomuls 60-40

Tallamide DEA [CAS #68155-20-4]
Synonyms: Diethanolamine tall oil acid amide; Tall oil acid diethanolamide
Monamine T-100; Schercomid TO-2

Tallamidopropyl dimethylamine
Schercodine T; Servamine KET 350

Tallamphopropionate [CAS #68991-88-8]
Miranol LM-SF Conc.; Monateric TA-35

Tall oil acid [CAS #61790-12-3]
Monomer C18

Tall oil amine
Lilamin 151

Tall oil aminoethyl imidazoline
Alkazine TO-A

Tall oil diethanolamide. See *Tallamide DEA*

Tall oil dioleate
Ninex TDO 5, 9, 14

Tall oil hydroxyethyl imidazoline [CAS #61791-39-7]
Synonyms: 1-Hydroxyethyl-2-tall oil imidazoline; Tall oil imidazoline
Alkazine TO; Amine T; Mazoline T; Miramine GT; Miramine TOC; Monazoline T; Textamine T-1; Unamine-T; Varine T

Tall oil oleate
Ninate TMO 5, 7, 9, 14

Tallowalkonium chloride
Synonyms: Tallow dimethyl benzyl ammonium chloride
Ammonyx 856; Kemamine BQ-9742C; Noramium S 75

Tallow amide [RD #977058-38-0]
Kemamide S-65

Tallowamide DEA [CAS #68140-08-9]
Synonyms: Tallow diethanolamide
Jordamide TC; Monamine 928; Product LT 19-14; Schercomid SO-T

Tallowamidopropylamine oxide [CAS #68647-77-8]
Ammonyx TDO

Tallowamidopropyl dimethylamine [CAS #68425-50-3]
Synonyms: Dimethylaminopropyl tallow amide
Incromine TB

Tallowamidopropyl hydroxysultaine
Mirataine TABS

Tallow amine [CAS #61790-33-8]
Amine BG; Armeen T, TD; Crodamine 1.T; Kemamine P-974, P-974D; Lilamin 170, 170 D; Noram S; Radiamine 6170, 6171; Tomah Tallow Amine

Tallow amine acetate
Acetamin T; Armac T; Kemamine A974; Lilamac 170; Noramac S; Radiamac 6179

Tallow amine oxide [RD #977058-39-1]
Synonyms: Tallow dimethylamine oxide
Conco XA-T

Tallowaminopropylamine [CAS #68439-73-6]
Synonyms: Tallow trimethylene diamine
Kemamine D-974

Tallow betaine
Alkateric TB

Tallow diamine
Lilamin 570; Radiamine 6570; Tomah Tallow Diamine

Tallow dihydroxyethyl betaine
Varion TEG

Tallow dimethyl trimethyl propylene diammonium chloride
Tomah Q-D-T

Tallowdimonium propyltrimonium dichloride [CAS #68607-29-4]
Synonyms: N,N,N′,N′,N′-Pentamethyl-N-tallow alkyl-1,3-propanediammonium dichloride; N-Tallow pentamethyl propane diammonium dichloride
Adogen 477

N-tallow dipropylene triamine
Trinoram S

Talloweth-11
Synonyms: PEG-11 tallow ether
Dehydol TA 11; Merpoxen T 110

Talloweth-18
Merpoxen T 180

Talloweth-20
Alkasurf TA-20

Talloweth-25
Merpoxen T 250; Rewopal TA 25/S

Talloweth-30
Alkasurf TA-30

Talloweth-40
Alkasurf TA-40

Talloweth-50
Alkasurf TA-50; Merpoxen T 500

Talloweth-80
Merpoxen T 800

Tallow glyceride [CAS #61789-13-7]
Synonyms: Tallow monoglyceride
Dimodan PM, PM 300, TH; Monomuls 90-20; Myverol 18-30

Tallow glycerides
Synonyms: Tallow mono, di and tri glycerides
Monomuls 60-20

Tallow hydroxyethyl imidazoline
Synonyms: Tallow imidazoline
Accosoft 808; Miramine TC

Tallow imidazolinium methosulfate
Incroquat S-75 CG; Incrosoft S-75, S-90, S-90M

Tallow nitrile
Arneel T

Tallow polypropylene polyamine
Polyram S

N-Tallow-1,3-propanediamine
Synonyms: Tallow-1,3-diaminopropane
Duomeen T, TX

N-Tallow-1,3-propanediamine diacetate
Duomac T

N-Tallow-1,3-propanediamine dioleate
Duomeen TDO

Tallow propylene diamine
Dinoram S; Lilamuls BG; Nissan Amine DT

Tallow propylene diamine acetate
Dinoramac S

N-Tallow 1,3-propylene diamine diacetate
Kemamine AD 974

Tallow propylene diamine PEG ammonium methosulfate
Rewoquat DQ 35

Tallow propylene diammonium caprylate
Rewocor TPAC 100

Tallow sulfosuccinate
Arylene TA

Tallow triamine
Tomah Tallow Triamine

Tallowtrimonium chloride [CAS #8030-78-2; 7491-05-2]
Synonyms: Tallow trimethyl ammonium chloride
Adogen 471, R-6; Arquad T-2C-50, T-27W, T-50; Jet Quat T-50; Quartamin TPR; Querton BGCL50; Radiaquat 6471

TEA-abietoyl hydrolyzed animal protein [CAS #68918-77-4]
Lamepon PA-TR

TEA-coco-hydrolyzed animal protein [CAS #68952-16-9]
Lamepon S-TR, ST40; Lexein S620S, S620TA; Maypon 4CT; Super Pro #5A, 100

TEA-cocoyl-glutamate [CAS #68187-29-1]
Acylglutamate CT-12; Amisoft C-T-12

TEA-cocoyl sarcosinate [CAS #68411-96-1; RD #977068-62-4]
Cycloryl TEALS

TEA-collagen polypeptide coco fatty acid condensate
Monteine LCT

TEA-dodecylbenzenesulfonate [CAS #27323-41-7]
Synonyms: Triethanolamine dodecylbenzene sulfonate
Alkasurf T; Arylan TE/C; Bio Soft N-300; Bio-Surf PBC-420; Calsoft T-60; Carsofoam T-60-L; Conco AAS-60S; Cycloryl DDB-60T; Elfan WAT; ESI-Terge T-60; Manro TDBS 60; Marlopon AT 50, CA; Merpisap AT 50, 90; Nansa TS 60; Norfox T-60; Polystep A-9; Puxol CB-22; Reworyl TKS 90/F; Richonate S-1280, TAB; Rueterg 97-T; Sandoz Sulfonate AAS 60S; Serdet DML 45; Steinaryl TKS 90/F; Sterling LA Oil; Surco 60T; Witconate 60T, 79S, TAB

TEA-hydrogenated tallow glutamate [RD #977068-63-5]
Synonyms: Triethanolamine hydrogenated tallowyl glutamate
Amisoft HT-12

TEA-isostearoyl hydrolyzed animal protein
Lexemul LPM

TEA-laureth sulfate [CAS #27028-82-6]
Synonyms: Triethanolamine lauryl ether sulfate
Cycloryl TD; Drewpon EDT; Empicol ETB; Montosol PL-14; Neopon LOT; Sactol 2 OT; Texapon NT; Tylorol LT50

TEA-lauroyl animal collagen amino acids
Synonyms: TEA-lauroyl collagen amino acids
Aminofoam C

TEA-lauroyl glutamate [CAS #53576-49-1]
Synonyms: Triethanolamine lauroyl glutamate
Amisoft LT-12

TEA-lauryl sulfate [CAS #139-96-8]
Synonyms: Triethanolamine lauryl sulfate; Triethanolammonium lauryl sulfate
Akyposal TLS 42; Alkasurf TLS; Avirol 300; Berol 480; Carsonol TLS; Cedepon LT-40; Conco Sulfate TL; Condanol TLS 40; Cosmopon TR; Cycloryl WAT; DeSonol T; Drewpon 40, 40LS; Elfan 240T, 240 T/S; Emal TD; Emersal 6434; Empicol TA40, TA40A; Empicol TC30/T, TC 34, TCR, TDL 75, TL 40, TL40/T, TLP, TLP/T, TLR, TLR/T; Equex STM, T; Incronol TLS; Jordanol TL-400; Lakeway 101-30; Lonzol LT-400; Manro TL 40; Maprofix TLS, TLS-65, TLS-106, TLS-500, TLS-513; Mars TLS-40; Merpinal LMT 40; Montovol RF-11, RL-10; Neopon LT; Nutrapon TLS-500; Perlankrol ATL40; Polyfac PCW-160, TLS-40; Polystep B-6; Rewopol TLS 40; Richonol T; Sactol 2 T; Sandoz Sulfate TL; Serdet DFL 40; Sipex TEA; Sipon LT-6; Standapol Conc. 7021, SHC-301, T, TS-100; Stepanol TLS-40, WAT; Sterling WAT; Sulfatol 33 TR; Sulfetal KT 400; Sulfotex WAT; Sulphonated Lorol Liquid TA, TN, TNR; Sunnol LST; Swascol 4L; Texapon N 42, T, T 35, T 42, TH; Witcolate T; Zoharpon LAT

TEA oleamide sulfosuccinate
Cyclopol SBG 280/T

TEA-oleamido PEG-2 sulfosuccinate
Synonyms: Triethanolamine monooleamido PEG-2 sulfosuccinate
Mackanate ODT; Standapol SH-300, SHC-301

TEA-oleth-2 phosphate
Crodafos O.2 TEA

TEA-PEG-3 cocamide sulfate
Genapol AMS

TEA-sodium dodecylbenzene sulfonate
Chimipon TSB

TEA tetrapropylene benzene sulfonate
Merpisap DT 90

Tetradecyl. See *Myristyl*

Tetraethylene glycol. See *PEG-4*

Tetrahydroxypropyl ethylenediamine [CAS #102-60-3]
Mazeen 173, 174-75

Tetramethyl decynediol [CAS #126-86-3]
Surfynol 104, 104A, 104E, 104H

Tetrapropylene benzene sulfonic acid
Merpisap TS 98

Tetrasodium dicarboxyethyl stearyl sulfosuccinamate [CAS #3401-73-8; 37767-39-8]
Synonyms: Tetrasodium dicarboxyethyl octadecyl sulfosuccinamate
Aerosol 22, 22N; Alconate 2CSA; Alcopol FL; Lankropol ATE; Monawet SNO-35; Rewopol B 2003

Thiodiglycol [CAS #111-48-8]
Synonyms: Thiodiethylene glycol; β-Bis-hydroxyethyl sulfide; Dihydroxyethyl sulfide
Kromfax

TIPA decyl diphenyloxide disulfonate
Synonyms: Triisopropanolamine decyl diphenyloxide disulfonate
Dowfax 3B-Triisopropanolamine

TIPA dodecyldiphenyloxide disulfonate
Synonyms: Triisopropanolamine dodecyldiphenyloxide disulfonate
Dowfax 2A-Triisopropanolamine

TIPA-lauryl sulfate [CAS #661-61-6]
Synonyms: Triisopropanolamine lauryl sulfate
Cycloryl WAT-1

Toluene sulfonic acid [CAS #104-15-4]
Synonyms: 4-Methylbenzenesulfonic acid
Eltesol TA, TA 65, TA 96, TA/F; Reworyl T 65; Steinaryl T 65; Sulframin TX

Triacetin [CAS #102-76-1]
Synonyms: Glyceryl triacetate; 1,2,3-Propanetriol triacetate; Triacetyl glycerol
Kessco Triacetin

Triamine dodecylbenzene sulfonate
Reworyl TKS 90/L

Triammonium dodecylbenzene sulfonate
Condasol TKS 90/F

Tri (caprylyl capryl) tert. amine
Lilamin 364

Tri (caprylyl/capryl/lauryl) tert. amine
Lilamin 368

Triceteareth-4 phosphate
Hostacerin CG; Hostaphat KW 340 N

Trideceth-2 phosphate
Synonyms: PEG-2 tridecyl ether phosphate
Crodafos T2 Acid

Trideceth-3 [CAS #4403-12-7; 24938-91-8 (generic); RD #977058-48-2]
Synonyms: PEG-3 tridecyl ether; PEG (3) tridecyl ether; POE (3) tridecyl ether
Chemal TDA-3; Emulphogene BC-420; Hetoxol TD-3; Iconol TDA-3; Lipal 3 TD; Lipocol TD-3; Macol TD-3; Siponic TD-3; Trycol TDA-3; Volpo T3

Trideceth-4
Macol TD-4; Merpoxen TR 40

Trideceth-5
Alkasurf TDA-5; Volpo T5

Trideceth-5 phosphate
Crodafos T5 Acid

Trideceth-6 [CAS #24938-91-8 (generic); RD #977058-49-3]
Ahcowet DQ-114; Alkasurf TDA-6; Chemal TDA-6; DeSonic 6T; Emulphogene BC-610; Ethosperse TDA-6; Hetoxol TD-6; Iconol TDA-6; Lipal 6 TD; Lipocol TD-6; Macol TD-6; Renex 36; Siponic TD-6; Trycol TDA-6

Trideceth-6 phosphate [CAS #9046-01-9 (generic)]
Gafac RS-610

Trideceth-7
Alkasurf TDA-7; Alkasurf TDA-7.5 (7.5 EO); Flo-Mo AJ-85, AJ-100 (7.5 EO); Merpoxen TR 70

Trideceth-7 carboxylic acid [CAS #56388-96-6 (generic); 66161-61-3; RD #977061-70-3]
Carsodet TD-7C; Rewopol CT

Trideceth-8
Alkasurf TDA-8.5 (8.5 EO); Iconol TDA-8, TDA-8-90%; Lipal 610; Macol TD-8, TD-610; Merpoxen TR 80; Sterox AJ (8.5 EO), AJ-100 (8.5 EO); Trycol TDA-8

Trideceth-9 [CAS #24938-91-8 (generic); RD #977062-57-9]
Chemal TDA-9; DeSonic 9T; Hetoxol TD-9; Iconol TDA-9; Siponic TD990; Teric 13A9; Trycol TDA-9

Trideceth-10 [CAS #24938-91-8 (generic); RD #977058-50-6]
Carsonon TD-10; DeSonic 10T; Emulphogene BC-720; Iconol TDA-10; Lipal 10TD; Macol TD-10; Merpoxen TR 100; Volpo T10

Trideceth-10 phosphate
Crodafos T10 Acid

Trideceth-11 [CAS #24938-91-8 (generic); RD #977067-68-7]
Carsonon TD-11, TD-11 70%; Trycol TDA-11

Trideceth-12 [CAS #24938-91-8 (generic); RD #977058-51-7]
Ahcowet DQ-145; Alkasurf TDA-12; Chemal TDA-12; DeSonic 12T; Hetoxol TD-12; Lipocol TD-12; Renex 30, 35; Siponic TD-12; Trycol TDA-12

Trideceth-15 [CAS #24938-91-8 (generic); RD #977067-69-8]
Alkasurf TDA-15; Chemal TDA-15; DeSonic 15T; Emulphogene BC-840; Renex 31; Volpo T15

Trideceth-18
Chemal TDA-18; Iconol TDA-18-80%; Trycol TDA-18

Trideceth-20
Volpo T20

Trideceth-25
Hetoxol TD-25

Trideceth-29
Iconol TDA-29-80%

Trideceth-100
Macol TD-100

Trideceth-150
T-Det TDA-150

Tridecylbenzene sulfonic acid [CAS #25496-01-9]
Arylan SO60 Acid; Conco ATR-98S; Nansa TDB

Tridecyloxypropylamine
Tomah PA-17

Tridecyloxypropyl dihydroxyethyl methyl ammonium chloride
Tomah Q-17-2

Tridecyl stearate [CAS #31556-45-3]
Grocor 5721

Triethanolamine [CAS #102-71-6]
Synonyms: TEA; 2,2′,2′′-Nitrilotris(ethanol); Trolamine; Tri(2-hydroxyethyl)amine
Surfonic HDL

Triethylamine decyl diphenyloxide disulfonate
Dowfax 3B-Triethylamine

Triethylamine dodecyl diphenyloxide disulfonate
Dowfax 2A-Triethylamine

Triethyl citrate [CAS #77-93-0]
Synonyms: Ethyl citrate
Hydagen DEO

Triethylene glycol. See *PEG-3*

Triglyceryl. See *Polyglyceryl-3*

Triisodecyl tert. amine
Lilamin 382

Triisooctyl tert. amine
Lilamin 381

Triisopropanolamine. See *TIPA*

Triisostearin
Synonyms: Glyceryl triisostearate
Cyclochem GTIS

Trilaureth-4 phosphate
Hostaphat KL340N

Trilaurin [CAS #538-24-9]
Synonyms: Glyceryl trilaurate
Cyclochem GTL; Softisan 100

Trilaurylamine [CAS #102-87-4]
Lilamin 363

Trilauryl citrate
Citrest LT

Trimethylolpropane oleate
Radiasurf 7372

Trimethylolpropane tri-C5/C9
Radiasyn 7364

Trimethylolpropane tri-C7
Radiasyn 7367

Trimethylolpropane tri-C8/C10
Radiasyn 7368

Trimethylolpropane trioleate
Radia 7370

Trioctyl phosphate
Servoxyl VPTZ 100

Triolein [CAS #122-32-7]
Synonyms: Glyceryl trioleate; Olein
Aldo TO; Alkamuls GTO; Emerest 2423; Grocor 1000, 1200; Juniorlan; Kemester 1000; Radia 7161, 7163, 7303, 7363

Triolein PEG-6 complex
Labrafil M 2735 CS

Trioleth-8 phosphate [RD #977058-53-9]
Hostaphat KO 380

Trioleyl phosphate [RD #977058-54-0]
Hostaphat KO 300

Tripropylene glycol. See *PPG-3*

Trisodium sulfosuccinate
T.S.S.

Tristearin [CAS #555-43-1]
Synonyms: Glyceryl tristearate
GTS-33; Neobee 62

Tristearyl citrate [CAS #7775-50-0]
Citrest ST

Tritridecyl tert. amine
Lilamin 383

Undecylenamide DEA [CAS #25377-64-1; 60239-68-1]
Synonyms: Undecylenoyl diethanolamide
Rewocid DU 185

Undecylenamide MEA [CAS #20545-92-0; 25377-63-3]
Synonyms: Undecylenoyl monoethanolamide
Incromide UM; Loramine U 185; Rewocid U 185; Rewomid U 185; Steinazid U 185

Undecylenamidopropyl trimonium methosulfate
Synonyms: Undecylenic acid propylamido trimethyl ammonium methosulfate
Rewocid UTM 185

Undecylenic sulfosuccinate
Mackanate UD

Undecylpentadecanol [CAS #68444-33-7]
Synonyms: Isohexacosanol
Standamul GTO-26

Urea [CAS #57-13-6]
Bio Soft LD-95; Renex 25, 35

Vegetable oil glyceride
Synonyms: Veg. oil monoglyceride
D-9; Dimodan O, PVP

Vegetable oil mono/diglycerides
Emuldan HV 40 K, HVF 52 K

Wheat germ glycerides [CAS #68990-07-8]
Synonyms: Wheat germ oil mono-, di-, and triglycerides
Wickenol 535

Wheat germ oil [CAS #8006-95-9]
Emcon W

Xanthan gum [CAS #11138-66-2]
Biozan, SPX 5423

Xylene sulfonic acid [CAS #25321-41-9]
Eltesol XA, XA 65, XA 90, XA/M65; Reworyl X 65

Zinc-bis (cocamidopropyl betaine)
Schercotaine CAB-Z

Zinc-bis (undecylenic amidopropyl dimethyl glycinate)
Schercotaine UAB-Z

Trademark Product Manufacturers Directory

[2] **Aarhus Oliefabrik A/S**, DK-1800, Aarhus C. Denmark
Skyllex; Sulfatol; Syntol

[4] **Aceto Chemical Co., Inc.**, 126-02 Northern Blvd., Flushing, NY 11368
Acetoquat; Ceramol

[8] **Air Products and Chemicals Inc.**, Box 538, Allentown, PA 18105
Surfynol

[9] **Ajinomoto USA, Inc.**, 9 West 57th St., #4625, New York, NY 10019
Acylglutamate; Ajicoat; Amifat; Amihope; Amisoft; Amiter; CAE; Pyroter

[10] **Akzo Chemie America/Armak Chemical**, 300 S. Riverside Dr., Chicago, IL 60606
[11] **Akzo Chemie B.V.**, Stationsstraat 48, POB 247, 3800 AE-Amersfoort, Netherlands
[12] **Akzo Chemie Italia SpA.**, Via Vismara, 20020 Arese, Milano, Italy
[10] to [12]: see also Armak; Armac; Armeen; Armotan; Armoteric; Aromox; Arquad; Dapral; Duomac; Duomeen; Duoquad; Elfacos; Elfan; Elfanol; Elfapur; Ethoduomeen; Lauridit

[15] **Albright & Wilson (Australia) Ltd.**, 610 St. Kilda Rd., PO Box 4544, Melbourne 3001, Australia
[16] **Albright & Wilson Inc.**, PO Box 26229, Richmond, VA 23260
[17] **Albright & Wilson Ltd./Detergents Div., Marchon Works**, PO Box 15, Whitehaven, Cumbria CA28 9QQ, UK
[18] **Albright & Wilson Ltd./Phosphates Div.**, PO Box 3, Hagley Rd. W., Oldbury, Warley, W. Midlands B68 0NN, UK
[15] to [18]: Aviscour; Eltesol; Empicol; Empicryl; Empigen; Empilan; Empimin; Empiwax; Gardilene; Gardinol; Gardiquat; Gardisperse; Laurex; Lorol; Nansa; Nansol; Nonion

[19] **Alco Chemical Corp.**, 909 Mueller Dr., Chattanooga, TN 37406
Astromid; Astrowet

[20] **Alcolac Inc.**, 3440 Fairfield Rd., Baltimore, MD 21226
[21] **Alcolac Ltd.**, 490 Dufferin St., Valleyfield, Quebec J6S 2B4 Canada

Abex; Akypo; Alacsan; Alcamate; Alconate; Emulsifier 83; Sipenol; Sipex; Sipomide; Sipon; Siponate; Siponic; Sipoteric

[22] **Alconox Inc.**, 215 Park Ave. So., New York, NY 10003

Alcojet; Alconox; Alcotab; Det-O-Jet; Terg-A-Zyme

[23] **Alkaril Chemicals (UK) Ltd.**, Southbank Daveylands, Wilmslow, Cheshire 5K9 2AG, UK
[24] **Alkaril Chemicals Inc.**, Industrial Pkwy., PO Box 1010, Winder, GA 30680
[25] **Alkaril Chemicals Ltd.**, 3265 Wolfedale Road, Mississauga, Ontario L5C 1V8, Canada

[23] to [25]: Alkadet; Alkameen; Alkamide; Alkamidox; Alkaminox; Alkamox; Alkamuls; Alkaphos; Alkapol; Alkaquat; Alkasurf; Alkateric; Alkatronic; Alkatrope; Alkazine; Indu-Sol

[27] **Allied Colloids Inc.**, One Robinson Lane, Ridgewood, NJ 07450

Collemul; Dispex

[28] **Allied Colloids Ltd.**, PO Box 38, Low Moor, Bradford, Yorkshire BD12 OJZ, UK

Alcopol; Imidrol

[32] **Alox Corp.**, 3943 Buffalo Ave., PO Box 517, Niagara Falls, NY 14302

Alox; Aqualox

[37] **Amerchol Corp./Unit of CPC International Inc.**, PO Box 4051, 136 Talmadge Road, Edison, NJ 08818
[38] **Amerchol Europe/Div. of Les Industries DuMaise S.A.**, Havenstraat 84, B-1800, Vilvoorde, Belgium

Alcolan; Amerchol; Amerlate; Ameroxol; Ceralan; Cetal; Glucam; Glucamate; Glucate; Lanamine; Lanogel; Lanogene; OHlan; Polylan; Solulan; Stearal

[40] **American Can Co./Lignin Chemicals**, Greenwich Office Park #8, Greenwich, CT 06830

Kelig; Marasperse; Norlig

[42] **American Color & Chemical Corp.**, 11400 Westinghouse Blvd., Charlotte, NC 28232

Amascour; Amasperse; Amawet; Aminogen; Depuma

[43] **American Cyanamid Co./Industrial Chem. Div.**, Berdan Ave., Wayne, NJ 07470

See also Cyanamid B.V. Aerosol; Complemix; Deceresol; Solusol

[43A] **American Enzyme,** PO Box 23441, Brown Deer, WI 53223
Amerzyme

[44] **American Hoechst Corp.,** Rt. 202-206 North, Somerville, NJ 08876
See also Hoechst AG. Arkopal; Bohrmittel Hoechst; Emulsogen; Genamine; Genapol; Hostacerin; Hostacor; Hostapal; Hostapur; Medialan

[45] **American Lecithin Co., Inc.,** 32-34 61st St., Woodside, NY 11377
Alcolec; Lexin; LKE

[50] **Anchor Chemical Ltd.,** Manchester M11 4SR, UK
Anchoid

[52] **Angus Chemical Co.,** 2211 Sanders Rd., Northbrook, IL 60062
AEPD; Alkaterge; AMP

[54] **Apex Chemical Co., Inc.,** 200 South First St., Elizabethport, NJ 07206
Apex; Apexomine; Apex Pentrapex; Apexscour; Apex Tarem

[58] **Ardmore Chemical Co.,** 29 Riverside Ave., Bldg. #14, Newark, NJ 07104
Ardet

[60] **Arizona Chemical Co.,** 16-00 Route 208, Fairlawn, NJ 07410
Arizona; Arkolene

[61] **Arjay Inc.,** PO Box 45045, Houston, TX 77045
See also Magna; Arsul

[62] **Arkansas Co., Inc.,** Box 210, Newark, NJ 07101
Nilofoam

[63] **Armak Co.,** 8401 West 47th St., McCook, IL 60525
[64] **Armak Industrial Chem. Div./Akzo Chemie America,** 300 S. Rivereside Dr., Chicago, IL 60606
[63] to [64]: See also Akzo Chemie. Armac; Armeen; Armid; Armoblen; Armogard; Armosoft; Arneel; Aromox; Arquad; Duomac; Duomeen; Ethoduomeen; Ethoduoquad; Ethofat; Ethomeen; Ethomid; Ethoquad

[66] **Armstrong Chemical Co., Inc.,** 1530 South Jackson St., Janesville, WI 53545
Accobetaine; Accomeen; Accomid; Acconon; Accoquat; Accosoft; Accosperse; Ampho; Sanac

[67] **Arol Chemical Products Co.,** 649 Ferry St., Newark, NJ 07105
Aremul; Arodet; Arofos; Arol; Arolterge; Arowet; Detergent CR

[69] **Ashland Chemical Co./Div. Ashland Oil, Inc.,** Box 2219, Columbus, OH 43216
Starfol

[71] **Atlas Chem.**
See also ICI Americas Inc. Renex; Tween

[72] **Atlas Refinery, Inc.,** 142 Lockwood St., Newark, NJ 07105
Atlas; Atlasoft; Atlasol; Atlox; Eureka

[73] **Atramax Inc.,** Box 278, Hawthorne, NJ 07507
Atranonic

[74] **Australian Bakels (Pty) Ltd.,** PO Box 147, Lidcombe N.S.W. 2141, Australia
Monomax; Swedex

[76] **AZS Chemical Co./Div. of AZS Corp.,** 762 Marietta Blvd. N.W., Atlanta, GA 30318
AZdry

[81] **BASF (UK) Ltd.,** PO Box 4, Earl Rd., Cheadle Hulme, Cheadle, Cheshire 5K8 60QG, UK
[82] **BASF AG,** Carl-Bosch-Strasse 38, D-6700 Ludwigshafen, F.R.G.
[83] **BASF Canada Ltd.,** PO Box 430, Montreal, Quebec H4L 4V8, Canada
[84] **BASF Espanola S.A.,** Apartado 762, Barcelona 8, Spain
[85] **BASF S.A., Compagnie Francaise,** MC-NT, 140, Rue Jules Guesde, 92303 Levallois-Perret, France
[86] **BASF Wyandotte Corp.,** 100 Cherry Hill Rd., Parsippany, NJ 07054

[81] to [86]: Basophor; Cremophor; Emulan; Icodimeen; Icomeen; Icomid; Iconol; Indoil; Industrol; Katioran; Laventin; Leophen; Lutensit; Lutensol; Nekal; Nekanil; OP; Pluracol; Pluradot; Plurafac; Pluraflo; Plurest; Pluriol; Pluronic; Polyol; Primasol; Separol; Tetronic; Twix; Uniperol; Wettol

[89] **Berol Chemicals, Inc.,** 274 Riverside Ave., Westport, CT 06880
[90] **Berol Kemi AB,** Box 851, S-444 01, Stenungsund 1, Sweden
Berol; Sulfamin; Sulfosoft

[92] **Bio-Lab Inc.,** 627 E. College Ave., Decatur, GA 30030
Bio-Dac; Bio-Quat; Bio-Surf

[95] **Bofors Lakeway,** 5025 Evanston Ave., PO Box 328, Muskegon, MI 49443
Lakeway

[97] **Borg-Warner Chemicals Inc.,** International Center, Parkersburg, WV 26102
Condensate; Pegafac; Peganol

[98] **Bostik South/Div. of USM Corp.,** PO Box 5695, Greenville, SC 29606
Atcopen; Atcowet

[109] **Canada Packers Inc./Edible Oils Div.**, 30 Weston Rd., Toronto, Ontario M6N 3P4, Canada

[110] **Canada Packers Ltd./Chemical Div.**, 5100 Timberlea Blvd., Mississauga, Ontario L4W 2S5, Canada

Canamulse; Sterling

[111] **Capital City Products Co./Div. of Stokely-Van Camp, Inc.**, PO Box 569, Columbus, OH 43216

Ampho; Caplube; Capmul; Caprol; Sanac

[113] **Carson Chemicals, Inc.** See Lonza Inc.

[114] **Carstab Corp.**, West Street, Reading, OH 45215

Base 11-T

[115] **CasChem Inc.**, 40 Avenue A, Bayonne, NJ 07002

ES; Naturechem; P-10 Acid

[116] **Catawba-Charlab, Inc.**, 5046 Pineville Rd., PO Box 240497, Charlotte, NC 28224

Charlab

[117] **Ceca SA**, 12/16 Allee des Vosges, Cedex 54-92062, La Defense 5, Paris, France

Amphoram; Dinoram; Dinoramac; Dinoramox; Noram; Noxamine; Noxamium; Polyram; Trinoram

[121] **Central Soya Co., Inc./Chemurgy Div.**, 1300 Ft. Wayne Bank Bldg., PO Box 1400, Fort Wayne, IN 46801

Actiflo; Centrocap; Centrol; Centrolene; Centrolex; Centromix; Centrophase; Centrophil

[122] **Chemax, Inc.**, PO Box 6067, Highway 25 South, Greenville, SC 29606

Chemal; Chemax; Chemazine; Chemeen; Chemfac; Chemquat; Chemsulf; Sorbax

[123] **Chemform Corp.**, 141 S.W. 8th St., Pompano Beach, FL 33061

Chemannate; Chemcol; Chemstick

[127] **Chemithon Corp.**, 5430 Marginal Way, S.W., Seattle, WA 98106

Detergent Slurry

[128] **Chemos Corp.**, 225-235 Emmet St., Newark, NJ 07114

Cresterge; Crestex; Crestoil; Crestol; Crestolan; Crestomite; Crestomul; Crestonol; Crestonon; Crestopen; Tetralex

[129] **Chemtall Inc.**, PO Box 250, Riceboro, GA 31323

Monomer

[130] **Chem-Y Fabriek Van Chemische Producten B.V.**, PO Box 50, 2410 AB, Bodegraven, Netherlands

Tephal Grunau; Wetting Agent 611, CG 16

[131] **ChemY GmbH**, Kupferstrasse 1, D-4240 Emmerich 1, West Germany

Akypo; Akypogene; Akypomine; Akyporox; Akyposal; Aminol; Letocil; Ogtac; Surfageen; Wibarco

[136] **Ciba-Geigy Corp.**, CH-4002, Basle, Switzerland
[137] **Ciba-Geigy Corp.**, PO Box 18300, Greensboro, NC 27419

Alrodyne; Alrosol; Alrosperse; Alrowet; Amine; Bina QAT; Erional; Irgalube; Irgasol; Kar-Kote; Levelene; Lostat; Nonisol; Quaternary; Reserve Salt Flake; Sarkosyl; Sorbit; Tinegal; Ultratex

[138] **Cincinnati Milacron Chem. Inc.**, West St., Reading, OH 45215

Ambidex; Base 11-T; Cardene; Pave

[139] **Cindet Chemicals, Inc.**, Greensboro, NC

Cindet

[140] **Cities Service Co.**, ICD, Cities Service Bldg., 3445 Peachtree Rd. N.E., Atlanta, GA 30326

AA-10; Sul-fon-ate

[143] **Cleary Chemical Corp., W.A.**, 1049 Somerset St., Somerset, NJ 08873

All Wet; Clearate

[144] **Clintwood Chemical Co.**, 4342 S. Wolcott Ave., Chicago, IL 60609

Clindrol; Product LT, WRS; Rodea

[145] **Clough Chemical Co., Ltd.**, 178 St. Pierre, St.-Jean, Quebec J3B 7B5, Canada

Castor Oil Sulfated; Closyl; Compene; Dousoft; Nutrapon; Nutrol; Orgozon; Pentine; Stamid; Sulfonated Neatsfoot 3089; Sulfonated Syn. Sperm Oil 2091; Thorowet

[146] **Coastal Industries**, 190 Jony Dr., Carlstadt, NJ 07072

Hi-Brite; Hi-Tex

[148] **Colloids Inc.**, 394 Frelinghuysen Ave., Newark, NJ 07114

Musloid; Pentex

[152] **Conoco Chemicals/Continental Oil Co.** See Vista Chemical Co.

[153] **Consos, Inc.**, PO Box 34186, Charlotte, NC 28234

Consamine; Consonyl; Consos; Consoscour; Consostat; Consotard; Pinamine

[154] **Continental Chemical Co.**, 270 Clifton Blvd., Clifton, NJ 07015

Conco; Concofac; Concopal; Condensate; STXS; SXS

[159] **Croda Chemicals Ltd.**, Cowick Hall, Snaith Goole, North Humberside DN14 9AA, UK

[160] **Croda Inc.**, 51 Madison Ave., New York, NY 10010

[161] **Croda Surfactants Inc.**, 183 Madison Ave., New York, NY 10016

[162] **Croda Universal Ltd.**, Oak Rd., Clough Rd., Hull, UK

[159] to [162]: Adinol; Aminofoam; Anfomul; Cetomacrogol; Cithrol; Corona; Coronet; Cosmowax; Crester; Crill; Crillet; Crillon; Crodafos; Crodalan; Crodamets; Crodamine; Crodamol; Crodasinic; Crodateric; Crodazoline; Croderol; Crodesta; Crodet; Crodex; Croduret; Crolactil; Datagel; Datem; Etocas; Fluilan; Fluilanol; GLS; Glycerox; Hartolan; Hartolite; Incrocas; Incrochem; Incromate; Incromectant; Incromide; Incromine; Incronam; Incronol; Incropearl; Incropol; Incroquat; Incrosoft; Incrosorb; Incrosul; Iscolan; Lanethyl; Lanex; Lanexol; Lanpol; Lanpolamide; Liquid Base; Novol; Paraffin Wax Emulsifier; Paralan; Polawax; Polychol; Procetyl; Prochols; Produkt B; Prostearyl; Provol; Satexlan; Satulan; Skliro; Solan; Solangel; Sorbeth; Spreading Agent ET0672; Squalane; Super Corona; Super Hartolan; Syncrowax; Usolan; Volpo; Vykacet; Vykamol; Vykasoid; White Swan; Yeoman

[163] **Crompton & Knowles Corp./Dyes & Chemical Div.**, Route 208, Fair Lawn, NJ 07410

Cenegen; Cenekol; Intralan; Intraphasol; Intratex; Intravon; Resogen; Retarder A

[164] **Crown Metro, Inc.**, PO Box 5695, Greenville, SC 29606

Atcowet

[165] **Crown Zellerbach Corp.**, 1 Bush St., San Francisco, CA 94104

Orzan

[167] **Cyanamid B.V.**, Postbus 1523, BM 3000, Rotterdam, The Netherlands

See also American Cyanamid Co. Aerosol; Complemix

[168] **Cyclo Chemical Corp.**, 7500 N.W. 66th St., Miami, FL 33166

Citrest; Cyclochem; Cyclogol; Cyclomide; Cyclomox; Cyclophos; Cyclopol; Cycloryl; Cyclosheen; Cycloteric; Cycloton

[172] **Dai-ichi Kogyo Seiyaku Co., Ltd.**, Miki Bldg., 3-12-1, Nihombashi, Chuo-ku, Tokyo, 103 Japan

Amiladin; Dianol; DK Ester; Lamigen; Monogen; Noigen; Sunny Safe

[173] **Daishowa Chemicals Inc.**, 81 Holly Hill Lane, Greenwich, CT 06830

Amberlig; Kelig; Lignosol; Maramul

[177] **Dee Zee Chemical Inc.**, 600 Mt. Ephraim Ave., Camden, NJ 08103

Zeebase; Zeeterge

[181] **DeSoto Inc./Chemical Specialties Div.**, PO Box 23523, Harahan, LA 70183

Armul; Arnox; DeSomeen; DeSonate; DeSonic; DeSonol; DeSophos; DeSotan; Flo-Mo; Morwet; Petrac; Petro; Petroscale; Udet

[182] **Devan Chemicals SA**, 34 St. Sauveur Str., B-9600 Ronse-Renaix, Belgium

Rychem; Rycofax

[183] **Dexter Chemical Corp.**, 845 Edgewater Rd., Bronx, NY 10474

Atolene; Barisol; Dextrol; Strodex

[185] **Diamond Shamrock (Australia) Pty.**, PO Box 66, Campbellfield, Victoria, Australia 4061

[186] **Diamond Shamrock Chemical Products AG**, Schutzenmatt Postfach 45 3280, Morat, Switzerland

[187] **Diamond Shamrock Corp./Process Chem. Div.**, 350 Mt. Kemble Ave., Morristown, NJ 07960

[188] **Diamond Shamrock Europe Corp.**, Chaussee dela Hulpe 185, B1170 Brussels, Belgium

[189] **Diamond Shamrock Eytesa S.A.**, Apartado de Correos 282, Terrasa (Barcelona), Spain

[190] **Diamond Shamrock India Ltd.**, Poonam Chambers, No. 702 B, 7th Fl., Shivsagar Estate, Dr. A.B. Rd., Worli, Bombay 400 018, India

[191] **Diamond Shamrock Italia SpA**, Via dei Giovi 6, 1-20032, Cormano, Milano, Italia

[192] **Diamond Shamrock Process Chem. Europe**, 185 Ave. de Fontainebleu, F. 77310, St.-Fargeau, Ponthierry, France

[193] **Diamond Shamrock Scandinavia A/S**, Korkeavuorenk 6A3, 00150 Helsinki 15, Finland

[194] **Diamond Shamrock Scandinavia AB**, Norra Gubberogt 23, S-416 63, Gothenburg, Sweden

[185] to [194]: Agrimul; Agriwet; Alphenate; Ampholan; Arylan; Delvet; Diazital; Doittol; Dyasulf; Dymsol; Emulgane; Foamaster; Hyonic; Intrapol; Kerapol; Kerasol; Kerinol; Lancare; Lankrolan; Lankro Mud; Lankropearl; Lomar; Lutopon; Lutostat; Modicol; Monosulph; Natrex; Nopalcol; Nopco; Nopcocastor; Nopcogen; Nopcolan; Nopcosant; Nopcosperse; Nopcosulf; Noramac; Noramium; Noramox; Phospholan; Sellogen; Sopralub; Steramine; Syntergent; Versilan

[196] **Dixo Co., Inc.**, 158 Central Ave., Rochelle Park, NJ 07662

CPA-Alpha

[199] **Domtar Inc./Organic Chemicals Div.**, 1136 Matheson Blvd., Mississauga, Ontario L4W 2V4, Canada

Cedemide; Cedepal; Cedepon

[201] **Dow Chemical Co.**, 1703 S. Saginaw Rd., Midland, MI 48640
[202] **Dow Chemical Europe S.A.**, Leland I. Doan Str. 3 CH-8810, Horgen, Switzerland

Dowfax; Methocel

[203] **Dow Corning Corp.**, Box 0994, Midland, MI 48640

Dow Corning; Dowfax

[205] **Drew Chem. Corp./Drew Industrial Div./Ashland**, One Drew Plaza, Boonton, NJ 07005

Drewfax

[206] **Drew Prudutos Quimicos Ltds.**, Rua Sampaio Viana, 425, 04004, Sao Paulo, Brazil

Drew; Drewfax; Drewmulse; Drewpon; Synotol

[207] **Duphar B.V.**, Postbus 900, 1380 DA Weesp, The Netherlands

Dulectin; Dusoran; Lanolin Fatty Acids G, O

[208] **E.I. DuPont de Nemours & Co.**, Nemours Bldg., Wilmington, DE 19898

Alkanol; Compound 8-S; Du Pont; Duponol; Merpol; Petrowet; Product BCO; Zelec; Zonyl

[210] **Durkee Industrial Foods/SCM Corp.**, 900 Union Commerce Bldg., Cleveland, OH 44115

CW; Dur-Em; Durfax; Durlac; Dur-Lo; Durpeg; Durpro; Durtan; EC; Ice; Kaorich; Santelle; Santone; Tally

[212] **Dutton & Reinisch Ltd.**, Crown House, London Road, Morden, Surrey SM45DU, UK

Condanol; Condasol; Loraderm; Loramine; Lorapal; Loraphat; Lorapon; Loraquat; Rewocid; Rewoderm; Rewolan; Rewomat; Rewomid; Rewomine; Rewominoxid; Rewopal; Rewophat; Rewopol; Rewopon; Reworyl; Steinamat; Steinamid; Steinamin; Steinaminoxid; Steinapal; Steinaphat; Steinapol; Steinapon; Steinaryl; Steinazid

[213] **Dynamit Nobel AG**, 2 Haberstrasse, D-5210 Troisdorf-Oberlar, West Germany
[214] **Dynamit Nobel Chem.**, 10 Link Dr., Rockleigh, NJ 07647

[213] to [214]: Dynacet; Imwitor; Softenol; Softigen; Softisan

[215] **Eastern Color and Chemical Co.**, 35 Livingston St., Providence, RI 02904

Eccoful; Eccolene; Eccoscour; Eccoterge; Eccotex; Eccowet

[216] **Eastman Chemical Products, Inc.**, PO Box 431, Kingsport, TN 37662

Myvacet; Myvaplex; Myvatem; Myvatex; Myverol

[217] **Emery Industries Inc.**, 1501 W. Elizabeth Ave., Linden, NJ 07036

Clearlan; Emerest; Emersal; Emersol; Emery; Emid; Emrite; Emsorb; Emthox;

Juniorlan; Lanacet; Lanfrax; Lanoquat; Lantrol; Nimco; Nimcolan; Propoxyl; Tergelan; Trycol; Trydet; Tryfac; Trylon; Trylox; Trymeen; Twitchell

[218] **Emkay Chemical Co.**, 319-325 Second St., Elizabeth, NJ 07206

Emkabase; Emkafix; Emkagen; Emkal; Emkalon; Emkane; Emkapon; Emkastat; Emkatex; Rexobase; Rexopene; Rexopon; Rexowet

[220] **Emulsion Systems Inc.**, 215 Kent Ave., Brooklyn, NY 11211

ESI-Terge

[224] **Essential Chemicals Corp.**, 28391 Essential Rd., Merton, WI 53056

Ecconol

[227] **Ethyl Corp./Chem. Group**, 451 Florida Blvd., Baton Rouge, LA 70801

Adma; ASA; Epal; Tetradecene

[228] **Evans Chemetics/W.R. Grace & Co.**, 90 Tokeneke Rd., Darien, CT 06820

Emulsifier K-700

[230] **Exxon Company U.S.A.**, PO Box 2180, Houston, TX 77001

Acto; Alkali Surfactant; Amphoteric

[232] **The Fanning Corp.**, 3117 North Clybourn Ave., Chicago, IL 60618

Emcon; Fancol; Fancor; Lan-Aqua-Sol; Lanotein; Vigilan

[237] **Finetex Inc.**, 418 Falmouth Ave., Elmwood Park, NJ 07407

Acrilev; Aminol; Cordon; Gemtex; Rueterg; Surfine; Tauranol

[242] **Formosa Chem. & Fibre Corp.**, 1 Nanking East Road, Sec. 2, Taipei, Taiwan, China

Tairygent

[245] **GAF Corp./Chemical Products**, 1361 Alps Rd., Wayne, NJ 07470

Alipal; Antara; Antaron; Antarox; Antifoam; Biopal; Blancol; Diazopon; Duowet; Emulphogene; Emulphor; Gafac; Gafamide; Gafterge; Gaftex; Ganex; Gantrez; Igepal; Igepon; Katapol; Monopol; Nekal; Pegafac; Pegameen; Peganate; Peganol; Pegeste; Pegol; Peregal

[247] **Gattefosse Corp.**, 3 Westchester Plaza, Elmsford, NY 10523
[248] **Gattefosse Etablissements**, 36 Chemin de Genas, 69800 Saint Priest, 92100-Boulogne, France

[247] to [248]: Compritol; Geleol; Gelot; Labrafac; Labrafil; Labrasol; Maisine; Monosteol; Olicine; Plurol; Polystate; Precirol; Sucro Ester; Tefose

[249] **General Electric Co./Silicone Prod. Div.**, Mechanicsville Rd., Waterford, NY 12188

AF-

[252] **Georgia-Pacific Corp.**, 300 W. Laurel St., Bellingham, WA 98225

CM; Dyqex; Ke-Mul; Lignosite

[253] **Geronazzo SpA**, Via Milano 78, 20021 Ospiate Di Bollate, Milano, Italy

Drynol; Geronol; Geropon; Soprofor

[256] **Glyco Inc.**, 51 Weaver St., PO Box 700, Greenwich, CT 06836

Aldo; Aldosperse; Ethosperse; Glycomul; Glycosperse; Pegosperse; Polyaldo

[258] **Th. Goldschmidt AG**, Goldschmidtstr. 100, Postfach 101461, D-4300 Essen 1, West Germany

[259] **Goldschmidt Chemical Corp.**, 914 Randolph Rd., Box 1299, Hopewell, VA 23860

[258] to [259]: Abil; Aminoxid; Axol; Datamuls; Protegin; Tagat; Tegacid; Tegin; Teginacid; Tego-Amid; Tego-Betaine; Tegomuls

[261] **B.F. Goodrich Co.**, 6100 Oak Tree Blvd., Cleveland, OH 44131

Carbopol; Carboset

[262] **W.R. Grace & Co./Organic Chem. Div.**, 55 Hayden Ave., Lexington, MA 02173

Hampfoam; Hamposyl

[263] **Graden Chemical Co., Inc.**, 426 Bryan St., Havertown, PA 19083

Denphos; Densol; Densulf; Denwet; Gradol; Gradonic; Haroil; Harol

[270] **Grindsted do Brazil**, Cx. Postal 20.999, 01000, Sao Paulo, S.P., Brazil

[271] **Grindsted France S.A.R.L.**, 30 Ave. Amiral Lemonnier, F 78160 Marly Le Roi, France

[272] **Grindsted Products A/S**, 38 Edwin Rahrs Vej, DK-8220 Brabrand, Denmark

[273] **Grindsted Products Inc.**, 201 Industrial Pkwy., PO Box 26, Industrial Airport, KS 66031

[274] **Grindsted Products Ltd.**, Northern Way, Bury St. Edmunds, Suffolk, IP32 6NP, UK

[275] **Grindstedvaerket GmbH**, Kellerbleek 3, D-2000 Hamburg 54, Germany

[270] to [275]: Acidan; Amidan; Artodan; Cetodan; Dimodan; Emuldan; Famodan; Grindtek; Homodan; Lactodan; Lecidan; Lipodan; Mayodan; Panodan; Promodan; Triodan

[276] **A. Gross & Co./Div. of Millmaster Onyx Corp.**, 652 Doremus Ave., PO Box 818, Newark, NJ 07101

Grocor; Grotex

[277] **Grünau, Chemische Fabrick Grünau GmbH**, Robert-Hansen Str. 1, Postfach 1063, D-7918 Jllertissen, Bavaria, West Germany

Arbyl; Arbylen; Lactomul; Lamacit; Lamecerin; Lamecreme; Lamefix; Lameform; Lamegin; Lamepon; Lamesoft; Monomuls; Osimol; Stabilizer T; Zoharsoft

[280] **C.P. Hall Co.**, 7300 South Central Ave., Chicago, IL 60638

CPH; Emulvis; Hallcomid

[282] **A. Harrison & Co. Inc.**, PO Box 494, Pawtucket, RI 02862

Iberpal; Iberpenetrant; Iberpol; Iberpon; Iberscour; Iberterge; Iberwet

[283] **Hart Chemicals Ltd.**, 256 Victoria Rd. South, Guelph, Ontario N1H 6K8 Canada

Acid Felt Scour; Aconol; Ammonium Stearate; Arylene; Duofol; Dyesperse; Foamer; Hartamide; Hartex; Hartolon; Hartonyl; Hartopol; Hartopon; Hartotrope; Hartowet; Hymolon; Nonasol; Nonex; Pogol; Polylube; Retarder CA; Rexol; Rexonic; Rexopal; Rexophos; Scourtex; Solvent Scour; Sulfonic Acid LS; Sulphonic Acid LS; Supratol; T.S.S.

[284] **Hart Products Corp.**, 173 Sussex St., Jersey City, NJ 07302

Alkylene; Duofol; Hartaine; Hartamide; Hartenol; Hartox; Penetron; Retarder N, N-85; Tergenol

[287] **Hefti Ltd. Chemical Products**, PO Box 1623, CH-8048, Zurich, Switzerland

AMS; CE; Cowax; DMO; DMS; DO; GMA; GMF; GMM; GMO; GMR; GMS; GTS; LA; ML; MO; MP; MS; NP; OL; OP; PGE; PMS; QO; RS; ST; Syntens; TO; TS

[288] **Henkel Argentina S.A.**, Avda. E. Madero Piso 14, 1106 Capital Federal, Argentina

[289] **Henkel Chem. (Canada) Ltd.**, 9550 Ray Lawson Blvd., Ville d'Anjou, Quebec H1J 1L3, Canada

[290] **Henkel Inc.**, 480 Alfred Ave., Teaneck, NJ 07666

[291] **Henkel KGaA**, Postfach 1100, D-4000, Dusseldorf 1, West Germany

[288] to [291]: Aethoxal; Amphocerin; Avirol; Cationic Guar; Comperlan; Cutina; Dehydag; Dehydol; Dehydrophen; Dehymuls; Dehypon; Dehyquart; Dehyton; Deriphat; Edenol; Emulgade; Eumulgin; Euperlan; Forlanit; Fosfamide; Fosterge; Fostex; Generol; Hydagen; Lanette; Maranil; Nitrene; Polyquart; Primarol; Quatrene; Standamid; Standamox; Standamul; Standapol; Standapon; Stantex; Sulfopon; Sulfotex; Supercol; Sykanol; Texapon; Texin; Textamine; Velvetex

[294] **Hercules B.V.**, 8 Veraartlaan, PO Box 5822, 2280 HV Rijswijk, The Netherlands

[295] **Hercules Inc.**, Hercules Plaza, Wilmington, DE 19894

[294] to [295]: Biozan; CMTC-T; Dresinate; Hercules; Surfactant AR 150

[296] **Heterene Chemical Co., Inc.**, PO Box 247, 792 21st Ave., Paterson, NJ 07513

Hetamide; Hetester; Hetoxamate; Hetoxamide; Hetoxamine; Hetoxide; Hetoxol; Hetsorb; Hetsulf

[297] **Hexcel Corp./Chemical Products Div.**, 215 N. Centennial St., Zeeland, MI 49464

Bromat; Cetats; Dibactol; Mytab; Stedbac

[298] **High Point Chemical Corp.**, 601 Taylor St., PO Box 2316, High Point, NC 27261

Hipochem

[299] **Hilton-Davis Chemical Co./Div. of Sterling Drug, Inc.**, 2235 Langdon Farm Rd., Cincinnati, OH 45237

Cyncal; Roccal; SCMS

[300] **Hodag Chemical Corp.**, 7247 N. Central Park Ave., Skokie, IL 60076

Nonionic; Sole-Onic; Sole-Terge

[301] **Hoechst AG**, Verhaufkanststoffe, D-6230, Frankfurt am Main 80, West Germany

See also American Hoecsht. Arkomon; Arkopal; Arkopon; Dispersogen; Dissolvan; Dodiflood; Emulsifier DMR; Emulsogen; Fluowet; Genagen; Genamin; Genaminox; Genapol; Hostacerin; Hostapal; Hostaphat; Hostapon; Hostapur; Humectol; Leomin; Leonil; Medialan; Peroxide Stabilizer; Phenyl Sulphonate; Prapagen; Sapogenat; Solegal; Stabilisol; Tylose

[305] **E.F. Houghton & Co.**, PO Box 930, Valley Forge, PA 19482

Cerfak; Cerfax; Surfax

[308] **Hüls AG, Chemische Werke**, Postfach 1320 D-4370, Marl 1, West Germany

Dionil; Marlamid; Marlazin; Marlican; Marlinat; Marlipal; Marlon; Marlophen; Marlophor; Marlopon; Marlosol; Marlowet; Marlox; Na-Cumene Sulfonate

[310] **Humko Chemical Div./Witco Corp.**, PO Box 125, 755 Crossover Lane, Memphis, TN 38101

Hystrene; Industrene; Intermediate; Kemamide; Kemamine; Kemester; Neustrene

[311] **Humko Products**, Westinghouse Rd., Trafford Park, Manchester M17 1DR, UK

Kemamine

[312] **Humko Sheffield Chem.** See Humko Chemical Div./Witco Corp.

[316] **ICI Australia Ltd.**, ICI House, 1 Nicholson St., Melbourne 3000 Australia

[317] **ICI Ltd./Organics Div., Chem. Auxiliaries Business Group**, Smith's Rd., Bolton, Lancs, BL3 2QJ, UK

[318] **ICI PLC/Petrochemicals & Plastics Div.**, PO Box 90, Wilton Middlesbrough, Cleveland TS6 8JE, UK

[319] **ICI Specialty Chemicals**, Everslann 45 B-3078, Kortenberg, Belgium

[320] **ICI Americas Inc.**, New Murphy Rd. & Concord Pike, Wilmington, DE 19897

[316] to [320]: Ahco; Ahcovel; Ahcowet; Alkadet; Arlacel; Arlamol; Arlasolve; Arlatone; Atlas; Atlox; Atmer; Atmos; Atmul; Atpet; Atplus; Atsurf; Brij; Calsolene; Cirrasol; Detergent ADC, RP; Detergyl; Lenetol; Lissapol; Lubrol; Matexil; Monaprin; Monflor; Myrj; Nylomine; Oxamin; Pycal; Querton; Renex; Span; Sunaptol; Synperonic; Synprol; Synprolam; Synthrapol; Tandem; Teric; Tridecanol; Tween; Unisol; Vantoc

[323] **IMC Chemical Corp./NP Div.**, 4415 W. Harrison St., Hillside, IL 60162
AEPD; Alkaterge; AMP; DMAMP

[325] **Inolex Chemical Co./Div. of American Can Co.**, Jackson & Swanson Sts., Philadelphia, PA 19148
Lexaine; Lexamine; Lexate; Lexein; Lexemul; Lexomul; Lexoteric

[329] **Interstab Chemicals, Inc.**, 500 Jersey Ave., New Brunswick, NJ 08903
Interwet

[330] **Intex Products, Inc.**, PO Box 6648, Greenville, SC 29606
Intex; Tex-Wet

[331] **Isochem Corp.**, 99 Cook St., Lincoln, RI 02865
Isokleen; Isomul; Isoplast; Isorezoff; Isoslushoff

[332] **ITT Rayonier Inc.**, 18000 Pacific Highway South, Suite 900, Seattle, WA 98188
Orzan; Raylig

[335] **Jefferson Chemical Co., Inc.**, PO Box 4128, Austin, TX 78765
Surfonic

[336] **Jetco Chemicals, Inc.**, PO Box 1898, Corsicana, TX 75110
Jet Amine; Jetco; Jet Quat

[339] **Jordan Chemical Co.**, 1830 Columbia Ave., Folcroft, PA 19032
Jorchem; Jordamide; Jordamine; Jordamox; Jordanol; Jordaphos; Jordapon; Jordaquat; Jordawet; Jorscour; Jortaine

[341] **Kao Corp.**, 14-10 Nihonbashi, Kayabacho 1-Chome, Chuo-ku, Tokyo 103, Japan
Acetamin; Coconad; Emal; Emanon; Emasol; Excel; Homotex; Kalcohl; Levenol; Loving; Neopelex; Pelex; Quartamin; Rheodol; Scourol; Softex

[344] **Kay-Fries, Inc./Dynamit Nobel of America, Chem. Div.**, 10 Link Dr., Rockleigh, NJ 07647
See also Dynamit Nobel; Imwitor; Softigen; Softisan

[345] **Keil Chemical Div./Ferro Corp.**, 3000 Sheffield Ave., Hammond, IN 46320
Base 75, 76, 7800, ML, MT; Inversol

[346] **Kelco Co./Div. Merck & Co., Inc.**, 8355 Aero Dr., San Diego, CA 92123
Dariloid; Kelacid; Kelcoloid; Kelco-Pac; Kelcosol; Kelgin; Kelmar; Kelset; Keltex; Keltone; Keltose; Kelvis

[347] **Kempen, Elektrochemische Fabrik Kempen GmbH,** Postfach 100 260, D-4152 Kempen 1, West Germany

Merpalyt; Merpasol; Merpemul; Merpinal; Merpinamid; Merpinol; Merpiphos; Merpiquat; Merpisap; Merpophal; Merpoxal; Merpoxen; Remanol; S-Oil

[348] **Kenobel S.A.,** Rue Gachard 88, Bte 9, B-1050 Bruxelles, Belgium

Amine; Diamine; Lilamuls; Querton

[350] **Knapp Products, Inc.,** PO Box 405, Lodi, NJ 07644

Polymulse; Polytex

[360] **Lankro Chemicals Ltd.,** PO Box 1, Eccles, Manchester M30 0BH, UK

Agrilan; Agrisol; Ampholan; Arylan; Ethylan; Landemul; Lankromul; Lankropol; Lankrosol; Lanquell; Levelan; Monolan; Perlankrol; Phospholan; Quadrilan

[367] **Lever Industriel,** 103 Rue DeParis, 9300 Bobigny, France

Sactol

[368] **Lilachim S.A.,** 37-33 Rue de la Loi, 1040 Brussels, Belgium

Lilamac; Lilamin; Lilaminox; Querton

[370] **Lion Corp.,** 2-22, 1-Chome, Yokoami, Sumida-ku, Tokyo, Japan

Dovanox; Lipolan; Lipomin; Lipon; Liponox; Liptol; Soft Detergent; Sunnol

[371] **Lipo Chemicals, Inc.,** 207 19th Ave., Paterson, NJ 07504

Lipamide; Lipo; Lipocol; Lipolan; Lipomulse; Liponic; Lipopeg; Lipophos; Liposorb; Lipowax

[372] **Lonza Inc.,** 22-10 Route 208, Fair Lawn, NJ 07410

Aldo; Alkawet; Amphoterge; Bardac; Barlox; Barquat; Carsamide; Carsamine; Carsodet; Carsofoam; Carsofos; Carsonam; Carsonol; Carsonon; Carsoquat; Carsosoft; Carsosulf; Lonzaine; Lonzest; Lonzol; Quad Opacifier; Unamide

[375] **Lucas Meyer GmbH.** See Meyer

[376] **Lyndal Chemical Co.,** PO Box 1740, Dalton, GA 30720

Karawet; Mollisan; Permalene; Phosfac; Progalan; Progasol; Velvamine

[377] **3M,** 3 M Center, St. Paul, MN 55144

Fluorad

[381] **Magna Corp./Arjay Chemicals Div.,** PO Box 33387, Houston, TX 77233

See also Arjay. Aramide; Arphos; Arsil; Arsul

[383] **Manchem Ltd.**, Aston New Rd., Manchester M11 4AT, UK

Manoxol

[384] **Manro Products Ltd.**, Bridge St., Stalybridge, Cheshire SK15 1PH, UK

Manro

[385] **Marathon Morco Co.**, 1502 Pine Ave., Dickinson, TX 77539

Morco

[386] **Marchon Espanola SA**, Carretera Montblanc Km 2, 4, Alcover (Tarragona), Spain
[387] **Marchon France SA**, BP 19, F-55300, St. Mihiel, France
[388] **Marchon Italiana SpA**, Casella Postale No. 30, 1-46043, Castiglione delle Stiviere, Italy

See Albright & Wilson/Marchon Div.

[391] **Marlowe-Van Loan Corp.**, PO Box 1851, High Point, NC 27261

Marvanol; Marvanscour; Marvantex

[392] **Mars Chemical Corp.**, 762 Marietta Blvd. N.W., Atlanta, GA 30318

Mars; Marsamid; Marsonol; Surfac; Surfam

[393] **Marubishi Oil Chemical Co., Ltd.**, 3-7-12 Tomobuchi-cho, Miyakojima-ku, Osaka, Japan

Babinar; Levelan; Value

[397] **Matsumoto Yushi-Seiyaku Co., Ltd.**, 1-3, 2-Chome, Shibukawa-cho, Yao City, Osaka, Japan

Effcol; Elimina; Marpol; Marvelin; Zontes

[400] **Mazer Chem., Inc.**, 3938 Porett Dr., Gurnee, IL 60031

Larosol; Larostat; Macol; Mafo; Mapeg; Maphos; Masil; Mazamide; Mazawet; Mazeen; Mazol; Mazoline; Mazon; M-Quat; S-Maz; T-Maz

[401] **McIntyre Chemical Co., Ltd.**, 4851 S. St. Louis Ave., Chicago, IL 60632

Mackadet; Mackam; Mackamide; Mackamine; Mackanate; Mackazoline; Mackester

[410] **Lucas Meyer GmbH & Co.**, Ausschlaeger Elbdeich 62, PO Box 280246, 2000 Hamburg 28, West Germany

Asol; Chocothin; Colorol; Emulbesto; Emulfluid; Emulpur; Emulthin; Magathin; M-C-Thin; Metarin; Topcithin

[411] **M. Michel & Co., Inc.**, 90 Broad St., New York, NY 10004

Michelene

[413] **Milliken Chemical/Div. of Milliken & Co.**, PO Box 1927, Spartanburg, SC 29304
Syn Fac; Syn Lube

[414] **Miranol Chemical Co., Inc.**, 68 Culver Rd., PO Box 411, Dayton, NJ 08810
Miramine; Miranate; Miranol; Mirapol; Mirapon; Mirataine; Mirawet

[416] **Mobay Chemical Corp.**, Penn Lincoln Parkway West, Pittsburgh, PA 15205
Emulvin

[418] **Mona Industries, Inc.**, PO Box 425, 76 E. 24th St., Paterson, NJ 07544
Detergent Concentrate 840; Monafax; Monamate; Monamid; Monamine; Monamulse; Monaquat; Monaterge; Monateric; Monawet; Monazoline

[419] **Monsanto Co.**, 800 N. Lindbergh Blvd., St. Louis, MO 63167
Alkylate; DDBSA; Scripset; Sterox

[422] **Morton Chemical/Div. of Morton-Norwich Prod., Inc.**, 1275 Lake Ave., Woodstock, IL 60098
Sotex

[424] **The Murphy-Phoenix Company**, PO Box 22930, Beachwood, OH 44122
Tallates

[427] **National Starch and Chemical Corp.**, 10 Finderne Ave., Bridgewater, NJ 08807
Emulsifying Oil C; Protamine; Protowax; Protowet; Sulfonated Castor Oil

[428] **Nease Chemical Co.**, PO Box 221, State College, PA 16801
Naxonate

[434] **Niacet Corp.**, 47th St. & Niagara Falls Blvd., Niagara Falls, NY 14304
Niaproof

[437] **Nikko Chemicals Co., Ltd.**, 1-4-8 Nihonbashi-Bakurocho, Chuo-ku, Tokyo 103, Japan
Deatron; Dispatex; LAN; Neocation; Neoprotex; Neorate; Nikkol; Sunsoflon; Sunsolt; Swanol; Vifcoll

[440] **Nippon Nyukazai Co., Ltd.**, 19-9, 3-Chome, Ginza, Chuo-ku, Tokyo 104, Japan
Minemal; Newcol; Paracol; Texnol

[441] **Nippon Oils & Fats Co., Ltd.**, No. 5-1 Chome, Yurakucho, Chiyoda-ku, Tokyo, Japan
Nissan; Stafoam

[442] **Nippon Senka Chemical Industries, Ltd.**, 17-34 Hanaten-Higashi, 1-Chome, Tsurumi-ku, Osaka, Japan
Migregal; Multinol

[446] **NL Industries/Industrial Chemicals Div.**, PO Box 700, Hightstown, NJ 08520
Flexricin; Solricin; Surfactol

[447] **NL Industries/NL Baroid**, PO Box 1675, 2404 Southwest Freeway, Houston, TX 77251
Aktaflo; Carbonox; CC; Driltreat; Invermul; Slik; Surflo

[451] **Norman, Fox & Co.**, 5511 S. Boyle Ave., PO Box 58727, Vernon, CA 90058
Norfox

[451a] **Noury Chem., Div. of Akzo Chemie**, 2153 Lockport-Olcott Rd., Burt, NY 14094
Armoslip

[454] **Oleofina S.A.**, 4, Rue Jacques de Lalaing, 1040 Brussels, Belgium
Radia; Radiamac; Radiamine; Radiamuls; Radianol; Radiaquat; Radiasurf; Radiasyn

[455] **Olin Chemicals**, 120 Long Ridge Rd., Stamford, CT 06904
Glycidol; Poly-G; Poly-Tergent; Wayhib; Wayplex

[456] **Onyx Chemical Co./Millmaster Onyx Group**, 190 Warren St., Jersey City, NJ 07302
Ammonyx; Anionyx; BTC; BTCO; Casul; Isothan; Maprofix; Maprolyte; Maprosyl; Neutronyx; Onamer; Onamine; Onyxide; Onyxol; Super Amide; Surco

[460] **Paniplus**, 100 Paniplus Rdwy., Olathe, KS 66061
Alpha; Elasdo; Panalite; PC; Soft Touch

[462] **Patco Products/Div. C.J. Patterson Co.**, 3947 Broadway, Kansas City, MO 64111
[463] **Patterson Co., C.J.**, 3947 Broadway, Kansas City, MO 64111
[462] to [463]: Alphadim; D-9; Emplex; Patco; Pationic; Patlac; Vanade; Vanall; Vanease; Verv

[467] **Penreco/Div. Pennzoil Co.**, 106 S. Main St., Butler, PA 16001
Petrosul

[469] **Petrochemicals Co., Inc.**, PO Box 2199, 2001 N. Grove, Ft. Worth, TX 76113
Petrac

[478] **Pilot Chemical Co.**, 11756 Burke St., Santa Fe Springs, CA 90670
Aristonate; Calamide; Calfoam; Calimulse; Calsoft; Calsuds; Emulsifier 30, 99; Pilot

[480] **Polymer Research Corp. of America**, 2186 Mill Ave., Brooklyn, NY 11234
Polylev

[482] **PPF International**, Lindtsedjik 8, 3336 LE, Zwijndrecht, The Netherlands
Admul; Hymono

[489] **Procter and Gamble Co./Industrial Chemicals Div**, 301 E. 6th St., PO Box 599, Cincinnati, OH 45201
AE; Equex; P & G

[490] **Proctor Chemical Co.**, PO Box 399, Salisbury, NC 28144
Condensate; Protodye; Protopan; Protosan; Protowet

[494] **Pulcra S.A.**, Farell 9, Apartado 7.096, 08014 Barcelona, Spain
Amfotex; Catinex; Crafol; Cralane; Crapol; Demelan; Depasol; Montosol; Montovol; Niox; Noiox; Puxol

[496] **PVO International, Inc.**, 416 Divison St., Boonton, NJ 07005
Compound 9264B; Drewlate; Drewlene; Drewmulse; Drewpol; Drewpone; Drewsorb; Drewsoft; Druspin; Lipal; Neobee; Synotol; Wecobee; Wecotop

[498] **Quad Chemical Corp.**, 2779 East El Presidio, Long Beach, CA 90810
Sorbitol

[500] **Quimigal-Quimica de Portugal E.P.**, Av. Infante Santo No. 2, 1300 Lisboa, Portugal
Quimipol

[502] **Reed Lignin Inc.** See Daishowa Chemicals Inc.

[503] **Reed Ltd./Chemical Div.-Lignin Products**, PO Box 2025, Quebec, P.Q., Canada
Lignosol; PB; Peltex

[505] **Reichhold Chemicals Inc.**, RCI Bldg., White Plains, NY 10603
Acofor

[506] **Reilly-Whiteman Inc.**, Washington & Righter Sts., Conshohocken, PA 19428
Carnapol; Emulsan; Infrasan; Petrosan; R3124 Ester; Solocod

[507] **Rewo Chemicals Inc.**, 107B Allen Blvd., E. Farmingdale, NY 11735
[508] **Rewo Chemicals Ltd.**, Crown House, London Rd., Morden, Surrey SM4 5DU, UK
[509] **Rewo Chemische Werke GmbH**, Postfach 1160, Industriegebiet West, D-6497 Steinau, West Germany
[510] **Rewo Quimica S.A.**, Barcelona, Spain
[507] to [510]: Condanol; Rewocid; Rewocor; Rewoderm; Rewolan; Rewolub; Rewomid; Rewominox; Rewomul; Rewopal; Rewophat; Rewopol; Rewopon; Rewoquat; Reworyl; Rewoteric; Rewowax

[514] **Rhone-Poulenc Industries,** 22 Ave. Montaigne, 75360 Paris Cedex 08, France

Lipoproteol

[515] **The Richardson Co.,** 2701 Lake St., Melrose Park, IL 60160

Chemadene; Chemadox; Richamate; Richamide; Richonate; Richonic; Richonol

[518] **Rilsan Corp./Subsid. of Atochem,** 139 Harristown Rd., Glen Rock, NJ 07452

Sorgen

[519] **RITA Corp.,** PO Box 556, Crystal Lake, IL 60014

Barre; Grilloten; Laneto; Pationic; Ritacetyl; Ritachol; Ritaderm; Ritahydrox; Ritalafa; Ritalan; Ritapro; Ritasol; Ritawax; Ritoleth; Rotolan; Super-Sat; Super Wet

[521] **Robeco Chemicals Inc.,** 99 Park Ave., New York, NY 10016

Cetina; Spermwax

[522] **Robinson-Wagner Co., Inc.,** 628 Waverly Ave., Mamaroneck, NY 10543

Lanogel; Promulgen

[523] **Rohm & Haas (Australia) Pty. Ltd.,** 969 Burke Rd., PO Box 115 Camberwell, Victoria 3124 Australia

[524] **Rohm & Haas Co.,** Independence Mall West, Philadelphia, PA 19105

[525] **Rohm and Haas Co. European Operations,** Chesterfield House, 15-19 Bloomsbury Way, London WC1A 2TP, UK

[523] to [525]: Hyamine; Triton

[527] **Ronsheim & Moore Ltd./Div. of Hickson & Welch Ltd.,** Ings Lane, Castleford, Yorkshire WF10 2JT, UK

Aremsol; Gardinol; Lanbritol; Lanette Wax; Sulphonated 'Lorol'

[533] **Ryco Inc./Div. Reilly-Whiteman Inc.,** 801 Washington St., Conshohocken, PA 19428

Rychem; Rycofax; Rycomid

[534] **Ryoto Co., Ltd.,** Sun Bldg., 13-12 Ginza 5-Chome, Chuo-ku, Tokyo 105, Japan

Ryoto

[536] **Sandoz Chemicals Corp.,** 4000 Monroe Rd., Charlotte, NC 28205

Algepon; Dergon; Keripon; Lyogen; Sandobet; Sandopan; Sandopure; Sandoxylate; Sandoz; Sandozin; Tetranol

[537] **Sandoz Inc./Colors & Chemicals Div.,** Route 10, East Hanover, NJ 07936

See Sandoz Chemicals Corp.

[538] **Sandoz Ltd./Chemicals Div.**, Lichstrasse 35, CH-4002 Basle, Switzerland

Cartaretin; Ceranine; Chemical Base; Dilasoft; Ekaline; Imerol; Mercerol; Nilo; Sandopan; Sandopur; Sandoz; Sandozin; Sandozol

[539] **Sanyo Chemical Industries, Ltd.**, 11-1 Ikkyo Nomoto-cho Higashiyama-ku, Kyoto 605, Japan

Eleminol; Ionet; Lebon; Newpol; Nonipol; Octapol; Profan; Sandet; Sanmorin; Sanstat

[541] **Scher Chemicals Inc.**, Industrial West & Styertowne Rd., Clifton, NJ 07012

Schercamox; Schercemol; Schercodine; Schercomid; S; Schercomul; Schercophos; Schercopol; Schercopon; Schercoquat; Schercotaine; Schercoterge; Schercoteric; Schercozoline

[546] **Seaboard Chemicals, Inc.**, 30 Foster St., PO Box 707, Salem, MA 01970

Seabomul; Seachem

[551] **Seppic**, 70 Champs Elysees, 75008 Paris, France

Alkylox; Amonyl; Delestat; Desadipol; Lanapol; Lanol; Lanoxal; Melanol; Montaline; Montane; Montanox; Montapol; Montegal; Monteine; Nonarox; Octaron; Oramide; Oramix; Orapol; Simulsol; Solamine; Supermontaline

[552] **Servo Chemische Fabriek B.V.**, PO Box 1, 7490 AA, Delden, Holland

Serdet; Serdolamide; Serdox; Sermul; Servamine; Servirox; Servo; Servoxyl; Serwet

[553] **Shamton Lubricants and Chem. Ltd.**, Tel Aviv, Israel

Emulton

[555] **Shell Chemical Co.**, One Shell Plaza, Houston, TX 77002
[556] **Shell Chemicals UK Ltd.**, 1 Northumberland Ave., London WC2N 5LA, UK
[557] **Shell Chimie France**, 27 Rue de Berri, 7539, Paris Cedex 08 France
[558] **Shell Nederland Chemie B.V.**, PO Box 187,2501 CD, The Hague, Netherlands

[555] to [558]: Dobane; Dobanic Acid; Dobanol; Neodene; Neodol; Teepol

[559] **Sherex Chemical Co., Inc./Subsid. of Schering A.G.**, 5777 Frantz Rd., PO Box 646, Dublin, OH 43017

Adogen; Adol; Arosurf; Hydrofol; Talofloc; Varamide; Varine; Varion; Variquat; Varisoft; Varonic; Varox; Varstat; Varsulf

[561] **Sinor Kao S.A.**, Calle Aragon, 383 Atico 08013, Barcelona, Spain

Acetamin; Amiet; Diamiet; Diamin; Farmin; Quartamin; SK Flot

[562] **Werner G. Smith Inc.**, 1730 Train Ave., Cleveland, OH 44113

Non Ionic Emulsifier

[565] **Soitem-Societa Italiana Emulsionanti Srl,** Via R. Cozzi, 34, 20125 Milano, Italy

Soil Mul; Soitem

[566] **Soluol Chemical Co.,** Green Hill & Market Sts., Box 112, W. Warwick, RI 02893

Examide

[568] **Southland Corp./Chem. Div.,** 7666 W. 63rd St., Summit, IL 60501

Actrabase; Actrafos; Actralube; Actramide; Actrasol; Actrol

[571] **Standard Chemical Co./Horsfield Bros. Ltd.,** Mill Lane, Cheadle, Cheshire SK8 2NX, UK

Atolex; Sovatex; Sufatol

[573] **Stauffer Chemical Co.,** Nyala Farm Rd., Westport, CT 06880

Drymet; Victawet

[573A] **Stave Chemical Co. Inc.,** 20 Marilyn St., Basking Ridge, NJ 07920

Stave

[575] **Stepan Co.,** Edens & Winnetka Roads, Northfield, IL 60093
[576] **Stepan Europe,** BP127, 38340 Voreppe, France

[575] to [576]: Agent; Amidox; Amphosol; Bio Soft; Bio Terge; Catisol; Emulsifier WHC; Lathanol; Makon; Maprofix; Maypon; Nacconol; Ninate; Ninex; Ninol; Ninox; Nipol; Onyxol; Pestilizer; Petrostep; Polypeptide; Polystep; Secolat; Secosol; Secoster; Secosyl; Steol; Stepan; Stepanate; Stepanol; Stepantan; Stepantex; Stepfac; Steposol; Super Pro; Toximul

[579] **Stokely Van Camp, Inc./Industrial Products Group,** PO Box 569, Columbus, OH 43216

Acconon

[584] **Swastik Household & Industrial Products Ltd.,** Shahibag House, PB362, 13 Walchand Hirachand Marg, Ballard Estate, Bombay 400 038, India

DET; DET-Washmatic; Endet; Idet; Swanic; Swascol; Swastik; Sway

[585] **Swift Chemical Co.,** 383 Orenda Road, Bramaleg, Ontario, Canada

Solar

[586] **SWS Silicones/Div. of Stauffer Chemical Co.,** PO Box 428, Sutton Road, Adrian, MI 49221

Antifoam Compound SWS; Antifoam Emulsion Q-, SWS-

[587] **Sybron Chemical Inc.,** 59 Lee Ave., Haledon, NJ 07508

Cellopal; Tanapal; Tanapon; Tanaquad; Tanaterge

[588] **Taiwan Surfactant Corp.**, 8-1 Floor, No. 106, Sec. 2, Changan East Road, Taipei, Taiwan, R.O.C.

Ablunol

[590] **Takemoto Oil & Fat Co., Ltd.**, No. 5, Sec. 2, Minato-Machi, Gamagori, Aichi, Japan 443

Chupol; Delion; Elecut; Honoralin; Newkalgen; Newlon; Polefine; Sofbon; Sofnon

[595] **Tennessee Chemical Co.**, 3475 Lenox Rd., N.E., Suite 670, Atlanta, GA 30326

Sulfonic

[598] **La Tessilchimica SpA**, Via Baertsch 1, 24100 Bergamo, Italy

Chimin; Chimipal; Chimipon; Cosmopon; Tequat; Tewax

[599] **Texaco Chemical Co./Div. Texaco Inc.**, PO Box 15730, Austin, TX 78761

Surfonic

[600] **Texo Corp.**, 2801 Highland Ave., Cincinnati, OH 45212

Pantex; Texo

[601] **Theobald Industries**, PO Box 72, Harrison, NJ 07029

Mercol

[603] **Thompson-Hayward Chem. Co./Specialty Chem. Div.**, PO Box 2383, 5200 Speaker Rd., Kansas City, KS 66110

AF; Amsul; Antifoam; Casul; T-9; T-Det; T-Mulz

[605] **Titan Chemical Products, Inc.**, PO Box 20, Short Hills, NJ 07078

Titadine; Titan; Titanole; Titanterge; Titapene; Titazole

[606] **Toho Chemical Industry Co., Ltd.**, 14-9, 1-Chome, Kakigara-cho, Nihonbashi, Chuo-ku, Tokyo 103, Japan

Airrol; Alscoap; Catinal; Demulfer; DRA; Emulbon; Neoscoa; Neospinol; Nonal; Nonipol; Obazoline; Pegnol; Pepol; Sorbon; Sorpol; Tohol

[607] **Tokai Seiyu Ind. Co., Ltd.**, Yamadahigashimachi 2-chome 67, Higashi-ku, Nagoya, Japan

Geitol; Ospin; Penestrol; Regitant; Softal

[608] **Tomah Products, Inc./Exxon Chemical Americas**, 1012 Terra Dr., PO Box 388, Milton, WI 53563

Acid Foamer; Acra; AO; Tomah

[610] **Thomas Triantaphyllou S.A.**, 405 Tatoiou Ave., TK 136 71, Acharnes, Athens, Greece

A.B.S. 87%; Loropan; Louryl; Tylorol

[611] **Troy Chemical Corp., Inc.**, One Avenue L, Newark, NJ 07105
Lecithin W.D.; Troykyd

[613] **Unger Fabrikker AS**, N-1601, Fredrikstad, Norway
Ufanon; Ufaryl; Ufasan; Ungerol

[615] **Union Camp Corp.**, 1600 Valley Rd., Wayne, NJ 07470
Acetamide

[616] **Union Carbide Corp.**, 39 Old Ridgebury Rd., Danbury, CT 06817
Tergitol; Ucar; Union Carbide

[618] **Uniroyal Inc./Chemical Group**, World Headquarters, Middlebury, CT 06749
Polywet

[620] **United States Borax & Chemical Corp.**, 3075 Wilshire Blvd., Los Angeles, CA 90010
Borax

[627] **Valchem (Australia) Pty. Ltd.**, PO Box 255, Wangaratta, Victoria 3677, Australia
[628] **Valchem/Chem. Div. of United Merchants & Mfger. Inc.**, PO Box 38, Langley, SC 29834
[627] to [628]: Valdet; Valscour; Valwet

[629] **Van Dyk & Co., Inc.**, 11 Williams St., Belleville, NJ 07109
Cerasynt

[630] **Van Schuppen Chemie**, Nieuweweg, Veenendaal, Netherlands
Dusoran; Lanolin Alcohols LG, LO, THG, THO; Lanolin Fatty Acids G, O; Solwax

[631] **R.T. Vanderbilt & Co., Inc.**, 30 Winfield St., Norwalk, CT 06855
Rheotol; Van Gel; Veegum

[633] **Vikon Chemical Co., Inc.**, PO Box 1520, Burlington, NC 27215
Vikosperse

[634] **Vista Chemical Co.**, 15990 N. Barkers Landing Rd., PO Box 19029, Houston, TX 77079
Alfol; Alfonic; AXS; Nalkylene; Novel; Vista

[639] **Westbrook Lanolin Co./Argonaut Works**, Laisterdyke, Bradford, Yorks BD4 8AU, UK
Albalan; Aqualose; Argobase; Argonol; Argowax; Golden Dawn; Golden Fleece

[642] **Westvaco Chemical Div.**, PO Box 70848, Charleston Hts., SC 29415

Indulin; Polyfac; Polyfon; Polytard; Reax; Westvaco

[646] **Wickhen Prod. Inc.**, Big Pond Rd., Huguenot, NY 12746

Wickenol

[650] **Witco Chemical Corp./Organics Div. & Sonnneborn Div.**, 277 Park Ave., New York, NY 10017
[651] **Witco Chemical Corp./Ultra Div.**, 2 Wood St., Paterson, NJ 07524
[652] **Witco Chemical Ltd. (Israel)**, PO Box 975, Haifa 31000, Israel
[653] **Witco Chemical Ltd. (UK)**, Union Lane, Droitwich, Worcester WR9 9BB, UK
[654] **Witco Chemical S.A.**, Rue Gravetel, Saint Pierre les Elbeuf 76320, France
[655] **Witco Chemical S.A./Cyclo Div.**, 10, Rue Cambaceres, 75008 Paris, France

[650] to [655]: Acetamide; Adsee; Alconate; Aluminum Stearate EA; Barium Petronate; Bubble Breaker; Calcium Petronate; Cetalox; Emargol; Emcol; Emphos; Emulpon; Finsolv; Hexaryl; Hybase; Kadif; Lanoxyl; Lauropal; Lauroxal; LSF; Neopon; Ocetox; Petromixes; Petronate; Pyronate; Sochamine; Sorban; Sorbanox; Sponto; Sulfosil; Sulframin; Synthionic; Syntofor; Syntophos; Syntopon; Ultra; Ultraphos; Witbreak; Witcamide; Witcamine; Witco; Witcolate; Witcomul; Witconate; Witconol; Witcor; Witcosperse

[658] **Yoshimura Oil Chemical Co., Ltd.**, Minami 5-Chrome-1-1 Honan-cho, Toyonaka-shi, Osaka 561, Japan

Permax

[659] **Zohar Detergent Factory**, Kibbutz Dalia, Israel 18920

Amido Betaine C; DDBS; Labs; Lauramide; Texamine; Textol; Zohar; Zoharex; Zoharlab; Zoharpon; Zoharteric

[660] **Zschimmer and Schwarz**, Postfach 2179, 4-5 Max-Schwarz-Strasse, D-5420 Lahnstein, West Germany

Alphoxat; Amphotensid; Lumo; Mulsifan; Oxetal; Oxypon; Phosfetal; Produkt RT; Propetal; Purton; Setacin; Skinotan; Sulfetal; Triumphnetzer; Zetesol; Zusolat; Zusomin

www.ingramcontent.com/pod-product-compliance
Lightning Source LLC
LaVergne TN
LVHW102102300926
849161LV00001B/4
* 9 7 8 0 8 2 0 6 0 3 3 5 3 *